MINITAB Commands Used Within the Text (continued)

C denotes a column, K denotes a constant

9. REGRESSION

REGRESS Y in C using K predictors in C,....,C

Subcommands: COEF into C

RESIDUALS into C

PREDICT for K,....,K

or PREDICT for C,....,C

Example: REGRESS C1 2 C2 C3 will regress Y in C1 using 2 predictors in C2 and C3.

BRIEF output at level K

STEPWISE regression of C using predictors C,....,C

Subcommands: FENTER = K

FREMOVE = K

FORCE C,....,C

ENTER C,....,C

REMOVE C,....,C

10. ANALYSIS OF VARIANCE

ONEWAY using data in C, levels in C

TWOWAY using data in C, levels, in C, blocks in C

(or TWOWAY using data in C, factor A levels in C, factor B levels in C)

11. NONPARAMETRIC STATISTICS

CHISQUARE using C,....,C

RUNS above and below K for data in C

MANN–WHITNEY [alternative = K] using C and C

WTEST using C

Subcommand: ALTERNATIVE = K

KRUSKAL–WALLIS test for data in C, levels in C

RANK the values in C, put ranks into C

12. TIME SERIES

ACF of C

DIFFERENCES [of lag K] using C, put into C

LAG [by K] using C, put into C

13. MISCELLANEOUS

IRANDOM K integers between K and K, put into C

ERASE C,....,C

14. SUBCOMMANDS

Some MINITAB commands have subcommands to convey additional information. To use a subcommand, include a semicolon at the end of the main command; the subcommands follow (one per line). End the final subcommand with a period.

Introduction to Business Statistics

A COMPUTER INTEGRATED APPROACH

Second Edition

Introduction to Business Statistics

A COMPUTER INTEGRATED APPROACH

Second Edition

Alan H. Kvanli
C. Stephen Guynes
Robert J. Pavur

UNIVERSITY OF NORTH TEXAS

WEST PUBLISHING COMPANY

ST. PAUL • NEW YORK • LOS ANGELES • SAN FRANCISCO

Copyediting: Linda Thompson
Composition: Syntax International
Cover Design: David Farr, Imagesmythe, Inc.
Cover Image: Melvin L. Prueitt, Computer Graphics Group, Los Alamos National Laboratories
Text Design and Production Coordination: Schneider & Company

COPYRIGHT © 1986 By WEST PUBLISHING COMPANY
COPYRIGHT © 1989 By WEST PUBLISHING COMPANY
50 W. Kellogg Boulevard
P.O. Box 64526
St. Paul, MN 55164-1003

Library of Congress Cataloging-in-Publication Data

Kvanli, Alan H.

Introduction to business statistics: a computer integrated approach/Alan H. Kvanli, C. Stephen Guynes, Robert J. Pavur.—2nd ed.

p.   cm.
Includes index.
ISBN 0-314-47148-0
1. Commercial statistics.   2. Statistics.   3. Commercial statistics—Data processing.   4. Statistics—Data processing.
I. Guynes, C. Stephen (Carl Stephen)   II. Pavur, Robert J.
III. Title.
HF1017.K83   1989
519.5′024658—dc19

88-28091

CIP

TO   Ann
      Holly, Stephen, and Darin
      Gail, Robert, and Michael

CONTENTS

Preface

As we approach the 1990s, the union of computers and statistics has moved from a casual relationship to a serious marriage, and all indications are that they'll be living happily ever after. With the mushrooming of microcomputer sales and capabilities, users of statistics are no longer required to wait all day for a megadollar number cruncher filling an entire room to give them a solution to their problem. As mentioned in the preface to the first edition, we feel that a statistics text that *fully* integrates the use of computers with statistics is a necessity in the marketplace. This edition has retained the "nonintimidating" approach to describing the concepts and applications of statistics while giving students the opportunity to observe (or actually carry out) computer generated solutions using a mainframe or microcomputer statistical package. The text has also been designed so that those requiring or desiring a more traditional calculator-based approach will find an abundance of exercises and examples that can be solved in this manner.

This edition expands the integration of computers to include microcomputer versions (along with earlier mainframe versions) of SAS, SPSS, and MINITAB. A new database and many new exercises have been added. Chapter material new to this edition includes a discussion of exploratory data analysis and a section on two-factor ANOVA designs. A section on multiple comparisons for both one-factor and two-factor designs has also been added. After much careful thought and helpful feedback, it was decided to delete the chapter on bivariate data from the first edition and move this material to the chapter dealing with simple linear regression. As a result, this edition contains one less chapter than the previous edition.

The text is intended to be an undergraduate or M.B.A. introduction to basic statistics. We assume that the student has a good understanding of basic algebra. Reference is made on a few occasions to calculus applications, but no calculus background is required to read the material. The writing style is interesting and easy-to-understand without sacrificing any credibility in the descriptive material. It is a nonmathematical, but not a "black box," approach to teaching the appreciation and application of statistics. We've included a large number of illustrative examples to better guide the student to an understanding of statistical concepts and applications.

To the Instructor

This text can be used for either a one- or two-semester introduction to business statistics. Suggested material to be covered in the first semester would be Chapters 1 through 8, in order, which concludes with an introduction to hypothesis testing. Chapters 9 and 10 could be included in a second-semester course, along with those remaining chapters that you feel are particularly relevant and of interest to your students.

The text has intentionally been written in a somewhat conversational style to make it less intimidating to the student. Our intent was for the student to read the text; not just use it as a source of homework exercises.

The text fully integrates three popular statistical packages: MINITAB, SAS, and SPSS. New to the second edition is the addition of program statements (and corresponding output) for use of the *microcomputer* versions of these packages. The featured package throughout all of the chapter examples is MINITAB, since these commands are simple English statements and are illustrative of computer capabilities whether or not you use the MINITAB package in your course instruction. MINITAB currently has a student version complete with diskettes and handbook, quite inexpensive and highly recommendable. Corresponding SAS and SPSS descriptions are contained at the ends of chapters—a feature unique to this text. We have fully integrated these packages throughout the text, making it possible for you to include computer usage as part of your course without having to spend a great deal of time explaining the mechanics of a particular package. An introduction to each of these three packages (both mainframe and micro versions) is presented at the end of the text. For instructors who wish to avoid computer usage, the text allows for a calculator-based approach—the exercises do not require a computer package and contain reasonably sized data sets.

Other features of the text include:

- a Look Back/Introduction at the start of each chapter to tie the chapter to the relevant material from the preceding chapters. Each chapter closes with a summary section containing the key words (in boldface print) introduced in the chapter.

- an abundance of exercises (over 1200) using realistic business situations. Each chapter also includes a case study containing an actual application of the chapter material and requiring an in-depth discussion.

- a full treatment of the use of p-values to make statistical decisions. These are derived and discussed throughout the entire text.

- three continuous distributions (normal, uniform, and exponential), along with three discrete distributions (binomial, hypergeometric, and Poisson).

- various sampling procedures, along with corresponding sample estimators and confidence intervals, as separate sections in two of the earlier chapters. In this way, the instructor is able to cover this often-neglected material without having to spend the time to cover an entire chapter.

- separate chapters for inference regarding normal parameters (μ, σ) and inference on a binomial parameter (p). Chapters 7, 8, and 9 are strictly devoted to normal inference, both one population (7 and 8) and two populations (9). Binomial inference (one and two populations) is covered in Chapter 10.

- an entire chapter devoted to forecasting using time series data. It includes several exponential smoothing models and discusses the pros and cons of using multiple regression versus time series modeling techniques for such data.

- an entire chapter on statistical decision theory. This chapter is placed near the end of the text (Chapter 17) but can be covered at any time, including the first semester, if desired.

- a large database (1140 observations) containing data on family income, family size, total indebtedness, monthly utility expenditures, and other variables. This is an end-of-text appendix and is available to adopters on a floppy disk.

- appendixes that provide an introduction to each of the three statistical packages utilized in the text.

New to the 2nd edition are:

- MINITAB/SAS/SPSS input and output for microcomputers.
- a second database containing 1000 observations selected from companies listed in the Moody's Investor Service Industrial Manual (also available on floppy disk).
- additional exercises within each chapter containing the interpretation of MINITAB output.

- additional exercises within each chapter requiring the use of a computer to answer questions related to both databases.

- additional exercises within each chapter using actual applications in a business setting (and the source of these applications).

- new material (in Chapter 3) on exploratory data analysis.

- new material (in Chapter 11) on designing an experiment, two-factor ANOVA and multiple comparisons.

- new material (in Chapter 18) on the nonparametric Friedman test.

- increased emphasis on the applications in the area of statistical quality control. There are additional examples and exercises in this area within many of the chapters. In particular, there are numerous case studies containing actual recorded applications of the chapter material in the field of quality control. These case studies are (1) Toward zero defects with statistical quality control at Hewlett-Packard (Chapter 5), (2) Optimizing a process with control charts: How X-bars saved National Semiconductor some gold (Chapter 7), (3) Monitoring the monitor: How the OC curve evaluates the sensitivity of the control chart (Chapter 8), (4) Quality improvement through process experimentation: An active use of statistics (Chapter 9), and (5) Whether to use 100 percent inspection: A decision theory analysis (Chapter 17). The following material is also available:

- an instructor's manual containing solutions to all exercises.

- a test bank containing true/false questions, completion exercises, and additional application problems. The test bank is also available on WESTEST, a computerized testing program for the IBM-PC and compatibles or the Apple II family of microcomputers.

- a student study guide written to put students at ease and guide them through applications of the chapter material. We certainly hope that this text will meet your classroom needs. If you care to offer comments and suggestions, we would like to hear from you. Address any correspondence to Al Kvanli, College of Business Administration, University of North Texas, Denton, Texas 76203.

To the Student

We believe you will find this text to be a readable, easily understood treatment of business statistics. Our intent is to carefully explain the various statistical concepts and strategies without getting bogged down in unnecessary mathematics. We have included many examples within each chapter to allow you to see how each procedure works. At the beginning of each chapter you will find a Look Back/Introduction section that will set up the chapter and tie it in with the previous chapters. At the end of each chapter is a summary containing all of the key definitions and concepts introduced within the chapter. At the end of the book you will find introductions to the three computer packages integrated into the text: MINITAB, SPSS, and SAS. Instructions for both *mainframe* and *microcomputer* versions of these packages are included.

As the old adage goes, "practice makes perfect," and mastering statistics is no exception. To this end, we have included a large number of exercises to help you along the road to perfection. Also, you will find the solutions to the odd-numbered exercises at the end of the text. A study guide, which contains additional examples along with their solutions, has also been prepared. These solutions take you step-by-step through the applications of the various statistical techniques with many blanks where you supply the missing number or word.

Acknowledgments

We are very much indebted to the people who helped in the production and preparation of this text. We especially want to thank Anis Kashani who not only helped prepare the first edition of this text, but had the patience to prepare the entire second edition. The editorial advice and assistance of Denise Simon and Theresa O'Dell were once again very timely, professional, and of great help. Special thanks to Beverly Kenney who typed the entire solutions manual and the many graduate students at the University of North Texas who supplied valuable input.

A heartfelt thanks goes to Jitendra Sarhad who has once again enriched our work by providing us with excellent case studies and many "real-world" exercises drawn from a vast variety of sources. His tireless efforts and amazing capacity for statistics and the English language are very much appreciated.

Wilke English has again authored a very helpful and entertaining study guide to accompany the text. We feel that his study guide has been (and will continue to be) a big plus for the textbook, and we are most appreciative for having his time and talent.

Last but certainly not least we would like to thank the reviewers who had a multitude of excellent suggestions for this edition. The following list contains the names and affiliations of these individuals:

Stan Bowen—Ryerson Polytechnical Institute
John R. Collins—University of Calgary
Robert Fountain—University of Texas at San Antonio
Richard G. Fritz—University of Central Florida
Geoffrey Jones—Seneca College of Applied Arts and Technology
Stephen K. Pollard—California State University, Los Angeles
Walter Pranger—DePaul University
W. Lee Schwendig—Idaho State University
LaVerne W. Stanton—California State University, Fullerton
Charles Zeis—University of South Colorado
Cathleen M. Zucco—LeMoyne College

Introduction to Business Statistics

A COMPUTER INTEGRATED APPROACH

Second Edition

■■■■

A First Look at Statistics

Definitions

A First Look at Statistics

Many people probably think a statistician is someone who helps figure batting averages during a baseball game broadcast. You might wonder how we can devote an entire textbook to compiling numbers and making simple calculations. Surely it cannot be that complicated!

Statistics is the science comprising rules and procedures for collecting, describing, analyzing, and interpreting numerical data. The applications of statistics are evident everywhere. Hardly a day goes by that we are not bombarded by such statements as:

Results show that Crest toothpaste helps prevent tooth decay.

The chances of a NASA rocket failure is higher than was quoted originally.

The state court has ruled that the XYZ Company is guilty of age discrimination in its termination procedure.

Or how about:

The surgeon general has determined that cigarette smoking is dangerous to your health.

Besides using statistics to inform the public, statisticians help businesses make forecasts for planning and decision making.

The use of statistics began as early as the first century A.D., when governments used a census of land and properties for tax purposes. Census taking was gradually extended to include such local events as births, deaths, and marriages. The *science* of statistics, which uses a sample to predict or estimate some characteristics of a population, began its development during the nineteenth century.

Use of statistical methods has undergone a dramatic change as computers have entered the research environment. Companies can store and manipulate large collections of data and once-formidable statistical calculations are reduced to a few key strokes. Sophisticated computer software allows users merely to specify the type of analysis desired and input the necessary data. This textbook concentrates on three of these statistical packages: MINITAB (a statistical computer package originally designed at Penn State University specifically for students), SAS (Statistical Analysis System), and SPSS (Statistical Package for the Social Sciences).

Although most statistical functions are performed by professional statisticians, you may have to draw a valid conclusion from a statistical report. Occasionally, however, statistics can obscure the truth or give an erroneous impression. Anyone who has ever changed plans due to a 90% chance of rainy weather only to sit home on a sunny day can attest to this fact. You often can avoid a bad decision by recognizing statistical errors and bias in the results that you review.

In addition, you may be asked to perform a statistical analysis. Although you may elect to obtain outside assistance, you will need to know when to consult a statistician and how to tell him or her what you need.

1.1 USES OF STATISTICS IN BUSINESS

Modern businesses have more need to predict future operations than did those in the past, when businesses were smaller. Small-business managers often can solve problems simply through personal contact. Managers in large corporations, however, must try to summarize and analyze the various data available to them. They do this by using modern statistical methods.

Areas of business that rely on statistical information and techniques include:

1. *Quality control.* Statistical quality-control procedures assure high product quality and enhance productivity.
2. *Product planning.* Statistical methods are used to analyze economic factors and business trends, and to prepare detailed sales budgets, inventory-control systems, and realistic sales quotas.
3. *Forecasting.* Statistics are used in business forecasting to predict sales, productivity, and employment trends.
4. *Yearly reports.* Annual stockholder reports are based on statistical treatment of the many cost and revenue factors analyzed by the business comptroller.
5. *Personnel management.* Statistical procedures are used in such areas as age- and sex-discrimination lawsuits, performance appraisals, and workforce-size planning.
6. *Market research.* Corporations that develop and market products or services use sophisticated statistical procedures to describe and analyze consumer purchasing behavior.

1.2 SOME BASIC DEFINITIONS

Statistics has specialized definitions for terms crucial to statistical reasoning. In **descriptive statistics**, you collect data and describe them. If you analyze and interpret the data, you are using **inferential statistics**.

Descriptive statistics are used to describe a large set of data. For example, you can reduce the set of data values to one or more single numbers, such as the average of 150 test scores, or you can construct a graph that represents some feature of the data.

You use inferential statistics to form conclusions about a large group—a population—by collecting a portion of it—a sample. Thus a **population** is the set of measurements (generally belonging to a group of people or objects) that is of interest. A **sample** is the portion of the population about which information is gathered.

FIGURE 1.1
Population versus a sample.

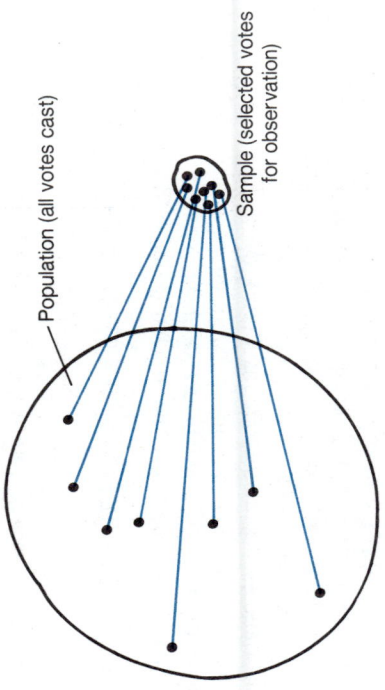

Population (all votes cast)

Sample (selected votes for observation)

The analyst decides what the population is. Typically, this population is so large that it would be nearly impossible to obtain information about every item in it. Instead, we obtain information about selected population members and attempt to draw a conclusion about all members. In other words, we attempt to infer something about the population using information about only some of the members of this population. For this procedure to be valid, the sample must be typical of the entire population.

To make an early prediction of the election results for governor of California, analysts could use a sample of voters leaving the voting booths, as illustrated in Figure 1.1. The population is all the votes cast in the election. To make a valid statistical inference using a sample, it is crucial that the sample **represent** the population. One way to do this is to collect a sample of size n, where each set of n people has the same chance of being selected for the sample. This is a **simple random sample** (Figure 1.1). It is akin to drawing names out of a hat; each name in the hat has the same chance of being pulled out. Thus, if our population is all votes cast on the day of the gubernatorial election, a sample of votes cast in only one city would not be representative. We would have no guarantee that these votes would represent the voting of the entire state. A random sample obtained across the entire state would better represent this population.

As another illustration, assume that Calcatron, a producer of electronic calculators, orders 50,000 components from GLC. Calcatron instructs GLC that they will accept the shipment if an outside laboratory that randomly selects 100 components from the batch finds that fewer than three are defective. Calcatron relies on inferential statistics; they infer that the population of components is of satisfactory quality if the sample is satisfactory. Note that it is possible that the sample will contain fewer than three defective components, whereas the population will contain, say, 80% defective parts. Whenever we attempt to infer something about a population from a sample, there is always a chance of drawing an incorrect conclusion. The only way of being 100% sure is to list the entire population. Such a sample is called a **census.**

Proper use of numerical data can be a great aid in making a critical decision. However, using an improper technique or "bad data" can lead you down the wrong path. Generally, the technique we use to analyze data in statistics depends on the nature of the data. We can distinguish between two types of numerical data.

How do the following two sets of numbers differ?

3, 5, 2, 1, 4, 4, 3, 5, 5, 1, 2, 4

4.31, 11.62, 5.37, 1.55, 3.71, 6.88, 7.23, 9.52, 2.36, 7.42, 6.11, 4.85

The primary difference is that the values in the first data set consist of *counting numbers*, or *integers*. Such data are **discrete**. For example, these data may be the coded responses from 12 people who answered a particular question in a marketing survey, where 1 = strongly agree, 2 = agree, 3 = uncertain, 4 = disagree, and 5 = strongly disagree. Note that discrete data may contain a decimal point. Nevertheless, such data have *gaps* in their possible values. For example, if you throw a single die twice and record the average of the two throws, the possible values are 1, 1.5, 2, 2.5, 3, 3.5, 4, 4.5, 5, 5.5, and 6.

Examples of discrete data that have integer values are the number of automobiles that arrive at a drive-up window over a 5-minute period, the number of children in your family, and the total of the two numbers appearing on a throw of two dice. Note that although the first two have infinite (theoretically, at least) possible values, the data are discrete. Your family cannot have 2.5 children.

Now consider the second data set in our original example. These data might represent the weights of 12 parcels received at a post office. A list of all the possible values of package weights would be long—if our scale were completely accurate, the list would be infinite and any value would be possible. Such data are **continuous**: *Any value* over some particular range is possible. There are no gaps in possible continuous data values. For example, although we may say Sandra is 5.5 feet tall, we mean her height is about 5.5 feet. In fact, this value may be 5.50372 feet. Height data are continuous. Or consider the contents of a coffee cup filled by a vending machine. Will the machine release exactly 6 ounces every time? Certainly not. In fact, if you were to observe the machine fill five such cups and measured the contents to the nearest .001 ounce, you might observe values of 6.031, 5.932, 5.871, 6.353, and 5.612 ounces. Here again, any value between, say, 5.5 ounces and 6.5 ounces is possible: these are continuous data. Data such as weights, heights, age (actual), and time are generally continuous data and will be used in the examples in the chapters to follow.

1.4 LEVEL OF MEASUREMENT FOR NUMERICAL DATA

As well as classifying numerical data as discrete or continuous, we can also label these data as to their level of measurement. We will discuss them in order of strength, beginning with the weakest. **Nominal data** are really not numerical at all, but are merely labels or assigned values. Examples include: sex (1 = male, 2 = female), manufacturer of automobile (1 = General Motors, 2 = Ford, 3 = Chrysler), or color of eyes (1 = blue, 2 = green, 3 = brown). Assigning a numerical code to such data is merely a convenience so that, for example, one can store the information in a computer. Therefore, it makes no sense to perform calculations with such numbers, such as finding their average. What would it mean to claim that 'the average eye color is 2.73?' This is a meaningless statement. Generally, we are interested in the **proportion** of such data in each category. Consider Calcatron's shipment, in which each component is either defective or not defective. We could assign the code 1 = defective, 0 = not defective. The value of interest here is p, where p = proportion of defective components in the population of 50,000 components. If Calcatron believes p is too large, they will not accept the shipment. We will consider what is "too large" in Chapter 10.

Ordinal data can be arranged in order, such as worst to best or F to A (grades on an exam). A classic example of ordinal data is the result of a cross-

FIGURE 1.2

Classifications of numerical data.

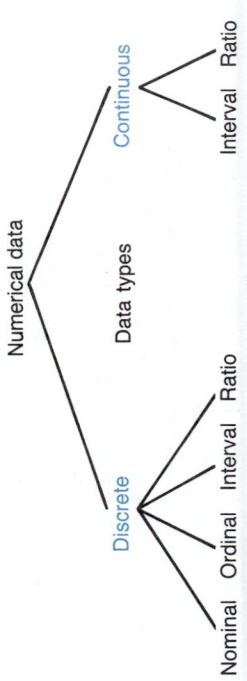

Numerical data — Data types — Discrete / Continuous

Discrete: Nominal Ordinal Interval Ratio

Continuous: Interval Ratio

country race, where ten people compete and 1 = the fastest (the winner), 2 = the runner-up, and so on, with 10 = the slowest. Here, the *order* of the values is important (3 finished before 4) but the *difference* of the values is not. For example, 2 − 1 = 1 and 10 − 9 = 1, but this does not imply that 1 and 2 were just as close in the final results as were 9 and 10.

The difference between values of **interval data** *does* have meaning. It is meaningful to add and average such data. The classic example is *temperature*, where it is true that the difference in heat between 60°F and 61°F is the same as that between 80°F and 81°F. Many of the techniques used to analyze data in statistics require data that are at least of this strength.

Ratio data differ from interval data in that there is a definite *zero point*. To decide if your data are interval or ratio, ask yourself whether twice the value is twice the strength. For example, is 100°F twice as hot as 50°F? The answer is no, so these data are interval. Is a 4-acre field twice as large as a 2-acre field? The answer is yes, so these are ratio data. Here the zero point is a field of 0 acres. Typically, data consisting of areas, counts, volumes, and weights are ratio data.

The techniques used in statistics generally do not distinguish between interval and ratio data. A summary of the various data classifications is shown in Figure 1.2. Notice that discrete data can have any of the four levels of measurement, whereas continuous data must be interval or ratio.

SUMMARY

Decision making using statistical procedures continues to grow in popularity, since calculators and computers make it easy to avoid "seat-of-the-pants" decisions by analyzing sample results in a scientific manner. Contemporary applications (such as, Should a particular company accept an outside shipment of components based upon a sample of these components?) covering a variety of business disciplines were mentioned.

The science of statistics comprises a set of rules and procedures used to describe numerical data or to make decisions using these data. The group of measurements that are of interest define the **population**. The portion of a population selected for observation is a **sample**. Most statistical methods assume that a **simple random sample** of size n has been collected, in which each set of n measurements has the same chance of being selected for the sample. A characteristic (such as average) of the population is referred to as a **parameter**, and the corresponding sample characteristic is a **statistic**. For example, the average age of a sample of 100 people passing the most recent CPA exam is a statistic. The average age of *all* people passing this exam is a parameter. A sample that contains the entire population is a **census**.

Descriptive statistics is concerned only with collecting and describing data. **Inferential statistics** is used when tentative conclusions about a population are drawn on the basis of data contained in a representative sample. The question of whether to accept the shipment of components (the population) based upon the sample of 100 components is an example of inferential statistics.

Numerical data are either discrete or continuous. **Continuous data** have limited, specific possible values. **Continuous data** can assume any value over some range. **Discrete data** have limited, specific possible values.

A further classification of data is their level of measurement. At the lowest level, **nominal data** are categorical data that are assigned numeric codes. **Ordinal data** are ranked—the order of the data values is meaningful. In **interval data**, both the order of the data and the difference between any two data values have meaning. Finally, **ratio data** have all the properties of interval data and also contain a definite zero point. Most statistical techniques do not distinguish between interval and ratio data but do require the data to have at least an interval level of measurement.

1.1 The manager of Computer Solutions, Inc., wants to determine the demand for the usage of computer consultants in small businesses in Los Angeles, California. The manager selects 50 small businesses at random in the Los Angeles metropolitan area. Analysis of the questionnaire sent to these small businesses revealed the extent of computerization and the willingness of the companies to utilize consulting services. Explain what group of people represent the population and what group of people represent the sample.

1.2 The manager of Easy Fly Airlines took a sample of 200 people who regularly fly on Easy Fly Airlines and collected information on the salaries of these people. The manager then graphed the data to get an idea of the distribution of the incomes. After studying the graph, the manager concluded that most Easy Fly Airlines customers are in the middle income bracket.

a. Explain how the manager used descriptive statistics.

b. Did the manager use inferential statistics?

1.3 Explain whether the following groups of people or objects represent a population or a sample.

a. A list of 500 employees of General Motors (*Hint:* Could this be either a sample or a population? Explain.)

b. Forty students who were randomly stopped and questioned on a university campus

c. A marketing questionnaire mailed to 200 people selected randomly from the telephone book

d. The list of all possible choices of 2 cards from a deck of 52 cards

1.4 Explain whether the following data are continuous or discrete.

a. The annual incomes of 20 executives

b. The number of long-distance calls made each month

c. The length of time for each long distance call for a particular month

d. The number of defective items in a manufacturing process

1.5 Explain why inferential statistics is not needed if data are collected by taking a census.

1.6 What is the lowest level of measurement for a set of data that would permit the valid calculation of a proportion?

1.7 Your student record contains information about your age, sex, race, current grade point average, and current classification (freshman, sophomore, . . .). State whether each of these data is nominal, ordinal, interval, or ratio.

1.8 Do you think nominal data would usually be continuous data or discrete data?

1.9 Give an example of ordinal data that would not be interval data.

1.10 What is the highest level of measurement for each of the following data sets?

a. The chain of command for officers in the army

b. The closing prices of the stocks in the Dow Jones Industrial Index

c. The temperatures in degrees Fahrenheit of several classrooms

d. The social security numbers of 12 randomly selected people

e. A listing of the college graduates and non–college graduates of a company

COMPUTER EXERCISES USING THE DATABASE

Exercise 1—Appendix H

For each of the variables defined in the database of household financial variables, determine if the corresponding data would be classified as discrete or continuous.

Exercise 2—Appendix H

For each of the variables in the database of household financial

variables, what is the highest level of measurement for the corresponding data?

Exercise 3—Appendix I

Answer Exercises 1 and 2 for each variable defined in the database using financial variables on companies.

Descriptive Graphs

Handwritten notes:

Frequency Distribution

Errors	f	f/n	By	Cf	cumulative prop.
0-4	7	.14	4.5	7	.14
5-9	9	.18	9.5	16	.32
7					
20-24	6				
	50				

A Look Back/Introduction

Chapter 1 introduced you to some of the basic terms used in statistics. One of the key concepts was the idea of acquiring data using a sample from a population. It was also emphasized that the proper use of statistics depends on the nature of the data involved. Are the data discrete or continuous? Are the values nominal, ordinal, interval, or ratio?

Once the data have been gathered, the problem becomes learning whatever we can from them. One method is to describe the data by means of a graph. A graph allows us to discuss intelligently the "shape" of the data.

Everyone has heard the expression that a picture is worth a thousand words or, more appropriately here, a thousand numbers. This is especially true in statistics, where it may be vital to reduce a large set of numbers to a graph (or picture) that illustrates the structure underlying your data. For example, in a business meeting a quick glance at a graph demonstrates a point much more easily than does a page filled with numbers and words.

To see why you may want to describe a data set by using a descriptive graph, suppose that the television department of Q-Mart sells color and black-and-white televisions and home videocassette recorders (VCRs). They decide to take a sample of 50 of their customers over a 3-month period. For each sale, they record (1) what the customer purchased, (2) the purchase price, and (3) the number of channels that each customer receives on his or her home set. The results are shown in Table 2.1.

How can you summarize and present these data in a form that is easily understood? There are many graphical methods to do just that, depending on the nature of the data and what you are trying to demonstrate about them. When presenting data graphically, the first step usually is to combine the data values into a frequency distribution.

TABLE 2.1

Data from Q-Mart Survey

ITEM	NO. OF SALES	PURCHASE PRICE/NO. OF CHANNELS						
Color TV	30	460.04	538.13	477.18	475.96	715.93	436.68	643.55
		3	9	6	12	5	8	9
		495.57	515.62	712.26	463.36	676.84	620.24	561.63
		18	6	15	10	8	12	4
		375.94	516.82	434.27	397.95	481.45	517.79	520.24
		16	10	5	13	8	7	12
		488.37	840.57	624.63	419.19	782.57	485.15	812.36
		8	11	20	15	6	9	8
		583.82	388.70					
Black-and-	6	345.88	255.46	295.77	318.91	362.81	405.16	
White TV		7	14	5	10	4	8	
VCR	14	478.03	715.71	450.36	488.34	582.36	657.41	684.71
		9	4	17	8	11	6	19
		631.78	521.48	515.61	540.44	528.57	564.16	745.28
		10	6	9	4	8	13	3

2.1

FREQUENCY DISTRIBUTIONS

We need to reduce a large set of data to a much smaller set of numbers that can be more easily comprehended. If you have recorded the population sizes of 500 randomly selected cities, there is no easy way to examine these 500 numbers visually and learn anything. It would be easier to examine a condensed version of this set of data, such as that presented in Table 2.2.

This type of summary, called a **frequency distribution**, consists of *classes* (such as "10,000 and under 15,000") and *frequencies* (the number of data values within each class). What do you gain using this procedure? You reduce 500 numbers to 10 classes and frequencies. You can study the frequency distribution in Table 2.2 and learn a great deal about the shape of this data set. For example, approximately 50% of the cities in your sample have a population between 20,000 and 35,000. Also, only 1% of the cities contain 50,000 people or more.

Frequency Distribution for Continuous Data

A frequency distribution is typically condensed from data having an interval or ratio level of measurement. When you construct a frequency distribution for

TABLE 2.2

Frequency Distribution of the Populations of 500 Cities

CLASS NUMBER	SIZE OF CITY	FREQUENCY
1	Under 10,000	4
2	10,000 and under 15,000	51
3	15,000 and under 20,000	77
4	20,000 and under 25,000	105
5	25,000 and under 30,000	84
6	30,000 and under 35,000	60
7	35,000 and under 40,000	45
8	40,000 and under 45,000	38
9	45,000 and under 50,000	31
10	50,000 and over	5
		500

continuous data, you need to decide how many classes to use (ten in Table 2.2) and the class width (5000 in Table 2.2).

There is no "correct" **number of classes (K)** to use in a frequency distribution. However, you can best condense a set of data using between 5 and 15 classes. The usual procedure is to choose what you think would be an adequate number of classes and to construct the resulting frequency distribution. A quick look at the resulting distribution will tell you if you have reduced the data too much (not enough classes; K is too small) or not enough (too many classes; K is too large). If you have a very large set of data, you can use a larger number of classes than you would for smaller data sets. Whenever you construct frequency distributions using a computer, select several different values of K and look at the effects of the different choices.

Having chosen a value for K, the next step is to examine

$$\frac{\text{range}}{\text{number of classes}} = \frac{H - L}{K}$$

where H = the highest value in your data and L = the lowest value in your data. Round the result to a value that provides an easy-to-interpret frequency distribution. This is the **class width (CW)**. The width of each class should be the same. Possible exceptions to this rule are for the first and last classes, which we will discuss later.

Suppose that, for a particular set of data, you have elected to use $K = 10$ classes in your frequency distribution. Also, $H = 106$ and $L = 10$, and so

$$\frac{H - L}{K} = \frac{106 - 10}{10} = 9.6$$

The desirable class width to use here is CW = 10.

Now let us use the 50 purchase prices in Table 2.1 to construct a frequency distribution of the purchase prices, using six classes. Your first step should be to arrange the data from smallest to largest. This is called an **ordered array**. Both the original data and the ordered data are **raw data**, since they are not grouped into classes. The ordered purchase prices are listed in Table 2.3. Using the ordered data, $H = 840.57$ and $L = 255.46$. Since $K = 6$, you compute CW:

$$\frac{840.57 - 255.46}{6} = 97.5$$

The best choice here is CW = 100.

There are two rules to remember in selecting the first class: This class must contain L, your lowest data value, and it should begin with a value that makes

TABLE 2.3

50 Purchase Prices from Q-Mart Survey Arranged as an Ordered Array

RAW DATA					ORDERED DATA				
460.04	463.36	520.24	345.88	582.36	255.46	434.27	488.34	538.13	657.41
538.13	676.84	488.37	255.46	657.41	295.77	436.68	488.37	540.44	676.84
477.18	620.24	840.57	295.77	684.71	318.91	450.36	495.57	561.63	684.71
475.96	561.63	624.63	318.91	631.78	345.88	460.04	515.61	564.16	712.26
715.93	375.94	419.19	362.81	521.48	362.81	463.36	515.62	582.36	715.71
436.68	516.82	782.57	405.16	515.61	375.94	475.96	516.82	583.82	715.93
643.55	434.27	485.15	478.03	540.44	388.70	477.18	517.79	620.24	745.28
495.57	397.95	812.36	715.71	528.57	397.95	478.03	520.24	624.63	782.57
515.62	481.45	583.82	450.36	564.16	405.16	481.45	521.48	631.78	812.36
712.26	517.79	388.70	488.34	745.28	419.19	485.15	528.57	643.55	840.57

TABLE 2.4
Frequency Distribution
of Purchase Prices Using
Six Classes

CLASS NUMBER	CLASS	FREQUENCY
1	250 and under 350	4
2	350 and under 450	8
3	450 and under 550	20
4	550 and under 650	8
5	650 and under 750	7
6	750 and under 850	3
		50

continuous

the frequency distribution easy to interpret. Because $L = 255.46$, this class should begin with either 200 or 250—we will use 250. The resulting frequency distribution is shown in Table 2.4.

Perhaps you think that six classes are not enough; that is, you have condensed this set of data too much. One indication of this would be that a large portion of your data (say, nearly 50%) lies in one class. Table 2.5 summarizes this set of data using $K = 10$ classes. Here, the class width chosen is $CW = 60$ because

discrete

TABLE 2.5
Frequency Distribution
of Purchase Prices Using
Ten Classes

CLASS NUMBER	CLASS	FREQUENCY	RELATIVE FREQUENCY
1	250 and under 310	2	.04
2	310 and under 370	3	.06
3	370 and under 430	5	.10
4	430 and under 490	12	.24
5	490 and under 550	10	.20
6	550 and under 610	4	.08
7	610 and under 670	5	.10
8	670 and under 730	5	.10
9	730 and under 790	2	.04
10	790 and under 850	2	.04
		50	

$$\frac{H - L}{10} = \frac{840.57 - 255.46}{10} = 58.5$$

As before, the first class begins at 250.

This table also contains each **relative frequency**, where

$$\text{relative frequency} = \frac{\text{frequency}}{\text{total number of values in data set}}$$

So, for example, in class 2 the relative frequency is .06; this class contains 3 out of the 50 values. The advantage of using relative frequencies is that the reader can tell immediately what percentage of the data values lies in each class.

Another alternative is to use $CW = 50$, because an increment of 50 produces classes easier to comprehend. This would produce 12 classes, as shown in Table 2.6. We could argue that 12 classes are too many, considering that the data set has only 50 values. Many classes contain only one or two data values.

The highest and lowest values in a class are the **class limits**. For example, in Table 2.4, the lower class limit of class 2 is 350, and the upper class limit is 450.

COMMENTS

TABLE 2.6

Frequency Distribution of Purchase Prices Using *CW* = 50. This Format is Used for *Continuous* Data

CLASS NUMBER	CLASS	FREQUENCY
1	250 and under 300	2
2	300 and under 350	2
3	350 and under 400	4
4	400 and under 450	4
5	450 and under 500	11
6	500 and under 550	9
7	550 and under 600	4
8	600 and under 650	4
9	650 and under 700	3
10	700 and under 750	4
11	750 and under 800	1
12	800 and under 850	2
		50

The **class midpoints** are those values in the center of the class.* Each midpoint in a sense "represents" its class. These values often are used in a statistical graph as well as for calculations performed on the information contained within a frequency distribution. The midpoint of class 2 in Table 2.4 is $(350 + 450)/2 = 400$.

Often, a set of data contains one or two very small or very large numbers quite unlike the remaining data values. Such values are called **outliers**. It is generally better to include these values in one or two **open-ended classes**. The distribution in Table 2.2 contains two open-ended classes: class 1 (under 10,000) and class 10 (50,000 and over). You may need an open-ended class if your data set includes one or more outliers or your present frequency distribution has too many empty classes on the low or high end.

CONSTRUCTING A FREQUENCY DISTRIBUTION

1. Gather the sample data.
2. Arrange the data in an ordered array.
3. Select the number of classes to be used.
4. Determine the class width.
5. Determine the class limits for each class by first selecting a lower class limit for the first class that provides a frequency distribution that will be easy to interpret.
6. Count the number of data values in each class (the class frequencies).
7. Summarize the class frequencies in a frequency distribution table.

Frequency Distribution for Discrete Data

When your data are discrete, the procedure is almost the same as when they are continuous, except (1) we define the class width CW to be the difference between the lower class limits and not the difference between an upper and lower limit (this will also work for continuous data) and (2) the description of each class is slightly different because we no longer use the "and under" definition of each class. Thus, if CW = 5 and *the data are continuous*, our classes might be 5 and under 10, 10 and under 15, and 15 and under 20. *If the data are discrete*, they might be 5 to 9, 10 to 14, and 15 to 19. Note that for the continuous data, the class midpoints are 7.5, 12.5, and 17.5. For the discrete data, however, the midpoints are 7, 12, and 17.

*Class midpoints are often referred to as *class marks*.

TABLE 2.7
Frequency Distribution of the Number of Channels Received. This Format is Used for *Discrete Data*.

CLASS NUMBER	CLASS	FREQUENCY	RELATIVE FREQUENCY
1	3–5	10	.20
2	6–8	15	.30
3	9–11	12	.24
4	12–14	6	.12
5	15–17	4	.08
6	18–20	3	.06
		50	1.0

Using the data in Table 2.1, we can construct a frequency distribution using six classes for the number of channels each customer receives on his or her television set. First we develop an ordered array:

3, 3, 4, 4, 4, 4, 5, 5, 5, 5, 6, 6, 6, 6, 6, 7, 7, 8, 8, 8, 8, 8, 8, 8, 8, 9,
9, 9, 9, 10, 10, 10, 10, 10, 11, 11, 11, 12, 12, 12, 12, 13, 13, 14, 15, 15, 15,
16, 17, 18, 19, 20

So, $H = 20$ and $L = 3$. Since

$$\frac{H - L}{K} = \frac{20 - 3}{6} = 2.83$$

we use CW = 3. The resulting frequency and relative frequency distribution is shown in Table 2.7.

EXERCISES

2.1 The following are the scores of the students of Oceanspray College on a statistics exam:

69, 47, 82, 73, 99, 97, 55, 18, 100, 85, 77, 80, 94, 79, 66, 81, 81, 88, 94, 70, 62, 58, 43, 21, 85, 68, 50, 43, 91, 85, 60, 45, 88, 95, 46, 59, 75, 80, 74, 71, 70

a. Convert the raw data into an ordered array.

b. If you have to transform the data into a frequency distribution, what value of K would you use? (K = number of classes)

c. Calculate the class width for the frequency distribution.

d. Present the data in the form of a frequency distribution.

e. Calculate relative frequencies of the scores of the college students.

f. Comment on the shape of the frequency distribution and on the shape of the relative frequency distribution.

2.2 The number of hours per day that the secretary of an accounting firm spends on the telephone is recorded. The following data are the number of hours per day over a 30-day period.

5.21, 2.12, 1.33, 7.10, 4.30, 4.20, 5.20, 2.50, 4.10, 1.50, 3.22, 3.51, 2.45, 2.54, 1.80, 1.70, 1.80, 7.10, 3.20, 2.50, 4.11, 6.20, 6.79, 1.67, 3.30, 2.90, 7.20, 5.00, 2.80, 3.90

a. Are these data discrete or continuous?

b. Construct a frequency and a relative frequency distribution.

c. What are the class midpoints?

2.3 A survey was conducted to find out how long each week homemakers spend shopping for groceries. One hundred homemakers were selected randomly in a telephone survey and asked to state the amount of time they spent in grocery food stores during the past week. The results were:

HOURS SHOPPING	FREQUENCY
0 and under 2	38
2 and under 4	31
4 and under 6	21
6 and under 8	6
8 and under 10	3
10 and over	1

a. Construct a relative frequency distribution.

b. What are the class limits?

c. What are the class midpoints?

d. Before the survey, the researchers believed that most homemakers spent no more than 2 hours per week shopping for groceries. Do the data for these 100 homemakers support that opinion?

2.4 A retail store charges its customers 15% of the value of a check on any check that bounces. In 1 week, 30 checks bounced and the following fees were collected (in dollars):

1.02, .50, 6.00, 7.21, 2.34, 2.51, 10.91, 5.95, 2.59, 4.31, 6.31, 8.30, 1.03, 8.62, 9.71, 5.21, 25.91, 10.91, 7.30, 12.51, 8.51, 6.51, 7.31, 1.19, 11.60, 12.51, 2.24, 5.41, 7.20, 8.51

a. Construct a frequency distribution using class intervals of $0 and under $2, $2 and under $4, and so on.

b. Can the value of $25.91 be considered to be an outlier?

2.5 Harberts Wholesale Plumbing Supply receives special orders for parts not in stock at local retail stores. The following data are the number of special orders received per day over a 30-day period.

2, 4, 10, 1, 12, 15, 5, 6, 9, 18, 14, 13, 19, 22, 18, 6, 4, 10, 20, 7, 15, 10, 6, 2, 7, 17, 23, 11, 16, 4

a. Are the data discrete or continuous?

b. Construct a relative frequency distribution.

2.6 Comment on the "correct" number of classes to be used in a frequency distribution.

2.7 The following are the unemployment rates (percent of available labor) for the year 1984 for selected northern cities in the United States:

6.8, 2.6, 4.6, 5.4, 6.8, 4.3, 17.1, 6.7, 6.2, 6.4, 4.7, 5.1, 2.7, 16.8, 3.4, 2.2, 4.1, 9.8, 10.4, 11.5

a. Calculate the class width for a frequency distribution for the data.

b. What do you hope to achieve by tabulating the data in the form of a frequency distribution?

c. Prepare a frequency distribution table.

2.2

HISTOGRAMS

After you complete a frequency distribution, your next step will be to construct a "picture" of these data values using a histogram. A **histogram** is a graphical representation of a frequency distribution. It describes the shape of the data. You can use it to answer quickly such questions as Are the data symmetric? and Where do most of the data values lie? For the frequency distribution in Table 2.4, the corresponding histogram is illustrated in Figure 2.1. The height of each bar represents the frequency of that particular class.

FIGURE 2.1

Frequency histogram for the frequency distribution shown in Table 2.4. 21 (out of 50) purchases were between $450 and $550, with 18 people spending $550 or more.

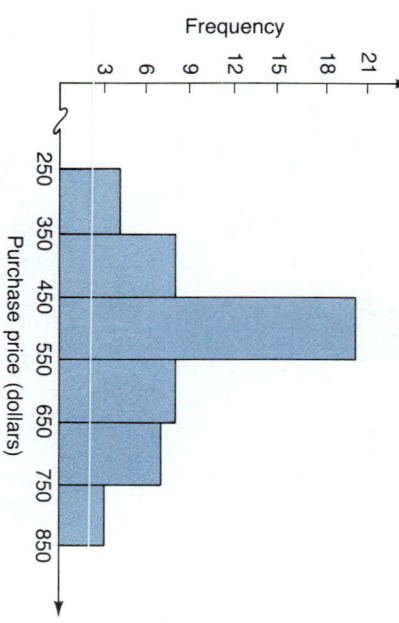

FIGURE 2.2

Relative frequency histogram of the frequency distribution in Table 2.4. 42% of the purchases were between $450 and $550; 36% of the purchases were $550 or more.

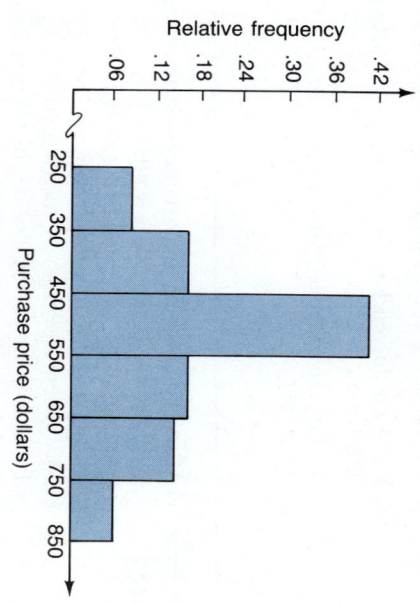

In a histogram, the bars must be adjoining (no gaps). For discrete data (such as that in Table 2.7), the right edge of each bar is midway between the upper limit of the class contained in this bar and the lower limit of the next class. For example, a histogram of Table 2.7 will contain a bar between 2.5 and 5.5 (with a height of 10), the next bar between 5.5 and 8.5 (height of 15), and so forth. The final bar (height of 3) will extend from 17.5 to 20.5.

Avoid constructing a "squashed" histogram by using the vertical axis wisely. The top of this axis (21 in Figure 2.1) should be a value close to your largest class frequency (20). Notice also that, for this example, you obtain a more concise picture by starting the horizontal axis at 250 rather than at zero.

A histogram can be constructed using the relative frequencies rather than the frequencies. **A relative frequency histogram** of the data in Table 2.4 is shown in Figure 2.2. Notice that the shape of a frequency histogram (Figure 2.1) and a relative frequency histogram (Figure 2.2) are the same. One advantage of using a relative frequency histogram is that the units on the vertical axis are always between zero and one, so the reader can tell at a glance what percentage of the data lies in each class.

Most standard statistical software packages will construct a histogram from your data. Using MINITAB, you can specify the class width and the starting class midpoint, or you can let MINITAB select these values. Your output will contain the frequency distribution as well as a graphical representation in the form of a histogram (without the bars). MINITAB will provide each class fre-

FIGURE 2.3

Histogram using MINITAB, where CW and the first class midpoint are not specified.

```
MTB > SET INTO C1
DATA> 460.04 538.13 477.18 475.96 715.93 436.68 643.55 495.57 515.62 712.26
DATA> 463.36 676.84 620.24 561.63 375.94 516.82 434.27 397.95 481.45 517.79
DATA> 520.24 488.37 840.57 624.63 419.19 782.57 485.15 812.36 583.82 388.70
DATA> 345.88 255.46 295.77 318.91 362.81 405.16 478.03 715.71 450.36 488.00
DATA> 582.36 657.41 684.71 631.78 521.48 515.61 540.44 528.57 564.16 745.28
DATA> END
MTB > PRINT C1

C1
460.04   538.13   477.18   475.96   715.93   436.68   643.55   495.57
515.62   712.26   463.36   676.84   620.24   561.63   375.94   516.82
434.27   397.95   481.45   517.79   520.24   488.37   840.57   624.63
419.19   782.57   485.15   812.36   583.82   388.70   345.88   255.46
295.77   318.91   362.81   405.16   478.03   715.71   450.36   488.00
582.36   657.41   684.71   631.78   521.48   515.61   540.44   528.57
564.16   745.28

MTB > HISTOGRAM OF C1

Histogram of C1    N = 50

Midpoint   Count
   250       1    *
   300       2    **
   350       2    **
   400       5    *****
   450       5    *****
   500      14    **************
   550       5    *****
   600       4    ****
   650       3    ***
   700       5    *****
   750       1    *
   800       2    **
   850       1    *
```

FIGURE 2.4

MINITAB histogram using specified classes. CW = 100, and the first midpoint is 300.

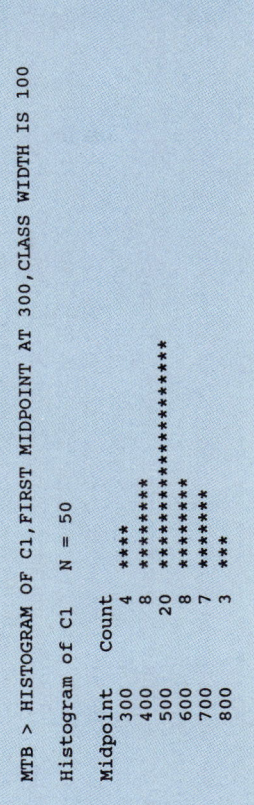

```
MTB > HISTOGRAM OF C1,FIRST MIDPOINT AT 300,CLASS WIDTH IS 100

Histogram of C1    N = 50

Midpoint   Count
   300        4    ****
   400        8    ********
   500       20    ********************
   600        8    ********
   700        7    *******
   800        3    ***
```

quency next to the corresponding class midpoint (not class limits). Figure 2.3 contains the necessary **MINITAB** statements and the resulting output, where the class width and the midpoint of the first class were not specified. Figure 2.4 specified CW = 100 and the first midpoint to be 300. We can use the output as it appears or use this information to construct Figure 2.1, which is a graphical representation of Table 2.4.

Although a histogram does demonstrate the shape of the data, perhaps a clearer method of illustrating this is to use a **frequency polygon**. Here, you merely connect the centers of the tops of the histogram bars (located at the class midpoints) with a series of straight lines. The resulting multisided figure is a frequency polygon. Figure 2.5 is an example; once again, the data in Table 2.4 were used.

2.3

FREQUENCY POLYGONS

FIGURE 2.5

Frequency polygon
using the frequency
distribution in Table 2.4.
21 (out of 50) purchases
were between $450 and
$550, with 18 people
spending $550 or more.

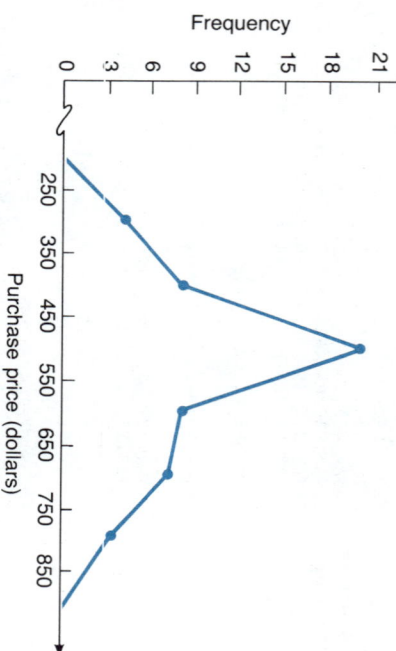

COMMENTS

The polygon can also be constructed from the relative frequency histogram. The
shape will not change, but the units on the vertical axis will now represent relative
frequencies.

The polygon must begin and end at zero frequency (as in Figure 2.5). To accomplish this, imagine a class at each end of the corresponding histogram that is empty
(contains no data values). Begin and end the polygon with the class midpoints of
these imaginary classes. Thus, your vertical axis *must* begin at zero. This need not
be true for the horizontal axis.

How do you handle an open-ended class? The easiest way is to construct a
frequency polygon of the closed classes and place a footnote at each open-ended
class location indicating the frequency of that particular class. Figure 2.6 demonstrates this, using the data from Table 2.2.

Frequency polygons are usually better than histograms for comparing the shape
of two (or more) different frequency distributions. For example, Figure 2.7 demonstrates at a glance that salaries are higher (for the most part) for management personnel at Texcom Electronics who have a college degree.

Both histograms and frequency polygons represent the actual number of data
values in each class. Suppose that your annual salary is one of the values contained in a sample of 250 salaries. One question of interest might be, What fraction
of the people in the sample have a salary *less than* mine? Such information can be
displayed using a statistical graph called an *ogive*.

FIGURE 2.6

Frequency polygon using
footnotes to handle
open-ended classes.
The data are from
Table 2.2.

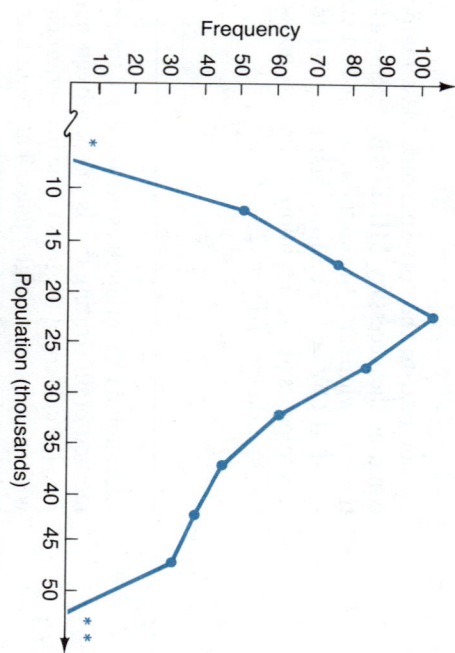

* 4 cities had populations of less than 10,000.
** 5 cities had populations of 50,000 or greater.

FIGURE 2.7

Frequency polygon showing annual salaries for Texcom Electronics management personnel. Higher salaries are observed in the college degree sample.

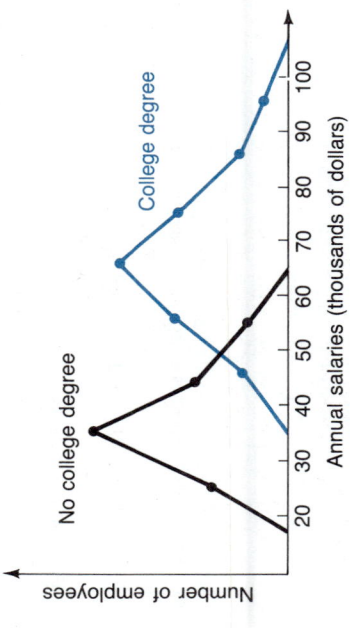

2.4

CUMULATIVE FREQUENCIES (OGIVES)

Another method of examining a frequency distribution is to list the number of observations (data values) that are *less than* each of the class limits rather than how many are *in* each of the classes. You are then determining **cumulative frequencies**. Take another look at the frequency distribution in Table 2.4. Table 2.8 shows the cumulative frequencies. Notice that you can determine cumulative frequencies (column 2) or cumulative relative frequencies (column 4). The results in Table 2.8 can be summarized more easily in a simple graph called an **ogive** (pronounced oh´-jive). The ogive is useful whenever you want to determine what percentage of your data lies *below* a certain value. Figure 2.8 is constructed by noting that

4 values (4/50 = .08) are less than 350

4 + 8 = 12 values (12/50 = .24) are less than 450

12 + 20 = 32 values (32/50 = .64) are less than 550, and so on.

TABLE 2.8

Purchase Prices from Table 2.3 with Cumulative Frequencies Analyzed

CLASS NUMBER	CLASS	FREQUENCY	CUMULATIVE FREQUENCY	RELATIVE FREQUENCY	CUMULATIVE RELATIVE FREQUENCY
1	250 and under 350	4	4	.08	.08
2	350 and under 450	8	12	.16	.24
3	450 and under 550	20	32	.40	.64
4	550 and under 650	8	40	.16	.80
5	650 and under 750	7	47	.14	.94
6	750 and under 850	3	50	.06	1.00
		50		1.0	

FIGURE 2.8

Ogive for cumulative relative frequencies using data from Table 2.8. One-half of the sample spent less than $515, and 80% of the sample spent less than $650.

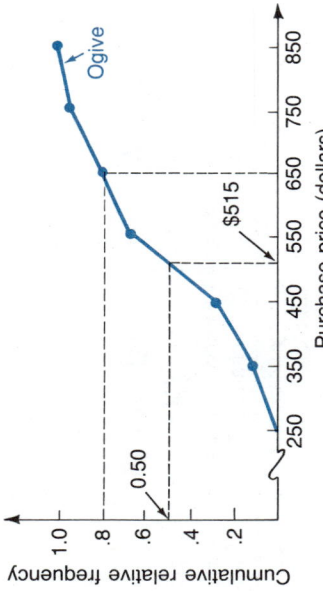

The ogive allows you to make such statements as Eighty percent of the purchase prices were less than $650 and Fifty percent of the purchase prices were under $515.

You always begin at the lower limit of the first class (250 here). The cumulative relative frequency at that point is always 0 because the number of data values less than this number is 0. You always end at the upper limit of the last class (850 here). The cumulative relative frequency here is always 1 because all the data values are less than this upper limit. This ogive value would be $n =$ the number of data values less than this upper limit. This ogive here is constructing a frequency ogive rather than a relative frequency ogive. However, the shape of the ogive is the same for both procedures.

The ogives discussed here are *less-than* ogives. Such an ogive always rises from left to right or is flat but never decreases. For example, the number of data values less than 450 must be at least as large as the number of values less than 350. A *greater-than* ogive plots the percentage of data values greater than each class limit. Such an ogive always decreases or remains flat if a particular class is empty.

EXERCISES

2.8   The following were the daily maximum temperatures in Dallas, Texas, for the month of June (in degrees Fahrenheit):

84, 84, 94, 97, 97, 89, 90, 95, 99, 94, 88, 91, 90, 97, 93, 91, 88, 89, 102, 100, 88, 85, 88, 106, 102, 86, 93, 90, 105, 99

a.   Convert the data into an ordered array.

b.   Present the data in the form of a frequency distribution, using six classes.

c.   Calculate the relative frequencies and the cumulative relative frequencies.

2.9   Construct a frequency histogram for the data in Exercise 2.2.

2.10   Draw a frequency polygon for the data in Exercise 2.3.

2.11   Construct the cumulative frequency distribution for the data in Exercise 2.4. Draw the ogive.

2.12   Does the shape of an ogive change if the cumulative relative frequencies are used instead of the cumulative frequencies?

2.13   The following is the distribution for a certain town of the population between ages 5 and 39 for the year 1988. ("Age" is defined as the age of the person at the person's last birthday.)

AGE	NUMBER
5–9	30,116
10–14	14,633
15–19	29,424
20–24	40,146
25–29	29,424
30–34	44,555
35–39	40,100

a.   Construct a frequency histogram.

b.   What does the shape of the histogram indicate?

c.   If a histogram was constructed from the relative frequency distribution, would the shape of the histogram change? Try it if you are not sure.

2.14   Draw a frequency histogram that indicates the scores of the students on a statistics exam given in Exercise 2.1.

2.15   The price:earnings (P-E) ratio is important for investors. The following is a list of P-E ratios for some of the major U.S. corporations in the year 1987.

COMPANY	P-E RATIO	COMPANY	P-E RATIO
Gulf & Western	12	JC Penney	10
Borden	14	Merrill Lynch	4
Anheuser-Busch, Inc.	15	Texaco	16
AT&T	21	Coca-Cola	14
International Paper	10	Walmart	25
Texas Instruments	18	IBM	15
McGraw-Hill	15	Ford	5
Polaroid	13	Exxon	12
Pennzoil	52	General Motors	7

(*Source: The Wall Street Journal*, January 21, 1988.)

a. Construct a frequency distribution.

b. Construct a frequency histogram for the P-E ratios.

c. If an investor wished to invest in stocks in the preceding list that had a P-E ratio below 12, what percentage of the stocks could the investor choose from?

2.16 For the data in Exercise 2.15, draw a relative frequency histogram showing the distribution of the P-E ratios. Also, draw a frequency polygon using the relative frequencies.

2.17 For the unemployment rate data given in Exercise 2.7:

a. Construct a frequency histogram.

b. Draw an ogive.

2.18 A county library's records show the following information regarding the number of patrons who used the library during the past 30 days.

100, 87, 44, 53, 17, 34, 88, 67, 31, 40, 98, 77, 55, 41, 73, 62, 88, 28, 70, 51, 82, 44, 32, 50, 33, 49, 59, 67, 79, 84

a. Construct a cumulative frequency distribution.

b. Convert the cumulative frequency distribution in part (a) into an ogive graph.

c. The number of patrons attending the library was less than what value 80% of the time?

2.19 The following is a frequency distribution of the number of daily automobile accidents reported for a month in Newark, New Jersey.

ACCIDENTS PER DAY	FREQUENCY
0–3	12
4–7	10
8–11	7
12–15	1
16–19	1

a. Construct a cumulative relative frequency distribution for the data.

b. What percentage of the time do eight or more daily accidents occur?

2.20 The profitability index is widely used by big corporations in making capital investment decisions. It is defined as the ratio of the present value of a project to its cost and should be at least equal to one. The following is a schedule of profitability indices developed by J. Conway, financial analyst of Control Systems:

PROFITABILITY INDEX	PROJECT NAME	PROFITABILITY INDEX	PROJECT NAME
1.70	A	3.50	I
.41	B	1.41	J
2.44	C	2.20	K
2.98	D	5.98	L
4.00	E	6.90	M
1.01	F	1.78	N
5.13	G	6.00	O
2.96	H		

a.   Construct a cumulative relative frequency distribution of the profitability indices.

b.   Construct a frequency histogram.

2.21   Price elasticity of electricity measures the responsiveness of consumers to changes in the price of electricity. It can be expressed as the percentage change in quantity demanded (of electricity) over percentage change in the price (of electricity). Because it is always a negative number, it is expressed only in absolute value. The following are the price elasticities (of electricity) for various utility companies in the country:

.40, .71, .33, .08, .14, .24, .38, .27, .44, .35, .22, .05, .39, .18, .22, .70, .52, .31,
.21, .36, .15, .38, .41, .23, .55, .61, .52, .35, .48, .62

a.   Construct a frequency polygon.

b.   Construct an ogive curve for the data.

2.22   An econometric model is a statistical model that predicts an econometric measure such as gross national product (GNP), unemployment rate, or inflation rate for a specific time span. The effectiveness of an econometric model is judged by the percentage of wrong predictions made when using the model to forecast. The following is a frequency distribution for a list of 25 econometric models and the percentage of errors (wrong predictions) created by them:

PERCENTAGE OF WRONG PREDICTIONS	NUMBER OF MODELS
1 and under 5	7
5 and under 10	10
10 and under 15	4
15 and under 20	2
20 and under 25	1
25 and under 30	1
	25

Construct a frequency polygon using this information.

2.23   The following are the test scores of freshmen on the first exam in an economics course at a local university:

62, 67, 74, 48, 100, 93, 49, 57, 77, 63, 82, 10, 78, 88, 99, 44, 51, 80, 71,
39, 58, 76, 89, 94, 70, 41, 66, 82, 18, 73

a.   Construct a relative frequency histogram.

b.   Draw an ogive curve.

c.   How do you interpret the distribution of the test scores?

2.24   David Bannerman, the president of Bannerman Automobile Manufacturing, has gathered the following data concerning the company's new sports car, the Chariot. The data show the numbers of cars (in hundreds) sold by the 22 top dealers during the past year. Transform the data into an appropriate graph to help David make management decisions in areas such as advertising expenditure and plant expansion. How would you describe the distribution of the data in your report to David Bannerman?

CARS SOLD	DEALERS
1 and under 5	4
5 and under 10	8
10 and under 15	2
15 and under 20	2
20 and under 25	3
25 and under 30	2
30 and under 35	1

2.25   Metro Power manufactures a high-powered copper coil to be used in giant power transformers. Tensile strength (given in thousands of pounds per square inch) is of critical importance in the manufacture of the copper coil. The following data are from a sample of copper coils tested for tensile strength:

COIL TENSILE STRENGTH	COIL TENSILE STRENGTH	COIL TENSILE STRENGTH	COIL TENSILE STRENGTH
5	18	7	11
8	21	18	15
12	24	10	9
10	7	6	22
15	26	4	10

a. Construct an ogive.

b. Find an appropriate value, X, in units of pounds per square inch such that more than one-half of the coils sampled have tensile strengths greater than X.

2.26 The following list summarizes the number and value of stocks (in dollars) that make up the investment portfolio of a mutual fund corporation:

VALUE OF STOCKS	NUMBER OF STOCKS
10,000–14,999	7
15,000–19,999	4
20,000–24,999	13
25,000–29,999	6
30,000–34,999	10

a. Draw a frequency histogram.

b. Construct an ogive.

2.27 Return on sales is a popular measure of corporate performance. It is expressed as a percentage and is equal to (net income/sales) × 100. The following are the returns on sales for 15 U.S. corporations for the year 1986.

COMPANY	RETURN ON SALES	COMPANY	RETURN ON SALES
Exxon	7.7	Occidental Petroleum	1.2
Texaco	2.3	General Electric	7.1
IBM	9.3	Xerox	5.0
Rockwell International	5.0	Pfizer	14.8
Ashland Oil	3.0	Johnson & Johnson	4.7
Aluminum Company of America	5.4	Time, Inc.	10.0
Westinghouse Electric	6.3	Merck	16.4
Monsanto	6.3		

(*Source:* Fortune 500, *Fortune*, April 27, 1987, p. 364.)

a. Construct a frequency histogram.

b. Construct a frequency polygon.

c. Construct an ogive.

d. What is a "typical" return on sales?

e. Are there companies with exceptionally high returns? What can you say about them?

2.5
BAR CHARTS

Histograms, frequency polygons, and ogives are used for data having an interval or ratio level of measurement. For data having a **nominal** level, we use a bar chart. For situations producing a sample of **ordinal** level data with a reasonable set of possible values (such as 1 = strongly agree, 2 = agree, ..., 5 = strongly disagree), a bar chart can be used to summarize the sample. A bar chart is similar to a histogram, in that the height of each bar is proportional

FIGURE 2.9

Bar chart for number of sales of each of three items. Data from Table 2.1. 60% of the purchases were for color TVs; the smallest number of sales was for black-and-white TVs (12%).

Q. If the price of natural gas goes down by 25% in the next few years, would you and your family use more or less?

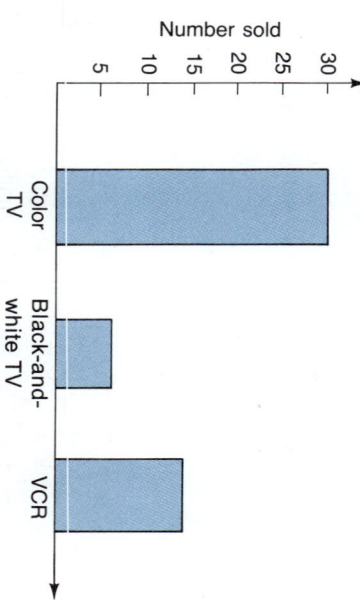

A.

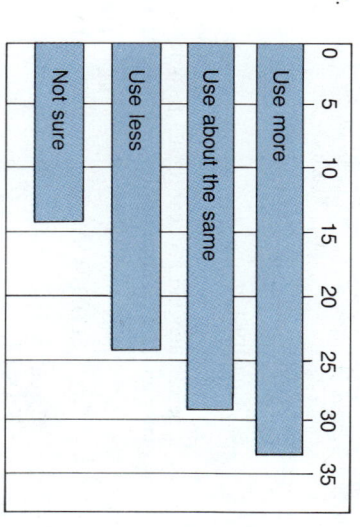

FIGURE 2.10

Bar chart drawn horizontally, which makes it easy to place labels within the boxes

to the frequency of that class. Such a graph is most helpful when you have many categories to represent.

Consider the data in Table 2.1. If you are interested in the number of sales for each of the three products (color televisions, black-and-white televisions, and VCRs), a bar chart will do a good job of summarizing this information (Figure 2.9). Notice that a gap is inserted between each of the bars in a bar chart. The data here are nominal, so the length of this gap is arbitrary.

Figure 2.10 is an example of a bar chart in which the bars are constructed horizontally rather than vertically. This enables you to label each category *within* the bar.

2.6   PIE CHARTS

A **pie chart** is used to split a particular quantity into its component pieces, typically at some specified point in time or over a specified time span. It is a convenient way of representing percentages or relative frequencies rather than frequencies. Figure 2.11 shows a pie chart of the 50 sales in Table 2.1. To construct a pie chart, draw a line from the center of the circle to the outer edge. Then construct the various pieces of the pie chart by drawing the corresponding angles. For example, the black-and-white televisions represent 12% of the total number of sales (6 out of 50). So angle *A* in Figure 2.11 is 12% of 360°, or 43.2°. Angle *B* is 28% of 360°, or 100.8°, and angle *C* is 60% of 360°, or 216°.

FIGURE 2.11

Pie chart of number of sales using data from Table 2.1. 60% of the purchases were for color TVs, the smallest number of sales was for black-and-white TVs (12%).

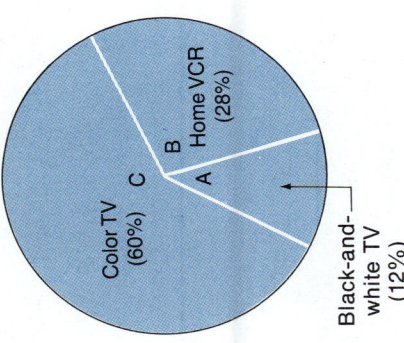

Color TV (60%)

C

B

A

Home VCR (28%)

Black-and-white TV (12%)

EXERCISES

2.28 Consumers spend their incomes on a vast array of goods and services. The following figures provide a quick summary of how the average consumer dollar is spent:

CATEGORY	PERCENT OF INCOME
Medical care	5
Clothing	5
Entertainment	4
Housing	46
Food	17
Transportation	19
Others	4
	100

a. Summarize the information in the form of a pie chart.

b. What area represents the largest piece of the pie? Is it very much larger than the next piece? How much?

2.29 Millions of business firms supply goods and services to us. Usually they are organized as proprietorships, partnerships, or giant corporations. In 1981, there are roughly 14.741 million firms producing goods and services, out of which 11.346 and 1.153 million were proprietorships and partnerships, respectively. The remainder were corporations. Present these data in the form of a pie diagram.

2.30 Using the data concerning econometric models in Exercise 2.22, construct a pie chart.

2.31 Reexamine the investment portfolio of the mutual fund corporation in Exercise 2.26. Present the information in the form of a pie chart.

2.32 The following are various current assets of nonfinancial corporations in the United States for the year 1985 (in billions of dollars):

CURRENT ASSETS	AMOUNT
Cash	189.2
U.S. Government securities	33.0
Notes and account receivables	671.5
Inventories	666.0
Other assets	224.9
Total	1784.6

(*Source: Economic Indicators*, September 1988, p. 29.)

Summarize the preceding information using a pie chart. Comment on the relative size of U.S. government securities as seen on the pie chart.

2.33 The following data indicate the electricity consumption (in kilowatt-hours) for 20 typical two-bedroom apartments in a major city:

10, 12, 17, 11, 12, 10, 9, 14, 12, 10, 14, 12, 8, 10, 8, 13, 15, 14, 12, 8

a. Construct a frequency distribution.

b. Construct a frequency histogram.

c. Is there much variation among the "typical" apartments?

2.34 The following data indicate the number of business starts for the year 1986 for selected states:

STATE	BUSINESS STARTS	STATE	BUSINESS STARTS
Alabama	3236	Massachusetts	6239
Arizona	4283	Minnesota	3555
Colorado	5292	North Carolina	5225
Connecticut	3822	Tennessee	4312
Georgia	7303	Washington	5238
Indiana	4023	Wisconsin	3448
Louisana	4141		

(*Source:* Adapted from *World Almanac and Book of Facts*, 1988, pp. 107–08.)

Construct a cumulative frequency distribution (ogive).

2.35 The number of new clients that a stock broker has opened a new account for over the past 2 years is summarized in years as follows:

AGE GROUP	NUMBER OF NEW CLIENTS	AGE GROUP	NUMBER OF NEW CLIENTS
15–19	8	40–44	12
20–24	19	45–49	17
25–29	12	50–54	8
30–34	9	55–59	4
35–39	6		

a. Construct a frequency histogram using the frequency distribution.

b. Draw a frequency polygon.

c. Comment on the shape of the distribution of the data.

2.36 A successful businessperson receives the following yearly incomes (in dollars) from seven business partnerships.

BUSINESS PARTNERSHIP	YEARLY INCOME
A	23,160
B	30,070
C	32,732
D	35,900
E	37,304
F	43,608
G	60,014
Total	262,788

Express the yearly incomes from each partnership as a percentage of the businessperson's total income and summarize this information using a pie chart.

2.37 The following data indicate the percentage of United States petroleum imports by source for the year 1985:

PERCENTAGE OF U.S. PETROLEUM IMPORTS

NATION	PERCENTAGE OF U.S. PETROLEUM IMPORTS
Algeria	3.8
Indonesia	6.1
Saudi Arabia	3.3
Iran	.5
Venezuela	12.1
United Arab Emirates	.9
Canada	15.2
Mexico	16.2
Virgin Islands	4.9
United Kingdom	6.2
Others	30.8
	100.0

(*Source: The World Almanac and Book of Facts, 1987, p. 148.*)

Construct a pie chart to illustrate these percentages.

2.38 The following table contains the total number of passengers arriving and departing at the ten busiest airports in the United States for 1986. The units are given in millions.

AIRPORT	PASSENGERS	AIRPORT	PASSENGERS
Chicago (O'Hare)	54.77	Newark	29.43
Atlanta	45.19	San Francisco	27.81
Los Angeles	41.42	New York (JFK)	27.22
Dallas/Fort Worth	39.95	New York (La Guardia)	22.19
Denver	34.69	Miami	21.95

(*Source: World Almanac, 1988, p. 152.*)

Present the data in the form of a bar graph.

2.39 The following are gross average weekly earnings of manufacturing workers in terms of current dollars and in terms of the buying power of 1977 dollars:

	CURRENT DOLLARS	1977 DOLLARS
1975	190.79	214.85
1980	288.62	212.06
1984	374.03	220.67
1985	385.97	219.93
1986	396.01	222.23
1987	401.47	222.30

(*Source: World Almanac, 1988, p. 89.*)

a. Are the data discrete or continuous?

b. Present the average weekly earnings in terms of current dollars in the form of a bar chart.

c. Repeat part (b) for 1977 dollars. Compare the two bar charts. What insight do we gain by comparing these two bar charts?

2.40 The amount of time it takes to order a special automotive part from the manufacturer is of great concern to a local automotive dealer. The following data are the average delivery time (in days) from 9 different stores for parts that were special ordered.

STORE	1	2	3	4	5	6	7	8	9
AVERAGE DELIVERY TIMES	2.5	4.5	3.0	6.0	3.0	7.0	4.0	3.5	4.0

a. Draw a bar chart and describe the shape of the chart.

b. How does a bar chart differ from a histogram?

2.7
DECEPTIVE GRAPHS

You might be tempted to be creative in your graphical displays by using, for example, a three-dimensional figure. Such originality is commendable, but does your graph accurately represent the situation? Consider Figure 2.12, which someone drew in an attempt to demonstrate that there are twice as many men as women in management positions. The artist constructed a box for the category *men* twice as high—but also twice as deep—as that for the category *women*. The result is a rectangular solid for men that is, in fact, four times the volume of the one for women. So the illustration is misleading; it appears that there are four times as many men as women in management.

When data values correspond to specific time periods—such as monthly sales or annual expenditures—the resulting data collection is a **time series**. A time series is represented graphically by using the horizontal axis for the time increments. For example, Figure 2.13 contains a return-on-investment time series for two mutual funds, plotted over a 6-year period. A glance at this figure might lead you to believe that mutual fund A is performing nearly twice as well as mutual fund B. A closer look, however, reveals that *the vertical axis did not start at zero,* which can seriously distort the information contained in such a graph. The 1985 return for fund A appears to be roughly twice that for fund B. However, the actual returns are 15.8% for fund A and 14.5% for fund B. Granted, fund A is outperforming fund B, but not nearly as dramatically as Figure 2.13 seems to indicate.

Such examples, and many others, are contained in an entertaining and en-

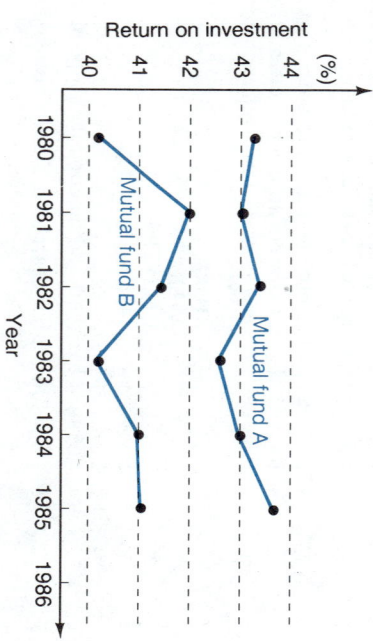

FIGURE 2.12

The illustrator wished to show that there are twice as many men as women in management positions. However, box B is twice the height *and* twice the depth of box A, and thus is four times the volume.

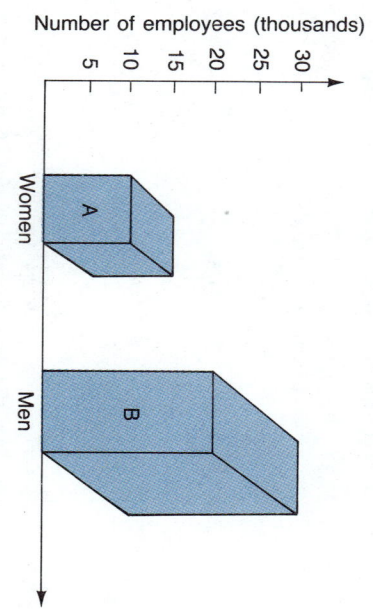

FIGURE 2.13

Time-series graph of the performance of two mutual funds. The graph is misleading because the vertical axis does not start at zero.

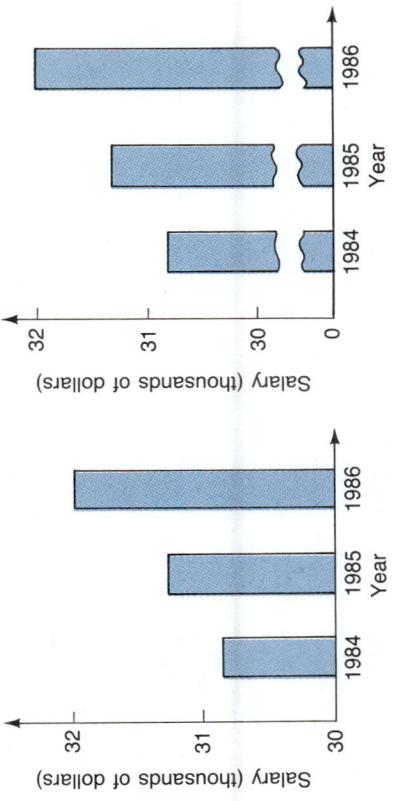

FIGURE 2.14

Two misleading bar charts. The vertical axis of the left chart does not begin at zero, and the bars in the right chart are chopped without a corresponding adjustment in the vertical axis.

lightening book by Darrell Huff entitled *How to Lie with Statistics*.* Other deceptive graphs mentioned by Huff include bar charts similar to those in Figure 2.14. Here, you may be tempted to conclude that there is a significant difference in bar heights, either because the vertical axis does not begin at zero (left side) or because the bars are chopped in the middle without a corresponding adjustment of the vertical axis (right side). As an observer, beware of such trickery. As an illustrator do not intentionally mislead your reader by disguising the results through the use of a misleading graph. It tends to give statisticians a bad name!

Now that you are ready to invest in graph paper, a straight edge, a protractor, and colored pens, you will be happy to learn that there is a much easier method of preparing professional-looking statistical graphs. There are programs available for practically all microcomputers that allow you to construct a variety of multicolored bar charts, pie charts, and so on. Figure 2.15 was drawn with a

2.8
COMPUTER GRAPHICS ON THE MICRO-COMPUTER

FIGURE 2.15

This graph was produced on an IBM PS-2 Color display 8513 (courtesy of International Business Machines Corporation).

* Darrell Huff, *How to Lie with Statistics* (New York: W. W. Norton, 1954 [and 1982 by Darrell Huff with Irving Geis, illustrator]). More recent discussions are included in *Statistics: Concepts and Controversies* by David S. Moore, 2d ed. (New York: Freeman, 1985) and "How to Display Data Badly" by Howard Wainer, *The American Statistician*, (The American Statistical Association: May 1984).

microcomputer package by using a few simple commands. If you think you will have to create many graphic summaries, try to obtain access to a computer graphics package and its output. No good report is complete without at least one such graph!

SUMMARY

This chapter examined methods of summarizing and presenting a large set of data using a graph. You begin by placing the sample data in order, from smallest to largest (an **ordered array**). The next step is to summarize the data in a **frequency distribution**, which consists of a number of classes (such as 150 and under 250) and corresponding frequencies.

The data summary can then be displayed using an appropriate graph. We discussed four kinds of graphs:

1. A **histogram**, or **frequency polygon**, is a graphical view of a frequency distribution.

2. A **bar chart** summarizes categorical (nominal) or ordinal data.

3. An **ogive** allows you to illustrate "less-than" percentages or frequencies.

4. A **pie chart** presents a percentage breakdown of a particular quantity.

A frequency distribution provides a summary of the data by placing them into groups, called **classes**. The number of values in each class is the **class frequency**. For example, there may be ten values in the class "150 and under 250." The numbers 150 and 250 here are the **class limits**, and the difference between consecutive lower class limits is the **class width**. The center of this class $[(150 + 250)/2 = 200]$ is the **class midpoint**. All classes should have the same width, except possibly the first and last class, which may be open-ended if you have a few outliers. For comparisons, the same data can be summarized using **relative frequencies**, which indicate the fraction of data values in each class rather than actual counts (**frequencies**).

A **histogram** is a graphical representation of a frequency distribution and is generally used for data having an interval or ratio level of measurement. When the data are nominal or ordinal, a **bar chart** provides a graphical summary. When constructing a bar chart, gaps are inserted between the bars due to the nature of this data type.

An excellent way to indicate the shape of the data values is to use a **frequency polygon**. This is constructed by replacing the bars in the histogram with straight lines connecting the midpoint of the top of each bar.

An **ogive** allows you to make such statements as 40% of the data values are less than 500. Like a frequency polygon, an ogive consists of many straight lines; in the ogive, the lines increase from 0 to 1 on the vertical axis. The final graph we discussed is a **pie chart**. This circular graph can be used to represent percentages (relative frequencies) at some point in time or over a certain time period.

What's Next?

A graph, such as a frequency polygon, is an excellent method of describing a set of data, but it does have its limitations. For example, examine Figure 2.5 and answer the following question: Where is the middle (center) of the data? One person might argue that it is "somewhere around 500," whereas someone else might decide that it is some value closer to 550. The point is that we need to define what the word *middle* means and define some method of calculating this value, so that we all get the *same* result. Such a value is called a *numerical measure*.

The next chapter examines a variety of such numerical measures. Rather than reducing a set of data to a graph we will reduce the data to one or more *numbers* that give us some information about the data.

REVIEW EXERCISES

2.41 Quick Checkout convenience store sells the following numbers of gallons of milk weekly over a 24-week period:

28, 36, 41, 23, 45, 23, 24, 45, 20, 26, 53, 54, 57, 43, 21, 29, 60, 42, 33, 28, 39, 44, 49, 52

a. Construct a frequency distribution using class intervals of 20–24, 25–30, and so on.

b. Construct a frequency histogram.

c. Draw the ogive.

2.42 An economist has a model to forecast the weekly money supply. The following values represent the difference between the forecasted money supply and the actual money-supply figures over a period of 25 weeks. Units are in hundreds of thousands of dollars.

11.4, 2.5, −50.5, −12.4, −5.1, 4.5, 13.6, 29.8, 51.6, −10.8, −17.8, 30.1, 33.8, 39.6, 44.7, −40.1, −35.6, −37.1, 46.7, 21.6, 18.2, −24.5, −20.5, −15.4, 53.4

a. Construct a frequency distribution.

b. Construct a cumulative relative frequency distribution.

c. Draw the ogive.

2.43 A large real-estate firm has 20 agents. The following data are the yearly salaries of each agent. Units are in thousands of dollars.

13.5, 19.6, 29.8, 43.4, 50.2, 18.7, 7.5, 24.6, 20.3, 27.4, 30.5, 34.6, 12.7, 31.7, 45.8, 41.4, 32.7, 22.6, 27.8, 20.1

a. Construct a frequency distribution.

b. Construct a cumulative relative frequency distribution.

c. Draw the ogive.

2.44 An investor owns several thousand shares of the stock Computer Graphics. Because the price of the stock is so volatile, the investor records the closing price of the stock every day to get an idea of the distribution of the price of the stock. The closing prices of this stock for 25 days are:

13.125, 13.5, 12.875, 12.25, 12.375, 13.00, 13.75, 13.375, 14.25, 15.00, 15.25, 15.375, 15.00, 14.75, 15.125, 15.375, 15.75, 16.125, 16.375, 16.50, 16.00, 15.50, 15.75, 16.25, 16.50

a. Construct a frequency distribution.

b. Construct a frequency histogram.

c. Construct a frequency polygon.

2.45 A mutual fund has its assets spread over seven sectors of the economy. The following data are the total value (in millions of dollars) of the stocks in which the fund is invested for each sector.

STOCK	VALUE
Electronics and electrical equipment	2.116
Aerospace and defense	10.375
Food and beverage	4.864
Utilities	2.713
Insurance and finance	6.538
Health care	3.675
Oil and gas	1.532

a.   Express the amount invested in each sector of the economy as a percent.

b.   Summarize the list in a pie chart.

2.46   The following data represent the scores on a computer-graded multiple-choice test. There are 20 questions, and each question is worth 5 points. Construct a frequency polygon for the grades of the 30 students.

95, 90, 80, 55, 90, 45, 50, 75, 75, 60, 55, 90, 85, 95, 95, 60, 50, 45, 95, 70, 60, 50, 85, 90, 55, 45, 95, 100, 90

2.47   An independent oil firm recently hired ten engineers, five geologists, three accountants, one statistician, four computer scientists, and one chemist. Present these data in the form of a pie chart.

2.48   The numbers of hours that a repair machine is used daily at Pat's Shoe Repair are listed below for 20 days.

3.2, 4.6, 3.1, 3.6, 2.5, 4.3, 2.1, 5.7, 6.1, 4.3, 3.0, 2.5, 1.3, 1.7, 2.4, 5.6, 5.1, 4.2, 1.9, 2.6

a.   Construct a frequency distribution.

b.   Construct a cumulative relative distribution.

c.   What can you say about machine usage (in hours per day)?

2.49   The ages (in years) of the 20 loan officers, four vice presidents, and the president of American Bank are

47, 52, 55, 65, 42, 37, 29, 52, 47, 36, 60, 50, 48, 42, 45, 35, 38, 45, 57, 43, 39, 41, 33, 58, 60

a.   Construct a frequency distribution.

b.   Construct a cumulative frequency distribution.

c.   Write a summary statement about the distribution of the ages of the loan officers.

2.50   A psychologist has designed a technique to improve a person's memory. Certain material is given to 30 people to memorize before they learn the technique. Similar material is given to the 30 people after the technique has been taught to them. The difference in the amount of time that it took to memorize the material (before–after) is given in minutes.

5, 10, 15, 11, 13, 20, 14, 5, 23, 18, 17, 4, 1, 5, 29, 18, 15, 21, 24, 16, 2, 15, 19, 30, 24, 21, 14, 18, 26, 10

a.   Construct a frequency distribution.

b.   Construct a frequency histogram.

c.   Construct a frequency polygon.

d.   Take one class interval and write out, in words, exactly what it tells you.

2.51   A manufacturing firm would like to find out the distribution of defective fuses in each package of fuses that it manufactures. Twenty boxes of 50 fuses were randomly selected and the following number of defective fuses were noted for each box:

3, 5, 10, 12, 0, 6, 17, 1, 0, 7, 3, 15, 21, 9, 13, 24, 12, 10, 6, 16

a.   Construct a frequency distribution.

b.   Construct a cumulative relative frequency distribution.

c.   Make several statements about the number of defectives usually found in a box of 50 fuses.

2.52   Custom House Products has seven stores. The following list indicates the total yearly sales for each store (in thousands of dollars):

STORE	SALES	STORE	SALES
1	60.5	5	88.7
2	70.3	6	142.6
3	44.6	7	104.2
4	59.8		

Draw a pie chart to represent the data.

2.53 The following are the number of workers absent each day recorded for 30 days in a steel factory:

10, 5, 2, 13, 17, 3, 16, 5, 7, 10, 3, 19, 22, 14, 11, 6, 9, 18, 23, 14, 7, 8, 20, 17, 13, 7, 24, 2, 6, 15

a. Construct a frequency distribution for the data.

b. Assume that the total work force is 100 people. If you were a supervisor, would you inform management of an absentee problem? If so, what would you say?

2.54 The number of cars rented daily at Rent-a-Cheapie is recorded over 40 days. Construct a frequency distribution and draw a frequency histogram for these data.

15, 20, 22, 10, 13, 24, 9, 8, 13, 21, 6, 19, 17, 13, 26, 21, 22, 19, 13, 11, 5, 26, 27, 19, 16, 10, 4, 18, 22, 14, 24, 31, 28, 18, 12, 19, 26, 7, 18, 30

2.55 In a research study activity primarily focused on human resource management issues in the hospitality industry, 26 hotel managers were randomly selected and their performances on implementing policy, making decisions, and delegating responsibility were evaluated on a scale from one to seven. The results are summarized next.

NUMERICAL RATINGS

	1	2	3	4	5	6	7
TOTAL NO. OF MANAGERS WITH EACH RATING	0	0	1	7	3	14	1

(*Source: International Journal of Hospitality Management* 5, no. 2, p. 60.)

a. Using intervals of .5 and less than 1.5, 1.5 and less than 2.5, and so on, construct a frequency histogram.

b. Make a statement which summarizes the results of the performance ratings of the hotel managers.

2.56 State gas tax in cents per gallon varies across the United States. At the beginning of 1987, the following state gas taxes were in effect for the 50 states and the District of Columbia. (Units are in cents.)

13, 8, 16, 13.5, 9, 18, 17, 13, 15.5, 9.7, 11, 14.5, 13, 14, 16, 11, 15, 16, 14, 13.5, 11, 15, 17, 9, 10, 17, 16.7, 13, 14, 8, 11, 8, 15.5, 13, 12, 10, 11, 12, 15, 13, 13, 17, 10, 14, 13, 11, 18, 10.5, 17.5, 8

(*Source: The World Almanac*, 1988, p. 143.)

a. Construct a frequency distribution.

b. Construct a cumulative frequency distribution.

c. In your judgment, what percentage of the states have an extremely high level of gas tax?

2.57 A total of 184 corporate members of the Forbes 500 chose to participate in a survey on computer crime. From the 184 firms, 219 incidents of different types of computer crime were reported. The following data are shown below for the 219 incidents.

CATEGORY	NUMBER OF INCIDENTS
Theft of computer hardware	75
Misuse of corporate computers for employee benefit	33
Theft of computer software	27
Destruction or alteration of corporate data	19
Embezzlement of corporate funds	15
Destruction of computer hardware	13
Destruction of computer software	11
Fraud against the corporation	10
Theft of output data	10
Extortion or blackmail	3
Theft of input data	3

(*Source: SIG Security Audit & Control Review* 5, no. 1, 1987.)

a. Summarize these results in a pie chart.

b. Compare the percentage of incidents related to the theft and destruction of computer hardware with the percentage of incidents related to the theft and destruction of computer software.

2.58 The numbers of business starts and failures for various parts of the United States for 1987 are given below. Draw a bar chart for the numbers of business starts by geographical region and also draw a separate bar chart for the numbers of business failures by geographical region. Compare these bar charts.

	BUSINESS STARTS	BUSINESS FAILURES
Pacific	26,530	8,676
Mountain West States	10,743	3,978
West North Central	8,852	3,902
West South Central	16,793	9,329
East North Central	22,005	6,645
East South Central	7,874	2,238
Middle Atlantic	27,550	2,775
South Atlantic	29,510	4,330
New England	8,654	716

(Source: Nation's Business, January 1988, p. 18.)

2.59 The 3-year return on an investment made in one of the various open-ended growth and income mutual funds was recorded. The 3-year returns (in percent) as of December 24, 1987, are given next.

63.1, 65.0, 55.5, 64.4, 66.8, 39.0, 63.8, 65.9, 43.3, 53.6, 36.9, 40.5, 75.1, 59.8,
59.5, 57.3, 24.4, 56.3, 41.3, 61.4, 53.4, 32.2, 68.7, 37.5, 35.6, 71.9, 61.9, 51.4,
73.4, 48.7, 25.2, 54.3, 49.0, 34.9, 54.6, 32.5, 64.4, 54.4, 60.7, 66.5, 64.2, 55.1,
58.1, 57.6, 26.6, 44.4, 66.1, 40.8, 55.0, 91.3, 82.6, 34.4, 54.9, 38.8, 43.5, 50.0,
26.7, 48.6, 64.9, 46.7, 66.7, 37.9, 61.1, 63.6, 63.3, 66.1, 52.5, 61.2, 60.8, 25.2,
41.6, 67.4, 66.1, 59.0, 67.1

(Source: Money Magazine, February 1988, pp. 151–68.)

a. Construct a frequency histogram using a statistical package.

b. Interpret the distribution. In your judgment, what percentage of growth and income mutual funds gave a poor return relative to a risk-free certificate of deposit at a bank? (Assume that one can receive a yield of 9% annually from a long-term certificate of deposit.)

2.60 The projected change in population from 1986 to 1991 for various cities in Texas is given below:

City	% CHANGE 1986–1991	City	% CHANGE 1986–1991
Abilene	6.7	Laredo	12.3
Amarillo	8.6	Longview-Marshall	7.8
Austin	17.4	Lubbock	2.0
Beaumont–Port Arthur	2.0	McAllen–Edinburgh Mission	15.8
Brazoria	12.5	Midland	17.6
Brownsville–Harlingen	11.2	Odessa	10.5
Bryan	16.0	San Angelo	8.9
Corpus Christi	6.7	San Antonio	10.2
Dallas	12.1	Sherman-Denison	5.7
El Paso	9.1	Tyler	13.0
Fort Worth–Arlington	15.9	Victoria	7.5
Galveston–Texas City	7.1	Waco	6.8
Houston	10.4	Wichita Falls	2.0
Killeen-Temple	7.6		

(Source: Survey of Buying Power, October 26, 1987, pp. 77–79.)

a. Construct a frequency distribution on the percentage change in total population for the given cities.

b. Construct a relative frequency distribution on the percentage change in total population for the given cities.

c. What percentage of the cities are expected to have growth of 10% or more?

2.61 A lawyer investigating the salary structure of a manufacturing firm is interested in the distribution of salaries for females and males. A random sample of 40 females and 40 males is selected from the blue-collar workers and managers. The salaries for the females are shown below in C1. The salaries for males are shown in C2. Describe the difference in the two histograms given in the following MINITAB output.

```
MTB > PRINT C1
C1
25389    21700    22000    23500    29000    33000    32800    27200    28450
24700    35500    29200    23400    23800    30250    37300    19450    31700
27300    28900    32600    30100    21900    25500    38450    19350    23800
28200    34800    23800    35800    22500    28450    18590    40250    29500
34900    38400    27700    26300

MTB > HISTOGRAM OF C1    N = 40

Histogram of C1

Midpoint   Count
  18000      1    *
  20000      2    **
  22000      4    ****
  24000      6    ******
  26000      3    ***
  28000      7    *******
  30000      5    *****
  32000      3    ***
  34000      3    ***
  36000      2    **
  38000      3    ***
  40000      1    *

MTB > PRINT C2
C2
23300    24400    34400    31800    25500    37150    16750     2950    19450
33800    19800    31800    24200    19750    34450    15950    22500    18500
27500    29500    25780    24350    28000    16800    34500    28500    20800
20700    26200    32500    21800    25750    24450    29950    31050    26500
20400    19200    19500    23800

MTB > HISTOGRAM OF C2    N = 40

Histogram of C2

Midpoint   Count
  16000      3    ***
  18000      1    *
  20000      8    ********
  22000      2    **
  24000      7    *******
  26000      5    *****
  28000      3    ***
  30000      2    **
  32000      4    ****
  34000      4    ****
  36000      0
  38000      1    *
```

2.62 The vice president of the Association for Manufacturing Excellence wished to determine the use of computerized manufacturing planning and control systems using just-in-time (JIT) manufacturing methods. A survey was sent to 100 factories in the industrialized Northeast part of the United States. Under PRINT C1 in the following MINITAB output, the percentage of parts manufactured with the use of the JIT method at each factory is given. Next, a frequency histogram of the values under PRINT C1 is given with the first midpoint of a class interval starting at 20. What additional insights does one gain about the frequency distribution from the frequency histogram, which has the first midpoint of a class interval starting at 15?

```
MTB > PRINT C1
C1
65  78  74  33  81  55  49  77  82  39  71  25  42
61  50  79  56  71  20  73  88  43  31  73  79  62
45  73  86  68  28  49  57  81  78  43  75  68  71
60  23  70  57  42  51  78  62  15  38  28  73  61
59  33  76  83  19  68  54  58  38  67  73  73  74
52  63  75  78  63  79  30  47  72  61  27  48
70  58  56  32  74  84  21  92  86  38  73  53  48
57  68  63  71  61  53  77  25  75  58  44  70

MTB > HISTOGRAM OF C1, FIRST MIDPOINT AT 20, CLASS WIDTH IS 10

Histogram of C1   N = 100

Midpoint   Count
20.0        5    *****
30.0       10    **********
40.0        9    *********
50.0       11    ***********
60.0       20    ********************
70.0       23    ***********************
80.0       18    ******************
90.0        4    ****

MTB > HISTOGRAM OF C1, FIRST MIDPOINT AT 15, CLASS WIDTH IS 10

Histogram of C1   N = 100

Midpoint   Count
15.0        2    **
25.0        8    ********
35.0        9    *********
45.0       10    **********
55.0       16    ****************
65.0       16    ****************
75.0       30    ******************************
85.0        8    ********
95.0        1    *
```

COMPUTER EXERCISES USING THE DATABASE

Exercise 1—Appendix H
Select 50 observations at random from the database. Using a convenient statistical package, construct a frequency histogram on the variable HPAYRENT (house payment or house/apartment rent). Also, using this same set of observations construct separate frequency histograms on variable HPAYRENT for those that own their place of residency, and for those that rent their place of residency. Comment on the shapes of the frequency histograms.

Exercise 2—Appendix H
Choose at random 30 observations from families that live in the NE sector and then choose another 30 observations at random from families that live in the SW sector. Using a convenient statistical computer package construct frequency histograms on variable INCOME1 (income of principal wage earner) for each group of 30 observations and comment on the frequency distribution of each.

Exercise 3—Appendix I
Select 100 observations at random from the database. Construct a frequency distribution and a histogram of the values of the variable ASSETS (current assets).

Exercise 4—Appendix I
Repeat Exercise 3 using the variable LIABIL (current liability).

CASE STUDY

A LOOK AT SOME AIRLINE PERFORMANCE DATA

Air traffic has increased dramatically in the wake of airline deregulation, the passenger total was about 420 million in 1986, 450 million in 1987, and is expected to exceed 650 million by 1997. Almost everybody is interested in flights being on time. Many a weary tale of woe could be told about flights delayed, connections missed, or baggage lost.

According to the Federal Aviation Administration (FAA), at least 60% of flight delays are caused by bad weather. Overcrowding of airspace, with flights bunching together at peak hours, contributes to delays. The National Transportation Safety Board (NTSB) has expressed criticism

about a shortage of air traffic controllers. The U.S. Department of Transportation (DOT) has looked into allegations that some airlines publish deceptive or unrealistic schedules to gain an advantage in computer reservation systems. Finally, a wave of airline mergers has disrupted organizations and affected the quality of service.

In response to all this, the U.S. Congress is trying to get airlines to make public their on-time, baggage handling, and other performance statistics so that consumers can make informed decisions about choosing airlines. Consider the following airline performance data. Table 2.9 shows the percentage of

TABLE 2.9 Percentage of On-Time Arrivals by Airline at the 27 Largest Airports

	AMERICAN	CONTINENTAL	DELTA	EASTERN	NORTHWEST	TRANS WORLD	UNITED	ON-TIME ARRIVALS
Atlanta	94	80	79	87	59	78	77	82.4
Boston	77	79	66	70	47	58	68	69.5
Charlotte	86	75	86	87	NS	75	86	85.1
Chicago	88	74	66	68	59	75	82	80.9
Denver	84	83	51	81	68	77	84	81.9
Dallas/Fort Worth	93	83	74	86	59	78	85	84.5
Detroit	81	76	68	87	67	72	67	69.3
Houston	89	84	68	82	64	75	76	80.6
Las Vegas	91	88	69	79	60	74	77	76.7
Los Angeles	75	66	63	72	51	66	81	70.4
Newark	79	80	67	74	62	76	75	76.2
NY-JFK	74	64	51	68	57	72	67	68.8
NY-La Guardia	80	76	70	77	50	81	79	75.1
Memphis	89	NS	80	NS	78	87	91	78.4
Miami	83	77	67	71	52	75	74	74.1
Minneapolis/St Paul	86	80	66	87	75	70	71	74.8
Orlando	82	74	72	78	60	76	71	74.0
Philadelphia	76	78	68	70	56	70	59	68.5
Phoenix	81	70	70	71	70	76	74	72.2
Pittsburgh	74	69	56	78	57	66	66	67.3
St Louis	92	81	68	94	71	83	83	82.0
Salt Lake	86	92	80	83	88	85	87	81.5
San Diego	77	73	57	76	64	65	71	71.7
San Francisco	63	67	58	63	46	72	71	65.3
Seattle	74	77	68	85	76	77	79	76.5
Tampa	88	79	74	79	66	81	76	76.7
Washington National	77	84	53	78	60	72	77	74.1

NS: Airline does not serve this city.

TABLE 2.10
Passenger Complaints
Against Airlines

| | MAY 1987 | | MAY 1986 | |
Airlines	Total	Per 100,000 Passengers	Total	Per 100,000 Passengers
Continental	793	21.39	50	3.08
Eastern	445	10.11	97	2.55
Trans World	194	8.68	57	4.10
Pan American	99	8.06	32	3.11
Northwest	283	8.01	16	1.11
Jet American	4	4.82	2	2.53
Transtar	10	4.69	3	1.60
Hawaiian	20	4.38	11	2.99
United	224	4.25	112	2.66
Midway	8	3.03	2	0.87
Braniff	6	2.52	3	1.24
American	107	2.37	50	1.24

(*Source:* Federal Aviation Administration, National Transportation Safety Board, and Department of Transportation, as reported by *Dallas Morning News*, June 22, 1987, p. 5A, and November 11, 1987, p. 12A.)

on-time arrivals for selected airlines at the 27 largest airports in the USA. Table 2.10 shows a comparison of passenger complaints against airlines for May 1986 and May 1987, both as absolute numbers and per 100,000 passengers.

Case Study Questions

1. From Table 2.9, use only the 27 observations in the last column (on-time arrivals for *all* airlines) to construct a frequency distribution. Next, construct another frequency distribution using the first seven data columns. (There are 27 observations for each airline, except where the airline does not serve that city.)

2. Prepare frequency histograms and frequency polygons from each of the above two frequency distributions. Does a comparison of these graphs give you any indication about whether the on-time performance of the second group (selected airlines) is similar to the overall performance of all airlines? Note that the number of data items in the first distribution is smaller. Does that make a difference?

3. Now prepare *relative* frequency histograms and *relative* frequency polygons, and address

the same issue raised in Question 2.

4. For a graphical comparison of the on-time arrivals of the seven selected airlines at Chicago versus Los Angeles, which of the following would be appropriate for this purpose: histograms, bar charts or pie charts?

5. From Table 2.10, use the absolute number of complaints (under the column heading of "Total") to prepare two pie charts, one each for 1986 and 1987, showing the share of complaints for each airline. What would the pie charts lead you to believe about the relative rankings of American Airlines and United Airlines with respect to passenger complaints?

6. Consider now the columns showing passenger complaints per 100,000 passengers. If pie charts were prepared using these figures, what happens to

the relative rankings for the above two airlines? Which of the pie charts are deceptive—these or the ones from question 5? Explain where the deception lies.

S P S S

MAINFRAME AND MICRO

EXAMPLE

CONSTRUCTING A HISTOGRAM

You can use SPSS to construct a histogram using the prices of television sets and VCRs purchased from the Q-Mart department store. The SPSS program listing in Figure 2.16 requests a histogram of the data in Table 2.1. As you can see, it is similar to the procedures in the SPSSX and SPSS/PC+ appendices at the end of the text. In this problem the SPSS commands are the same for both the mainframe and PC versions. **(Remember to end each command line with a period when using the PC version.)**

The TITLE command names the SPSS run.

The DATA LIST command gives each variable a name and describes the data as being in free form.

The BEGIN DATA command indicates to SPSS that the input data immediately follow.

The next 10 lines contain the data values, which represent the prices of the items in the sample.

The END DATA statement indicates the end of the data entry.

The FREQUENCIES statement specifies the variable from which we wish to produce a histogram, with the HISTOGRAM statement generating the actual graph.

Figure 2.17 shows the SPSSX output obtained by executing the listing in Figure 2.16, whereas Figure 2.18 shows the SPSS/PC+ output.

FIGURE 2.16

Input for SPSSX or SPSS/PC+. All commands for SPSS/PC+ should end with a period.

```
TITLE    Q-MART PURCHASE PRICE
DATA LIST FREE   /TVDAT
BEGIN DATA
460.04  463.36  520.24  345.88  582.36
538.13  676.84  488.37  255.46  657.41
477.18  620.24  840.57  295.77  684.71
475.96  561.63  624.63  318.91  631.78
715.93  375.94  419.19  362.81  521.48
436.68  516.82  782.57  405.16  515.61
643.55  434.27  485.15  478.03  540.44
495.57  397.95  812.36  715.71  528.57
515.62  481.45  58.82   450.36  564.16
712.26  517.79  388.70  488.34  745.28
END DATA
FREQUENCIES VAR=TVDAT/HISTOGRAM
```

FIGURE 2.17

SPSSX output.

```
COUNT   MIDPOINT   ONE SYMBOL EQUALS APPROXIMATELY   .20 OCCURRENCES

1         268      *****
1         296      *****
1         324      *****
2         352      **********
2         380      **********
3         408      ***************
3         436      ***************
5         464      *************************
6         492      ******************************
7         520      ***********************************
3         548      ***************
3         576      ***************
4         604      ********************
1         632      *****
4         660      ********************
2         688      **********
3         716      ***************
1         744      *****
1         772      *****
1         800      *****
1         828      *****
                   I........+....I....+....I....+....I....+....I....+....I
                   0         2         4         6         8        10
                              HISTOGRAM FREQUENCY

VALID CASES    50    MISSING CASES    0
```

FIGURE 2.18
SPSS/PC+ output.

```
                    TITLE    Q-MART PURCHASE PRICE

TVDAT
   Count  Midpoint
     1      268    ——
     2      303    ——
     1      338    ——
     3      373    ———
     3      408    ———
     4      443    ————
     4      478    ————
     8      513    ————————
     8      548    ————————
     4      583    ————
     2      618    ——
     3      653    ———
     2      688    ——
     2      723    ——
     3      758    ———
     1      793    —
     1      828    —
     2

            I.....I.....I.....I.....I.....I
            0     2     4     6     8    10
                   Histogram Frequency
```

MAINFRAME AND MICRO

[S A S]

EXAMPLE

CONSTRUCTING A HISTOGRAM

You can use SAS to construct a histogram using the prices of television sets and VCRs purchased from the Q-mart department store. The SAS program listing in Figure 2.19 requests a histogram of the data in Table 2.1. As you can see, it is similar to the procedures in the SAS and SAS/PC appendices at the end of the text. In this problem the SAS commands are the same for both the mainframe and PC versions.

The TITLE command names the SAS run (enclose in single quotes).

The DATA command gives the data a name.

The INPUT command names and gives the correct order for the different fields on the data lines.

The CARDS command indicates to SAS that the input data immediately follow.

The next 50 lines contain the data. The first line, for example, represents the price of the first item in the sample. The remaining lines indicate the prices of the other 49 items.

The PROC CHART command requests a SAS procedure to print a histogram. VBAR PRICE generates a histogram of the variable PRICE. The resulting output contains the class midpoints and frequencies.

Figure 2.20 shows the SAS mainframe output obtained by executing the listing in Figure 2.19, whereas Figure 2.21 shows the SAS/PC output.

FIGURE 2.19

Input for SAS (mainframe or micro version).

```
TITLE    'Q-MART PURCHASE PRICES'';
DATA TVDAT;
INPUT PRICE;
CARDS;
460.04
538.13
477.18
475.96
715.93
436.68
643.55
495.57
515.62
712.26
463.36
676.84
620.24
561.63
375.94
516.82
434.27
397.95
481.45
517.79
520.24
488.37
840.57
624.63
419.19
782.57
485.15
812.36
583.82
388.70
345.88
255.46
295.77
318.91
362.81
405.16
478.03
715.71
450.36
488.34
582.36
657.41
684.71
631.78
521.48
515.61
540.44
528.57
564.16
745.28
PROC CHART;
VBAR PRICE;
```

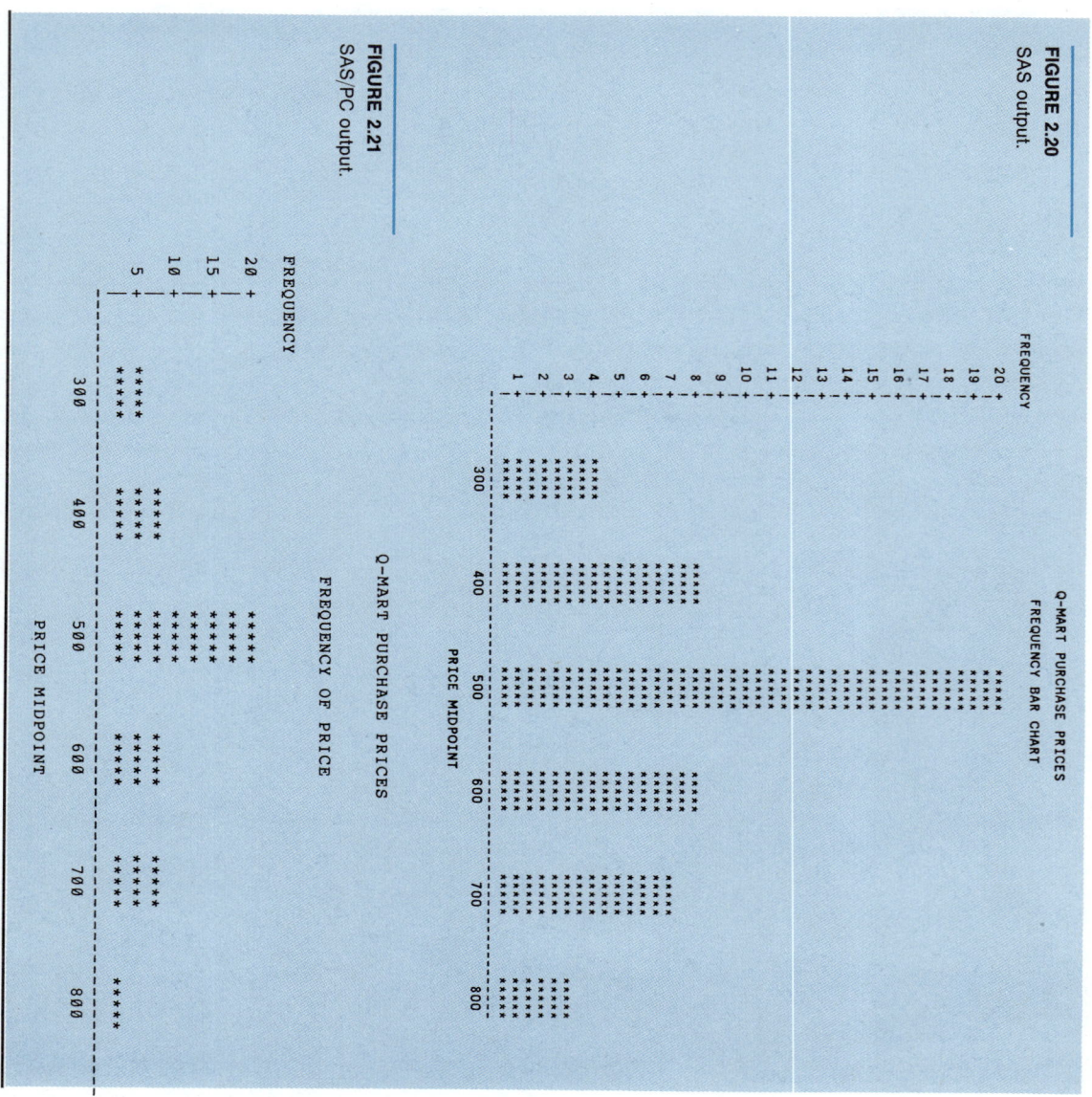

FIGURE 2.20
SAS output.

FIGURE 2.21
SAS/PC output.

CHAPTER

Descriptive Measures

A Look Back/Introduction

The first two chapters focused on different types of numerical sample data and methods of summarizing and presenting data. A frequency distribution is used to condense data from a sample into groups (called classes). Different types of statistical graphs can be used to illustrate sample data in different ways. The types of graphs we have discussed so far include the histogram, bar chart, ogive, frequency polygon, and pie chart. The purpose of these graphs is to convey information at a glance about the distribution of the values in your sample.

Every sample data set is a small part of a much larger population. Even if we don't always mention the word population, it is always there, since the sample values were selected from this group of interest. Every population has properties (called parameters) that describe it. By collecting a set of sample data, we can then estimate these properties by computing statistics and making graphs.

We have seen how to reduce a set of sample data to a graph. It also is helpful to reduce data to one or more numbers (such as an average). Such a number is called a descriptive measure. Because this number is derived from a sample, it also can be called a sample statistic. We discuss the commonly used descriptive measures and explain what you can expect to learn from each one. In later chapters we discuss how you can use many of these sample statistics to estimate the corresponding population parameters.

3.1

VARIOUS TYPES OF DESCRIPTIVE MEASURES

A **descriptive measure** is a *single* number that provides information about a set of sample data. The class of descriptive measures described here consists of four types. Which one you select depends on what you want to measure. These types are:

1. *Measures of central tendency.* These answer the questions Where is the "middle" of my data? and Which data value occurs most often?

2. *Measures of dispersion.* These answer the questions How spread out are my data values? and How much do the data values jump around?

3. *Measures of position.* These answer the questions How does my value (score on an exam, for example) compare with those of everyone else? and Which data value was exceeded by 75% of the data values? by 50%? by 25%?

4. *Measures of shape.* These answer the questions Are my data values symmetric? and If not symmetric, just how nonsymmetric (skewed) are the data?

3.2

MEASURES OF CENTRAL TENDENCY

The purpose of a **measure of central tendency** is to determine the "center" of your data values or possibly the "most typical" data value. Some measures of central tendency are the mean, median, midrange, and mode. We will illustrate each of these measures using as data the number of accidents (monthly) reported over a particular 5-month period:

accident data: 6, 9, 7, 23, 5

The Mean

The **mean** is the most popular measure of central tendency. It is merely the average of the data. The mean is easy to obtain and explain, and it has several mathematical properties that make it more advantageous to use than the other three measures of central tendency.

Business managers often use a mean to represent a set of values. They select one value as typical of the whole set of values, such as average sales, average price, average salary, or average production per hour. In economics, the term *per capita* is a measure of central tendency. The income per capita of a certain district, the number of clothes washers per capita, and the number of televisions per capita are all examples of a mean.

The sample mean, $\bar{x}$ (read "x bar"), is equal to the sum of the data values divided by the number of data values. For the accident data set,

$$\bar{x} = \frac{6 + 9 + 7 + 23 + 5}{5} = 10.0$$

In general, let an arbitrary data set be represented as:

$$x_1, x_2, x_3, \ldots, x_n$$

where n is the number of data values. (In the accident data set, $x_1 = 6$, $x_2 = 9$, $x_3 = 7$, $x_4 = 23$, $x_5 = 5$, and n is 5.) Then,

$$\bar{x} = \frac{x_1 + x_2 + \cdots + x_n}{n} = \frac{\Sigma x}{n} \qquad (3.1)$$

$\mu = \dfrac{\Sigma f(x)}{N} = \dfrac{60}{10} = 6$ *mean*

The symbol Σ (sigma) means "the sum of." In this case, the sample mean, $\bar{x}$, is the sum of the x values divided by n.* When dealing with discrete data, the sample mean is very often *not* an integer (such as 10, here) and should *not* be rounded to an integer. For example, remove the last value (5) from the accident data set. The sample mean is now

$\bar{X}$ - sample mean

$$\bar{x} = \frac{6 + 9 + 7 + 23}{4} = 11.25$$

In subsequent chapters, we will be concerned with the mean of the *population*. The symbol for the population mean is μ (mu). For a population consisting of N elements, denoted by

$$x_1, x_2, x_3, \ldots, x_N$$

the population mean is defined to be

$$\mu = \frac{x_1 + x_2 + \cdots + x_N}{N} = \frac{\Sigma x}{N} \qquad (3.2)$$

- Population: $\quad x_1, x_2, \ldots, x_N$

- Population mean $= \mu = \dfrac{x_1 + x_2 + \cdots + x_N}{N} = \dfrac{\Sigma x}{N}$

- Sample values (selected from the population): $\quad x_1, x_2, \ldots, x_n$, where $n \leq N$

- Sample mean $= \bar{x} = \dfrac{x_1 + x_2 + \cdots + x_n}{n} = \dfrac{\Sigma x}{n}$

The Median — positional average | 50%, 50% | Median

The **median** of a set of data is the value in the center of the data values when they are arranged from smallest to largest. Consequently, it is in the center of the ordered array.

Using the accident data set, the median **Md** is found by first constructing an ordered array:

$.50(5) + .50 = $ 3rd position

$$5, 6, 7, 9, 23$$

The value that has an equal number of items to the right and the left is the median. Thus, $\underline{Md = 7.}$

---

* In another application of this symbol, we square each of the sample values and sum these values. For the accident data, this operation would be written as

$$\Sigma x^2 = 5^2 + 6^2 + 7^2 + 9^2 + 23^2$$
$$= 25 + 36 + 49 + 81 + 529$$
$$= 720$$

For these data, then, $\Sigma x = 50$ and $\Sigma x^2 = 720.$

f	x	$f(x)$
1	2	2
2	4	8
3	6	18
4	8	32
10		60

In general, if n is *odd*, Md is the center data value of the ordered set:

$$Md = \left(\frac{n+1}{2}\right) \text{st ordered value}$$

Here, the median is the $(5 + 1)/2 = $ 3rd value in the ordered array. Note that for these data, the *position* of the median is 3, and the *value* of the median is 7. If n is *even*, Md is the average of the two center values of the ordered set. Thus, the median of the array 3, 8, 12, 14 is $(8 + 12)/2 = 10.0$.

In our accident data set, one of the five values (23) is much larger than the remaining values—it is an outlier. Notice that the median (Md = 7) was much less affected by this value than was the mean ($\bar{x} = 10$). When dealing with data that are likely to contain outliers (for example, personal incomes or prices of residential housing), the median usually is preferred to the mean as a measure of central tendency, since the median provides a more "typical" or "representative" value for these situations.

The Midrange

Although less popular than the mean and median, the **midrange (Mr)** provides an easy-to-grasp measure of central tendency. Notice that it also is severely affected (even more than $\bar{x}$) by the presence of an outlier in the data. In general:

$$Mr = \frac{(\text{smallest value}) + (\text{largest value})}{2} \tag{3.3}$$

Using the accident data set,

$$Mr = \frac{5 + 23}{2} = 14.0$$

Compare this to $\bar{x} = 10$ and Md = 7.

The Mode

The **mode (Mo)** of a data set is the value that occurs the most often. The mode is not always a measure of central tendency; this value need not occur in the "center" of your data. One situation in which the mode is the value of interest is the manufacturing of clothing. The *most common* hat size is what you would like to know, not the *average* hat size. Can you think of other applications where the mode would provide useful information?

Note that there is no mode for our accident data set because all values occur only once. Instead, consider the data set

4, 8, 7, 6, 9, 8, 10, 5, 8

Mo = 8 (occurs three times).

There may be more than one mode if several numbers occur the same (and the largest) number of times.

A sample of ten was taken to determine the typical completion time (in months) for the construction of a particular model of Brockwood Homes:

4.1, 3.2, 2.8, 2.6, 3.7, 3.1, 9.4, 2.5, 3.5, 3.8

EXAMPLE 3.1

FIGURE 3.1

Dot array diagram of the measure of central tendency for a sample of ten housing construction times. See text for explanation.

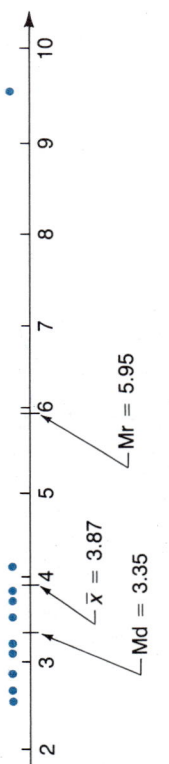

Construction time (months)

We find the average completion time as follows:

$$\bar{x} = \frac{4.1 + 3.2 + \cdots + 3.8}{10} = \frac{38.7}{10} = 3.87 \text{ months}$$

Notice that there is an outlier in the data, namely, 9.4 months. To be safe, you should double-check this figure to make sure that it is, in fact, correct; that is, that there was no mistake in recording or transcribing this value. In the presence of one or two outliers, the median generally provides a more reliable measure of central tendency, so we construct an ordered array:

2.5, 2.6, 2.8, 3.1, **3.2**, **3.5**, 3.7, 3.8, 4.1, 9.4

Consequently,

$$Md = \frac{3.2 + 3.5}{2} = 3.35 \text{ months}$$

Also, the midrange is given by

$$Mr = \frac{2.5 + 9.4}{2} = 5.95 \text{ months}$$

This value is severely affected by the presence of the outlier; the midrange of nearly 6 months is a poor measure of central tendency for this application.

Finally, no mode exists because there are no repeats in the data values. These results are summarized in the graph in Figure 3.1, a **dot array diagram**. Each data value is represented as a dot on the horizontal line.

EXERCISES

3.1 The appraised value of 20 selected new homes, all of equal square footage, in a suburb of Birmingham, Alabama, is given in thousands of dollars:

115, 121, 118, 113, 115, 117, 120, 112, 116, 119, 118, 116, 119, 118, 115, 118, 120, 118, 119, 116

Compute the mean, median, mode, and midrange for the appraised value of the 20 homes. Should the values of the mean, median, mode, and midrange be very close?

3.2 Compute the mean, median, mode, and midrange for the price:earnings-ratio data in Exercise 2.15.

3.3 Compute the mean, median, mode, and midrange for the profitability-index data in Exercise 2.20.

3.4 Compute the mean, median, mode, and midrange for the price elasticity of electricity in Exercise 2.21.

3.5 The distribution of family income (in dollars) in the United States in 1985 was as follows:

INCOME RANGE	PERCENT DISTRIBUTION OF FAMILIES
under 5,000	8.3
5,000 and under 10,000	17.0
10,000 and under 15,000	14.9
15,000 and under 20,000	12.1
20,000 and under 25,000	11.3
25,000 and under 35,000	16.0
35,000 and under 50,000	12.5
50,000 and over	8.1
Total	100.0

(*Source: Statistical Abstract of the United States 1987*, U.S. Department of Commerce. p. 436.)

Estimate the median. Give an interpretation of the value of the median.

3.6   Ratings measure a percentage of the audience tuned to a particular station and are used by advertisers when they place orders for commercials. The following data, released by Arbitron Rating Co., pertains to the period April 3, 1986, to June 25, 1986, for radio stations in a major metropolitan area:

NUMBER	STATION	RATING
1	KVIL-FM	8.8
2	KKDA-FM	8.5
3	KPLX-FM	7.3
4	WBAP-AM	6.9
5	KRLD-AM	6.3
6	KMEZ-FM	6.3
7	KSCS-FM	5.7
8	KEGL-FM	5.2
9	KTXQ-FM	5.0
10	KQZY-FM	3.8

(*Source: Dallas Morning News*, July 22, 1986.)

Compute the mean, median, and mode for the ratings.

3.7   Compute the mean, median, and mode for the return-on-sales (ROS) data in Exercise 2.27.

3.8   Give an appropriate measure of central tendency for the following data on the monthly commissions (in hundreds of dollars) for eight salespersons:

0.5, 12.3, 15.9, 16.1, 16.2, 16.3, 16.4, 18.6

3.9   Compute the mean, median, and mode for the daily advertising expense of a car dealer using the following data. The daily advertising expense in dollars is given for 20 days. Which measure of central tendency is most appropriate? Why?

38, 60, 20, 130, 55, 150, 47, 35, 86, 95, 31, 46, 112, 130, 55, 42, 130, 35, 60, 130

3.10   Moody's bond yield average on medium grade bonds is given below for the 12 months of 1986 as a percentage:

MONTH	YIELD ON MEDIUM GRADE BONDS	MONTH	YIELD ON MEDIUM GRADE BONDS
Jan	10.79	July	9.39
Feb	10.47	Aug	9.28
Mar	9.86	Sept	9.30
Apr	9.69	Oct	9.38
May	9.62	Nov	9.28
Jun	9.57	Dec	9.15

(*Source: Moody's Bank and Finance Manual*, 1987, p. a13.)

Calculate the mean and median of the yields. Interpret the meaning of these two statistics.

3.11 Outstanding external debt of Third World nations totaled over $1 trillion at the end of 1986, according to a survey by the World Bank. The following are the leading debtor nations at the beginning of 1987.

COUNTRY	DEBT (in billions of dollars)
Brazil	107.8
Mexico	102.0
Argentina	53.0
Indonesia	43.9
Venezuela	34.1
Philippines	28.1

COUNTRY	DEBT (in billions of dollars)
Nigeria	22.1
Chile	21.2
Peru	14.7
Colombia	14.7
Ecuador	9.1

(*Source: World Almanac and Book of Facts, 1988, p. 95.*)

a. Calculate the mean and median of the debt of the 11 countries.

b. Omit Brazil's debt and recalculate the mean and median of the remaining ten nations.

c. Which value, the mean or median, is affected more by the omission of Brazil's debt? Which measure do you think will be more easily affected by the omission of other debt values?

3.3
MEASURES OF DISPERSION

A measure of central tendency, such as the mean, is certainly useful. However, the use of any single value to describe a complete distribution fails to reveal important facts.

The more homogenous a set of data is, the better the mean will represent a "typical value." **Dispersion** is the tendency of data values to scatter about the mean, $\bar{x}$. If all the data values in a sample are identical, then the mean provides perfect information, and the dispersion is zero. This is rarely the case, however, so we need a measure of this dispersion that will increase as the scatter of the data values about $\bar{x}$ increases.

Knowledge of dispersion can sometimes be used to control the variability of your data values in the future. Industrial production operations maintain quality control by observing and measuring the dispersion of the units produced. If there is too much variation in the production process, the causes are determined and corrected using an inspection control procedure.

Some measures of dispersion are the range, mean absolute deviation, variance, standard deviation, and coefficient of variation. To illustrate the various dispersion measures, we will use the accident data from the previous section: 6, 9, 7, 23, 5.

The Range

The simplest measure of dispersion is the **range** of the data, which is the numerical difference between the largest value and the smallest value. For the accident data,

$$\text{range} = 23 - 5 = 18$$

The range is a rather crude measure of dispersion, but it is an easy number to calculate and contains valuable information for many situations. Stock reports generally give prices in terms of their ranges, citing the high and low prices of the day. The value of the range is strongly influenced by an outlier in the sample data.

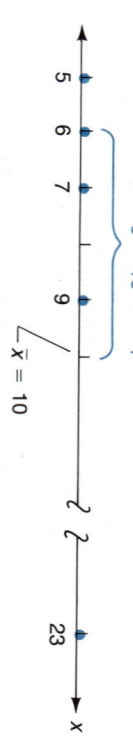

$$x - \bar{x} = 6 - 10 = -4$$

$\bar{x} = 10$

FIGURE 3.2

This presentation of the accident data shows their variation.

Mean Absolute Deviation (MAD)

The purpose of a measure of dispersion is to determine the variability of your data. The more variation there is in your data, the larger this measure should become. Take a look at the accident data illustrated in Figure 3.2. To measure the variation about the sample mean, $\bar{x}$, consider the distance from each data value to $\bar{x}$ (that is, $x - \bar{x}$) and its absolute value:

DATA VALUE (x)	$x - \bar{x}$	$|x - \bar{x}|$
5	−5	5
6	−4	4
7	−3	3
9	−1	1
23	13	13
	$\Sigma(x - \bar{x}) = 0$	$\Sigma|x - \bar{x}| = 26$

As a possible measure, consider the average of the $(x - \bar{x})$ values:

$$\frac{\Sigma(x - \bar{x})}{5} = \frac{0}{5} = 0$$

This value is *always* zero for any set of data because the positive deviations from the sample mean always balance out the negative ones. To overcome this, use the actual distance from each data value to the sample mean, paying no attention to the side of the mean on which it lies, by taking the **absolute value** of each deviation. The **mean absolute deviation (MAD)** is the average of these distances.

$$\boxed{\text{MAD} = \frac{\Sigma|x - \bar{x}|}{n}} \qquad (3.4)$$

Using the accident data,

$$\text{MAD} = \frac{5 + 4 + 3 + 1 + 13}{5} = \frac{26}{5} = 5.2$$

What is the MAD without the value 23?

$$\bar{x} = \frac{5 + 6 + 7 + 9}{4} = 6.75$$

and so

$$\text{MAD} = \frac{|5 - 6.75| + |6 - 6.75| + |7 - 6.75| + |9 - 6.75|}{4}$$

$$= \frac{1.75 + .75 + .25 + 2.25}{4} = 1.25$$

Here the MAD is much lower than before. This indicates that the smaller data set has much less variation than does the one containing the outlier.

The Variance and Standard Deviation

By far the most widely used measures of dispersion are the **variance** and **standard deviation**. They resemble the MAD in that they are based on deviations of all the values from the sample mean, $\bar{x}$. The problem encountered earlier in examining the sum of each $(x - \bar{x})$ was that the negative deviations balanced out the positive ones. The MAD handled this situation by taking the absolute value of each deviation. Another possibility is to *square* each of these deviations, thereby removing all the negative signs. We can illustrate this using our example data. Recall that $\bar{x} = 10$.

DATA VALUE (x)	$(x - \bar{x})$	$(x - \bar{x})^2$
5	-5	25
6	-4	16
7	-3	9
9	-1	1
23	13	169
	$\Sigma(x - \bar{x}) = 0$	$\Sigma(x - \bar{x})^2 = 220$

So, $\Sigma(x - \bar{x})^2 = 220$.

The obvious thing to do next would be to find the average of these squared deviations:

$$(1/n) \cdot \Sigma(x - \bar{x})^2$$

One use of this particular statistic in subsequent chapters is as an *estimator*. In particular, we will need to estimate the variation within an entire population, using sample data collected from the population. However, a better estimator is obtained by dividing the sum of the squared deviations by $n - 1$ rather than by n. This leads to the **sample variance, s^2**. In general,

$$s^2 = \frac{\Sigma(x - \bar{x})^2}{n - 1} \qquad (3.5)$$

Using the accident data,

$$s^2 = \frac{220}{5 - 1} = \frac{220}{4} = 55.0$$

The square root of the variance is referred to as the sample standard deviation, s. In general,

$$s = \sqrt{\frac{\Sigma(x - \bar{x})^2}{n - 1}} \qquad (3.6)$$

Using the accident data,

$$s = \sqrt{55.0} = 7.416$$

As previously mentioned, the sample variance, s^2, is used to estimate the variance of the entire population. The symbol for the population variance is σ^2 (read as sigma squared). For a population consisting of N elements,

$$x_1, x_2, x_3, \ldots, x_N$$

the population variance is defined to be

$$\sigma^2 = \frac{\Sigma(x - \mu)^2}{N} \tag{3.7}$$

where μ is the population mean, defined in equation 3.2.

As we saw, the *population* variance can be obtained by dividing the sum of the squared deviations about μ by the population size N. The *sample* variance is calculated by dividing the sum of the squared deviations about $\bar{x}$ by the sample size (n) minus one. Had we chosen to divide by n rather than by $n - 1$, the resulting estimator would (on the average) underestimate σ^2. For this reason, we use $n - 1$ in the denominator of s^2.

- Population: $x_1, x_2, \ldots, x_N$

- Population variance $= \sigma^2 = \dfrac{(x_1 - \mu)^2 + \cdots + (x_N - \mu)^2}{N}$

$$= \frac{\Sigma(x - \mu)^2}{N}$$

- Population standard deviation $= \sigma = \sqrt{\dfrac{\Sigma(x - \mu)^2}{N}}$

- Sample values (selected from the population): $x_1, x_2, \ldots, x_n$, where $n \le N$

- Sample variance $= s^2 = \dfrac{(x_1 - \bar{x})^2 + \cdots + (x_n - \bar{x})^2}{n - 1}$

$$= \frac{\Sigma(x - \bar{x})^2}{n - 1}$$

- Sample standard deviation $= s = \sqrt{\dfrac{\Sigma(x - \bar{x})^2}{n - 1}}$

Now consider what the units of measurement are for s and s^2. The units on s are the same as the units on the data. If the data are measured in pounds, the units on s are pounds. Consequently, the units on the variance, s^2, would be (pounds)^2—a rather difficult unit to grasp at best.

For the accident data,

$$s = 7.416 \text{ accidents}$$
$$s^2 = 55 \text{ (accidents)}^2$$

For this reason, s (rather than s^2) is typically the preferred measure of dispersion.

There is another way to compute the sample variance. Using equation 3.5 to compute the value of s^2 may have appeared easy enough, but this was helped in part by the fact that the sample mean, $\bar{x}$, was an integer (10). When $\bar{x}$ is not an integer, this is not the easiest way to find s^2. Instead, use

$$s^2 = \frac{\Sigma x^2 - (\Sigma x)^2/n}{n - 1} \tag{3.8}$$

As before, the standard deviation is the square root of the variance. To illustrate the use of equation 3.8, consider the accident data:

x	x^2
5	25
6	36
7	49
9	81
23	529
$\overline{50}$	$\overline{720}$

So, $n = 5$, $\Sigma x = 50$, and $\Sigma x^2 = 720$. Consequently, using equation 3.8:

$$s^2 = \frac{720 - (50)^2/5}{5 - 1}$$

$$= \frac{720 - 500}{4} = 55.0 \quad \text{(as before)}$$

Also

$$s = \sqrt{55.0} = 7.416 \quad \text{(as before)}$$

Finally, you may wish to interpret the magnitude of the value of s or s^2—that is, whether your value of s (or s^2) is large or not. This is difficult to determine because the values of s and s^2 depend on the magnitude of the data values. In other words, large data values generally lead to large values of s. For example, which of the following two data sets exhibits more variation?

data set 1:   5, 6, 7, 9, 23   (accident reports)

data set 2:   5000, 6000, 7000, 9000, 23,000

As we have already seen, for data set 1, $\overline{x} = 10.0$ and $s = 7.416$. For data set 2, $\overline{x} = 10,000$ and $s = 7416$ (we will discuss this later).

Does this mean that data set 2 has a great deal more variation, given that its standard deviation is 1000 times that of data set 1? Another look at the values reveals that the large value of s for data set 2 is due to the large values within this set. In fact, considering the size of the numbers within each data set, the *relative* variation within each group of values is the same. So comparing the standard deviations or variances of two data sets is not a good idea unless you know that their mean values ($\overline{x}$) are approximately equal. The next section deals with another statistical measure that will allow you to compare the relative variation within two data sets.

The Coefficient of Variation

Consider again our two data sets that appear to have the same variation (relative to the size of the data values) yet have vastly different standard deviations. These data sets are

data set 1:   5, 6, 7, 9, 23   ($\overline{x} = 10$, $s = 7.416$)

data set 2:   5000, 6000, 7000, 9000, 23,000   ($\overline{x} = 10,000$, $s = 7416$)

To compare their variation, we need a measure of dispersion that will produce the same value for both of them. The solution here is to measure the standard deviation in terms of the mean; that is, what percentage of $\overline{x}$ is s? This is the

coefficient of variation, CV. In general,

$$CV = \frac{s}{\bar{x}} \cdot 100 \qquad (3.9)$$

For our example data sets:

data set 1: $\quad CV = \dfrac{7{,}416}{10} \cdot 100 = 74.16$

data set 2: $\quad CV = \dfrac{7{,}416}{10{,}000} \cdot 100 = 74.16$

So our conclusion here is that both data sets exhibit the same relative variation; s is 74.16% of the mean for both sets.

EXAMPLE 3.2

To review the various measures of dispersion, use the data on housing construction time that you used in Example 3.1.

First, compute the range:

completion time: 4.1, 3.2, 2.8, 2.6, 3.7, 3.1, 9.4, 2.5, 3.5, 3.8 (months)

(largest value) − (smallest value) = 9.4 − 2.5 = 6.9 months

To determine the MAD, recall that $\bar{x}$ is 3.87 months:

$$\begin{aligned} MAD &= (1/10)(|4.1 - 3.87| + |3.2 - 3.87| + \cdots + |3.8 - 3.87|) \\ &= (1/10)(11.52) = 1.152 \text{ months} \end{aligned}$$

Now find the variance and the standard deviation:

$$\sum x = 4.1 + 3.2 + \cdots + 3.8 = 38.7$$

and

$$\sum x^2 = (4.1)^2 + (3.2)^2 + \cdots + (3.8)^2 = 186.25$$

Hence,

$$\begin{aligned} s^2 &= \frac{186.25 - (38.7)^2/10}{10 - 1} \\ &= \frac{186.25 - 149.77}{9} = 4.05 \text{ (months)}^2 \end{aligned}$$

and

$$s = \sqrt{4.05} = 2.01 \text{ months}$$

To calculate the coefficient of variation, use the previously obtained values of s and $\bar{x}$, where

$$CV = \frac{2.01}{3.87} \cdot 100 = 51.9$$

The standard deviation is 51.9% of the sample mean.

So far, you can reduce a set of sample data to a number that indicates a typical or average value (a measure of central tendency) or one that describes the amount of variation within the data values (a measure of dispersion). The next section examines yet another set of statistics—measures of position.

EXERCISES

3.12 When making capital-investment decisions, firms frequently consider the dispersion of the estimated future cash flows. Usually, the project with lesser dispersion is preferred to the one with more dispersion. Given the estimated cash flows (after tax) for projects A and B, which one would you prefer? Why?

MONTH	PROJECT A CASH FLOW	PROJECT B CASH FLOW	MONTH	PROJECT A CASH FLOW	PROJECT B CASH FLOW
January	4,000	700	July	10,800	710
February	7,200	1,100	August	9,100	450
March	8,800	600	September	2,000	580
April	2,400	1,300	October	14,000	640
May	7,400	800	November	7,700	330
June	4,100	650	December	3,900	210

3.13 The following are the scores made on an aptitude test by a group of job applicants:

53, 55, 43, 14, 64, 39, 65, 22, 17, 74, 36, 24, 13, 28, 40, 96, 92, 32, 92, 36, 18, 100, 84, 65

Calculate the:

a. Range.

b. Mean absolute deviation.

c. Variance.

d. Standard deviation.

e. Coefficient of variation.

3.14 The surplus or deficit in the U.S. trade with each of its top trading partners in manufactured goods is given in billions of dollars for the year 1985.

COUNTRY	SURPLUS OR DEFICIT	COUNTRY	SURPLUS OR DEFICIT
Canada	-8.2	Saudi Arabia	+3.7
Japan	-59.2	South Korea	-6.9
Mexico	+1.7	Italy	-5.9
United Kingdom	-1.1	Singapore	-.8
West Germany	-12.8	Taiwan	-14.4
France	-3.4	Brazil	-2.5
Australia	+4.3	Hong Kong	-6.8
The Netherlands	+1.4		

(*Source:* Bryan Berry, "U.S. Battles the World With More Exports and Offshore Manufacturing," *Iron Age 3,* no. 2 (February 1987): 17–23.)

a. Calculate the mean surplus or deficit for the 15 countries given.

b. Recalculate the mean surplus or deficit for the same groups of countries, but omit Japan's deficit. Compare this value to the mean in part (a).

c. Calculate the standard deviation of the surplus or deficit figure given for the 15 countries.

d. Omit the value for Japan's deficit and calculate the standard deviation of the remaining 14 countries.

e. Compare your answers in parts (c) and (d) and comment on the difference.

3.15 For the price:earnings-ratio data in Exercise 2.15, calculate the variance and standard deviation. Is the standard deviation always smaller than the variance?

3.16 The following values represent years of experience of 12 auditors in a small accounting firm.

2.3, 4.1, 7.3, 5.2, 5.8, 3.6, 11.6, 5.0, 4.7, 3.2, 5.5, 4.5

a.   Calculate the standard deviation.

b.   Which one value in the data set, if omitted, would change the value of standard deviation the most?

c.   Which one value in the data set, if omitted, would change the value of standard deviation the least?

d.   Compare your answers in parts (b) and (c) and comment on their values.

3.17   Show that

is equivalent to

$$s^2 = \frac{\sum x^2 - (\sum x)^2/n}{n-1}$$

3.18   If a data set consists of five observations and if the variance of the observations in the data set is zero, what can we say about the value of each of the five observations?

3.19   The game of cricket, which originated in England, is popular in Australia, India, and the West Indies. The following are the runs (strikes) scored by two players, A and B, in various innings:

PLAYER A	PLAYER B
47	66
0	10
14	11
33	22
101	88
68	32
87	40
14	38
22	18
46	41
Total 432	366

a.   Who is the better player? On what basis?

b.   Who is more consistent? Why?

3.20   The yield on a stock is equal to dividend per share/market price × 100. It is one of the popular measures used by individual investors when they make investment decisions. The following are the yield figures (in percent) for two stocks, A and B, for the past 10 years:

STOCK A	STOCK B	STOCK A	STOCK B
7	12	18	9
9	13	7	18
14	13	11	12
13	17	12	17
11	11	10	13

Which is a more stable stock in terms of yields? Why?

3.21   Calculate the variance for the daily advertising expense of the car dealer in Exercise 3.9.

3.4
MEASURES
OF POSITION

Suppose that you think you are drastically underpaid as compared with other people with similar experience and performance. One way to attack the problem is to obtain the salaries of these other employees and demonstrate that *comparatively* you are way down the list. To evaluate your salary as compared with

TABLE 3.1

Ordered Array of
Aptitude Test Scores
for 50 Applicants
($\bar{x} = 60.36$, $s = 18.61$)

22	44	56	68	78
25	44	57	68	78
28	46	59	69	80
31	48	60	71	82
34	49	61	72	83
35	51	63	72	85
39	53	63	74	88
39	53	63	75	90
40	55	65	75	92
42	55	66	76	96

the entire group, you would use a measure of position. **Measures of position** are indicators of how a particular value fits in with all the other data values. Two commonly used measures of position are (1) a percentile (and quartile), and (2) a Z score.

To illustrate these measures, we suppose that the personnel manager of Texon Industries has administered an aptitude test to 50 applicants. The ordered data are shown in Table 3.1. The mean of the data is $\bar{x} = 60.36$, and the standard deviation is $s = 18.61$. Ms. Jenson received the score of 83. She wishes to measure her performance in relation to all the applicant scores. We will return to this illustration in Example 3.3.

Percentiles

A **percentile** is the most common measure of position. The value of, for example, the 35th percentile is essentially the value that exceeds 35% of all the data values. More precisely, the 35th percentile is that value (say, P_{35}) such that at most 35% of the data values are less than P_{35} and at most 65% of the data values are greater than P_{35}. We will use the Texon Industries applicant data to determine the 35th percentile. Which data value is 35% of the way between the smallest and largest value? Here the number of data values is $n = 50$ and the percentile is $P = 35$. We define the *position* of the 35th percentile as follows:

$$n \cdot \frac{P}{100} = 50 \cdot .35 = 17.5$$

To satisfy the more precise definition of a percentile, whenever $n \cdot P/100$ is *not* a counting number, it should be rounded *up* to the next counting number. So, 17.5 is rounded up to 18, and the 35th percentile is the 18th value *of the ordered values.* Referring to Table 3.1, the 35th percentile is $P_{35} = 53$.

In general, to find the **location** of the Pth percentile, determine $n \cdot P/100$ and use one of the following two location rules.

Location rule 1. If $n \cdot P/100$ *is not* a counting number, round it up, and the Pth percentile will be the value in this position of the ordered data.

Location rule 2. If $n \cdot P/100$ *is* a counting number, the Pth percentile is the average of the number in this location (of the ordered data) and the number in the next largest location.

Now we can use the applicant data to determine the 40th percentile. Here $n \cdot P/100 = (50)(.4) = 20$. Then, using the second rule,

$$P_{40} = \text{40th percentile} = \frac{(\text{20th value}) + (\text{21st value})}{2}$$

$$= \frac{55 + 56}{2} = 55.5$$

EXAMPLE 3.3

SOLUTION

Recall that Ms. Jenson received a score of 83. What is her percentile value? Her value is the 45th largest value (out of a total of 50). An initial guess of the percentile here would be:

$$P = \frac{45}{50} \cdot 100 = 90$$

However, due to the percentile rules used here, this may be slightly incorrect. Your next step should be to examine this value of P, along with the next two smaller values. The following calculations of $P = 88$, $P = 89$, and $P = 90$ reveal that Ms. Jenson's score is the 89th percentile.

P	$n \cdot P/100$	Pth PERCENTILE
88	$50 \cdot .88 = 44$	$(82 + 83)/2 = 82.5$
89	$50 \cdot .89 = 44.5$	45th value = 83
90	$50 \cdot .90 = 45$	$(83 + 85)/2 = 84$

EXAMPLE 3.4

What is the 50th percentile for the applicant data in Table 3.1?

SOLUTION

Here, $n \cdot P/100 = 50 \cdot .5 = 25$. The 50th percentile is an average of the 25th and 26th ordered data values:

$$P_{50} = 50\text{th percentile} = \frac{61 + 63}{2} = 62$$

Quartiles

Quartiles are merely particular percentiles that divide the data into quarters, namely:

$Q_1 = $ 1st quartile = 25th percentile (P_{25})

$Q_2 = $ 2nd quartile = 50th percentile = median (P_{50})

$Q_3 = $ 3rd quartile = 75th percentile (P_{75})

They are used as benchmarks, much like the use of A, B, C, D, and F on examination grades. Using the applicant data in Table 3.1, we can determine the first quartile by first calculating:

$$n \cdot \frac{P}{100} = 50 \cdot .25 = 12.5$$

This is rounded up to 13, and $Q_1 = $ 13th ordered value = 46.

$$Q_2 = \text{median} = 62$$

from Example 3.4. Finally,

$$n \cdot \frac{P}{100} = 50 \cdot .75 = 37.5$$

This is rounded up to 38, and $Q_3 = $ 38th ordered value = 75.

Another measure commonly used in conjunction with quartiles is the **inter-quartile range (IQR)**, defined as

$$IQR = Q_3 - Q_1$$

In the applicant data, the interquartile range is

$$IQR = 75 - 46 = 29$$

Consequently, the middle 50% of the data are between 29 and 75.

Strictly speaking, the interquartile range is a measure of dispersion since it can be expected to increase as the data become more "spread out." It is not a commonly used measure of dispersion, although it is certainly easy to compute (much like the range of a sample data set). Its primary disadvantage is that it measures the spread within the middle of the data, not within the entire data set. A graphical illustration of the interquartile range is contained in a simple graph, called a box and a whisker plot, discussed in Section 3.8.

Z Scores

Another measure of position is a sample **Z score**, which is based on the mean ($\bar{x}$) and standard deviation of your data set. As with percentiles, a Z score determines the relative position of any particular data value x expressed in terms of the number of standard deviations above or below the mean. The Z score of x is defined as

$$Z = \frac{x - \bar{x}}{s} \tag{3.10}$$

Recall from Example 3.3 that Ms. Jenson had a score of 83 on the test. For this data set, $\bar{x} = 60.36$ and $s = 18.61$. Her score of 83 is in the 89th percentile. The corresponding Z score is

$$Z = \frac{83 - 60.36}{18.61} = 1.22$$

This means that Ms. Jenson's score of 83 is 1.22 standard deviations to the *right* of the mean, or above the group's average. Thus, if Z is positive, it indicates how many standard deviations x is to the right of the mean.

A negative value implies that x is to the *left* of the mean. Again referring to Table 3.1, what is the Z score for the individual who obtained a total of 35 on the aptitude examination?

$$Z = \frac{35 - 60.36}{18.61} = -1.36$$

This individual's score is 1.36 standard deviations to the left of the mean, or below the group's average.

When you derive sample statistics, you have essentially two options: Use a calculator or use a computer. Calculators work well for small data sets but involve too much time (and opportunity for error) for moderate or large sample sizes. Practically all statistical computer packages will provide you with the basic sample statistics (mean, median, variance, and so on) in response to only a few commands once the data have been read in. Figure 3.3 contains the MINITAB commands (along with the output) necessary to derive the basic statistics for the data in Table 3.1. The appendices at the end of the chapter demonstrate this procedure using SPSS and SAS.

FIGURE 3.3

MINITAB procedure for describing sample data

```
MTB > SET INTO C1
DATA> 22 25 28 31 34 35 39 39 40 42 44 44 46 48 49
DATA> 51 53 53 55 55 55 56 57 59 60 61 63 63 65 66
DATA> 68 68 69 71 72 72 74 75 75 76 78 78 80 82 83
DATA> 85 88 90 92 96
DATA> END
MTB > DESCRIBE C1
```

```
                      x̄                               s
C1     N     MEAN     MEDIAN    TRMEAN    STDEV    SEMEAN
       50    60.36    62.00     60.57     18.61    2.63

       MIN      MAX      Q1       Q3
C1     22.00    96.00    45.50    75.00
```
⌐ 1st quartile   └── 3rd quartile

3.5

MEASURES OF SHAPE

A basic question in many applications is whether your data exhibit a **symmetric** pattern. **Measures of shape** determine skewness and kurtosis.

Skewness

The histogram in Figure 3.4 demonstrates a perfectly symmetric distribution. When the data are symmetric, the sample mean, x̄, the sample median, Md, and the sample mode, Mo, are the same. As the data tend toward a nonsymmetric distribution, referred to as **skewed**, the mean and median drift apart. The easiest method of determining the degree of skewness present in your sample data is to calculate a measure referred to as the **Pearsonian coefficient of skewness, Sk.** Its value is given by:

$$Sk = \frac{3(\bar{x} - Md)}{s} \qquad (3.11)$$

where s is the standard deviation of the sample data.

The value of Sk ranges from −3 to 3. If the data are perfectly symmetric (a rare event), Sk = 0, because x̄ = Md. For Figure 3.4, Sk is zero. If Sk is positive, the mean is larger than the median, and we say that the data are *skewed right*. This merely means the data exhibit a pattern with a right tail, as illustrated in Figure 3.5. We know the mean is affected by extreme values, so we would expect the mean to move toward the right tail, above the median. This results in a positive value of Sk. Similarly, if Sk is negative, the data are *skewed left* and the mean is smaller than the median. Figure 3.6 shows a data distribution exhibiting a left tail and negative skew.

Using the aptitude examination scores in Table 3.1, we have x̄ = 60.36, s = 18.61, and Md = 62.

$$Sk = \frac{3(60.36 - 62)}{18.61} = -.26$$

Consequently, a histogram of these data should be just slightly skewed left.

Kurtosis

Sk measures the tendency of a distribution to stretch out in a particular direction. Another measure of shape, referred to as the **kurtosis,** measures the *peakedness* of your distribution. The calculation of this measure is a bit cumbersome,

3(Mean − Median)
Stand. Dev.

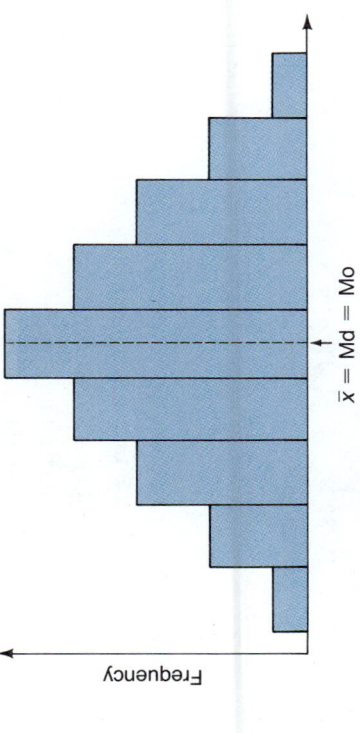

FIGURE 3.4

Histogram constructed with symmetric data. The mean, median, and mode are equal.

$\bar{x} = Md = Mo$

Frequency

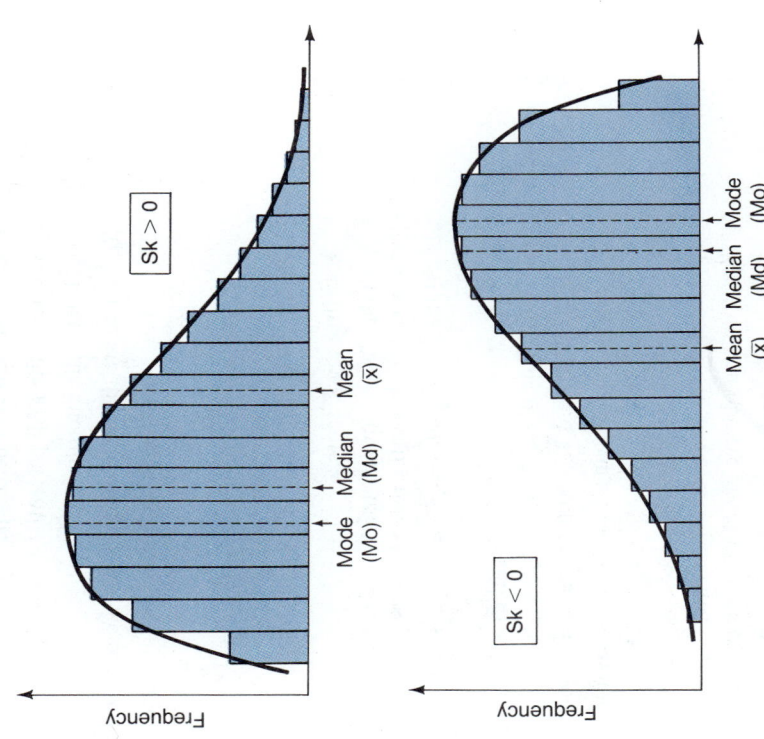

FIGURE 3.5

Histogram showing right (positive) skew

$Sk > 0$

Mean
$(\bar{x})$

Median
(Md)

Mode
(Mo)

Frequency

FIGURE 3.6

Histogram showing left (negative) skew

$Sk < 0$

Mean Median Mode
$(\bar{x})$ (Md) (Mo)

Frequency

and the kurtosis value is not needed in the remaining text material.* Briefly, this value is small if the frequency of observations close to the mean is high and the frequency of observations far from the mean is low.

EXERCISES

3.22 The following table indicates the scores of the students at Hillside College on a statistics examination:

18, 58, 71, 83, 89, 96, 21, 62, 74, 84, 90, 97, 43, 66, 75, 86, 92, 98, 47, 66,
77, 86, 94, 100, 55, 68, 78, 88, 95, 100

* The following texts contain an alternate method of computing sample skewness as well as a procedure for computing the sample kurtosis; L. Ott and D. K. Hildebrand, *Statistical Thinking for Managers*, 2d ed. (Boston: Duxbury Press, 1987); C. L. Olsen, and M. J. Picconi, *Statistics for Business Decision Making* (Glenview, Ill.: Scott, Foresman, 1983).

a. Calculate the 40th percentile

b. Calculate the 77th percentile.

c. Interpret the meaning of the numbers calculated for the two percentiles above.

d. Calculate the interquartile range and explain what the value means.

3.23 In Exercise 3.22, how would you evaluate the performance of a student who scored 74 on the exam?

3.24 Using the data in Exercise 3.22, calculate the first and third quartiles. State the result in words.

3.25 Consider the following data:

7, 8, 8, 9, 11, 14, 15, 16, 18, 19, 21 27, 28, 30, 32, 35

a. Calculate the 90th percentile.

b. Calculate the 58th percentile.

c. Calculate the interquartile range.

3.26 Calculate the three quartiles for the data in Exercise 3.25.

3.27 Assume a particular set of sample data such that

$$\bar{\bar{x}} = 49$$
$$s = 18$$

Consider one particular value of $x = 63$.

a. Calculate the Z score.

b. Interpret the Z score.

3.28 If the Z score for an observation is 1.50, the standard deviation is 14, and the observation is 32, what is the mean?

3.29 If the mean is 40, $x = 35$, and its Z score is -2, what is the value of the standard deviation?

3.30 If the Z score for an observation is -1.22, $\bar{x} = 83$ and the variance is 84, what is the value of the observation?

3.31 If the Z score is 1.78 and the variance is 64, what is $x - \bar{x}$?

3.32 Assume the following about a set of sample data:

$$s = 14$$
$$\bar{x} = 21$$
$$Md = 18.5$$

a. What do you observe about the pattern of the data?

b. Calculate the coefficient of skewness. What does this value suggest?

3.33 The GNP is the market value of all final goods and services produced by an economy in a given time period. It is probably the most important indicator of economic health of a country. The GNPs (in billions of constant 1982 dollars) of the United States for the years 1970 through 1986 were as follows:

YEAR	GNP	YEAR	GNP
1970	2416.2	1979	3192.4
1971	2484.8	1980	3187.1
1972	2608.5	1981	3248.8
1973	2744.1	1982	3166.0
1974	2729.3	1983	3279.1
1975	2695.0	1984	3489.9
1976	2826.7	1985	3585.2
1977	2958.6	1986	3676.5
1978	3115.2		

(Source: Economic Report of the President, January 1987.)

a. Calculate the mean, median, variance, and standard deviation of the GNP values.
b. Calculate the coefficient of skewness. Are the data skewed to the left?

3.6

INTERPRETING $\bar{x}$ AND s

Now that you have gone through several pencils determining the sample mean and standard deviation, what can you learn from these values? The type of question that you can answer is How many of the data values are within two standard deviations of the mean?

Take a look at the aptitude test scores in Table 3.1. Here, $\bar{x} = 60.36$ and $s = 18.61$, and so we obtain

$$\bar{x} - s = 60.36 - 18.61 \qquad \bar{x} + s = 60.36 + 18.61$$
$$= 41.75 \qquad\qquad = 78.97$$

$$\bar{x} - 2s = 60.36 - 37.22 \qquad \bar{x} + 2s = 60.36 + 37.22$$
$$= 23.14 \qquad\qquad = 97.58$$

$$\bar{x} - 3s = 60.36 - 55.83 \qquad \bar{x} + 3s = 60.36 + 55.83$$
$$= 4.53 \qquad\qquad = 116.19$$

Examine these data and observe that (1) 33 out of the 50 values (66%) lie between $\bar{x} - s$ and $\bar{x} + s$; (2) 49 out of the 50 values (98%) lie between $\bar{x} - 2s$ and $\bar{x} + 2s$; and (3) 50 out of the 50 values (100%) lie between $\bar{x} - 3s$ and $\bar{x} + 3s$. Or, put another way: (1) 66% of the data values have a Z score between -1 and 1; (2) 98% have a Z score between -2 and 2, and (3) 100% have a Z score between -3 and 3.

What can we say in general for any data set? There are two types of statements we can make. One of these, *Chebyshev's inequality*, is usually conservative but makes *no assumption* about the population from which you obtained your data. Following are the components of Chebyshev's inequality.

CHEBYSHEV'S INEQUALITY

1. At least 75% of the data values are between $\bar{x} - 2s$ and $\bar{x} + 2s$.
2. At least 89% of the data values are between $\bar{x} - 3s$ and $\bar{x} + 3s$.
3. In general, there are at least $(1 - 1/k^2)$ of your data values between $\bar{x} - ks$ and $\bar{x} + ks$, for $k \geq 1$.

Note that if $k = 1$, $1 - 1/k^2 = 0$; so Chebyshev's inequality provides no information on the number of data values to expect between $\bar{x} - s$ and $\bar{x} + s$.

The other type of statement is called the **empirical rule**. We make a key assumption here, namely, that the population from which you obtain your sample has a *bell-shaped distribution*; that is, it is symmetric and tapers off smoothly into each tail. Such a population is called a **normal population** and is illustrated in Figure 3.7. Thus, the data set should have a skewness measure, Sk, near zero and a histogram similar to that in Figure 3.4. However, the empirical rule is still quite accurate even if your distribution is not exactly bell-shaped. Following are the components of the empirical rule.

FIGURE 3.7
A bell-shaped (normal) population.

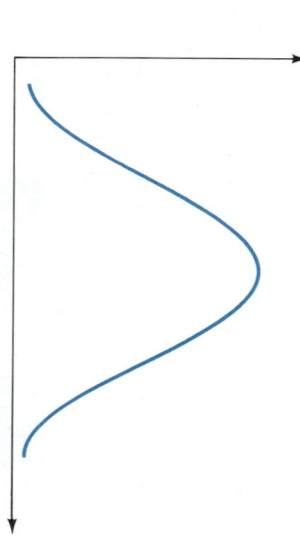

EMPIRICAL RULE

Under the assumption of a bell-shaped population:

1. Approximately 68% (large majority) of the data values lie between $\bar{x} - s$ and $\bar{x} + s$.
2. Approximately 95% (19 out of 20) of the data values lie between $\bar{x} - 2s$ and $\bar{x} + 2s$.
3. Approximately 99.7% (nearly all) of the data values lie between $\bar{x} - 3s$ and $\bar{x} + 3s$.

Returning to Table 3.1, we can summarize our previous results along with the information provided by Chebyshev's inequality and the empirical rule. The actual percentage of the sample values in each interval, as well as the percentage specified using each of the two rules, are shown in Table 3.2.

As you can see, Chebyshev's inequality is very conservative, but it always works. The empirical rule predicted results close to what was observed. This is not surprising because the skewness measure is only slightly different from zero ($Sk = -.26$).

EXAMPLE 3.5

In a random sample of 200 automotive insurance claims obtained from Pearson Insurance Company, $\bar{x} = \$615$ and $s = \$135$.

1. What statement can you make using Chebyshev's inequality?
2. If you have reason to believe that the population of all insurance claims is bell-shaped (normal), what does the empirical rule say about these 200 values?

SOLUTION 1

Chebyshev's inequality provides information regarding the number of sample values within a specified number of standard deviations of the mean. For $k = 2$, we have:

$$\bar{x} - 2s = 615 - 2(135) = \$345$$
$$\bar{x} + 2s = 615 + 2(135) = \$885$$

TABLE 3.2
Summary of Percentages of Sample Values by Interval, Using Data from Table 3.1

BETWEEN	ACTUAL PERCENTAGE	CHEBYSHEV INEQUALITY PERCENTAGE	EMPIRICAL RULE PERCENTAGE
$\bar{x} - s$ and $\bar{x} + s$	66% (33 out of 50)	—	≈68%
$\bar{x} - 2s$ and $\bar{x} + 2s$	98% (49 out of 50)	≥75%	≈95%
$\bar{x} - 3s$ and $\bar{x} + 3s$	100% (50 out of 50)	≥89%	≈100%

We conclude that at least 75% of the sample values lie between $345 and $885. Because $.75 \cdot 200 = 150$, this implies that at least 150 of the claims are between $345 and $885.

For $k = 3$,

$$\bar{x} - 3s = 615 - 3(135) = \$210$$
$$\bar{x} + 3s = 615 + 3(135) = \$1020$$

and we conclude that at least 8/9 (89%) of the data values are between $210 and $1020. Here, $8/9 \cdot 200 = 177.8$, and so at least 178 of the claims are between $210 and $1020.

SOLUTION 2

If the distribution of automotive claims at Pearson Insurance Company is believed to be bell-shaped, the empirical rule allows us to draw stronger conclusions. In particular, for $k = 1$, we have

$$\bar{x} - s = 615 - 135 = \$480$$
$$\bar{x} + s = 615 + 135 = \$750$$

and we conclude that approximately 68% of the data values ($.68 \cdot 200 = 136$) are between $480 and $750.

For $k = 2$,

$$\bar{x} - 2s = \$345$$
$$\bar{x} + 2s = \$885$$

and we conclude that approximately 95% of the data values ($.95 \cdot 200 = 190$) will lie between $345 and $885. ∎

EXERCISES

3.34 Use the mean, median, and standard deviation obtained for the data in exercise 3.33. Using Chebyshev's inequality, find the range of GNP that will include 89% of the data.

3.35 The mean of the wages of a sample of production workers in a company is $18,600, and the standard deviation is $445. Assuming the corresponding population is normally distributed, estimate the range of wages within which about 95% of the employees in the sample are expected to lie.

3.36 In the annual spaghetti-eating contest of a small town, 100 people participated. The contestants' spaghetti consumption averaged 120 feet with a standard deviation of 14 feet.

a. At least how many contestants ate between 92 and 148 feet of spaghetti?

b. At least how many contestants ate between 78 and 162 feet of spaghetti?

c. Did you have to make any assumptions about the data to answer parts (a) and (b)?

3.37 What do you think will be the value of the coefficient of skewness for a symmetric distribution? Will the mean and median be equal?

3.38 Refer to the data on exam scores in Exercise 2.1. Using Chebyshev's inequality, estimate the values that should contain at least 75% of the data.

3.39 At the University of Nebraska, a study found that financial performance at financial institutions is related to the match between environmental perceptions and environmental analysis techniques. Data were collected by a mailed questionnaire from 246 financial institutions. Perceived environmental complexity and change were measured by asking subjects to indicate, on a 5-point scale, the extent of agreement with 13 statements. The means and standard deviations of the responses pertaining to each variable are given below.

VARIABLE	MEAN	STANDARD DEVIATION
Perceived Environmental		
Complexity	3.51	.72
Change in Competition	3.85	.91
Change in Market	3.48	.58
Change in Information	3.45	.76

(*Source:* P. H. Specht and L. Trussell, "Perceived Environmental Characteristics, Environmental Analysis For Strategic Planning, and Small Firm Financial Performance," *1987 Proceedings of the Annual Meeting of Decision Sciences Institute,* pp. 1136–38.)

a. Calculate the coefficient of variation for each variable. Do the data sets for each variable exhibit the same relative variation?

b. Under the assumption that the population from which the samples are obtained has a bell-shaped distribution, between what two values will approximately 68% of the responses lie for each variable above?

3.40 The mean score of students in an accounting class was 78.5, and the standard deviation was 22. Using the empirical rule, estimate the range of scores within which about 95% of the data values are expected to lie.

3.41 If the mean of a sample is 28.2 with a standard deviation of 6.2, what data values would lie within two standard deviations of the mean?

3.42 For the P:E ratio data in Exercise 2.15:

a. Calculate the coefficient of skewness.

b. Using the empirical rule, find the range of values within which approximately 100% of the values are expected to lie.

3.43 Using Chebyshev's inequality, find the scores on the aptitude test in Exercise 3.13 between which at least 89% of the data will fall.

3.44 The average number of visitors to the local museum is 48 per day and the standard deviation was 22. Using the empirical rule, estimate the range of scores within which about 95% of the data values are expected to lie.

3.45 A shoe salesperson computes that his mean daily sales of shoes is $280 with a standard deviation of $40. Give two bounds within which the shoe salesperson can expect daily sales to lie at least 75% of the time.

3.7
GROUPED DATA

You may have to work with data in the form of a frequency distribution, called **grouped data,** when the raw data are not available. This situation can arise when we find a histogram or frequency distribution in a magazine or newspaper article in which the actual raw data used to construct the histogram are not included. We do not have the data values used to make up this frequency distribution, so we are forced to approximate the sample statistics, in particular the mean, median, and standard deviation.

Approximating the Sample Mean, $\bar{x}$

Assume we obtain the frequency distribution shown in Table 3.3, containing the ages of 36 individuals who recently passed a CPA examination. The 36 data values are not available, so we cannot add them up. A procedure that works well for estimating $\bar{x}$ is simply to pretend that the 36 data values are equal to their respective class midpoints. Consequently, there are

5 values at $(20 + 30)/2 = 25$

14 values at $(30 + 40)/2 = 35$

$\vdots$

2 values at $(60 + 70)/2 = 65$

TABLE 3.3

Age of 36 Individuals Who Recently Passed A CPA Examination

CLASS NUMBER	CLASS (age in years)	FREQUENCY
1	20 and under 30	5
2	30 and under 40	14
3	40 and under 50	9
4	50 and under 60	6
5	60 and under 70	2
		36

The value of $\bar{x}$ is estimated by finding what it is approximately equal to ($\cong$).

$$\bar{x} \cong \frac{(25 + 25 + 25 + 25) + \cdots + (65 + 65)}{35}$$

$$= \frac{(5)(25) + (14)(35) + (9)(45) + (6)(55) + (2)(65)}{35}$$

$$= \frac{1480}{36} = 41.1$$

Our estimate of the average age of these 36 individuals is

$$\bar{x} \cong 41.1 \text{ years}$$

In general,

$$\bar{x} \cong \frac{\sum f \cdot m}{n} \qquad (3.12)$$

where n = sample size, f = frequency of each class, and m = midpoint of each class.

Approximating the Sample Standard Deviation, s

Using the same fictitious data set at the various class midpoints, the variance, s^2, can be found in the usual way, using equation 3.8.

$$s^2 = \frac{\sum(\text{each data value})^2 - [\sum(\text{each data value})]^2/n}{n-1}$$

$$\sum(\text{each data value})^2 = \overbrace{(25^2 + 25^2 + \cdots + 25^2)}^{5 \text{ times}}$$
$$+ \overbrace{(35^2 + 35^2 + \cdots + 35^2)}^{14 \text{ times}} + \cdots$$
$$+ (65^2 + 65^2)$$
$$= (5)(25^2) + (14)(35^2) + (9)(45^2) + (6)(55^2) + (2)(65^2)$$
$$= 65,100$$

Also, $\sum(\text{each data value}) = 1480$. This was determined previously when approximating $\bar{x}$.

$$s^2 \cong \frac{65,100 - (1480)^2/36}{35} = \frac{4255.56}{35} = 121.59$$

TABLE 3.4
Summary of Calculations
for Grouped Data

CLASS NUMBER	CLASS	f	m	$f \cdot m$	$f \cdot m^2$
1	20 and under 30	5	25	125	3,125
2	30 and under 40	14	35	490	17,150
3	40 and under 50	9	45	405	18,225
4	50 and under 60	6	55	330	18,150
5	60 and under 70	2	65	130	8,450
		36		$\Sigma f \cdot m = 1{,}480$	$\Sigma f \cdot m^2 = 65{,}100$

and

$$s \cong \sqrt{121.59} = 11.03$$

In general,

$$s^2 \cong \frac{\Sigma f \cdot m^2 - (\Sigma f \cdot m)^2 / n}{n - 1} \qquad (3.13)$$

where f, m, and n are as defined in equation 3.12.

The calculations necessary to approximate $\bar{x}$ and s are more easily performed using a table similar to Table 3.4.

Approximating the Sample Median, Md

The sample median can best be approximated by approximating the $(n/2)$th ordered value. Using the previous example, we have

$$\text{Md} \cong \left(\frac{36}{2}\right) \text{th ordered value} = 18\text{th ordered value}$$

Where is this value in the frequency distribution? The first class contains the five smallest values, and the first two classes contain the first 19 ordered values $(5 + 14 = 19)$. So the 18th value is in the second class.

We can better approximate the median by assuming that the values in this class (and all classes) are spread *evenly* between the lower and upper limits. Because the first class contains five values, the median is 13 (18 − 5) values into the second class. This class begins at 30, has a width of 10, and has 14 values in it. So we want to go 13 values into a class of width 10 containing 14 values. The resulting estimate of the median is

$$\text{Md} \cong 30 + \frac{13}{14}(10) = 39.3$$

In general,

$$\text{Md} \cong L + \frac{k}{f} \cdot W \qquad (3.14)$$

where L = lower limit of the class containing the median (called the **median class**); $k = n/2 -$ (the number of data values preceding the median class); f = frequency of the median class; and W = class width.

In the previous example, $L = 30$, $f = 14$, $W = 10$, and thus

$$k = \frac{36}{2} - 5 = 13$$

If n is odd (say, $n = 25$), then $n/2 = 12.5$, and you need to estimate the 12.5th ordered value. This is halfway between the 12th and the 13th value. The procedure to follow here is exactly the same, except k will not be a counting number.

Remember that these procedures for approximating the sample statistics are used only when the raw data are not available and your only information is a frequency distribution or corresponding histogram. If the actual data values are available, these statistics can be determined exactly, and the approximation procedures described in this section should not be used.

EXERCISES

3.46 The following data are the per capita taxes for the 50 states in the United States for the fiscal year 1985.

PER CAPITA TAXES	NUMBER OF STATES
$300 and under $600	2
$600 and under $900	28
$900 and under $1200	15
$1200 and under $1500	3
$1500 and under $2000	1
$2000 and above	1

(*Source: The World Almanac and Book of Facts*, 1988, p. 107.)

Compute the median. Is the median an appropriate summary number for these data? Why or why not?

3.47 The following is the summarized information concerning family income (in dollars) for a specific neighborhood of a major city.

INCOME LEVEL	PERCENT OF FAMILIES
0 and under 10,000	11
10,000 and under 20,000	19
20,000 and under 30,000	30
30,000 and under 40,000	15
40,000 and under 50,000	10
50,000 and under 60,000	15

Find the median family income.

3.48 Advertising expenditures constitute one of the important components of the cost of goods sold. From the following data concerning advertising expenditures (in millions of dollars) of 50 companies, find the median advertising expenditure.

ADVERTISING EXPENDITURE	NUMBER OF COMPANIES
25 and under 35	5
35 and under 45	11
45 and under 55	18
55 and under 65	6
65 and under 75	10
	50

3.49 Calculate the median age of the U.S. population for the year 1988 using the data in Exercise 2.13.

3.50 Glen's Mufflers and Exhaust has been experiencing a steady growth in sales. The following grouped data show a breakdown of the frequency of sales volume (in hundreds of dollars):

DAILY SALES	NUMBER OF DAYS
.5 and under 1	2
1 and under 1.5	2
1.5 and under 2	3
2 and under 2.5	11
2.5 and under 3	20
3 and under 3.5	12
3.5 and under 4	5
4 and under 4.5	3
4.5 and under 5	2

a. Calculate the mean daily sales.

b. Calculate the standard deviation of the mean daily sales.

3.51 The following is the final distribution of grades in an introductory economics course at a local university:

GRADE	NUMBER OF STUDENTS
90–99	5
80–89	14
70–79	11
60–69	4
50–59	7
	41

a. Calculate the mean score.

b. Calculate the standard deviation.

c. Interpret the value of the mean and standard deviation.

3.52 The industrial engineer of the Bright Light company is interested in examining the average burning hours (life) of the 100-watt bulbs manufactured. Using the following data, determine the mean burning hours of a 100-watt bulb:

BURNING HOURS	NUMBER OF BULBS
0 and under 40	231
40 and under 50	168
50 and under 60	244
60 and under 70	300
70 and under 80	111
80 and under 90	48
90 and under 100	98
	1200

3.8 EXPLORATORY DATA ANALYSIS

Exploratory data analysis (EDA) is a recently developed technique that provides easy-to-construct pictures that summarize and describe a set of sample data. Two popular diagrams that fall under this category are *stem and leaf diagrams* and *box and whisker plots*. Stem and leaf diagrams are like histograms, since both allow you to see the "shape" of the data. Box and whisker plots are graphical representations of the quartile measures of position, discussed in Section 3.4.

Stem and Leaf Diagrams

Stem and leaf diagrams were originally developed by John Tukey (pronounced Too'key) of Princeton University. They are extremely useful in summarizing reasonably sized data sets (under 100 values as a general rule) and, unlike his-

FIGURE 3.8

Stem and leaf diagram for aftertax profits

```
2 | .3  .7  .4
3 | .4  .8  .4  .6
4 | .5  .7  .1
5 | .9  .1
```

tograms, result in no loss of information. By this we mean that it is possible to construct the original data set from a stem and leaf diagram, which is not the case when using a histogram. Consequently, some of the information in the original data is lost when constructing a histogram. Nevertheless, histograms provide the best alternative when attempting to summarize a large set of sample data.

To illustrate the construction of a stem and leaf diagram, suppose that a study contains the after-tax profits for 12 selected companies. The profits (recorded as cents per dollar revenue) are as follows:

3.4   4.5   2.3   2.7   3.8   5.9   3.4   4.7   2.4   4.1   3.6   5.1

The stem and leaf diagram for these data is shown in Figure 3.8. Each observation is represented by a **stem** to the left of the vertical line and a **leaf** to the right of the vertical line. For example, the stems and leaves for the first and last observation would be:

Stem	Leaf		Stem	Leaf
3	.4		5	.1

In a stem and leaf diagram, the stems are put *in order* to the left of the vertical line. The leaf for each observation is generally the last digit (or possibly the last two digits) of the data value, with the stem consisting of the remaining first digits. The value 562 could be represented as 5|62 or as 56|2 in a stem and leaf diagram, depending upon the range of the sample data. Whether or not the raw data are ordered, the stem and leaf diagram provides at least a partial ordering of the data. If the diagram is rotated counterclockwise, it has the appearance of a histogram and clearly describes the shape of the sample data.

The aptitude scores of the 50 job applicants contained in Table 3.1 are represented by a stem and leaf diagram in Figure 3.9. From this diagram we observe that the minimum score is 22, the maximum score is 96, and the largest group of scores is between 60 and 69. Also, the five leaves in stem row 3 indicate that five people scored at least 30 but less than 40. The three leaves in stem row 9 tells us at a glance that 3 people scored 90 or better.

For larger data sets, you may want to consider spreading out the stem column by repeating the stem value two or three times. To illustrate, the 50 applicant test scores are put into another stem and leaf diagram in Figure 3.10, where each stem value is repeated twice. The first stem value contains leaves between 0 and 4, the second stem contains leaves between 5 and 9. A MINITAB version of the stem and leaf diagram using these data is shown in Figure 3.11. Notice that MINITAB used the double stems as in Figure 3.10.

FIGURE 3.9

Stem and leaf diagram for aptitude test scores

```
2 | 2 5 8
3 | 1 4 5 9 9
4 | 0 2 4 4 6 8 9
5 | 1 3 3 5 5 6 7 9
6 | 0 1 3 3 5 6 8 8 9
7 | 1 2 2 4 5 5 6 8 8
8 | 0 2 3 5 8
9 | 0 2 6
```

FIGURE 3.10
Stem and leaf diagram of aptitude test scores (Table 3.1) using MINITAB

```
2 | 2
2 | 5 8
3 | 1 4
3 | 5 9
4 | 0 2 4 4
4 | 6 8 9
5 | 1 3 3
5 | 5 6 7 9
6 | 0 1 3 3
6 | 5 6 8 9
7 | 1 2 4
7 | 5 6 8 8
8 | 0 2 3
8 | 5 8 8
9 | 0 2
9 | 6
```

FIGURE 3.11
Stem and leaf diagram for aptitude test scores using repeated stems

```
MTB > SET INTO C1
DATA> 22 25 28 31 34 35 39 39 40 42
DATA> 44 44 46 48 49 51 53 53 55 55
DATA> 56 57 59 60 61 63 63 63 65 66
DATA> 68 68 69 71 72 72 74 75 75 76
DATA> 78 78 80 82 83 85 88 90 92 96
DATA> END
MTB > STEM AND LEAF USING C1

Stem-and-leaf of C1        N = 50
Leaf Unit = 1.0

   1    2  2
   3    2  58
   5    3  14
   8    3  599
  12    4  0244
  15    4  689
  18    5  133
  23    5  55679
  (5)   6  01333
  22    6  56889
  17    7  1224
  13    7  55688
   8    8  023
   5    8  58
   3    9  02
   1    9  6
```

FIGURE 3.12
Stem and leaf diagram for production amounts

```
5 | 21  50  36  72
6 | 71  33  47  62  55
7 | 83  62  11  31  40  21  57
8 | 44  16  35  92
```

In many situations the leaves in your diagram may consist of a *pair* of digits. In Figure 3.12, the tons of chemical produced by Sarhad Industries are illustrated for 20 randomly selected days. Here each stem consists of a single digit and each leaf represents the last two digits of the production amount.

Box and Whisker Plots

A **box and whisker plot** is a graphical representation of a set of sample data that illustrates the lowest data value (L), the first quartile (Q_1), the median (Q_2, Md), the third quartile (Q_3), the interquartile range (IQR), and the highest data value (H).

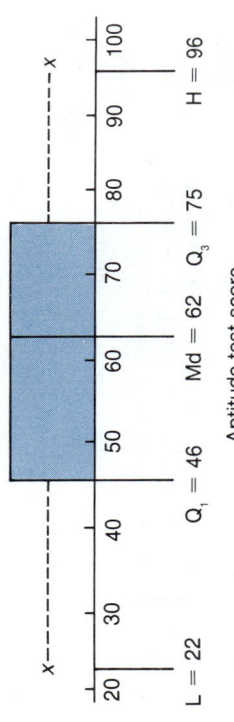

Aptitude test score

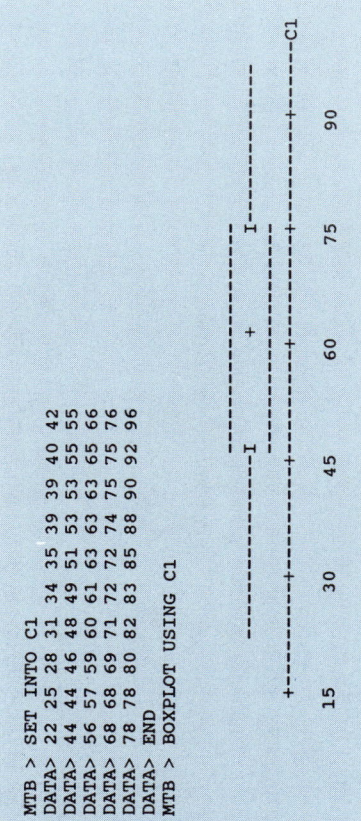

```
MTB > SET INTO C1
DATA> 22 25 28 31 34 35 39 39 40 42
DATA> 44 44 46 48 49 51 53 53 55 55
DATA> 56 57 59 60 61 63 63 65 66
DATA> 68 68 69 71 72 72 74 75 75 76
DATA> 78 78 80 82 83 85 88 90 92 96
DATA> END
MTB > BOXPLOT USING C1
```

In Section 3.4, the following values were determined for the aptitude test scores in Table 3.1.

$$L = 22$$
$$Q_1 = 46$$
$$Q_2 = Md = 62$$
$$Q_3 = 75$$
$$IQR = 75 - 46 = 29$$
$$H = 96$$

A box and whisker plot of these values is shown in Figure 3.13. The ends of the *box* are located at the first and third quartile with a vertical bar inserted at the median. Consequently, the length of the box is the interquartile range. The dotted lines are the *whiskers* and connect the highest and lowest data values to the ends of the box. This means that approximately 25% of the data values will lie in each whisker and in each portion of the box. If the data are symmetric, the median bar should be located at the center of the box. Consequently, the bar location informs you as to the *skewness* of the data; if located in the left half of the box, the data are skewed right and if located in the right half, the data are skewed left.

According to Figure 3.13, the data appear to be nearly symmetric (with a slight left skew), which is supported using the stem and leaf diagram in Figure 3.9. A box and whisker plot using MINITAB is shown in Figure 3.14.

EXERCISES

3.53 The average monthly expenditures on newspaper advertising by 36 real estate offices in the New Orleans metropolitan area are as follows:

590, 593, 651, 765, 503, 921, 833, 580, 804, 841, 865, 905, 881, 845, 807, 593, 551, 845, 650, 884, 920, 596, 630, 838, 889, 655, 585, 751, 885, 908, 586, 843, 654, 583, 847, 589,

a. List the stem possibilities, first two digits, in order.

b. Form a stem and leaf display by attaching the appropriate leaves to stems.

c. Comment on the shape of the distribution of the data.

3.54 The number of no-shows for four daily commuter flights between Boston's Logan Airport and JFK Airport in New York City is given below for 20 randomly selected business days.

4, 15, 11, 16, 24, 31, 12, 5, 11, 20, 21, 13, 7, 12, 15, 9, 13, 14, 22, 16

a. Construct a stem and leaf display of the data.

b. Comment on the shape of the distribution of the no-shows.

3.55 Forty samples of boxes of electronic resistors were checked for defective resistors. Each box contains 200 resistors. The number of defectives in each box is given below.

11, 13, 8, 12, 9, 12, 11, 37, 13, 48, 14, 19, 12, 14, 15, 13, 19, 14, 12, 17, 12, 16, 15, 33, 18, 11, 13, 15, 26, 56, 22, 14, 17, 18, 12, 7, 15, 12, 14, 23

a. Construct a stem and leaf display and use the first digit of each number as the stem.

b. What number of defectives are most frequently found?

c. Construct a box and whisker plot of this data. What can you say about the data based on this plot?

3.56 The manager of a small restaurant wished to determine how long the average customer had to wait to be served during the lunch hour. At the lunch hour on a particular "typical" day, 20 customers were served with the following waiting times given in minutes.

5.5, 10.3, 7.5, 8.1, 6.8, 11.0, 10.2, 9.0, 7.0, 5.8, 12.5, 7.5, 6.0, 13.7, 5.5, 14.0, 7.0, 6.9, 6.3, 7.4

a. Construct a box and whisker plot of the data.

b. Should the manager feel comfortable in advertising that meals are served in 10 min or less, or else the customer eats for free?

3.57 The yields in percent on the 15 utility stocks that are included in the Dow Jones Utility Average are given next for March 16, 1988.

8.0, 15.9, 3.6, 10.6, 7.2, 4.5, 12.1, 9.2, 8.7, 11.7, 7.7, 8.5, 11.4, 8.3, 7.5

(Source: *Wall Street Journal*, March 17, 1988.)

a. Which measure of central tendency do you think is most appropriate for the preceding data set? Compute it. If the yield of 15.9 was considered an outlier, what would be the value of this measure of central tendency after eliminating this yield from the data set?

b. Compute the standard deviation and the mean absolute deviation. Compare these two values.

c. Construct a stem and leaf display of the yields.

d. Construct a box and whisker plot of the yields.

e. What do the stem and leaf display in part (c) and the box and whisker plot in part (d) tell you about the shape of the distribution?

3.58 The following data reflect the number of consecutive hours that a sample of 20 medical interns worked in a local emergency room:

14, 21, 35, 11, 18, 28, 26, 38, 21, 19, 17, 27, 18, 22, 16, 30, 17, 26, 20, 25

Construct a stem and leaf diagram of these data.

3.59 The scores on a newly devised instrument for measuring life satisfaction could range from a minimum of 100 to a maximum of 800. A sample of 25 scores contained the following values:

512, 587, 563, 542, 556, 577, 515, 520, 532, 526, 560, 584, 530, 593, 542, 518,
570, 588, 534, 542, 557, 538, 562, 548, 525

The researcher noticed that all the scores were between 500 and 600, not a good situation due to the lack of variation in these scores. Illustrate this by constructing a stem and leaf diagram of the instrument scores.

3.60   The ages of 40 individuals working at a telephone crisis center were recorded as follows:

33, 40, 35, 24, 31, 32, 45, 23, 45, 34, 34, 54, 25, 36, 38, 44, 62, 31, 56, 47,
30, 42, 43, 52, 48, 56, 64, 40, 61, 39, 26, 34, 37, 43, 47, 32, 45, 27, 35, 29

a.   Construct a stem and leaf diagram of the data. What conclusions can you make from this picture?

b.   Construct a box and whisker plot of these data. What can you say about the data based on this plot?

3.9

CALCULATING DESCRIPTIVE STATISTICS BY CODING

When you use a calculator to determine the sample mean or standard deviation, one problem that can occur is that the data values are too large or too small to "fit" into your calculator. To avoid having the calculator self-destruct in your hands, a procedure referred to as **data adjusting**, or **data coding**, allows you to derive these statistics using more reasonable data values. You can then work backward to get the desired statistics. To code, or adjust, the data you subtract (or add) or divide (or multiply) your original data set by a fixed amount. Figure 3,15 demonstrates this procedure using data sets containing several large values. To adjust the data when subtracting (adding) a positive constant to each data value,

actual $\bar{x}$ = adjusted $\bar{x}$ plus (minus) the constant

actual s = adjusted s (no change)

When dividing (multiplying) by a positive constant,

actual $\bar{x}$ = adjusted $\bar{x}$ times (divided by) the constant

actual s = adjusted s times (divided by) the constant

FIGURE 3.15   Data coding

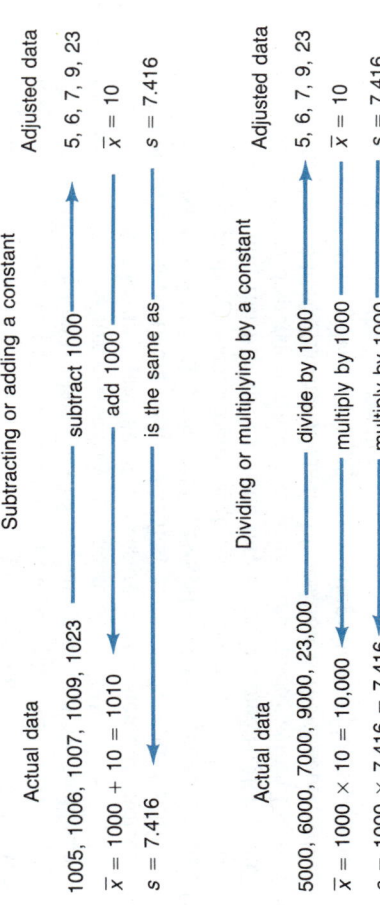

Subtracting or adding a constant

Actual data   Adjusted data

1005, 1006, 1007, 1009, 1023   → subtract 1000 →   5, 6, 7, 9, 23

$\bar{x}$ = 1000 + 10 = 1010   ← add 1000 $\bar{x}$ = 10

s = 7.416   is the same as s = 7.416

Dividing or multiplying by a constant

Actual data   Adjusted data

5000, 6000, 7000, 9000, 23,000   → divide by 1000 →   5, 6, 7, 9, 23

$\bar{x}$ = 1000 × 10 = 10,000   ← multiply by 1000 $\bar{x}$ = 10

s = 1000 × 7.416 = 7,416   ← multiply by 1000 s = 7.416

EXERCISES

3.61 Using the subtraction rule, adjust the following data and calculate the mean and standard deviation.

413, 407, 411, 402, 425, 408, 410, 421

3.62 Using the multiplication rule, adjust the following data and calculate the mean and standard deviation:

0.00119, 0.00101, 0.00121, 0.00108, 0.0010, 0.00114, 0.00117, 0.00104, 0.00123, 0.00124

3.63 Using the division rule, calculate the mean and variance for the following data:

200, 600, 800, 1000, 400, 1200, 10,000, 1400, 1600, 400

3.64 Using the division rule, calculate the mean and standard deviation for the following data:

500, 3000, 600, 1000, 400, 300, 100, 2500, 700, 300

3.65 Using the multiplication rule, adjust the following data and calculate the mean and standard deviation:

0.001182, 0.001104, 0.001270, 0.001251, 0.001407, 0.001553, 0.001177, 0.001333, 0.001489, 0.001505

3.66 Using the subtraction rule, adjust the following data and calculate the mean:

1013, 1007, 1011, 1102, 1025, 1008, 1110, 1021, 1111, 1009

SUMMARY

The purpose of analyzing or describing sample data is to learn more about the population from which it was obtained. Every population has properties that describe it. These properties are referred to as **parameters**. We can estimate these parameters by obtaining a sample and deriving the corresponding sample statistic, which is a particular **descriptive measure**.

This chapter has introduced you to some of the more popular descriptive measures used to describe a set of sample values. **Measures of central tendency** are used to describe a typical value within the sample. They are the **mean** (the average of the sample data), the **median** (the value in the center of the ordered data), the **mode** (that data value occurring the most often), and the **midrange** (an average of the lowest and highest data values). To measure the variation within a set of sample data, we use **measures of dispersion**: the **range** (difference between the highest and lowest data values), the **mean absolute deviation** (an average of the absolute deviations from the sample mean), the **variance** (sum of the squares of the deviations from the sample mean, divided by $n - 1$), the **standard deviation** (square root of the variance), and the **coefficient of variation** (standard deviation divided by the mean, times 100).

Percentiles and **quartiles** are **measures of position** and indicate the relative position of a particular value. The first quartile (Q_1) and the third quartile (Q_3) are the 25th and 75th percentiles, respectively. The second quartile (Q_2) is the 50th percentile, which is identical to the sample median. The difference between the first and third quartiles is the **interquartile range** (IQR), which is another measure of dispersion, since it measures the spread within the middle 50% of the data values. Another measure of position is the **Z score**, which is derived for a particular observation by subtracting the sample mean and dividing by the sample standard deviation.

Finally, the shape of a data set can be described using various **measures of shape**. Two such measures are the sample **skewness** (the degree of symmetry in the data) and **kurtosis** (the tendency of a distribution to stretch out in a particular direction).

The two most commonly used measures are the sample mean and standard deviation. These two statistics can be used together to describe the sample data by applying **Chebyshev's inequality** or the **empirical rule.** The latter procedure draws a stronger conclusion about the concentration of the data values but assumes that the population of interest is bell-shaped (normal).

We examined how to estimate the sample mean and standard deviation when the only information available is a frequency distribution, or **grouped data.** **Data coding** can be used to calculate these two measures more easily when you encounter data sets containing extremely large or small values.

Stem and leaf diagrams and **box and whisker plots** are easy-to-construct pictures that illustrate the shape of the sample data. A stem and leaf diagram is similar to a histogram but results in no loss of information. A box and whisker box plot summarizes the lowest and highest data values, along with the three quartiles, in a simple diagram.

REVIEW EXERCISES

3.67 The unemployment rate (U) represents the percent of the civilian labor force that is unemployed. The following are the U.S. unemployment rates for the years 1951 through 1985.

YEAR	U	YEAR	U	YEAR	U
1951	3.3	1963	5.6	1975	8.5
1952	3.0	1964	5.2	1976	7.7
1953	2.9	1965	4.5	1977	7.0
1954	5.6	1966	3.8	1978	6.0
1955	4.4	1967	3.9	1979	5.8
1956	4.1	1968	3.6	1980	7.1
1957	4.3	1969	3.5	1981	7.6
1958	6.8	1970	4.9	1982	9.7
1959	5.5	1971	5.9	1983	9.6
1960	5.5	1972	5.6	1984	7.5
1961	6.7	1973	4.9	1985	7.2
1962	5.6	1974	5.6		

(*Source: Statistical Abstract of the United States 1987*, U.S. Department of Commerce.)

Calculate the following statistics and explain how these statistics help to describe the distribution of the data:

a. Midrange   **b.** Mean

c. Median   **d.** Mode

e. Standard deviation   **f.** Coefficient of variation

g. First quartile   **h.** Ninetieth percentile

i. Interquartile range

3.68 The productivity of the U.S. working person is important in keeping U.S. goods and services competitive with world markets. The annual percent change in output per

YEAR	PERCENT CHANGE IN PRODUCTIVITY	YEAR	PERCENT CHANGE IN PRODUCTIVITY
1975	2.5	1981	2.2
1976	4.6	1982	2.2
1977	3.0	1983	5.8
1978	1.5	1984	5.5
1979	−.1	1985	5.1
1980	.0	1986	3.7

(*Source: World Almanac and Book of Facts*, 1988.)

hour for all persons in the manufacturing sector in the United States is given for the years 1975–1986.

a.   Calculate the mean, median, and mode for the given data set. Which measure do you think is most appropriate as a measure of central tendency?

b.   From the value of the mean and the value of median calculated in part (b), what can you say about the skewness of the data? Calculate the coefficient of skewness and interpret its value.

c.   Calculate the standard deviation for the data set.

d.   Find the number of values that are between $\bar{x} - 3s$ and $\bar{x} + 2s$ and also the number of values between $\bar{x} - 2s$ and $\bar{x} + 3s$. Are these numbers consistent with Chebyshev's inequality?

e.   Calculate the midrange and the interquartile range. Which is more easily affected by an outlier?

3.69   With the following data set, answer the following questions.

2.0, 1.1, −1.5, .3, 1.9, −2.3, −1.2, .7, −.5, 3.1, −.2, 1.0, −.6, 1.3, −.9, −.1, .8, −1.1, 2.4, .9, −.1, −1.5

a.   Do the data appear to be bell-shaped?

b.   Calculate the coefficient of skewness.

c.   Using the empirical rule, estimate the range of values within which about 68% of the data values are expected to lie.

3.70   The following is the distribution of the annual incomes (in thousands of dollars) of the households in a neighborhood:

ANNUAL INCOME	NUMBER OF HOUSEHOLDS
0 and under 10	21
10 and under 20	11
20 and under 30	9
30 and under 40	13
40 and under 50	17
50 and under 60	20
60 and under 70	14
70 and under 80	20
80 and under 90	7
90 and under 100	2
	$\overline{134}$

a.   Calculate the mean annual income.

b.   Calculate the variance.

c.   Estimate the interval within which at least 75% of the data values are expected to fall.

d.   Calculate the coefficient of skewness.

3.71   The mean rate charged by the CPAs in a certain city is about $75 per hour, with a standard deviation of $15. Assuming that the data came from a normal population, estimate the range of rates within which about 95% of the charges by the CPAs are expected to lie.

3.72   The Z score is −1.50, the mean is 45, and $x = 15$. What is the value of the variance?

3.73   The mean is 81, the standard deviation is 9, and one particular x value is 45.

a.   Calculate the Z score.

b.   Interpret the Z score.

3.74   The mean GMAT score of the 65 applicants who were accepted into the MBA program of Xavier Business School was 520 with a standard deviation of 25. About how many applicants scored between 470 and 570 on the GMAT?

3.75 Calculate the mean and standard deviation for these data:

50.2, 53.8, 51.4, 52.2, 50.8, 59.1, 52.8, 57.7, 51.1, 54.3, 55.5, 52.1, 57.6, 55.9, 50.9, 54.7

3.76 Calculate the mean and standard deviation for the following data:

1000, 700, 400, 100, 800, 20,000, 4000, 300, 900, 600, 200, 500, 2000, 700, 2500, 5500

3.77 The industrial production index (IPI) is one of the useful indicators that helps to monitor the economy. Simply stated, it indicates the gross value of production in a given sector of an industry. Because it is expressed relative to a base year (such as 1967 = 100), it facilitates comparison over a period of time. The following are the IPIs for the month of July 1984 for various industries.

INDUSTRY	IPI	INDUSTRY	IPI
Foods	164.9	Furniture and fixtures	192.6
Tobacco products	115.1	Clay, glass, and stone products	160.9
Textile mills	139.8	Fabricated metal products	140.6
Paper and products	176.7	Electrical instruments	221.5
Printing and publishing	172.6	Motor vehicles and parts	169.0
Chemicals	232.0	Oil and gas extraction	122.8
Petroleum products	124.7	Coal mining	176.5
Rubber and plastics	341.4	Utilities	181.8
Leather products	60.6	Stone and earth minerals	147.9
Lumber products	146.0		

(*Source: Federal Reserve Bulletin* 71, no. 2, February 1985, Washington, D.C.: Board of Governors of the Federal Reserve System.)

Calculate the values for parts (a) through (l).

a. Mean
b. Median
c. Mode
d. Variance
e. Standard deviation
f. Mean absolute deviation
g. Range
h. Coefficient of variation
i. Coefficient of skewness
j. First quartile
k. Eighty-fifth percentile
l. Z score for lumber products.

m. Using Chebyshev's inequality, find the values of the IPI between which at least 75% of the data values will fall.

n. Calculate the proportion of the data values that will have a Z score between −1 and +1, assuming that the empirical rule holds.

3.78 The 15 industrial corporations in the United States with the largest sales in 1986 are as follows:

	SALES (in billions of dollars)		SALES (in billions of dollars)
General Motors	102	DuPont	27
Exxon	69	Chevron	24
Ford	62	Chrysler	22
IBM	51	Philip Morris	20
Mobil	44	Amoco	18
General Electric	35	RJR Nabisco	17
AT&T	34	Shell Oil	16
Texaco	31		

(*Source: World Almanac and Book of Facts,* 1988, p. 104.)

A MINITAB computer printout is given next on the sales data. From the printout,

```
MTB > PRINT C1
C1
   102    69    62    51    44    35    34    31    27    24    22
    20    18    17    16

MTB > STEM AND LEAF USING C1

Stem-and-leaf of C1      N = 15
Leaf Unit = 1.0

    3     1  678
    7     2  0247
   (3)    3  145
    5     4  4
    4     5  1
    3     6  29
    1     7
    1     8
    1     9
    1    10  2

MTB > DESCRIBE C1

C1       N      MEAN    MEDIAN    TRMEAN    STDEV    SEMEAN
         15     38.13   31.00     34.92     24.10    6.22

C1       MIN     MAX       Q1       Q3
         16.00   102.00    20.00    51.00
```

a. Summarize the shape of the distribution of the data.

b. Construct a box and whisker plot.

3.79 The insurance agent at Central Insurance Company wished to determine the distribution of the number of minutes that the company's secretary was spending on phone calls about insurance rates. The times for 30 randomly selected calls were analyzed in the MINITAB computer printout given below.

a. Compute the coefficient of skewness and comment on the shape of the distribution.

b. What is the range, midrange, and interquartile range? How do these values help describe the distribution?

c. Within what limits would you expect most of the times per phone call to fall?

```
MTB > PRINT C1
C1
  17.8   3.8   1.5   6.5   15.1   3.4   13.0   6.5   13.4   3.2   9.8
   9.4   2.8   8.0   7.1    2.4   7.4    5.3   2.1   19.3   5.1   2.2
   7.3   4.3   1.7   6.8    4.1   2.7    7.4   4.8

MTB > DESCRIBE C1

C1       N      MEAN     MEDIAN    TRMEAN    STDEV     SEMEAN
         30     6.807    5.900     6.304     4.743     0.866

C1       MIN      MAX       Q1       Q3
         1.500    19.300    3.100    8.350

MTB > HISTOGRAM OF C1

Histogram of C1    N = 30

Midpoint    Count
     2        7    *******
     4        6    ******
     6        5    *****
     8        5    *****
    10        2    **
    12        0
    14        2    **
    16        1    *
    18        1    *
    20        1    *
```

COMPUTER EXERCISES USING THE DATABASE

Exercise 1—Appendix H

Randomly select 100 observations of variable INCOME1 (income of principal wage earner) from the database.

a. Use a convenient statistical computer package to find the various descriptive measures to describe the distribution of INCOME1.

b. What are the actual proportions of observations that are between ±2.0 and ±3.0 standard deviations of the mean of the data set? Are these proportions consistent with:
(i) Chebyshev's inequality?
(ii) The empirical rule?

Exercise 2—Appendix I

Randomly select 100 observations of the variable SALES from the database. Use a convenient statistical computer package to find the mean, median, range, variance, coefficient of variation, and coefficient of skewness for this variable.

CASE STUDY

WOULD YOU RELOCATE YOUR BUSINESS FOR LOWER TAXES?

Business promoters often argue that a low tax burden attracts new industry. Some corporate relocation experts say taxes are no longer as important as other factors such as location, size of city, availability of transportation, and quality of educational facilities. Whatever the case, taxation is certainly a factor that will continue to be analyzed.

Suppose that you were a corporate relocation consultant. Like it or not, you would probably have to analyze taxation levels for your clients. The accompanying data (Table 3.5) on per capita state taxes and the questions that follow should be of interest to you. Of course, in an actual situation, the depth of analysis might be somewhat more sophisticated. What is illustrated here is some simple number crunching to elucidate the use of descriptive statistical measures.

Case Study Questions

1. Compute the mean and median values for the sales tax, personal income tax, and corporate income tax.

2. Compute the variance and standard deviations for the three types of taxes in Question 1.

3. Which of these three types of taxes tends to vary more among the states?

4. (Do this part assuming you have a roughly symmetrical bell-shaped distribution.)
   a. Compute the Z score for a per capita personal income tax of $250.

   b. What are the two per-capita personal income tax values that correspond to $Z = -2.0$ and $Z = 2.0$?

   c. What percentage of the states should have per-capita personal income tax figures between the two values mentioned in (b)? How many actually do?

5. If you did not have a bell-shaped curve, how does that affect your answer for Question 4(c)?

TABLE 3.5 Per Capita State Taxes for Fiscal '85

	TOTAL	RANK	SALES	LICENSES	PERS. INCOME	CORP. INCOME	SEVERANCE	OTHER
Alabama	727	41	401	55	177	53	22	19
Alaska	3620	1	217	109	2	393	2667	232
Arizona	924	19	568	62	191	63	0	40
Arkansas	740	38	419	48	200	55	0	7
California	1098	9	459	37	408	139	1	54
Colorado	707	42	335	41	281	31	9	10
Connecticut	1102	8	750	52	92	154	0	54
Delaware	1312	3	158	384	588	124	0	58
Florida	694	44	535	45	0	40	15	59
Georgia	757	37	369	23	288	70	0	7
Hawaii	1293	4	808	18	407	46	0	14
Idaho	730	40	360	68	257	42	1	2
Illinois	800	32	429	61	225	61	0	24
Indiana	789	33	479	28	234	32	0	16
Iowa	800	31	371	68	286	54	0	21
Kansas	782	35	352	49	246	65	46	24
Kentucky	809	28	363	42	208	57	61	78
Louisiana	860	23	434	68	118	66	166	8
Maine	864	22	473	66	255	46	0	24
Maryland	984	16	444	35	403	56	0	46
Massachusetts	1137	7	382	34	543	146	0	32
Michigan	956	18	387	45	335	153	8	28
Minnesota	1247	5	516	76	533	91	19	12
Mississippi	693	45	455	58	99	41	36	4
Missouri	667	46	364	55	210	32	0	6
Montana	776	36	169	68	219	76	182	62
Nebraska	648	47	352	56	199	30	3	8
Nevada	1005	14	859	112	0	0	0	34
New Hampshire	435	50	201	63	25	96	0	50
New Jersey	1021	12	536	71	256	122	0	36
New Mexico	993	15	565	48	59	44	270	7
New York	1164	6	389	43	585	105	0	42
North Carolina	831	26	346	58	323	78	0	26
North Dakota	1011	13	429	84	111	123	257	7
Ohio	805	29	431	70	259	41	1	3
Oklahoma	903	20	340	82	220	32	215	16
Oregon	738	39	88	82	488	57	12	11
Pennsylvania	857	24	427	87	218	80	0	45
Rhode Island	891	21	469	34	291	73	0	24
South Carolina	816	27	448	41	254	60	0	13
South Dakota	502	49	414	43	0	24	6	15
Tennessee	630	48	481	64	13	54	1	17
Texas	705	43	485	98	0	0	133	9
Utah	805	30	443	35	262	32	30	3
Vermont	857	25	432	73	271	65	0	16
Virginia	783	34	323	47	341	50	0	22
Washington	1040	11	782	56	0	0	8	194
West Virginia	958	17	587	47	260	51	0	13
Wisconsin	1061	10	461	44	421	87	0	48
Wyoming	1584	2	452	120	0	0	794	218
Average	902	—	441	57	267	74	30	33

(*Source:* U.S. Census Bureau.)

SAS

MAINFRAME AND MICRO

SOLUTION

TABLE 3 .1

We can use SPSS for the aptitude test scores in Table 3.1. The SPSS program listing in Figure 3.16 was used to request the calculation of the mean, standard deviation, and other descriptive statistics. The mainframe and micro commands are the same. **(Remember to end each command line with a period when using the PC version.)**

The TITLE command names the SPSS run.

The DATA LIST command gives each variable a name and describes the data as being in free form. The variable name is APTEST.

The BEGIN DATA command indicates to SPSS that the input data immediately follow.

The next 10 lines contain the data values. Each line represents the test score of five of the 50 applicants.

The END DATA statement indicates the end of the data entry.

The CONDESCRIPTIVE statement requests an SPSS procedure to compute simple descriptive statistics for the variable(s) in the applicant data set.

Figure 3.17 shows the SPSSX output obtained by executing the listing in Figure 3.16, whereas Figure 3.18 shows the SPSS/PC+ output.

FIGURE 3.16

Input for SPSSX or SPSS/PC+. All commands for SPSS/PC+ should end with a period.

```
TITLE       DESCRIPTIVE STATISTICS
DATA LIST FREE   /APTEST
BEGIN DATA
22  44  56  68  78
25  44  57  68  78
28  46  59  69  80
31  48  60  71  82
34  49  61  72  83
35  51  63  72  85
39  53  63  74  88
39  53  63  75  90
40  55  65  75  92
42  55  66  76  96
END DATA
CONDESCRIPTIVE APTEST
```

FIGURE 3.17

SPSSX output.

```
NUMBER OF VALID OBSERVATIONS (LISTWISE) =       50.00
    VARIABLE    MEAN    STD DEV   MINIMUM   MAXIMUM  VALID N   LABEL
    APTEST     60.360   18.605    22.00     96.00     50
```

FIGURE 3.18

SPSS/PC+ output.

```
            DESCRIPTIVE STATISTICS

Number of Valid Observations (Listwise) =       50.00

Variable    Mean    Std Dev   Minimum   Maximum   N  Label

APTEST     60.36    18.61     22.00     96.00     50
```

MAINFRAME AND MICRO

SOLUTION

TABLE 3.1

We can use SAS for the aptitude test scores in Table 3.1. The SAS program listing in Figure 3.19 was used to request the calculation of the mean, standard deviation, and other descriptive statistics. The mainframe and micro commands are the same.

FIGURE 3.19
Input for SAS
(mainframe or micro
version).

```
TITLE   'DESCRIPTIVE STATISTICS';
DATA APTDAT;
INPUT APTEST;
CARDS;
22
25
28
31
34
35
39
40
42
44
44
46
48
49
51
53
53
55
55
56
57
59
60
61
63
63
65
65
66
68
68
69
71
72
72
74
75
75
76
78
78
80
82
83
85
88
90
92
96
PROC MEANS;
```

The TITLE command names the SAS run (enclose in single quotes).

The DATA command gives the data a name.

The INPUT command names and gives the correct order for the different fields on the data lines. The variable name is APTEST.

The CARDS command indicates to SAS that the input data immediately follow.

The next 50 lines contain the data. Each line represents one applicant's aptitude test score.

The PROC MEANS command requests a SAS procedure to print simple descriptive statistics for the variable APTEST.

Figure 3.20 shows the SAS mainframe output obtained by executing the listing in Figure 3.19, whereas Figure 3.21 shows the SAS/PC output.

FIGURE 3.20 SAS output.

```
                              DESCRIPTIVE STATISTICS

                                  MINIMUM      MAXIMUM    STD ERROR
VARIABLE   N    MEAN            STANDARD        VALUE       VALUE      OF MEAN          SUM          VARIANCE       C.V.
                                DEVIATION

APTEST     50   60.36000000   18.60520006   22.00000000  96.00000000  2.63117263   3018.0000000   346.15346939   30.824
```

FIGURE 3.21
SAS/PC output.

```
                          DESCRIPTIVE STATISTICS

Analysis Variable : APTEST

N Obs   N     Minimum        Maximum          Mean          Std Dev
-----------------------------------------------------------------------
  50   50   22.0000000    96.0000000    60.3600000    18.6052001
-----------------------------------------------------------------------
```

Probability Concepts

A Look Back/Introduction

You use descriptive statistics to summarize or present data that consist of observations that have already occurred. If these data are drawn from a population, then you describe your sample in some way. If you wish to infer something about the population using the smaller sample, you must deal with uncertainty. To measure the chance that something will occur, you use its **probability**. The concepts of probability form the foundation of all decision making in statistics. By using probabilities, you are able to deal with uncertainty because you are able, at least, to measure it.

To illustrate this idea, suppose that a recent report contained the results of a random sample of 100 homes within a large metropolitan city and stated that the average electric bill was $185. However, the electric company claims that the average bill for all of its customers is $110. Is something wrong here? Do you believe that the *population* mean is $110 based on the fact that the *sample* mean is $185? Here we need a probability, in particular the probability of observing a sample mean at least this large (that is, $185 or more), assuming that the electric company is correct in their claim. If we decide that this probability is extremely small, we can infer that the *population* claim is incorrect, based on the *sample* results.

As mentioned in Chapter 1, there is always the possibility of arriving at the wrong decision (maybe the electric company *is* correct) when using sample results to infer something concerning a population. This particular type of question will be addressed in Chapter 8. We begin the journey into probability in this chapter by introducing some basic concepts and discussing various ways of determining probabilities.

4.1

EVENTS AND PROBABILITY

An activity for which the outcome is uncertain is an **experiment**. An experiment need not involve mixing chemicals in the laboratory; it could be as simple as throwing two dice and observing the total of the faces turned up. At the completion of an experiment, a measurement of some kind is obtained. An **event** consists of one or more possible outcomes of the experiment; it is usually denoted by a capital letter.* The following are examples of experiments and some corresponding events:

1. *Experiment:* Rolling two dice; *events:* A = rolling a total of 7, B = rolling a total greater than 8, C = rolling two 4s.

2. *Experiment:* Taking a CPA exam; *events:* A = pass, B = fail.

3. *Experiment:* Observing the number of arrivals at a drive-up window over a 5-minute period; *events:* A_0 = no arrivals, A_1 = one arrival, A_2 = two arrivals, and so on.

When you estimate a probability, you are estimating the probability *of an event.* For example, when rolling two dice, the probability that you will roll a total of 7 (event A) is the probability that event A occurs. It is written $P(A)$. The probability of any event is always between 0 and 1, inclusive.

NOTATION

$P(A)$ = probability that event A occurs

Classical Definition of Probability

Suppose a particular experiment has n possible outcomes and event A occurs in m of the n outcomes. The *classical definition* of the probability that event A will occur is

$$P(A) = m/n \qquad (4.1)$$

This definition assumes that all n possible outcomes have the same chance of occurring. Such outcomes (events) are said to be **equally likely,** and each has probability $1/n$ of occurring. If this is not the case, the classical definition does not apply.

Consider the experiment of tossing a nickel and a dime into the air and observing how they fall. Event A is observing one head and one tail. The possible outcomes are (H = head, T = tail):

NICKEL	DIME
H	H
H	T
T	H
T	T

Thus, there are two (m = 2) outcomes that constitute event A of the four possible outcomes (n = 4). These four outcomes are equally likely, so each oc-

---

* The set of all possible outcomes of an experiment is often referred to as the *sample space.*

curs with probability 1/4. Consequently,

$$P(A) = 2/4 = .5$$

Relative Frequency Approach

Another method of estimating a probability is referred to as the **relative frequency** approach. This is based on observing the experiment n times and counting the number of times an event (say, A) occurs. If event A occurs m times, your estimate of the probability that A will occur in the future is

$$P(A) = m/n \qquad \text{(4.2)}$$

Suppose that a particular production process has been in operation for 250 days; 220 days have been accident-free. Let A = a randomly chosen day in the future is free of accidents. Using the relative frequency definition, then

$$P(A) = 220/250 = .88$$

Subjective Probability

Another type of probability is **subjective probability**. This is a measure (between 0 and 1) of your belief that a particular event will occur. A value of one indicates that you believe this event will occur with complete certainty. Examples of situations requiring a subjective probability are:

The probability that the Dow Jones closing index will be below 1800 at some time during the next 6 months.

The probability that your newly introduced product will capture at least 10% of the market.

The probability that an audited voucher will contain an error.

The probability that your recently married cousin, divorced five times already, will once again go down alimony lane.

Although no two people may agree on a particular subjective probability, these probabilities are governed by the same rules of probability, which are developed later in the chapter.

Datacomp has recently conducted a survey of 200 selected purchasers of their new microcomputer to obtain a sex-and-age profile of their customers. The results obtained are shown in Table 4.1, which is a **contingency**, or **cross-tab table**. Such tables are a popular method of summarizing a group by means of

TABLE 4.1
Datacomp Survey of Microcomputer Purchasers

| SEX | AGE (YEARS) | | | |
	<30 (U)	30–45 (B)	>45 (O)	TOTAL
Male (M)	60	20	40	120
Female (F)	40	30	10	80
Total	100	50	50	200

two categories—in this case, age and sex. The numbers within the table represent the frequency, or number of individuals, within each pair of subcategories and so the contingency table allows you to see how these two categories interact.

There are 60 purchasers who are male *and* under 30; 10 purchasers are female *and* over 45. One person from the total group of 200 is to be selected at random to receive a free software package. We can define the following events:

M = a male is selected

F = a female is selected

U = the person selected is under 30

B = the person selected is between 30 and 45

O = the person selected is over 45

Because there are 200 people, there are 200 possible outcomes to this experiment. All 200 outcomes are equally likely (the person is randomly selected), so the classical definition provides an easy way of determining probabilities.

The probability of any one single event used to define the contingency table is a **marginal probability**. When you use a contingency table, you can obtain the marginal probabilities by merely counting. For example, of the 200 purchasers, 120 are males. So the probability of selecting a male is

$$P(M) = 120/200 = .6$$

Similarly,

$$P(F) = 80/200 = .4$$
$$P(U) = .5$$
$$P(B) = .25$$
$$P(O) = .25$$

Notice that $P(O) = 50/200 = .25$, which implies that (1) if you repeatedly selected a person at random from this group, 25% of the time the person selected would be over 45 years of age and (2) 25% of the people in this group are over 45 years old. So, a probability here is simply a **proportion**.

Complement of an Event The **complement** of an event A is the event that A does *not* occur. This event is denoted by $\bar{A}$. For example, A = it rains tomorrow, $\bar{A}$ = it does not rain tomorrow; or A = stock market rises tomorrow, $\bar{A}$ = stock market does not rise tomorrow.

In our Datacomp survey, $P(M) = .6$ and so

$$P(\bar{M}) = P(F) = .4$$

Notice that $P(M) + P(\bar{M}) = .6 + .4 = 1.0$. In general, for any event A, either A or $\bar{A}$ must occur. Consequently,

$$P(A) + P(\bar{A}) = 1$$

and so

$$P(\bar{A}) = 1 - P(A)$$

Written another way,

$$P(A) = 1 - P(\bar{A})$$

How can we determine what proportion of the purchasers are age 45 or younger?

$$P(\bar{O}) = 1 - P(O) = 1 - .25 = .75$$

Joint Probability

What if we wish to know the probability of selecting a purchaser who is female *and* under age 30? Such a person is selected if events *F and U* occur. This probability is written $P(F$ and $U)$ and is referred to as a **joint probability**.* There are 40 purchasers who are female and under 30, so

$$P(F \text{ and } U) = 40/200 = .2$$

What proportion are males between 30 and 45? This is the same as

$$P(M \text{ and } B) = 20/200 = .1$$

because 20 out of 200 satisfy both requirements.

Probability of A or B

In addition to calculating joint probabilities involving two events, we can also determine the probability that *either* of the two events will occur. In our discussion, "either *A* or *B*" will refer to the event that *A* occurred, *B* occurred, or both occurred. This probability is written as

$$P(A \text{ or } B)$$

for any two events *A* and *B*.†

Now we will calculate the probability of selecting someone who is a male *or* under 30 years of age. This is $P(M$ or $U)$. How many people qualify? There are 120 males and there are 100 people under 30. Is the answer $(120 + 100)/200 = 1.1$? You should realize that this is not correct because *a probability is never greater than 1*. What is the mistake here? The problem is that the 60 males under age 30 were counted *twice*. How many purchasers are male or under 30? The answer is the 120 males plus the 40 females under age 30. So

$$P(M \text{ or } U) = (120 + 40)/200 = .8$$

What is $P(F$ or $B)$? The people in the shaded area in Table 4.1 qualify. So,

$$P(F \text{ or } B) = (80 + 20)/200 = .5$$

Conditional Probability

Suppose that someone has some inside information about who has been selected from the group of 200 purchasers. This person informs you that the selected individual is under 30 years of age; that is, event *U* occurred. Armed with this information, we can calculate the probability that the selected person is a male. Given that event *U* occurred, we have immediately narrowed the number of possible outcomes from 200 to the 100 people under age 30. Each of these 100 people is equally likely to be chosen, and 60 of them are male. So the answer is $60/100 = .6$.

Whenever you are given information and are asked to find a probability based on this information, the result is a **conditional probability**. This probability is written as

$$P(A|B)$$

* The joint probability of events *A* and *B* is often written as $P(A \cap B)$, read as "the probability of *A* intersect *B*."

† The probability $P(A$ or $B)$ can be written as $P(A \cup B)$, read as "the probability of *A* union *B*."

where B is the event that you know occurred and A is the uncertain event whose probability you need, given that event B has occurred. The vertical line indicates that the occurrence of event B is given, so the expression is read as the "probability of A given B." In the example, $P(M|U)$.

Suppose that you were given *no information* about U and were asked to find the probability that a male is selected. This is a marginal probability. We earlier determined that $P(M) = .6$. For our example, note that

$$P(M) = P(M|U) = .6$$

This means that being given the information that the person selected is under 30 has *no effect* on the probability that a male is selected. In other words, whether U happens has no effect on whether M occurs. Such events are said to be independent. Thus, events A and B are **independent** if the probability of event A is unaffected by the occurrence or nonoccurrence of event B.

There are a number of ways to demonstrate that any two events A and B are independent.

DEFINITION

Events A and B are **independent** if and only if:

1. $P(A|B) = P(A)$ (assuming $P(B) \neq 0$), or   **(4.3)**
2. $P(B|A) = P(B)$ (assuming $P(A) \neq 0$), or   **(4.4)**
3. $P(A \text{ and } B) = P(A) \cdot P(B)$.   **(4.5)**

You need not demonstrate all three conditions. If one of the equations is true, they are all true; if one is false, they are all false (in which case A and B are not independent). Events that are not independent are **dependent** events.

In our example, are events F and O independent? We previously showed that

$$P(O) = 50/200 = .25$$

Since $P(O|F) = 10/80 = .125$, then $P(O) \neq P(O|F)$, and these events are dependent. Put another way, if someone informs you that event F (a female) has occurred, this *does* have an effect on whether or not the person selected is over 45 years of age. If you are told that F occurred, the probability that the selected person is over 45 *drops* from .25 to .125. These events do affect each other and so are dependent events.

We could also approach this by showing that $P(F|O)$ is not the same as $P(F)$:

$$P(F|O) = 10/50 = .2$$
$$P(F) = 80/200 = .4$$

These are not the same values, so events F and O are not independent.

The final option is to show that $P(F \text{ and } O)$ is not the same as $P(F) \cdot P(O)$. This follows since

$$P(F \text{ and } O) = 10/200 = .05$$
$$P(F) \cdot P(O) = (.4)(.25) = .1$$

In our discussion of joint probabilities, we showed that

$$P(F \text{ and } U) = 40/200 = .2$$

Consequently, events F and U *can both occur* because their joint probability is not zero.

How would you calculate $P(F$ and $M)$? One cannot be both a male and a female, so $P(F$ and $M) = 0$. Because events M and F cannot both occur, these events are said to be mutually exclusive.

DEFINITION

Events A and B are **mutually exclusive** if A and B cannot both occur simultaneously. To demonstrate that two events A and B are mutually exclusive, you must show that their joint probability is zero: $P(A$ and $B) = 0$.

EXAMPLE 4.1

The quality-control department of Lectron has selected ten devices for testing purposes. Which of these outcomes are mutually exclusive?

A = exactly one device is defective

B = more than two devices are defective

C = fewer than four devices are defective

SOLUTION

A and B are mutually exclusive events—they cannot both occur.

A and C are *not* mutually exclusive—if A occurs, so does event C.

B and C are *not* mutually exclusive—if three devices are defective, both events B and C will occur.

Note: By "not mutually exclusive" we do not mean that both of these events *must* occur, only that both *could* occur.

■

SUMMARY OF PROBABILITY DEFINITIONS

1. **Experiment:** An experiment is any process that yields a measurement (observation).
2. **Outcome:** An outcome is any particular result of an experiment.
3. **Event:** An event consists of one or more possible outcomes of an experiment.
4. **Complement:** The complement of event A is the event that A does not occur. This is written $\bar{A}$.
5. **Mutually exclusive events:** Two events are mutually exclusive if they cannot both occur simultaneously.
6. **Independent events:** Two events are independent if the probability of one event occurring is unaffected by the occurrence or nonoccurrence of the other.
7. **Probability:** A probability is a measure of the likelihood that an event will occur when the experiment is performed.
8. **Marginal probability:** A marginal probability is the probability that any one single event used to define a contingency table will occur.
9. **Joint probability:** The joint probability of events A and B is the probability that both A and B will occur. This is written as $P(A$ and $B)$.
10. **Conditional probability:** The conditional probability of A given B is the probability that event A occurs given that event B occurs. This is written $P(A|B)$.

EXERCISES

4.1 Which of the following values cannot be a probability? Why?

a. .02   **b.** 0   **c.** 5/4   **d.** 985/1051

4.2 Explain what the following terminologies mean about two events A and B.

a. Probability of A and B

b. Probability of A or B

c. Probability of the complement of A

4.3 If there are 20 sophomores, 10 juniors, and 5 seniors in a classroom, what is the probability of choosing a junior at random? Is this the relative frequency approach to estimating a probability?

4.4 Assume that 20 doctors are chosen at random from the Houston telephone directory. Of these 20, there are 15 who recommend Little's pills and 5 who do not. If a doctor was chosen at random from the city of Houston, estimate the probability that this doctor would recommend Little's pills.

4.5 Let A and B be two mutually exclusive events. Are the complement of A and the event B mutually exclusive?

4.6 Four hundred randomly sampled automobile owners were asked whether they selected the particular make and model of their present car mainly because of its appearance or because of its performance. The results were as follows:

OWNER	APPEARANCE	PERFORMANCE	TOTALS
Male	95	55	150
Female	85	165	250
Both	180	220	400

a. What is the probability that an automobile owner buys a car mainly because of its appearance?

b. What is the probability that an automobile owner buys a car mainly because of its appearance and that the automobile owner is a male?

c. What is the probability that a female automobile owner is a male?

d. What is the probability that a female automobile owner purchases the car mainly because of its appearance?

4.7 A large sports chain wants to know whether it should concentrate its advertising on the serious athlete or on the "weekend" athlete. The sports store also wants to know which sports are the most popular. The marketing department gathered the following information on 500 randomly selected customers:

ATHLETE	TENNIS	RUNNING	BASKETBALL	SWIMMING	SOCCER	RACQUETBALL	TOTAL
Serious	46	17	60	43	59	50	275
Weekend	54	63	20	37	11	40	225

a. What is the probability that a customer's favorite sport is basketball?

b. What is the probability that a customer is a weekend athlete?

c. What is the probability either that a customer is a serious athlete or that a customer's favorite sport is running?

d. What is the probability that a customer's favorite sport is not swimming?

4.8 The employment center at a university wanted to know the proportion of students who worked and also the proportion of those students who lived in the dorm. The following data were collected:

LIVING ARRANGEMENTS	WORK FULL TIME	WORK PART TIME	DO NOT WORK	TOTAL
In dorm	19	22	20	61
Not in dorm	25	9	5	39
				100

a. What is the probability of selecting a student at random who works either full or part time?

b. What is the probability that a student who works lives in the dorm?

c. What is the probability that a student either works full time or else does not live in the dorm?

d. Is the event that a student lives in the dorm independent of the event that a student works full time? Discuss what your answer means.

4.9 An investment-newsletter writer wanted to know in which investment areas her subscribers were most interested. A questionnaire was sent to 331 randomly selected professional clients, with the following results:

BUSINESS	STOCKS	BONDS	COMMERCIAL PAPER	COMMODITIES	STOCK OPTIONS	TOTAL
Doctors	30	25	15	2	0	72
Lawyers	29	34	12	0	5	80
Bankers	50	35	29	5	10	129
Others	21	14	10	3	2	50
						331

a. What is the probability that an investment client is neither a doctor nor a lawyer?

b. What is the probability that an investment client is a banker and that the investment client's main investment interest is in commodities?

c. If an investment client's main investment interest is commodities, what is the probability that he or she is a banker?

d. What is the probability that an investment client's main investment interest is not in stock options?

e. Let A be the event that an investment client is a lawyer. Let B be the event that an investment client's main investment interest is in commodities. Are the events A and B mutually exclusive?

4.10 If events A and B are mutually exclusive, is the occurrence of event A affected by the occurrence of event B? Can one say that if two events are mutually exclusive, they are not independent?

4.11 A large supermarket has 67 employees classified by job and by number of years of schooling. The following contingency table gives the categories:

JOB	≤8	9–10	11–12	13–14	15–16	TOTAL
Stocker	1	5	8	1	0	15
Checker	0	5	6	3	0	14
Meat cutter	1	3	7	1	0	12
Cashier	0	0	4	10	5	19
Manager	0	0	1	3	3	7
						67

a. What is the probability that an employee selected at random has 11 or more years of schooling?

b. What is the probability that an employee is either a manager or has 13 or more years of schooling?

c. What is the probability that a cashier has 15 to 16 years of schooling?

d. Let A be the event that an employee is a meat cutter. Let B be the event that an employee has 15 to 16 years of schooling. Are the events A and B mutually exclusive? Are the events A and B independent?

4.12 A statistics instructor wishes to find out the relationship between the classification of a student and the student's grade in the course. The following is a breakdown of three sections of an introductory statistics course:

GRADE	FRESHMAN	SOPHOMORE	JUNIOR	SENIOR
A	0	7	9	10
B	0	6	8	11
C	1	7	9	12
D	2	4	1	4
F	0	6	2	1
Total	3	30	29	38

a. Suppose that one student is randomly selected. What is the probability that the student is a junior and makes at least a B in the course?

b. What is the probability that a senior does not make an A in the course?

c. What is the probability that the student makes a D or F in the course?

d. Let A be the event that a sophomore is taking the course. Let B be the event that the student makes a C in the course. Are the events A and B independent? Are the events A and B mutually exclusive?

4.13 A local bank has 5276 accounts cross-classified by type of account and average account balance. The summarized results are (in dollars):

ACCOUNT BALANCE	CHECKING ACCOUNT	SAVINGS ACCOUNT	NEW ACCOUNT	MONEY-MARKET ACCOUNT	TOTAL
<500	1020	803	21	90	1934
500–1000	640	774	452	112	1978
>1000	51	659	538	116	1364
				Total	5276

a. What is the probability that an account does not have over $1000 in it and that the account is not a money-market account?

b. What is the probability that a new account's balance is between $500 and $1000?

c. What is the probability that an account has less than $500 in it or that the account is a savings account?

d. Given that an account is not a savings account, what is the probability that the account has $1000 or less in it?

4.14 If the probability that it is going to rain today is .3, what can you say about the probability that it is not going to rain today?

4.15 Give an example of two events that are mutually exclusive. Explain why they are mutually exclusive. Give an example of two events that are independent. Explain why they are independent.

4.3 GOING BEYOND THE CONTINGENCY TABLE

Our Datacomp survey served as an intuitive introduction to probability definitions. The classical approach was used to derive probabilities by dividing the number of outcomes favorable to an event by the total number of (equally likely) outcomes. Not all probability problems, however, are concerned with randomly selecting an individual from a contingency table.

When dealing with two or more events in general, one approach is to illustrate these events by means of a **Venn diagram.** A Venn diagram representing any two events A and B is shown in Figure 4.1.

In a Venn diagram, the probability of an event occurring is its corresponding area. This may sound complicated, but it really is not. The Venn diagram for $P(A) = .4$ is shown in Figure 4.2. The area of the rectangle is 1; it represents all possible outcomes. The shaded area is the complement of A, namely, $\bar{A}$. Here, $P(\bar{A}) = 1 - P(A) = 1 - .4 = .6$. No effort is made to construct

FIGURE 4.1
Venn diagram for events
A and B. The rectangle
represents all possible
outcomes of an
experiment.

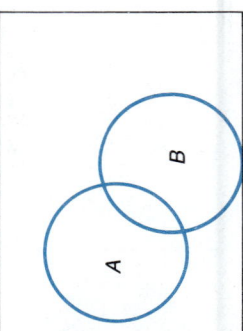

FIGURE 4.2
Venn diagram for
$P(A) = .4$.

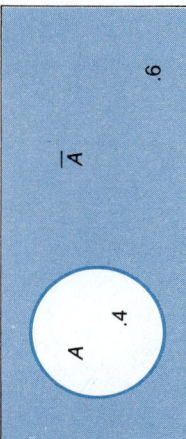

FIGURE 4.3
$P(A$ and $B)$. The points
in the shaded area are
in A and B.

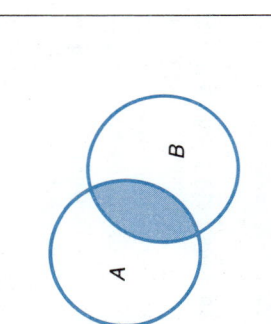

FIGURE 4.4
$P(A$ or $B)$. The points in
the shaded area are in
A or B.

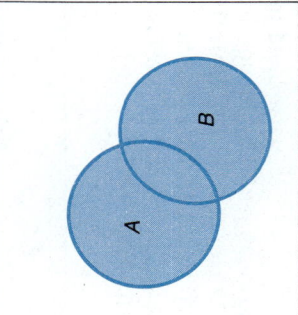

a circle with an area of .4; it is simply labeled .4. The shaded area then represents $\overline{A}$, and the corresponding area must be .6.

Figure 4.3 shows $P(A$ and $B)$, and Figure 4.4 shows $P(A$ or $B)$.

If A and B are mutually exclusive (they cannot both occur), $P(A$ and $B) = 0$. For example, an auto dealer has data that indicate that 20% of all new cars ordered contain a red interior, whereas 25% have a blue interior. Only one color is allowed when selecting an interior color. Let A be the event that a red interior is selected and B be the event that a blue interior is selected. A Venn diagram for this situation is shown in Figure 4.5.

Each person can select only one color, so events A and B are mutually

FIGURE 4.5
Venn diagram of mutually exclusive events. $P(A$ and $B) = 0$. $P(A$ or $B) = P(A) + P(B) = .2 + .25 = .45$.

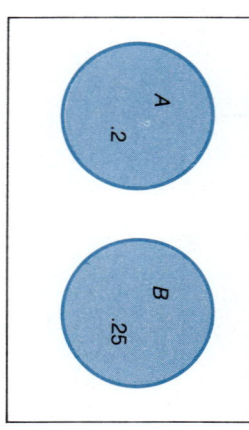

□ = $P(A$ and $B)$

▨ = $P(A$ or $B)$

FIGURE 4.6
A Venn diagram illustrating $P(A$ or $B)$ and $P(A$ and $B)$.

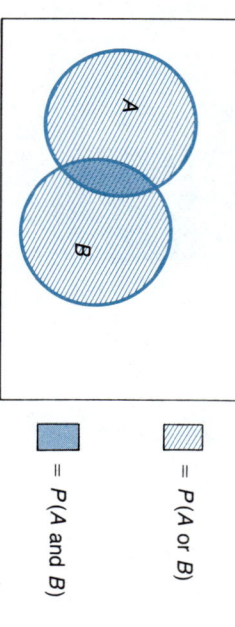

exclusive, and the resulting circles do not overlap in the Venn diagram. What is the probability that a person selects red or blue? This is $P(A$ or $B)$ and is represented by the shaded area in the circles in Figure 4.5. The Venn diagram allows us to see clearly that this shaded area is $P(A) + P(B) = .2 + .25 = .45$. In other words, 45% of the people will purchase either red or blue interiors. We thus have the following rule.

RULE

If events A and B are mutually exclusive, then

$$P(A \text{ or } B) = P(A) + P(B) \qquad (4.6)$$

This rule does *not* work when A and B can both occur, but there is an easy way to devise another solution. Look at the Venn diagram for this situation, shown in Figure 4.6. By adding $P(A) + P(B)$, we do not obtain $P(A$ or $B)$ because we have counted $P(A$ and $B)$ *twice*. So we need to subtract $P(A$ and $B)$ to obtain the actual area corresponding to $P(A$ or $B)$. This is the **additive rule of probability**.

ADDITIVE RULE

For any two events, A and B,

$$P(A \text{ or } B) = P(A) + P(B) - P(A \text{ and } B) \qquad (4.7)$$

Notice that if A and B are mutually exclusive, then $P(A$ and $B) = 0$, and we obtain the previous rule; namely, that $P(A$ or $B) = P(A) + P(B)$.

EXAMPLE 4.2

Draw a single card from a deck of 52 playing cards. Let S be the event that the card is a seven and H be the event that the card is a heart.

FIGURE 4.7
$P(S) = 4/52$;
$P(H) = 13/52$.

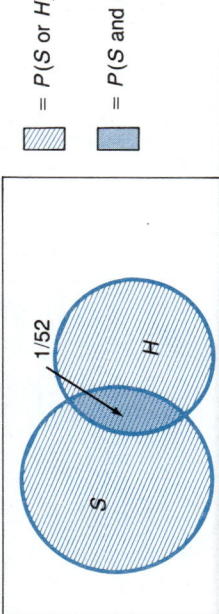

▨ = $P(S$ or $H)$

▨ = $P(S$ and $H)$

SOLUTION 1

1. Are these events mutually exclusive?
2. What is $P(S$ and $H)$?

Both S and H can occur. A seven of hearts is a possible outcome here. S and H are *not* mutually exclusive. There are (1) 52 equally likely outcomes (cards), (2) 4 sevens, and (3) 13 hearts, so

$$P(S) = 4/52$$
$$P(H) = 13/52$$

SOLUTION 2

$P(S$ and $H)$ is the probability of selecting a seven of hearts from the deck. There is only one such card, so

$$P(S \text{ and } H) = 1/52$$

A Venn diagram for this situation is shown in Figure 4.7. Using the additive rule, the proportion of draws (probability) that a seven *or* a heart will be selected from the deck is

$$P(S \text{ or } H) = P(S) + P(H) - P(S \text{ and } H)$$
$$= 4/52 + 13/52 - 1/52$$
$$= 16/52$$

Refer back to the Datacomp survey data in Table 4.1. Does the additive rule work here also? It does—this rule works for *any* two events—but it certainly is a hard way to solve this problem. Suppose we want to find the probability (from our previous example) that the person selected is a male or is under age 30. By inspection, we previously found that

$$P(M \text{ or } U) = 160/200 = .8$$

Using the additive rule, we obtain the same result:

$$P(M \text{ or } U) = P(M) + P(U) - P(M \text{ and } U)$$
$$= 120/200 + 100/200 - 60/200$$
$$= 160/200$$
$$= .8$$

■

Conditional Probabilities

Using the Datacomp survey data, we found that the probability the person selected is a male (M), given the information that the person selected is under 30 (U), was $P(M|U) = .6$. Our reasoning here was: (1) There are 100 people under 30 years of age, (2) 60 of them are male, (3) each of these 100 people is equally likely to be selected, and so (4) the result is 60/100 = .6. Notice that

$$P(U) = 100/200 = .5$$
$$P(M \text{ and } U) = 60/200 = .3$$
$$P(M|U) = P(M \text{ and } U)/P(U) = .3/.5 = .6$$

FIGURE 4.8
A Venn diagram illustrating a conditional probability.

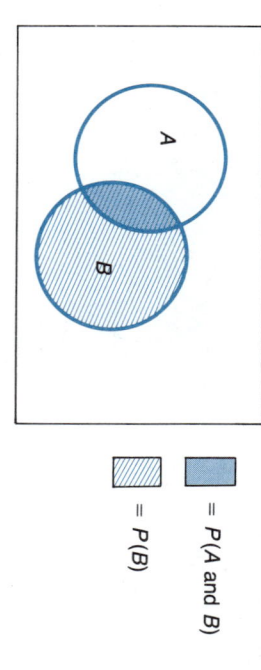

= $P(A \text{ and } B)$

= $P(B)$

This procedure for finding a conditional probability applies to *any* two events. Use the Venn diagram in Figure 4.8 to determine $P(A|B)$. Given the information that event B occurred, we are immediately restricted to the lined area (B). What is the probability that a point in B is also in A (that is, event A occurs) given this information? A point is also in A if it lies in the shaded area, and

$$P(A|B) = \frac{\text{shaded area}}{\text{striped area}}$$

$$= \frac{P(A \text{ and } B)}{P(B)}$$

This is the rule for conditional probabilities.

RULE FOR CONDITIONAL PROBABILITIES

For any two events, A and B,

$$P(A|B) = \frac{P(A \text{ and } B)}{P(B)} \qquad (P(B) \neq 0) \qquad \textbf{(4.8)}$$

and

$$P(B|A) = \frac{P(A \text{ and } B)}{P(A)} \qquad (P(A) \neq 0) \qquad \textbf{(4.9)}$$

Independent Events

In the discussion of the Datacomp survey example, a summary of how to demonstrate that two events are independent was provided in equations 4.3, 4.4, and 4.5. One need demonstrate only that one of these equations holds to verify independence. These three methods of proving independence apply to *any two events*, not just to contingency table applications.

To summarize, events A and B are **independent** if any of the following statements can be verified:

$$P(A|B) = P(A)$$
$$P(B|A) = P(B)$$
$$P(A \text{ and } B) = P(A) \cdot P(B)$$

In many situations, it is unnecessary (or impossible) to prove independence of two events. However, one can often argue convincingly that two events are independent or dependent without resorting to a mathematical proof. Consider these events:

> A = Procter and Gamble's new laundry detergent will capture at least 5% of the market next year

and

> B = General Motors will introduce a new line of compact automobiles next year

Whether event B happens should have no effect on whether event A occurs. So $P(A|B) = P(A)$, and these events are independent. Next, change event A to be: Toyota automobile sales will drop next year. Now whether event B occurs could very well have an effect on whether event A occurs. So it is not safe to assume that $P(A|B) = P(A)$ because it seems reasonable that $P(A|B)$ is *larger* than $P(A)$. Notice that we have not discussed the values of $P(A)$ and $P(A|B)$. The probability values are not necessary to show that the events are dependent. The important thing is that $P(A|B) \neq P(A)$, so these events are clearly dependent events.

Joint Probabilities

The rule for conditional probabilities in equations 4.8 and 4.9 can be rewritten as

MULTIPLICATIVE RULE

For any two events A and B,

$$P(A \text{ and } B) = P(A|B) \cdot P(B) \quad (4.10)$$
$$= P(B|A) \cdot P(A) \quad (4.11)$$

This is the **multiplicative rule of probability**. Using equation 4.5, we also have the following rule for two independent events.

RULE

For any two independent events A and B,

$$P(A \text{ and } B) = P(A) \cdot P(B) \quad (4.12)$$

You may be wondering how we can use the same equation to define the rule for $P(A|B)$ (equation 4.8) and the rule for $P(A \text{ and } B)$ (equation 4.10). This is not a bad question! It appears that we have used the same rule twice to make two different statements—and in fact we have. However, for any application you encounter, either $P(A|B)$ or $P(A \text{ and } B)$ must be provided or can be determined without resorting to formulas. We can clarify this using our card-drawing example:

> S = select a seven
>
> H = select a heart

Here $P(S \text{ and } H)$ (the probability of selecting a seven of hearts) is 1/52. No formulas were necessary to determine this, only a little head scratching.

Now, what is $P(S|H)$? Using equation 4.8,

$$P(S|H) = P(S \text{ and } H)/P(H)$$
$$= (1/52)/(13/52)$$
$$= 1/13$$

Assume that you select a card from a deck, examine it, and then discard it. You then select another card. This is called **sampling without replacement**. Let

A = selecting a seven on the first draw

B = selecting a seven on the second draw

What is the probability of drawing two sevens [$P(A \text{ and } B)$]? If you selected a seven on the first draw, then, of the 51 cards remaining, three are sevens. So $P(B|A) = 3/51$. Again, we used no formulas.

Next, we use the multiplicative rule, equation 4.11:

$$P(A \text{ and } B) = P(B|A) \cdot P(A)$$
$$= \left(\frac{3}{51}\right) \cdot \left(\frac{4}{52}\right) \approx .0045$$

Notice that $P(A) = 4/52$ because there are four sevens available on the first draw. So you would expect to draw two sevens from a card deck about 45 times out of 10,000, if you are drawing without replacement.

Now suppose you select a card from a deck but replace it before selecting the second card. This is called **sampling with replacement**. What is $P(B|A)$? There are still 52 cards in the deck when you select your second card, and four of these are sevens. So

$$P(B|A) = 4/52 = P(B)$$

If event A occurs, the probability of a seven on the second draw is unaffected. This probability is 4/52 *whether or not A occurs*; these events are now independent. For this situation,

$$P(A \text{ and } B) = P(A|B) \cdot P(B)$$
$$= P(A) \cdot P(B) \quad \text{(since they are independent)}$$
$$= 4/52 \cdot 4/52 = .0059$$

The probability of getting two sevens is higher when drawing cards with replacement—not a surprising result.

4.4

APPLYING THE CONCEPTS

EXAMPLE 4.3

SOLUTION

In a particular city, 20% of the people subscribe to the morning newspaper, 30% subscribe to the evening newspaper, and 10% subscribe to both. Determine the probability that an individual from this city subscribes to the morning newspaper, the evening newspaper, or both.

The most important thing when solving a wordy probability problem is to set up the problem correctly. Your first step when solving any probability application should always be to *define* the events clearly using capital letters. Your initial step should be to define

M = person subscribes to the morning newspaper

E = person subscribes to the evening newspaper

We do not need to define another event for a person subscribing to both newspapers, as we shall see.

We now have

$$P(M) = .2$$
$$P(E) = .3$$

The probability that a selected individual subscribes to the morning *and* the evening newspaper is given as .10. This is a *joint* probability:

$$P(M \text{ and } E) = .1$$

We want to find the probability of M or E. Using the additive rule,

$$P(M \text{ or } E) = P(M) + P(E) - P(M \text{ and } E)$$
$$= .2 + .3 - .1$$
$$= .4$$

So 40% of the people in this city subscribe to at least one of the two newspapers.

Suppose we also know that 1/3 of the evening newspaper subscribers are also morning newspaper subscribers. How can you translate this statement into a probability? Another way of stating the preceding sentence is, "Given that you subscribe to the evening newspaper, the probability that you also subscribe to the morning newspaper is 1/3." In other words, this is a *conditional* probability:

$$P(M|E) = 1/3$$

■

EXAMPLE 4.4

Referring to the subscription data in Example 4.3, what percentage of the evening subscribers do not subscribe to the morning newspaper?

SOLUTION

A Venn diagram for this problem is shown in Figure 4.9. Notice that M (the morning subscribers) is made up of two components, (1) those people in E (the evening subscribers) and (2) those not in E. Since $P(M \text{ and } E) = .1$, the area of M that is striped is

$$P(M) - P(M \text{ and } E) = P(M \text{ and } \bar{E})$$
$$= .2 - .1 = .1$$

Similarly, the area of E that is striped is

$$P(E) - P(M \text{ and } E) = P(E \text{ and } \bar{M})$$
$$= .3 - .1 = .2$$

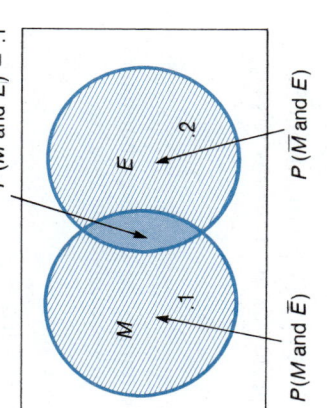

FIGURE 4.9
Venn diagram for
Example 4.4.

$P(M \text{ and } \bar{E})$

$P(\bar{M} \text{ and } E)$

$P(M \text{ and } E) = .1$

Our question could be stated, "Given that a person subscribes to the evening newspaper, what is the probability that this person does not subscribe to the morning newspaper?" This is the *conditional* probability

$$P(\overline{M}|E)$$

Look at the Venn diagram. You know that E occurred, so the outcome is in the E circle. What is the probability that the outcome is not in M? We know that the total area of E is .3 and that the area that is not in M but is in E is .2. So

$$P(\overline{M}|E)$$

Another approach here is to utilize the formulas in the section by noting that given event E has occurred, either event M occurs or it doesn't. Consequently,

$$P(\overline{M}|E) = .2/.3 = 2/3$$

$$
\begin{aligned}
P(\overline{M}|E) &= 1 - P(M|E) \\
&= 1 - [P(M \text{ and } E)]/P(E) \\
&= 1 - (.1/.3) = 2/3
\end{aligned}
$$

■

HELPFUL HINTS FOR PROBABILITY APPLICATIONS

1. Define each event using capital letters.
2. Translate each statement into a probability. Is a particular statement telling you $P(A)$? $P(B)$? $P(A$ and $B)$? $P(A$ or $B)$? $P(A|B)$? $P(B|A)$?
3. Determine the answer by identifying the probability rule that applies and by using a Venn diagram. Using both allows you to check your logic and your arithmetic.

EXAMPLE 4.5

In a certain northeastern state going through financial difficulties, it is believed that 5% of the banks will fail. It is known that, within this state, 90% of the banks have deposits insured by the Federal Depository Insurance Company (FDIC). It is also believed, from past experience, that 3% of the banks protected by FDIC will fail. A bank examiner employed by the federal government would like to know:

1. What is the probability that, for a randomly chosen bank, the bank has deposits protected by FDIC and the bank will fail?
2. What is the probability that, for a randomly chosen bank, the bank has deposits covered by FDIC or the bank will fail?
3. What percentage of the banks that go under have deposits protected by FDIC?

SOLUTION 1

The first step is to define appropriate events.

$$A = \text{bank has deposits protected by FDIC}$$
$$B = \text{bank will fail}$$

We now translate each of the statements into a probability. We have the following marginal probabilities:

$$P(A) = .90$$
$$P(B) = .05$$

The last statement in the problem can be written as, "Given that a bank has accounts protected by FDIC, the probability that the bank will fail is .03." So this is a conditional probability, namely,

$$P(B|A) = .03$$

What is question 1 asking for? $P(A$ or $B)$? $P(A|B)$? $P(A$ and $B)$? The examiner wishes to know the probability that a bank is protected by FDIC *and* will fail. This is $P(A$ and $B)$. Using the multiplicative rule,

$$P(A \text{ and } B) = P(B|A) \cdot P(A)$$
$$= (.03)(.90) = .027$$

Thus, 92.3% of the banks covered by the FDIC, will fail, or both.

SOLUTION 2

For question 2, we wish to know the probability that A or B occurs. By the additive rule,

$$P(A \text{ or } B) = P(A) + P(B) - P(A \text{ and } B)$$
$$= .90 + .05 - .027 = .923$$

SOLUTION 3

Question 3 can be phrased as, "Given that a bank has failed, what is the probability that this bank has deposits protected by FDIC?" This is $P(A|B)$.

$$P(A|B) = [P(A \text{ and } B)]/P(B) = .027/.05 = .54$$

Therefore, 54% of those banks that fail have deposits protected by FDIC. ▪

EXERCISES

4.16 If $P(A) = .5$, $P(B) = .3$, and $P(A|B) = .4$, what is $P(B|A)$?

4.17 Let $P(A) = .7$, $P(B) = .3$, and $P(A$ and $B) = .2$. Find the following probabilities:

  a. $P(\overline{A})$ **b.** $P(A$ or $B)$ **c.** $P(B|A)$

  d. $P(A$ and $\overline{B})$ **e.** $P(A|\overline{B})$ **f.** $P(\overline{A}$ and $\overline{B})$

  g. $P(\overline{A \text{ and } B})$

4.18 If $P(A) = .4$, $P(B) = .5$, and $P(A$ or $B) = .8$, what is $P(A$ and $B)$?

4.19 If $P(A|B) = .8$ and $P(A$ and $B) = .6$, what is the $P(B)$?

4.20 If $P(A) = .5$, $P(B) = .2$, and $P(A$ or $B) = .7$, are the events A and B mutually exclusive? Explain.

4.21 If $P(A) = .5$ and $P(B) = .6$, are the events A and B mutually exclusive? Explain.

4.22 If $P(A) = .4$, $P(B) = .3$, and $P(A$ and $B) = .12$, what is the probability of A given B? Are the events A and B independent?

4.23 If a penny, a nickel, and a dime are flipped, what is the probability of getting three heads, given that the flip of the penny resulted in a head?

4.24 Suppose one card is randomly picked from a deck of 52 playing cards. Event A is the occurrence of a king. Event B is the occurrence of a spade.

  a. What is the probability of A and B?

  b. What is the probability of A or B?

  c. What is the probability of A given B?

4.25 A manufacturer of widgets historically has produced 80 good widgets out of every 100 widgets. If two widgets are randomly selected off the assembly line, what is the probability that both widgets will be nondefective? What is the probability that two randomly selected widgets will be defective?

4.26 A manufacturer claims that a customer has a 30% chance of noticing a particular flaw in a dress it makes. Two dresses, one without the flaw and one with the flaw, are given to two customers to see whether they can recognize the dress with the flaw.

a. What is the probability that a customer will notice the flaw if each of the two customers examines a different dress?

b. What is the probability that a customer will notice the flaw if the two customers examine both dresses?

c. Assume that a customer randomly selects one of the two dresses and then examines the selected dress. What is the probability that the customer will find a flaw?

d. If both customers select a dress at random and examine the dress in the manner described in part (c), what is the probability that both customers will find a flaw?

4.27 If the probability that a person orders the morning newspaper is .5 and the probability that a person orders the evening newspaper is .3, and if the probability that a person orders at least one of the two newspapers is .7, then what is the probability that a person orders both the morning and evening newspapers?

4.28 At a certain university, 30% of the students major in mathematics. Of the students majoring in mathematics, 60% are males. Of all the students at the university, 70% are males.

a. What is the probability that a student selected at random in the university is a male majoring in mathematics?

b. What is the probability that a student selected at random in the university is a male or is majoring in mathematics?

c. What proportion of the males are majoring in mathematics?

4.29 At a semiconductor plant, 60% of the workers are skilled and 80% of the workers are full time. Ninety percent of the skilled workers are full-time.

a. What is the probability that an employee, selected at random, is a skilled full-time employee?

b. What is the probability that an employee, selected at random, is a skilled worker or a full-time worker?

c. What percentage of the full-time workers are skilled?

4.30 A supermarket has 40% of its merchandise on sale. Twenty percent of its merchandise consists of nonedible items. Fifty percent of the sale items consists of nonedible items.

a. What is the probability that an item, selected at random in the supermarket, is nonedible and on sale?

b. What is the probability that an item, selected at random, is either nonedible or on sale?

c. What proportion of nonedible items are on sale?

4.31 For every person who visits a leasing office of an apartment community near a certain university, there is an 80% chance that the person will lease an apartment if the person is a student and a 50% chance that the person will lease an apartment if the person is not a student. If two people, one of whom is a student and the other is not, enter the office, what are the chances of leasing an apartment to at least one of the two people? What assumption did you have to make here?

4.32 An independent oil-drilling company drills wildcat oil wells. So far, 60% of the wells have been oil-producing and 40% have been dry. A private investor wishes to go into a partnership with the oil company in two wells. Assuming that the outcome of one well does not affect the outcome of the other well, what is the probability that both of the oil wells in the partnership will produce oil? What is the probability that at least one of the two wells will produce oil?

4.33 Use the additive rule for the data in Exercise 4.7(c) to find the probability that a customer is either a serious athlete or that a customer's favorite sport is running.

4.34 In an article describing the participation of business on state policy-making about medical care in the 1980s, 18 of the 50 states in the United States were classified as having high business participation. States with high business participation were more likely to have more active alternative delivery systems such as health maintenance organizations (HMOs) and preferred provider organizations (PPOs). In 17 of the states with high business participation, PPOs were either starting up or highly active.

(Source: Linda A. Bergthold, "Business and the Pushcart Vendors in an Age of Supermarkets," *International Journal of Health Services* 17, no. 1: 7–17.)

a. What is the probability that a state in the United States is not among the states with high business participation?

b. What is the probability a state in the United States has high business participation, but does not have any PPOs starting up or active?

c. What is the probability that a state with high business participation does not have any PPOs starting or active?

4.35 In 1986, 14 of the 50 states in the United States had a state per capita income over $15,000. Let A and B be two states that are sampled at random from the 50 states of the United States without replacement. What is the probability that either state A or B has a state per capita income over $15,000.

(Source: The World Almanac and Book of Facts, 1988, p. 84.)

We illustrate what happens when you encounter more than two events by using three events, *A*, *B*, and *C*. The following rules can easily be extended to any finite number of events. In the applications of probability in the chapters that follow, we usually will be dealing with multiple events that are either mutually exclusive or independent.

Mutually Exclusive Events

Events *A*, *B*, and *C* are mutually exclusive if no two events can occur simultaneously. A Venn diagram of this situation is shown in Figure 4.10. When dealing with mutually exclusive events, we usually will be interested in the probability that *one* of these events will occur—that is *P(A or B or C)*. We can use a simple rule here:

> For mutually exclusive events, *A*, *B*, and *C*,
>
> $$P(A \text{ or } B \text{ or } C) = P(A) + P(B) + P(C) \qquad (4.13)$$

Thus, to determine "or" probabilities when the events are mutually exclusive, you add the respective probabilities.

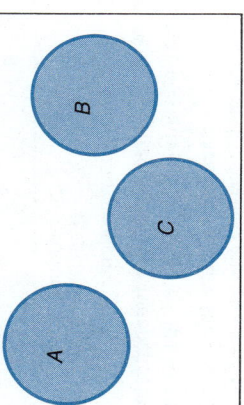

FIGURE 4.10
Three mutually exclusive events.

Independent Events

Events A, B, and C are independent if all the following are true:

$$P(A \text{ and } B) = P(A) \cdot P(B)$$
$$P(A \text{ and } C) = P(A) \cdot P(C)$$
$$P(B \text{ and } C) = P(B) \cdot P(C)$$
$$P(A \text{ and } B \text{ and } C) = P(A) \cdot P(B) \cdot P(C)$$

Thus the events are independent if the "and" probability for *any* subset of the events (including the set containing all the events) is equal to the corresponding product of marginal probabilities. When dealing with independent events, the probability of interest usually is that *all* of the events occur—that is, $P(A$ and B and $C)$. Using the fourth condition above, we can make the following statement.

For independent events, A, B, and C,

$$P(A \text{ and } B \text{ and } C) = P(A) \cdot P(B) \cdot P(C) \qquad (4.14)$$

Thus, to determine "and" probabilities when the events are independent, you multiply the respective probabilities.

EXAMPLE 4.6

Dellex Industries makes memory units for a microcomputer. Dellex customers have agreed to select three units randomly from each shipment and test them. If none of the units is defective, the customer will accept the shipment.

Usually, 2% of all Dellex units are defective. Determine the probability that a shipment will be accepted—that is, that all three units tested will be nondefective.

SOLUTION

Let

A = first unit is nondefective
B = second unit is nondefective
C = third unit is nondefective

We know that 2% of the units produced are defective, so 98% of them are not defective; consequently, $P(A) = P(B) = P(C) = .98$.

We want to find $P(A$ and B and $C)$. Are the events independent? There is no need to use fancy formulas here. The answer is yes, simply because the units are selected randomly from the entire shipment. Therefore,

$$P(A \text{ and } B \text{ and } C) = P(A) \cdot P(B) \cdot P(C)$$
$$= .98 \cdot .98 \cdot .98 = .94$$

So 94% of shipments will be accepted. ■

EXERCISES

4.36 Let $P(A) = .3$, $P(B) = .4$, $P(C) = .5$, and $P(A$ and B and $C) = .06$. Are the events A, B, and C independent?

4.37 If three events are independent, are the three events mutually exclusive? If the three events are mutually exclusive, then are the three events independent?

4.38 In Exercise 4.12, are the events "receiving the grade of A," "receiving the grade of B," and "receiving the grade of C or better" mutually exclusive? What is the probability that at least one of these events will occur?

4.39 In Exercise 4.9, let A be the event that a subscriber is a banker. Let B be the event that a subscriber is a lawyer. Let C be the event that the subscriber is a doctor.

a. Are the events A, B, and C mutually exclusive?

b. What is $P(A$ or B or $C)$?

c. What is $P(A$ and B and $C)$?

4.40 In Exercise 4.11, let A be the event that an employee has 8 years or less of schooling. Let B be the event that the employee has 9–10 years of schooling. Let C be the event that an employee has 15–16 years of schooling.

a. Are the events A, B, and C mutually exclusive?

b. What is $P(A$ or B or $C)$?

c. What is $P(A$ and B and $C)$?

4.41 Three cards are picked with replacement from a deck of 52 playing cards. What is the probability that the first card will be a queen, that the second will be a spade, and that the third will be a king?

4.42 At a university, 40% of the accounting majors, 20% of the marketing majors, and 15% of the finance majors are from out of state. If a student is selected randomly from each of these three majors, what is the probability that all three are from out of state?

4.43 A retailer receives, on the average, one defective calculator out of every 20. What is the probability that three calculators, randomly selected by the retailer, are nondefective?

4.6

COUNTING RULES (OPTIONAL)

Counting rules determine the number of possible outcomes that exist for a certain broad range of experiments. They can be extremely useful in determining probabilities. For instance, consider an experiment that has 200 possible outcomes, all of which are equally likely to occur. The probability of any one such outcome is $1/200 = .005$.

The question we wish to answer here is, For a particular experiment, how many possible outcomes are there? No set of rules applies to all situations, but we will consider three very popular counting procedures: (1) filling slots, (2) permutations (a special case of filling slots), and (3) combinations.

Filling Slots

We use **counting rule 1** to fill k different slots. Let

n_1 = the number of ways of filling the first slot

n_2 = the number of ways of filling the second slot *after* the first slot is filled.

n_3 = the number of ways of filling the third slot *after* the first two slots are filled

$\vdots$

n_k = the number of ways of filling the kth slot *after* filling slots 1 through $k-1$.

the number of ways of filling all the k slots is

$$n_1 \cdot n_2 \cdot n_3 \cdot \cdots \cdot n_k$$

EXAMPLE 4.7 When ordering a new car, you have a choice of eight interior colors, ten exterior colors, and four roof colors. How many possible color schemes are there?

SOLUTION There are three slots to fill here, (eight) interior color, (ten) exterior color, and

EXAMPLE 4.8

(four) roof color. To answer the question, you simply *multiply* the number of ways of filling each slot. So the answer is $8 \cdot 10 \cdot 4 = 320$ different color schemes.

The order in which you fill the slots is unimportant. So $n_2 = 10$, regardless of whether or not you have filled the first slot. For some applications, this is not the case. Consider the following example.

A local PTA group is selecting their officers for the current year. There are 15 individuals in the group, whom we label as $I_1, I_2, \ldots, I_{15}$. They need to select a president, vice president, secretary, and treasurer. How many possible groups of officers are there?

SOLUTION

We have four slots to fill here, president (n_1), vice president (n_2), secretary (n_3), and treasurer (n_4). We know that n_1 is 15. After a president is elected, only 14 people remain, so $n_2 = 14$. By a similar argument, $n_3 = 13$ and $n_4 = 12$. The answer is $15 \cdot 14 \cdot 13 \cdot 12 = 32{,}760$ different slates of officers.   ■

Permutations

Example 4.8 is a counting situation in which you select people *without replacement*. If a particular person, say I_3, is elected president, then I_3 is not available to fill the remaining slots. Another way of stating the result is that there are 32,760 ways of selecting four people out of 15, where the **order of selection** is important. For example,

$$I_2 = \text{president}$$
$$I_6 = \text{vice president}$$
$$I_{12} = \text{secretary}$$
$$I_7 = \text{treasurer}$$

is not the same slate of officers as

$$I_7 = \text{president}$$
$$I_{12} = \text{vice president}$$
$$I_2 = \text{secretary}$$
$$I_6 = \text{treasurer}$$

even though the same four people are involved.

The number of ways of selecting k objects (or people) from a group of n distinct objects, where the order of selection is important, is referred to as the number of permutations of n objects using k at a time. This is written

$$_nP_k$$

In Example 4.8, $_{15}P_4 = 32{,}760$. Determining the number of permutations is just a special case of counting rule 1; this is also a slot-filling application.

The symbol $n!$ is read as "*n* factorial." Its value is determined by multiplying n by all the positive integers smaller than n.

For example,

$$n! = (n)(n-1)(n-2)\cdots(2)(1) \tag{4.15}$$

$$5! = (5)(4)(3)(2)(1) = 120$$
$$1! = 1$$
$$0! = 1 \quad \text{(by definition)}$$

Notice that $n!$ is the number of ways of filling n slots using n objects. There are

n ways of filling the first slot, $(n - 1)$ ways of filling the second slot, $(n - 2)$ ways for the third slot, and so on.

In Example 4.8, the result was obtained by finding $15 \cdot 14 \cdot 13 \cdot 12 = 32{,}760$. This also can be written as

$$\frac{15 \cdot 14 \cdot 13 \cdot 12 \cdot \cancel{11} \cdot \cancel{10} \cdot \cancel{9} \cdot \cancel{8} \cdot \cancel{7} \cdot \cancel{6} \cdot \cancel{5} \cdot \cancel{4} \cdot \cancel{3} \cdot \cancel{2} \cdot \cancel{1}}{\cancel{11} \cdot \cancel{10} \cdot \cancel{9} \cdot \cancel{8} \cdot \cancel{7} \cdot \cancel{6} \cdot \cancel{5} \cdot \cancel{4} \cdot \cancel{3} \cdot \cancel{2} \cdot \cancel{1}} = 32{,}760$$

This is an application of **counting rule 2:** The number of **permutations** of n objects using k objects at a time is

$$_nP_k = \frac{n!}{(n - k)!} = (n)(n - 1) \cdots (n - k + 1) \qquad (4.16)$$

EXAMPLE 4.9

How many two-digit numbers can you construct using the digits 1, 2, 3, and 4, without repeating any digit?

SOLUTION

The order of selection is certainly important here—the number 42 is not the same as 24. The answer is $_4P_2$, where

$$_4P_2 = \frac{4!}{(4 - 2)!} = \frac{(4)(3)(\cancel{2!})}{\cancel{2!}} = 12$$

These 12 permutations are

12	21	31	41
13	23	32	42
14	24	34	43

■

Combinations

Take another look at Example 4.8, where we selected 4 people from a group of 15. This time, however, choose a committee of 4 people from a group of 15 where the order of selection does not matter. Each such committee is one *combination* of the 15 people, using four at a time. For example,

$$I_2 \quad I_6 \quad I_{12} \quad I_7$$

and

$$I_7 \quad I_{12} \quad I_2 \quad I_6$$

are different permutations but are the same combination. These two arrangements are made up of the same individuals; hence they form the same committee or combination.

Clearly, there are not as many combinations (using I_2, I_6, I_{12}, I_7) as there are permutations. The two preceding permutations form the same combination. There are 24 possible permutations of this combination $(4 \cdot 3 \cdot 2 \cdot 1 = 24)$. Now we wish to determine how many possible committees (combinations) of four there are for the group of 15. This is written as

$$_{15}C_4$$

Each combination has 24 permutations, so

$$_{15}C_4 = \frac{_{15}P_4}{24} = \frac{32{,}760}{24} = 1365$$

There are 1365 possible committee combinations. Notice that 24 is the number of *permutations* of these four numbers (2, 6, 7, and 12); that is, $24 = {}_4P_4 = 4!$

Counting rule 3 is used to count the number of possible combinations. The number of **combinations** of n objects using k at a time is

$$_nC_k = \frac{_nP_k}{k!} = \frac{n!}{k!(n-k)!} \qquad (4.17)$$

EXAMPLE 4.10

A company must select 5 employees from a department of 40 people to attend a national conference. How many possible delegations are there?

SOLUTION

The order of selection is not a factor here, so this is a combination problem rather than a permutation problem. The answer is

$$_{40}C_5 = \frac{40!}{5!35!}$$

$$= \frac{(40)(39)(38)(37)(36)(35!)}{(5!)(35!)} = \frac{\overset{8}{(40)}\overset{13}{(39)}\overset{19}{(38)}(37)(36)}{(5)(4)(3)(2)(1)} = 658{,}008 \qquad ∎$$

EXERCISES

4.44 A cafeteria serves four different vegetables, five different main dishes consisting of either fish or meat, and three different desserts. If a customer chooses one serving from each of these three categories, how many different combinations of vegetables, main dishes, and desserts are possible?

4.45 How many different three-digit numbers can be constructed using the digits 3, 5, 7, and 9 if no digit can be repeated?

4.46 How many different ways can you select four playing cards from a deck of 52 such that the first is a heart, the second is a diamond, the third is a club, and the fourth is a spade?

4.47 Five offices are available for five recently hired junior executives. How many different ways can the five junior executives be assigned to the offices?

4.48 Seven assistant professors have applied for tenure at a university. However, only two assistant professors can be granted tenure. How many different sets of two assistant professors can be selected to receive tenure?

4.49 A shoe salesperson has ten different western boots to display in her showcase window. She can display only four at one time. How many different sets of four western boots can the salesperson select?

4.50 Six chairs are available for the six typists at a certain firm. How many seating arrangements are possible?

4.51 A committee of ten people needs to select a committee chairperson and committee secretary. How many different ways can the chairperson and secretary be selected from this committee?

4.52 A builder has five different house styles and three lots on which to build. If each lot has a different style of house on it, how many sets of the five house styles are possible on these three lots?

4.53 A firm has 100 laborers, 20 salespersons, and 10 executives. If an employee is chosen from each of these categories, how many different sets of three employees are possible?

4.54 To cut down on overhead, a small business decides to reduce the number of keypunchers employed. How many different sets of five people can be selected from 25 keypunchers if the firm decides to reduce the number of keypunchers by 5?

4.55 A person has six different-colored shirts and ten different-colored trousers. How many color schemes are possible if one shirt and one pair of trousers are chosen?

4.56 A company is trying to encourage women to fill executive positions. For the latest batch of executive trainees, it wishes to fill seven vacancies with 5 women and 2 men. The company has 7 women and 8 men, making a total of 15 finalists for these seven vacancies.

a. In how many ways can the 7 vacancies be filled if sex is disregarded?

b. In how many ways can the 7 vacancies be filled if the company insists on 5 women and 2 men?

c. What is the probability that the company gets the combination it wants if positions are filled randomly?

4.57 Ten employees have the option of selecting the day shift or the night shift. How many different sets of employees can be found such that only two choose the night shift?

4.58 Initial computer screening has given the IRS auditor 12 income tax returns to work on. The auditor intends to select 5 returns randomly from these 12. How many different sets of 5 returns are possible? If John, Mary, Tom, Ahmed, and Lim are among the 12 originally screened by the computer, what is the probability that all five of them will be selected by this auditor?

SIMPLE RANDOM SAMPLES (OPTIONAL)

Practically all the applications in later chapters that use probabilities derived from sample results to make a decision concerning the population are based on the assumption of a simple random sample or, more simply, a random sample.

We introduced this concept in Chapter 1. A sample of size n, selected from a population of size N, constitutes a *simple random sample* if every possible sample of size n has the same probability of being selected.

In Example 4.10, we determined that there were 658,008 possible delegations when selecting 5 people from a group of 40. If we view this group as the population of interest, then $n = 5$ and $N = 40$. Our concern here is to determine the probability that any specified group of five individuals will be the designated delegation if simple random sampling is used.

Notice here that we do not allow any one person to be selected more than once (that is, all five members of the delegation are different people). In effect, we randomly select the first individual from the group of 40, randomly select the second individual from the remaining group of 39, randomly select the third individual from the remaining group of 38, and so on. Consequently, at each step we do *not* replace the people previously selected for the sample when selecting the next individual. Such a sampling procedure is called **sampling without replacement**. This procedure can be simplified in practice by randomly selecting 5 different individuals *at one time* from the group of 40.

When the sample data are obtained one at a time by returning a person or object to the population prior to selection of the next sample value, this procedure is termed **sampling with replacement**. Using this scheme, a particular person (or object) can be selected *more than once* in the sample. These sampling procedures will be discussed further in Chapter 7.

For this illustration there are $_{40}C_5 = 658,008$ possible delegations when selecting without replacement, and each has the same probability of being selected. Therefore, the probability that any one combination of people will be picked is $1/658,008 = .000002$.

In general, when employing a simple random sample of size n, selected without replacement from a population of size N, the total number of possible random samples is

$$\frac{1}{{}_N C_n}$$

Also, each of these samples has a probability of being selected equal to

$$\frac{1}{{}_N C_n}$$

EXAMPLE 4.11

Your task is to obtain a random sample of two individuals selected without replacement from a group of five employees (E_1, E_2, E_3, E_4, and E_5). What is the probability that you select E_2 and E_5 as your sample?

SOLUTION

There are

$${}_5 C_2 = \frac{5!}{2!3!} = 10$$

possible random samples. They are:

SAMPLE NUMBER	SAMPLE
1	E_1, E_2
2	E_1, E_3
3	E_1, E_4
4	E_1, E_5
5	E_2, E_3
6	E_2, E_4
7	E_2, E_5
8	E_3, E_4
9	E_3, E_5
10	E_4, E_5

Each random sample (including the one containing E_2 and E_5) has a probability of being selected of $1/10 = .1$. ∎

Obtaining a Random Sample (Without Replacement)

When N is small, we can put the N names in a hat and pick out n of them for our sample. This will constitute a random sample (if you do not have a hat, improvise). When N is moderately large (for example, 10,000), we need a more practical method of selecting n items from this population. A common procedure is to select n **random numbers** between 1 and 10,000 using a table of random numbers or a computer-generated list of n random numbers. A list of random numbers is provided in Table A.13 at the end of the text. To generate a list of random numbers between 1 and 10,000, one procedure you could use is to:

1. Start in any arbitrary position, such as row 5 of column 3.
2. Select a list of random numbers by reading either across or down the table.
3. For each five-digit number selected, place a decimal before the final digit and round this value to the nearest counting number; for example, 24127 would become 2412.7, which is then rounded to 2413.

A computer-generated list of random numbers is easiest to use. Figure 4.11 contains the instructions for generating 100 random numbers between 1 and 10,000 using MINITAB. We give you the necessary commands in SAS and SPSS at the end of the chapter.

Using either procedure, let us assume that the resulting set of random numbers is 415, 6962, and 4815 (for $n = 3$). Then you must (1) code your population from 1 to 10,000 in some manner and (2) select individuals 415, 4815, and 6962 for your random sample. This topic and extensions of simple random sampling are further discussed in Chapter 7.

FIGURE 4.11

Generating 100 random numbers using MINITAB.

```
MTB > RANDOM 100 VALUES INTO C1;          ← Input: command to generate
SUBC> UNIFORM A=1 AND B=10000.              100 random numbers
MTB > PRINT C1                              between 1 and 10,000

C1
```

7478.70	4806.13	702.13	7650.34	6264.10	2966.48	722.85
241.21	29.66	3584.07	7929.06	1107.40	8314.07	9237.93
4732.05	1441.28	54.54	6692.89	6570.06	1215.40	1815.62
6850.35	6254.82	1806.54	5715.44	4377.34	7098.00	7213.67
1674.91	9259.29	7402.37	5264.13	7956.73	4565.76	653.35
1552.95	4014.14	1693.66	1604.75	490.24	1161.95	5133.46
1622.26	2678.70	4746.88	3294.51	1730.14	6164.95	571.34
1300.24	2421.46	2588.30	3445.97	664.86	2990.92	3778.03
2176.35	1947.26	3307.84	3397.42	4595.07	4316.88	9539.29
2406.14	673.40	4058.78	7273.44	9145.90	3227.31	3330.20
6192.00	3952.82	4027.23	3329.65	6122.72	5291.39	1365.13
534.43	6686.30	5746.26	8229.06	8610.19	6256.28	1989.36
8569.81	1209.32	1055.74	1855.96	1893.85	6629.71	8671.19
3882.93	5290.70	1279.76	9860.55	2567.28	818.11	2150.02
8654.63	1812.48					

Output: 100 random numbers

MINITAB, SAS, and SPSS all have options available that allow you to sample randomly from a stored data set and save the results for further analysis. You can find the necessary commands to carry out this procedure in the appropriate user's manual.

To obtain a representative sample when N is extremely large or unknown requires good judgment. Stopping the first ten people you meet on the street is a very poor way of sampling your population.

You sometimes may be forced to select items for the sample that represent the population as accurately as possible, realizing that a poorly gathered sample can easily lead to an incorrect decision of significant importance. Accountants often encounter this problem when performing a statistical audit. However, when such a sample is *not* a random sample, it is not correct to use probability theory in your analysis.

EXERCISES

4.59 An instructor requires each of 35 students in a class to write a term paper comparing nonparametric with parametric statistics. Two students are selected to present their term papers in class. What is the probability that Georgia and Fred (two students in the class) will be picked?

4.60 There are eight identically shaped objects in a jar. Two objects are selected randomly without replacement. If the objects are numbered one through eight, what is the probability that a six and a seven are drawn?

4.61 A firm has 12 skilled technicians, 8 of whom have college degrees. If a group of 4 technicians is chosen randomly to form a production team, what is the probability that none of the 4 will have a college degree? Do the 4 selected at random constitute a simple random sample?

4.62 Explain how to select a simple random sample for a population of 100,000 using the computer-generated random numbers in Table A.13.

4.63 The school-newspaper photographer takes ten different pictures of the homecoming queen at the school's football game. All ten pictures are excellent, so the photographer chooses two at random to place in the school newspaper. What is the probability that the first two pictures taken will be selected?

SUMMARY

This chapter has examined methods of dealing with uncertainty by applying the concept of **probability**. An activity that results in an uncertain outcome is called an **experiment**; the possible outcomes are **events**. Uncertainty is measured in terms of the probabilities of events. To determine the value of a particular probability, we used the classical approach, the relative frequency method, and the subjective probability approach. The **classical** definition for the probability of event *A* occurring assumes that the experiment has *n* equally likely outcomes and event *A* occurs in *m* of the *n* outcomes with a resulting probability of occurring equal to *m/n*. When using the **relative frequency** approach, the experiment is observed *n* times. Letting *m* represent the number of times event *A* occurs out of the *n* times, then the resulting probability of event *A* occurring in the future is *m/n*. A **subjective probability** is a measure of your belief that a particular event will occur, and like all probabilities, ranges from zero to one, inclusive.

When examining more than one event, say *A* and *B*, several types of probabilities can be derived. The probability of *A* and *B*. The probability of *A* and *B* occurring is a **joint probability** and is written *P(A* and *B*). The **multiplicative rule** is a method of determining a joint probability. The probability of *A* or *B* (or both) occurring is written *P(A* or *B*) and can be obtained using the **additive rule**. When asked to find a probability given particular information about events, you determine a **conditional probability**. For example, the probability that *B* occurs given that *A* has occurred is a conditional probability, written *P(B|A*). A variation of the multiplicative rule provides a method of determining a conditional probability. The probability of a single event, such as *P*(the person selected is a female) or *P*(an individual subscribes to the *Wall Street Journal*), is a **marginal probability**.

An effective method of determining a probability in complicated situations is to use a **Venn diagram**. When you represent the various events visually, you can often obtain a seemingly complex probability easily.

Two events are said to be **independent** if the occurrence of the one event has no effect on the probability that the other event occurs. Do not confuse "no effect" with "never occur simultaneously." If two events can never occur simultaneously, they are **mutually exclusive**. For example, these two events are certainly independent but are not mutually exclusive (since both events could occur): *A*, the stock market drops more than two points during a particular week, and *B*, your company's copying machine breaks down during the same week.

We discussed various counting rules, including **permutations** and **combinations**. These rules are used to count the number of possible outcomes for experiments that select a certain number of people or objects (*k*) from a large group of *n* such objects. When determining the corresponding number of permutations (written ₙ*P*ₖ), the order of selection is considered. The number of combinations for this situation (written ₙ*C*ₖ) ignores the order of selection and counts only the number of groups that can be obtained.

We also discussed the number of random samples that exists when the population size is known, and we examined methods of obtaining such a sample. *In the chapters to follow, any results using a statistical sample assume that the sample is obtained randomly.*

4.64 Assume events *A* and *B* are mutually exclusive. Find the following probabilities where *P(A)* = 0.4 and *P(B)* = 0.15:

a. *P(A* or *B*)   b. *P(A* or *B̄*)   c. *P(Ā* or *B̄*)
d. *P(Ā* and *B*)   e. *P(A|B*)   f. *P(A|B̄*)

4.65 Consider the experiment in which a single die is tossed.
a. What is the probability that an even number occurs?

b. What is the probability that an even number occurs, given that the number is greater than three?

c. What is the probability that an even number occurs or that a number greater than three occurs?

4.66 If you are selecting playing cards at random without replacement from a deck of 52 and you have already drawn a king of spades, queen of spades, ten of spades, and nine of spades, then what is the probability of drawing a jack of spades?

4.67 A marketing-research group conducted a survey to find out where people did their holiday shopping. Out of a group of 110 randomly selected shoppers, 70 said that they shopped exclusively at the local mall, 30 said that they shopped exclusively in the downtown area, and 10 said that they shopped at both the local mall and the downtown area.

a. What is the probability that a customer shops in both the local mall and the downtown area?

b. What proportion of customers who shop at the local mall also shop in the downtown area?

c. What is the probability that a customer shops downtown but not at the local mall?

4.68 An electronics firm decides to market three different software packages for its personal computers. The marketing analyst gives each of the three packages an 80% chance of success. The outcomes for each of the software packages are independent.

a. What is the probability that all three will be a success?

b. What is the probability that only two of the packages will be a success?

c. What is the probability that none will be successful?

4.69 A payroll record with an error in it is placed in a filing cabinet with six error-free payroll records. Two payroll records are randomly selected, without replacement, by an auditor.

a. What is the probability of drawing the payroll record with the error on the first draw?

b. What is the probability of drawing the payroll record with the error on the second draw?

c. What is the probability of drawing the payroll record with the error on the first or second draw?

d. What is the probability of drawing an error-free payroll record on the first or second draw?

4.70 A survey of 715 television ads over three major television networks produced the following breakdown of ads with nonelderly people and ads with elderly people.

NETWORK	ADS WITH NONELDERLY PEOPLE	ADS WITH ELDERLY PEOPLE
ABC	263	20
CBS	208	16
NBC	194	14
Total	665	50

(*Source: Journal of Advertising* 16, no. 1, 1987, p. 50.)

a. What is the probability that an ad uses an elderly person?

b. What is the probability that one of the 715 television ads does not use an elderly person and is not from the CBS network?

c. What is the probability that an ad with elderly people is from the ABC network?

d. What is the probability that an ad from the 715 television ads is an ad with non-elderly people from the ABC network or an ad with elderly people from the NBC network?

4.71 For a marketing survey, 200 customers were classified according to their age (in years) and their favorite type of donut.

AGE OF CUSTOMER	GLAZED	CHOCOLATE-COVERED	CREME-FILLED	CAKE
<21	3	25	10	7
21–30	5	23	26	10
31–45	15	12	3	20
>45	29	5	1	6
Total	52	65	40	43 (200)

a. What is the probability that a person prefers creme-filled donuts and is age 45 years or less?

b. What is the probability that a person's favorite donut is not glazed or that the person is less than 21 years of age?

c. What is the probability that a person is between 21 and 30 years of age if that person favors chocolate-covered donuts?

d. Are age and favorite donut independent variables?

4.72 The probability that a person buys a car after receiving a sales pitch is .10. After a customer decides to buy a car, the probability that the customer will arrange financing through the dealer is .75. What is the probability that a customer who hears a sales pitch will buy a car and arrange financing through the dealer?

4.73 An instructor has 40 questions from which she will draw to make up a 30-question test. How many different tests can the instructor design?

4.74 A busy executive has to meet with five production managers during the day. The executive needs to decide in which order to see the managers. How many different orderings can the executive choose? What is the probability of the executive's choosing any one ordering, if the choice is random?

4.75 A defective tape recorder is inspected by two service representatives. If one representative has a 50% chance of finding the defect, and the other has a 60% chance, then what is the probability that at least one will find the defect if both check the tape recorder independently? What is the probability that neither will spot the defect?

4.76 A student forgot the combination for his bike lock. The combination consists of a sequence of three numbers and each number can range from zero to nine. How many different sequences are possible?

4.77 In a survey of some of the largest corporations in the United States, the number of employees known or assumed to have been involved in incidents of computer crime within a corporation was tabulated. A classification of the 124 employees responsible for incidents of computer crime is given next.

NUMBER OF INDIVIDUALS

Employees not directly involved in the operation of computers	62
Employees directly involved in the operation of computers	45
Unknown individuals, believed to be employees	17
Total	124

(Source: J. O'Donoghue, "Strategies found to be effective in the control of computer crime in the Forbes 500 corporations," SIG Security Audit and Control 5, no. 1, 1987.)

a. Are the three preceding classifications mutually exclusive?

b. If employee A is selected at random from the 124 employees involved in computer crime, what is the probability that employee A was a known individual either directly or indirectly involved in the operation of computers?

c. Suppose employee A is selected at random from the 124 employees and suppose that employee B is also selected at random from the full set of 124 employees. What is the probability that both employees A and B were known individuals either directly or indirectly involved in the operation of computers?

4.78 Forty percent of the students in an economics class major in business and 70% are from St. Louis, Missouri. Also, 20% are neither business majors nor from St. Louis. What is the probability that a student in the economics class selected at random is a business major from St. Louis?

4.79 A 1987 Gallup survey for the American Society of Quality Control asked 615 executives the following questions:

Question A: Based on your experience, does increasing quality lead to greater profits?

Question B: Based on your experience, does increasing quality lead to cost reduction?

Two-thirds of the executives answered yes to question A, whereas 43% said yes to question B. Just 6% disagreed with both statements.

(*Source: Training*, February 1988, p. 84.)

In finding the probabilities asked for in the following questions, assume that an executive is randomly selected from the respondents to the survey.

a. What is the probability that an executive answered yes to question B or disagreed with both questions A and B?

b. Is the set of executives who answered yes to question A mutually exclusive from the set of executives who answered yes to question B? Justify your answer from the probabilities that are given.

c. Suppose that it could be assumed that those executives who answered yes on question B also answered yes on question A. What is the probability that an executive answered yes on question A or question B?

d. Using the assumption made in part (c), what is the probability that an executive answered yes on question A and no on question B?

e. Using the assumption made in part (c), what is the probability that an executive, who answered yes to question A, answered no to question B?

4.80 In a study titled "Merger, is it the answer?", a questionnaire was mailed to 450 CPAs who have experienced business mergers. Overall, 97% of the participants stated that their mergers were successful. Seventy-nine percent of the participants also mentioned that they would engage in a merger again, given the right opportunity.

(*Source: The CPA Journal*, February 1988, pp. 80–85.)

a. What would one expect the probability to be that three different mergers are considered successful? (Assume that separate mergers are independent.)

b. Consider two different mergers. What is the probability that one merger is successful and the other merger is not successful?

c. What additional information would one need to calculate the probability that a participant either considered their merger as successful or would engage in a merger again?

d. Suppose one could assume that the 79% of the participants who would engage in a merger again also considered their merger as being successful. Find the probability that a participant who considered their company's merger to be successful would not engage in a merger again.

4.81 An estimated 61,236 businesses failed in the United States in 1987. A breakdown by industry is as follows:

Services	24,029
Retail trade	12,185
Construction	6,724
Manufacturing	4,317
Wholesale trade	4,304
Agriculture, forestry, and fishing	3,783
Finance, insurance, and real estate	2,492
Other	3,402

(*Source:* "Business Failures," *Wall Street Journal,* March 3, 1988, p. 19)

a. What is the probability that a business selected at random from the businesses that failed in the United States in 1987 is in the services industry or the wholesale trade industry?

b. What is the probability that a business selected at random from the businesses that failed in the United States in 1987 is not in the construction industry and not in the finance, insurance, and real estate industry?

c. What is the probability that a business selected at random from the businesses that are not in the services industry and that failed in the United States in 1987 is in the retail trade industry?

4.82 A computer-generated list of numbers between 0.5 and 6.5 is given in the MINITAB printout. Round these numbers off into integers. Using the relative frequency approach to finding a probability value, find the probability of an odd number. In your judgment, does this value tend to support the assumption that the numbers are truly random?

```
MTB > random 100 values into c1;
SUBC> uniform A=.5 and B=6.5.
MTB > PRINT C1
C1
3.18409   6.01078   5.34752   6.43991   4.98901   3.62637   1.29673
4.09175   5.46874   3.59305   3.13126   5.40799   1.99838   1.79702
0.62798   4.49777   2.22069   5.58658   6.32294   2.36695   5.86879
5.59827   1.78393   4.99184   3.97935   3.41907   5.38338   4.92242
1.30241   4.80093   4.11574   2.46772   6.46524   2.15536   4.91979
5.47431   4.28828   6.03483   2.35320   4.15030   0.78688   6.36033
1.04070   2.08737   0.92095   5.11850   1.81216   2.52028   1.03504
1.38061   2.57570   1.96239   3.29816   2.26940   5.67520   5.39965
0.95661   3.57579   0.97396   5.74543   2.17879   6.34877   5.59579
1.47414   2.26745   5.43085   4.85600   5.00017   5.02085   1.60677
0.84645   1.80587   1.73359   4.69857   3.32107   5.13373   3.71653
0.56630   2.78776   4.46958   4.69761   3.20144   2.18042   0.55235
1.04325   2.40601   4.75175   3.96839   2.04846   2.05730   3.16237
3.29612   2.01459   3.82397   1.99612   1.51512   1.39047   3.80817
6.02175   0.71840
```

4.83 The following data given in the MINITAB printout represent the number of years of managerial experience and the ages of 20 midlevel managers of a department store chain. The plot with the shaded areas can be considered a Venn diagram of the set *A* of midlevel managers between 32 and 40 years of age, inclusively, and the set *B* of midlevel managers with greater than 5 years of managerial experience.

a. What is the probability that a midlevel manager is in set *A* or set *B*?

b. What is the probability that a midlevel manager is in set *A* and not in set *B*?

c. What is the probability that a midlevel manager is not in set *A* and not in set *B*?

```
MTB > print c1 c2
ROW   age   exper.
  1    35    5
  2    33    4
  3    30    2
  4    41    6
  5    42    7
  6    45    5
  7    29    2
  8    28    1
  9    38    4
 10    40    5
 11    43    4
 12    38    2
 13    35    8
 14    34    6
 15    38    3
 16    40    5
 17    35    5
 18    45    7
 19    40    3
 20    38    6

MTB > plot c2 c1;
SUBC> xincrement = 1;
SUBC> xstart at 25;
SUBC> yincrement = 1;
SUBC> ystart at 0.
* Increment increased to cover range
* Increment increased to cover range
```

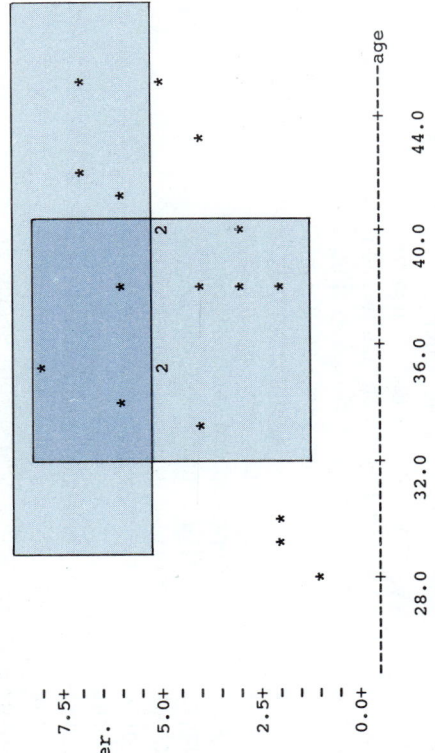

```
exper. -
     7.5+
       -
       -
       -
     5.0+
       -
       -
       -
     2.5+
       -
       -
       -
     0.0+
        --------+---------+---------+---------+---------+------age
             28.0      32.0      36.0      40.0      44.0
```

COMPUTER EXERCISES USING THE DATABASE

Exercise 1—Appendix H

From the following MINITAB computer printout of random numbers between 1 and 1140, select a random sample of 100 observations from the database.

```
MTB > RANDOM 100 VALUES INTO C1;
SUBC> UNIFORM A=1 AND B=1140.
MTB > PRINT C1
C1
```

219.92	29.97	205.64	523.37	374.75	21.34	265.15
1127.27	687.22	354.07	852.81	550.19	309.99	1038.31
957.40	1095.47	129.98	177.44	415.22	523.41	378.88
536.87	923.31	250.19	396.48	459.26	333.99	620.59
1136.57	710.82	1025.30	470.51	601.22	993.71	1077.17
119.91	57.82	269.25	500.65	950.78	266.90	206.90
681.06	723.10	282.46	1012.94	64.99	26.69	934.81
549.24	191.51	1034.75	512.59	165.82	101.75	66.18
175.63	188.43	650.22	284.70	153.94	894.24	34.29
745.52	806.84	498.98	741.91	356.07	1102.66	1028.57
878.92	397.14	541.65	381.00	802.61	1109.45	738.77
1101.68	906.32	404.38	307.24	694.51	126.23	848.36
1132.92	255.11	1011.61	1037.69	880.72	622.30	211.16
74.56	83.69	86.26	406.88	620.13	1079.34	390.94
905.89	351.37					

Using the relative frequency approach of finding a probability value, find the probability that a family owns its home.

Exercise 2—Appendix I

Generate 200 random numbers from a uniform distribution to select 200 observations from the database. Using the relative frequency approach, find the probability that a company has an A bond rating. Also find the probability that a company has a B bond rating.

CASE STUDY

ANALYSIS OF CHARACTER FREQUENCIES TO COMPRESS FILES FOR DATA COMMUNICATION

Two major technologies have been converging at an accelerating pace in the last decade: telecommunications and data processing. So powerful is the convergence that AT&T surrendered its monopoly in telecommunications to enter the field of computers and data processing. As technology marches on, the demarcation between voice communication and data communication is being blurred. Desktop computer workstations, facsimile (FAX) machines, electronic mail facilities, and local area networks (LANs) are very common today.

Salespersons can upload daily sales data to their company's mainframe computer via a modem; reservations clerks can access remote databases of flight bookings; copies of new advertising layouts can be sent to clients via FAX (long-distance photocopying!)—the common element is the use of data communications technology. However, as the volume of data being transferred increases, so does the cost of the transfer.

One way to cut costs is to sacrifice speed and go for cheaper, low-speed channel links. Another approach is to reduce the quantity of data by compressing it. For example, in files containing text, there is a lot of "white space," that is, spaces and blank lines. From the point of view of the computer, a space is just another character, such as A, B, or C. Instead of sending 50 character codes for a space, a special code could be sent, followed by the number 50, which would have the meaning 50 blank spaces. Fifty characters would thus be compressed into two or three characters. At the receiving side, the compressed characters would be expanded back to their original form.

Held and Marshall (1987) discuss various techniques of data compression. Not all techniques are universally applicable. For example, a method called relative encoding is efficient when applied to telemetry data or facsimile digital scan codes. To look for character patterns in the data to decide which compression technique will work best, Held and Marshall used a computer program called DATANALYSIS on various data files. Some of the results pertaining to one file are given in Tables 4.2–4.3. Our focus, of course, will not be on data compression techniques, but on the probabilistic aspects of how the characters are distributed in the file.

Table 4.2 lists the frequency of occurrence of individual characters. The total number of characters in the file analyzed was 99,132. Note that SP refers to the space (blank) character. Character codes denoted as SH, EX, SX, ET, EQ, DL, and so on, are so-called control characters and are not generally printable. Table 4.3 gives the frequency analysis for 138 of the most common pair sequences of characters, that is, two characters occurring together in sequence. Counting multiple occurrences of these pairs, a total of 4528 such combinations were found. When answering the questions that follow, consider carefully which table you should use.

TABLE 4.2
System Standard
Frequency of Occurrence
—Table of Characters
Found in Sysout File

CHAR.	COUNT	%	CHAR.	COUNT	%	CHAR.	COUNT	%
SH	0.	0.	,	1327.	1.34	W	197.	0.20
EX	0.	0.	–	135.	0.14	X	196.	0.20
SX	0.	0.	.	135.	0.14	Y	190.	0.19
ET	0.	0.	/	60.	0.06	Z	62.	0.06
EQ	0.	0.	0	1066.	1.08	[	1.	0.00
AK	0.	0.	1	905.	0.91	\	1.	0.00
BL	0.	0.	2	1427.	1.44	]	1.	0.00
BS	0.	0.	3	405.	0.41	^	1.	0.00
HT	0.	0.	4	354.	0.36	_	53.	0.05
LF	0.	0.	5	353.	0.36	`	0.	0.
VT	0.	0.	6	248.	0.25	a	0.	0.
FF	0.	0.	7	281.	0.28	b	0.	0.
CR	0.	0.	8	312.	0.31	c	0.	0.
SO	0.	0.	9	242.	0.24	d	0.	0.
SI	0.	0.	:	21.	0.02	e	0.	0.
DE	0.	0.	;	4.	0.00	f	0.	0.
D1	0.	0.	<	1.	0.00	g	0.	0.
D2	0.	0.	=	152.	0.15	h	0.	0.
D3	0.	0.	>	1.	0.00	i	0.	0.
D4	0.	0.	?	5.	0.01	j	0.	0.
NK	0.	0.	@	2.	0.00	k	0.	0.
SY	0.	0.	A	713.	0.72	l	0.	0.
EB	0.	0.	B	252.	0.25	m	0.	0.
CN	0.	0.	C	427.	0.43	n	0.	0.
EM	0.	0.	D	290.	0.29	o	0.	0.
SB	0.	0.	E	928.	0.94	p	0.	0.
EC	0.	0.	F	279.	0.28	q	0.	0.
FS	0.	0.	G	192.	0.19	r	0.	0.
GS	0.	0.	H	1095.	1.10	s	0.	0.
RS	0.	0.	I	1223.	1.23	t	0.	0.
US	0.	0.	J	97.	0.10	u	0.	0.
SP	73509.	74.15	K	121.	0.12	v	0.	0.
EP	13.	0.01	L	355.	0.36	w	0.	0.
'	201.	0.20	M	332.	0.33	x	0.	0.
#	2.	0.00	N	644.	0.65	y	0.	0.
$	2.	0.00	O	694.	0.70	z	0.	0.
%	7.	0.01	P	432.	0.44	{	0.	0.
&	105.	0.11	Q	98.	0.10	|	0.	0.
,	1.	0.00	R	813.	0.82	}	0.	0.
(	542.	0.55	S	573.	0.58	~	0.	0.
)	542.	0.55	T	1022.	1.03	DL	0.	0.
*	1055.	1.06	U	511.	0.52	TOTAL	99132.	100.00
+	74.	0.07	V	95.	0.10			

** Denotes total characters in file

TABLE 4.3
Paired Character
Compression Analysis

PAIR/COUNT		PAIR/COUNT		PAIR/COUNT		PAIR/COUNT	
E——I	156	RO	20	MA	60	——A	21
E——	80	LY	19	FO	53	AB	20
RE	68	HG	18	——D	46	YS	19
——D	55	HU	18	AR	42	HO	18
ON	50	HY	17	——S	37	——B	17
EN	45	CU	16	GO	37	UE	16
NT	39	IO	16	IO	33	MI	16
IT	34	EG	16	TO	29	RR	15
RA	30	EL	15	IM	26	UR	14
UB	28	HB	14	NE	24	HF	13
AC	25	OS	14	PR	23	——W	85
LI	24	RI	110	NA	21	——F	68
TA	22	IN	75	HS	20	O——	56
NG	21	AT	66	HC	19	SE	52
CE	19	IR	53	HI	18	S——	45
RD	18	R——	46	HQ	17	——T	39
IP	18	CO	43	HM	17	AN	35
HW	17	SI	38	G——	16	HE	31
PI	16	PA	33	CS	16	FI	18
AP	16	IA	30	PE	14	DI	17
HN	15	EQ	26	HV	14	——N	16
TS	14	CH	24	HR	13	BE	18
ES	14	OM	23	PU	96	NO	19
TE	149	DA	22	OR	70	AI	21
TI	75	MP	20	ER	59	IL	22
N——	66	RS	19	T——	52	NU	24
AL	55	IX	18	HA	46	OU	25
L——	48	ST	17	IS	40	ME	28
SU	45	HK	16	TH	37	UN	16
LA	39	HD	16	WE	33	ED	15
——E	33	WO	16	H——R	29	TR	14
RM	30	HH	14	H——O	25	——G	14
LE	27	X——	14				
GE	24	EP	14				
CT	23	UT	108	ND	22	HT	13

TOTAL COMBINATIONS FOUND: 4528

Note: The underline —— represents a space in Table 4.3.

(*Source* for Tables 4.2 and 4.3: Gilbert Held and Thomas R. Marshall, *Data Compression: Techniques and Applications*, 2d ed., New York: John Wiley, 1987, Tables 3.3, 3.5, 3.8, pp. 128, 130–31, 137. Reprinted by permission of John Wiley & Sons, Ltd.)

Case Study Questions

1. Which is the most frequently occurring individual character? What proportion of the file does it occupy?

2. If *one* character is selected at random from this file, what is the probability that it is:
   a. A space (blank)
   b. An uppercase letter (*A* to *Z*)
   c. A lowercase letter (*a* to *z*)
   d. *Not* a number (0 to 9)
   e. Alphanumeric (upper- or lowercase letter or number)

3. If *two* characters, not necessarily sequential pairs, are randomly and independently selected from this file, what is the probability that:
   a. Both are blank spaces
   b. One is an upper- or lower-case letter and the other is a number
   c. Neither one is a blank space

4. In computing the above probabilities, which view of probability are you assuming: the

classical view, the relative frequency approach, or subjective probability?

5. If a pair of characters occurring in sequence is randomly selected from this file, what is the probability that it will be one of the pairs shown in Table 4.3?
(*Hint:* Determine how many *sequential* pair combinations are possible in this file.)

SOLUTION

RANDOM NUMBER GENERATION

You can use SPSSX to generate random numbers. The SPSSX program listing in Figure 4.12 was used to request the generation of 100 random numbers between 1 and 10,000.

The TITLE command names the SPSSX run.

The INPUT PROGRAM statement allows you to build your own subprograms to either input or generate data.

The LOOP statement sets up a loop that is terminated by an END LOOP statement. In this example, we are looping 100 times to COMPUTE 100 different random numbers.

The RND (UNIFORM (10000)) statement generates a random number from a uniform distribution between 1 and 10,000 and rounds this value to the nearest integer.

The END CASE statement passes control of the loop to the END LOOP statement, which passes control to the loop until 100 random numbers have been generated.

The END FILE statement terminates loop processing.

The END INPUT PROGRAM statement terminates the INPUT PROGRAM.

The PRINT statement sets up the 100 random numbers that were generated for printing.

The EXECUTE command causes the actual printing to occur.

Figure 4.13 shows the output obtained by executing the program listing in Figure 4.12.

FIGURE 4.12 Input for SPSSX program to generate 100 random numbers between 1 and 10,000.

```
TITLE     RANDOM NUMBERS
INPUT PROGRAM
LOOP #1=1 TO 100
COMPUTE RANNUM = RND(UNIFORM(10000))
END CASE
END LOOP
END FILE
END INPUT PROGRAM
PRINT    /RANNUM
EXECUTE
```

FIGURE 4.13 SPSSX output.

1396.00	1006.00	2060.00	3339.00	2286.00
4313.00	7326.00	7734.00	6208.00	7262.00
6122.00	2254.00	669.00	6411.00	8825.00
2908.00	2214.00	5271.00	7521.00	1822.00
1557.00	6101.00	6618.00	2050.00	2797.00
6995.00	9540.00	9358.00	1433.00	6955.00
3463.00	2369.00	3157.00	4459.00	9026.00
4456.00	3385.00	5227.00	3691.00	1228.00
524.00	5228.00	7658.00	4853.00	5077.00
1032.00	7383.00	1271.00	993.00	1173.00
1412.00	1198.00	5594.00	7257.00	9247.00
429.00	589.00	9826.00	373.00	5173.00
6217.00	1880.00	3253.00	1747.00	2714.00
1536.00	996.00	5355.00	2884.00	6446.00
7152.00	6759.00	5077.00	6689.00	6438.00
9283.00	6927.00	2820.00	7612.00	8528.00
5781.00	8668.00	3472.00	8031.00	652.00
2619.00	3512.00	9211.00	7029.00	4068.00
7248.00	4964.00	4081.00	479.00	5667.00
371.00	7020.00	957.00	3943.00	8433.00

SOLUTION

RANDOM NUMBER GENERATION

You can use SPSS/PC+ to generate random numbers. The program listing in Figure 4.14 was used to request the generation of 100 random numbers between 1 and 10,000.

The TITLE command names the SPSS/PC+ run.

The DATA LIST FREE / A, BEGIN DATA, the 100 values of 1, and the END DATA statements will provide 100 random numbers when the COMPUTE RANNUM = RND (UNIFORM (10000)) command is executed. These 100 values are from a uniform distribution between 1 and 10,000 and are rounded to the nearest integer. If 250 random numbers are desired, for instance, the only change necessary is to replace the 100 values of 1 with 250 such values.

The LIST RANNUM command will output the 100 random numbers.

Figure 4.15 shows the output obtained by executing the program listing in Figure 4.14.

FIGURE 4.14

Input for SPSS/PC+ program to generate 100 random numbers between 1 and 10,000.

```
TITLE    RANDOM NUMBERS.
DATA LIST FREE / A.
BEGIN DATA.
1 1 1 1 1 1 1 1 1 1 1 1 1 1 1 1 1 1 1 1
1 1 1 1 1 1 1 1 1 1 1 1 1 1 1 1 1 1 1 1
1 1 1 1 1 1 1 1 1 1 1 1 1 1 1 1 1 1 1 1
1 1 1 1 1 1 1 1 1 1 1 1 1 1 1 1 1 1 1 1
1 1 1 1 1 1 1 1 1 1 1 1 1 1 1 1 1 1 1 1
END DATA.
COMPUTE RANNUM = RND (UNIFORM(10000)).
LIST RANNUM.
```

FIGURE 4.15

SPSS/PC+ output

RANNUM	RANNUM	RANNUM	RANNUM	RANNUM
6869.00	2399.00	6442.00	5024.00	6567.00
8789.00	8900.00	6185.00	813.00	4369.00
5913.00	7730.00	8040.00	64.00	4669.00
4493.00	8383.00	5211.00	9654.00	5088.00
2233.00	8792.00	7102.00	629.00	5436.00
9899.00	898.00	1050.00	8939.00	6770.00
7468.00	3662.00	751.00	9873.00	8592.00
5298.00	2419.00	8831.00	1399.00	1226.00
9864.00	3249.00	9210.00	1613.00	1944.00
3033.00	5306.00	1648.00	8405.00	8948.00
5299.00	6427.00	1230.00	7982.00	8986.00
4199.00	681.00	263.00	4641.00	1015.00
2266.00	8944.00	3728.00	9700.00	4393.00
1062.00	8037.00	453.00	3477.00	5104.00
9689.00	4803.00	5606.00	5888.00	7718.00
2885.00	4599.00	1125.00	7894.00	4078.00
5582.00	797.00	8520.00	8083.00	2273.00
9943.00	221.00	8530.00	9206.00	8384.00
6331.00	4882.00	8701.00	9178.00	8885.00
8699.00	2426.00	4898.00	3741.00	4865.00

SOLUTION

RANDOM NUMBER GENERATION

You can generate random numbers using SAS. The SAS program listing in Figure 4.16 was used to request the generation of 100 random numbers between 1 and 10,000.

The TITLE command names the SAS run (enclose in single quotes).

The PROC MATRIX command allows the user to build a table to hold the generated random numbers.

FIGURE 4.16 Input for SAS (mainframe version).

```
TITLE  'RANDOM NUMBERS';
PROC MATRIX;
X=UNIFORM (J(20,5,0));
X = X * 10000;
PRINT X FORMAT=F6.0;
```

FIGURE 4.17 SAS output.

RANDOM NUMBERS

X	COL1	COL2	COL3	COL4	COL5
ROW1	5212	3656	5943	9183	2727
ROW2	6840	8385	8551	6154	6223
ROW3	2645	475	5029	1314	2976
ROW4	6313	6059	6093	7290	1754
ROW5	8773	4584	5711	1772	5024
ROW6	1893	4824	3906	1010	3421
ROW7	7204	9976	4315	9087	5948
ROW8	2805	5934	2545	4208	7932
ROW9	7307	5403	9720	337	2351
ROW10	6442	4337	6159	4486	5835
ROW11	8559	9409	8674	4364	1064
ROW12	3695	8445	7642	3241	8939
ROW13	7928	5329	8721	3093	8886
ROW14	5350	7905	8863	2745	3118
ROW15	7500	5421	7625	5749	5046
ROW16	5131	369	1282	8451	5262
ROW17	3581	2896	520	4710	7650
ROW18	8133	8485	5715	7787	3112
ROW19	8542	5582	6051	1752	5373
ROW20	3397	8383	9986	6686	6746

The X = UNIFORM (J(20, 5, 0)) command calls 100 random numbers and stores them in a table of 20 rows by 5 columns.

The X = X * 10000 statement scales the numbers up from a decimal number to an integer between 1 and 10,000.

The PRINT × FORMAT = F6.0 statement establishes the size of the numbers to be printed, and then prints the 20-by-5 table.

Figure 4.17 shows the output obtained by executing the program listing in Figure 4.16.

SOLUTION

RANDOM NUMBER GENERATION

You can generate random numbers using SAS/PC. The program listing in Figure 4.18 was used to request the generation of 100 random numbers between 1 and 10,000.

The TITLE command names the SAS run (enclose in single quotes).

The DATA command gives the data a name.

The DO statement sets up a loop that is terminated by the END statement. In this example, we are looping 100 times to compute 100 different random numbers.

The Y = UNIFORM (0) * 10000 statement generates a random number from a uniform distribution between 1 and 10,000. The zero in parentheses acts as a seed for the random number generator. This value can be changed to obtain a different set of random numbers.

FIGURE 4.18 Input for SAS (micro version).

```
TITLE  'RANDOM NUMBERS';
DATA NEW;
DO I = 1 TO 100;
Y = UNIFORM (0) * 10000;
X = ROUND (Y);
OUTPUT;
END;
PROC PRINT;
    VAR X;
```

The X = ROUND (Y) statement rounds the random number to the nearest integer.

The OUTPUT statement stores the X and Y values.

The END statement marks the end of the loop used to generate the random numbers.

The PROC PRINT command prints the (rounded) random numbers as specified by the VAR X statement.

Figure 4.19 shows the output obtained by executing the program listing in Figure 4.18.

FIGURE 4.19
SAS/PC output.

RANDOM NUMBERS

OBS	X	OBS	X	OBS	X
1	1460	33	5247	65	5524
2	713	34	4390	66	8343
3	7630	35	89	67	5005
4	432	36	8902	68	4607
5	4590	37	190	69	2311
6	4212	38	4845	70	2993
7	5812	39	7647	71	1879
8	1865	40	7055	72	8198
9	191	41	6901	73	6498
10	6181	42	8582	74	9424
11	5783	43	734	75	4631
12	7328	44	273	76	1640
13	8235	45	1861	77	7699
14	9646	46	3643	78	5769
15	5332	47	6368	79	8988
16	7992	48	9822	80	6261
17	1829	49	5443	81	7217
18	7817	50	7356	82	2818
19	2694	51	5719	83	476
20	5480	52	3996	84	6193
21	8239	53	4298	85	298
22	5678	54	9543	86	2016
23	9402	55	9156	87	794
24	2347	56	3195	88	7253
25	4168	57	491	89	1178
26	2280	58	5923	90	7370
27	7831	59	3458	91	1696
28	6157	60	8896	92	6551
29	5997	61	5087	93	1690
30	570	62	9680	94	8321
31	5863	63	9806	95	6710
32	4549	64	524	96	7337
				97	7688
				98	6457
				99	9782
				100	766

Discrete
Probability
Distributions

The early chapters were concerned with describing sample data that had been gathered from a previous experiment, a printed report, or some other source. The data were summarized using one or more numerical measures (for example, a sample mean, variance, or correlation) or using a statistical graph (such as a histogram, bar chart, or scatter diagram).

Chapter 4 introduced you to methods of dealing with uncertainty by using a probability to measure the chance of a particular event occurring. Rules were defined that enable you to compute the various probabilities of interest, such as a conditional or a joint probability. However, so far we have defined only the probability of a certain event happening.

Whenever an experiment results in a numerical outcome, such as the total value of two dice, we can represent the various possible outcomes and their corresponding probabilities much more conveniently by using a **random variable**, the topic of this chapter. Suppose that your company manufactures a product that is sometimes defective and is returned for repair in 10% of the cases. An excellent way of describing the chance that 3 of 20 products will be returned before the warranty runs out is to use the concept of a random variable.

Random variables can be classified into two categories: discrete and continuous. This chapter introduces both types but concentrates on the discrete type of random variable. Several commonly used discrete random variables will be discussed, as will methods of describing and applying them.

A Look Back/Introduction

DEFINITION

A **random variable** is a function that assigns a numerical value to each outcome of an experiment.

5.1

RANDOM
VARIABLES

Discrete Random Variables

The probability laws developed in the previous chapter provide a framework for the discussion of random variables. We will still be concerned about the probability of a particular event; often, however, some aspect of the experiment can be easily represented using a random variable. The result of a simple experiment can sometimes be summarized concisely by defining a discrete random variable to describe the possible outcomes.

Flip a coin three times. The possible outcomes for each flip are heads (H) and tails (T). According to counting rule 1 from Chapter 4, there are $2 \cdot 2 \cdot 2 = 8$ possible results. These are TTT, TTH, THT, HTT, HHT, THH, and HHH. Let

A = event of observing 0 heads in 3 flips (TTT)
B = event of observing 1 head in 3 flips (TTH, THT, HTT)
C = event of observing 2 heads in 3 flips (HHT, HTH, THH)
D = event of observing 3 heads in 3 flips (HHH)

We wish to find $P(A)$, $P(B)$, $P(C)$, and $P(D)$.

Consider one outcome, say, HTH. The coin flips are independent, so we use equation 4.14:

$$P(\text{HTH}) = (\text{probability of H on 1st flip}) \cdot (\text{probability of T}$$
$$\text{on 2nd flip}) \cdot (\text{probability of H on 3rd flip})$$

$$= (1/2) \cdot (1/2) \cdot (1/2) = 1/8$$

This same argument applies to all eight outcomes. These outcomes are all equally likely, and each occurs with probability 1/8.

Event A occurs only if you observe TTT. This has the probability of occurring one time out of eight:

$$P(A) = 1/8$$

Event B will occur if you observe HTT, TTH, or THT. It would be impossible for HTT and TTH *both* to occur, so $P(\text{HTT and TTH}) = 0$. This is true for any combination here, so these three events are all mutually exclusive. Consequently, according to equation 4.13,

$$P(B) = P(\text{HTT or TTH or THT})$$
$$= P(\text{HTT}) + P(\text{TTH}) + P(\text{THT})$$
$$= 1/8 + 1/8 + 1/8 = 3/8$$

By a similar argument,

$$P(C) = 3/8 \quad \text{(using HHT, HTH, THH)}$$
$$P(D) = 1/8 \quad \text{(using HHH)}$$

The variable of interest in this example is X, defined as

$$X = \text{number of heads out of three flips}$$

We defined all the possible outcomes of X by defining the four events A, B, C, and D. This works but is cumbersome. Consider having to do this for 100 flips of a coin! A more convenient way to represent probabilities is to examine the value of X for each possible outcome.

OUTCOME	VALUE OF X	
TTT	0	1 outcome
THT	1	
TTH	1	3 outcomes
HTT	1	
HHT	2	
HTH	2	3 outcomes
THH	2	
HHH	3	1 outcome

Each outcome has probability 1/8, so the probability that X will be 0 is 1/8, written:

$$P(X = 0) = P(0) = 1/8$$

The probability that X will be 1 is 3/8, written:

$$P(X = 1) = P(1) = 3/8$$

The probability that X will be 2 is 3/8, written:

$$P(X = 2) = P(2) = 3/8$$

The probability that X will be 3 is 1/8, written:

$$P(X = 3) = P(3) = 1/8$$

Notice that

$$P(X = 0) + P(X = 1) + P(X = 2) + P(X = 3) = 1/8 + 3/8 + 3/8 + 1/8$$
$$= 1$$

because 0, 1, 2, and 3 represent *all the possible values* of X.

The values and probabilities for this random variable can be summarized by listing each value and its probability of occurring.

$$X = \begin{cases} 0 & \text{with probability } 1/8 \\ 1 & \text{with probability } 3/8 \\ 2 & \text{with probability } 3/8 \\ 3 & \text{with probability } 1/8 \end{cases}$$

This list of possible values of X and the corresponding probabilities is a **probability distribution**.

In any such formulation of a problem, the variable X is a **random variable**. Its value is not known in advance, but there is a probability associated with each possible value of X. Whenever you have a random variable of the form

$$X = \begin{cases} x_1 & \text{with probability } p_1 \\ x_2 & \text{with probability } p_2 \\ x_3 & \text{with probability } p_3 \\ \vdots \\ x_n & \text{with probability } p_n \end{cases}$$

where $x_1, \ldots, x_n$ is the set of possible values of X, then X is a **discrete random variable**. In the coin-flipping example, $x_1 = 0$ and $p_1 = 1/8$; $x_2 = 1$ and $p_2 = 3/8$; $x_3 = 2$ and $p_3 = 3/8$, and $x_4 = 3$ and $p_4 = 1/8$.

Other examples of a discrete random variable include:

X = the number of cars that drive up to a bank within a 5-minute period ($X = 0, 1, 2, 3, \ldots$).

X = the number of people out of a group of 50 who will suffer a fatal accident within the next 10 years ($X = 0, 1, 2, \ldots, 50$).

X = the number of people out of 200 who make an airline reservation and then fail to show up ($X = 0, 1, 2, \ldots, 200$).

X = the number of calls arriving at a telephone switchboard over a 2-minute period ($X = 0, 1, 2, 3, \ldots$).

Notice that, for each example, the discrete random variable is a *count* of the number of people, calls, accidents, and so on that can occur.

EXAMPLE 5.1

You roll two dice, a red die and a blue die. What is a possible random variable X for this situation? What are its possible values and corresponding probabilities? (*Hint:* Roll the dice and observe a particular number. This number is your value of the random variable, X. What observations are possible from the roll of two dice?)

SOLUTION

There are many possibilities here, including

X = total of the two dice

X = average of the two dice

X = the higher of the two numbers that appear (possible values: 1, 2, 3, 4, 5, 6)

X = the number of dice with 3 appearing (possible values: 0, 1, 2)

Suppose that the random variable X equals the total of the two dice. The next step is to determine the possible values of X and the corresponding probabilities. When you roll the two colored dice, there are $6 \cdot 6 = 36$ possible outcomes, using counting rule 1 from Chapter 4.

OUTCOME	RED DIE	BLUE DIE	VALUE OF X
1	1	1	2
2	1	2	3
3	1	3	4
4	1	4	5
5	1	5	6
6	1	6	7
7	2	1	**3**
8	2	2	4
9	2	3	5
⋮	⋮	⋮	⋮
34	6	4	10
35	6	5	11
36	6	6	12

$P(X = 3) = 2/36$

The 36 outcomes are equally likely because the number appearing on each die (1, 2, 3, 4, 5, or 6) has the same chance of appearing. Notice that we are *not* saying that each value of X is equally likely, as the following discussion will make clear. Each of the above 36 outcomes has probability 1/36 of occurring. If you write down all 36 outcomes and note what can happen to X, your random variable, you will observe:

VALUE OF X	NUMBER OF POSSIBLE OUTCOMES	
2	1	(rolling a 1, 1)
3	2	(rolling a 1, 2, or 2, 1)
4	3	(rolling a 1, 3 or 3, 1 or 2, 2)
5	4	(and so on)
6	5	
7	6	
8	5	
9	4	
10	3	
11	2	
12	1	

Consequently,

$$
X = \begin{cases}
2 \text{ with probability } \dfrac{1}{36} \\[4pt]
3 \text{ with probability } \dfrac{2}{36} \\[4pt]
4 \text{ with probability } \dfrac{3}{36} \\[4pt]
5 \text{ with probability } \dfrac{4}{36} \\[4pt]
6 \text{ with probability } \dfrac{5}{36} \\[4pt]
7 \text{ with probability } \dfrac{6}{36} \\[4pt]
8 \text{ with probability } \dfrac{5}{36} \\[4pt]
9 \text{ with probability } \dfrac{4}{36} \\[4pt]
10 \text{ with probability } \dfrac{3}{36} \\[4pt]
11 \text{ with probability } \dfrac{2}{36} \\[4pt]
12 \text{ with probability } \dfrac{1}{36} \\[4pt]
\hline
\phantom{12 \text{ with probability }} 1.0
\end{cases}
$$

Because 2 through 12 represent all possible values of X, the total of all probabilities is equal to 1.

Suppose instead X is defined to be the *average* of the two dice, rather than the total. Now the possible values of X are 1 (with probability 1/36), 1.5 (with probability 2/36), ..., 5.5 (with probability 2/36), and 6 (with probability 1/36). Notice that X is still a discrete random variable, since there are gaps in the possible values (a value of 4.2 is not possible, for example). However, the possible values of X are *not* all counting numbers. In general, the possible values of a discrete random variable need not be positive integers but generally are since the discrete random variable typically counts the number of occurrences of a particular event.

■

Continuous Random Variables

The previous section introduced you to the discrete random variable, where the possible values of X can be listed along with corresponding probabilities. Characteristic of this type of random variable is the presence of *gaps* in the list of possible values. For example, when throwing two dice, a total of 8.5 cannot occur.

The other type of random variable is the **continuous random variable**, for which *any* value is possible over some range of values. For a random variable of this type, there are no gaps in the set of possible values. As a simple example, consider two random variables: X is the number of days that it rained in Boston during any particular month and Y is the amount of rainfall during this month. X is a *discrete* random variable, because it counts the number of days, and consequently there are gaps in the possible values (7.4, for example, is not possible). Y, on the other hand, is a *continuous* random variable because at least in principle, the amount of rainfall could be any nonnegative value.

Suppose the heights of all adult males in the United States range from 3 feet to 7.5 feet. Your task is to describe these heights using such statements as

X = height of a randomly selected adult male in the United States

We first define the random variable

X = height of a randomly selected adult male in the United States

Figure 5.1 shows the range of X.

We are unable to list all possible values of X, since *any* height is possible over this range. However, we can still discuss probabilities associated with X. For example, the two preceding statements can be described by using the probability statements

$$P(X < 5.5) = .15$$
$$P(X \text{ is between 5 ft and 6 ft}) = P(5 < X < 6) = .88$$

For this situation, X is a continuous random variable. Probabilities for continuous random variables can be found only for *intervals.* (Probabilities of exact values are meaningful only for discrete random variables.) Determining probabilities for a continuous random variable is discussed in Chapter 6.

In Chapter 1, we discussed discrete and continuous data. They are directly related to our present discussion. When you observe a discrete random variable, you obtain discrete data. When you observe a continuous random variable (such as 100 heights), you obtain continuous data.

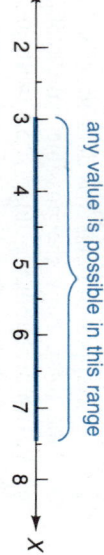

FIGURE 5.1
Example of a continuous random variable.
X = height in feet of a randomly selected adult male in the United States.

any value is possible in this range

2   3   4   5   6   7   8   X

EXERCISES

5.1   If a bank is interested in the business received from customers daily, what random variables would be of interest to the bank?

5.2   If a student is taking a multiple-choice test with ten questions and is interested in the final score, what random variable would be of interest to the student?

5.3   Classify the following random variables as discrete or continuous.

a.   The number of pages in a statistics book.

b.   The daily number of passengers on an airline flight.

c.   The time that business professionals spend in teleconferences.

d.   The life of a fluorescent tube.

e.   The amount in dollars of loans defaulted on new car loans.

5.4   Consider an experiment in which two dice are rolled. Let X be the total of the numbers on the two dice. What is the probability that X is equal to two or four?

5.5   Consider an experiment in which a coin is tossed and a die is rolled. Let X be the number observed from rolling the die. Let Y be the value one if a head appears and zero if a tail appears. List the values that the random variables X and Y can have, along with the corresponding probabilities.

5.6   Consider an experiment in which a coin is tossed four times. List all possible outcomes. Let X be the number of heads in each outcome. What is the value of $P(X = 2)$? of $P(X = 3)$?

5.7   If the random variable X can take on the values of 2, 3, 4, and 5 with equal probability, what is the probability that X is equal to 3? Assume that X cannot take on any other values.

5.8   Consider an experiment in which two dice are rolled. Let X take on the value of 1 if both dice have the same number and of 0 otherwise. What is the probability that X is equal to 1?

5.2

REPRESENTING PROBABILITY DISTRIBUTIONS FOR DISCRETE RANDOM VARIABLES

There are three popular methods of describing the probabilities associated with a discrete random variable, X. They are:

List each value of X and its corresponding probability.

Use a histogram to convey the probabilities corresponding to the various values of X.

Use a function that assigns a probability to each value of X.

Remember our coin-flipping example, in which $X =$ number of heads in three flips of a coin. We can list each value and probability.

$$X = \begin{cases} 0 \text{ with probability } 1/8 \\ 1 \text{ with probability } 3/8 \\ 2 \text{ with probability } 3/8 \\ 3 \text{ with probability } 1/8 \end{cases}$$

This works well when there are only a small number of possible values for X; it would not work well for 100 flips of a coin.

Using a histogram also is a convenient way to represent the shape of a discrete distribution having a small number of possible values. For this situation, you construct a histogram in which the height of each bar is the probability of observing that value of X (Figure 5.2). It is easier to determine the

FIGURE 5.2

A histogram representation of a discrete random variable, where X = number of heads in three coin flips.

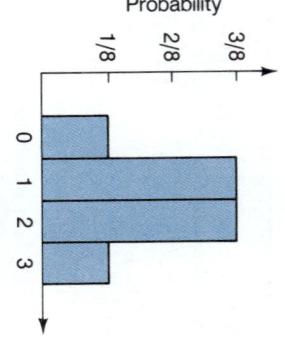

shape of the probability distribution by using such a chart. The distribution in Figure 5.2 is clearly symmetric and concentrated in the middle values.

Using a function (that is, an algebraic formula) to assign probabilities is the most convenient method of describing the probability distribution for a discrete random variable. For any given application of such a random variable, however, this function may or may not be known. Later in the chapter we identify certain useful discrete random variables, each of which has a corresponding function that assigns these probabilities.

The function that assigns a probability to each value of X is called a **probability mass function (PMF)**. Denoting a particular value of X as x, this function is of the form

$$P(X = x) = \text{some expression (usually containing } x \text{) that}$$
$$\text{produces the probability of observing } x$$

$$= P(x)$$

Not every function can serve as a PMF. The requirements for a PMF function are:

1. $P(x)$ is between 0 and 1 (inclusively) for each x
2. $\Sigma P(x) = 1$

EXAMPLE 5.2

Consider a random variable X having possible values of 1, 2, or 3. The corresponding probability for each value is:

$$X = \begin{cases} 1 \text{ with probability } 1/6 \\ 2 \text{ with probability } 1/3 \\ 3 \text{ with probability } 1/2 \end{cases}$$

Determine an expression for the PMF.

SOLUTION

Consider the function

$$P(X = x) = P(x) = x/6 \qquad \text{for } x = 1, 2, 3$$

This function provides the probabilities

$$\begin{aligned} P(X = 1) &= P(1) = 1/6 &\text{(OK)} \\ P(X = 2) &= P(2) = 2/6 = 1/3 &\text{(OK)} \\ P(X = 3) &= P(3) = 3/6 = 1/2 &\text{(OK)} \end{aligned}$$

This function satisfies the requirements for a PMF: Each probability is between 0 and 1, and $P(1) + P(2) + P(3) = 1/6 + 1/3 + 1/2 = 1$. Consequently, the function

$$P(x) = x/6 \qquad \text{for } x = 1, 2, 3 \qquad \text{(and zero elsewhere)}$$

is the PMF for this discrete random variable.

EXAMPLE 5.3

Consider Example 5.1, where X is the total of two dice. Determine the PMF for this discrete random variable.

SOLUTION

Consider the expression

$$P(x) = \frac{x - 1}{36} \qquad \text{for } x = 2, 3, 4, \ldots, 12 \qquad \text{(and zero elsewhere)}$$

If this is the proper PMF, then, for example,

$$P(2) = P(X = 2) = \frac{2 - 1}{36} = 1/36$$

This does appear to be correct, so far. Also,

$$P(5) = P(X = 5) = \frac{5 - 1}{36} = 4/36$$

This also is correct. But now consider

$$P(10) = P(X = 10) = \frac{10 - 1}{36} = 9/36$$

According to our previous solution, we know that $P(10) = 3/36$, not $9/36$. So this particular function is not a bona fide PMF for this example (it is). Do not worry about where this expression came from, but do verify that it works. The truth of the matter is that often PMFs are derived by trial and error.

Consider the expression

$$P(x) = \frac{6 - |x - 7|}{36} \qquad \text{for } x = 2, 3, \ldots, 12 \qquad \text{(and zero elsewhere)}$$

where $|\ |$ represents the absolute value of a number. See if you can demonstrate that this function is a bona fide PMF for this random variable; the PMF must work for *all* values of X.

Notice that a probability mass function provides a theoretical "model" of the population by describing the chance of observing any particular value of the random variable. You can view the population as what you would obtain if you observed the corresponding random variable indefinitely.

EXERCISES

5.9 Let X be the value observed from rolling a die.

a. What is the probability mass function of X?

b. Construct a histogram in which the height of each bar is the probability of X.

5.10 A salesperson calls on three customers who request to see his product. The salesperson has a probability of 50% of selling his product. What is the probability mass function of X, where X is the number of customers (out of three) who buy his product?

5.11 Five copies of the minutes of a business meeting are available for circulation.

Suppose two of the five copies have a blurred spot over an important dollar figure. The secretary chooses one of the copies at random with replacement each time an employee requests to read a copy. Let X be the number of times that a blurred copy is chosen for two requests. What is the probability distribution of X?

5.12 Is the following function a probability mass function? Why or why not?

$$P(X = x) = (x - 2)/6 \quad \text{for } x = 1, 4, 7 \quad \text{(and zero elsewhere)}$$

5.13 Is the following function a probability mass function? Why or why not?

$$P(X = x) = x^2/10 \quad \text{for } x = -2, -1, 1, 2 \quad \text{(and zero elsewhere)}$$

5.14 Let X be equal to the number of heads from tossing a coin twice. Verify that the following is the probability mass function of X.

$$P(X = x) = \frac{1}{2(2 - x)! \, x!} \quad \text{for } x = 0, 1, 2 \quad \text{(and zero elsewhere)}$$

5.15 Do you recognize the following probability mass function? What is it?

$$P(X = x) = 1/6 \quad \text{if } x = 1, 2, 3, 4, 5, 6 \quad \text{(and zero elsewhere)}$$

5.16 A real estate broker needs to advertise two townhouses, two duplexes, and two single family homes. However, the broker decides to choose at random only one of the six properties for open house on a certain weekend. Let the random variable X take on the value 1 if a townhouse is chosen, 2 if a duplex is chosen, and 3 if a single family home is chosen. Write the probability mass function of X.

5.17 Suppose that in Exercise 5.16, the random variable X is assigned the value of 2 if a townhouse is chosen, 4 if a duplex is chosen, and 6 if a single family home is chosen. Write the probability mass function of X.

5.18 Suppose a probability mass function is defined to be nonzero at three points, $X = 1, 2,$ and 3. If $P(X = 1) = .2$ and $P(X = 2) = .3$, what is $P(X = 3)$?

5.19 A quality-control inspector is inspecting incoming lots of materials to check for excessive numbers of defective items. The inspector needs to choose at random one of six incoming lots, and then needs to inspect only that lot. Suppose one lot has zero defective items, two of the lots have three defective items, and the remaining three each have ten defective items. Let the random variable X be equal to 1 if the lot with zero defective items is drawn, 2 if a lot with three defective items is drawn, and 3 if a lot with ten defective items is drawn. Verify that the following function is the probability mass function of X.

$$P(X = x) = x/6 \quad \text{if } x = 1, 2, 3 \quad \text{(and zero elsewhere)}$$

5.3 MEAN AND VARIANCE OF DISCRETE RANDOM VARIABLES

Mean of Discrete Random Variables

Chapter 3 introduced you to the mean and variance of a set of sample data consisting of n values. Suppose that these values were obtained by observing a particular random variable n times. The sample mean, $\bar{X}$, represents the *average* value of the sample data. In this section, we determine a similar value, the **mean of a discrete random variable**, written as μ. The value of μ represents the average value of the random variable if you were to observe this variable over an indefinite period of time.

Reconsider our coin-flipping example, where X is the number of heads in three flips of a coin. Suppose you flip the coin three times, record the value of X, flip the coin three times again, record the value of X, and repeat this process ten times. Now you have ten observations of X. Suppose they are

2, 1, 1, 0, 2, 3, 2, 1, 1, 3

The mean of these data is the *statistic* $\bar{X}$, where

$$\bar{x} = \frac{2 + 1 + 1 + \cdots + 1 + 3}{10}$$

$$= 1.6 \text{ heads}$$

If you observed X *indefinitely*, what would X be on the average?

$$X = \begin{cases} 0 \text{ with probability } 1/8 \\ 1 \text{ with probability } 3/8 \\ 2 \text{ with probability } 3/8 \\ 3 \text{ with probability } 1/8 \end{cases}$$

So 1/8 of the time you should observe the value 0, 3/8 of the time, the value 1, 3/8 of the time, the value 2, and 1/8 of the time, the value 3. In a sense, each probability represents the *relative frequency* for that particular value of X. So the average value of X is

$$(0)(1/8) + (1)(3/8) + (2)(3/8) + (3)(1/8) = 1.5 \text{ heads}$$

Notice that X cannot be 1.5; this is merely the value of X on the average.

DEFINITION

The average value of the discrete random variable X (if observed indefinitely) is the mean of X. The symbol for this parameter is μ.

We found that $\mu = 1.5$ by multiplying each value of X by its corresponding probability and summing the results:

$$\mu = 1.5 = 0 \cdot P(0) + 1 \cdot P(1) + 2 \cdot P(2) + 3 \cdot P(3)$$

This procedure applies to any discrete random variable, and so we define

$$\mu = \Sigma \, xP(x) \qquad\qquad (5.1)$$

EXAMPLE 5.4

A personnel manager in a large production facility is investigating the number of reported on-the-job accidents over a period of 1 month. We define the random variable

$$X = \text{number of reported accidents per month}$$

Based on past records, she has derived the following probability distribution for X:

$$X = \begin{cases} 0 \text{ with probability } .50 \\ 1 \text{ with probability } .25 \\ 2 \text{ with probability } .10 \\ 3 \text{ with probability } .10 \\ 4 \text{ with probability } \underline{.05} \\ \phantom{4 \text{ with probability }} 1.0 \end{cases}$$

During 50% of the months there were no reported accidents, 25% of the months had one accident, and so on. (Notice that deriving an algebraic expression for the PMF for this distribution would be extremely difficult, if not impossible. This poses no problem, however.)

What is the mean (average value) of X?

Using equation 5.1,

$$\mu = (0)(.5) + (1)(.25) + (2)(.1) + (3)(.1) + (4)(.05)$$
$$= .95$$

There is .95 (nearly 1) accident reported on the average per month. ■

Variance of Discrete Random Variables

We previously considered ten observations of the random variable that counted the number of heads in three flips of a coin. These data were 2, 1, 1, 0, 2, 3, 2, 1, 1, 3. We used the notation from Chapter 3 to define the mean of these data, and we obtained $\bar{x} = 1.6$. The variance of these data, using equation 3.8, is $s^2 = .933$. Since s^2 describes a sample, it is a statistic.

Once again, consider observing X indefinitely. For this situation, the average value of X is defined as the mean of X, μ. When we observe X indefinitely, this particular variance is defined to be the variance of the random variable, X, and is written σ^2 (read as "sigma squared").

$$\sigma^2 = \text{variance of the discrete random variable, } X$$

The **variance of a discrete random variable**, X, is a parameter describing the corresponding population and can be obtained by using one of the following expressions, which are mathematically equivalent:

$$\sigma^2 = \Sigma(x - \mu)^2 \cdot P(x) \qquad (5.2)$$
$$\sigma^2 = \Sigma x^2 P(x) - \mu^2 \qquad (5.3)$$

Equation 5.3 generally provides an easier method of determining the variance and will be used in all of the examples to follow. For the coin-flipping example,

$$\sigma^2 = \Sigma x^2 P(x) - \mu^2$$
$$= [(0)^2 \cdot (1/8) + (1)^2 \cdot (3/8) + (2)^2 \cdot (3/8) + (3)^2 \cdot (1/8)] - (1.5)^2$$
$$= 3 - 2.25 = .75$$

So our final results would be:

USING THE SAMPLE OF TEN OBSERVATIONS	FOR THE RANDOM VARIABLE, X (INDEFINITE # OF OBSERVATIONS)
$\bar{x} = 1.6$ $s^2 = .933$	$\mu = 1.5$ $\sigma^2 = .75$
Mean Variance	Mean Variance
Statistics	Parameters

In Chapter 3, the square root of the variance, s, was defined to be the standard deviation of the data. The same definition applies to a random variable. The **standard deviation of a discrete random variable**, X, is denoted σ, where:

$$\sigma = \sqrt{\Sigma(x - \mu)^2 \cdot P(x)} \qquad (5.4)$$
$$\sigma = \sqrt{\Sigma x^2 P(x) - \mu^2} \qquad (5.5)$$

EXAMPLE 5.5

Determine the variance and standard deviation of the random variable described in Example 5.4 concerning on-the-job accidents.

SOLUTION

A convenient method of determining both the mean and variance of a discrete random variable is to summarize the calculations in tabular form:

x	$P(x)$	$x \cdot P(x)$	$x^2 \cdot P(x)$
0	.5	0	0
1	.25	.25	.25
2	.1	.2	.4
3	.1	.3	.9
4	.05	.2	.8
	1.00	.95	2.35

So,

$$\mu = \Sigma x P(x) = .95 \text{ accident}$$

and

$$\sigma^2 = \Sigma x^2 P(x) - \mu^2 = 2.35 - (.95)^2$$
$$= 1.45$$

Also

$$\sigma = \sqrt{1.45} = 1.20 \text{ accidents}$$

∎

EXERCISES

5.20 Goodwin and Sanati (1986) describe a new approach to the teaching of a beginning course in Pascal computer programming using a new learning package called PASLAB. There were 322 students taught by the traditional approach in 1984 and 296 students participating under the new conditions in 1985. The following results were reported:

FINAL GRADES UNDER TRADITIONAL (AB84) AND NEW (CA85) CONDITIONS

DEFINITION OF GRADE	X = GRADE	PERCENT OF STUDENTS (AB84)	PERCENT OF STUDENTS (CA85)
Did not pass; completed 1–2 assignments/exams	0	2	4
Did not pass; completed half of assignments/exams	1	6	3
Did not pass; completed almost all assignments/exams	2	15	4
Passed courses with grade of "Acceptable"	3	55	68
Passed courses with grade of "Distinction"	4	22	21
		100	100

(*Source:* Leonard Goodwin and Mohammad Sanati, "Learning computer programming through dynamic representation of computer functioning: Evaluation of a new learning package for Pascal." *Int. J. Man-Machine Studies* 25 (1986): 327–341.)

a. What is the average grade for the traditional students?

b. What is the average grade for the new students?

c. Compute the variance of the distribution for each of the two groups.

5.21 Several students in a finance class subscribe to the *Wall Street Journal*. If two students are chosen at random, the probability of choosing no students who subscribe is .81. The probability of choosing one student who subscribes and one who does not subscribe is .18, and the probability of choosing two students who subscribe is .01. X is a random variable equal to the number of students who subscribe from the two chosen at random. Find the mean value of X and the variance of X.

5.22 Find the mean and variance of a random variable X, the probability mass function of which is as follows:

X	$P(x)$
-2	.12
-1	.3
0	.1
1	.3
2	.18

5.23 Suppose that a coin is flipped three times. Define the random variable X to be equal to twice the number of heads that appear. Determine the mean and variance of X.

5.24 Show that $\Sigma(x - \mu)P(x) = 0$, where the summation is over all outcomes of X, for any discrete random variable.

5.25 Determine the mean and standard deviation of the random variable for which the probability mass function is defined as follows:

$$P(X = x) = (x - 2)/30 \quad \text{if } x = 3, 12, 21 \quad \text{(and zero elsewhere)}$$

5.26 Determine the mean and standard deviation of the random variable for which the probability mass function is defined as follows:

$$P(X = x) = 1/5 \quad \text{if } x = 1, 2, 3, 4, 5 \quad \text{(and zero elsewhere)}$$

5.27 Pierson and Shorter (1987) conducted a survey of business schools to determine the hardware and software used by students enrolled in introductory database courses. They found a greater prevalence of relational database management systems (DBMS) as opposed to network and hierarchical DBMS. Among other things, they reported the following distribution of DBMS usage:

NUMBER OF DBMS PER SCHOOL USED FOR INSTRUCTIONAL PURPOSES

X = NUMBER OF DBMS	NUMBER OF SCHOOLS	PERCENTAGE OF SCHOOLS
1	34	42.0
2	22	27.1
3	17	21.0
4	5	6.2
5	3	3.7
		100.00

(*Source:* J. K. Pierson and J. D. Shorter, "Study reveals emphasis of relational DBMS in introductory database courses." *Decision Sciences: Theory and Applications;* Proceedings of the Eighteenth Annual Conference of the Southwest Region Decision Sciences Institute, Houston, Texas, March 10–14, 1987.)

a. Without any computations, what do you think the mean and standard deviation of the random variable X might be?

b. Compute the mean and standard deviation of the random variable X.

5.4

BINOMIAL RANDOM VARIABLES

The random variable X representing the number of heads in three flips of a coin is a special type of discrete random variable, a **binomial** random variable. We next list the conditions for a binomial random variable in general and as applied to our coin-flipping example.

A BINOMIAL SITUATION

1. Your experiment consists of n repetitions, called **trials**.

2. Each trial has two mutually exclusive possible outcomes, (or can be considered as having two outcomes), referred to as **success** and **failure**.

3. The n trials are *independent*.

4. The probability of a success for each trial is denoted p; the value of p remains the same for each trial.

5. The random variable X is the number of *successes* out of n trials.

FOR EXAMPLE 5.1

1. n = three flips of a coin

2. Success = head, failure = tail (this is arbitrary).

3. The results on one coin flip do not affect the results on another flip.

4. p = the probability of flipping a head on a particular trial = 1/2.

5. X = the number of heads out of three flips.

You encounter a binomial random variable when a certain experiment is repeated many times (n trials), the trials are independent, and each experiment results in one of two mutually exclusive outcomes. For example, a randomly selected individual is either male or female, is on welfare or is not, will vote Republican or will not, and so on.

The two outcomes for each experiment are labeled as *success* or *failure*. A success need not be considered "good" or "desirable." Instead, it depends on what you are counting at the completion of the n trials. If, for example, the object of the experiment is to determine the probability that 3 people out of 20 randomly selected individuals *are* on welfare, then a success on each of the $n = 20$ trials is the event that the person selected on each trial *is* on welfare.

EXAMPLE 5.6

In Example 4.3, it was noted that 30% of the people in a particular city read the evening newspaper. Select four people at random from this city. Consider the number of people out of these four that read the evening paper. Does this satisfy the requirements of a binomial situation? What is your random variable here?

SOLUTION

Refer to conditions 1 through 5 in our list for a binomial situation.

1. There are $n = 4$ trials, where each trial consists of selecting one individual from this city.

2. There are two outcomes for each trial. We are interested in counting the number of people, out of the four selected, who *do* read the evening paper, so define

success = read the evening newspaper

failure = do not read the evening newspaper

3. The trials are independent since the people are selected randomly.

4. p = probability of a success on each trial = .3.

5. The random variable here is X, where

X = number of successes in n trials

X = number of people (out of four) that read the evening newspaper

All the requirements are satisfied. Thus, X is a binomial random variable (it is also discrete). ∎

Counting Successes for a Binomial Situation

How many ways are there of getting two heads out of four flips of a coin? There are six: HHTT, HTHT, HTTH, THHT, THTH, and TTHH. How many ways can you select two people from a group of four people, where the order of selection is unimportant (say you are selecting a two-person committee)? Label the individuals as I_1, I_2, I_3, and I_4. You want to find the number of combinations of four people using two at a time:

$$_4C_2 = \frac{4!}{2! \, 2!} = 6$$

Put these results side by side. The scheme for matching the two results is to select I_1 if H appears on the first flip, select I_2 if H appears on the second flip, and so on.

TWO HEADS OUT OF FOUR FLIPS	TWO PEOPLE FROM A GROUP OF FOUR
HHTT	I_1, I_2
HTHT	I_1, I_3
HTTH	I_1, I_4
THHT	I_2, I_3
THTH	I_2, I_4
TTHH	I_3, I_4

You should see a direct correspondence between the two solutions. Our conclusion is that the number of ways of getting two heads out of four flips of a coin is $_4C_2$. Extending this to any number of flips of a coin, the number of ways of getting k heads out of n flips of a coin is $_nC_k$. Finally, for any binomial situation, the number of ways of getting k successes out of n trials is $_nC_k$. We are thus able to determine the probability mass function (PMF) for any binomial random variable.

Once again, let X equal the number of heads out of three flips. Here X is a binomial random variable, with $p = .5$. Consider any value of X, say, $X = 1$. Then (1) 1/8 is the probability of any one outcome where $X = 1$, such as HTT, and (2) $_3C_1 = 3$ is the number of ways of getting one head (success) out of three flips (trials). Consequently, the probability that X will be one is:

$$P(1) = {_3C_1}(1/8) = 3/8$$

The resulting PMF for this situation can be written as

$$P(x) = {_3C_x} \cdot (1/8) \qquad \text{for } x = 0, 1, 2, 3 \qquad \text{(and zero elsewhere)}$$

Using this function, we obtain the same results as before:

$$P(0) = {_3C_0}(1/8) = 1 \cdot (1/8) = 1/8$$
$$P(1) = {_3C_1}(1/8) = 3 \cdot (1/8) = 3/8$$
$$P(2) = {_3C_2}(1/8) = 3 \cdot (1/8) = 3/8$$
$$P(3) = {_3C_3}(1/8) = 1 \cdot (1/8) = \frac{1/8}{1}$$

EXAMPLE 5.7

In Example 5.6, the binomial random variable X is the number of people (out of four) who read the evening newspaper. Also, there are $n = 4$ trials (people) with $p = .3$ (30% of the people read the evening newspaper). Let S denote a success and F a failure. Then define:

S = a person reads the evening newspaper

F = a person does not read the evening newspaper

What is the probability that exactly two people (out of four) will read the evening paper?

SOLUTION

This is $P(X = 2)$, or $P(2)$. Consider any one result where $X = 2$, such as SFSF. The probability of this result, using equation 4.14, is (probability of S on first trial) · (probability of F on second trial) · (probability of S on third trial) · (probability of F on fourth trial), which is

$$(.3)(.7)(.3)(.7) = (.3)^2(.7)^2$$

Also note that the probability of *each* result with two S's and two F's ($X = 2$) also is $(.3)^2(.7)^2 = p^2(1 - p)^2$. How many ways can we get two successes out of four trials? This is:

$$_4C_2 = \frac{4!}{2!2!} = 6$$

So, the final result here is

$P(2) =$ (number of ways of getting $X = 2$)(probability of each one)

$$= {}_4C_2(.3)^2(.7)^2$$

$$= (6)(.09)(.49) = .265$$

So, 26.5% of the time, exactly two people out of four will read the evening newspaper. ■

We can extend the results of Example 5.7 to obtain the PMF for a binomial random variable:

$$P(x) = {}_nC_x p^x (1 - p)^{n-x} \qquad \text{for } x = 0, 1, 2, \ldots, n \qquad \text{(and zero elsewhere)} \qquad \textbf{(5.6)}$$

For the newspaper example, $x = 2$, $n = 4$, and $p = .3$. The complete list of probabilities for this example is:

$$X = \begin{cases} 0 \text{ with probability } {}_4C_0 (.3)^0 (.7)^4 = .240 \\ 1 \text{ with probability } {}_4C_1 (.3)^1 (.7)^3 = .412 \\ 2 \text{ with probability } {}_4C_2 (.3)^2 (.7)^2 = .265 \\ 3 \text{ with probability } {}_4C_3 (.3)^3 (.7)^1 = .076 \\ 4 \text{ with probability } {}_4C_4 (.3)^4 (.7)^0 = \underline{.008} \\ \phantom{4 \text{ with probability } {}_4C_4 (.3)^4 (.7)^0 = } 1.001 \end{cases}$$

Note that the total value may be slightly greater or less than 1.0, due to rounding. A graphical representation of this PMF is shown in Figure 5.3.

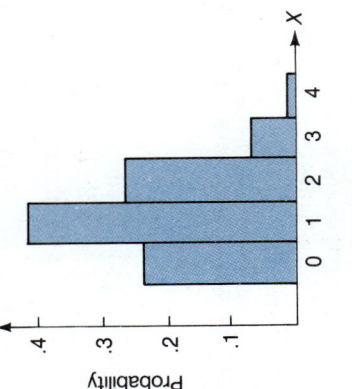

FIGURE 5.3

Probability mass function for $n = 4$, $p = .3$.

Using the Binomial Table

The binomial PMFs have been tabulated in Table A.1 for various values of n and p. The maximum number of trials in this table is $n = 20$. For binomial situations where $n > 20$, one alternative is to use an approximation to a binomial probability. This is considered in the next section and in Chapter 6.

For the evening newspaper illustration in Example 5.6, $n = 4$ and $p = .3$. To find $P(2)$, locate $n = 4$ and $x = 2$. Go across the table to $p = .3$ and you will find the corresponding probability (after inserting the decimal in front of the number). This probability is .265. Similarly, $P(0) = .240$, $P(1) = .412$, $P(3) = .076$, and $P(4) = .008$, as before.

The probability that no more than two people will read the evening paper is written $P(X \le 2)$, where

$$P(X \le 2) = P(X = 0) + P(X = 1) + P(X = 2)$$
$$= P(0) + P(1) + P(2)$$
$$= .240 + .412 + .265$$
$$= .917$$

This is a **cumulative probability** and is obtained by summing $P(x)$ over the appropriate values of X.

Shape of the Binomial Distribution

Figure 5.4 contains a graphical representation of four binomial distributions. In particular, notice that:

1. When $p = .5$, the shape is perfectly *symmetrical* and resembles a bell-shaped (normal) curve.

2. When $p = .2$, the distribution is *skewed right*. This skewness increases as p decreases.

3. For $p = .8$, the distribution is *skewed left*. As p approaches 1, the amount of skewness increases.

Compare Figure 5.4(c) and (d). Notice that, in both cases, p is .2; however, the number of trials increased from $n = 10$ (in c) to $n = 20$ (in d). For the larger value of n, the shape of this distribution is nearly bell-shaped, *despite the small value of p*. This implies that, regardless of the value of p, the shape of a binomial distribution approaches a bell-shaped distribution as the number of trials (n) increases. We will use this fact in the next chapter, when we demonstrate an approximation to the binomial distribution using a bell-shaped (normal) curve.

In summary, the shape of a binomial distribution is:

1. Skewed left for $p > 1/2$ and small n.
2. Skewed right for $p < 1/2$ and small n.
3. Approximately bell-shaped (symmetric) if p is near 1/2 or if the number of trials is large.

Mean and Variance of Binomial Random Variables

In Example 5.6, we examined the binomial random variable X representing the number of people (out of four) who read the evening newspaper. If you select four people, observe X, select four more people, observe X, and repeat this procedure indefinitely, what will X be on the average? This is the mean of X,

FIGURE 5.4
Shape of the binomial distribution.

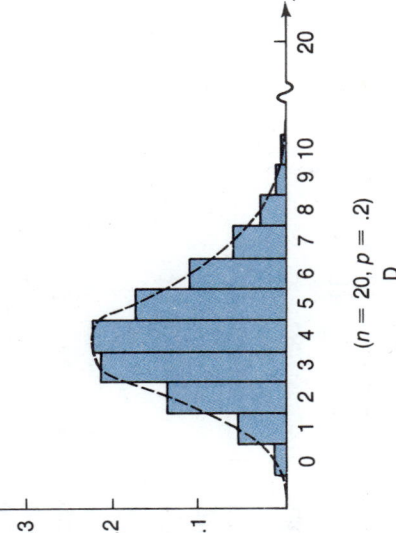

$(n = 10, p = .5)$
A

$(n = 10, p = .8)$
B

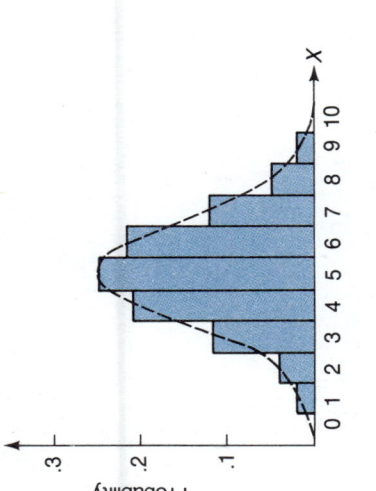

$(n = 10, p = .2)$
C

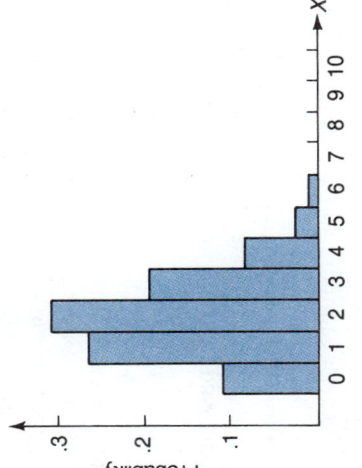

$(n = 20, p = .2)$
D

where, using equation 5.1,

$$\mu = \Sigma \, xP(x)$$
$$= (0)(.240) + (1)(.412) + (2)(.265) + (3)(.076) + (4)(.008)$$
$$= 1.2 \text{ people}$$

Also, using equation 5.3, the variance of X is

$$\sigma^2 = \Sigma \, x^2 P(x) - \mu^2$$
$$= [(0)^2(.240) + (1)^2(.412) + (2)^2(.265) + (3)^2(.076) + (4)^2(.008)] - (1.2)^2$$
$$= 2.28 - 1.44 = .84$$

and so $\sigma =$ standard deviation of $X = \sqrt{.84} = .92$ people. (Watch the units.)

The good news is that there is a convenient shortcut for finding the mean and variance of a binomial random variable. For this situation, you need not use equations 5.1 and 5.3. Instead, for any binomial random variable,

$$\mu = np \qquad \qquad (5.7)$$
$$\sigma^2 = np(1 - p) \qquad \qquad (5.8)$$

How these expressions were derived is certainly not obvious, but let us verify that they work for Example 5.6. Here $n = 4$ and $p = .3$, so

$$\mu = (4)(.3) = 1.2 \quad (\text{OK})$$
$$\sigma^2 = (4)(.3)(.7) = .84 \quad (\text{OK})$$

EXAMPLE 5.8

If you repeat Example 5.6 using $n = 50$ people (rather than $n = 4$ people), how many evening newspaper readers will you observe on the average?

SOLUTION

Now, X is the number of people (out of 50) who read the evening paper. Consequently,

$$\mu = np = (50)(.3) = 15$$

So, on the average, X will be 15 people. For this situation, the variance of X is

$$\sigma^2 = np(1 - p) = (50)(.3)(.7) = 10.5$$

Also,

$$\sigma = \sqrt{10.5} = 3.24 \text{ people}$$

■

EXAMPLE 5.9

Airline overbooking is a common practice. Many people make reservations on several flights due to uncertain plans and then cancel at the last minute or simply fail to show up. Eagle Air is a small commuter airline. Their planes hold only 15 people. Past records indicate that 20% of the people making a reservation do not show up for the flight.

Suppose that Eagle Air decides to book 18 people for each flight.

1. Determine the probability that on any given flight, at least one passenger holding a reservation will not have a seat.
2. What is the probability that there will be one or more empty seats for any one flight?
3. Determine the mean and standard deviation for this random variable.

SOLUTION 1

The binomial random variable for this situation is $X =$ the number of people (out of 18) who book a flight and actually do appear. For this binomial situation, $n = 18$ (18 reservations are made) and $p = 1 - .2 = .8$ (the probability that any one person will show up). At least one passenger will have no place to sit if X is 16 or more. Using Table A.1,

$$P(X \geq 16) = P(X = 16) + P(X = 17) + P(X = 18)$$
$$= .172 + .081 + .018 = .271$$

We see that if the airline follows this policy, 27% of the time one or more passengers will be deprived of a seat—not a good situation.

SOLUTION 2

We want to find the probability that the number of people who actually arrive (X) is 14 or less. Using Table A.1 (where $n = 18$, $p = .8$),

$$P(X \leq 14) = .215 + .151 + \cdots + .003 + .001 = .50$$

(Notice that the four remaining probabilities are nearly zero.) With this booking policy, the airline will have flights with one or more empty seats approximately one-half of the time.

SOLUTION 3

The mean of X is

$$\mu = np = (18)(.8) = 14.4 \text{ people}$$

which implies that the average number of people that book a flight and do appear is 14.4.

The standard deviation of X is

$$\sigma = \sqrt{np(1-p)} = \sqrt{(18)(.8)(.2)} = 1.70 \text{ people}$$

■

EXAMPLE 5.10

It is estimated that one out of ten vouchers examined by the audit staff employed by a branch of the Department of Health and Human Services will contain an error. Define X to be the number of vouchers in error out of 20 randomly selected vouchers.

1. What is the probability that at least three vouchers will contain an error?
2. What is the probability that no more than one contains an error?
3. Determine the mean and standard deviation of X.

SOLUTION 1

The random variable X satisfies the requirements for a binomial random variable with $n = 20$ and $p = .1$. For this situation, a "success" is defined to be that a voucher contains an error. The probability that at least three vouchers will contain an error is the probability that X is *3 or more*. This is

$$P(X \geq 3) = 1 - P(X < 3)$$
$$= 1 - P(X \leq 2)$$
$$= 1 - [P(0) + P(1) + P(2)]$$
$$= 1 - (.122 + .270 + .285)$$
$$= .323$$

Consequently, the probability that at least three vouchers will contain an error is .323.

SOLUTION 2

The chance that no more than one voucher is in error is the probability that X is *1 or less*. This is

$$P(X \leq 1) = P(0) + P(1) = .122 + .270 = .392$$

So, this event will occur with probability .392.

SOLUTION 3

The mean of the random variable X is

$$\mu = np = (20)(.1) = 2 \text{ vouchers}$$

and the standard deviation of X is

$$\sigma = \sqrt{np(1-p)} = \sqrt{(20)(.1)(.9)} = 1.34 \text{ vouchers}$$

This implies that, on the average, the audit staff will encounter 2 vouchers containing an error (out of 20 randomly selected vouchers).

■

EXAMPLE 5.11

A situation that requires the use of a binomial random variable is **lot acceptance sampling**, where you decide to accept or send back a lot (batch) consisting of many electrical components, machine parts, or whatever.

A shipment of 500 calculator chips arrives at Cassidy Electronics. The contract specifies that Cassidy will accept this lot if a sample size of ten from the shipment has no more than one defective chip. What is the probability of accepting the lot if, in fact, 10% of the lot (50 chips) are defective? If 20% are defective?

SOLUTION

This is approximately a binomial situation where:

1. There are $n = 10$ trials.
2. Each trial has two outcomes:

   success = chip is defective

   failure = chip is not defective

(*Note:* Since the object is to count the number of *defective* chips in the shipment, a success on each trial (chip) will be that the chip is defective. As mentioned earlier, a success need not be a desirable event.)

3. p = probability of a success = .10.*

4.   The random variable here is X = number of *defective* chips in the number of defective chips out of ten. Cassidy accepts the lot of chips if X is 0 or 1. The corresponding probability is a cumulative probability:

$$P(\text{accept}) = P(X \le 1)$$
$$= P(0) + P(1)$$

Using Table A.1 (for $n = 10$, $p = .10$), you obtain

$$P(0) = .349 \quad \text{and} \quad P(1) = .387$$

The resulting probability of accepting the lot is $.349 + .387 = .736$. This means that such a sampling procedure will result in Cassidy accepting the entire batch of chips 73.6% of the time.

If $p = .20$, then $P(0) = .107$ and $P(1) = .268$, again using Table A.1; Cassidy now accepts the lot with probability $.107 + .268 = .375$. ■

The concept of lot acceptance sampling was originally presented in Chapter 1 to illustrate the distinction between a population and a sample. It also serves as a brief introduction to the area of inferential statistics, discussed at length in Chapter 7. In Example 5.11, we inferred something about a population (the lot of 500 chips) using a sample (the ten chips selected for testing). The sample does not include all elements of the population, so there is a risk of making an incorrect decision, such as (1) accepting the lot of chips when in fact Cassidy should not have, or (2) rejecting the lot of chips when in fact it was satisfactory. Such possibilities for error *always* exist when a statistical sample is used as a basis for an assertion about a population.

5.28   An investment advisor predicts that five stocks will grow over the next 18 months. From the advisor's records, 40% of the stocks she recommends are profitable.

a.   What is the probability that exactly two of the five stocks are profitable?

b.   What is the probability that at least three of the five stocks are profitable?

5.29   If four trials are independently conducted and each trial results in a success with probability one-third, what is the probability that exactly two of the four trials result in successes?

5.30   A survey reveals that 60% of the eligible people in a certain county vote during a county election.

a.   If 20 people in the county who are eligible to vote are chosen at random, what is the probability that exactly 12 vote during the next county election?

b.   What is the probability that exactly 10 people in the county vote during this election out of the 20 chosen at random?

* If the lot size is large (500 here) and the sample size is small (10 here), then the value of p is nearly, although not completely, unaffected by the previous trials. For example, if 10% of the chips are defective, then on the first trial, p is 50/500 = .10. On the second trial, p is either 50/499 = .1002 (if the first chip was nondefective) or 49/499 = .098 (if the first chip was defective). We typically ignore this minor problem in lot sampling. Situations in which the value of p is severely affected by what occurred on previous trials will be dealt with in the next section, where we discuss the hypergeometric distribution.

5.31 The vice president of a business firm has reviewed the records of the firm's personnel and has found that 70% of the employees read the *Wall Street Journal*. If the vice president was to choose 12 employees at random, what is the probability that the number of these employees who read the *Wall Street Journal* is the following?

a. At least equal to five.

b. Between four and ten, inclusive.

c. No more then seven.

5.32 A lawyer estimates that 40% of the cases in which she represented the defendant were won. If the lawyer is presently representing ten defendants in different cases, what is the probability that at least five of the cases will be won? What are you assuming here?

5.33 A market-research firm has discovered that 30% of the people who earn between $25,000 and $50,000 per year have bought a new car within the past two years. In a sample of 12 people earning between $25,000 and $50,000 per year, what is the probability that between four and ten people, inclusive, have bought a new car within the past two years?

5.34 A newsstand owner has calculated that 80% of the midday newspapers are sold. If the owner orders 25 midday newspapers daily, what is the probability that on any day 23 or more of the newspapers will be sold?

5.35 The sales manager of an insurance company knows that the company's best salesperson can sell an insurance policy 60% of the time. If this salesperson were to make 15 calls to sell insurance, what is the probability that at least ten insurance policies would be sold?

5.36 Let the random variable X represent the number of loans that have gone into default from a sample of eight loans made 5 years ago. The probability of a loan going into default within 5 years is equal to .15.

a. What is the mean and standard deviation of the random variable X?

b. What is $P(X = 2)$?

5.37 Let the random variable X represent the number of correct responses on a multiple-choice test that has 15 questions. Each question has five multiple-choice answers.

a. What is the probability that the random variable X is greater than 8 if the person taking the test randomly guesses?

b. What is the mean value of X if the person randomly guesses?

c. What is the standard deviation of X if the person randomly guesses?

d. Estimate the probability that X will fall within the limits $\mu \pm 2\sigma$.

5.38 The *Professional Technician* recommends stocks each month. If 40% of the stocks recommended advance at least 20%, what is the probability that of the five stocks most recently recommended, at least three will advance at least 20%?

5.39 The manager of a retail store knows that 10% of all checks written are "hot" checks. Of the next 25 checks written at the retail store, what is the probability that no more than 3 checks are hot?

5.40 There has not been a strong need for market segmentation sophistication in the insurance industry. Insurance companies deal with the consumer typically only through an agent. The success of the company depends on how successful the agents are in selling. A survey by Prudential Insurance Company (reported in the *Journal of Consumer Marketing* 4, no. 1, p. 52, 1987) revealed that 75% of the recent purchasers of insurance policies did not shop around.

a. If a random sample of eight recent purchasers of insurance policies were selected, how many would you expect shopped around before buying?

b. What is the probability that all eight did not shop around before buying in part (a)? Would this outcome be considered a rare event?

c. Within what limits would you expect the number of purchasers who did not shop around to fall at least 75% of the time, if repeated samples of size 8 were taken? (*Hint:* Use Chebyshev's inequality.)

5.41 Competition and change in the corporate world can create critical levels of stress, which can eventually affect one's job. Mental experts estimate that as many as 15% of executives and managers suffer from depression and critical levels of stress. Let X represent the number of executives and managers that suffer from depression and critical levels of stress from a random sample of five executives and managers.

(*Source:* "Stress: The Test Americans Are Failing," *Business Week*, April 18, 1988, pp. 74–76.)

a. What is the probability that X is no more than one?

b. What are the mean and standard deviation of X?

5.5
THE HYPER-GEOMETRIC DISTRIBUTION (OPTIONAL)

Another type of discrete random variable that fits many sampling situations is the **hypergeometric random variable**. It bears a strong resemblance to the binomial random variable since the experiment once again consists of n trials, with each trial having two possible outcomes (success or failure).

The conditions for a hypergeometric random variable are:

1. Population size = N. In this population, k members are S (successes) and $N - k$ are F (failures).

2. Sample size = n trials, obtained *without replacement*.

3. X = the number of successes out of n trials (a hypergeometric random variable).

The main distinction between a hypergeometric and a binomial situation is that the trials in the former *are not independent*. As a result, the probability of a success on each trial is affected by the results of the previous trials. This occurs when sampling *without replacement* from a *finite* population.

The situation surrounding a hypergeometric random variable is similar to the binomial situation in that you count "successes" in both cases. However, for the hypergeometric situation, you have a *finite* population (of size N) and you know the number of successes (k) and failures ($N - k$) that make up this population. For example, you might select a random sample of $n = 8$ from a group of $N = 30$ unionized workers, of which $k = 20$ are in favor of a strike and $N - k = 10$ are not. For this situation, the hypergeometric random variable is X = the number of workers (of the 8) who favor the strike.

We can repeat Example 5.11 using 50 chips (instead of 500), 10 of which are selected for testing. Suppose that 10% of these chips (5 chips) are defective. As before, define

$$S = \text{success} = \text{chip is defective}$$
$$F = \text{failure} = \text{chip is not defective}$$

In Example 5.11, we used $p = P(S) = .10$ for each trial. Here, out of the 50 chips, five are defective. So

$$P(S \text{ on first trial}) = 5/50 = .10$$

The conditional probability of S on the second trial is:

$$5/49 = .102 \text{ if first chip was not defective}$$
$$4/49 = .082 \text{ if first chip was defective}$$

The probability of a success on the second trial is affected by what occurred on the first trial; this is a hypergeometric situation.

The PMF for the hypergeometric random variable is:

$$P(x) = \frac{_kC_x \cdot _{N-k}C_{n-x}}{_NC_n} \quad (5.9)$$

for $x = a, a+1, a+2, \ldots, b$, where a is the maximum of 0 and $n+k-N$ and b is the minimum of k and n. The value of $P(x)$ is zero for all other values of X.

EXAMPLE 5.12

SOLUTION

Determine the probability of observing exactly one defective chip out of a sample of size ten.

Imagine two containers (the population). One contains 5 S's and the other has 45 F's. The sample consists of 10 chips, randomly selected from these two containers. If x chips are selected from the success container, then $10-x$ chips are selected from the failure container. For this situation, $N = 50$, $k = 5$ and $n = 10$. The possible values for X are from $a =$ maximum of 0 and -35 (0) to $b =$ minimum of 5 and 10 (5). The probability of obtaining one S and nine F's in your sample is

$$P(X = 1) = P(1) = \frac{_5C_1 \cdot _{45}C_9}{_{50}C_{10}}$$

As you will quickly see, the term $_NC_n$ gets very large—in fact, it becomes too large for many calculators. The only practical way to evaluate a hypergeometric probability, short of relying on a computer, is to cancel as many terms as possible in the expression.

The final result here is $P(1) = .431$: 43% of the time, you will obtain exactly 1 defective chip in your sample of size 10. ∎

EXAMPLE 5.13

SOLUTION

A local group of 30 unionized workers contains 20 people who are in favor of a strike and 10 who are not. Determine the probability that a random sample of 8 workers contains 5 individuals who favor the strike and 3 that are opposed.

This situation fits the requirements for a hypergeometric random variable, where X is the number of workers (out of 8) that favor a strike, $n = 8$, $N = 30$, and $k = 20$. Consequently,

$$P(X = 5) = P(5) = \frac{_{20}C_5 \cdot _{10}C_3}{_{30}C_8}$$

$$= \frac{\frac{20!}{5!15!} \cdot \frac{10!}{3!7!}}{\frac{30!}{8!22!}}$$

$$= \frac{(15,504)(120)}{5,852,925} = .318$$

Approximately 32% of the time, in a sample of size 8 from this group, 5 people would favor a strike. ∎

Mean and Variance of a Hypergeometric Random Variable

As we did with the binomial random variable, we could use the definition of the mean and variance of a discrete random variable contained in equations 5.1 and 5.3. For example,

$$\mu = \Sigma \, xP(x)$$

where $P(x)$ is the PMF given in equation 5.9.

As in the binomial situation, simpler expressions exist for both the mean and the variance of the hypergeometric random variable. These are:

$$\mu = \Sigma \, xP(x) = \frac{nk}{N} \qquad (5.10)$$

and

$$\sigma^2 = \Sigma \, x^2 P(x) - \mu^2$$
$$= \frac{k(N-k)n(N-n)}{N^2(N-1)} = \left[n \left(\frac{k}{N} \right) \left(1 - \frac{k}{N} \right) \right] \left(\frac{N-n}{N-1} \right) \qquad (5.11)$$

For Example 5.12, $N = 50$, $k = 5$, $n = 10$. Consequently,

$$\mu = \frac{(10)(5)}{50} = 1 \text{ chip}$$

$$\sigma^2 = \frac{(5)(45)(10)(40)}{(50)^2(49)} = .735$$

and so

$$\sigma = \sqrt{.735} = .857 \text{ chip}$$

This means that if we observed this process of sampling 10 chips out of a batch of 50 indefinitely, we would obtain one ($= \mu$) defective chip on the average. Also, $\sigma = .857$ (or $\sigma^2 = .735$) is our measure of the variation in the observations of this random variable if we observe it over an indefinite period.

Using the Binomial to Approximate the Hypergeometric

Whenever $n/N < .05$, the binomial distribution will provide a good approximation to the hypergeometric distribution. Here, define

$$p = \frac{\text{number of successes in population}}{\text{size of population}} = \frac{k}{N}$$

Then X is the number of successes in the sample. X is approximately a binomial random variable with n trials and probability of success p. Briefly, the binomial approximation works well if your sample size is *less than* 5% of your population size. This was the case in Example 5.11, where $n/N = 10/500 = .02$.

What probability would you obtain had you treated Example 5.12 as a binomial situation, where $p = k/N = 5/50 = .10$? Here you have a binomial situation with $n = 10$ and $p = .10$. Using equation 5.6,

$$P(1) = {}_{10}C_1 (.1)^1 (.9)^9 = (10)(.1)(.387) = .387$$

The same result is obtained using Table A.1.

For this example, .431 is the *exact* probability using the hypergeometric distribution and .387 is the *approximate* probability using the binomial distribution. We did not obtain a very good approximation here. The problem is that the population size is $N = 50$ and the sample size is $n = 10$, which is 20% of the population size.

The Poisson distribution, named after the French mathematician Simeon Poisson, is useful for counting the number of times a particular event occurs over a specified period of time. It also can be used for counting the number of times an event (such as a manufacturing defect) occurs over a specified area (such as a square yard of sheet metal) or in a specified volume. We will restrict our discussion to counting over time, although any unit of measurement is permissible.

The random variable X for this situation is the number of occurrences of a particular event over a specified period of time. The possible values are 0, 1, 2, 3, ... For X to be a **Poisson random variable**, the occurrences of this event need to occur *randomly*, as summarized by the following three conditions:

1. The number of occurrences in one interval of time is unaffected by (statistically independent of) the number of occurrences in any other non-overlapping time interval. For example, what took place between 3:00 and 3:20 P.M. is unaffected by what took place between 9:00 and 10:00 A.M.

2. The expected (or average) number of occurrences over any time period is proportional to the size of this time interval. For example, we would expect half as many occurrences between 3:00 and 3:30 P.M. as between 3:00 and 4:00 P.M.

   This also implies that the probability of an occurrence must be constant over any intervals of the same length. A situation in which this is usually *not* true is at a restaurant from 12:00 noon to 12:10 P.M. and 2:00 to 2:10 P.M. Due to the differences in traffic flow for these two intervals, we would not expect the arrivals between, say, 11:30 A.M. and 2:30 P.M. to satisfy the requirements of a Poisson situation.

3. Events cannot occur exactly at the same time. More precisely, there is a unit of time sufficiently small (such as one second) during which no more than one occurrence of an event is possible.

Four situations that usually meet these conditions are:

The number of arrivals at a local bank over a 5-minute interval.

The number of telephone calls arriving at a switchboard over a 1-minute interval.

The number of daily accidents reported along a 20-mile stretch of an inter-city toll road.

The number of trucks from a fleet that break down over a 1-month period.

For each situation, the (discrete) random variable X is the number of occurrences over the time period T. If all the assumptions are satisfied, then X is a Poisson random variable. Define μ to be the expected (or average) number of occurrences over this period of time.* For any application, the value of μ must be specified or estimated in some manner. The Poisson PMF for X follows.

---

* The symbol λ (lambda) often is used to denote this parameter.

POISSON PROBABILITY MASS FUNCTION

X = number of occurrences over time period T.

$$P(x) = \frac{\mu^x e^{-\mu}}{x!} \qquad \text{for } x = 0, 1, 2, 3, \ldots \qquad (5.12)$$

where μ = expected number of occurrences over T.

Equation 5.12 contains the number e. This is an interesting and useful number in mathematics and statistics. To get an idea how this number is derived, consider the following sequence:

$(1 + 1/2)^2 = 2.25$

$(1 + 1/3)^3 = 2.37$

$(1 + 1/4)^4 = 2.44$

$(1 + 1/5)^5 = 2.49$

...

$(1 + 1/100)^{100} = 2.705$

...

$(1 + 1/1000)^{1000} = 2.717$

...

This sequence of numbers is approaching e. The actual value is

$$e = 2.71828 \ldots$$

We use e again in Chapter 6.

One interesting application of the number e occurs when calculating compound interest. For example, if you invest $100 at 12% compounded annually, then at the end of the year you will have $112. However, if your interest is compounded not monthly, not daily, but continuously, the amount in your account will be $(100)(e^{.12}) = (100)(1.1275) = 112.75. The difference in these amounts is not as large as you might expect!

Mean and Variance of a Poisson Random Variable

Once again, we could use the definition of the mean and variance of a discrete random variable in equations 5.1 and 5.3. However, this is not necessary. It is fairly easy to show that, using equation 5.12,

$$\text{mean of } X = \sum x P(x)$$

$$= \mu$$

This is hardly a surprising result, given how μ was originally defined. Also,

$$\text{variance of } X = \sigma^2$$

$$= \sum x^2 P(x) - \mu^2$$

$$= \mu$$

So, both the mean and the variance of the Poisson random variable X are equal to μ. Recall that the Poisson random variable is the number of occurrences of a particular event (such as a traffic accident) over a given time period (such as 1 hour). If the time period is doubled to 2 hours, then the mean of the "new" Poisson random variable is twice the original mean; if the time period is halved to 30 minutes, the corresponding mean is halved, and so on. This is illustrated in the next two examples.

Applications of a Poisson Random Variable

EXAMPLE 5.14

Handy Home Center specializes in building materials for home improvements. They recently constructed an information booth in the center of the store. Define X to be the number of customers who arrive at the booth over a 5-minute period. Assume that the conditions for a Poisson situation are satisfied with

$$\mu = 4 \text{ customers over a 5-minute period}$$

A graph of the Poisson probabilities for $\mu = 4$ is contained in Figure 5.5.

1. What is the probability that over any 5-minute interval, exactly four people arrive at the information booth?
2. What is the probability that more than one person will arrive?
3. What is the probability that you observe exactly six people over a 10-minute period?

SOLUTION 1

First, this probability is not 1 because $\mu = 4$ is the *average* number of arrivals over this time period. The actual number of arrivals over some 5-minute period may be fewer than four, more than four, or exactly four. The fraction of time that you observe exactly four people is, using Table A.3,

$$P(4) = \frac{4^4 e^{-4}}{4!} = .1954$$

If you stand in the booth for many 5-minute periods, 19.5% of the time you will observe four people arrive.

SOLUTION 2

This is $P(X > 1) = P(X \geq 2)$. We could try

$$P(X \geq 2) = P(X = 2) + P(X = 3) + \cdots$$
$$= P(2) + P(3) + \cdots$$
$$= .1465 + .1954 + \cdots$$

There is an infinite number of terms here, however, so this is *not* the way to find this probability. A much better way is to use the fact that these probabilities

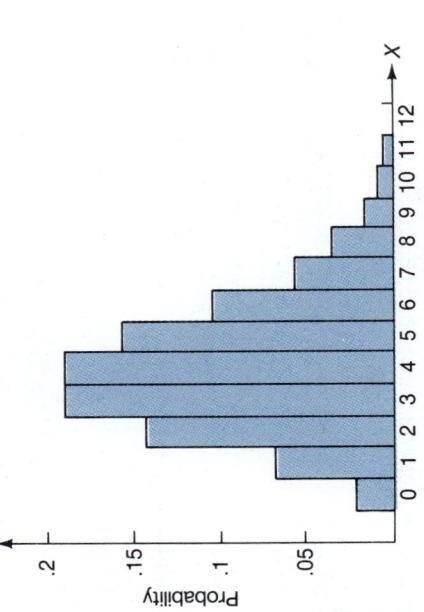

FIGURE 5.5
Poisson probabilities for $\mu = 4$.

sum to 1. Consequently,

$$P(X \geq 2) = 1 - P(X < 2)$$
$$= 1 - P(X \leq 1)$$
$$= 1 - [P(0) + P(1)]$$
$$= 1 - \left[\frac{4^0 e^{-4}}{0!} + \frac{4^1 e^{-4}}{1!} \right]$$
$$= 1 - [.0183 + .0733] = .9084$$

SOLUTION 3

For this time interval,

μ = expected (average) number of people over a 10-minute
time period

$\mu = 8$ (we expect four people over a 5-minute period)

Therefore, the probability of observing six people over a 10-minute period is

$$\frac{8^6 e^{-8}}{6!} = .1221$$

using Table A.3.   ∎

EXAMPLE 5.15

During the mid to late 1980s, the number of Texas savings and loan (S and L) institutions that failed reached alarming proportions due to poor oil and real estate investments. Suppose that the number of S and L failures follows a Poisson distribution with an average of two failures per month. Determine the probability that:

1. There will be no S and L failures during any particular month.
2. There will be more than eight failures during a randomly chosen 3-month period.

SOLUTION 1

The Poisson random variable X for this situation is the number of S and L failures. The average number of failures over a 1-month period is two, so

$$P(X = 0) = \frac{2^0 e^{-2}}{0!} = .1353$$

using Table A.3. This means that, 13.5% of the time, there will be no S and L failures over a randomly chosen 1-month period.

SOLUTION 2

The average number of failures over a 3-month period is six, given that the average is two over a 1-month period. Therefore, using Table A.3 with $\mu = 6$:

$$P(X > 8) = 1 - P(X \leq 8)$$
$$= 1 - \left[\frac{6^0 e^{-6}}{0!} + \frac{6^1 e^{-6}}{1!} + \cdots + \frac{6^7 e^{-6}}{7!} + \frac{6^8 e^{-6}}{8!} \right]$$
$$= 1 - (.0025 + .0149 + \cdots + .1377 + .1033) = .153$$

We can expect more than eight savings and loan institutions to fail over a 3-month period 15.3% of the time.   ∎

Poisson Approximation to the Binomial

There will be many times when you are in a binomial situation but n is too large to be tabulated. For such situations, you can use a computer. If that is not convenient, there are methods of *approximating* these probabilities without

FIGURE 5.6

Poisson distributions provide a good approximation of binomial probabilities where $n > 20$ and $np \leq 7$. Here, $n = 20$, $p = .10$.

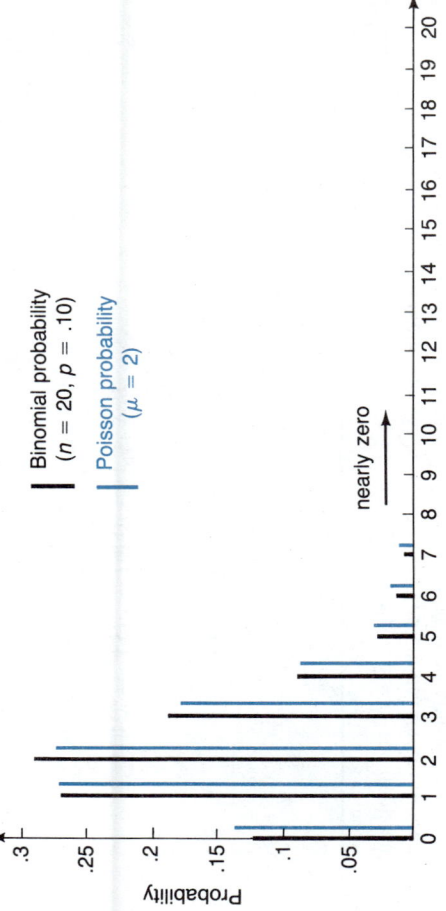

sacrificing much accuracy. One method is to pretend that your binomial random variable, X, is a Poisson random variable having the same mean. The corresponding Poisson probability may be much simpler to derive and will serve as an excellent approximation to the binomial probability.

A good approximation to a binomial probability is obtained using the Poisson distribution if n is large and p is small. For most situations, you can trust this approximation if $n > 20$ and $np \leq 7$. An illustration using $n = 20$ and $p = .10$ is shown in Figure 5.6. The binomial probabilities are from Table A.1 and the Poisson probabilities are from Table A.3.

EXAMPLE 5.16

In Example 5.11, Cassidy Electronics received a batch (lot) of 500 calculator chips, 10 of which they sampled. Suppose instead that they receive a batch of 2500 chips and test 100 of them. They will accept the lot if the sample contains no more than one defective chip. If we assume that 5% of the chips are defective, what is the probability that they will accept the lot?

SOLUTION

We can treat this as a binomial situation (rather than the more complicated hypergeometric situation) because

$$\frac{n}{N} = \frac{100}{2500} = .04 < .05$$

This is approximately a binomial situation with $n = 100$ trials and $p = .05$. The binomial random variable X here is the number of defective chips out of 100. So

$$P(\text{accept}) = P(X \leq 1) = P(0) + P(1)$$

Using the PMF in equation 5.6:

$$P(\text{accept}) = {}_{100}C_0 \cdot (.05)^0 (.95)^{100} + {}_{100}C_1 \cdot (.05)^1 (.95)^{99}$$

Table A.1 does not contain values of n larger than 20, so you need another means of determining these probabilities. A computer or a calculator will allow you to determine these exactly. The exact answer here is $P(\text{accept}) = .037$.

The other alternative is to pretend that X is a Poisson random variable with the same mean as the actual binomial random variable; that is,

$$\mu = np = (100)(.05) = 5$$

The approximation should work quite well because $n > 20$ and np is ≤ 7.

Using Table A.3, with $\mu = 5$,

$$P(0) = .0067$$

and

$$P(1) = .0337$$

Therefore,

$$P(X \leq 1) = .0067 + .0337 = .0404$$

Using the Poisson approximation,

$$P(\text{accept}) \cong .0404$$

which is quite close to the binomial value of .037.

◼

EXERCISES

5.42 Ten people apply for a job as a bookkeeper. Six of the applicants have college degrees and the remainder do not. If four of the applicants are randomly selected for the job, what is the probability that exactly three have college degrees?

5.43 Six vegetables are available at a cafeteria. Four vegetables are green and the other two are not green. If three different vegetables are ordered at random, what is the probability that at least two of the vegetables are green?

5.44 A population consists of eight round and seven square objects. Let the random variable X be equal to the number of round objects selected randomly without replacement from a sample of nine objects.

a. Find $P(2 \leq X \leq 5)$.

b. Find $P(X > 4)$.

c. Find the mean of the random variable X.

d. Find the standard deviation of the random variable X.

5.45 A batch of 350 resistors is to be shipped if a random sample of 15 resistors has two or fewer defective resistors. If it is known that there are 50 defective resistors in the batch, what is the probability that two or fewer of the sample of 15 resistors will be defective?

5.46 The Good Olde Boys used-car lot has 20 cars for sale. It is known that 8 of the cars get over 28 miles per gallon on the highway and 12 do not. Let X be the random variable equal to the number of cars sold that get over 28 miles per gallon out of the next five cars sold. Assume that each car is equally likely to sell.

a. What is $P(X \leq 2)$?

b. What is $P(1 \leq X \leq 3)$?

c. Find the mean and variance of X.

5.47 In a sample of ten men, it is found that six are physically fit. If four men are randomly selected from this sample of ten, what is the probability that no more than three are physically fit?

5.48 A box contains eight golf balls. Four of these balls are not perfectly round. If three balls are randomly selected without replacement from the eight golf balls, what is the probability that at least one is not perfectly round?

5.49 A factory manufactures rubber grommets to be placed on the stick shift of a car. A sample of ten grommets is chosen from a box of 200. Let the random variable X be the number of defective rubber grommets. Assume that it is known that there are ten defective grommets in the box.

a. What is the probability that X is greater than two?

b. What is the probability that X is equal to zero?

c. What is the standard deviation of X?

5.50 A textbook copy editor is reviewing a manuscript for grammatical errors. Let the random variable X represent the number of grammatical errors made in a particular chapter. Assume that the conditions of a Poisson distribution are satisfied with an average of ten grammatical errors per chapter.

a. What is the probability that X is less than seven?

b. What is the mean value of X?

c. What is the standard deviation of X?

5.51 Let the random variable X be binomially distributed with $n = 60$ and $p = 0.05$. Use the Poisson distribution to approximate the probability that X is greater than or equal to three.

5.52 A police officer writes an average of two speeding tickets per hour. What is the probability that in 1 hour, the police officer writes no more than one speeding ticket? What assumptions need to be made?

5.53 The auto parts department of an automotive dealership sends out an average of eight special orders daily. The number of special orders is assumed to follow a Poisson distribution.

a. What is the probability that for any day, the number of special orders sent out will be more than four?

b. What is the standard deviation of the number of special orders sent out daily?

5.54 A survey indicates that 10% of the people who earn less than $20,000 per year are homeowners.

a. If a sample of 40 people who earn less than $20,000 per year is randomly selected, what is the probability that more than four people are homeowners?

b. What is the probability that exactly four people are homeowners from the sample of 40 people who earn less than $20,000 per year?

5.55 In a study on "How Important Is Computer Service In Retail Branch Banking?" (authored by M. Maggard and S. Globerson and reported in *The Proceedings of the Decision Sciences Institute,* 1987: 906–8), a survey of over 50 branch bank managers revealed that an "acceptable" average waiting time for a customer to see a teller ranged from a low of .75 minute to a high of 10 minutes. Suppose that a certain manager believes that if the mean number of customers arriving over a 10-minute interval is a Poisson random variable with mean 3, then the average waiting time for a customer will be acceptable.

a. What percentage of the time will the number of arrivals over a 10-minute interval exceed three, assuming the manager's belief is correct?

b. Can the event of seven or more arrivals over a 10-minute interval be considered a relatively rare event under the manager's assumption?

5.56 According to a survey commissioned by Arthur Young on a large sample of the Times 1000 companies, accountancy firms are not attractive to potential investors. The survey revealed that only 6% of top management would consider investing in an accountancy business. If 22 companies from the Times 1000 companies were randomly selected by a separate small survey, approximately what is the probability that more than one company in the survey would consider investing in an accountancy business?

(*Source:* "Briefing: Accountancy Profession," *The Accountant's Magazine* (March 1988): 5.)

SUMMARY

When an experiment results in a numerical outcome, a convenient way of representing the possible values and corresponding probabilities is to use a random variable. A **random variable** takes on a numerical value for each outcome of an experiment. If the possible values of this variable can be listed along with the probability for each value, this variable is said to be a **discrete random variable.**

Conversely, if any value of this variable can occur over a specific range, then it is a **continuous random variable**. This chapter concentrated on the discrete type, whereas Chapter 6 discusses the continuous random variable.

For a discrete random variable, the set of possible values and corresponding probabilities is a **probability distribution**. There are several ways of representing such a distribution, including a list of each value and its probability, a histogram, or an expression called a **probability mass function (PMF)**, which is a numerical function that assigns a probability to each value of the random variable.

If you could observe a random variable indefinitely, you would obtain the corresponding *population*. A *sample* then consists of a finite number of random variable observations. In Chapter 3, we introduced ways of describing a set of sample data using various statistics, including the sample mean and variance. Similarly, we can describe a random variable using its mean and variance. Since they describe the population, they are parameters. The **mean** of a discrete random variable, μ, is the average value of this variable if observed over an indefinite period. The mean is found by summing the product of each value and its probability of occurring. The **variance** of a discrete random variable, σ^2, is a measure of the variation for this variable. The **standard deviation**, σ, also measures this variation and is the square root of the variance.

Three popular discrete random variables used in a business setting are the binomial, hypergeometric, and Poisson random variables. A **binomial** random variable counts the number of successes out of n independent trials. When the trials are dependent and the population size (N) as well as the number of successes in the population (k) are known, the **hypergeometric** random variable can be used to describe the number of successes out of the n trials. The **Poisson** random variable is used for situations where you are observing the number of occurrences of a particular event over a specified period of time or space. A table of binomial probabilities is contained at the back of the text in Table A.1. Tables A.2 and A.3 can be used for Poisson probabilities. For these three distributions, shortcut formulas exist for deriving the mean and variance of the random variable. Often, the probabilities for one of these discrete distributions are difficult to calculate due to the magnitude of the numbers involved. In many situations, you can use one discrete distribution to *approximate* the probability for another. Figure 5.7 summarizes how this is done.

Summary of the Three Most Commonly Used Discrete Random Variables

Binomial Distribution

1. X denotes the number of successes out of n independent trials. Each trial results in a success (with probability p) or a failure (with probability $1 - p$).
2. PMF is $P(X = x) = P(x) = {}_nC_x p^x(1 - p)^{n-x}$ for $x = 0, 1, \ldots, n$.
3. Mean $= \mu = np$.
4. Variance $= \sigma^2 = np(1 - p)$ and standard deviation $= \sigma = \sqrt{np(1 - p)}$.
5. Probabilities for the binomial random variable are provided in Table A.1.

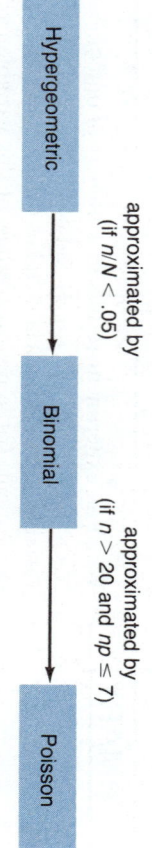

FIGURE 5.7

Summary of how the three most common types of discrete random variables can be used to approximate values for one another.

Hypergeometric $\longrightarrow$ approximated by (if $n/N < .05$) $\longrightarrow$ Binomial $\longrightarrow$ approximated by (if $n > 20$ and $np \leq 7$) $\longrightarrow$ Poisson

Hypergeometric Distribution

1. X represents the number of successes in a sample of size n when selecting from a population of size N containing k successes and $N - k$ failures.
2. PMF is $P(x) = ({}_kC_x \cdot {}_{N-k}C_{n-x})/{}_NC_n)$ for $x = a,\ a+1, \ldots, b$, where $a =$ maximum $\{0, n + k - N\}$ and $b =$ minimum $\{k, n\}$.
3. Mean $= \mu = n(k/N)$.
4. Variance $= \sigma^2 = [n(k/N)(1 - k/N)][(N - n)/(N - 1)]$.

Poisson Distribution

1. $X =$ the number of occurrences of a particular event over a certain unit of time, length, area, or volume.
2. PMF is $P(x) = (\mu^x e^{-\mu})/(x!)$ for $x = 0, 1, 2, \ldots$.
3. Mean $= \mu$.
4. Variance $= \mu$.
5. Probabilities for the Poisson random variable are provided in Table A.3.

REVIEW EXERCISES

5.57 Is the following function a probability mass function? Why or why not?

$$P(X = x) = x^3/153 \quad \text{for } x = 1, 3, 5 \quad \text{(and zero elsewhere)}$$

5.58 Assume that a fair die has one blue face, two white faces, and three black faces. Define the random variable X as follows:

$$X = \begin{cases} 1 & \text{if blue} \\ 2 & \text{if white} \\ 3 & \text{if black} \end{cases}$$

Find the probability mass function of X. Construct a histogram in which the height of each bar is the probability of X.

5.59 Find the mean and variance of the following random variable X with probability mass function $P(x)$. Can a value of X be negative as shown in the table?

X	P(x)
−3	.2
0	.1
3	.2
5	.3
10	.2

5.60 A bakery knows that historically the number of cakes sold daily has the following probability distribution:

X: NUMBER OF CAKES DAILY	P(x)
0	.40
1	.30
2	.15
3	.10
4	.05

a. Find the probability that at least two cakes are sold daily.

b. Find the mean and standard deviation of the number of cakes sold daily.

5.61 For a binomially distributed random variable X with 12 trials and with the probability of a success equal to .3, find the following.

a. $P(X = 7)$.   **b.** $P(4 < X \le 6)$.

c. $P(X > 5)$.   **d.** $P(X < 2)$.

5.62 Let the variable X be equal to minus one if stock XYZ declines, zero if stock

XYZ remains unchanged, and one if stock XYZ increases in price. If $P(X = x)$ is equal to $(x + 2)/6$, what are the mean and standard deviation of X?

5.63 A manager has ten research projects to assign to either engineer 1 or 2. If each research project is randomly assigned to either one of the two engineers, what is the probability that engineer 1 will be assigned no more than five research projects?

5.64 An average of five books per week are returned to a bookstore. Assume that the number of returned books is Poisson distributed.

a.   What is the probability that less then four books will be returned in one week?

b.   What is the standard deviation of the distribution of the number of books returned in one week?

5.65 The supervisor of the employees who solder resistors on certain electrical components would like to know what the average number of absentees is daily and also what the standard deviation is of the daily employee absentee rate. Find these two values from the following probability mass function, which was constructed from historical data of the company:

X: NUMBER OF DAILY ABSENTEES	P(x)
0	.50
1	.23
2	.12
3	.10
4	.02
5	.02
6	.01

5.66 Ten employees are being reviewed for promotion. Four of the employees are females. If each employee is equally likely to get promoted, what is the probability that two females and three males will be promoted if a total of five promotions are given?

5.67 There are 90 drill bits in a box at a machine shop. Fifty of the drill bits are 3/8-inch diameter, and 40 are 7/16-inch diameter. If four drill bits are selected at random, what is the probability that two drill bits of 3/8-inch diameter and two drill bits of 7/16-inch diameter will be chosen?

5.68 A population consists of 15 employees, 6 of whom have less than 2 years experience. Let X be equal to the number of employees with less than 2 years experience from a sample of eight employees drawn from this population.

a.   Find $P(X = 3)$.

b.   Find $P(X \leq 2)$.

c.   Find the average value of X.

d.   Find the standard deviation of X.

5.69 Blair's Moving Company loads an average of three boxes of damaged merchandise daily. What is the probability that exactly three boxes of damaged merchandise are shipped daily? What is the standard deviation of the number of boxes of damaged merchandise that are shipped daily? Assume a Poisson distribution.

5.70 A person has written seven songs, three of which are ballads. If the songwriter chooses two at random, what is the probability of the following?

a.   Exactly one is a ballad.

b.   None are ballads.

c.   Both are ballads.

5.71 The owner of Fashion Designs knows that only six customers can be handled effectively in a 15-minute period. If the average number of customers in a 15-minute interval is five, what is the probability that more than six customers will arrive in a 15-minute interval? What distribution is assumed of the number of customers in a 15-minute period?

5.72 A survey was conducted on consumer responses to implied superiority claims made in slogans used to advertise products. The results of the survey are given in the February/March 1987 issue of the *Journal of Advertising Research*. When asked whether the slogan "No leading brand gets rid of dandruff better than *Selsun Blue*," meant that the product was equal to or better than the competitive brands, approximately 55% indicated that the claim was for superiority.

a. If 5 consumers were selected from the 187 consumers responding in the survey, then could the number in the sample of size 5 who believed the slogan meant superiority be considered a binomial random variable? Why?

b. Would it be unreasonable to expect all 5 consumers in this sample to believe that the slogan meant superiority? Why?

5.73 The number of independent television stations has mushroomed over the past decade. Three major networks (NBC, CBS, and ABC) attracted more than 90% of the prime-time viewers before the arrival of cable television. In 1987, these three major networks' share of the prime-time audience had shrunk to 76% of the prime-time viewers.

(*Source:* Tom Graver, "Technology Changing Media Markets," *Leisure-Time*, July 30, 1987, Section 3, pp. 1–6.)

a. If a random sample of 20 prime-time viewers were selected randomly across the country before the arrival of cable television, what is the probability that at least 18 viewers were watching one of the three major networks?

b. Answer part (a), but assume that the sample is taken in 1987, when the three major networks' share of prime-time viewers has shrunk to 76% of the prime-time viewers.

5.74 The state of Delaware is the corporate home of 43% of the companies listed on the New York Stock Exchange. If a stock broker chooses three companies at random from the companies listed on the New York Stock Exchange, what is the probability that at least one company does not have its corporate home in Delaware?

(*Source:* "Oh, Come On, Delaware," *Business Month* (March 1988): 12.)

5.75 According to an article in the *Business Monthly*, there is no strategic advantage in knowing what happened yesterday. If a banker wants to read tomorrow's *Wall Street Journal* today, the banker needs DJ/NR (Dow Jones/News Retrieval), an electronic service. In late 1987, DJ/NR was serving the needs of 77 of the nation's 100 largest banks. Determine the probability of observing exactly 3 banks with DJ/NR out of a sample (without replacement) of 5 of the nation's 100 largest banks.

(*Source:* "Friendly Software For The Bank CEO," by Michael Violano, *Bankers Monthly* (May 1988): 44–48.)

5.76 According to an article referring to consumers over 60 years of age, by the year 2000, the most important demographic market will not be yuppies but *grampies*, Growing Retired Active Married People In an Excellent State. By 1990, one in 11 consumers will be over 60. By the year 2000, one-tenth of the world's market will be over 60 years of age. Suppose that these percentages will hold for a large department store chain in the United States in future years. If 36 consumers were randomly surveyed in 1990 by the department store chain, what is the approximate probability that no more than 2 consumers are over 60 years of age? What would the probability be if the survey was taken in the year 2000?

(*Source:* Sandra van der Merve, "Grampies: A New Breed of Consumers Comes of Age," *Business Horizons* (November-December 1987).)

5.77 The following MINITAB computer printout displays the results of randomly generating 500 observations from a population with a binomial distribution with $n = 17$ and $p = .50$.

a. Find the relative frequency at each possible value of a binomial random variable with $n = 17$ and $p = .50$. How do these compare to the actual probabilities given in the binomial table in the appendix?

b. Calculate the mean of the generated observations by using the formula $\Sigma f_i x_i / 500$ where f_i is the frequency. Compare this value to the population mean.

c. Calculate the variance of the generated observations by using the formula $[\Sigma f_i x_i^2 - (\Sigma f_i x_i)^2 / 500)] / (500 - 1)$. Compare this value to the population variance.

```
MTB > random 500 c1;
SUBC> binomial n=17 and p=.50.
MTB > tally c1

        C1    COUNT
         3        6
         4       11
         5       25
         6       43
         7       72
         8       93
         9       87
        10       69
        11       56
        12       26
        13        8
        14        4
                500
       N=      500
```

5.78 The following MINITAB computer printout displays the results of randomly generating 500 observations from a population with a Poisson distribution with a mean of 3.

a. Find the relative frequencies at each possible value in the population. Compare these relative frequencies to the probabilities in the Poisson table in the appendix.

b. Answer part (b) given in Exercise 5.77 with respect to the randomly generated data given.

c. Answer part (c) given in Exercise 5.77 with respect to the data given.

```
MTB > random 500 c1;
SUBC> poisson mean = 3.
MTB > tally c1

        C1    COUNT
         0       25
         1       83
         2      100
         3      106
         4       91
         5       57
         6       17
         7       11
         8        8
         9        2
                500
       N=      500
```

COMPUTER EXERCISES USING THE DATABASE

Exercise 1—Appendix H

Generate 100 random numbers and select a sample of 100 observations from the database. Consider the variable FAMLSIZE (family size). Let p represent the proportion of observations in your set of 100 observations in which the family size is no greater than 2. If you randomly select 10 observations (with replacement) from the 100 possible, what is $P[X \leq 5]$, where

X is the number of observations (out of 10) in which the family size is no greater than 2? What type of random variable is X?

Exercise 2—Appendix H

Repeat Exercise 1, where 10 observations are selected without replacement.

Exercise 3—Appendix H

Referring to Exercise 2, if you were to obtain samples of size 10 indefinitely, what would X be on the average? What is the standard deviation of this random variable?

Exercise 4—Appendix H

Estimate the proportion of homeowners in a randomly selected set of 100 observations. From this set of 100 observations, select with replacement a random set of ten observations. Can this be considered a binomial experiment with $n = 10$ and p equal to the proportion of homeowners in the set of 100 observations? Estimate the probability that the number of homeowners in the set of 10 observations is greater than or equal to 5.

Exercise 5—Appendix H

Randomly select 200 observations from the database. From this set, estimate the proportion of observations in which the location of residence is in one of the northern sectors. From this set of observations, randomly select without replacement eight observations. Find the prob-

ability that, in this sample of size eight, the residences of four or less observations are in the northern sectors. Can the binomial approximation be used? Why?

Exercise 6—Appendix I

Select 200 observations at random from the database. Let p be the proportion of companies with a positive net income. If you were to randomly select (with replacement) 15 observations from the 200 possible, what is the probability of selecting at least 9 companies with a positive net income?

Exercise 7—Appendix I

For the 20 observations, in Exercise 6, let p be equal to the proportion of companies in which the number of employees exceeds 10,000. If you were randomly to select 15 observations (with replacement) from the 200 possible, what is the probability of selecting at least 7 companies in which the number of employees exceeds 10,000?

TOWARD ZERO DEFECTS WITH STATISTICAL QUALITY CONTROL AT HEWLETT-PACKARD

Japan's challenge to American business is partly built on a reputation for quality and reliability. It is somewhat ironic that this reputation comes from applying statistical quality control techniques pioneered by American authorities such as W. Edward Deming and Joseph M. Juran. In 1949, Deming visited Japan and introduced his ideas. The Japanese applied them with fervor and improved on them. It is a testament to Deming's influence that in Japan, the Deming prize for the best quality improvement effort in industry is regarded as equivalent to getting a Nobel prize for quality assurance (QA).

However, U.S. industry is catching up, and today's buzzwords in industry are SQC (statistical quality control), SPC (statistical process control), TQC (total quality control) and similar variations. One also hears about the Taguchi

method, Ishikawa's cause-and-effect diagrams, and the 80/20 rule of Pareto analysis. One of the ideas in modern QA is that managers need to move away from the inspection philosophy. A saying is often quoted: "You can't inspect quality into a product, you have to make it right the first time." The focus should be on quality *improvement*. To do this, it is first necessary to understand the process. Statistical methods can provide certain important insights about a process. For now, let us look at a successful use of SQC.

Hewlett-Packard (HP) is a worldwide electronics and computer manufacturer based in California. Sepehri and Walleigh (1986) have described the use of SQC at HP. Prior to 1982, HP's Computer Systems Division suffered from a large number of backorders and inflated inventory. These matters also influenced quality problems. Various techniques were employed, including SQC and later TQC, to improve the situation. The company emphasized to its employees that it believed that *defects were caused by problems inherent in the process itself* and were not the fault of the operator. This ensured the proper attitude and cooperation from employees.

In the following questions, we focus on the probabilistic behavior of defect rates and how these

Case Study Questions

1. One item being made at HP was IC boards using the wave solder process, which solders a whole bunch of integrated circuits at once. The solder joint failure rate was 5000 ppm, i.e., .5%, in July 1982. About a year later, it was down to 10 ppm, i.e., .001%. The failure rate had been as high as 18,000 ppm (1.8%). Defects versus no defects is clearly a binomial situation. However, with defect rates measured in parts per million, the computation of binomial probabilities gets to be messy. For example, a random sample of 20 circuit boards, with just 200 solder joints per board, represents 4000 solder joints. Suppose you wished to compute the probability that more than 2 defects will be seen when the known defect rate is 1.8%. Explain the difficulty involved in calculating this.

2. Another way of measuring solder-related defects was to note that these declined from an average of 5 defects per 100 boards to 1 defect per 100 boards.

   a. Could the Poisson distribution be a possible model for defects per 100 boards?

   b. At an average of 5 defects

per 100 boards, what is the probability that in any randomly selected lot of 100 boards, the number of defects is more than 2?

   c. What happens to the probability computed in 2(b) when the defect rate drops to 1 defect per 100 boards? Does it drop to one-fifth of what it was?

3. Another area of improvement was kitting errors, where material discrepancies in an assembly kit lead to returning excess parts to stock and exchanging parts to stock and exchanging wrong parts. The material discrepancy rate was initially 1.3 errors per kit, on the average. This was lowered to .2 errors per kit after SQC was implemented. Assuming a Poisson distribution of errors per kit, compute the probability of observing between 1 to 3 errors per kit in any selected kit when the error rate was 1.3 errors per kit. Repeat this computation based on .2 errors per kit.

4. Make some generalizations about how probabilities changed as the defect rates went down. Do these suggest progress toward the goal of zero defects?

change as improvements in QC are gained. At HP, failure information was entered into a LOTUS program on a PC, and the defect rate in parts per million (ppm) was plotted over time for various processes. The goal at HP was to achieve zero defects. This elusive target was, in fact, briefly achieved in December 1984 for the first time. However, it is not easy to consistently maintain a zero defect process for an extended length of time.

(*Source:* Mehran Sepehri and Richard C. Walleigh, "Quality and Inventory Control Go Hand In Hand At Hewlett-Packard's Computer Systems Division," *Industrial Engineering* (February 1986): 44–51. Copyright February 1986 Institute of Industrial Engineers, 25 Technology Park/Atlanta, Norcross, GA 30092.)

Continuous Probability Distributions

A Look Back/Introduction

After we described the use of descriptive statistics, we introduced you to the area of uncertainty by using probability concepts and random variables. Random variables offer you a convenient method of describing the various outcomes of an experiment and their corresponding probabilities.

When each value of the random variable as well as its probability of occurring can be listed, the random variable is discrete. The other type, a continuous random variable, can assume any value over a particular range. This includes such variables as X = height, X = weight, and X = time. For this kind of situation, it is impossible to list all values of X, yet you can still make probability statements regarding X if you can make certain assumptions about the type of population.

Making decisions from sample information in statistics is called **statistical inference**. In subsequent chapters, we will develop a formal set of rules to offer you a guide in making statistical decisions. The making of such a decision typically involves one or more assumptions about the population from which the sample was obtained. One such assumption, widely used in statistics, is that the data came from a normal population, which means that you are dealing with a normal random variable.

The concept of a continuous random variable was introduced in Chapter 5. What distinguishes a discrete random variable from one that is continuous is the presence of *gaps* in the possible values for a discrete random variable. To illustrate, X = total of two dice is a discrete random variable; there are many gaps over the range of possible values, and a value of 10.4, for example, is not possible. One can list the possible values of a discrete variable, along with the probability that each value will occur.

Determining probabilities for a continuous random variable is quite different. For such a variable, any value is possible over a specific range. This means that we are unable to list all the possible values of this variable. Probability statements for a continuous random variable (such as X) are not concerned

FIGURE 6.1

Finding a probability for a continuous random variable.

with specific values of X (such as the probability that X will equal 50) but rather deal with probabilities over a range of values, such as the probability that X is *between* 40 and 50, *greater than* 65, or *less than* 20, for example.

Such probabilities can be determined by first making an assumption regarding the nature of the population involved. We assume that the population can be described by a curve having a particular shape—such as normal, uniform, or exponential. Once this curve is specified, a probability can be determined by finding the corresponding area under this curve. As an illustration, Figure 6.1 demonstrates a particular curve (called the normal curve) for which the probability of observing a value of X between 20 and 60 is the area under this curve between these two values. The entire range of probability is covered using such a curve, since, for any continuous random variable, the total area under the curve is equal to 1.

The following sections examine the normal, uniform, and exponential distributions, since these are the most widely encountered random variables in practice. The graphs and descriptive statistics described in the previous chapters can help determine if one of these random variables might be appropriate for a particular situation. If a histogram of the sample data appears nearly flat, the population might be represented by a uniform random variable. If the histogram is symmetric with decreasing tails in each end, a normal random variable may be in order. If the sample histogram steadily decreases from left to right, the population of all possible values perhaps can be described using an exponential random variable. In the first two cases, the mean and median should be nearly equal (the population is symmetric) providing a skewness measure near zero. For the exponential case, the median should be less than the mean with a corresponding positive measure of skewness.

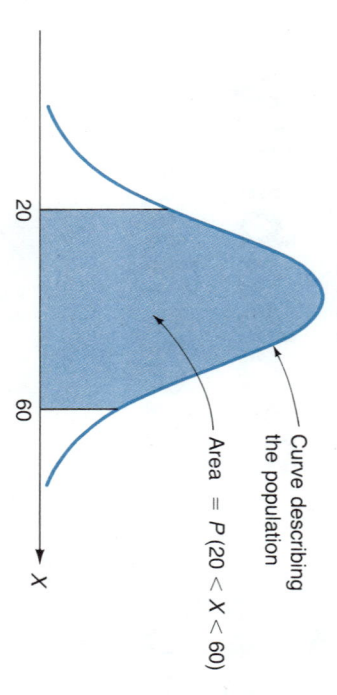

Area = $P(20 < X < 60)$

Curve describing the population

The normal distribution is the most important of all the continuous distributions. You will find that this distribution plays a key role in the application of many statistical techniques. When attempting to make an assertion about a population by using sample information, a major assumption often is that the population has a normal distribution.

When discussing measurements such as height, weight, thickness, or time, the resulting population of all measurements often can be assumed to have a probability distribution that is normal.

A histogram constructed from a large *sample* of such measurements can help determine whether this assumption is realistic. Assume, for example, that data were collected on the length of life of 200 Everglo light bulbs. Let X represent the length of life (in hours) of an Everglo bulb. One thing we are

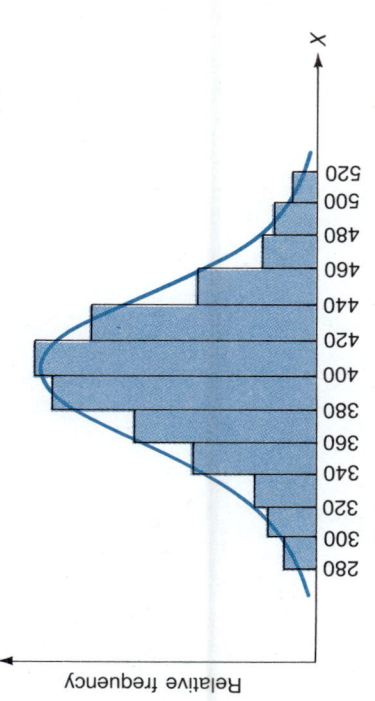

FIGURE 6.2

Histogram of 200
Everglo light bulb
lifetimes (in hours). The
curve represents all
possible values
(population). The
histogram represents
the sample (200 values).

interested in is the *shape* of the distribution of the 200 lifetimes. Where are they
centered? Are they symmetric? The easiest way to approach such questions is
to construct a histogram of the 200 values, as illustrated in Figure 6.2. This
histogram indicates that the data are nearly symmetric and are centered at
approximately 400.

The curve in Figure 6.2 is said to be a **normal curve** because of its shape.
A normal curve is characterized by a **symmetric, bell-shaped appearance**, with
tails that "die out" rather quickly. We use such curves to represent the **assumed
population** of all possible values. This example contained 200 values observed
in a *sample*. Consequently:

1. A histogram represents the shape of the sample data.
2. A smooth curve represents the assumed shape (distribution) of the popu-
    lation.

If all possible values of a variable X follow an assumed normal curve,
then X is said to be a **normal random variable**, and the population is **normally
distributed**.

When you assume that a particular population follows a normal distribution, you assume that X, an observation randomly obtained from this population, is a normal random variable. Based on the histogram in Figure 6.2, it
appears to be a reasonable assumption that the smooth curve describing the
population of *all* Everglo bulbs can be approximated using a normal curve
centered at 400 hours. Therefore, we will assume that X is a normal random
variable, centered at 400 hours.

There are two numbers used to describe a normal curve (distribution);
where the curve is centered and how wide it is. The **center** of a normal curve
is called the *mean* and is represented by the symbol μ (mu). The **width** of a
normal curve can be described using the *standard deviation*, represented by the
symbol σ (sigma).

These are illustrated in Figure 6.3, which shows the normal curve representing the lifetime of Everglo light bulbs. Another way of stating this situation
is: X is a normal random variable with $\mu = 400$ hours and $\sigma = 50$ hours. Notice
that the units of μ and σ are the same as the units of the data (hours).

In Figure 6.3, there is a point P on the normal curve. Above, this point P,
the curve resembles a bowl that is upside down, and below P the curve is "right
side up." In calculus, this point is referred to as an **inflection point**. The distance
between vertical lines through μ and P is the value of σ.

Because μ and σ represent the location and spread of the normal distribution, they are called **parameters**. The parameters are used to define the distribution completely. The values of μ and σ of a normal population are all you

FIGURE 6.3

Distribution of the lifetime of Everglo bulbs showing the mean ($\mu = 400$), the standard deviation ($\sigma = 50$), and the inflection point (P).

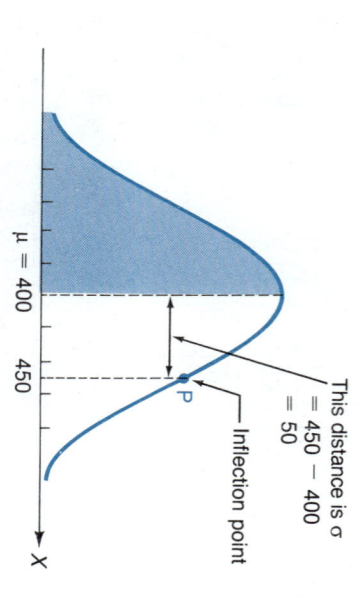

This distance is σ
$= 450 - 400$
$= 50$

Inflection point

$\mu = 400$

450

X

need to separate it from all other normal populations that have the same bell shape but different location and/or variability. The values of the parameters must be specified in order to make probability statements regarding X. As a result, there are infinitely many normal curves (populations), one for each pair of values of μ and σ.

In Chapter 5, we discussed the mean of (say) ten observations of the random variable X, written as $\bar{X}$. If you were to observe X indefinitely, then you could obtain the mean of the population, μ. The same concept applies to continuous random variables, where, for the Everglo example, $\bar{X}$ represents the mean of the 200 bulbs (the sample) and μ is the mean of all Everglo bulbs (the population).

MEAN	
SAMPLE	**POPULATION**
$\bar{X}$	μ
the average	the average
of the	of the
sample	population

STANDARD DEVIATION	
SAMPLE	**POPULATION**
s	σ
the standard	the standard
deviation of	deviation of
the sample	the population

In our Everglo example, the average lifetime of all bulbs is *assumed* to be $\mu = 400$ hours. The standard deviation of the population, σ, just like s, is a measure of **variability**. The larger σ is, the more variation (jumping around) we would see if X were observed indefinitely. For both the sample and the population, the square of the standard deviation is referred to as the variance. It is another measure of the variability of X. The **variance** of a random variable, X, is represented by σ^2.

Consider whether the sample average ($\bar{X}$) of the 200 values in our example is the *same* as μ. It is not. Do not confuse the average lifetime of all light bulbs (μ) with the average lifetime of just 200 bulbs ($\bar{X}$). This is an important distinction in statistics. However, if our assumed normal distribution (with $\mu = 400$ and $\sigma = 50$) is correct, then $\bar{X}$ most often will be "close to" μ. We examine this again in Chapter 7.

The curve in Figure 6.3 is an illustration of a normal random variable with a mean of 400 hours and a standard deviation of 50 hours. We can compare normal curves that may differ in mean, standard deviation, or both. The normal curves in Figure 6.4 indicate that, on the average, males are taller than females. The mean of the male curve is to the right of the mean of the female curve. The male heights "jump around" about as much as female heights. In other words, there is about the same amount of *variation* in male and female heights. This is because the standard deviation of each curve is the same; that is, each curve is equally wide.

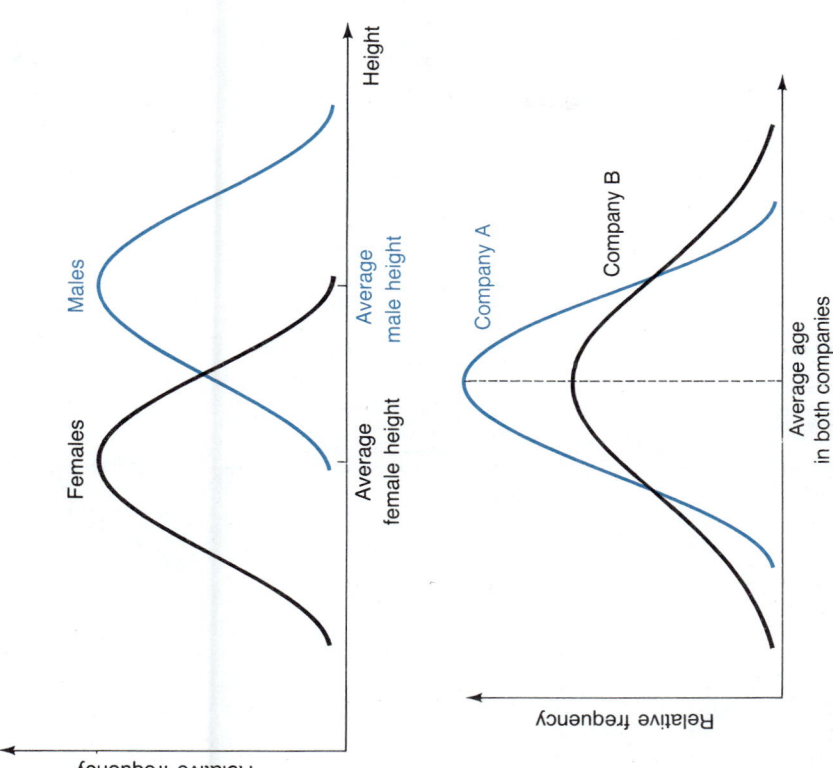

FIGURE 6.4

Two normal curves with unequal means and equal standard deviations.

FIGURE 6.5

Two normal curves with equal means and unequal standard deviations.

In Figure 6.5, the two normal curves represent the ages of the employees at two large companies. It appears that:

1. The average age of employees for the two companies is the same.
2. The ages in Company B have more variability. This simply means that there are more old people and more young people in Company B than in Company A.

6.2

DETERMINING A PROBABILITY FOR A NORMAL RANDOM VARIABLE

So you have assumed that the lifetime of an Everglo light bulb is a normal random variable with $\mu = 400$ and $\sigma = 50$. Now what? This brings us back to the subject of probability. Before we describe probabilities for a normal random variable, consider one important property of *any* normal curve (or of any curve representing a continuous random variable, for that matter), namely, that the total area under the curve is 1 (see Figure 6.6). When we described the normal curve as bell-shaped, we also determined that it was symmetrical. If the halves are identical, then the probability above the mean (μ) is equal to .5 and is the same as the probability below the mean. Thus, in Figure 6.3 the shaded area is equal to the nonshaded area under the curve.

Returning to the Everglo bulb example, what percentage of the time will the burnout time, X, be less than 360? This probability is written as

$$P(X < 360)$$

FIGURE 6.6

Area under a normal curve.

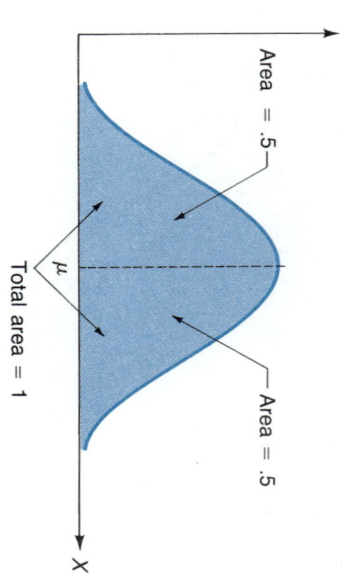

Total area = 1

Area = .5

Area = .5

FIGURE 6.7

Normal curve for Everglo light bulbs showing $P(X < 360)$. The shaded area is the percentage of time that X will be less than 360. (X = lifetime of Everglo bulb.)

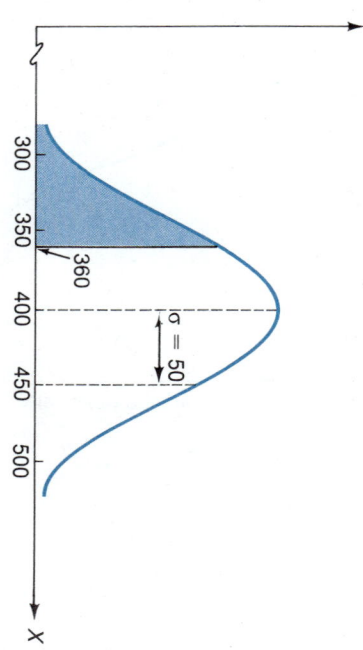

We discuss how to determine this area (a simple procedure) later in the chapter, but for now, just remember that when dealing with a normal random variable, a **probability** is represented by an **area** under the corresponding normal curve. The value of $P(X < 360)$ is illustrated in Figure 6.7. It appears that roughly 20% of the total area has been shaded, so we conclude that (1) roughly 20% of the Everglo bulbs will burn out in less than 360 hours, and (2) the probability that X is less than 360 is approximately .2.

6.3   FINDING AREAS UNDER A NORMAL CURVE

Areas under the Standard Normal Curve

We begin our discussion by finding the area under a special normal curve—namely, one that is centered at zero ($\mu = 0$) and has a standard deviation of one ($\sigma = 1$). This random variable is typically represented by the letter Z and is referred to as the **standard normal random variable**. As Figure 6.8 demonstrates, Z is as likely to be negative as positive; that is, $P(Z \leq 0) = P(Z \geq 0) = .5$. Although you probably never will observe a random variable like Z in practice,

FIGURE 6.8

Standard normal curve.

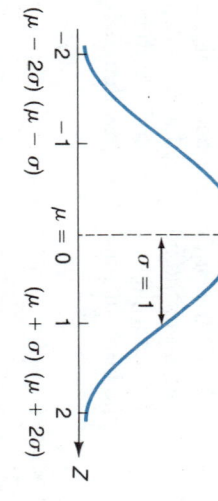

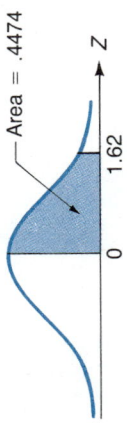

FIGURE 6.9
Shaded area = .4474, from Table A.4.

it is a useful normal random variable. In fact, an area under *any* normal curve (as in Figure 6.7) can be determined by finding the corresponding area under the standard normal curve.

To derive the area under the standard normal curve requires the use of integral calculus. Unfortunately, the integral of the function describing the standard normal curve does not have a simple (closed form) expression. By using excellent approximations of this integral, however, we can tabulate these areas—see Table A.4 and Figure 6.9.

For example, say that we want to determine the probability that a standard normal random variable will be between 0 and 1.62. This is written as

$$P(0 < Z < 1.62)$$

The value of this probability is obtained from Table A.4 by noting that it contains the area under the curve between the mean of zero and the particular value of Z. The far left column of Table A.4 identifies the first decimal place for Z, and you read across the table to obtain the second decimal place.

In our example, we find the intersection between 1.6 on the left and .02 on the top, because $Z = 1.62$. Look at Table A.4; the value .4474 is the *area* between 0 and 1.62. In other words, the probability that Z will lie between 0 and 1.62 is .4474.

You can begin to see why it is a good idea to sketch the curve and shade in the area when dealing with normal random variables. It gives you a clear picture of what the question is asking and cuts down on mistakes.

EXAMPLE 6.1

SOLUTION

What is the probability that Z will be greater than 1.62?

We wish to find $P(Z > 1.62)$. Examine Figure 6.10. The area under the right half of the Z curve is .5, so, using our value from Table A.4, the desired area here is

$$.5 - .4474 = .0526$$

So the probability that Z will exceed 1.62 is .05.

What if we wish to know the probability that Z is equal to a particular value, such as $P(Z = 1.62)$? There is no area under the curve corresponding to $Z = 1.62$, so

$$P(Z = 1.62) = 0$$

In fact,

$$P(Z = \text{any value}) = 0$$

One nice thing about this fact is that $P(Z \geq 1.62)$ is the *same* as $P(Z > 1.62)$ (that is, .0526). So putting the equal sign on the inequality ($\geq$ or $\leq$) has *no* effect on the resulting probability.

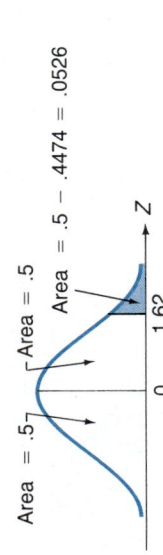

FIGURE 6.10
The shaded area represents the probability that Z will be greater than 1.62 $[P(Z > 1.62)]$.

FIGURE 6.11

Area under the Z curve for $P(Z < 1.62)$.

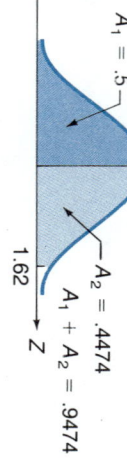

FIGURE 6.12

Area under the Z curve for $P(1.0 < Z < 2.0)$.

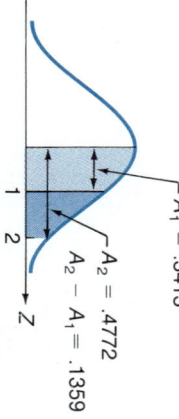

FIGURE 6.13

Area under the Z curve for $P(-1.25 < Z < 1.15)$.

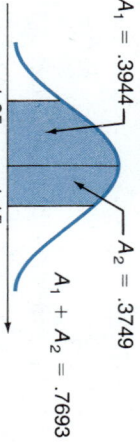

FIGURE 6.14

Z curve for $P(0 < Z < -1.25) = P(-1.25 < Z < 0)$.

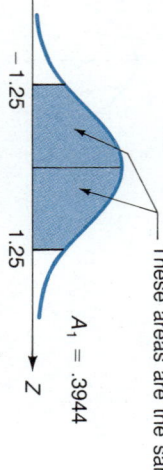

By looking at the Z curve in Figure 6.11, you can see that

$$P(Z < 1.62) = .5 + .4474 = .9474$$

As before, this also is $P(Z \leq 1.62)$.

Figure 6.12 shows $P(1.0 < Z < 2.0)$ (areas from Table A.4). We see that

$$P(1.0 < Z < 2.0) = P(0 < Z < 2.0) - P(0 < Z < 1.0)$$
$$= .4772 - .3413$$
$$= .1359$$

By subtracting the two areas, we find that the probability that Z will lie between 1.0 and 2.0 is .1359.

We use Figure 6.13 and Table A.4 to determine $P(-1.25 < Z < 1.15)$:

$$P(-1.25 < Z < 1.15) = P(-1.25 < Z < 0) + P(0 < Z < 1.15)$$
$$= A_1 + A_2.$$

Using the symmetry of the Z curve and Figure 6.14, the area of A_1 is the same as $P(0 < Z < 1.25)$ and thus is .3944. The area of A_2, from Table A.4, is .3749. So we add A_1 and A_2:

$$.3944 + .3749 = .7693$$

Finally, we can determine $P(Z < -1.45)$ using Figure 6.15. This can be written as $P(Z < 0) - P(-1.45 < Z < 0)$. Using the discussion from Figure 6.14, the area between zero and -1.45 is .4265 (from Table A.4). As a result, Z will be less than (or equal to) -1.45 approximately 7.35% of the time.

FIGURE 6.15

Area under the Z curve
for $P(Z < -1.45)$.

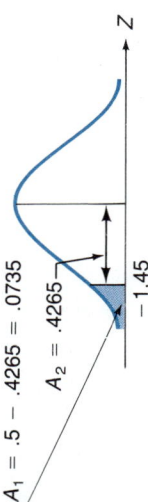

$A_1 = .5 - .4265 = .0735$

$A_2 = .4265$

EXERCISES

6.1 Explain how the parameters μ and σ determine the graph of a normal distribution.

6.2 Find the area under the standard normal curve for the following Z values. Sketch the corresponding area.

a. $Z \leq 0$ **b.** $Z \leq 1.0$

c. $Z \geq 1.0$ **d.** $Z \leq -1.0$

6.3 Find the area under the standard normal curve bounded by the following Z values. Sketch the corresponding area.

a. $Z = 0$ to 1.0   **b.** $Z = 1.0$ to 1.5

c. $Z = -1.0$ to 1.0   **d.** $Z = -2.5$ to -1.5

6.4 Find the following probabilities. Sketch the corresponding area.

a. $P(Z \leq 1.75)$ **b.** $P(Z \geq 1.96)$

c. $P(-1.0 \leq Z \leq 2.5)$ **d.** $P(-.5 \leq Z \leq .5)$

6.5 Find the probability that an observation taken from a standard normal population will be

a. Between -3 and 1.6   **b.** Less than -2.1

c. Between .76 and 1.96   **d.** Between -1.65 and 1.65

6.6 Find the value of z for the following probability statements and sketch the corresponding area.

a. $P(Z \leq z) = .95$ **b.** $P(Z \leq z) = .10$

c. $P(Z \geq z) = .025$ **d.** $P(Z \geq z) = .55$

6.7 Find the value of z for the following probability statements and sketch the corresponding area.

a. $P(-1.8 \leq Z \leq z) = .6$ **b.** $P(0 \leq Z \leq z) = .25$

c. $P(1.0 \leq Z \leq z) = .1$ **d.** $P(-2.8 \leq Z \leq z) = .05$

6.8 Find the two Z values such that

a. The area bounded by them is equal to the middle 40% of the standard normal distribution.

b. The area bounded by them is equal to the middle 80% of the standard normal distribution.

6.9 Find the Z values such that the area under the standard normal curve between the Z value and $Z = 1.0$ is equal to .10. Find both Z values that make this possible.

6.10 The output from a monitor that measures the amperage of an electronic circuit follows a normal distribution with mean 0 and variance 1. What proportion of the data would be outside the interval from -2 to 2?

Areas under Any Normal Curve

Take another look at the histogram of the 200 Everglo light bulb lifetimes in Figure 6.2. A normal curve with $\mu = 400$ hours and $\sigma = 50$ hours was used to describe the population of *all* Everglo lifetimes. So, X = Everglo lifetime is a normal random variable with $\mu = 400$ and $\sigma = 50$.

FIGURE 6.16

Histogram obtained by subtracting $\mu = 400$ (compare with Figure 6.2).

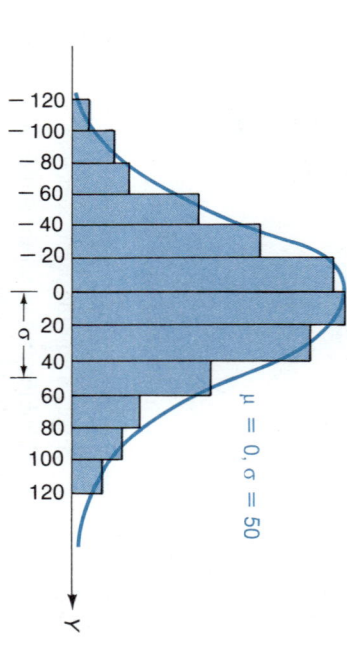

$\mu = 0, \sigma = 50$

What happens to the shape of the data if we take each of the 200 lifetimes in this example and subtract 400 (that is, subtract μ)? As you can see in Figure 6.16, the histogram (and corresponding normal curve) is merely "shifted" to the left by 400. It resembles the normal curve for X, except the "new" mean is zero. The random variable defined by $Y = X - 400$:

1. Is a normal random variable;
2. Has a mean equal to zero;
3. Has a standard deviation equal to that of X; that is, 50.

Figure 6.17 shows what happens to the shape of 200 Y values if each of them is *divided* by 50 (that is, by σ). Notice the horizontal axis in the histogram and the corresponding normal curve. The resulting normal curve resembles a normal curve with a mean of zero and a standard deviation equal to 1.

Thus, if X is a normal random variable with mean 400 and standard deviation 50, then the random variable defined by

$$Z = \frac{X - 400}{50},$$

1. Is a normal random variable;
2. Has a mean equal to zero;
3. Has a standard deviation equal to 1.

This means that, in general, for *any normal* random variable X,

$$Z = \frac{X - \mu}{\sigma}$$

FIGURE 6.17

Histogram obtained by subtracting μ and dividing by σ (compare with Figures 6.2 and 6.16).

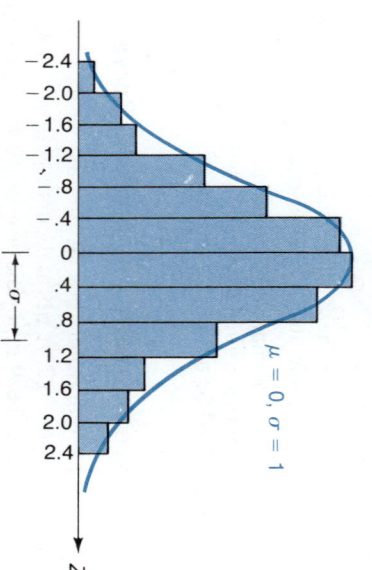

$\mu = 0, \sigma = 1$

is a **standard normal random variable**. This procedure of subtracting μ and dividing by σ is referred to as **standardizing** the normal random variable X. It allows us to determine probabilities for any normal random variable by first standardizing it and then using Table A.4. So the standard normal distribution turns out to be much more important than you might have expected!

EXAMPLE 6.2

The normal curve in Figure 6.7 represented the lifetime of all Everglo bulbs, with $\mu = 400$ hours and $\sigma = 50$ hours. What percentage of the bulbs will burn out in less than (or equal to) 360 hours? Or, put another way, what is the probability that any particular bulb will last less than 360 hours?

SOLUTION

This is a probability and is written as

$$P(X < 360)$$

This random variable is continuous, so $P(X < 360) = P(X \leq 360)$. To determine the probability, you need to standardize this variable:

$$P(X < 360) = P\left(\frac{X - 400}{50} < \frac{360 - 400}{50}\right)$$

$$= P(Z < -.8)$$

where $Z = (X - 400)/50$ (Figure 6.18).

Earlier, by examining Figure 6.7, we estimated this area to be roughly 20%. The actual area, from Figure 6.18, is .2119; that is, it is 21.19% of the total area. The conclusion here is that

$$P(X < 360) = .2119$$

and so 21% of all Everglo bulbs will have a lifetime of less than 360 hours. ■

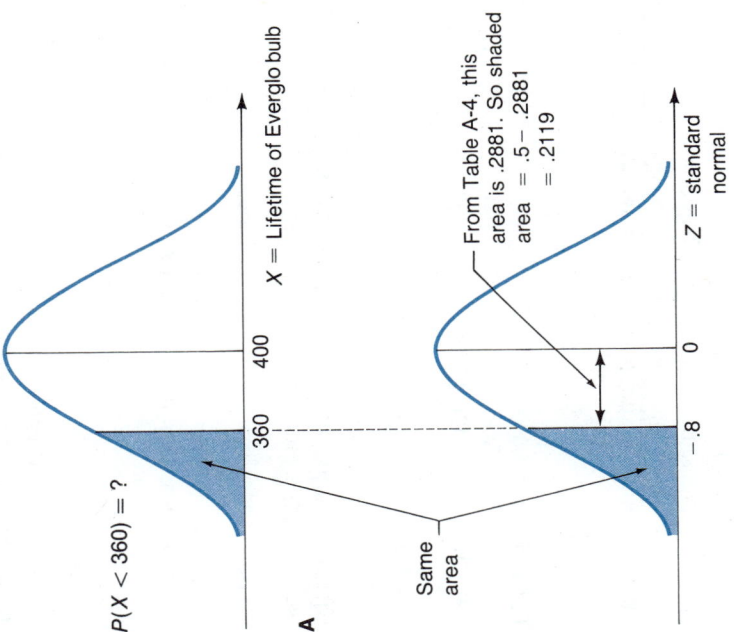

FIGURE 6.18

Compare the areas for the X (**A**) and Z (**B**) normal curves to find $P(X < 360)$.

Interpreting Z

What does a Z value of $-.8$ imply in Example 6.2? It simply means that 360 is .8 standard deviations to the left (Z is negative) of the mean. So,

$$\mu - .8(\sigma) = 400 - .8(50) = 360$$

Recall that a Z score was defined in exactly the same way in Chapter 3 using a sample mean ($\bar{X}$) and standard deviation (s). In this chapter, we use the population mean (μ) and standard deviation (σ). In general:

1. A *positive* value of Z designates how many standard deviations (σ) X is to the *right* of the mean (μ).

2. A *negative* value of Z designates how many standard deviations X is to the *left* of the mean.

EXAMPLE 6.3

A large midwestern bank has reason to believe that average monthly account balances of their bank card accounts follow a normal distribution with a mean of \$650 and a standard deviation of \$100. If an account is selected at random, what is the probability that the account balance is between \$450 and \$600?

SOLUTION

This probability can be written as

$$P(450 < X < 600)$$

Using the standardizing procedure,

$$P(450 < X < 600) = P\left[\frac{450 - 650}{100} < \frac{X - 650}{100} < \frac{600 - 650}{100}\right]$$

$$= P(-2.0 < Z < -.5)$$

where Z once again represents the *standardized* normal random variable, which, for this example, is defined by

$$Z = \frac{X - 650}{100}$$

Refer to Table A.4 and Figure 6.19. Comparing Figures 6.19(a) and (b), the areas are equal:

$$.4772 - .1915 = .2857$$

Thus, 28.57% of the accounts will have a balance between \$450 and \$600, which implies that the probability of a randomly selected account having a balance between these two amounts is .2857.

1. 450 is two standard deviations to the left of the mean: $Z = -2$ and $450 = 650 - 2(100)$.
2. 600 is .5 standard deviation to the left of the mean: $Z = -.5$ and $600 = 650 - .5(100)$.
3. $P(X = 450) = P(X = 600) = 0$, so $P(450 < X < 600) = P(450 \leq X \leq 600) = .2857$. ■

EXAMPLE 6.4

Actuarial scientists in an insurance company formulate insurance policies that will be both profitable and marketable. For a particular policy, the lifetimes of the policyholders follow a normal distribution with $\mu = 66.2$ years and $\sigma = 4.4$ years. One of the options with this policy is to receive a payment following the 65th birthday and a payment every 5 years thereafter.

FIGURE 6.19

A: The probability that X is between $450 and $600. **B:** The probability that Z is between −2.0 and −.5.

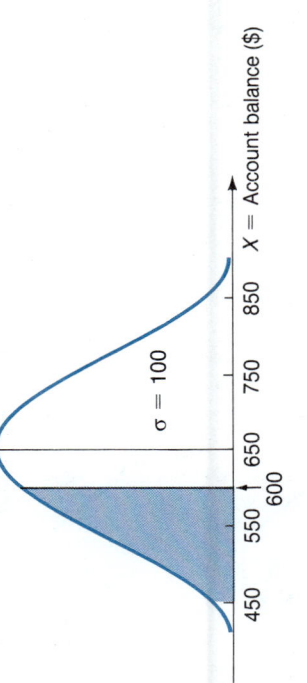

A

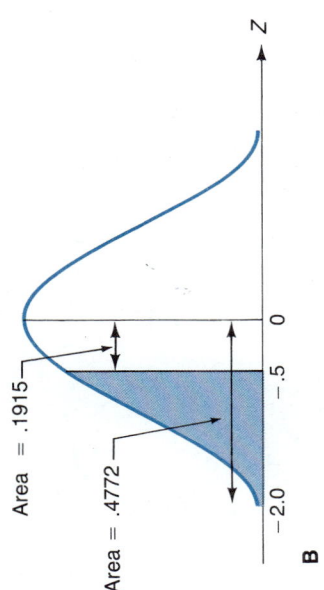

B

1. What percentage of policy holders will receive at least one payment using this option?
2. What percentage will receive two or more payments?
3. What percentage will receive exactly two payments?

SOLUTION 1

The normal curve for the policyholder lifetimes is shown in Figure 6.20. To receive at least one payment, the policy holder must live beyond 65 years of age. So we need to determine (see Figure 6.21):

$$P(X > 65) = P[(X - 66.2)/4.4 > (65 - 66.2)/4.4]$$
$$= P(Z > -.27) = .1064 + .5$$
$$= .6064$$

So nearly 61% of the policy holders will receive at least one payment.

SOLUTION 2

Because the policy holder receives a payment every 5 years, he or she will receive two or more payments provided he or she lives to be older than 70 years of age. This means that the probability of two or more payments is determined

FIGURE 6.20

The normal curve for policy-holder lifetimes. X = age at death (in years).

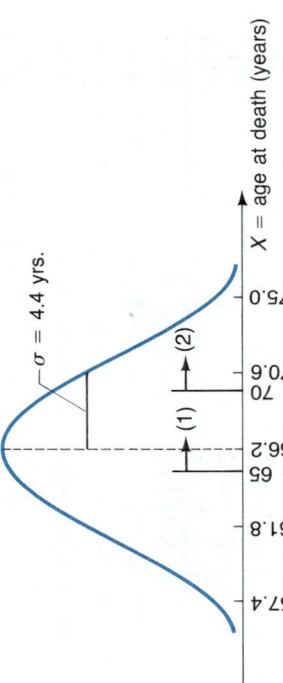

FIGURE 6.21

Z curve for $P(Z > -.27)$.

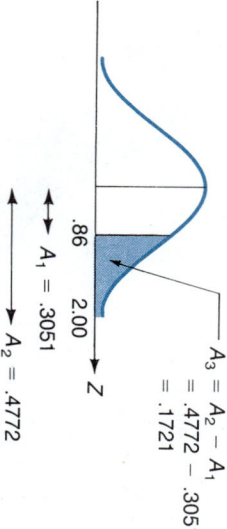

$$A_1 = .1064$$
$$A_2 = .5$$

$-.27$

$A_1 + A_2 = P(Z > -.27)$
$= .1064 + .5$
$= .6064$

FIGURE 6.22

Z curve for $P(Z > .86)$.

$A_1 = .3051$ (Table A-4)

$A_2 = .5 - .3051$
$= .1949$

.86

FIGURE 6.23

Z curve for
$P(.86 < Z < 2.00)$.

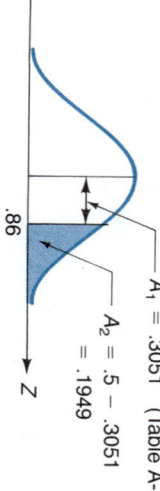

$A_3 = A_2 - A_1$
$= .4772 - .3051$
$= .1721$

.86   2.00

$A_1 = .3051$

$A_2 = .4772$

SOLUTION 3

To receive exactly two payments, the policy holder must live longer than 70 years and less than 75 years. This probability is

$$P(70 < X < 75)$$

Using the same standardization procedure (see Figure 6.23):

$$P(70 < X < 75) = P[(70 - 66.2)/4.4 < (X - 66.2)/4.4 < (75 - 66.2)/4.4]$$
$$= P(.86 < Z < 2.00)$$
$$= .4772 - .3051 = .1721$$

So 17.21% of the policy holders will receive exactly two payments. ∎

by (see Figure 6.22):

$$P(X > 70) = P[(X - 66.2)/4.4 > (70 - 66.2)/4.4]$$
$$= P(Z > .86) = .5 - .3051$$
$$= .1949$$

Thus, 19.5% of the policy holders will survive long enough to collect two payments.

6.4

APPLICATIONS WHERE THE AREA UNDER A NORMAL CURVE IS PROVIDED

Another twist to dealing with normal random variables is a situation where you are given the area under the normal curve and asked to determine the corresponding value of the variable. This is a common application of a normal random variable. For example, the manufacturer of a product may want to determine a warranty period during which the product will be replaced if it becomes defective, so that at most 5% of the items are returned during this period. Or, in a grocery store on any given day, the demand for a freshly made food item may or may not exceed the supply. The owner may want to determine

FIGURE 6.24

A: $P(X < X_0) = .8.$
B: $P(Z < .84) = .8.$

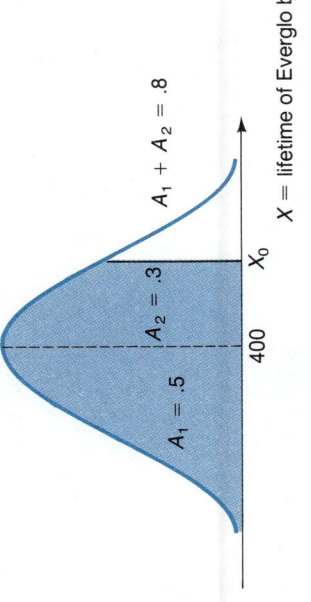

A

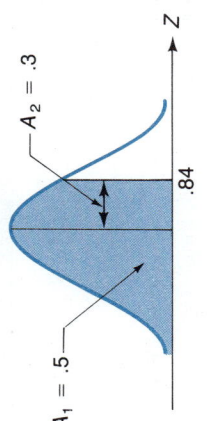

B

how much to supply each day, such that the demand (a normal random variable) will exceed this value 10% of the time (in other words, the customers will be disappointed no more than 10% of the time).

EXAMPLE 6.5

Referring to Example 6.2, 80% of the Everglo bulbs will burn out before what period of time? Recall that $\mu = 400$ and $\sigma = 50$.

SOLUTION

The first step here is to sketch this curve (Figure 6.24(a)) and estimate the value of X (say X_0) so that

$$P(X < X_0) = .8$$

Because .8 is larger than .5, X_0 must lie to the *right* of 400.

Next, find the point on a standard normal (Z) curve such that the area to the left is also .8 (Figure 6.24(b)). Using Table A.4, the area between 0 and .84 is .2995. This means that

$$P(Z < .84) = .5 + .2995$$
$$= .7995$$
$$= .8 \quad \text{(approximately)}$$

By standardizing X, we conclude that

$$\frac{X_0 - 400}{50} = .84$$

$$X_0 - 400 = 42$$

$$X_0 = 400 + 42 = 442$$

So 80% of the Everglo bulbs will burn out within 442 hours. ■

EXAMPLE 6.6

A bakery shop sells loaves of freshly made French bread. Any unsold loaves at the end of the day are either discarded or sold elsewhere at a loss. The demand for this bread has followed a normal distribution with $\mu = 35$ loaves and $\sigma = 8$ loaves.

FIGURE 6.25

A: $P(X \leq X_0) = .90.$
B: $P(Z \leq 1.28) = .90.$

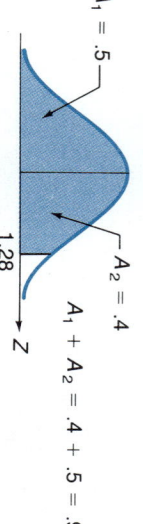

Area = .9

$\mu = 35$
$\sigma = 8$

35   X_0 $X =$ demand for French bread (loaves)

A

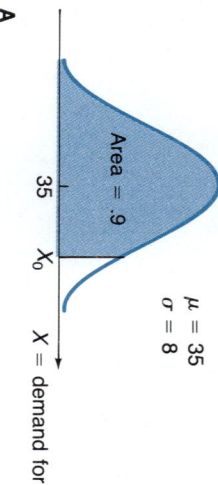

$A_1 = .5$ $A_2 = .4$

$A_1 + A_2 = .4 + .5 = .9$

1.28 Z

B

How many loaves should the bakery make each day so that they can meet the demand 90% of the time?

SOLUTION

The normal random variable X here is the demand for French bread (measured in loaves) (Figure 6.25(a)). To meet the demand 90% of the time, the bakery must determine an amount, say X_0 loaves, such that:

$$P(X \leq X_0) = .90$$

Proceeding as before, examine a Z curve having an area to the *left* $= .90$ (Figure 6.25(b)). Using Table A.4,

$$P(0 \leq Z \leq 1.28) = .4 \quad \text{(more accurately, .3997)}$$

which means that

$$P(X \leq X_0) = .90$$

So

$$P(Z \leq 1.28) = .4 + .5 = .9$$

and

$$\frac{X_0 - 35}{8} = 1.28$$

So

$$X_0 = 35 + (1.28)(8) = 45.24$$

To be conservative, round this value up to 46 loaves. By stocking 46 loaves each day, the bakery will meet the demand for this product 90% of the time. ■

6.5
ANOTHER LOOK
AT THE EMPIRI-
CAL RULE

In Chapter 3, the empirical rule specified that when sampling from a bell-shaped distribution (which means a normal distribution):

1. Approximately 68% of the data values should lie between $\bar{X} - s$ and $\bar{X} + s.$
2. Approximately 95% of them should lie between $\bar{X} - 2s$ and $\bar{X} + 2s.$
3. Approximately 99.7% of them should lie between $\bar{X} - 3s$ and $\bar{X} + 3s.$

Nothing was said at that time about the origin of these numbers. They ac-

FIGURE 6.26
Z curve for
$P(-1 < Z < 1) \cong .68$.

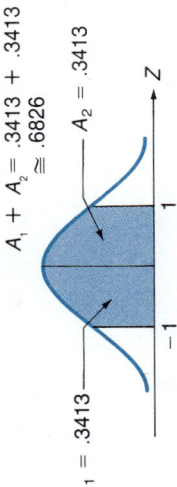

$A_1 + A_2 = .3413 + .3413$
$\cong .6826$

$A_2 = .3413$

$A_1 = .3413$

tually came directly from Table A.4. To see this, consider Figure 6.26, in which

$$P(-1 < Z < 1) = .68$$

This implies that, for any normal random variable X,

$$P[-1 < (X - \mu)/\sigma < 1] = .68$$

That is,

$$P[(\mu - \sigma) < X < (\mu + \sigma)] = .68$$

Thus, for a set of data from a normal population, where $\bar{X}$ is the sample mean and s is the sample standard deviation, approximately 68% of the data will be between $\bar{X} - s$ and $\bar{X} + s$.

Similarly, $P(-2 < Z < 2) = .4772 + .4772 = .9544$, so you can expect (approximately) 95% of the data points from a normal (bell-shaped) population to lie between $\bar{X} - 2s$ and $\bar{X} + 2s$.

Finally, $P(-3 < Z < 3) = .4987 + .4987 = .9974$, which leads to the third conclusion of the empirical rule.

EXERCISES

6.11 Let the random variable X be normally distributed with mean 5 and variance 4. Find the following probabilities.

a. $P(X \geq 5.7)$ **b.** $P(X \leq 3.4)$

c. $P(2.8 \leq X \leq 5.1)$ **d.** $P(5.7 \leq X \leq 6.8)$

6.12 Find the value of x if the random variable X is normally distributed with mean 10 and variance 9.

a. $P(X \leq x) = .51$ **b.** $P(X \geq x) = .805$

c. $P(10 \leq X \leq x) = .05$ **d.** $P(8 \leq X \leq x) = .13$

6.13 High-Tech, Inc. produces an electronic component, GX-7, that has an average life span of 4500 hours. The life span is normally distributed with a standard deviation of 500 hours. The company is considering a 3800 hours warranty on GX-7. If this warranty policy is adopted, what proportion of GX-7 components should High-Tech expect to replace under warranty?

6.14 The estimated miles-per-gallon (on the highway) ratings of a class of trucks are normally distributed with a mean of 12.8 and a standard deviation of 3.2. What is the probability that one of these trucks selected at random would get

a. Between 13 and 15 miles-per-gallon?

b. Between 10 and 12 miles-per-gallon?

6.15 The yearly cost of dental claims for the employees of D. S. Inc. is normally distributed with a mean of $75 and a standard deviation of $30. At least what yearly cost would be expected for 40 percent of the employees?

6.16 The diameter of $\frac{1}{2}$-inch bolts produced by a workshop is normally distributed with a mean of .5 inch and a standard deviation of .04 inch. What is the probability that a bolt selected at random will fit in a hole whose diameter is between 0.475 and 0.525 inch?

6.17 To become a member of MENSA, the nationwide organization for people with high I.Q.'s, one has to pass the qualifying examination. If the scores on the exam are normally distributed with a mean of 80 and a standard deviation of 25 and if only 20% of the people taking this exam are admitted to the organization, what is the passing score?

6.18 Harvard University has an extremely large endowment fund ($2.7 billion). However, on a per-student basis, Princeton is tops (with $250,530 per student), Harvard is second ($174,080), Cal Tech is third ($161,080), and Rice University is fourth ($151,210). Assume that endowment dollars per student for major universities across the nation is a normally distributed variable with a mean of $80,000 and a standard deviation of $30,000.

a. Convert the figure for Rice to a Z value.

b. Find the endowment per student such that 30% of universities will have endowments less than this amount.

c. For any randomly selected university, what is the probability that the endowment per student is greater than $100,000?

6.19 A recent article in *Bankers Monthly* investigated why some of the largest financial institutions in the United States have yet to fully automate their branch systems. One reason cited was that the annual cost to maintain an automated system at a 200-branch bank would average $5 million per year. Suppose that from past experience that for a 200-branch bank the cost to maintain an automated system is normally distributed with a mean of $5 million and a standard deviation of $.5 million.

(*Source:* "Why Big Banks Balk at Branch Automation," Michael Violano, *Bankers Monthly* (February 1988): 36–41.)

a. Find the probability that a randomly selected bank with 200 branches will have a cost greater than $4 million to maintain an automated system.

b. If a randomly selected bank with 200 branches had a cost greater than $6.5 million to maintain an automated system, would this be considered unusual? Why?

6.20 The vice president of Offshore Oil and Gas, a consulting firm, notices that the average length of time that a consultant spends on the telephone with a client at any one time is 40 minutes with a standard deviation of 18 minutes. Assuming that the length of time a consultant talks is normally distributed, what percent of the time would a consultant spend longer than 50 minutes on the phone?

6.21 As part of an experiment conducted for a graduate class in organizational behavior, the time taken to complete an assembly task was measured for two groups of workers. For the first group, the mean time was 10 minutes with a standard deviation of 1.5 minutes. For the second group, the mean time was 11.5 minutes with a standard deviation of 2.0 minutes. Assume that the times for both groups follow normal distributions. A worker from each group is randomly selected. What is the probability that the assembly time for this worker is less than 9 minutes, if the worker is from

a. The first group?   **b.** The second group?

6.22 Find the value of k such that $P(\mu \le X \le \mu + k\sigma) = .251$, for a random variable X having a normal distribution with mean μ and standard deviation σ.

6.23 If X is a normally distributed random variable with standard deviation of 10, find the mean μ given that $P(X \le .35) = .182$.

6.24 If X is normally distributed with a mean of 100, find the standard deviation given that $P(X \ge 110) = .123$.

6.25 If X is a normally distributed random variable with $P(X \ge 2) = .1$ and $P(X \le 1) = .3$, find both the mean and standard deviation.

6.6

NORMAL APPROXIMA-TION TO THE BINOMIAL

The binomial random variable was introduced in Chapter 5. It is a discrete random variable used to count the number of successes in a binomial situation.

CHARACTERISTICS OF A BINOMIAL SITUATION

1. You have n independent identical trials.
2. Each trial is a success (with probability p) or a failure (with probability $1 - p$).
3. The binomial random variable X is the number of successes out of n trials.
4. The mean of X is $\mu = np$, and the standard deviation of X is $\sigma = \sqrt{np(1 - p)}$.

Examples included:

X = the number of heads (successes) out of three flips (trials) of a coin

X = the number of people that read the evening newspaper (successes) out of a sample of 50 people (trials)

X = the number of defectives (successes) out of a sample of ten electrical components (trials)

Table A.1 contains values of n (the number of trials) only up to $n = 20$. In Chapter 5, we used the Poisson approximation to determine binomial probabilities for values of $n > 20$. In other words, we pretend that X is a Poisson random variable *having the same mean* as the actual binomial random variable. This is a good approximation, provided n is large (>20) and p is small ($np \leq 7$). We can also use the **normal approximation** to the binomial random variable. Here you pretend that X is a normal random variable *having the same mean and standard deviation* as the actual binomial random variable. This approximation works well when p is near .5 and in general offers a good estimate when both $np > 5$ and $n(1 - p) > 5$.

APPROXIMATIONS TO THE BINOMIAL

- Poisson approximation: Use when $n > 20$ and $np \leq 7$.
- Normal approximation: Use when $np > 5$ and $n(1 - p) > 5$.

Consider 12 flips of a coin. We want to determine (1) the probability of observing no more than 4 heads and (2) the probability of observing more than 5 heads. First, notice that a normal approximation is not necessary here. This is a binomial situation with $n = 12$ and $p = .5$, and Table A.1 does contain probabilities for this set of values. We chose this illustration to compare the actual binomial probability to the approximated probability using the normal distribution. Look at Figure 6.27, which demonstrates how we estimate binomial probabilities using a normal curve.

To solve Question 1, let X = the number of heads in 12 flips, so X is a binomial random variable. We want to determine $P(X \leq 4)$. We can obtain an exact solution using Table A.1:

$$P(X \leq 4) = P(0) + P(1) + P(2) + P(3) + P(4)$$
$$= 0 + .003 + .016 + .054 + .121$$
$$= .194$$

FIGURE 6.27

Approximating binomial probabilities using a normal curve.

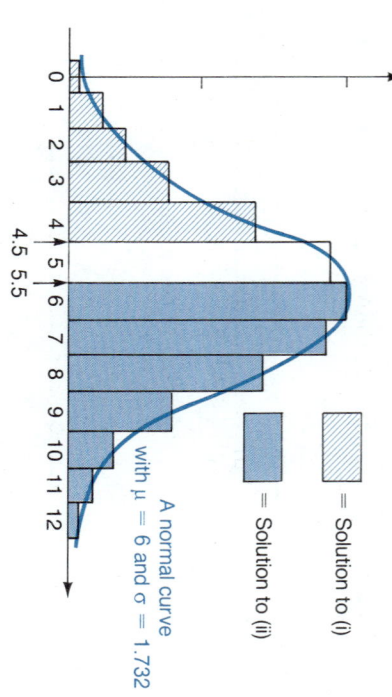

A normal curve
with $\mu = 6$ and $\sigma = 1.732$

= Solution to (ii)

= Solution to (i)

In Figure 6.27, this value is the sum of the areas of the boxes corresponding to $X = 0, 1, 2, 3,$ and 4. Note that the width of each box is one and so the height of the box (the probability) is the same as the area of the box. As a result, the total area of the boxes is 1.

We can also obtain an approximate solution. For this binomial random variable,

$$\mu = np = (12)(.5) = 6$$

and

$$\sigma = \sqrt{np(1-p)}$$
$$= \sqrt{3} = 1.732$$

To obtain an approximation, treat X as a normal random variable with $\mu = 6$ and $\sigma = 1.732$, illustrated in Figure 6.27. Note that both the total area of the boxes and the total area under the normal curve are 1. The area under the normal curve that approximates $P(X \le 4)$ is the area to the left of 4.5. So we obtain a better approximation here if we find the area under the normal curve to the left of 4.5, not 4.0. This .5 adjustment is referred to as an **adjustment for continuity**. This adjustment is necessary whenever you approximate a *discrete* random variable (such as binomial) using a *continuous* distribution (such as normal). Remember that the discrete distribution has gaps, whereas the continuous does not, so we must assign a portion of the space (probability) between 4 and 5 when we use a continuous distribution to approximate a discrete one. Using Table A.4,

Binomial	**Normal**
($n = 12, p = .5$)	($\mu = 6, \sigma = 1.732$)

$$P(X \le 4) \cong P(X \le 4.5)$$
$$= P\left[Z \le \frac{4.5 - 6}{1.732} \right]$$
$$= P(Z \le -.87) = .1922$$

Notice that the approximate solution of .1922 is very close to the actual probability of .194. This is helped in part by the fact that $p = .5$ for this situation, which means that the binomial distribution is perfectly symmetric. As the value of p moves away from .5, larger values of n are necessary to achieve an approximation this good.

Now consider Question 2, the probability of observing more than 5 heads in 12 flips, or $P(X > 5) = P(X \ge 6)$. Using Table A.1, we can obtain an *exact* solution:

$$P(X \geq 6) = P(6) + P(7) + \cdots + P(11) + P(12)$$

$$= .226 + .193 + \cdots + .003 + 0 = .613$$

We can also obtain an approximate solution. Using Figure 6.27, the area under the normal curve that corresponds to the lined area representing the exact solution is the area to the right of 5.5. So, using Table A.4:

Binomial   **Normal**

$(n = 12, p = .5)$ $(\mu = 6, \sigma = 1.732)$

$$P(X \geq 6) \cong P(X \geq 5.5)$$

$$= P\left(Z \geq \frac{5.5 - 6}{1.732}\right)$$

$$= P(Z \geq -.29) = .6141$$

Again, we obtain a very good approximation, helped by the fact that we are using a perfectly symmetrical binomial distribution.

HOW TO ADJUST FOR CONTINUITY

If X is a binomial random variable with n trials and probability of success $= p$, then:

1. $P(X \leq b) \cong P\left(Z \leq \dfrac{b + .5 - \mu}{\sigma}\right)$

2. $P(X \geq a) \cong P\left(Z \geq \dfrac{a - .5 - \mu}{\sigma}\right)$

3. $P(a \leq X \leq b) \cong P\left(\dfrac{a - .5 - \mu}{\sigma} \leq Z \leq \dfrac{b + .5 - \mu}{\sigma}\right)$

where

$$\mu = np, \qquad \sigma = \sqrt{np(1 - p)}$$

and Z is a standard normal random variable

4. Be sure to convert a $<$ probability to a $\leq$, and convert a $>$ probability to a $\geq$ before switching to the normal approximation.

EXAMPLE 6.7

In Chapter 5, we discussed a binomial situation (approximated by the Poisson) in which we had a sample of 100 chips to be tested. Each chip was either defective (a success) or not defective (a failure). Therefore, X was the number of defective chips (out of 100) and was a binomial random variable. We assumed that $p = .05$, which resulted in a very good Poisson approximation because n was large and p was small. Suppose, instead, that 10% of these chips are defective; that is, $p = .10$. Now, $np = 10$ and, because this is greater than 7, the Poisson distribution cannot be expected to provide a good approximation. However, we can obtain a good normal approximation here because $np = 10$ and $n(1 - p) = 90$, both of which are > 5.

For this situation, what is the probability that you observe one or fewer defective chips in a sample of 100, in which case the lot of chips is accepted?

SOLUTION

X is a binomial random variable with

$$\mu = np = (100)(.10) = 10$$

and

$$\sigma = \sqrt{np(1 - p)} = \sqrt{9} = 3$$

Therefore, using Table A.4:

Binomial

($n = 100, p = .10$)

$P(X \leq 1) \cong P(X \leq 1.5)$

Normal

($\mu = 10, \sigma = 3$)

$$= P\left[Z \leq \frac{1.5 - 10}{3} \right]$$

$$= P(Z \leq -2.83) = .0023$$

Consequently, there is a very small chance of accepting the lot, since you can expect to accept it only 23 times out of 10,000 using this procedure. ∎

EXAMPLE 6.8

In Chapter 5, we discussed a binomial situation in which Eagle Air was intentionally overbooking their flights. On a particular flight from Dallas to El Paso, they use a much larger aircraft that holds 200 people. As in our previous example, 20% of the people do not show up for a reserved flight. If Eagle Air accepts 235 reservations, what is the probability that at least one passenger will end up without a seat on this flight?

SOLUTION

The binomial random variable X here is the number of people (out of 235) who show up for the flight. For this situation, $n = 235$, and $p = .8$ represents the probability that any one passenger *will* show up. The mean of this random variable is

$$\mu = (235)(.8) = 188$$

and the standard deviation is

$$\sigma = \sqrt{(235)(.8)(.2)} = 6.13$$

At least one person holding a reservation will be deprived of a seat if $X \geq 201$ because the plane holds only 200 people. Once again, we use the normal approximation (Table A.4) to obtain the following probability:

Binomial

($n = 235, p = .8$)

$P(X \geq 201) \cong P(X \geq 200.5)$

Normal

($\mu = 188, \sigma = 6.13$)

$$= P\left[Z \geq \frac{200.5 - 188}{6.13} \right]$$

$$= P(Z \geq 2.04)$$

$$= .5 - .4793$$

$$= .0207$$

So on approximately 2 flights out of 100, at least one person will be unable to secure a seat. ∎

EXERCISES

6.26 A random variable X has a binomial distribution with the probability of a success, p, equal to .25.

a. Would it be appropriate to use the normal approximation to the binomial if $n = 30$? if $n = 15$?

b. With $n = 40$, use the normal approximation to find $P(2 \leq X \leq 10)$.

c. What is the smallest value that n can be and still have the normal distribution to be appropriate for approximating the binomial distribution?

6.27 Let the random variable X indicate the number of female students chosen (with replacement) in a sample of 15 from a student body with 40% female students.

a. Using the binomial table, find the probability that X is greater than 4 and less than 9.

b. Use the normal approximation to answer part (a).

c. Compare the answers in parts (a) and (b).

6.28 Thirty percent of the computer programmers who are hired to work for Techtronics do not have work experience in programming. If a random sample of 35 computer programmers is selected, what is the probability that fewer than 20 have had experience in computer programming before being hired by Techtronics?

6.29 A travel agency promotes vacation packages by phoning households at random in the evening hours. Historically, only 65% of heads of households are at home when the agency phones. If 30 households are phoned on a given evening, what is the probability that the agency will find between 15 and 25 households, inclusively, with the head of the household at home?

6.30 Many fast-food restaurants target their advertising toward the age group below 30 years. It is estimated that 70% of the people between ages 18 and 24 visit fast-food restaurants at least once a week.

(Source: USA Today, May 24, 1988, Section A, p. 1.)

a. In a random sample of 100 people between the ages 18 and 24, what is the probability that between 50 to 70 people, inclusively, visit fast-food restaurants at least once a week.

b. In a random sample of 100 people between the ages 18 and 24, what is the probability that less than 60 visit fast-food restaurants at least once a week?

6.31 The percentage of cars sold at Lance Holey's used-car lot that required financing is 58%. If 30 car buyers at this lot are randomly selected, what is the probability that between 15 and 25 buyers (inclusive) financed their car?

6.32 A study sponsored by Allstate Insurance Company and Fortune Magazine examined the views of United States corporate executives on major problems facing the United States. It was found that approximately 45% of the executives chose AIDS (Acquired Immune Deficiency Syndrome) as one of the three major problems facing the United States.

(Source: Alison Kittrell, "Employers Lack AIDS Strategy: A Study," Business Insurance, February 1988.)

a. If 100 corporate executives were selected at random, what is the probability that between 30 and 60 executives, inclusively, consider AIDS as one of three major problems facing the United States?

b. Suppose that in the sample of 100 executives in part (a) it was found that either less than 20 or more than 80 executives considered AIDS as one of three major problems facing the United States. Would this make one question whether the 45% figure given in the survey was accurate? Why?

6.33 If a pair of fair dice is rolled 70 times, what is the probability that a pair of snake eyes (a one on each die) will appear between five and ten times, inclusively? Is the normal approximation appropriate here? What other approximation should work well?

OTHER CONTINUOUS DISTRIBUTIONS (OPTIONAL)

The normal distribution is one example of a continuous distribution. A normal random variable X is a continuous random variable. This simply means that over some specific range, *any* value of X is possible. We used X to represent the lifetime of an Everglo bulb to illustrate a continuous random variable because any value between 280 hours and 520 hours (see Figure 6.2) is possible. In fact, any value less than 280 or more than 520 is also possible, although not likely to occur.

For the Everglo example, a normal distribution seemed appropriate because the histogram of 200 sample bulbs in Figure 6.2 revealed a concentration of burnout times in the "middle" and not nearly as many burnout times around 300 or 500. This why the normal curve has a "mound" in the center and "tails" on each end.

There are many continuous distributions that do not resemble a normal curve in appearance. For example, consider these two situations in which a random variable, X, ranges from one to ten.

Situation 1

the chance that X is between 1.0 and 1.5

= the chance that X is between 1.5 and 2.0

= the chance that X is between 2.0 and 2.5

⋮

= the chance that X is between 9.0 and 9.5

= the chance that X is between 9.5 and 10.0

Situation 2

The larger X is, the less likely it is to occur. Thus, the chance that X is between 1.0 and 1.5

> the chance that X is between 1.5 and 2.0

> the chance that X is between 2.0 and 2.5

⋮

> the chance that X is between 9.0 and 9.5

> the chance that X is between 9.5 and 10.0

These two cases can be represented by two other popular continuous distributions. Situation 1 can be represented by a uniform random variable, whereas situation 2 could be described using an exponential random variable.

Although there are other random variables that apply to these two situations, the uniform and exponential distributions most often fit the applications encountered in business.

The Uniform Distribution

Consider spinning the minute hand on a clock face. Define a random variable X to be the stopping point of the minute hand. It seems reasonable to assume that, for example, the probability that X is between two and four is *twice* the probability of observing a value of X between eight and nine. In other words, the probability that X is in any particular interval is *proportional* to the width of that interval.

A random variable of this nature is a **uniform random variable.** The values of such a variable are evenly distributed over some interval because the random variable occurs *randomly* over this interval. Unlike the normal random variable, values of the uniform random variable do not tend to be concentrated about the mean.

Assume that the manager of Dixie Beverage Service is concerned about the amount of soda that is released by the dispensing machine that the company is now using. She is considering the purchase of a new machine that electronically controls the cutoff time and is supposed to be very accurate. The present ma-

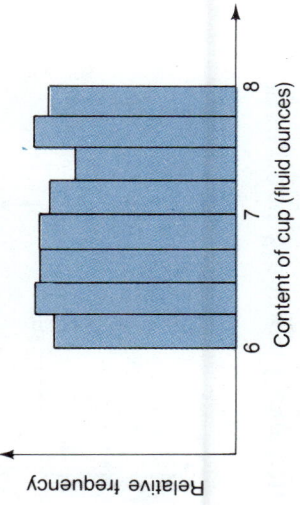

FIGURE 6.28

Relative frequency histogram of a sample of 150 cups of soda.

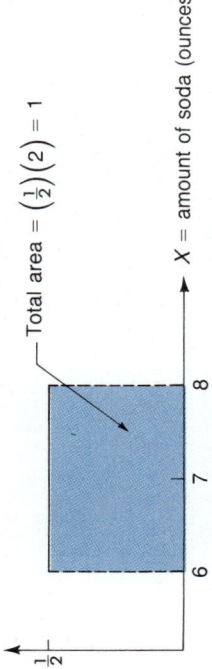

FIGURE 6.29

Uniform distribution for X = soda content (compare with Figure 6.28).

chine cuts off mechanically, and she suspects that the device shuts off the fluid flow *randomly* at anywhere between 6 and 8 ounces. To test the present system, a sample of 150 cups is taken from the machine, and the amount of soda released into each cup is recorded. The relative frequency histogram made from these 150 observations is shown in Figure 6.28.

Would you be tempted to describe the population of *all* cup contents using a normal curve? We hope not, because there is no evidence of a declining number of observations in the tails. As a word of warning here, we often have a tendency to think of all continuous random variables as being normally distributed. As this application demonstrates, this is certainly not the case. Instead, this distribution is a flat or uniform distribution. The random variable, X = content of soda, is a uniform random variable. The corresponding smooth curve describing the population is shown in Figure 6.29.

Notice that the total area here is given by a rectangle, and, as is true of all continuous random variables, this total area must be 1. The area of a rectangle is given by (width) · (height). By making the height of this curve (a straight line, actually) equal to .5, the total area is

$$(8 - 6)(.5) = 1.0$$

In general, the curve defining the probability distribution for a uniform random variable is shown in Figure 6.30. The total area is

$$(b - a) \left[\frac{1}{b - a} \right] = 1.0$$

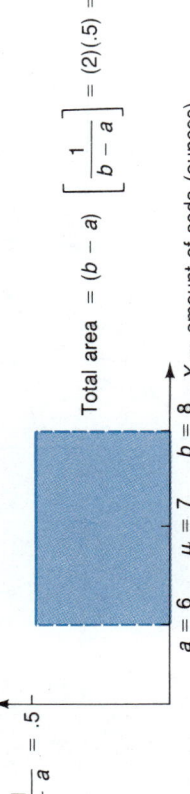

FIGURE 6.30

Total area for a uniform distribution.

FIGURE 6.31

The probability that X exceeds 7.5. The shaded area represents the percentage of cups containing more than 7.5 ounces.

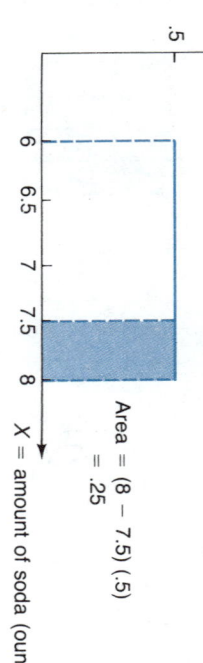

Mean and Standard Deviation

Refer to Figure 6.30. The mean (μ) of X is the value midway between a and b, namely,

$$\mu = \frac{a+b}{2}$$

The standard deviation (σ) of X is, as before, a measure of how much variation there would be in X if you were to observe it indefinitely. Unlike when using the normal distribution, σ is hard to represent graphically here as a particular distance on the probability curve. Its value, however, is given by

$$\sigma = \frac{b-a}{\sqrt{12}}$$

Determining Probabilities

As for all continuous random variables, a probability using a uniform random variable is determined by finding an area under a curve. Suppose, for example, the manager of Dixie Beverage Service would like to know what percentage of the cups will contain more than 7.5 ounces, using the present machines. In Figure 6.31, the shaded area is a rectangle, so its area is easy to find:

$$\text{Area} = (\text{width}) \cdot (\text{height}) = (8 - 7.5) \cdot .5 = .25$$

So 25% of the cups will contain more than 7.5 ounces.

EXAMPLE 6.9

What is the probability that a cup will contain between 6.5 and 7.5 ounces? What is the average content?

SOLUTION

The first result is the same as the percentage of cups containing between 6.5 and 7.5 ounces. Based on Figure 6.32, we conclude that

$$P(6.5 < X < 7.5) = .5$$

The average cup content (mean of X) is

$$\mu = \frac{6+8}{2} = 7 \text{ ounces}$$

FIGURE 6.32

The probability that X is between 6.5 and 7.5.

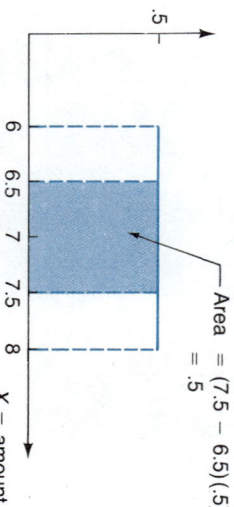

The standard deviation of X is

$$\sigma = \frac{8 - 6}{\sqrt{12}} = 0.58 \text{ ounce}$$

Notice that, as with the normal random variable, the probability that X is equal to any particular value is zero. So,

$$P(X = 6.5) = P(X = 7.5) = 0$$

As a result,

$$P(6.5 \leq X \leq 7.5) = P(6.5 < X < 7.5) = .5$$

Simulation is an area of statistics that relies heavily on the uniform distribution. In fact, this distribution is the underlying mechanism for this often complex procedure. So, although not as many "real-world" populations resemble this distribution as they do the normal one, the uniform distribution is extremely important in the application of statistics.

The Exponential Distribution

The final continuous distribution we will discuss is the **exponential distribution**. Similar to the uniform random variable, the exponential random variable is used in a variety of applications in statistics. One application is observing the time between arrivals at, for example, a drive-up bank teller. Another situation that often fits the exponential distribution is observing the lifetime of certain components in a machine.

Chapter 5 discussed the Poisson random variable, which often is used to describe the *number* of arrivals over a specified time period. If the random variable Y, representing the number of arrivals over time period T, follows a Poisson distribution, then X, representing the *time between* successive arrivals, will be an **exponential random variable**. The exponential random variable has many applications when describing any situation in which people or objects have to wait in line. This line is called a **queue**. People, machines, or telephone calls may wait in a queue.

The Exponential Random Variable The shape of the exponential distribution is represented by a curve that steadily decreases as the value of the random variable, X, increases. Thus, the larger X is, the probability of observing a value of X at least this large decreases exponentially. This type of curve is illustrated in Figure 6.33.

Determining Probabilities Determining areas for exponential random variables is not as simple as for uniform ones, but it is easier than for normal random variables because exponential probabilities can be derived on a calculator. Table A.2 also can be used to determine the probability for an exponential random variable.

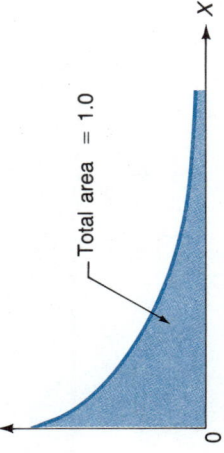

Total area = 1.0

FIGURE 6.33
Curve showing the distribution of an exponential random variable.

FIGURE 6.34

Curve used for
determining a probability
for an exponential
random variable.

As Figure 6.34 illustrates, for an exponential random variable, X, the probability that X exceeds or is equal to a specific value, X_0, is

$$P(X \geq X_0) = e^{-A \cdot X_0}$$

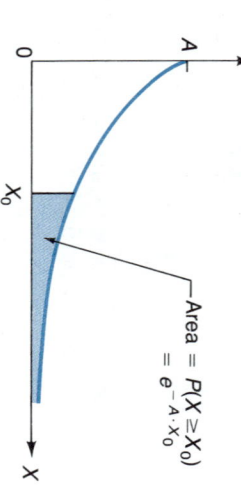

Area $= P(X \geq X_0)$
$= e^{-A \cdot X_0}$

The parameter A is related to the Poisson random variable we used when
discussing arrivals. In fact, the Poisson distribution for arrivals per unit time
and the exponential distribution for time *between* arrivals provide two alternative ways of describing the same thing. For example, if the number of arrivals
per unit time follows a Poisson distribution with an average of $A = 6$ per hour,
then an alternate way of describing this situation is to say that the time between
arrivals is exponentially distributed with mean time between arrivals equal to
$1/A = 1/6$ hour (10 minutes).

In general, $1/A$ is the average (mean) value of the exponential random variable, X. It is also equal to the standard deviation of X. So,

$$\mu = 1/A$$
$$\sigma = 1/A$$

In applications using this distribution, the value of A either will be given
or can be estimated in some way.

EXAMPLE 6.10

A manufacturer of color televisions has determined that the lifetime of the
picture tube follows an exponential distribution with an average lifetime of 10
years. Determine the fraction of picture tubes that;

1. Fail after 15 years
2. Fail before the warranty period of 2 years

SOLUTION 1

Define X to be the lifetime of a picture tube. Since $\mu = 10$ years and $\mu = 1/A$,
then $A = 1/\mu = .1$. We want to determine $P(X > 15)$, which is illustrated in
Figure 6.35.

We see that $X_0 = 15$ and $A = .1$. Values of e^{-x} are contained in Table A.2.
Using this table or your calculator,

$$P(X > 15) = P(X \geq 15)$$
$$= e^{-A \cdot X_0} = e^{-(.1)(15)}$$
$$= e^{-1.5} = .22$$

FIGURE 6.35

Curve showing the
probability that X
exceeds 15 [$P(X > 15)$].

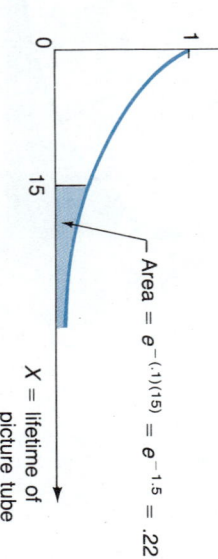

Area $= e^{-(.1)(15)} = e^{-1.5} = .22$

$X =$ lifetime of
picture tube

FIGURE 6.36

Curve showing the probability that X is less than $2[P(X < 2)]$.

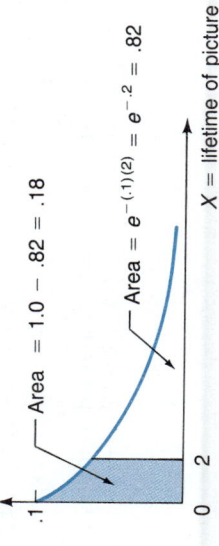

Area $= 1.0 - .82 = .18$

Area $= e^{-(.1)(2)} = e^{-.2} = .82$

$X =$ lifetime of picture tube

Thus, 22% of the television picture tubes will survive longer than 15 years.

SOLUTION 2

Here the problem is to find $P(X < 2) = P(X \le 2)$, which is $1 - P(X > 2)$. Using Table A.2 and Figure 6.36,

$$P(X > 2) = e^{-(.1)(2)}$$
$$= e^{-.2}$$
$$= .82$$

The total area under the curve is 1.0, so

$$P(X < 2) = P(X \le 2) = 1 - .82 = .18$$

The manufacturer will be forced to replace 18% of the tubes during the 2-year warranty period. ■

EXAMPLE 6.11

The owner of the Downtown Haircut Emporium believes the best way to run his barbershop is to rely on walk-in customers and not schedule appointments. From past experience, the arrival of customers follows a Poisson distribution with an average arrival rate of $\lambda = 4$ customers per hour.

1. If the owner just witnessed the arrival of a customer, what is the probability that a new arrival will occur within 30 minutes?
2. If X represents the time between successive arrivals, what are the mean and standard deviation of X?

SOLUTION 1

To determine this probability, we must first convert 30 minutes to .5 hour, since the arrival rate is 4 **per hour**. The desired probability then is $P(X \le .5)$. Referring to Figure 6.37, the probability that X *exceeds* .5 is

$$P(X > .5) = P(X \ge .5)$$
$$= e^{-(4)(.5)}$$
$$= e^{-2}$$
$$= .135$$

FIGURE 6.37

Curve showing the probability that X exceeds $.5[P(X > .5)]$.

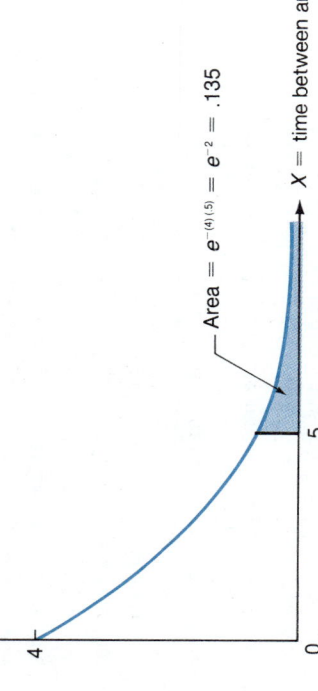

Area $= e^{-(4)(.5)} = e^{-2} = .135$

$X =$ time between arrivals

SOLUTION 2

Consequently, $P(X \leq .5) = 1 - .135 = .865$, and so 86.5% of the time, the time between successive arrivals will not exceed 30 minutes.

Both the mean and standard deviation of X (the time between successive arrivals) are $1/\lambda = 1/4$ hour (15 minutes). ∎

EXERCISES

6.34   A random variable X has a uniform distribution between the values 0 and 4.

a.   What is the mean of X?

b.   What is the standard deviation of X?

c.   What is the height of the probability distribution of X?

d.   What is the probability that X is greater than 1.23?

6.35   The errors from a forecasting technique appear to be uniformly distributed between -3 and 3.

a.   Find the probability that the errors deviate by no more than 1.5 from the mean.

b.   Find the value x such that 60% of the errors occur between $-x$ and x.

6.36   The temperature of a warming tray is uniformly distributed between the values of 100°F and 104°F.

a.   What percent of the time is the warming tray temperature less than 101.5°?

b.   What is the mean temperature of the warming tray?

c.   What is the standard deviation of the temperature of the warming tray?

6.37   The rate at which a swimming pool is filled is uniformly distributed between 20 and 26.3 gallons per minute.

a.   What is the probability that the rate at which the swimming pool is filled at any one time is between 21.3 and 24.6 gallons per minute?

b.   What is the mean rate at which the swimming pool is filled?

c.   What is the standard deviation of the rate at which the swimming pool is filled?

6.38   A quality-control engineer records that a certain machine uniformly produces between 10 and 15 precision ball bearings per hour. At least how many precision ball bearings are produced per hour 75% of the time?

6.39   If the amount of time spent by visitors in a certain zoo follows an exponential distribution and if it is known that the average visitor spends 1.9 hours at the zoo, calculate the probability that a given visitor will spend at least 1.5 hours at the zoo?

6.40   If the amount of time that a customer spends in Ricky's Hide-Away restaurant follows an exponential distribution and if the average time spent by a customer is 0.75 hours, what is the probability that a customer will spend more than an hour in the restaurant? What is the standard deviation of the amount of time spent by a customer in the restaurant?

6.41   Yellow Rose taxi company estimates that it makes an average of $415 in profits per day. Assuming that the daily profit follows an exponential distribution, what is the probability that on a given day at least $500 in profits will be made?

6.42   The president of Bright-Light Candles estimates that the average burning time of their "medium-K" candles is 40 hours. Assuming that burning time follows an exponential distribution, calculate the probability that a given medium-K candle will burn for at least 50 hours?

6.43   If the amount of time ships spend at the Philadelphia dockyard follows an exponential distribution and if the average ship spends 3.1 days there, what is the probability that a given ship spends no more than 1.5 days at the dockyard?

6.44   The Mylapore County fire department has determined that the amount of time per month spent fighting fires follows an exponential distribution. If the average fire-fighting time per month is 10.4 hours, what is the probability that in a given month no more than 15 hours will be spent fighting fires?

6.45 Use a convenient computer package to generate randomly 50 uniformly distributed observations between 2.0 and 4.0. Find the mean and standard deviation of this sample of 50 observations. How closely does the sample mean and sample standard deviation agree with the true population mean and standard deviation? Repeat this procedure with 100 uniformly distributed observations. In MINITAB, the format is as follows:

```
MTB  >   RANDOM    50 observations, put in C1;
SUBC >   UNIFORM with continuous uniform on 2.0 to 4.0 .
MTB  >   Describe C1
```

SUMMARY

A random variable that can assume any value over a specific range is a **continuous random variable**. Many business applications have continuous probability distributions that can be approximated using a normal, uniform, or exponential random variable. Each of these distributions has a unique curve that can be used to determine probabilities by finding the corresponding area under this curve. The **normal** distribution is characterized by a bell-shaped curve with values concentrated about the mean. The **uniform** distribution (curve) is flat; values of this random variable are evenly distributed over a specified range. The **exponential** distribution has a shape that steadily decreases as the value of the random variable increases. Table A.2 (or a good calculator) can be used to derive probabilities for the exponential distribution.

We discussed examples illustrating the shape of each distribution. The exact curve for a particular random variable is specified using one or two *parameters* that describe the corresponding population. As in the case of a discrete random variable, the population consists of what you would obtain if the random variable was observed indefinitely. The resulting average value and standard deviation represent the **mean** and **standard deviation** of the random variable and corresponding population.

There are infinitely many normal random variables, one for each mean (μ) and positive standard deviation (σ). If $\mu = 0$ and $\sigma = 1$, this normal random variable is the **standard normal** random variable, Z. Consequently, there is only *one* normal random variable of this type. Table A.4 gives the probabilities (areas) under the standard normal curve. You can also use this table to determine a probability for any normal random variable if you first **standardize** the variable by defining $Z = (X - \mu)/\sigma$. For this situation, Z represents the number of standard deviations that X is to the right (Z is positive) or left (Z is negative) of the mean.

The normal distribution can be used to approximate binomial probabilities for a large number of trials, n. Because the normal distribution is continuous and the binomial is discrete, the approximation can be significantly improved by **adjusting for continuity** before applying the normal approximation.

REVIEW EXERCISES

6.46 Determine each of the following for a standard normal curve. Sketch the corresponding area.

a. $P(0 < Z < 1.5)$

b. $P(Z > -3)$

c. $P(Z < -1.88)$

d. $P(-2.5 < Z < 2.5)$

6.47 Calculate and sketch the area under the standard normal curve between the following Z values.

a. 2.2 and 3.25

b. −1.5 and 1.5

c. −.75 and 0

6.48 A commodities broker has a record of being correct 30% of the time in transactions which the broker solicits. From a random sample of 35 different recommendations to clients, what is the probability that less than 11 of the recommendations by the broker are profitable?

6.49 A quality control engineer noted that about 2% of all smoke detectors do not go off when a fire is present. Out of a sample of 600 smoke detectors that were in homes that caught fire, what is the probability that more than eight smoke detectors did not sound an alarm?

6.50 The mean length of certain gauges manufactured by a firm is 20 inches with a standard deviation of 0.44 inches. A random sample of 100 gauges was taken. Assuming that the length of gauges manufactured is approximately normally distributed, what percentage of these gauges measured less than 20 inches in length?

6.51 Let X be a normally distributed random variable. Find the values of X that bound the middle 50% of the distribution of X if the mean is 5 and the variance is 9.

6.52 Scores on the English screening exam for international students are distributed normally with a mean of 68 and a standard deviation of 11. Calculate the following.

a. The percentage of scores between 70 and 80

b. The percentage of scores that are less than 60

6.53 The examination committee of the Institute of Chartered Accountants passes only 20% of those who take the examination. If the scores follow a normal distribution with an average of 72 and a standard deviation of 18, what is the passing score?

6.54 The shelf life of cookies made by a firm is considered to be exponentially distributed with a mean equal to 3 days. What percentage of the boxes of cookies placed on the shelf today would still be considered marketable after 2.75 days?

6.55 The time that a certain drug has an effect on a normal human being is considered to be exponentially distributed when a standard dose is taken. If the average length of time that the drug has an effect is 30 hours, what is the probability that any given normal person will be affected by the drug for at least 32 hours? What is the standard deviation for the length of time that the drug affects a person?

6.56 Accidents such as the 1987 Amtrak-Conrail disaster in Maryland, in which 16 people died, have heightened employers' awareness of their legal liabilities of personal and property losses caused by workers under the influence of drugs and alcohol. Federal experts estimate that 10% to 23% of all American workers use drugs on the job. Suppose that a certain firm has over 100 manufacturing plants across the country. Let X be the percentage of employees who use drugs on the job at a particular plant. Assume that past data indicate that X has a normal distribution with mean 16% and standard deviation 4%.

a. What is the probability that X is greater than 10%?

b. What is the probability that X is less than 5%, which may be considered an acceptable figure with regard to risk taken by the employer?

(Source: "Drug Testing—Walking a Legal Tightrope" Robert J. Aalberts, *Business* (January–March 1988): 52–56.)

6.57 Clearvision Company manufactures picture tubes for color television sets and claims that the life spans of their tubes are exponentially distributed with a mean of 1800 hours. What percentage of the picture tubes will last no more than 1600 hours?

6.58 The amount of time each day that the copying machine is used at a certain business is approximately exponentially distributed with a mean of 3.5 hours. What is the probability that the copying machine will be used at least 2 hours a day?

6.59 The diameter of a special aluminum pipe made by Everything Aluminum Inc. is normally distributed with a mean of 3.00 centimeters and a standard deviation of 0.1 centimeter. Calculate the proportion of pipes whose diameters are more than 3.15 centimeters.

6.60 The Defense Contract Audit Agency (DCAA), which audits defense contractors, has a backlog of unaudited contracts in any given year. Suppose that the dollar amounts of unaudited contracts (in billions of dollars per annum) is normally distributed, with a mean of $74.5 billion and a standard deviation of $11.6 billion. What is the probability that in any given year, the backlog is less than $50 billion?

6.61 A manufacturer of heating elements for hot water heaters ships boxes that contain 100 elements. A quality-control inspector randomly selects a box in each shipment and accepts the shipment if there are 5 or less defective heating elements in the box. Assuming that the manufacturer has had a rate of 6% defective items, what is the probability that a shipment of heating elements will pass the inspection?

6.62 A recent poll of the chief financial officers (CFOs) at 100 major financial institutions in the United States (reported in the "Best Brokers Survey" of *Financial World*, January 1988) revealed that 53% of the CFOs favored abolishing the Glass-Steagall law, which separates commercial banking from investment banking.

a. If 20 CFOs were chosen at random from any of the major financial institutions in the United States, what is approximately the probability that less than 18 CFOs will favor abolishing the Glass-Steagall law? Assume that the percentage 53% can be used as a close approximation of the population proportion.

b. If in part (a), 20 CFOs were chosen at random and more than 15 CFOs favored abolishing the Glass-Steagall law, would this be considered an unusual sample? Why?

6.63 A paint sprayer coats a metal surface with a layer of paint between 0.5 and 1.5 millimeters thick. The thickness of the coat of paint is approximately uniformly distributed.

a. What is the mean and standard deviation of the thickness of the coat of paint on the metal surface?

b. What is the probability that paint from this sprayer on any given metal surface will be between 1.0 and 1.3 mm thick?

6.64 The rate at which a sack of soybeans is filled varies uniformly from 50 pounds per hour to 65 pounds per hour. What percent of the time is the rate greater than 55 pounds per hour?

6.65 If X is a uniform random variable that represents the percentage of time each day that a machine does not work, what is the probability that X is greater than the mean percentage of time that the machine does not work?

6.66 If the random variable X has a uniform distribution between -10 and 10, find the value of x such that $P[X \geq x] = .25$.

6.67 The marketing division of Goodlife Tires determined the average (mean) life of tires to be 30,000 miles with a standard deviation of 5,000 miles. Given that tire life is normally distributed random variable, find the following.

a. The probability that tires last between 25,000 and 35,000 miles

b. The probability that tires last between 28,000 and 33,000 miles

c. The probability that tires last less than 28,000 miles

d. The probability that tires last more than 35,000 miles

6.68 The random variable X is normally distributed with mean μ and variance σ^2. Find k if $P(\mu - k\sigma \leq X \leq \mu + k\sigma) = .67$.

6.69 If the random variable X is normally distributed with mean 25, find the variance if $P(X \geq 29) = .27$.

6.70 The random variable X is normally distributed such that $P(X \leq 10) = .12$ and $P(X \geq 15) = .4$. Find the mean and variance of the random variable X.

6.71 For the 12-month period ending March 31, 1988, Chrysler Corporation noted that consumers were shunning extended-term car loans despite the lower monthly payments. During this time period, only about 7% of all Chrysler minivan buyers opted for the 6-year loans on the minivans. Suppose that an owner of a Chrysler dealership wished to know if it would be considered unusual for more than 11 buyers to opt for

6-year loans in a random sample of 100 Chrysler minivan buyers over this time period. Find the probability of this event happening.

(Source: Melinda Grenier Gruler, "A Cold Shoulder to Longer-Term Loans," *Wall Street Journal,* May 12, 1988, p. 25.)

6.72 The mechanics at Quick Brown Fox can tune up a car in an average of 30 minutes with a standard deviation of 5 minutes. If a car arrives for a tune-up 25 minutes before closing, what is the probability that the car will be serviced by closing, assuming that the time it takes for a tune-up is normally distributed.

6.73 Portfolio insurance can protect a portfolio from losing value during a declining market. However, an insured portfolio will not have as much upside potential as an un-insured portfolio. In a study by Clarke and Arnott in 1987, return characteristics were studied for insured portfolios with varying minimum returns. One such portfolio was considered to have a mean return of 13% with standard deviation of 14.2%. However with insurance the minimum return would be 0% and thus no loss would occur. Assume that the return on this portfolio can be approximated by a normal distribution.

a. Find the probability that the return would be less than 0% without insurance. Do you think one should strongly consider insuring the portfolio?

b. Find the probability that the return will be greater than 5%.

(Source: "The Cost of Portfolio Insurance: Tradeoffs and Choices" R. G. Clarke and R. D. Arnott, *Financial Analyst Journal,* 1987.)

6.74 There are certain situations where both $np > 5$ and $n(1 - p) > 5$ do not hold but the normal approximation may be a reasonable approximation. The following MINITAB

```
MTB > BINO N=20 P = .2 PUT IN C2

BINOMIAL PROBABILITIES FOR N =   20   AND P = 0.200000

       K    P( X = K )              P(X LESS OR = K)
       0      0.0115                      0.0115
       1      0.0576                      0.0692
       2      0.1369                      0.2061
       3      0.2054                      0.4114
       4      0.2182                      0.6296
       5      0.1746                      0.8042
       6      0.1091                      0.9133
       7      0.0545                      0.9679
       8      0.0222                      0.9900
       9      0.0074                      0.9974
      10      0.0020                      0.9994
      11      0.0005                      0.9999
      12      0.0001                      1.0000

MTB > PLOT C2 C1

      C2
   0.280+
        -
        -                                              *
        -
   0.210+                                          *       *
        -
        -                                      *             *
        -
   0.140+                                   *                   *
        -
        -                               *                          *
        -
   0.070+                            *                                *
        -                         *                                     *
        -                      *                                           *  *
   0.000+                   *                                                    *  *  *  *  *  *  *  *  *  *
        +---------+---------+---------+---------+---------+---------+---------+---------C1
       -0.0       4.0       8.0      12.0      16.0      20.0
```

computer printout shows the distribution of a binomial random variable X with $n = 20$ and $p = .20$. Examine the graph and describe how well a bell-shaped curve could fit this distribution? Calculate the probability that $P[X \leq 3]$ and $P[X \geq 5]$ by using both the normal approximation and the binomial table. Would you say that the normal approximation is reasonably close?

6.75 The following MINITAB computer printout generates 60 random observations from a uniform distribution between 0 and 1. Find the percentage of observations that fall between $\bar{x} - s$ and $\bar{x} + s$ and also between $\bar{x} - 2s$ and $\bar{x} + 2s$. Show that $\mu - 2\sigma$ and $\mu + 2\sigma$ for the uniform distribution will always contain all data points.

```
MTB > RANDOM 60 C1;
SUBC> UNIFORM CONTINUOUS ON 0 TO 1.0.
MTB > DESCRIBE C1
```

	N	MEAN	MEDIAN	TRMEAN	STDEV	SEMEAN
C1	60	0.4599	0.4077	0.4574	0.3027	0.0391

	MIN	MAX	Q1	Q3
C1	0.0034	0.9542	0.2017	0.7854

```
MTB > PRINT C1
C1
```

0.178674	0.334305	0.788096	0.512027	0.003356	0.419453	0.43159
0.949007	0.625865	0.233143	0.142913	0.864091	0.011382	0.42273
0.841612	0.201545	0.193092	0.136525	0.065687	0.210867	0.35833
0.791633	0.954189	0.273619	0.202318	0.289786	0.223228	0.90355
0.944187	0.023396	0.924546	0.568281	0.035168	0.396006	0.50074
0.592827	0.103347	0.918322	0.790217	0.777163	0.145397	0.17460
0.825177	0.147107	0.388373	0.546681	0.335123	0.890333	0.29163
0.454211	0.776412	0.051564	0.445526	0.690789	0.348568	0.57100
0.375630	0.953717	0.214655	0.831832			

COMPUTER EXERCISES USING THE DATABASE

Exercise 1—Appendix H

Select 100 observations at random from the database and use a convenient statistical computer package to estimate the mean and standard deviation of the variable HPAYRENT (house payments or apartment/house rents). Find the percentage of the observations between $\bar{x} \pm s$, $\bar{x} \pm 2s$, and $\bar{x} \pm 3s$. Comment on whether these percentages support the conclusion that the data come from a normally distributed population.

Exercise 2—Appendix H

Select 150 observations at random from the database and, with reference to the variable OWNORENT, calculate the proportion of those observations that indicate the house is owned rather than rented. If a random sample of 20 observations

were chosen from this set of 150 observations with replacement, what is the probability that more than half of the homes in the sample of 20 are owned by households? Is the normal approximation appropriate for this situation?

Exercise 3—Appendix I

Select 100 observations at random from the database on the variable EMPLOYEES (number of employees). Use a convenient statistical computer package to construct a histogram. What type of distribution does the histogram approximate? Normal? Uniform? Exponential? None of these?

Exercise 4—Appendix I

Repeat Exercise 3 using the variable SALES.

CASE STUDY

WHAT DOES IT
MEAN WHEN A
NORMAL CURVE
SHIFTS OF
FLATTENS OUT?

R&D magazine conducts an annual survey of salaries of professionals in the research and development field. According to editor Robert R. Jones, in most recent years the frequency histogram of salary levels produced "a fairly nice bell curve, with a distinct peak and almost-matching halves," Each year, as general salary levels were increased to keep pace with inflation, the bell shape shifted a little to the right, toward the higher-income side, but essentially retaining its shape.

A trend that had been developing over the past 3 years became especially noticeable after the 1988 *R&D* opinion poll of 1900 respondents. For the third year in a row, the peak of the curve has not shifted to the left or right. The salary range of $40,000 to $44,999 has represented the modal class, but the size of this group has shrunk from 16.5% of the total respondents in 1986 to 14.5% in 1988. Expressed graphically, the height of the plotted peak and the peak's shoulders also are lower, as if someone were pressing down on the hump from above.

Do you think this would cause the curve to flatten out? Well, yes

and no. According to *R&D* magazine, those who were "pushed out" from under the peak are now to be found along a flattened right-hand slope, representing the higher-income portion of the curve. In 1986, 60.8% of *R&D* workers fell into the salary range of $35,000 to $59,999. In 1988, the figure for this group declined to 58.7%. Over the same period, those in the group receiving a salary above $59,999 increased from 8.7% to 14.4%, even while the percentage earning above $80,000 remained the same for 1987 and 1988.

Consider the histogram in Figure 6.38. Salary is a continuous variable, but since the chart shows 3 years side by side, gaps have been left between the bars for the sake of readability. The shape seems to follow a normal distribution. The *R&D* article did not provide the mean and standard deviation. This is typical of much data published in magazines and newspapers, so we have to learn how to come up with a good estimate ourselves. Statistics is not only the science of numbers, but also the art of approximation.

Case Study Questions

1. Which class contains the median? Would the midpoint of this class be a good approximation of the mean? Could the upper class limit be a better approximation?

2. Estimate the percentages in each class from the above preceding discussion and from Figure 6.38. Use this to estimate the standard deviation. Your answer may differ from someone else's, depending on how you argue your case.

3. Assume a mean of $45,000 and a standard deviation of $10,000 for the year 1988.

   a. What should be the proportion of individuals in the *R&D* field earning $85,000 or more? How well does this agree with Figure 6.38?

   b. What is the probability that an *R&D* professional selected at random earns between $30,000 and $49,999 annually? Again, how well does this agree with Figure 6.38?

4. Use the concept of skewness (discussed in Chapter 3) to explain the differences between the theoretical proportions indicated by the ideal normal

CASE STUDY

FIGURE 6.38 Salary levels for all R&D (3 years compared).

Salary levels for all R&D (3 years compared)

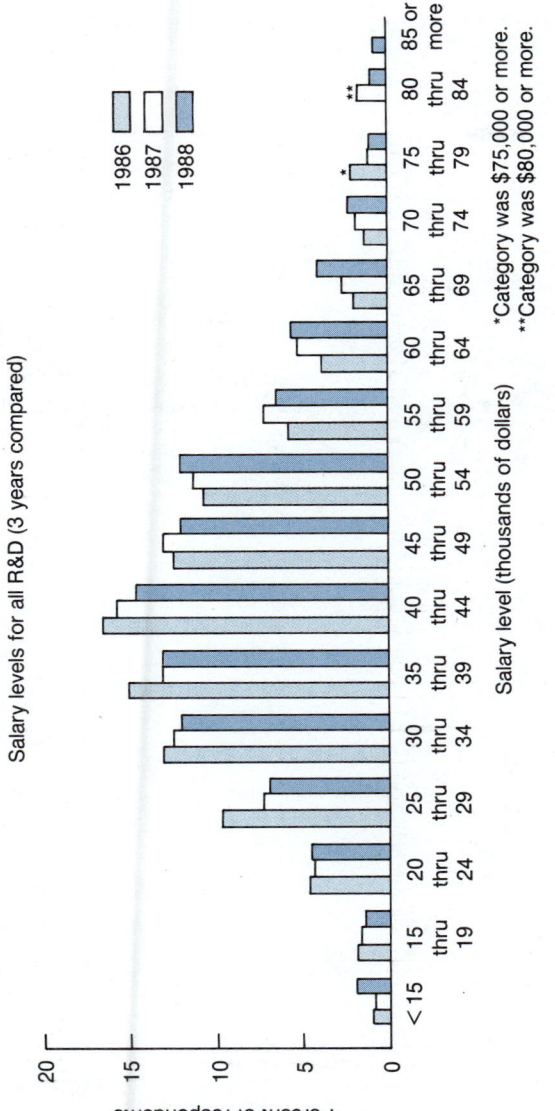

*Category was $75,000 or more.
**Category was $80,000 or more.

curve, and those indicated by Figure 6.38.

5. *R&D* began their article with the words: "The *R&D* salary bell is going flat. This is not to

say that it makes a sour note. In fact, there are many people . . . . who find it especially attractive this way." To what was the editor referring?

(*Source:* Robert R. Jones, editor, *Research & Development* (March 1988): 57–60.)

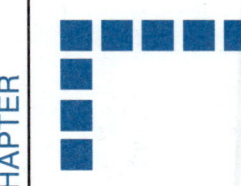

Statistical Inference and Sampling

A Look Back/Introduction

The previous three chapters laid the foundation for using statistical methods in decision making. Any such decision will have uncertainty associated with it, but we can attempt to measure this uncertain outcome using a probability. Random variables (both discrete and continuous) allow you conveniently to represent certain outcomes of an experiment and their corresponding probabilities. If the experiment fits a particular discrete situation (such as binomial), you can easily determine the probability of certain events or determine the mean (average) value of the related distribution.

If the random variable of interest is continuous, you can make probability statements after assuming the probability distribution involved (such as normal, exponential, uniform, or others not discussed). Both discrete and continuous random variables come into play in all areas of decision making. They allow us to make decisions concerning a large population using the information contained in a much smaller sample.

This is the area of **statistical inference**, which this chapter introduces by demonstrating how to estimate something about the population (such as the average value, μ) by using the corresponding value from a sample (such as the sample average, $\bar{X}$). Recall that μ (belonging to the population) is a parameter and $\bar{X}$ (belonging to the sample) is a statistic. When dealing with a normal population, for example, what does one do if the population mean, μ, is unknown? So far in the text, this value has been specified for you. In this chapter, we discuss methods of estimating population parameters using sample statistics along with several methods of gathering your sample data.

7.1

RANDOM SAMPLING AND THE DISTRIBU-TION OF THE SAMPLE MEAN

In Chapter 3, you learned how to calculate the mean of a sample, $\bar{X}$. This sample is drawn from a population having a particular distribution, such as normal, exponential, or uniform. If you were to obtain another sample (you probably will not, as most decisions are made from just one sample), would you get the same value of $\bar{X}$? Assuming that the new sample was made up of different individuals than was the first sample, then almost certainly the two $\bar{X}$'s would not be the same. So, $\bar{X}$ itself is a random variable. We will demonstrate that if a sample is large enough, $\bar{X}$ is very nearly *normally* distributed regardless of the shape of the sampled population. That is, if you were to obtain many samples, calculate the resulting $\bar{X}$'s, and then make a histogram of these $\bar{X}$'s, this histogram would always approximately resemble a bell-shaped (normal) curve.

Simple Random Samples

In Chapter 4, the concept of a simple random sample was introduced. The mechanics of obtaining a random sample range from drawing names out of a hat to using a computer to generate lists of random numbers. For extremely large populations, one is often forced to select individuals (elements) from the population in a *nearly* random manner.

The underlying assumption behind a random sample of size n is that any sample of size n has the same chance (probability) of being selected. To be completely assured of obtaining a random sample from a *finite* population, you should number the members of the population from 1 to N (the population size) and, using a set of n random numbers, select the corresponding sample of n population elements for your sample.

This procedure was described in Chapter 4 and is often used in practice, particularly when you have a sampling situation that needs to be legally defensible. Such is the case in many statistical audits. However, for situations in which the population is extremely large, this strategy may be impractical, and instead you can use a sampling plan that is nearly random. Several other sampling procedures are discussed in the last section of this chapter.

The main point of all this lengthy discussion is that practically all the procedures presented in subsequent chapters relating to decision making and estimation assume that you are using a random sample. In the chapters that follow, the word *sample* will mean *simple random sample*.

Estimation

The idea behind statistical inference has two components:

1. The *population* consists of everyone of interest. By "everyone" we mean all people, machine parts, daily sales, or whatever else you are interested in measuring or observing. The mean value (for example, average height, average income) of everyone in this population is μ and generally is not known.

2. The *sample* is randomly drawn from this population. Elements of the sample thus are part of the population—but certainly not all of it. The exception to this is a *census*, a sample that consists of the entire population.

The sample values should be selected randomly, one at a time, from the entire population. Figure 7.1 emphasizes our central point—namely, an unknown population **parameter** (such as μ = the mean value for the entire population) can be **estimated** using the corresponding sample **statistic** (such as $\bar{X}$ = the mean of your sample).

FIGURE 7.1

The sample mean, $\bar{X}$, is used to estimate the population mean, μ. In general, sample statistics are used to estimate population parameters.

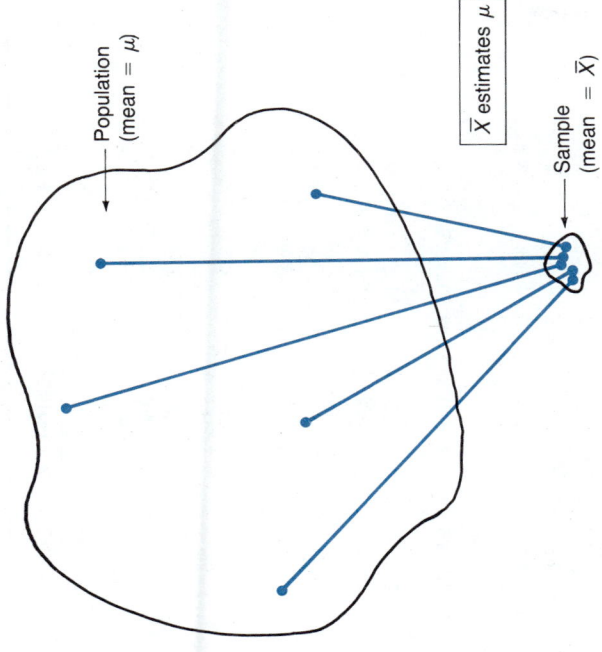

Population
(mean = μ)

$\bar{X}$ estimates μ

Sample
(mean = $\bar{X}$)

It makes sense, doesn't it? It would be most desirable to know the average value for everyone in the population, but in practice this is nearly always impossible. It may take too much time or money, we may not be able to obtain values for them all even if we want to, or the process of measuring the individual items may destroy them (such as measuring the lifetime of a light bulb). In many instances, estimating the population value using a sample estimate is the best we can do.

EXAMPLE 7.1

In Chapter 6, the monthly account balance of bank card accounts at a midwestern bank was assumed to follow a normal distribution with a mean of $\mu = \$650$ and a standard deviation of $\sigma = \$100$. There is no way of *knowing* that $\mu = \$650$ unless all such accounts are examined. Assume that

$$X = \text{monthly account balance}$$

is a normal random variable, but do not assume anything about the mean and standard deviation. Ignoring the standard deviation, estimating μ involves obtaining a random sample of accounts and recording their balances. Suppose you obtain a sample of size $n = 10$, with the following results, in dollars:

535.65, 641.67, 663.18, 512.40, 600.51, 784.92, 587.37, 650.14, 572.21, 725.35

What is the estimate of μ, based on these values?

SOLUTION

The sample mean would be $\bar{x} = \$627.34$. Thus, based on ten sample values, our best estimate of μ is $\bar{x} = \$627.34$.* ■

Distribution of $\bar{X}$

Referring to Example 7.1 the value of $\bar{X}$ would almost certainly change if you were to obtain another sample. The question of interest here is, if we *were* to obtain many values of $\bar{X}$, how would they behave? If we observed values of $\bar{X}$

* The notation $\hat{\mu}$ is commonly used (in place of $\bar{x}$) to denote an *estimate* of μ. For this example, the estimate of μ is $\hat{\mu} = \$627.34$.

TABLE 7.1
20 Samples of 10
Everglo Bulbs

SAMPLE 1	SAMPLE 2	SAMPLE 3	SAMPLE 4	SAMPLE 5
308	431	416	373	354
419	448	361	451	385
389	380	389	329	449
432	371	497	460	419
362	387	400	481	483
302	410	489	350	396
440	400	406	431	317
430	426	333	356	457
375	381	307	410	404
383	361	375	353	480
$\bar{X} = 384.0$	$\bar{X} = 399.5$	$\bar{X} = 397.3$	$\bar{X} = 399.4$	$\bar{X} = 414.4$
$s = 49.30$	$s = 28.54$	$s = 60.51$	$s = 53.99$	$s = 54.25$

SAMPLE 6	SAMPLE 7	SAMPLE 8	SAMPLE 9	SAMPLE 10
404	372	449	403	354
390	404	389	350	446
390	493	397	565	343
454	344	428	354	458
386	396	374	358	404
385	441	502	412	468
384	373	365	441	416
351	438	402	340	340
392	360	416	359	409
396	367	316	446	408
$\bar{X} = 393.2$	$\bar{X} = 398.8$	$\bar{X} = 403.8$	$\bar{X} = 402.8$	$\bar{X} = 404.6$
$s = 25.45$	$s = 46.10$	$s = 50.32$	$s = 68.93$	$s = 46.28$

SAMPLE 11	SAMPLE 12	SAMPLE 13	SAMPLE 14	SAMPLE 15
329	429	461	448	457
473	286	399	386	432
336	382	416	375	425
356	380	378	488	391
385	423	359	447	429
365	388	408	429	448
419	329	393	377	416
448	438	374	380	429
459	423	440	372	414
449	378	454	408	315
$\bar{X} = 401.9$	$\bar{X} = 385.6$	$\bar{X} = 408.2$	$\bar{X} = 411.0$	$\bar{X} = 415.6$
$s = 54.12$	$s = 47.91$	$s = 34.60$	$s = 40.12$	$s = 39.73$

SAMPLE 16	SAMPLE 17	SAMPLE 18	SAMPLE 19	SAMPLE 20
491	439	331	418	428
353	336	427	422	368
375	425	445	341	445
536	419	420	485	429
447	346	401	442	475
415	408	389	470	437
322	392	363	404	475
350	409	439	370	458
453	313	352	539	308
343	334	346	435	408
$\bar{X} = 408.5$	$\bar{X} = 382.1$	$\bar{X} = 391.3$	$\bar{X} = 432.6$	$\bar{X} = 423.1$
$s = 71.46$	$s = 45.28$	$s = 41.35$	$s = 56.78$	$s = 51.48$

FIGURE 7.2
Assumed distribution of
Everglo bulbs.

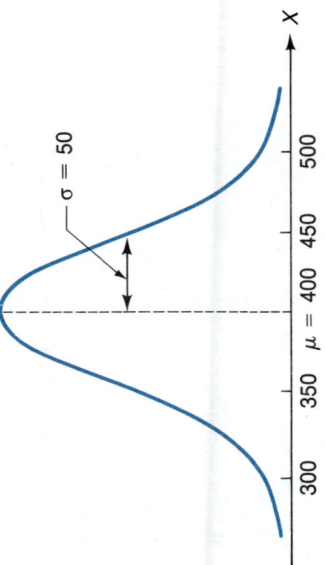

indefinitely, where would they center; that is, what is the **mean** of the distribution for the random variable, $\bar{X}$? Is the variation of the $\bar{X}$ values more, less, or the same as the variation of individual observations? This is measured by the **standard deviation** of the distribution for $\bar{X}$.

In Example 6.2, it was assumed that the average lifetime of an Everglo light bulb was $\mu = 400$ hours, with a population standard deviation of $\sigma = 50$ hours. This does not imply that if you obtain a random sample of these bulbs, the resulting sample mean, $\bar{X}$, always will be 400. Rather, a little head scratching should convince you that $\bar{X}$ will not be exactly 400, but $\bar{X}$ should be *approximately* 400.

Twenty samples of 10 bulbs each and the calculated $\bar{X}$ for each sample are shown in Table 7.1. We will assume for now that the population parameters are $\mu = 400$ hours and $\sigma = 50$ hours (Figure 7.2).

The 20 values of $\bar{X}$ are:

384.0, 399.5, 397.3, 399.4, 414.4, 393.2, 398.8, 403.8, 402.8, 404.6, 401.9, 385.6, 408.2, 411.0, 415.6, 408.5, 382.1, 391.3, 428.6, 423.1

They are not each 400, but they are all close to 400. Using a calculator or computer, you would also find that (1) the average (mean) of these 20 values is 402.88 (this is close to $\mu = 400$) and (2) the standard deviation of these 20 values is 12.78 (this is *much smaller* than $\sigma = 50$).

The $\bar{X}$ values appear to be centered at $\mu = 400$ hours but have *much less variation* than the individual observations in each of the samples. A histogram of these 20 values generated by MINITAB is contained in Figure 7.3. Based on the shape of this histogram, it seems reasonable to assume that the values

FIGURE 7.3
Histogram of 20 sample
means generated by
MINITAB. Compare with
Figure 7.2.

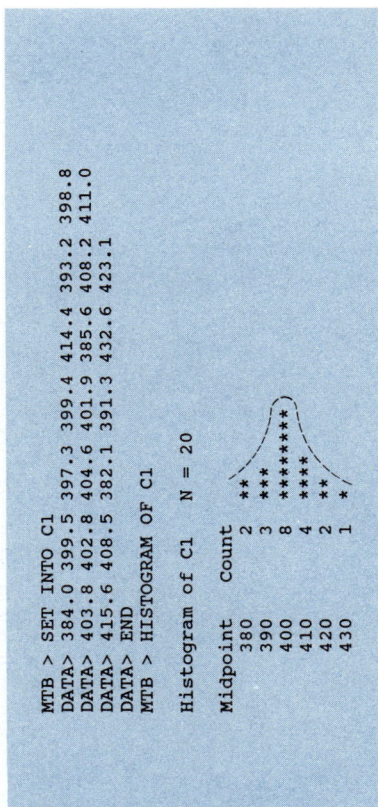

of $\bar{X}$ follow a normal distribution, but one that is much *narrower* than the population of individual lifetimes in Figure 7.2.

7.2 THE CENTRAL LIMIT THEOREM

Our last example illustrates a useful result, the Central Limit Theorem.

CENTRAL LIMIT THEOREM

When using a random sample of size n from a population with mean μ and standard deviation σ, the resulting sample mean, $\bar{X}$, has a normal distribution with mean μ and standard deviation $\sigma/\sqrt{n}$. This is true for any sample size, n, if the underlying population is normally distributed, and it is approximately true for large sample sizes (generally $n > 30$) obtained from any population.

In other words, the distribution of all possible $\bar{X}$ values has an exact or approximate normal distribution with mean μ and standard deviation $\sigma/\sqrt{n}$.

COMMENTS

more variation in dispersion, greater the S.D
AVG of all $\bar{X}$ = M

The second part of the Central Limit Theorem is an extremely strong result; it says that you can assume that $\bar{X}$ follows an approximate normal distribution *regardless* of the shape of the population from which the sample was obtained if the sample size (n) is large. For example, if you repeatedly sampled from a population with an exponential distribution, the resulting $\bar{X}$'s would follow a *normal* (not an exponential) curve.

In Table 7.1, 20 samples of size ten were obtained and the corresponding values of $\bar{X}$ were determined. Suppose samples of size ten were obtained *indefinitely* and we wished to describe the shape of the resulting $\bar{X}$'s. According to the Central Limit Theorem, $\bar{X}$ will be a normal random variable. We are assuming that the individual lifetimes follow a normal curve (see Figure 7.2), so this will be true for any sample size—in particular, $n = 10$. So the resulting $\bar{X}$'s will describe a normal curve similar to the curve in Figure 7.3.

Where is the curve centered? According to the Central Limit Theorem, the mean of this normal random variable is the *same* as that in Figure 7.2; that is, it is the mean of the population from which you are sampling. This value is $\mu = 400$, and so, on the average, the value of $\bar{X}$ is $\mu_{\bar{X}} = 400$ hours. Notice that the average of the 20 values of $\bar{X}$ that we did observe was 402.88. This value will get closer to, or **tend toward**, 400 as we take more samples of size ten.

What is the standard deviation of the normal curve for $\bar{X}$? As we noted earlier, the 20 values of $\bar{X}$ jump around (vary) much less than do the individual observations in each of the samples. Consequently, the standard deviation of the $\bar{X}$ normal curve will be much less than that of the population curve (describing individual lifetimes) in Figure 7.2. In fact, according to the Central Limit Theorem, this will be

$$\sigma_{\bar{X}} = \frac{\sigma}{\sqrt{n}} \qquad \textbf{(7.1)}$$

where σ is the standard deviation of the population ($\sigma = 50$ in Figure 7.2). Consequently,

$$\sigma_{\bar{X}} = \frac{50}{\sqrt{10}} = 15.81$$

FIGURE 7.4

Normal curves for
population and sample
mean.

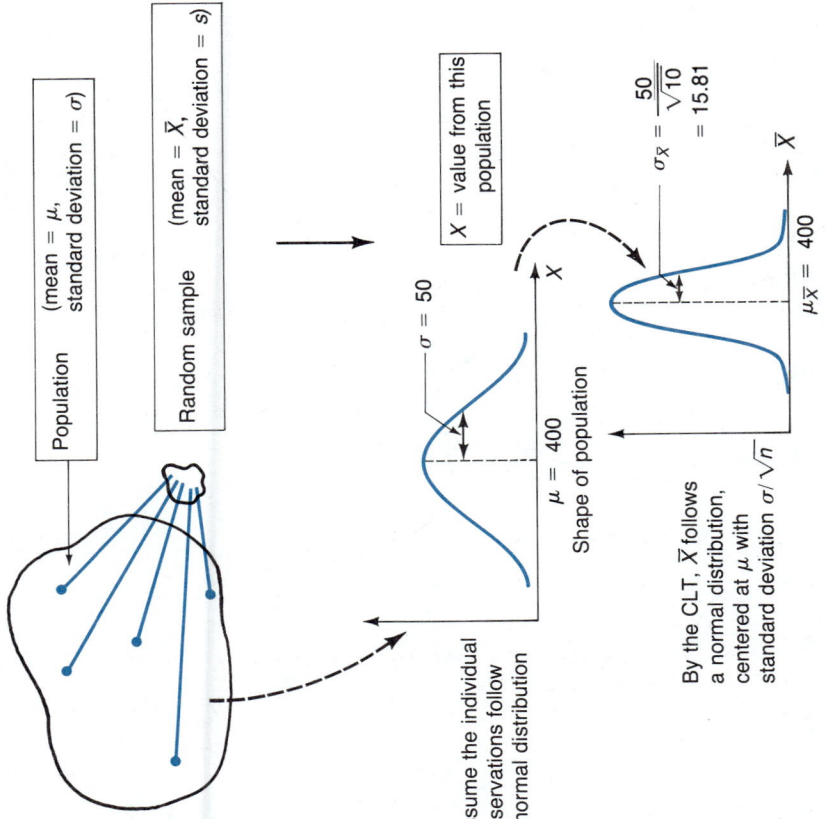

Recall that the standard deviation of the 20 observed $\bar{X}$ values was 12.78. This value will tend toward 15.81 if we take more samples of size ten. These results are summarized in Figure 7.4, where $\mu_{\bar{x}} = 400$ and $\sigma_{\bar{x}} = 15.81$.

Basically, the Central Limit Theorem says that the normal curve (distribution) for $\bar{X}$ is centered at the same value as the population distribution but has a much smaller standard deviation. Notice that as the sample size, n, increases, $\sigma/\sqrt{n}$ decreases, and so the spread relative to the mean of the $\bar{X}$ curve (that is, the variation in the $\bar{X}$ values) decreases. If we repeatedly obtained samples of size 100 (rather than 10), the corresponding $\bar{X}$ values would lie even closer to μ. If $n = 400$ because now $\sigma_{\bar{x}}$ would equal $50/\sqrt{100} = 5$. This is illustrated in Figure 7.5.

For the 20 values of $\bar{X}$ in Table 7.1, it was assumed that the population mean was *known* to be $\mu = 400$, so each of the $\bar{X}$ values estimates μ with a certain amount of error. The more variation in the $\bar{X}$ values, the more error we encounter using $\bar{X}$ as an estimate of μ. Consequently, the standard deviation of $\bar{X}$ also serves as a measure of the error that will be encountered using a sample mean to estimate a population mean. The standard deviation of the $\bar{X}$ distribution is often referred to as the **standard error** of $\bar{X}$.

Standard error of $\bar{X}$ = standard deviation of the probability distribution for $\bar{X}$

$$= \frac{\sigma}{\sqrt{n}}$$

FIGURE 7.5
Normal curves for the sample mean ($n = 10$, 20, 50, 100).

Population

$\mu = 400$

$\sigma = 50$

X

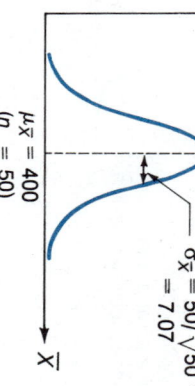

$\mu_{\bar{X}} = 400$
($n = 10$)

$\sigma_{\bar{X}} = 50/\sqrt{10}$
$= 15.81$

$\bar{X}$

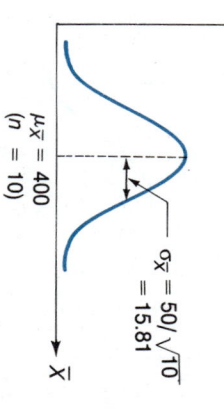

$\mu_{\bar{X}} = 400$
($n = 20$)

$\sigma_{\bar{X}} = 50/\sqrt{20}$
$= 11.18$

$\bar{X}$

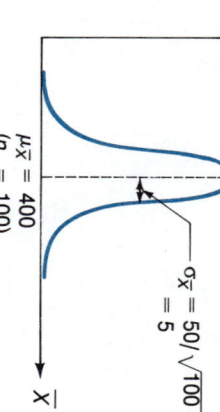

$\mu_{\bar{X}} = 400$
($n = 50$)

$\sigma_{\bar{X}} = 50/\sqrt{50}$
$= 7.07$

$\bar{X}$

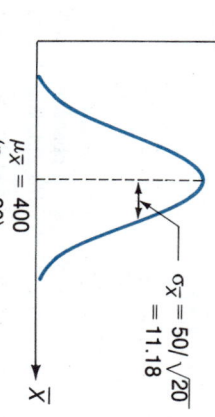

$\mu_{\bar{X}} = 400$
($n = 100$)

$\sigma_{\bar{X}} = 50/\sqrt{100}$
$= 5$

$\bar{X}$

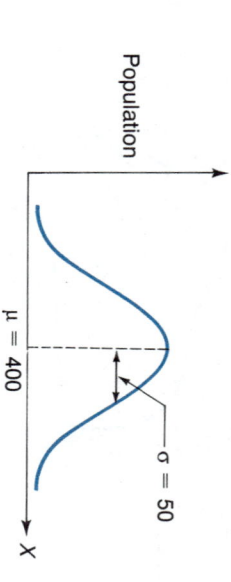

EXAMPLE 7.2

Electricalc has determined that the assembly time for a particular electrical component is normally distributed with a mean of 20 minutes and a standard deviation of 3 minutes.

1. What is the probability that an employee in the assembly division takes longer than 22 minutes to assemble one of these components?

2. What is the probability that the average assembly time for 15 such employees exceeds 22 minutes?

3. What is the probability that the average assembly time for 15 employees is between 19 and 21 minutes?

SOLUTION 1

The random variable X here is the assembly time for a component. This was assumed to be a normal random variable, with $\mu = 20$ minutes and $\sigma = 3$ minutes (Figure 7.6). We wish to determine $P(X > 22)$. Standardizing this variable and using Table A.4, we obtain

$$P(X > 22) = P\left[\frac{X - 20}{3} > \frac{22 - 20}{3}\right]$$

$$= P(Z > .67)$$

$$= .5 - .2486 = .2514$$

Therefore, a randomly chosen employee will require longer than 22 minutes to assemble the component with probability .25.

SOLUTION 2

Figure 7.6 does *not* apply to this question because we are concerned with the

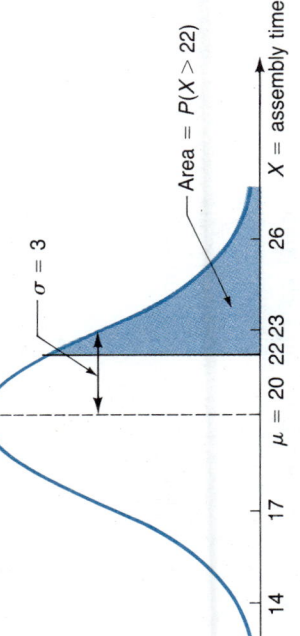

FIGURE 7.6

Assembly time for electrical components. See Example 7.2.

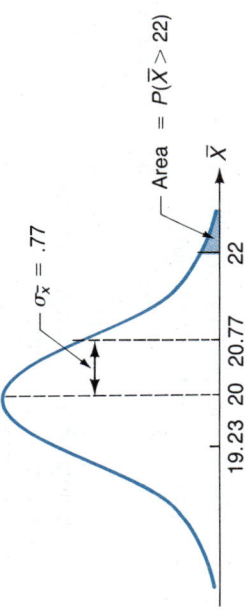

FIGURE 7.7

Curve for $\bar{X}$ = average of 15 employees' assembly times. Shaded area shows $P(\bar{X} > 22)$.

average time for 15 employees, not an individual employee. Using the Central Limit Theorem, we know that the curve describing $\bar{X}$ (an average of 15 employees) is normal with

$$\text{mean} = \mu_{\bar{X}} = \mu = 20 \text{ minutes}$$

$$\text{standard deviation (standard error)} = \sigma_{\bar{X}} = \sigma/\sqrt{n}$$

$$= 3/\sqrt{15} = .77 \text{ minutes}$$

(See Figure 7.7.)

The procedure is the same as in Solution 1, except now the standard deviation of this curve is .77 rather than 3:

$$P(\bar{X} > 22) = P\left[\frac{\bar{X} - 20}{.77} > \frac{22 - 20}{.77}\right]$$

$$= P(Z > 2.60)$$

$$= .5 - .4953 = .0047$$

So an average assembly time for a sample of 15 employees will be more than 22 minutes with less than 1% probability; that is, it is very unlikely that an average of 15 assembly times will exceed 22 minutes.

SOLUTION 3

The curve for this solution is shown in Figure 7.8. We wish to find $P(19 < \bar{X} < 21)$.

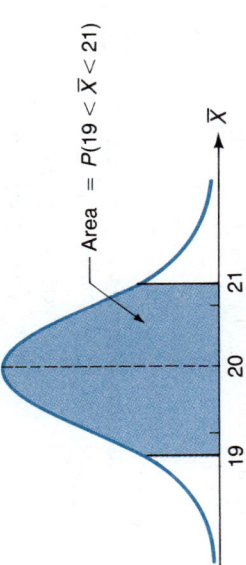

FIGURE 7.8

Curve for average assembly time of 15 employees. Shaded area shows $P(19 < \bar{X} < 21)$.

$$P(19 < \bar{X} < 21) = P\left[\frac{19 - 20}{.77} < \frac{\bar{X} - 20}{.77} < \frac{21 - 20}{.77}\right]$$
$$= P(-1.30 < Z < 1.30)$$
$$= .4032 + .4032 = .8064$$

Thus, a sample of 15 employees will produce an average assembly time between 19 and 21 minutes with probability about .81. ∎

EXAMPLE 7.3

The price-earnings (P-E) ratio of a stock is usually considered by analysts who put together financial portfolios. Suppose a population of all P-E ratios has a mean of 10.5 and a standard deviation of 4.5.

1. What is the probability that a sample of 40 stocks will have an average P-E ratio less than nine?

2. What assumptions are necessary about the population of all P-E ratios in your answer to Question 1?

SOLUTION 1

By the Central Limit Theorem, $\bar{X}$ is approximately a normal random variable with mean $= \mu = 10.5$ and standard deviation $= \sigma/\sqrt{n} = 4.5/\sqrt{40} = .71$. So

$$Z = \frac{\bar{X} - 10.5}{.71}$$

is approximately a standard normal random variable, and consequently

$$P(\bar{X} < 9) = P\left[\frac{\bar{X} - 10.5}{.71} < \frac{9 - 10.5}{.71}\right]$$
$$= P(Z < -2.11) = .0174$$

SOLUTION 2

No assumptions regarding the shape of the P-E ratio population are necessary. This population may be normal or it may not be—it simply does not matter because we are using a fairly large sample ($n = 40$). The distribution of $\bar{X}$ is approximately normal, regardless of the shape of the population of all P-E ratios. Our only assumptions in Solution 1 were that $\mu = 10.5$ and $\sigma = 4.5$. ∎

EXERCISES

7.1 Let $\bar{X}$ be the average of a sample of size 18 from a normally distributed population with mean 37 and variance 16. Find the following probabilities.

a. $P(\bar{X} \leq 35)$
b. $P(\bar{X} \geq 38)$
c. $P(34 \leq \bar{X} \leq 36.5)$
d. $P(36 \leq \bar{X} \leq 38)$

7.2 The manager of Homer and Gordon Realty finds that their four realtors have sold 0, 1, 3, and 4 homes, respectively, in the past month.

a. List the number of homes sold by two realtors selected randomly with replacement for all possible samples of size two.

b. Calculate the sample mean for each sample of size two. Construct the probability distribution for the sample mean.

In Example 7.2, it was assumed that the individual assembly times followed a normal distribution. However, remember that the strength of the Central Limit Theorem is that this assumption is not necessary for large samples. We can answer Questions 2 and 3 for *any* population whose mean is 20 minutes and standard deviation is 3 minutes, provided we take a *large* sample ($n > 30$). In this case, the normal distribution of $\bar{X}$ is not exact, but it provides a very good approximation.

c. Draw a histogram showing the distribution of the sample mean ($\bar{X}$).

7.3 A southwestern bank issues traveler's checks in denominations of $10, $20, $50, $100, and $500. All five amounts have occurred with equal probability.

a. List all possible samples of three from these five denominations. (Denominations may not be repeated.)

b. Calculate the sample mean for each sample of size three.

c. Construct the probability distribution of the sample mean.

d. Draw a histogram showing the distribution of the sample mean ($\bar{X}$).

7.4 Five machines produce electronic components. The number of components produced per hour is normally distributed with a mean of 25 and a standard deviation of 4.

a. What percentage of the time does a machine produce more than 27 components per hour?

b. What percentage of the time is the average rate of output of the five machines more than 27 components per hour?

7.5 A fireworks manufacturer has found that the height reached by his rockets follows a normal distribution, with a mean of 200 feet and a standard deviation of 12 feet.

a. What percentage of the individual rockets achieve a height exceeding 210 feet?

b. What percentage of the time will groups of one dozen rockets have an average height exceeding 210 feet?

7.6 The average length of actual running time (excluding advertisements) for television feature films is 1 hour and 40 minutes, with a standard deviation of 15 minutes. If a sample of 49 TV feature films is taken at random, what is the probability that the average running time for this group is 1 hour and 45 minutes or more?

7.7 In an article in *Business Horizons*, it is reported that although 4 of the 800 leading Chief Executive Officers (CEOs) in the United States are under 40 years of age, the median age is a mature 59.

(*Source:* Louis E. Boone and David L. Kurtz, "CEOs: A Group Profile," *Business Horizons* 31, no. 4 (July–August, 1988): 38–42.)

a. Assume that the distribution of ages of CEOs is approximately normally distributed with a mean of 59 and a standard deviation of 6 years. Find the probability that the mean age of 35 CEOs chosen at random is between 57 and 60 years of age.

b. Find the probability in part (a), assuming that the standard deviation of the ages is 8 years instead of 6 years.

7.8 In a study to determine the annualized return on stocks that have significant repurchase programs, it was found that over the period 1974–1983, the average annualized return for 600 companies that engaged in significant repurchase programs was 22.6%. Over the same period the Standard and Poor's 500's return averaged only 14.1%. Assume that an investor chose a random sample of 35 stocks from the group of 600 companies engaged in repurchasing and that these stocks were held from 1974 to 1983.

(*Source:* "The Stock Market Reaction To Significant Share Repurchases," by W. Davidson, S. Garrison, and D. Worrell, *Proceedings of the Decision Sciences Institute* (1986): 194–6.)

a. Is it possible that some of the stocks with repurchasing programs could have showed a loss over this time period? Explain.

b. For the sample of 35 stocks chosen by the investor, what is the probability that the investor's annualized return will be greater than 18%? Assume a standard deviation of 9% for the return from the 600 companies engaged in repurchasing programs.

c. Assume that the standard deviation is 25% for the return from the 600 companies engaged in repurchasing programs. Answer part (b) using this standard deviation.

Applying the Central Limit Theorem to Normal Populations

The Central Limit Theorem tells us that $\bar{X}$ tends toward a normal distribution as the sample size increases. If you are dealing with a population that has an assumed normal distribution (as in Example 7.2), then $\bar{X}$ is normal regardless of the sample size. However, as the sample size increases, the variability of $\bar{X}$ decreases, as is illustrated in Figure 7.5.

This means that for large sample sizes, if you were to get many samples and corresponding values of $\bar{X}$, these values of $\bar{X}$ would be more concentrated around the middle, with very few extremely large or extremely small values.

Look at Figure 7.5, which illustrates the assumed normal distribution of all Everglo bulbs. We know that (using Table A.4) 95% of a normal curve is contained within 1.96 standard deviations of the mean. For a sample size of $n = 10$ from a normal population with $\mu = 400$ and $\sigma = 50$, $\sigma_{\bar{X}} = 15.81$. Now,

$$\mu_{\bar{X}} + 1.96\sigma_{\bar{X}} = 400 + 1.96(15.81) = 431.0$$

and

$$\mu_{\bar{X}} - 1.96\sigma_{\bar{X}} = 400 - 1.96(15.81) = 369.0$$

Thus, if we repeatedly obtain samples of size ten, 95% of the resulting $\bar{X}$ values will lie between 369.0 and 431.0.

This result and the corresponding results using $n = 20$, 50, and 100 are contained in Table 7.2. This reemphasizes that for larger samples, you are much more likely to get a value of $\bar{X}$ that is close to $\mu = 400$. In practice, you typically do not know the value of μ. However, by using a larger sample size, you are more apt to obtain an $\bar{X}$ that is a good estimate of the unknown μ.

Applying the Central Limit Theorem to Nonnormal Populations

The real strength of the Central Limit Theorem is that $\bar{X}$ will tend toward a normal random variable regardless of the shape of your population. You need a large sample ($n > 30$) to obtain a nearly normal distribution for $\bar{X}$. The Central Limit Theorem also holds when sampling from a discrete population.

Figures 7.9, 7.10, and 7.11 illustrate the distribution of $\bar{X}$ for three nonnormal populations. Notice that the uniform population (Figure 7.9) is at least symmetric about the mean, so the distribution of the sample mean, $\bar{X}$, tends toward a normal distribution for much smaller sample sizes. The U-shaped distribution (Figure 7.11) is another continuous distribution. It is characterized by many small and large values, with few values in the middle. This distribution is symmetric about the mean, but its shape is opposite to that of a normal distribution. Here, $\bar{X}$ requires a large sample ($n > 30$) to attain a normal distribution.

TABLE 7.2

Sampling from a normal population with $\mu = 400$ and $\sigma = 50$; 95% of the time, the value of $\bar{X}$ will be between $\mu_{\bar{X}} - 1.96\sigma_{\bar{X}}$ and $\mu_{\bar{X}} + 1.96\sigma_{\bar{X}}$. Refer to Figure 7.5 for the values of $\sigma_{\bar{X}}$.

SAMPLE SIZE	$\sigma_{\bar{X}}$	$\mu_{\bar{X}} - 1.96\sigma_{\bar{X}}$	$\mu_{\bar{X}} + 1.96\sigma_{\bar{X}}$	CONCLUSION
$n = 10$	15.81	369.0	431.0	95% of the time, the value of $\bar{X}$ will be between 369.0 and 431.0
$n = 20$	11.18	378.1	421.9	95% of the time, the value of $\bar{X}$ will be between 378.1 and 421.9
$n = 50$	7.07	386.1	413.9	95% of the time, the value of $\bar{X}$ will be between 386.1 and 413.9
$n = 100$	5	390.2	409.8	95% of the time, the value of $\bar{X}$ will be between 390.2 and 409.8

FIGURE 7.9

Distribution of $\bar{X}$ for a uniform population.

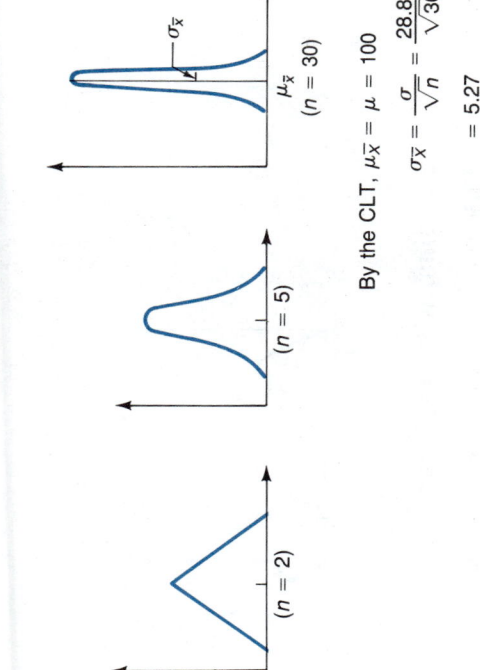

$$\mu = \frac{a+b}{2} = 100$$

$$\sigma = \frac{b-a}{\sqrt{12}} = 28.87$$

$a = 50 \quad \mu = 100 \quad b = 150$

Uniform population

$(n = 2)$

$(n = 5)$

$\mu_{\bar{x}}$ $(n = 30)$

$\sigma_{\bar{x}}$

By the CLT, $\mu_{\bar{x}} = \mu = 100$

$$\sigma_{\bar{x}} = \frac{\sigma}{\sqrt{n}} = \frac{28.87}{\sqrt{30}}$$

$$= 5.27$$

FIGURE 7.10

Distribution of $\bar{X}$ for an exponential population.

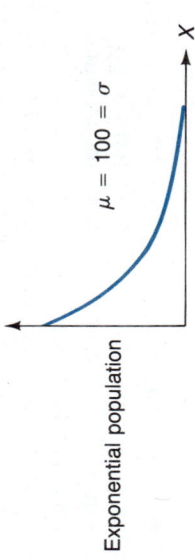
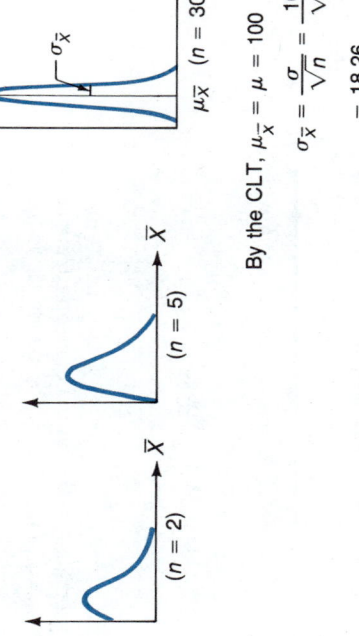

$\mu = 100 = \sigma$

Exponential population

X

$(n = 2)$

$\bar{X}$

$(n = 5)$

$\bar{X}$

$\mu_{\bar{X}}$ $(n = 30)$

$\sigma_{\bar{x}}$

$\bar{X}$

By the CLT, $\mu_{\bar{x}} = \mu = 100$

$$\sigma_{\bar{x}} = \frac{\sigma}{\sqrt{n}} = \frac{100}{\sqrt{30}}$$

$$= 18.26$$

Sampling from a Finite Population

In the previous discussion, we assumed that the population was large enough that the sample was extremely small by comparison. We will now consider whether our results, including the Central Limit Theorem, apply when the exact size of the population is known and the sample is a large portion of the population.

FIGURE 7.11

Distribution of $\bar{X}$ for a U-shaped population.

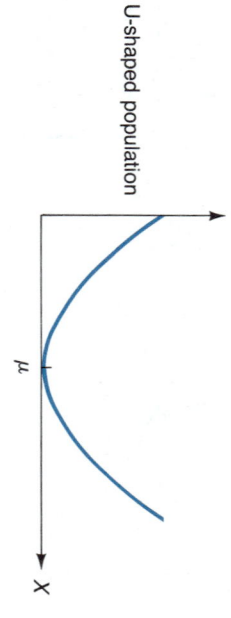

U-shaped population

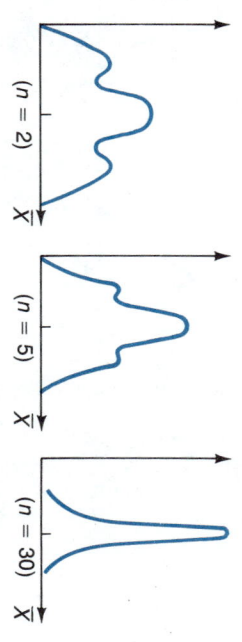

($n = 2$)   ($n = 5$)   ($n = 30$)

Sampling with Replacement

When you return each element of the sample to the population before taking the next sample element, you are sampling with replacement. This is not a common sampling procedure; people generally obtain their sample all at once, which makes it impossible to sample with replacement. When sampling with replacement, it is possible to obtain the same element more than once. When sampling with replacement, the same person could be chosen all three times in a sample of size $n = 3$. For example, the same person could be chosen all three times in a sample of size $n = 3$.

When sampling with replacement, the Central Limit Theorem applies exactly as before, without any adjustments necessary.

CENTRAL LIMIT THEOREM: SAMPLING WITH REPLACEMENT FROM A FINITE POPULATION

When sampling with replacement from a finite population with mean μ and standard deviation σ, the sample mean $\bar{X}$ tends toward a normal distribution with

$$\text{mean} = \mu_{\bar{x}} = \mu$$

$$\text{standard deviation (standard error)} = \sigma_{\bar{x}} = \frac{\sigma}{\sqrt{n}}$$

(7.2)

where n = sample size.

Sampling without Replacement

We first encountered the problem of sampling without replacement from a finite population in Chapter 5, where the hypergeometric distribution considered the population size (N) and the binomial distribution did not. It is easy to show that, for this situation,

$$\begin{bmatrix} \text{variance of hypergeometric} \\ \text{random variable} \end{bmatrix} = \begin{bmatrix} \text{variance of corresponding} \\ \text{binomial random variable} \end{bmatrix} \cdot \begin{bmatrix} N - n \\ \overline{N - 1} \end{bmatrix}$$

because

$$\frac{k(N-k)n(N-n)}{N^2(N-1)} = n\frac{k}{N}\left[1 - \frac{k}{N}\right]\cdot\left[\frac{N-n}{N-1}\right]$$

$$= np(1-p)\cdot\left[\frac{N-n}{N-1}\right]$$

where $p = k/N$. Here, $(N-n)/(N-1)$ is called the **finite population correction (fpc) factor**. When the sample size, n, is very small as compared with the population size, N, the fpc factor is nearly 1 and can be ignored. In fact, as discussed in Chapter 5, the binomial distribution serves as a good approximation to the hypergeometric whenever $n/N < .05$. The same result applies to sampling situations as well. We can express this as a rule: The fpc can be ignored whenever $n/N < .05$.

We can also use the Central Limit Theorem in this situation.

CENTRAL LIMIT THEOREM: SAMPLING WITHOUT REPLACEMENT FROM A FINITE POPULATION

When sampling without replacement from a finite population (of size N), with mean μ and standard deviation σ, the sample mean $\bar{X}$ tends toward a normal distribution with

$$\text{mean} = \mu_{\bar{X}} = \mu$$

$$\text{standard deviation (standard error)} = \sigma_{\bar{X}} = \frac{\sigma}{\sqrt{n}}\cdot\sqrt{\frac{N-n}{N-1}} \qquad (7.3)$$

where n = sample size.

EXAMPLE 7.4

A group of women managers at Compumart are considering filing a sex-discrimination suit. A recent report stated that the average annual income of all employees in middle management positions at Compumart is $48,000 and the standard deviation is $8500. A random sample of 45 women in these positions taken from a population of 350 female middle managers at Compumart had an average income of $\bar{x} = \$43,900$. If the population of all female incomes at this level is assumed to have the same mean ($48,000) and standard deviation ($8500) as the distribution of incomes for all employees, what is the probability of observing a value of $\bar{X}$ this low?

SOLUTION

Because we have a large sample, we can assume (using the Central Limit Theorem) that the curve describing $\bar{X}$ is normal, as shown in Figure 7.12. Here, $n = 45$ and $N = 350$. We need to find $P(\bar{X} \leq 43,900)$. Standardizing and using

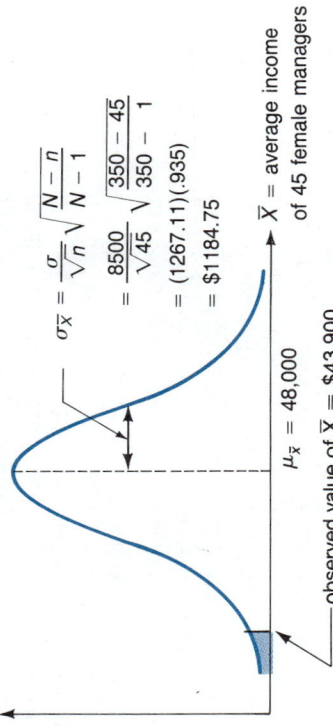

$$\sigma_{\bar{X}} = \frac{\sigma}{\sqrt{n}}\sqrt{\frac{N-n}{N-1}}$$

$$= \frac{8500}{\sqrt{45}}\sqrt{\frac{350-45}{350-1}}$$

$$= (1267.11)(.935)$$

$$= \$1184.75$$

$\mu_{\bar{x}} = 48,000$

$\bar{X}$ = average income
of 45 female managers

observed value of $\bar{X} = \$43,900$

FIGURE 7.12

Distribution of sample mean of annual salaries (assuming $\mu = \$48,000$, $\sigma = \$8500$). The shaded area represents the solution to Example 7.4, $P(\bar{X} \leq 43,900)$.

Table A.4, we find that

$$P(\bar{X} \le 43,900) = P\left[Z \le \frac{43,900 - 48,000}{1184.75}\right]$$
$$= P(Z \le -3.46) = .0003$$

So, if the female population has an average salary of $48,000 (and standard deviation of $8500), then the chance of obtaining an $\bar{X}$ as low as $43,900 is extremely small. If we assume that the standard deviation is correct, then, based strictly on this set of data, our conclusion would be that the average salary for women at this level is not $48,000 but is less than $48,000.

With the type of question asked in Example 7.4, there is always the chance that we will reach an incorrect decision using the sample data; there is always the chance of error due to sampling. This possible error will be a concern whenever you test a hypothesis. For now, remember that, when dealing with sample data, statistics never *prove* anything. They do, however, *support* or *fail to support* a claim (such as $\mu < $48,000$).

■

EXERCISES

7.9 From a finite population of size 300 that is approximately normally distributed with a mean of 50 and a standard deviation of 10, what is the probability that a random sample of size 30 without replacement will yield a sample mean larger than 55? What is the probability that a random sample of size 30 with replacement will yield a sample mean larger than 55?

7.10 Advanced Machinery manufactured 1584 diagnostic machines. The machines can pinpoint electrical problems in a certain type of machinery in 5 minutes on the average with a standard deviation of 2 minutes. If a random sample of 300 diagnostic machines are selected without replacement, what is the probability that the sample mean of the time it takes to pinpoint the electrical problems is greater than 6 minutes?

7.11 The electric bill for 250 households in a small midwestern town was found to have a mean of $120 with a standard deviation of $25 for the month of November. If ten households are selected at random from the 250 households, what is the probability that the sample mean will be between $110 and $130? What are you assuming about the population?

7.12 General Appliances has 70 microwave ovens that need repair. The mean cost of repair for the 70 microwaves is $80. The standard deviation of the cost is $35. The cost can be considered to be approximately normally distributed.

a. If a sample of 10 from the 70 microwaves is selected without replacement, what is the probability that the mean cost for the sample is greater than $100?

b. If a sample of 10 from the 70 microwaves is selected with replacement, what is the probability that the mean cost for the sample is greater than $100?

7.13 The mean daily time spent on the telephone by the 60 personnel managers of Retail Products is 1.25 hours; the standard deviation is .62 hours. Assuming that the time spent on the telephone is approximately normally distributed, what is the probability that the mean daily time spent on the telephone by 10 different personnel managers selected at random is greater than 1.5 hours?

7.14 As the finite population size gets large for a fixed sample size, explain how the finite population correction factor is affected.

7.15 National Distributing employs 500 salespersons in 200 territories throughout the United States. The average yearly commission earned by a salesperson is $8540. If a random sample of 60 salespersons is selected, what is the probability that the sample mean of their yearly commission is less than $45,000?

7.16 An aptitude test on the theory of electronics was given to all the 275 repair-people of A.N.P. Micronics. The mean score on the test was 112 and the standard deviation of the test scores was 18.6. Twenty repairpeople were selected at random. What is the probability that the sample mean would be between 110 and 120? What are you assuming about the distribution of the exam scores?

7.17 Legislators are allowed certain postal privileges to communicate with their constituency. Suppose that for 435 members of Congress, the mean amount of annual postal charges used is $1630 with a standard deviation of $170. If a sample of 40 members of Congress is obtained, what is the probability that the average annual postal charge for this group of 40 is $1600 or more? If a sample of 20 members of Congress is taken, what is the probability that the average annual charge for this group of 20 is $1600 or more? What must you assume to answer the latter question?

7.18 Two hundred boxes of meat patties are loaded into a truck. The mean weight of the 200 boxes is 8.2 pounds. The standard deviation of the weight is .5 pounds. If a fast-food restaurant randomly selects 40 boxes of meat patties, what is the probability that the mean weight of the 40 boxes is less than 8.0 pounds?

7.3

CONFIDENCE INTERVALS FOR THE MEAN OF A NORMAL POPULATION (σ KNOWN)

Return to the situation where we have obtained a sample from a normal population with unknown mean, μ. We first consider a case in which we know the variability of the normal random variable, the value of σ (Figure 7.13). (The situation where both μ and σ are unknown is dealt with in the next section.)

We know that to estimate μ, the average of the entire population, we obtain a sample from this population and calculate $\bar{X}$, the average of the sample. The sample mean, $\bar{X}$, is the estimate of μ and is also called a **point estimate** because it consists of a single number.

In Example 7.2, it was assumed that the assembly time for a particular electrical component followed a normal distribution, with $\mu = 20$ minutes and $\sigma = 3$ minutes. What if μ is not known for *all* workers? A random sample of 25 workers' assembly times was obtained with the following results (in minutes):

22.8, 29.3, 27.2, 30.2, 24.0, 23.2, 22.9, 30.3, 27.1, 31.2, 27.0, 32.0, 28.6, 24.1, 28.9, 26.8, 26.6, 23.4, 25.1, 26.6, 25.7, 28.1, 31.5, 24.8, 25.2

Based on these data,

estimate of μ = sample mean, $\bar{X}$

$$= \frac{22.8 + 29.3 + \cdots + 25.2}{25} = 26.9 \text{ minutes}$$

Is this large value of $\bar{X}$ ($= 26.9$) due to random chance? We know that 50% of the samples drawn will have $\bar{X}$ larger than 20, even if $\mu = 20$ (Figure 7.14). Or is this value large because μ is a value larger than 20? In other words, does this value of $\bar{X}$ provide just cause for concluding that μ is larger than 20? We tackle this type of question in Chapter 8.

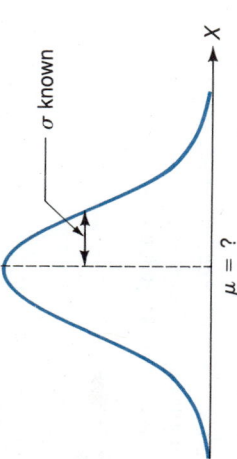

FIGURE 7.13

An example where the standard deviation σ is known, but the mean μ is unknown.

FIGURE 7.14

Distribution of $\bar{X}$ if $\mu = 20$ minutes.

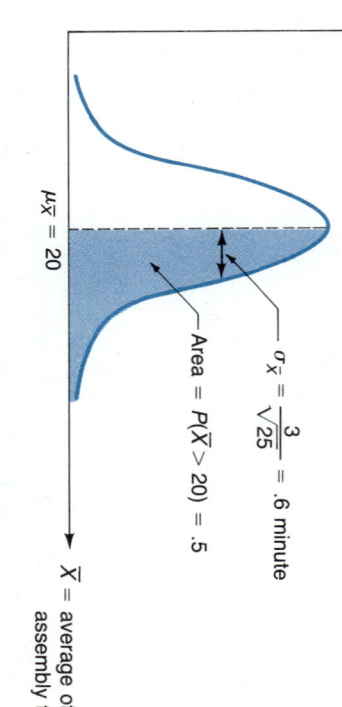

How accurate is a derived estimate of the population mean, μ? This depends, for one thing, on the sample size. We can measure the precision of this estimate by constructing a **confidence interval (CI)**. By providing the confidence interval, one can make such statements as "I am 95% confident that the average assembly time, μ, is between 25.7 minutes and 28.1 minutes." For this illustration, (25.7, 28.1) is called a 95% confidence interval for μ. The following discussion demonstrates how to construct such a confidence interval.

Using the Central Limit Theorem, we know that $\bar{X}$ is approximately a normal random variable with*

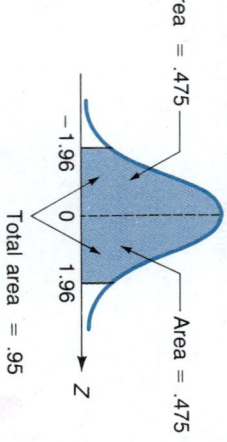

$$\mu_{\bar{X}} = \mu$$
$$\sigma_{\bar{X}} = \frac{\sigma}{\sqrt{n}}$$

where μ and σ represent the mean and standard deviation of the population. To standardize $\bar{X}$, you subtract the mean (μ) of $\bar{X}$ and divide by the standard deviation ($\sigma/\sqrt{n}$) of $\bar{X}$. Consequently,

$$Z = \frac{\bar{X} - \mu}{\sigma/\sqrt{n}}$$

is a standard normal random variable. Consider the following statement and refer to Figure 7.15:

$$P(-1.96 \leq Z \leq 1.96) = .95$$

so

$$P\left(-1.96 \leq \frac{\bar{X} - \mu}{\sigma/\sqrt{n}} \leq 1.96\right) = .95$$

FIGURE 7.15

$P(-1.96 \leq Z \leq 1.96) = $ .95.

* This discussion ignores the finite population correction (fpc) factor defined in the previous section. For the case of sampling without replacement, where the population size (N) is known, see the discussion of simple random sampling in Section 7.6.

After some algebra and rearrangement of terms, we get

$$P\left(\bar{X} - 1.96 \frac{\sigma}{\sqrt{n}} \le \mu \le \bar{X} + 1.96 \frac{\sigma}{\sqrt{n}}\right) = .95$$

How does the last statement apply to a *particular* sample mean, $\bar{x}$? Consider the interval

$$\left(\bar{x} - 1.96 \frac{\sigma}{\sqrt{n}}, \bar{x} + 1.96 \frac{\sigma}{\sqrt{n}}\right) \qquad (7.4)$$

Using the values from our assembly-time example, we have $\bar{x} = 26.9$, $\sigma = 3$, and $n = 25$. The resulting 95% confidence interval is

$$\left(26.9 - 1.96 \cdot \frac{3}{\sqrt{25}}, 26.9 + 1.96 \cdot \frac{3}{\sqrt{25}}\right)$$

or

$$(25.72, 28.08)$$

Since μ is unknown, it is unknown whether μ lies between 25.72 and 28.08. However, if you were to obtain random samples repeatedly, calculate $\bar{x}$, and determine the intervals defined in formula 7.4, then 95% of these intervals would contain μ and 5% would not. For this reason, formula 7.4 is called a **95% confidence interval** for μ. Using our assembly-time illustration, we are 95% confident that the average assembly time, μ, lies between 25.72 and 28.08.

Notation Let Z_a denote the value of Z with an area *to the right* of this value equal to a. How can we determine $Z_{.025}$, $Z_{.05}$, and $Z_{.1}$ (Figure 7.16)? Using Table A.4, $Z_{.025} = 1.96$, $Z_{.05} = 1.645$, and $Z_{.1} = 1.28$.

When defining a confidence interval for μ, we can define a 99% confidence interval, a 95% confidence interval, a 90% confidence interval, or whatever. The specific percentage represents the **confidence level**. The *higher* the confidence level, the *wider* the confidence interval. The confidence level is written as $(1 - \alpha) \cdot 100\%$, where $\alpha = .01$ for a 99% confidence interval, $\alpha = .05$ for a 95% confidence interval, and so on. Thus, a $(1 - \alpha) \cdot 100\%$ confidence interval for the mean of a normal population, μ, is

$$\left[\bar{x} - Z_{\alpha/2}\left(\frac{\sigma}{\sqrt{n}}\right), \bar{x} + Z_{\alpha/2}\left(\frac{\sigma}{\sqrt{n}}\right)\right] \qquad (7.5)$$

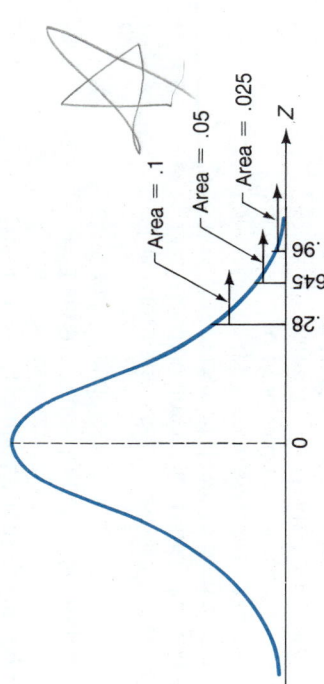

FIGURE 7.16

$1.28 = Z_{.1}$, $1.645 = Z_{.05}$, and $1.96 = Z_{.025}$.

According to the Central Limit Theorem, formula 7.5 provides an *approximate* confidence interval for the mean of any population, provided the sample size, n, is large ($n > 30$).

EXAMPLE 7.5

Determine a 90% and a 99% confidence interval for the average assembly time of all workers using the 25 observations given on page 223.

the 25 observations given on page 223.

SOLUTION

The sample mean here was $\bar{x} = 26.9$. The population standard deviation is assumed to be 3 minutes. The resulting 90% confidence interval for the population mean μ is

$$26.9 - Z_{.05}\left(\frac{3}{\sqrt{25}}\right) \quad \text{to} \quad 26.9 + Z_{.05}\left(\frac{3}{\sqrt{25}}\right)$$

$$= 26.9 - 1.645\left(\frac{3}{\sqrt{25}}\right) \quad \text{to} \quad 26.9 + 1.645\left(\frac{3}{\sqrt{25}}\right)$$

$$= 26.9 - .99 \quad \text{to} \quad 26.5 + .99$$

$$= 25.91 \text{ minutes} \quad \text{to} \quad 27.89 \text{ minutes}$$

The 99% confidence interval for μ is

$$26.9 - Z_{.005}\left(\frac{3}{\sqrt{25}}\right) \quad \text{to} \quad 26.9 + Z_{.005}\left(\frac{3}{\sqrt{25}}\right)$$

$$= 26.9 - 2.575\left(\frac{3}{\sqrt{25}}\right) \quad \text{to} \quad 26.9 + 2.575\left(\frac{3}{\sqrt{25}}\right)$$

$$= 26.9 - 1.54 \quad \text{to} \quad 26.9 + 1.54$$

$$= 25.36 \text{ minutes} \quad \text{to} \quad 28.44 \text{ minutes}$$

Consequently, we are 90% confident that the mean assembly time for all workers is between 25.91 and 27.89 minutes. We are also 99% confident that this parameter is between 25.36 and 28.44 minutes based on the results of this sample. Notice that the width of the interval increases as the confidence level increases when using the same sample data.   ■

Discussing a Confidence Interval

The narrower your confidence interval, the better, for the same level of confidence. Suppose Electricalc spent $50,000 investigating the average time necessary to assemble their electrical components. Part of this study included obtaining a confidence interval for the average assembly time, μ. Which statement would they prefer to see?

1. I am 95% confident that the average assembly time is between 2 minutes and 50 minutes.

2. I am 95% confident that the average assembly time is between 25 minutes and 27 minutes.

The information contained in the first statement is practically worthless, and that's $50,000 down the drain. The second statement contains useful information; μ is narrowed down to a much smaller range.

Given the second statement, can you tell what the corresponding value of $\bar{X}$ was that produced this confidence interval? For any confidence interval for μ, $\bar{X}$ (the estimate of μ) is always *in the center*. So $\bar{X}$ must have been 26 minutes. For the 90% confidence interval in Example 7.5, the following conclusions are valid:

1. I am 90% confident that the average assembly time lies between 25.91 and 27.89 minutes.

2. If I repeatedly obtained samples of size 25, then 90% of the resulting confidence intervals would contain μ and 10% would not. (Question from the audience: Does this confidence interval [25.91, 27.89] contain μ? Your response: I don't know. All I can say is that this procedure leads to an interval containing μ 90% of the time.)

3. I am 90% confident that my estimate of μ (namely, $\bar{x} = 26.9$) is within .99 minute of the actual value of μ.

Here .99 is equal to $1.645 \cdot (\sigma/\sqrt{n})$. This is referred to as the **maximum error, E**.

$$E = \text{maximum error} = Z_{\alpha/2}\left(\frac{\sigma}{\sqrt{n}}\right) \qquad (7.6)$$

Be careful! The following statement is *not* correct: The probability that μ lies between 25.91 and 27.89 is .90. What is the probability that the number 27 lies in this confidence interval? How about 24? The answer to the first question is 1, and to the second, 0, because 27 lies in the confidence interval and 24 does not. So what is the probability that μ lies in the confidence interval? Remember that μ is a fixed number; we just do not know what its value is. It is *not a random variable*, unlike its estimator, $\bar{X}$. As a result, this probability is either 0 or 1, not .90. Therefore, remember that, once you have inserted your sample results into formula 7.5 to obtain your confidence interval, the word *probability* can no longer be used to describe the resulting confidence interval.

EXAMPLE 7.6

Refer to the 20 samples of Everglo bulbs in Table 7.1. Using sample 1, what is the resulting 95% confidence interval for the population mean, μ? Assume that σ is 50 hours.

SOLUTION

Here, $n = 10$ and $\bar{x} = 384.0$. The confidence level is 95%, so $Z_{\alpha/2} = Z_{.025} = 1.96$ (from Table A.4). Therefore, the resulting 95% confidence interval for μ is

$$384.0 - 1.96\left(\frac{50}{\sqrt{10}}\right) \quad \text{to} \quad 384.0 + 1.96\left(\frac{50}{\sqrt{10}}\right)$$

$$= 384.0 - 31.0 \quad \text{to} \quad 384.0 + 31.0$$

$$= 353.0 \quad \text{to} \quad 415.0$$

So we are 95% confident that μ lies between 353 and 415 hours. Also, we are 95% confident that our estimate of μ ($\bar{x} = 384.0$) is within 31.0 hours of the actual value. A confidence interval constructed using MINITAB is shown in Figure 7.17. ■

FIGURE 7.17
MINITAB solution to Example 7.6.

```
MTB > SET INTO C1
DATA> 308 419 389 432 362 302 440 430 375 383
MTB > END
MTB > ZINTERVAL USING 95%, SIGMA = 50, DATA IN C1

    THE ASSUMED SIGMA = 50.0

          N      MEAN     STDEV   SE MEAN    95.0 PERCENT C.I.
    C1    10     384.0    49.3     15.8    (  353.0,    415.0)
```

FIGURE 7.18 Confidence intervals constructed using 20 samples in Table 7.1

Sample	Lower limit	Upper limit
1	353.0	415.0
2	368.5	430.5
3	366.3	428.3
4	368.4	430.4
5	383.4	445.4
6	362.2	424.2
7	367.8	429.8
8	372.8	434.8
9	371.8	433.8
10	373.6	435.6
11	370.9	432.9
12	354.6	416.6
13	377.2	439.2
14	380.0	442.0
15	384.6	446.6
16	377.5	439.5
17	351.1	413.1
18	360.3	422.3
19	401.6	463.6
20	392.1	454.1

To illustrate the nature of a confidence interval, take a closer look at the 20 samples in Table 7.1. By repeating the procedure in Example 7.6 for the remaining 19 samples, Figure 7.18 can be constructed. If this procedure of obtaining samples of size 10 were repeated indefinitely, we would expect 95% of the resulting confidence intervals to contain the population mean (known to be $\mu = 400$ here), and 5% would not. For these 20 samples, in fact, we observe that all but one (sample 19) resulted in a confidence interval containing $\mu = 400$.

EXERCISES

7.19 A random sample of 125 observations is obtained from a normally distributed population with a standard deviation of five. Given that the sample mean is 20.6, construct a 90% confidence interval for the mean of the population.

7.20 The following data are the values of a random sample from a normally distributed population. Assume that the population variance is 10.2. Construct a 99% confidence interval for the mean of the population.

50.6, 52.3, 48.6, 45.3, 51.8, 50.8, 46.7, 56.1, 47.7, 49.3, 44.9, 57.0, 50.7, 42.6,
49.8, 46.1, 48.7, 51.8, 54.3, 48.4, 50.5

7.21 A random sample of size 60 from a normally distributed population yields a mean of 100. The standard deviation of the population is 30. Construct a 95% confidence interval for the mean of the population.

7.22 The monthly advertising expenditure of Discount Hardware Store is normally distributed with a standard deviation of $100. If a sample of 10 randomly selected months yields a mean advertising expenditure of $380 monthly, what is a 90% confidence interval for the mean of the store's monthly advertising expenditure?

7.23 A vending machine containing laundry detergent produces a total profit of $1800 over 7 randomly selected months. If the monthly profit from the vending machine is considered to be normally distributed with a standard deviation of $300, what would be a 90% confidence interval for the mean monthly profit from this machine?

7.24 The perfectionist owner of Kwik Kar Kare has reduced an oil-change job to a science and wants to keep it that way. The owner constantly monitors the performance of the staff. This week, 15 oil-change jobs performed were sampled with a sample mean of 9.8 minutes per job. From experience, the times follow a normal distribution and the standard deviation of the population is known to be 1.2 minutes. Based on this week's sample, construct a 90% confidence interval for the population mean (average time for an oil-change job).

7.25 A manufacturer of ten-speed racing bicycles believes that the average weight of the bicycle is normally distributed with a mean of 22 pounds and a standard deviation of 1.5 pounds. A random sample of 30 bicycles is selected. If the mean from this sample is 22.8, what is a 96% confidence interval for the mean weight of the bicycle?

7.26 A quality-control engineer is concerned about the breaking strength of a metal wire manufactured to stringent specifications. A sample of size 25 is randomly obtained and the breaking strengths are recorded. The breaking strength of the wire is considered to be normally distributed with a standard deviation of three. Find a 95% confidence interval for the mean breaking strength of the wire.

26, 27, 18, 23, 24, 20, 21, 24, 19, 27, 25, 20, 24, 21, 26, 19, 21, 20, 25, 20, 23, 25, 21, 20, 21

7.27 An investor would like to bid on a tract of forest land and then clear the land selectively to market the timber. To arrive at an estimate of the total weight of the lumber, a random sample of 50 trees is selected and their diameters are measured. The sample yields a mean diameter of 13.2 inches. Find a 90% confidence interval for the mean diameter of the trees if the diameter of the trees on the tract of land is considered to be normally distributed with a variance of 4.3 inches squared.

7.28 As the sample size increases, would a confidence interval given by equation 7.5 get smaller or larger? For a given random sample, would the confidence interval given by equation 7.5 for a 90% confidence interval be larger or smaller than that for an 80% confidence interval?

7.29 A medical researcher would like to obtain a 99% confidence interval for the mean length of time that a particular sedative is effective. Thirty subjects are randomly selected. The mean length of time that the sedative was effective is found to be 8.3 hours for the sample. Find the 99% confidence interval, assuming that the length of time that the sedative is effective is considered to be approximately normally distributed with a standard deviation of .93 hours.

7.30 A safety council is interested in the age at which a person first obtains his or her driver's license. If the ages of people who obtain their first driver's license is considered to be normally distributed with a standard deviation of 2.5 years, what is an 80% confidence interval for the mean age, given that a random sample of 20 new drivers yields a mean age of 19.3 years?

7.4

CONFIDENCE INTERVALS FOR THE MEAN OF A NORMAL POPULATION (σ UNKNOWN)

If σ is unknown, it is impossible to determine a confidence interval for μ using formula 7.5 because we are unable to evaluate the standard error $\sigma/\sqrt{n}$. Let us take another look at how we estimate the parameters of a normal population.

When a population mean is unknown, we can estimate it using the sample mean. The logical thing to do if σ is unknown is to replace it by its estimate, the standard deviation of the sample, s. But consider what happens when

$$\frac{\bar{X} - \mu}{\sigma/\sqrt{n}}$$

is replaced by

$$\frac{\bar{X} - \mu}{s/\sqrt{n}}$$

This is no longer a standard normal random variable, Z. However, it does follow another identifiable distribution, the **t distribution**. Its complete name is *Student's t distribution*, named after W. S. Gosset, a statistician in a Guinness brewery who used the pen name Student. The distribution of

$$\frac{\bar{X} - \mu}{s / \sqrt{n}}$$

will follow a *t* distribution, *provided* the population from which you are obtaining the sample is normally distributed.

The *t* distribution is similar in appearance to the standard normal (Z) distribution in that it is symmetric about zero. Unlike the Z distribution, however, its shape depends on the sample size, *n*. Consequently, when you use the *t* distribution, you must take into account the sample size. This is accomplished by using **degrees of freedom**. For this application using the *t* distribution,

$$\text{degrees of freedom} = \text{df} = n - 1$$

The value of df = $n - 1$ can be explained by observing that, for a given value of $\bar{X}$, only $n - 1$ of the sample values are free to vary. For example, in a sample of size $n = 3$, if $\bar{x} = 5.0$, $x_1 = 2$, and $x_2 = 7$, then x_3 must be 6 because this is the only value providing a sample mean equal to 5.0.

Two *t* distributions are illustrated in Figure 7.19. Notice that the *t* distributions are symmetrically distributed about zero but have wider tails than does the standard normal, Z. Observe that as *n* increases, the *t* distribution tends toward the standard normal, Z. In fact, for $n > 30$, there is little difference between these two distributions. Areas under a *t* curve are provided in Table A.5 for various df. So, for large samples ($n > 30$), it does not matter whether σ is known (Z distribution, Table A.4) or σ is unknown (*t* distribution, Table A.5) because the *t* and Z curves are practically the same. For this reason, the *t* distribution often is referred to as the **small-sample distribution** for $\bar{X}$. The Z table can be used as an approximation even if σ is unknown provided *n* is larger than 30.

Using the *t* distribution, then, a $(1 - \alpha) \cdot 100\%$ confidence interval for μ is

$$\bar{x} - t_{\alpha/2, n-1}\left(\frac{s}{\sqrt{n}}\right) \quad \text{to} \quad \bar{x} + t_{\alpha/2, n-1}\left(\frac{s}{\sqrt{n}}\right) \tag{7.7}$$

where $t_{\alpha/2, n-1}$ denotes the *t* value from Table A.5 using a *t* curve with $n - 1$ df and a right-tail area of $\alpha/2$.

Do you remember our sample of 25 assembly times that produced a point estimate for μ having a value of $\bar{x} = 26.9$ minutes? This estimate was used in Example 7.5, where it was assumed that the population standard deviation was

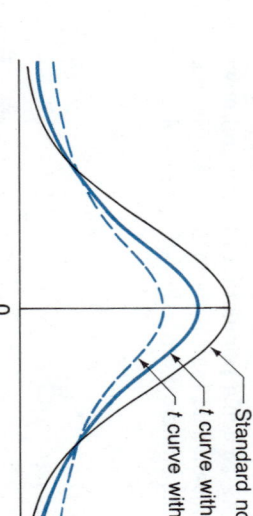

Standard normal, Z
t curve with 20 df
t curve with 10 df

FIGURE 7.19
The *t* distribution.

$\sigma = 3$, in constructing a confidence interval for μ. Furthermore, the assembly times were assumed to follow a *normal* distribution.

Suppose that we do not know σ, either. Then the point estimate of the population standard deviation is

$$s = \sqrt{\dfrac{22.8^2 + 29.3^2 + \cdots + 25.2^2) - (22.8 + 29.3 + \cdots + 25.2)^2/25}{24}}$$

$$= \sqrt{\dfrac{18{,}285.14 - (672.6)^2/25}{24}}$$

$$= \sqrt{7.896} = 2.81 \text{ minutes}$$

Using Table A.5 to find a 90% confidence interval for μ, you first determine that

$$t_{\alpha/2,\, n-1} = t_{.05,\, 24} = 1.711$$

The resulting 90% confidence interval is

$$26.9 - 1.711\left(\dfrac{2.81}{\sqrt{25}}\right) \quad \text{to} \quad 26.9 + 1.711\left(\dfrac{2.81}{\sqrt{25}}\right)$$

$$= 26.9 - .96 \quad \text{to} \quad 26.9 + .96$$

$$= 25.94 \quad \text{to} \quad 27.86$$

Using these data, we are 90% confident that the estimate for the mean of this normal population ($\bar{x} = 26.9$) is within .96 minutes of the actual value. Comparing this result with Example 7.5, we notice little difference in the two 90% confidence intervals. This is due mostly to the fact that the estimate of σ ($s = 2.81$) is very close to the assumed value of $\sigma = 3$.

EXAMPLE 7.7

Metro Moving Company is considering the purchase of a new, large moving van. The sales agency has agreed to lease the truck to Metro for 3 weeks (18 working days) on a trial basis. Of primary concern to Metro is the miles per gallon (mpg) that the van obtains on a typical moving day. The mpg values for the 18 trial days are

9.4, 8.6, 11.1, 7.5, 9.1, 10.4, 8.8, 10.8, 11.4, 6.8, 7.8, 11.8, 12.1, 10.7, 8.4, 9.5, 9.3, 10.1

What is the 95% confidence interval for the average mpg for this van, assuming that the daily mpg follows a normal distribution?

SOLUTION

Your (point) estimate of σ is $s = 1.52$ mpg. Also, your point estimate of μ is $\bar{x} = 9.64$ mpg. A 95% confidence interval for the average mpg of the new van (if daily mpg were recorded indefinitely) is

$$9.64 - t_{.025,\, 17}\left(\dfrac{1.52}{\sqrt{18}}\right) \quad \text{to} \quad 9.64 + t_{.025,\, 17}\left(\dfrac{1.52}{\sqrt{18}}\right)$$

$$= 9.64 - 2.11\left(\dfrac{1.52}{\sqrt{18}}\right) \quad \text{to} \quad 9.64 + 2.11\left(\dfrac{1.52}{\sqrt{18}}\right)$$

$$= 9.64 - .756 \quad \text{to} \quad 9.64 + .756$$

$$= 8.88 \quad \text{to} \quad 10.40 \text{ mpg}$$

We are thus 95% confident that the daily mpg of the van is between 8.88 and

FIGURE 7.20

MINITAB solution to
Example 7.7.

```
MTB > SET INTO C1
DATA> 9.4 8.6 11.1 7.5 9.1 10.4 8.8 10.8 11.4 6.8 7.8
DATA> 11.8 12.1 10.7 8.4 9.5 9.3 10.1
DATA> END
MTB > TINTERVAL WITH 95 PERCENT CONFIDENCE USING C1

        N     MEAN   STDEV  SE MEAN    95.0 PERCENT C.I.
C1     18    9.644   1.515    0.357  (  8.891,  10.398)
```

10.40 mpg. Notice here that the maximum error is

$$E = 2.11 \left(\frac{1.52}{\sqrt{18}} \right) = .756 \text{ mpg}$$

which implies that we are 95% confident that $\bar{X}$ is within .756 mpg of the actual average mpg. The MINITAB solution for this example is shown in Figure 7.20.

■

EXERCISES

7.31 Find the t values for the following α levels and degrees of freedom.

a. $t_{.10,29}$ **b.** $t_{.025,13}$ **c.** $t_{.05,18}$

d. $t_{.90,20}$ **e.** $t_{.95,25}$ **f.** $t_{.10,40}$

7.32 For a sample of size 21 used to test a hypothesis about a sample mean, what is the t value from a t distribution such that the following are true?

a. Ninety percent of the area under the t distribution is to the right of the t value.

b. Ten percent of the area under the t distribution is to the right of the t value.

c. Five percent of the area under the t distribution is to the left of the t value.

7.33 A random sample of size 15 is selected from a normally distributed population. The sample mean is 30 and the sample variance is 16. Find a 95% confidence interval for the population mean.

7.34 The price per month of back orders at Harlington Industries is considered to be normally distributed. A random sample of back orders for 12 randomly selected months yields a mean of $115,320 with a standard deviation of $35,000. Construct a 90% confidence interval for the mean price of back orders per month at Harlington Industries.

7.35 The mean monthly expenditure on gasoline per household in Middletown is determined by selecting a random sample of 36 households. The sample mean is $68 with a sample standard deviation of $17.

a. What is a 95% confidence interval for the mean monthly expenditure per household on gasoline in Middletown?

b. What is a 90% confidence interval for the mean monthly expenditure per household on gasoline in Middletown?

7.36 Second Federal Savings and Loan would like to estimate the mean number of years in which 30-year mortgages are paid off. Eighteen paid-off 30-year mortgages are randomly selected and the numbers of years in which the loans were paid in full are

19.6, 20.8, 29.6, 6.3, 3.1, 10.6, 30.0, 21.7, 10.5, 26.3, 10.7, 6.1, 7.3, 12.6, 9.8, 27.4, 20.1, 10.8

Assuming that the number of years in which 30-year mortgages are paid off is normally distributed, construct an 80% confidence interval for the mean number of years in which the mortgages are paid off.

7.37 An apartment-finder service would like to estimate the average cost of a one-bedroom apartment in Kansas City. A random sample of 41 apartment complexes yielded a mean of $310 with a standard deviation of $29. Construct a 90% confidence interval for the mean cost of one-bedroom apartments in Kansas City.

7.38 In a study carried out to see how people learn about operating a complex programmable device without a user's guide or other assistance ("instructionless learning"), Shrager and Klahr (1986) provided the following data on gross performance of 7 subjects during a stage called the systematic investigation phase:

SUBJECT	PHASE DURATION	KEY PRESSES	WORDS PER MINUTE	KEYSTROKE RATE/MINUTE
1	29.3	423	150	14.5
2	22.3	470	70	21.1
3	25.7	362	71	14.1
4	22.3	319	60	14.3
5	23.3	543	125	23.3
6	18.7	391	101	21.0
7	18.8	345	92	17.4
Mean	23.0	408	96	18.0
Std deviation	3.6	78	32	3.8

(*Source:* Jeff Shrager and David Klahr, "Instructionless Learning about a Complex Device: The Paradigm and Observations. *International Journal Man-Machine Studies* 25, 160 (1986): Table 2.)

Construct the following confidence intervals using the preceding data:

a. The 90% confidence interval for the mean phase duration in minutes

b. The 95% confidence interval for the mean number of key presses

c. The 99% confidence interval for the mean keystroke rate per minute

7.39 An investment advisor believes that the return on interest-sensitive stocks is approximately normally distributed. A sample of 24 interest-sensitive stocks was selected and their yearly return (including dividends and capital appreciation) was as follows (in percentages):

11.1, 12.5, 13.6, 9.1, 8.7, 10.6, 12.5, 15.6, 13.8, 8.0, 10.9, 7.6, 5.2, 1.2, 12.8, 16.7, 13.9, 10.1, 9.6, 10.8, 11.6, 12.3, 12.9, 11.6

Find a 90% confidence interval for the mean yearly return on interest-sensitive stocks.

7.40 The president of Secure Savings and Loan Association would like to estimate the average salaries of vice presidents of savings and loan associations. After selecting a random sample of 41 vice presidents, the following statistics were calculated regarding annual salaries:

$$\bar{x} = 52,100$$

$$s = 10,350$$

Construct a 95% confidence interval for the mean annual salaries of vice presidents.

7.41 A quality-control engineer conducted a test of the tensile strength of 20 aluminum wires. The coded data represent the tensile strength of 20 aluminum wires selected at random. Find a 90% confidence interval for the mean value of the coded data. What are you assuming about the distribution of the tensile strengths?

105, 113, 95, 90, 112, 93, 106, 80, 95, 90, 88, 101, 93, 91, 86, 107, 103, 93, 84, 87

7.42 MAXIX has been selling automatic toll-collection systems in the United States at various prices depending on the competition. It is believed that the price at which the systems are sold can be approximated by a normal distribution. A sample of 15 systems was sold at the following values (in dollars):

16,500, 13,200, 14,560, 12,320, 13,640, 12,980, 13,350, 12,130, 11,980, 13,590, 15,670, 16,350, 11,860, 13,400, 13,860

Find a 90% confidence interval for the mean price at which the automatic toll-collection systems are sold.

7.5

SELECTING THE
NECESSARY
SAMPLE SIZE

Sample Size for Known σ

How large a sample do you need? This is often difficult to determine, although a carefully chosen *large* sample generally provides a better representation of the population than does a smaller sample. Acquiring large samples can be costly and time-consuming; why obtain a sample of size $n = 1000$ if a sample size of $n = 500$ will provide sufficient accuracy for estimating a population mean? This section will show you how to determine what sample size is necessary when the maximum error, E, is specified in advance.

In Example 7.6, we assumed that the lifetime of Everglo bulbs is normally distributed with standard deviation $\sigma = 50$ hours but unknown mean μ. Based on the results of sample 1 from Table 7.1, the conclusion was that we were 95% confident that the estimate of μ ($\bar{x} = 384.0$) was within 31.0 hours of the actual value of μ for $n = 10$. How large a sample is necessary if we want our point estimator ($\bar{X}$) to be within 10 hours of the actual value of μ, with 95% confidence? The value 10 here is the maximum error, E, defined in equation 7.6. We would like the estimate of μ (that is, $\bar{X}$) to be within 10 of the actual value, so

$$E = 10 = Z_{\alpha/2}\left(\frac{50}{\sqrt{n}}\right)$$

Because the confidence level is 95%, $Z_{\alpha/2} = Z_{.025} = 1.96$. Consequently,

$$10 = (1.96)\frac{50}{\sqrt{n}}$$

$$\sqrt{n} = \frac{(1.96)(50)}{10} = 9.8$$

Squaring both sides of this statement produces

$$n = (9.8)^2 = 96.04$$

Rounding this value *up* (always), then a sample size of $n = 97$ will produce a confidence interval with $E \leq 10$ hours. As a result, your point estimate of μ, $\bar{X}$, will be within 10 hours of the actual value, with 95% confidence.

This sequence of steps can be summarized using the following expression:

$$n = \left[\frac{Z_{\alpha/2} \cdot \sigma}{E}\right]^2 \qquad (7.8)$$

Sample Size for Unknown σ

Equation 7.8 works if σ is known but does not apply to situations where both μ and σ are unknown. There are two approaches to the latter situation.

A Preliminary Sampling

If you have already obtained a small sample, you have an estimate of σ, namely, the sample standard deviation, s. Replacing σ by s in equation 7.8 gives you the desired sample size, n. Assuming that the resulting value of n is greater than

30, the $Z_{\alpha/2}$ notation in equation 7.8 is still valid because the actual t distribution here will be closely approximated by the standard normal.

When you do obtain the confidence interval using the larger sample, the resulting maximum error, E, may not be exactly what you originally specified because the new sample standard deviation will not be the same as that belonging to the smaller original sample.

EXAMPLE 7.8

In Example 7.7, Metro Moving Company obtained 18 observations consisting of daily miles per gallon (mpg) on a large moving van. For these data,

$$\bar{x} = 9.64 \text{ mpg}$$

$$s = 1.52 \text{ mpg}$$

How large a sample would they need for $\bar{X}$ to be within .5 mpg of the actual average mpg, with 95% confidence?

SOLUTION

Based on the results of the original sample, $s = 1.52$, so

$$n = \left[\frac{Z_{\alpha/2} \cdot s}{E} \right]^2 = \left[\frac{(1.96)(1.52)}{.5} \right]^2$$

$$= 35.5$$

We round this up to 36. Assuming a six-day workweek, Metro would need six weeks of driving time to make a statement with this much precision, that is, within .5 mpg. Of course, they already have three weeks of data that can be included in the larger sample. ∎

Obtaining a Rough Approximation of σ

We know from the empirical rule and Table A.4 that 95.4% of the population will lie between $\mu - 2\sigma$ and $\mu + 2\sigma$. Because $(\mu + 2\sigma) - (\mu - 2\sigma) = 4\sigma$, this is a span of four standard deviations. One method of obtaining an estimate of σ is to ask a person who is familiar with the data to be collected these questions:

1. What do you think will be the highest value in the sample (H)?
2. What will be the lowest value (L)?

The approximation of σ is then obtained by assuming that $\mu + 2\sigma = H$ and $\mu - 2\sigma = L$, so

$$H - L = (\mu + 2\sigma) - (\mu - 2\sigma) = 4\sigma$$

Consequently,

$$\sigma \cong \frac{H - L}{4} \qquad (7.9)$$

We can use this estimate of σ in equation 7.8 to determine the necessary sample size, n.

EXAMPLE 7.9

The quality-control manager of a division that produces hair dryers is interested in the average number of switches that can be tested by the division's employees. Assuming that the number of switches that are tested each hour by an employee follows a normal distribution (centered at μ), the manager wants to estimate μ with 90% confidence. Also, this estimate must be within one unit (switch) of μ.

The manager estimates that H is 45 switches and L is 25 switches. How large a sample will be necessary?

SOLUTION

Based on $H = 45$ switches and $L = 25$ switches,

$$\sigma \cong \frac{45 - 25}{4} = 5 \text{ switches}$$

The sample size necessary to obtain a maximum error of $E = 1$ is

$$n = \left[\frac{(1.645)(5)}{1} \right]^2 = 67.7$$

Thus, a sample size of 68 should produce a value of E close to one switch. The value will not be exactly one because the sample standard deviation, s, probably will not be exactly five. Estimating σ in this manner, however, produces a value that is "in the neighborhood" of σ.

■

EXERCISES

7.43 A 99% confidence interval is to be constructed such that $\bar{X}$ is within 1.5 units of the mean of a normal population. Assuming that the population variance is 30, what sample size would be necessary to achieve this maximum error?

7.44 To be 95% confident that $\bar{X}$ is within .65 of the actual mean of a normal population with a standard deviation of 2.5, what sample size would be necessary?

7.45 The Chamber of Commerce of Tampa, Florida, would like to estimate the mean amount of money spent by a tourist to within $100 with 95% confidence. If the amount of money spent by tourists is considered to be normally distributed with a standard deviation of $200, what sample size would be necessary for the Chamber of Commerce to meet their objective in estimating this mean amount?

7.46 Security Savings and Loan Association's manager would like to estimate the mean deposit by a customer into a savings account to within $500. If the deposits into savings accounts are considered to be normally distributed with a standard deviation of $1250, what sample size would be necessary to be 90% confident?

7.47 If a sample size of 70 was necessary to estimate the mean of a normal population to within 1.2 with 90% confidence, what is the approximate value of the standard deviation of the population?

7.48 The marketing agency for computer software of Personal Micro Systems would like to estimate with 95% confidence the mean time that it takes for a beginner to learn to use a standard software package. Past data indicate that the learning time can be approximated by a normal distribution with a standard deviation of 20 minutes. How large a sample size should the marketing agency choose if the mean time to learn to use the software package is to be estimated within 8 minutes with 90% confidence?

7.49 Past data indicate that the distribution of the daily price-earnings ratio of National Health and Medical Services can be approximated by a normal distribution with a variance of 17. How large a sample size would be necessary to estimate the mean price-earnings ratio of National Health and Medical Services to within 2 units with 98% confidence?

7.50 A chemist at International Chemical would like to measure the adhesiveness of a new wood glue. From past experiments, a measure used to indicate adhesiveness has ranged from 7.3 to 11.1 units. To be 98% confident, how large a sample would be necessary to estimate the mean adhesiveness to within .5 units?

7.51 An investor would like to obtain an idea of the profitability of a soft-drink vending machine by taking a random sample of several days and recording the daily number of soft drinks sold. A preliminary sample shows that the highest number of drinks sold daily is 106 and the least sold in a day is 36. What sample size would be necessary to estimate the mean number of soft drinks sold to within six soft drinks with 99% confidence?

7.52   Federal law requires companies producing wastewater that includes toxic chemicals to treat and dilute these chemicals before the wastewater reaches municipal sewage treatment plants. Violations are punishable by fines up to $2000 per day in Dallas. If one can assume that the most significant violator in the Dallas metropolitan area paid $10,158 in 1987, find the sample size necessary to estimate the mean amount in fines per company to within $800 with 95% confidence.

(*Source:* "Dallas Businesses Dumped Array of Toxic Pollutants," *Dallas Morning News*, p. 6A, March 7, 1988.)

7.53   An economist would like to estimate the rise in personal income for a particular quarter. A preliminary study shows that the rise is between 1% and 6%. What sample size would be necessary to obtain a 95% confidence interval to estimate the mean rise in personal income for the quarter to within .1%?

7.6
OTHER SAMPLING PROCEDURES (OPTIONAL)

To discuss methods of sampling other than simple random sampling, we need to define several terms. These definitions also apply to simple random sampling.

1.   **Population.** As before, this refers to the collection of people or objects of interest about which we desire to learn something. It may be as large as the set of all voting adults in the United States or as small or smaller than the set of all top-level managers in a particular company. In this section, we will assume that we are sampling from a *finite* population.

2.   **Sampling unit.** This is a collection of elements or an individual element selected from the population. Elements within one sampling unit must not overlap with the elements in other sampling units.

3.   **Cluster.** This is a sampling unit that is a group of elements from the population, such as all adults in a particular city block.

4.   **Sampling frame.** This is a list of population elements from which the sample is to be selected. Ideally, the sampling frame should be identical to the population. In many situations, however, this is impossible, in which case the frame must be *representative* of the population.

5.   **Strata.** These are nonoverlapping subpopulations. For example, the population of all cigarette smokers can be split into two strata—men and women. You can then use **stratified sampling,** in which your total sample consists of a sample selected from each individual stratum.

6.   **Sampling design.** This is a plan that specifies the manner in which the sampling units are to be selected for your sample. Examples include simple random sampling, systematic sampling, stratified sampling, and cluster sampling.

Simple Random Sampling

The results obtained when using a simple random sample were presented earlier and are summarized here for the usual case of sampling without replacement, where every sample of n elements (from a population of size N) has an equal chance of being selected.

According to the Central Limit Theorem, for large samples the distribution of the sample mean, $\bar{X}$, is approximately normal, without making any assumptions concerning the shape of the population being sampled. The resulting confidence interval for the population mean, μ, is an *approximate* confidence interval for this parameter. If you assume that the population has a normal distribution with mean μ, the confidence interval is exact.

SIMPLE RANDOM SAMPLING

- Population mean: μ
- Estimator:

$$\bar{X} = \frac{\Sigma x}{n}$$

(7.10)

- Variance of the estimator:

$$\sigma_{\bar{X}}^2 = \frac{\sigma^2}{n} \cdot \frac{N-n}{N-1}$$

(7.11)

- Approximate confidence interval: $\bar{X} \pm Z_{\alpha/2}\sigma_{\bar{X}}$

The fpc is $(N-n)/(N-1)$ and can be ignored if $n/N < .05$.

Systematic Sampling

For large populations, obtaining a random sample can be quite cumbersome. Perhaps you have just informed a group of bank tellers that you need a random sample of their customers over the next few days. For them to select people randomly would be nearly impossible. A much easier scheme would be to have them select, say, every tenth customer to be included in the sample. This is systematic sampling.

Other situations where systematic sampling is advantageous include:

1. The population consists of N records on a magnetic tape or disk. The sample of n is obtained by sampling every k th record, where k is an integer approximately equal to N/n . For example, if there are $N = 9435$ records and you need a sample of size $n = 100$, selecting every $9435/100 \cong 94$ th record would result in a systematic sample. Typically, a random starting point (record) is determined, and then every k th record is selected for your sample.

2. The population consists of a collection of files stored consecutively by date of birth. A quick (although not necessarily reliable) method of obtaining a "nearly random" sample is to select every k th file for your sample. (What could cause the sample selected from such a list not to be random?)

There are many situations in which it is dangerous to use systematic sampling. If there are obvious patterns contained in the sample frame listing, your sample may be far from random. If elements are stored according to days, for example, your sample could consist of data that all belong to Tuesday. If the data are cyclic, your sample might consist of all the peaks or all the valleys of the population. Basically, systematic sampling works best when the order of your population is fairly random with respect to the measurement of interest.

Despite its dangers, systematic sampling can provide an easy method of obtaining a representative sample. If the order of your population is in fact random (no cycles, no obvious patterns of any kind), a systematic sample can be analyzed as though it were a simple random sample.

Stratified Sampling

Suppose that you own a chain of four tire stores in four different cities and you are interested in the average amount due on delinquent accounts. The population sizes of these four cities differ considerably, ranging from a small store in an east Texas town to a large store in downtown Houston. To obtain a random sample, you could combine the delinquent accounts from all four stores into one large population and obtain your random sample from this group of accounts. On the other hand, because of the different sizes, locations, and credit policies of the stores, you might want to sample the stores individually. You could obtain the largest sample from the Houston store and smaller samples from the smaller stores. This is proportional stratified sampling.

Stratified sampling is used when the population can be physically or geographically separated into two or more groups (strata), where the variation within the strata is less than the variation within the entire population. The cost of obtaining the stratified sample may be less than that of collecting a random sample of the same size, especially if the sampling units are determined geographically.

The advantages of stratified sampling are:

1. By stratifying, we can obtain more information from the sample because data are more homogeneous within each stratum; consequently, confidence intervals are narrower than those obtained through random sampling.
2. We do obtain a cross section of the entire population.
3. We do obtain an estimate of the mean within each stratum as well as an estimate of μ for the entire population. We use the following notation:

n_i = sample size in stratum i

N_i = number of elements in stratum i

N = total population size = ΣN_i

n = total sample size = Σn_i

$\bar{X}_i$ = sample mean in stratum i

s_i = sample standard deviation in stratum i

STRATIFIED SAMPLING

- Population mean: μ
- Estimator:

$$\bar{X}_s = \frac{\Sigma N_i \bar{X}_i}{N}$$

(7.12)

- Variance of the estimator:

$$\sigma^2_{\bar{X}_s} = \frac{\Sigma N_i^2 \left(\dfrac{N_i - n_i}{N_i - 1} \right) \dfrac{s_i^2}{n_i}}{N^2}$$

(7.13)

- Approximate confidence interval: $\bar{X}_s \pm Z_{\alpha/2} \sigma_{\bar{X}_s}$

One method often used to determine the strata sample sizes, n_i, is to select each sample size proportional to stratum size. Consequently,

$$n_i = n \left(\frac{N_i}{N} \right)$$

In this way, you obtain larger samples from the larger strata.

Because you desire an estimator with *small* variance (that is, one that will not drastically vary from one data set to the next), you should attempt to create strata such that the individual variances, s_i^2, are as small as possible.

Assume you would like to obtain a sample of size 20 from the chain of four tire stores. You want to use a stratified sample with proportional sample sizes because each of the stores has a different volume of customers, credit policy, and credit ceiling. Here, N_i is the number of delinquent accounts at each store; $N_1 = 72$, $N_2 = 39$, $N_3 = 25$, and $N_4 = 44$. So

$$N = 72 + 39 + 25 + 44 = 180 \quad \text{delinquent accounts}$$

Your sample sizes are:

$$n_1 = 20\left(\frac{72}{180}\right) \qquad n_3 = 20\left(\frac{25}{180}\right)$$

$$\cong 8 \qquad \cong 3$$

$$n_2 = 20\left(\frac{39}{180}\right) \qquad n_4 = 20\left(\frac{44}{180}\right)$$

$$\cong 4 \qquad \cong 5$$

The randomly selected accounts are analyzed to find the dollar amounts due on delinquent accounts. The sample results are:

	STORE 1	STORE 2	STORE 3	STORE 4
	$150	$ 82	$186	$321
	175	106	162	285
	216	98	174	306
	205	110		356
	182			332
	240			
	195			
	213			
N_i	72	39	25	44
n_i	8	4	3	5
$\bar{X}_i$	$197.00	$ 99.00	$174.00	$320.00
s_i	$ 27.91	$ 12.38	$ 12.00	$ 26.75

To estimate μ from these data,

$$\bar{X}_s = \frac{(72)(197) + (39)(99) + (25)(174) + (44)(320)}{180}$$

$$= \$202.64$$

Also

$$\sigma_{\bar{X}_s}^2 = \left[(72)^2\left(\frac{72 - 8}{72 - 1}\right)\left(\frac{(27.91)^2}{8}\right) + (39)^2\left(\frac{39 - 4}{39 - 1}\right)\left(\frac{(12.38)^2}{4}\right) \right.$$

$$\left. + (25)^2\left(\frac{25 - 3}{25 - 1}\right)\left(\frac{(12.00)^2}{3}\right) + (44)^2\left(\frac{44 - 5}{44 - 1}\right)\left(\frac{(26.75)^2}{5}\right) \right] \div (180)^2$$

$$= 787,475.21 \div 32,400 = 24.305$$

Consequently,

$$\sigma_{\bar{X}_s} = \sqrt{24.305} = \$4.93$$

The corresponding approximate 95% confidence interval for the average overdue amount, μ, is

$$\$202.64 - (1.96)(4.93) \quad \text{to} \quad \$202.64 + (1.96)(4.93) = \$192.98 \quad \text{to} \quad \$212.30$$

So we are 95% confident that the average delinquent amount for the four stores is between $192.98 and $212.30.

Cluster Sampling

We can sample clusters (groups) within the population rather than collecting individual elements one at a time. For example, to determine the opinions of the members of a particular labor union, you might interview everyone attending several of the local meetings. Of course, the danger here is that possibly (1) the people attending the local meetings that were sampled (clusters) do not represent the population of all voting members, and (2) the people attending the local meetings do not provide an adequate representation of the local members. As a general rule, it is advisable to select many small clusters rather than a few large clusters to obtain a more accurate representation of your population.

Cluster sampling is preferred to (and less costly than) random and stratified sampling when:

1. The only sampling frame that can be constructed consists of clusters (for example, all people in a particular household, city block, or Zip code).
2. The population is extremely spread out, or it is impossible to obtain data on all the individual members.

When using cluster sampling, you should *randomly* select a set of clusters (once they have been clearly defined) for sampling. You can then include all individuals within each cluster selected for the sample **(single-stage cluster sampling)** or randomly select individuals from the sampled clusters to be included in the sample **(two-stage cluster sampling).**

We use the following notation:

M = total number of clusters in the population

m = number of clusters randomly selected for the sample

n_i = number of elements in sample cluster i

$\bar{n}$ = average cluster size of the sampled clusters ($\bar{n} = \Sigma n_i/m$)

N = total population size (N = total of all M cluster sizes that make up the population)

$\bar{N}$ = average cluster size for the population ($\bar{N} = N/M$)

T_i = total of all observations within cluster i (required for the sampled clusters only)

CLUSTER SAMPLING (SINGLE STAGE)

- Population mean: μ
- Estimator:

$$\bar{X}_c = \frac{\Sigma T_i}{\Sigma n_i} \qquad (7.14)$$

- Variance of estimator:

$$\sigma^2_{\bar{X}_c} = \left(\frac{M - m}{mM\bar{N}^2}\right) \frac{\Sigma (T_i - \bar{X}_c n_i)^2}{m - 1} \qquad (7.15)$$

If $\bar{N}$ is unknown, this can be replaced by its estimate, $\bar{n}$.

- Approximate confidence interval: $\bar{X}_c \pm Z_{\alpha/2}\sigma_{\bar{X}_c}$

As marketing director for a cable-television company in a large city, you are trying to decide whether to begin a major advertising campaign to reach tenants in local high-rise apartment buildings. Your staff disagree about whether or not this is a good idea. One group of your employees feels that people living in high-rise apartments are always on the go and are not likely to spend much time watching television—cable or network. The others tend to believe that such tenants have very little grass to mow and few leaves to rake and so have a great deal of time to spend watching television.

Rather than drawing a sample from all high-rise tenants, you construct a sampling frame consisting of all 18 ($= M$) high-rise apartment complexes. From these, you randomly select a sample of $m = 4$ complexes (clusters). Each tenant in these four complexes is then asked how many hours per week he or she watches television. You obtain the following results:

	COMPLEX 1	COMPLEX 2	COMPLEX 3	COMPLEX 4
Number of units (n_i)	260	220	310	274
Total number of hours per cluster (complex)	2475	2750	3160	4110

You begin by noting

N = the total number of units in the 18 high-rise complexes (population) = 4590

$$\bar{N} = \frac{4590}{18} = 255$$

$$\Sigma(T_i - \bar{X}_c n_i)^2 = \Sigma T_i^2 - 2\bar{X}_c \Sigma T_i n_i + \bar{X}_c^2 \Sigma n_i^2 \qquad (7.16)$$

Using the sample data,

$$\Sigma T_i = 2475 + \cdots + 4110 = 12{,}495$$
$$\Sigma n_i = 260 + \cdots + 274 = 1064$$
$$\Sigma T_i^2 = (2475)^2 + \cdots + (4110)^2 = 40{,}565{,}825$$
$$\Sigma T_i n_i = (2475)(260) + \cdots + (4110)(274) = 3{,}354{,}240$$
$$\Sigma n_i^2 = (260)^2 + \cdots + (274)^2 = 287{,}176$$

As a result,

$$\bar{X}_c = \frac{\Sigma T_i}{\Sigma n_i} = \frac{12{,}495}{1064} = 11.743$$

Also, using equation 7.16,

$$\frac{\Sigma(T_i - \bar{X}_c n_i)^2}{m-1} = [40{,}565{,}825 - (2)(11.743)(3{,}354{,}240) + (11.743)^2(287{,}176)] \div 3$$
$$= 463{,}051.49$$

Consequently,

$$\sigma_{\bar{X}_c}^2 = \frac{18-4}{(4)(18)(255)^2} \cdot 463{,}051.49 = 1.385$$

and so

$$\sigma_{\bar{X}_c} = \sqrt{1.385} = 1.177$$

The resulting approximate 95% confidence interval for the average number of television hours (for all 18 complexes) is

$$11.743 - 1.96(1.177) \quad \text{to} \quad 11.743 + 1.96(1.177) = 9.44 \text{ hours} \quad \text{to} \quad 14.05 \text{ hours}$$

Therefore, we are 95% confident that μ lies between 9.44 and 14.05 hours and that we have estimated μ to within 2.3 hours. This example has illustrated single-stage cluster sampling where everyone in the selected clusters (apartment complexes) was used in the sample. Another look at this example indicates that a two-stage cluster procedure might be more practical where each of the four sample clusters is also sampled to obtain the final sample.

EXERCISES

7.54 A real estate agent would like to estimate the average price of a home in the suburbs of a major metropolitan city. The agent decides to use stratified random sampling. The population of homes is stratified into the five major suburbs. The results of the stratified sample yield the following statistics. Construct a 95% confidence interval for the mean price of a home in units of one thousand.

STRATUM	N_i: NUMBER OF HOUSES	n_i: SAMPLE SIZE	$\bar{X}_i$ (IN THOUSANDS)	s_i^2
Suburb 1	150	22	101.2	64.2
Suburb 2	220	33	80.7	24.3
Suburb 3	140	21	61.4	20.8
Suburb 4	70	11	139.6	53.5
Suburb 5	90	13	76.8	30.1
	670	100		

7.55 An advertising firm would like to estimate the amount of money spent per month on advertising by certain retail stores in an industrial sector of northeast New Jersey. Three sizes of retail stores were chosen—small, medium, and large. The random sample for each stratum yielded the following values. Construct a 90% confidence interval for the mean monthly advertising expenditure of retail stores.

STRATUM	N_i: NUMBER OF STORES	n_i: SAMPLE SIZE	MONTHLY ADVERTISING EXPENDITURE (IN THOUSANDS)
Large	40	8	2.1 1.6 1.8 1.2 0.7 2.6 0.9 0.8
Medium	112	22	0.5 0.7 0.9 1.1 0.6 1.4 1.7 0.4
			0.8 0.7 0.9 0.7 0.9 1.3 1.1 1.2
			0.4 0.3 0.8 0.6 0.9 1.1
Small		15	0.3 0.4 0.3 0.6 0.5 0.4 0.3 0.4
	80		0.1 0.3 0.2 0.4 0.5 0.2 0.7
	232	45	

7.56 Basic Microcomputers would like to market its version of the professional computer. To price the professional computer and its peripheral equipment properly, a survey is taken among the lower-middle, upper-middle, and high income groups to find out what a businessperson would be willing to pay. The survey was restricted to a certain city in an industrial area. A stratified sample among these three groups yielded the following statistics. Construct a 90% confidence interval for the mean price that a professional businessperson would be willing to pay, in units of one thousand.

INCOME LEVEL	N_i	n_i	$\bar{X}_i$	s_i^2
Lower-middle	8,641	56	2.3	1.6
Upper-middle	14,683	95	4.6	1.9
High	7,457	49	4.8	1.4
	30,781	200		

7.57 A market-research firm would like to estimate the average number of hours that a householder spends shopping each week. Four neighborhoods were selected from a total of 24 neighborhoods for sampling purposes. Find a 90% confidence interval for the mean number of hours that a householder spends shopping each week from the following data (units are in hours per week):

1	2	3	4
2.3	5.4	1.6	5.6
1.1	4.2	0.9	4.1
4.3	3.6	4.6	2.3
0.5	7.2	5.4	7.3
3.7	8.4	5.6	6.1
4.6	11.8	4.6	4.7
10.1	2.1	7.1	5.8
6.3	1.5	3.2	4.8
7.8	8.1	4.5	5.3
8.4	4.1	3.1	1.9
7.9	3.4	2.6	8.4
10.6			

7.58 In what situation is the use of systematic sampling appropriate? Explain how a systematic sample would be taken from a file of students listed by social security number.

7.59 The administration of Digital Systems would like to obtain an estimate of the amount of time workers spend on physical exercise. Five departments out of 20 in Digital Systems were selected for sampling purposes. Find a 95% confidence interval for the mean time that an employee spends on physical fitness per week given the following data (units are in hours per week):

1	2	3	4	5
1.1	2.3	3.5	7.9	0.1
0.2	0.4	4.6	1.3	0.0
2.3	0.3	1.5	2.5	7.6
4.6	1.0	0.7	5.7	5.1
0.1	4.6	3.6	7.8	4.0
0.0	8.3	9.5	10.3	3.0
2.6	7.1	0.8	0.6	6.5
6.8	0.2			
1.1	2.7			

SUMMARY

This chapter introduced you to **statistical inference,** an extremely important area of statistics. Inference procedures were used to estimate a certain unknown *parameter* (such as the mean, μ, or the standard deviation, σ) of a population by using the corresponding sample *statistic* (such as the sample mean, $\bar{X}$, or the sample standard deviation, s).

The **Central Limit Theorem** states that for large samples, the sample mean $\bar{X}$ always follows an approximate normal distribution. If, in addition, you assume that the population is normally distributed, then $\bar{X}$ will follow an exact normal distribution. The strength of the Central Limit Theorem is that no assumptions need be made concerning the shape of the population, provided the sample is large ($n > 30$). The Central Limit Theorem allows you to make probability statements concerning $\bar{X}$, such as $P(\bar{X} < 150)$. When sampling without replacement from a finite population, the standard deviation of the normal distribution for $\bar{X}$ (the **standard error**) is obtained by including a **finite population correction factor (fpc),** which adjusts the standard error by including the effect of the known population size, N.

FIGURE 7.21

The correct table to use for constructing a confidence interval for a population mean.

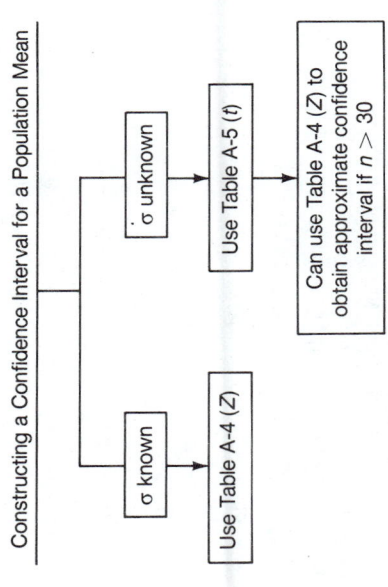

Constructing a Confidence Interval for a Population Mean

The sample mean, $\bar{X}$, provides a **point estimate** of μ because it estimates this parameter using a single number. A **confidence interval** for μ measures the precision of the point estimate. If the population standard deviation σ is known, the standard normal table (Table A.4) is used to derive the confidence interval. If σ is unknown, it can be replaced by its estimate—the sample standard deviation, s. This provided an introduction to the **t distribution**. The corresponding confidence interval for μ is constructed using the t table (Table A.5) and assumes that the sampled population is normally distributed (that is, that μ is the mean of a normal population). For sample sizes (n) greater than 30, the standard normal table can be used to construct an approximate confidence interval for the population mean when σ is unknown. A summary of this procedure is contained in Figure 7.21.

For many applications, the precision of the point estimate, $\bar{X}$, is specified using the **maximum error**, E. When constructing a confidence interval, E is the amount that is added to and subtracted from the point estimate to obtain the endpoints of the desired interval. The sample size n necessary to achieve a desired accuracy can be obtained using a specified value of E. The population standard deviation can be estimated from a preliminary sample, or a rough approximation procedure can be used.

Simple random sampling is employed when every sample of n elements has an equal chance of being selected. Nonrandom sampling procedures can often be used to obtain a more precise estimate of the population mean, μ, providing confidence intervals for this parameter that are much narrower. These sampling techniques include systematic, stratified, and cluster sampling. **Systematic** sampling selects a random starting point and then selects every kth value for some counting number $k > 0$. This procedure assumes that the population is stored sequentially in some manner such as a computer file. **Stratified** sampling is used whenever the population can be physically or geographically separated into two or more groups (strata) where the variation within the strata is less than the variation within the entire population. When sampling groups of people (clusters) within the population, rather than selecting individual elements one at a time, **cluster** sampling is used.

REVIEW EXERCISES

7.60 Medical Products Consolidated wishes to estimate the yearly maintenance costs of the mechanical ventilators that hospitals buy from them. Fifteen hospitals were randomly chosen from a total of 45. From the data, construct a 90% confidence interval for the mean yearly maintenance costs (in thousands of dollars) of the ventilators.

HOSPITAL	NUMBER OF VENTILATORS	MAINTENANCE COST	HOSPITAL	NUMBER OF VENTILATORS	MAINTENANCE COST
1	2	1.2	9	5	2.3
2	5	2.3	10	4	1.9
3	2	0.6	11	5	2.8
4	7	4.1	12	7	3.8
5	6	3.0	13	6	3.4
6	5	2.0	14	4	1.8
7	4	1.3	15	5	2.7
8	2	0.9			

7.61 An accounting firm has a large pool of secretaries. It is assumed that the time it takes a secretary to type a certain legal document is normally distributed with a mean time of 30 minutes and a standard deviation of 4 minutes.

a. What is the probability that a secretary will spend less than 27 minutes typing the legal document?

b. In a randomly selected sample of six secretaries, what is the probability that the average time that it takes them to type the legal document is less than 27 minutes?

7.62 At a manufacturing firm approximately 300 similar machines are used in a production process for manufacturing certain steel products. It is discovered that the number of times per day that a machine needs to be readjusted for going out of control has a Poisson distribution with mean one.

a. Describe the distribution of the average number of times a daily readjustment is necessary per machine.

b. Find the probability that the average number of readjustments per machine is less than 1.1 on a certain day.

7.63 The ages of applicants for the position of manager of a particular fast-food restaurant are considered to be approximately normally distributed. A sample of the ages of 36 applicants selected at random yielded a sample mean of 29.4 years with a sample standard deviation of 2.1 years. Construct a 95% confidence interval for the mean age of applicants for this position.

7.64 A local merchant would like to estimate the mean amount of money that a family spends at the state fair to within \$15. If the amount spent by a family is considered to be normally distributed with a standard deviation of \$27, what sample size would be necessary to be 90% confident?

7.65 A random variable is found to range from a high of 50 to a low of 25 from past data. The distribution of the random variable can be approximated by a normal distribution. To estimate the mean of the random variable to within 2.1, what sample size would be necessary in selecting a random sample to achieve a 99% confidence level?

7.66 A wholesale furniture store has 160 dining tables that have a mean weight of 47.3 pounds with a standard deviation of 9.8 pounds. If 15 tables are randomly selected, what is the probability that the average weight of the 15 tables will be between 41 and 56 pounds? What are you assuming about the distribution of the table weights?

7.67 The research and development department of a large oil company employs 253 engineers who have an average of 6.2 years of practical experience with a standard deviation of 2.1 years.

a. If a sample of 35 engineers is selected randomly without replacement, what is the probability that the average number of years of experience of the sample will be greater than 6.8 years?

b. If a sample of 35 engineers is selected randomly with replacement, what is the probability that the average number of years of experience will be greater than 6.8 years?

7.68 A confidence interval for the mean of a normally distributed population is found to range from 70.1 to 80.2. What is the level of confidence for the confidence interval if the sample size is 36 and the population standard deviation is 13.2?

7.69 The tensile strength of a high-powered copper coil used in giant power transformers is believed to follow a normal distribution. A sample of 14 high-powered copper coils yields the following tensile strength (in units of thousands of pounds per square inch). Construct a 90% confidence interval for the mean tensile strength of copper coils.

6.1, 2.6, 3.5, 4.3, 3.1, 5.2, 3.6, 3.5, 5.4, 4.2, 3.2, 2.8, 4.0, 3.7

7.70 The mean number of defectives found in a box of electrical resistors is 10 with a population standard deviation of 4. If 11 boxes are selected at random, what is the probability that the average number of defectives will be between 9 and 12? What are you assuming about the probability distribution here?

7.71 Roadway Express was the biggest LTL (less-than-truckload) carrier in the United States in 1987. Roadway's director of safety firmly believes that goods must be delivered undamaged and on time. The company averages about 2 accidents every one million miles on the highway. Assume that the number of accidents every one million highway miles can be approximated by a Poisson distribution. Suppose a random sample of 35 million highway miles are selected at random. Let the random variable X represent the number of accidents per million highway miles.

(*Source:* "Roadway Plays It Safe," *Purchasing* (November 1987): 21–2.)

a. Describe the sampling distribution of $\bar{X}$. What are the mean and variance of $\bar{X}$?

b. Find the probability that the average number of accidents per one million highway miles in the sample of 35 is between 1.5 and 2.0.

7.72 A personnel administrator for Teltronix would like to estimate the amount of term life insurance that an employee carries. Three strata are used for finding a stratified random sample of all employees. From the data, construct a 95% confidence interval for the mean amount of term life insurance that an employee carries (given in units of one thousand dollars).

STRATUM	TOTAL NUMBER IN COMPANY N_i	n_i	$\bar{X}_i$	s_i
Employees paid by the hour	350	67	28.5	5.7
Engineers and technicians	112	22	80.6	10.3
Management	57	11	125.2	13.6
	519	100		

7.73 A machine at a manufacturing plant fills sacks with 10 pounds of oats. Each case contains ten sacks of oats. The quality-control engineer would like to find a 99% confidence interval for the mean weight per sack of oats. Using cases as clusters, construct the confidence interval if eight cases are chosen at random. Assume that there is a total of 50 cases from which to choose.

CASES	SACKS PER CASE	WEIGHT OF CASE	CASES	SACKS PER CASE	WEIGHT OF CASE
1	10	96.7	5	10	110.8
2	10	99.8	6	10	104.6
3	10	103.5	7	10	93.5
4	10	92.7	8	10	112.3

7.74 The average weekly wage for workers in the manufacturing industry in April 1988 was $414.92. Suppose that the population standard deviation of the wage in the manufacturing industry is $100. What is the probability that the average weekly wage of 35 workers in the manufacturing industry is less than $395 for April 1988?

(*Source: USA Today,* May 23, 1988, Section B, p. 1.)

7.75 A binomial random variable can be thought of as a sum of 1s and 0s, where 1 represents a success on a trial and 0 represents a failure on a trial. Dividing the binomial random variable by n gives an estimate of the proportion of successes. This estimate of the proportion is equivalent to a sample mean of 1s and 0s. Show that the central limit theorem can be applied to the estimate of the proportion of successes in a binomial experiment.

7.76 As described in Exercise 7.75, the estimate of the proportion of successes in a binomial experiment can be treated the same as $\bar{X}$. In the following MINITAB computer printout, 1000 random observations from a binomial distribution with $n = 35$ and $p = .40$ are generated. A histogram is printed to represent the distribution of the estimates of the proportion—that is, the distribution of the values of the binomial random variables divided by $n = 35$. Does the distribution of the estimate of the proportion appear to comply with the empirical rule? Approximately what percentage of the observations are within one standard deviation of the true proportion, two standard deviations of the true proportion, and three standard deviations of the true proportion, respectively? Assume that the population standard deviation of the estimate of the proportion is $\sqrt{p(1-p)/n}$.

```
MTB > random 1000 c1;
SUBC> binomial n = 35 p = .40.
MTB > let c2 = c1/35
MTB > erase c1
MTB > histogram c2

Histogram of C2   N = 1000
Each * represents 5 obs.

Midpoint   Count
   0.10       1   *
   0.15       2   *
   0.20      10   **
   0.25      51   ***********
   0.30     123   *************************
   0.35     246   **************************************************
   0.40     138   ****************************
   0.45     223   *********************************************
   0.50     144   *****************************
   0.55      52   ***********
   0.60       5   *
   0.65       4   *
   0.70       1   *
```

7.77 The following MINITAB computer printout shows the results of simulating 50 times a sample of 13 observations from a uniform distribution over the interval 0 to 20. According to the central limit theorem, the distribution of the sample mean should be approximately normal for large sample sizes. However, the distribution of the sample mean from a population with a uniform distribution approaches a normal distribution

```
MTB > let k2 = 1
MTB > store
MTB > random 13 observations, put into c1;
SUBC> uniform with continuous uniform on 0 to 20.
MTB > mean data in c1, put in k1
MTB > let c2(k2) = k1
MTB > let k2 = k2 + 1
MTB > end
MTB > execute 50 times
MTB > random 13 observations, put into c1;
SUBC> uniform with continuous uniform on 0 to 20.
  .
  .
  .
MTB > histogram of c2

Histogram of C2   N = 50

Midpoint   Count
    6        1   *
    7        2   **
    8        6   ******
    9        7   *******
   10       13   *************
   11       10   **********
   12        8   ********
   13        1   *
   14        1   *
   15        1   *
```

for fairly small sample sizes. Comment on the shape of the histogram of the sample mean. Repeat this simulation experiment with sample sizes of $n = 18$ and 25. Comment on the shapes of the histogram of the sample mean.

COMPUTER EXERCISES USING THE DATABASE

Exercise 1—Appendix H

Select the first 400 observations of the database as the population of interest. Estimate the mean of HPAYRENT (house payment or house/apartment rent) for this population by taking a simple random sample of size 40. Also estimate the mean by taking a stratified sample proportional to the size of the two strata. Let stratum one be the group of observations in which a secondary wage earner (variable INCOME2) has a positive income and let stratum two be the group of observations in which there is no secondary wage earner. Compare the confidence intervals on mean house payment/rent (HPAYRENT) for both the simple random sample and the stratified random sample.

Exercise 2—Appendix H

Select a simple random sample of 32 observations from the database. Calculate a 95% confidence interval on the mean income of the principal wage earner (variable INCOME1). Select another simple random sample of size 60 from the database in Appendix H. Calculate a 95% confidence interval on the mean income of the principal wage earner. Comment on the widths of these two confidence intervals.

Exercise 3—Appendix I

Generate 20 random samples of size 10 using the data for the variable ASSETS (current assets). For each sample determine the sample mean. Construct a histogram of these 20 sample means. Also find the mean and the variance of these 20 values. Repeat this procedure for samples of size 25 and 40. Does the Central Limit Theorem appear to be operating correctly here? Discuss.

CASE STUDY

OPTIMIZING A PROCESS WITH CONTROL CHARTS—HOW X-BARS SAVED NATIONAL SEMICONDUCTOR SOME GOLD BARS

Control charts are one of the most widely used tools of statistical process control (SPC) and statistical quality control (SQC). Introduced by Walter A. Shewhart in 1925, they are often called Shewhart charts. Many variations have been developed, but the most well known are probably the $\bar{x}$ (x-bar) chart and the R chart. These are known as *control charts for variables*, the latter in this case being the mean and range of observations obtained from sampling a process.

In this case study, we will look at the $\bar{x}$ chart. First, we consider a bit of theory. The central idea behind all control charts is that variation is inevitable and inherent in any production process. If any process were "frozen" at any instant, we would get a single observation that would tell us very little. But if groups of observations are taken and this sampling of the process is repeated at regular intervals, then statistical theory tells us that certain predictable patterns can be expected. This should remind you of the sampling distribution of $\bar{x}$ and the Central Limit Theorem. It is precisely the predictability associated with repeated sampling that is exploited in control charts.

In a typical application, a mean (process norm) is established, either from specifications known beforehand, or by observation of a process while it is "in control," i.e, stable and behaving properly. Three standard deviations above and below the mean, an upper control limit (UCL) and a lower control limit (LCL), respectively, are established. Subsequently, as the process is run, samples are taken at regular intervals ($n = 5$ is very common). The mean of each such sample is plotted over time, resulting in an $\bar{x}$ chart, where a central line is the (assumed) process mean, and the control limits are $\bar{x} \pm 3\sigma_{\bar{x}}$ usually. (The corresponding R chart, representing range = highest − lowest for each sample, is plotted below the $\bar{x}$ chart.)

Interpretation of such a chart is simple. By virtue of the Central Limit Theorem, sample averages above or below $3\sigma_{\bar{x}}$ are very unusual. So long as sample averages that exceed the limits are like red flags, alerting the operator that the process may be unstable or "out of control." As opposed to random influences, the causes of instability are often referred to as *assignable causes.*

The preceding description represents the classical theory of control charts. We now look at how managerial discretion can extend the classical theory, by shifting the focus from *controlling* to *optimizing.* We introduce now Mr. M. N. Bhatt, Director of Quality Assurance at Dyna-Craft, Inc., a wholly owned subsidiary of National Semiconductor (NS), a worldwide manufacturer of integrated chips. Dyna-Craft is the stamping and plating arm of NS.

Mr. Bhatt argues, "The classical theory assumes that the process is well defined, whereas in actual fact we are often fine-tuning it. You should interrogate your process, *discover* something about your process."

Our example involves a part called a lead frame, which is used in IC assembly. Large reels of the raw material are fed into a process, each reel producing 30,000 to 50,000 parts. A couple of reels may be used per day. A crucial part of the process involves selectively plating a part with gold by electro-deposition. A minimum plating thickness of 50 microinches of gold has been specified. A microinch is one thousandth of an inch. As you know, gold is extremely expensive to use in a mass produced item. If Dyna-Craft used too little gold, the plating would be too thin and would not meet specifications; if they made the plating too thick, this would push the cost up. Thickness was determined by the amount of electric current and the length of time it was applied; this was the process variable to be fine-tuned.

A process has to start somewhere. Initial settings are based on established engineering theory, general definitions of each type of process and customer specifications. These are determined by the engineering personnel. At the beginning and the end of a reel, five observations are taken. Since the process is continuous, the end of one reel is the beginning of the other. This means that five observations per reel are obtained. Data are entered manually into a computer, which computes the sample mean and range automatically. Table 7.3 provides the readings from May 2, 1988 to May 24, 1988 for the $\bar{x}$ and R (range).

Mr. Bhatt refers to the control limits of $\mu \pm 3\sigma_{\bar{x}}$ as the "process window" within which the process operates. None of the reels had a mean outside the control chart limits. Also, none of the days had a reading below specifications. Nevertheless, Mr. Bhatt decided to shift the process window upward by increasing the plating time and, thereby, the plating thickness. Table 7.4

TABLE 7.3

REEL	DAY	x̄	R	REEL	DAY	x̄	R
1	May 2	61.4	14	24	May 9	61.6	11
2	May 3	53.8	6	25	May 10	62.2	9
3		58.0	11	26		60.0	12
4		55.8	5	27		59.4	7
5		59.6	10	28		65.4	9
6	May 4	74.2	11	29	May 11	53.6	7
7		71.2	8	30		60.6	13
8		69.4	10	31		58.2	11
9		81.8	7	32		59.4	11
10		78.6	21	33	May 12	60.4	14
11	May 5	70.8	13	34		62.2	14
12		58.8	12	35		60.6	12
13		61.2	11	36		60.4	8
14		57.0	8	37		57.8	8
15		61.6	9	38		61.2	10
16	May 6	62.4	10	39	May 13	58.2	11
17		55.2	7	40		60.0	8
18		56.2	6	41		55.6	9
19		58.0	7	42	May 16	56.0	10
20	May 7	54.8	8	43	May 24	59.8	15
21	May 9	59.2	7	44		59.0	12
22		52.0	3				
23		59.0	7				

TABLE 7.4

REEL*	DAY	x̄	R	REEL	DAY	x̄	R
15	May 25	64.0	13	35	Jun 2	76.2	9
16		65.0	11	36		73.6	16
17	May 27	87.0	14	37		92.4	2
18		65.6	9	38		80.4	18
19		93.8	19	39		74.0	13
20	May 31	74.6	11	40	Jun 3	70.8	5
21		78.6	43	41		68.8	11
22		63.2	6	42		65.0	4
23		73.4	15	43		66.6	11
24	Jun 1	60.8	9	44		73.6	11
25		76.0	12	45		76.6	16
26		66.0	14	46	Jun 4	85.2	16
27		75.2	11	47		85.2	16
28		80.8	21	48		83.6	5
29		69.6	12	49		86.0	12
30		69.0	5	50		65.8	4
31		80.4	6	51	Jun 6	59.8	23
32		70.8	10	52		60.6	6
33		74.0	7	53		76.2	6
34		80.0	7	54		77.4	9

(*Source*: Data supplied by Mr. M. N. Bhatt, Director of Quality Assurance, Dyna-Craft, Inc., 2919 San Ysidro Way, Santa Clara, CA 95051. Dyna-Craft is a wholly owned subsidiary of National Semiconductor. Mr. Bhatt's contribution to this case study is gratefully acknowledged.)

*Reels 1—14 are unrelated to this process and are hence omitted, but original numbering is retained.

provides the readings after this shift in the process was effected, from May 25, 1988, to June 6, 1988. This decision might seem surprising, since it meant increasing the cost, but Mr. Bhatt had to weigh that against the demands of quality assurance. Without the use of statistics, Dyna-Craft would have had to "play safe" and actually use a thicker gold plating

Case Study Questions

1. For the data in Table 7.3, compute the grand mean (average of the means) and the upper and lower control limits based on a process window of $\pm 3\sigma_{\bar{x}}$. The grand mean is often written by SQC practitioners as $\bar{\bar{x}}$ (x-double bar). Following the notation used in this book, which parameter does it estimate?

2. What is the probability that the $\bar{x}$ for any individual reel falls
   (a) below the lower control limit,
   (b) above the upper control limit, and (c) outside the process window?

3. Since the *minimum* thickness has to be 50 microinches, this should be the (ideal) lower limit for control purposes; batches below this thickness would then have a very low probability of occurring. Consider now the process described in Table 7.3. What is the probability that in this process a sample mean lower than 50 microinches will occur? Is this acceptable for our process window? Justify Mr. Bhatt's decision to shift the process mean upwards.

4. After the process mean had been shifted up, the LCL was still below 50 microinches. The standard deviation had increased a bit, but more importantly, the deviations were coming more from the higher side. Classical theory assumes a symmetrical normal curve, but this process distribution was skewed. At this point managerial discretion was invoked. Although the LCL was still less than 50, it was felt that there was no need to adjust the process upward anymore. Comment on this line of reasoning.

5. Suppose you wish to raise the process window a little more, such that the LCL is just about 50 microinches. Assuming a standard deviation of 9 microinches, what process mean should you aim for, and what will be the UCL?

6. Try to relate the discussion on confidence intervals in this chapter to the upper and lower control limits of the control chart.

or risk a serious violation of the specifications, whereas by monitoring the process with control charts, they were able to move the process mean as *low* as it could safely get. As Mr. Bhatt put it, they were *optimizing* the process. In the long run, the x-bars would save Dyna-Craft and NS a few gold bars!

Hypothesis Testing for the Mean and Variance of a Population

A Look Back/Introduction

We have seen that statistical inference is used to estimate a population parameter using a sample statistic. For the rest of this book, the mean (μ) and standard deviation (σ) of the parent population will be unknown and will have to be estimated from the sample. Do not forget that even though you have estimated μ or σ, these values still are unknown and will forever remain unknown.

As a measure of how reliable your point estimate of the population mean μ really is, you can determine a confidence interval for this parameter. For a given confidence level, the narrower your resulting confidence interval is, the more faith you can have in the ability of your sample mean, $\bar{X}$, to provide an accurate estimate of the population mean. Also, when the Central Limit Theorem is applicable, you need not worry about the shape of the parent population (normal, exponential, and so on), when making probability statements regarding $\bar{X}$, provided you have a large sample (generally, $n > 30$). When you do have a large sample, the distribution of $\bar{X}$ closely approximates the normal. This allows you to construct confidence intervals for population means without worrying about the nature of the parent population, simply because it doesn't matter.

Next, we turn to the situation in which someone makes a claim regarding the value of the population mean, μ. For example, when dealing with the lifetime of Everglo light bulbs in Chapter 6, we assumed that the population average of *all* bulbs was $\mu = 400$ hours. Where did this value come from? Suppose that Everglo advertisements claim that the average lifetime of the bulbs is 400 hours. By testing a sample of bulbs, can we prove this statement? The answer is an emphatic no; the only way to know the value of μ exactly is to obtain data for *all* Everglo bulbs; that is, obtain the entire population.

The sample, however, may allow us to reject the claim that μ is 400 hours. However, since the sample is only a portion of the population, this conclusion may be incorrect. Such is the nature of hypothesis testing.

8.1
HYPOTHESIS TESTING ON THE MEAN OF A POPULATION: LARGE SAMPLE

A newspaper article claims that the average height of adult males in the United States is not the same as it was 50 years ago; it claims the average height is now 5.9 feet (approximately 5'11"). Your firm manufactures clothing, so the value of this population mean is of vital interest to you. To investigate the article's claim, you randomly select 75 males and measure their heights.* Your results for $n = 75$ are $\bar{x} = 5.76$ feet and $s = .48$ feet.

Let μ represent the population average (mean) of all U.S. male heights. We do have a point estimate of μ, $\bar{x} = 5.76$ feet is an estimate of μ. Keep in mind that the actual value of μ is unknown (although it *does exist*) and will remain that way. What we can do is estimate μ using the sample data. This situation can be summarized by considering the following pair of hypotheses.

Null hypothesis:

H_0: $\mu = 5.9$

Alternative hypothesis:

H_a: $\mu \neq 5.9$

H_0 asserts that the value of μ that has been claimed to be correct is in fact correct. H_a asserts that μ is some value other than 5.9 feet. The alternative hypothesis typically contains the conclusion that the researcher is attempting to demonstrate using the sample data. In our height example, if you do not believe that the average height is 5.9 feet and you expect the data to demonstrate that μ has some other value, H_a is $\mu \neq 5.9$.

The task of all tests of hypothesis is to **reject** H_0 or **fail to reject** H_0. Notice that we do not say "reject H_0 or accept H_0." This is an important distinction.

In our study of male heights, the (point) estimate of μ is $\bar{x} = 5.76$ feet. Should we reject H_0, given that it claims that μ is 5.9? First, we need not worry about the shape of the underlying population of male heights because, by the Central Limit Theorem, $\bar{X}$ is approximately normally distributed for large samples, regardless of the shape of this population. So, $\bar{X}$ is approximately a normal (and thus continuous) random variable. What is the probability that *any* continuous random variable is equal to a certain value? In particular, what is the probability that $\bar{X}$ is exactly equal to 5.9 feet? The answer to both questions is zero. Thus we see that we cannot reject H_0 simply because $\bar{X}$ is not equal to 5.9 feet. What we do is to allow H_0 to stand, provided $\bar{X}$ is "close to" 5.9 feet, and reject H_0 otherwise. To define what "close" means, we need to take an in-depth look at what happens when you test hypotheses.

Type I and Type II Errors

Because the sample does not consist of the entire population, there always is the possibility of drawing an incorrect conclusion when inferring the value of a population parameter using a sample statistic. When testing hypotheses, there are two types of possible error:

Type I error A Type I error occurs if you rejected H_0 when in fact it is true. For example, this would occur if you were to reject the claim (hypothesis) that the population mean is 5.9 feet when in fact it really is true.

---

* The size of this sample is unrealistically small (yet large, statistically).

Type II error A Type II error occurs if you fail to reject H_0 when in fact H_0 is not true. For example, a Type II error is encountered if you were to fail to reject the hypothesis that the population mean is 5.9 feet when in fact the mean is *not* 5.9 feet.

CONCLUSION	ACTUAL SITUATION	
	H_0 True	H_0 False
FAIL TO REJECT H_0	Correct decision	Type II error
REJECT H_0	Type I error	Correct decision

For any test of hypothesis, define

α = the probability of rejecting H_0 when H_0 is true

= P(Type I error)

β = the probability of failing to reject H_0 when H_0 is false

= P(Type II error)

For any test of hypothesis, you would like to have control over n (the sample size), α (the probability of a Type I error), and β (the probability of a Type II error). However, in reality, you can control only two of these: n and α, n and β, or α and β. In other words, for a *fixed* sample size, you cannot control both α and β.

Suppose you decide to set $\alpha = .02$. This means that the procedure you use to test H_0 versus H_a will reject H_0 when it is true with a probability of .02. You may wonder why we do not set $\alpha = 0$, so we would never have a Type I error. The thought of never rejecting a correct H_0 sounds appealing, but the bad news is that β (the probability of a Type II error) is then equal to 1; that is, you will *always* fail to reject H_0 when it is false. If we set $\alpha = 0$, then the resulting test of H_0 versus H_a will automatically fail to reject H_0: $\mu = 5.9$ whenever μ is, in fact, any value other than 5.9 feet. If, for example, μ is 7.5 feet (hardly the case, but interesting), we would still fail to reject H_0—not a good situation at all. We therefore need a value of α that offers a better compromise between the two types of error probabilities. (Note that for the situation where $\alpha = 0$ and $\beta = 1$, $\alpha + \beta = 1$. As later examples will demonstrate, this is *not true* in general.)

The value of α you select depends on the relative importance of the two types of error. For example, consider the following hypotheses and decide if the Type I error or the Type II error is the more serious.

You have just been examined by a physician using a sophisticated medical device, where the hypotheses under consideration are:

H_0: you do not have a particular serious disease

H_a: you do have the disease

$\alpha = P$(rejecting H_0 when it is true)

= P(device indicates that you have the disease when you do not have it)

$\beta = P$(fail to reject H_0 when in fact it is false)

= P(device indicates that you do not have the disease when you do have it)

For this situation, the Type I error (measured by α) is not nearly as serious as the Type II error (measured by β). Provided the treatment for the disease does you no serious harm if you are well, the Type I error is not serious. But the Type II error means you fail to receive the treatment even though you are ill.

We never set β in advance, only α. This will allow us to carry out a test of H_0 versus H_a. The smaller α is, the larger β is. Consequently, if you want β to be small, you choose a large value of α. For most situations, the range of acceptable α values is .01 to .1.

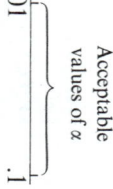

For the medical-device problem, you could choose a value of α near .1 or possibly larger, due to the seriousness of a Type II error. On the other hand, if you are more worried about Type I errors for a particular test (such as rejecting an expensive manufactured part that really is good), a small value of α is in order. What if there is no basic difference in the effect of these two errors? If there is no significant difference between the effects of a Type I error versus a Type II error, researchers often choose $\alpha = .05$.

Performing a Statistical Test

The claim that the average adult male height is 5.9 feet resulted in the following pair of hypotheses:

$$H_0: \mu = 5.9$$
$$H_a: \mu \neq 5.9$$

We decide to use a test that carries a 5% risk of rejecting H_0 when it is correct; that is, $\alpha = .05$. In hypothesis testing, α is referred to as the **significance level** of your test. Using $n = 75$, $\bar{x} = 5.76$ ft, and $s = .48$ ft, we wish to carry out the resulting statistical test of H_0 versus H_a. We decided to let H_0 stand (not reject it) if $\bar{X}$ was "close to" 5.9 feet. In other words, we will reject H_0 if $\bar{X}$ is "too far away" from 5.9 feet. We write this as follows:

$$\text{reject } H_0 \text{ if } |\bar{X} - 5.9| \text{ is "too large"}$$

or, by standardizing $\bar{X}$, we can

$$\text{reject } H_0 \text{ if } \left| \frac{\bar{X} - 5.9}{s/\sqrt{n}} \right| \text{ is "too large"}$$

We rewrite the last statement as

$$\text{reject } H_0 \text{ if } \left| \frac{\bar{X} - 5.9}{s/\sqrt{n}} \right| > k, \text{ for some } k$$

What is the value of k? Here is where the value of α has an effect. If H_0 is true and the sample size is large, then using the Central Limit Theorem, $\bar{X}$ is approximately a normal random variable with

$$\text{mean} = \mu = 5.9 \quad \text{and} \quad \text{standard deviation} \cong \frac{s}{\sqrt{n}}$$

So, if H_0 is true, $(\bar{X} - 5.9)/(s/\sqrt{n})$ is approximately a standard normal random variable, Z, for large samples.[*] In this case, we reject H_0 if $|Z| > k$, for approximately a normal random variable.

---

[*] In Chapter 7, we mentioned that $(\bar{X} - 5.9)/(s/\sqrt{n})$ actually follows a t distribution, but, for large sample sizes, it can be approximated well using the standard normal (Z) distribution. This section deals with large samples, so we will use the Z notation to represent this random variable.

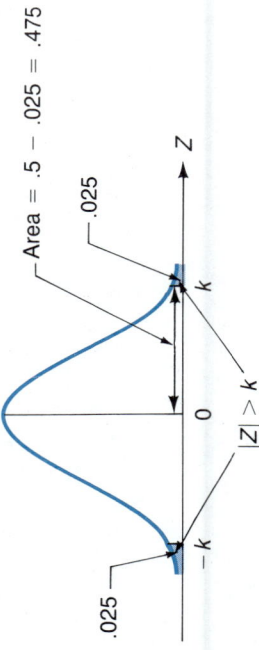

FIGURE 8.1
The shaded area represents the significance level, α.

some k. Suppose $\alpha = .05$. Then,

$$.05 = \alpha = P(\text{rejecting } H_0 \text{ when it is true})$$

$$= P\left(\left|\frac{\bar{X} - 5.9}{s/\sqrt{n}}\right| > k, \text{ when } \mu = 5.9\right)$$

$$= P(|Z| > k)$$

To find the value of k that satisfies this statement, consider Figure 8.1. When $|Z| > k$, either $Z > k$ or $Z < -k$, as illustrated. Since $P(|Z| > k) = .05$, the total shaded area is .05, with .025 in each tail due to the symmetry of this curve. Consequently, the area between 0 and k is .475, and, using Table A.4, $k = 1.96$. So our test of H_0 versus H_a is

$$\text{reject } H_0 \text{ if } \left|\frac{\bar{X} - 5.9}{s/\sqrt{n}}\right| > 1.96$$

and fail to reject H_0 otherwise. So,

$$\text{reject } H_0 \text{ if } \frac{\bar{X} - 5.9}{s/\sqrt{n}} > 1.96$$

or

$$\text{reject } H_0 \text{ if } \frac{\bar{X} - 5.9}{s/\sqrt{n}} < -1.96$$

This test will reject H_0 when it is true 5% of the time. This means that there is a 5% risk of making a Type I error.

Using the sample data, we obtained $n = 75$, with $\bar{x} = 5.76$ feet and $s = .48$ feet. Is $\bar{x} = 5.76$ feet far enough away from 5.9 feet for us to reject H_0? This was not at all obvious at first glance; it may have seemed that this value of $\bar{X}$ is "close enough to" 5.9 for us not to reject H_0. Such is not the case, however, because

$$Z = \frac{\bar{X} - 5.9}{s/\sqrt{n}} = \frac{5.76 - 5.9}{.48/\sqrt{75}} = -2.53 = Z^*$$

where Z^* is the **computed value** of Z. Because $-2.53 < -1.96$, we reject H_0. We thus conclude that based on the sample results and a value of $\alpha = .05$, the average population male height (μ) is not equal to 5.9 feet.

Another way of phrasing this result is to say that if H_0 is true (that is, if $\mu = 5.9$ feet), the value of $\bar{X}$ obtained from the sample (5.76 feet) is 2.53 standard deviations to the left of the mean using the normal curve for $\bar{X}$ (Figure 8.2). Because a value of $\bar{X}$ this far away from the mean is very unlikely (that is, with probability less than $\alpha = .05$), our conclusion is that H_0 is not true, and so we reject it.

When testing $\mu = $ (some value) versus $\mu \neq $ (some value), the null hypothesis, H_0, always contains the $=$, and the alternative hypothesis, H_a, always contains

FIGURE 8.2

Distribution of $\bar{X}$ if H_0 is true (H_0: $\mu = 5.9'$).

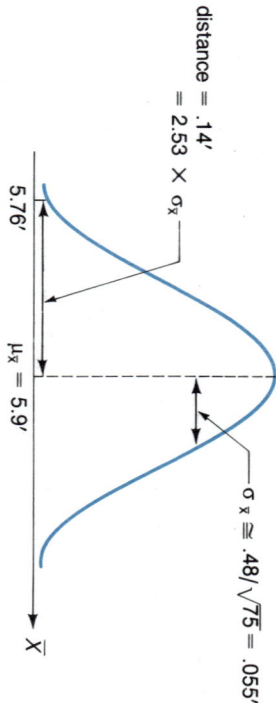

distance = .14'
= 2.53 × $\sigma_{\bar{x}}$

5.76' $\mu_{\bar{x}} = 5.9'$

$\sigma_{\bar{x}} \cong .48/\sqrt{75} = .055'$

$\bar{X}$

---

the $\neq$. In our example, this resulted in splitting the significance level, α, in half and including one-half in each tail of the test statistic, Z. Consequently, when testing H_0: $\mu = $ (some value) versus H_a: $\mu \neq$ (some value), we refer to this as a **two-tailed test**.

EXAMPLE 8.1

Using the data from our example of male heights, what would be the conclusion using a significance level α of .01?

SOLUTION

The only thing that we need to change from our previous solution is the value of k. Now,

$$P(|Z| > k) = \alpha = .01$$

as shown in Figure 8.3. Using Table A.4, $k = 2.575$, and the test is (see Figure 8.4):

$$\text{reject } H_0 \text{ if } Z > 2.575 \quad \text{or} \quad Z < -2.575$$

What is the value of $(\bar{X} - 5.9)/(s/\sqrt{n})$? Our data values have not changed, so the value of this expression is the same: $Z^* = -2.53$.

The region defined by values of Z to the right of 2.575 and to the left of -2.575 in Figure 8.4 is the **rejection region**. The value of k (2.575) defining this region is the **critical value**. Z^* fails to fall in this region, so we fail to reject H_0. In other words, for $\alpha = .01$, the value of $\bar{X}$ is "close enough" to 5.9 to let

FIGURE 8.3

The shaded area is $\alpha = .01$.

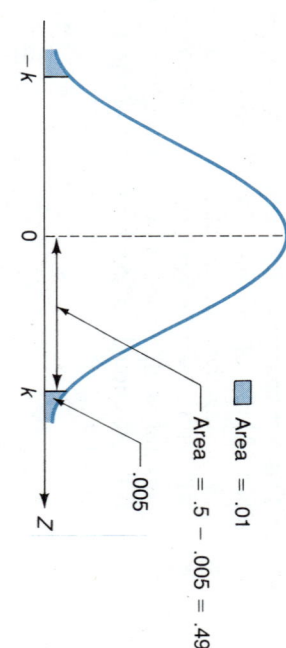

Area = .01

Area = .5 − .005 = .495

.005

Area = .01

$-k$ 0 k Z

FIGURE 8.4

We reject H_0 if Z^* falls within either tail—the rejection region for $\alpha = .01$.

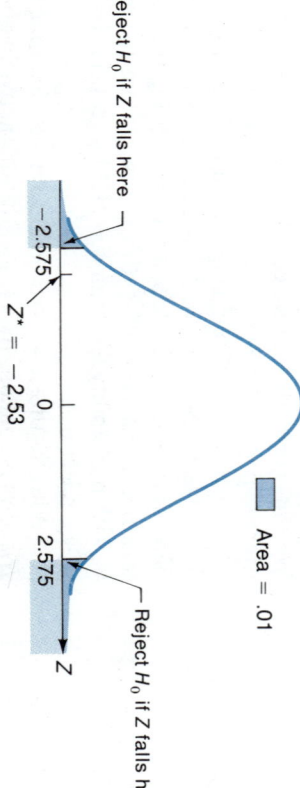

Reject H_0 if Z falls here

Reject H_0 if Z falls here

-2.575 0   2.575   Z

$Z^* = -2.53$

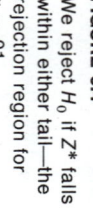

H_0 stand; there is insufficient evidence to conclude that μ is different from 5.9 feet.

Accepting H_0 or Failing to Reject

It may appear that there is no difference between "accepting" and "failing to reject" a null hypothesis, but there *is* a difference between these two statements. When you test a hypothesis, H_0 is *presumed innocent* until it is demonstrated to be guilty. In Example 8.1, using $\alpha = .01$ we failed to reject H_0. Now, how certain are we that μ is *exactly* 5.9 feet? After all, our estimate of μ is 5.76 feet. Clearly, we do not believe that μ is precisely 5.9 feet. There simply was not enough evidence to *reject* the claim that $\mu = 5.9$ feet.

For any hypothesis-testing application, the only hypothesis that can be *accepted* is the alternative hypothesis, H_a. Either there is sufficient evidence to *support* H_a (we reject H_0) or there is not (we fail to reject H_0). The focus of our attention is whether there is sufficient evidence within the sample data to conclude that H_a is correct. By failing to reject H_0, we are simply saying that the data do not allow us to support the claim made in H_a (such as $\mu \neq 5.9$ feet) and not that we accept the statement made in H_0 (such as $\mu = 5.9$ feet).

THE FIVE-STEP PROCEDURE FOR HYPOTHESIS TESTING

The discussion up to this point has concentrated on hypothesis testing on the unknown mean of a particular population. We want to emphasize that the shape of the parent population is not important, provided you have a large sample. In other words, the population may be a normal (bell-shaped) one or it may not—it simply does not matter for large samples. Once the level of significance (α) has been determined, the steps carried out when attempting to reject or failing to reject a claim regarding the population mean μ are:

Step 1. *Set up the null hypothesis, H_0, and the alternative hypothesis, H_a.* If the purpose of the hypothesis test is to test whether the population mean is equal to a particular value (say, μ_0), the "equal hypothesis" always is stated in H_0 and the "unequal hypothesis" always is stated in H_a.

Step 2. *Define the test statistic.* This is evaluated, using the sample data, to determine if the data are compatible with the null hypothesis. For tests regarding the mean of a population using a large sample, the test statistic is approximately a standard normal random variable given by the equation

$$Z = \frac{\bar{X} - \mu_0}{s/\sqrt{n}} \qquad (8.1)$$

where μ_0 is the value of μ specified in H_0.

Step 3. *Define a rejection region,* having determined a value for α, the significance level. In this region the value of the test statistic will result in rejecting H_0.

Step 4. *Calculate the value of the test statistic, and carry out the test.* State your decision: to reject H_0 or to fail to reject H_0.

Step 5. *Give a conclusion in the terms of the original problem or question.* This statement should be free of statistical jargon and should merely summarize the results of the analysis.

Steps 1 through 5 apply to all tests of hypothesis in this and subsequent chapters. The form of the test statistic and rejection region change for different applications, but the sequence of steps always is the same.

EXAMPLE 8.2

Remember that Everglo light bulbs are advertised as lasting 400 hours on the average. As manager of the quality-control department, you need to examine this claim closely. If the average lifetime is, in fact, less than 400 hours, you can expect at least a half-dozen government watchdog agencies knocking on your door. If the light bulbs last longer than the 400 hours (on the average) claimed, you want to revise your advertising accordingly. To check this claim, you have tested the lifetimes of 100 bulbs, each under the same circumstances (power load, room temperature, and so on). The results of this sample are $n = 100$, $\bar{x} = 411$ hours, and $s = 42.5$ hours. What conclusion would you reach using a significance level of .1?

SOLUTION

Step 1. *Define the hypotheses.* We will test $H_0: \mu = 400$ versus $H_a: \mu \neq 400$.

Step 2. *Define the test statistic.* The proper test statistic for this problem is

$$ Z = \frac{\bar{X} - 400}{s/\sqrt{n}} $$

Step 3. *Define the rejection region.* The steps for finding the rejection region are shown in Figure 8.5. We conclude:

reject H_0 if $Z > 1.645$ or $Z < -1.645$

Step 4. *Calculate the value of the test statistic and carry out the test.* The computed value of Z is

$$ Z^* = \frac{411 - 400}{42.5/\sqrt{100}} = \frac{11}{4.25} = 2.59 $$

Since $2.59 > 1.645$, our decision is to reject H_0. In Figure 8.5, Z^* falls in the rejection region.

Step 5. *State a conclusion.* Based on the sample data, there is sufficient evidence to conclude that the average lifetime of Everglo bulbs is not 400 hours. ■

COMMENTS

In Example 8.2, $\bar{X}$ was "far enough away from" 400 for us to reject the claim that the average lifetime is 400 hours (H_0). However, remember that you cannot decide what is "far enough away from" without also considering the value of the standard deviation ($s = 42.5$ hours in Example 8.2). This is why the value of s (or σ, if it is known) is a vital part of the test statistic.

Examine the test statistic in Example 8.2. Observe that for *small* s, it is "easier" to reject H_0. As s becomes smaller, the absolute value of the test statistic, Z, becomes larger, and the test statistic is more likely to be in the rejection region for a given value of α.

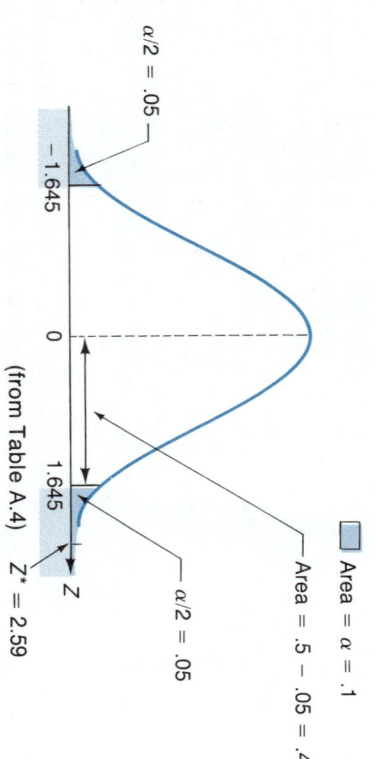

FIGURE 8.5
See Example 8.2; the rejection region is $|Z| >$ 1.645.

Confidence Intervals and Hypothesis Testing

What is the relationship, if any, between a 95% confidence interval and performing a *two-tailed* test using $\alpha = .05$? There is a very simple relationship here: When testing $H_0: \mu = \mu_0$ versus $H_a: \mu \neq \mu_0$ using the five-step procedure and a significance level, α, H_0 will be rejected if and only if μ_0 does *not* lie in the $(1 - \alpha) \cdot 100\%$ confidence interval for μ.

The five-step procedure and the confidence interval procedure always lead to the same result. In fact, you can think of a confidence interval as that set of values of μ_0 that would not be rejected by a two-tailed test of hypothesis.

In our example involving heights of U.S. males, a sample of 75 heights produced $\bar{x} = 5.76$ feet and $s = .48$ feet. The resulting 95% confidence interval for μ is

$$\bar{X} - k\left[\frac{s}{\sqrt{n}}\right] \quad \text{to} \quad \bar{X} + k\left[\frac{s}{\sqrt{n}}\right]$$

What is the value of k? The population standard deviation (σ) is unknown, so we need to use the t table (Table A.5). We do have a large sample, however, so the t value will be closely approximated by the corresponding Z value (Table A.4). Keep in mind that when dealing with large samples, it really does not matter if σ is known or replaced by s. In either case, the standard normal table (Table A.4) gives us the probability points we need.

The value of k that provides a 95% confidence interval here is the *same value of k that provides a two-tailed area under the Z curve equal to $1 - .95 = .05$. In other words, we use the same k value that we used in a two-tailed test of H_0 versus H_a—namely, $k = 1.96$. So the 95% confidence interval for μ is

$$\bar{X} - 1.96\left(\frac{s}{\sqrt{n}}\right) \quad \text{to} \quad \bar{X} + 1.96\left(\frac{s}{\sqrt{n}}\right)$$

$$= 5.76 - 1.96\left(\frac{.48}{\sqrt{75}}\right) \quad \text{to} \quad 5.76 + 1.96\left(\frac{.48}{\sqrt{75}}\right)$$

$$= 5.76 - .11 \quad \text{to} \quad 5.76 + .11$$

$$= 5.65 \quad \text{to} \quad 5.87$$

The value of μ we are investigating here is $\mu = 5.9$ feet, and the corresponding hypotheses are $H_0: \mu = 5.9$ and $H_a: \mu \neq 5.9$. For $\alpha = .05$, our result using the two-tailed test was to reject H_0. Using the confidence interval procedure, we obtain the same result because 5.9 does not lie in the 95% confidence interval.

Thus, if you already have computed a confidence interval for μ, you can tell at a glance whether to reject H_0 for a two-tailed test, provided the significance level, α, for the hypothesis test and the confidence level, $(1 - \alpha) \cdot 100\%$, match up.

EXAMPLE 8.3

Repeat the heights of U.S. males example using a 99% confidence interval. Is the result the same as in Example 8.1, where we failed to reject $H_0: \mu = 5.9$ using $\alpha = .01$?

SOLUTION

Using $\alpha = .01$, we failed to reject H_0 because the absolute value of the test statistic did not exceed the critical value of $k = 2.575$. The corresponding 99%

confidence interval for μ is

$$\bar{X} - 2.575\left(\frac{s}{\sqrt{n}}\right) \quad \text{to} \quad \bar{X} + 2.575\left(\frac{s}{\sqrt{n}}\right)$$

$$= 5.76 - 2.575\left(\frac{.48}{\sqrt{75}}\right) \quad \text{to} \quad 5.76 + 2.575\left(\frac{.48}{\sqrt{75}}\right)$$

$$= 5.76 - .143 \quad \text{to} \quad 5.76 + .143$$

$$= 5.617 \quad \text{to} \quad 5.903$$

Because 5.9 does (barely) lie in this confidence interval, our decision is to fail to reject H_0—the same conclusion reached in Example 8.1.   ■

The Power of a Statistical Test

Up to this point, the probability of a Type II error, β, has remained a phantom—we know it is there, but we don't know what it is. One thing we can say is that a *wide* confidence interval for μ means that the corresponding two-tailed test of H_0 versus H_a has a *large* chance of failing to reject a false H_0; that is, β is large. Now,

$$\beta = P(\text{fail to reject } H_0 \text{ when } H_0 \text{ is false})$$

which means that

$$1 - \beta = P(\text{rejecting } H_0 \text{ when } H_0 \text{ is false})$$

The value of $1 - \beta$ is referred to as the **power** of the test. Since we like β to be small, we prefer the power of the test to be large. Notice that $1 - \beta$ represents the probability of making a *correct* decision in the event that H_0 is false, because in this case we *should* reject it. The more powerful your test is, the better.

Determining the power of your test (hence, β) is not difficult. We will illustrate this procedure for the previous two-tailed test of H_0: $\mu = \mu_0$ versus H_a: $\mu \neq \mu_0$, for some μ_0. We will first consider the case where σ is known and then discuss the situation where σ is unknown.

Power of the Test σ Known   In Example 8.2 we looked at the data on Everglo light bulbs, where the hypotheses were H_0: $\mu = 400$ hours and H_a: $\mu \neq 400$ hours. Assume that the actual population standard deviation is known to be $\sigma = 50$ hours. For this situation, our test statistic is (using a sample size of $n = 100$):

$$Z = \frac{\bar{X} - 400}{\sigma / \sqrt{n}} = \frac{\bar{X} - 400}{50 / \sqrt{100}}$$

$$= \frac{\bar{X} - 400}{5}$$

Proceeding as in Example 8.2, using $\alpha = .10$, we reject H_0 if $Z > 1.645$ or $Z < -1.645$; that is, $|Z| > 1.645$. So, reject H_0 if $(\bar{X} - 400)/5 > 1.645$ (same as $\bar{X} > 400 + (1.645)(5) = 408.225$) or if $(\bar{X} - 400)/5 < -1.645$ (same as $\bar{X} < 400 - (1.645)(5) = 391.775$). This way of representing the rejection region is illustrated in Figure 8.6, using the shaded area under curve A. The power of this test is

$$1 - \beta = P(\text{rejecting } H_0 \text{ if } H_0 \text{ is false})$$

$$= P(\text{rejecting } H_0 \text{ if } \mu \neq 400)$$

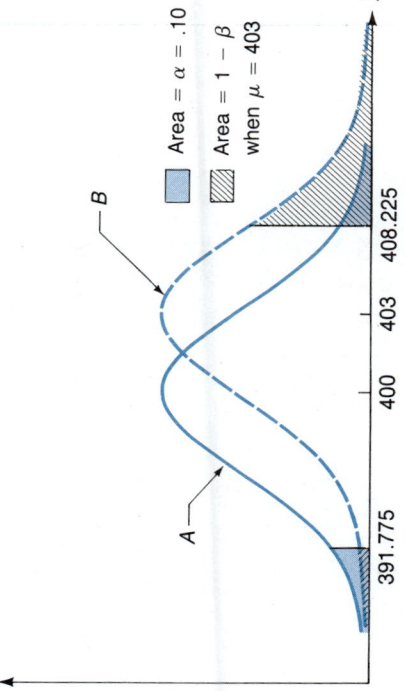

FIGURE 8.6

The shaded area is the probability of rejecting H_0 if $\mu = 400$ (that is, $\alpha = .10$), and the striped area is the probability of rejecting H_0 if $\mu = 403$ (that is, the power of the test $1 - \beta$ when $\mu = 403$).

▢ Area $= \alpha = .10$

▨ Area $= 1 - \beta$
  when $\mu = 403$

What is the power of this test if μ is not 400 but is 403? What you have here is a value of $1 - \beta$ for *each* value of $\mu \neq 400$.

Recall that we reject H_0 if $\bar{X} > 408.225$ or $\bar{X} < 391.775$. The probability of this occurring if $\mu = 403$ is illustrated as the lined area under curve B in Figure 8.6. Now, if $\mu = 403$ and $\sigma = 50$ (assumed), then

$$Z = \frac{\bar{X} - 403}{50/\sqrt{n}} \qquad \frac{\bar{X} - 403}{5}$$

is a standard normal random variable. So, in Figure 8.6, the striped area to the right of 408.225 is

$$
\begin{aligned}
P(\bar{X} > 408.225) &= P\left[\frac{\bar{X} - 403}{5} > \frac{408.225 - 403}{5}\right] \\
&= P\left[Z > \frac{5.225}{5}\right] \\
&= P(Z > 1.04) \\
&= .5 - .3508 \\
&= .1492
\end{aligned}
$$

Also, the striped area to the left of 391.775 is

$$
\begin{aligned}
P(\bar{X} < 391.775) &= P\left[\frac{\bar{X} - 403}{5} < \frac{391.775 - 403}{5}\right] \\
&= P(Z < -2.24) \\
&= .5 - .4875 \\
&= .0125
\end{aligned}
$$

Adding these two areas, we find that, if $\mu = 403$, the power of the test of H_0: $\mu = 400$ versus H_a: $\mu \neq 400$ is

$$1 - \beta = .1492 + .0125 = .1617$$

This means that if $\mu = 403$, the probability of making a Type II error (not rejecting H_0) is $\beta = 1 - .1617 = .8383$ (rather high).

This procedure is summarized in the following box. Notice that in the previous discussion, $Z_{\alpha/2} = Z_{.05} = 1.645$, $z_1 = 1.645 - (403 - 400)\sqrt{100}/50 = 1.04$, and $z_2 = -1.645 - (403 - 400)\sqrt{100}/50 = -2.24$.

POWER OF TEST FOR $H_0: \mu = \mu_0$ versus $H_a: \mu \neq \mu_0$

1. Determine

$$z_1 = Z_{\alpha/2} - \frac{(\mu - \mu_0)\sqrt{n}}{\sigma}$$

and

$$z_2 = -Z_{\alpha/2} - \frac{(\mu - \mu_0)\sqrt{n}}{\sigma}$$

where $Z_{\alpha/2}$ is the value of Z from Table A.4 having a right-tailed area of $\alpha/2$ and μ_0 is the specific value of $\mu(= 403$ in Figure 8.6).

2. Power of test $= P(Z > z_1) + P(Z < z_2)$.

The power of your test increases (β decreases) as μ moves away from 400. This is illustrated in Figure 8.7. Using the five-step procedure, which uses the test statistic $Z = (\bar{X} - 400)/(\sigma/\sqrt{n})$, the resulting power curve is the solid line curve in Figure 8.7. It is symmetric, and its lowest point is located at $\mu = 400$. For this value of μ, H_0 is actually true, so that a Type II error was not committed. Nevertheless, the value on the power curve corresponding to $\mu = 400$ is always

$$P(\text{rejecting } H_0 \text{ if } \mu = 400) = \alpha = .10 \text{ (for this example)}$$

The *steeper* your power curve is, the better. You are more apt to reject H_0 as μ moves away from 400—certainly a nice property. If we assume that the sampled population is normally distributed, Figure 8.7 illustrates that the power curve using the five-step procedure lies above (is steeper than) the power curve for any other testing procedure. To illustrate briefly another testing procedure, rather than basing the test statistic on the sample mean $\bar{X}$, we could derive a test statistic using the sample *median*. The resulting power curve for this procedure would lie *below* the one using $\bar{X}$, indicating that the test using the sample median is less powerful and thus inferior. So, in this sense, the five-step procedure defines the best (most powerful) test of $H_0: \mu = \mu_0$ versus $H_a: \mu \neq \mu_0$.

Power of the Test: σ Unknown When σ is unknown, we are forced to *approximate* the power of the test by replacing σ with the sample estimate, s. We are dealing with large samples, so we can use Table A.4 (the Z table).

In our discussion of the power of our test for Everglo bulb lifetimes, we treated the population standard deviation, σ, as known. If we make no assumptions about this parameter, we need to approximate the power of the test for $\mu = 403$. We assume $s = 42.5$ hours (as before) and use $\alpha = .10$.

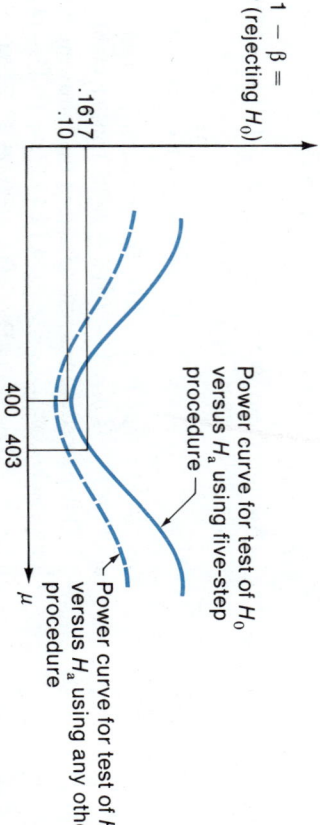

FIGURE 8.7
Power curve for $H_0: \mu = 400$ versus $H_a: \mu \neq 400$.

$1 - \beta = P$ (rejecting H_0)

.1617
.10

400 403

Power curve for test of H_0 versus H_a using five-step procedure

Power curve for test of H_0 versus H_a using any other procedure

We now reject H_0 if

$$\bar{X} > 400 + 1.645\left[\frac{s}{\sqrt{n}}\right] \quad \text{or} \quad \bar{X} < 400 - 1.645\left[\frac{s}{\sqrt{n}}\right]$$

So, H_0 is rejected, provided

$$\bar{X} > 400 + 1.645\left[\frac{42.5}{\sqrt{100}}\right] = 406.99$$

or

$$\bar{X} < 400 - 1.645\left[\frac{42.5}{\sqrt{100}}\right] = 393.01$$

The resulting power of the test for $\mu = 403$ is approximately equal to

$$P(\bar{X} > 406.99 \text{ if } \mu = 403) + P(\bar{X} < 393.01 \text{ if } \mu = 403)$$

$$= P\left[Z > \frac{406.99 - 403}{42.5/\sqrt{100}}\right] + P\left[Z < \frac{393.01 - 403}{42.5/\sqrt{100}}\right]$$

$$= P(Z > .94) + P(Z < -2.35)$$

$$= (.5 - .3264) + (.5 - .4906) = .1736 + .0094$$

$$= .183$$

EXERCISES

8.1 The vice president of Metropolitan Bank must decide whether to grant a large loan to an independent energy-exploration company. Consider the null hypothesis: the energy-exploration company will pay back the entire loan.

a. Describe the four possible outcomes from deciding either to fail to reject the null hypothesis or to reject the null hypothesis.

b. Which of the two errors, Type I or Type II, is more serious?

c. If the energy-exploration company does not qualify for the loan, does this "prove" that the energy-exploration company would not pay back the entire loan?

8.2 State what type of error can be made in the following situations:

a. The conclusion is to reject the null hypothesis.

b. The conclusion is to fail to reject the null hypothesis.

c. The calculated value of the test statistic does not fall in the rejection region.

8.3 Explain why the following statements are true or false.

a. The probability of the Type I error and the probability of the Type II error always add to 1.

b. Increasing the value of α increases the value of β.

c. A large value for the power at a specified value of the alternative hypothesis indicates a small value for the probability of a type II error given the specified value stated in the alternative hypothesis.

d. The smaller the specified value of α is, the larger the rejection region.

8.4 Hallman Industrial is interested in testing the null hypothesis that a particular applicant is qualified for the position of marketing strategist.

a. Explain what the Type I and Type II errors are for this situation.

b. Which of the two errors in part (a) is more serious?

8.5 The mean of a normally distributed population is believed to be equal to 50.1. A sample of 36 observations is taken and the sample mean is found to be 53.2. The alternative hypothesis is that the population mean is not equal to 50.1. Complete the hypothesis test, assuming that the population standard deviation is equal to 4. Use a .05 significance level.

8.6   The average life expectancy of males in a developing nation was believed to be 62.5 years. However, the belief was based on data that might be considerably out-of-date. A random survey of 250 deaths in that country revealed an average life span of 64.2 years with a standard deviation of 8.8 years. Does the sample evidence indicate that the mean life expectancy differs from 62.5 years? Use a significance level of .05.

8.7   The weights of fish in a certain pond that is regularly stocked are considered to be normally distributed with a mean of 3.1 pounds and a standard deviation of 1.1 pounds. A random sample of size 30 is selected from the pond and the sample mean is found to be 2.4 pounds. Is there sufficient evidence to indicate that the mean weight of the fish differs from 3.1 pounds? Use a 10% significance level.

8.8   A crime reporter was told that, on the average, 3000 burglaries per month occurred in his city. The reporter examined past data, which was used to compute a 95% confidence interval for the number of burglaries per month. The confidence interval was from 2176 to 2784. At a 5% level of significance, do these data tend to support the alternative hypothesis, H_a: $\mu \neq 3000$?

8.9   A 95% confidence interval for the mean time that it takes a city bus to complete its route is 2.2 hours to 2.6 hours. The time that it takes the bus to complete its route is normally distributed. Is there sufficient evidence to indicate that the mean time to complete the route is different from 2.0 hours? Use a 5% significance level.

8.10   The life span of an electronic chip used in a high-powered microcomputer is estimated to be 625.35 hours from a random sample of 40 chips. The life of an electronic chip is considered to be normally distributed with a population variance of 400 hours.

a.   Find a 90% confidence interval for the mean life of the electronic chips.

b.   Is the true mean life of the electronic chips different from 633 hours? Use a 10% significance level.

8.11   The manufacturer of a special-purpose industrial pipe is interested in testing the hypothesis that the mean diameter of the pipes is 12.75 inches. A sample of 100 pipes was randomly selected and the diameters were measured. The sample mean was found to be 12.73 inches and the sample standard deviation was found to be .01.

a.   Find a 99% confidence interval for the mean diameter of the pipes.

b.   Is there evidence that the mean diameter of the pipes is different from 12.75 inches? Use a 1% significance level.

8.12   The hypotheses for a situation are

$$H_0: \mu = 20$$
$$H_a: \mu \neq 20$$

If the population of interest is normally distributed, what is the power of the test for the mean if μ is actually equal to 22? Assume that a sample of size 49 is used and the sample standard deviation is 4.2. Use a significance level of .05.

8.13   Find the power of the test for the mean for the following situations if the true population mean is 30 and the population variance is 25. Use a 10% significance level.

a. $H_0: \mu = 26$, $H_a: \mu \neq 26$, $n = 20$.

b. $H_0: \mu = 36$, $H_a: \mu \neq 36$, $n = 25$.

c. $H_0: \mu = 33$, $H_a: \mu \neq 33$, $n = 25$.

8.14   An electro-optical firm currently uses a laser component in producing sophisticated graphic designs. The time it takes to produce a certain design with the current laser component is 70 seconds, with a standard deviation of 8 seconds. A new laser component is bought by the firm because it is believed that the time it takes this laser to produce the same design is not equal to 70 seconds, and has a standard deviation of 8 seconds. The research-and-development department is interested in constructing the power curve for testing the claim that the time it takes to produce the same design by the new laser component is not equal to 70 seconds. Graph the power function for a sample of size 25 and a significance level of .05.

8.15 Fermet's Soup is interested in knowing how much the average homemaker spends on soup and ingredients to make soup per month. The company's marketing analyst takes a sample of 100 homemakers from a certain city and finds the standard deviation of the amount spent monthly on soup to be \$1.50. What would be the power of the test for the hypothesis that the monthly expenditure on soup is equal to \$8 if the true monthly expenditure on soup was \$10? Assume a significance level of .05.

8.16 Explain why the sample mean, rather than the sample median, is used as a basis for testing the hypothetical mean of a normally distributed population.

8.2

ONE-TAILED TEST FOR THE MEAN OF A POPULATION: LARGE SAMPLE

There are many situations in which you are interested in demonstrating that the mean of a population is *larger* or *smaller* than some specified value. For example, as a member of a consumer-advocate group, you may be attempting to demonstrate that the average weight of a bag of sugar for a particular brand is not 10 pounds (as specified on the bag) but is in fact less than 10 pounds. Because the situation that you are attempting to demonstrate goes into the alternative hypothesis, the resulting hypotheses would be $H_0: \mu \geq 10$ and $H_a: \mu < 10$. Remember that we said it is standard practice always to put the equal sign in the null hypothesis. In the testing procedure only the **boundary value** is important, and so the hypotheses may be written as

$$H_0: \mu = 10$$
$$H_a: \mu < 10$$

In this way, we can identify the distribution of $\bar{X}$ when H_0 is true—namely, $\bar{X}$ is a normal random variable centered at ten with standard deviation $s/\sqrt{n}$ (or $\sigma/\sqrt{n}$ if σ is known). Because the focus of our attention is on H_a (can we support it or not?), which of the two ways you use to write H_0 is not an important issue. The procedure for testing H_0 versus H_a is the same regardless of how you state H_0.

The resulting test is referred to as a **one-tailed test**, and it uses the same five-step procedure as the two-tailed test. The only change we make is to modify the rejection region: All the error is in a single tail.

EXAMPLE 8.4

A foreign car manufacturer advertises that its newest model, the Bullet, rarely stops at gas stations. In fact, they claim its EPA rating for highway driving is at least 32.5 mpg. However, the results of a recent independent study determined the mpg for 50 identical models of the Bullet, with these results: $n = 50$, $\bar{x} = 30.4$ mpg, and $s = 5.3$ mpg. This report failed to offer any conclusion, and you have been asked to interpret these results by someone who has always felt that the 32.5 figure is too high. What would be your conclusion using a significance level of $\alpha = .05$?

SOLUTION

Step 1. An important point to be made here is that H_0 and H_a (as well as α) must be defined *before* you observe any data. In other words, *do not let the data dictate your hypotheses*; this would introduce a serious bias into your final outcome. For this application, we want to demonstrate that the population mean, μ, is less than 32.5 mpg, and so this goes into H_a. The appropriate hypotheses then are $H_0: \mu \geq 32.5$ and $H_a: \mu < 32.5$.

Step 2. The test statistic for a one-tailed test is the same as that for a two-tailed

FIGURE 8.8

The one-tailed rejection region is Z > 1.645. We reject H_0 if $Z = (\bar{X} - 32.5)/(s/\sqrt{n}) < -1.645$.

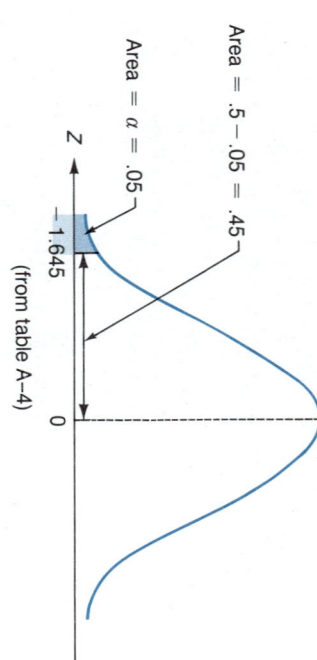

Area = .5 − .05 = .45

Area = α = .05

-1.645 0   Z

(from table A–4)

test, namely,

$$Z = \frac{\bar{X} - \mu_0}{s/\sqrt{n}} \quad \text{or, if } \sigma \text{ is known,} \quad \frac{\bar{X} - \mu_0}{\sigma/\sqrt{n}};$$

$$= \frac{\bar{X} - 32.5}{s/\sqrt{n}}$$

Step 3. What happens to Z when H_a is true? Here we would expect $\bar{X}$ to be <32.5 (because μ is), so the value of Z should be negative. Consequently, our procedure will be to reject H_0 if Z lies "too far to the left" of zero; that is,

$$\text{reject } H_0 \text{ if } Z = \frac{\bar{X} - 32.5}{s/\sqrt{n}} < k \text{ for some } k < 0$$

Since $\alpha = .05$, we will choose a value of k (the critical value) such that the resulting test will reject H_0 (shoot down the mpg claim) when it is true, with a 5% risk of an incorrect decision. This amounts to defining a rejection region in the *left tail* of the Z curve, the shaded area in Figure 8.8. Using Table A.4, we see that the critical value is $k = -1.645$, and the resulting test of H_0 versus H_a is

$$\text{reject } H_0 \text{ if } Z = \frac{\bar{X} - 32.5}{s/\sqrt{n}} < -1.645$$

Step 4. Using the sample results, the value of the test statistic is

$$Z^* = \frac{30.4 - 32.5}{5.3/\sqrt{50}} = -2.80$$

Because $-2.80 < -1.645$, the decision is to reject H_0.

Step 5. The results of this study support the claim that the average mileage for the Bullet is *less than* 32.5 mpg. This would provide just cause for claiming false advertising by the auto manufacturer.   ■

One-Tailed Test or Two-Tailed Test?

The decision to use a one-tailed test or a two-tailed test depends on what you are attempting to demonstrate. For example, when the quality-control department of a manufacturing facility receives a shipment from one of its vendors and wants to determine if the product meets minimal specifications, a one-tailed test is appropriate. If the product does not meet specifications, it will be rejected. This problem was first encountered in Chapter 5, where we examined lot acceptance sampling. Here, the product is *not* checked to see whether it *exceeds* specifications because any product that exceeds specifications is acceptable.

On the other hand, the vendors who supply the products would generally run two-tailed tests to determine two things. First, they must know if the product meets the minimal specifications of their customers before they ship it. Second, they must determine whether the product greatly exceeds specifications because this can be very costly in production. If they are making a product that in effect is too well built, this costs them extra money.

The testing of electric fuses is a classic example of a two-tailed test. A fuse must break when it reaches the prescribed temperature or a fire will result. However, the fuse must not break before it reaches the prescribed temperature or it will shut off the electricity when there is no need to do so. Therefore, the quality-control procedures for testing fuses must be two-tailed.

EXAMPLE 8.5

The mean consumption of electricity for the month of June at the Southern States Power Company (SSPC) historically has been 918 kilowatt-hours per residential customer. As part of its request for a rate increase, SSPC is arguing that the power consumption for June of the current year is substantially higher. To demonstrate this, they hired an independent consulting firm to examine a random sample of customer accounts. The results of the sample were $n = 60$ customers, $\bar{x} = 952.36$ kilowatt-hours, and $s = 173.92$ kilowatt-hours. Can you conclude that the average consumption for all users during June of this year (denoted by μ) is larger than 918? Use $\alpha = .01$.

SOLUTION

Step 1. The hypotheses here are $H_0: \mu \leq 918$ and $H_a: \mu > 918$.

Step 2. The correct test statistic is

$$Z = \frac{\bar{X} - 918}{s/\sqrt{n}}$$

Step 3. For this situation, what happens to Z if H_a is true? The value of $\bar{X}$ should then be *larger* than 918 (on the average), resulting in a positive value of Z. So we

$$\text{reject } H_0 \text{ if } Z = \frac{\bar{X} - 918}{s/\sqrt{n}} > k \text{ for some } k > 0$$

Examine the standard normal curve in Figure 8.9, where the area corresponding to α is the shaded part of the *right tail*; using Table A.4, the critical value is $k = 2.33$. The test of H_0 versus H_a will be

$$\text{reject } H_0 \text{ if } Z > 2.33$$

Step 4. The value of your test statistic is

$$Z^* = \frac{952.36 - 918}{173.92/\sqrt{60}} = 1.53$$

Because $1.53 < 2.33$, the decision is to fail to reject H_0.

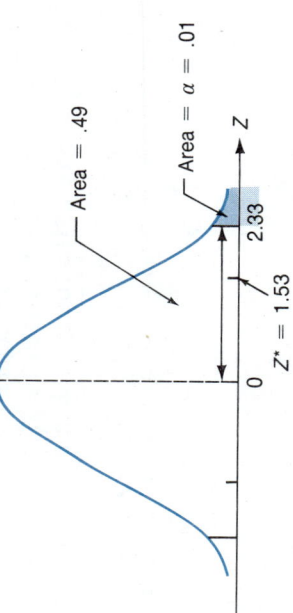

FIGURE 8.9

One-tailed rejection region; reject H_0 if $Z > 2.33$.

Area = .49

Area = α = .01

z

2.33

0

$Z^* = 1.53$

Step 5. Using this value of α, there is insufficient evidence to support the power company's claim that the power consumption for June has increased. ∎

This result is very much tied to the value of α. Using $\alpha = .10$ in Example 8.5, we would obtain the *opposite* conclusion—which you may find somewhat disturbing. You often hear the expression that "statistics lie." This is not true—statistics are merely mistreated, either intentionally or accidentally. One can often obtain the desired conclusion by choosing the value of α that produces a desired conclusion. We therefore reemphasize that you must choose α by weighing the seriousness of a Type I versus a Type II error *before* seeing the data. A partial remedy for this dilemma is discussed in Section 8.3.

Large Samples Taken from a Finite Population

For applications in which we take a large sample from a finite population, we make a slight adjustment to the standard error of $\bar{X}$ by including the finite population correction (fpc) factor.

For the finite population case, the standard error (standard deviation) of $\bar{X}$ is not $s/\sqrt{n}$ but instead

$$\text{standard error of } \bar{X} = s_{\bar{X}} = \frac{s}{\sqrt{n}} \sqrt{\frac{N-n}{N-1}} \qquad (8.2)$$

Once again using the results of the Central Limit Theorem, the test statistic is an *approximate* standard normal random variable, given by

$$Z = \frac{\bar{X} - \mu_0}{s_{\bar{X}}}$$

As a result, the five-step procedure can be carried out exactly as before.

EXAMPLE 8.6

In Example 7.4 we considered a sample of 45 incomes from a group of female managers at Compumart. The women wished to demonstrate that the average income of the population of 350 female middle managers was less than $48,000. For this illustration, it was assumed that the population standard deviation was known. Because this assumption is *not necessary* and perhaps incorrect, a safer procedure would be to use the sample estimate, s. The results of the sample were $n = 45$, $\bar{x} = \$43,900$, and $s = \$7140$.

What would be your conclusion from these results, using a significance level of $\alpha = .05$?

SOLUTION

Step 1. The hypotheses are $H_0: \mu \geq 48,000$ and $H_a: \mu < 48,000$, where $\mu =$ the average annual income for all females in middle-management positions at Compumart.

Step 2. The corresponding test statistic here is

$$Z = \frac{\bar{X} - 48,000}{s_{\bar{X}}}$$

where

$$s_{\bar{X}} = \frac{s}{\sqrt{n}} \sqrt{\frac{N-n}{N-1}}$$

Step 3. Using Figure 8.8 the rejection region is:

$$\text{reject } H_0 \text{ if } Z < -1.645$$

Step 4. Here,

$$s_{\bar{X}} = \frac{7140}{\sqrt{45}} \sqrt{\frac{350 - 45}{350 - 1}} = 995.01$$

so our computed test statistic is

$$Z^* = \frac{43,900 - 48,000}{995.01} = -4.12$$

Step 5. Since $-4.12 < -1.645$, we (strongly) reject H_0 in favor of H_a. The sample results strongly support the assertion that the female middle managers are underpaid. We reached the same conclusion in Example 7.4, where we based this decision on the extremely small probability of observing a value of $\bar{X}$ this small if μ was in fact \$48,000. ∎

COMMENTS

1. As mentioned in Chapter 7, the fpc factor of $(N - n)/(N - 1)$ can be ignored whenever your sample size is less than 5% of the population size—that is, when $n/N < .05$. Such is also the case when using the fpc in hypothesis testing. In the preceding example, $n/N = 45/350 = .129$. Consequently, ignoring the fpc would have produced a much smaller (and less accurate) value of Z^*.

2. To calculate the power of a one-sided test, refer to the box on page 264. We modify this procedure for a one-sided test by determining $z_1 = Z_\alpha - (\mu - \mu_0)\sqrt{n}/\sigma$ for H_a: $\mu > \mu_0$ or $z_2 = -Z_\alpha - (\mu - \mu_0)\sqrt{n}/\sigma$ for H_a: $\mu < \mu_0$. The resulting power is $P(Z > z_1)$ for H_a: $\mu > \mu_0$ or $P(Z < z_2)$ for H_a: $\mu < \mu_0$.

LARGE-SAMPLE TESTS ON A POPULATION MEAN

TWO-TAILED TEST

$$H_0: \mu = \mu_0$$
$$H_a: \mu \neq \mu_0$$

reject H_0 if $|Z| > Z_{\alpha/2}$

$$(Z_{\alpha/2} = 1.96 \text{ for } \alpha = .05)$$

ONE-TAILED TEST

$H_0: \mu \leq \mu_0$	$H_0: \mu \geq \mu_0$
$H_a: \mu > \mu_0$	$H_a: \mu < \mu_0$
reject H_0 if $Z > Z_\alpha$	reject H_0 if $Z < -Z_\alpha$
$(Z_\alpha = 1.645 \text{ for } \alpha = .05)$	$(-Z_\alpha = -1.645 \text{ for } \alpha = .05)$

where $Z = (\bar{X} - \mu_0)/s_{\bar{X}}$

For a finite population with $n/N > .05$,

$$s_{\bar{X}} = \frac{s}{\sqrt{n}} \sqrt{\frac{N - n}{N - 1}}$$

Otherwise,

$$s_{\bar{X}} = \frac{s}{\sqrt{n}}$$

EXERCISES

8.17 Find the rejection region of the Z-statistic in a hypothesis test of the population mean for the following situations:

a. It is believed that the mean monthly advertising expenditure for a company was greater than $2000. A significance level of .05 is used.

b. It is believed that the average length of sick time taken by an employee for firm XYZ is equal to 5.2 days per year. A significance level of .10 is used.

c. It is believed that the mean age of an applicant applying for a particular job is less than 25 years. A significance level of .01 is used.

8.18 A sample of size 20 is drawn from a finite population of size 225. The finite population can be approximated by a normal distribution. The sample of size 20 yields a sample mean of 75.8. The population variance is equal to 16. Is there sufficient evidence to indicate that the mean of the population is different from 82.5? Use a 10% significant level.

8.19 Carry out the hypothesis test for the mean of the normally distributed population given the following information:

$$H_0: \mu \geq 4.5 \qquad \bar{x} = 3.9$$
$$H_a: \mu < 4.5 \qquad \sigma = 1.12$$
$$n = 30 \qquad \alpha = .07$$

8.20 Bobby Marks is seriously considering investing in the grocery business in the southeast United States. He believes that the industry's average return on sales (ROS) is less than 5%. A random sample of 46 such businesses in various sectors of the southeast United States revealed that

$$\bar{x} = 4.6\% \qquad s = 1.2\%$$

Test Marks' belief concerning the ROS of the grocery business in the southeast United States, using a 5% significance level.

8.21 A delivery company claims that the mean time it takes to deliver frozen food between two particular cities is less than 3.7 hours. A random sample of 50 deliveries yielded a sample mean of 3.3 hours with a sample standard deviation of .2 hours. Do the data support the claim? Use a 10% significance level.

8.22 An auditing firm would like to test the belief that the average customer of a small town's utility service pays the utility bill in less than 15 days after receipt of the bill. The town has only 12,352 customers. A sample of size 1325 yields a sample mean of 14.6 days in which a customer paid the utility bill. The sample standard deviation is 6 days. Do the data support the belief? Use a 5% significance level.

8.23 Two hundred fifty applicants apply for the same position at an assembly plant. A random sample of 25 applicants is reviewed carefully. The average experience of the 25 applicants is 3.4 years. Can it be concluded that the mean experience of the 250 applicants is greater than 2.5 years at a significance level of .05? Assume that the population standard deviation of the experience of the 250 applicants is 1.3 years.

8.24 The quality-control engineer of a battery-manufacturing firm has been asked to verify the marketing department's claim that the mean life of the multipurpose battery made by the firm is greater than 47 hours. The quality-control engineer takes a random sample of 80 batteries and finds the sample mean to be 47.5 hours with a sample standard deviation of 1.6 hours. Do the data support the marketing department's claim? Use a .05 significance level.

8.25 The average hourly earnings for manufacturing production workers across the United States in January 1987 was $9.84. Assume that the vice president at a manufacturing plant in Indiana believes that the average hourly earnings for production workers in Indiana are above the national average. To verify his belief a random sample of 150 manufacturing production workers is taken from across the state of Indiana with the following results:

$$\bar{x} = \$10.20$$
$$s = \$ 1.70$$

(Source: The World Almanac and Book of Facts, 1988, p. 131.)

a. Do the data support the vice president's belief that the average hourly earnings for production workers in Indiana are greater than the national average? Use a .05 significance level.

b. Find a 99% confidence interval for the mean hourly earnings for production workers in Indiana. Interpret the confidence interval.

8.26 Is the average rent in housing estates built by the Liberian National Housing Authority (NHA) more than $50 per month? Use the following data to address this issue:

HOUSING ESTATES BUILT BY NHA (1970–1980)

ESTATE	NUMBER OF UNITS	AVERAGE MONTHLY RENT
Cabral	72	$38.61
Goodridge	576	$56.00
New Georgia	226	$43.96

(Source: Republic of Liberia 1980, National Housing Authority Annual Report 1980. Adapted from L. Lacey and S. E. Owusu, "Self-help shelter and related programs in Liberia." J. Am. Planning Assoc., 53, no. 2 (Spring 1987: 209, Table 1.)

a. Compute the overall average monthly rent for the 874 units.

b. Although this is not strictly a random sample of 874 units, you may assume that for this exercise. Given a standard deviation in monthly rent of $22.00, conduct a hypothesis test at a 5% significance level to decide if the average monthly rent is greater than $50. What is your conclusion?

8.3

REPORTING TESTING RESULTS USING A *p*-VALUE

In Example 8.1, we noted that for one value of α we rejected H_0, and for another (seemingly reasonable) value of α we failed to reject H_0. Is there a way of summarizing the results of a test of hypothesis that allows you to determine whether these results are barely significant (or insignificant) or overwhelmingly significant (or insignificant)? Did we barely reject H_0, or did H_0 go down in flames?

A convenient way to summarize your results is to use a *p*-value, often called the *observed α* or *observed significance level*.

> The **p-value** is the value of α at which the hypothesis test procedure changes conclusions. It is the smallest value of α for which you can reject H_0 (that is, at which the test is significant).

Consequently, the *p*-value is the point at which the five-step procedure leads us to switch from rejecting H_0 to failing to reject H_0.

Determining the *p*-Value

The *p*-value for *any* test is determined by replacing the area corresponding to α by the area corresponding to the *computed* value of the test statistic. In our discussion and Example 8.1, using $\alpha = .05$ you reject H_0 and using $\alpha = .01$ you fail to reject H_0. We know that the *p*-value here is between .01 and .05. For

FIGURE 8.10

Rejection regions for $\alpha = .01, .05$.

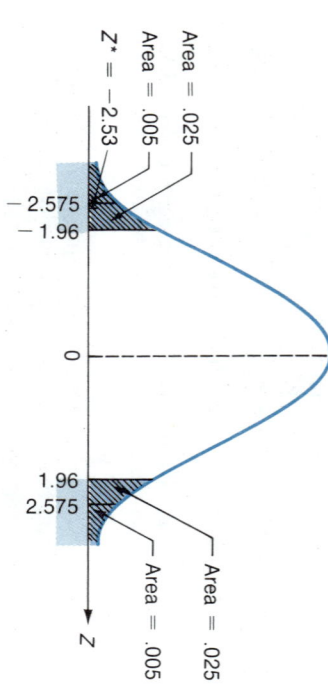

Area = .025
Area = .005
$Z^* = -2.53$

-2.575
-1.96

Area = .5 − .4943
= .0057

Area = .4943 (Table A−4)

Area = p value

Area = .005
Area = .025

1.96
2.575

FIGURE 8.11

p-value is determined by replacing the area corresponding to α (see Figure 8.10) by the area corresponding to Z^*. Here $Z^* = -2.53$, and the p-value $= 2 \cdot .0057 = .0114$ (total shaded area).

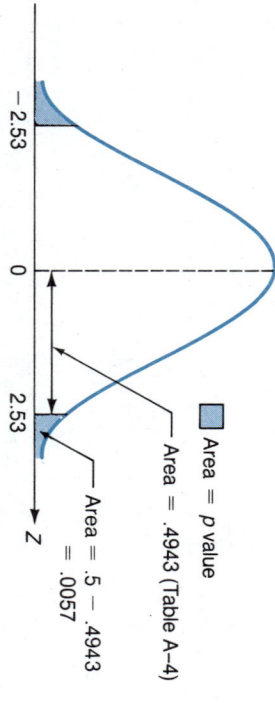

-2.53

0

2.53

this example, the computed value of the test statistic was $Z^* = -2.53$, where the hypotheses are $H_0: \mu = 5.9$ feet and $H_a: \mu \neq 5.9$ feet. The Z curve for this situation is shown in Figure 8.10.

For which value of α does the testing procedure change the conclusions here? In Figure 8.10, if you were using a predetermined significance level α, you would split α in half and put $\alpha/2$ into each tail. So the total tail area represents α. Using Figure 8.11, we reverse this procedure by finding the *total* tail area corresponding to a two-tailed test with $Z^* = -2.53$; we add the area to the left of -2.53 (.0057) to that to the right of 2.53 (also .0057). This total area is .0114, which is the p-value for this application. Thus, if you choose a value of $\alpha > .0114$ (such as .05), you will reject H_0. If you choose a value of $\alpha < .0114$ (such as .01), you will fail to reject H_0.

PROCEDURE FOR FINDING THE p-VALUE

1. For $H_a: \mu \neq \mu_0$

   $$p = 2 \cdot (\text{area outside of } Z^*)$$

   Reason: When using a significance level α, the value of α represents a two-tailed area.

2. For $H_a: \mu > \mu_0$

   $$p = \text{area to the right of } Z^*$$

   Reason: When using a significance level α, the value of α represents a right-tailed area.

3. For $H_a: \mu < \mu_0$

   $$p = \text{area to the left of } Z^*$$

   Reason: When using a significance level α, the value of α represents a left-tailed area.

EXAMPLE 8.7

What is the p-value for Example 8.5?

SOLUTION

The results of the sample were $n = 60$, $x = 952.36$ kilowatt-hours, and $s =$

FIGURE 8.12
p-value for $Z^* = 1.53$.

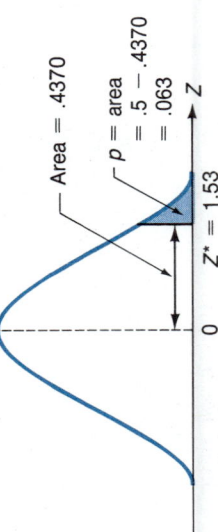

173.92 kilowatt-hours. The corresponding value of the test statistic was

$$Z^* = \frac{952.36 - 918}{173.92/\sqrt{60}} = 1.53$$

The alternative hypothesis is $H_a: \mu > 918$, so the *p*-value will be the area to the *right* of the computed value, 1.53, as illustrated in Figure 8.12. Notice that the inequality in H_a determines the *direction* of the tail area to be found. The *p*-value here is .063, which is consistent with the results of Example 8.5, where we concluded that for $\alpha = .01$, you fail to reject H_0 and for $\alpha = .10$, you reject H_0. That is, the *p*-value is between .01 and .10.

Most statistical computer packages will provide you with the computed *p*-value when testing the mean of a population. The MINITAB solution to Example 8.5 is provided in Figure 8.13. This procedure assumes that the population standard deviation (σ) is unknown, and so it uses the command *TTEST* (as in *t*-test). The *p*-value in Figure 8.13 is slightly different than the value obtained in Example 8.7, since MINITAB uses the *t* distribution to obtain this value. We discuss this point further in Section 8.4, but for now remember that the *t* random variable is closely approximated by the standard normal, Z, when using a large sample.

Interpreting the *p*-Value

We will consider two ways of using the *p*-value to arrive at a conclusion. The first is the **classical approach** that we have used up to this point: We choose a value for α and base our decision on this value. When using a *p*-value in this manner, the procedure is:

reject H_0 if *p*-value $< \alpha$

fail to reject H_0 if *p*-value $\geq \alpha$

FIGURE 8.13
MINITAB solution for
Example 8.5.

```
MTB > SET INTO C1
DATA> 950.14 1006.79 119.31 1115.08 ...
   . .    (the 60 data values)

DATA> END
MTB > AVERAGE OF C1
    MEAN   =    952.36
MTB > STDEV OF C1
    ST.DEV. =    173.92
MTB > TTEST OF MU=918 USING C1;
SUBC> ALTERNATIVE=1.

TEST OF MU = 918.0 VS MU G.T. 918.0

         N      MEAN     STDEV    SE MEAN       T     P VALUE
   C1    60     952.4    173.9      22.4       1.53    0.066
```

= 1 for $H_a: \mu > \mu_0$
= −1 for $H_a: \mu < \mu_0$
This step is not necessary
for $H_a: \mu \neq \mu_0$

The second approach is a **general rule of thumb** that applies to most applications of hypothesis testing on μ. We previously stated that typical values of α range from .01 to .10. This implies that for most applications we will not see values of α smaller than .01 or larger than .1. With this in mind, the following rule can be defined:

reject H_0 if the p-value is small ($p < .01$)

fail to reject H_0 if the p-value is large ($p > .1$)

Consequently, if $.01 \leq p\text{-value} \leq .1$, the data are *inconclusive*.

The advantage of this approach is that you avoid having to choose a value of α; the disadvantage is that you may arrive at an inconclusive result.

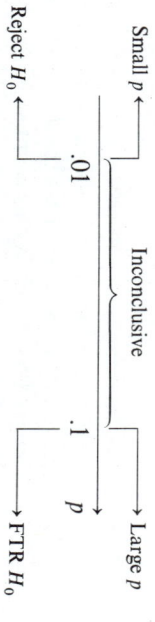

Now for a brief disclaimer: This rule does not apply to all situations. If a Type I error would be extremely serious and you prefer a very small value of α using the classical approach, then you can lower the .01 limit. Similarly, you might raise the .1 limit if the Type II error is extremely critical and you prefer a large value for α. However, this gives a working procedure for most applications in business.

What can you conclude if the p-value is $p = .0001$? This value is extremely small as compared with *any* reasonable value of α. So we would strongly reject H_0. Consequently, if you are making an investment decision based on these results, for example, you can breathe a little easier. This data set supports H_a overwhelmingly. On the other hand, if $p = .65$, this value is large as compared with any reasonable value of α. Without question, we would fail to reject H_0.

There is yet one other interpretation of the p-value, summarized in the following box.

ANOTHER INTERPRETATION OF THE p-VALUE

1. For a two-tailed test where H_a: $\mu \neq \mu_0$, the p-value is the probability that the value of the test statistic, Z^*, will be at least as large (in absolute value) as the observed Z^*, if μ is in fact equal to μ_0.

2. For a one-tailed test where H_a: $\mu > \mu_0$, the p-value is the probability that the value of the test statistic, Z^*, will be at least as large as the observed Z^*, if μ is in fact equal to μ_0.

3. For a one-tailed test where H_a: $\mu < \mu_0$, the p-value is the probability that the value of the test statistic, Z^*, will be at least as small as the observed Z^*, if μ is in fact equal to μ_0.

In Example 8.7, we determined the p-value to be .063; the computed value of the test statistic was $Z^* = 1.53$; the hypotheses were H_0: $\mu \leq 918$ and H_a: $\mu > 918$. So the probability of observing a value of Z^* as large as 1.53 (that is, $Z \geq 1.53$) if μ is 918 is $p = .063$.

Based on this description of the p-value, if p is small, conclude that H_0 is not true and reject it. We obtain precisely the same result using the classical and rule-of-thumb options of the p-value. Small values of p favor H_a, and large values favor H_0.

EXAMPLE 8.8

In Example 8.6 we performed a one-tailed test of H_0: $\mu \geq 48{,}000$ and H_a: $\mu < 48{,}000$. The sample results were $n = 45$, $x = \$43{,}900$, and $s = \$7140$. The calculated value of the test statistic was

$$Z^* = \frac{43{,}900 - 48{,}000}{s_{\bar{x}}}$$

where

$$s_{\bar{x}} = \frac{s}{\sqrt{n}} \sqrt{\frac{N-n}{N-1}}$$

$$= \frac{7140}{\sqrt{45}} \sqrt{\frac{350-45}{350-1}} = 995.01$$

so $Z^* = -4.12$.

1. What is your conclusion based on the corresponding *p*-value, using $\alpha = .05$?
2. Without specifying a value of α, what would be your conclusion based on the calculated *p*-value?
3. Interpret the *p*-value for this application.

SOLUTION 1

The *p*-value is illustrated in Figure 8.14. We are unable to determine the *p*-value exactly using Table A.4; however, this area is roughly the same as the area to the left of -4.0 under the Z curve—namely, $.5 - .49997 = .00003$. So $p \cong .00003$. Because *p* is less than $\alpha = .05$, we reject H_0. Our conclusion is the same as that of Example 8.6 (which also used $\alpha = .05$), where we concluded that the female managers were underpaid.

SOLUTION 2

We use the general rule of thumb for interpreting the *p*-value. Since $p \cong .00003$, it is extremely small, so we strongly reject H_0 (same conclusion as Solution 1).

SOLUTION 3

We can make the following statements:

1. The significance level at which the conclusion indicated by the testing procedure changes is $\alpha = .00003$.
2. The smallest significance level for which you can reject the null hypothesis is $\alpha = .00003$.
3. The probability of observing a value of the test statistic as small as the one obtained (≤ -4.12) is .00003 if, in fact, the population mean is \$48,000. ∎

Practical Versus Statistical Significance

Researchers often calculate what appears to be a conclusive result without considering the practical significance of their findings. For example, consider a situation similar to the one described in Example 8.4; this time, a sample of 1000 Bullets, tested under normal highway conditions, results in a sample average of $\bar{x} = 32.32$ mpg, with a standard deviation of $s = 2.15$ mpg. Advertising for

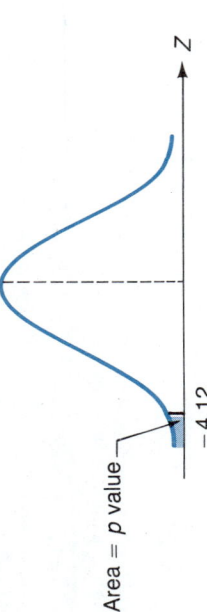

FIGURE 8.14
Illustration of the *p*-value for Example 8.8.

FIGURE 8.15

p-value for Z* = −2.65.

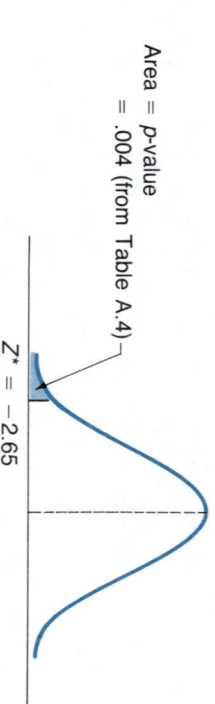

Area = p-value
= .004 (from Table A.4)

Z* = −2.65

this car claims that the mpg under test conditions is at least 32.5 mpg. Is there sufficient evidence to reject this claim?

The hypotheses are $H_0: \mu \geq 32.5$ and $H_a: \mu < 32.5$. The value of the test statistic is

$$Z^* = \frac{\bar{X} - 32.5}{s/\sqrt{n}} = \frac{32.32 - 32.5}{2.15/\sqrt{1000}} = -2.65$$

The p-value here is the area to the left of −2.65 under the Z curve, as illustrated in Figure 8.15. This value (from Table A.4) is .004. Based on this small p-value, we reject H_0 and conclude (as we did in Example 8.4) that the mpg for these cars under normal highway conditions is less than 32.5. Statistically speaking, this is correct, and the data do provide sufficient evidence to support the statement that their mpg claim is overstated. As a consumer, however, how concerned would you be that the sample average ($\bar{x} = 32.32$) is (only) .18 mpg under the advertised level? In other words, in a practical sense, how misleading is the Bullet advertising?

What we have seen is that $\bar{X}$ is far enough away from 32.5 (in a statistical sense) to conclude that μ is less than 32.5 mpg. However, perhaps in the eyes of a consumer about to invest $15,000 in a new car, this value of $\bar{X}$ is really "close enough" to 32.5.

Moral: It is possible for a statistically significant result to be of no particular practical significance, depending on the context of the analysis.

EXERCISES

8.27 State whether you would reject or fail to reject the null hypothesis in each of the following cases.

a. $p = .12, \alpha = .05$.

b. $p = .03, \alpha = .05$.

c. $p = .001, \alpha = .01$.

d. $p = .01, \alpha = .001$.

8.28 Using the rule-of-thumb option (not selecting a value of α) in the interpretation of the p-value, state whether the test statistic would be statistically significant in the following situations.

a. $p = .57$.

b. $p = .008$.

c. $p = .12$.

d. $p = .04$.

8.29 Explain the difference between "significance" in a statistical sense and "significance" in a practical sense.

8.30 Find p-values for the following situations with calculated test statistics given by Z^*.

a. $H_0: \mu = 30, H_a: \mu \neq 30, Z^* = 2.38$

b. $H_0: \mu \leq 20, H_a: \mu > 20, Z^* = 1.645$

c. $H_0: \mu \geq 15, H_a: \mu < 15, Z^* = -2.54$

d. $H_0: \mu = 50, H_a: \mu \neq 50, Z^* = -1.85$

8.31 Test the belief that the mean of a normally distributed population exceeds 20, assuming that a sample of size 60 yields the following statistics:

$$\bar{x} = 20.4 \qquad s = 3.0$$

Use the p-value criteria.

8.32 The producer of Take-a-Bite, a snack food, claims that each package weighs 175 grams. A representative of a consumer advocate group selected a random sample of 70 packages. From this sample, the mean and standard deviation were found to be 172 grams and 8 grams, respectively.

a. Find the *p*-value for testing the claim that the mean weight of Take-a-Bite is less than 175 grams.

b. Interpret the *p*-value in part (a).

8.33 A marketing-research analyst is interested in examining the statement made by the makers that brand A cigarettes contain less than 3 milligrams of tar. The marketing-research analyst randomly selected 60 cigarettes and found the mean amount of tar to be 2.75 milligrams with a standard deviation of 1.5 milligrams. Do the data support the claim? Find the *p*-value.

8.34 The Association of Independent Commercial Producers enlisted the Television Bureau of Advertising in 1987 to get estimates on the average cost of a 30-second TV spot. From surveying 60 production houses, the Television Bureau of Advertising found that on the average, it would cost approximately $50,000 to shoot a 30-second TV spot. Assume that a group of advertisers wished to verify this claim and that a separate random sample of 55 production houses produced a sample mean of $57,386 with a standard deviation of $10,112.

(*Source:* "Spot Discrepancies," *Sales and Marketing Management* (January 1988); 27.)

a. Does the sample evidence indicate that the mean cost to shoot a 30-second TV spot is not $50,000? Interpret the *p*-value for the test.

b. From the *p*-value given in part (a), would you expect a 99% confidence interval for the mean cost to contain $50,000? Find a 99% confidence interval for the mean cost of shooting a 30-second TV spot.

8.35 Find the *p*-value for the test conducted in Exercise 8.24.

8.36 A recruiter from a large recruiting firm wishes to determine if the mean starting salaries for students with an MBA and no experience is greater than $34,000 for a certain metropolitan area. From a random sample of 50 starting salaries for MBA's without experience, the mean and standard deviation were found to be $34,715 and $2960. Do the data support the belief that the mean starting salary of MBAs without experience is greater than $34,000? Use the *p*-value criteria.

8.4
HYPOTHESIS TESTING ON THE MEAN OF A NORMAL POPULATION: SMALL SAMPLE

Our approach to hypothesis testing with small samples when the standard deviation, σ, is unknown uses the same technique we used for dealing with confidence intervals on the mean of a population: We switch from the standard normal distribution, Z, to the t distribution. However, we need to examine the distribution of the population when the sample is small—the population distribution determines the procedure that we use. In this section, we have reason to believe that the population has a normal distribution. When it does not, we use a nonparametric procedure, which is discussed in Chapter 18.

Certain variations from a normal population *are* permissible with the small-sample test. If a test of hypothesis is still reliable when slight departures from the assumptions are encountered, this test is said to be **robust**. If you believe the parent population to be reasonably symmetric, the level of your confidence interval and Type I error (α) will be quite accurate, even if the population has heavy tails (unlike the normal distribution), as shown in Figure 8.16(A). However, when using small samples, the small-sample test is *not* robust for populations that are heavily skewed (see Figure 8.16(B)). A nonparametric procedure offers a much better solution for this situation. For larger sizes, a histogram of your data often can detect whether a population is heavily skewed in one direction.

FIGURE 8.16
A: Small-sample test is valid.
B: Small-sample test is not valid.

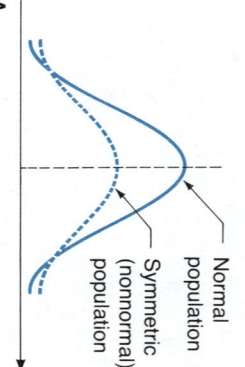

A

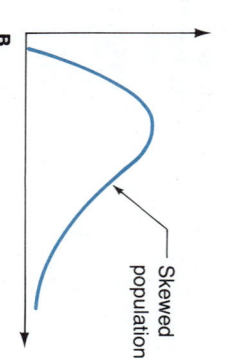

B

To reemphasize, the discussion in this section assumes a normal population. In other words, if X is an observation from this population, then $\bar{X}$ is a normal random variable with unknown mean μ. Also, we assume that σ is unknown. (If σ is known, the resulting test statistic is $Z = (\bar{X} - \mu_0)/(\sigma/\sqrt{n})$, and the five-step procedure of Section 8.1 allows you to do hypothesis testing on μ.)

The only distinction between using a small and a large sample is the form of the test statistic. Using the discussion from Chapter 7, if we define the test statistic as

$$t = \frac{\bar{X} - \mu_0}{s/\sqrt{n}} \qquad (8.3)$$

we now have a t distribution with $n - 1$ degrees of freedom (df). The procedure to use for testing $H_0: \mu = \mu_0$ and $H_a: \mu \neq \mu_0$ is the same five-step procedure, except that the rejection region is defined using the t table (Table A.5) rather than the Z table (Table A.4). This also applies to a one-tailed test. Because we are looking at very small samples (typically $n < 30$), we can ignore the finite population correction factor.

EXAMPLE 8.9

You may recall from Example 7.7 that Metro Moving Company is considering the purchase of a new moving van. They will purchase the van if it can be demonstrated that its average miles per gallon is greater than 9. Using the $n = 18$ data values from Example 7.7, how would you advise Metro? The daily mpg values are believed to follow a normal distribution.

SOLUTION

Step 1. What you are attempting to demonstrate goes into the alternative hypothesis, H_a, so the hypotheses are $H_0: \mu \leq 9$ and $H_a: \mu > 9$.

Step 2. The test statistic here is

$$t = \frac{\bar{X} - 9}{s/\sqrt{n}}$$

Step 3. The implications of making a Type I error (rejecting a correct H_0) and a Type II error (fail to reject an incorrect H_0) appear to be the same, so you decide on a significance level of $\alpha = .05$.

As before, we will reject H_0 when the value of the test statistic lies in the right tail (Figure 8.17):

$$\text{reject } H_0 \text{ if } t > t_{.05, 17} = 1.74$$

because df $= n - 1 = 17$.

Step 4. For this data set, $n = 18$, $\bar{x} = 9.64$ mpg, and $s = 1.52$ mpg. The value of

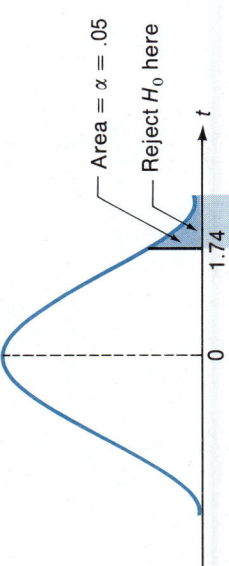

FIGURE 8.17

t distribution; the rejection region is the lightly shaded area to the right of 1.74, for Example 8.9.

the test statistic is

$$t^* = \frac{9.64 - 9}{1.52/\sqrt{18}} = 1.79$$

Because $1.79 > 1.74$, we reject H_0.

Step 5. The average daily mpg of this van is larger than 9. ■

What is the p-value in Example 8.9, and what can we conclude based on this value? We run into a slight snag when dealing with the t distribution because we are not able to determine precisely the p-value. You can see this in Figure 8.18, using Table A.5 (17 df). The p-value is the area to the right of $t^* = 1.79$. The best we can do here is to say that p is *between* .025 and .05. (*Note:* A reliable computer package will provide the exact p-value. Using MINITAB, this value is $p = .0456$.)

Using the classical approach and $\alpha = .05$ we *can* say that p is less than .05, despite not knowing p exactly. Consequently, we reject H_0. This procedure *always* produces the same result as the five-step procedure. Notice that this conclusion does not tell us how strongly we reject H_0 but simply that H_0 is rejected at this significance level.

Suppose we choose not to select a significance level (α) but prefer to base our conclusion strictly on the calculated p-value. We use the rule of thumb and decide whether p is small ($<.01$), large ($>.1$), or in between. Despite not having an exact value of p, we can say that this p-value falls in the inconclusive range. These data values do not provide us with any strong conclusion. Our advice to Metro would be to obtain some additional data.

EXAMPLE 8.10

An auditing firm was hired to determine if a particular defense plant was overstating the value of their inventory items. It was decided that 15 items would be randomly selected. For each item, the recorded amount, the audited (exact) amount, and the difference between these two amounts (recorded − audited) were determined. Of particular interest was whether it could be demonstrated that the average difference exceeds $25, in which case the defense plant would be subject to a loss of contract and financial penalties. The following 15 differences were obtained (in dollars):

17, 35, 31, 22, 50, 42, 56, 23, 27, 38, 20, 25, 43, 45, 21

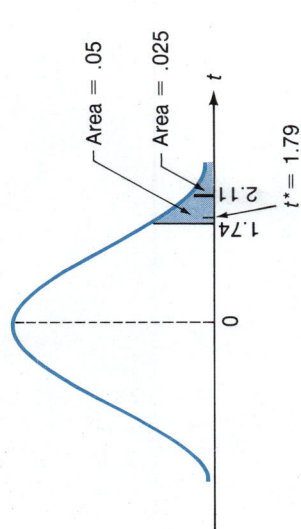

FIGURE 8.18

t curve with 17 df. The p-value is the area to the right of $t^* = 1.79$, so we can say only that it is between .025 and .05.

So $n = 15$, $\bar{x} = \$33.00$, and $s = \$12.15$. Set up the appropriate hypotheses and test them using a significance level of $\alpha = .05$. The population of differences is believed to be normally distributed.

SOLUTION

Step 1. The hypotheses are H_0: $\mu \leq 25$ and H_a: $\mu > 25$, where μ is the average difference between the recorded and audited amounts for *all* the inventory items.

Steps 2, 3.

$$\text{reject } H_0 \text{ if } t = \frac{\bar{X} - 25}{s/\sqrt{n}} > t_{.05, 14} = 1.761,$$

where the df $= n - 1 = 14$.

Step 4. The calculated t is

$$t^* = \frac{33 - 25}{12.15/\sqrt{15}} = 2.55$$

Because 2.55 exceeds the tabulated value of 1.761, we reject H_0. Also, the p-value (using Table A.5 and 14 df) is the area to the right of 2.55. This is between .01 and .025, which is less than $\alpha = .05$, and so (as before) we reject H_0.

Step 5. These data indicate that the defense plant is overstating the value of their inventory items by more than \$25. ∎

A MINITAB solution for this example is contained in Figure 8.19. Note that the calculated (exact) p-value is .012.

SMALL-SAMPLE TESTS ON A NORMAL POPULATION MEAN

TWO-TAILED TEST

H_0: $\mu = \mu_0$
H_a: $\mu \neq \mu_0$

reject H_0 if $|t| > t_{\alpha/2, n-1}$

where $n =$ sample size and $t = \dfrac{\bar{X} - \mu_0}{s/\sqrt{n}}$

ONE-TAILED TEST

H_0: $\mu \leq \mu_0$ H_0: $\mu \geq \mu_0$
H_a: $\mu > \mu_0$ H_a: $\mu < \mu_0$

reject H_0 if $t > t_{\alpha, n-1}$ reject H_0 if $t < -t_{\alpha, n-1}$

FIGURE 8.19
MINITAB solution for Example 8.10.

```
MTB > SET INTO C1
DATA> 17 35 31 22 50 42 6 23 27 38 20 25 43 45 21
DATA> END
MTB > TTEST OF MU=25 USING C1;
SUBC> ALTERNATIVE=1.

TEST OF MU = 25.000 VS MU G.T. 25.000

           N     MEAN   STDEV   SE MEAN      T   P VALUE
C1        15   33.000  12.148     3.137   2.55     0.012
```

EXERCISES

8.37 Find the rejection region of the *t*-test used to test the following situations for a normally distributed population:

a. Twenty observations are randomly selected to test the claim that mean yearly maintenance expense on a certain type of lawn mower is less than $28 per year. A significance level of .05 is used.

b. Twenty-five observations are randomly selected to test the claim that managers of convenience stores have an annual income of more than $30,000. A significance level of .10 is used.

c. Fifteen observations are randomly selected to test the claim that the tensile strength of steel rods is equal to the tensile strength specified by the firm ordering the steel rods. A significance level of .05 is used.

8.38 Find the *p*-value for the following situations with calculated test statistics given by t^*.

a. $H_0: \mu = 40$, $H_a: \mu \neq 40$, $t^* = 2.30$, $n = 12$.

b. $H_0: \mu \leq 13.6$, $H_a: \mu > 13.6$, $t^* = 2.73$, $n = 19$.

c. $H_0: \mu \geq 100.80$, $H_a: \mu < 100.80$, $t^* = 1.25$, $n = 20$.

d. $H_0: \mu = 35.6$, $H_a: \mu \neq 35.6$, $t^* = 1.57$, $n = 11$.

8.39 Carry out the hypothesis test for the mean of a normally distributed population given the following information:

$$H_0: \mu \leq 1.6$$
$$H_a: \mu > 1.6$$

$$n = 15 \qquad \bar{x} = 1.8 \qquad s^2 = 1.7 \qquad \alpha = .10$$

8.40 A sample of size 12 is drawn from a finite population of size 300. The finite population can be approximated by a normal distribution. The sample mean and sample standard deviation are 100.6 and 3.7, respectively. Is there evidence to indicate that the mean of the population is different from 107 at the 10% significance level?

8.41 Refer to Exercise 8.40. Is there evidence to indicate that the mean of the population is different from 107, if it is known that the population standard deviation is equal to 4?

8.42 The senior executive of a publishing firm would like to train employees to read faster than 1000 words per minute. A random sample of 21 employees underwent a special speed-reading course. This sample yielded a mean of 1018 words per minute with a standard deviation of 30 words per minute. Do the data support the belief that the speed-reading course will enable the employees to read more than 1000 words per minute at a significance level of .05? Assume that the reading speeds of persons who have taken the course are normally distributed.

8.43 It is believed that the mean score on an aptitude test of engineers graduating from Safire University is greater than 180. Assuming the scores are normally distributed, do the data support the belief if a random sample of 26 engineers yielded a mean score of 186 with a standard deviation of 10.2? Use the *p*-value.

8.44 Gopal and Krause, an investment firm, have made public a new growth and income mutual fund. To enjoy the fruits of this well-managed fund, an initial deposit of $10,000 is required. After the account is opened, the balance can fall as low as $5000. Gopal and Krause believe that the account balances in this mutual fund are normally distributed with a mean greater than $10,000. To test this belief, a random sample of 23 accounts is selected. The sample mean is $10,963 and the sample standard deviation is $446. Do the data support the claim that the mean account balance in this fund is greater than $10,000? Use a .05 significance level.

8.45 The comptroller of National Insurance Company states that the average claim against the company for an automobile accident is less than $4500. A random sample of 14 claims yielded a mean amount of $4200 with a sample standard deviation of $171. It is believed that the claims are normally distributed. Use the *p*-value criteria to determine if the data supports the comptroller's statement.

8.46 The Chevrolet Sprint Metro is one of the most fuel-efficient automobiles on the market with 57 miles per gallon in the city. The actual miles per gallon can vary according to conditions that may be less than ideal. Assume that a consumer advocacy group was interested in the average miles per gallon for the Chevrolet Sprint Metro for a typical worker in Houston, Texas, in which the car was used in city driving. Fifteen Chevrolet Sprint Metros in top condition were each used to drive 100 miles in city traffic. The miles per gallon for the fifteen cars were

52, 53, 50, 53, 46, 45, 51, 53, 43, 49, 52, 56, 55, 53, 54

(Source: World Almanac and Book of Facts, 1988, p. 802.)

a. Do the data provide sufficient evidence to conclude that the mean miles per gallon is significantly less than 57 miles per gallon? Use a .05 significance level.

b. Find the *p*-value and interpret this value.

8.5

INFERENCE FOR THE VARIANCE AND STANDARD DEVIATION OF A NORMAL POPULATION (OPTIONAL)

Our discussion in Chapters 7 and 8 has been concerned with the mean of a particular random variable or population. In other words, we are trying to decide or estimate what is occurring *on the average*. Suppose someone involved with a production process that is manufacturing 2-inch bolts has just been informed that, without a doubt, these bolts are 2 inches long, on the average. Is there anything else this person might like to know about the production process? Suppose that one-half of the bolts produced are 1 inch long and the other half are 3 inches. The report was accurate—on the average, they *are* 2 inches long.* However, such a production process certainly will not satisfy the customers, and this company soon will be out of the bolt business.

What was missing in the report was the amount of *variation* in this production process. If the variation was zero, every bolt would be exactly 2 inches long—an ideal situation. In practice, there always will be a certain amount of variation in any mechanical or production process. So we are concerned about not only the mean length μ of the population of bolts but also the variance σ^2 or standard deviation σ of the lengths of these bolts. If the variance is *too large*, the process is not operating correctly and needs adjustment.

The variance of a population also is of vital interest to someone making investment decisions. Here the *risk* of a venture (or portfolio) often is measured by the variance of the return paid by the venture in the past. Often, financial analysts prefer a financial package with a relatively small average return (based on past history) that appears to be low risk on the basis of only small fluctuations in its past performance.

In the inference procedures for a population variance (and standard deviation) to follow, we will assume that the population of interest is normally distributed. Unlike the *t* test, the hypothesis testing procedures and confidence intervals for the variance are very sensitive to departures from the normal population—notably, heavy tails in the distribution or heavy skewness will have a large effect. In other words, the following tests of hypothesis are less robust than are those we discussed earlier.

Confidence Interval for the Variance and Standard Deviation

The point estimate of a population variance is the obvious one—namely, the sample variance. This was discussed in Chapter 7, where we used the variance, s^2, of a sample to estimate the variance, σ^2, of the much larger population.[†]

---

* A statistician often is described as someone who thinks that if one-half of you is in an oven and the other half is in a deep freeze, on the average you are very comfortable.

† The notation $\hat{\sigma}^2$ is often used to represent an estimate of σ^2. Consequently, $s^2 = \hat{\sigma}^2$.

FIGURE 8.20

Shape of a chi-square distribution.

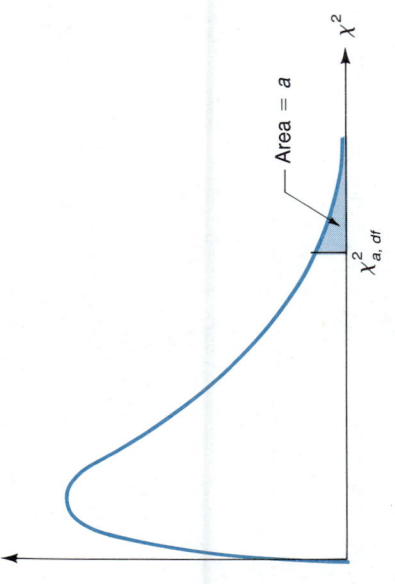

When constructing a confidence interval for μ using a small sample, we used the t distribution. Such a distribution is referred to as a **derived distribution** because it was derived to describe the behavior of a particular test statistic. This type of distribution is not used to describe a population, as is the normal distribution in many applications. For example, you will *not* hear a statement such as: Assume that these data follow a t distribution—normal, exponential, uniform, maybe, but not a t distribution. The t random variable merely offers us a method of testing and constructing confidence intervals for the mean of a *normal* population when the standard deviation is unknown and is replaced by its estimate.

Another such continuous derived distribution allows us to determine confidence intervals and perform tests of hypothesis on the variance and standard deviation of a normal population. This is the **chi-square** (pronounced ky) distribution, written as χ^2. The shape of this distribution is illustrated in Figure 8.20. Notice that, unlike the Z and t curves, the χ^2 distribution is not symmetric and is definitely skewed right.

For chi-square, as with all continuous distributions, a probability corresponds to an area under a curve. Also, the shape of the chi-square curve, like that of its cousin the t distribution, depends on the sample size n. As before, this will be specified by the corresponding degrees of freedom (df).

When using the χ^2 distribution to construct a confidence interval or perform a test of hypothesis on a population variance or standard deviation, the degrees of freedom are given by

$$df = n - 1$$

Let $\chi^2_{a, df}$ be the χ^2 value whose area to the right is a, using the proper df.

EXAMPLE 8.11

Using a chi-square curve with 12 df, determine $P(\chi^2 > 18.5494)$ and $P(\chi^2 < 6.30380)$.

SOLUTION

Tabulated values for the χ^2 distribution are contained in Table A.6. This table contains *right-tailed* areas (probabilities). Based on this table (see Figure 8.21),

$$P(\chi^2 > 18.5494) = .1$$

This can be written as

$$\chi^2_{.1, 12} = 18.5494$$

For $\chi^2 = 6.30380$, Table A.6 informs us that the area to the right of 6.30380 is .900. Because the total area is 1, the area to the left of 6.30380 is $1 - .900 = .1$,

FIGURE 8.21

χ^2 curve with 12 df. The shaded area represents $P(\chi^2 > 18.5494)$.

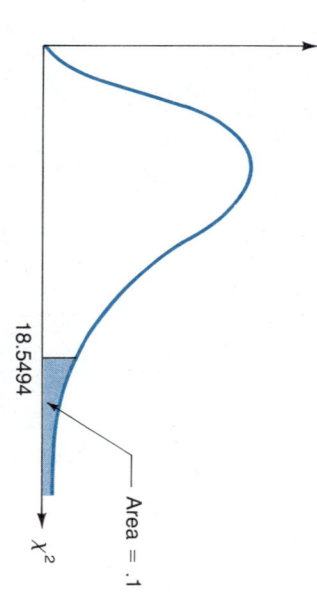

Area = .1

18.5494 χ^2

and so $P(\chi^2 < 6.30380) = .1$. As a result, we can say that

$$P(6.30380 \leq \chi^2 \leq 18.5494) = 1 - .1 - .1 = .8$$

That is, 80% of the time a χ^2 value (with 12 df) will be between 6.30380 and 18.5494. ∎

EXAMPLE 8.12

Using Example 8.11, determine a and b that satisfy

$$P(a < \chi^2 < b) = .95, \qquad \text{with df} = 12$$

Choose a and b so that an equal area occurs in each tail.

SOLUTION

Figure 8.22 shows the areas for a and b. Using Table A.6,

a = the χ^2 value whose left-tailed area is .025

= the χ^2 value whose area to the right is .975

= 4.40

and

b = the χ^2 value whose right-tailed area is .025

= 23.3 ∎

To derive a confidence interval for σ^2, we need to examine the sampling distribution of s^2. If we repeatedly obtained a random sample from a normal population with mean μ and variance σ^2, calculated the sample variance s^2, and made a histogram of these s^2 values, what would be the resulting shape of this histogram? It can be shown that the shape will depend on the sample size n and the value of σ^2 but *not* on the value of the population mean μ. In fact, the values of n and σ^2, along with the random variable s^2, can be combined to define a chi-square random variable, given by

$$\chi^2 = \frac{(n-1)s^2}{\sigma^2} \qquad (8.4)$$

FIGURE 8.22

χ^2 curve with 12 df.

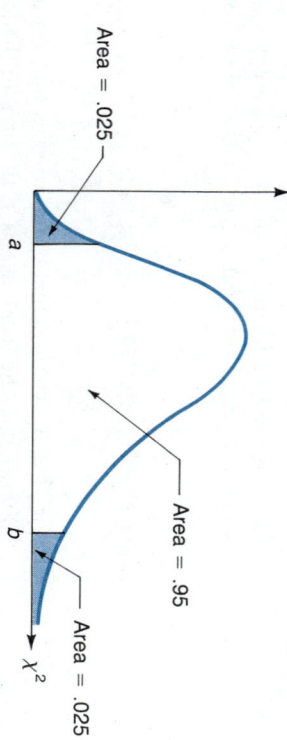

Area = .025

Area = .95

Area = .025

a b χ^2

having a chi-square distribution with $n - 1$ df. Therefore, the sampling distribution for s^2 can be defined using the chi-square distribution in equation 8.4.

For example, a sample size of $n = 13$ results in 12 df. From Example 8.12, it follows that

$$P(4.40 < \chi^2 < 23.3) = .95$$

So,

$$P\left[4.40 < \frac{12s^2}{\sigma^2} < 23.3\right] = .95 \qquad \text{using equation 8.4}$$

or

$$P\left[\frac{12s^2}{23.3} < \sigma^2 < \frac{12s^2}{4.40}\right] = .95$$

As in all confidence interval constructions, the parameter (σ^2) is bounded between two limits defined by a random variable (s^2). This means that a 95% confidence interval for σ^2 is

$$\frac{12s^2}{23.3} \quad \text{to} \quad \frac{12s^2}{4.40}$$

In general, the following procedure can be used to construct a confidence interval for σ^2 or σ. A $(1 - \alpha) \cdot 100\%$ confidence interval for σ^2 is

$$\frac{(n-1)s^2}{\chi^2_{\alpha/2,\,n-1}} \quad \text{to} \quad \frac{(n-1)s^2}{\chi^2_{1-\alpha/2,\,n-1}} \qquad \textbf{(8.5)}$$

The corresponding confidence interval for σ is

$$\sqrt{\frac{(n-1)s^2}{\chi^2_{\alpha/2,\,n-1}}} \quad \text{to} \quad \sqrt{\frac{(n-1)s^2}{\chi^2_{1-\alpha/2,\,n-1}}} \qquad \textbf{(8.6)}$$

EXAMPLE 8.13

Vitamix Dog Chow comes in 10-, 25-, and 50-pound bags. The owners are concerned about the variation in the weight of the 50-pound bags because they have recently acquired a new mechanical packaging device. A random sample of the weights of 15 bags (in pounds) was obtained, with the following results:

51.2, 47.5, 50.8, 51.5, 49.5, 51.1, 51.3, 50.7, 46.7, 49.2, 52.1, 48.3, 51.6, 49.2, 51.5

For these data, $\bar{x} = 50.15$ pounds and $s = 1.65$ pounds. Determine a 90% confidence interval for σ^2 and for σ. The bag weights are believed to come from a normal population.

SOLUTION

The corresponding 90% confidence interval for σ^2 is

$$\frac{(15-1)(1.65)^2}{\chi^2_{.05,14}} \quad \text{to} \quad \frac{(15-1)(1.65)^2}{\chi^2_{.95,14}} = \frac{(14)(1.65)^2}{23.7} \quad \text{to} \quad \frac{(14)(1.65)^2}{6.57}$$

$$= 1.61 \quad \text{to} \quad 5.80$$

The 90% confidence interval for σ would be

$$\sqrt{1.61} \quad \text{to} \quad \sqrt{5.80}$$

that is, 1.27 pounds to 2.41 pounds.

Hypothesis Testing for the Variance and Standard Deviation

For many applications, we are concerned that the standard deviation or variance of our population may be exceeding some specified value. If this claim is supported, then, for example, we may wish to shut down a production process and make adjustments that will reduce this excessive variation. As you could with the tests of hypothesis examined so far, you can (although this is not the usual case) perform a two-tailed test where either too much variation or too little variation is the topic of concern.

HYPOTHESIS TESTING ON σ^2

TWO-TAILED TEST

$H_0: \sigma^2 = \sigma_0^2$
$H_a: \sigma^2 \neq \sigma_0^2$

reject H_0 if $\chi^2 > \chi_{\alpha/2, n-1}^2$ or   if $\chi^2 < \chi_{1-\alpha/2, n-1}^2$

test statistic: $\chi^2 = \dfrac{(n-1)s^2}{\sigma_0^2}$

ONE-TAILED TEST

$H_0: \sigma^2 \leq \sigma_0^2$ $H_0: \sigma^2 \geq \sigma_0^2$
$H_a: \sigma^2 > \sigma_0^2$ $H_a: \sigma^2 < \sigma_0^2$

reject H_0 if $\chi^2 > \chi_{\alpha, n-1}^2$ reject H_0 if $\chi^2 < \chi_{1-\alpha, n-1}^2$

EXAMPLE 8.14

Example 8.13 was concerned with the variation of the actual weight of a (supposedly) 50-pound bag of Vitamix Dog Chow. Based on earlier production tests, management is convinced that the average weight of all bags being produced is, in fact, 50 pounds. However, the production supervisor has been informed that at least 95% of the bags produced *must* be within 1 pound of the specified weight (50 pounds). Using a significance level of $\alpha = .1$, what can we conclude? Assume a normal distribution for the bag weights.

What is the supervisor being told about σ? Remember that for a normal population, 95% of the observations will lie within two standard deviations of the mean (empirical rule, Chapter 3). So, if two standard deviations are the same as 1 pound, then the supervisor is being told that σ must be no more than .5 pound. Is there any evidence to conclude that this is not the case—that is, that σ is larger than .5 pound? Let's investigate.

SOLUTION

Step 1. The appropriate hypotheses are $H_0: \sigma \leq .5$ and $H_a: \sigma > .5$ (production is not meeting required standards).

(Note that these hypotheses are precisely the same as $H_0: \sigma^2 \leq .25$ and $H_a: \sigma^2 > .25$. Whether you write H_0 and H_a in terms of σ or σ^2 does not matter; the testing procedure is the same in either case.)

Step 2. The test statistic is

$$\chi^2 = \frac{(15-1)s^2}{(.5)^2} = \frac{14s^2}{.25}$$

which has a chi-square distribution with 14 df.

Step 3. Using $\alpha = .1$ and Table A.6, the rejection region for this test is

reject H_0 if $\chi^2 > 21.1$

FIGURE 8.23

Illustration for the p-value for Example 8.14.

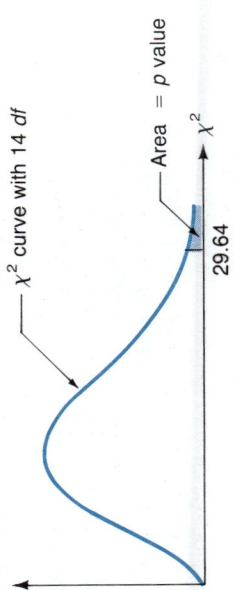

χ^2 curve with 14 df

Area = p value

29.64

χ^2

Step 4. The computed value using the sample data is

$$\chi^{2*} = \frac{(15-1)(1.65)^2}{(.5)^2} = 152.5$$

Since $152.5 > 21.1$, we reject H_0. This is hardly a surprising result; the point estimate of σ is $s = 1.65$, quite a bit larger than .5.

Step 5. We conclude rather convincingly that σ is larger than .5 pound. The bagging procedure has far too much variation in the weight of the bags produced. ∎

Note that the p-value for the test of hypothesis in Example 8.14 is the area to the right of 152.5 under the χ^2 curve with 14 df (illustrated in Figure 8.23). All we are able to determine about this value using Table A.6 is that it is much smaller than .005 (the smallest tabulated value). Using this information, we arrive at the same decision—namely, reject H_0—because (1) using the classical approach, p is less than $\alpha = .10$, or (2) the p-value is extremely small ($<.01$) by the general rule of thumb described in Section 8.3.

EXERCISES

8.47 From the tabulated values for the chi-square distribution, find the following values and indicate graphically where the values fall with respect to other values of the chi-square distribution.

a. $\chi^2_{.10, 10}$.

b. $\chi^2_{.025, 30}$.

c. $\chi^2_{.95, 15}$.

d. $\chi^2_{.01, 26}$.

8.48 A sample of size 25 from a normally distributed population yields a sample standard deviation of 12.8. At the 10% significance level, determine if there is sufficient evidence to indicate that the population standard deviation is greater than 11.3.

8.49 A sample of size 15 from a normally distributed population yields the sample statistic

$$\Sigma(x - \bar{x})^2 = 180.3$$

a. Construct a 90% confidence interval for the population variance.

b. Construct a 90% confidence interval for the population standard deviation.

c. Do the data indicate that the population variance differs from 10? Use a 10% significance level.

8.50 A sample of size five from a normally distributed population yields the following sample statistics:

$$\Sigma x^2 = 135 \qquad \Sigma x = 23$$

a. Construct a 95% confidence interval for the population variance.

b. Construct a 95% confidence interval for the population standard deviation.

c. Is there sufficient evidence to indicate that the population standard deviation differs from 2.8? Use a 5% significance level.

8.51 The production manager of Crystal-Clear Picture Tubes believes that the life of the picture tubes is 25,000 hours. However, to maintain the company's reputation for quality, the manager would like to keep the standard deviation of the life span of the picture tubes to less than 1000 hours. A sample of 24 picture tubes was randomly selected and the sample standard deviation was found to be 928 hours. Do the data indicate that the population standard deviation is less than 1000 hours? Use a 10% significance level. What assumption must be made about the distribution of the life span of the picture tubes?

8.52 For a random variable χ^2, which has chi-square distribution with 18 df, determine the values of a and b such that $P(a \leq \chi^2 \leq b) = .90$ and such that the areas in each tail are equal, that is, $P(0 \leq \chi^2 < a) = P(b < \chi^2)$.

8.53 A production manager in charge of manufacturing plastic discs must maintain a standard deviation less than 2 millimeters for the diameter of the disc. A sample of 26 plastic discs randomly selected reveals a standard deviation of 1.85 millimeters. Assuming that the diameters of the disc are normally distributed, do the data indicate that the standard deviation of the disc is less than 2 millimeters? Use the p-value criteria.

8.54 The salaries for mathematics teachers in secondary schools in Connecticut are believed to be normally distributed with a variance greater than 3000. Test this belief using the following sample statistics:

$$\Sigma(x - \bar{x})^2 = 45,130$$

$$n = 14$$

where X represents a math teacher's salary. Use a 1% significance level.

8.55 With the widespread use of computer terminals in American society, numerous studies have investigated the interaction between human beings and machines. One study looked at (among other things) the degree of flicker and fuzziness that a human operator can tolerate for characters on the video screen. A scale of measure for flicker and fuzziness was devised by the researchers, who concluded that a variance of .25 was the optimum level. A manufacturer of video terminals claimed that his machines have precisely this level of variance (i.e., .25). A sample of 40 terminals from this manufacturer had a variance of .29. At a 1% significance level, do the data indicate that the true population variance may differ from .25?

8.56 A real estate agent believes that the standard deviation of the prices of homes in Bloomington Hill Estates is less than $6000. A random sample of 25 homes in Bloomington Hill Estates yields a sample standard deviation of $5030. Do the data support the agent's belief? Use the p-value criteria.

In Chapter 7 you were introduced to the topic of **statistical inference** by discussing the concept of estimating a population parameter (such as μ or σ) by using a corresponding sample estimate. The reliability of using the sample mean to estimate μ was measured using a confidence interval. This chapter presented the other side of statistical inference—**hypothesis testing** regarding these two population parameters, along with a method of deriving a confidence interval for the population standard deviation or variance.

For testing against a hypothetical value of the population mean (μ), we introduced a procedure that used the standard normal (Z) distribution for large samples ($n > 30$) and the t distribution for small samples. For small samples, the hypotheses are concerned with the mean of a normal population. However, we are able to discuss the mean of any continuous population when we have large samples by using the Central Limit Theorem.

The two hypotheses under investigation are the **null hypothesis**, H_0, and the **alternative hypothesis**, H_a. Typically, a claim that one is attempting to demonstrate goes into the alternative hypothesis.

Since any test of hypothesis uses a sample to infer something about a population, errors can result. Two specific errors are of great concern when you use the hypothesis-testing procedure. A **Type I error** occurs in the event you reject a null hypothesis when in fact it is true; a **Type II error** occurs when you fail to reject a null hypothesis when in fact it is not true.

The probability of a Type I error is the **significance level** of the test and is written as α. The probability of a Type II error is β; large values of β are associated with small values of α and vice versa. To define a test of hypothesis, you **select a value of** α that considers the cost of rejecting a correct H_0 and failing to reject an incorrect H_0. Typical values of α range from .01 to .1.

The **power** of a statistical test is defined as $1 - \beta$ and is equal to the probability of rejecting H_0 when it is in fact false. The value of β (and so $1 - \beta$) depends on the actual value of the parameter under investigation, and so the power of the test can be obtained for each possible value of this parameter. The resulting set of power values defines a **power curve** for this test of hypothesis.

A five-step procedure was defined for any test of hypothesis:

Step 1. Set up H_0 and H_a.
Step 2. Define the **test statistic**, which is evaluated using the sample data.
Step 3. Define a **rejection region**, using the value of α, stating which values of the test statistic will result in rejecting H_0.
Step 4. Calculate the value of the test statistic from the sample data and carry out the test. This will result in rejecting H_0 or failing to reject H_0.
Step 5. Give a conclusion in the language of the problem.

A test such as H_0: $\mu = 50$ versus H_a: $\mu \neq 50$ is called a **two-tailed test** because we reject H_0 whenever the sample estimate of μ ($\bar{X}$) is either too large (test statistic is in the right tail) or too small (test statistic is in the left tail). Similarly, a test on the population variance (or standard deviation) such as H_0: $\sigma^2 = .2$ versus H_a: $\sigma^2 \neq .2$ also is a two-tailed test.

$$H_0: \mu \leq 50 \text{ versus } H_a: \mu > 50 \quad \text{or} \quad H_0: \mu \geq 50 \text{ versus } H_a: \mu < 50$$

or

$$H_0: \sigma^2 \leq .2 \text{ versus } H_a: \sigma^2 > .2 \quad \text{or} \quad H_0: \sigma^2 \geq .2 \text{ versus } H_a: \sigma^2 < .2$$

are all examples of **one-tailed tests** of hypothesis, since the rejection region lies in either the left tail or the right tail.

The tests on a population variance introduced the **chi-square distribution**, χ^2. This distribution was used to construct confidence intervals for σ^2 and σ as well as to define a distribution for the test statistic when performing a test of hypothesis on the variance or standard deviation.

Finally, we discussed why you should always include a **p-value** in the results of any hypothesis test. This value measures the strength of your point estimate (such as $\bar{X}$ or s^2). When using a predetermined significance level, α, you reject H_0 whenever the p-value is less than α and fail to reject H_0 otherwise. Another option you can use is not to select the somewhat arbitrary value of α but simply to reject H_0 whenever the p-value is "small" (say, $<.01$), fail to reject H_0 if it is "large" (say, $>.1$), or decide that the data are inconclusive if the p-value lies between these two values. You can also use the p-value to measure the enthusiasm (p-value very small) with which you reject H_0 or the authority (p-value quite large) with which you fail to reject H_0.

REVIEW
EXERCISES

8.57   Explain how changes in the α level affect the following.

a.   The rejection region.

b.   The Type II error.

8.58   The manager of Jack-Be-Nimble candle company would like to claim that a certain type of their candles burns more than 14 hours. To test this claim, the manager randomly selects 50 candles and finds that the sample mean is equal to 14.75 hours with a standard deviation of 1.8 hours.

a.   What are the null and alternative hypotheses?

b.   Which error would you consider to be more serious, Type I or Type II?

c.   At a significance level of .05, what is your conclusion?

8.59   Given the following statistics from a normally distributed population, is there sufficient evidence to support the claim that the mean of the population differs from 235.6?

$$n = 21$$
$$\bar{x} = 234.1$$
$$\sigma = 2.3$$

Use the p-value to draw your conclusion.

8.60   The manager of the Train Depot Restaurant believes that the average time customers wait before being served is 10 minutes. To test the belief, the manager selects 50 customers at random and records that the average waiting time is 11.9 minutes with a standard deviation of 1.4 minutes.

a.   Find a 95% confidence interval for the mean waiting time of a customer.

b.   Do the data indicate that the mean waiting time differs from 10 minutes, at a 5% significance level?

8.61   Calculate the power of the test for the mean of a normally distributed population with known population variance for the following situations, assuming that the true population mean is 10 and the known population standard deviation is 3.1. Use a significance level of .05.

a. H_0: $\mu = 11$, H_a: $\mu \neq 11$, $n = 14$.

b. H_0: $\mu = 9.5$, H_a: $\mu \neq 9.5$, $n = 25$.

c. H_0: $\mu = 8$, H_a: $\mu \neq 8$, $n = 40$.

8.62   The federal government provides an enormous amount of funding to the states for research and development (R&D), according to a federal official interviewed by a journalist. These funding amounts are believed to follow a normal distribution with a standard deviation of $45 million. The official said that each state, on the average, gets $150 million or more per annum. The journalist thought this figure might be too high and felt it was less than $150 million. Ten states were selected at random. The average amount of federal funding for R&D received per state was $120 million. Does this provide enough evidence to reject the federal official's claim, using a 5% significance level? Determine the p value for your test. (*Hint:* Use $N = 50$ in the finite population correction factor.)

8.63   A manufacturer of drugs and medical products claims that a new anti-inflammatory drug will be effective for 4 hours after the drug is administered in the prescribed dosage. A random sample of 50 volunteers demonstrated that the average effective time is 3.70 hours with a sample standard deviation of .606 hours. Use the p-value criteria to determine if there is sufficient evidence to support that the mean effective time of the drug differs from 4 hours?

8.64   Indicate what the p-values are for the following situations in which the mean of a normally distributed population is being tested.

a. H_0: $\mu = 31.6$, H_a: $\mu \neq 31.6$ (population variance is known), $Z^* = 2.16$.

b. $H_0: \mu = 4.07$, $H_a: \mu \neq 4.07$ (population variance is known), $Z^* = -1.35$.

c. $H_0: \mu = 87.6$, $H_a: \mu \neq 87.6$ (population variance is unknown), $t^* = 2.51$, $n = 15$.

d. $H_0: \mu = 195.3$, $H_a: \mu \neq 195.3$ (population variance is unknown), $t^* = -1.71$, $n = 25$.

8.65 Using the following information, perform the hypothesis test for the mean of a normally distributed population:

$$H_0: \mu \geq 7.19$$
$$H_a: \mu < 7.19$$
$$\bar{x} = 6.21$$
$$s^2 = .26$$
$$n = 23$$
$$\alpha = .10$$

8.66 The vice president of academic affairs at a small private college believes that the average full-time student who lives off campus spends about $300 per month for housing. A random sample of 200 full-time students living off campus spent an average of $305 per month with a standard deviation of $70 a month.

a. Find the p-value to determine whether there is sufficient evidence to indicate that a full-time student spends more than $300 per month on housing.

b. Would you reject the null hypothesis for the test in question a if $\alpha = .01$? if $\alpha = .05$? if $\alpha = .10$?

8.67 A marketing analyst is looking at the feasibility of opening a new movie theater in a small town. The town currently has only two movie theaters. The movie theater would be a practical investment if the average family in the town spends at least 14 hours at the movies each year. A random sample of 80 households yielded a sample mean of 14.5 hours per year with a standard deviation of 1.4.

a. Find the 95% confidence interval for the mean time that a family spends per year at the movies.

b. Is there sufficient evidence to indicate that the mean time that a family spends at the movies is greater than 14 hours per year? Use a .05 significance level.

8.68 From a finite population of size 425, a random sample of size 20 is drawn. The finite population can be approximated by a normal distribution. From the sample, it is found that

$$\bar{x} = 43.7 \qquad s^2 = 6.7$$

Do the data support the statement that the mean of the population is greater than 40? Use a 1% significance level.

8.69 The manager of a real-estate firm is concerned about the yearly vacancy rate of apartments in a large city. There are 325 apartment complexes in the city. A random sample of 25 apartment complexes reveals that the average yearly vacancy rate is 11.5%. The vacancy rate for the apartment complexes is considered to follow a normal distribution. The sample standard deviation for the vacancy rate is 1.9%.

a. Is there sufficient evidence to indicate that the vacancy rate is greater than 10% for apartment complexes in the city? Use a 1% significance level.

b. Find the p-value for the test in part (a).

8.70 The owners of a shopping center are contemplating increasing the parking space in front of the shopping center. The owners would like to demonstrate that the average driver parks for more than .75 hours. The length of time parked is considered to be normally distributed. A random sample of 45 parked cars is observed; the average time parked was .80 hours with a standard deviation of .12. Do the data support the idea that the average driver parks for more than .75 hours? Use a 10% significance level.

8.71 There are 420 persons attending a conference on the Strategic Defense Initiative (SDI). A reporter randomly selected 50 persons from those attending, to determine the average income of the participants. From this sample, a mean of 38.6 (thousand dollars) and a standard deviation of 6.7 (thousand dollars) were obtained.

a. Construct the 95% confidence interval for the mean income of the conference participants.

b. Do the data indicate that the average income is less than 40 (thousand dollars)? Use a 5% significance level.

c. Determine the p-value.

8.72 From a normally distributed population, a random sample of size 22 yields the following statistic:

$$\Sigma(x - \bar{x})^2 = 1.67$$

a. Find a 95% confidence interval for the population variance.

b. Find a 95% confidence interval for the population standard deviation.

c. Test the null hypothesis that the population variance is equal to .07. Use a two-tailed test and a .05 significance level.

8.73 Using a significance level of .05, perform the hypothesis test for the standard deviation of a normally distributed population given the following information:

$$H_0: \sigma \geq 20.6$$
$$H_a: \sigma < 20.6$$
$$\Sigma(x - \bar{x})^2 = 6100$$
$$n = 18$$

8.74 Eastern State Bank currently operates five drive-in teller windows. Management is concerned about the variability of the time spent by a customer using the windows. A sample of 24 customers was taken and the sample standard deviation was found to be 4.7 minutes. Management would like to keep the standard deviation below 4 minutes and may consider adding another drive-in teller window.

a. Test the null hypothesis that the standard deviation of the waiting time by a customer is less than or equal to 4 minutes. Use a 10% significance level.

b. What assumption should be made about the distribution of the waiting time of a customer who uses the drive-in teller windows?

8.75 An investment counselor would like to know how much variability there is in the yield of money-market funds. The yields of the funds can be considered to be approximately normally distributed for the time frame of interest. A sample of 21 money-market funds yields a sample standard deviation of .7%. At the .05 significance level, is there sufficient evidence to indicate that the standard deviation of the yields of money-market funds is greater than .6%?

8.76 A large university had recently converted to a computerized registration system for enrollment. After the first semester, administrators found that the average time spent registering per student was quite satisfactory, yet there still continued to be substantial complaints and dissatisfaction among students. Further study indicated that although the *average* time might seem satisfactory, there might be too much *variation* in the registration times. It was decided to study the situation during the next semester's registration period. If the standard deviation was greater than 20 minutes, six additional computer terminals would be installed; otherwise, two new computer terminals would be installed. From a random sample, the following data were obtained:

$$\Sigma(x - \bar{x})^2 = 6900, \qquad n = 18$$

Assume the population is normally distributed.

a. Is there sufficient evidence to indicate that the population standard deviation exceeds 20 minutes? Use a 5% significance level.

b. What is the decision indicated by the test: install six new terminals or two new terminals?

c. State the p-value for the test.

8.77 Microcomputers have proliferated in business schools throughout the United States in the past 8 years. A study in 1984 of 210 business schools and a study in 1986 of 160 business schools, with each set of schools being randomly selected, revealed that the average number of laboratory-based micros on campus increased from 30.7 to 53.8. This study projected an average of 85.5 micros in laboratories at universities which have a school of business in 1988. Assume that the dean at a growing college of business believed that this projection was overly optimistic. Suppose that in 1988 the dean was able to get a random sample of 20 colleges of business and found the results to be a sample mean of 78. Assume that the standard deviation of the number of micros in business schools in 1988 is 15.

(*Source:* B. Render and R. Stair, "Trends In The Use of Personal Computers In Business Schools", *The Proceedings of the Decision Sciences Institute* (1986; 899–901.)

a. Do the data provide sufficient evidence to conclude that the projection of the mean number of micros in laboratories at schools of business is too optimistic? Use a significance level of .05.

b. What assumptions are necessary for the statistical test in part (a) to be valid?

8.78 Economists at the Federal Reserve Bank of Chicago have compiled an index to show that the Midwest did benefit from the surge in exports in 1987. Using factory hours worked and electricity usage as a way to gauge output, the economists found that factories in Illinois, Indiana, Iowa, Michigan, and Wisconsin boosted production by 4.2% in 1987. Assume that an economist was interested in whether the manufacturer in the Midsouth shared in the same production increase. A sample of 50 manufacturers in the Midsouth were selected and change in production was recorded. The results can be summarized as follows:

$$\bar{x} = 3.8\%$$
$$s^2 = 0.81$$

(*Source:* "Why Mid-Western Manufacturers Outpace The Nation," *Business Week*, April, 11, 1988, p. 27.)

a. Is there sufficient evidence to claim that the mean production increase in the Midsouth is less than 4.2% at the .01 significance level?

b. Suppose it is known that the standard deviation of the production increase for manufacturers in the Midwest is believed to be equal to 1. Does the sample evidence indicate that the standard deviation for the production increase in the Midsouth differs significantly from 1? Use a .05 significance level.

8.79 Expert systems are special-purpose computer programs, which help to solve problems in the same manner as human experts. For 50 different situations involving various production planning problems, the amount of time spent for a manager to interact with a particular expert system on these production planning problems are given below in units of hours. The following MINITAB computer printout displays a 95% confidence interval.

1.2	1.4	0.9	1.8	0.7	1.1	1.5	0.6	0.8	1.2	1.3
1.7	0.7	0.9	1.2	1.5	0.7	1.0	1.3	1.1	1.4	1.8
1.2	1.1	0.4	.3	1.8	0.7	1.9	0.9	1.0	0.9	0.7
1.5	0.8	1.1	.7	1.9	1.2	0.7	0.6	1.0	1.3	1.4
0.7	1.9	1.1	1.2	0.7	1.1	1.2

```
MTB > describe c1

        N      MEAN    MEDIAN    TRMEAN    STDEV    SEMEAN
C1     51    1.1333    1.1000    1.1289    0.4068    0.0570

       MIN       MAX        Q1        Q3
C1    0.3000    1.9000    0.8000    1.4000

MTB > zint with 95 percent, sample stdev = .4068, data in c1

THE ASSUMED SIGMA =0.407

        N      MEAN     STDEV    SE MEAN    95.0 PERCENT C.I.
C1     51    1.1333    0.4068    0.0570   ( 1.0215,  1.2451)
```

a. At the .05 significance level, can one conclude from the computer printout that the data support the belief that that the mean time spent by the manager with the expert system is not equal to one hour?

b. If a significance level of .10 was used, can the question in part (a) be answered using the same computer printout? Why?

c. What justifies the use of the confidence interval with the normal distribution instead of the t distribution in part (a)?

8.80   The manager at A & A Plumbing and Air-conditioning Company is concerned about the time that it takes a serviceperson to repair a defective switch in a high-efficiency heat pump at local residences. The manager believes that the average repair time is greater than 30 minutes. Past data shown in the histogram below give the following times in minutes for a serviceperson to service a defective switch.

```
MTB > describe c1

        N     MEAN   MEDIAN  TRMEAN   STDEV  SEMEAN
C1      25   31.348  31.200  31.352   4.278  0.856

        MIN    MAX     Q1      Q3
C1     19.800 42.800 29.150  33.350

MTB > histogram of c1

Histogram of C1   N = 25

Midpoint   Count
20       1   *
22       0
24       0
26       1   *
28       4   ****
30       6   ******
32       6   ******
34       4   ****
36       1   *
38       1   *
40       0
42       1   *

MTB > ttest mu = 30, data in c1

TEST OF MU = 30.000 VS MU N.E. 30.000

        N     MEAN   STDEV  SE MEAN    T    P VALUE
C1      25   31.348  4.278   0.856   1.58    0.13

MTB > ttest mu = 30, data in c1;
SUBC> alternate = 1.

TEST OF MU = 30.000 VS MU G.T. 30.000

        N     MEAN   STDEV  SE MEAN    T    P VALUE
C1      25   31.348  4.278   0.856   1.58    0.064

MTB > tinterval with 90 percent confidence, data in c1

        N     MEAN   STDEV  SE MEAN    90.0 PERCENT C.I.
C1      25   31.348  4.278   0.856   ( 29.884, 32.812)
```

a. Do the assumptions needed to perform a t test appear to hold?

b. Do the data support the manager's claim? At what significance level? (Note that the MINITAB printout with "Alternate = 1" indicates a right-tailed test. "Alternate = 0" is the default value if no value is assigned to the alternate.)

c. Should the same conclusion hold in testing the manager's belief for a one-tailed test as for a two-tailed test? What is the conclusion from the two-tailed test? Does the confidence interval support this conclusion at a significance level of 10%.

8.81   To understand better the difference between the t distribution and the normal distribution, a MINITAB computer printout is given to show the output of 1000 observations randomly generated from a population with a normal distribution with mean 0

```
MTB > random 1000 observations, put into c1;
SUBC> normal mean = 0, sigma = 1.
MTB > histogram c1

Histogram of C1    N = 1000
Each * represents 5 obs.

Midpoint   Count
  -3.5        1    *
  -3.0        3    *
  -2.5        5    *
  -2.0       28    ******
  -1.5       65    *************
  -1.0      112    ***********************
  -0.5      191    **************************************
   0.0      191    **************************************
   0.5      177    ************************************
   1.0      111    ***********************
   1.5       72    ***************
   2.0       31    *******
   2.5        8    **
   3.0        2    *
   3.5        2    *
   4.0        1    *
```

```
MTB > random 1000 observations, put into c1;
SUBC> t degrees of freedom = 7.
MTB > histogram of c1

Histogram of C1    N = 1000
Each * represents 10 obs.

Midpoint   Count
  -4         7    *
  -3        18    **
  -2        69    *******
  -1       212    **********************
   0       379    **************************************
   1       225    ***********************
   2        67    *******
   3        18    **
   4         1    *
   5         3    *
   6         0
   7         1    *
```

and variance 1 and also from a population with a t distribution with degrees of freedom equal to 7. Comment on the differences between the two histograms.

8.82 Use a convenient statistical computer package to generate 24 observations from a normal population with mean 0 and standard deviation 4 and then find a 90% confidence interval of the mean using the t distribution. Execute this program 200 times. Find the number of times that the confidence intervals contained the true mean. Should the confidence interval always contain the true mean? What does the 90% confidence level indicate? The needed program using MINITAB commands is given. Simply change the command execute 1 to execute 200.

```
MTB > store
MTB > random 24 observations, put into c1;
MTB > normal with mean = 0, sigma = 4.
MTB > tinterval with confidence level = 90,data in c1
MTB > end
MTB > execute 1
```

COMPUTER EXERCISES USING THE DATABASE

Exercise 1—Appendix H

Randomly select 100 observations from the database. Use a convenient statistical computer package to determine if the sample evidence indicates that the mean of the variable TOTLDEBT (total indebtedness) exceeds $10,000 at the .05 significance level. Also determine if the sample evidence indicates that the standard deviation of the variable TOTLDEBT exceeds $3000 at the .05 significance level.

Exercise 2—Appendix H

Randomly select 30 observations from the database in which the location of residence is in the NE sector and also randomly select another 30 observations in which the location of residence is in the NW sector. Find separate 90% confidence intervals on the mean of the variable INCOME1 (income of principal wage earner) from each of

these two sets of data. Comment on the difference in the confidence intervals.

Exercise 3—Appendix I

Randomly select 30 observations from the database from companies with a bond rating of A and 30 observations from companies with a bond rating of C. Find separate 95% confidence intervals on the mean of sales minus cost of sales for each of these two random samples. Compare and comment on the differences.

CASE STUDY

THE OC CURVE AND CONTROL CHART SENSITIVITY

The control chart is a powerful statistical tool that monitors a process and detects shifts in μ, the process mean. Actually, the $\bar{x}$ control chart is just a graphical way of conducting a hypothesis test on μ. Thus, it is subject to the Type I and Type II errors. This raises the question of sensitivity, i.e., how well a control chart detects shifts in the mean. For statisticians the answer is expressed as a probability. The instrument that provides this information is known as the **operating characteristic curve,** or OC curve. To see how the OC curve evaluates the sensitivity of a control chart, we must first understand how a control chart really represents a hypothesis test.

Assume a process defined to be in control when the process mean is μ, and the process fluctuates within $\pm 3\sigma$ units from μ. You do not really *know* that the process mean is actually μ, you merely *hypothesize* that this is so. Suppose a magnetic oxide coating process is stable when the mean thickness is 5 microns with $\sigma = .2$ microns. This could be stated formally as a hypothesis test:

$H_0: \mu = 5$ microns (the process is in control)

$H_a: \mu \neq 5$ microns (the process is not in control)

For samples of size 4, the standard error, $\sigma_{\bar{x}}$, is $.2/\sqrt{4} = .1$, so the control limits for $\bar{x}$ are $5 \pm 3(.1) = 4.7$ and 5.3. The areas above and below these values are the rejection region in a two-tailed hypothesis test. Note that the limits are in terms of $\sigma_{\bar{x}}$ rather than σ because although *individual* observations of $\pm 3\sigma$ are permissible, *sample means* must have a narrower distribution. Now, a sample with $\bar{x} = 5.25$ microns would lead to a decision to not reject H_0 (the process is assumed to be in control), whereas $\bar{x} = 5.60$ microns leads to a decision to reject H_0 (the process could be out of control). These are precisely the same decisions that are made with the control chart.

Now consider the two types of errors in hypothesis testing. Type I error, α, is the probability of rejecting H_0 even when it is true. In a control chart, this is the situation where the process mean has not actually shifted, but some extreme observations lead us to believe (erroneously) it has. Things have not changed, but we wrongly conclude they have. So even though the goods being produced are within specifications, they are treated as if they violated specifications. In acceptance sampling, this amounts to rejecting a good lot;

so the Type I error is called the *producer's risk.*

The Type II error, β, is the probability of "accepting" (not rejecting) H_0 even when it is false. In a control chart, this translates to the situation where the process mean has actually shifted, but we have not yet obtained extreme observations that would reveal this. Instead, the observations appear to be within control limits. Figure 8.24 shows graphically, for an upward shift of the mean, that this can happen because there is a certain amount of overlap between the distribution describing the "old" control mean and the "new"

out-of-control mean; β is the area on the new curve that corresponds to the acceptance region of the old curve. Within the overlapped region, observations from the new distribution may be wrongly interpreted as coming from the old distribution. Things have changed, but we complacently believe they have not. Even though the goods being produced may be violating specifications, they are treated as if they did not. In acceptance sampling, this is the situation where a bad lot gets accepted, hence the Type II error is called the *consumer's risk.*

Just as β = the probability of not rejecting a false H_0, $1 - \beta$ =

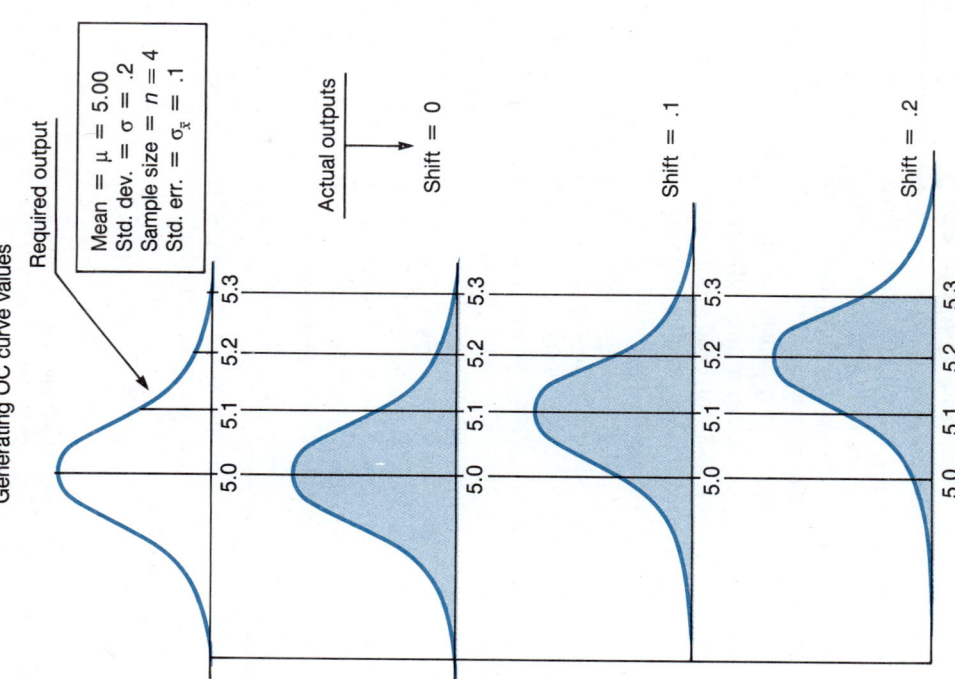

Required output

| Mean = μ = 5.00 |
| Std. dev. = σ = .2 |
| Sample size = n = 4 |
| Std. err. = $\sigma_{\bar{x}}$ = .1 |

Actual outputs $\longrightarrow$ Shift = 0

Shift = .1

Shift = .2

Generating OC curve values

FIGURE 8.24
Determining β for various upward shifts of the mean.

(*Source:* Peter Rob and Elias R. Callahan, Jr., "Detecting Shifts in Mean Value for Quality Control," *Industrial Engineering* (December 1982): 18–21, Figure 2. Copyright December 1982 Institute of Industrial Engineers, 25 Technology Park/Atlanta, Norcross, GA 30092.)

FIGURE 8.25

OC curves for various sample sizes.

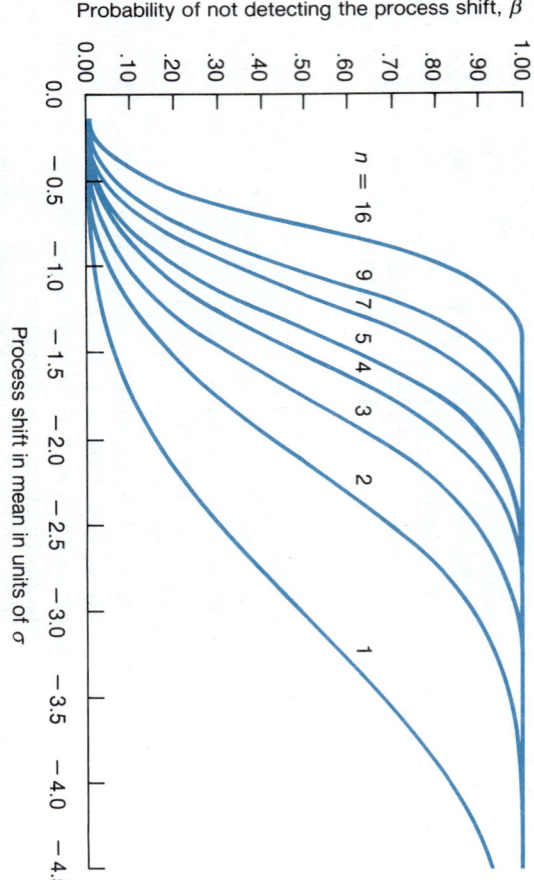

OC curves for x̄ control charts

(*Source:* H. M. Wadsworth, Jr., K. S. Stephens, and A. B. Godfrey, *Modern Methods for Quality Control and Improvement*, New York: John Wiley, 1986. Chart adapted from Figure 7-7, p. 219)

the probability of rejecting a false H_0. The better the test is at doing the latter, the more powerful it is. Thus $1 - \beta$ is the power of a test. A plot of $1 - \beta$ against different possible shifts of the mean is known as a **power curve**, as explained in the text. The OC curve is simply the complement of the power curve. Instead of plotting $1 - \beta$, the OC curve plots β against various shifts of the mean. OC curves may be obtained for different sample sizes. Figure 8.25 shows the OC curves generated for various sample sizes, with $n = 1$ shown as a basis for comparison. Showing shifts of the mean in units of standard deviations is the convention. The vertical axis shows β, the probability of not detecting a shift in the mean.

The elongated S shapes in Figure 8.25 are typical of OC curves generated for left-sided hypothesis tests. Those for right-sided tests would be a mirror image

of Figure 8.25, with elongated backward S shapes, whereas for two-tailed tests, the OC curves would be a composite of Figure 8.25 and its mirror image, i.e., hump-shaped curves. The steeper the curve, the more sensitive it is at detecting shifts in the mean. Note that sample sizes larger than 1 are more discriminating than individual observations ($n = 1$), and sensitivity increases with sample size. Remember from the Central Limit Theorem that as n increases, $\sigma_{\bar{x}}$ gets smaller, and the sampling distribution gets tighter.

Of course, larger samples provide more information but cost more to obtain, while smaller samples are less sensitive but cost less. It is the classic tradeoff: cost versus benefit. Having a set of OC curves as in Figure 8.25 enables us to fine-tune our selection of the sample size that gives us the level of sensitivity we require.

Case Study Questions

1. The control chart was described as a hypothesis-testing device, with critical values (control

limits) placed at $\pm 3\sigma_{\bar{x}}$. What is the level of significance at which the control chart performs this two-tailed hypothesis test?

2. For the magnetic oxide coating process discussed earlier, a sample average of 5.25 microns led to a decision not to reject H_0. Suppose the sample size had been 16 instead of 4.

  a. Conduct a hypothesis test to decide if the process may be assumed to be in control. Use a .01 significance level.

  b. What is the value of the test statistic and the *p*-value associated with it?

  c. If H_0 is rejected at the .01 significance level, does that automatically mean that H_0 is also rejected in terms of the control limits of $\pm 3\sigma_{\bar{x}}$?

  d. If H_0 is not rejected at the .01 significance level, does that automatically mean that H_0 is also not rejected in terms of the control limits of $\pm 3\sigma_{\bar{x}}$?

  e. Each time a sample mean signals a process out of control, the process has to be stopped for readjustment. Leaving aside the cost of

larger samples, do you think it is advisable to take very large samples to obtain a super-sensitive control chart that can detect even very small shifts in the process mean?

3. [Optional] Obtain the power curve and the OC curve for the oxide coating process to detect *upward* shifts of .05 microns, using a sample size of 4. You should first construct a table with four columns: amount of shift in microns (5.00, 5.05, 5.10, . . . , up to 5.50); amount of shift in units of standard deviations; the OC curve probabilities (β); and the power curve probabilities ($1 - \beta$). Then plot the standard error units on the horizontal axis, and the β values on the vertical axis to get the OC curve. Do the same thing with the $1 - \beta$ values to get the power curve. The power curve should look like an elongated *S* shape. What is the shape of the OC curve?

MAINFRAME AND MICRO

SOLUTION

EXAMPLE 8.10

Example 8.10 was concerned with the computation of means and a *t* statistic to examine average differences between the recorded and audited values of inventory items. You can use SPSS to solve this problem by borrowing a technique that actually belongs in Chapter 9 (and is used again in that chapter). The program listing in Figure 8.26 was used to request the calculation of the *t* statistic and the corresponding *p*-value. Note that SPSS automatically assumes a two-tailed test; this was a one-tailed test, so the calculated *p*-value needs to be divided by two. In this problem the SPSS commands are the same for both the mainframe and PC versions. **(Remember to end each command line with a period when using the PC version.)**

The TITLE command names the SPSS run.

The DATA LIST command gives each variable a name and describes the data as being in free form. The variable AMT (abbreviated form of "amount") is defined as the value of the mean contained in the hypotheses. Actually, AMT is a constant but is treated as a variable in the analysis.

The BEGIN DATA command indicates to SPSS that the input data immediately follow.

FIGURE 8.26 Input for SPSSX or SPSS/PC+. All command lines should end with a period when the PC version is used.

```
TITLE     DEFENSE
DATA LIST FREE   / DIFFER, AMT
BEGIN DATA
17 25
35 25
31 25
22 25
50 25
42 25
56 25
23 25
27 25
38 25
20 25
25 25
43 25
45 25
21 25
END DATA
T-TEST PAIRS = DIFFER AMT
```

The next 15 lines contain the data values. Each line contains the difference (DIFFER) between the stated value and the audited value of one inventory item, and the constant 25 (AMT).

The END DATA statement indicates the end of the data.

The T-TEST PAIRS = DIFFER AMT command computes the t statistic using the variable DIFFER and the corresponding p-value.

Figure 8.27 shows the SPSSX output obtained by executing the listing in Figure 8.26, whereas Figure 8.28 shows the SPSS/PC+ output.

FIGURE 8.27 SPSSX output.

```
- - - - - - - - - - - - - - - - - - - - - - - - - - - - T - T E S T - - - - - - - - - - - - - - - - - - - - - - - - - - - - -

VARIABLE   NUMBER              STANDARD   STANDARD  *(DIFFERENCE) STANDARD   STANDARD  *        2-TAIL *   T    DEGREES OF 2-TAIL
           OF CASES    MEAN   DEVIATION     ERROR   *     MEAN    DEVIATION    ERROR   * CORR.  PROB.  * VALUE    FREEDOM   PROB.
-------------------------------------------------------------------------------------------------------------------------------
DIFFER                                              *                                  *               *
                     33.0000    12.148      3.137   *                                  *               *
              15                                    *   8.0000     12.148      3.137   *99.00099.000   *  2.55      14      0.023
                     25.0000     0.000      0.000   *                                  *               *
AMT                                                 *                                  *               *
-------------------------------------------------------------------------------------------------------------------------------
```

value of t statistic

p-value

for a one-tailed test, the actual p-value = .023/2 = .012

FIGURE 8.28 SPSS/PC+ output.

```
Paired samples t-test:  DIFFER
                        AMT

Variable     Number                Standard    Standard
           of Cases       Mean    Deviation       Error

DIFFER         15       33.0000     12.148       3.137
AMT            15       25.0000       .000        .000

(Difference) Standard    Standard |       2-Tail  |     t     Degrees of   2-Tail
    Mean     Deviation     Error  |  Corr. Prob.  |   Value    Freedom      Prob.

   8.0000     12.148       3.137  | 99.00099.000  |   2.55       14         .023
------------------------------------------------------------------------------------
```

for a one-tailed test the actual p-value = .023/2 = .012

value of t-statistic

p-value

MAINFRAME AND MICRO

SOLUTION

EXAMPLE 8.10

Example 8.10 was concerned with the computation of means and a *t*-statistic to examine average differences between the recorded and audited values of inventory items. You can use SAS to solve this problem. The program listing in Figure 8.29 was used to request the calculation of the *t*-statistic and the resulting *p*-value. Note that SAS automatically assumes a two-tailed test; this was a one-tailed test, so the calculated *p*-value needs to be divided by two. In this problem the SAS commands are the same for both the mainframe and PC versions.

FIGURE 8.29
Input for SAS (mainframe or micro version).

```
TITLE    'DEFENSE';
DATA DIFFDATA;
  INPUT DIFFER;
  H01=DIFFER-25.0;
  CARDS;
17.0
35.0
31.0
22.0
50.0
42.0
56.0
23.0
27.0
38.0
20.0
25.0
43.0
45.0
21.0
PROC MEANS N MEAN T PRT;
  VAR H01;
  TITLE 'ONE-TAILED TEST';
```

The TITLE command names the SAS run (enclose in single quotes).

The DATA command gives the data a name.

The INPUT command gives the variable a name.

The H01 = DIFFER − 25.0. statement is used to compute a new variable, H01, which is the difference between the variable DIFFER and the constant 25.0.

The CARDS command indicates to SAS that the input data immediately follow.

The next 15 lines are the data values, representing the differences between the stated value and the audited value of the inventory items along with the constant 25.

The PROC MEANS command requests a SAS procedure to print the number of observations, the mean, the *t*-statistic, and the *p*-value.

The VAR statement specifies that the variable H01 is the variable to be used in computing the statistics.

The TITLE statement specifies the heading for the printout.

Figure 8.30 shows the SAS output obtained by executing the listing in Figure 8.29, and Figure 8.31 shows the SAS/PC output.

FIGURE 8.30
SAS output.

```
                          ONE-TAILED TEST

VARIABLE        N      MEAN          T        PR>!T!
HO1            15   8.00000000     2.55       0.0231
```

Value of *t*-statistic → $\quad$ *p*-value →

for a one-tailed test, the actual
p-value = .023/2 = .012

FIGURE 8.31
SAS/PC output.

```
                     ONE-TAILED TEST

Analysis Variable : HO1

N Obs   N       Mean          T     Prob>|T|
----------------------------------------------
 15    15    8.0000000     2.5505537    0.0231
----------------------------------------------
```

for a one-tailed test, the actual
p-value = .023/2 = .012

Inference Procedures for Two Populations

A Look Back/Introduction

We have learned to describe and summarize data from a single population using a statistic (such as the sample mean, $\bar{X}$), or a graph (such as a histogram). Chapters 7 and 8 introduced you to statistical inference, where we (1) attempted to estimate a parameter (such as the mean, μ) from this population by using the corresponding sample statistic and (2) arrived at a conclusion about this parameter (such as $\mu > 5.9$ ft) by performing a test of hypothesis. The concept behind hypothesis testing was described, and we paid special attention to the errors (Type I and Type II) that can occur when we use a sample to infer something about a population.

Next we learn how to compare two populations. Questions of interest here include:

1. Are the values in population 1 larger, on the average, than those in population 2? (For example, are men taller, on the average, than women?)
2. Do the values in population 1 exhibit more variation than those in population 2? (For example, do male heights vary more than female heights?)

The two populations under observation may or may not be normally distributed. When comparing two population means using large samples, once again using the Central Limit Theorem, it simply does not matter. For small samples, we need to examine the distribution of the populations so we can use the proper procedure to construct confidence intervals and perform tests of hypothesis.

This chapter discusses two different sampling situations. In the first, random samples from two populations are obtained *independently* of each other; in the second, corresponding data values from the two samples are matched up, or paired. Paired samples are *dependent*.

9.1

INDEPENDENT VERSUS DEPENDENT SAMPLES

When making comparisons between the means of two populations, we need to pay particular attention to how we intend to collect sample data. For example, how would you determine if tire brand A lasts longer than brand B? You might decide to put one of each brand on the rear wheels of ten cars and measure the tires' wear. Or you might randomly select ten brand A and ten brand B tires, attach them to a machine that wears them down for a certain time, and then measure the resulting tire wear. If you use the first procedure (putting both brands of tire on the same car), you obtain *dependent* samples; in the latter situation, you obtain *independent* samples.

Consider another situation. Suppose you are interested in male heights as compared with female heights. You obtain a sample of $n_1 = 50$ male heights and $n_2 = 50$ female heights. You obtain these data:

OBSERVATION	MALE HEIGHTS	OBSERVATION	FEMALE HEIGHTS
1	5.92 ft	1	5.36 ft
2	6.13 ft	2	5.64 ft
3	5.78 ft	3	5.44 ft
⋮	⋮	⋮	⋮
50	5.81 ft	50	5.52 ft

Is there any need to match up 5.92 with 5.36, 6.13 with 5.64, 5.78 with 5.44, and so on? The male heights were randomly selected and the female heights were obtained independently, so there is no reason to match up the first male height with the first female height, the second male height with the second female height, and so on. Nothing relates male 1 with female 1 other than the accident of their being selected first—these are **independent samples**.

What if you wish to know whether husbands are taller than their wives? To collect data, you select 50 married couples. Suppose you obtain the 100 observations from the previous male and female height example. Now, is there a reason to compare the first male height with the first female height, the second with the second, and so on? The answer is a definite yes since each pair of heights belongs to a married couple. The resulting two samples are **dependent, or paired, samples**.

In summary,

1. If there is a definite reason for pairing (matching) corresponding data values, the two samples are **dependent** samples.

2. If the two samples were obtained independently and there is no reason for pairing the data values, the resulting samples are **independent** samples.

Why does this distinction matter? If you are trying to decide whether male heights are, on the average, greater than female heights, the procedure that you use for testing this depends on whether the samples are obtained independently.

Applications of dependent samples in a business setting include data from the following situations.

1. Comparisons of *before versus after*. Sample 1: person's weight before a diet plan is begun. Sample 2: person's weight 6 months after starting the diet. Why pair the data? Each pair of observations belongs to the same person.

2. Comparisons of people with *matching characteristics*. Sample 1: salary for a male employee at Company ABC. Sample 2: salary for a female employee at Company ABC, where the woman has education and job experience equal to the man's. Why pair the data? The two paired employees are identical in their job qualifications.

3. Comparisons of observations *matched by location.* Sample 1: sales of brand A tires for a group of *n* stores. Sample 2: sales of brand B tires for the same group of stores. Why pair the data? Both observations were obtained from the same store. Your data consist of sales (weekly, monthly, and so on) from a sample of stores selling these two brands.

4. Comparisons of observations *matched by time.* Sample 1: sales of restaurant A during a particular week. Sample 2: sales of restaurant B during this week. Why pair the data? Each pair of observations corresponds to the same week of the year.

EXERCISES

9.1 For each of the following claims determine whether the paired samples or the independent samples procedure would be appropriate.

a. The mean of the scores attained by students before a tutorial session is less than the mean of the scores attained by the same students after the tutorial session.

b. There is no difference between the mean grade point averages of females and males in the MBA program.

c. The average wage for Japanese auto workers is less than that for European auto workers.

9.2 Two private colleges decided to compare the mean SAT scores of their incoming freshmen. One college gathered 98 scores and the other took a sample of 52 scores. Do the two sets of data represent dependent or independent samples?

9.3 A medical institution is examining the effectiveness of a newly developed drug. The drug was administered to 18 patients whose health condition before and after taking the drug was recorded. Is this a case of dependent or independent samples?

9.4 The advertising division of a chemical company would like to see how two different dishwashing detergents are rated by homemakers. Homemakers are chosen at random. They assign a value from zero to ten to each product. They assign a value of zero to the product if they think the detergent is worthless and ten to the detergent if they believe it is the best on market. How can dependent samples be chosen? How can independent samples be chosen?

9.5 The career placement center at Safire University conducts a survey of beginning salaries for MBAs with no on-the-job experience. Ten pairs of men and women are chosen randomly such that each pair of one man and one woman have nearly identical qualifications. Can the sample of observations from men be independent of the sample of observations from women?

9.6 A retail store would like to compare sales from two different arrangements of displaying its merchandise. Sales are recorded for a 30-day period with one arrangement and then sales are recorded for another 30-day period for the alternative arrangement. Can the data for each of the two 30-day periods be independent or dependent?

9.7 Fifty people were randomly selected to rate particular brands of soft drink on a scale from one to ten, with ten being the highest rating. If 25 people rated brand A and the other 25 people rated brand B, would the samples from these two groups be independent or dependent?

9.8 In Exercise 9.7 suppose the 50 people were each asked to rate both brand A and brand B. Would the sample of 50 observations of brand A be independent of the sample of 50 observations of brand B?

9.2

COMPARING TWO MEANS USING TWO LARGE INDEPENDENT SAMPLES

When comparing the means of two independent samples from different populations, we can use Figure 9.1 to help visualize the situation. The two populations are shown to be normally distributed, but, because we will be using large samples from these populations, this is *not* a necessary assumption. For these populations,

μ_1 = mean of population 1

μ_2 = mean of population 2

σ_1 = standard deviation of population 1

σ_2 = standard deviation of population 2

For example, if we wished to compare U.S. adult male and female heights:

μ_1 = average of all female heights

μ_2 = average of all male heights

σ_1 = standard deviation of all female heights

σ_2 = standard deviation of all male heights

The point estimates discussed in earlier chapters apply here as well—we simply have two of everything because we are dealing with two populations.

The procedure we follow is to obtain a random sample of size n_1 from population 1 and then obtain another sample of size n_2, completely independent of the first sample, from population 2. So, $\bar{X}_1$ is our best (point) estimate of μ_1. Likewise, $\bar{X}_2$ estimates μ_2. The sample standard deviations (s_1 and s_2) provide the best estimates of the population standard deviations (σ_1 and σ_2).

Constructing a Confidence Interval for $\mu_1 - \mu_2$

Ace Delivery Service operates a fleet of delivery vans in the Houston area. They prefer to have all their drivers charge their gasoline using the same brand of credit card. Presently, they all use a Texgas credit card. Ace management has decided that perhaps Quik-Chek, a chain of convenience stores also selling gasoline but not accepting credit cards, is worth investigating. A random sample of gas prices at 35 Texgas stations and 40 Quik-Chek stores in the Houston area is obtained. The cost of 1 gallon of regular gasoline is recorded; the data are summarized:

Sample 1 (Texgas)	Sample 2 (Quik-Chek)
$n_1 = 35$	$n_2 = 40$
$\bar{x}_1 = \$1.48$	$\bar{x}_2 = \$1.39$
$s_1 = \$.12$	$s_2 = \$.10$

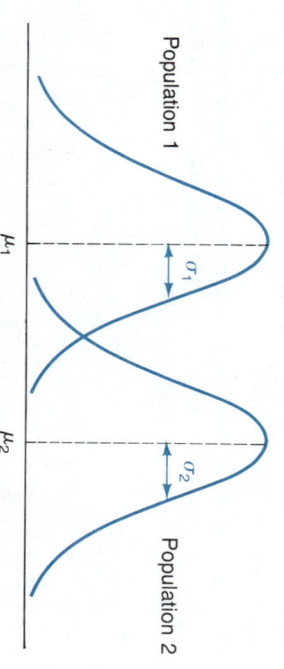

FIGURE 9.1

Example of two populations. Is $\mu_1 = \mu_2$?

Let μ_1 be the average price of regular gasoline at *all* Texgas stations in the Houston area, and let μ_2 be the average price of regular gasoline at all Quik-Chek stores in the Houston area.

When dealing with these two populations, the parameter of interest is $\mu_1 - \mu_2$ rather than the individual values of μ_1 and μ_2. Here, $\mu_1 - \mu_2$ represents the difference between the average gasoline prices at the Texgas stations and Quik-Chek stores. If we conclude that $\mu_1 - \mu_2 > 0$, then this means that $\mu_1 > \mu_2$. In this case, the gasoline *is* more expensive at the Texgas stations.

The point estimator of $\mu_1 - \mu_2$ is the obvious one: $\bar{X}_1 - \bar{X}_2$. For our data, the (point) estimate of $\mu_1 - \mu_2$ is $\bar{x}_1 - \bar{x}_2 = 1.48 - 1.39 = .09$. How much more expensive is the gasoline from all of the Texgas stations, on the average? We do not know because this is $\mu_1 - \mu_2$, but we *do* have an estimate of this value—namely, 9¢.

What kind of random variable is $\bar{X}_1 - \bar{X}_2$? First, because the samples are moderately large, we know by using the Central Limit Theorem that $\bar{X}_1$ is approximately a normal random variable with mean μ_1 and variance σ_1^2/n_1 and that $\bar{X}_2$ is approximately a normal random variable with mean μ_2 and variance σ_2^2/n_2. Because these are two independent samples, it follows that $\bar{X}_1 - \bar{X}_2$ is also approximately a normal random variable with mean $\mu_1 - \mu_2$ and variance $(\sigma_1^2/n_1) + (\sigma_2^2/n_2)$. Note that the variance of $\bar{X}_1 - \bar{X}_2$ is obtained by *adding* the variances for $\bar{X}_1$ and $\bar{X}_2$.

By standardizing this normal distribution, we obtain an approximate standard normal random variable defined by

$$Z = \frac{(\bar{X}_1 - \bar{X}_2) - (\mu_1 - \mu_2)}{\sqrt{\dfrac{\sigma_1^2}{n_1} + \dfrac{\sigma_2^2}{n_2}}} \qquad (9.1)$$

We do not need normal populations. The results of equation 9.1 are approximately valid *regardless* of the shape of the two populations, provided both samples are large (from the Central Limit Theorem). We pointed out that the two populations illustrated in Figure 9.1 need not follow a normal distribution. In fact, they can have any shape, such as exponential, uniform, or possibly a discrete distribution of some sort.

This enables us to derive a confidence interval for $\mu_1 - \mu_2$. By using Table A.4, we know that for the standard normal random variable Z,

$$P(-1.96 < Z < 1.96) = .95$$

Using equation 9.1 (after rearranging the inequalities), we can make the following statement about a random interval prior to obtaining the sample data:

$$P\left[(\bar{X}_1 - \bar{X}_2) - 1.96\sqrt{\frac{\sigma_1^2}{n_1} + \frac{\sigma_2^2}{n_2}} < \mu_1 - \mu_2 < (\bar{X}_1 - \bar{X}_2) + 1.96\sqrt{\frac{\sigma_1^2}{n_1} + \frac{\sigma_2^2}{n_2}}\right] = .95$$

This produces a $(1 - \alpha) \cdot 100\%$ confidence interval for $\mu_1 - \mu_2$ (large samples), where σ_1 and σ_2 are *known*, of:

$$(\bar{X}_1 - \bar{X}_2) - Z_{\alpha/2}\sqrt{\frac{\sigma_1^2}{n_1} + \frac{\sigma_2^2}{n_2}} \quad \text{to} \quad (\bar{X}_1 - \bar{X}_2) + Z_{\alpha/2}\sqrt{\frac{\sigma_1^2}{n_1} + \frac{\sigma_2^2}{n_2}} \qquad (9.2)$$

If σ_1 and σ_2 are *unknown*, we have:

$$(\bar{X}_1 - \bar{X}_2) - Z_{\alpha/2}\sqrt{\frac{s_1^2}{n_1} + \frac{s_2^2}{n_2}} \quad \text{to} \quad (\bar{X}_1 - \bar{X}_2) + Z_{\alpha/2}\sqrt{\frac{s_1^2}{n_1} + \frac{s_2^2}{n_2}} \quad \textbf{(9.3)}$$

Notice that this interval is very similar to the confidence interval for a single population mean using a large sample, namely,

$$(\text{point estimate}) \pm Z_{\alpha/2} \cdot (\text{standard deviation of the point estimator})$$

To construct the confidence interval if σ_1 and σ_2 are unknown (the usual case), you simply substitute the sample estimates in their place *provided you have large samples* ($n_1 > 30$ and $n_2 > 30$). Consequently, the confidence interval in equation 9.2 is exact (σ_1, σ_2 known) and the confidence interval in equation 9.3 is approximate (σ_1, σ_2 unknown).

EXAMPLE 9.1

Using the data from the two gas-price samples, construct a 90% confidence interval for $\mu_1 - \mu_2$.

SOLUTION

To begin with, the estimate of μ_1 is $\bar{x}_1 = \$1.48$, and the estimate of μ_2 is $\bar{x}_2 = \$1.39$. We are constructing a 90% confidence interval, so (using Table A.4) we find that $Z_{.05} = 1.645$ (Figure 9.2). The resulting 90% confidence interval for $\mu_1 - \mu_2$ is

$$(\bar{X}_1 - \bar{X}_2) - 1.645\sqrt{\frac{s_1^2}{n_1} + \frac{s_2^2}{n_2}} \quad \text{to} \quad (\bar{X}_1 - \bar{X}_2) + 1.645\sqrt{\frac{s_1^2}{n_1} + \frac{s_2^2}{n_2}}$$

$$= (1.48 - 1.39) - 1.645\sqrt{\frac{(.12)^2}{35} + \frac{(.10)^2}{40}} \quad \text{to} \quad (1.48 - 1.39) + 1.645\sqrt{\frac{(.12)^2}{35} + \frac{(.10)^2}{40}}$$

$$= .09 - (1.645)(.0257) \quad \text{to} \quad .09 + (1.645)(.0257)$$

$$= .09 - .042 \quad \text{to} \quad .09 + .042$$

$$= .048 \quad \text{to} \quad .132$$

We can summarize this result in several ways:

1. We are 90% confident that $\mu_1 - \mu_2$ lies between .048 and .132.
2. We are 90% confident that the average price of Texgas regular gasoline is between 4.8¢ and 13.2¢ higher than the regular gasoline at Quik-Chek.
3. We are 90% confident that our estimate of $\mu_1 - \mu_2$ ($\bar{X}_1 - \bar{X}_2 = .09$) is within 4.2¢ of the actual value.

The confidence intervals defined in equations 9.2 and 9.3 will contain $\mu_1 - \mu_2$ 90% of the time. In other words, if you repeatedly obtained indepen- ∎

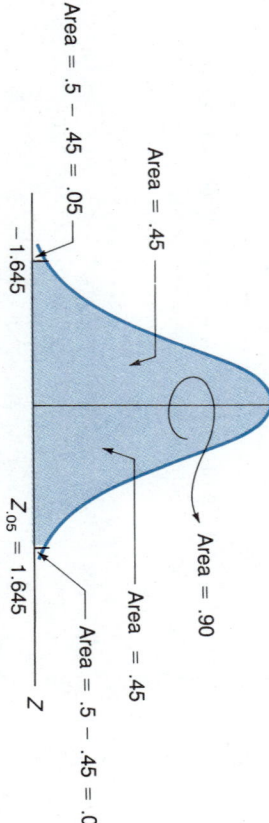

FIGURE 9.2
Finding the pair of Z values containing 90% of the area under the curve. The values are −1.645 and 1.645.

Area = .45

Area = .90

Area = .45

Area = .5 − .45 = .05

−1.645

Area = .5 − .45 = .05

$Z_{.05} = 1.645$

dent samples and repeated the procedure in Example 9.1, 90% of the corresponding confidence intervals would contain the unknown value of $\mu_1 - \mu_2$, and 10% of them would not.

Sample Sizes

The amount that you add to and subtract from your point estimate to obtain the confidence interval is the **maximum error, E.** For Example 9.1, this value is $E = .042$ (4.2¢). If you think that E is too large and you would like it to be smaller, one recourse is to *obtain larger samples* from your two populations. To determine how large a sample you need, one procedure is to select equal sample sizes. Consider the illustration in Example 9.1, where in this case you want large enough sample means so that the difference in sample means is within 2¢ (rather than 4.2¢ in Example 9.1) of the difference in population means, with 90% confidence. So, $E = .02$. By insisting on equal sample sizes, where $n_1 = n_2 = n$ (say), then

$$.02 = 1.645 \sqrt{\frac{(.12)^2}{n} + \frac{(.10)^2}{n}}$$

After some algebraic manipulation, we have

$$n = \frac{(1.645)^2[(.12)^2 + (.10)^2]}{(.02)^2} \simeq 166 \qquad \text{(by rounding up)}$$

In general, this value is

$$n = \frac{Z_{\alpha/2}^2(s_1^2 + s_2^2)}{E^2} \tag{9.4}$$

In this illustration, the total sample size is $n = n_1 + n_2 = 166 + 166 = 332$. A better way to proceed here is to find the values of n_1 and n_2 that **minimize the total sample size, n.** The values of n_1 and n_2 that accomplish this are

$$n_1 = \frac{Z_{\alpha/2}^2 s_1 (s_1 + s_2)}{E^2} \tag{9.5}$$

$$n_2 = \frac{Z_{\alpha/2}^2 s_2 (s_1 + s_2)}{E^2} \tag{9.6}$$

For this illustration, $Z_{\alpha/2} = Z_{.05} = 1.645$, $s_1 = .12$, $s_2 = .10$, and $E = .02$. Consequently,

$$n_1 = \frac{(1.645)^2(.12)(.22)}{(.02)^2} \simeq 179$$

$$n_2 = \frac{(1.645)^2(.10)(.22)}{(.02)^2} \simeq 149$$

and the total sample size is $n = 179 + 149 = 328$.

A derivation of this result is contained in Appendix B. Keep in mind that when you use these values of n_1 and n_2, the resulting value of E may not be exactly what you previously specified—because the values of s_1 and s_2 in the new samples will change. If no prior estimates of σ_1 and σ_2 are available, each can be roughly estimated using the high/low procedure discussed in Chapter 8.

Using equations 9.5 and 9.6, observe that if $s_1 = s_2$, your total sample size $(n_1 + n_2)$ will be the smallest when $n_1 = n_2$. If $s_1 > s_2$, you will select $n_1 > n_2$,

FIGURE 9.3
Hypothesis testing for two populations. Sample 1: size, n_1, mean, $\bar{X}_1$, and standard deviation, s_1. Sample 2: size, n_2, mean, $\bar{X}_2$, and standard deviation, s_2. Is $\mu_2 > \mu_1$?

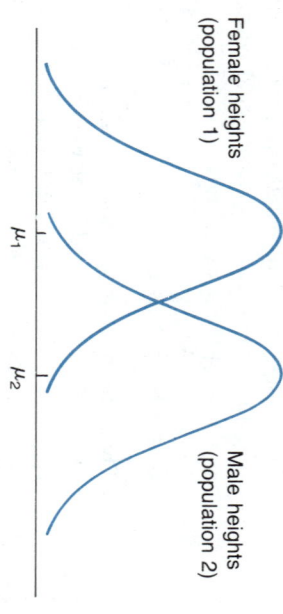

Female heights (population 1)

Male heights (population 2)

μ_1 μ_2

and if $s_1 < s_2$, you will select $n_1 < n_2$. Finally, note that the ratio of the sample sizes (n_1/n_2) is the same as the ratio of the estimated standard deviations (s_1/s_2).

Hypothesis Testing for μ_1 and μ_2 (Large Samples)

Are men on the average taller than women? How do you answer such a question? We know that we can start by getting a sample of male heights and independently obtain a sample of female heights. Figure 9.3 shows two such samples.

We proceed as before and put the claim that we are trying to demonstrate into the *alternative* hypothesis. The resulting hypotheses are

$$H_0: \mu_1 \geq \mu_2 \qquad \text{(men are not taller, on the average)}$$
$$H_a: \mu_1 < \mu_2 \qquad \text{(men are taller, on the average)}$$

We have estimators of μ_1 and μ_2, namely, $\bar{X}_1$ and $\bar{X}_2$. A sensible thing to do would be to reject H_0 if $\bar{X}_2$ is "significantly larger" than $\bar{X}_1$. In this case, the obvious conclusion is that μ_2 (the average of all male heights in your population) is larger than μ_1 (for female heights).

To define significantly larger, we need to know what chance we are willing to take in rejecting H_0 when in fact it is true. This is α (the significance level) and, as before, is determined prior to seeing any data. Typical values range from .01 to .1, with $\alpha = .05$ generally providing a good tradeoff between the Type I and Type II errors. The test statistic here is the same as the one used to derive a confidence interval for $\mu_1 - \mu_2$. We are dealing with large samples ($n_1 > 30$ and $n_2 > 30$), so this is approximately a standard normal random variable, defined by

$$Z = \frac{\bar{X}_1 - \bar{X}_2}{\sqrt{\dfrac{s_1^2}{n_1} + \dfrac{s_2^2}{n_2}}} \qquad (9.7)$$

EXAMPLE 9.2

The Ace Delivery people suspected that the gasoline at the Quik-Chek stores was less expensive than that at Texgas before they obtained any data. (*Note:* This is important! Do not let the data dictate your hypotheses for you. If you do, you introduce a serious bias into your testing procedure, and the "true" significance level may no longer be the predetermined α.) Here, μ_1 represents the average price at all of the Texgas stations and μ_2 is the average price at the Quik-Chek stores in the area. Is $\mu_2 < \mu_1$? Or, put another way, is $\mu_1 > \mu_2$? Use a significance level of .05.

SOLUTION

Step 1. *Define the hypotheses.* The question is whether or not the data support

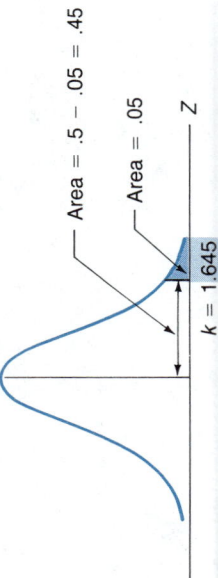

FIGURE 9.4

Z curve showing rejection region for Example 9.2.

the claim that $\mu_1 > \mu_2$, so we put this statement in the alternative hypothesis.

$$H_0: \mu_1 \leq \mu_2 \qquad \text{(Texgas is less expensive or the same)}$$
$$H_a: \mu_1 > \mu_2 \qquad \text{(Quik-Chek is less expensive)}$$

As in Chapter 8, when defining a one-tailed test the equal sign goes into H_0. In other words, the case where $\mu_1 = \mu_2$ is contained in the null hypothesis.

Step 2. *Define the test statistic.* This is the statistic that you evaluate using the sample data. Its value will either support the alternative hypothesis or it will not. The test statistic for this situation is equation 9.7:

$$Z = \frac{\bar{X}_1 - \bar{X}_2}{\sqrt{\dfrac{s_1^2}{n_1} + \dfrac{s_2^2}{n_2}}}$$

Step 3. *Define the rejection region.* In Figure 9.4, where should the null hypothesis H_0 be rejected? We simply ask, what happens to Z when H_a is true? In this case ($\mu_1 > \mu_2$), we *should see* $\bar{X}_1 > \bar{X}_2$. In other words, Z will be positive. So we reject H_0 if Z is "too large"; that is,

$$\text{reject } H_0 \text{ if } Z > k \text{ for some } k > 0$$

Using $\alpha = .05$, we use Table A.4 to find the corresponding value of Z (that is, k). In Figure 9.4, $k = 1.645$. This is the same value and rejection region we obtained in Chapter 8 when using Z for a one-tailed test in the right tail. The test is

$$\text{reject } H_0 \text{ if } Z > 1.645$$

Step 4. *Evaluate the test statistic and carry out the test.* The data collected showed $n_1 = 35$, $\bar{x}_1 = 1.48$, $s_1 = .12$ (from the Texgas sample) and $n_2 = 40$, $\bar{x}_2 = 1.39$, $s_2 = .10$ (from the Quik-Chek sample). Based on these sample results, can we conclude that $\bar{x}_1 = 1.48$ is significantly *larger* than $\bar{x}_2 = 1.39$? If we can, the decision will be to reject H_0. The following value of the test statistic will answer our question.

$$Z = \frac{\bar{X}_1 - \bar{X}_2}{\sqrt{\dfrac{s_1^2}{n_1} + \dfrac{s_2^2}{n_2}}} = \frac{1.48 - 1.39}{\sqrt{\dfrac{(.12)^2}{35} + \dfrac{(.10)^2}{40}}}$$

$$= \frac{.09}{.0257} = 3.50 = Z^*$$

Because $3.50 > 1.645$, we reject H_0; $\bar{x}_1$ is significantly larger than $\bar{x}_2$. Therefore, we claim that $\mu_1 > \mu_2$.

Step 5. *State a conclusion.* We conclude that the Quik-Chek stores *do* charge less

FIGURE 9.5
Z curve showing
p-value for $Z^* = 3.50$.

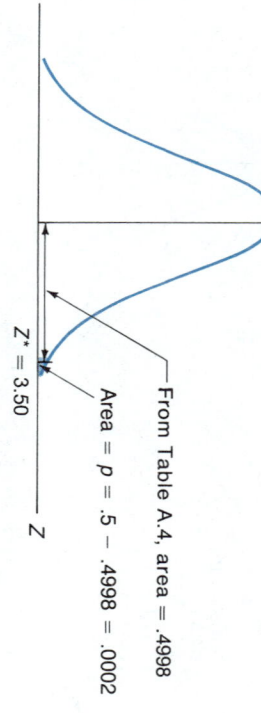

From Table A.4, area = .4998

Area = p = .5 − .4998 = .0002

$Z^* = 3.50$

Z

for gasoline (on the average) than do the Texgas stations. If the locations of these stores are equally convenient to Ace Delivery Service, buying gas from Quik-Chek appears to be a money-saving alternative. ■

Using the corresponding *p*-value for the data in Example 9.2, what would you conclude using the classical approach (with $\alpha = .05$)? For this example, the *p*-value will be the area under the Z curve (Z is our test statistic) to the right (we reject H_0 in the right tail for this example) of the calculated test statistic, $Z^* = 3.50$. In general,

$$p = p\text{-value} = \begin{cases} \text{area to the right of } Z^* \text{ for } H_a: \mu_1 > \mu_2 \\ \text{area to the left of } Z^* \text{ for } H_a: \mu_1 < \mu_2 \\ 2 \cdot (\text{tail area of } Z^*) \text{ for } H_a: \mu_1 \neq \mu_2 \end{cases} \quad (9.8)$$

These three alternative hypotheses are your choices for this situation. Once again, $H_0: \mu_1 = \mu_2$ versus $H_a: \mu_1 \neq \mu_2$ is a two-tailed test, and the first two alternative hypotheses represent one-tailed tests.

Returning to our example, we can see from Figure 9.5 that the resulting *p*-value is $p = .0002$ (very small). Using the classical approach, because $p <$ the significance level of .05, we reject H_0—the same conclusion as before. In fact, this procedure *always* leads to the same conclusion as the five-step solution, as we saw in Chapter 8.

If we elect not to select a significance level α and instead use only the *p*-value to make a decision, we proceed as before:

Reject H_0 if p is small ($p < .01$).

Fail to reject H_0 if p is large ($p > .1$).

Data are inconclusive if p is neither small nor large ($.01 \leq p \leq .1$).

For this example, $p = .0002$ is clearly small, and so we again reject H_0. The Quik-Chek gasoline definitely appears to be less expensive than the Texgas gasoline. As was pointed out in the previous chapter, you often encounter a result that is *statistically* significant but not significant in a *practical* sense. To illustrate, suppose that the *p*-value of .0002 was the result of two very large samples and that the difference in gasoline price for the two samples was $\bar{x}_1 - \bar{x}_2 = .008$. You might not view this difference (less than 1¢) as being worth the inconvenience of having to pay cash for all gasoline purchases.

COMMENT

There may well be situations where the severity of the Type I error requires a significance level smaller than .01 on the low end, or the impact of a Type II error dictates a significance level larger than .1 on the upper end. This rule is thus only a general yardstick that applies to most, but certainly not all, business applications.

LARGE-SAMPLE TESTS FOR μ_1 AND μ_2

TWO-TAILED TEST

$H_0: \mu_1 = \mu_2$

$H_a: \mu_1 \neq \mu_2$

reject H_0 if $|Z| > Z_{\alpha/2}$

where

$$Z = \frac{\bar{X}_1 - \bar{X}_2}{\sqrt{\dfrac{s_1^2}{n_1} + \dfrac{s_2^2}{n_2}}}$$

ONE-TAILED TEST

$H_0: \mu_1 \leq \mu_2$ $H_0: \mu_1 \geq \mu_2$

$H_a: \mu_1 > \mu_2$ $H_a: \mu_1 < \mu_2$

reject H_0 if $Z > Z_\alpha$ reject H_0 if $Z < -Z_\alpha$

Two-Sample Procedure for any Specified Value of $\mu_1 - \mu_2$

The two-tailed hypotheses for large sample tests for μ_1 and μ_2 can be written as

$$H_0: \mu_1 - \mu_2 = 0$$
$$H_a: \mu_1 - \mu_2 \neq 0$$

The right-sided one-tailed hypotheses are

$$H_0: \mu_1 - \mu_2 \leq 0$$
$$H_a: \mu_1 - \mu_2 > 0$$

The left-tailed hypotheses can be written in a similar manner. The point is that H_0 (so far) claims that $\mu_1 - \mu_2$ is equal to *zero* or lies to one side of zero (the one-tailed tests).

Suppose the claim is that $\mu_1 - \mu_2$ is more than ten. To demonstrate that this is true, we must make our alternative hypothesis $H_a: \mu_1 - \mu_2 > 10$; the corresponding null hypothesis is $H_0: \mu_1 - \mu_2 \leq 10$.

In general, to test that $\mu_1 - \mu_2 =$ (some specified value, say D_0), the five-step procedure still applies, except the test statistic is now

$$Z = \frac{(\bar{X}_1 - \bar{X}_2) - D_0}{\sqrt{\dfrac{s_1^2}{n_1} + \dfrac{s_2^2}{n_2}}} \qquad (9.9)$$

Equation 9.9 applies to both one-tailed and two-tailed tests. It can be used to compare two means directly (for example, $H_0: \mu_1 = \mu_2$ versus $H_a: \mu_1 \neq \mu_2$) by setting $D_0 = 0$, as in Example 9.2.

In Example 9.2, we decided that Ace Delivery Service would save money if they purchased their gasoline from Quik-Chek because that store's average gasoline price appeared to be less than that of the Texgas stations. Because Quik-Chek

EXAMPLE 9.3

does not accept credit cards, the owner of Ace is willing to purchase their gasoline only if their average price is more than 6¢ per gallon less than Texgas's. Do the data indicate that it is? (α is still .05.)

SOLUTION

The question now is whether the data support the claim that the difference between the two means (Texgas and Quik-Chek) is larger than 6¢. So the hypotheses are $H_0: \mu_1 - \mu_2 \leq .06$ and $H_a: \mu_1 - \mu_2 > .06$, where μ_1 is Texgas's mean and μ_2 is Quik-Chek's mean.

The test statistic is

$$Z = \frac{(\bar{X}_1 - \bar{X}_2) - .06}{\sqrt{\frac{s_1^2}{n_1} + \frac{s_2^2}{n_2}}}$$

The computed value of Z is

$$Z^* = \frac{(1.48 - 1.39) - .06}{\sqrt{\frac{(.12)^2}{35} + \frac{(.10)^2}{40}}} = \frac{.03}{.0257} = 1.17$$

The testing procedure is exactly as it was previously—reject H_0 if $Z^* > 1.645$. Because $1.17 < 1.645$, we fail to reject H_0. The difference between the two sample means (9¢) was *not* significantly larger than the hypothesized value of 6¢.

These data provide insufficient evidence to conclude that Quik-Chek is more than 6¢ less expensive (on the average) than Texgas. If Ace's owner thinks that not using credit cards would be too much trouble for less than a savings of 6¢ per gallon, Ace should use the Texgas gasoline. ■

EXERCISES

9.9 Determine the value of the test statistic and the *p*-value that would result from the hypothesis test in each of the following cases:

a. $H_0: \mu_1 - \mu_2 \leq 100$, $H_a: \mu_1 - \mu_2 > 100$, $n_1 = 31$, $n_2 = 34$, $\bar{x}_1 = 190$, $\bar{x}_2 = 80$, $s_1 = 25.1$, $s_2 = 20$.

b. $H_0: \mu_1 - \mu_2 = 0$, $H_a: \mu_1 - \mu_2 \neq 0$, $n_1 = 40$, $n_2 = 64$, $\bar{x}_1 = 4.5$, $\bar{x}_2 = 5.8$, $\sigma_1 = 19.1$, $\sigma_2 = 49.3$.

c. $H_0: \mu_1 - \mu_2 \geq 406$, $H_a: \mu_1 - \mu_2 < 406$, $n_1 = 100$, $n_2 = 100$, $\bar{x}_1 = 1050$, $\bar{x}_2 = 650$, $s_1 = 900$, $s_2 = 330$.

9.10 The computer center managers in Becker Industries would like to know if there is a difference in the weekly computer time (in seconds) used by the employees within the financial-planning department and that used by those in the legal-aid department. Conduct a test of hypothesis to determine if there is a significant difference in computer usage for the two departments. Let the significance level be .05. Use the sample statistics that follow, in which 40 weeks were randomly selected for each department.

STATISTIC	n	$\bar{x}$	$\Sigma(x - \bar{x})^2$
Financial Planning	40	2503	2180.4
Legal	40	2510	2291.6

9.11 Fort Worth and Dallas are two large cities that, despite their geographical closeness, have somewhat different economies. To find out the difference in the amount of unemployment, a statistician randomly interviewed 120 unemployed workers from Fort Worth and 150 unemployed workers from Dallas. These people indicated the number of weeks they had been out of work over the past 52 weeks. The data are as follows. Total number of weeks: Fort Worth, 2031; Dallas, 3713. Standard deviation: Fort Worth, 1.3; Dallas, 2.1. Find a 90% confidence interval for the mean difference in the number of weeks unemployed for the work forces of Fort Worth and Dallas.

9.12 First National Bank and City National Bank are competing for customers who would like to open IRAs (individual retirement accounts). Thirty-two weeks are randomly selected for First National Bank and another 32 weeks are randomly selected for City National. The total amount of deposits from IRAs is noted for each week. A summary of data (deposits in thousands of dollars) from the survey is as follows. First National: $\bar{x} = 4.1$, $s = 3.5$, $s = 1.2$. City National: $\bar{x} = 3.5$, $s = 0.9$. Use a 98% confidence interval to estimate the difference in the mean weekly deposits from IRAs for each bank.

9.13 Two discount stores in a popular shopping mall have their merchandise laid out differently. Both stores claim that the arrangement of goods in their store makes the customer buy more on impulse. A survey of 100 customers from each store is taken. Each customer is asked how much money he or she spent on merchandise he or she did not originally intend to buy before walking into the store. The results are as follows. Discount store 1, $\bar{x} = \$15.50$, $s = \$3.20$. Discount store 2: $\bar{x} = \$19.40$, $s = \$4.80$. Find a 90% confidence interval for the difference in the mean amount of cash spent per customer on impulse buying for the two different stores. Is layout affecting impulse buying? How do you know?

9.14 If a 99% confidence interval is found in Exercise 9.13 instead of a 90% confidence interval, would the former confidence interval be larger or smaller than the latter? Explain.

9.15 A personnel director wants to know if the mean length of employment in years with the company is about the same for assembly and clerical workers. A sample 35 employees was randomly drawn from each of these two groups of workers. Conduct a test of hypothesis to determine if there is a significant difference in the mean length of employment with the company for the two groups. Use a significance level of .03.

Assembly workers: $n = 35$, $\bar{x} = 4.1$, $s^2 = 30.2$.

Clerical workers: $n = 35$, $\bar{x} = 3.2$, $s^2 = 28.1$.

9.16 Construct the 90% confidence interval for the difference in the mean length of employment from the two groups of workers in Exercise 9.15.

9.17 Consider the following two independent random samples: $n_1 = 37$, $s_1 = 210$, $\bar{x}_1 = 1360$, and $n_2 = 64$, $s_2 = 108$, $\bar{x}_2 = 1270$. Test the hypotheses: H_0: $\mu_1 - \mu_2 \leq 75$ and H_a: $\mu_1 - \mu_2 > 75$. Use a 10% significance level.

9.18 Alan Nakkupuchi, a designer for high-quality stereo systems, is interested in constructing a 95% confidence interval for the difference between the signal/noise ratios for the two models he developed. Data were obtained from each model as follows. Model **A**: $\bar{x} = 54$, $n = 45$, $s = 8$. Model B: $\bar{x} = 62$, $n = 60$, $s = 17$. Find the 95% confidence interval.

9.19 The average annual pay in 1986 in the state of Rhode Island was \$17,733 and in the state of Connecticut was \$22,516. This is a difference of \$4783. Suppose that a statistician believes that the difference is much less for employees in the manufacturing industry. Suppose that an independent random sample of employees in the manufacturing industry is taken in each state and the results are as follows:

STATE	$\bar{x}$	s	n
Connecticut	21,525	2831	180
Rhode Island	17,630	2600	200

(*Source: The World Almanac and Book of Facts,* 1988, p. 89).

At the .05 significance level, do the data support the statistician's belief that for employees in the manufacturing industry, the mean annual salary in Connecticut differs from the mean annual salary in Rhode Island by less than \$4783? Report the *p*-value and interpret its value.

9.20 The systems manager of Ace Manufacturing is about to purchase a microcomputer and has narrowed her choices to the Alpha and Gemini models. She is seriously concerned about the cost of maintenance of these two models. After interviewing 40 experts on the Alpha model and a different 40 experts on the Gemini model, she obtained the following information on the cost of maintenance. Average annual cost of maintenance: Alpha, $46.50; Gemini, $37.20. Standard deviation: Alpha, 4.20; Gemini, 6.10. Do the data indicate a significant difference in the cost of maintenance for the two models? Use the p-value to justify your answer.

9.21 The financial analyst of Hogan Securities believes that there is no difference in the annual average returns for steel industry stocks and mineral industry stocks. Using the following information, test the hypothesis that there is no significant difference in the average returns for these two types of stocks. Steel industry stocks: $\bar{x} = 9\%$, $n = 33$, $s = 2.4\%$. Mineral industry stocks: $\bar{x} = 11\%$, $n = 41$, $s = 4\%$. Use a 10% significance level.

9.22 From an initial study of a sample of lengths of time that it takes for a package to be delivered through two different express mail companies, it was found that the standard deviation of times to send a package is 1.5 days for company A and 2.3 days for company B. Let E be the maximum error of estimating the mean difference in the times with a 90% confidence interval. If E is taken to be .6, find the values of n_1 and n_2 that minimize the total sample size.

9.23 An education analyst is studying the performance of high-school seniors on the SAT examination. He is specifically testing whether there is any difference between the mean SAT scores of seniors who attended public schools and those who attended private schools. He believes that private schools should score more than 50 points higher than public schools. Using the following information, what would be your conclusion at a significance level of .05? Do you agree with the analyst?

Seniors—public schools: $\bar{x} = 590$, $s = 67$, $n = 40$

Seniors—private schools: $\bar{x} = 680$, $s = 110$, $n = 55$

9.24 The department of management is interested in comparing the utilization of two student laboratories. A crude measure of utilization was used, namely, time between sign-in and sign-out, in hours. Attendance records for the two labs from the previous semester were obtained and 40 students were randomly selected for each lab. The results are summarized as follows:

LAB	n	$\bar{x}$	s
1	40	1.5	0.9
2	40	1.3	1.1

a. Construct a 90% confidence interval for the true population difference $(\mu_1 - \mu_2)$ in the utilization of the two labs.

b. If a 99% confidence interval is wanted but the maximum error E has to remain the same as obtained in part (a), what are the optimum sample sizes n_1 and n_2?

c. Conduct a hypothesis test to determine if there is any difference in the utilization rate of the two labs. Use $\alpha = .05$.

d. Indicate the p-value for the test.

9.3

COMPARING TWO NORMAL POPULATION MEANS USING TWO SMALL INDEPENDENT SAMPLES

When dealing with *small* samples from two populations, we need to consider the assumed distribution of the populations because the Central Limit Theorem no longer applies. This section is concerned with comparing two population means when two small independent random samples are used. It differs from the previous section in two respects:

1. We are dealing with *small samples*.

2. We have reason to believe that the two populations of interest are *normal* populations. In Figures 9.1 and 9.3, where we had large sample sizes, this was not a necessary assumption. When you use small samples from two populations, one or both of which appear to be *not* normally distributed, a nonparametric procedure is the proper method for analyzing such data. This is discussed in Chapter 18.

In Chapter 8, we showed that when going from large samples to small samples from normal populations, the confidence interval and hypothesis-testing procedures both remained exactly the same, except we use the *t* distribution rather than the *Z* distribution to describe the test statistic. We will use the same approach for small samples from two populations.

Confidence Interval for $\mu_1 - \mu_2$ (Small Independent Samples)

When using large samples from the two populations to compare μ_1 and μ_2, we used the *Z*-statistic defined by

$$Z = \frac{\bar{X}_1 - \bar{X}_2}{\sqrt{\frac{s_1^2}{n_1} + \frac{s_2^2}{n_2}}}$$

When using small samples ($n_1 < 30$ or $n_2 < 30$), this statistic no longer approximates the standard normal. To make matters more complicated, it is not a *t* random variable either. However, this expression is *approximately* a *t* random variable with a somewhat complicated expression used to derive the degrees of freedom (df). So we define

$$t' = \frac{\bar{X}_1 - \bar{X}_2}{\sqrt{\frac{s_1^2}{n_1} + \frac{s_2^2}{n_2}}} \qquad \textbf{(9.10)}$$

This statistic approximately follows a *t* distribution with df given by

$$\text{df for } t' = \frac{\left[\frac{s_1^2}{n_1} + \frac{s_2^2}{n_2}\right]^2}{\frac{\left(\frac{s_1^2}{n_1}\right)^2}{n_1 - 1} + \frac{\left(\frac{s_2^2}{n_2}\right)^2}{n_2 - 1}} \qquad \textbf{(9.11)}$$

Admittedly, equation 9.11 is a bit messy, but a good calculator or computer package makes this calculation relatively painless. To be on the conservative side, if df as calculated is not an integer ($1, 2, 3, \ldots$), it should be rounded *down* to the next integer. As a check of your calculations, the df should be between A and B, where A is the smaller of $(n_1 - 1)$ and $(n_2 - 1)$ and B is $(n_1 - 1) + (n_2 - 1)$.

When finding the df, you can scale *both* s_1 and s_2 any way you wish, provided you scale them both the same way. By scaling, we mean that you can use s_1 and s_2 as is, or you can move the decimal point to the right or left. The resulting df will be the same *regardless* of the scaling used. However, when you evaluate the test statistic, t', or later perform a test of hypothesis, you must return to the *original* values of s_1 and s_2.

To derive an approximate confidence interval for $\mu_1 - \mu_2$, we use the same logic as in the previous (large samples) procedure. Thus, a $(1 - \alpha) \cdot 100\%$

confidence interval for $\mu_1 - \mu_2$ (small samples) is:

$$(\bar{X}_1 - \bar{X}_2) - t_{\alpha/2,\text{df}} \sqrt{\frac{s_1^2}{n_1} + \frac{s_2^2}{n_2}} \quad \text{to} \quad (\bar{X}_1 - \bar{X}_2) + t_{\alpha/2,\text{df}} \sqrt{\frac{s_1^2}{n_1} + \frac{s_2^2}{n_2}} \quad (9.12)$$

where df is specified in equation 9.11. If df is not an integer, round this value *down* to the next integer.

EXAMPLE 9.4

Checkers Cab Company is trying to decide which brand of tires to use for the coming year. Based on current price and prior experience, they have narrowed their choice to two brands, Beltex and Roadmaster. A recent study examined the durability of these tires by using a machine with a metallic device that wore down the tires. The time it took (in hours) for the tire to blow out was recorded.

Because the test for each tire took a great deal of time and the tire itself was ruined by the test, small samples of 15 of each brand were used. Notice that these are *independent* samples; there is no reason to match up the first Beltex tire with the first Roadmaster tire in the sample, the second Beltex with the second Roadmaster, and so on. As discussed in Section 9.1, they would be *dependent* samples if the tires were tested by putting one of each brand on the rear wheels of 15 different cars.

The blowout times (hours) were as follows:

BELTEX	ROADMASTER	BELTEX	ROADMASTER
3.82	4.16	2.84	3.65
3.11	3.92	3.26	3.82
4.21	3.94	3.74	4.55
2.64	4.22	3.04	3.82
4.16	4.15	2.56	3.85
3.91	3.62	2.58	3.62
2.44	4.11	3.15	4.88
4.52	3.45		

SOLUTION

Construct a 90% confidence interval for $\mu_1 - \mu_2$, letting μ_1 be the average blowout time for *all* Beltex tires and μ_2 be the average blowout time for *all* Roadmaster tires.

Here is a summary of the data from these two samples.

Sample 1 (Beltex)	Sample 2 (Roadmaster)
$n_1 = 15$	$n_2 = 15$
$\bar{x}_1 = 3.33$ hours	$\bar{x}_2 = 3.98$ hours
$s_1 = .68$ hours	$s_2 = .38$ hours

Your next step is to get a t-value from Table A.5. To do this, you first must calculate the correct df using equation 9.11. This is

$$\text{df} = \frac{\left[\frac{(.68)^2}{15} + \frac{(.38)^2}{15}\right]^2}{\frac{\left(\frac{(.68)^2}{15}\right)^2}{14} + \frac{\left(\frac{(.38)^2}{15}\right)^2}{14}}$$

$$= \frac{(.0404)^2}{.0000679 + .00000662} = 21.9$$

Rounding down, we use df = 21. Using Table A.5:

$$t_{.10/2, 21} = t_{.05, 21} = 1.721$$

The resulting 90% confidence interval for $\mu_1 - \mu_2$ is

$$(\bar{X}_1 - \bar{X}_2) - t_{.05,21}\sqrt{\frac{s_1^2}{n_1} + \frac{s_2^2}{n_2}} \quad \text{to} \quad (\bar{X}_1 - \bar{X}_2) + t_{.05,21}\sqrt{\frac{s_1^2}{n_1} + \frac{s_2^2}{n_2}}$$

$$= (3.33 - 3.98) - 1.721\sqrt{\frac{(.68)^2}{15} + \frac{(.38)^2}{15}} \quad \text{to} \quad (3.33 - 3.98) + 1.721\sqrt{\frac{(.68)^2}{15} + \frac{(.38)^2}{15}}$$

$$= -.65 - .35 \quad \text{to} \quad -.65 + .35$$

$$= -1.00 \text{ hr} \quad \text{to} \quad -.30 \text{ hr}$$

So we are 90% confident that the average blowout time for the Beltex tires is between 18 minutes (.3 hours) and 1 hour *less* than the average for the Roadmaster tires. Based on these results, Roadmaster appears to be the better (longer-wearing) tire. ∎

Hypothesis Testing for μ_1 and μ_2 (Small Independent Samples)

The five-step procedure for testing hypotheses concerning μ_1 and μ_2 with large samples also applies to the small-sample situation. The only difference is that Table A.5 is used (rather than Table A.4) to define the rejection region.

EXAMPLE 9.5

In Example 9.4 a confidence interval was constructed for the difference between average blowout times for Beltex and Roadmaster tires. Can we conclude that these average blowout times are in fact not the same? Use a significance level of .10.

SOLUTION

Step 1. We are testing for a difference between the two means (not that Roadmaster is longer-wearing than Beltex or vice versa). The corresponding appropriate hypotheses are H_0: $\mu_1 = \mu_2$ and H_a: $\mu_1 \neq \mu_2$.

Step 2. The test statistic is

$$t' = \frac{\bar{X}_1 - \bar{X}_2}{\sqrt{\frac{s_1^2}{n_1} + \frac{s_2^2}{n_2}}}$$

which approximately follows a t distribution with df given by equation 9.11.

Step 3. You next need the df in order to determine your rejection region. In Example 9.4 we found that df = 21. Because H_a: $\mu_1 \neq \mu_2$, we will reject H_0 if t' is too large ($\bar{X}_1$ is significantly *larger* than $\bar{X}_2$) or if t' is too small ($\bar{X}_1$ is significantly *smaller* than $\bar{X}_2$). As in previous two-tailed tests using the Z- or t-statistic, H_0 is rejected if the absolute value of t exceeds the value from the table corresponding to $\alpha/2$. Using Table A.5, the rejection region for this situation will be

reject H_0 if $|t'| > t_{\alpha/2,\text{df}} = t_{.05,21} = 1.721$

Step 4. The value of the test statistic is

$$t^* = \frac{3.33 - 3.98}{\sqrt{\frac{(.68)^2}{15} + \frac{(.38)^2}{15}}} = \frac{-.65}{.20} = -3.25$$

Step 5. Because $|t^*| = 3.25 > 1.721$, we reject H_0. Consequently, the difference between the sample means (−.65) *is* significantly large (in absolute value), which leads to a rejection of the null hypothesis. There *is* a significant difference in the average blowout times for the two brands. ∎

FIGURE 9.6
MINITAB solution to
Example 9.5

```
MTB > SET INTO C1
DATA> 3.82 3.11 4.21 2.64 4.16 3.91 2.44 4.52 2.84
DATA> 3.26 3.74 3.04 2.56 2.58 3.15
DATA> END
MTB > SET INTO C2
DATA> 4.16 3.92 3.94 4.22 4.15 3.62 4.11 3.45 3.65
DATA> 3.82 4.55 3.82 3.85 3.62 4.88
DATA> END
MTB > TWOSAMPLE T WITH 90% CONFIDENCE USING C1 AND C2 ↑

                                            No subcommands
                                            are necessary
TWOSAMPLE T FOR C1 VS C2                     for a two-tailed test.
        N     MEAN    STDEV   SE MEAN
C1     15    3.332    0.679    0.18
C2     15    3.984    0.377    0.097

90 PCT CI FOR MU C1 = MU C2: (-1.00, -0.307)
                                            t'
TTEST MU C1 = MU C2 (VS NE): T= -3.25  P=0.0038  DF= 21
                                                 p-value
```

COMMENTS

The hypotheses in Example 9.4 could be written as $H_0: \mu_1 - \mu_2 = 0$ and $H_a: \mu_1 - \mu_2 \neq 0$. Having already determined a 90% confidence interval for $\mu_1 - \mu_2$, a much simpler way to perform this two-tailed test (using $\alpha = .10$) would be to reject H_0 if 0 does not lie in the confidence interval for $\mu_1 - \mu_2$ and fail to reject H_0 otherwise. The confidence interval according to Example 9.4 is $(-1.00, -.30)$, which does not contain zero, and so we reject H_0 (as before).

This alternate method of testing H_0 versus H_a holds only for a two-tailed test in which the significance level of the test, α, and the confidence level $[(1 - \alpha) \cdot 100\%]$ of the confidence interval "match up." For example, a significance level of $\alpha = .05$ would correspond to a 95% confidence interval, a value of $\alpha = .10$ would correspond to a 90% confidence interval, and so on.

A MINITAB solution to Example 9.5 is provided in Figure 9.6. The calculated p-value is $p = .0038$. Based on this extremely small value, we again reject H_0.

Notice that the procedure in this section for testing μ_1 versus μ_2 and constructing confidence intervals for $\mu_1 - \mu_2$ made no mention as to whether the population variances (or standard deviations) were equal or not. In fact, we can say that this procedure did not assume that $\sigma_1 = \sigma_2$; it also did not assume that $\sigma_1 \neq \sigma_2$. Next, we will examine a special case where we have reason to believe that the standard deviations are equal. For this situation, we will define another t test to detect any difference between the population means.

Special Case of Equal Variances

There are some situations in which we are willing to assume that the population variances (σ_1^2 and σ_2^2) are equal. This is common in many long-running production processes for which, based on past experience, you are convinced that the variation within population 1 is the same as the variation within population 2.

Another situation in which we may assume $\sigma_1 = \sigma_2$ arises when we obtain two *additional* samples from the two populations, which we use strictly to determine if the population standard deviations are equal. If there is not sufficient evidence to indicate that $\sigma_1 \neq \sigma_2$, then there is no harm in assuming that $\sigma_1 = \sigma_2$. A procedure for comparing the population standard deviations is discussed in Section 9.4.

Why make the assumption that $\sigma_1 = \sigma_2$? Remember, we are still interested in the means, μ_1 and μ_2. As before, we would like to obtain a confidence interval for $\mu_1 - \mu_2$ and to perform a test of hypothesis. If, in fact, σ_1 *is* equal to σ_2, then we can construct a slightly stronger test of μ_1 versus μ_2. By stronger, we mean that we are *more likely* to reject H_0 when it is actually false. This test is said to be more **powerful**.

For this case, because we believe that $\sigma_1^2 = \sigma_2^2 = \sigma^2$ (say), it makes sense to combine—or **pool**—our estimate of σ_1^2 (s_1^2) with the estimate of σ_2^2 (s_2^2) into one estimate of this common variance (σ^2). The resulting estimate of σ^2 is called the **pooled sample variance** and is written s_p^2. This estimate is merely a *weighted average* of s_1^2 and s_2^2, defined by

$$s_p^2 = \frac{(n_1 - 1)s_1^2 + (n_2 - 1)s_2^2}{n_1 + n_2 - 2} \qquad (9.13)$$

Constructing Confidence Intervals for $\mu_1 - \mu_2$

To construct the confidence interval, we make two changes in the previous procedure. First, t' is replaced by

$$t = \frac{\bar{X}_1 - \bar{X}_2}{\sqrt{\dfrac{s_p^2}{n_1} + \dfrac{s_p^2}{n_2}}} \qquad (9.14)$$

$$= \frac{\bar{X}_1 - \bar{X}_2}{s_p\sqrt{\dfrac{1}{n_1} + \dfrac{1}{n_2}}} \qquad (9.15)$$

Here (unlike the previous test statistic), t exactly follows a t distribution (assuming the two populations follow normal distributions).

Second, the df for t are much easier to derive:

$$df = n_1 + n_2 - 2$$

So you avoid the difficult df calculation in equation 9.11, but you need to derive the pooled variance, s_p^2, using the individual sample variances, s_1^2 and s_2^2.

As a check, your resulting pooled value for s_p^2 should be between s_1^2 and s_2^2, since it is a weighted average of these two values.

Hypothesis Testing for μ_1 and μ_2

In hypothesis testing for $\mu_1 - \mu_2$, the previous procedure applies except that t' is replaced by t, where the df used in Table A.5 is $df = n_1 + n_2 - 2$ rather than $df = $ value from equation 9.11.

In Examples 9.4 and 9.5, we examined the blowout times for two brands of tires as measured by a machine performing a stress test of the sampled tires. Assume we have determined from previous tests that the *variation* of the blowout times is not affected by the tire brand. Assuming that σ_1^2 (Beltex) $= \sigma_2^2$ (Roadmaster), how can we construct a 90% confidence interval for $\mu_1 - \mu_2$ and determine whether there is a difference in the mean blowout times?

Sample 1 (Beltex)	Sample 2 (Roadmaster)
$n_1 = 15$	$n_2 = 15$
$\bar{x}_1 = 3.33$ hr	$\bar{x}_2 = 3.98$ hr
$s_1 = .68$ hr	$s_2 = .38$ hr

Our first step is to pool the sample variances:

$$s_p^2 = \frac{(15-1)(.68)^2 + (15-1)(.38)^2}{15+15-2} = \frac{(14)(.4624) + (14)(.1444)}{28}$$

$$= \frac{8.495}{28} = .303$$

$$s_p = \sqrt{.303} = .55 \text{ hr}$$

Is .303 between .1444 and .4624? Yes. Consequently, $s_p^2 = .303$ is our estimate of the common variance (σ^2) of the two tire populations. To find the 90% confidence interval for $\mu_1 - \mu_2$, we use

$$(\bar{X}_1 - \bar{X}_2) - t_{\alpha/2,\,\mathrm{df}} \sqrt{\frac{s_p^2}{n_1} + \frac{s_p^2}{n_2}} \quad \text{to} \quad (\bar{X}_1 - \bar{X}_2) + t_{\alpha/2,\,\mathrm{df}} \sqrt{\frac{s_p^2}{n_1} + \frac{s_p^2}{n_2}} \quad \textbf{(9.16)}$$

where df $= n_1 + n_2 - 2$ and $\alpha = .10$.

Because $n_1 + n_2 - 2 = 28$, we find (from Table A.5) that $t_{.05,\,28} = 1.701$. Next,

$$\sqrt{\frac{s_p^2}{n_1} + \frac{s_p^2}{n_2}} = s_p \sqrt{\frac{1}{n_1} + \frac{1}{n_2}} = .55 \sqrt{\frac{1}{15} + \frac{1}{15}} = .20$$

The resulting confidence interval is

$$(3.33 - 3.98) - (1.701)(.20) \quad \text{to} \quad (3.33 - 3.98) + (1.701)(.20)$$

$$= -.65 - .34 \quad \text{to} \quad -.65 + .34$$

$$= -.99 \quad \text{to} \quad -.31$$

Comparing this result to the confidence interval in Example 9.4, you see little difference in the two confidence intervals although the interval using the pooled variance is a bit narrower. Oftentimes these intervals can differ considerably, depending on the relative sizes of n_1 and n_2 as well as the relative values of s_1^2 and s_2^2.

Now we wish to test H_0: $\mu_1 = \mu_2$ versus H_a: $\mu_1 \neq \mu_2$. For this particular example, we can, as noted earlier, reject H_0 (using $\alpha = .10$) because zero does not lie in the previously derived confidence interval for $\mu_1 - \mu_2$. When using the five-step procedure, there are only two changes we need to make when using the pooled sample variances. First, when defining our rejection region, we use $n_1 + n_2 - 2 = 28$ df. From Table A.5, the test procedure is to

$$\text{reject } H_0 \text{ if } |t| > t_{\alpha/2,\,\mathrm{df}}$$

where $t_{.05,\,28} = 1.701$.

Second, the value of our test statistic is now

$$t = \frac{\bar{X}_1 - \bar{X}_2}{s_p \sqrt{\dfrac{1}{n_1} + \dfrac{1}{n_2}}} \qquad \textbf{(9.17)}$$

Here,

$$t = \frac{3.33 - 3.98}{.55 \sqrt{\dfrac{1}{15} + \dfrac{1}{15}}} = \frac{-.65}{.20} = -3.25$$

FIGURE 9.7
MINITAB pooled variances solution using data from Example 9.4.

```
MTB > TWOSAMPLE TEST WITH 90% CONFIDENCE USING C1 AND C2;
SUBC> POOLED.
                                   This subcommand
TWOSAMPLE T FOR C1 VS C2           is necessary when
                                   you assume σ₁ = σ₂
       N    MEAN    STDEV   SE MEAN
C1    15   3.332   0.679    0.18
C2    15   3.984   0.377    0.097

90 PCT CI FOR MU C1 - MU C2: (-0.99, -0.311)

TTEST MU C1 = MU C2 (VS NE): T= -3.25  P=0.0030  DF= 28

POOLED STDEV =    0.549
                              t'          p-value
```

Because $|-3.25| = 3.25 > 1.701$, we reject H_0; once again the two sample means are significantly different. We conclude that there is a difference in the population mean blowout times for the two brands of tires.

A MINITAB solution for this example is provided in Figure 9.7. As in Example 9.5 (Figure 9.6), we obtain a very small p-value when pooling the sample variances. For this particular example, we observe little difference in the two solutions.

To Pool or Not to Pool? You might think, based on the previous examples, that it really does not matter whether you assume $\sigma_1 = \sigma_2$ or not. The two confidence intervals were nearly the same and the tests of hypothesis results were very close, differing only in their df for the test statistic. However, this is not always the case. Unless you have strong evidence that the variances are the same, *we suggest you not pool the sample variances* and use the test statistic defined in equation 9.10. If you assume that $\sigma_1 = \sigma_2$ and use the t test statistic in equation 9.17 but in fact $\sigma_1 \neq \sigma_2$, your results will be unreliable. This test is quite sensitive to this particular assumption. Also, if σ_1 and σ_2 *are* the same, we would expect s_1 and s_2 to be nearly the same. If, in addition, $n_1 = n_2$ (or nearly so), then the computed values of t' and t will be practically identical (including the df). What this means is that you have little to gain by pooling the variances (and using t) but a great deal to lose if your assumption is incorrect.

We will show you in the next section how to use two samples to test the hypothesis that $\sigma_1 = \sigma_2$. With those results in hand, one possible procedure to use when testing the *means* would be: (1) if you reject $H_0: \sigma_1 = \sigma_2$, then use t' to test $H_0: \mu_1 = \mu_2$, and (2) if you fail to reject $H_0: \sigma_1 = \sigma_2$, then use t to test $H_0: \mu_1 = \mu_2$.

At first glance this may appear to be statistically sound, but it has some problems. The main one is that these two tests use the same data, and so the tests are not performed independently of one another. Also, your actual significance level may not be the α that you had previously chosen before you saw any data. This *can* be a valid procedure if you obtain separate samples—one to test the σ values and the other to test the μ values. Again however, caution is in order, since the test of $H_0: \sigma_1 = \sigma_2$ is very sensitive to the assumption of normal populations and so using small samples to carry out this test can be unreliable. Consequently, if there is reason to believe that the standard deviations might not be equal, a safe procedure is to proceed as if they weren't.

The next section provides a procedure for testing the standard deviations from two normal populations using independent samples. By comparing the standard deviations using separate data from the two populations, one can decide whether the pooling procedure should be used when using additional data to test μ_1 versus μ_2. If you reject $H_0: \sigma_1 = \sigma_2$, then the t'-statistic in equation 9.10 is the proper test statistic to use on a test for the means because it does *not* assume that the population standard deviations are equal. On the other hand, if you fail to reject H_0, then the t-statistic in equation 9.17, which *does* assume that $\sigma_1 = \sigma_2$, is the recommended test statistic for testing μ_1 versus μ_2.

SMALL-SAMPLE TESTS FOR μ_1 AND μ_2

TWO-TAILED TEST

$$H_0: \mu_1 - \mu_2 = D_0$$
$$H_a: \mu_1 - \mu_2 \neq D_0$$
$$(D_0 = 0 \text{ for } H_0: \mu_1 = \mu_2)$$
reject H_0 if $|T| > t_{\alpha/2, \text{df}}$

where, not assuming $\sigma_1 = \sigma_2$:

$$T = t' = \frac{(\bar{X}_1 - \bar{X}_2) - D_0}{\sqrt{\dfrac{s_1^2}{n_1} + \dfrac{s_2^2}{n_2}}}$$

Or, assuming $\sigma_1 = \sigma_2$:

$$T = t = \frac{(\bar{X}_1 - \bar{X}_2) - D_0}{s_p \sqrt{\dfrac{1}{n_1} + \dfrac{1}{n_2}}}$$

and, for t':

$$\text{df} = \frac{\left[\dfrac{s_1^2}{n_1} + \dfrac{s_2^2}{n_2}\right]^2}{\dfrac{\left(\dfrac{s_1^2}{n_1}\right)^2}{n_1 - 1} + \dfrac{\left(\dfrac{s_2^2}{n_2}\right)^2}{n_2 - 1}}$$

and, for t:

$$\text{df} = n_1 + n_2 - 2$$

where

$$s_p = \sqrt{\frac{(n_1 - 1)s_1^2 + (n_2 - 1)s_2^2}{n_1 + n_2 - 2}}$$

ONE-TAILED TEST

$H_0: \mu_1 - \mu_2 \leq D_0$	$H_0: \mu_1 - \mu_2 \geq D_0$
$H_a: \mu_1 - \mu_2 > D_0$	$H_a: \mu_1 - \mu_2 < D_0$
$(D_0 = 0 \text{ for } H_0: \mu_1 \leq \mu_2)$	$(D_0 = 0 \text{ for } H_0: \mu_1 \geq \mu_2)$
reject H_0 if $T > t_{\alpha, \text{df}}$	reject H_0 if $T < -t_{\alpha, \text{df}}$

9.25   Achieving a high score on the LSAT examination is a prerequisite to acceptance to law school. Scores on the LSAT are considered to be normally distributed. Two law schools decided to compare the mean scores on the LSAT for students enrolled in their schools. Is there sufficient evidence to indicate that the average scores differ between the two schools? Law school 1: $\bar{x} = 680$, $s = 84$, $n = 15$. Law school 2: $\bar{x} = 634$, $s = 92$, $n = 21$. Use a 1% significance level. Assume that the population variances are equal for law school 1 and law school 2.

9.26 Construct a 95% confidence interval for $\mu_1 - \mu_2$ in Exercise 9.25. Assume that the population variances are equal for the two law schools.

9.27 The president of a personnel agency is interested in examining the annual mean salary differences between vice presidents of banks and vice presidents of savings and loan institutions. A random sample of eight of each kind of vice president was selected. Their annual salaries (in dollars) were as follows:

n	BANKS	SAVINGS AND LOAN INSTITUTIONS	n	BANKS	SAVINGS AND LOAN INSTITUTIONS
1	84,320	73,420	5	48,940	88,670
2	67,440	49,580	6	56,790	59,640
3	98,590	58,750	7	77,610	65,590
4	111,780	101,400	8	62,000	74,810

Conduct a test of hypothesis to determine if there is a significant difference in the average salary for the two vice president groups. The salaries for both groups are considered to be approximately normally distributed. Use a significance level of .05. Do not assume that the population variances are equal.

9.28 Construct a 90% confidence interval for the difference in the means of the salaries for vice presidents in the banking industry and for vice presidents of savings and loan institutions for Exercise 9.27. Do not assume that the population variances are equal.

9.29 Using the data in Exercise 9.27, test the same hypothesis, but assume that the population variances *are* equal.

9.30 The production supervisor of Dow Plast is conducting a test of the tensile strengths of two types of copper coils. The relevant data are as follows. Coil A: $\bar{x} = 118$, $s = 17$, $n = 9$. Coil B: $\bar{x} = 143$, $s = 24$, $n = 16$. The tensile strengths for the two types of copper coils are approximately normally distributed. Based on the p-value, do the data support the conclusion that the mean tensile strengths of the two coils are different at a significance level of 7%? Do not assume that the population variances are equal.

9.31 Construct a 99% confidence interval for $\mu_A - \mu_B$ in Exercise 9.30. Do not assume that the population variances are equal.

9.32 Using a pooled estimate of the variance, perform the test of hypothesis in Exercise 9.30. Compare the two answers.

9.33 Are you easily distracted when taking a test? It is generally assumed that a quiet, distraction-free situation is necessary for testing. However, some research does not support the general assumption that distractions in the testing situation influence test scores. Many studies have been conducted on this issue. Trentham (1975) randomly selected 72 students from the sixth-grade classes of three elementary schools in Paducah, Kentucky. These 72 students were randomly divided into experimental and control groups. The "Torrance Test of Creative Thinking" (TTCT) was chosen as the instrument for this study. The experimental group was subjected to distractions like an alarm bell going off, blinding lights, books being dropped, and various other interruptions. Conditions were kept as "ideal" as possible for the control group. The t scores were calculated for verbal and figural totals on the TTCT using pooled samples, i.e., assuming equal variances. The following table also provides the computed t-statistic for comparing two means:

TTCT SUBTEST	DISTRACTION	NONDISTRACTION	t-STATISTIC
Verbal scores	46.80	53.95	2.95
Figural scores	48.68	52.36	1.63

Source: L. L. Trentham, The Effect of Distractions on Sixth-grade Students in a Testing Situation. *Journal Education Measurement* 12, no. 1 (Spring 1975): 13–17.

a. What are the degrees of freedom for the tests?

b. Given the computed t-statistic, is there a significant difference between the TTCT verbal scores of the distraction and nondistraction groups? Use a significance level of .05.

c. Did the distractions imposed during the test have a significant influence on the TTCT figural scores? Use a significance level of .05.

9.34 A machine operator is interested in whether there is a significant difference in the time to produce a particular item of output between machine 1 and machine 2. The time that it takes to output an item is normally distributed. Ten items produced by machine 1 and then another 10 items produced by machine 2 were recorded. The resulting times in minutes were:

MACHINE 1		MACHINE 2	
40.3	39.7	43.7	41.6
35.6	40.2	42.1	42.3
42.7	38.2	41.8	40.9
41.9	39.6	42.8	43.8
38.6	40.3	40.2	42.7

Without assuming equal population variances, perform the test of hypothesis to determine if there is a significant difference in the mean time for machine 1 and machine 2 to produce 1 item of output. Use a significance level of .05.

9.35 Construct a 95% confidence interval for $\mu_1 - \mu_2$ for Exercise 9.34.

9.36 Using the pooled estimate of the variance, perform the test of hypothesis in Exercise 9.34. Calculate the p-value.

9.4
COMPARING THE VARI-ANCES OF TWO NORMAL POPULATIONS USING INDEPENDENT SAMPLES

Once again we concentrate on independent samples from two normal populations, only this time we focus our attention on the *variation* of these populations rather than on their averages. This is illustrated in Figure 9.8. When estimating and testing σ_1 versus σ_2, we will not be concerned about μ_1 and μ_2. They may be equal, or they may not—it simply does not matter for this test procedure.

In business applications, you may want to compare the variation of two different production processes or compare the risk involved with two proposed investment portfolios. As mentioned previously, when testing for population *means* using small independent samples, you must pay attention to the population standard deviations (variances). Based on your belief that σ_1 does or does not equal σ_2, you select your corresponding test statistic for testing the means, μ_1 and μ_2. As a reminder, it is *not* a safe procedure to use the *same data set* to test both $\sigma_1 = \sigma_2$ and $\mu_1 = \mu_2$. A proper procedure would be to test σ_1 and σ_2

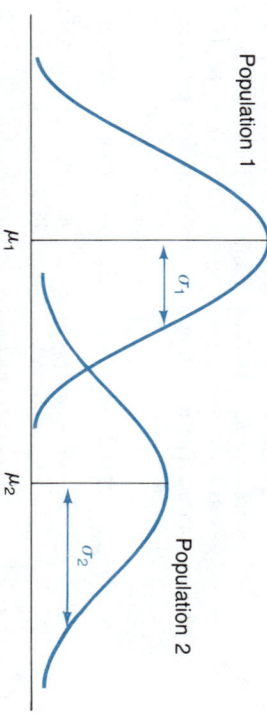

FIGURE 9.8
Comparing two standard deviations. Is $\sigma_1 = \sigma_2$?

Population 1

Population 2

μ_1

μ_2

σ_1

σ_2

Comparing two standard deviations. Is $\sigma_1 = \sigma_2$?

FIGURE 9.9

Shape of the F
distribution.

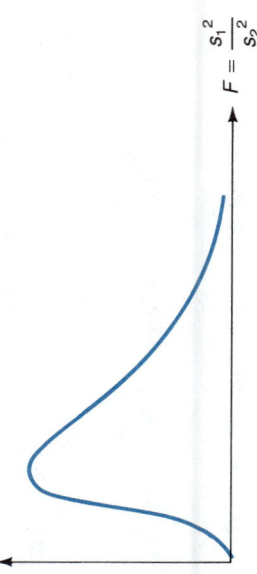

$$F = \frac{s_1^2}{s_2^2}$$

using one set of samples (as outlined in this section) and to obtain another set of samples *independently* of the first to test the means.

In the previous section, when trying to decide if $\mu_1 = \mu_2$, we examined the *difference* between the point estimators, $\bar{X}_1 - \bar{X}_2$. If $\bar{X}_1 - \bar{X}_2$ was large enough (in absolute value), we rejected H_0: $\mu_1 = \mu_2$. When looking at the variances, we use the **ratio of the sample variances**, s_1^2 and s_2^2, to derive a test of hypothesis and construct confidence intervals. We do this because the distribution of $s_1^2 - s_2^2$ is difficult to describe mathematically, but s_1^2/s_2^2 does have a recognizable distribution when in fact σ_1^2 and σ_2^2 are equal. So we define

$$F = \frac{s_1^2}{s_2^2} \qquad (9.18)$$

If you were to obtain sets of two samples repeatedly, calculate s_1^2/s_2^2 for each set, and make a histogram of these ratios, the shape of this histogram would resemble the curve in Figure 9.9. This is the **F distribution**. Its shape resembles the chi-square curve—it is nonsymmetric, skewed right (right-tailed), and the corresponding random variable is never negative. There are many F curves, depending on the sample sizes, n_1 and n_2. The shape of the F curve becomes more symmetric as the sample sizes, n_1 and n_2, increase. As later chapters will demonstrate, the F distribution has a large variety of applications in statistics. Right-tail areas for this random variable have been tabulated in Table A.7. As a final note here, the F-statistic in equation 9.18 is highly sensitive to the assumption of normal populations. For larger data sets, it is recommended that you examine the shape of the sample data when using this particular F-statistic.

When using the t- and χ^2-statistics, we needed a way to specify the sample size(s) because the shape of these curves changes as the sample size changes. The same applies to the F distribution. There are two samples here, one from each population, and we need to specify *both* sample sizes. As before, we use the degrees of freedom (df) to accomplish this, where

$$v_1 = \text{df for numerator} = n_1 - 1$$
$$v_2 = \text{df for denominator} = n_2 - 1$$

So, the F-statistic shown in Figure 9.9 follows an F distribution with v_1 and v_2 df provided $\sigma_1^2 = \sigma_2^2$ ($\sigma_1 = \sigma_2$). What happens to F when $\sigma_1 \neq \sigma_2$? Suppose that $\sigma_1 > \sigma_2$? Then we would expect s_1 (the estimate of σ_1) to be larger than s_2 (the estimate of σ_2); we should see

$$s_1^2 > s_2^2$$

or

$$F = \frac{s_1^2}{s_2^2} > 1$$

Similarly, if $\sigma_1 < \sigma_2$, then we expect an F-value < 1. We will use this reasoning to define a test of hypothesis for σ_1 versus σ_2.

Hypothesis Testing for $\sigma_1 = \sigma_2$

Is $\sigma_1 = \sigma_2$? We use the usual five-step procedure for testing a hypothesis concerning the two variances. Your choice of hypotheses is (as usual) a two-tailed test or a one-tailed test. For the two-tailed test the hypotheses are $H_0: \sigma_1 = \sigma_2$ $(\sigma_1^2 = \sigma_2^2)$ and $H_a: \sigma_1 \neq \sigma_2$ $(\sigma_1^2 \neq \sigma_2^2)$. For the one-tailed test the hypotheses are $H_0: \sigma_1 \leq \sigma_2$ and $H_a: \sigma_1 > \sigma_2$ (Figure 9.10(A)) or $H_0: \sigma_1 \geq \sigma_2$ and $H_a: \sigma_1 < \sigma_2$ (Figure 9.10(B)).

Notice that the hypotheses can be written in terms of the standard deviations (σ_1 and σ_2) or the variances (σ_1^2 and σ_2^2); if $\sigma_1 > \sigma_2$, then $\sigma_1^2 > \sigma_2^2$.

Right-tail areas under an F curve are provided in Table A.7. Notice that we have a table for areas of .1 (Table A.7(a)), .05 (Table A.7(b)), .025 (Table A.7(c)), and .01 (Table A.7(d)). These are the most commonly used values. For each table, the df for the numerator (v_1) run across the top, and the df for the denominator (v_2) run down the left margin.

Suppose we want to know which F-value has a right-tail area of .05, using 10 and 12 df. Let the F-value whose right-tail area is α, where the df are v_1 and v_2, be

$$F_{\alpha, v_1, v_2}$$

For example, $F_{.05, 10, 12} = 2.75$ (Figure 9.11).

Notice that Table A.7 contains *right-tail* areas only. Later, we will show you how to find *left-tail* areas. We can, however, define each of our tests of hypothesis as a right-tailed test by simply and arbitrarily putting the larger sample variance in the numerator for a two-tailed test. Then F always will be greater than or equal to 1. This procedure is summarized in the accompanying box.

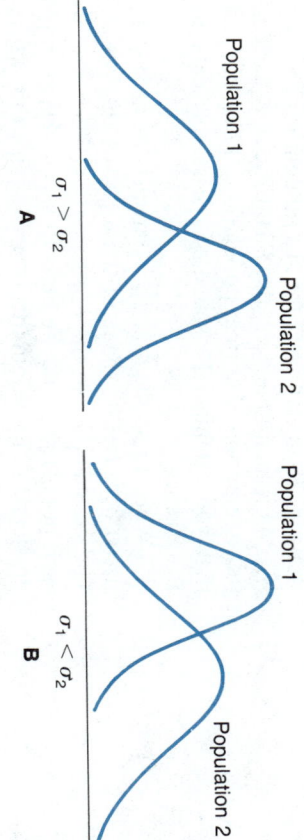

FIGURE 9.10
Unequal population variances.

Population 1   Population 2   Population 1   Population 2

$\sigma_1 > \sigma_2$ $\sigma_1 < \sigma_2$

A   **B**

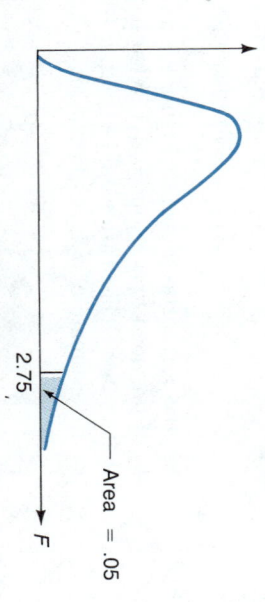

FIGURE 9.11
F curve with 10 and 12 df for probability that F exceeds 2.75 (2.75 is from Table A.7b).

Area = .05

2.75   F

HYPOTHESIS TESTS FOR σ_1 AND σ_2

TWO-TAILED TEST

$$H_0: \sigma_1 = \sigma_2$$
$$H_a: \sigma_1 \neq \sigma_2$$

$$F = \frac{\text{larger of } s_1^2 \text{ and } s_2^2}{\text{smaller of } s_1^2 \text{ and } s_2^2}$$

reject H_0 if $F > F_{\alpha/2, v_1, v_2}$

where

$$v_1 = \begin{cases} n_1 - 1 & \text{if } s_1^2 \geq s_2^2 \\ n_2 - 1 & \text{if } s_1^2 < s_2^2 \end{cases}$$

and

$$v_2 = \begin{cases} n_2 - 1 & \text{if } s_1^2 \geq s_2^2 \\ n_1 - 1 & \text{if } s_1^2 < s_2^2 \end{cases}$$

ONE-TAILED TEST

$$H_0: \sigma_1 \leq \sigma_2$$
$$H_a: \sigma_1 > \sigma_2$$

$$F = \frac{s_1^2}{s_2^2}$$

reject H_0 if $F > F_{\alpha, v_1, v_2}$

where

$$v_1 = n_1 - 1$$

and

$$v_2 = n_2 - 1$$

$$H_0: \sigma_1 \geq \sigma_2$$
$$H_a: \sigma_1 < \sigma_2$$

$$F = \frac{s_2^2}{s_1^2}$$

reject H_0 if $F > F_{\alpha, v_2, v_1}$
(be careful about order) where

$$v_2 = n_2 - 1$$

and

$$v_1 = n_1 - 1$$

EXAMPLE 9.6

The management of Case Automotive Products is considering the purchase of some new equipment that will fill 1-quart containers with a recently introduced radiator additive. They have narrowed their choice of brand of filling machine to brand 1 and brand 2. Although brand 1 is considerably less expensive than brand 2, they suspect that the contents delivered by the brand 1 machine will have *more variation* than that obtained using brand 2. In other words, brand 1 is more apt to slightly (or severely) overfill or underfill containers. (The Case people realize that they must use a container slightly larger than 1 quart in any event, to allow for heat expansion and overfill of their product.)

The Case production department was able to obtain data on the performance of both brands for a sample of 25 containers using brand 1 and 20 containers using brand 2. Using their summary information, can you confirm Case's suspicions? Use $\alpha = .05$. All mean and standard deviation measures are fluid ounces.

Brand 1	Brand 2
$n_1 = 25$	$n_2 = 20$
$\bar{x}_1 = 31.8$	$\bar{x}_2 = 32.1$
$s_1 = 1.21$	$s_2 = .72$

SOLUTION

Step 1. The purpose of the test is to determine if one standard deviation (or variance) is *larger* than the other; this calls for a one-tailed test. The suspicion is that σ_1 is larger than σ_2, so this statement is put in the alternative hypothesis. The resulting hypotheses are

$$H_0: \sigma_1 \leq \sigma_2 \qquad H_a: \sigma_1 > \sigma_2$$

Step 2. The appropriate test statistic is

$$F = \frac{s_1^2}{s_2^2}$$

Step 3. Because the df are $v_1 = 25 - 1 = 24$ and $v_2 = 20 - 1 = 19$, we find $F_{.05, 24, 19} = 2.11$. The test of H_0 versus H_a will be to

reject H_0 if $F > 2.11$

Step 4. The computed F-value is

$$F^* = \frac{(1.21)^2}{(.72)^2} = 2.82$$

Because $2.82 > 2.11$, we reject H_0.

Step 5. On the basis of these data and this significance level, Case is correct in its belief that the variation in the containers filled by brand 1 exceeds that of the containers filled by brand 2. ■

EXAMPLE 9.7

Using the blowout times data from Example 9.4, examine the variances *only*. Can you conclude that there is a difference in the two population variances, using a significance level of .05?

SOLUTION

Step 1. We are trying to detect a *difference* in the two variances (not whether one exceeds the other); a two-tailed test should be used. We define

$$H_0: \sigma_1 = \sigma_2 \text{ (or } \sigma_1^2 = \sigma_2^2\text{)}$$
$$H_a: \sigma_1 \neq \sigma_2 \text{ (or } \sigma_1^2 \neq \sigma_2^2\text{)}$$

Step 2. The test statistic here is

$$F = \frac{\text{larger of } s_1^2 \text{ and } s_2^2}{\text{smaller of } s_1^2 \text{ and } s_2^2}$$

From these data, we determined that $s_1 = .68$ hour and $s_2 = .38$ hour, with $n_1 = n_2 = 15$.

Step 3. We need $F_{.025, 14, 14}$ from Table A.7(c). Unfortunately, it is not there; this table contains only selected df. In this situation, we pick the nearest df, which, for this example, is $F_{.025, 15, 14} = 2.95$. So our test of H_0 versus H_a is to

reject H_0 if $F > 2.95$

Step 4. Since $s_1^2 = (.68)^2$ is larger than $s_2^2 = (.38)^2$, the computed value of the test statistic is

$$F^* = \frac{(.68)^2}{(.38)^2} = 3.20 > 2.95$$

Therefore, reject H_0 in favor of H_a.

Step 5. There *is* sufficient evidence to conclude that the two variances are unequal. If additional data are obtained to test the population *means*

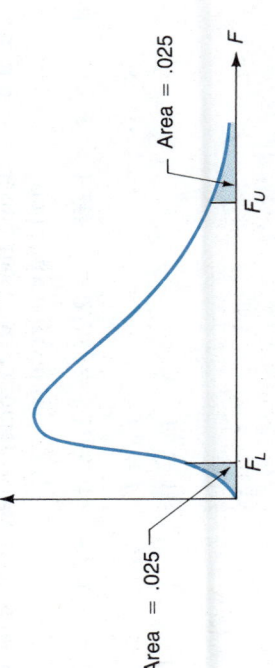

FIGURE 9.12

F curve with v_1 and v_2 df showing F values used for a 95% confidence interval.

(such as $H_0: \mu_1 \geq \mu_2$ versus $H_a: \mu_1 < \mu_2$) using small samples, the correct procedure would be to use the t' statistic described earlier, which does not assume that σ_1 and σ_2 are equal, provided both populations are believed to be normally distributed. ■

Confidence Interval for σ_1^2/σ_2^2

Consider an F curve with v_1 and v_2 df. To construct a 95% confidence interval for σ_1^2/σ_2^2, you first need to find the values of F_L and F_U, where (Figure 9.12)

$$F_L \text{ has a left-tail area} = .025$$
$$F_U \text{ has a right-tail area} = .025$$

F_U can be found directly from Table A.7(c). It is F_L that poses a problem, however, because Table A.7 contains only right-tail areas and the F distribution is *not symmetric*. However,

$$F_L = \frac{1}{F_{.025, v_2, v_1}} \qquad (9.19)$$

Notice that we *switched* the df used when finding F_U because F_U can be written

$$F_U = F_{.025, v_1, v_2}$$

The confidence interval for σ_1^2/σ_2^2 is then

$$\frac{s_1^2/s_2^2}{F_U} \quad \text{to} \quad \frac{s_1^2/s_2^2}{F_L}$$

In general, we have a $(1 - \alpha) \cdot 100\%$ confidence interval for σ_1^2/σ_2^2 (independent samples):

$$\frac{s_1^2/s_2^2}{F_U} \quad \text{to} \quad \frac{s_1^2/s_2^2}{F_L} \qquad (9.20)$$

where

$$F_U = F_{\alpha/2, v_1, v_2}$$
$$F_L = 1/F_{\alpha/2, v_2, v_1}$$
$$v_1 = n_1 - 1$$
$$v_2 = n_2 - 1$$

EXAMPLE 9.8

Using the Case Automotive Products data in Example 9.6, determine a 95% confidence interval for σ_1^2/σ_2^2.

SOLUTION

Here, $n_1 = 25$, $s_1 = 1.21$, and $n_2 = 20$, $s_2 = .72$. So we need

$$F_U = F_{.025, 24, 19} = 2.45$$
$$F_L = 1/F_{.025, 19, 24}$$
$$\cong 1/2.33 \qquad (\text{using } F_{.025, 20, 24})$$
$$= .43$$

The 95% confidence interval for σ_1^2/σ_2^2 is

$$\frac{(1.21)^2/(.72)^2}{2.45} \quad \text{to} \quad \frac{(1.21)^2/(.72)^2}{.43} = 1.15 \quad \text{to} \quad 6.57$$

As a result, we are 95% confident that σ_1^2/σ_2^2 is between 1.15 and 6.57. This means that we are 95% confident that σ_1^2 is between 1.15 and 6.57 *times as large as* σ_2^2. ■

EXAMPLE 9.9

For the Case Automotive Products data in Example 9.6, determine a 95% confidence interval for σ_1/σ_2. Use the results of Example 9.8.

SOLUTION

This is obtained simply by finding the *square root* of each endpoint of the confidence interval for σ_1^2/σ_2^2. Your 95% confidence interval for σ_1/σ_2 will be

$$\sqrt{1.15} \quad \text{to} \quad \sqrt{6.57} = 1.07 \quad \text{to} \quad 2.56 \text{ (fluid ounces)}$$ ■

EXERCISES

9.37 In evaluating capital-investment projects, the variability of the cash flows of the returns is carefully assessed. The higher the variability, the higher the risk associated with that project. Boone Enterprises is currently evaluating two projects. The mean expected net cash flow for the next 11 years for project 1 is $134,000, as against $166,000 for project 2 for the next eight years. The standard deviations of the net cash flows are $28,000 and $37,000 for project 1 and project 2, respectively. Assume that cash flows are normally distributed. Do these sample standard deviations present sufficient evidence to indicate that project 1 and project 2 are not equally risky? Use a significance level of 5%.

9.38 A computer program generates data that are approximately normal. Two independent samples are generated:

SAMPLE 1		SAMPLE 2	
17	26	24	29
25	34	32	34
30	19	29	26
28	25	25	30

Using a significance level of .05, would you reject the null hypothesis that the variance of the population from which Sample 1 was taken is less than or equal to the variance of the population from which Sample 2 was taken?

9.39 Construct a 95% confidence interval for the ratio of the population variances in Exercise 9.38.

9.40 The Water Pollution Prevention Council (WPPC) had recommended that the discharge of industrial waste and effluents into rivers in the district should be done at a slow and steady rate of 100 pound/hour (i.e., 2400 pound/day). Industrial plants tended to concentrate their effluent discharge activity in the night shift. The WPPC found that although companies might technically achieve an average discharge rate of 2400 pound/day, the rivers could not cope with the erratic rate of discharge. The effluents needed to be released throughout the day, rather than all at night. The following was recommended: The true variance of the discharge rate should not exceed 600. The following results were obtained from samples of 21 observations each.

factory A: variance 585
factory B: variance 618

At a 5% significance level, is there sufficient evidence to conclude that the variance for factory A is less than that for factory B?

9.41 The transportation of frozen food by trucks requires that the temperature be maintained within a narrow range. If the temperature is kept too low, extra fuel is consumed and unnecessary costs are incurred. If the temperature is too high, health standards would be violated, and there is the danger of food spoilage and bacterial contamination. Two models of refrigeration units were being compared for their variation from a given temperature setting. A special sensor attached to the units took readings every hour. The variance of temperatures for model 1 was $1.44°^2$C. The variance for model 2 was $2.15°^2$C. In both cases, the sample consisted of 16 readings for each model.

a. Construct the 95% confidence interval for the ratio of the two variances.

b. At a 5% significance level, is the variation of temperature in model 2 significantly greater than that for model 1?

9.42 The statistical quality-control department of a company that manufactures wall clocks is studying the variability of two types of wall clocks that have been recently developed. Using the following information, test the hypothesis that $H_0: \sigma_1 = \sigma_2$, using a significance level of .05. Assume that the samples are taken from populations that are approximately normally distributed. Clock 1: $n = 25$, $s = 1.8$. Clock 2: $n = 21$, $s = 1.39$.

9.43 Construct a 95% confidence interval for σ_1^2/σ_2^2 using the data in Exercise 9.42.

9.44 The following is a summary of the mean annual return ($\bar{X}$) and variance (s^2) of the annual return of common stocks for three different industries. Computer industry: $n = 16$, $\bar{X} = 14.3\%$, $s^2 = 5.6$. Steel industry: $n = 9$, $\bar{X} = 8.5\%$, $s^2 = 11.2$. Oil-and-gas industry: $n = 13$, $\bar{X} = 11.8\%$, $s^2 = 16.4$. Using these data, can we conclude that computer stocks are less risky than oil and gas stocks? Use a significance level of .05. Assume that the mean annual return for the industries are approximately normally distributed.

9.45 Using the data in Exercise 9.44 test the hypothesis that the steel industry's stocks are just as risky as the computer industry's stocks. Use a significance level of .05.

9.46 The manager of a vending-machine company decided to buy one of two types of dispenser to put in her vending machines. Both dispensers claim to dispense, on the average, 6 ounces of fluid in a plastic cup when used. The amount dispensed is approximately normally distributed. However, to test this claim, the manager would first like to know whether the variability in the amount of fluid dispensed is the same for both dispensers. Using a significance level of .05, is the manager justified in using a pooled estimate of the population variance, if 16 replications on each dispenser give the following results? Dispenser 1: $s = 1.5$, $n = 16$. Dispenser 2: $s = 3.4$, $n = 16$. If so, derive the pooled estimate.

COMPARING THE MEANS OF TWO NORMAL POPULATIONS USING PAIRED SAMPLES

This final section examines the situation in which the two samples are *not* obtained independently. All discussion up to this point has assumed that the two samples *are* independent. By not independent, we mean that the corresponding elements from the two samples are *paired*. Perhaps each pair of observations corresponds to the same city, the same week, the same married couple, or even the same person. This discussion focuses on comparing the two population means for this situation where two *dependent* samples are obtained from the two populations.

When attempting to estimate or test for the difference between two population means, your first question always should be, is there any natural reason to pair the first observation from sample 1 with the first observation from sample 2, the second with the second, and so on? If there is no reason to pair these data and the samples were obtained independently, the previous methods for finding confidence intervals and testing μ_1 versus μ_2 apply. If the data were gathered such that pairing the values is necessary, then it is *extremely* important that you recognize this and treat the data in a different manner. We can still determine confidence intervals and perform a test of hypothesis, but the procedure is different.

We will assume here that we have reason to believe the populations follow *normal* distributions (as we did in Sections 9.3 and 9.4). As a result, we need not worry about large samples versus small samples because we will use the t distribution for our confidence intervals and test of hypothesis, regardless of the sample sizes. Of course, if the samples are large ($n_1 > 30$ and $n_2 > 30$), this distribution is very closely approximated by the standard normal distribution.

If you have reason to suspect that your two populations are *not* normally distributed, then one alternative is to use a nonparametric procedure—in particular, the Wilcoxon signed rank test (discussed in Chapter 18.)

Assume that the city council of a large western city is taking a close look at the number of people who visit two local museums, one displaying the history of humanity and the other containing space-exploration exhibits. They believe that the space museum is attracting more people, even though the history museum has a national reputation. If this is correct, the space museum will receive additional funding for the coming year. The following data were gathered based on the number of adult admissions (in thousands) for 12 randomly selected weeks during the past year. Sample 1 is the space museum; sample 2 is the history museum. The letter d represents the *difference* of each pair of values.

WEEK	1	2	3	4	5	6	7	8	9	10	11	12
Space museum	.6	.8	.7	1.2	1.4	2.3	3.8	4.4	1.5	1.3	1.1	.8
History museum	.5	1.0	.5	.8	1.2	2.5	2.8	3.5	1.2	1.4	.8	.6
d	.1	−.2	.2	.4	.2	−.2	1.0	.9	.3	−.1	.3	.2
d^2	.01	.04	.04	.16	.04	.04	1.0	.81	.09	.01	.09	.04

$\Sigma d^2 = .01 + .04 + \cdots + .04 = 2.37$
$\Sigma d = .1 + (-.2) + \cdots + .2 = 3.1$

Confidence Interval for μ_d Using Paired Samples

The statistic used to derive a confidence interval for μ_d and perform a test of hypothesis using *dependent* samples is

$$t_D = \frac{\bar{X}_1 - \bar{X}_2}{s_d/\sqrt{n}} = \frac{\bar{d}}{s_d/\sqrt{n}} \qquad (9.21)$$

Each pair of data values was collected during the same week, so these data values clearly need to be paired—these are dependent samples. It seems reasonable to examine the difference of the two values for each week, so these differences (d), along with the d^2 values, are also shown. We have thus reduced the problem from two sets of values to a single new set. The parameter of interest here is the **difference** of the population means, μ_d. Put another way, μ_d is the **mean of the population differences.** The following discussion will demonstrate how to use the sample differences to construct a confidence interval for μ_d or perform a test of hypothesis.

where

n = the number of pairs of observations

s_d = the standard deviation of the n differences

$$= \sqrt{\frac{\sum d^2 - (\sum d)^2/n}{n-1}}$$

df for $t_D = n - 1$

This is a t *random variable* with $n - 1$ df. Notice that the numerator of t_D is the same as before—namely, $\bar{X}_1 - \bar{X}_2$, which is also represented by $\bar{d} = \sum d/n$, the mean of the differences. The mean of the differences $\bar{d}$ always is equal to $\bar{X}_1 - \bar{X}_2$. Notice that this can help you in checking your arithmetic when computing the d's.

Using this t-statistic, we obtain a $(1 - \alpha) \cdot 100\%$ confidence interval for μ_d:

$$\bar{d} - t_{\alpha/2, n-1} \frac{s_d}{\sqrt{n}} \quad \text{to} \quad \bar{d} + t_{\alpha/2, n-1} \frac{s_d}{\sqrt{n}} \qquad (9.22)$$

EXAMPLE 9.10

Using the data from the history and space museums, derive a 95% confidence interval for μ_d, where

μ_d = average weekly difference in attendance

SOLUTION

We have

$$\bar{d} = \frac{\sum d}{n} = \frac{3.1}{12} = .258$$

Notice that

$$\bar{x}_1 = \frac{.6 + .8 + \cdots + .8}{12} = 1.658$$

and

$$\bar{x}_2 = \frac{.5 + 1.0 + \cdots + .6}{12} = 1.4$$

so $\bar{d} = \bar{x}_1 - \bar{x}_2 = .258$. It checks! Also,

$$s_d = \sqrt{\frac{\sum d^2 - (\sum d)^2/n}{n-1}} = \sqrt{\frac{2.37 - (3.1)^2/12}{11}}$$

$$= \sqrt{\frac{1.569}{11}} = \sqrt{.143} = .378$$

The resulting 95% confidence interval for μ_d is

$$\bar{d} - t_{.025, 11} \frac{s_d}{\sqrt{n}} \quad \text{to} \quad \bar{d} + t_{.025, 11} \frac{s_d}{\sqrt{n}}$$

$$= .258 - 2.201 \frac{.378}{\sqrt{12}} \quad \text{to} \quad .258 + 2.201 \frac{.378}{\sqrt{12}}$$

$$= .258 - .240 \quad \text{to} \quad .258 + .240$$

$$= .018 \quad \text{to} \quad .498$$

Based on these data, we are 95% confident that the number of admissions per week to the space museum is between 18 and 498 *more* than that to the history museum.

∎

Hypothesis Testing Using Paired Samples

The test statistic for testing the means is

$$t_D = \frac{\bar{d} - D_0}{s_d / \sqrt{n}} \tag{9.23}$$

where D_0 is the hypothesized value of μ_d. When testing $H_0: \mu_d = D_0$ versus $H_a: \mu_d \neq D_0$, reject H_0 if $|t_D| > t_{\alpha/2, n-1}$. Here, $t_{\alpha/2, n-1}$ is obtained from Table A.5 using $n-1$ df. One-tailed tests are performed in a similar manner by placing α in either the right tail ($H_a: \mu_d > D_0$) or in the left tail ($H_a: \mu_d < D_0$). A summary is provided in the box on paired sample tests for μ_d and D_0 on page 341.

EXAMPLE 9.11

Consider the data on the admissions per week at the space and history museums. Can you confirm the suspicion that the average difference in attendance (space museum − history museum) is positive? Use a significance level of $\alpha = .05$.

SOLUTION

Step 1. We are attempting to demonstrate that the average difference in attendance is positive; this claim goes into the alternative hypothesis. The resulting hypotheses are

$$H_0: \mu_d \leq 0$$
$$H_a: \mu_d > 0$$

Step 2. We are dealing with paired data, so the correct test statistic is

$$t_D = \frac{\bar{d}}{s_d / \sqrt{n}}$$

What is k? As before, this depends on α and, in the usual manner, we have

$$\text{reject } H_0 \text{ if } t_D > t_{\alpha, n-1}$$

where $t_{\alpha, n-1}$ is obtained from Table A.5. For this situation, $t_{.05, 11} = 1.796$, and so we

$$\text{reject } H_0 \text{ if } t_D > 1.796$$

Step 3. What happens to t_D when H_a is true? If $\mu_d > 0$, then we would expect $\bar{d}$ to be *positive*. So, the test procedure is to

$$\text{reject } H_0 \text{ if } t_D > k, \text{ for some } k > 0$$

Step 4. Using the sample data,

$$t_D^* = \frac{.258}{.378 / \sqrt{12}} = 2.36$$

Because $2.36 > 1.796$, we reject H_0.

FIGURE 9.13

MINITAB solution to Examples 9.10 and 9.11. This is the correct way to analyze these data.

```
MTB > SET INTO C1
DATA> 0.6 0.8 0.7 1.2 1.4 2.3 3.8 4.4 1.5 1.3 1.1 0.8
DATA> END
MTB > SET INTO C2
DATA> 0.5 1.0 0.5 0.8 1.2 2.5 2.8 3.5 1.2 1.4 0.8 0.6
DATA> END
MTB > SUBTRACT C2 FROM C1, PUT INTO C3
MTB > TINTERVAL WITH 95% CONFIDENCE USING C3

        N    MEAN   STDEV  SE MEAN      95.0 PERCENT C.I.
C3      12   0.258  0.378   0.109     ( 0.018,   0.498)

MTB > TTEST OF MU=0 USING C3;
SUBC> ALTERNATIVE=1.
```

$= 1$ for $H_a: \mu_d > D_0$
$= -1$ for $H_a: \mu_d < D_0$
(not necessary for $H_a: \mu_d \neq D_0$)

```
TEST OF MU = 0.000 VS MU G.T. 0.000

        N    MEAN   STDEV  SE MEAN    T
C3      12   0.258  0.378   0.109    2.37
```

$\bar{d}$ s_d t_D^*

P VALUE
0.019

FIGURE 9.14

MINITAB solution to Example 9.11. This is the incorrect way to analyze these data. Compare with Figure 9.13.

```
MTB > TWOSAMPLE TEST USING C1 AND C2;
SUBC> ALTERNATIVE=1.

TWOSAMPLE T FOR C1 VS C2
        N    MEAN   STDEV  SE MEAN
C1      12   1.66   1.23    0.36
C2      12   1.400  0.991   0.29

95 PCT CI FOR MU C1 - MU C2: (-0.69, 1.21)

TTEST MU C1 = MU C2 (VS GT): T= 0.57  P=0.29  DF= 21
```

p-value

Step 5. The average attendance at the space museum *is* higher than that at the history museum. ■

A MINITAB solution to Example 9.11 is provided in Figure 9.13. Notice that the differences are first derived, and then a standard t test (described in section 8.4) is used to test that the mean difference μ_d is ≤ 0 versus the alternative $H_a: \mu_d > 0$. The resulting *p*-value is $p = .019$, which, using $\alpha = .05$, again results in rejecting H_0 because $p < \alpha$.

What happens if you fail to pair these observations and perform a regular two-sample t test, as we did in Section 9.3 for small *independent* samples? The results are summarized in Figure 9.14, where we observe an interesting result. The t value (using the test statistic from equation 9.10) now is .57, with a corresponding *p*-value of $p = .29$. This means that, using this test, we now *fail to reject* H_0. We are unable to demonstrate a difference between the average weekly attendances, which, according to Figure 9.13, is *not* a correct conclusion. Figure 9.14 shows convincingly that failing to pair the observations when you should can cause you to obtain an incorrect result. More importantly, there is nothing to warn you that this has occurred.

EXAMPLE 9.12

The market research staff at Allied Foods is considering two different packaging designs for an instant breakfast cereal that Allied is about to introduce. The

SOLUTION

first type of container under consideration is a rectangular box, whereas the second container type has a cylindrical shape.

They decide to conduct a pilot study by placing the product in both containers at opposite ends of the breakfast cereal section in ten different supermarkets. All the containers are placed at eye level to remove any effect due to the height of the display. The main question under consideration is whether there is any difference in sales of the two types of container. Using these data, can you conclude that there is a difference in sales for the rectangular and cylindrical containers. Use $\alpha = .05$ to define your test.

SUPERMARKET	1	2	3	4	5	6	7	8	9	10
Rectangular	194	152	160	172	118	110	137	126	176	145
Cylindrical	184	161	153	184	105	123	155	111	156	129

The data were gathered by collecting a pair of observations from each supermarket, so this is a clear-cut case of dependent sampling. Your next step should be to determine the paired differences. Define d to be the rectangular box sales minus the cylindrical box sales.

Step 1. We are attempting to detect a difference in the two means: a two-tailed test in order. Let

μ_d = average difference in sales for the two container types.

The correct hypotheses are

$$H_0: \mu_d = 0$$
$$H_a: \mu_d \neq 0$$

Steps 2, 3. Using the t_D test statistic, the test will be to

reject H_0 if $|t_D| > t_{\alpha/2, \, n-1}$

where $t_{\alpha/2, \, n-1} = t_{.025, \, 9} = 2.262$.

Step 4. Using the sample data,

d	10	-9	7	-12	13	-13	-18	15	20	16	**TOTAL** 29
d^2	100	81	49	144	169	169	324	225	400	256	1917

From these values we obtain

$$\bar{d} = \frac{\Sigma d}{n} = \frac{29}{10} = 2.9$$

$$s_d = \sqrt{\frac{\Sigma d^2 - (\Sigma d)^2/n}{n-1}}$$

$$= \sqrt{\frac{1917 - (29)^2/10}{9}}$$

$$= \sqrt{\frac{1832.9}{9}} = 14.271$$

Step 5. Based on these data, there is *insufficient evidence* to conclude that the container type has an effect upon sales.

$$t_D^* = \frac{\bar{d}}{s_d/\sqrt{n}} = \frac{2.9}{14.271/\sqrt{10}} = .643$$

Because $.643 < 2.262$, we fail to reject H_0.

PAIRED SAMPLE TESTS FOR μ_1 AND μ_2

TWO-TAILED TEST

$H_0: \mu_d = D_0$

$H_a: \mu_d \neq D_0$

reject H_0 if $|t_D| > t_{\alpha/2, n-1}$

where

1. Each difference, d, is (sample 1 value) − (sample 2 value)

2. $t_D = \dfrac{\bar{d} - D_0}{s_d/\sqrt{n}}$

3. $\bar{d} = \bar{x}_1 - \bar{x}_2 = \dfrac{\Sigma d}{n}$

4. $s_d = \sqrt{\dfrac{\Sigma d^2 - (\Sigma d)^2/n}{n - 1}}$

5. df for $t_D = n - 1$

ONE-TAILED TEST

$H_0: \mu_d \leq D_0$ $\qquad$ $H_0: \mu_d \geq D_0$

$H_a: \mu_d > D_0$ $\qquad$ $H_a: \mu_d < D_0$

reject H_0 if $t_D > t_{\alpha, n-1}$ $\qquad$ reject H_0 if $t_D < -t_{\alpha, n-1}$

EXERCISES

9.47 A hospital is experimenting with the effectiveness of a newly developed drug that controls blood pressure. The blood-pressure level is measured using a sphygmometer before and after administration of the drug to a sample of hypertensive patients with a history of elevated blood pressure. The question is whether there is a measured decrease in systolic blood pressure (in mm Hg) after administration of the drug. The difference in blood pressure before and after is believed to be approximately normally distributed.

PATIENT	BEFORE DRUG	AFTER DRUG	PATIENT	BEFORE DRUG	AFTER DRUG
1	110	94	6	82	85
2	88	81	7	96	77
3	84	82	8	97	89
4	94	88	9	134	110
5	108	97			

a. Using a significance level of .10, can you conclude that the blood-pressure level is lower after the drug is administered?

b. Should you use an independent or dependent sample t statistic to analyze this experiment?

9.48 Construct a 99% confidence interval for μ_d using the data in Exercise 9.47 ($d =$ before − after).

9.49 Suppose that in Exercise 9.47, the manufacturer of the drug claims that the new drug is effective in reducing the average blood-pressure scores by more than 8 mm Hg. Would you support that statement at a significance level of .01?

9.50 The controller of a fast-food chain is interested in determining whether there is any difference in the weekly sales of restaurant 1 and restaurant 2. The weekly sales are approximately normally distributed. The sales, in dollars, for seven randomly selected weeks are:

WEEK	RESTAURANT 1	RESTAURANT 2
1	4100	3800
2	1800	4600
3	2200	5100
4	3400	3050
5	3100	2800
6	1100	1950
7	2200	3400

a. Should this problem be analyzed using an independent or dependent samples t statistic?

b. Using a significance level of .01, is there evidence to support the conclusion that there is a significant difference in the weekly sales of the two restaurants?

9.51 Calculate the p-value for Exercise 9.50 and interpret it.

9.52 Conduct a 95% confidence interval for the mean difference in the weekly sales of the two restaurants in Exercise 9.50.

9.53 Smart Look, an exercise program developed by Joni Beauty consultants, is claimed to be effective in reducing the weight of a typical overweight woman by more than 17 pounds. In order to examine the validity of this hypothesis, the program was tried on a group of middle-aged women, and their weights (in pounds) were recorded before and after completion of the exercise program. Assume that the difference in a woman's weight after the program is approximately normally distributed.

WOMAN	BEFORE	AFTER	WOMAN	BEFORE	AFTER
1	140	115	5	175	165
2	160	130	6	145	125
3	110	100	7	115	101
4	132	109	8	122	105

a. Using a significance level of .10, what would be your conclusion?

b. Why did you select the particular test statistic you used to analyze this problem?

9.54 Using the data from Exercise 9.53, test the hypothesis that $H_0: \mu_d = 0$ versus $H_a: \mu_d \neq 0$ at a .01 significance level (d = before − after).

9.55 Ten commonly bought automotive parts that are available at both a dealership and at a local automotive parts shop are randomly selected to determine if the mean price at the dealership is significantly higher than the mean price at the local shop. Prices are as follows:

AUTO PARTS	PRICE AT LOCAL SHOP (IN DOLLARS)	PRICE AT DEALERSHIP (IN DOLLARS)	AUTO PARTS	PRICE AT LOCAL SHOP (IN DOLLARS)	PRICE AT DEALERSHIP (IN DOLLARS)
1	31	38	6	45	43
2	25	44	7	21	26
3	36	47	8	69	64
4	51	56	9	58	73
5	42	53	10	34	33

a. At the .01 significance level, do the data support the belief that the mean price at the dealership is higher than the mean price at the local automotive parts shop?

b. What assumptions are necessary to ensure the validity of the test procedure?

9.56 Some researchers on stock security returns believe that there is a "neglected firm effect" in terms of superior performance for less researched companies. A study was conducted to examine the effect of visibility on security returns. Visibility of a company may result from either advertising or media exposure. In this study, highly visible firms were paired with less visible firms. For example, Gillette Company (highly visible) was paired with Avon Products, Inc. (less visible). Random samples of pairs of companies were chosen over 3 time periods. The results are given in the following table where n represents the sample size and $\bar{d}$ represents the mean difference (with d equal to return on highly visible company minus return on less visible company).

TEST RESULTS

STUDY PERIOD	n	$\bar{d}$	MEAN STD ERROR	t
1	27	-.0146	.0058	-2.52
2	20	-.0154	.0057	-2.71
3	29	.0086	.0053	1.62

(*Source:* Robert Johnson and Gerald Jension, "An Empirical Examination of the Effect of Visibility on Security Returns," *Proceedings of the Annual Meeting of the Decision Sciences Institute* (1986): 218–20.)

At the .05 significance level, for which study period is there sufficient evidence to conclude that less visible firms tend to exhibit higher rates of returns than the highly visible firms?

SUMMARY

This chapter has presented an introduction to **statistical inference for two populations**. We examined tests of hypothesis and confidence intervals for the means and variances (for example, whether they are equal) of the two populations, using both independent and dependent samples.

When we used large **independent samples** to test the population means, we defined a test statistic having an approximate standard normal distribution, which we also used to define a confidence interval for $\mu_1 - \mu_2$. For small independent samples ($n_1 < 30$ or $n_2 < 30$), hypothesis testing on μ_1 versus μ_2 is concerned with means from two normal populations. For this situation, although we are concerned with the means, we must pay special attention to whether we also have reason to believe that the population standard deviations (σ_1 and σ_2) are equal.

If we do not assume that the σ values are equal, we use a test statistic for μ_1 versus μ_2 having an *approximate t* distribution. This statistic also results in an approximate confidence interval for $\mu_1 - \mu_2$. If we assume that the σ values are equal, we use a procedure that pools the sample variances and results in a test statistic having an *exact t* distribution. We also derived confidence intervals for $\mu_1 - \mu_2$ for this situation.

To determine whether two population variances (or standard deviations) are the same, we introduced the **F distribution**. This distribution is nonsymmetric (right skew) and assumes that two independent samples were obtained from normal populations. Probabilities (areas under the curve) for the F random variable are contained in Table A.7. Using this distribution, we can perform two-tailed tests (such as $H_a: \sigma_1 \neq \sigma_2$) or one-tailed tests (such as $H_a: \sigma_1 > \sigma_2$) on the two standard deviations. We also use it to construct a confidence interval for σ_1^2/σ_2^2 or σ_1/σ_2.

When two samples are obtained such that corresponding observations are paired (matched), the resulting samples are **dependent** or **paired**. When using two such samples, we defined a *t*-statistic to test the mean of the population differences, μ_d, and to construct a confidence interval for μ_d. We need not be concerned about whether the population standard deviations are equal for this situation because the test statistic uses the differences between the paired observations, a new variable.

REVIEW EXERCISES

9.57 To evaluate the expected life of two types of tires, a car manufacturer decided to use a randomly selected set of 20 similar cars for testing the mean difference in the amount of wear (in thousandths of an inch) for the two brands of tires after 10,000 miles. The manufacturer placed two tires of the first brand and two tires of the second brand on each car. Will the resulting samples be independent samples or dependent samples? Discuss.

9.58 A sandwich shop wishes to test the effectiveness of its coupons. The manager believes that the business brought in by the responses to the coupon in the *Highland Village Daily* is equal to the business brought in by the responses to the coupon placed in the *Green Sheet*. The amount spent by each customer using a coupon is recorded (in dollars) and can be considered to be normally distributed. Test the manager's belief with a significance level of .01. *Highland Village Daily*: $n = 32$, $\bar{x} = 9.50$, $s = 26.3$. *Green Sheet*: $n = 39$, $\bar{x} = 11.80$, $s = 29.4$.

9.59 Dairy Castle would like to boost the sales of their "Country Baskets." They think that it might be helpful to hang posters that picture the item. They recorded the number of Country Baskets sold during lunch time for one week at its various stores. They repeated the sampling for another week when the poster advertising was used. Assume that weekly sales are normally distributed. Is there sufficient evidence to say that hanging the posters improved sales of the Country Baskets? Use a .05 significance level.

STORE	BEFORE	AFTER	STORE	BEFORE	AFTER
1218	215	240	1270	201	220
1224	180	220	1282	207	215
1236	150	190	1292	195	219
1252	180	175	1304	180	195

9.60 Denver Hydro-Mulch Company helps lawns grow by spraying a prepared mixture on top of each lawn. A chemical company sales representative would like to convince Denver Hydro-Mulch that her company has a better fertilizer mixture. She has agreed to give the company enough fertilizer mixture to spray on eight randomly selected lawns. An additional set of eight randomly selected lawns are sprayed with the fertilizer mix that the company currently is using. At the end of 4 weeks, the eight lawns prepared with the new mixture had an average growth of 32 cm and a standard deviation of 7.8 cm. The eight lawns sprayed with the fertilizer mixture that the company is currently using had an average growth of 25 cm and a standard deviation of 6 cm. The growth of the grass at the end of the 4 week period is considered to be normally distributed. Test the claim that the new fertilizer mixture is superior to the current one. Use a 10% significance level. Do not assume that the population variances are equal.

9.61 The suggested 1988 retail prices on a Ford Escort and on a Chevrolet Cavalier are $6895 and $7395, respectively, according to *U.S. News and World Report* (April 1988, p. 83). A manager at a Ford automotive dealership believes that the mean advertised price of the Chevrolet Cavalier exceeds the mean advertised price of the Ford Escort at automotive dealerships in the Chicago metropolitan area. Assume that a random sample of seven Ford dealerships yielded a mean of $6291 with a sample standard deviation of $300. Also assume that a random sample of seven Chevrolet dealerships yielded a mean of $6981 with a sample standard deviation of $380. At the .10 level of significance, is there sufficient evidence to conclude that the advertised price of the Chevrolet Cavalier exceeds the advertised price of the Ford Escort by $500? Assume equal population variances, and that the data come from populations that are normally distributed.

9.62 An insurance company wants to compare the amount of damage (dollar value) from a rear-end collision of cars equipped with 5-mph bumpers to those equipped with 1983 2.5-mph bumpers. Twenty cars are tested, 10 with the 5-mph bumpers and the other 10 with 2.5-mph bumpers. The cars are put through 10 different tests. Is there sufficient evidence to indicate a significant difference in the dollar value of damage between the two types of bumpers? Use a .05 significance level. What assumption needs to be made about the distribution of the damage from a rear-end collision?

COLLISION	2.5-MPH BUMPER	5.0-MPH BUMPER	COLLISION	2.5-MPH BUMPER	5.0-MPH BUMPER
1	750	435	6	650	700
2	675	600	7	575	405
3	825	739	8	450	350
4	439	325	9	580	470
5	980	650	10	625	485

9.63 Researchers at Stanton University have said that while older workers were often more productive than their younger counterparts, supervisors tended to rate the older workers lower. Suppose we have the following data after sampling at random the ratings of 8 older workers and 8 younger workers.

RATINGS

Younger Workers	82	91	50	82	75	89	90	76
Older Workers	77	80	65	80	70	83	79	75

Do the data support the contention the mean rating for younger workers is higher than the mean rating for older workers? Assume the data came from normally distributed populations with equal population variances. Use a significance level of .05.

9.64 A financial analyst measures the risk in investing in a particular type of mutual fund by the variance in the rate of return for funds with similar goals. The distribution of the rate of return for the funds is considered to be approximately normally distributed. Based on the following data, in which 31 maximum capital gain funds and 41 long-term growth funds were sampled, can you conclude that the long-term growth funds are riskier at the .05 level of significance? Maximum capital gains: $n = 31$, $s = 112$. Long-term growth: $n = 41$, $s = 209$.

9.65 A high rate of worker turnover is a major problem in the food-service industry. Although this is widely accepted as an unavoidable fact of life, Mok and Finley (1986) argue that high turnover may be related to factors over which management does have control. They investigated the relationship between job satisfaction and labor turnover at three first-class chain hotels in Hong Kong. A fairly reliable instrument called the Job Descriptive Index (JDI) was used to measure job satisfaction on five components. They first surveyed 373 food-service workers at three hotels. Six months later, turnover data were obtained from the hotels. They found that 39 workers had voluntarily left their jobs. For each of these 39 terminators, they selected 2 stayers who were matched for age, marital status, sex, and length of employment. The JDI component mean scores for the 39 terminators and 78 stayers are as follows:

JDI MEAN SATISFACTION SCORES OF TERMINATORS AND STAYERS

JDI VARIABLE	TERMINATORS ($n = 39$)		STAYERS ($n = 78$)		t-TEST STATISTIC
	Mean	S.D.	Mean	S.D.	
Work	21.10	9.67	25.88	10.51	2.38
Supervision	28.07	11.83	32.98	11.45	2.16
Pay	11.15	5.73	11.17	5.63	0.02
Promotion	8.64	6.42	10.89	6.55	1.77
Co-Workers	30.20	13.79	35.43	11.21	2.20

(*Source:* C. Mok and D. Finley, Job Satisfaction and Its Relationship to Demographics and Turnover of Hotel Food-service Workers in Hong Kong, *International Journal Hospitality Management* 5, no. 2 (1986): 71–8.

a. Using the preceding t-statistic (based on the pooled variance method) and $\alpha = .05$, compare satisfaction scores of terminators versus stayers on each of the five JDI components. Is there a significant difference?

b. Find a 95% confidence interval for the mean difference on each of the five JDI components for the two groups.

9.66 Determine which of the following sets of hypotheses are equivalent.

a. $H_0: \sigma_1^2/\sigma_2^2 \leq 1$ and $H_a: \sigma_1^2/\sigma_2^2 > 1$.

b. $H_0: \sigma_2^2/\sigma_1^2 \geq 1$ and $H_a: \sigma_2^2/\sigma_1^2 < 1$.

c. $H_0: \sigma_2^2 \geq \sigma_1^2$ and $H_a: \sigma_2^2 < \sigma_1^2$.

d. $H_0: \sigma_1 \leq \sigma_2$ and $H_a: \sigma_1 > \sigma_2$.

9.67 A study is designed to determine the effect of an office-training course on typing productivity. Ten typists are randomly selected and are asked to type 15 pages of equally difficult text before and after completing the training course. Their productivity is measured by the total number of errors made.

TYPIST	BEFORE	AFTER	TYPIST	BEFORE	AFTER
1	30	27	6	33	31
2	19	14	7	28	22
3	36	31	8	30	25
4	42	37	9	27	30
5	35	29	10	34	33

Assume that the total number of errors can be approximated by a normal distribution. Test the claim that taking the office-training course leads to a reduction in the average number of errors made by a typist. Use a significance level of .05.

9.68 Suppose that a sample of size 16 is chosen from population 1 and a sample of size 26 is drawn from population 2. Assume that both populations are normally distributed. If a 90% confidence interval for the ratio of the variance of population 1 to the variance of population 2 is .367 to 1.753, what is the point estimate of the ratio of the two population variances?

9.69 A number of institutions of higher education around the country televise engineering education. In 1983, the Virginia legislature created a cooperative graduate engineering program, wherein engineering classes were broadcast live from special classrooms at the University of Virginia and other institutions via the public television network. The on-campus professor had regular campus students in front of him but also televised lectures to students at remote locations. The remote classroom was equipped with TV receivers and audio facilities, which permitted remote students to ask questions and participate in discussions. The same admission standards and examinations were applied to on-campus and off-campus (TV) students. An evaluation committee, representing participating institutions and the State Department of Information Technology, was formed to assess the program. One basic question raised was: Do the TV students perform academically on a par with their on-campus counterparts? Wergin et al. report the following results for the 1983–1984 academic year:

GRADE POINT AVERAGE, ON-CAMPUS VS. TV STUDENTS, BY COURSE AND TERM*

COURSE	FALL			SPRING		
	n	x̄	t	n	x̄	t
Electrical Engg.						
On-campus	36	3.40	1.94	13	3.41	2.30
TV	19	3.07		9	2.59	
Industrial Engg.						
On-campus	13	3.49	.16	11	3.60	3.10
TV	18	3.45		12	2.79	
Civil Engg.						
On-campus	10	3.20	.19	14	3.36	.17
TV	12	3.11		9	3.39	
Material Science						
On-campus	14	3.59	1.12	11	3.39	6.00
TV	17	3.40		18	2.95	

* n = number of students, x̄ = mean GPA, t = t-test results, pooled sample.
(Source: J. F. Wergin, D. Boland, and T. W. Hass. Televising Graduate Engineering Courses: Results of an Instructional Experiment. Engineering Education 110 (November 1986).)

REVIEW EXERCISES

a. In the fall term, was there a significant difference in course grades between on-campus and TV students, for any of the four engineering courses offered? Use a .05 significance level.

b. In the spring term, did on-campus students perform significantly better than TV students? Use a .05 significance level.

c. Without conducting any formal tests, but just by examining the figures in the table and taking into account (a) and (b), comment on whether you would agree with the statement: The trend toward increasing differences in GPA between on-campus and TV students during the year was due primarily to the declining performance of TV students (except in civil engineering) compared to the relatively steady performance of on-campus students throughout the year.

9.70 The sales of two Stop-N-Go convenience stores are compared for 12 randomly selected weeks. The MINITAB computer printout gives a printout of store 1 in C1 and store 2 in C2. The units are in thousands of dollars.

```
MTB > print c1
C1
  16.8   17.3   18.2   17.0   16.2   16.8   15.7   18.3   17.5
  15.0   16.3   17.3

MTB > print c2
C2
  17.2   17.1   18.5   17.9   15.7   17.7   15.9   19.7   18.2
  15.1   16.2   17.7

MTB > let c3 = c2 - c1
MTB > print c3
C3
  0.40000  -0.20000   0.30000   0.90000  -0.50000   0.90000
  0.20000   1.40000   0.70000   0.10000  -0.10000   0.40000

MTB > ttest of mu = 0 using c3;
SUBC> alternative = 0.

TEST OF MU = 0.000 VS MU N.E. 0.000

         N      MEAN     STDEV    SE MEAN       T    P VALUE
C3      12     0.375     0.534      0.154     2.43      0.033
```

a. Is there sufficient evidence to conclude that the two convenience stores' sales differ significantly? Use the p-value to justify your conclusion.

b. Construct a 99% confidence interval on the difference in the mean sales for the two stores. Interpret the confidence interval.

c. What assumptions are necessary for the test procedure in part (a) to be valid?

9.71 Twelve stock market analysts randomly selected from the Marty Sinch brokerage firm and ten stock market analysts randomly selected from the E. P. Sutton brokerage firm were asked to forecast the percentage change in Standard and Poor's 500 index after one year. The MINITAB computer printout gives several confidence intervals comparing the means of the two groups. The values of the random sample from the Marty Sinch brokerage firm are in column C1 and the values of the random sample from the E. P. Sutton brokerage firm are in C2. The figures are in percentages.

```
MTB > PRINT C1
SINCH
   3.4   10.5   20.6   18.5   11.3   12.6   14.7   10.9    9.6
  15.7   25.6    9.5

MTB > PRINT C2
SUTTON
  16.5   11.8   17.3   18.8   11.6    6.1   12.6   22.8   20.6
  14.8

MTB > TWOSAMPLE TEST WITH 99% CONFIDENCE USING C1 AND C2;
SUBC> ALTERNATIVE = 0;
SUBC> POOL.
```

```
TWOSAMPLE T FOR SINCH VS SUTTON
          N    MEAN    STDEV   SE MEAN
SINCH     12   13.57   5.90    1.7
SUTTON    10   15.29   4.95    1.6

MTB > TWOSAMPLE TEST WITH 95% CONFIDENCE USING C1 AND C2;
SUBC> ALTERNATIVE = 0;
SUBC> POOL.

99 PCT CI FOR MU SINCH - MU SUTTON: (-8.4, 5.0)
TTEST MU SINCH = MU SUTTON (VS NE): T=-0.73 P=0.47 DF=20.0
```

```
TWOSAMPLE T FOR SINCH VS SUTTON
          N    MEAN    STDEV   SE MEAN
SINCH     12   13.57   5.90    1.7
SUTTON    10   15.29   4.95    1.6

MTB > TWOSAMPLE TEST WITH 90% CONFIDENCE USING C1 AND C2;
SUBC> ALTERNATIVE = 0;
SUBC> POOL.

95 PCT CI FOR MU SINCH - MU SUTTON: (-6.6, 3.2)
TTEST MU SINCH = MU SUTTON (VS NE): T=-0.73 P=0.47 DF=20.0
```

```
TWOSAMPLE T FOR SINCH VS SUTTON
          N    MEAN    STDEV   SE MEAN
SINCH     12   13.57   5.90    1.7
SUTTON    10   15.29   4.95    1.6

90 PCT CI FOR MU SINCH - MU SUTTON: (-5.8, 2.3)
TTEST MU SINCH = MU SUTTON (VS NE): T=-0.73 P=0.47 DF=20.0
```

a. Interpret the three confidence intervals. Explain why the width of the confidence intervals changes for each of the confidence levels?

b. What are the necessary assumptions for the confidence intervals to be valid?

9.72 A manager at a manufacturing plant was interested in whether the job proficiency of workers with exactly 3 years' experience was significantly different from the job proficiency of workers with exactly 5 years' experience. A random sample of 12 workers is selected from each of these two groups. A job proficiency test is administered to each worker and a score obtained. A MINITAB printout gives an analysis of the data. The printout shows the results of pooling and not pooling in testing that $H_0: \mu_1 - \mu_2 = 0$ versus $H_a: \mu_1 - \mu_2 \neq 0$.

```
MTB > name c1 '3 years'
MTB > name c2 '5 years'
MTB > print c1

3 years
92   64   79   51   95   83   76   87   58   86
68   88

MTB > print c2

5 years
85   86   89   82   91   83   86   83   87   85
90   84

MTB > twosample test with 95% confidence using c1 and c2;
SUBC> alternative = 0.

TWOSAMPLE T FOR 3 years VS 5 years
           N    MEAN    STDEV   SE MEAN
3 years    12   77.3    14.1    4.1
5 years    12   85.92   2.87    0.83

95 PCT CI FOR MU 3 years - MU 5 years: (-17.8, 0.46)
TTEST MU 3 years = MU 5 years (VS NE): T=-2.09 P=0.061 DF=11.9

MTB > twosample test with 95% confidence using c1 and c2;
SUBC> alternative = 0;
SUBC> pool.

TWOSAMPLE T FOR 3 years VS 5 years
           N    MEAN    STDEV   SE MEAN
3 years    12   77.3    14.1    4.1
5 years    12   85.92   2.87    0.83

95 PCT CI FOR MU 3 years - MU 5 years: (-17.3, -0.07)
TTEST MU 3 years = MU 5 years (VS NE): T=-2.09 P=0.048 DF=22.0
```

a. What conclusion would one make for the test procedure with pooling and without pooling at the .05 significance level?

b. Intuitively does it appear that the standard deviations of the two groups differ significantly? At the .05 significance level, do the data provide sufficient evidence to conclude that the population standard deviation for the workers with 3 years' experience differs from the population standard deviation of the workers with 5 years' experience? What conclusion would you draw from the MINITAB printout?

COMPUTER EXERCISES USING THE DATABASE

Exercise 1—Appendix H

Choose at random ten observations from the database in which the family owns their home and ten observations in which the family rents their home. (Refer to the variable OWNORENT.) Do the data support the conclusion that the home payment for homeowners is larger than the home payment for renters? Use a .05 significance level. What assumptions are necessary to ensure that the test procedure is valid? Assume unequal population variances.

Exercise 2—Appendix H

Choose at random ten observations from the database from a family of size 2 and ten observations from a family of size 4. Do the data support the conclusion that the monthly utility expenditure (variable UTILITY) is larger for a family of size 4? Use a .05 significance level. Assume unequal population variances.

Exercise 3—Appendix I

Choose a random sample of 12 companies from the database with an A bond rating and another random sample of 12 companies with a C bond rating. Do the data support the conclusion that the net income of companies with a C bond rating is less than the net income of companies with an A rating? Use a .05 significance level. Assume unequal population variances.

Exercise 4—Appendix I

Choose a random sample of 15 companies from the database with a B bond rating and another random sample of 15 companies with a C bond rating. Do the data support the conclusion that the variances of the current assets of the companies with B bond ratings and C bond ratings differ significantly at the .05 level?

passive uses of statistics, analogous to a physician listening to a patient. Besides being a good listener, a doctor must also actively question a patient and perform necessary tests to diagnose and improve the patient's condition. Similarly, quality managers must also interrogate their procedures with the *active* use of statistics in experiments with process variables.

Hunter makes an important point:

"Statistics is much more than a collection of tools and computing protocols . . . Statistics is a lan-

CASE STUDY

QUALITY
IMPROVEMENT
THROUGH
PROCESS
EXPERIMENTATION

Evolutionary operation (EVOP) is a formalized statistical procedure for production process experimentation introduced by G. E. P. Box. Standard statistical methods of experimental design are applied to the task of eliciting information about quality improvement in production processes, with floor personnel acting as experimenters.

Hunter (1985) sees it as an issue of active versus passive statistics. Inspection methods, analysis of past data, and control charts, according to Hunter, are all

guage ... for ... quantitative concepts and ideas. Managers insist that their staff speak and write succinctly. Numerical literacy ... is no less important. *Since most quality problems are characterized in numerical terms, their description and analysis requires ... the language of statistics.* The ability to employ both the language and tools of statistics on behalf of product quality and productivity will identify tomorrow's "quality" manager. [Emphasis added.] (*Source:* Hunter, 1985, p. 14.)

An example where the active use of statistics is useful is the study of the influence of some factor on a process variable. It is necessary to distinguish between two types of factors: the *studied factor* whose influence one wishes to study, and the *blocking factor*, which is an

extraneous distorting influence that is present and needs to be accounted for. (A later chapter will cover this topic in greater depth.)

Two different types of NO_x (nitrous oxide) sensors are being compared. Ten sensors of type A and 10 of type B are placed in the exhaust system of twenty different automobiles. The resulting measurements of emissions of NO_x in parts per million (ppm) are displayed in Table 9.1. No attempt is made to account for the influence of differences in the combustion systems of different cars.

In a second experiment, each of the 10 cars is equipped with both sensors. To dramatize the difference between the two experiments, suppose that we obtain the difference shown in Table 9.2.

TABLE 9.1 Comparing Two Sensors: NO_x Sensor Experiment I (NO_x in ppm)

NO_x sensor experiment I
NO_x in ppm

Type A	Type B
72.1	74.0
68.2	68.8
70.9	71.2
74.3	74.2
70.7	71.8
66.6	66.4
69.5	69.8
70.8	71.3
68.8	69.3
73.3	73.6

$n_A = 10$ $n_B = 10$

$\bar{x}_A = 70.63$ $\bar{x}_B = 71.04$

$\bar{x}_A - \bar{x}_B = -0.41$

NO_x (ppm)

65   70   75

o = Type A
x = Type B

(*Source:* J. Stuart Hunter, "The Technology of Quality," *RCA Engineer* 30, no. 3 (May/June 1985): 8–15. Copyright RCA Corporation.)

TABLE 9.2 Comparing Two Sensors with Blocking: NO_x Sensor Experiment II (NO_x in ppm)

NO_x sensor experiment II
NO_x in ppm

Car	Type A	Type B	A − B
1	72.1	74.0	−0.8
2	68.2	68.8	−0.6
3	70.9	71.2	−0.3
4	74.3	74.2	+0.1
5	70.7	71.8	−1.1
6	66.6	66.4	+0.2
7	69.5	71.8	−0.3
8	70.8	71.3	−0.5
9	68.8	69.8	−0.5
10	73.3	73.6	−0.3

n = 10 pairs

NO_x (ppm)

65   70   75

o = Type A
x = Type B

Case Study Questions

1. Using Table 9.1, perform a hypothesis test at a .05 significance level to decide if there is a difference between the two types of NO_x sensors. (Assume equal variances for the two groups.)

2. Repeat the above test for the data in Table 9.2.

3. Could these two statistical tests arrive at different conclusions, even with identical data? Explain how.

4. Why is the experiment shown in Table 9.1 badly designed? How does the use of paired data improve the second experiment?

5. Identify the studied factor and the blocking factor.

6. State the p-value for the test in question 2. What is the usefulness of knowing this value over a prespecified significance level?

▦ ▦ ▦ ▦ ▦

MAINFRAME AND MICRO

SOLUTION

EXAMPLE 9.4

Example 9.4 was concerned with a t-test for two independent samples to compare two population means. The problem was to determine if the average blowout time using the test apparatus was different for BELTEX and ROADMASTER tires. (H_0: $\mu_1 = \mu_2$)

The SPSS program listing in Figure 9.15 was used to test the variances and means for two independent samples. For the test on the means, the value of the test statistic and the p-value are provided for the case where the population variances are assumed equal and the case where it is not assumed that the variances are equal.

In this problem the SPSS commands are the same for both the mainframe and PC versions. **(Remember to end each command line with a period when using the PC version.)**

The TITLE command names the SPSS run.

The DATA LIST command gives each variable a name and describes the data as being in free form.

The BEGIN DATA command indicates to SPSS that the input data immediately follow.

The next 30 lines contain the data values, with each line representing the brand (Roadmaster or Beltex) and the wear factor. The first line, for example, represents brand B (BELTEX) and a wear factor of 3.82.

The END DATA statement indicates the end of the data.

FIGURE 9.15

Input for SPSSX or SPSS/PC+. All command lines should end in a period when the PC version is used.

```
TITLE   BELTEX-ROADMASTER TIRES
DATA LIST FREE/WEAR BRAND
BEGIN DATA
3.82 0
3.11 0
4.21 0
2.64 0
   :
3.82 1
3.85 1
3.62 1
4.88 1
END DATA
T-TEST GROUPS=BRAND(0,1)  /  VARIABLES=WEAR
```

FIGURE 9.16 SPSSX output.

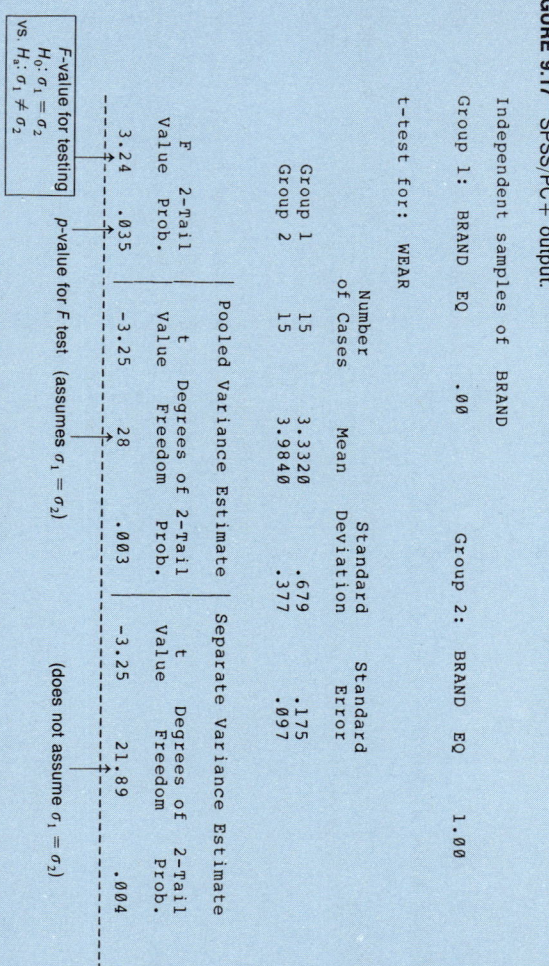

F-value for testing $H_0: \sigma_1 = \sigma_2$ vs. $H_a: \sigma_1 \neq \sigma_2$ p-value for F-test   value of t value of t'

VARIABLE WEAR

VARIABLE	NUMBER OF CASES	MEAN	STANDARD DEVIATION	STANDARD ERROR	F VALUE	2-TAIL PROB.
GROUP 1	15	3.3320	0.679	0.175	3.24	0.035
GROUP 2	15	3.9840	0.377	0.097		

	POOLED VARIANCE ESTIMATE			SEPARATE VARIANCE ESTIMATE		
	T VALUE	DEGREES OF FREEDOM	2-TAIL PROB.	T VALUE	DEGREES OF FREEDOM	2-TAIL PROB.
	-3.25	28	0.003	-3.25	21.89	0.004

assumes $\sigma_1 = \sigma_2$ does not assume $\sigma_1 = \sigma_2$

The TTEST command compares two sample means. The GROUPS and VARIABLES subcommands divide the cases into two groups for a comparison of sample means.

Figure 9.16 shows the SPSSX output obtained by executing the listing in Figure 9.15, whereas Figure 9.17 shows the SPSS/PC+ output.

FIGURE 9.17 SPSS/PC+ output.

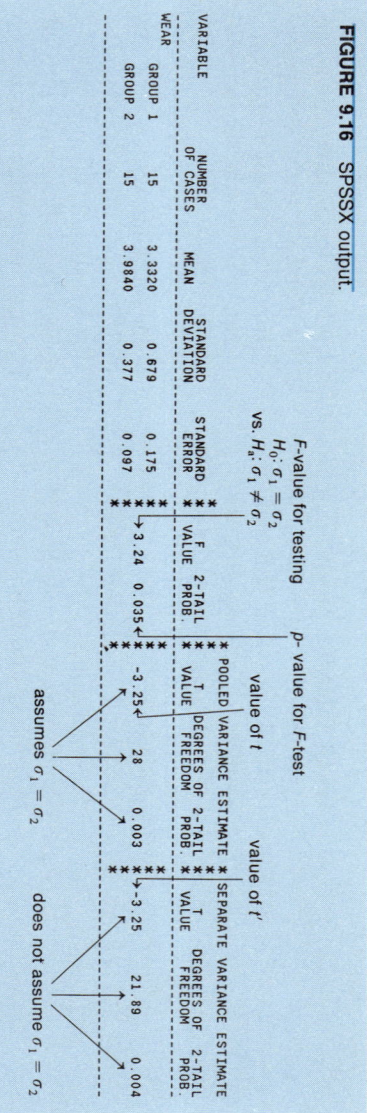

Independent samples of BRAND

Group 1: BRAND EQ .00   Group 2: BRAND EQ 1.00

t-test for: WEAR

	Number of Cases	Mean	Standard Deviation	Standard Error
Group 1	15	3.3320	.679	.175
Group 2	15	3.9840	.377	.097

		Pooled Variance Estimate			Separate Variance Estimate		
F Value	2-Tail Prob.	t Value	Degrees of Freedom	2-Tail Prob.	t Value	Degrees of Freedom	2-Tail Prob.
3.24	.035	-3.25	28	.003	-3.25	21.89	.004

F-value for testing $H_0: \sigma_1 = \sigma_2$ vs. $H_a: \sigma_1 \neq \sigma_2$

p-value for F test (assumes $\sigma_1 = \sigma_2$)   (does not assume $\sigma_1 = \sigma_2$)

MAINFRAME AND MICRO

S P S S

SOLUTION

EXAMPLE 9.11

Example 9.11 was concerned with the computation of the *t*-statistic for the means of two populations using paired (dependent) samples. The problem was to determine whether the average difference in weekly attendance (space museum — history museum) is positive. ($H_a : \mu_d > 0$).

The SPSS program listing in Figure 9.18 was used to request a mean, *t*-score and *p*-value. Note that SPSS assumes a two-tailed test. This was a one-tailed test, so the calculated *p*-value must be divided by two.

In this problem the SPSS commands are the same for both the mainframe and PC versions. (**Remember to end each command line with a period when using the PC version.**)

FIGURE 9.18

Input for SPSSX or SPSS/PC+. All command lines should end in a period when the PC version is used.

```
TITLE   MUSEUM ATTENDANCES
DATA LIST FREE/SPACE HISTORY
BEGIN DATA
0.6 0.5
0.8 1.0
0.7 0.5
1.2 0.8
1.4 1.2
2.3 2.5
3.8 2.8
4.4 3.5
1.5 1.2
1.3 1.4
1.1 0.8
0.8 0.6
END DATA
T-TEST PAIRS=SPACE HISTORY
```

The TITLE command names the SPSS run.

The DATA LIST command gives each variable a name, and describes the data as being in free form.

The BEGIN DATA command indicates to SPSS that the input data immediately follow.

The next 12 lines contain the data values, which represent the attendance in thousands at the space and history museums respectively. The first line, for example, implies that there were 600 guests at the space museum and 500 guests at the history museum.

The END DATA statement indicates the end of the data.

The T-TEST command compares two sample means. The PAIRS subcommand names the variables being compared.

Figure 9.19 shows the SPSSX output obtained by executing the listing in Figure 9.18, whereas Figure 9.20 shows the SPSS/PC+ output.

FIGURE 9.19 SPSSX output.

```
-  -  -  -  -  -  -  -  -  -  -  -  -  -  -  -  -  T - T E S T  -  -  -  -  -  -  -  -  -  -  -  -  -  -

VARIABLE  NUMBER           STANDARD  STANDARD  *(DIFFERENCE) STANDARD  STANDARD  *            2-TAIL  *  T      DEGREES OF  2-TAIL
          OF CASES  MEAN   DEVIATION  ERROR    *  MEAN       DEVIATION  ERROR    * CORR.  PROB.*  VALUE   FREEDOM     PROB.

SPACE        12    1.6583   1.235     0.356    *                                *            *    *      *       *      *
                                               *   0.2583    0.378     0.109     * 0.966  0.000 *  2.37     11       0.037
HISTORY      12    1.4000   0.991     0.286    *                                *            *    *      *       *      *
```

t_D^* ↑
for a one-tailed
test, the actual
p-value $= .037/2$
$= .019$

FIGURE 9.20 SPSS/PC+ output.

```
            MUSEUM ATTENDANCES

Paired samples t-test:  SPACE
                        HISTORY

                                    Standard   Standard
Variable   Number            Mean   Deviation  Error
           of Cases

SPACE         12           1.6583   1.235      .356
HISTORY       12           1.4000    .991      .286

(Difference) Standard  Standard  |          2-Tail   |   t     Degrees of   2-Tail
   Mean      Deviation   Error   | Corr.   Prob.     | Value   Freedom      Prob.

   .2583      .378       .109    |  .966    .000     |  2.37      11          .037
```

t_D^* ↑
for a one-tailed
test, the actual
p-value $= .037/2 = .019$

MAINFRAME AND MICRO

S A S

EXAMPLE 9.4

SOLUTION

Example 9.4 was concerned with a t-test for two independent samples to compare two population means. The problem was to determine if the average blowout time using the test apparatus was different for BELTEX and ROADMASTER tires. ($H_0: \mu_1 = \mu_2$).

The SAS program listing in Figure 9.21 was used to request the variances and the means for two independent samples. For the test on the means, the value of the test statistic and the p-value are provided for the case where the population variances are assumed equal and the case where it is not assumed that the variances are equal.

In this problem the SAS commands are the same for both the mainframe and PC versions.

The TITLE command names the SAS run (enclose in single quotes).

The DATA command gives the data a name.

The INPUT command names and gives the correct order for the different fields on the data lines. The $ implies that BRAND contains character data.

The CARDS command indicates to SAS that the input data immediately follow.

The next 30 lines contain the data values, with each line representing the brand (Roadmaster or Beltex) and the wear factor. The first line for example represents brand B and a wear factor of 3.82.

The PROC TTEST command compares the means of two groups of observations. The subcommand CLASS identifies BRAND as the variable to be classified in this example. The subcommand TITLE provides a report heading for the output.

FIGURE 9.21
Input for SAS (mainframe or micro version).

```
TITLE    'BELTEX-ROADMASTER TIRES';
DATA BLOWOUT;
INPUT BRAND $ WEAR;
CARDS;
B 3.82
B 3.82
B 3.11
B 4.21
B 2.64
   :
R 3.82
R 3.85
R 3.62
R 4.88
PROC TTEST;
CLASS BRAND;
TITLE 'INDEPENDENT SAMPLES TTEST';
```

FIGURE 9.22
SAS output.

```
                    INDEPENDENT SAMPLES TTEST

                          TTEST PROCEDURE

VARIABLE: WEAR

BRAND   N   MEAN         STD DEV      STD ERROR    MINIMUM      MAXIMUM
B       15  3.33200000   0.67915073   0.17535596   2.44000000   4.52000000
R       15  3.98400000   0.37727974   0.09741321   3.45000000   4.88000000

VARIANCES     T        DF     PROB > |T|
UNEQUAL       -3.2503   21.9   0.0037
EQUAL         -3.2503   28.0   0.0030

FOR H0: VARIANCES ARE EQUAL, F' = 3.24 WITH 14 AND 14 DF    PROB > F'  0.0354
```

F-value for testing
$H_0: \sigma_1 = \sigma_2$
vs. $H_a: \sigma_1 \neq \sigma_2$

value of t

value of t

p-values (2-tailed)

p-value for F-test

EQUAL assumes $\sigma_1 = \sigma_2$

UNEQUAL does not assume $\sigma_1 = \sigma_2$

FIGURE 9.23 SAS/PC output.

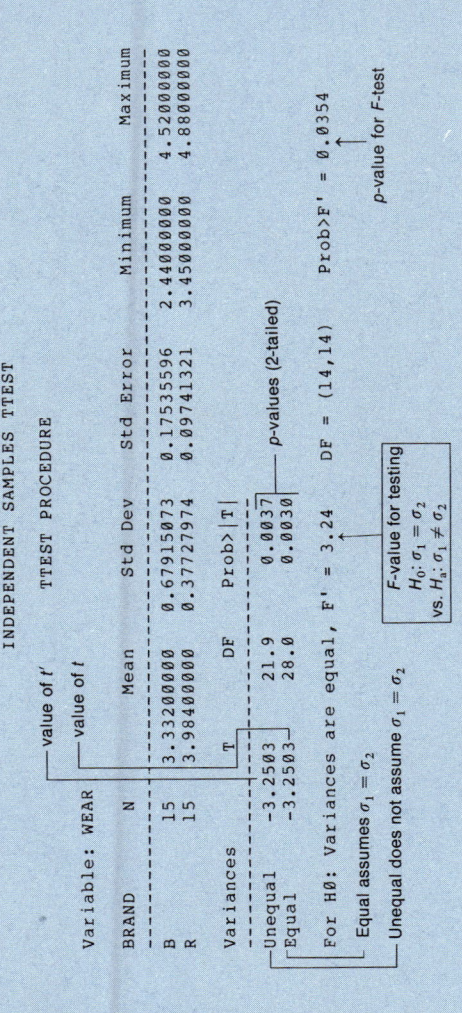

```
                        INDEPENDENT SAMPLES TTEST

Variable: WEAR                 TTEST PROCEDURE

BRAND    N    Mean         Std Dev      Std Error     Minimum       Maximum
B       15    3.33200000   0.67915073   0.17535596    2.44000000    4.52000000
R       15    3.98400000   0.37727974   0.09741321    3.45000000    4.88000000

             T       DF     Prob>|T|
Variances
Unequal   -3.2503   21.9    0.0037
Equal     -3.2503   28.0    0.0030

For H0: Variances are equal, F' = 3.24   DF = (14,14)   Prob>F' = 0.0354
```

- value of t' — value of t
- p-values (2-tailed)
- p-value for F-test
- F-value for testing $H_0: \sigma_1 = \sigma_2$ vs. $H_a: \sigma_1 \neq \sigma_2$
- Equal assumes $\sigma_1 = \sigma_2$
- Unequal does not assume $\sigma_1 = \sigma_2$

Figure 9.22 shows the SAS output obtained by executing the listing in Figure 9.21, whereas Figure 9.23 shows the SAS/PC output.

▮▮▮▮▮▮▮▮▮▮ S A S ▮▮▮▮▮▮▮▮▮▮

MAINFRAME AND MICRO

SOLUTION

EXAMPLE 9.11

Example 9.11 was concerned with the computation of the t-statistic for the means of two populations using paired (dependent) samples. The problem was to determine if the average difference in weekly attendance (space museum – history museum) is positive. ($H_a: \mu_d > 0$)

The SAS program listing in Figure 9.24 was used to request a mean, t-score, and p-value. Note that SAS assumes a two-tailed test. This was a one-tailed test, so the calculated p-value must be divided by two.

In this problem the SAS commands are the same for both the mainframe and PC versions.

The TITLE command names the SAS run (enclose in single quotes). The DATA command gives the data a name.

FIGURE 9.24
Input for SAS (mainframe or micro version).

```
TITLE 'MUSEUM ATTENDANCE';
DATA MUSEUM;
INPUT SPACE HISTORY;
DIFF=SPACE-HISTORY;
CARDS;
0.6 0.5
0.8 1.0
0.7 0.5
1.2 0.8
     . . .
1.5 1.2
1.3 1.4
1.1 0.8
0.8 0.6
PROC MEANS N MEAN T PRT;
VAR DIFF;
TITLE 'TWO DEPENDENT SAMPLES';
```

The INPUT command names and gives the correct order for the different fields on the data cards.

The DIFF = SPACE – HISTORY statement is used to compute a new variable, DIFF, which is the difference between the value of SPACE and the value of HISTORY.

The CARDS command indicates to SAS that the input data immediately follow.

The next 12 lines contain the data values. Each line represents the attendance in thousands at the space and history museums respectively. The first line for example implies that there were 600 guests at the space museum and 500 guests at the history museum.

The PROC MEANS command requests an SAS procedure to print simple descriptive statistics for the variable in the following subcommand VAR DIFF. The TITLE subcommand names the output.

Figure 9.25 shows the SAS output obtained by executing the listing in Figure 9.24, whereas Figure 9.26 shows the SAS/PC output.

FIGURE 9.25
SAS output

```
                    TWO DEPENDENT SAMPLES

VARIABLE    N         MEAN              T        PR>|T|
DIFF        12        0.2583333        2.37      0.0372
                                        │          │
                                       t_b*        │
                                      → 2.37      0.0372 ←
                                      for a one-tailed
                                      test, the actual
                                      p-value = .037/2 = .019
```

FIGURE 9.26
SAS/PC output

```
                    TWO DEPENDENT SAMPLES

            Analysis Variable : DIFF

N Obs   N       Mean            T       Prob>|T|
12      12      0.2583333       2.3693700    0.0372
                                    │          │
                                   t_b*        │
                                 → 2.3693700  0.0372 ←
                                 for a one-tailed test,
                                 the actual p-value = .019
```

Estimation and Testing for Population Proportions

A Look Back/Introduction

By now you should be comfortable with the concepts of estimation and hypothesis testing. If, for example, you have reason to believe that the population is normally distributed, you can then estimate the necessary population parameters (such as the mean and standard deviation) using the corresponding sample statistics. You should be well aware that there always is the risk of arriving at an incorrect conclusion when using sample information to infer something about an entire population. Due to the Central Limit Theorem, the assumptions regarding normality can be relaxed when making inferences about population means if large samples are used.

Chapters 7, 8, and 9 have concentrated primarily on normal populations. We provided you with confidence intervals for the mean and the variance of a single normal population. We examined how to check a statement regarding one of these parameters (such as $\mu < 100$ or $\sigma > .5$) using a test of hypothesis. Then this concept was extended to comparing the means or variances of two normal populations.

Now we return to the *binomial* situation, in which we are interested in the *proportion* of your population that has a certain attribute. This includes personal attributes, such as willingness to buy a product or being in favor of a proposed labor contract. We can also examine proportions as they relate to a particular physical attribute, such as the proportion of defective components in a batch.

We are interested in a single parameter, referred to as **p**, which is the **proportion** of the population having this attribute. For example, suppose that a recent report claims that only 10% of all registered voters in a certain area are in favor of forced busing for school children ($p = .10$). Or suppose it has been reported that a lower proportion of families with children favor busing than do those without children. How can we estimate the actual proportions here and test these claims?

10.1
ESTIMATION AND CONFIDENCE INTERVALS FOR A POPULATION PROPORTION

A test for a population *proportion* deals with a binomial situation. Using the definitions from Chapter 5, each member of your population is either a *success* or a *failure*. These words can be misleading; it is necessary only that each person (or object) in your population either have a certain attribute (a success) or not have it (a failure). So we define *p* to be the proportion of successes in the population—that is, the proportion that have a certain attribute.

Do not confuse the notation *p* = a population proportion with the previously used shorthand for a *p*-value. They do not mean the same thing. We hope that the context will make it clear which of the two *p*'s is being described.

In Chapter 5 we assumed that *p* is known. For any binomial situation, perhaps *p* is known, or more likely it was estimated in some way. This chapter examines how you can estimate *p* by using a sample from the population. Also, we can support (or fail to support) claims concerning the value of *p*. The final section in this chapter compares two samples from two separate populations.

Point Estimate for a Population Proportion

Suppose that the management of Cassidy Electronics, a manufacturer of calculators and microcomputers, is considering offering a dental plan to their employees. Because the monthly premium will be deducted from employee paychecks, perhaps not all employees will wish to join the plan. The insurance company is interested in the proportion of employees who will want to join. A random sample of 200 employees was interviewed. Of these, 137 said they would purchase the dental insurance if it were offered. What can you say about the proportion (*p*) of all employees who wish to join?

We view this problem as a binomial situation and define success as a person who will sign up for the dental insurance and failure as a person who will not sign up for the dental insurance. Consequently, *p* is the proportion of successes in the population (proportion of all employees who favor the dental insurance). Remember that *p*, like μ and σ previously, will remain *unknown* forever. To *estimate p*, we obtain a random sample and observe the proportion of successes in our sample. We use $\hat{p}$ (read "*p* hat") to denote the estimate of *p*, which is the proportion of successes in the sample. Here, $\hat{p}$ = proportion of employees in the sample who will sign up for the dental insurance, so $\hat{p}$ = 137/200 = .685.

In general,

$$\begin{aligned} \hat{p} &= \text{estimate of } p \\ &= \text{proportion of sample having a specified attribute} \\ &= \frac{x}{n} \end{aligned}$$

(10.1)

where *n* = sample size and *x* = the number of sample observations having this attribute.

The symbol ˆ is used to denote an *estimate*. Distinguish between $\hat{p}$ obtained from sample information with *p*, the population proportion being estimated by $\hat{p}$. This is the same type of difference that we previously recognized between a sample mean, $\bar{X}$ (often referred to as $\hat{\mu}$), and population mean, μ.

Confidence Intervals for a Population Proportion (Using a Small Sample)

The calculations involved in determining a confidence interval for p using a small sample are fairly complex. To make them easier, we have listed 90% and 95% confidence intervals for sample sizes of $n = 5, 6, \ldots, 20$ in Table A.8. For sample sizes other than these, you can (1) use the large sample confidence interval (described next) or (2) extend Table A.8 by consulting your local statistician. Or you can use a computer subroutine to derive additional values for this table. An explanation of the method used to generate these confidence limits is included at the end of the table.

Using Table A.8 is much like using Table A.1, the table of binomial probabilities. Let n = sample size and x = the observed number of successes in your sample. Based on these values, the confidence interval (p_L, p_U) can be obtained directly from the table.

EXAMPLE 10.1

A private company is considering the purchase of 200 Beagle microcomputers to monitor seismic activity. These computers will be placed in outdoor stations where they must be able to operate in extremely cold weather. If the computers will operate in temperatures as low as $-10°F$, the company will purchase them. Beagle, anxious to demonstrate the reliability of their system, has agreed to subject 15 computers to a "cold test." Let p = proportion of *all* Beagle computers that will function at $-10°F$.

Of the 15 sample computers, three of them stopped operating at or above $-10°F$. What can you say about p? Construct a 95% confidence interval for p.

SOLUTION

Let a success be that a computer *survives* the cold test (still functions at $-10°F$). We observe 12 successes out of 15 in the sample. So,

$$\hat{p} = \frac{12}{15} = .8$$

Using Table A.8 for $n = 15$, $x = 12$, and $\alpha = .05$, we find $p_L = .519$ and $p_U = .957$. The corresponding 95% confidence interval for p is

$$p_L \quad \text{to} \quad p_U = .519 \quad \text{to} \quad .957$$

So we are 95% confident that the actual (population) percentage of Beagle computers that can function at $-10°F$ is between 51.9% and 95.7%. ∎

Confidence Intervals for a Population Proportion (Using a Large Sample)

When dealing with large samples, the Central Limit Theorem once again provides us with a reliable method of determining approximate confidence intervals for a population proportion. For each element in your sample, assign a value of 1 if this observation is a success (has the attribute) or 0 if this observation is a failure (does not have the attribute). Using the dental-plan example to illustrate, for *each* person in the sample we assign 1 if this person wants the dental insurance and 0 if this person does not want the dental insurance. So what is $\hat{p}$? We can write this as

$$\hat{p} = \frac{\overbrace{1+1+\cdots+1}^{137 \text{ times}} + \overbrace{0+0+\cdots+0}^{63 \text{ times}}}{200} = \frac{137}{200} = .685$$

In this sense, then, $\hat{p}$ is a **sample average**: it is an average of 0s and 1s. As a result, we can apply the Central Limit Theorem to $\hat{p}$ and conclude that $\hat{p}$ is (approximately) a *normal random variable* for large samples. This work reasonably well provided np and $n(1 - p)$ are both greater than 5. So the distribution of $\hat{p}$ [large sample; $np > 5$ and $n(1 - p) > 5$] can be summarized: $\hat{p}$ is (approximately) a normal random variable with

$$\text{mean} = p$$

$$\text{standard deviation (standard error)} = \sqrt{\frac{p(1 - p)}{n}}$$

$$\tag{10.2}$$

By standardizing this result, we have

$$Z = \frac{\hat{p} - p}{\sqrt{\dfrac{p(1 - p)}{n}}}$$

which is approximately a standard normal random variable. This allows us to use Table A.4 to construct a confidence interval for p. This confidence interval is obtained in an identical manner used to construct previous confidence intervals with the standard normal distribution, namely,

$$(\text{point estimate}) \pm Z_{\alpha/2} \cdot (\text{standard deviation of point estimator}) \tag{10.3}$$

Thus, a $(1 - \alpha) \cdot 100\%$ confidence interval for p (large sample; np and $n(1 - p) > 5$) is

$$\hat{p} - Z_{\alpha/2}\sqrt{\frac{\hat{p}(1 - \hat{p})}{n}} \quad \text{to} \quad \hat{p} + Z_{\alpha/2}\sqrt{\frac{\hat{p}(1 - \hat{p})}{n}} \tag{10.4}$$

where $\hat{p}$ = sample proportion. Notice that we have used $\hat{p}$ and $1 - \hat{p}$ under the square root in equation 10.4 rather than p and $1 - p$. This was necessary because p is *unknown* and must be replaced by its estimate, $\hat{p}$. As we observed in previous chapters, replacing an unknown parameter by its estimate works well provided our sample is large enough. For this situation, both np and $n(1 - p)$ should be greater than 5.

The mean of the random variable $\hat{p}$ is the (unknown) value of p. In other words, the average value of $\hat{p}$ is the parameter it is estimating. Such an estimator is said to be **unbiased**. If we obtained random samples indefinitely, the resulting $\hat{p}$'s—on the average—will equal p. This is a desirable property for a sample estimator to have. We have actually discussed two other unbiased estimators previously; $\bar{X}$ is an unbiased estimator of a population mean (μ) and s^2 is an unbiased estimator of a population variance (σ^2).

EXAMPLE 10.2   Using the data regarding employees' desire to join the dental plan, what is a 90% confidence interval for the proportion of all employees who would participate in the dental insurance program?

SOLUTION   Using Table A.4, $Z_{\alpha/2} = Z_{.05} = 1.645$. Also, $\hat{p} = 137/200 = .685$. So the 90%

confidence interval for p is

$$.685 - 1.645 \sqrt{\frac{(.685)(.315)}{200}} \quad \text{to} \quad .685 + 1.645 \sqrt{\frac{(.685)(.315)}{200}}$$

$$= .685 - .054 \quad \text{to} \quad .685 + .054$$

$$= .631 \quad \text{to} \quad .739$$

Based on the sample data, we are 90% confident that the percentage of employees who would purchase the dental insurance is between 63.1% and 73.9%. ■

EXAMPLE 10.3

Remember that in lot acceptance sampling, we either accept or reject a batch (lot) of components, parts, or assembled products based on tests using a random sample drawn from the lot.

Suppose we draw a sample of size 150 from a lot of calculators. We test each of the sampled calculators and find 13 defectives. Determine a 95% confidence interval for the proportion of defectives in the entire batch.

SOLUTION

Let p = proportion of defective calculators in the batch. Based on the sample of 150 calculators, we have

$$\hat{p} = \frac{13}{150} = .0867$$

Because $Z_{.025} = 1.96$, the 95% confidence interval for p is

$$.0867 - 1.96 \sqrt{\frac{(.0867)(.9133)}{150}} \quad \text{to} \quad .0867 + 1.96 \sqrt{\frac{(.0867)(.9133)}{150}}$$

$$= .0867 - .045 \quad \text{to} \quad .0867 + .045$$

$$= .042 \quad \text{to} \quad .132$$

Consequently, we are 95% confident that our estimate $\hat{p} = .0867$ is within .045 of the actual value of p. In other words, this sample estimates the actual percentage of defective calculators to within 4.5%, with 95% confidence. ■

Choosing the Sample Size (One Population)

Suppose that you want your point estimate, $\hat{p}$, to be within a certain amount of the actual proportion, p. In Example 10.3 the *maximum error*, E, was $E = .045$, that is, 4.5%. What if the buyer's specifications necessitate that we estimate the parameter p to within 2% with 95% confidence? Now,

$$E = 1.96 \sqrt{\frac{p(1-p)}{n}} \qquad (10.5)$$

We have an earlier estimate of p ($\hat{p} = .0867$) using the sample of size 150; this can be used in equation 10.5. The purpose is to extend this sample in order to obtain this specific maximum error, E. The specified value of E is .02, so

$$E = .02 = 1.96 \sqrt{\frac{(.0867)(.9133)}{n}}$$

Therefore,

$$\sqrt{\frac{(.0867)(.9133)}{n}} = \frac{.02}{1.96}$$

Squaring both sides and rearranging leads to

$$n = \frac{(1.96)^2(.0867)(.9133)}{(.02)^2} = 760.5$$

Rounding up (*always*), we come to the conclusion that a sample of size $n = 761$ calculators will be necessary to estimate p to within 2%.

In general, the following equation provides the necessary sample size to estimate p with a specified maximum error, E, and confidence level $(1 - \alpha) \cdot 100\%$:

$$n = \frac{Z_{\alpha/2}^2 \hat{p}(1 - \hat{p})}{E^2} \qquad \textbf{(10.6)}$$

In this illustration, we used an estimate of p from a prior sample to determine the necessary sample size using equation 10.6. If the sample of size n based on this equation is our first and only sample, then we *have no estimate* $\hat{p}$. There is a conservative procedure we can follow here that will guarantee the accuracy (E) that we require. Look at the curve of different values of $\hat{p}(1 - \hat{p})$ in Figure 10.1. Consider these values:

$\hat{p}$	$\hat{p}(1 - \hat{p})$
.2	.16
.4	.24
.5	.25
.7	.21
.9	.09

Note that the largest value of $\hat{p}(1 - \hat{p})$ is .25.

If we make $\hat{p}(1 - \hat{p})$ in equation 10.6 *as large as possible*, this will provide a value of n that will result in a maximum error that is sure to be less than the specified value. So we can formulate this rule: If no prior estimate of p is available, a conservative procedure to determine the necessary sample size from equation 10.6 is to use $\hat{p} = .5$.

EXAMPLE 10.4

Suppose that the insurance company underwriting the dental plan wishes to obtain a single sample that will estimate, to within 2% with 90% confidence, the proportion (p) of employees who would purchase the dental insurance. They have *no prior knowledge* of this proportion. Their intent is to obtain a large enough sample the first time so that they can estimate the population proportion with this much accuracy. How large a sample is required?

SOLUTION

We have no prior knowledge of p, so we use $\hat{p} = .5$ in equation 10.6 to obtain

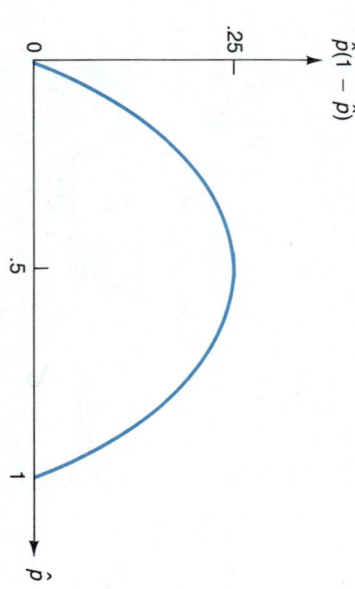

FIGURE 10.1
Curve of values of
$\hat{p}(1 - \hat{p})$.

a sample size of

$$n = \frac{(1.645)^2(.5)(.5)}{(.02)^2} = 1691.3$$

To obtain an estimate of p with a maximum error of $E = .02$, we will need a sample size of $n = 1692$ employees. With a sample of this size, we can safely say that the point estimate, $\hat{p}$, will be within 2% of the actual value of p, with 90% confidence (however, this is a very large sample).

EXERCISES

10.1 Mary Pharmaceuticals is interested in estimating the proportion (p) of its employees who would accept a substantial increase in benefits instead of an annual raise in salary for a particular year. A random sample of 20 employees was obtained, and 12 welcomed the idea. How can you estimate the proportion in the population who would welcome this plan? Construct a 90% confidence interval for p. Use Table A.8.

10.2 In Exercise 10.1, let the maximum error of the estimating proportion be 4%. Estimate the necessary sample size for a 95% confidence interval. Use the estimate of p from Exercise 10.1 for the value of p.

10.3 In Exercise 10.2, assume that you have no prior knowledge of p. Estimate the sample size that is required to estimate the proportion (p) of employees who would accept the plan with a 95% confidence level, using the same maximum error of estimate for the proportion.

10.4 A company wishes to estimate the proportion, p, of its employees who went on sick leave during the past six months. A random sample of 18 employees was taken; of them, ten went on sick leave. Construct a 90% confidence interval for p. Use Table A.8.

10.5 An investment firm surveyed 18 randomly selected economists. The survey found that 13 of them felt that the economy of the United States would not slip into a recession for at least a year. Use Table A.8 to construct 90% and 95% confidence intervals for the percentage of economists who believe that a recession will not occur in the United States for at least a year.

10.6 A math workshop will be offered only if the student demand is sufficiently high. What is the required sample size necessary to estimate with 90% confidence the proportion of students who would register for the workshop if we specify a value of .03 for the maximum error, E?

10.7 In Exercise 10.6, if a previous study indicated that the proportion of students who would register for the workshop was .68, estimate the necessary sample size for a maximum error, E, of 3%.

10.8 Winthrop Boat Lines is exploring the possibility of offering a ferry service between the cities of Patna and Madura, provided there is sufficient demand to make it feasible. The firm randomly interviewed 210 commuters from the two cities, and 146 of them indicated they would patronize the ferry service instead of the present bus service. Estimate the population proportion p of commuters from the two cities who prefer the ferry service. Construct a 95% confidence interval for p.

10.9 In Exercise 10.8, if the maximum error is $E = .01$, estimate the necessary sample size at the 95% confidence level. Use the value obtained in Exercise 10.8 for the estimate of p.

10.10 Suppose a small sample ($n = 12$) was drawn from a lot of electric bulbs. Each bulb was tested; four defectives were found. Determine the 90% confidence interval for the proportion of defectives in the entire batch.

10.11 A manufacturer of microcomputers purchases electronic chips from a supplier that claims its chips are defective only 5% of the time. Determine the sample size that would be required to estimate the true proportion of defective chips if we wanted our estimate, $\hat{p}$, to be within 1.25% of the true proportion, with 99% confidence.

10.12 MasterCard was the leading credit card in 1973. Recently it appears that Visa has become the choice among credit card holders of both cards. In a study on the changes in the use of Visa and MasterCard, it was found that 55% of the households across the United States had a Visa card and 43% had a MasterCard. Suppose that a local merchant wanted to estimate the proportion of Visa credit-card holders in a certain metropolitan area to within 0.10 with 95% confidence. Using the initial estimate of the proportion of Visa credit-card holders locally to be equal to the national proportion, find the sample size necessary to estimate the proportion of Visa credit-card holders in the local metropolitan area.

(*Source*: Douglas K. Hawes, "Profiling Visa and MasterCard Holders: An Overview of Changes—1973 to 1984, and Some Thoughts for Future Research," *Journal of the Academy of Marketing Sciences* 15, no. 1 (1987): 62–9.)

10.13 Blackburn Candies is considering the withdrawal from the market of its product Nutty Bar if Nutty Bar has not captured at least 5% of the candy bar market. A random sample of 115 candy-bar buyers was taken; four bought the Nutty Bar. Find a 95% confidence interval for the proportion, p, of the population of candy-bar buyers who choose Nutty Bar.

10.14 Using the data in Exercise 10.13, what is the required sample size necessary to estimate with 95% confidence, and to be within .03, the proportion of candy-bar buyers who would choose the candy bar? Assume that we have no prior knowledge of p.

10.15 Japan is considered to be one of the major competitors of the United States in the world of electronics. Japanese companies generally hire a worker for life, and Japanese employees are extremely loyal to their employers. However, according to *Japan Electronics Update*, a publication of the American Electronics Association, this attitude may be changing. A survey of 830 workers by a Japanese newspaper indicated that workers may not be so reluctant to switch jobs. In response to the question "Have you ever considered changing your job?" 30% of those surveyed said they had changed jobs, and another 31% reported that they had been contacted by an executive search firm (a relatively new trend in Japan) but had not changed jobs.

a. Construct the 90% confidence interval for the percentage of Japanese workers who have changed jobs.

b. Construct the 95% confidence interval for the percentage of Japanese workers who have neither changed jobs nor been approached by an executive search firm.

10.2   HYPOTHESIS TESTING FOR A POPULATION PROPORTION

How can you statistically reject a statement such as, at least 60% of all heavy smokers will contract a serious lung or heart ailment before age 65? Perhaps someone merely took a wild guess at the value of 60%, and it is your job to gather evidence that will either shoot down this claim or let it stand if there is insufficient evidence to conclude that this percentage actually is less than 60%. We set up hypotheses and test them much like before, only now we are concerned about a population proportion, p, rather than the mean or standard deviation of a particular population.

Hypothesis Testing Using a Small Sample

Because confidence intervals can be used to perform a test of hypothesis, we will use Table A.8 to conduct such a test. Table A.8 contains sample sizes of $n = 5$ to 20 and $\alpha = .05$ and .10. If the sample size exceeds 20 and np and $n(1 - p)$ are both greater than 5, the large-sample approximation will provide an accurate test. For sample sizes contained in Table A.8, use the procedure outlined in the accompanying box.

HYPOTHESIS TESTING (SMALL SAMPLE; n IS BETWEEN 5 AND 20)

TWO-TAILED TEST

$$H_0: p = p_0$$
$$H_a: p \neq p_0$$

1. Obtain the $(1 - \alpha) \cdot 100\%$ confidence interval from Table A.8; that is, (p_L, p_U), using $x =$ the observed number of successes.
2. Reject H_0 if p_0 does not lie between p_L and p_U.
3. Fail to reject H_0 if $p_L \leq p_0 \leq p_U$.

ONE-TAILED TEST

$$H_0: p \leq p_0 \qquad\qquad H_0: p \geq p_0$$
$$H_a: p > p_0 \qquad\qquad H_a: p < p_0$$

1. Obtain the $(1 - 2\alpha) \cdot 100\%$ confidence interval from Table A.8; that is, (p_L, p_U), using $x =$ the observed number of successes.
2. Reject H_0 if $p_0 < p_L$.
3. Fail to reject H_0 if $p_0 \geq p_L$.

1. Obtain the $(1 - 2\alpha) \cdot 100\%$ confidence interval from Table A.8; that is, (p_L, p_U), using $x =$ the observed number of successes.
2. Reject H_0 if $p_0 > p_U$.
3. Fail to reject H_0 if $p_0 \leq p_U$.

Notice that for a one-tailed test, we *double* α when finding the confidence interval for p from Table A.8. For example, if $\alpha = .05$, then $2\alpha = .10$, and so we retrieve a 90% confidence interval from the table. As a result, this particular binomial table can be used only when $\alpha = .025$ or $.05$ for a one-tailed test.

EXAMPLE 10.5

In Example 10.1 suppose that the company interested in the Beagle microcomputers will purchase them if Beagle's claim that the proportion, p, of all Beagle computers that can survive these cold temperatures is greater than .75 (75%) can be shown to be true. Do the data support this claim using $\alpha = .05$?

SOLUTION

The claim under investigation goes into the alternative hypothesis. The appropriate hypotheses are

$$H_0: p \leq .75 \quad \text{and} \quad H_a: p > .75$$

We observed, in the sample of 15 computers, $x = 12$ successes (computers that survived). Because $\alpha = .05$, we double this ($2\alpha = .10$) and refer to Table A.8 for a 90% confidence interval for p when $n = 15$, $x = 12$. This is

$$(p_L, p_U) = (.560, .943)$$

We will reject H_0 provided p_L lies to the right of $p_0 = .75$. Because $p_L = .560$ is less than $p_0 = .75$, we fail to reject H_0.

Based on the evidence gathered from this sample, we cannot demonstrate that p is greater than the required 75%. Notice that we are not *accepting* H_0 — we simply *fail to reject* it. This means that the point estimate $\hat{p} = 12/15 = .8$ is not enough larger than .75 to justify the claim made in H_a. The fact that $\hat{p}$ exceeds .75 may be due to the sampling error that is possible when using a sample statistic ($\hat{p}$) to infer something about a population parameter (p). ∎

Hypothesis Testing Using a Large Sample

The standard five-step procedure is outlined for testing H_0 versus H_a when attempting to support a claim regarding a binomial parameter, p, using a large sample. To define a test statistic for this situation, the approximate standard normal random variable contained in equation 10.2 is used.

The rejection region for this test is defined by determining the distribution of the test statistic, given that H_0 is true. This means that the unknown value of p in equation 10.2 is replaced by the value of p specified in H_0 (say, p_0). For a one-tailed test, the boundary value of p in H_0 is used. This procedure is summarized in the following box.

HYPOTHESIS TESTING (LARGE SAMPLE; np_0 AND $n(1 - p_0)$ BOTH GREATER THAN 5)

TWO-TAILED TEST

$H_0: p = p_0$
$H_a: p \neq p_0$

reject H_0 if $|Z| > Z_{\alpha/2}$

where

$$Z = \frac{\hat{p} - p_0}{\sqrt{\dfrac{p_0(1 - p_0)}{n}}}$$

ONE-TAILED TEST

$H_0: p \leq p_0$ $H_0: p \geq p_0$
$H_a: p > p_0$ $H_a: p < p_0$

reject H_0 if $Z > Z_\alpha$ reject H_0 if $Z < -Z_\alpha$

Notice that the form of the test statistic is that used in many of the previous large sample test statistics, namely,

$$Z = \frac{\text{(point estimate)} - \text{(hypothesized value)}}{\text{(standard deviation of point estimator)}} \qquad (10.7)$$

EXAMPLE 10.6   In Example 10.2, we estimated the proportion of employees at Cassidy Electronics who would sign up for the dental insurance. The insurance company is not willing to offer such a plan unless more than 60% of the employees will participate. Using the sample of 200 employees, can you conclude that this percentage is greater than the required 60%? Use a significance level of $\alpha = .10$.

SOLUTION

Step 1.   Your hypotheses should be

$$H_0: p \leq .6$$
$$H_a: p > .6$$

Step 2.   Since $np_0 = (200)(.6) = 120$ and $n(1 - p_0) = (200)(.4) = 80$ are   both greater than 5, the large-sample test statistic can be used, namely,

$$Z = \frac{\hat{p} - p_0}{\sqrt{\frac{p_0(1 - p_0)}{n}}} = \frac{\hat{p} - .6}{\sqrt{\frac{(.6)(.4)}{200}}}$$

Step 3. The testing procedure, using $\alpha = .10$, will be to

reject H_0 if $Z > Z_{.10} = 1.28$

Step 4. Using the sample data, $\hat{p} = 137/200 = .685$, so

$$Z^* = \frac{.685 - .6}{\sqrt{\frac{(.6)(.4)}{200}}} = \frac{.085}{.0346} = 2.45$$

Because $2.45 > 1.28$, we reject H_0 in favor of H_a.

Step 5. This sample indicates that the population proportion of employees who would participate in the dental insurance plan *is greater than* 60%. ■

In Example 10.6, the computed test statistic *was* $Z^* = 2.45$. Figure 10.2 shows the Z curve and the calculated *p*-value, which is .0071. Using the classical approach, because $.0071 < \alpha = .10$, we reject H_0. If we choose to base our conclusion strictly on the *p*-value (without choosing a significance level, α), this value would be classified as *small*—it is less than .01. This means that (if we were to use the classical approach) we would reject H_0 for any α greater than or equal to .01. Consequently, using this procedure, we once again reject H_0.

EXAMPLE 10.7

In Example 10.3, we estimated the proportion of calculators that were defective in a batch (lot). The company has determined that a good target for this defective percentage is 4%. The sample of 150 had 13 defectives. Can we conclude that the actual proportion of defective calculators is different from 4%? Use $\alpha = .05$.

SOLUTION

Step 1. We wish to see if p is *different* from 4%, so we should use a two-tailed test with hypotheses

$$H_0: p = .04$$
$$H_a: p \neq .04$$

Step 2. Here $np_0 = (150)(.04) = 6$ and $n(1 - p_0) = (150)(.96) = 144$. Both are >5, so the appropriate test statistic is

$$Z = \frac{\hat{p} - p_0}{\sqrt{\frac{p_0(1 - p_0)}{n}}} = \frac{\hat{p} - .04}{\sqrt{\frac{(.04)(.96)}{150}}}$$

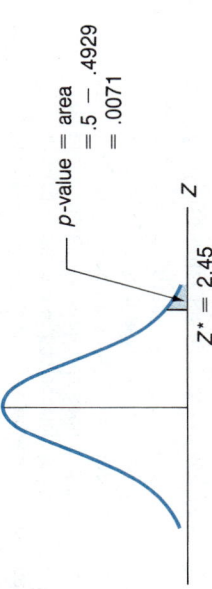

p-value = area
= .5 − .4929
= .0071

$Z^* = 2.45$

FIGURE 10.2
Z curve showing *p*-value for Example 10.6.

FIGURE 10.3

Z curve showing p-value
(twice the shaded area)
for Example 10.7.

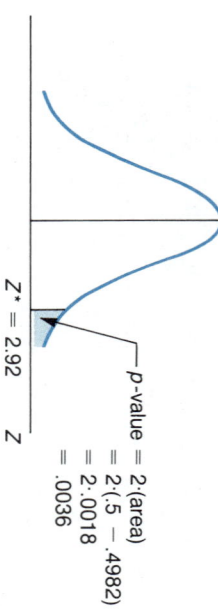

$Z^* = 2.92$

p-value $= 2\cdot(\text{area})$
$= 2\cdot(.5 - .4982)$
$= 2\cdot.0018$
$= .0036$

Step 3. With $\alpha = .05$, the test procedure of H_0 versus H_a will be to

$$\text{reject } H_0 \text{ if } |Z| > 1.96$$

Step 4. Using $\hat{p} = 13/150 = .0867$,

$$Z^* = \frac{.0867 - .04}{\sqrt{\dfrac{(.04)(.96)}{150}}} = \frac{.0467}{.016} = 2.92$$

Because $2.92 > 1.96$, we reject H_0.

Step 5. The company is *not* meeting their target percentage of defectives. As a reminder, because $\alpha = .05$, 5% of the time this particular test will reject H_0 when in fact it is true.

In Example 10.7, $Z^* = 2.92$. What is the p-value? This is a two-tailed test, so we need to *double* the right-tail area, as illustrated in Figure 10.3. So $p = 2\cdot.0018 = .0036$. Thus, using either the classical procedure (comparing the p-value to $\alpha = .05$) or basing our decision strictly on the p-value, we reject H_0 because of this extremely small p-value.

■

EXERCISES

10.16 Using the data in Exercise 10.13, test the hypothesis that Nutty Bar captures less than 5% of the market using a .05 level of significance.

10.17 Calculate the p-value for Exercise 10.16. Based on the p-value, would you reject the null hypothesis at the .05 level?

10.18 An official for a computer firm was told by an independent source that 20% of the employees of the computer firm perceived that there was sex discrimination in the salary structure of the company. A quick random survey by the official of 18 employees found that three of the employees thought there was sex discrimination in the salary structure of the company. Is there sufficient evidence by the official's survey to indicate that the figure of 20% given by the independent source is in error? Conduct a hypothesis test using a significance level of .10 and using Table A.8.

10.19 Using the data in Exercise 10.10, test the hypothesis that the defective rate is more than 10%, using a .05 level of significance. Use Table A.8.

10.20 There are about 87 million households in the United States. In 1970, about 71% of the households were occupied by married couples. In the last 15 years, the trend toward living alone has accelerated. One market researcher believes that at present the proportion of households occupied by married couples is 58%. A random sample of 20 households reveals that 12 are occupied by married couples. At a significance level of 10%, are you in a position to contradict the market researcher and say that this person is probably wrong?

10.21 In order for $np > 5$ and $n(1 - p) > 5$, how large must n be if $p = .03$?

10.22 A significant trend in higher education has been the increasing enrollments in business programs. Accompanying this increase has been a shortage of doctorally qualified faculty. Dolecheck (1987) sent a questionnaire to 369 MBA graduates of Northeast Louisiana University (NLU) who had received their MBA degree between May 1965 and May 1984. Research of this sort generally does not obtain a 100% reply rate. Of the 175 usable replies to the questionnaire, 33 respondents had opted to pursue a doctorate and teach, while 147 had chosen to work in business or government. Those who had opted to pursue a doctorate and teach were asked at what point in their lives they made their decision. The percentage distribution of their responses is given below:

POINT IN TIME DECISION TO PURSUE A DOCTORATE WAS MADE ($n = 33$)

POINT IN TIME	PERCENT
Before entering undergraduate program	0.0
While pursuing undergraduate degree	3.0
After undergrad. degree but before entering MBA	9.1
While pursuing MBA degree	36.4
After receiving MBA degree	51.5
	100.0

(*Source:* M. Dolecheck, "University Teaching: Perceptions of MBAs Who Earned the Doctorate and Those Who Did Not." *Proceedings of the 18th Annual Conference of the Southwest Region, Decision Sciences Institute,* Houston, Texas, March 10–14, 1987, Table IV.)

a. Construct a 95% confidence interval for the proportion of persons who decide to pursue a doctorate while pursuing the MBA.

b. Does the above data support the contention that more than half of those who decide to pursue a doctorate do so after receiving their MBA degree? Use a 5% significance level for the test, and state the *p*-value.

10.23 The editor of a famous weekly magazine is concerned about typographical errors and believes that about 1.5% of the number of lines printed have at least one error. An examination of 650 different lines revealed 11 lines that had at least one error. Do the data support the editor's belief? Use a significance level of .10.

10.24 A manager at National Insurance believes that out of the total number of automobile-accident claims settled in a particular month, there are more claims related to speeding by the driver than there are claims that are not related to speeding. From a random sample of 75 claims, 40 were found to be associated with speeding. Test the manager's belief. Use a significance level of .05.

10.25 Calculate the *p*-value for Exercise 10.24 and interpret it.

10.26 An instructor believes that of the students who take a certain course, there are more students who have not taken the prerequisites for the course than students who have taken the prerequisites. The instructor randomly selected 70 students and found that only 30 students had taken the prerequisites for the course. Do these data support the instructor's belief? Use a significance level of .10.

10.27 Based on the *p*-value for Exercise 10.26, would you reject the null hypothesis at the .01 level?

10.28 From 1983 to 1987 diversified mutual funds gained on the average five percentage points less than the rise in the Standard and Poor's 500 (S & P 500). However, in the first quarter of 1988, diversified mutual funds were leaving the market indexes in the dust. According to *Business Week* magazine, seven of the ten largest stock funds beat the S & P 500 in the first quarter of 1988. Suppose that an investor wished to determine if this proportion held true for diversified mutual funds in general. Assume that a random sample of 40 diversified mutual funds are selected and 24 funds beat the S & P 500 index. Is there sufficient evidence to conclude that the proportion of diversified mutual funds that outperformed the S & P 500 differs from .7? Use the *p*-value to determine the conclusion.

(*Source:* "Mutual Funds Are Riding High—Thanks To Little Companies," *Business Week* (April 1988): 92–4.)

10.29   KNNN, a television news channel, claimed that more than 65% of its subscribers had an annual income of \$40,000 or more. A random sample of 160 subscribers was interviewed; 71% of them had incomes of \$40,000 or more. Does this information support KNNN's claim? Use a significance level of .05.

10.30   Calculate and interpret the *p*-value for Exercise 10.29.

10.3
COMPARING TWO POPULATION PROPORTIONS (LARGE INDEPENDENT SAMPLES)

Consider the following questions:

Is the divorce rate higher in California than it is in New York?

Is there any difference in the proportion of cars manufactured by Henry Motor Company requiring an engine overhaul before 100,000 miles and the proportion of General Auto (GA) automobiles requiring one?

Is there a higher rate of lung cancer among cigarette smokers than there is among nonsmokers?

These questions are concerned with proportions from *two* populations. Our method of estimating these proportions will be exactly as it was for one population. We simply have two of everything—two populations, two samples, two estimates, and so on. In this section, it is assumed that the two samples are obtained *independently*.

For example, consider the question concerning the proportion of cars requiring an engine overhaul. Population 1 is all Henry cars, with p_1 = proportion of Henry cars requiring an engine overhaul before 100,000 miles, n_1 = Henry sample size, and x_1 = number of Henry cars requiring an overhaul before 100,000 miles. Population 2 is all GA cars, with p_2 = proportion of GA cars requiring an engine overhaul before 100,000 miles, n_2 = GA sample size, and x_2 = number of GA cars requiring an overhaul before 100,000 miles.

Define a success to be that a car requires an overhaul before 100,000 miles. (Keep in mind that "success" is merely a label for the trait you are interested in. It need not be a desirable trait.) Our unbiased point estimator of p_1 will be as before:

$$\hat{p}_1 = \frac{\text{observed number of successes in the sample}}{\text{sample size}}$$

$$= \frac{\text{number of cars in the Henry sample requiring an overhaul}}{n_1}$$

That is, the unbiased point estimator of p_1 is

$$\hat{p}_1 = \frac{x_1}{n_1} \qquad (10.8)$$

Similarly, the unbiased point estimator of p_2, obtained from the second sample, is

$$\hat{p}_2 = \frac{x_2}{n_2} \qquad (10.9)$$

For the two-population case, the parameter of interest will be the *difference* between the two population proportions, $p_1 - p_2$. The next section discusses a method of estimating $p_1 - p_2$ by using a point estimate along with a corresponding confidence interval.

Confidence Interval for $p_1 - p_2$ (Large Independent Samples)

The logical estimator of $p_1 - p_2$ is $\hat{p}_1 - \hat{p}_2$, the difference between the sample estimators. What kind of random variable is $\hat{p}_1 - \hat{p}_2$? We are dealing with large independent samples (where $n_1\hat{p}_1$, $n_1(1 - \hat{p}_1)$, $n_2\hat{p}_2$, and $n_2(1 - \hat{p}_2)$ are each larger than five) so it follows that $\hat{p}_1 - \hat{p}_2$ is (approximately) a normal random variable with

$$\text{mean} = p_1 - p_2$$

and

$$\text{standard deviation} = \sqrt{\frac{p_1(1 - p_1)}{n_1} + \frac{p_2(1 - p_2)}{n_2}}$$

In a previous section, we observed that $\hat{p}_1$ is a sample mean, where the sample consists of observations that are either a 1 (a particular event occurred) or a 0 (this event did not occur). Because the two samples are obtained independently, the results extend to this situation, leading to the approximate normal distribution for $\hat{p}_1 - \hat{p}_2$. Since $\hat{p}_1$ and $\hat{p}_2$ are unbiased estimators of p_1 and p_2, respectively, the mean of the estimator $\hat{p}_1 - \hat{p}_2$ is $p_1 - p_2$; that is, $\hat{p}_1 - \hat{p}_2$ is an unbiased estimator of $p_1 - p_2$. Notice that the variance of $\hat{p}_1 - \hat{p}_2$ is obtained by *adding* the variance of $\hat{p}_1$, or $p_1(1 - p_1)/n_1$, to the variance of $\hat{p}_2$, or $p_2(1 - p_2)/n_2$.

To evaluate the confidence interval, we are forced to approximate the confidence limits by replacing p_1 by $\hat{p}_1$ and p_2 by $\hat{p}_2$ under the square root sign. This approximation works well provided both sample sizes are large. So we derive a $(1 - \alpha) \cdot 100\%$ confidence interval for $p_1 - p_2$ (large independent samples; $n_1\hat{p}_1$, $n_1(1 - \hat{p}_1)$, $n_2\hat{p}_2$, and $n_2(1 - \hat{p}_2)$ are each greater than 5):

$$(\hat{p}_1 - \hat{p}_2) - Z_{\alpha/2}\sqrt{\frac{\hat{p}_1(1 - \hat{p}_1)}{n_1} + \frac{\hat{p}_2(1 - \hat{p}_2)}{n_2}}$$

$$\text{to} \quad (\hat{p}_1 - \hat{p}_2) + Z_{\alpha/2}\sqrt{\frac{\hat{p}_1(1 - \hat{p}_1)}{n_1} + \frac{\hat{p}_2(1 - \hat{p}_2)}{n_2}} \quad \text{(10-10)}$$

where $\hat{p}_1 = x_1/n_1$ and $\hat{p}_2 = x_2/n_2$ are the sample proportions. Observe that the construction of this confidence interval was the "usual" procedure employing Table A.4 described in equation 10.3.

EXAMPLE 10.8

Of a random sample of 100 cars manufactured by Henry (population 1), 28 needed an engine overhaul before reaching 100,000 miles. A second sample, obtained independently of the first, consisted of 150 cars produced by GA; 48 of them required an engine overhaul before 100,000 miles. Both sets of cars were subjected to the same weather conditions, maintenance program, and driving conditions. Construct a 99% confidence interval for $p_1 - p_2$.

SOLUTION

We have $\hat{p}_1 = 28/100 = .28$ and $\hat{p}_2 = 48/150 = .32$. Also, $Z_{\alpha/2} = Z_{.005} = 2.575$, using Table A.4. The resulting confidence interval for $p_1 - p_2$ is

$$(.28 - .32) - 2.575 \sqrt{\frac{(.28)(.72)}{100} + \frac{(.32)(.68)}{150}}$$

$$\text{to} \quad (.28 - .32) + 2.575 \sqrt{\frac{(.28)(.72)}{100} + \frac{(.32)(.68)}{150}}$$

$$= -.04 - .15 \quad \text{to} \quad -.04 + .15$$

$$= -.19 \quad \text{to} \quad .11$$

This confidence interval leaves us unable to conclude that either manufacturer produces a better engine. We are 99% confident that the percentage of Henry engines requiring an overhaul before 100,000 miles is between 19% *lower* to 11% *higher* than for the GA engines.

◼

EXAMPLE 10.9

Boone Advertising Agency handles the advertising for Slick cigarettes. They recently completed a 6-month advertising campaign in an attempt to increase the market share for Slick. A private marketing consulting firm was chosen to estimate the market share before and after the campaign. A set of vendors who supply cigarettes nationwide was used to determine market share. Prior to the campaign, a random sample of 1200 cartons supplied by these vendors was selected. Following the campaign, a second random sample of 1200 cartons was selected (independently of the first sample). The results were as follows. After the advertising campaign: $n_1 = 1200$ cartons and $x_1 =$ number of Slick cartons = 90. Before the advertising campaign: $n_2 = 1200$ cartons and $x_2 =$ number of Slick cartons = 54. Let $p_1 =$ proportion of cartons sold *after* the advertising campaign that were Slick cigarettes and $p_2 =$ proportion of cartons sold *before* the campaign. Determine a 95% confidence interval for $p_1 - p_2$.

SOLUTION

Our proportion estimates are

$$\hat{p}_1 = \frac{90}{1200} = .075$$

$$\hat{p}_2 = \frac{54}{1200} = .045$$

The 95% confidence interval for $p_1 - p_2$ is

$$(.075 - .045) - 1.96 \sqrt{\frac{(.075)(.925)}{1200} + \frac{(.045)(.955)}{1200}}$$

$$\text{to} \quad (.075 - .045) + 1.96 \sqrt{\frac{(.075)(.925)}{1200} + \frac{(.045)(.955)}{1200}}$$

$$= .03 - .019 \quad \text{to} \quad .03 + .019$$

$$= .011 \quad \text{to} \quad .049$$

So we are 95% confident that (1) our estimate of the difference in proportion (after minus before), namely $\hat{p}_1 - \hat{p}_2 = .03$, is within 1.9% of the actual difference, and (2) the proportion increase after the advertising campaign is between 1.1% and 4.9%.

◼

Choosing the Sample Sizes (Two Populations)

In Chapter 9, we discussed how to select samples from two populations when the desired accuracy of the point estimate of the difference between two population means is specified—this is the maximum error, E. If E is 10 pounds, for instance, then what sample sizes (n_1 and n_2) are necessary for the point estimate of $\mu_1 - \mu_2$ (namely, $\bar{X}_1 - \bar{X}_2$) to be within 10 pounds of the actual value, with 95% (or whatever) confidence? Using the results contained in Appendix B at the end of the text, values of n_1 and n_2 were provided in Chapter 9 that minimized the total sample size, $n_1 + n_2$, for this specific value of E.

We encounter a similar situation when dealing with two population proportions, p_1 and p_2. If a maximum error of $E = .10$, for instance, is specified, then the question of interest is: What sample sizes (n_1 and n_2) are necessary for the point estimate of $p_1 - p_2$ (namely, $\hat{p}_1 - \hat{p}_2$) to be within .10 of the actual value, with 95% (or whatever) confidence?

The maximum error, E, always is the amount that you *add to* and *subtract from* the point estimate when determining a confidence interval. When dealing with two proportions, this is

$$E = Z_{\alpha/2} \sqrt{\frac{p_1(1 - p_1)}{n_1} + \frac{p_2(1 - p_2)}{n_2}} \qquad (10.11)$$

To evaluate this expression, you will need estimates of p_1 and p_2. You have two options. If you have previously obtained small samples from these two populations, you can use the resulting sample estimates $\hat{p}_1$ and $\hat{p}_2$. The purpose then will be to extend these samples to obtain better accuracy in the point estimate, $\hat{p}_1 - \hat{p}_2$. If no information regarding p_1 and p_2 is available, then you can use the conservative approach discussed in Section 10.1 by letting $\hat{p}_1 = \hat{p}_2 = .5$.

By applying the results of Appendix B to this situation, the sample sizes n_1 and n_2 that minimize the total sample size $n_1 + n_2$ are given by

$$n_1 = \frac{Z_{\alpha/2}^2(A + B)}{E^2} \qquad (10.12)$$

$$n_2 = \frac{Z_{\alpha/2}^2(C + B)}{E^2} \qquad (10.13)$$

where

$$A = p_1(1 - p_1)$$
$$B = \sqrt{p_1 p_2(1 - p_1)(1 - p_2)}$$
$$C = p_2(1 - p_2)$$

To determine A, B, and C, estimates of p_1 and p_2 should be substituted for p_1 and p_2 by using one of the two options described.

EXAMPLE 10.10

Using the engine overhaul data from Example 10.8, determine what sample sizes are necessary for the estimate of the difference between the proportion of cars needing an overhaul before 100,000 miles to be within .10 of the actual value, with 99% confidence, if (1) the results from Example 10.8 are available and (2) no sample information is available.

SOLUTION 1

The specified maximum error is $E = .10$. Sample data have been collected regarding these proportions, so we use the corresponding estimates to determine the sample sizes necessary to obtain this degree of accuracy. Using Table A.4,

$Z_{\alpha/2} = Z_{.005} = 2.575$. Here, $\hat{p}_1 = .28$ and $\hat{p}_2 = .32$. Consequently,

$$A = \hat{p}_1(1 - \hat{p}_1)$$
$$= .2016$$

$$B = \sqrt{\hat{p}_1\hat{p}_2(1 - \hat{p}_1)(1 - \hat{p}_2)}$$
$$= .2094$$

$$C = \hat{p}_2(1 - \hat{p}_2)$$
$$= .2176$$

To obtain the *smallest possible* total sample size for the required accuracy, the two sample sizes should be

$$n_1 = \frac{(2.575)^2(.2016 + .2094)}{(.10)^2} \cong 273$$

(remember—always round up) and

$$n_2 = \frac{(2.575)^2(.2176 + .2094)}{(.10)^2} \cong 284$$

providing a total sample size of $n_1 + n_2 = 557$ cars.

If no prior estimates of p_1 and p_2 are available, using $\hat{p}_1 = \hat{p}_2 = .5$ will result in sample sizes n_1 and n_2 that, when obtained, will provide a maximum error *no larger than* the specified value of $E = .10$. Here, $A = (.5)(.5) = .25$. Similarly, $B = C = .25$, so

$$n_1 = n_2 = \frac{(2.575)^2(.25 + .25)}{(.10)^2} \cong 332$$

Consequently, a total sample size of $n_1 + n_2 = 664$ cars will be necessary for $\hat{p}_1 - \hat{p}_2$ to be within .10 of the actual value of $p_1 - p_2$, with 99% confidence. ∎

Hypothesis Testing for p_1 and p_2 (Large Independent Samples)

Suppose that a recent report stated that, based on a sample of 500 people, 35% of all cigarette smokers had at some time in their life developed a particular fatal disease. On the other hand, 25% of the nonsmokers in the sample acquired the disease. Can we conclude from this sample that, because $\hat{p}_1 = .35 > \hat{p}_2 = .25$, the proportion ($p_1$) of all smokers who will acquire the disease exceeds the proportion (p_2) for nonsmokers? In other words, is $\hat{p}_1$ *significantly* larger than $\hat{p}_2$? After all, even if $p_1 = p_2$, there is a 50–50 chance that $\hat{p}_1$ will be larger than $\hat{p}_2$ because for large samples, the distribution of $\hat{p}_1 - \hat{p}_2$ is approximately a bell-shaped (normal) curve centered at $p_1 - p_2$, which, if $p_1 = p_2$, would be zero.

Are the results of the sample significant, or are they due simply to the sampling error that is always possible when estimating from a sample? Your alternative hypothesis can be that two proportions are *different* (a two-tailed test) or that one *exceeds* the other (a one-tailed test). As before, we will assume that the two random samples are obtained *independently*. The possible hypotheses are these: For a two tailed test,

$$H_0: p_1 = p_2$$
$$H_a: p_1 \neq p_2$$

and for a one-tailed test,

$$H_0: p_1 \leq p_2$$
$$H_a: p_1 > p_2$$

or

$$H_0: p_1 \geq p_2$$
$$H_a: p_1 < p_2$$

One possible test statistic to use here would be the standard normal (Z) statistic that was used to derive a confidence interval for $p_1 - p_2$, namely,

$$Z = \frac{\hat{p}_1 - \hat{p}_2}{\sqrt{\dfrac{\hat{p}_1(1 - \hat{p}_1)}{n_1} + \dfrac{\hat{p}_2(1 - \hat{p}_2)}{n_2}}} \qquad (10.14)$$

In all previous tests of hypothesis, we always examined the distribution of the test statistic when H_0 was *true*. For a one-tailed test, we assumed the boundary condition of H_0, which in this case is $p_1 = p_2$. Because of this, whenever we obtained a value of the test statistic in one of the tails, our decision was to reject H_0 because this value would be very unusual if H_0 were true. This reasoning was used for test statistics that followed a Z, t, χ^2, or F distribution.

We use the same approach here. If $p_1 = p_2 = p$ (for example), we can improve the test statistic in equation 10.14. For this situation, p is the proportion of successes in the combined population. Our best estimate of p is the proportion of successes in the *combined sample*. So define

$$\bar{p} = \frac{x_1 + x_2}{n_1 + n_2}$$

Thus, assuming $p_1 = p_2$, $\hat{p}_1 - \hat{p}_2$ is approximately a normal random variable with

$$\text{mean} = p_1 - p_2 = 0$$

and

$$\text{standard deviation} = \sqrt{\frac{p_1(1 - p_1)}{n_1} + \frac{p_2(1 - p_2)}{n_2}}$$
$$\cong \sqrt{\frac{\bar{p}(1 - \bar{p})}{n_1} + \frac{\bar{p}(1 - \bar{p})}{n_2}}$$

The resulting test statistic for p_1 versus p_2 (large independent samples; $n_1\hat{p}_1$, $n_1(1 - \hat{p}_1)$, $n_2\hat{p}_2$, and $n_2(1 - \hat{p}_2)$ are each greater than 5) is

$$Z = \frac{\hat{p}_1 - \hat{p}_2}{\sqrt{\dfrac{\bar{p}(1 - \bar{p})}{n_1} + \dfrac{\bar{p}(1 - \bar{p})}{n_2}}} \qquad (10.15)$$

where

$$\hat{p}_1 = x_1/n_1$$
$$\hat{p}_2 = x_2/n_2$$
$$\bar{p} = \frac{x_1 + x_2}{n_1 + n_2}$$

Observe that the form of this test statistic is the same as for the single-population case described in equation 10.7. The test procedure is the standard routine when using the Z distribution. For a two-tailed test,

$$H_0: p_1 = p_2$$
$$H_a: p \neq p_2$$
$$\text{reject } H_0 \text{ if } |Z| > Z_{\alpha/2}$$

where Z is defined in equation 10.15. For a one-tailed test,

$$H_0: p_1 \leq p_2$$
$$H_a: p_1 > p_2$$
$$\text{reject } H_0 \text{ if } Z > Z_\alpha$$

or

$$H_0: p_1 \geq p_2$$
$$H_a: p_1 < p_2$$
$$\text{reject } H_0 \text{ if } Z < -Z_\alpha$$

EXAMPLE 10.11   Use the engine overhaul data from Example 10.8 and determine whether there is any difference between the proportion of Henry cars and that of GA cars that required an engine overhaul before 100,000 miles. Let $\alpha = .01$.

SOLUTION   The five-step procedure is the correct one. The confidence interval derived in Example 10.8 would produce the same result as the five-step procedure *if* the test statistic were the one defined in Equation 10.14. The correct procedure here is to use the Z-statistic in equation 10.15 as your test statistic.

Step 1. Since we are looking for a difference between p_1 and p_2, define

$$H_0: p_1 = p_2$$
$$H_a: p_1 \neq p_2$$

Step 2. The test statistic is

$$Z = \frac{\hat{p}_1 - \hat{p}_2}{\sqrt{\dfrac{\bar{p}(1 - \bar{p})}{n_1} + \dfrac{\bar{p}(1 - \bar{p})}{n_2}}}$$

Step 3. Using $\alpha = .01$, then $Z_{\alpha/2} = Z_{.005} = 2.575$. The test procedure will be to

$$\text{reject } H_0 \text{ if } |Z| > 2.575$$

Step 4. Since $n_1 = 100$, $x_1 = 28$ and $n_2 = 150$, $x_2 = 48$, then

$$\bar{p} = \frac{x_1 + x_2}{n_1 + n_2} = \frac{76}{250} = .304$$

Therefore, our estimate of the proportion of cars needing an overhaul in the combined population (if $p_1 = p_2$) is $\bar{p} = .304$ (30.4%). Also, $\hat{p}_1 = 28/100 = .28$, and $\hat{p}_2 = 48/150 = .32$. The value of the test statistic is

$$Z^* = \frac{.28 - .32}{\sqrt{\dfrac{(.304)(.696)}{100} + \dfrac{(.304)(.696)}{150}}} = \frac{-.04}{.059} = -.68$$

Because $|Z^*| = 68 < 2.575$, we fail to reject H_0.

FIGURE 10.4

Z curve showing p-value
(twice the shaded area)
for Example 10.11.

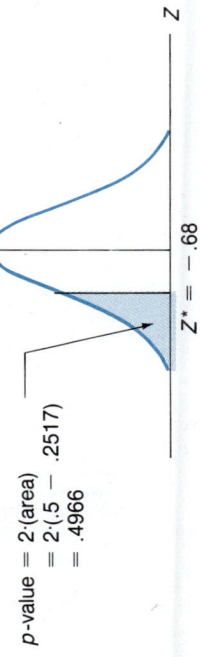

$p\text{-value} = 2(\text{area})$
$= 2(.5 - .2517)$
$= .4966$

$Z^* = -.68$

Z

Step 5. There is *insufficient evidence* to conclude that a difference exists between the Henry and GA cars as far as engine durability is concerned.

The Z curve and calculated p-value for Example 10.11 are shown in Figure 10.4. The p-value is twice the shaded area (this was a two-tailed test) and is .4966, which is extremely large. Using the classical approach, because $.4966 > \alpha = .01$, we fail to reject H_0—there is insufficient evidence to indicate a difference in engine durability. As a reminder, this reasoning *always* leads to the same conclusion as the five-step procedure. Because .4966 exceeds *any* reasonable value of α, we fail to reject H_0 quite strongly for this application.

EXAMPLE 10.12

In Example 10.9 we derived a confidence interval for Slick cigarettes' proportion sold after the campaign (p_1) and their proportion sold before the campaign (p_2). Based on these data, can you conclude that the proportion sold increased as a result of this campaign? Use a significance level of $\alpha = .05$.

SOLUTION

Step 1. We wish to know whether the data warrant the conclusion that p_1 is *larger* than p_2. Placing this in the alternative hypothesis leads to

$$H_0: p_1 \leq p_2$$
$$H_a: p_1 > p_2$$

Steps 2, 3. Using the test statistic in equation 10.15, the resulting one-tailed test procedure would be to

reject H_0 if $Z > Z_{.05} = 1.645$

Step 4. We have

$$\hat{p}_1 = \frac{90}{1200} = .075$$

and

$$\hat{p}_2 = \frac{54}{1200} = .045$$

Also,

$$\bar{p} = \frac{(90 + 54)}{(1200 + 1200)} = \frac{144}{2400} = .06$$

Consequently,

$$Z^* = \frac{.075 - .045}{\sqrt{\frac{(.06)(.94)}{1200} + \frac{(.06)(.94)}{1200}}} = \frac{.03}{.0097} = 3.09$$

Because $3.09 > 1.645$, we reject H_0.

FIGURE 10.5
Z curve showing the
calculated *p*-value for
Example 10.12.

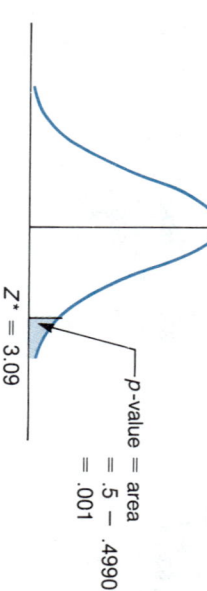

p-value = area
= .5 − .4990
= .001

$Z^* = 3.09$

Step 5. There *is* evidence of an increase in the proportion of Slick cigarettes sold following the advertising campaign. ■

The Z curve and calculated *p*-value for Example 10.12 (a one-tailed test) are shown in Figure 10.5. The *p*-value is .001. This is definitely a very small *p*-value and (as before) leads to rejecting H_0 using a significance level of .05. Based on this *p*-value alone, we arrive at the same conclusion—namely, that the market share *did* increase following the 6-month advertising campaign.

EXERCISES

10.31 A manufacturer of storm windows sampled 250 new (less than 5 years of age) homes and found that 142 of them had storm windows. Another sample of size 320 of older (at least five years of age) homes was taken; 150 of them had storm windows. The manufacturer believes that the proportion of older homes that have storm windows is larger than the proportion of new homes that have storm windows. Do the sample statistics support the manufacturer's claim at the .05 significance level?

10.32 Using the data in Exercise 10.31, construct a 90% confidence interval for the difference between the proportion of new homes with storm windows and the proportion of older homes with storm windows.

10.33 How does the value of the test statistic given in equation 10.15 change if $\hat{p}_1$, the proportion of successes for sample 1, is replaced by the proportion of failures for sample 1 and if $\hat{p}_2$, the proportion of successes for sample 2, is replaced by the proportion of failures for sample 2?

10.34 Corporations in America have a long tradition of making charitable contributions to support the fine arts. A 1987 Lou Harris poll showed that though Americans seem to want more arts (such as music, ballet, and so on) they seem to be enjoying them less. In 1984, 35% of those surveyed said they attended either an opera or a musical theater, whereas in 1987, the percentage had fallen to 27%. Find a 99% confidence interval on the difference in the proportion of people who attended either an opera or a musical theater in 1984 and in 1987. Assume that 2000 observations were used in each of the surveys in 1984 and 1987. If you were a corporate donor supporting the arts for many years with charitable contributions, would you say that your donations provided support during a time of declining audiences?

(Source: "So Much To Do, So Little Time," *Dallas Morning News,* April 17, 1988, p. 12C)

10.35 The owner of two hotels in Atlanta is interested in the proportion of "no-shows" on Friday night. The manager believes that the proportion of no-shows does not differ significantly between the two hotels. Assume that preliminary estimates of the proportion of no-shows at hotel A is 15% and at hotel B is 19%. How many random Friday nights would the owner need to select from each hotel to estimate the difference in the proportions of no-shows to within .20 with 95% confidence so that the total sample size is minimized?

10.36 A random sample of 125 manufacturing firms showed that 64% of them spent more than 75% of their total revenue on salaries and wages. A random sample of 100 wholesale firms showed that 57% of them spent more than 75% of their total revenue on salaries and wages. Let $\alpha = .05$ and test $H_0: p_1 \leq p_2$ and $H_a: p_1 > p_2$ where p_1 and p_2

are the proportions of manufacturing and wholesale firms, respectively, that spent more than 75% of their total revenue on salaries and wages.

10.37 Construct a 99% confidence interval for $p_1 - p_2$ using the information in Exercise 10.36.

10.38 A financial analyst compared the performances of individual stocks with the performance of the industry average (in terms of rate of return). The industry average is the average of all stocks that belong to the same industry and are listed on the New York Stock Exchange. The analyst believed that the proportion of stocks that perform better than the industry average is the same for both the oil and the steel industry. In a random sample of 37 oil-industry stocks, 17 performed better than the industry average. Similarly, in a random sample of 30 steel-industry stocks, 11 did better than the industry average. Using a .05 significance level, test the validity of the financial analyst's belief.

10.39 Two manufacturers supply rebuilt motors to an air-conditioning repair company. A preliminary study showed that the proportion of defective motors from one manufacturer was 10% and from the other manufacturer was 16%. Determine what sample sizes are necessary for the estimate of the difference between the proportions of defective motors from the two manufacturers to be within .15 with 95% confidence such that the total sample size is minimized.

SUMMARY

You will often encounter a situation in which you are concerned with a population **proportion** rather than the mean or variance. For example, the parameter of interest might be the proportion (p) of executives earning more than $100,000 annually rather than the average salary (μ) or the standard deviation (σ) of the salaries. The usual procedure of estimating a population parameter using the sample estimator, $\hat{p}$, allows us to derive a point estimate and construct a confidence interval for p. When the sample is small, Table A.8 provides an exact confidence interval for p. For large samples, the Central Limit Theorem can be applied to determine an approximate confidence interval, provided that *both* np and $n(1 - p)$ are greater than 5.

When the desired accuracy of the point estimator, $\hat{p}$, is specified in advance, you can determine the sample size necessary to obtain this degree of accuracy for a certain confidence level. To derive this sample size, an estimate of p is necessary. You can calculate this value using a previous sample estimate or, if no information is available, using a conservative procedure and making $\hat{p} = .5$.

When you investigate a statement concerning a population proportion, you can use a statistical test of hypothesis. For small samples, the confidence interval from Table A.8 provides an exact procedure for either a one- or two-tailed test. For tests of a hypothesis when a large sample is used, a test statistic having an approximate standard normal distribution can be used.

To compare **two population proportions** (p_1 and p_2), two *independent* random samples are obtained, one from each population. Procedures for large independent samples generally provide an accurate confidence interval or test of hypothesis whenever $n_1\hat{p}_1$, $n_1(1 - \hat{p}_1)$, $n_2\hat{p}_2$, and $n_2(1 - \hat{p}_2)$ each exceeds 5. Using a standard normal approximation, we can construct a confidence interval for $p_1 - p_2$. If the accuracy of this estimate is specified, the sample sizes necessary to obtain this level of accuracy as well as to minimize the total sample size $n_1 + n_2$ can be obtained.

Two population proportions can be compared by using two large independent samples to evaluate a test statistic having an approximate standard normal distribution. We examined procedures for a one-tailed test (for example, $H_a: p_1 > p_2$) or a two-tailed test ($H_a: p_1 \neq p_2$). The rejection regions for these tests are defined using the areas from Table A.4.

REVIEW
EXERCISES

10.40 Ten of the 17 employees who took an in-house speed-reading course can show that the course has substantially increased their efficiency on the job.

a. Are there significantly more employees who have benefited from the course than have not at the .05 significance level? Use Table A.8.

b. Find a 95% confidence interval for the proportion of employees who have benefited from the course. Use Table A.8.

10.41 An advertising agent for Computerized Telephone Systems claims that the proportion of installed telephone systems that have maintenance problems during the first 3 years is less than 10%. A random sample of 19 computerized telephone systems that were installed within the last 3 years was taken, and one of the telephone systems was found to have needed repairs.

a. Test the advertising agent's claim at the .05 significance level. Use Table A.8.

b. Find a 90% confidence interval for the true proportion of installed telephone systems that have maintenance problems.

10.42 Fifteen male customers were asked which of two electric shavers, brand 1 or brand 2, they preferred. Nine of them preferred brand 1.

a. At the .05 level of significance, can it be concluded that brand 1 was preferred to brand 2 by male shoppers? Use Table A.8.

b. Find a 95% confidence interval for the proportion of male shoppers who preferred brand 1 over brand 2. Use Table A.8.

c. Assume you have no prior knowledge of p (the proportion of males who preferred brand 1 over brand 2). Estimate the sample size that is required to estimate p, with 90% confidence, assuming a maximum error of .08.

10.43 A small sample ($n = 15$) is drawn from a lot of dry-cell batteries. Each of the batteries was tested; seven were defective. Determine a 95% confidence interval for the proportion of defectives in the entire batch. Use Table A.8.

10.44 William's Packaging is interested in estimating the proportion of its employees who would attend an alcohol-awareness program. In a random sample of 70 employees, 39 said that they would attend the program. Calculate the estimate of the proportion of all employees who would attend the program. Find a 90% confidence interval for this proportion.

10.45 Using the data from Exercise 10.44, if the maximum error of the estimate for the proportion is 4%, estimate the sample size needed for a 95% confidence level. Use the value of $\hat{p}$ from Exercise 10.44.

10.46 Harry Truman once said, "The buck stops here." However, presidents are not the only ones who are passed the buck. The following interesting data were published in an article by Jill Kiecolt (1987) of Louisiana State University:

WHO IS TO BLAME FOR NATIONAL
ECONOMIC PROBLEMS

WHO IS TO BLAME?	NUMBER	%
Congress	181	18.7
President	110	11.3
Labor unions	313	32.3
Big business	328	33.8
Big business and labor	10	1.0
Other combinations	28	2.9
Total sample:	970	100.0

(*Source:* K. J. Kiecolt, Group Consciousness and the Attribution of Blame for National Economic Problems. *Am. Politics Quar.* **15**, no. 2 (April 1987): 214, Table 1.)

a. Construct a 90% confidence interval for the proportion of people who blame Congress for national economic problems.

b. Construct a 95% confidence interval for the proportion of people who did not blame the President for national economic problems.

c. Construct a 99% confidence interval for the proportion of people who blame big business for national economic problems.

10.47 *People's Choice*, a monthly magazine, claimed that more than 40% of its subscribers had an annual income of $50,000 or more. In a random sample of 62 subscribers, 30 had incomes of $50,000 or more. Does this information substantiate the magazine's claim? Use a significance level of .10.

10.48 Calculate and interpret the *p*-value for Exercise 10.47.

10.49 A statistician reported to a car insurance company a confidence interval for the proportion of convertible cars that had been involved in major accidents during the past year. The 95% confidence interval for *p* was reported to be the interval from .10 to .36.

a. What is the statistician's estimate of *p*?

b. What is the maximum error of estimate (E) of the proportion for this confidence interval?

c. Approximately what sample size did the statistician use?

10.50 A confidence interval is reported for the proportion of male YLU students who belong to the $\alpha\sigma\mu$ fraternity. The confidence interval is based on a sample of 100 students and is given to be .22 to .44. What level of confidence was used in obtaining this interval?

10.51 Must a confidence interval for a proportion contain the true proportion of the population? Explain what the "level of confidence" means for a confidence interval.

10.52 A market-research firm believed that the proportion of households with more than four family members in county 1 was greater than the proportion of households with more than four family members in county 2. The firm gathered random samples of 180 and 155 from counties 1 and 2, respectively. The number of households with more than four members were 74 from county 1 and 61 from county 2. From these data, can we conclude that the proportion of households with more than four members is higher in county 1 than in county 2? Use a significance level of .01.

10.53 Calculate the *p*-value for exercise 10.52. Using the *p*-value, would you reject the null hypothesis at the .05 level?

10.54 Construct a 95% confidence interval for the difference of the two proportions in Exercise 10.52.

10.55 A market-research firm is interested in testing the hypothesis that the proportion of students who own a car is the same for the local state university campus and a local private college. They interviewed 240 students from the state university campus and a local private college. The number of students who did not own a car was 78 at the state university and 82 at the private college. Using a .02 significance level, test the hypothesis.

10.56 For Exercise 10.55, construct a 95% confidence interval for the difference of the true proportions of students who own cars at the two campuses.

10.57 A recent item in *Working Women* magazine stated that U.S. businesses make about 350 billion photocopies a year, of which 130 billion end up in the trash. Also, 29% of the "good" copies generally are filed and not used again, while 8% of the copies are made for employees' personal use. A female executive decided to investigate her own office. She obtained a random sample of 200 photocopies made and determined that 30 ended up in the trash, 80 "good" copies were filed, 70 good copies were mailed out, and 20 were made for the personal use of employees. Conduct three hypothesis tests, using $\alpha = .05$, to decide if her office is different from the national figures given in *Working Women* magazine, in the following categories:

a. The proportion of photocopies trashed.

b. The proportion of photocopies filed.

c. The proportion of photocopies made for personal use.

d. State the *p*-value in each case.

10.58 ABC's Monday Night Football had a rating of 19.7 and a share of 32 in 1985.

"Rating" is the percentage of homes with television sets that are tuned to a particular program out of all the homes that have television sets in the United States. "Share" is the percentage of sets actually in use that are tuned to a particular program. Although the actual process of determining these figures is fairly sophisticated, let us assume, for the sake of simplicity, that the "share" figure of 32 is computed from a sample of 1000 homes.

a. What is the 95% confidence interval for ABC's 1985 share for Monday Night Football?

b. What is the maximum error of the estimate in part (a)?

10.59 A credit union randomly selected 110 savings-account customers and found that 85 of them also had checking accounts with the union. Construct a 95% confidence interval for the true proportion of savings-account customers who also have checking accounts.

10.60 A labor-union leader stated that at least 40% of the members of the union had a college degree. A random sample of 440 members revealed that only 170 had a college degree. Test this hypothesis if H_0 is H_0: the proportion of union members with a college degree is greater than or equal to 40%. Use a .05 level of significance.

10.61 Many industries are responding to the need to increase the quality of their goods in order to stay competitive in international trade. In the 1980s, Ford Motor Company turned back shipments from Bethlehem Steel Corp. 8% of the time. In 1987, this percentage dropped to 1% of the time. Suppose that in a random sample of 600 shipments, 6 were turned back. Find a 90% confidence interval on the true proportion of shipments that were acceptable.

(*Source:* "The Push For Quality," *Business Week*, June 1987, pp. 131–5.)

10.62 The manager at a manufacturing firm is interested in the proportion of the quality problems that originate with the worker rather than with the system. In a random sample of 200 quality problems, it was found that 90 of the quality problems originated with the workers. From the following MINITAB computer printout, is there sufficient evidence for the manager to conclude that less than 50% of the quality problems originate with the worker? At what significance level? (Note that 90 ones and 110 zeros are placed in C1, and sigma is taken to be the square root of p times $1 - p$, where p is the hypothesized proportion.)

```
MTB > set c1
DATA> 90(1) 110(0)
DATA> end
MTB > let k1 = sqrt (.5 * .5)
MTB > ztest, mu = .5, alternative hypothesis = -1, sigma = k1, data in c1

TEST OF MU = 0.5000 VS MU L.T. 0.5000
THE ASSUMED SIGMA =0.500

        N    MEAN    STDEV   SE MEAN    Z    P VALUE
C1    200  0.4500  0.4987   0.0354  -1.41   0.079
```

10.63 An industrial psychologist conducted a survey of 150 randomly selected middle-level managers 40 years or older and of 150 randomly selected middle-level managers younger than 40. The industrial psychologist recorded that 81 of the managers 40 years of age or older did not think they would achieve a top-level management position in their firms. In the group of younger managers, 70 of the managers did not think they would achieve a top-level management position in their firm.

A MINITAB computer printout performs the statistical test procedure to determine if the proportions of the two groups are significantly different. What conclusion can be drawn from the analysis? Using the statistics in the computer printout, construct a 95% confidence interval on the difference of the proportions of the two groups. (Note that 81 ones and 69 zeros are placed in C1 and 70 ones and 80 zeros are placed in C2. Note that sigma is taken to be equal to $\sqrt{[2\bar{p}(1 - \bar{p}]}$. Using the MINITAB commands to perform the Z-test in this fashion requires that n_1 and n_2 be equal.)

```
MTB > set c1
DATA> 81(1) 69(0)
DATA> end
MTB > set c2
DATA> 70(1) 80(0)
DATA> end
MTB > let c3 = c1 - c2
MTB > let k2 = (81 + 70) / (150 + 150)
MTB > let k1 = sqrt(2*k2*(1-k2))
MTB > ztest, mu = 0, alternative hypothesis = 0, sigma = k1, data in c3
```

TEST OF MU = 0.0000 VS MU N.E. 0.0000
THE ASSUMED SIGMA =0.707

	N	MEAN	STDEV	SE MEAN	Z	P VALUE
C3	150	0.0733	0.2616	0.0577	1.27	0.20

COMPUTER EXERCISES USING THE DATABASE

Exercise 1—Appendix H

Randomly select 100 observations from the database. Find a 95% confidence interval on the proportion of households that own their homes. (Refer to the variable OWNORENT.)

Exercise 2—Appendix H

Randomly select 100 observations from the database. Estimate the proportion of observations from the NE sector in which the households own their homes. (Refer to the variables LOCATION and OWNORENT.) Also estimate the proportion of observations from the NW sector in which the households own their homes. Find a 95% confidence interval on the difference of the proportions of house owners for the two sectors.

Exercise 3—Appendix I

Randomly select 100 observations from the database. Find a 95% confidence interval on the proportion of companies with a positive net income. (Refer to the variable NETINC.)

Exercise 4—Appendix I

Randomly select 100 observations from the database. Estimate the proportion of observations from companies with an A bond rating that have a positive net income. (Refer to the variables BONDRATE and NETINC.) Also estimate the proportion of observations from companies with a B bond rating that have a positive net income. Find a 95% confidence interval on the difference of the proportions of companies with positive net income from those with A bond ratings and those with B bond ratings.

CASE STUDY

A LOOK AT AMERICAN AND JAPANESE MANAGEMENT STYLES

Japanese management practices and quality control techniques have been the subject of much discussion and research during the past two decades. Sullivan and Nonaka addressed the question: Just how different is Japanese management from American/European practices? They theorized that a greater Japanese commitment to an organizational learning perspective contributes to differences in the strategy-formulating behavior of Japanese and American executives.

The purpose of their research was to explore the organizational learning practices of Japanese managers as compared to a similar group of American managers. Senior American managers were considered to be company presidents since they are the individuals who carry out the theory of action in American firms. However, in

Japanese firms, company presidents typically are not active managers. Consequently, the Japanese sample mostly focused on *bucho*, those managers who are above section chiefs, since they carry out the same senior functions as American company presidents.

Responses to a questionnaire were received from 75 American company presidents, and 75

responses were randomly selected from 422 Japanese managers who participated in the study. The English questionnaire was translated into Japanese and then translated back into English by an American who did not have access to the original. Minor discrepancies were identified and resolved. A portion of the results are summarized in the following table.

Information
Management
Approaches of Senior
American ($n_1 = 75$)
and Japanese
($n_2 = 75$) Managers

	SENIOR AMERICAN MANAGERS PREFER	SENIOR JAPANESE MANAGERS PREFER
1. To communicate realistic rather than idealistic goals.	73%	65%
2. To inspire employees by talking about values rather than rewards.	52%	95%
3. To assign tasks which challenge employees to use greater effort and ability than they have ever done before rather than tasks which require routine or average effort and ability.	58%	73%
4. To stress employee task rotation rather than specialization.	60%	75%
5. To do some employee training themselves.	68%	62%
6. To stress learning from mistakes rather than avoiding mistakes.	81%	89%
7. To base employee knowledge on experience rather than on rules, manuals or information systems.	61%	78%
8. To encourage overlapping, loosely defined projects to foster information sharing.	70%	51%
9. To give general, rather than specific directions to project teams to foster freedom in their deciding how to proceed.	72%	89%

(*Source:* Sullivan, J. J. and Nonaka, I., *J. of International Business Studies*, **XVII** no. 3 (Fall 1986): 129–142. Reprinted with permission of Hong Kong Baptist College, Hong Kong.)

Case Study Questions

1. Do you agree with the authors' procedure of randomly selecting the 75 Japanese responses from

   A study of the table reveals some apparently significant differences between the proportions of American and Japanese managers who prefer a particular management style. For example, 52% of the American managers prefer to inspire

   employees by talking about values rather than rewards, as compared to 95% for the Japanese managers. Are such sample percentages statistically significant? The following questions will take a closer look at these percentages.

   the 422 managers who participated in the study? Why not simply select the first 75 responses?

2. Give the point estimate for the number of American managers who stress task rotation rather than specialization.

3. Construct a 95% confidence interval for the parameter in question 2.

4. Give the point estimate for the difference between the proportions of American managers (p_1) and Japanese managers (p_2) who prefer to do some employee training themselves. (Use $p_1 - p_2$.)

5. Construct a 99% confidence interval for question 4.

6. Do the data support the hypothesis that more than half of the American managers inspire employees by talking about values rather than

rewards? Conduct a hypothesis test at a 5% significance level, and state the p-value.

7. Is there evidence to indicate that a higher percentage of American managers (p_1) prefer to communicate realistic rather than idealistic goals than do the Japanese managers (p_2)? Conduct a hypothesis test at the 1% level of significance, and state the p-value.

8. Is there evidence to indicate that a higher percentage of American managers prefer to encourage overlapping, loosely defined projects to foster information sharing? Conduct a test of hypothesis at the 5% level of significance, and state the p-value.

TEST STATISTIC
F RATIO

AVG. VAR "Between" means
"   "   within Samples

Analysis of Variance

A Look Back/Introduction

In Chapter 9, we considered a question of the type, Do men have the same heights as women? By this we mean, Is the *average* height of males equal to the average height of females? We were interested in the means of two populations and performed a test of hypothesis, using, for example, $H_0: \mu_M = \mu_F$ and $H_a: \mu_M \neq \mu_F$. This works well when dealing with two populations, but how can we compare the means of *more than two* populations? For example, we might wish to examine the average sales of five different training programs to see whether they are the same. Our hypotheses become

$$H_0: \mu_1 = \mu_2 = \mu_3 = \mu_4 = \mu_5$$

$$H_a: \text{not all } \mu\text{'s are equal}$$

We test such a hypothesis by first collecting five samples, one from each of the training programs (populations). We will see that to compare these five means one pair at a time is *not* the correct approach. This results in ten different pairwise tests, and what was intended to be a testing procedure with, say, a .05 significance level results in a much higher significance level. In other words, the overall significance level, α, is *larger* than the predetermined value. The correct procedure for this situation is to examine the *variation* of the sales values, both (1) within each of the samples (examining the variability of each sample alone) and (2) among the five samples (for example, are the values in sample 1 larger or smaller, on the average, than the values in the other samples?).

In Chapter 9, we saw that when trying to decide if $\bar{X}_1$ is "significantly different" from $\bar{X}_2$, a key part of the answer rested on the values of s_1 and s_2, the variation *within* the two samples. Both s_1 and s_2 affect the width of the confidence interval for $\mu_1 - \mu_2$. Consequently, we infer something about the *means* of several populations by utilizing the *variation* of the resulting samples. Hence the term *analysis of variance*—our next topic.

11.1

COMPARING
TWO MEANS:
ANOTHER
LOOK

We begin with an example. The manager of a convenience store wants to know whether the sales of two particular brands of cigarettes (brand 1 and brand 2) are the same. Based on past experience, he believes that weekly sales follow a *normal* distribution. By using past sales records, he obtains that weekly sales follow a *normal* distribution. By using past sales records, he obtains the number of cartons sold per week for brand 1 using a randomly selected 5-week period. The sales for brand 1, brand 2 are randomly obtained for a *different* 5-week period, with the following results (in numbers of cartons):

BRAND 1	BRAND 2
43	30
48	26
38	37
41	31
51	34

Let μ_1 be the average weekly sales (if observed indefinitely) for brand 1 and μ_2, be the average for brand 2. We wish to determine whether the data allow us to conclude that $\mu_1 \neq \mu_2$, using $\alpha = .10$.

We examined the same type of question in Chapter 9; we are dealing with two small independent samples. In Chapter 9 we advised against assuming that σ_1 was equal to σ_2. As a result, we generally used a t-test that did *not* pool the sample variances. However, when examining more than two normal populations (the main concern of this chapter), the following testing procedure for detecting a difference in the population means requires that the populations have the *same* distribution. This distribution can be used only when we are willing to assume that the population means are equal. Consequently, it can be used only when we are willing to assume that the population means *are equal* (or approximately equal). The analysis of variance procedure is *not* extremely sensitive to departures from this assumption, especially if equal-sized samples are obtained from each population. A procedure for verifying this assumption (similar to the F-test used to compare two variances in Chapter 9) is discussed in this chapter.

As a result, we will assume that we have reason to believe that the variation of the brand 1 sales is the same as for brand 2 sales; that is, $\sigma_1 = \sigma_2$. Using the approach discussed in Chapter 9, we first find

$$s_p^2 = \text{pooled variance} = \frac{(n_1 - 1)s_1^2 + (n_2 - 1)s_2^2}{n_1 + n_2 - 2}$$

where n_1, n_2 = sample sizes for brand 1, brand 2 and s_1^2, s_2^2 = sample variances for brand 1, brand 2. Using the sample data,

BRAND 1	BRAND 2
$n_1 = 5$	$n_2 = 5$
$\bar{x}_1 = 44.2$	$\bar{x}_2 = 31.6$
$s_1 = 5.263$	$s_2 = 4.159$

Consequently,

$$s_p^2 = \frac{(4)(5.263)^2 + (4)(4.159)^2}{8}$$

$$= \frac{180.0}{8}$$

$$= 22.5$$

and so

$$s_p = \sqrt{22.5} = 4.74$$

The appropriate hypotheses are $H_0: \mu_1 = \mu_2$ and $H_a: \mu_1 \neq \mu_2$. The resulting test statistic is

$$t = \frac{\bar{X}_1 - \bar{X}_2}{s_p \sqrt{\dfrac{1}{n_1} + \dfrac{1}{n_2}}}$$

$$= \frac{44.2 - 31.6}{4.74 \sqrt{\dfrac{1}{5} + \dfrac{1}{5}}} = \frac{12.6}{2.998}$$

$$= 4.20$$

That is, $t^* = 4.20$.

We are dealing with a two-tailed test using a t-statistic with $(n_1 - 1) + (n_2 - 1) = 4 + 4 = 8$ df, so the test procedure is to

$$\text{reject } H_0 \text{ if } |t^*| > t_{\alpha/2, \text{df}} = t_{.05, 8} = 1.86$$

Comparing $t^* = 4.20$ to 1.86, we reject H_0 and conclude that the mean sales for the two brands are not the same. Looking at the sample data, we can say that $\bar{x}_1 = 44.2$ is significantly different from $\bar{x}_2 = 31.6$.

The Analysis of Variance Approach

We need to introduce two new terms. The previous example examined the effect of one **factor** (brand), consisting of two **levels** (brand 1 and brand 2). If you want to extend this to four brands (say, brands 1, 2, 3, and 4), then you still have *one* factor but you now have four levels.

The purpose of **analysis of variance (ANOVA)** is to determine if this factor has *a significant effect* on the variable being measured (sales, in our example). If, for instance, the brand factor *is* significant, the mean sales for the different brands will not be equal. Consequently, testing for equal means among the various brands is the same as attempting to answer the question, Is there a significant effect on sales due to this factor?

This section examines the effect of a single factor on the variable being measured, **one-factor ANOVA**. Extensions of this technique include ANOVA procedures that determine the effect of two or more factors operating simultaneously. These factors may be *qualitative* (such as brand in the previous illustration) or *quantitative* (such as several levels of advertising expenditure).

All ten values in the cigarette-sales example are different, and we observe a variation in these values. We will look at two *sources of variation*: (1) variation *within* the samples (levels) and (2) variation *between* the samples.

Within-Sample Variation

When you obtain a sample, you usually obtain different values for each observation. The five sample values for brand 1 vary about the mean $\bar{x}_1 = 44.2$ cartons, as measured by $s_1 = 5.26$ cartons. Likewise, the five values in the second sample also exhibit some variation ($s_2 = 4.16$) about $\bar{x}_2 = 31.6$ cartons. These are the **within-sample variations**. They are used when estimating the common population variance, say σ^2. This procedure provides an accurate estimate of σ^2, whether or not the sample means are equal.

Between-Sample Variation

When you compare the two samples, you observe that the values for brand 1 are *larger*, on the average, than are those for brand 2. This is summarized in the sample means, where $\bar{x}_1 = 44.2$ appears to be considerably larger than $\bar{x}_2 = 31.6$. So there is a variation in the ten values due to the *brand*; that is, due to the factor. This is **between-sample variation**. In general, if this variation is large, we expect considerable variation among the sample means. The between-sample variation is also used in another estimate of the common variance, σ^2, *provided the population means are equal*. In other words, if the means are equal, the between-sample and within-sample estimates of σ^2 should be nearly the same. As we will see later in this section, we can derive a test of hypothesis procedure for determining whether the means are equal by comparing these two estimates.

Measuring Variation

When using the ANOVA approach, we measure these two sources of variation by calculating various **sums of squares, SS**. We determine

SS(factor), which measures between-sample variation (also called SS(between))

SS(error), which measures within-sample variation (also called SS(within))

SS(total) = SS(between) + SS(within) = SS(factor) + SS(error)

In addition, each of the first two sums of squares will have corresponding degrees of freedom, df, which are determined from the number of terms that make up this particular SS. The df are given by

$$\text{df for factor} = (\text{number of levels}) - 1$$
$$= (\text{number of brands}) - 1$$
$$= 2 - 1 = 1$$

$$\text{df for error} = (n_1 - 1) + (n_2 - 1)$$
$$= n_1 + n_2 - 2 = 5 + 5 - 2 = 8$$

We will show how to determine these sums of squares and how we combine them, and their df, into another test statistic for testing $H_0 : \mu_1 = \mu_2$ against $H_a : \mu_1 \neq \mu_2$. The beauty of this approach is that it extends nicely to the situation in which you wish to compare more than two means using a *single* test.

Determining SS(factor)   SS(factor) is the sum of squares that determines whether the values in one sample are larger or smaller on the average than the values in the second sample.

$$SS(\text{factor}) = n_1(\bar{x}_1 - \bar{\bar{x}})^2 + n_2(\bar{x}_2 - \bar{\bar{x}})^2 \qquad (11.1)$$

where $\bar{x}_1, \bar{x}_2$ are the two sample means and

$$\bar{\bar{x}} = \frac{\sum (\text{all data values})}{n} = \frac{n_1\bar{x}_1 + n_2\bar{x}_2}{n_1 + n_2}$$

and $n = n_1 + n_2 = $ total sample size.

A method of determining this sum of squares that is much easier using a calculator is

$$SS(\text{factor}) = \left[\frac{T_1^2}{n_1} + \frac{T_2^2}{n_2}\right] - \frac{T^2}{n} \qquad (11.2)$$

where T_1 = total of the sample 1 observations, T_2 = total of the sample 2 observations, and T = grand total = $T_1 + T_2$.

Determining SS(total) SS(total) is a measure of the variation in all $n = n_1 + n_2$ data values. You obtain its value as though you were finding the *variance of* these n values, except that you do not divide by $n - 1$. So,

$$SS(\text{total}) = \Sigma(x - \bar{x})^2 \qquad (11.3)$$

or (after some algebra similar to that used in Chapter 3),

$$SS(\text{total}) = \Sigma x^2 - \frac{(\Sigma x)^2}{n} = \Sigma x^2 - \frac{T^2}{n} \qquad (11.4)$$

Determining SS(error) SS(error) is the measure of the variation *within* each of the samples. Its value simply is the *numerator of the pooled variance, s_p^2,* obtained using the previous *t*-test. Thus,

$$SS(\text{error}) = \underbrace{\Sigma(x - \bar{x}_1)^2}_{\text{first sample}} + \underbrace{\Sigma(x - \bar{x}_2)^2}_{\text{second sample}} \qquad (11.5)$$

and therefore,

$$SS(\text{error}) = \Sigma x^2 - \left[\frac{T_1^2}{n_1} + \frac{T_2^2}{n_2}\right] \qquad (11.6)$$

Given that

$$SS(\text{total}) = SS(\text{factor}) + SS(\text{error})$$

a much easier way to find this value is

$$SS(\text{error}) = SS(\text{total}) - SS(\text{factor}) \qquad (11.7)$$

Let us return to the cigarette sales example. To find the SS(factor) here, we first determine

$$T_1 = 43 + 48 + 38 + 41 + 51$$
$$= 221$$

$$T_2 = 30 + 26 + 37 + 31 + 34$$
$$= 158$$

$$T = T_1 + T_2 = 221 + 158 = 379$$

So, using equation 11.2,

$$SS(factor) = \frac{221^2}{5} + \frac{158^2}{5} - \frac{379^2}{10}$$
$$= 14,761 - 14,364.1$$
$$= 396.9$$

To find SS(total), the only new term we need to evaluate is

$$\Sigma x^2 = \text{sum of each data value squared}$$
$$= 43^2 + 48^2 + \cdots + 31^2 + 34^2$$
$$= 14,941$$

So, using equation 11.4 (the value 14,364.1 was obtained in SS(factor)),

$$SS(total) = \Sigma x^2 - \frac{T^2}{n}$$
$$= 14,941 - 14,364.1$$
$$= 576.9$$

Finally, we find SS(error) by subtraction:

$$SS(error) = SS(total) - SS(factor)$$
$$= 576.9 - 396.9$$
$$= 180.0$$

ANOVA TEST for H_0: $\mu_1 = \mu_2$ Versus H_a: $\mu_1 \neq \mu_2$

To begin with, the procedure we are about to define is valid for a *two-tailed test only*. In other words, the alternative hypothesis must be that the two means differ, not that one is larger than the other (a one-tailed test). (When examining more than two means, the alternative hypothesis will be that *at least* two of the means are unequal and H_0 will be that all the means are equal.) The next step, when using the ANOVA procedure, is to determine something resembling an "average" sum of squares, referred to as a **mean square**. We compute a mean square for only SS(factor) and SS(error), not for SS(total).

$$MS(factor) = \frac{SS(factor)}{\text{df for factor}} = SS (factor)/1 \qquad (11.8)$$

Note that the df for this term always is (number of levels) − 1. In this section, we are dealing with two levels (populations), and so here df is 1.

$$MS(error) = \frac{SS(error)}{\text{df for error}} = \frac{SS \text{ (error)}}{(n_1 + n_2 - 2)} \qquad (11.9)$$

We denote the common variance of the two normal populations as σ^2. So, $\sigma^2 = \sigma_1^2 = \sigma_2^2$. If the null hypothesis—H_0: the means are equal—is true, then, because the populations have identical means and variances, this implies that under H_0 we are dealing with a *single population*. The ANOVA procedure is based on a comparison between two separate estimates of the variance, σ^2. The first estimate is derived using the variation among the sample means (only two in the previous example). The other estimate is determined using the variation *within* each of the samples.

The ANOVA procedure is based on a comparison of these two estimates of σ^2 because they should be approximately equal *provided H_0 is true.* We have derived these two estimates:

MS(factor) = estimate of σ^2 based on the variation among the sample means

MS(error) = estimate of σ^2 based on the variation within each of the samples

Our new test statistic for testing $H_0: \mu_1 = \mu_2$ versus $H_a: \mu_1 \neq \mu_2$ is the *ratio* of these two estimates:

$$F = \frac{\left(\begin{array}{c}\text{estimated population variance based on the variation} \\ \text{among the sample means}\end{array}\right)}{\left(\begin{array}{c}\text{estimated population variance based on the variation} \\ \text{within each of the samples}\end{array}\right)}$$

$$= \frac{\text{MS(factor)}}{\text{MS(error)}} \tag{11.10}$$

This test statistic follows an F distribution, which was first introduced in Chapter 9 as a ratio of two variance estimates. The degrees of freedom (df) for the F-statistic in equation 11.10 are the df for factor and the df for error; that is, in our present examples, the df for F are 1 and $(n_1 + n_2 - 2)$. Because the F-statistic is based on a comparison of two variance estimates, this technique is called analysis of variance.

This is our second encounter with the F distribution. In Chapter 9, we used this distribution to compare two population variances (σ_1^2 and σ_2^2). The shape of this distribution is illustrated in Figure 11.1 and is tabulated in Table A.7. Remember that the shape of the F curve is affected by both the df for the numerator (1 here) and the df for the denominator ($n_1 + n_2 - 2$ here).

Defining the Rejection Region

What happens to the F statistic when H_a is true; that is, when $\mu_1 \neq \mu_2$? In this case, we would expect $\bar{X}_1$ and $\bar{X}_2$ to be "far apart." As a result, the estimate of the variance σ^2 using the *between-sample* variation (measured by MS(factor)) will be *larger* than the estimate of σ^2 based on the *within-sample* variation (measured by MS(error)). This implies that we should reject H_0 in favor of H_a whenever the ratio of these two estimates is large—in which case the computed F-value is in the right tail. Consequently, the test procedure will be to

reject H_0 if $F^* > F_{\alpha, v_1, v_2}$

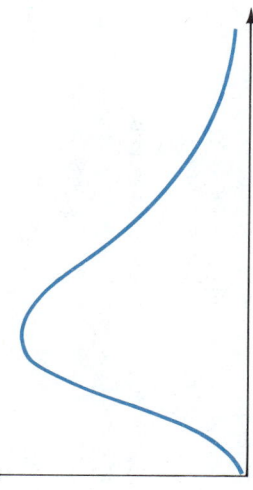

FIGURE 11.1

Shape of the F distribution shown by F curve with 1 and $n_1 + n_2 - 2$ df.

where $v_1 = $ df for numerator $= $ (number of levels) $- 1 = 1$, $v_2 = $ df for denominator $= n_1 + n_2 - 2$, and F_{α, v_1, v_2} is obtained from Table A.7 with a right-tail area $= \alpha$.

EXAMPLE 11.1

Using the data from the cigarette sales example and the previously calculated sums of squares, test H_0: $\mu_1 = \mu_2$ versus H_a: $\mu_1 \neq \mu_2$, where $\mu_1 = $ average weekly number of cartons sold for brand 1, if observed indefinitely, and $\mu_2 = $ average for brand 2. Use a significance level of $\alpha = .10$.

SOLUTION

Step 1. The hypotheses are as defined—H_0: $\mu_1 = \mu_2$ and H_a: $\mu_1 \neq \mu_2$.

Step 2. The test statistic is

$$F = \frac{\text{MS(factor)}}{\text{MS(error)}}$$

Step 3. The rejection region (using Table A.7(a)) is

reject H_0 if $F > F_{.10, 1, 8} = 3.46$

Step 4. From the previous calculations, SS(factor) $= 396.9$ and SS(error) $= 180$. So,

$$\text{MS(factor)} = \text{SS(factor)}/1 = 396.9$$

and

$$\text{MS(error)} = \frac{\text{SS(error)}}{(n_1 + n_2 - 2)} = 180.0/8 = 22.5$$

The resulting value of the test statistic is

$$F^* = \frac{396.9}{22.5} = 17.64$$

Because $17.64 > 3.46$, we reject H_0.

Step 5. These data indicate that the mean sales for brand 1 and brand 2 are *not* the same. ∎

COMMENTS

Compare our first treatment of the cigarette sales problem with Example 11.1. Both solutions led to the same conclusion, namely, that the two average sales are not the same. In fact, both solutions *always* lead to the same conclusion when comparing two means. Furthermore, the p-values for both solutions *are the same*, as illustrated in Figure 11.2. The values were obtained using a computer program (available in many statistical packages) that provides an exact p-value for a t- or F-statistic, given the computed value and corresponding degrees of freedom.

The computed value of the F statistic is equal to the square of the computed value of the t-statistic because $17.64 = (4.20)^2$. This is true whenever you have an F statistic or table value with 1 df in the numerator. So,

$$F^* = (t^*)^2$$

Furthermore, the table values satisfy the same relationship, namely,

$$F_{.10, 1, 8} = 3.46 = (1.86)^2 = [t_{.05, 8}]^2$$

We see that the two tests are *identical*; they produce the same conclusion and p-value. Furthermore, the computed value and the table value for the F-statistic are the squares of the corresponding values using the t-statistic. This comparison applies *only* when the F-statistic has 1 df in the numerator—that is, when there are two factor levels (as in this illustration). As mentioned previously, the advantage of the ANOVA approach is that it extends very easily to the situation of comparing means for more than two populations (covered in the next section).

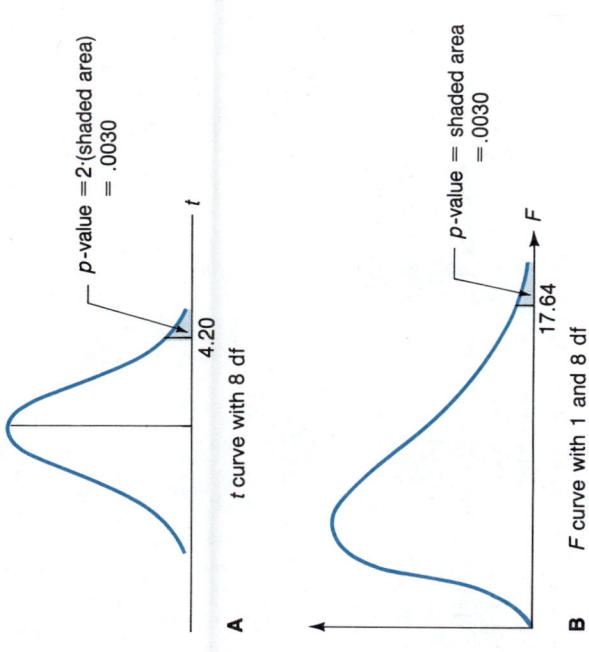

FIGURE 11.2

p-values for the solution to the cigarette sales example. **A**: Solution using pooled variance t test. **B**: Solution using ANOVA (see Example 11.1).

The ANOVA Table

Rather than carrying out the five-step procedure using the F-statistic, an easier method is to use an **ANOVA table** of the various sums of squares. The format of this table is as follows:[†]

SOURCE	df	SS	MS	F
Factor	1	SS(factor)	$MS(factor) = \dfrac{SS(factor)}{1}$	$\dfrac{MS(factor)}{MS(error)}$
Error	$n-2$	SS(error)	$MS(error) = \dfrac{SS(error)}{n-2}$	
Total	$n-1$	SS(total)		

To fill in this table, you compute the necessary sums of squares along with the mean squares and insert them. Notice that $n = n_1 + n_2 =$ total sample size and that column 3 (MS) = column 2 (SS) divided by column 1 (df).

The ANOVA table for Example 11.1 follows.

SOURCE	df	SS	MS	F
Factor	1	396.9	396.9	17.64
Error	8	180.0	22.5	
Total	9	576.9		

Summary of the ANOVA Approach for One-Factor Tests

In Example 11.1 we concluded that a difference existed between the two *means* because the variation *between* the two samples (measured by MS(factor)) was much greater than the variation *within* the samples (measured by MS(error)). Thus, the ratio of these values was very large and $F*$ fell in the rejection region.

[†] The headings under the "Source" column will vary, depending on the computer package. SS(factor) often is labeled "between groups" (SAS) or "among groups", SS(error) often is labeled "within groups" (SAS), "residual" (SPSS), or "error" (MINITAB).

FIGURE 11.3

Dot-array diagram of replicates in Example 11.1.

FIGURE 11.4

Dot-array diagram where between-sample and within-sample variations are nearly the same. The F statistic would not lie in the rejection region.

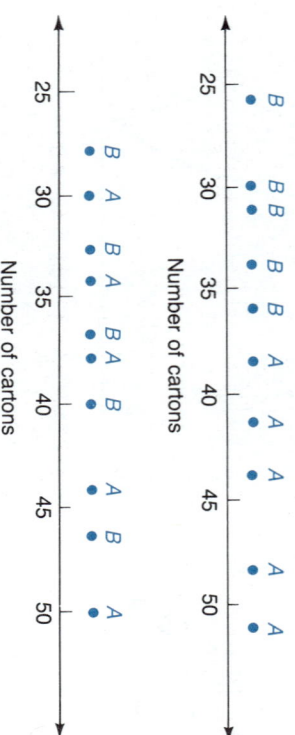

Number of cartons

Number of cartons

Consequently, we rejected $H_0: \mu_1 = \mu_2$. What this means in the language of ANOVA is that there *is* a significant effect on sales due to the brand factor.

To carry out the F-test, we first randomly obtain observations, called **replicates**, from each population. Example 11.1 used five replicates (monthly sales) from each of the two cigarette populations. It is *not necessary* to obtain the same number of replicates from each population.

Figure 11.3 is a dot-array diagram of the data in Example 11.1, where the symbol A represents a value from brand 1 and B represents brand 2. You do not need to be an expert statistician to observe that a clear difference exists between the sales of the two brands of cigarettes. The variation within the A's alone and the B's alone is the within-sample variation. Because the distances from the A values to the B values are much larger than the distances among the A values alone, the between-sample variation is quite large, as we have already observed.

Suppose instead that your dot-array diagram looks like Figure 11.4. Now the two sources of variation appear to be nearly the same and there is no obvious difference between the two brands. The resulting F-statistic here would not lie within the rejection region and we would not be able to demonstrate, using the ANOVA approach, a difference between the two mean sales.

11.2

ONE-FACTOR ANOVA COMPARING MORE THAN TWO MEANS

In the previous section, we examined a single *factor* with two *levels*. Our concern was whether there was any difference between the two levels of this factor. This amounted to performing a test of hypothesis on the means of two populations. Because we were dealing with the effect of a single factor, this was a one-factor (or one-way) ANOVA.

In general, one-factor ANOVA techniques can be used to study the effect of any single factor on sales, performance, and the like. This factor can consist of any number of levels—say, k levels. To determine if the levels of this factor affect our measured observations, we examine the hypotheses

$$H_0: \mu_1 = \mu_2 = \cdots = \mu_k$$
$$H_a: \text{not all } \mu\text{'s are equal}$$

Suppose we are interested in the average sales for not two but five brands of cigarettes. Is there any difference in these five mean sales? To answer this question, we test

$$H_0: \mu_1 = \mu_2 = \mu_3 = \mu_4 = \mu_5$$
$$H_a: \text{not all } \mu\text{'s are equal}$$

We have a single factor (brand) consisting of five levels (brand 1, brand 2, ..., brand 5). One possibility is to examine these samples one pair at a time using the t-statistic discussed in the previous section. This appears to be a safe way to proceed here, although there are $_5C_2 = 10$ such pairs of tests to perform this way. The main problem with this approach when performing many tests of this nature is determining the probability of making an incorrect decision. In particular, what value does α have, where α is the probability of rejecting H_0: all μ's are equal, when in fact it is true? You set α in advance but, after performing ten of these pairwise tests ($\mu_1 = \mu_2, \mu_1 = \mu_3, \ldots$), for instance, what is your *overall* probability of concluding that at least one pair of means are not equal when they actually are? This is a difficult question. The overall probability is not the significance level, α, with which you started for just one pair. So we need an approach that will test for the equality of these five means using a single test. This is what the ANOVA approach does.

Assumptions Behind the ANOVA Analysis

When using the ANOVA procedure, there are three key assumptions that must be satisfied. They are basically the same assumptions that were necessary when testing two means using small independent samples and the pooled variance approach. These assumptions are:

1. The replicates are obtained *independently* and *randomly* from each of the populations. The value of one observation has no effect on any other replicates within the same sample or within the other samples.
2. The observations (replicates) from each population follow (approximately) a *normal* distribution.
3. The normal populations all have a *common variance*, σ^2. We expect the values in each sample to vary about the same amount. The ANOVA procedure will be much less sensitive to violations of this assumption when we obtain samples of equal size from each population.

Deriving the Sum of Squares

When examining k populations, for example, the data will be configured somewhat like this:

Level 1	Level 2	$\cdots$	Level k
$\vdots$	$\vdots$		$\vdots$
n_1 replicates	n_2 replicates	$\cdots$	n_k replicates
$\vdots$	$\vdots$		$\vdots$
Totals $\overline{\quad T_1 \quad}$	$\overline{\quad T_2 \quad}$	$\cdots$	$\overline{\quad T_k \quad}$

This resembles the data from Example 11.1, where $k = 2$ and $n_1 = n_2 = 5$ replicates. To derive the sum of squares for this situation, we extend the results in equations 11.2, 11.4, and 11.6 to

$$\text{SS(factor)} = \left[\frac{T_1^2}{n_1} + \frac{T_2^2}{n_2} + \cdots + \frac{T_k^2}{n_k} \right] - \frac{T^2}{n} \quad (11.11)$$

$$\text{SS(total)} = \Sigma x^2 - \frac{T^2}{n} \quad (11.12)$$

$$\text{SS(error)} = \Sigma x^2 - \left[\frac{T_1^2}{n_1} + \frac{T_2^2}{n_2} + \cdots + \frac{T_k^2}{n_k} \right] \quad (11.13)$$

$$= \text{SS(total)} - \text{SS(factor)} \quad (11.14)$$

Here, n = the total number of observations = $n_1 + n_2 + \cdots + n_k$, and T = Σx = the sum of all n observations = $T_1 + T_2 \cdots + T_k$. Also, to find Σx^2, you square each of the n observations and sum the results.

The ANOVA Table

The good news is that the format of the ANOVA table is the same regardless of the number of populations (levels), k. The only change from the two-population case is that

$$\text{df for factor} = k - 1$$
$$\text{df for error} = n - k$$

As before, the total df are $n - 1$. The resulting ANOVA table follows.

SOURCE	df	SS	MS	F
Factor	$k - 1$	SS(factor)	$\text{MS(factor)} = \dfrac{\text{SS(factor)}}{k - 1}$	$\dfrac{\text{MS(factor)}}{\text{MS(error)}}$
Error	$n - k$	SS(error)	$\text{MS(error)} = \dfrac{\text{SS(error)}}{n - k}$	
Total	$n - 1$	SS(total)		

Note that

$$\text{MS(factor)} = \frac{\text{SS(factor)}}{\text{df for factor}}$$
$$= \frac{\text{SS(factor)}}{k - 1}$$

$$\text{MS(error)} = \frac{\text{SS(error)}}{\text{df for error}}$$
$$= \frac{\text{SS(error)}}{n - k}$$

(11.15)

The test statistic for testing $H_0: \mu_1 = \mu_2 = \cdots = \mu_k$ versus H_a: not all μ's are equal is

$$F = \frac{\text{MS(factor)}}{\text{MS(error)}}$$

(11.16)

which has an F distribution with $k - 1$ and $n - k$ df.

As in the two-sample case, the procedure is to reject H_0 when the variation among the sample means (measured by MS(factor)) is *large* compared to the variation within the samples (measured by MS(error)). Consequently, the test will be to reject H_0 whenever F lies in the *right-tailed* rejection region, defined by the significance level, α.

EXAMPLE 11.2

A convenience-store manager has asked you to examine the sales for the four brands of ice cream that she carries. Space in her refrigerated storage area is limited, so she would like to eliminate one or more of the brands that sell less well. She wants to know whether there is any difference among the average sales of these four brands. Past sales records showed data for the sales from six randomly selected weeks for *each* brand (gallons sold per week). Overall, the sales from 24 randomly selected weeks were used. Past experience has indicated that sales follow a normal distribution.

	BRAND 1	BRAND 2	BRAND 3	BRAND 4
	41	32	35	33
	35	37	30	27
	48	46	24	36
	40	53	26	35
	45	41	28	27
	52	43	31	25
Total (T)	261	252	174	183
Average ($\bar{X}$)	43.5	42.0	29.0	30.5
Variance (s^2)	37.1	52.8	15.2	22.3

The four sample averages are $\bar{x}_1 = 43.5$, $\bar{x}_2 = 42.0$, $\bar{x}_3 = 29.0$, and $\bar{x}_4 = 30.5$. Brands 1 and 2 appear to be outselling brands 3 and 4. In other words, it appears that there is a significant *between-group variation*. But do these sample means provide sufficient evidence to reject H_0: $\mu_1 = \mu_2 = \mu_3 = \mu_4$, where each μ_i represents the average of *all* weekly sales of brand i? Use the ANOVA procedure to answer this question with $\alpha = .05$.

The assumptions behind this analysis are that (1) the samples were obtained randomly and independently from each of the four populations and (2) the number of gallons of each brand that is sold each week follow a *normal* distribution, with a *common variance*, say, σ^2.

$$SS(\text{factor}) = \left[\frac{T_1^2}{n_1} + \frac{T_2^2}{n_2} + \frac{T_3^2}{n_3} + \frac{T_4^2}{n_4} \right] - \frac{T^2}{n}$$

So $n = n_1 + n_2 + n_3 + n_4 = 24$, and

$$T = \Sigma x = T_1 + T_2 + T_3 + T_4$$
$$= 261 + 252 + 174 + 183$$
$$= 870$$

Therefore,

$$SS(\text{factor}) = \frac{261^2}{6} + \frac{252^2}{6} + \frac{174^2}{6} + \frac{183^2}{6} - \frac{870^2}{24}$$
$$= 32,565 - 31,537.5$$
$$= 1027.5$$

$$SS(\text{total}) = \Sigma x^2 - \frac{T^2}{n}$$
$$= [41^2 + 35^2 + \cdots + 27^2 + 25^2] - \frac{870^2}{24}$$
$$= 33,202 - 31,537.5$$
$$= 1664.5$$

$$SS(\text{error}) = SS(\text{total}) - SS(\text{factor})$$
$$= 1664.5 - 1027.5$$
$$= 637$$

The ANOVA table for this analysis follows.

SOURCE	df	SS	MS	F
Factor	$k - 1 = 3$	1027.5	$1027.5/3 = 342.5$	$342.5/31.85 = 10.75$
Error	$n - k = 20$	637	$637/20 = 31.85$	
Total	23	1664.5		

The computed F value using the ANOVA table is $F^* = 10.75$. Since $\alpha = .05$, we use Table A.7 to find that $F_{.05, 3, 20} = 3.10$. Comparing these two values, $F^* = 10.75 > 3.10$, so we reject H_0.

We conclude that the average sales for the four brands are not the same. This confirms our earlier suspicion based on the variation among the four sample means. Our results indicate that the brand factor *does* have a significant effect on sales.

■

The Assumptions Behind ANOVA and a Test for Equal Variances

Use of independent random samples is of extreme importance when using the ANOVA procedure. The F-test used for comparing the population means in the ANOVA table is very sensitive to departures from this assumption, so the safest way to guard against incorrect conclusions is to use random sampling techniques. In many situations, however, such as when using the same set of people for before-after experiments, this may be difficult or impossible. One solution to this problem is to modify your study design, such as by using a randomized block design, discussed later in this chapter.

Lack of **normality** within the populations is not a critical matter provided the departure is not too extreme. The F-test used to test the means is not severely affected by populations that are somewhat nonnormal in nature. One way of making the ANOVA procedure even less sensitive to this assumption is to use *large samples*.

If the *variances* of the population *are not equal*, the F-test used in the ANOVA procedure for testing the means is only slightly affected, provided the *sample sizes are equal* (or nearly so). However, for this case, there is a very simple test of hypothesis for verifying this assumption.

In Chapter 9 an F-test was defined for determining whether two normal population variances (or standard deviations) are equal. A similar test is used when you are comparing more than two normal population variances, provided the sample sizes are equal.[†] Here the hypotheses are

$$H_0: \sigma_1^2 = \sigma_2^2 = \cdots = \sigma_k^2$$
$$H_a: \text{at least two variances are unequal (the } k \text{ variances are not the same)}$$

We warned you in Chapter 9 about the dangers of using the same data to test both the variances *and* the means. This argument applies to tests of more than two populations. A better procedure is to use a different data set for testing H_0: the variances are equal. This requires a much larger data set than is necessary if you use the same data for both tests. The test for equal variances

---

[†] When the sample sizes are unequal, a computationally more difficult test for equal variances can be performed, derived by M. S. Bartlett. For details, see J. Neter, W. Wasserman, and M. Kutner, *Applied Linear Statistical Models*, 2d ed. (Homewood, Ill.: Richard D. Irwin, 1985), pp. 618–22.

is the Hartley test; the test statistic is defined to be

$$H = \frac{\text{maximum } s^2}{\text{minimum } s^2} \qquad (11.17)$$

which is simply the ratio of the largest sample variance divided by the smallest of these k variances.

If H_0 is false, the test statistic will be "large," so the testing procedure is to reject H_0 if the computed value of H lies in the right tail. The rejection region for a 5% level of significance can be obtained from Table A.14. This region depends on the number (k) of populations or levels and the number of observations in *each* sample.

Suppose we use the ice cream sales data from Example 11.2 only for testing the hypotheses

$$H_0: \sigma_1^2 = \sigma_2^2 = \sigma_3^2 = \sigma_4^2$$
$$H_a: \text{at least two variances are unequal}$$

Using $\alpha = .05$ and Table A.14, because $k = 4$ and there are six observations in each sample, the test is to

$$\text{reject } H_0 \text{ if } H > 13.7$$

Using the data summary in Example 11.2, the minimum s^2 is 15.2 and the maximum s^2 is 52.8. Consequently,

$$H = \frac{52.8}{15.2} = 3.47$$

which is less than 13.7, and so the conclusion is that we have no reason to suspect unequal variances for this situation.

If other data are available for testing the means, the assumption of equal variances behind the ANOVA procedure appears to be safe.

Confidence Intervals in One-Factor ANOVA

When we deal with normal populations, as we do here, we can supply:

1. A point estimate of each mean, μ_i; for example, an estimate of μ_2 is $\bar{X}_2$.
2. A point estimate of each mean difference, $\mu_i - \mu_j$; for example, an estimate of $\mu_1 - \mu_3$ is $\bar{X}_1 - \bar{X}_3$.

When using the ANOVA procedure, the populations are believed to have a common variance, say σ^2. To estimate this variance, we use an estimate of σ^2 that does not depend on whether the population means are equal. This is the *within*-sample variation, measured by MS(error). The point estimate of σ^2 is

$$s_p^2 = \text{pooled variance}$$
$$= \text{MS(error)}$$

where MS(error) is defined in equation 11.16.

In previous chapters, we always supplied a confidence interval along with a point estimate to provide a measure of how reliable this estimate really is. The narrower the confidence interval, the more faith you have in your point estimate.

A $(1 - \alpha) \cdot 100\%$ confidence interval for μ_i is

$$\bar{X}_i - t_{\alpha/2,\, n-k}\, s_p \sqrt{\frac{1}{n_i}} \quad \text{to} \quad \bar{X}_i + t_{\alpha/2,\, n-k}\, s_p \sqrt{\frac{1}{n_i}} \qquad \textbf{(11.18)}$$

where

$$k = \text{number of populations (levels)}$$
$$n_i = \text{number of replicates in the } i\text{th sample}$$
$$n = \text{total number of observations}$$
$$s_p = \sqrt{\text{MS(error)}}$$

$t_{\alpha/2,\, df}$ is the value from Table A.5 with df = df for error = $n - k$, and right-tail area = $\alpha/2$.

A $(1 - \alpha) \cdot 100\%$ confidence interval for $\mu_i - \mu_j$ is

$$(\bar{X}_i - \bar{X}_j) - t_{\alpha/2,\, n-k}\, s_p \sqrt{\frac{1}{n_i} + \frac{1}{n_j}} \quad \text{to} \quad (\bar{X}_i - \bar{X}_j) + t_{\alpha/2,\, n-k}\, s_p \sqrt{\frac{1}{n_i} + \frac{1}{n_j}} \qquad \textbf{(11.19)}$$

EXAMPLE 11.3

Using the ice cream sales data from Example 11.2, construct a 95% confidence interval for the average sales of brand 1. Also determine a 95% confidence interval for the difference between the average sales of brands 1 and 3.

SOLUTION

First, your point estimate of μ_1 is $\bar{x}_1 = 43.5$. Using the ANOVA table from Example 11.2,

$$s_p^2 = \text{MS(error)} = 31.85$$

and so

$$s_p = \sqrt{31.85} = 5.64$$

Because $n = 24$ and $k = 4$, the resulting 95% confidence interval for μ_1 is

$$43.5 - t_{.025,\,20}(5.64)\sqrt{\frac{1}{6}} \quad \text{to} \quad 43.5 + t_{.025,\,20}(5.64)\sqrt{\frac{1}{6}}$$
$$= 43.5 - (2.086)(5.64)(.408) \quad \text{to} \quad 43.5 + (2.086)(5.64)(.408)$$
$$= 43.5 - 4.80 \quad \text{to} \quad 43.5 + 4.80$$
$$= 38.7 \quad \text{to} \quad 48.3$$

As a result, we are 95% confident that the average sales of the brand 1 ice cream is between 38.7 and 48.3 gallons per week.

The 95% confidence interval for $\mu_1 - \mu_3$ is

$$(\bar{X}_1 - \bar{X}_3) - t_{.025,\,20}\, s_p \sqrt{\frac{1}{n_1} + \frac{1}{n_3}} \quad \text{to} \quad (\bar{X}_1 - \bar{X}_3) + t_{.025,\,20}\, s_p \sqrt{\frac{1}{n_1} + \frac{1}{n_3}}$$

$$= (43.5 - 29.0) - (2.086)(5.64)\sqrt{\frac{1}{6} + \frac{1}{6}} \quad \text{to} \quad (43.5 - 29.0) + (2.086)(5.64)\sqrt{\frac{1}{6} + \frac{1}{6}}$$

$$= 14.5 - (2.086)(5.64)(.577) \quad \text{to} \quad 14.5 + (2.086)(5.64)(.577)$$

$$= 14.5 - 6.79 \quad \text{to} \quad 14.5 + 6.79$$

$$= 7.71 \quad \text{to} \quad 21.29$$

Based on this confidence interval, we are 95% confident that the average sales for brand 1 are between 7.71 and 21.29 gallons per week *higher* than the average sales for brand 3.

■

A Word of Warning

The procedure we used in Example 11.3 for determining confidence intervals is reliable, providing you decide which intervals you want computed *before* you observe your data. For example, constructing a confidence interval for the difference of two population means having the corresponding largest and smallest sample means is not an accurate procedure. If you do this, you let the data dictate which confidence interval you determine.

When using the procedure in Example 11.3 to construct confidence intervals for the difference of two population means, it is important to keep the number of such intervals as small as possible. This is because the probability of any one interval containing the true population difference is $1 - \alpha$, but the probability that *all* the intervals contain their respective population differences is not $1 - \alpha$. In other words, if $\alpha = .05$, the overall confidence level of this procedure is not 95%; it is something much less than 95%. To compare all possible pairs of means effectively, you will need to use a technique that will allow you to make all possible comparisons between population means while maintaining the Type I error rate at α. This is called a **multiple comparison** procedure; one such procedure is discussed following Example 11.5.

EXAMPLE 11.4

Comptek, a computer software development firm, is interested in the effect of educational level on the job knowledge of the company's employees. They administer an exam to a randomly selected group of people having various educational backgrounds.

In the sample of 15 employees, 6 have only high school diplomas, 5 have only bachelor's degrees, and 4 have master's degrees. The exam scores are:

HIGH SCHOOL DIPLOMA	BACHELOR'S DEGREE	MASTER'S DEGREE
81	94	88
84	83	89
69	86	78
85	81	85
84	78	
95		
Total (T) 498	422	340
Average ($\bar{X}$) 83.0	84.4	85.0

What would be your conclusion using a significance level of .10?

SOLUTION

Examining the sample means, you might be tempted to conclude that the higher a person's level of education, the higher their score on the exam. But is there a significant difference among these three means? An ANOVA analysis will clarify this.

The assumptions necessary here are:

1. The scores were obtained randomly and independently from each of the three populations.
2. The exam scores for each of the three populations follow a normal distribution, with means μ_1, μ_2, and μ_3. The scores in each of the samples are assumed to have the same amount of variation.

Because the sample sizes are not the same, the Hartley test for equal variances cannot be used here. As discussed earlier, we prefer not to use the same data for testing both the means and variances, and so a better procedure would be to obtain additional data (with equal sample sizes) for testing the equality of these three variances. We begin by calculating the necessary sum of squares.

$$SS(\text{factor}) = \left[\frac{T_1^2}{n_1} + \frac{T_2^2}{n_2} + \frac{T_3^2}{n_3}\right] - \frac{T^2}{n}$$

where

$$T = \Sigma x = T_1 + T_2 + T_3$$
$$= 498 + 422 + 340 = 1260$$
$$n = n_1 + n_2 + n_3$$
$$= 6 + 5 + 4 = 15$$

So,

$$SS(\text{factor}) = \frac{498^2}{6} + \frac{422^2}{5} + \frac{340^2}{4} - \frac{1260^2}{15}$$
$$= 105,850.8 - 105,840.0$$
$$= 10.8$$

$$SS(\text{total}) = \Sigma x^2 - \frac{T^2}{n}$$
$$= [81^2 + 84^2 + \cdots + 78^2 + 85^2] - \frac{1260^2}{15}$$
$$= 106,424 - 105,840$$
$$= 584$$

Finally, because $k = 3$ and $n = 15$,

$$SS(\text{error}) = 584 - 10.8 = 573.2$$

df for factor $= k - 1 = 2$
df for error $= n - k = 12$
df for total $= n - 1 = 14$

The resulting ANOVA table is

SOURCE	df	SS	MS	F
Factor	2	10.8	5.4	.11
Error	12	573.2	47.8	
Total	14	584		

The hypotheses are

$$H_0: \mu_1 = \mu_2 = \mu_3$$
$$H_a: \text{not all } \mu\text{'s are equal}$$

where each μ_i represents the average score of *all* employees having this particular educational level at Comptek.

We will reject H_0 if

$$F^* > F_{.10, 2, 12} = 2.81$$

Because $.11 < 2.81$, we fail to reject H_0.

We conclude that there is not sufficient evidence to indicate that the average performance on the exam is different among the three groups. As usual, we do not *accept* H_0; that is, we do *not* conclude that these three means *are* equal. There is simply not enough evidence to support the claim that employees with a higher educational level perform better at this particular company.

■

The factor in Example 11.4 was the educational level of the employee; it had three levels. The results show that we are unable to demonstrate that this factor has a significant effect on exam performance.

EXAMPLE 11.5

In Example 11.4, before the exam was given, Comptek decided to construct a 95% confidence interval for the average exam score of all people holding a master's degree and the difference between the average exam scores for personnel with a master's degree and those with only a high school diploma. What are the confidence intervals?

SOLUTION

The point estimates are

for μ_3: $\bar{x}_3 = 85.0$

for $\mu_3 - \mu_1$: $\bar{x}_3 - \bar{x}_1 = 85.0 - 83.0 = 2.0$

To construct the confidence intervals, you first need an estimate of the common variance of these three populations. Based on the results of Example 11.4, this is

$$s_p^2 = MS(error) = 47.8$$

so

$$s_p = \sqrt{47.8} = 6.91$$

Because $n = 15$, $k = 3$, and $n_3 = 4$, the 95% confidence interval for μ_3 is

$$\bar{X}_3 - t_{.025,12}\, s_p \sqrt{\frac{1}{n_3}} \quad \text{to} \quad \bar{X}_3 + t_{.025,12}\, s_p \sqrt{\frac{1}{n_3}}$$

$$= 85.0 - (2.179)(6.91)(.5) \quad \text{to} \quad 85.0 + (2.179)(6.91)(.5)$$

$$= 85.0 - 7.53 \quad \text{to} \quad 85.0 + 7.53$$

$$= 77.47 \quad \text{to} \quad 92.53$$

The 95% confidence interval for $\mu_3 - \mu_1$ is

$$(\bar{X}_3 - \bar{X}_1) - t_{.025,12}\, s_p \sqrt{\frac{1}{n_3} + \frac{1}{n_1}} \quad \text{to} \quad (\bar{X}_3 - \bar{X}_1) + t_{.025,12}\, s_p \sqrt{\frac{1}{n_3} + \frac{1}{n_1}}$$

$$= (85.0 - 83.0) - (2.179)(6.91)(.645) \quad \text{to} \quad (85.0 - 83.0) + (2.179)(6.91)(.645)$$

$$= 2.0 - 9.71 \quad \text{to} \quad 2.0 + 9.71$$

$$= -7.71 \quad \text{to} \quad 11.71$$

Consequently, we are 95% confident that the average exam score of all employees with master's degrees is between 7.71 *lower* to 11.71 *higher* than those with a high school diploma only. This implies that the data do *not* allow us to say that the employees with master's degrees performed better on the exam than those with high school degrees.

A MINITAB solution to this example is shown in Figure 11.5. The output contains summary information for each sample, the ANOVA table, and a graphical representation of the confidence interval for each population mean. ■

Multiple Comparisons: A Follow-up to the One-Factor ANOVA Procedure

If the one-factor ANOVA procedure leads to a rejection of H_0: all population means are equal, a logical question would be, Which means do differ? In other words, rejecting the ANOVA null hypotheses informs us that the means are not all the same but provides no clue as to which of the population means are

FIGURE 11.5

MINITAB solution for Examples 11.4 and 11.5.

```
MTB > READ INTO C1 C2 ────── Factor level: 1 = H. S. diploma
DATA> 81 1                                  2 = Bachelor's degree
DATA> 84 1                                  3 = Master's degree
DATA> 69 1
DATA> 85 1
DATA> 84 1
DATA> 95 1
DATA> 94 2
DATA> 83 2
DATA> 86 2
DATA> 81 2
DATA> 78 2
DATA> 88 3
DATA> 89 3
DATA> 78 3
DATA> 85 3
DATA> END
      15 ROWS READ
MTB > ONEWAY USING DATA IN C1, LEVELS IN C2

ANALYSIS OF VARIANCE ON C1
SOURCE     DF       SS       MS       F       P
C2          2     10.8      5.4    0.11   0.894
ERROR      12    573.2     47.8
TOTAL      14    584.0

                                  INDIVIDUAL 95 PCT CI'S FOR MEAN
                                  BASED ON POOLED STDEV
LEVEL    N      MEAN    STDEV   ----+---------+---------+---------+---
1        6    83.000    8.367       (-----------*-----------)
2        5    84.400    6.107        (------------*------------)
3        4    85.000    4.967          (-------------*-------------)
                                  ----+---------+---------+---------+---
POOLED STDEV =   6.911          80.0      85.0      90.0
```

different. As was discussed prior to Example 11.4, performing a series of t-tests to compare all possible pairs of means is not a good idea, since the chances of making at least one Type I error (concluding that a difference exists between two population means when in fact they are the same) using such a procedure is much larger than the predetermined α used for each of the t-tests.

What is needed is a technique that compares all possible pairs of means in such a way that the probability of making **one or more** Type I errors is α. This is a *multiple comparison procedure*. There are several methods available for making multiple comparisons; the one presented here is **Tukey's** test for multiple comparisons. (Tukey is pronounced too'-key.)

Tukey's procedure is based on a statistical test that uses the largest and smallest sample means. The form of this statistic is

$$Q = \frac{\text{maximum }(\bar{X}_i) - \text{minimum }(\bar{X}_i)}{\sqrt{\text{MS(error)}/n_r}} \qquad (11.20)$$

where

1. maximum $(\bar{X}_i)$ and minimum $(\bar{X}_i)$ are the largest and smallest sample means, respectively.
2. MS(error) is the sample variance.
3. n_r is the number of replicates in each sample.

Notice that Tukey's procedure assumes that each sample contains the same number (n_r) of replicates. Critical values of the Q-statistic are contained in Table A.16. Define

$Q_{\alpha,k,v} = $ critical value of the Q-statistic from Table A.16, using a significance level of α; k is the number of sample means (groups), and v is the df freedom associated with MS(error)

MULTIPLE COMPARISON PROCEDURE

1. Find $Q_{\alpha, k, v}$ using Table A.16.
2. Determine

$$D = Q_{\alpha, k, v} \cdot \sqrt{\frac{MS(error)}{n_r}}$$

where MS(error) is the sample variance and n_r is the number of replicates in each sample. For one-factor ANOVA, MS(error) is the same as s_p^2.

3. Place the sample means in order, from smallest to largest.
4. If two sample means differ by more than D, the conclusion is that the corresponding population means are unequal. In other words, if $|\bar{X}_i - \bar{X}_j| > D$, this implies that $\mu_i \neq \mu_j$.

To illustrate this procedure, reconsider Example 11.2. Here we concluded that the average number of gallons sold per week was not the same for the four brands of ice cream. The four sample means were

$$\begin{align}
\text{Brand 1:} \quad & (\bar{x}_1 = 43.5) \\
\text{Brand 2:} \quad & (\bar{x}_2 = 42.0) \\
\text{Brand 3:} \quad & (\bar{x}_3 = 29.0) \\
\text{Brand 4:} \quad & (\bar{x}_4 = 30.5)
\end{align}$$

For this study, there were $n_r = 6$ replicates in each sample, with a resulting pooled standard deviation of $s_p^2 = MS(error) = 31.85$. The study contained $k = 4$ groups and the df for the error sum of squares was $v = n - k = 24 - 4 = 20$. Using a significance level of .05, we begin by finding $Q_{.05, 4, 20}$ in Table A.16. This value is 3.96. Next we determine

$$D = Q_{.05, 4, 20} \cdot \sqrt{\frac{MS(error)}{n_r}}$$
$$= (3.96)\sqrt{31.85/6}$$
$$= 9.12$$

The sample means, in order, are

$$\underline{29.0, \ 30.5, \ 42.0, \ 43.5}$$

Two sample means are significantly different using the Tukey procedure if they differ by an amount greater than $D = 9.12$. Here there are four significant differences, namely,

$$\begin{align}
\bar{x}_1 - \bar{x}_3 &= 43.5 - 29.0 = 14.5 > 9.12 \\
\bar{x}_2 - \bar{x}_3 &= 42.0 - 29.0 = 13.0 > 9.12 \\
\bar{x}_1 - \bar{x}_4 &= 43.5 - 30.5 = 13.0 > 9.12 \\
\bar{x}_2 - \bar{x}_4 &= 42.0 - 30.5 = 11.5 > 9.12
\end{align}$$

The conclusion from the multiple comparison analysis is that $\mu_1 \neq \mu_3$, $\mu_1 \neq \mu_4$, $\mu_2 \neq \mu_3$, and $\mu_2 \neq \mu_4$. There is no evidence of a difference between the brand 1 and the brand 2 populations or between the brand 3 and the brand 4 populations. This is indicated by the two overbars connecting these two pairs of sample means. In general, there is no evidence to indicate a difference in the population means for any group of sample means under such a bar.

ONE-FACTOR ANOVA PROCEDURE

ASSUMPTIONS

1. The replicates are obtained *independently* and *randomly* from each of the populations. The value of one observation has no effect on any other replicates within the same sample or within the other samples.

2. The observations (replicates) from each population follow (approximately) a *normal* distribution.

3. The normal populations all have a common variance, σ^2. We expect the values in each sample to vary about the same amount. The ANOVA procedure will be much less sensitive to this assumption when we obtain samples of equal size from each population.

HYPOTHESES

$$H_0: \mu_1 = \mu_2 = \cdots = \mu_k$$
$$H_a: \text{not all } \mu\text{'s are equal}$$

Note that H_a is not the same as H_a': all μ's are unequal; H_a states that *at least two* of the μ's are different.

SUM OF SQUARES

$$SS(\text{factor}) = \left[\frac{T_1^2}{n_1} + \frac{T_2^2}{n_2} + \cdots + \frac{T_k^2}{n_k} \right] - \frac{T^2}{n}$$

where $n = n_1 + n_2 + \cdots + n_k$ and $T = \Sigma x = T_1 + T_2 + \cdots + T_k$.

$$SS(\text{total}) = \Sigma x^2 - \frac{T^2}{n}$$

$$SS(\text{error}) = SS(\text{total}) - SS(\text{factor})$$

$$= \Sigma x^2 - \left[\frac{T_1^2}{n_1} + \frac{T_2^2}{n_2} + \cdots + \frac{T_k^2}{n_k} \right]$$

DEGREES OF FREEDOM

df for factor $= k - 1$

df for error $= n - k$

df for total $= n - 1$

Note that $(k - 1) + (n - k) = n - 1$.

ANOVA TABLE

SOURCE	df	SS	MS	F
Factor	$k - 1$	SS(factor)	MS(factor) $= \dfrac{SS(\text{factor})}{k-1}$	MS(factor)/MS(error)
Error	$n - k$	SS(error)	MS(error) $= \dfrac{SS(\text{error})}{n-k}$	
Total	$n - 1$	SS(total)		

where MS = mean square = SS/df.

TESTING PROCEDURE

reject H_0 if $F^* > F_{\alpha, k-1, n-k}$

where $F_{\alpha, k-1, n-k}$ is obtained from Table A.7.

EXERCISES

11.1 A shoe manufacturer wanted to test whether there is a difference in the amount of wear on three different designs of rubber soles for a particular jogging shoe. Eighteen joggers were selected for the experiment. Each type of design was randomly assigned to six joggers. After running 200 miles, the joggers turned in their shoes. The manufacturer used an index to indicate the amount of rubber left on the sole. The measures obtained were:

Design 1	3.2	4.1	6.2	5.3	4.9	3.5
Design 2	4.7	6.3	4.0	5.4	7.1	4.5
Design 3	3.9	3.9	6.0	5.5	4.2	3.1	5.1

a. State the null and alternative hypotheses.

b. What assumptions are necessary to use the ANOVA procedure on these data?

c. What are the point estimates of the mean wear for each of the three designs?

d. What is the within-groups mean square?

e. Set up an ANOVA table and state the conclusion. Use a significance level of .05.

f. Using a significance level of .05, perform a multiple comparisons procedure, if appropriate.

11.2 In Exercise 11.1, subtract 3.0 from each of the observations in the table. Perform the ANOVA procedure. Are the sum of squares the same for the coded data as for the original data? Why or why not? What happens when any set of data is coded by adding or subtracting the same number to each observation value in terms of the sum of squares?

11.3 A manufacturer introduces a new car that gets 40 mpg with a standard deviation of 3. The manufacturer's competitor introduces a similar economy car and claims it also gets the same mpg with the same standard deviation. A random sample of 15 observations of mpg is taken for each manufacturer's car. Is there any difference in the mean mpg for these two cars? Use the ANOVA procedure.

Manufacturer	41	40	40	39	36	41	40	42	42	39	40	41	39	38	41
Competitor	38	39	37	40	42	43	41	39	38	37	37	38	39	39	38

11.4 Use the two-sample t-test for the data in Exercise 11.3 to test for any difference in the mean mpg for these two cars. What is the relationship between the t-test of this exercise and the F-test of Exercise 11.3?

11.5 The science of ergonomics studies the influence of "human factors" in technology, i.e., how human beings relate to and work with machines. With the widespread use of computers for data processing, computer scientists and psychologists are getting together to study human factors. One typical study investigated the productivity of secretaries with different word processing programs. An identical task was given to 18 secretaries, randomly allocated to three groups. Group 1 used a primarily menu-driven program, Group 2 used a command-driven program, and Group 3 was a mixture of both approaches. The secretaries all had about the same level of experience, typing speed, and computer skills. The time (in minutes) taken to complete the task was observed. The results are as follows:

GROUP 1 (MENU-DRIVEN)	GROUP 2 (COMMAND-DRIVEN)	GROUP 3 (MIXED)
12	14	10
15	11	8
11	13	9
12	12	10
10	11	7
13	14	8

a. Do the necessary calculations to construct an ANOVA table, and test the hypothesis that there is no difference between the three types of word-processing programs (i.e., on the average, the time taken to complete the task is about the same). Use $\alpha = .05$.

b. State the p-value for the test.

c. Does the type of word-processing software used affect the performance of the secretaries?

d. If the secretaries had different levels of experience, typing speed, and computer skills, how would it affect the data? (Would it be an extraneous source of variation, or "noise"? Would it tend to increase the "within-sample" variation, or the "between-sample" variation, or both, or neither?)

e. Using a significance level of .05, perform a multiple comparisons procedure (if appropriate).

11.6 A small engine-repair shop can special-order parts from any one of three different warehouses and receive a substantial discount on the price. The manager of the shop is concerned with the length of time that it takes to special order a part from one of the warehouses. The number of days it takes to special-order a part is recorded for 15 randomly selected orders from each of the three warehouses, as shown in the following table. Do the data indicate that there is a difference in the mean times that it takes to special-order a part from a warehouse? Use a .05 significance level. State the *p*-value.

WAREHOUSE

A	13	17	14	10	9	15	18	11	13	18	16	13	15	12	16
B	7	12	8	15	6	10	12	10	8	14	10	6	9	13	11
C	10	12	18	19	9	15	20	11	15	13	17	13	10	14	16

11.7 A sales manager wanted to know whether there was a significant difference in the monthly sales of her three sales representatives. John is strictly on commission, Randy is on commission and a small salary, and Ted is on a small commission and a salary. Eight months were chosen at random. The data represent monthly sales.

John	969	905	801	850	910	1030	780	810
Randy	738	773	738	805	850	800	690	720
Ted	751	764	701	810	840	790	720	735

a. Using a significance level of .05, test the hypothesis that there is no difference in the mean monthly sales. (Coding the data may make the computations easier.)

b. What is the *p*-value?

c. Using a significance level of .05, perform a multiple comparisons procedure, if appropriate.

11.8 An instructor wanted to test whether there was a difference in effectiveness of four different teaching techniques. Four groups of students were taught using one of the four teaching techniques. If the instructor examined the groups for mean differences one pair at a time, how many *t*-tests would have to be performed? What is the advantage of using an ANOVA procedure instead?

11.9 Astral Airlines recently introduced a nonstop flight between Houston and Chicago. The vice president of marketing for Astral decided to run a test to see whether Astral's passenger load was similar to that of its two major competitors. Ten daytime flights were picked at random from each of the three airlines and the percent of unfilled seats on each flight was as follows:

Astral	10	14	12	10	8	13	11	8	12	9
Competitor 1	12	9	8	9	10	12	7	11	10
Competitor 2	15	10	15	8	14	9	8	11	10	12

Use a significance level of .05 and perform an ANOVA procedure. Find the *p*-value.

11.10 What assumptions do the data need to satisfy in Exercise 11.9 to ensure that the ANOVA procedure is valid?

11.11 After performing an ANOVA procedure on three groups, an analyst found that a significant difference existed at the .01 significance level. From the following statistics, perform a multiple comparison procedure, using a significance level of .01:

$$\bar{x}_1 = 18.21, \quad \bar{x}_2 = 19.14, \quad \bar{x}_3 = 14.97, \quad n_1 = 7, \quad n_2 = 7, \quad n_3 = 7, \quad MSE = 2.8$$

11.12 The workers at a calculator assembly plant wish to bargain for more breaks during the work day. The manager believes that increasing the number of 15-minute breaks will affect productivity. The workers currently receive three breaks during the 8-hour work day. The manager decides to run a test by choosing four groups of five workers each and giving one group three breaks, the next group four breaks, and so on. The number of calculators assembled per day is recorded for five days. Test the manager's claim using an ANOVA procedure with a .10 significance level. Find the p-value. Using a significance level of .05, perform a multiple comparisons procedure, if appropriate.

3 breaks	200	205	197	210	205
4 breaks	210	203	201	197	199
5 breaks	198	190	185	188	180
6 breaks	197	180	190	192	175

11.13 A study was conducted to determine if the structure of strategic decision making in not-for-profit hospitals significantly constrained or promoted the development of internal politics. A questionnaire was developed to measure internal politics in hospitals with integrated strategic management structures and with unintegrated strategic structures. The results of the survey are shown in the following ANOVA table:

SOURCE	df	SS	MS	F
Factor	1	1.6913		
Error				
Total	37	21.5202		

(*Source:* Robert Jones and Karen Fowler, "The Relationships Between The Structures of Strategic Decision Making And Internal Politics In Not-For-Profit Hospitals: An Empirical Test," *Proceedings of 1986 Annual Meeting of the Decision Sciences Institute*, pp. 1255–7).

a. Complete the ANOVA table. How many total observations were used in this study?

b. Can one conclude that not-for-profit hospitals with strategic management structures were significantly different in terms of internal politics with the not-for-profit hospitals without strategic management structures? At what significance level?

11.14 Independent samples of size 16 are drawn from each of four normally distributed populations. The resulting sample standard deviations are: $s_1 = 2.0$, $s_2 = 2.5$, $s_3 = 2.2$, $s_4 = 2.5$. Do the data provide sufficient evidence that a significant difference exists in the population standard deviations of the four populations? Use a .05 significance level.

11.15 Three machines package 50 pound sacks of pinto beans. A preliminary test is performed using data from a pilot study to determine whether a significant difference exists among the variances of the amount of beans packaged for each machine. Use a 5% significance level in conducting a test for equality of variances for each of the machines from the sample data, in pounds:

MACHINE 1	MACHINE 2	MACHINE 3
52	50	48
51	49	46
48	51	51
50	50	50
46	52	52
55	53	50
53	55	51

11.16 A sales manager would like to determine whether there is a significant difference in the variance of sales of three salespersons. Three independent samples of daily sales (in hundreds of dollars) are collected. Using a 5% significance level, determine whether the data indicate a difference in the variance of the sales of the three salespersons.

MACHINE 1	MACHINE 2	MACHINE 3
49	50	52
56	49	50
51	51	51
56	50	49
45	49	50
50	48	50

SALESPERSON 1	SALESPERSON 2	SALESPERSON 3	SALESPERSON 1	SALESPERSON 2	SALESPERSON 3
1.2	2.5	1.4	1.0	2.2	1.7
1.1	2.1	1.8	1.3	2.3	1.3
1.4	2.3	1.5	1.8	2.4	1.2
1.6	2.0	1.6	1.4	2.3	1.5
1.4	2.3	1.4	1.5	2.5	1.6

11.17 A target-shooting club performed an experiment on a randomly selected group of 21 beginning shooters to determine whether shooting accuracy is affected by the method of sighting: right eye only open, left eye only open, or both eyes open. The 21 beginners were randomly divided into three groups of seven each. Each group went through the same training and practicing procedures with one exception—the use of eyes for sighting. The scores from each shooter were coded and are:

Right Eye	2	3	0	8	2	5	3	1
Left Eye	3	6	0	9	0	4	2
Both Eyes	1	7	2	7	1	5	2

a. Do the data provide sufficient evidence to indicate that a difference exists among the mean scores of the three methods? Use a .05 significance level. Find the p-value.

b. Find a 95% confidence interval to estimate the mean score of the coded data for each method.

c. Using a significance level of .05, perform a multiple comparisons procedure, if appropriate.

11.18 In a small company, upper management wants to know if there is a difference in the three types of methods used to train its machine operators. One method uses a hands-on approach but is very expensive. A second method uses a combination of classroom instruction and some on-the-job training. The third method is the least expensive and is confined completely to the classroom. Eight trainees are assigned to each training technique. The following table gives the results of a test administered after completion of the training. Do the data provide sufficient evidence to indicate a difference in the methods of training at the .01 level of significance?

METHOD 1	METHOD 2	METHOD 3	METHOD 1	METHOD 2	METHOD 3
95	85	88	81	93	81
100	90	94	85	86	84
90	95	90	96	94	90
91	88	80	95	95	87

11.3 DESIGNING AN EXPERIMENT

The previous section introduced you to one-factor (or one-way) ANOVA. In this type of analysis, you randomly obtain samples from each of the k populations (levels) describing a single factor. The variable that is being measured (such as weekly sales) is referred to as the **dependent** variable. Since replicates (repeat observations) are obtained in a completely random manner from each population, this type of sampling plan is called a **completely randomized design**. This section discusses other experimental designs, including the randomized block design and the two-way factorial design.

Suppose that the personnel director at Blackburn Industries is interested in examining the amount of dental claims filed by Blackburn employees. Let us consider using the amount of these claims to examine various group differences and what type of design would be appropriate for each situation.

Situation 1: The Completely Randomized Design

One question of interest to the personnel director is whether the average annual amount claimed on the dental insurance plan differs among the four employee classifications. These classifications range from category 1 (consisting of production line workers) to category 4 (consisting of upper-level management). Replicates are obtained randomly within each population (category) and the four samples are not related in any way. This illustration consists of one factor (employee classification) consisting of four levels. The question of interest here is: Is there a difference in the average annual claims among the four types of employees? The corresponding null hypothesis is

$$H_0: \mu_1 = \mu_2 = \mu_3 = \mu_4$$

Essentially, this type of analysis (called one-way ANOVA) will fail to reject H_0 if the sample means are "close together" and reject this hypothesis otherwise.

Situation 2: The Randomized Block Design

Suppose instead that the personnel director at Blackburn Industries wished to investigate family dental claims. In particular, she wished to know if there was a difference in the amounts claimed (1) by the husband, (2) by the wife, and (3) per child in the family. For the study she randomly selected 15 (or however many) families having at least one child and recorded these three amounts for each family. This is an example of a **randomized block design**. The configuration of the sample results would resemble the following scheme, where each x represents a dollar amount.

FAMILY	HUSBAND	WIFE	PER CHILD	
1	x	x	x	(1st block)
2	x	x	x	(2nd block)
⋮				
15	x	x	x	(15th block)

This design consists of one *factor* with three levels (husband/wife/per child). Unlike the completely randomized design, the three samples *are not independent*, since the data are grouped (blocked) by family. For example, the first husband value is not independent of the first wife value, since they both belong to the same family. We encountered this very same design in Chapter 9, where we compared two population means using paired (that is, blocked) samples. When using the randomized block design, you can compare the means of more than two populations using a blocking strategy to gather your data.

The question of interest here is: Is there a difference in the husband, wife, and per-child claims? Similar to the completely randomized design, the question of interest is whether the factor of interest (family member type, here) has a significant effect on the value of the dependent variable (amount of the annual claim, here). The difference between this and the completely randomized situation is that here we use a blocking strategy rather than independent samples to obtain a more precise test for examining differences in the factor level means. The null hypothesis for this illustration is that the group (factor level) means are identical, that is,

$$H_0: \mu_H = \mu_W = \mu_C$$

where μ_H is the average annual amount claimed by the husband, μ_W is the average amount for the wife, and μ_C is the average amount claimed per child.

The analysis for the randomized block design is discussed in the next section, but essentially this procedure removes the effects of the blocks (families, here) before testing for a difference between the factor level means. Consequently, this design removes the block effect from the error sum of squares in the completely randomized design. Several examples in the next section illustrate this technique.

Situation 3: The Two-Way Factorial Design

The **two-way factorial design** is very similar to situation 1, the one-way analysis of variance, except now *two* factors are of interest to the individual conducting the study. Suppose that the personnel director at Blackburn Industries decides to examine the dental claims for all the unmarried employees. She wants to investigate the effect of sex (factor A) and employee classification (factor **B**) on the amount of the annual claims. As before, employee classification ranges from category 1 (production line workers) to category 4 (upper-level management). The previous one-way ANOVA illustration examined only the effect of employee classification on the amount of dental claims. The inclusion of the sex factor accomplishes two things: first, you can determine if the sex of the employee has an effect on the amount of the annual claim, and second, you can investigate whether the relationship between employee classification and the amount of the annual claim is different for male and female employees. In other words, whether factor B relates to the dependent variable (amount of annual dental claim) depends on the level of factor A. This type of effect is called **interaction** between factors A and B. This differs from the randomized block design, where it is assumed that *no interaction is present between the factor of interest and the blocks.*

Consequently, there are three sets of hypotheses that can be tested using the two-way factorial design. The corresponding null hypotheses are:

$H_{0,A}$: factor A (sex) is not significant

$H_{0,B}$: factor B (employee classification) is not significant

$H_{0,AB}$: there is no interaction between factor A and factor B

The first two hypotheses are similar to those tested in one-way ANOVA, but the third hypothesis is unique to the two-way (or higher) factorial design.

To collect data for this design, the personnel director would obtain an amount for *every* combination of a factor A level and a factor B level. This particular illustration consists of two levels for factor A (male/female) and four levels for factor B (category 1/.../category 4), and is called a **2 × 4 factorial design.** Consequently, data are collected for eight possible factor level combinations, referred to as **treatments.** Data values within the same treatment are termed **replicates.** Furthermore, it is necessary when using this type of design *to obtain more than one replicate for each treatment.* An illustration using three replicates (two would be sufficient) is shown in Figure 11.6. Each *x* represents the amount of the annual dental claim. The actual two-way analysis of variance for this illustration is demonstrated in Section 11.5.

		Factor B		
	Category 1	Category 2	Category 3	Category 4
Factor A Male	x,x,x	x,x,x	x,x,x	x,x,x
Female	x,x,x	x,x,x	x,x,x	x,x,x

EXERCISES

11.19 a. Name the three types of experimental design discussed in a preceding section.

b. Which design or designs does not involve the necessity of having replicates?

c. What is "interaction" between factors and which design permits testing for interaction?

d. In each of the three designs, how many dependent variables are there?

e. If a one-way (one-factor) ANOVA design has four levels of the factor and six replicates at each level, how many treatments are being considered and how many total number of observations are made?

f. If a 4×6 two-way factorial design is chosen for an ANOVA-based experiment, what are the number of treatments being considered and the *minimum* number of total observations necessary?

g. What is the advantage of blocking and when is it necessary?

h. What are some potential problems that can arise with the randomized block design?

11.20 An appraisal firm is interested in the amount of time it takes appraisers to appraise fairly new homes with a market value between $100,000 and $150,000. An experiment is set up in which the factor is experience of the appraiser. Three levels are used: less than 1 year of experience, 1 year to 3 years of experience, and over 3 years of experience.

a. To test if there is a significant difference in the levels of experience with respect to the amount of time that it takes an appraiser to appraise a house, which design is appropriate? Construct a diagram to show how the data would look.

b. Suppose that a second factor was added. Assume that the firm is interested in whether the low and high-priced homes affect the appraised time differently. Two levels are considered: houses with market value between $50,000 and $100,000 and houses with market value between $100,000 and $150,000. Construct a diagram to show the setup of the data. What is the appropriate experimental design if interaction exists between the factors?

11.21 Suppose a supervisor is interested in the productivity of three different machines used to assemble electronic components.

a. What design would be appropriate to test if there is a significant difference in the productivity of the machines?

b. Suppose that 20 operators use the machines over three shifts to assemble the components. How can a randomized block design be used?

c. How can one be assured that observations within each block are randomized?

11.22 Explain what observations are independent in a completely randomized design. Are all observations independent in a randomized block design?

11.4
RANDOMIZED BLOCK DESIGN

The previous section described the difference between the randomized block and completely randomized designs. Rather than obtaining independent samples from the k populations, the data for a randomized block design are organized into homogeneous units, referred to as **blocks**. *Within* each block, any predictable difference in the observations is due to the effect of the factor of interest, such as sex or employee classification.

Consider a situation where the editors of ten automotive magazines are asked to evaluate the new-car warranty for Henry and GA automobiles. Several criteria are included, leading to a composite score ranging from 0 to 100. A higher score indicates the editor thinks it is a better warranty. The scores are:

Once again, there is a single factor of interest, namely, *brand* of automobile. But are the 20 sample observations in fact replicates? Replicates (by definition) are obtained under (nearly) identical circumstances, so any variation in the values within any one sample is due strictly to chance. For this situation, the person from whom the values were obtained heavily influences each sample value. Also, each pair of values is supplied by the same person, and so these samples are *not independent*, violating a key assumption of the completely randomized design.

PERSON	HENRY	GA		PERSON	HENRY	GA
1	68	72		6	80	91
2	40	43		7	47	58
3	82	89		8	55	68
4	56	60		9	78	77
5	70	75		10	53	65

This situation fits the *paired sample design* discussed in Section 9.5, provided we assume that both populations are normally distributed, as we have been assuming throughout this chapter. Our discussion now becomes an extension of Section 9.5, except that we can now consider more than two populations.

To determine if there is a factor (brand) effect in these 20 ratings, we must first account for the block (person) effect. If we ignore this effect, we could easily come to an incorrect conclusion regarding a brand difference. This same point was made in Chapter 9, where a crucial question was whether to pair (block) the data. Figures 9.13 and 9.14 illustrated how one can arrive at an incorrect conclusion by failing to block the data.

Now assume that the warranty analysis was extended to three brands of automobiles by also including the warranty provided by the manufacturer of Roadster automobiles, using the same ten people. The results were:

PERSON	HENRY	GA	ROADSTER		PERSON	HENRY	GA	ROADSTER
1	68	72	65		6	80	91	86
2	40	43	42		7	47	58	50
3	82	89	84		8	55	68	52
4	56	60	50		9	78	77	75
5	70	75	68		10	53	65	60

The data from the warranty example constitute a randomized block design with a single factor (brand) containing three levels (Henry, GA, Roadster) as well as ten blocks (person 1, . . . , person 10). The general appearance of such a design is shown in Table 11.1.

When using the randomized block design, the various levels should be applied in a *random* manner within each block. In our warranty example, each

TABLE 11.1
The Randomized Block Design

	FACTOR LEVEL (POPULATION)					
Block	1	2	3	...	k	Total
1	x	x	x	...	x	S_1
2	x	x	x	...	x	S_2
3	x	x	x	...	x	S_3
⋮						
b	x	x	x	...	x	S_b
Total	T_1	T_2	T_3	...	T_k	T
Sample Mean	$\bar X_1$	$\bar X_2$	$\bar X_3$	...	$\bar X_k$	$\bar X$

person should *not* always examine the Henry warranty first, the GA warranty second, and the Roadster warranty last. Instead, these three companies should be assigned in a randomized order to each person—hence the name *randomized block design*.

The assumptions for the randomized block design are:

1. The observations within each factor level/block combination are obtained from a normal population.

2. These normal populations have a common variance, σ^2.

Futhermore, we assume that the factor effects are the same within each block; that is, there is no interaction effect between the factor and the blocks.

The analysis using the randomized block design is similar to that for the one-factor ANOVA, except that the total sum of squares (SS(total)) has an additional component. Now,

$$SS(total) = SS(factor) + SS(block) + SS(error)$$

where SS(block) measures the variation due to the blocks. Consequently, this design extracts the block effect, as measured by SS(blocks), from the error of sum of squares in the completely randomized design.

If you use the randomized block design when blocking is not necessary, SS(block) will be very small in comparison to the other sums of squares. Referring to Table 11.1, this will occur when $S_1, S_2, \ldots, S_b$ are nearly the same. If all the S_i's are equal, then SS(block) = 0. The effect of the blocks will be significant whenever you observe a lot of variation in these block totals.

The sum of squares for the randomized block design is thus

$$SS(factor) = \frac{1}{b}\left[T_1^2 + T_2^2 + \cdots + T_k^2\right] - \frac{T^2}{bk}$$

where

n = number of observations = bk

$T_1, T_2, \ldots, T_k$ represent the totals for the k factor levels

$S_1, S_2, \ldots, S_b$ are the totals for the b blocks

$T = T_1 + T_2 + \cdots + T_k$

$\quad = S_1 + S_2 + \cdots + S_b$ = total of all observations.

$$SS(blocks) = \frac{1}{k}\left[S_1^2 + S_2^2 + \cdots + S_b^2\right] - \frac{T^2}{bk}$$

$$SS(total) = \Sigma x^2 - \frac{T^2}{bk}$$

where Σx^2 = sum of the squares for each of the n ($=bk$) observations.

$$SS(error) = SS(total) - SS(factor) - SS(blocks)$$

The degrees of freedom are

$\quad$ df for factor = $k - 1$

$\quad$ df for blocks = $b - 1$

$\quad$ df for error = $(k - 1)(b - 1)$

$\quad$ df for total = $bk - 1$

The ANOVA table for a blocked design is very similar to the one-factor ANOVA table. There is one additional row because you now include the effect of the various blocks in your design:

SOURCE	df	SS	MS	F
Factor	$k-1$	SS(factor)	MS(factor) = $\dfrac{\text{SS(factor)}}{k-1}$	F_1 = MS(factor)/MS(error)
Blocks	$b-1$	SS(blocks)	MS(blocks) = $\dfrac{\text{SS(blocks)}}{b-1}$	F_2 = MS(blocks)/MS(error)
Error	$(k-1)(b-1)$	SS(error)	MS(error) = $\dfrac{\text{SS(error)}}{(k-1)(b-1)}$	
Total	$bk-1$	SS(total)		

where

$$\text{MS (factor)} = \frac{\text{SS(factor)}}{(k-1)}$$

$$\text{MS(blocks)} = \frac{\text{SS(blocks)}}{(b-1)}$$

$$\text{MS(error)} = \frac{\text{SS(error)}}{[(k-1)(b-1)]}$$

Hypothesis Testing

Is there a difference in the average rating of the three warranties in the previous illustration? In other words, does the brand have a significant effect on the warranty ratings? The hypotheses for this situation are $H_0 : \mu_1 = \mu_2 = \mu_3$ and H_a: not all the means are equal. We determine the test statistic exactly as we did for the one-factor ANOVA:

$$F_1 = \frac{\text{MS(factor)}}{\text{MS(error)}} \qquad (11.21)$$

where the mean square values are obtained from the ANOVA table. Notice that the MS(factor) value is *the same* regardless of whether or not you block. However, when you use the block effect in the design, the SS(error) is smaller than the SS(error) value obtained using the completely randomized design. The df in the error term are different for the two designs, so it does not necessarily follow that the MS(error) is smaller in the randomized block design. If there is considerable variation among the block totals, then quite likely MS(error) *will* be smaller in the randomized block design. You are thus more likely to detect a difference in these k means when a difference does exist. Had you not included the block effect, the block variation would have been included in SS(error), resulting in a smaller F-value. This value often becomes small enough not to fall in the rejection region, leading you to conclude that no difference exists. But perhaps there *is* a difference among these means (the factor *does* have a significant effect) that will go undetected if an incorrect experimental design is used.

For the randomized block design, the test will be to

$$\text{reject } H_0 \text{ if } F_1 > F_{\alpha, v_1, v_2}$$

where $v_1 = k - 1$ and $v_2 = (k - 1)(b - 1)$. So, once again, we reject H_0 if the F-statistic falls in the right-tail rejection region, this time using Table A.7 with $v_1 = k - 1$, the df for factor, and $v_2 = (k - 1)(b - 1)$, the df for error.

Now suppose we wish to determine whether the effect of the person evaluating the warranties is significant. We are attempting to determine whether there is a block effect, so the hypotheses are

H_0': there is no effect due to the evaluators (blocks)
(the block means are equal)

H_a': there is an effect due to the evaluators (the block means are not all equal)

The corresponding test uses the "other" F-statistic from the randomized block ANOVA table, namely,

$$F_2 = \frac{MS(\text{blocks})}{MS(\text{error})} \qquad (11.22)$$

and the test procedure is to

reject H_0' if $F_2 > F_{\alpha, v_1', v_2'}$

where $v_1' = b - 1$ and $v_2' = (k - 1)(b - 1)$.

Let us reexamine our data. We will use $\alpha = .05$. Here, $k = 3$ levels (brands), $b = 10$ blocks (people), and $n = bk = 30$ observations.

PERSON	HENRY	GA	ROADSTER	TOTALS
1	68	72	65	205
2	40	43	42	125
3	82	89	84	255
4	56	60	50	166
5	70	75	68	213
6	80	91	86	257
7	47	58	50	155
8	55	68	52	175
9	78	77	75	230
10	53	65	60	178
Total	629	698	632	1959
$\bar{X}$	62.9	69.8	63.2	

$$SS(\text{factor}) = \frac{1}{10}\left[629^2 + 698^2 + 632^2\right] - \frac{1959^2}{30}$$
$$= 128,226.9 - 127,922.7$$
$$= 304.2$$

$$SS(\text{blocks}) = \frac{1}{3}\left[205^2 + 125^2 + \cdots + 178^2\right] - \frac{1959^2}{30}$$
$$= 133,627.7 - 127,922.7$$
$$= 5705.0$$

$$SS(\text{total}) = \left[68^2 + 40^2 + \cdots + 75^2 + 60^2\right] - \frac{1959^2}{30}$$
$$= 134,107 - 127,922.7$$
$$= 6184.3$$

$$SS(\text{error}) = SS(\text{total}) - SS(\text{factor}) - SS(\text{blocks})$$
$$= 6184.3 - 304.2 - 5705.0$$
$$= 175.1$$

So

$$MS(factor) = \frac{SS(factor)}{(k-1)}$$
$$= \frac{304.2}{2} = 152.1$$

$$MS(blocks) = \frac{SS(blocks)}{(b-1)}$$
$$= \frac{5705.0}{9} = 633.9$$

$$MS(error) = \frac{SS(error)}{[(k-1)(b-1)]}$$
$$= \frac{175.1}{18} = 9.73$$

The resulting ANOVA table is

SOURCE	df	SS	MS	F
Factor	2	304.2	152.1	$152.1/9.73 = 15.63$ (F_1)
Blocks	9	5705.0	633.9	$633.9/9.73 = 65.15$ (F_2)
Error	18	175.1	9.73	
Total	29	6184.3		

We first consider the hypotheses

$$H_0: \mu_1 = \mu_2 = \mu_3$$
$$H_a: \text{not all } \mu\text{'s are equal}$$

where

μ_1 = average rating of Henry warranty (estimate is $\bar{x}_1 = 62.9$)

μ_2 = average rating of GA warranty (estimate is $\bar{x}_2 = 69.8$)

μ_3 = average rating of Roadster warranty (estimate is $\bar{x}_3 = 63.2$)

Because $F_1 = 15.63 > F_{.05, 2, 18} = 3.55$, we reject H_0 and conclude that there *is* a difference in the perceived quality of the three warranties. This is not a surprising result because it appears that the GA warranty scored much higher than Henry and Roadster warranties. This means that the factor (car brand) *does* have a significant effect on the warranty rating.

We also wish to test

$$H_0': \text{there is no block effect}$$
$$H_a': \text{there is a block effect}$$

$S_1, S_2, \ldots, S_{10}$ appear to contain considerable variation, so our initial guess is that there is a block effect. Carrying out the statistical test, we see that

$$F_2 = 65.15 > F_{.05, 9, 18} = 2.46$$

Consequently, we strongly reject H_0 in favor of H_a'. The effect of the person doing the three evaluations *is* significant.

A MINITAB solution for this problem is shown in Figure 11.7.

What would the result have been had we treated these 30 observations as replicates, ten from each of the three brands? In other words, what would happen if we failed to recognize that blocking was necessary and we incorrectly

FIGURE 11.7

MINITAB solution for car warranties data.

```
MTB > READ INTO C1-C3
DATA> 68 1 1
DATA> 40 1 2
DATA> 82 1 3
DATA> 56 1 4
DATA> 70 1 5
DATA> 80 1 6
DATA> 47 1 7
DATA> 55 1 8
DATA> 78 1 9
DATA> 53 1 10
DATA> 72 2 1
DATA> 43 2 2
DATA> 89 2 3
DATA> 60 2 4
DATA> 75 2 5
DATA> 91 2 6
DATA> 58 2 7
DATA> 68 2 8
DATA> 77 2 9
DATA> 65 2 10
DATA> 65 3 1
DATA> 42 3 2
DATA> 84 3 3
DATA> 50 3 4
DATA> 68 3 5
DATA> 86 3 6
DATA> 50 3 7
DATA> 52 3 8
DATA> 75 3 9
DATA> 60 3 10
DATA> END
  30 ROWS READ
MTB > TWOWAY USING DATA IN C1, LEVELS IN C2, BLOCKS IN C3

ANALYSIS OF VARIANCE  C1
```

C_3 contains block values: 1 = person 1.
2 = person 2, etc.

C_2 contains factor (brand) levels
1 = Henry
2 = GA
3 = Roadster

SOURCE	DF	SS	MS
C2	2	304.20	152.10
C3	9	5704.97	633.89
ERROR	18	175.13	9.73
TOTAL	29	6184.30	

F value not provided
$F_1 = 152.10/9.73 = 15.63$ (factor)
$F_2 = 633.89/9.73 = 65.15$ (block)

used the one-factor ANOVA? Because both SS(factor) and SS(total) do not change, the only difference is a new SS(error). Therefore,

$$SS(error) = SS(total) - SS(factor)$$
$$= 6184.3 - 304.2$$
$$= 5880.1$$

Also, in the one-factor ANOVA design,

$$\text{df for total} = (\text{df for factor}) + (\text{df for error})$$

So,

$$\text{df for error} = (\text{df for total}) - (\text{df for factor})$$
$$= 29 - 2 = 27$$

The resulting F-value will be

$$F = \frac{MS(factor)}{MS(error)} = \frac{304.2/2}{5880.1/27} = .70$$

Because $F^* = .70$ is *much less* than $F_{.05,\, 2,\, 27} = 3.35$, *we fail to detect a difference in the three means*, μ_1, μ_2, and μ_3. This is the effect of assuming independence among the samples when it does not exist. This emphasizes that failing to recognize the need for a randomized block design can have serious consequences!

EXAMPLE 11.6

The personnel director at Blackburn Industries is investigating dental claims submitted by married employees having at least one child. Of interest is whether the average annual dollar amounts of dental work claimed by the husband, by the wife, and per child are the same. Data were collected by randomly selecting 15 families and recording these three dollar amounts (total claims for the year by the husband, by the wife, and per child). The results of the sample are shown next. Can the personnel director conclude that there is a difference in the three population means using a significance level of .05?

FAMILY	HUSBAND	WIFE	PER CHILD	TOTAL
1	78	84	112	274
2	105	80	274	459
3	95	184	305	584
4	85	158	280	523
5	148	180	263	591
6	284	208	145	637
7	124	145	340	609
8	118	75	130	323
9	153	112	239	504
10	143	204	262	609
11	84	110	182	376
12	106	172	248	526
13	218	185	320	723
14	145	90	226	461
15	175	304	152	631
Total	2061	2291	3478	7830
$\bar{X}$	$ 137.40	$ 152.73	$ 231.87	

SOLUTION

The hypotheses for this situation are

$$H_0: \mu_H = \mu_W = \mu_C$$
$$H_a: \text{not all three means are equal}$$

where μ_H is the average annual amount claimed by the husband, μ_W is the average for the wife, and μ_C is the average amount per child.

The various block and factor totals shown are obtained by summing across and down the array of data.

$$SS(\text{factor}) = \frac{1}{15}[(2061)^2 + (2291)^2 + (3478)^2] - \frac{7830^2}{45}$$

$$= 1,439,525.73 - 1,362,420 = 77,105.73$$

$$SS(\text{blocks}) = \frac{1}{3}[(274)^2 + (459)^2 + \cdots + (631)^2] - \frac{7830^2}{45}$$

$$= 1,435,654 - 1,362,420 = 73,234$$

$$SS(\text{total}) = [(78)^2 + (105)^2 + \cdots + (226)^2 + (152)^2] - \frac{7830^2}{45}$$

$$= 1,610,430 - 1,362,420 = 248,010$$

$$SS(\text{error}) = SS(\text{total}) - SS(\text{factor}) - SS(\text{blocks})$$

$$= 97,670.27$$

This leads to the following ANOVA table. Note that the df for the factor are $3 - 1 = 2$, for blocks are $15 - 1 = 14$, and for total are $45 - 1 = 44$, leaving $44 - 2 - 14 = 28$ df for error.

SOURCE	df	SS	MS	F
Factor	2	77,105.73	38,552.87	$F_1 = 38552.87/3488.22 = 11.05$
Blocks	14	73,234	5,231	$F_2 = 5231/3488.22 = 1.50$
Error	28	97,670.27	3,488.22	
Total	44	248,010		

To test $H_0: \mu_H = \mu_W = \mu_C$, we use $F_1 = 11.05$. Since $11.05 > F_{.05, 2, 28} = 3.34$, we reject H_0 and conclude that the three average claim amounts are not equal. By observing the sample means, we notice that the claims per child are considerably higher than those for the husband and wife.

As a final note, the block (family) effect is *not* significant here, since $F_2 = 1.50 < F_{.05, 14, 28} \cong 2.04$ (using 15 and 28 df). This does not mean that including the block effect in the analysis was a mistake, since the samples were clearly *not* independent. Furthermore, there is no guarantee that for the next set of data in this situation, the block effect will once again turn out to be insignificant.

■

Constructing a Confidence Interval for the Difference Between Two Population Means

We can construct a confidence interval for the difference between any pair of means, $\mu_i - \mu_j$. Remember, however, that we must determine which confidence intervals we will construct *before* observing the data. Do not fall into the trap of letting the data dictate which confidence intervals you construct.

When using the randomized block design, our estimate of the common variance, σ^2, is now

$$s^2 = \text{estimate of } \sigma^2$$
$$= \text{MS(error)} = \frac{\text{SS(error)}}{(k-1)(b-1)} \qquad (11.23)$$

Thus, a $(1 - \alpha) \cdot 100\%$ confidence interval for $\mu_i - \mu_j$ is

$$(\bar{X}_i - \bar{X}_j) - t_{\alpha/2, \text{df}} \cdot s \cdot \sqrt{\frac{1}{b} + \frac{1}{b}} \quad \text{to} \quad (\bar{X}_i - \bar{X}_j) + t_{\alpha/2, \text{df}} \cdot s \cdot \sqrt{\frac{1}{b} + \frac{1}{b}}$$
$$(11.24)$$

where df = degrees of freedom for the t-statistic (Table A.5) = $(k-1)(b-1)$; b = number of blocks; k = number of factor levels; and s is determined from equation 11.23.

EXAMPLE 11.7

Assume you have not yet observed the dental claim data in Example 11.6, and you decided to construct a 95% confidence interval for the difference between the average annual claim for the wife and the average annual claim per child. What does it tell you?

SOLUTION

Using the ANOVA table for these data,

$$s^2 = \text{MS(error)} = 3488.22$$

and so $s = 59.06$. Also, $t_{.025, 28} = 2.048$ using Table A.5. The resulting 95% confidence interval for $\mu_C - \mu_W$ is

$$(\bar{X}_C - \bar{X}_W) - t_{.025, 28}\sqrt{\frac{1}{15} + \frac{1}{15}} \quad \text{to} \quad (\bar{X}_C - \bar{X}_W) + t_{.025, 28}\sqrt{\frac{1}{15} + \frac{1}{15}}$$

$$= (231.87 - 152.73) - (2.048)(59.06)(.365) \quad \text{to}$$
$$(231.87 - 152.73) + (2.048)(59.06)(.365)$$
$$= 79.14 - 44.15 \quad \text{to} \quad 79.14 + 44.15$$
$$= 34.99 \quad \text{to} \quad 123.29$$

We are thus 95% confident that the average annual claim per child is between $34.99 and $123.29 *higher* than the average annual claim for the wife.

■

EXERCISES

11.23 A real estate firm used two independent property appraisers. The firm wanted to know whether the two appraisers were consistent in determining the market value of local buildings. The appraisers each appraised 11 buildings and the following data were collected (values in dollars):

BUILDING	APPRAISER 1	APPRAISER 2	BUILDING	APPRAISER 1	APPRAISER 2
1	25,000	25,500	7	29,950	30,590
2	28,000	30,000	8	38,100	39,500
3	41,200	41,100	9	31,350	32,750
4	48,300	47,600	10	25,890	24,900
5	51,350	50,100	11	48,500	47,300
6	32,450	34,125			

a. Use a paired t-test to test for differences due to the appraisers. Use a .05 significance level.

b. Use the F-test in the randomized block design to test for no differences in appraisers. Use a .05 significance level.

c. What is the relationship between the t-test in part (a) and the F-test in part (b)?

11.24 A study compared the price of regular gas at Exgas stations and at Argas stations. Ten locations were randomly chosen in which both Exgas and Argas service stations were located. The price per gallon (in dollars) was:

LOCATION	1	2	3	4	5	6	7	8	9	10
Exgas	1.08	1.05	1.09	1.04	1.10	1.09	1.05	1.06	1.09	1.10
Argas	1.07	1.04	1.04	1.07	1.08	1.10	1.05	1.03	1.06	1.07

a. Use a paired t-test to test for differences in the mean price of regular gas of the two companies. Use a .01 significance level.

b. Use the F-test in the randomized block design to test for the difference in the mean price of regular gas at the Exgas and Argas service stations. Use a .01 significance level. Is the F-value equal to the square of the t-value in part (a)?

11.25 A particular application contains four blocks and four levels for the factor of interest. The totals for each of the four blocks are given as: $S_1 = 170, S_2 = 184, S_3 = 182, S_4 = 240$, and the totals for each of four levels of the factor are given as: $T_1 = 120, T_2 = 240, T_3 = 210, T_4 = 206$. Construct the ANOVA table for the randomized block design and assume that the total sum of squares is 2836.

11.26 Complete the following ANOVA table for a randomized block design. Find the p-value.

SOURCE	df	SS	MS	F
Factor	7	105.6		
Blocks	3	90.8		
Error				
Total		336.5		

11.27 The study in Exercise 11.5 was modified such that only six secretaries were used. Each secretary had a different typing speed. Each secretary tested all three word-processing software packages. The same task could not be used for testing all three packages, since the "learning effect" would come into play, so each secretary performed three separate tasks. However, the tasks were of essentially the same length and difficulty level. Furthermore, which task was assigned to which word processor was randomly determined, and the order in which the three word processors were tested was also randomly decided. The secretaries relaxed between tasks to avoid "fatigue effects." Thus, a randomized block design was achieved, with secretaries constituting blocks and the three observations (levels of the factor) within each block being randomized. The following data were obtained (the secretary's typing speed in words per minute is given in parentheses for reference purposes, and the body of the table contains time taken to complete the tasks):

SECRETARY	GROUP 1 (MENU)	GROUP 2 (COMMAND)	GROUP 3 (MIXED)
1 (75 wpm)	9	10	7
2 (65 wpm)	12	11	9
3 (55 wpm)	12	14	11
4 (50 wpm)	13	13	11
5 (45 wpm)	16	15	13
6 (30 wpm)	18	16	15

a. Compute the ANOVA table for the preceding data.

b. Conduct a hypothesis test to address the question: Is there a significant difference between the three word processors (as measured by the performance of the secretaries)? Use $\alpha = .10$.

c. Determine the p-value. Does the conclusion change at $\alpha = .05$ and at $\alpha = .01$?

d. Is the block (secretary's) effect significant at $\alpha = .01$?

11.28 An analyst with a marketing firm wished to know if there was a difference in the number of responses from advertising a certain product at three different times on television. Ten days were randomly selected to run a commercial with a call-in phone number at each of the three times. The number of responses was recorded for 16 products:

PRODUCT	NOON	5:00 P.M.	10:30 P.M.	PRODUCT	NOON	5:00 P.M.	10:30 P.M.
1	12	18	14	9	7	5	8
2	12	30	22	10	35	39	37
3	5	4	3	11	17	15	16
4	21	20	24	12	31	45	40
5	13	19	14	13	15	25	21
6	17	16	15	14	18	29	18
7	35	37	33	15	7	10	6
8	20	29	20	16	20	17	18

a. At the .05 significance level, is there sufficient evidence to conclude that the mean number of responses at noon, 5:00 P.M., and 10:30 P.M. are different?

b. Find a 90% confidence interval for the difference in the mean number of responses for noon and 5:00 P.M.

11.29 In Exercise 11.28, subtract 3.0 from each of the observations in the table. Perform the ANOVA procedure at the .05 significance level and test the hypothesis of no effect due to the three different times of day. Is the sum of squares the same for the coded data as for the original data? If any set of data is coded by adding or subtracting the same number to the value of each observation, how will the sum of squares be affected?

11.30 A computer firm compared the performance of four of its compilers. Five different programs were tested. The time required to compile each of the five programs was recorded. The observations are given in seconds. Do the data provide sufficient evidence to indicate that there is a difference in the performance of the compilers at the .05 significance level? Is there a significant difference due to blocks?

	1	2	3	4
Program 1	31.103	24.315	33.058	22.013
Program 2	30.111	25.216	34.698	21.001
Program 3	29.903	25.347	33.872	20.314
Program 4	30.013	24.136	35.671	24.316
Program 5	31.981	24.977	34.751	22.591

11.31 In a randomized block design, if the df for the factor sum of squares is given as 12 and the df for the error sum of squares is 3, can you find the number of blocks used in the experiment? If yes, how many were used? If the total sum of squares is given as 520, the error sum of squares as 110, and the sum of squares due to blocks as 280, can you find the F test for this experiment? If yes, what is it?

11.32 A machine-shop supervisor is interested in knowing whether there is a significant difference among the production times of three machines running six different jobs.

	1	2	3	4	5	6
Machine 1	4.2	2.1	1.3	7.1	6.0	3.4
Machine 2	6.1	2.9	2.0	7.8	6.8	4.3
Machine 3	5.3	2.1	1.4	7.3	5.8	3.1

a. At the .01 level, test the hypothesis that there is no mean differences in the production time for each of the three machines.

b. Is the test for blocks at the .05 level significant?

c. Find a 90% confidence interval for the difference in mean production time for machine 1 and machine 2.

11.33 The Green Thumb lawn-care company is testing three different formulas for a fertilizer especially designed for lawns in Denton County. To adjust for variation in the soil, the formulas are tested in 13 locations. The growth rate of the grass is recorded. Do the coded results indicate a difference in the lawn growth due to the formulas at the .05 level of significance?

LOCATION	FORMULA 1	FORMULA 2	FORMULA 3
1	3.1	3.0	2.7
2	2.6	2.4	2.5
3	2.9	2.1	2.3
4	3.5	3.4	3.1
5	3.8	3.7	3.2
6	2.9	2.5	2.6
7	3.1	3.3	3.2

LOCATION	FORMULA 1	FORMULA 2	FORMULA 3
8	3.4	2.9	3.1
9	3.1	3.2	3.2
10	3.3	3.0	2.9
11	2.7	2.4	2.2
12	2.8	2.3	2.1
13	3.4	3.0	3.1

11.34 Linoleum Unlimited is experimenting with three types of adhesives for laying linoleum. Each glue is tested on five different surfaces. The adhesiveness of the glue is measured and the coded results are as follows. Construct the ANOVA table and test the hypothesis that there is no difference in the three types of adhesives at the .10 level of significance.

SURFACE	ADHESIVE 1	ADHESIVE 2	ADHESIVE 3
1	1.5	2.1	2.4
2	1.6	1.8	1.9
3	2.4	2.5	2.4
4	3.1	3.4	3.1
5	4.5	4.2	4.0

11.35 Suppose the following results were given from a computer printout: SS(factor) = 293.1, SS(blocks) = 5160.2, and SS(error) = 170.2. Assume four factor levels and ten blocks were used. Develop the ANOVA table and test for the effect of the factor and also for the effect of blocks. Use a significance level of .05.

11.36 What assumptions need to be made about the distribution of the population from which data are obtained in a randomized block design?

11.37 A consumer wished to know whether three leading brands of bread were priced approximately the same. Twelve supermarkets were selected as blocks. The following data were collected:

SUPERMARKET	BRAND 1	BRAND 2	BRAND 3
1	45	40	38
2	40	39	38
3	41	40	37
4	42	41	37
5	40	40	40
6	39	40	40

SUPERMARKET	BRAND 1	BRAND 2	BRAND 3
7	46	42	38
8	43	40	39
9	42	40	40
10	43	41	36
11	40	38	37
12	39	38	37

Test the hypothesis $H_0: \mu_1 = \mu_2 = \mu_3$ at the .01 significance level. Is the test for blocks significant at the .01 level?

11.5
THE TWO-WAY FACTORIAL DESIGN

The two-way factorial design was introduced in Section 11.3. For this type of experiment, the researcher is considering two factors of interest, say factor A and factor B. Of concern will be whether the individual factors have a significant effect on the observed variable (called the dependent variable) as well as the combined effect of the two factors.

Consider a simple example where the dependent variable is the score on a test designed to measure assertiveness and managerial potential. The factors are sex and marital status (single or married). Each of these two factors consists of two levels. Suppose we observed a significant difference between the male and female scores. Thus we would conclude that factor A (sex) is significant. The analysis procedure to investigate this hypothesis is described in this section. If a significant difference between the scores of the single and married subjects is observed, then we would conclude that factor B (marital status) is also significant.

Suppose that a closer look at the scores revealed that the married males and single females scored high, but the single males and married females scored low on the test, as illustrated in Figure 11.8.

Consequently, the relationship between sex and the dependent variable (exam score) *depends upon the marital status*, since this relationship is different for the single and married groups. Similarly, the relationship between marital status and the dependent variable depends upon the particular level of factor A (sex). This is an illustration of **interaction** between factors A and B. A method of detecting interaction using a simple graph, along with a statistical test of hypothesis, is explained.

FIGURE 11.8
Scores on assertiveness/ managerial potential exam.

	Single	Married
Male	Low	High
Female	High	Low

FIGURE 11.9
Layout for two-way
factorial design.

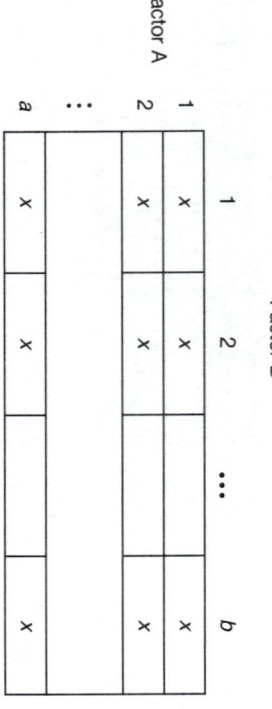

	Factor B			
Factor A	1	2	...	b
1	x	x		x
2	x	x		x
...				
a	x	x		x

Degrees of Freedom

In a two-way factorial design, each level of factor A is combined with each level of factor B when obtaining the sample data. Suppose that factor A has a levels and factor B has b levels, as shown in Figure 11.9. Each x represents a test score.

If we record one observation for each factor A and factor B combination (referred to as a **treatment**), then we have $n = ab$ total observations. The df for each factor is one less than the number of levels and the df for the interaction term is the product of the factor A df and the factor B df. Consequently,

$$\text{df for factor A} = a - 1$$
$$\text{df for factor B} = b - 1$$
$$\text{df for interaction} = (a - 1)(b - 1)$$
$$\text{df for total} = n - 1 = ab - 1$$

We have a bit of a problem here. This design, like all experimental designs, must contain a source of variation due to error, that is, the unexplained variation. Suppose $a = 4$ and $b = 3$. Then the remaining df for error is (df for total) − (df for factor A) − (df for factor B) − (df for interaction), which in this case is $11 - 3 - 2 - 6 = 0$. It can be shown that the error df is zero *regardless* of the values of a and b. Since it will be necessary to measure this unexplained variation, this design requires that you obtain repeat observations (**replicates**) for each treatment. An illustration (using two replicates) including the various totals needed to carry out the analysis is shown in Figure 11.10. In general, you will need two or more replicates at each treatment. The number of replicates at each treatment need not be the same, but we consider here only the case where there are r replicates at each treatment.

In the replicated design, the degrees of freedom are

$$\text{df for factor A} = a - 1$$
$$\text{df for factor B} = b - 1$$
$$\text{df for interaction} = (a - 1)(b - 1)$$
$$\text{df for total} = (\text{number of observations} - 1)$$
$$= abr - 1$$
$$\text{df for error} = (abr - 1) - (a - 1) - (b - 1) - (a - 1)(b - 1)$$
$$= ab(r - 1)$$

Sum of Squares and Mean Squares

The necessary sums of squares can be computed in a manner similar to that used in the previous designs. Using Figure 11.10, the following expressions can be used to find the corresponding sums of squares.

FIGURE 11.10

Illustration of two replicates in a two-way factorial design ($r = 2$).

Factor B

Factor A	1	2	$\cdots$	b	Totals
1	x, x (total = R_{11})	x, x (total = R_{12})		x, x (total = R_{1b})	T_1
2	x, x (total = R_{21})	x, x (total = R_{22})		x, x (total = R_{2b})	T_2
$\cdots$					
a	x, x (total = R_{a1})	x, x (total = R_{a2})		x, x (total = R_{ab})	T_a
Totals	S_1	S_2		S_b	

factor A:
$$SSA = \frac{1}{br}\left[T_1^2 + T_2^2 + \cdots + T_a^2\right] - \frac{T^2}{abr} \qquad (11.25)$$

where T = total of all n observations (that is, $T = T_1 + T_2 + \cdots + T_a$).

factor B:
$$SSB = \frac{1}{ar}\left[S_1^2 + S_2^2 + \cdots + S_b^2\right] - \frac{T^2}{abr} \qquad (11.26)$$

interaction:
$$SSAB = \frac{1}{r}\left[\sum R^2\right] - SSA - SSB - \frac{T^2}{abr} \qquad (11.27)$$

where the sum in the brackets is the sum of all the squares of the replicate totals, illustrated in Figure 11.10.

total:
$$SS(\text{total}) = \sum x^2 - \frac{T^2}{abr} \qquad (11.28)$$

where $\sum x^2$ is the sum of the squares for each of the $n = abr$ observations. By subtraction,

$$SS(\text{error}) = SS(\text{total}) - SSA - SSB - SSAB \qquad (11.29)$$

The corresponding mean squares can be obtained by dividing each sum of squares by the corresponding degrees of freedom. Thus we have

$$MSA = \frac{SSA}{(a-1)} \qquad (11.30)$$

$$MSB = \frac{SSB}{(b-1)} \qquad (11.31)$$

$$MSAB = \frac{SSAB}{(a-1)(b-1)} \qquad (11.32)$$

$$MS(\text{error}) = \frac{SS(\text{error})}{ab(r-1)} \qquad (11.33)$$

This analysis can be summarized in the following ANOVA table:

SOURCE	df	SS	MS	F
Factor A	$a-1$	SSA	$MSA = \dfrac{SSA}{a-1}$	$F_1 = MSA/MS(error)$
Factor B	$b-1$	SSB	$MSB = \dfrac{SSB}{b-1}$	$F_2 = MSB/MS(error)$
Interaction	$(a-1)(b-1)$	SSAB	$MSAB = \dfrac{SSAB}{(a-1)(b-1)}$	$F_3 = MSAB/MS(error)$
Error	$ab(r-1)$	SS(error)	$MS(error) = \dfrac{SS(error)}{ab(r-1)}$	
Total	$abr-1$	SS(total)		

The personnel director at Blackburn Industries is interested in examining the effect of sex and employee classification on the annual amount of dental claims for unmarried employees at Blackburn. Employee classification ranges from category 1 (production line workers) to category 4 (upper-level management). By utilizing a two-way factorial design, she can study the effect of sex (factor A) and employee classification (factor B), as well as the interaction effect between sex and employee classification, on the amount of the annual dental claims. This results in a 2×4 factorial design, since factor A consists of two levels and factor B has four levels. She decided to use three replicates for each of the eight treatment combinations, requiring annual claims from 24 different employees. The sample results are as follows, where the values in parentheses are the replicate totals for each of the treatments.

EMPLOYEE CLASSIFICATION (FACTOR B)

SEX (FACTOR A)	CATEGORY 1	CATEGORY 2	CATEGORY 3	CATEGORY 4	TOTAL	AVERAGE
Male	190,225,200 (615)	135,180,100 (415)	260,330,350 (940)	305,275,240 (820)	2790	232.50
Female	235,190,270 (695)	275,305,285 (865)	160,205,140 (505)	155,110,75 (340)	2405	200.42
Total	1310	1280	1445	1160	5195	
Average	218.33	213.33	240.83	193.33		

Using the previous discussion, the necessary sums of squares can be derived.

$$SSA = \frac{1}{(4)(3)}(2790^2 + 2405^2) - \frac{5195^2}{24} = 6176.04$$

$$SSB = \frac{1}{(2)(3)}(1310^2 + 1280^2 + 1445^2 + 1160^2) - \frac{5195^2}{24} = 6853.12$$

$$SSAB = \frac{1}{3}(615^2 + 415^2 + 940^2 + 820^2 + 695^2 + 865^2 + 505^2 + 340^2)$$
$$- 6176.04 - 6853.12 - \frac{5195^2}{24} = 98578.13$$

$$SS(total) = (190^2 + 225^2 + 200^2 + \cdots + 155^2 + 110^2 + 75^2) - \frac{5195^2}{24}$$
$$= 131173.96$$

Consequently,

$$SS(error) = 131173.96 - 6176.04 - 6853.12 - 98578.13 = 19566.67$$

The degrees of freedom here will be:

Sex factor: $a - 1 = 2 - 1 = 1$

Employee classification factor: $b - 1 = 4 - 1 = 3$

Interaction: $(a - 1)(b - 1) = (1)(3) = 3$

Error: $ab(r - 1) = (2)(4)(3 - 1) = 16$

Total: $abr - 1 = (2)(4)(3) - 1 = 24 - 1 = 23$

These calculations and the resulting mean squares can be summarized in the following ANOVA table.

SOURCE	df	SS	MS	F
Sex	1	6176.04	6176.04	$F_1 = 6176.04/1222.92 = 5.05$
Employee classification	3	6853.12	2284.37	$F_2 = 2284.37/1222.92 = 1.87$
Interaction	3	98578.13	32859.38	$F_3 = 32859.38/1222.92 = 26.87$
Error	16	19566.67	1222.92	
Total	23	131173.96		

Hypothesis Testing   When using a two-way factorial design, you can test for the significance of factor A, factor B, and the interaction of the two factors. For factor A, the null hypothesis is that the means are equal across the factor A levels. Written another way, we can define the following hypotheses:

$H_{0,A}$: factor A is not significant $(\mu_M = \mu_F)$

$H_{a,A}$: factor A is significant $(\mu_M \neq \mu_F)$

The corresponding test statistic is

$$F_1 = \frac{MSA}{MS(error)}$$

   (11.34)

and the testing procedure is to reject $H_{0,A}$ if

$$F_1 > F_{\alpha, v_1, v_2}$$

where F_{α, v_1, v_2} is from Table A.7, $v_1 =$ df for factor A $= a - 1$, and $v_2 =$ df for error $= ab(r - 1)$.

Similarly, to test for equal means of the factor B levels, we can define the hypotheses:

$H_{0,B}$: factor B is not significant $(\mu_1 = \mu_2 = \mu_3 = \mu_4)$

$H_{a,B}$: factor B is significant (not all μ_i's are equal)

The test statistic for determining the factor B effect is

$$F_2 = \frac{MSB}{MS(error)}$$

   (11.35)

and factor B is significant ($H_{0,B}$ is rejected) if

$$F_2 > F_{\alpha, v_1, v_2}$$

where $v_1 =$ df for factor B $= b - 1$, and $v_2 =$ df for error $= ab(r - 1)$.

The final set of hypotheses is concerned with the interaction effect between the two factors. The hypotheses for this procedure can be stated

$H_{0, AB}$: there is no significant interaction between factor A and factor B

$H_{a, AB}$: there is significant interaction between factor A and factor B

The test statistic is the remaining F-statistic in the ANOVA table, namely,

$$F_3 = \frac{MSAB}{MS(error)} \qquad (11.36)$$

and the test procedure is to reject $H_{0, AB}$ if

$$F_3 > F_{\alpha, v_1, v_2}$$

where $v_1 = $ df for interaction $= (a-1)(b-1)$ and $v_2 = $ df for error $= ab(r-1)$.

Multiple Comparisons

The method of multiple comparisons discussed in Section 11.2 for the one-factor ANOVA procedure (Tukey's method) can be used to examine pairwise differences between the various treatment means in two-way factorial designs. Since factor A has a levels and factor B has b levels, there are ab such means that can be compared, one pair at a time.

For the two-way factorial design, we use Table A.16 to find $Q_{\alpha, k, v}$, where α is the desired experimentwise significance level, $k = ab$ is the number of treatment means, and v is the degrees of freedom associated with MS(error). Any two (sample) treatment means differing by more than

$$D = Q_{\alpha, k, v} \cdot \sqrt{\frac{MS(error)}{r}}$$

will imply that the corresponding population means are unequal. This procedure is illustrated in the next example.

EXAMPLE 11.8

Using the previous ANOVA table constructed using the dental claims and the two factors sex (factor A) and employee classification (factor B), determine whether (1) factor A is significant, (2) factor B is significant, (3) there is significant interaction between sex and employee classification, and (4) which pairs of the eight (population) treatment means are unequal. Use a significance level of .05.

SOLUTION 1

The df for the F-statistic are $v_1 = 1$ and $v_2 = 16$. Using Table A.7, $F_{.05, 1, 16} = 4.49$, and the test is to reject $H_{0, A}$ if $F_1 > 4.49$. Since $F_1 = 5.05 > 4.49$, we conclude that the sex factor is significant. Examining the raw data, we observe that the sample mean for the males is $2790/12 = 232.50$, and the female average is $2405/12 = 200.42$. Thus we conclude that the difference between these sample means is significant, with higher dental claims occurring in the male population.

SOLUTION 2

For the employee classification factor, the df for the F-statistic are $v_1 = 3$ and $v_2 = 16$, with a corresponding table value of $F_{.05, 3, 16} = 3.24$. Since $F_2 = 1.87 < 3.24$, the employee classification factor is **not significant**. Taking a closer look, we observe that the high (low) female values were balanced by the low (high) male values within each employee classification category. Consequently, there is insignificant variation in the means for the four employee classification groups, leading to the "fail to reject $H_{0, B}$" conclusion.

FIGURE 11.11 Illustration of interaction effect. **A:** Interaction effect in Example 11.8. **B:** Hypothetical situation containing no significant interaction between sex and employee classification.

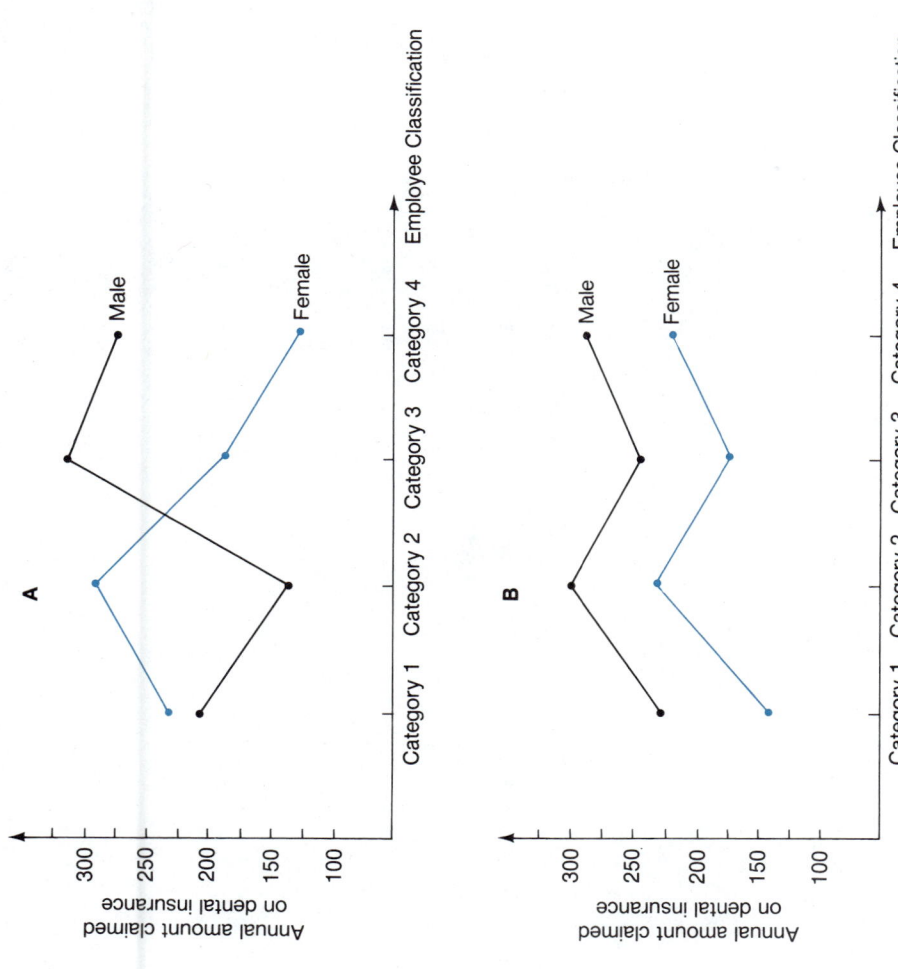

SOLUTION 3

The discussion in the solution to part 2 indicates the presence of interaction between the two factors. This means that the relationship between employee classification and the amount of the claim is not the same for males and females. The four male means are $615/3 = 205$ (category 1), $415/3 = 138.33$ (category 2), $940/3 = 313.33$ (category 3), and $820/3 = 273.33$ (category 4). The corresponding means for the female sample are 231.67, 288.33, 168.33, and 113.33. These means are shown in Figure 11.11(A), where interaction effect is very apparent, since the male and female lines **are not parallel**. When no interaction exists between the two factors, this graph should contain lines that are **nearly parallel**, as illustrated in Figure 11.11(B).

The statistical test here supports this conclusion, since there is significant interaction provided F_3 is larger than $F_{.05, 3, 16} = 3.24$. Here, $F_3 = 26.87$, and so we once again conclude that there is significant interaction between sex and employee classification for this population.

The MINITAB solution for this example is contained in Figure 11.12. Notice that the same TWOWAY command used for the randomized block design is used for the two-way factorial design. Due to the presence of replications, MINITAB assumes a possible interaction effect and includes this line in the ANOVA table.

FIGURE 11.12

MINITAB solution for two-way factorial design in Example 11.8.

```
MTB > READ INTO C1-C3
DATA> 190 1 1
DATA> 225 1 1
DATA> 200 1 1
DATA> 135 1 2
DATA> 180 1 2
DATA> 100 1 2
DATA> 260 1 3
DATA> 330 1 3
DATA> 350 1 3
DATA> 305 1 4
DATA> 275 1 4
DATA> 240 1 4
DATA> 235 2 1
DATA> 190 2 1
DATA> 270 2 1
DATA> 275 2 2
DATA> 305 2 2
DATA> 285 2 2
DATA> 160 2 3
DATA> 205 2 3
DATA> 140 2 3
DATA> 155 2 4
DATA> 110 2 4
DATA> 75 2 4
DATA> END
     24 ROWS READ
MTB > TWOWAY USING DATA IN C1, A LEVELS IN C2, B LEVELS IN C3

ANALYSIS OF VARIANCE C1

SOURCE        DF      SS       MS
C2             1    6176     6176
C3             3    6853     2284
INTERACTION    3   98578    32859
ERROR         16   19567     1223
TOTAL         23  131174
```

SOLUTION 4

A multiple comparisons analysis will determine if the average annual amount for category 1 males is the same as for category 4 males, the average for category 2 males is the same for category 3 females, and so forth. There are $2 \cdot 4 = 8$ means here, providing $_8C_2 = 28$ possible pairwise comparisons. In general, for a two-way factorial design there are $_{ab}C_2$ possible pairs of means that can be compared using the multiple comparison procedure.

The critical value corresponding to $\alpha = .05$, $k = 8$, and $v = 16$ (the df associated with the error sum of squares) from Table A.16 is $Q_{.05, 8, 16} = 4.90$. Since MS(error) $= 1222.92$ and there are $r = 3$ replicates at each treatment level, we next determine

$$D = Q_{\alpha, k, v} \cdot \sqrt{\frac{MS(error)}{r}}$$

$$= (4.90) \sqrt{\frac{1222.92}{3}}$$

$$= 98.93$$

Consequently any pair of sample treatment means differing by more than 98.93 will imply that the corresponding population means are unequal.

The eight sample means are obtained by dividing the corresponding replicate totals (R) by $r = 3$. Placing them in order, we obtain

113.33	138.33	168.33	205.00	231.67	273.33	288.33	313.33
(F, 4)	(M, 2)	(F, 3)	(M, 1)	(F, 1)	(M, 4)	(F, 2)	(M, 3)

Here, M and F represent the sex (factor A) and 1, 2, 3, and 4 represent the employee classification (factor B).

Since $205.00 - 138.33 = 66.67 < 98.93$, we *cannot* conclude that $\mu_{M,1} \neq \mu_{M,2}$. Consider category 4 males and category 4 females. Here $273.33 - 113.33 = 160 > 98.93$, and so we conclude that there *is* a difference in the average amounts for these two groups; that is, $\mu_{M,4} \neq \mu_{F,4}$. This can be observed in the preceding sample means, since there is no overbar connecting these two means. Continuing this procedure, we arrive at the following summary for the multiple comparison analysis:

males only: $\quad \mu_{M,2} \neq \mu_{M,4}, \mu_{M,2} \neq \mu_{M,3}, \mu_{M,1} \neq \mu_{M,3}$

females only: $\quad \mu_{F,4} \neq \mu_{F,1}, \mu_{F,4} \neq \mu_{F,2}, \mu_{F,3} \neq \mu_{F,2}$

males and females: $\quad \mu_{F,4} \neq \mu_{M,4}, \mu_{F,4} \neq \mu_{M,3}, \mu_{M,2} \neq \mu_{F,2},$
$\mu_{F,3} \neq \mu_{M,4}, \mu_{F,3} \neq \mu_{M,3}$

Consequently, we observe three significant differences in each of the male and female populations and five significant differences in the amount of annual dental claims when comparing employee classifications across both sexes. ■

EXERCISES

11.38 A two-way factorial experiment has five observations for each treatment. Three levels of factor A are used and two levels of factor B are used. The sample means for each treatment and the sum of squares are as follows.

		LEVELS (FACTOR A)		
		1	2	3
LEVELS (FACTOR B)	1	21.6	22.2	20.6
	2	23.0	23.6	22.8

$SSA = 7.20, \quad SSB = 20.83, \quad SSAB = 1.07, \quad SS(TOTAL) = 236.30$

a. From viewing the sample means of each treatment, would you consider any of the factors or interaction to be significant?

b. Complete an ANOVA table. Find the corresponding p-values. Interpret the results.

c. Which pairs of the six treatment means are significantly different at the .05 significance level?

11.39 The comparative study of word-processing software in Exercise 11.27 was modified to take into account different types of keyboards: enhanced keyboard, modified keyboard, and standard keyboard. Keyboard layout and type of software could not be assumed to be independent, because it was possible that a certain type of software might actually be enhanced by a certain type of keyboard (e.g., one with special function keys). In other words, interaction between factors was possible. Therefore, a 3×3 factorial design was implemented. The following table gives the completion time in minutes, with three observations for each treatment "cell."

	SOFTWARE TYPE		
KEYBOARD TYPE	GROUP 1 (MENU)	GROUP 2 (COMMAND)	GROUP 3 (MIXED)
Enhanced	9, 8, 10	8, 7, 7	8, 10, 10
Modified	14, 14, 13	10, 14, 12	12, 10, 14
Standard	15, 18, 17	18, 16, 15	15, 15, 14

a. Calculate the ANOVA table for the preceding experiment.

b. Assume that the assumptions of an ANOVA have been satisfied. Is there a significant difference in the three word processors, as measured by the productivity of the secretaries? Use $\alpha = .05$.

c. Do the different keyboards seem to affect productivity, as measured by completion times? Use $\alpha = .05$.

d. Is there a significant interaction between software type and keyboard type, at $\alpha = .05$?

e. For parts (b), (c), and (d), find the corresponding p-value for each test.

11.40 A manufacturer is interested in reducing the number of defective components produced by its employees. A consultant recommends that each employee follow one of three proposed systematic procedures. To determine if there was a difference in the three procedures, an experiment was conducted with one factor having three levels: one level for less than a year experience, one level for 1 to 4 years experience, and one level for over 4 years experience. The second factor was the systematic procedure, one level for each of the three proposed procedures. Data from the experiment, with three replications per cell, is given next. The average number of nondefective components produced per day over a week was recorded for each employee.

EXPERIENCE	PROCEDURE 1	2	3
Less than a year	12.6, 15.7, 10.5	8.6, 9.89, 11.2	13.6, 12.8, 10.2
Between 1 and 4 years	13.7, 14.2, 15.8	9.2, 12.6, 13.1	12.1, 11.8, 14.1
Over 4 years	17.5, 19.8, 20.4	16.4, 17.1, 14.2	16.5, 18.7, 17.0

At the 10 significance level, is there a difference in results of the three systematic procedures?

11.41 A study employed a 2×2 experimental design to study the effect of diversity on the performance of firms. One factor was broad spectrum diversity (BSD), which is defined to be the number of two-digit SIC (standard industrial classification) categories in which a firm concurrently operates. The second factor was mean narrow spectrum diversity (MNSD), which is defined to be the number of four-digit SIC categories in which a firm operates divided by the number of two-digit SIC categories in which the firm operates. High and low levels were chosen to be the two levels for each of these two factors. The dependent variable was the return on equity for each firm. The F ratios for the factors are as follows:

SOURCE	F RATIO
MNSD	2.87
BSD	5.85
Interaction	.06

(*Source:* "Diversification and Performance: A Reexamination Using a New Two-Dimensional Conceptualization of Diversity in Firms," by R. "Rajan" Varadarajan and V. Ramanujam, *Academy of Management Journal* (1987): 380–89.)

Test for significance of each F ratio, assuming that the second degrees of freedom can be taken to be infinity. Use a .05 significance level. What are your conclusions?

11.42 A study was conducted to examine the influence of group makeup of males and females on the conformity of individuals in a business setting. One factor, labeled "gender context," has eight levels, with each level representing a group of individuals with a certain proportion of males and females. Another factor was the sex of each individual in the experiment. An objective measure of the conformity of each individual was obtained. The results of the experiment are summarized in the following ANOVA table. What conclusions can be drawn based on the p-values given in the ANOVA table?

SOURCE OF VARIATION	F	p-VALUE
Gender context	6.93	.0002
Sex	.02	.89
Gender context and sex	.20	.90

(Source: "The Impact of Gender Context On Conformity to Co-worker Pressure," by Joy A. Schneer, Proceedings of the 1986 Annual Meeting of the Decision Sciences Institute, 1986, pp. 1003–5.)

11.43 A program to train middle-level managers is being experimented with to evaluate the overall training. One factor that is considered has two levels: computer-assisted training and no computer-assisted training. A second factor also has two levels: group instruction or self-paced program. A 2×2 factorial design is implemented. A training score on a 100 point scale is computed for each manager at the end of the training session.

	GROUP INSTRUCTION	SELF-PACED PROGRAM
Computer-assisted training	90, 93, 84, 94, 87	96, 92, 97, 99, 96
No computer used	87, 89, 98, 93, 94	97, 96, 91, 96, 95

a. How many replicates are there for each treatment combination?

b. Compute the ANOVA table for the preceding design.

c. Interpret the results of the ANOVA table. Are the factors or interaction significant at the .05 significance level?

d. Which pairs of the four treatment means are significantly different at the .05 level?

e. What assumptions are necessary to ensure that the ANOVA procedure is valid?

SUMMARY

The **analysis of variance (ANOVA)** procedure is a method of detecting differences between the means of two or more normal populations. The various populations represent the *levels* of a *factor* under observation. The factor might consist of, for example, different locations (Does the crime rate differ among five cities?), brands (Does one brand outsell the others?), or time periods (is average attendance the same during each day of the week?).

Samples for this analysis must be obtained independently of each other. The ANOVA technique measures sources of variation among the sample data by computing various **sums of squares**. The variation from one level (population) to the next is measured by the factor sum of squares (SS(factor)), which is large when there is great variation among the sample means. The variation *within* the samples is measured by the error sum of squares (SS(error)). Each of these SS has a corresponding df, which is divided into the SS to produce a **mean square, MS**.

The ratio of MS(factor) to MS(error) produces an *F*-statistic that is used to test for equal means within the various populations. If the *F*-value is large (significant), we conclude that the means are not all the same, which implies that the factor of interest *does* have a significant effect on the variable under observation, called the **dependent variable**.

When we analyze the effect of a single factor, we perform a one-factor ANOVA and use a **completely randomized design**. The results of this analysis, including the various sums of squares, mean squares, and df, are summarized in an **ANOVA table**. If the ANOVA procedure concludes that the population means are not the same, a follow-up analysis can be conducted to determine which of the population means are unequal. This analysis is a **multiple comparisons** procedure and should only be performed in the event that the ANOVA null hypothesis of equal means is rejected.

When samples are not obtained in an independent manner, a **randomized block design** often can be used to test for differences in the population means. Again, there is a single factor of interest, but, to determine the effect of this factor, the sample data are organized into *blocks*. For this situation, the samples are not independently obtained, but data within the same block may be gathered from the same city or person or at the same point in time. By including a block effect in the ANOVA procedure, we can analyze the factor of interest (the population means) using an *F*-test. In addition, another *F*-statistic can be used for determining whether there is a significant block effect within the sample data.

The other experimental design that was discussed was the **two-way factorial design**, where the effect of two factors can be investigated. Observations are obtained for each combination of factor levels, called **treatments**. For such a design, it is necessary to obtain two or more independent replicates for each treatment. The two-way factorial design allows the researcher to investigate the effect of each factor individually, as well as the combined effect of the two factors, referred to as the **interaction** effect.

11.44 Research conducted by social scientists in the Northeast suggests that although older workers are often more productive than their younger counterparts, supervisors tend to rate the older workers lower. Consider the following experimental setup, where the rating is shown on a scale of 1 to 10. The observations are random and independent.

AGE GROUP   **RATINGS BY SUPERVISORS**

AGE GROUP							
<30 years	7.2	5.5	8.0	7.5	6.3	9.0	6.6	7.1
31–45 years	6.1	7.9	5.8	8.0	6.8	7.3	8.2	7.7
>45 years	5.6	6.0	4.9	6.8	5.3	7.0	5.9	5.8

a. State the number of factor levels, and the number of replicates at each level.

b. If the assumptions of an ANOVA design are satisfied, what is the 95% confidence interval for:

(i) The average rating for those less than 30 years of age?

(ii) The average difference between ratings for the "less than 30 years" and the "more than 45 years" groups?

c. Construct an ANOVA table and test the hypothesis that there is no significant difference among the three age groups. At a 10% significance level, is there sufficient evidence to say that age level seems to affect the worker's rating?

d. State the *p*-value for the test.

e. Using a significance level of .05, perform a multiple comparisons procedure (if appropriate).

11.45 To compare the effectiveness of three motivational lectures, 21 employees hired in the past seven months were randomly divided into three groups. Each group heard one lecture. The increase in productivity of the 21 employees was measured over the two weeks following the lectures. The coded values for the increase in productivity were:

Lecture 1	5	6	7	4	6	5	6
Lecture 2	3	4	8	3	5	4	4
Lecture 3	1	2	6	7	2	4	3

a. State the null and alternative hypotheses for this experiment.

b. What is the between-sample variation?

c. What is the within-sample variation?

d. Test the null hypothesis at the .05 significance level and find the p-value.

11.46 The results of a one-factor ANOVA for four groups with six replicates per group yielded the following statistics:

$$\bar{x}_1 = 85, \quad \bar{x}_2 = 90, \quad \bar{x}_3 = 58, \quad \bar{x}_4 = 53, \quad MSE = 198.73$$

Perform a multiple comparison procedure at the .05 significance level.

11.47 The following data give scores obtained by respondents from an index to measure leadership ability, for three levels of management. It is assumed that the populations are normal and independent and the observations are randomly obtained. High scores represent greater leadership ability.

SUPERVISOR	MIDDLE-LEVEL MANAGER	UPPER-LEVEL MANAGER
18	36	55
21	60	42
16	21	68
45	31	33
20	40	48

a. Compute the ANOVA table.

b. Is there significant difference, on the average, between the leadership scores of the three groups? Use $\alpha = .05$. Find the p-value.

c. What is the 95% confidence interval for the mean leadership score of middle-level managers?

d. What is the 99% confidence interval for the difference between mean scores for the upper-level managers and the middle-level managers?

e. Using a significance level of .05, perform a multiple comparisons procedure, if appropriate.

11.48 The study in Exercise 11.47 has been modified to cover the same manager's leadership scores from the post of supervisor to upper-level manager. Fifteen persons were initially chosen, but only 10 actually went all the way to upper-level managerial positions. Since the same managers were used, a "randomized block" design was obtained, with three scores for each manager. Assume that the populations are normally distributed. The following table lists the leadership scores for the 10 managers who completed the study.

RESPONDENT	SUPERVISOR	MIDDLE-LEVEL MANAGER	UPPER-LEVEL MANAGER
1	25	30	30
2	16	35	48
3	17	18	20
4	30	25	20
5	35	30	32
6	28	29	28
7	29	30	35
8	30	40	48
9	27	29	35
10	40	30	32

a. Compute the ANOVA table.

b. Test the hypothesis that the means for the three classes are equal, at $\alpha = .05$. Can you conclude that, on the average, leadership scores remain stable, or do they tend to change as the persons move up the managerial scale?

c. Is there a significant difference among the managers' mean leadership scores? Use $\alpha = .05$.

11.49 Exercise 11.47 was further modified to take into account the influence of sex. Thus, a 3×2 two-way factorial design was implemented. It was decided to have 3 replicates for each cell. The leadership ability scores are given in the following table.

	SEX	SUPERVISOR	MIDDLE-LEVEL MANAGER	UPPER-LEVEL MANAGER
	Male	16, 17, 25	18, 25, 30	20, 30, 42
	Female	18, 20, 28	20, 28, 30	30, 41, 55

a. Compute the ANOVA table.

b. At $\alpha = .10$, test for a significant difference in leadership scores among the three groups of managers.

c. At $\alpha = .10$, is there a difference between the leadership scores of males and females?

d. Test for interaction between managerial level and sex at $\alpha = .10$.

e. Find the p-values for the three preceding hypothesis tests. Do the conclusions change at $\alpha = .05$ or $\alpha = .01$?

11.50 A factorial experiment with two levels for factor A and 4 levels for factor B is conducted. Five observations per treatment are used. The sample means for each treatment and the sums of squares are given.

LEVELS (FACTOR A)		LEVELS (FACTOR B)			
		1	**2**	**3**	**4**
	1	34.8	35.8	31.0	32.0
	2	34.8	32.0	31.2	31.0

$$SSA = 13.2, \quad SSB = 97.9, \quad SS(\text{error}) = 453.2, \quad SS(\text{total}) = 589.80$$

a. Conduct the ANOVA table with interaction.

b. What conclusions can be drawn from the ANOVA using a .05 significance level?

c. Which pairs of the eight treatment means are significantly different at the .05 significance level?

11.51 Assume four samples of size ten are taken from four normally distributed populations with a common variance. The total sum of squares is 5083, and five blocks and four factor levels are used in the experiment. What would be the value of the F-test for testing the hypothesis that there is no difference in the mean levels of the factor?

11.52 Assume that in Exercise 11.51, you are given the additional information that the sample means of populations 1 and 2 are 33.5 and 37.9, respectively. Find the 95% confidence interval for the difference in the means of populations 1 and 2.

11.53 Complete the following table for a randomized block design.

SOURCE	df	SS	MS	F
Factor	9			
Blocks		250		
Error	18	149		
Total		589		

11.54 Suppose it is known in a randomized block design that the mean square for blocks is 75, the error mean square is 291, the total sum of squares is 221.6 and the error sum of squares is 3.7. Test the null hypothesis that there is no difference in the mean of each population at the .05 level. Find the p-value.

11.55 The block totals (S's) and totals (T's) for each level of the factor for a randomized block design are given as follows:

$S_1 = 30,$	$S_2 = 46,$	$S_3 = 38,$	$S_4 = 40$
$S_5 = 40,$	$S_6 = 48,$	$S_7 = 38,$	$S_8 = 36$
$T_1 = 102,$	$T_2 = 92,$	$T_3 = 122$

Construct the ANOVA table for the randomized block design and assume that the total sum of squares is 274.

11.56 Fifteen university campuses of similar size were selected to determine which of three methods of advertising a blood-donation drive was most effective. Five randomly selected campuses advertised in the university newspaper (method 1). Another five advertised only by posters and signs around campus (method 2). The remaining five had each professor credit five points to the student's last test if the student contributed (method 3). The table gives the percentage of the student body that contributed.

Method 1	.10	.15	.19	.21	.25
Method 2	.20	.18	.20	.23	.19
Method 3	.29	.20	.25	.30	.25

Do these data provide sufficient evidence at the .01 level of significance to reject the null hypothesis that there is no difference in the effectiveness of the three methods of advertising?

11.57 Sample data from three normally distributed populations were generated by a computer program for a simulation study. Do the data provide evidence to indicate that at least two of the population variances are not equal? Use a .05 significance level.

SAMPLE 1	SAMPLE 2	SAMPLE 3	SAMPLE 1	SAMPLE 2	SAMPLE 3
38	45	28	33	47	30
37	47	27	32	45	31
35	43	31	39	43	32
40	44	30	37	42	31
39	42	31	36	44	29
35	44	32	35	45	30
34	45	29			

11.58 Three different investment advisers were asked to give a performance rating from 0 to 100 on the risk-adjusted performances of 12 randomly selected aggressive growth mutual funds. From the following data, is there sufficient evidence to conclude that the three investment advisers differ significantly in their mean performance ratings? Use a .05 significance level.

MUTUAL FUNDS	INVESTMENT ADVISORS		
	A	B	C
1	81	76	70
2	83	72	75
3	51	51	43
4	96	92	90
5	67	70	68
6	71	64	71
7	88	75	83
8	51	55	53
9	41	37	45
10	88	90	87
11	59	61	60
12	78	73	74

11.59 In a 3 × 3 factorial experiment with two replicates per treatment, how many error df are there? If a one-way ANOVA was used with three levels and six observations per level, how many error df are there? If the error df are very small for a two-way factorial experiment, how can one increase the error df?

11.60 According to an article in the Journal of Advertising, puffery in advertising claims is defined as "advertising or other sales representations which praise the item to be sold with subjective opinion, superlatives, or exaggerations, vaguely and generally, stating no specific facts." From a major west coast university, students were randomly selected to participate in an experiment. These students were randomly assigned to one of four treatment groups. Each group was exposed to one of two types of advertising appeals combined with one of two levels of puffery (low versus high). In the experiment each subject was asked, "How truthful did you believe the advertiser was being with you?" The ANOVA results are as follows.

	F
Appeal (A)	4.53
Puffery (B)	3.83
Interaction (A × B)	1.34

(*Source*: Michael Kamino and Lawrence Marks, "Advertising Puffery: The Impact of Using Two-sided Claims on Product Attitude and Purchase Intention," *Journal of Advertising* 16, no. 4 (1987): 6–15).

Suppose that you could assume that 21 students were assigned to each treatment combination. Find the degrees of freedom associated with each F-statistic. At what significance level are each of the F-statistics significant?

11.61 A study was conducted to detect the difference in advertising attitude of single women. Two factors were considered. One factor consisted of three levels: never married, divorced, and widowed. The other factor consisted of seven age groups. The attitudes of the single women randomly selected for the experiment are recorded from various questions. Two questions are given, along with the F-statistics for the factors (MS = marital status, A = age).

ATTITUDINAL ITEM	MS	F A	F MS × A
Advertising insults my intelligence.	.45	2.22	1.03
I don't believe a company's ad when it claims test results show its product to be better than competitive products.	6.78	1.52	.44

(*Source*: Alan Buch and John Burnett, "Assessing the Homogeneity of Single Females in Respect to Advertising, Media, and Technology," *Journal of Advertising* 16, no. 3 (1987): 31–8.)

a. Without knowing the denominator degrees of freedom, which F-statistics can you say are significant at the .05 significance level? (Use the F table for considering rejection regions.)

b. If it was known that the error df was greater than 4, would the F-value of 6.78 be considered significant at the .05 level?

c. Can one always consider an F-value of less than one to be nonsignificant regardless of the degrees of freedom at significance levels of .10, .05, and .01?

11.62 A corporate educator was interested in whether there was a significant difference in the time that it takes an individual to complete each of three computer-aided instruction courses. Ten assistant managers were randomly selected to take each of the three courses. The time for completion of each course was recorded in hours.

PERSON	COURSE 1	COURSE 2	COURSE 3	PERSON	COURSE 1	COURSE 2	COURSE 3
1	2.5	2.8	2.0	6	2.5	2.4	2.1
2	2.9	2.7	2.6	7	2.9	2.8	2.5
3	3.5	3.0	2.9	8	3.8	4.0	3.5
4	2.4	2.8	2.4	9	2.6	2.2	2.3
5	3.8	3.5	3.1	10	2.7	2.4	2.2

Is there sufficient evidence that the mean times for completing each computer-aided instruction course differ? Use a .05 significance level.

11.63 The supervisor at an assembly plant can set a machine at three different speeds—fast, normal, and slow. The supervisor is interested in whether the mean number of times that the machine goes out of control differs at each of the speeds. The machine is run on eight randomly selected days at each of the three speeds. The MINITAB printout shows the data, with level 1 representing the fast speed, level 2 representing the normal speed, and level 3 representing the slow speed. From the printout, what statistical conclusions can the supervisor make, assuming a .10 significance level?

```
MTB > READ INTO C1 C2
DATA> 12  1
DATA> 10  1
DATA> 18  1
DATA>  7  1
DATA> 20  1
DATA> 14  1
DATA> 20  1
DATA> 15  1
DATA> 14  2
DATA>  7  2
DATA> 11  2
DATA> 15  2
DATA> 16  2
DATA> 18  2
DATA> 15  2
DATA> 13  2
DATA> 10  3
DATA> 11  3
DATA> 12  3
DATA>  6  3
DATA>  5  3
DATA> 13  3
DATA> 15  3
DATA> 14  3
DATA> END
      24 ROWS READ
MTB > ONEWAY ANOVA, DATA IN C1, LEVELS IN C2
```

```
ANALYSIS OF VARIANCE ON C1
SOURCE    DF      SS       MS        F
C2         2    61.6     30.8     1.98
ERROR     21   327.4     15.6
TOTAL     23   389.0
```

```
                                        INDIVIDUAL 95 PCT CI'S FOR MEAN
                                        BASED ON POOLED STDEV
LEVEL    N     MEAN    STDEV       --+---------+---------+---------+----
1        8   14.500   4.721                    (---------*---------)
2        8   13.625   3.378               (---------*---------)
3        8   10.750   3.615       (---------*---------)
                                  --+---------+---------+---------+----
POOLED STDEV =   3.948            9.0      12.0      15.0      18.0
```

11.64 Workers at a production plant are paid three separate ways: (1) salary, (2) by the hour at $11 per hour, and (3) by the hour at $9 per hour but with a bonus at the end of each month for acceptable productivity levels. Let plans A, B, and C represent each of these pay plans, respectively. Each of eight managers randomly selected three groups of ten workers for each pay plan. Also, each manager recorded the total percent of increase or decrease in productivity for a 6-month period for each group that they selected.

MANAGER	PLAN A	PLAN B	PLAN C
1	2.1	1.1	7.3
2	5.3	2.1	5.1
3	3.6	-1.6	4.3
4	-2.1	-3.5	1.1
5	-3.5	-4.1	.9
6	4.1	3.0	5.1
7	2.8	2.1	3.1
8	3.1	1.1	4.7

a. From the MINITAB computer printout, what conclusions can be drawn? Use a .01 significance level.

b. Is the block effect, that is, the effect of different managers, significant at the .05 level?

c. Construct a 95% confidence interval for the difference in the mean change in productivity between plan A and plan B.

11.65 An experiment is set up to determine if there is a significant difference between three methods of debugging various programs. The first two methods are specific procedures for a programmer to follow, whereas the third method gives the programmer no instructions on how to debug the programs. Three types of difficulty levels are used

```
MTB > name c2 'manager'
MTB > name c3 'pay-plan'
MTB > print c1 c2 c3

ROW    C1    manager    pay-plan
  1    2.1      1           1
  2    1.1      1           2
  3    7.3      1           3
  4    5.3      2           1
  5    2.1      2           2
  6    5.1      2           3
  7    3.6      3           1
  8   -1.6      3           2
  9    4.3      3           3
 10   -2.1      4           1
 11   -3.5      4           2
 12    1.1      4           3
 13   -3.5      5           1
 14   -4.1      5           2
 15    0.9      5           3
 16    4.1      6           1
 17    3.0      6           2
 18    5.1      6           3
 19    2.8      7           1
 20    2.1      7           2
 21    3.1      7           3
 22    3.1      8           1
 23    1.1      8           2
 24    4.7      8           3

MTB > twoway anova,data in c1,pay-plan in c3,manager in c2

ANALYSIS OF VARIANCE C1

SOURCE      DF      SS       MS
pay-plan     2    61.64    30.82
manager      7   128.30    18.33
ERROR       14    22.99     1.64
TOTAL       23   212.93
```

```
MTB > name c1 'time'
MTB > name c2 'program'
MTB > name c3 'method'
MTB > print c1-c3

ROW   time   program   method
  1    21       1         1
  2    19       1         1
  3    21       1         2
  4    17       1         2
  5    18       1         3
  6    20       1         3
  7    21       2         1
  8    17       2         1
  9    19       2         2
 10    20       2         2
 11    19       2         3
 12    19       2         3
 13    20       3         1
 14    18       3         1
 15    19       3         2
 16    20       3         2
 17    24       3         3
 18    26       3         3

MTB > twoway anova data in c1, levels in c2, c3

ANALYSIS OF VARIANCE   time

SOURCE        DF      SS      MS
program        2    14.78    7.39
method         2    11.11    5.56
INTERACTION    4    34.89    8.72
ERROR          9    25.00    2.78
TOTAL         17    85.78
```

for the programs needing debugging. Thus one factor is labeled "method" and the other factor is labeled "program." Two inexperienced programmers are randomly assigned to each combination of debugging method and program difficulty level. The time (in minutes) it takes to debug each program is recorded. From the MINITAB computer printout, what conclusions can be drawn? Use a .10 significance level.

COMPUTER EXERCISES USING THE DATABASE

Exercise 1—Appendix H

From the database, select 10 observations each at random from the NE sector, the NW sector, and SE sector (variable LOCATION). Using a .05 significance, is there sufficient evidence to conclude that the mean house payment or apartment/house rent (variable HPAYRENT) is significantly different for the three locations? Include a multiple comparisons procedure, if appropriate.

Exercise 2—Appendix H

Select at random 12 observations for each level of two factors from the database. Factor A has two levels: a nonzero income from the secondary wage earner, or no secondary income (variable INCOME2). Factor B has two levels: own or rent one's residence (variable OWNORENT). Determine

the effect of these two factors on house payment or house/apartment rent (variable HPAYRENT). Set up an ANOVA table that includes interaction. What conclusions can be drawn at the .05 significance level?

Exercise 3—Appendix I

From the database, select at random six observations from each bond rating and region combination. (Refer to variables BONDRATE and REGION.) Determine the effect of bond rating and region on the assets (variable ASSETS). Use a .05 significance level. Discuss the effect of interaction.

Exercise 4—Appendix I

Repeat Exercise 3, but determine the effect of bond rating (BONDRATE) and region (REGION) on the net income (NETINC) instead of assets.

CASE STUDY

CAN AUDITORS SERVE THE ROLE OF CORPORATE POLICE OFFICERS?

The late 1980s have seen a host of business scandals, such as check floating scams, blatant insider trading, nepotism in bank loans, and so on. However, corporate irregularity is not a recent phenomenon, and public concern about corporate accountability has existed for some time. Recent legislation has pushed managers and auditors into having greater responsibility for preventing and detecting corporate hanky-panky. In this context, one can easily visualize auditors as playing the role of corporate policemen and acting as a deterrent to such behavior. One might also assume

that the more aggressive the auditor is perceived to be, the greater will be the deterrent effect. But is this actually the case?

A study by Uecker, Brief, and Kinney (1981) addressed precisely this issue. Middle- and upper-level managers located throughout the United States were selected, through the researchers' own contacts with business and industry. A packet of "in-basket exercises" and "manipulation check questionnaires" was mailed to each of 143 managers; 104 were received back. Of these 104, 18 were invalidated, leaving 86 usable responses. Although the researchers reported their analysis based on these 86 responses, for purposes of this case study we shall work with a slightly reduced set of 72 observations. (The reason for this is to make certain calculations simpler and more straightforward, without detracting from the essence of the experiment.)

The experimental design chosen by the researchers was a 2×2 factorial analysis of variance, which permits an assessment of main effects and interaction effects of the independent variables. The independent variables (i.e., factors or treatments) were perceived "aggressiveness" of the internal auditor (IA) and the external auditor (EA). The perception of high or low aggressiveness was created by means of certain communications and memos included in the in-basket exercises, and was further validated by a seven-point Likert scale in the manipulations check questionnaire. The purpose of the questionnaire was to check the internal validity of the experiment.

The researchers postulated the following hypotheses:

H_1: increasing the perceived "aggressiveness" of internal auditing activities in the organization decreases the occurrence of corporate irregularities.

H_2: increasing the perceived "aggressiveness" of external auditing activities in the client organization decreases the occurrence of corporate irregularities.

With the manipulation of the independent variables set up (as high versus low aggressiveness of IA and EA), it was necessary now

to measure the response of the dependent variable, which had to capture the occurrence of corporate irregularity. This was a major purpose of the in-basket exercises, so called because they created a very realistic simulation of an executive's in-basket. For this experiment, the in-basket put the manager in a sort of dilemma.

Acting in the capacity of president, he or she had to decide on the size of allowance for writing off a bad debt relating to a loan of $84,000 made by the company to a senior corporate officer, the marketing vice president. The most "honest" estimate was to write off at least $75,600, since the officer was not likely to be able to repay more than one-tenth of the original loan. The problem was that a large allowance for the write-off would reduce the net income before taxes (NIBT). Without any write-off, the projected NIBT was $755,000 versus a target of $742,500, so there was only $12,500 room to "play with." If more than $12,500 was written off, the NIBT would not be met, the president would lose a promised bonus of $20,000, and face problems with stockholders. If less than $75,600 was written off, the president would be guilty of fraudulently overstating the NIBT.

Faced with this dilemma, would the decision-maker be influenced by the chances of getting caught, that is, by the perception of high or low aggressiveness on the part of the internal and external auditors? The table on the next page summarizes the results. Note that if more than one person chose a specific amount as a write-off allowance, this is indicated in the frequency column.

Case Study Questions

1. What is the dependent variable used to measure corporate irregularity?

2. How many replicates are there in each cell?

3. If the data had been collected

Summary Statistics for Amount of Allowance for Doubtful Collections for Each Experimental Condition

			INTERNAL AUDITOR (IA)			
		LOW AGGRESSIVENESS $\bar{x} = \$56,450$		HIGH AGGRESSIVENESS $\bar{x} = \$54,228$		
		Frequency	Allowance	Frequency	Allowance	
	Low Aggressiveness	5	$ 0	5	$ 0	
		1	31,500	1	12,500	
		1	60,000	1	40,000	
		1	75,600	1	75,600	
		1	76,000	1	76,000	
		7	84,000	8	84,000	
		1	85,000	1	100,000	
		1	100,000			
		LOW AGGRESSIVENESS $\bar{x} = \$47,311$		HIGH AGGRESSIVENESS $\bar{x} = \$63,033$		
		Frequency	Allowance	Frequency	Allowance	
External Auditor (EA)	High Aggressiveness	6	$ 0	4	$ 0	
		1	6,000	1	60,000	
		1	40,000	1	75,000	
		1	74,000	1	75,600	
		1	75,600	2	80,000	
		1	76,000	7	84,000	
		2	80,000	2	88,000	
		5	84,000			

(*Source*: Wilfred C. Uecker. Arthur P. Brief, and William R. Kinney, Jr. "Perception of the Internal and External Auditor as a Deterrent to Corporate Irregularities. *Accounting Review* LVI, no. 3 (July 1981): 465–478, Table 5. Slightly modified to achieve the same number of replicates in each treatment cell. Reprinted with permission.)

only on the basis of EA alone (or IA alone), what kind of experimental design is that?

4. Consider hypothesis H_1 formulated by the researchers. Do the results suggest that perceived aggressiveness of the internal auditor affects the amount of the write-off allowance, at 5% significance? Report the *p*-value for the test.

5. Consider hypothesis H_2 formulated by the researchers. What do the results indicate about the influence of the external auditor? Use $\alpha = .05$ for the test. Report the *p*-value.

6. Viewing the internal and external auditors as "police officers in corporate society," do they act as a deterrent to managers contemplating irregular corporate activities, on the basis of the preceding study?

MAINFRAME AND MICRO

SOLUTION

EXAMPLE 11.4

Example 11.4 was concerned with using one-factor ANOVA procedures when testing for a difference in two or more population means. The purpose was to test for a difference in job knowledge among three groups. The groups were (1) employees with a high school diploma only, (2) employees with a bachelor's degree only, and (3) employees with a master's degree. The SPSS program listing in Figure 11.13 was used to request the ANOVA table and, in particular, the *F*-value and the *p*-value for *F*.

In this problem the SPSS commands are the same for both the mainframe and PC versions. (**Remember to end each command line with a period when using the PC version.**)

FIGURE 11.13

Input for SPSSX or SPSS/PC+. All command lines should end with a period when the PC version is used.

```
TITLE   JOB KNOWLEDGE
DATA LIST FREE/EDUC GRADE
BEGIN DATA
1 81
1 84
1 69
1 85
  .
  .
3 88
3 89
3 78
3 85
END DATA
ONEWAY GRADE BY EDUC(1,3)
```

The TITLE command names the SPSS run.

The DATA LIST command gives each variable a name and describes the data as being in free form.

The BEGIN DATA command indicates to SPSS that the input data immediately follow.

The next 15 lines contain the data values, which represent the level of education and the respective grade on the test. The first line implies that the individual had a level 1 (high school) education, and scored 81 on the exam.

The END DATA statement indicates the end of the data.

The ONEWAY command specifies a one-way analysis of variance model. GRADE is the dependent variable and EDUC is the independent variable. EDUC has minimum and maximum values of 1 and 3, respectively.

Figure 11.14 shows the SPSSX output obtained by executing the listing in Figure 11.13, whereas Figure 11.15 shows the SPSS/PC+ output.

FIGURE 11.14 SPSSX output.

```
- - - - - - - - - - - - - - - O N E W A Y - - - - - - - - - - - - - -
    Variable   GRADE
    By Variable   EDUC
```

ANALYSIS OF VARIANCE

SOURCE	D.F.	SUM OF SQUARES	MEAN SQUARES	F RATIO	F PROB.
BETWEEN GROUPS	2	10.8000	5.4000	.1130	.8940
WITHIN GROUPS	12	573.2000	47.7667		
TOTAL	14	584.0000			
				F^*	p-value

FIGURE 11.15 SPSS/PC+ output.

```
- - - - - - - - - - - - - O N E W A Y - - - - - - - - - - - - -
    Variable   GRADE
    By Variable   EDUC

JOB KNOWLEDGE
```

Analysis of Variance

Source	D.F.	Sum of Squares	Mean Squares	F Ratio	F Prob.
Between Groups	2	10.8000	5.4000	.1130	.8940
Within Groups	12	573.2000	47.7667		
Total	14	584.0000			
				F^*	p-value

MAINFRAME AND MICRO

SOLUTION

SECTION 11.4

The car-warranty example was based on a randomized block design. The purpose was to analyze ten individual assessments of new-car warranties for three brands of automobiles. The three brands of automobiles represented three factor levels; the ten individuals represented ten blocks. Each observation consisted of an individual's assessment of the new car warranty for a particular brand. The SPSS program listing in Figure 11.16 was used to request the ANOVA table and in particular the sum of squares for error, block, factor, and total. We were also interested in the F-values for the brands and individuals, and their respective p-values.

FIGURE 11.16

Input for SPSSX or SPSS/PC+. All command lines should end with a period when the PC version is used.

```
TITLE   WARRANTY ASSESMENTS
DATA LIST FREE/PERSON BRAND SCORES
BEGIN DATA
1 1 68
1 2 72
1 3 65
2 1 40
2 2 43
    :
9 2 77
9 3 75
10 1 53
10 2 65
10 3 60
END DATA
ANOVA SCORES BY BRAND(1,3) PERSON(1,10)
OPTIONS 3
```

In this problem the SPSS commands are the same for both the mainframe and PC versions. **(Remember to end each command line with a period when using the PC version.)**

The TITLE command names the SPSS run.

The DATA LIST command gives each variable a name and describes the data as being in free form.

The BEGIN DATA command indicates to SPSS that the input data immediately follow.

The next 30 lines contain the data values. In this example each line of data represents one rater's scoring of the warranty of one brand of automobile. For example, in line one the first 1 represents the first rater, the second 1 represents the first brand, and the 68 represents the warranty rating or score.

The END DATA statement indicates the end of the data.

The ANOVA command specifies the ANOVA model. SCORES is the dependent variable, BRAND is the factor of interest (ranging from 1 to 3), and PERSON represents the blocking variable (ranging from 1 to 10).

The OPTIONS 3 statement specifies that there is no interaction between the factor and blocking variables. (Use /OPTIONS 3 with the PC version.)

Figure 11.17 shows the SPSSX output obtained by executing the listing in Figure 11.16, whereas Figure 11.18 shows the SPSS/PC+ output.

FIGURE 11.17 SPSSX output.

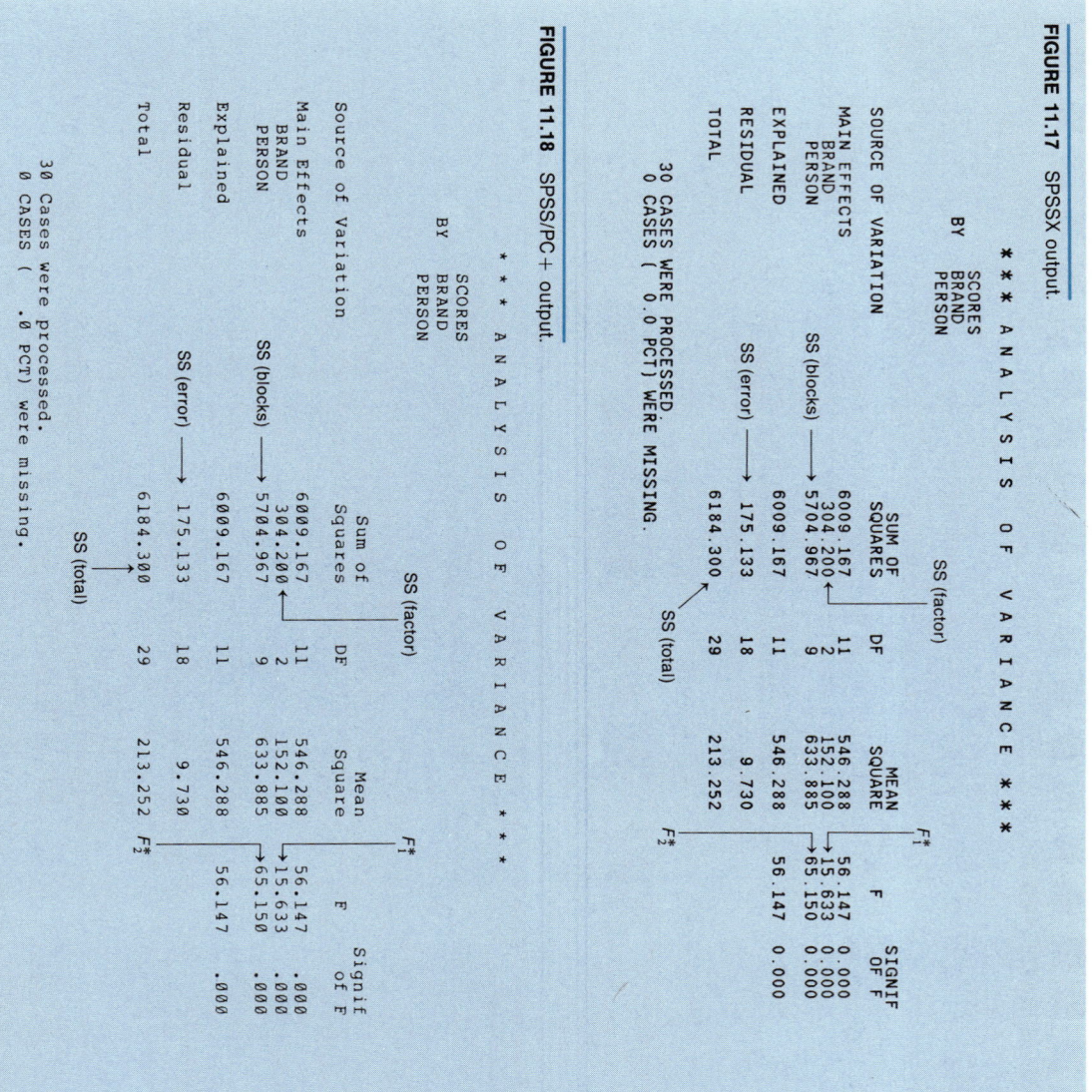

```
* * * A N A L Y S I S   O F   V A R I A N C E * * *

              SCORES
         BY   BRAND
              PERSON

                                                                                      SS (factor)

                                        SUM OF              MEAN
SOURCE OF VARIATION                     SQUARES      DF      SQUARE           F       SIGNIF
                                                                                      OF F
MAIN EFFECTS                            6009.167     11
  BRAND                                  304.200      2      152.100      15.633      0.000
  PERSON                                5704.967      9      633.885      65.150      0.000

EXPLAINED                               6009.167     11      546.288      56.147      0.000

RESIDUAL                                 175.133     18        9.730

TOTAL                                   6184.300     29      213.252

30 CASES WERE PROCESSED.
 0 CASES ( 0.0 PCT) WERE MISSING.
```

SS (blocks) →   SS (error) →   SS (total) →

F_1^* F_2^*

FIGURE 11.18 SPSS/PC+ output.

```
* * * A N A L Y S I S   O F   V A R I A N C E * * *

              SCORES
         BY   BRAND
              PERSON

                                                                                      SS (factor)

                            Sum of                 Mean
Source of Variation         Squares      DF        Square          F        Signif
                                                                            of F
Main Effects               6009.167      11
  BRAND                     304.200       2        152.100      15.633      .000
  PERSON                   5704.967       9        633.885      65.150      .000

Explained                  6009.167      11        546.288      56.147      .000

Residual                    175.133      18          9.730

Total                      6184.300      29        213.252
```

SS (blocks) →   SS (error) →   SS (total) →

F_1^* F_2^*

30 Cases were processed.
0 CASES (.0 PCT) were missing.

MAINFRAME AND MICRO

§ P § §

EXAMPLE 11.8

SOLUTION

The dental claims example was based on a two-way factorial design. The purpose was to determine the effects of sex and employee classification on level of dental claims by an employee. The purpose was to determine the effects of these two factors as well as test for a possible interaction effect between these two variables. The SPSS program listing in Figure 11.19 was used to request the ANOVA table, F-values, and their respective p-values.

In this problem the SPSS commands are the same for both the mainframe and PC versions. **(Remember to end each command line with a period when using the PC version.)**

SPSS

FIGURE 11.19

Input for SPSSX or SPSSX/PC+. All command lines should end with a period when the PC version is used.

```
TITLE   DENTAL CLAIMS ANALYSIS
DATA LIST FREE/SEX EMPCLASS CLAIMS
BEGIN DATA
1 1 190
1 1 225
1 1 200
1 2 135
 :
2 3 140
2 4 155
2 4 110
2 4 075
END DATA
ANOVA CLAIMS BY SEX(1,2)  EMPCLASS(1,4)
```

The TITLE command names the SPSS run.

The DATA LIST command gives each variable a name and describes the data as being in free form.

The BEGIN DATA command indicates to SPSS that the input data immediately follow.

The next 24 lines contain the data values. For example, the first line of data represents one employee's sex (1 or 2), their job classification (1, 2, 3, or 4), and their level of dental claims (190).

The END DATA statement indicates the end of the data.

The ANOVA command specifies the ANOVA model. CLAIMS is the dependent variable, SEX (ranging from 1 to 2) and EMPCLASS (ranging from 1 to 4) are the two factors of interest.

Figure 11.20 shows the SPSS output obtained by executing the listing in Figure 11.19, whereas Figure 11.21 shows the SPSS/PC+ output.

FIGURE 11.20 SPSSX output.

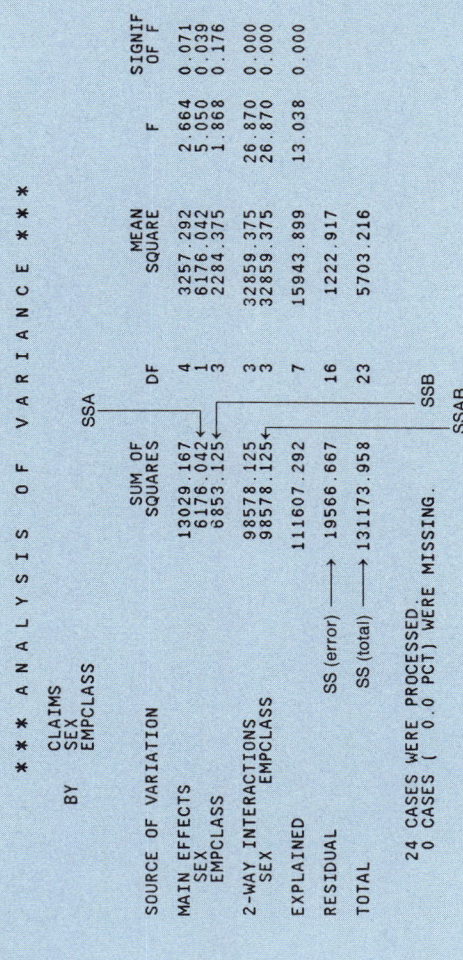

* * * A N A L Y S I S   O F   V A R I A N C E * * *

CLAIMS
BY   SEX
        EMPCLASS

SOURCE OF VARIATION	SUM OF SQUARES	DF	MEAN SQUARE	F	SIGNIF OF F
MAIN EFFECTS	13029.167	4	3257.292	2.664	0.071
SEX	6176.042	1	6176.042	5.050	0.039
EMPCLASS	6853.125	3	2284.375	1.868	0.176
2-WAY INTERACTIONS	98578.125	3	32859.375	26.870	0.000
SEX EMPCLASS	98578.125	3	32859.375	26.870	0.000
EXPLAINED	111607.292	7	15943.899	13.038	0.000
RESIDUAL	19566.667	16	1222.917		
TOTAL	131173.958	23	5703.216		

SSA

SSB

SSAB

SS (error)

SS (total)

24 CASES WERE PROCESSED
0 CASES (0.0 PCT) WERE MISSING.

FIGURE 11.21 SPSS/PC+ output.

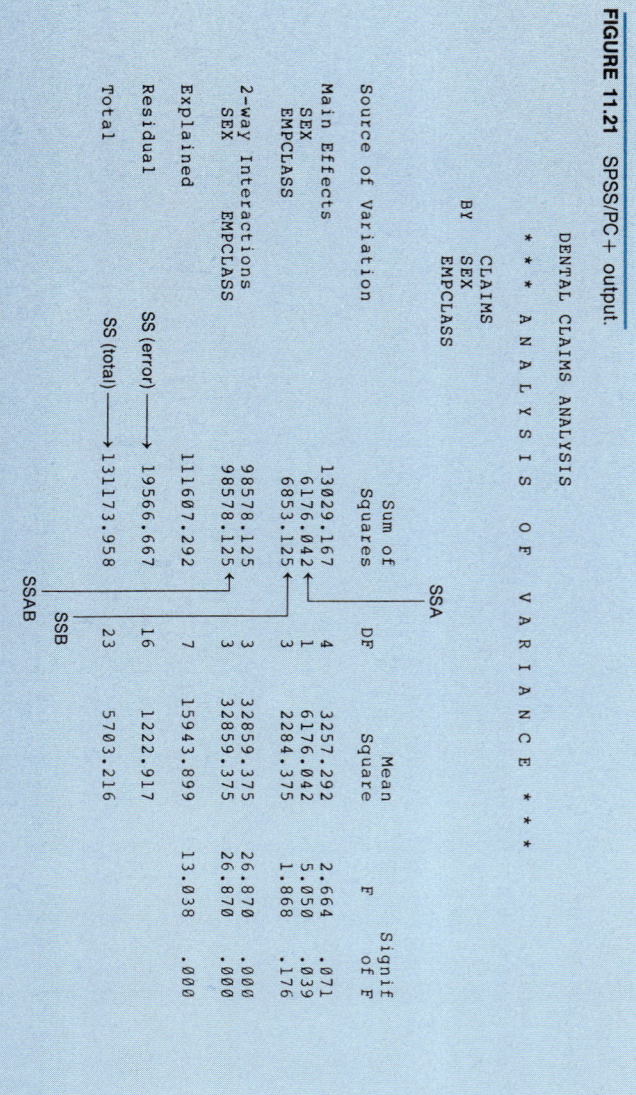

```
DENTAL CLAIMS ANALYSIS
* * *  A N A L Y S I S   O F   V A R I A N C E  * * *

         CLAIMS
     BY  SEX
         EMPCLASS
```

Source of Variation	Sum of Squares	DF	Mean Square	F	Signif of F
Main Effects	13029.167	4	3257.292	2.664	.071
SEX	6176.042	1	6176.042	5.050	.039
EMPCLASS	6853.125	3	2284.375	1.868	.176
2-way Interactions	98578.125	3	32859.375	26.870	.000
SEX EMPCLASS	98578.125	3	32859.375	26.870	.000
Explained	111607.292	7	15943.899	13.038	.000
Residual	19566.667	16	1222.917		
Total	131173.958	23	5703.216		

SSA → SEX, EMPCLASS main effects
SSB → SEX EMPCLASS
SSAB
SS(error) → Residual
SS(total) → Total

MAINFRAME AMD MICRO

S A S

EXAMPLE 11.4

SOLUTION

Example 11.4 was concerned with using one-factor ANOVA procedures when testing for a difference in two or more population means. The purpose was to test for a difference in job knowledge among three groups. The groups were (1) employees with a high school diploma only, (2) employees with a bachelor's degree only, and (3) employees with a master's degree. The SAS program listing in Figure 11.22 was used to request the ANOVA table and, in particular, the F-value and the p-value for F.

In this problem the SAS commands are the same for both the mainframe and PC versions.

The TITLE command names the SAS run.

The DATA command gives the data a name.

FIGURE 11.22

Input for SAS (mainframe or micro version).

```
TITLE  'JOB KNOWLEDGE';
DATA EDUCEXP;
INPUT EDUC GRADE;
CARDS;
1 81
1 84
1 69
1 85
  .
  .
  .
3 88
3 89
3 78
3 85
PROC ANOVA;
CLASS EDUC;
MODEL GRADE=EDUC;
```

SAS

FIGURE 11.23 SAS output.

```
                        JOB KNOWLEDGE
              ANALYSIS OF VARIANCE PROCEDURE
                  CLASS LEVEL INFORMATION
                  CLASS    LEVELS    VALUES
                  EDUC     3         1 2 3

          NUMBER OF OBSERVATIONS IN DATA SET = 15
```

```
                        JOB KNOWLEDGE
              ANALYSIS OF VARIANCE PROCEDURE
```

DEPENDENT VARIABLE: GRADE

SOURCE	DF	SUM OF SQUARES	MEAN SQUARE	F VALUE	PR > F	R-SQUARE	C.V.
MODEL	2	10.80000000	5.40000000	0.11	0.8940	0.018493	8.2278
ERROR	12	573.20000000	47.7666667			ROOT MSE	GRADE MEAN
CORRECTED TOTAL	14	584.00000000				6.9113433	84.00000000

SOURCE	DF	ANOVA SS	F VALUE	PR > F
EDUC	2	10.80000000	0.11	0.8940

F^* ← 0.11 p-value ← 0.8940

FIGURE 11.24 SAS output.

```
              JOB KNOWLEDGE                          (1)

          Analysis of Variance Procedure
              Class Level Information

          Class    Levels    Values
          EDUC     3         1 2 3

      Number of observations in data set = 15
```

```
              JOB KNOWLEDGE                          (2)

          Analysis of Variance Procedure
```

Dependent Variable: GRADE

Source	DF	Sum of Squares	Mean Square	F Value	Pr > F
Model	2	10.80000000	5.40000000	0.11	0.8940
Error	12	573.20000000	47.7666667		
Corrected Total	14	584.00000000			

F^* ← 0.11 p-value ← 0.8940

R-Square	C.V.	Root MSE	GRADE Mean
0.018493	8.227790	6.911343	84.0000000

```
              JOB KNOWLEDGE                          (3)

          Analysis of Variance Procedure
```

Dependent Variable: GRADE

Source	DF	Anova SS	Mean Square	F Value	Pr > F
EDUC	2	10.80000000	5.40000000	0.11	0.8940

The INPUT command names and gives the correct order for the different fields on the data lines.

The CARDS command indicates to SAS that the input data immediately follow.

The next 15 lines contain the data values. Each line represents the amount of education and the respective grade on the test. The first line implies that the individual had a level 1 (high school) education and scored 81 on the exam.

The PROC ANOVA command requests the ANOVA analysis procedure. The CLASS subcommand identifies EDUC as the factor of interest. The MODEL subcommand indicates that the exam grade is the dependent variable and the education level is the independent variable.

Figure 11.23 shows the SAS output obtained by executing the listing in Figure 11.22, whereas Figure 11.24 shows the SAS/PC output.

<hr>

MAINFRAME AND MICRO

■ ■ ■ ■ ■ ■ ■ ■   S A S   ■ ■ ■ ■ ■ ■ ■ ■

SOLUTION
Section 11.4

The car-warranty example was based on a randomized block design. The purpose was to analyze ten individual assessments of new car warranties for three brands of automobiles. The three factor levels; the ten individuals represented three factor levels; the ten individuals represented ten blocks. Each observation consisted of an individual's assessment of the new car warranty for a particular brand. The SAS program listing in Figure 11.25 was used to request the ANOVA table and in particular the sum of squares for error, block, factor, and total. We were also interested in the F-values for the brands and individuals, and their respective p-values.

In this problem the SAS commands are the same for both the mainframe and PC versions.

The TITLE command names the SAS run.

The DATA command gives the data a name.

The INPUT command names the variables and specifies the correct order during input.

The CARDS command indicates to SAS that the input data immediately follow.

FIGURE 11.25
Input for SAS (mainframe or micro version).

```
TITLE    'WARRANTY ASSESSMENTS';
DATA WARRANTY;
INPUT PERSON BRAND SCORES;
CARDS;
1 1 68
1 2 72
1 3 65
2 1 40
    :
    :
9 3 75
10 1 53
10 2 65
10 3 60
PROC ANOVA;
CLASS BRAND PERSON;
MODEL SCORES = BRAND PERSON;
```

FIGURE 11.26 SAS output.

WARRANTY ASSESSMENTS
ANALYSIS OF VARIANCE PROCEDURE

DEPENDENT VARIABLE: SCORES

SOURCE	DF	SUM OF SQUARES	MEAN SQUARE	F VALUE	PR > F	R-SQUARE	C.V.
MODEL	11	6009.16666667	546.28787879	56.15	0.0001	0.971681	4.7768
ERROR	18	175.13333333	9.72962963			ROOT MSE	SCORES MEAN
CORRECTED TOTAL	29	6184.30000000				3.11923542	65.30000000

SS (total)   SS (error)

SOURCE	DF	ANOVA SS	F VALUE	PR > F
BRAND	2	304.20000000	15.63	0.0001
PERSON	9	5704.96666667	65.15	0.0001

SS (blocks)   SS (factor)

F_2^* F_1^* p-values

FIGURE 11.27 SAS/PC output.

WARRANTY ASSESSMENTS

Analysis of Variance Procedure
Class Level Information

Class	Levels	Values
BRAND	3	1 2 3
PERSON	10	1 2 3 4 5 6 7 8 9 10

Number of observations in data set = 30

1

WARRANTY ASSESSMENTS

Analysis of Variance Procedure

Dependent Variable: SCORES

Source	DF	Sum of Squares	Mean Square	F Value	Pr > F
Model	11	6009.166667	546.287879	56.15	0.0001
Error	18	175.133333	9.729630		
Corrected Total	29	6184.300000			

SS (error)   SS (total)

R-Square	C.V.	Root MSE	SCORES Mean
0.971681	4.776777	3.119235	65.3000000

2

WARRANTY ASSESSMENTS

Analysis of Variance Procedure

Dependent Variable: SCORES

Source	DF	Anova SS	Mean Square	F Value	Pr > F
BRAND	2	304.200000	152.100000	15.63	0.0001
PERSON	9	5704.966667	633.885185	65.15	0.0001

SS (blocks)   SS (factor)

F_2^* F_1^* p-values

3

The next 30 lines contain the data values. In this example, each line of data represents one rater's scoring of the warranty of one brand of automobile. For example, in line one the first 1 represents the first rater, the second 1 represents first brand, and the 68 represents the warranty rating or score.

The PROC ANOVA command requests the ANOVA analysis procedure. The CLASS subcommand identifies brand and person as variables to be classified in our study. The MODEL subcommand indicates that SCORES is the dependent variable while PERSON and BRAND are the independent variables.

Figure 11.26 shows the SAS output obtained by executing the listing in Figure 11.25, whereas Figure 11.27 shows the SAS/PC outout.

■ ■ ■ ■ ■ ■ ■ ■ ■ ■ ■ ■ ■ S A S ■ ■ ■ ■ ■ ■ ■ ■ ■ ■ ■ ■ ■

MAINFRAME AND MICRO

EXAMPLE 11.8

The dental claims example was based on a two-way factorial design. The purpose was to determine the effects of sex and employee classification on level of dental claims by an employee. The purpose was to determine the effects of these two factors as well as test for a possible interaction effect between these two variables. The SAS program listing in Figure 11.28 was used to request ANOVA table, F-values, and their respective p-values.

SOLUTION

In this problem the SAS commands are the same for both the mainframe and PC versions.

The TITLE command names the SAS run.

The DATA command gives the data a name.

The INPUT command names the variables and specifies the correct order during input.

The CARDS command indicates to SAS that the input data immediately follow.

The next 24 lines contain the data values. For example the first line of data represents one employee's sex (1 or 2), their job classification (1, 2, 3, or 4) and their level of dental claims (190).

The PROC ANOVA command requests the ANOVA analysis procedure. The CLASS subcommand identifies sex and empclass as variables

FIGURE 11.28

Input for SAS (mainframe or micro version).

```
TITLE  'DENTAL CLAIMS ANALYSIS';
DATA DENTAL;
INPUT SEX EMPCLASS CLAIMS;
CARDS;
1 1 190
1 1 225
1 1 200
1 2 135
          .
          .
          .
2 3 140
2 4 155
2 4 110
2 4 075
PROC ANOVA;
CLASS SEX EMPCLASS;
MODEL CLAIMS = SEX EMPCLASS SEX*EMPCLASS;
```

to be classified in our study. The MODEL subcommand indicates that CLAIMS is the dependent variable and that the analysis will consider the effects of SEX, EMPCLASS, and the interaction between them (SEX*EMPCLASS).

Figure 11.29 shows the SAS output obtained by executing the listing in Figure 11.28, whereas Figure 11.30 shows the SAS/PC output.

FIGURE 11.29 SAS output.

DENTAL CLAIMS ANALYSIS

ANALYSIS OF VARIANCE PROCEDURE

DEPENDENT VARIABLE: CLAIMS

SOURCE	DF	SUM OF SQUARES	MEAN SQUARE	F VALUE	PR > F	R-SQUARE	C.V.
MODEL	7	111607.29166667	15943.89880952	13.04	0.0001	0.850834	16.1556
ERROR	16	19566.66666667	1222.91666667			ROOT MSE	CLAIMS MEAN
CORRECTED TOTAL	23	131173.95833333				34.97022543	216.45833333

SS (total) — SS (error)

SOURCE	DF	ANOVA SS	F VALUE	PR > F
SEX	1	6176.04166667	5.05	0.0391
EMPCLASS	3	6853.12500000	1.87	0.1757
SEX*EMPCLASS	3	98578.12500000	26.87	0.0001

SSB · SSA · SSAB · p-values

FIGURE 11.30 SAS/PC output.

DENTAL CLAIMS ANALYSIS

Analysis of Variance Procedure

Dependent Variable: CLAIMS

Source	DF	Sum of Squares	Mean Square	F Value	Pr > F
Model	7	111607.2917	15943.8988	13.04	0.0001
Error	16	19566.6667	1222.9167		
Corrected Total	23	131173.9583			

SS (error) · SS (total)

	R-Square	C.V.	Root MSE	CLAIMS Mean
	0.850834	16.15564	34.97023	216.458333

DENTAL CLAIMS ANALYSIS

Analysis of Variance Procedure

Dependent Variable: CLAIMS

Source	DF	Anova SS	Mean Square	F Value	Pr > F
SEX	1	6176.04167	6176.04167	5.05	0.0391
EMPCLASS	3	6853.12500	2284.37500	1.87	0.1757
SEX*EMPCLASS	3	98578.12500	32859.37500	26.87	0.0001

SSAB · SSB · SSA · p-value

CHAPTER

Applications of the Chi-Square Statistic

A Look Back/Introduction

We have examined several topics in both *descriptive* and *inferential* statistics. The descriptive area introduced you to the numeric (for example, mean, median, and variance) and the graphic (for example, histogram and scatter diagram) methods of describing data. In inferential statistics, we discussed point estimation, confidence intervals, and tests of hypothesis. In the remaining chapters, we turn our attention to other applications of the material from these earlier chapters.

In Chapter 8, we introduced the chi-square (χ^2) distribution. We used this distribution to test that the variance of a normal distribution was equal to a specified value. The shape of the chi-square distribution is skewed right (with a right tail) and is nonnegative. The shape of a chi-square curve and areas (probabilities) under such a curve are contained in Table A.6. The test statistic in Chapter 8 had a chi-square distribution. This chapter introduces you to additional applications of statistics by using the chi-square distribution to answer such questions as:

Do reported percentages of market share accurately describe the product mix for the new cars sold this past year in Minneapolis, Minnesota?

Does a person's age have an influence on buying behavior?

12.1 CHI-SQUARE GOODNESS-OF-FIT TESTS

The Binomial Situation

The binomial situation was introduced in Chapter 5, where the three following conditions had to be satisfied:

1. The experiment consists of n repetitions, called *trials*.
2. The trials are *independent*.
3. Each trial has two (and only two) outcomes, referred to as *success* and *failure*.
4. The probability of a success for each trial is p, where p remains the same for each trial. For a large finite population, p is the *proportion of successes* in this population.

Consequently, the binomial distribution applies to applications where there are only two possible outcomes, such as:

The person selected is a male or a female.

The product tested is either defective or it is not.

A new-car buyer buys either an American-made car or a foreign-made car.

Inferences for the Binomial Situation

Estimating the binomial parameter, p, was covered in Chapter 10. We obtained a random sample of size n and observed the number of successes, x. The estimator of p, the proportion of successes in the *population*, was $\hat{p} = x/n =$ the proportion of successes in the *sample*. We also discussed hypothesis testing for p. For example, we discussed a binomial situation in which a calculator was either defective (with probability p) or was not defective. The hypothetical value of p was .04, and we determined whether the results of the sample (13 defectives out of 150) indicated a departure from this percentage. Here $\hat{p} = 13/150 = .0867$. So, 8.67% of the sampled calculators were defective. Is this a large-enough percentage for us to conclude that p is different from .04, or is this large value of $\hat{p}$ just due to the fact that we tested a sample and not the entire population—that is, is this sampling error?

The resulting value of the test statistic was

$$Z^* = \frac{\hat{p} - .04}{\sqrt{\dfrac{(.04)(.96)}{150}}} = \frac{.0867 - .04}{.016} = 2.92$$

By comparing $Z^* = 2.92$ with the value 1.96 in Table A.4, we rejected H_0 using $\alpha = .05$; that is, the proportion of defective calculators was not 4%. The corresponding p-value was .0036.

Another Test for H_0: $p = p_0$ versus H_a: $p \neq p_0$

There is another test for a *two-tailed* test on p. This new test extends easily to a situation in which there are *more than two possible outcomes* for each trial: the **multinomial situation**.

To demonstrate this new testing procedure, a **chi-square goodness-of-fit** test, look at the lot sampling example. Note that the population consists of two **categories**—defective (category 1) and nondefective (category 2). Let $p_1 =$ the proportion of defectives in the population and $p_2 =$ the proportion of non-defectives in the population. In the previous solution, $p_1 = p$ and $p_2 = 1 - p$. We *observed* 13 sample values in category 1 (defective) and 137 in category 2. So define

$$O_1 = 13$$
$$O_2 = 137$$

How many units do we *expect* to see in each category if H_0 is *true*? The hy-potheses here can be written

$$H_0: p_1 = .04, p_2 = .96$$
$$H_a: p_1 \neq .04, p_2 \neq .96$$

This means that if H_0 is true, then, on the average, 4% of the sample values should be defective (category 1) and 96% should be nondefective (category 2). Define

$E_1 =$ expected number of sample values in category 1 if H_0 is true
$$= (150)(.04) = 6$$

$E_2 =$ expected number of sample values in category 2 if H_0 is true
$$= (150)(.96) = 144$$

We next define a test statistic that has an approximate chi-square distribution:

$$\chi^2 = \sum \frac{(O - E)^2}{E} \qquad (12.1)$$

where the summation is over all categories ($= 2$ here). In previous uses of this distribution, its shape depended on the sample size, specified by the degrees of freedom (df). Now the shape depends on the number of categories, where

$$df = \text{number of categories} - 1$$

For the binomial situation this is

$$df = 2 - 1 = 1$$

Therefore, for any *binomial* application, the test statistic in equation 12.1 has a *chi-square distribution with 1 df.*

EXAMPLE 12.1

Analyze the lot sampling data using the chi-square test statistic and a signifi-cance level of $\alpha = .05$.

SOLUTION

Step 1. The hypotheses are

$$H_0: p_1 = .04, p_2 = .96$$
$$H_a: p_1 \neq .04, p_2 \neq .96$$

Step 2. The test statistic is

$$\chi^2 = \sum \frac{(O - E)^2}{E}$$
$$= \frac{(O_1 - E_1)^2}{E_1} + \frac{(O_2 - E_2)^2}{E_2}$$

Step 3. If H_0 is not true (H_a is true), we expect the observed values to be different from the expected values. This would result in a *large* value for χ^2, so the procedure is to reject H_0 if the chi-square test statistic lies in the *right* tail. Consequently, the test procedure is to

$$\text{reject } H_0 \text{ if } \chi^2 > \chi^2_{.05,1}$$

where $\chi^2_{.05,1}$ is the χ^2 value having a right-tail area of .05 with 1 df. Using Table A.6, this is 3.84. Therefore,

$$\text{reject } H_0 \text{ if } \chi^2 > 3.84$$

Step 4. We have

$$O_1 = 13 \qquad E_1 = 6$$
$$O_2 = 137 \qquad E_2 = 144$$

(Note that $O_1 + O_2 = E_1 + E_2 = n = 150$.) The calculated value of the test statistic is

$$\chi^{2*} = \frac{(13 - 6)^2}{6} + \frac{(137 - 144)^2}{144}$$
$$= 8.17 + .34$$
$$= 8.51$$

Step 5. We conclude, as before, that the proportion of defectives (p_1) is not .04.

This value is larger than 3.84, so we reject H_0.

The *p*-value for Example 12.1 using the chi-square analysis is shown in Figure 12.1; it is the shaded area to the right of 8.51. Using Table A.6, all we can say is that this value is less than .005. The actual value is .0036 (calculated using a statistical software package). This is the *same p*-value as obtained when Z^* was used to perform this test of hypothesis.

In the lot sampling examples, we observe some quite fascinating (would you believe mildly interesting?) parallels with Chapter 11. In Chapter 11, we noted that when using the F-test from the ANOVA procedure to test H_0: $\mu_1 = \mu_2$, we obtained an F-value that was the *square* of the t value obtained using the corresponding t-test. Also, the value from the F table used to define the rejection region was the square of the corresponding t-value. This relationship held only when testing the equality of two means. Finally, the *p*-values from the two tests were identical; it made no difference which test we used for a two-tailed test on two means because the results were the same for both procedures. However, the ANOVA technique also could be used for comparing the means of more than two populations.

Using the results from the lot sampling example in this chapter, we again

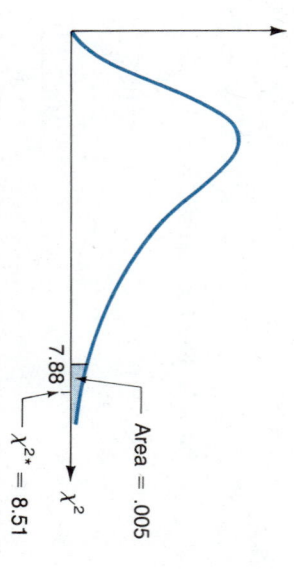

FIGURE 12.1
p-value for Example 12.1 using the chi-square analysis.

find that:

1. $\chi^{2*} = 8.51 = (2.92)^2 = (Z*)^2$.
2. The table values for the rejection region for Z test $= 1.96$ and for χ^2 test $= 3.84 = (1.96)^2$.
3. The p-value for each test was the same.

So again we have two testing procedures that produce identical conclusions. The chi-square test, however, extends easily to the multinomial situation. The chi-square goodness-of-fit test is an *extension* of the Z test used to test a binomial parameter. Furthermore, there is a definite relationship between the standard normal distribution (Z) and the chi-square distribution: The *square* of Z always is a chi-square random variable with 1 df.

TESTING $H_0: p = p_0$ VERSUS $H_a: p \neq p_0$

USING Z TEST

Test statistic:

$$Z = \frac{\hat{p} - p_0}{\sqrt{\dfrac{p_0(1 - p_0)}{n}}}$$

Rejection region:

reject H_0 if $|Z| > Z_{\alpha/2}$

(Use Table A.4.)

USING χ^2 TEST

Test statistic:

$$\chi^2 = \frac{(O_1 - E_1)^2}{E_1} + \frac{(O_2 - E_2)^2}{E_2}$$
$$= \sum \frac{(O - E)^2}{E}$$

Rejection region:

reject H_0 if $\chi^2 > \chi^2_{\alpha, 1}$

(Use Table A.6.)

The Multinomial Situation

The multinomial situation is identical to the binomial situation, except that there are k possible outcomes on each trial rather than two. Here k is any integer that is at least 2.

Suppose that a recent survey indicated the following percent of market share for U.S. auto manufacturers:

GA	62.5
K	22.6
L	11.5
M	1.4
Other	2.0
	100

An executive at GA questions whether these percentages apply to the new cars sold during the past year in Minneapolis, Minnesota. Obtaining a random sample of 500 new car registrations from that year, she observes the following frequencies of cars sold:

GA	290
K	125
L	65
M	10
Other	10
	500

The assumptions necessary for a multinomial experiment are:

1. The experiment consists of n independent repetitions (trials).
2. Each trial outcome falls in exactly one of k categories.
3. The probabilities of the k outcomes are denoted by $p_1, p_2, \ldots, p_k$, where these probabilities (proportions) remain the same on each trial. Also, $p_1 + p_2 + \cdots + p_k = 1$.

For this situation, we can define k random variables as the k observed values, where

$$O_1 = \text{the observed number of sample values in category 1}$$
$$O_2 = \text{the observed number of sample values in category 2}$$
$$\vdots$$
$$O_k = \text{the observed number of sample values in category } k$$

For this example, $n = 500$ trials, where each trial consists of obtaining a new car registration and observing in which of the $k = 5$ categories this new car lies. Assuming these registrations are obtained in a *random* manner (not the first 500 cars in May, for example), then these trials are independent. Also,

$p_1 = $ the proportion of cars sold in Minneapolis that were GA cars for that year

$p_2 = $ the proportion of cars sold in Minneapolis that were K cars for that year

$$\vdots$$

The five random variables here are

$O_1 = $ the number of GA cars in the sample
$O_2 = $ the number of K cars in the sample

$$\vdots$$

Thus, this example fits the assumptions for the multinomial situation.

Hypothesis Testing for a Multinomial Situation

The hypotheses for the Minneapolis market share example are

$$H_0: p_1 = .625, p_2 = .226, p_3 = .115, p_4 = .014, p_5 = .02$$
$$H_a: \text{at least one of the } p_i\text{'s is incorrect}$$

Notice that H_a is *not* $p_1 \neq .625, p_2 \neq .226, \ldots, p_5 \neq .02$. This is too strong and is not the opposite of H_0.

Let $p_{1,0}$ be any specified value of $p_1, p_{2,0}$ any specified value of p_2, and so on. The multinomial goodness-of-fit hypotheses are

$$H_0: p_1 = p_{1,0}, p_2 = p_{2,0}, \ldots, p_k = p_{k,0}$$
$$H_a: \text{at least one of the } p_i\text{'s is incorrect}$$

Using the observed values $(O_1, O_2, \ldots)$, the point estimates here are

$$\hat{p}_1 = \text{estimate of } p_1 = O_1/n$$
$$\hat{p}_2 = \text{estimate of } p_2 = O_2/n$$
$$\vdots$$

To test H_0 versus H_a, we use the previously stated chi-square statistic. To define the rejection region, notice that when H_a is true, we would expect the O's and E's to be "far apart" because the E's are determined by assuming that H_0 is true. In other words, if H_a is true, the chi-square test statistic should be large. Consequently, we always reject H_0 when χ^{2*} lies in the *right tail* when using this particular statistic.

To test H_0 versus H_a, compute

$$\chi^2 = \sum \frac{(O - E)^2}{E} \qquad (12.2)$$

where

1. The summation is across all categories (outcomes).
2. The O's are the *observed* frequencies in each category using the sample.
3. The E's are the *expected* frequencies in each category if H_0 is true, so

$$E_1 = np_{1,0}$$
$$E_2 = np_{2,0}$$
$$E_3 = np_{3,0}$$
$$\vdots$$

4. The df for the chi-square statistic are $k - 1$, where k is the number of categories.

To carry out the test,

reject H_0 if $\chi^2 > \chi^2_{\alpha,\,df}$

Notice that the hypothetical proportions (probabilities) for each of the categories are specified in H_0. Consequently, we will complete the analysis by concluding that at least one of the proportions is incorrect (we reject H_0) or that there is not enough evidence to conclude that these proportions are incorrect (we fail to reject H_0). We do not *accept* H_0; we never conclude that these specified proportions *are* correct. We act like the juror who acquits a defendant not because he or she is convinced that this person is innocent but rather because there was not sufficient evidence for conviction.

When we introduced the ANOVA technique, we mentioned that this procedure allowed us to determine whether many population means were equal using a *single* test. This was preferable to using many t tests to test the equality of two means, one pair at a time; these tests are not independent, and the overall significance level is difficult to determine. We encounter the same situation here. It is much better to use a chi-square goodness-of-fit test to test *all* of the proportions at once rather than using many Z tests to test the individual proportions.

EXAMPLE 12.2

What do the observed number of cars sold in our market share example tell us about the mix of new car sales in Minneapolis for that year? Do they conform to the percentages for all U.S. auto sales? Use a significance level of $\alpha = .05$.

SOLUTION

Step 1. Let p_1 = proportion of all Minneapolis new car sales that are GA, p_2 are K, p_3 are L, p_4 are M, and p_5 are all other new (U.S. made) cars.
The hypotheses under investigation are

$H_0: p_1 = .625, p_2 = .226, p_3 = .115, p_4 = .014, p_5 = .02$

$H_a:$ at least one of these p_i's is incorrect

Step 2. The test statistic is

$$\chi^2 = \sum \frac{(O - E)^2}{E}$$

where the summation is over the five categories.

Step 3. Your test procedure here is to

$$\text{reject } H_0 \text{ if } \chi^2 > \chi^2_{\alpha,\,df}$$

The df are (number of categories) $- 1$, so df $= 5 - 1 = 4$. The chi-square value from Table A.6 is $\chi^2_{.05,\,4} = 9.49$, and we

$$\text{reject } H_0 \text{ if } \chi^2 > 9.49$$

Step 4. The observed values are

$$O_1 = 290, \quad O_2 = 125, \quad O_3 = 65, \quad O_4 = 10, \quad O_5 = 10$$

The expected values when H_0 is true are obtained by multiplying $n = 500$ by each of the proportions in H_0. So,

$$
\begin{aligned}
E_1 &= (500)(.625) = 312.5 \\
E_2 &= (500)(.226) = 113 \\
E_3 &= (500)(.115) = 57.5 \\
E_4 &= (500)(.014) = 7 \\
E_5 &= (500)(.02) = \underline{10} \\
&= 500
\end{aligned}
$$

Note that we do not round the expected values because they are *averages*.

The computed value of the chi-square test statistic is

$$\chi^{2*} = \frac{(290 - 312.5)^2}{312.5} + \frac{(125 - 113)^2}{113} + \frac{(65 - 57.5)^2}{57.5} + \frac{(10 - 7)^2}{7} + \frac{(10 - 10)^2}{10}$$

$$= 5.16$$

Step 5. There is insufficient evidence to suggest that the Minneapolis car sales differ from the U.S. mixture. In other words, the observed values were "close enough" to the expected values under H_0 to let this hypothesis stand.

Because 5.16 does not exceed 9.49, we fail to reject H_0.

In Example 12.2, the proportions under investigation were directly specified. We can also use the chi-square statistic when the proportions are implied. ∎

EXAMPLE 12.3

Allied Health Corporation owns and operates hospitals in the southeast. Three of their hospitals were recently audited by federal auditors to determine if the three hospitals were in compliance with Medicare billing regulations. According to the auditors, an audited billing had an equal probability of being selected from each of the three hospitals. A random sample of the audited billings revealed the following number of billings selected from each hospital: Hospital 1: 485, Hospital 2: 405, Hospital 3: 310; total = 1200. What can you conclude using a significance level of .01?

SOLUTION

Step 1. Let $p_1 =$ proportion of audited billings from hospital 1, $p_2 =$ proportion from hospital 2, and $p_3 =$ proportion from hospital 3. If an audited billing has an equal probability of belonging to each hospital, then each of these proportions will be 1/3. This provides the values for H_0:

$$H_0: p_1 = 1/3, \ p_2 = 1/3, \ p_3 = 1/3$$

$$H_a: \text{at least one of these proportions is incorrect}$$

Steps 2, 3. The test procedure will be to

reject H_0 if $\chi^2 > \chi^2_{.01, 2} = 9.21$

because df $= k - 1 = 3 - 1 = 2$. The value of χ^2 is determined from equation 12.2.

Step 4. The observed and expected values are

O	E (if H_0 is true)	$\hat{p}$
$O_1 = 485$	$E_1 = (1200)(1/3) = 400$	$485/1200 = .404$
$O_2 = 405$	$E_2 = (1200)(1/3) = 400$	$405/1200 = .338$
$O_3 = 310$	$E_3 = (1200)(1/3) = 400$	$310/1200 = .258$
$\overline{1200}$	$\overline{1200}$	$\overline{1.0}$

$$\chi^{2*} = \frac{(485 - 400)^2}{400} + \frac{(405 - 400)^2}{400} + \frac{(310 - 400)^2}{400}$$

$$= 18.06 + .06 + 20.25$$

$$= 38.4$$

So we reject H_0 because $38.4 > 9.21$.

Step 5. There *is* an unequal distribution of hospital selection here. Your next step should be to examine the three values making up this large χ^2 value. This large value is due to the 18.06 (the number of billings from hospital 1 was larger than expected under H_0) and the 20.25 (the number of billings from hospital 3 was smaller than expected under H_0). This can also be seen in the $\hat{p}$ values from step 4.

■

The *p*-value for the results in Example 12.3 is shown in Figure 12.2. Using Table A.6 and 2 df, the largest value here is 10.6, with a corresponding right-tail area of .005. All you can say using this table is that the *p*-value is less than .005. At any rate, it is small and would lead you to reject H_0 for the most common values of α.

Pooling Categories

When using the chi-square procedure of comparing observed and expected values, we determine the difference between these two values for each category, square it, and *divide by the expected value, E.* If the value of E is very small (say, less than 5), then this produces an extremely *large* contribution to the final χ^2 value from this category. In other words, this small expected value produces

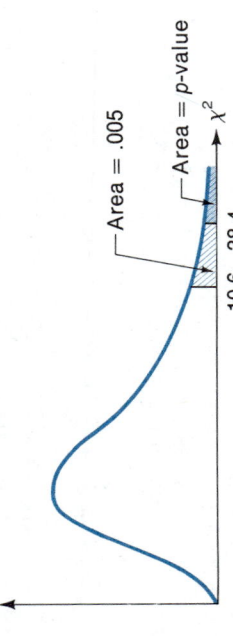

FIGURE 12.2
Shaded area is *p*-value for Example 12.3.

an inflated chi-square value, with the result that we reject H_0 when perhaps we should not have. To prevent this from occurring, we use the following rule: When using equation 12.2, each expected value, E, should be at least 5.[†]

If you encounter an application where one or more of the expected values is less than 5, you can handle this situation by *pooling* your categories such that each of the new categories has an expected value that is at least 5.

EXAMPLE 12.4

The analysis in Example 12.2 was repeated for a much smaller community using 200 observations rather than $n = 500$. The following mixture was observed (number of cars sold):

GA 95
K 50
L 41
M 5
Other 9
 200

Do the data from this community appear to fit the U.S. proportions of 62.5% for GA, 22.6% for K, and so on? Use $\alpha = .05$.

SOLUTION

CATEGORY	OBSERVED (O)	EXPECTED (E), IF H_0 IS TRUE
GA	$O_1 = 95$	$E_1 = (200)(.625) = 125$
K	$O_2 = 50$	$E_2 = (200)(.226) = 45.2$
L	$O_3 = 41$	$E_3 = (200)(.115) = 23$
M	$O_4 = 5$	$E_4 = (200)(.014) = 2.8$
Other	$O_5 = 9$	$E_5 = (200)(0.02) = 4$
	200	200

Notice that the last two expected values are less than 5. These two categories need to be pooled (combined) into a new category, which we will label category 4: other. The new summary is

CATEGORY	OBSERVED (O)	EXPECTED (E), IF H_0 IS TRUE
GA	95	125
K	50	45.2
L	41	23
Other	14 $(= 5 + 9)$	6.8 $(= 2.8 + 4)$

Now each of the expected values is at least 5, and we can continue the analysis. The hypotheses using the four categories are

H_0: $p_1 = .625$, $p_2 = .226$, $p_3 = .115$, $p_4 = .034$
H_a: at least one of the proportions is incorrect

The value of p_4 represents the proportion of all other cars (including those from M) sold in this community. The hypothetical value of p_4 becomes $.014 + .020 = .034$.

The computed chi-square value is now

$$\chi^{2*} = \frac{(95 - 125)^2}{125} + \frac{(50 - 45.2)^2}{45.2} + \frac{(41 - 23)^2}{23} + \frac{(14 - 6.8)^2}{6.8}$$

$$= 7.2 + .51 + 14.09 + 7.62 = 29.4$$

[†] This rule is somewhat arbitrary but commonly used. Another procedure used for pooling requires that all the expected values be 3 or more, while yet another procedure requires that no more than 20% of all the expected values be less than 5 with none less than 1.

This exceeds the Table A.6 value of $\chi^2_{.05,3} = 7.81$, so we reject H_0 and conclude that we do have a significant departure from the U.S. percentages in this community. The largest contributor to this chi-square value is from the L category, which exceeds expectations by 18 cars (78% more than expected if H_0 is true).

Testing a Hypothesis about a Distributional Form

In this discussion of the goodness-of-fit test, we examine such questions as:

Is it true that these data came from a binomial distribution with probability of success, p, equal to .2?

Does this particular set of data violate the assumption that the number of defects in this product follow a Poisson distribution?

Is there any reason to doubt the assumption that the weights of all Rice Krinkle cereal boxes follow a normal distribution using a recently obtained sample of boxes?

The first two questions concern *discrete* distributions (binomial and Poisson). The final question is concerned with whether the data came from a particular *continuous* (in this case, normal) distribution. We illustrate the chi-square technique using a goodness-of-fit test for a discrete situation. (Goodness-of-fit tests for the normal distribution are illustrated in Exercises 12.13, 12.14, and 12.15.)

Suppose that Blitz laundry detergent is well known for its obnoxious commercials, which advertise that 20% of all Blitz boxes contain a valuable discount coupon. A recent study obtained a random sample of ten Blitz boxes from each of 100 different stores. The results were

OF 10 BOXES, NUMBER CONTAINING COUPONS	NUMBER OF STORES
0	9
1	31
2	29
3	18
>3	13
	$\overline{100}$

We wish to know whether these data appear to come from a binomial distribution with $p = .2$, using $\alpha = .05$.

Your immediate reaction may well be that this problem does not fit a multinomial situation. However, there are 100 independent trials, each trial consisting of randomly selecting ten boxes of Blitz detergent. Also, we can set up five categories here, namely:

Category 1: Observe 0 coupons in the ten boxes (probability p_1)
Category 2: Observe 1 coupon in the ten boxes (probability p_2)
Category 3: Observe 2 coupons in the ten boxes (probability p_3)
Category 4: Observe 3 coupons in the ten boxes (probability p_4)
Category 5: Observe >3 coupons in the ten boxes (probability p_5)

So we *do* have a multinomial situation here. The hypotheses can be stated as

H_0: the data follow a binomial distribution with $p = .2$

H_a: the data do not follow a binomial distribution with $p = .2$

If H_0 is true, how often should we observe zero coupons in ten boxes? Each multinomial trial fits a binomial situation, where a success consists of a box containing a coupon. We repeat the trial ten times and count the number of successes (coupons). According to Table A.1, with $n = 10$ and probability of success $p = .2$, you should observe zero coupons out of ten boxes 10.7% of the time. So, if H_0 is true, $p_1 = .107$. Similarly, if H_0 is true, we should see one success out of ten trials 26.8% of the time. In other words, $p_2 = .268$. Therefore, another way to state your hypotheses is

$$H_0: p_1 = .107, p_2 = .268, p_3 = .302, p_4 = .201, p_5 = .122$$
$$H_a: \text{at least one of these } p_i\text{'s is incorrect}$$

So this is a multinomial test of hypothesis in disguise. Next, we compute the expected values, E.

We obtain p_5 by finding the probability of more than three successes in ten trials. This is

$$1 - (\text{probability of 3 or less}) = 1 - (p_1 + p_2 + p_3 + p_4)$$
$$= 1 - .878 = .122$$

CATEGORY	OBSERVED (O)	EXPECTED (E), IF H_0 IS TRUE
0 boxes	9	$(100)(.107) = 10.7$
1 box	31	$(100)(.268) = 26.8$
2 boxes	29	$(100)(.302) = 30.2$
3 boxes	18	$(100)(.201) = 20.1$
>3 boxes	13	$(100)(.122) = 12.2$
	100	100

Make sure that all the E's are greater than 5, so no pooling of categories is necessary.

To define the rejection region, we notice that there are $k = $ five categories. So the df in the chi-square statistic are df $= k - 1 = 4$. Also, $\chi^2_{.05, 4} = 9.49$, and so the test procedure will be to

$$\text{reject } H_0 \text{ if } \chi^2 > 9.49$$

The value of our test statistic is

$$\chi^{2*} = \frac{(9 - 10.7)^2}{10.7} + \frac{(31 - 26.8)^2}{26.8} + \cdots + \frac{(13 - 12.2)^2}{12.2}$$
$$= 1.25$$

Because 1.25 is less than 9.49, we fail to reject H_0 and conclude that there is no evidence to suggest that these data have violated the binomial assumption. These 100 observations suggest that we have no reason to accuse Blitz of false advertising for their claim that 20% of their boxes contain coupons.

In summary, suppose you are trying to perform a test of hypothesis on a binomial parameter, p—say, $H_0: p = .2$.

1. If we are using a *single* sample of size n, then the results of Chapter 10 apply; we are dealing with a binomial experiment.
2. If we are using *many* samples of size n, then the results of this section apply; this problem can be expressed as a multinomial experiment. The chi-square goodness-of-fit procedure allows us to perform a test on the population proportion, p, as well as determine if the population follows a binomial distribution. The previous example provides an illustration of this type of situation.

Distributional Form with Unknown Parameters

In the Blitz cereal example, H_0 not only stated that the data followed a binomial distribution, it also specified a value of the binomial parameter p (namely, $p = .2$). Your only concern often is whether the data follow a particular distribution (such as binomial, Poisson, or normal), and the value of the corresponding parameters is not preset.

For example, say that the manager of Case Electronics has always assumed that the weekly sales of his top-of-the-line telephone answering machine followed a Poisson distribution. Data from a 50-week period were gathered, with the following results:

UNITS SOLD	NUMBER OF WEEKS		UNITS SOLD	NUMBER OF WEEKS
0	1		6	5
1	3		7	3
2	6		8	4
3	11		>8	0
4	10			$\overline{50}$
5	7			

How can we test the hypothesis that the number of units sold follows a Poisson distribution, using $\alpha = .1$?
The correct hypotheses here are

H_0: weekly sales follow a Poisson distribution

H_a: weekly sales do not follow a Poisson distribution

The probability function for the Poisson distribution has one parameter, μ, because this function (from equation 5.12) is given by

$$P(X = x) = \frac{\mu^x e^{-\mu}}{x!}$$

for $x = 0, 1, 2, \ldots$, where $x =$ the number of units sold during a particular week.

However, the value of μ was not specified in H_0. In this case, we estimate any unknown parameter (μ, here) from the sample information and replace each parameter by its estimate in the probability function. In this way, we can estimate all the expected frequencies ($E_1, E_2, E_3, \ldots$).

Whenever you estimate unknown parameters for use with the chi-square test, you need to *adjust the corresponding degrees of freedom*, df. In general, the df for the chi-square goodness-of-fit statistic is given by

$$df = (number of classes) - 1 - (number of estimated parameters)$$

For the Poisson situation, you are estimating only one parameter, μ, and so the df = (number of classes) - 1 - 1 = (number of classes) - 2. The same argument holds true for a test of hypothesis on a binomial distribution where the single parameter, p, is unspecified in H_0 and is instead estimated from the sample information.

Estimating μ

Because μ is the mean of the telephone answering machine sales population, we estimate it using the average (mean) of the sample. In the sample, we observe 1 value of zero, 3 values of one, 6 values of two, and so on for all 50 values. This means that the sample average, our estimate of μ, will be

$$\hat{\mu} = \frac{(0)(1) + (1)(3) + (2)(6) + \cdots + (8)(4)}{50}$$

$$= \frac{206}{50} = 4.12$$

Rounding this to $\hat{\mu} = 4.1$, the estimated probability function is

$$P(X = x) = \frac{(4.1)^x e^{-4.1}}{x!}$$

for $x = 0, 1, 2, \ldots$.

We can now use Table A.3 (the Poisson table) to estimate the expected number of weeks with zero sales, with one sale, and so on. We are *estimating* each expected value, so we denote each of them as $\hat{E}$.

X	P(X = x)		$\hat{E}$		O
0	.0166	(.0166)(50) =	.83		1
1	.0679	(.0679)(50) =	3.39		3
2	.1393	(.1393)(50) =	6.97		6
3	.1904	(.1904)(50) =	9.52		11
4	.1951	(.1951)(50) =	9.76		10
5	.1600	(.1600)(50) =	8.00		7
6	.1093	(.1093)(50) =	5.46		5
7	.0640	(.0640)(50) =	3.20		3
8	.0328	(.0328)(50) =	1.64		4
>8	.0246	(.0246)(50) =	1.23		0
	1		50		50

Notice that, for the category $X > 8$, the corresponding probability is $1 - (.0166 + .0679 + \cdots + .0640 + .0328) = .0246$.

The next step always is to check your expected frequencies ($\hat{E}$) to see if pooling is necessary. Each $\hat{E}$ value must be at least 5, so it is necessary to pool the first three classes (.83 + 3.39 + 6.97) and the last three classes (3.20 + 1.64 + 1.23). Now you can evaluate the chi-square statistic.

X	$\hat{E}$	O	$(O - \hat{E})$	$(O - \hat{E})^2/\hat{E}$
≤2	11.19	10	−1.19	.127
3	9.52	11	1.48	.230
4	9.76	10	.24	.006
5	8.00	7	−1.00	.125
6	5.46	5	− .46	.039
≥7	6.07	7	.93	.142
	50	50	0	.669

$$\chi^2 = \sum \frac{(O - \hat{E})^2}{\hat{E}} = .669$$

(check)

The degrees of freedom for the corresponding test are

df = (number of classes) − 1 − (number of estimated parameters)

= 6 − 1 − 1 = 4

The resulting test, using Table A.6 and $\alpha = .1$, is

reject H_0 if $\chi^2 > 7.779$

Because $.669 < 7.779$, we fail to reject H_0 and conclude that there is not enough evidence to indicate that the Poisson distribution assumption is incorrect.

EXERCISES

12.1 A researcher in the marketing department of an investment firm believes that the proportion of full-time workers that are over 40 years of age in a certain locality is equal to 35%. A random sample of 100 observations was taken and the estimate of the proportion of full-time workers that are over 40 years of age was .31.

a. Is there sufficient evidence that the proportion of full-time workers that are over 40 years of age is not equal to 35%? Use the Z-test from Section 12.1. Let the significance level be .05.

b. Use the chi-square goodness-of-fit test to test the hypothesis in (a). Let the significance level be .05.

c. What is the relationship between the test statistics in (a) and (b)?

12.2 The owner of a car-insurance company believes that 20% of drivers under 25 years of age have been in exactly one automobile accident in the past 2 years. She also believes that 15% of the drivers under 25 have been in exactly two automobile accidents in the past 2 years. Finally, she believes that 10% of the drivers under 25 have been in more than two automobile accidents in the past 2 years. A survey of 300 randomly selected drivers under 25 years of age was taken. Test the beliefs using the following data and letting the significance level be .10.

NO ACCIDENT	1 ACCIDENT	2 ACCIDENTS	>2 ACCIDENTS
153	68	51	28

12.3 In exercise 12.2, the equality of proportions for any two categories can be tested using the Z test in section 12.1. Is there any difference in testing the equality of two proportions one pair at a time and testing all the proportions at once using the chi-square goodness-of-fit test? Is the overall significance level the same in both cases?

12.4 Barton's Food Store carries three brands of milk. Recently, Barton had been getting numerous complaints about the milk being spoiled after being sold to the customer. Barton decided to categorize 34 randomly selected complaints to see whether they were equally divided among the three brands of milk. Using the accompanying data, test that the complaints are equally divided among the three brands. Let the significance level be .05.

BRAND A	BRAND B	BRAND C
7	13	14

12.5 A stockbroker believes that when too many of the stockmarket newsletters are bullish on the market (that is, they predict that stock prices will go higher), the stockmarket will most likely fall. Thirty-two randomly selected stockmarket newsletters were each placed in one of three categories:

BEARISH ON STOCKMARKET	NEUTRAL ON STOCKMARKET	BULLISH ON STOCKMARKET
9	10	13

Test the null hypothesis that the newsletters are equally divided among the three categories. Use a .05 significance level.

12.6 At a major stereo and video exhibition that ran for 3 days in the convention center of a major New England city, 200 individuals were asked if they considered an American or Japanese videocassette recorder to be a better product. The results of the survey are as follows.

ANSWER	RESPONSES
1. American	37
2. Japanese	141
3. No preference	22

(*Source:* Kip Becker and Haluk Bekiroglu, "International Competitive Strategies For Positioning Against Nations With Favored Perceived Quality," *Proceedings of the 1986 Annual Meeting of the Decision Sciences Institute,* 1986, pp. 509–11.)

Using a .05 significance level, can the null hypothesis of no difference among respondents' selection of answers 1, 2, or 3 be rejected?

12.7   A soft-drink company believes that people are particular about the type of sweetener used in the soft drinks. The manager of the marketing department believes that 50% of the people prefer sugar, 35% prefer aspartame, 10% prefer saccharin, and 5% have no preference. Thirty people who regularly drank sweetened soft drinks were randomly selected. Using the following data and a significance level of 5%, test the manager's claim. (Do any of the categories need to be pooled?)

SUGAR	ASPARTAME	SACCHARIN	NO PREFERENCE
12	11	5	2

12.8   Electrical fuses are packaged in lots of 20. The quality-control department claims that only about 10% of the fuses are defective on the average for each package of 20 fuses. A random sample of 40 packages was selected and the results were:

DEFECTIVE FUSES	PACKAGES
0	7
1	12
2	10
3	7
4	1
5	1
>5	2

Do these data appear to have come from a binomial distribution with $p = .10$? Use a significance level of .05.

12.9   A large stock brokerage firm is considering introducing an international bond mutual fund. A study was first conducted at 100 branch brokerage firms across the United States. Twenty active investors at each brokerage firm were asked if they would invest in the bond mutual fund when the bond fund was available to the general public.

X = NUMBER OF ACTIVE INVESTORS WILLING TO INVEST IN FUND	FREQUENCY WITH WHICH X WAS OBSERVED AT 100 BROKERAGE FIRMS
0	7
1	8
2	9
3	10
4	11
5	12 or more

X = NUMBER OF ACTIVE INVESTORS WILLING TO INVEST IN FUND	FREQUENCY WITH WHICH X WAS OBSERVED AT 100 BROKERAGE FIRMS
0	20
1	10
2	3
3	2
4	1
5	0
6	

Do these data appear to come from a binomial distribution with $p = .30$? Use a significance level of .05.

12.10   An auditor believes that the number of errors per 25 invoices contained in the records of a discount furniture store chain follow a Poisson distribution with a mean of 2.2. To test the auditor's belief, 25 stores were randomly selected and the number of errors were tabulated.

ERRORS PER 25 INVOICES	STORES	ERRORS PER 25 INVOICES	STORES
0	3	4	2
1	5	5	2
2	7	6	1
3	4	>6	1

Do these data appear to have come from a Poisson distribution with a mean of 2.2? Use a .10 significance level.

12.11 On each flight of Astral Airways, 12 randomly selected passengers are asked if they would be willing to pay a 5% air fare increase to fly on an airline that had an open bar in the airplane. Results of the survey from 50 different flights were as follows:

YES ANSWERS	FLIGHTS	YES ANSWERS	FLIGHTS
0	3	4	7
1	10	5	3
2	13	6	2
3	12	>6	0

Use a chi-square goodness-of-fit test to determine whether these data came from a binomial distribution. Let the significance level be .05. [*Hint: p* must be estimated using (total number of people who said yes)/(total number asked).]

12.12 The Cranberry Mountain Visitor Center in the Monongahela National Forest in Richwood, West Virginia, includes 53,000 acres of back country, 36,000 of which are part of the National Wilderness Preservation System, and the 750-acre Cranberry Glades Botanical Area, which contains the largest bogs in the state. The Visitor Center is looked upon as being the pulse of the Monongahela National Forest. In a study of population density and crowding, Burrus-Bammel and Bammel (1986) noted the following pattern of arrivals of visitors at the Cranberry Mountain Visitor Center between July 24 and August 5, 1983, where "party size" is the number of persons per vehicle and "number" is the number of such vehicle arrivals.

PARTY SIZE	NUMBER	PERCENTAGE
1	25	6.5%
2	165	43.0
3	62	16.1
4	83	21.6
5	21	5.5
6	14	3.6
7	3	0.8
8	4	1.0
9	4	1.0
10	1	0.3
11	2	0.6
Total	384	100.0%

(*Source:* L. L. Burrus-Bammel and G. Bammel, Visiting patterns and effects of density at a Visitors' Center. *Journal of Environmental Education* 18, no. 1 (Fall 1986): 8, Table 2.)

a. Test whether the above data on arrival patterns follows a Poisson distribution, at a 5% significance level.

b. State your conclusion, and report the *p*-value for the test.

12.13 To perform certain statistical tests on a set of data, the assumption of normality is required. It is thought that the percentage gain over the past 3 years in mutual funds that have balanced portfolios of both long-term-growth stocks and income-oriented stocks is normally distributed, with a mean of 35% and a standard deviation of 10%. A sample of 75 mutual funds of this type is selected. To test this assumption of normality, a chi-square goodness-of-fit test can be used. For the intervals listed, probabilities can be found from the normal table (Table A.4). To find the expected frequencies in the third column, the sample size is multiplied by each probability. If the differences between the observed and expected frequencies are large, then the chi-square statistic based on the observed and expected frequencies would be large and would cause the null hypothesis, which is that the data was sampled from a normally distributed population with mean $= 35$ and standard deviation $= 10$, to be rejected.

INTERVAL	PROBABILITY	EXPECTED FREQUENCY	OBSERVED FREQUENCY
less than 20	.0668	5.01	7
20 and less than 30	.2417	18.1275	15
30 and less than 40	.3830	28.725	26
40 and less than 50	.2417	18.1275	21
50 or more	.0668	5.01	6

At the 5% significance level, complete the chi-square goodness-of-fit test by calculating the chi-square statistic presented in the chapter and by using a tabulated chi-square value for the critical value of the rejection region. The degrees of freedom is taken to be equal to the number of intervals minus one.

12.14 The monthly maintenance time on a particular machine at a manufacturing plant is believed to be normally distributed with a mean of 6 hours and a standard deviation of 1.5 hours. A sample of 45 months is selected and the maintenance time is recorded (in hours). From the following data, use the chi-square goodness-of-fit procedure in Exercise 12.13 to test whether there is enough evidence to support the hypothesis that the maintenance time is not normally distributed with a mean of 6 and a standard deviation of 1.5. Use a 10% significance level.

INTERVAL	OBSERVED FREQUENCY
less than 5	10
5 hours but less than 6	6
6 hours but less than 6.5	5
6.5 hours but less than 7	7
7 hours but less than 8	6
8 hours or more	11

12.15 The weekly traffic flow between 8 A.M. and 6 P.M. at a certain intersection in Oklahoma City is believed to be normally distributed, with a mean of 65,000 cars and a standard deviation of 12,000. Fifty weeks are randomly selected over the past 3 years. Using the following data and the chi-square goodness-of-fit test procedure in Exercise 12.13, is there sufficient evidence to conclude that the traffic flow (in thousands of cars) is not normally distributed with a mean of 65 (thousand) and a standard deviation of 12 (thousand)? Use a .10 significance level.

TRAFFIC FLOW	OBSERVED FREQUENCY
less than 50	9
50 and less than 60	10
60 and less than 70	11
70 and less than 80	10
80 or more	10

12.2
CHI-SQUARE TESTS OF INDEPENDENCE

In the previous section, we classified each member of a population into one of many categories. This was a one-dimensional situation because each member was classified using only *one* criterion (brand, color, and so on). In this section, we extend this idea to a two-dimensional situation, in which each element in the population is classified according to two criteria, such as sex and income level (high, medium, or low). The question of interest is, Are these two variables (classifications) *independent*? For example, if sex and income level are not independent, perhaps sex discrimination is present in the salary structure of a company. If a person's salary is not related to sex, these two classifications *would* be independent.

In Chapter 4, we examined a survey concerned with the age and sex of the purchasers of a recently released microcomputer. The results were summarized in a *contingency* (or *cross-tab*) table. This table consisted of **cells**, where each cell contains the **frequency** of people in the sample that satisfy each of the various cross-classifications.

SEX	AGE <30	AGE 30–45	AGE >45	TOTAL
Male	60	20	40	120
Female	40	30	10	80
Total	100	50	50	200

This is a 2×3 contingency table. It shows that there were 60 people who were both male *and* under 30. In Chapter 4, we selected a person at random from this group of 200 and determined various probabilities, such as the probability that this person is both a male and over 45 years. Here we do not select a person at random. Instead, we view these data as the results of a particular experiment (survey) and attempt to determine whether the variables—age and sex—are independent for this application. Put another way, is the age structure of the male buyers the same as that for the female purchasers? The hypotheses are

H_0: the classifications (age and sex) are independent

H_a: the classifications are dependent

This problem can also be viewed as a multinomial experiment containing 200 trials and $(2)(3) = 6$ possible categories for each trial outcome.

Deriving a Test of Hypothesis for Independent Classifications

Calculating the Expected Values   We want to decide whether the data about the purchasers exhibit random variation or a pattern of some type due to a dependency between age and sex. If these classifications *are* independent (H_0 is true), how many people would you expect in each cell? Consider the upper right cell, which shows males over 45 years. The expected number of sample observations in this cell is $200 \cdot P$(sampled purchaser is a male and over 45). Assuming independence, this is $200 \cdot P$(sampled purchaser is a male) $\cdot P$(sampled purchaser is over 45) using the multiplicative rule for independent events discussed in Chapter 4.

What is P(sampled purchaser is a male)? We do not know, because we do not have enough information to determine what percentage of *all* purchasers are male. However, from these data, we can *estimate* this probability using the percentage of males in the sample. This is $120/200 = .6$.

Similarly, P(sampled purchaser is over 45) can be estimated by the fraction of people over 45 in the sample—namely, $50/200$. So, our estimate of the expected number of observations for this cell is

$$\hat{E} = 200 \cdot \frac{120}{200} \cdot \frac{50}{200} = \frac{(120)(50)}{200} = 30$$

So, for this cell, the observed frequency is $O = 40$, and our estimate of the expected frequency (if H_0 is true) is $\hat{E} = 30$. In general,

$$\hat{E} = \frac{\text{(row total for this cell)} \cdot \text{(column total for this cell)}}{n}$$

where n = total sample size. A summary of the calculations can be tabulated as follows.

SEX	AGE	OBSERVED (O)	EXPECTED ($\hat{E}$) IF H_0 IS TRUE
Male	<30	60	(120)(100)/200 = 60
	30–45	20	(120)(50)/200 = 30
	>45	40	(120)(50)/200 = 30
Female	<30	40	(80)(100)/200 = 40
	30–45	30	(80)(50)/200 = 20
	>45	10	(80)(50)/200 = 20
		$\overline{200}$	$\overline{200}$

The easiest way to represent these 12 values is to place the expected value in parentheses alongside the observed value in each cell:

SEX	AGE <30	AGE 30–45	AGE >45	TOTAL
Male	60 (60)	20 (30)	40 (30)	120
Female	40 (40)	30 (20)	10 (20)	80
Total	100	50	50	200

The Test Statistic

The test statistic for testing H_0: the classifications are independent versus H_a: the classifications are dependent is the usual chi-square statistic, which in this case compares each *observed* frequency with the corresponding *expected* frequency estimate.

$$\chi^2 = \sum \frac{(O - \hat{E})^2}{\hat{E}} \qquad (12.3)$$

where the summation is over all cells of the contingency table.

Degrees of Freedom

For the multinomial situation, the degrees of freedom for the chi-square statistic were $k - 1$, where k = the number of categories (outcomes). For this situation, there were k values of $(O - \hat{E})$. However, because the sum of the observed frequencies is the same as the sum of the expected frequencies, the sum of the k values of $(O - \hat{E})$ is *always zero*. This means that, of these k values, only $k - 1$ are free to vary. This resulted in $k - 1$ df for the chi-square statistic.

Take a close look at the observed and expected frequencies in the contingency table for age and sex of purchasers. Notice that (1) for each row, sum of O's = sum of $\hat{E}$'s and (2) for each column, sum of O's = sum of $\hat{E}$'s. In general, if classification 1 has c categories and classification 2 has r categories, you

Pooling

At this point, you need to check your expected values. If any one of them is less than 5, you need to combine the column (or row) in which this small value occurs with another column (or row). This is to comply with the earlier assumption that all expected values in the chi-square statistic are at least 5. The observed and expected values for this new column (row) are obtained by summing the values for the two columns (rows).

FIGURE 12.3

Expected value estimates for an $r \times c$ contingency table.

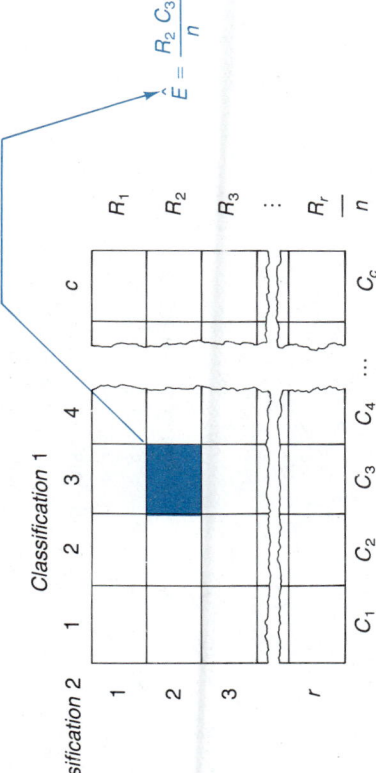

construct an **$r \times c$ contingency table** (Figure 12.3). Of the c values of $(O - \hat{E})$ in each row, only $c - 1$ are free to vary. Similarly, only $r - 1$ of the values in each column are free to assume any value. So, for this contingency table, only $(r - 1)(c - 1)$ values are free to vary. Therefore, for the chi-square test of independence,

$$\text{df} = (r - 1)(c - 1) \qquad (12.4)$$

where $\text{df} = (r - 1)(c - 1)$.

In summary, the chi-square test for independence hypotheses are

H_0: the row and column classifications are independent (not related)

H_a: the classifications are dependent (related or associated in some way)

The test statistic is

$$\chi^2 = \sum \frac{(O - \hat{E})^2}{\hat{E}}$$

where

1. The summation is over all cells of the contingency table consisting of r rows and c columns.
2. O is the observed frequency in this cell.
3. $\hat{E}$ is the estimated expected frequency for this cell.

$$\hat{E} = \frac{\left(\begin{array}{c} \text{total of row in} \\ \text{which the cell lies} \end{array} \right) \cdot \left(\begin{array}{c} \text{total of column in} \\ \text{which the cell lies} \end{array} \right)}{\text{(total of all cells)}}$$

4. The degrees of freedom for the chi-square statistic are $\text{df} = (r - 1)(c - 1)$.

The test procedure is (using Table A.6):

reject H_0 if $\chi^2 > \chi^2_{\alpha, \text{df}}$

Testing Procedure

When H_0 is not true, the expected frequencies and observed frequencies will be very different, producing a large χ^2 value. We again reject H_0 if the value of the test statistic falls in the *right-tail* rejection region, so we

reject H_0 if $\chi^2 > \chi^2_{\alpha, \text{df}}$

We can now return to our question of whether age and sex of purchasers are independent. Step 1 (statement of hypotheses) and step 2 (definition of test statistic) of our five-step procedure have been discussed already. Assume that a significance level of $\alpha = .1$ was specified. For step 3, the df are $(2 - 1)(3 - 1) = 2$. Using Table A.6, $\chi^2_{.1, 2} = 4.61$. So we will reject H_0 if $\chi^2 > 4.61$. For step 4, referring to the contingency table,

$$\chi^{2*} = \frac{(60 - 60)^2}{60} + \frac{(20 - 30)^2}{30} + \frac{(40 - 30)^2}{30} + \frac{(40 - 40)^2}{40}$$
$$+ \frac{(30 - 20)^2}{20} + \frac{(10 - 20)^2}{20}$$
$$= 0 + 3.33 + 3.33 + 0 + 5 + 5$$
$$= 16.66$$

This exceeds the table value of 4.61, so we reject H_0. We thus conclude that the age and sex classifications are *not* independent.

If the results of the chi-square test lead to a conclusion that the classifications are not independent, a closer look at the individual terms in the chi-square statistic can often reveal what the relationship is between these two variables. Examining the six terms, we observe four large values, namely, 3.33 (male/age 30–45), 3.33 (male/age over 45), 5 (female/age 30–45), and 5 (female/age over 45). We obtained more men (and fewer women) over 45 years than we would expect if there was no dependency. Similarly, there were fewer men (and more women) between 30 and 45 years.

We can find the p-value for this also, given $\chi^{2*} = 16.66$. Using a χ^2 curve with 2 df, the area to the right of 16.66, using Table A.6, is $< .005$. The p-value indicates the **strength** of the dependency between two classifications. The *smaller* the p-value is, the more you tend to support the alternative hypothesis, which indicates a *stronger* dependency between the two variables. For the age and sex illustration, $p < .005$, so we conclude that the age and sex of these purchasers are strongly related.

It is worth mentioning at this point that it is possible that examining one category (such as sex) can fail to show any differences among subcategories (male versus female), but when the category is examined along with another category (such as age classification), patterns can emerge. Such a technique is often useful in detecting job discrimination within companies. For example, no sex discrimination may be evident in a sample, but when it is examined along with race or age categories, certain discriminatory practices can be identified.

EXAMPLE 12.5

In example 11.4, a personnel director attempted to determine whether an employee's educational level had an effect on his or her job performance. An exam was given to a sample of the employees, and we used the ANOVA procedure to test for a difference among the three groups: (1) those with a high school diploma only, (2) those with a bachelor's degree only, and (3) those with a master's degree.

The director decided to expand this procedure by testing 120 employees; rather than recording the exam scores, she rated each person's exam performance as high, average, or low. The results of this study are:

	HIGH	AVERAGE	LOW	TOTAL
Master's degree	4	20	11	35
Bachelor's degree	12	18	15	45
High school diploma	9	22	9	40
Total	25	60	35	120

Does job performance as measured by the exam appear to be related to the level of an employee's education, at this particular firm? Use $\alpha = .05$.

SOLUTION

Step 1. This calls for a chi-square test of independence, with hypotheses

H_0: exam performance is independent of educational level

H_a: these classifications are dependent

Steps 2, 3. Your test statistic is the chi-square statistic in equation 12.3. The table of frequencies here is a 3×3 contingency table, which means that the degrees of freedom are df $= (3 - 1)(3 - 1) = 4$. From Table A.6, we determine that $\chi^2_{.05,4} = 9.49$, so the testing procedure is to

reject H_0 if $\chi^2 > 9.49$

Step 4. Computing the expected frequency estimates in the usual way, we arrive at the following table:

	HIGH	AVERAGE	LOW	TOTAL
Master's degree	4 (7.29)	20 (17.5)	11 (10.21)	35
Bachelor's degree	12 (9.38)	18 (22.5)	15 (13.12)	45
High school diploma	9 (8.33)	22 (20.0)	9 (11.67)	40
Total	25	60	35	120

To illustrate the calculations, the 11.67 in the lower right cell is $(40 \cdot 35)/120$. The computed chi-square value is

$$\chi^{2*} = \frac{(4 - 7.29)^2}{7.29} + \frac{(20 - 17.5)^2}{17.5} + \cdots + \frac{(9 - 11.67)^2}{11.67} = 4.67$$

This value is <9.49, and so we fail to reject H_0.

Step 5. We see no evidence of a relationship between job performance and level of education.

We do not conclude that these data demonstrate that the two classifications are clearly *independent* because this amounts to accepting H_0. We are simply unable to demonstrate that a relationship exists. ■

A MINITAB solution to Example 12.5 is contained in Figure 12.4. Notice

FIGURE 12.4

MINITAB solution to Example 12.5 (test for independence).

```
MTB > READ INTO C1-C3
DATA> 4 20 11
DATA> 12 18 15
DATA> 9 22 9
DATA> END
    3 ROWS READ
MTB > CHISQUARE USING C1-C3

Expected counts are printed below observed counts

                C1      C2      C3      Total
                                            35
    1            4      20      11              Observed (O)
              7.29   17.50   10.21               Expected (E)

    2           12      18      15              45
              9.37   22.50   13.12

    3            9      22       9              40
              8.33   20.00   11.67

    Total       25      60      35      120

    Chisq =  1.486 +  0.357 +  0.061 +
             0.735 +  0.900 +  0.268 +
             0.053 +  0.200 +  0.610 = 4.670

    df = 4
```

that the format of this table is similar to that of the one we constructed, with the expected value (assuming H_0) and the observed value shown for each cell.

In Example 12.5, the personnel director recorded the exam performance as high, average, or low rather than listing the actual exam score. Why would anyone take *interval/ratio* data (the exam scores) and convert them to seemingly weaker *ordinal* data (the exam performance classifications)? Do you lose useful information by doing this? When using the ANOVA procedure, we were forced to assume that these data came from *normal* populations with equal variances. In this chapter, aside from the randomness of the sample, *no* assumptions regarding the populations were necessary. So, by converting the exam scores to a form suitable for a contingency table and using the chi-square test of independence, we can avoid the assumptions of normality and equal variances.

This introduces **nonparametric statistics**, often called *distribution-free* statistics. The beauty of these procedures is that they require only very weak assumptions regarding the populations. However, if the data *do* satisfy the requirements of the ANOVA procedure (or nearly so), the nonparametric test is less sensitive to differences among the populations (such as educational level) and so is less *powerful* than the ANOVA *F*-test. Additional nonparametric tests of hypothesis are discussed in Chapter 18.

Test of Independence with Fixed Marginal Totals (Test of Homogeneity)

A slightly different interpretation of the previous chi-square procedure occurs when we determine *in advance* the number of observations to be sampled within each column (or row). In the previous discussion, the row and column totals were random variables because we had no way of knowing what they would be before the sample was obtained. In this discussion, the contingency table is the same, except that the column (or row) totals are predetermined.

Assume Lextron International, a manufacturer of electronic components, has facilities located in Dallas, Boston, Seattle, and Denver. Over the years, Lextron has gone to great lengths to discourage the formation of labor unions at these plants, including constructing employee recreational centers and offering better-than-average employee benefits. Management suspects, however, that there is growing interest among the employees in forming a union. Of particular interest is whether employee interest in a union differs among the four plants.

The Dallas and Denver plants are considerably larger than the other two, so Lextron obtains a random sample of 200 employees from each of these two plants and of 100 from each of the two smaller facilities. The results of the survey are:

	DALLAS	BOSTON	SEATTLE	DENVER	TOTAL
Interested	120	41	45	112	318
Not interested	35	38	40	36	149
Indifferent	45	21	15	52	133
Total	200	100	100	200	600

In the previous tests of independence, we had a *single* population, where each member was classified according to two criteria, such as age and sex. Now we have four distinct populations, namely, the Lextron employees in each of the four cities. Consequently, we obtained a random sample from each one. The column totals (sample sizes) were determined in advance. This differs from our previous examples, where we had no idea what the row or column totals would be before the sample was obtained.

The question of interest here becomes, Is interest in a labor union the same in each of the four cities? In other words, we are trying to determine whether these four populations can be viewed as belonging to the *same* population (in terms of this criterion). Identical populations are said to be **homogeneous**. Consequently, the test of hypothesis here is a **test of homogeneity** as well as a test for independence. The null hypothesis can be written as

H_0: the four populations are homogeneous in their interest in a union

or as

H_0: plant location and employee interest are independent classifications

The procedure for analyzing a contingency table is the *same* whether or not the column (or row) totals are fixed in advance.

A MINITAB solution using $\alpha = .05$ is provided in Figure 12.5. The expected cell frequencies are computed by finding

$$\hat{E} = \frac{\text{(row total)(column total)}}{600}$$

The computed chi-square value is

$$\chi^{2*} = \frac{(120 - 106.0)^2}{106} + \frac{(41 - 53.0)^2}{53} + \cdots + \frac{(52 - 44.3)^2}{44.3} = 34.16$$

The degrees of freedom here are $(3 - 1)(4 - 1) = 6$. This means that we reject H_0 if $\chi^2 > 12.59$, where 12.59 is $\chi^2_{.05, 6}$. The computed value (34.16) exceeds the tabled value, so we reject H_0. We conclude that these four populations are *not* homogeneous. The employee interest in a labor union is not identical at each of the four locations. We can also say that the location and union interest classifications are not independent.

Examining the individual terms of the chi-square value in Figure 12.5, we note that the larger plants (Dallas and Denver) had a higher proportion of employees interested in forming a union. In Dallas, for example, if these classifications were independent, we would expect 106 employees to be interested; instead, we observed 120. The same argument applies to the Denver plant.

FIGURE 12.5

MINITAB solution to test H_0: Plant location and employee interest are independent classifications (test of homogeneity).

```
MTB > READ INTO C1-C4
DATA> 120 41 45 112
DATA> 35 38 40 36
DATA> 45 21 15 52
DATA> END
     3 ROWS READ
MTB > CHISQUARE USING C1-C4

Expected counts are printed below observed counts

              C1       C2       C3       C4    Total
  1          120       41       45      112      318
          106.00    53.00    53.00   106.00

  2           35       38       40       36      149
           49.67    24.83    24.83    49.67

  3           45       21       15       52      133
           44.33    22.17    22.17    44.33

Total        200      100      100      200      600

ChiSq =    1.849 +   2.717 +   1.208 +   0.340 +
           4.331 +   6.981 +   9.263 +   3.761 +
           0.010 +   0.061 +   2.317 +   1.326 = 34.163

df = 6
```

EXERCISES

12.16 Suppose you are interested in determining whether there is a relationship between one's educational preference (major) and one's sex. A random sample of 172 students at Hamilton College yields the following data:

MAJOR	FEMALE	MALE	TOTAL
Liberal arts	35	25	60
Home economics	6	9	15
Physics	18	21	39
Business	26	32	58
Total	85	87	172

a. Formulate the necessary hypotheses.

b. Using the chi-square test, test the hypotheses to determine whether educational preference is independent of sex, using a significance level of .10.

c. Find and interpret the p-value for the chi-square test.

12.17 A lawn-equipment shop is considering adding a brand of lawnmowers to its merchandise. The manager of the shop believes that the highest-quality lawnmowers are Trooper, Lawneater, and Nipper and needs to decide whether it makes a difference which of these three the shop will add to its existing merchandise. Twenty owners of each of these three types of lawnmowers are randomly sampled and asked how satisfied they are with their lawnmowers.

LAWNMOWER	VERY SATISFIED	SATISFIED	SATISFIED	NOT SATISFIED	TOTAL
Trooper	11	6	1	3	20
Lawneater	13	4	3	3	20
Nipper	13	6	1	1	20

Are the owners of the lawnmowers homogeneous in their response to the survey? Use a 5% significance level.

12.18 A real-estate firm wanted to know whether the type of house purchased is associated with the amount of education of the head of the household. Fox and Jones Construction builds four styles of homes. A random sample of 175 homeowners who own a Fox and Jones house was taken and the education level of the household head was noted.

TYPE OF HOUSE	NO COLLEGE DEGREE	BACHELOR'S DEGREE	MASTER'S DEGREE	DOCTORAL DEGREE	TOTAL
1	12	5	1	0	18
2	13	10	8	2	33
3	10	20	25	10	65
4	2	18	30	9	59

Do the data provide sufficient evidence to indicate that the type of house owned is related to the education level of the head of the household? Use a .05 level of significance.

12.19 At the University of Tennessee at Chattanooga, 1207 students received grades in the course Introduction to Business Statistics. Students who withdrew from the course are not included. The following table gives a listing of grades by the sex of the student.

From Table A.6, the p-value here is less than .005. Because of this extremely small value, we can conclude that employees at these four plants have considerably different views in regard to the formation of a labor union. The small p-value also implies that there is an extremely strong dependence between the two classifications.

GRADE	MALE	FEMALE
A	151	145
B	181	168
C	163	92
D	77	53
F	110	67

Source: Farhad Raiszadeh and Mohammad Ahmadi, "Students' Race and Gender in Introductory Business Statistics," *Journal of Education for Business* 63, no. 1 (1987): 20–23.

Use the chi-square statistic to test the hypothesis that grades in the first course in statistics are independent of the sex of the students. Calculate the *p*-value and interpret it.

12.20 An insurance company claims that full-size cars are more prone to automobile accidents and hence should be subject to a higher insurance premium. To test the validity of the claim, an auto firm gathered a random sample.

CAR SIZE	AT LEAST ONE ACCIDENT	NO ACCIDENTS
Full size	24	13
Compact	36	117
Small	108	214

Formulate the necessary hypotheses and test at the 10% significance level. Also calculate the *p*-value. Based on this value, would you reject the null hypothesis that car size and occurrence of accidents are independent at the 1% level?

12.21 The Meyers-Briggs Type Indicator (MBTI) is a personality scale that can be used to classify qualities like Extrovert (E), Introvert (I), Intuitive (N), Feeling (F), Sensing (S), Thinking (T), Judging (J), Perceptive (P). Thus, EN means extrovert-intuitive, SP means sensing-perceptive. Consider the following hypothetical frequencies for a cross-tabulation of four types of personality using the MBTI against profession.

PROFESSION	\multicolumn{4}{c|}{PERSONALITY (MBTI)}	
	EN	IF	SP	JT	TOTAL
Computer programmer	4	6	5	6	21
Accountant	3	7	5	5	20
Marketer	9	3	7	4	23
Educator	5	5	5	5	20
Total	21	21	22	20	84

a. From the above table, can you conclude at a 1% significance level, whether personality type and profession are related?

b. State the *p* value for your test.

12.22 A study was undertaken to examine the composition of corporate board committees. A random sample of small companies was taken. In this sample 3075 directors were categorized by membership on the audit committees and by tenure in years on the audit committees. Tenure was considered an important characteristic which distinguished major committee members from nonmembers. Previous researchers have argued that it takes at least three to five years for directors to gain an adequate understanding of a firm.

COMMITTEE MEMBERSHIP	\multicolumn{3}{c}{TENURE IN YEARS}	
	0–5	6–10	>10
Major member	645	557	932
Not a major member	270	310	361

Source: Idalene F. Kesner, "Directors' Characteristics and Committee Membership: An Investigation of Type, Occupation, Tenure, and Gender," *Academy of Management Journal* 31, no. 1 (1988): 66–84.

Do the data provide sufficient evidence to indicate that type of membership and tenure are related? Base your conclusion on the p-value.

12.23 A research team conducts a study to see whether voting for the candidates in the recent local election is homogeneous within age groups (given in years). One-hundred voters were randomly selected from each of five age classifications. Do the data indicate that voting is homogeneous with respect to age group? Use a 5% significance level.

AGE	CANDIDATE A	CANDIDATE B	CANDIDATE C	TOTAL
Less than 25	48	22	30	100
25 and less than 35	55	20	25	100
35 and less than 45	50	28	22	100
45 and less than 55	45	21	34	100
Over 55	49	21	30	100

SUMMARY

When performing a two-tailed test of hypothesis on a binomial parameter (for example, $p = .75$) we can use a chi-square test statistic. The advantage of this approach is that it extends easily to the **multinomial situation**, where each trial can result in any specified number of outcomes. An example is the roll of a single die, which has six possible outcomes on each roll.

For the multinomial situation, the probability of observing each possible outcome may be specified (such as 1/6 for each outcome in the single die illustration). To test the hypothesis, a random sample of observations is obtained, and a chi-square test statistic is evaluated either to reject or to fail to reject this set of probabilities (percentages). Such a test is referred to as a **chi-square goodness-of-fit test**. The form of this chi-square test statistic is

$$\chi^2 = \sum \frac{(O - E)^2}{E}$$

where

1. O represents the *observed* frequency of observations in a particular category (such as the observed number of 3s in 60 rolls of a single die).

2. E is the *expected* frequency for this category. For example, we would expect to see $60 \cdot \frac{1}{6} = 10$ values of 3 in the die illustration.

3. The chi-square value is obtained by summing over all categories of the multinomial random variable.

4. Categories must be combined (pooled) together whenever an expected value (E) for a particular category is less than 5.

The chi-square goodness-of-fit procedure can be used to determine whether a certain set of sample data came from a specified probability distribution. For example, you might attempt to determine whether the number of defects in a particular product follows a Poisson distribution. By collecting a random sample and counting the number of defects in each product, you can compare the observed values (how many zeroes, how many ones, and so on) with what you would expect if the null hypothesis—H_0: the data are from a Poisson distribution—is true. If the calculated chi-square value is significantly large (in the right tail), this hypothesis will be rejected. Whenever any of the parameters for this distribution are unknown (such as μ for the Poisson illustration), they can be estimated using the sample data. The degrees of freedom of the chi-square test statistic are reduced by one for each estimated parameter.

Finally, this chi-square statistic can be used to test whether two classifications (such as age and performance) used to define a contingency table are independent. This is the **chi-square test of independence**. The expected value within each cell of the contingency table is determined under the assumption that H_0 is true, where H_0: the row and column classifications are independent. This also leads to a right-tailed rejection region using the chi-square statistic.

This procedure can be used as a **test for homogeneity** when fixed sample sizes are used for each row or column of the table. If the column totals are fixed in advance, this test will determine whether the populations defined by the column categories are homogeneous (identical) with respect to the variable defining the rows. A similar argument applies when the row totals are predetermined. The test statistic used for a test of homogeneity is the same chi-square statistic used in the test of independence.

REVIEW EXERCISES

12.24 The manager of the Grandiose Hotel guarantees that a customer's room will be ready at 6:00 P.M. if a reservation is made. Otherwise, the customer stays at the hotel for free. The manager believes that this policy should be continued; he believes that a room is not available on time only 5% of the time. A random sample of 200 past reservations was selected and the estimate of the proportion of times when a room was not available on time was .065.

a. Letting the significance level be .05, test the belief that the proportion of occurrences when a room is not available on time is .05. Use the Z-test from Section 10.2.

b. Use the chi-square goodness-of-fit test instead of the Z-test in part (a).

c. What is the relationship between the test statistics in parts (a) and (b)?

12.25 A student wanted to know whether the answers a, b, c, d, and e on the standardized departmental multiple choice test occurred equally as often. Several old departmental tests were randomly selected and the occurrences of each answer were tabulated.

ANSWER	TIMES USED AS ANSWER
a	39
b	26
c	43
d	42
e	25

Test the belief that all answer choices occur equally as often. Use a significance level of .10.

12.26 A manufacturer of clothes dryers believes that historically 40% of its sales are for the basic 18-pound-capacity clothes dryer, 35% are for the 20-pound-capacity dryer and 25% are for the 22-pound-capacity dryer. A random sample of 200 clothes dryers sold during the past 6 months was obtained, with the following results (capacity in pounds):

CAPACITY	NUMBER SOLD
18	72
20	68
22	60

Using a significance level of .10, is there sufficient evidence to indicate that the manufacturer was wrong in its statement of these percentages?

12.27 A construction company sells and installs solid vinyl siding. Siding comes in five basic colors: white, brown, avocado, reddish-tan and yellow. Historically, the percentages of sales for each color are 50%, 27%, 12%, 8% and 3%, respectively. A random sample of 100 recent sales gives the following data:

COLOR	NUMBER SOLD
White	43
Brown	29
Avocado	14
Reddish-tan	13
Yellow	1

Does the sales distribution of these colors appear to be the same as the historical distribution? Test at the 1% significance level.

12.28   A large department store in New York City has five entrances and exits. It is believed that the proportion of shoppers entering or leaving the store is approximately the same for each of the five doorways on any single day. The number of customers entering or leaving the store is tallied at each doorway for three randomly selected days.

DOORWAYS	CUSTOMERS
1	150
2	123
3	126
4	163
5	152

Do the data justify the statement that all five entrances and exits are used equally often? Use a 5% significance level.

12.29   The manager of a news and magazine store in a metropolitan area with four newspapers believes that the proportions of customers preferring newspapers A, B, C, and D are 30%, 40%, 15%, and 15%. One hundred randomly selected customers were chosen, and the newspaper they bought was recorded.

NEWSPAPER	NUMBER OF CUSTOMERS BUYING THE NEWSPAPER
A	26
B	44
C	12
D	18

Do the data justify the percentages used by the managers of the news and magazine store? Use a 1% significance level.

12.30   A car-rental company has 15 cars to rent. The owner believes that the number of cars rented daily is binomially distributed. He also believes that each car has a 30% chance of being rented each day. Forty-five randomly selected days and the corresponding number of cars rented are indicated. From the data, test the hypothesis that the daily rental of cars is binomially distributed with $p = .30$. Use a 5% significance level.

CARS RENTED	DAYS OCCURRED	CARS RENTED	DAYS OCCURRED
0	0	5	12
1	3	6	6
2	3	7	3
3	6	≥8	3
4	9		

12.31   An advertising firm believes that the number of daily responses to an advertisement in the *Wall Street Journal* follows a Poisson distribution. Forty days were randomly selected and the following data were collected:

RESPONSES	DAYS	RESPONSES	DAYS
0	0	4	6
1	8	5	6
2	8	6	2
3	10		

Can you conclude that the data did come from a Poisson distribution? Use a 1% level of significance.

12.32   The assistant dean of the College of Business at Oceanside University believes that the number of students dropping a class is Poisson distributed. Fifty classes, all containing the same number of students, were randomly selected. The number of withdrawals from the classes was recorded. Based on the following data, what conclusion can be drawn about whether or not these data come from a Poisson distribution? Use a significance level of .05 to justify your conclusion.

DROPS	CLASSES	DROPS	CLASSES
0	0	5	4
1	2	6	6
2	6	7	4
3	10	>7	0
4	18		

12.33 A temporary-help employment agency regularly has ten workers who are ready to work. An employer telephones the agency when he or she needs temporary help. The owner of the agency believes that the number of workers used each day by the employers in the community is binomially distributed. Sixty days of operation were used to compile the following:

WORKERS (OUT OF 10) USED	DAYS USED
≤4	0
5	12
6	12
7	12
8	14
9	8
10	2
Total	60

With a significance level of .10, test the goodness-of-fit of the data to a binomial distribution.

12.34 A computer generates 100 observations from a normally distributed population with mean 35 and standard deviation 2. The results of the 100 observations generated are:

INTERVAL	OBSERVED FREQUENCY	INTERVAL	OBSERVED FREQUENCY
less than 32	6	35 but less than 36	19
32 but less than 33	9	36 but less than 37	15
33 but less than 34	12	37 but less than 38	11
34 but less than 35	23	38 or more	5

Use the chi-square goodness-of-fit procedure in Exercise 12.13 to test that there is enough evidence to support the conclusion that the generated numbers did not come from a normally distributed population with mean = 35 and standard deviation = 2. Use a 1% significance level.

12.35 The vice president of a national firm wants to know the response of workers at a certain plant to a proposal to relocate the plant. Forty workers were randomly selected from each of the five divisions at the plant. The workers were asked if they favored a relocation of the plant.

DIVISION	FAVORED	DO NOT FAVOR	TOTAL
A	15	25	40
B	18	22	40
C	24	16	40
D	17	23	40
E	20	20	40

Do the data indicate that the divisions are not homogeneous with respect to the proportion of workers who favor a relocation of the plant? Use a .05 significance level.

12.36 A manufacturer of bleach sold the same bleach to three different companies, placing their respective brand names on the bottle. A marketing-research firm wanted to know whether the same amount of bleach would be sold from each of the three brands, provided the price was the same for each brand. Supplies of the three brands of bleach were placed on a supermarket shelf. By the end of the month, 335 bottles of bleach were sold, as follows:

	BLEACH X	BLEACH Y	BLEACH Z
	105	133	97

Test the null hypothesis that each brand sells equally well. Use a significance level of .10.

12.37 An immigration attorney was investigating which industries to target for obtaining new clients who might have problems with changes in the immigration laws. Five industries were selected. Twenty workers were chosen in each industry, and their visa status was verified. The data are summarized as follows:

	INDUSTRY					
VISA STATUS	A	B	C	D	E	TOTAL
Illegal alien	8	10	5	10	1	34
Legal resident	4	2	6	4	9	25
U.S. citizen	8	8	9	6	10	41
Total	20	20	20	20	20	100

a. Are the five industries homogeneous with respect to the visa status of its workers? Use $\alpha = .05$.

b. State the p-value.

12.38 An analyst for a marketing firm believes that the number of shoppers at a shopping mall who come from a distance of over 10 miles is equal to 25%. A random sample of 200 shoppers was selected and the estimate of the proportion of shoppers who traveled over 10 miles was .22.

a. Is there sufficient evidence to conclude that the percentage of shoppers who come from a distance over 10 miles is not equal to 25%? Use a significance level of .05 and the Z test from Section 12.1.

b. Use the chi-square goodness-of-fit test instead of the Z test in part (a). Let the significance level be .05.

c. What is the relationship between the test statistic in parts (a) and (b)?

12.39 Microtron, a maker of semiconductors, has plants in four states: Texas, Georgia, California, and New York. Management would like to evaluate the effect of a mandatory retirement age of 60. The largest plants are in California and New York, so they sample 300 employees from each of these states and 200 employees each in Texas and Georgia.

RESPONSE	TEXAS	GEORGIA	CALIFORNIA	NEW YORK
Favor	81	85	122	128
Against	119	115	178	172

Do the data support the hypothesis that employees within different states have different opinions about the mandatory retirement age? Use a 5% significance level.

12.40 A marketing survey was taken of 277 frequent flyers by recording their socioeconomic class and airline preference. Do the sample data provide sufficient evidence to reject the null hypothesis of independence of airline preference and social class at the .05 significance level?

SOCIOECONOMIC CLASS	DELTA	SOUTHWEST	AMERICAN	OTHER	TOTAL
Low	20	45	23	4	92
Middle	25	40	20	20	105
Upper	18	15	30	17	80

12.41 A record company wanted to survey its customers regarding music preferences. A random sample of 258 frequent customers of the record company was taken and information was gathered on their music preference and job classification. From the following data, can the null hypothesis of independence between type of music preferred and working status be rejected at the 10% significance level?

JOB CLASSIFICATION	COUNTRY AND WESTERN	ROCK	CLASSICAL	JAZZ	TOTAL
Clerical	25	40	17	5	87
Managerial	21	25	29	15	90
Blue collar	27	33	14	7	81
Total	73	98	60	27	258

12.42 The personnel department of a particular firm wants to know if an employee's age is associated with productivity (given in items per hour). The manager of the personnel department draws a random sample of 60 employees from each of the age classifications listed. Do the data support the hypothesis that the five age categories are not homogeneous with respect to productivity? Use a 10% significance level.

AGE	4–5 ITEMS	6–7 ITEMS	≥ 8 ITEMS	TOTAL
20 and under 30	15	25	20	60
30 and under 40	13	29	18	60
40 and under 50	16	26	18	60
50 and under 60	19	26	15	60
60 and under 70	22	24	14	60

12.43 An aspiring politician decided to sample 300 citizens from each of two major cities to find out whether the two populations were homogeneous with regard to their opinion on gun control. Do the following data indicate a lack of homogeneity? Use a 10% significance level.

CITY	FAVOR GUN CONTROL	AGAINST GUN CONTROL
A	126	174
B	148	152

12.44 Axiom Market Research published the following data concerning education level and attendance at "regular" theater performances. A sample size of 950 was selected. Do the data indicate, at the .05 significance level, a relationship between level of education and regular theater attendance?

EDUCATION	ATTEND MORE THAN ONCE PER YEAR	ATTEND NO MORE THAN ONCE PER YEAR	TOTAL
College graduate	82	120	202
Some college	75	131	206
High school graduate	106	215	321
Not a high school graduate	51	170	221

12.45 A sample of households classified as having incomes below the poverty level revealed the following distribution of persons by age:

AGE	FREQUENCY
0 to <5	27
5 to <18	53
18 to <22	16
22 to <45	60
45 to <65	26
65 to <72	18
Total	200

a. Compute the mean and standard deviation for the above distribution.

b. Using a 10% significance level, determine with a chi-square test whether the data fit a normal population.

c. Find the p-value for the test.

d. Does your conclusion change if $\alpha = .05$ or $\alpha = .01$?

12.46 In an effort to monitor the service of its employees, a parcel-delivery firm keeps a tally of the number of packages misrouted each week at each of its 25 distribution centers, with the following results:

MISROUTED PACKAGES	DISTRIBUTION CENTERS
0	5
1	6
2	8
3	6
>3	0

Do these data appear to come from a Poisson distribution? Use a .05 significance level.

12.47 In the field of organizational psychology, extensive study has been made of different leadership styles. One researcher refers to two extremes as authoritarian versus democratic; another refers to task-oriented versus people-oriented; yet others have their own labels for these qualities. Whatever the label, do these different styles affect the morale of the subordinates? To address this issue, a researcher established a ranking scale for worker morale, based on interviews, and grouped the workers into low, acceptable, and high morale categories. These were cross-classified against the leadership style of the supervisor. The following contingency table summarizes the results.

		LEADERSHIP STYLE	
WORKER MORALE	AUTHORITARIAN	DEMOCRATIC	TOTAL
Low	10	5	15
Acceptable	8	12	20
High	6	9	15
Total	24	26	50

a. Apply the chi-square test of independence on these data, at a 5% significance level.

b. State the p-value for your test.

c. Is worker morale related to the supervisor's leadership style, or are these qualities independent?

12.48 Kingston Pencils is considering a new bonus plan. Under the current bonus plan, the amount of bonus is not linked to the production but only linked to the profits. According to the proposed bonus plan, the amount of bonus will be linked to the quantity produced but will be subject to the amount of profits. The controller of Kingston is interested in examining whether employee opinion of the bonus plan is independent of job classification.

EMPLOYEE	FAVORABLE	UNFAVORABLE
White Collar	67	28
Blue Collar	43	19

Calculate the p-value and interpret it.

12.49 The need for managers to be directly involved and to participate in the design of management information or decision support systems is broadly accepted as one of the essential principles of information systems development. A random sample of managers were selected to examine the claim: When top management makes a clear long-term financial commitment to management information systems development, user management involvement and participation is encouraged. Results of the survey are given in the table below.

	DID YOU PARTICIPATE IN THE DESIGN OF THE SYSTEMS YOU ARE USING?	
	YES	NO
YES	55	24
NO	40	33

Has top management made a long-term commitment to provide stable funding for systems development?

(Source: William J. Doll, "Encouraging User Management Participation in Systems Design," Information and Management, 1987 13, no. 2: 25–32.)

At a .05 significance level, what conclusion can be drawn from these data?

12.50 The regulatory environment affecting financial institutions and markets has undergone substantial change during the 1980s. A study involving the states of Arkansas, Kansas, Missouri, and Oklahoma was conducted to examine attitudes of both urban and rural consumers toward the deregulation of the financial institutions. There were 101 respondents from the rural areas and 130 respondents from the urban areas. The following table gives the results of the survey for the question, Do you desire to see banking facilities owned and operated within major retail stores?

RESPONSE	URBAN	RURAL
Yes	24	9
No	88	78
Don't know	18	14

(*Source:* Donald J. Brown, "A Comparison of Urban vs. Rural Attitudes Concerning Interest Rates, Products, Services and Competition, of Financial Institutions," *Central State Business Review* 6, no. 2 (1987): 15–19.)

Is there sufficient evidence that the response to this question depends on whether the respondent lives in an urban area or rural area? Use a 10% significance level.

12.51 A bank offers three types of money market accounts. The vice president is interested in whether there is a relationship between the account balances and the type of money market account. A random sample of 2500 accounts are selected, with the results given in the following table. What conclusions can be drawn from the MINITAB computer printout? Base your decision on the *p*-value.

ACCOUNT BALANCE	MONEY MARKET ACCOUNT I	MONEY MARKET ACCOUNT II	MONEY MARKET ACCOUNT III
<500	194	190	215
500 and <1000	220	225	231
1000 and <5000	239	215	250
≥5000	178	172	171

```
MTB > CHISQUARE USING C1-C3

Expected counts are printed below observed counts

          ACCT I   ACCT II   ACCT III    Total
    1        194      190       215        599
           199.1    192.2     207.7

    2        220      225       231        676
           224.7    216.9     234.4

    3        239      215       250        704
           234.0    225.8     244.1

    4        178      172       171        521
           173.2    167.1     180.7

Total        831      802       867       2500

ChiSq =    0.13 +   0.02 +    0.25 +
           0.10 +   0.31 +    0.05 +
           0.11 +   0.52 +    0.14 +
           0.13 +   0.14 +    0.52 = 2.43

df = 6
```

12.52 In a survey, consumers are questioned about how often they purchase each of three products. One hundred consumers are randomly selected to participate in the survey. From the MINITAB computer printout, is there sufficient evidence to indicate that there is a relationship between the products purchased and how often the products are purchased at a significance level of .05? At a significance level of .01? Each row represents how often a product is purchased (frequently, occasionally, or never) and each column represents one of the three products.

```
MTB > NAME C1 'ITEM A'
MTB > NAME C2 'ITEM B'
MTB > NAME C3 'ITEM C'
MTB > PRINT C1-C3
ROW  ITEM A  ITEM B  ITEM C
1    30      10      8
2    8       6       11
3    9       7       11

MTB > CHISQUARE USING C1-C3

Expected counts are printed below observed counts

         ITEM A   ITEM B   ITEM C   Total
1        30       10       8        48
         22.6     11.0     14.4

2        8        11       11       25
         11.8     5.7      7.5

3        9        7        11       27
         12.7     6.2      8.1

Total    47       23       30       100

Chisq =  2.45 +  0.10 +  2.84 +
         1.20 +  0.01 +  1.63 +
         1.07 +  0.10 +  1.04 = 10.45

df = 4
```

COMPUTER EXERCISES USING THE DATABASE

Exercise 1—Appendix H

Randomly select 100 observations from the database. Determine whether the variable family size (FAMLSIZE) has a distribution that is significantly different from the binomial distribution. Use a .05 significance level.

Exercise 2—Appendix H

Randomly select 50 observations from the database. Are the categories own or rent one's residence (variable OWNORENT) and family size (variable FAMLSIZE) independent? Use a .05 significance level.

Exercise 3—Appendix I

Randomly select 100 observations from the database. Determine whether the total asset value (variable TOTAL) has a distribution which is significantly different from the normal distribution. Use a .05 significance level. (*Hint:* See Exercise 12.13.)

Exercise 4—Appendix I

Randomly select 100 observations from the database. Are the categories of bond rating (BONDRATE) and positive or negative net income (NETINC) independent? Use a .05 significance level.

CASE STUDY

COMPETITIVE STRATEGIES IN THE HOTEL AND MOTEL INDUSTRY

Competitive strategy is an abstract management concept that has been defined in many ways. One simple view of it is that it is the means through which an organization carries out its objectives by linking with, responding to, integrating with, or exploiting its environment. Research across many industries suggests a limited number of "generic" strategies (archetypes). Also, the nature of an industry will affect the configuration of competitive strategies actually used in that industry.

Schaffer (1987) conducted an empirical study of competitive strategies in the lodging industry, which is segmented into four sectors: transient hotels, resort hotels, motels with restaurants, and motels without restaurants. In order to identify the competitive strategies used by these sectors, a questionnaire was mailed to 386 lodging organizations in North America. Top management executives were asked to specify the relative importance to their firm of 26 strategic methods and characteristics identified by previous research. These 26 strategic approaches were derivatives of archetypes such as defenders, prospectors, analyzers, reactors, or differentiation versus focus versus cost leadership, and other strategies discussed in Schaffer's article.

A total of 101 firms responded to the survey. Their responses were translated from archetypes to actual strategies used by the lodging industry by using advanced statistical techniques such as principal component factor analysis and *K*-means cluster analysis. Schaffer found five distinct strategic configurations.

1. *Do-it-all differential* types emphasize uniqueness and innovation as well as efficiency and quality control.

2. *Internalized resource-controller* types emphasize control of resources and an internalized focus on channels of distribution, raw material purchases, and inventory levels.

3. *Narrow-focused marketing-innovator* types place heavy emphasis on innovative marketing techniques with a narrow product-market focus, deemphasizing quality control, operating efficiency, customer service, etc.

4. *Efficiency/quality-controller* types represent a defensive, cost-leadership strategy with an aversion toward innovation.

5. *Geographically-focused price-leadership* types emphasize serving and becoming known in a limited geographic market.

The distribution of these five competitive profiles cross-classified against the four market segments of the industry are given in the following table.

SEGMENT OF INDUSTRY IN WHICH ORGANIZATION PRIMARILY COMPETES

Strategic Types	Cluster	Transient hotels 1	Resort hotels 2	Motels with restaurants 3	Motels without restaurants 4	Total
Do-it-all differentiation	1	15	6	9	1	31
Internalized resource	2	9	7	10	4	30
Narrow-focused marketing innovator	3	5	0	5	7	17
Efficiency/quality controller	4	2	1	2	7	12
Geographically-focused price leader	5	1	1	1	2	5
Total		32	15	27	21	95

(*Source:* Jeffrey D. Schaffer, "Competitive Strategies in the Lodging Industry," *International Journal of Hospitality Management* 6, no. 1 (1987): 33–42.)

Case Study Questions

1. Strategy theorists have said that the nature of an industry will influence the configuration of competitive strategies in that industry. Thus, having identified five distinct strategic profiles, the researcher was interested in whether a firm's choice of strategy was related to the segment of the industry to which it belonged. Which chi-square test is appropriate for this purpose?

2. Conduct a chi-square test to decide this issue. State the proper hypotheses and your conclusion. Use a 5% level of significance.

3. Does the conclusion change at a 1% level of significance? What is the smallest significance level at which the null hypothesis in your test could be rejected?

4. Which strategy or strategies are most common in the industry as a whole? Which strategies are dominant among resort hotels?

5. The author stated in his article: "The lodging industry is a service industry where customers must come to specially designed facilities to receive the service . . . [L]odging units must be geographically dispersed . . . along travel routes so as to be positioned to be able to service customer travel needs . . . [and] must also attempt to have representation at many destinations and along many travel routes . . . [as] evidenced by the great importance attached to centralized reservation system [in] the lodging industry." What was it about the data that probably led the author to make this observation?

MAINFRAME AND MICRO

■ ■ ■ ■ ■ ■ ■ **S P S S** ■ ■ ■ ■ ■ ■ ■ ■

EXAMPLE 12.4

SOLUTION

Example 12.4 was concerned with a multinomial goodness-of-fit test. One of the expected values was less than 5, and it was necessary to pool the categories and create a new category with an expected value of at least 5. The problem was to determine whether the percentage of sales in the sample community fit the U.S. proportions. The percentages in the null hypothesis were .625, .226, .115, and .034. The SPSS program listing in Figure 12.6 was used to request a computed chi-square value when the categories were pooled.

In this problem the SPSS commands are the same for both the mainframe and PC versions. **(Remember to end each command line with a period when using the PC version.)**

```
TITLE    NEW CAR SALES MIXTURE
DATA LIST FREE/TYPE CARSSOLD
BEGIN DATA
1 95
2 50
3 41
4 14
END DATA
VALUE LABELS TYPE 1 'GA' 2 'K' 3 'L' 4 'OTHER'
WEIGHT BY CARSSOLD
NPAR TESTS CHISQUARE=TYPE/EXPECTED=0.625,0.226,0.115,0.034
```

The TITLE command names the SPSS run.

The DATA LIST command gives each variable a name and describes the data as being in free form.

The BEGIN DATA command indicates to SPSS that the input data immediately follow.

The next four lines contain the data values, with each line representing a type (1 = GA, etc.) and the observed number of cars sold.

The END DATA statement indicates the end of the data.

The VALUE LABELS statement assigns the labels GA to type 1 records, K to type 2 records, L to type 3 records, and OTHER to type 4 records.

The WEIGHT command is used to weight our cases by the number of observed cars sold.

FIGURE 12.6

Input for SPSSX or SPSS/PC+. All command lines should end with a period when the PC version is used.

The NPAR TESTS CHISQUARE = statement requests a chi-square test between the observed and expected sales. For instance, .625 is the expected market share for the variable labeled GA.

Figure 12.7 shows the SPSSX output obtained by executing the listing in Figure 12.6, whereas Figure 12.8 shows the SPSS/PC+ output.

FIGURE 12.7
SPSSX output.

```
- - - - CHI-SQUARE TEST

   TYPE
                          O        E        O - E

                        CASES
         CATEGORY     OBSERVED   EXPECTED   RESIDUAL

GA          1.00          95      125.00     -30.00
K           2.00          50       45.20       4.80
L           3.00          41       23.00      18.00
OTHER       4.00          14        6.80       7.20
                        -----
            TOTAL        200

         CHI-SQUARE      D.F.     SIGNIFICANCE
           29.420          3         0.000
```

$$\chi^2 = \sum \frac{(O-E)^2}{E}$$

FIGURE 12.8
SPSS/PC+ output.

```
                  NEW CAR SALES MIXTURE
- - - - Chi-square Test

   TYPE
                          O        E        O - E

                         Cases
         Category      Observed   Expected   Residual

GA          1.00          95      125.00     -30.00
K           2.00          50       45.20       4.80
L           3.00          41       23.00      18.00
OTHER       4.00          14        6.80       7.20
                        -----
            Total        200

         Chi-Square      D.F.     Significance
           29.420          3         .000
```

$$\chi^2 = \sum \frac{(O-E)^2}{E}$$

MAINFRAME AND MICRO

SOLUTION

EXAMPLE 12.5

Example 12.5 was concerned with a chi-square test of independence. The purpose was to determine whether an employee's educational level had an effect on job performance, as measured by an exam. The null hypothesis was that there is no relationship between educational level and job performance. The SPSS program listing in Figure 12.9 requests the chi-square and p-value statistics from the data obtained by testing 120 employees.

In this problem the SPSS commands are the same for both the mainframe and PC versions. **(Remember to end each command line with a period when using the PC version.)**

The TITLE command names the SPSS run.

The DATA LIST command gives each variable a name and describes the data as being in free form.

FIGURE 12.9

Input for SPSSX or SPSS/PC+. All command lines should end with a period when the PC version is used.

```
TITLE   EDUCATIONAL LEVEL VERSUS JOB KNOWLEDGE
DATA LIST FREE/LEVEL PERFORM COUNT
BEGIN DATA
1 1 4
1 2 20
1 3 11
2 1 12
2 2 18
2 3 15
3 1 9
3 2 22
3 3 9
END DATA
VALUE LABELS LEVEL 1 'MASTERS' 2 'BACHELOR' 3 'HIGH SCHOOL'/
             PERFORM 1 'HIGH' 2 'AVERAGE' 3 'LOW'
WEIGHT BY COUNT
CROSSTABS TABLES=LEVEL BY PERFORM
STATISTICS 1
OPTIONS 14
```

The BEGIN DATA command indicates to SPSS that the input data immediately follow.

The next nine lines contain the data values, representing an educational level code, a performance level code, and the number of observations that comprise the two adjacent categories.

The END DATA statement indicates the end of the data.

The VALUE LABELS statement assigns codes to different categories (LEVEL and PERFORM) of data. These are positional categories. The first position of the input data stream is the LEVEL category, while the second position is the PERFORM category. A 1 in the first position of the data stream indicates that the employee has a master's degree, a 2 indicates a bachelor's degree, and a 3 indicates a high school degree. For the second position of the data stream, a 1 indicates that the employee scored high on the test for job knowledge, 2 indicates average, and 3 indicates low.

The WEIGHT command requests that the data be weighted by the variable COUNT. This variable indicates the total number of employees who satisfy the criteria in the adjacent LEVEL and PERFORM categories. Looking at the first row of data, four employees had master's degrees and scored high on the exam.

FIGURE 12.10   SPSSX output.

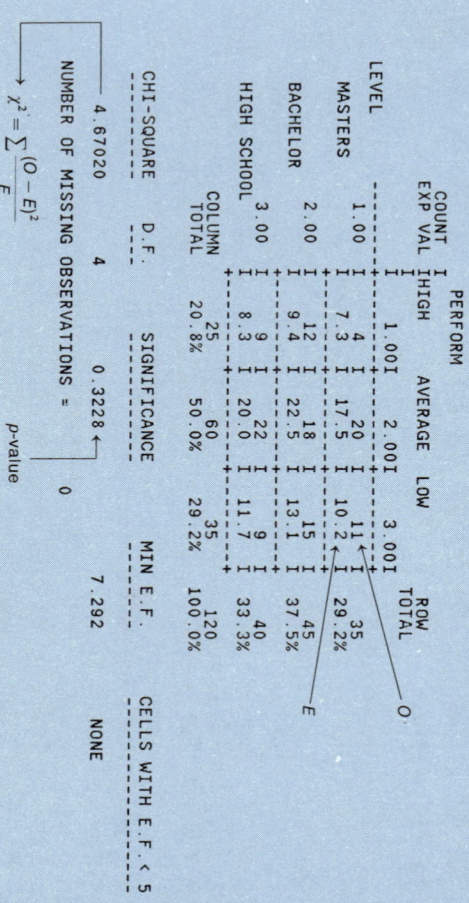

| | | PERFORM | | | |
LEVEL	COUNT EXP VAL	HIGH 1.00	AVERAGE 2.00	LOW 3.00	ROW TOTAL
MASTERS 1.00		4 / 7.3	20 / 17.5	11 / 10.2	35 / 29.2%
BACHELOR 2.00		12 / 9.4	18 / 22.5	15 / 13.1	45 / 37.5%
HIGH SCHOOL 3.00		9 / 8.3	22 / 20.0	9 / 11.7	40 / 33.3%
COLUMN TOTAL		25 / 20.8%	60 / 50.0%	35 / 29.2%	120 / 100.0%

CHI-SQUARE	D.F.	SIGNIFICANCE	MIN E.F.	CELLS WITH E.F. < 5
4.67020	4	0.3228	7.292	NONE

NUMBER OF MISSING OBSERVATIONS = 0

$$\chi^2 = \sum \frac{(O - E)^2}{E}$$

p-value

FIGURE 12.11 SPSS/PC+ output.

```
            EDUCATIONAL LEVEL VERSUS JOB KNOWLEDGE

Crosstabulation:    LEVEL
                 By PERFORM

          Count   HIGH    AVERAGE  LOW        Row
PERFORM->  Exp Val  1.00    2.00    3.00     Total
LEVEL     --------+--------+--------+--------+
  1.00    |    4   |   20   |   11   |   35
MASTERS   |   7.3  |  17.5  |  10.2  |  29.2%
          --------+--------+--------+--------+
  2.00    |   12   |   18   |   15   |   45
BACHELOR  |   9.4  |  22.5  |  13.1  |  37.5%
          --------+--------+--------+--------+
  3.00    |    9   |   22   |    9   |   40
HIGH SCHOOL|  8.3  |  20.0  |  11.7  |  33.3%
          --------+--------+--------+--------+
          Column      25      60      35      120
          Total     20.8%   50.0%   29.2%   100.0%

Chi-Square      D.F.    Significance    Min E.F.   Cells with E.F.< 5
----------      ----    ------------    --------   ------------------
 4.67020          4        .3228          7.292          None

Number of Missing Observations =    0
```

$$\chi^2 = \sum \frac{(O - E)^2}{E}$$

The CROSSTABS command produces a cross-tabulation of the variables LEVEL and PERFORM.

The STATISTICS 1 command requests the chi-square test. (Use /STATISTICS 1 for the PC version.) The OPTIONS 14 command requests the expected frequencies be printed. (Use /OPTIONS 14 for the PC version.)

Figure 12.10 shows the SPSSX output obtained by executing the listing in Figure 12.9, whereas Figure 12.11 shows the SPSS/PC+ output.

▪▪▪▪▪▪▪ S A S ▪▪▪▪▪▪▪▪▪▪

MAINFRAME AND MICRO

SOLUTION

EXAMPLE 12.5

Example 12.5 was concerned with a chi-square test of independence. The purpose was to determine whether an employee's educational level had an effect on job performance, as measured by an exam. The null hypothesis was that there is no relationship between educational level and job performance. The SAS program listing in Figure 12.12 requests the chi-square and *p*-value statistics from the data obtained by testing 120 employees.

FIGURE 12.12
Input for SAS (mainframe or micro version).

```
TITLE 'EDUCATIONAL LEVEL VERSUS JOB KNOWLEDGE';
DATA EXAM PERFORM;
    INPUT LEVEL $ PERFORM $ COUNT@@;
CARDS;
MASTERS HIGH 4 MASTERS AVERAGE 20 MASTERS LOW 11
BACHELORS HIGH 12 BACHELORS AVERAGE 18 BACHELORS LOW 15
HIGHSCHOOL HIGH 9 HIGHSCHOOL AVERAGE 22 HIGHSCHOOL LOW 9
PROC FREQ;
    WEIGHT COUNT;
    TABLES LEVEL*PERFORM/CHISQ;
```

In this problem the SAS commands are the same for both the mainframe and PC versions.

The TITLE command names the SAS run.

The DATA command gives the data a name.

The INPUT command names and gives the correct order for the different fields on the data lines. The $ implies that both LEVEL and PERFORM are character data. The @@ signs indicate that each data line contains two additional sets of data.

The CARDS command indicates to SAS that the input data immediately follow.

The next three lines are data values. The first line, for example, indicates that 4 employees have a master's degree and scored high on the exam, 20 with master's degrees scored average, and 11 with master's degrees scored low.

The PROC FREQ command and WEIGHT COUNT subcommand specify that the values of the variable COUNT are relative weights for the observations.

The TABLES subcommand produces a cross-tabulation of the variables LEVEL and PERFORM.

The CHISQ command generates chi-square statistics.

Figure 12.13 shows the SAS output obtained by executing the listing in Figure 12.12, whereas Figure 12.14 shows the SAS/PC output.

FIGURE 12.13

SAS output.

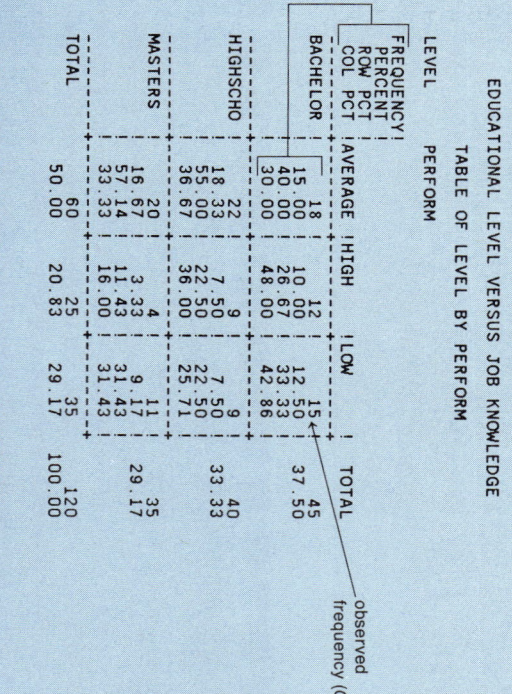

```
EDUCATIONAL LEVEL VERSUS JOB KNOWLEDGE
           TABLE OF LEVEL BY PERFORM

LEVEL      PERFORM

FREQUENCY|
PERCENT  |
ROW PCT  |
COL PCT  |AVERAGE |HIGH    |LOW     |  TOTAL
---------+--------+--------+--------+
BACHELOR |     18 |     12 |     15 |     45
         |  15.00 |  10.00 |  12.50 |  37.50
         |  40.00 |  26.67 |  33.33 |
         |  30.00 |  48.00 |  42.86 |
---------+--------+--------+--------+
HIGHSCHO |     22 |      9 |      9 |     40
         |  18.33 |   7.50 |   7.50 |  33.33
         |  55.00 |  22.50 |  22.50 |
         |  36.67 |  36.00 |  25.71 |
---------+--------+--------+--------+
MASTERS  |     20 |      4 |     11 |     35
         |  16.67 |   3.33 |   9.17 |  29.17
         |  57.14 |  11.43 |  31.43 |
         |  33.33 |  16.00 |  31.43 |
---------+--------+--------+--------+
TOTAL          60       25       35      120
            50.00    20.83    29.17   100.00
```

observed frequency (O)

```
STATISTICS FOR TABLE OF LEVEL BY PERFORM

STATISTIC                        DF      VALUE      PROB
----------------------------------------------------------
CHI-SQUARE                        4      4.670      0.323
LIKELIHOOD RATIO CHI-SQUARE       4      4.986      0.289
MANTEL-HAENSZEL CHI-SQUARE        1      1.094      0.296
PHI                                      0.197
CONTINGENCY COEFFICIENT                  0.194
CRAMER'S V                               0.139

SAMPLE SIZE = 120
```

computed value of chi-square statistic

p-value

SAS

FIGURE 12.14
SAS/PC output.

EDUCATIONAL LEVEL VERSUS JOB KNOWLEDGE

TABLE OF LEVEL BY PERFORM

LEVEL   PERFORM

Frequency Percent Row Pct Col Pct	AVERAGE	HIGH	LOW	Total
BACHELOR	18 15.00 40.00 30.00	12 10.00 26.67 48.00	15 12.50 33.33 42.86	45 37.50
Total	60 50.00	25 20.83	35 29.17	120 100.00

observed
frequency (O) → 15

EDUCATIONAL LEVEL VERSUS JOB KNOWLEDGE

TABLE OF LEVEL BY PERFORM

LEVEL   PERFORM

Frequency Percent Row Pct Col Pct	AVERAGE	HIGH	LOW	Total
HIGHSCHO	22 18.33 55.00 36.67	9 7.50 22.50 36.00	9 7.50 22.50 25.71	40 33.33
Total	60 50.00	25 20.83	35 29.17	120 100.00

EDUCATIONAL LEVEL VERSUS JOB KNOWLEDGE

TABLE OF LEVEL BY PERFORM

LEVEL   PERFORM

Frequency Percent Row Pct Col Pct	AVERAGE	HIGH	LOW	Total
MASTERS	20 16.67 57.14 33.33	4 3.33 11.43 16.00	11 9.17 31.43 31.43	35 29.17
Total	60 50.00	25 20.83	35 29.17	120 100.00

EDUCATIONAL LEVEL VERSUS JOB KNOWLEDGE

STATISTICS FOR TABLE OF LEVEL BY PERFORM

Statistic	DF	Value	Prob
Chi-Square	4	4.670	0.323
Likelihood Ratio Chi-Square	4	4.986	0.289
Mantel-Haenszel Chi-Square	1	1.094	0.296
Phi Coefficient		0.197	
Contingency Coefficient		0.194	
Cramer's V		0.139	

Sample Size = 120

→ p-value

computed value of
chi-square statistic

Simple Linear Regression

A Look Back/Introduction

The early chapters discussed methods of reducing a set of values of one variable to a graph (such as a histogram) or a numerical measure (such as a mean). A **variable** here is the characteristic of the population being measured or observed. For example, variables of interest might be an individual's height or income. The sample then consists of random observations of the variable describing a given population.

In this chapter we discuss the situation in which the population and sample consist of measurements not on *one* variable but on *two*. As a result, we cannot only describe each variable individually but can describe how the two variables are related. The relationship between the two variables can be described using a simple graph or a numerical measure (statistic). We can then use the sample results to form a conclusion about the population from which the sample was obtained. If we believe that a linear relationship exists, the next step is to construct the "best-fitting" line through the points defined by the sample bivariate data.

Finally, we turn our attention to the question of what we are estimating when using a bivariate sample. How can we determine whether a significant linear relationship exists? We answer this by introducing the concept of a **statistical model** and the assumptions behind it. Various tests of hypothesis examine the adequacy of this model (Is it a good one?), and an assortment of confidence intervals measure the reliability of the corresponding estimates using this model.

13.1

BIVARIATE DATA AND CORRELATION

In bivariate data, each observation consists of data on two variables. For example, you obtain a sample of people and record their ages (X) and liquid assets (Y). Or, for each month, you record the average interest rate (X) and the number of new housing starts (Y). These data are *paired*.

Suppose that a real-estate developer is interested in determining the relationship between family income (X, in thousands of dollars) of the local residents and the square footage of their homes (Y, in hundreds of square feet). A random sample of ten families is obtained with the following results:

Income (X)	22	26	45	37	28	50	56	34	60	40
Square footage (Y)	16	17	26	24	22	21	32	18	30	20

Bivariate data can be represented graphically using a **scatter diagram**. In this graph, each observation is represented by a point, where the X axis is always horizontal and the Y axis is vertical. A scatter diagram of the real-estate data is shown in Figure 13.1a. The underlying pattern here appears to be that larger incomes (X) are associated with larger home sizes (Y). This means that X and Y have a **positive relationship**. A **negative relationship** occurs when Y decreases as X increases—for example, when Y is the demand for a particular consumer product and X is the selling price.

We next try to determine whether we can estimate this relationship by means of a straight line. One possible line is sketched in Figure 13.1b, which passes among these points and has a positive slope. To measure the strength of the linear relationship between these two variables, we determine the coefficient of correlation.

Coefficient of Correlation

It is often difficult to determine whether a *significant* linear relationship exists between X and Y by inspecting a scatter diagram of the data. A second procedure is to include a *measure* of this linearity—the sample coefficient of correlation. It is computed from the sample data by combining these pairs of values into a single number, written as r. Thus, the sample **coefficient of correlation**, r, measures the strength of the linear relationship that exists within a sample of

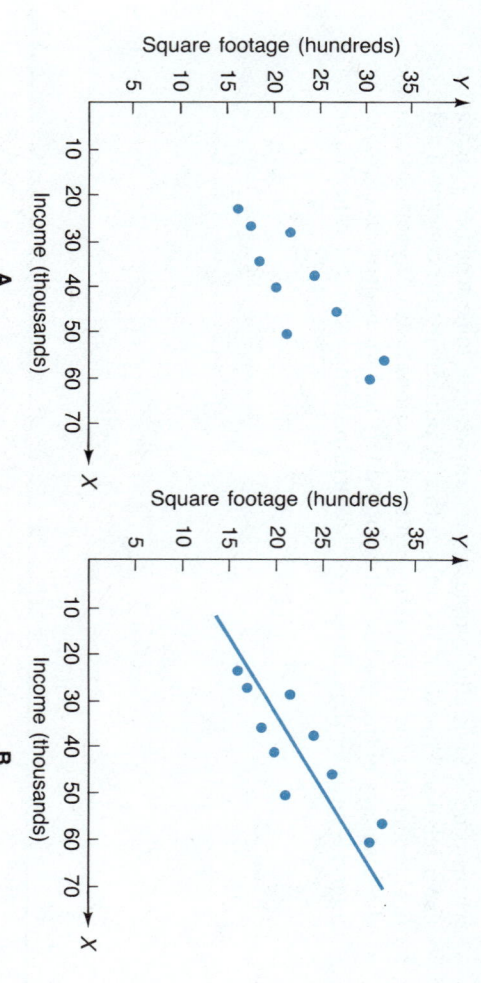

FIGURE 13.1 Scatter diagram of real-estate data. **A:** Scatter diagram of sample data. **B:** Line through sample data.

(Handwritten notes in margin:)

$\hat{Y} = b_0 + b_1 X$

Slope

b_0 = Y intercept

$\sum (y - \hat{y})^2 = \text{min}$

$E(y - \hat{y})^2 = \text{min}$

I. $\sum y = n b_0 + b_1 \sum x$

II. $\sum xy = b_0 \sum x + b_1 \sum x^2$

$b_1 = \dfrac{n \sum xy - (\sum x)(\sum y)}{n \sum x^2 - (\sum x)^2}$

$b_0 = \bar{Y} - [b_1] \bar{X}$

n bivariate data. Its value is given by

$$r = \frac{\sum(x - \bar{x})(y - \bar{y})}{\sqrt{\sum(x - \bar{x})^2}\sqrt{\sum(y - \bar{y})^2}} \qquad (13.1)$$

$$= \frac{\sum xy - (\sum x)(\sum y)/n}{\sqrt{\sum x^2 - (\sum x)^2/n}\sqrt{\sum y^2 - (\sum y)^2/n}} \qquad (13.2)$$

where $\sum x$ = sum of X values, $\sum x^2$ = sum of X^2 values, $\sum y$ = sum of Y values, $\sum y^2$ = sum of Y^2 values, $\sum xy$ = sum of XY values, $\bar{x} = \sum x/n$, and $\bar{y} = \sum y/n$. When using a calculator to determine a coefficient of correlation, equation 13.2 provides a computationally easier procedure. Notice that the summations in the denominator of equation 13.1 are the numerators for the sample variances of X and Y.

Sum of Squares

We will introduce a shorthand notation at this point, related to the notation in Chapter 11 for ANOVA. Let

$$\begin{aligned} SS_X &= \text{sum of squares for } X \\ &= \sum(x - \bar{x})^2 \\ &= \sum x^2 - (\sum x)^2/n \end{aligned} \qquad (13.3)$$

$$\begin{aligned} SS_Y &= \text{sum of squares for } Y \\ &= \sum(y - \bar{y})^2 \\ &= \sum y^2 - (\sum y)^2/n \end{aligned} \qquad (13.4)$$

$$\begin{aligned} SS_{XY} &= \text{sum of squares for } XY \\ &= \sum(x - \bar{x})(y - \bar{y}) \\ &= \sum xy - (\sum x)(\sum y)/n \end{aligned} \qquad (13.5)$$

Using this notation, we can write r as

$$r = \frac{SS_{XY}}{\sqrt{SS_X}\sqrt{SS_Y}} \qquad (13.6)$$

Properties of the Sample Correlation Coefficient The following are some important properties of the sample correlation coefficient, r.

1. r ranges from -1.0 to 1.0.
2. The larger $|r|$ (absolute value of r) is, the stronger is the linear relationship.
3. r near zero indicates that there is no linear relationship between X and Y, and the scatter diagram typically appears to have a shotgun effect (Figure 13.2a). Here, X and Y are uncorrelated.
4. $r = 1$ or $r = -1$ implies that a perfect linear pattern exists between the two variables in the sample, that is, a single line will go *through* each point. Here we say that X and Y are **perfectly correlated** (Figure 13.2b and c).
5. Values of $r = 0$, 1, or -1 are rare in practice. Several other values of the correlation coefficient are illustrated in Figure 13.2d, e, and f.

$SST = 40 = \sum(y - \bar{y})^2$

$\sum(y - \hat{y})^2 = SSE$
unexplained

SSR = Regression Sum of Sq. Explained

Add x's.

$r^2 = 1 - \frac{SSE}{SST}$

coefficient of determination

FIGURE 13.2

Scatter diagrams for various values of the sample correlation coefficient.

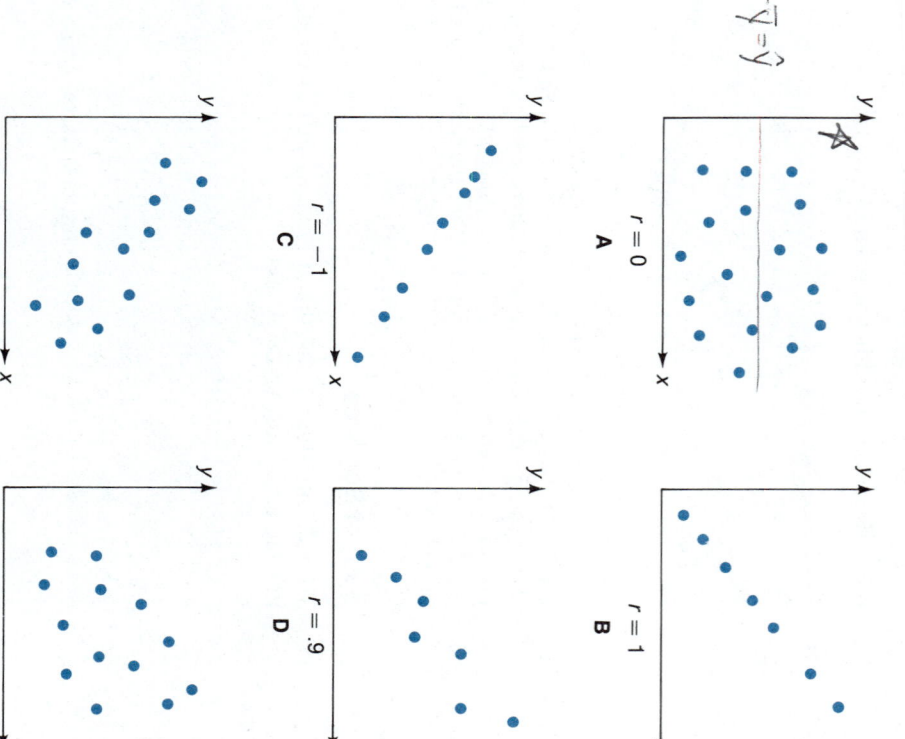

6. The sign of r tells you whether the relationship between X and Y is a positive (direct) or a negative (inverse) one.

7. The value of r tells you nothing about slope of the line through these points (except for the sign of r). If r is positive, the line through these points has positive slope, and, similarly, this line will have negative slope if r is negative. However, a set of data with $r = .9$ will not necessarily have a steeper line passing through it than will a set of data with $r = .4$. All you will observe in the first data set is a set of points that is very close to some straight line with positive slope, but you know nothing (except for the sign) about the slope of this line. See Figure 13.3, where both sets of data provide an r value of .9.

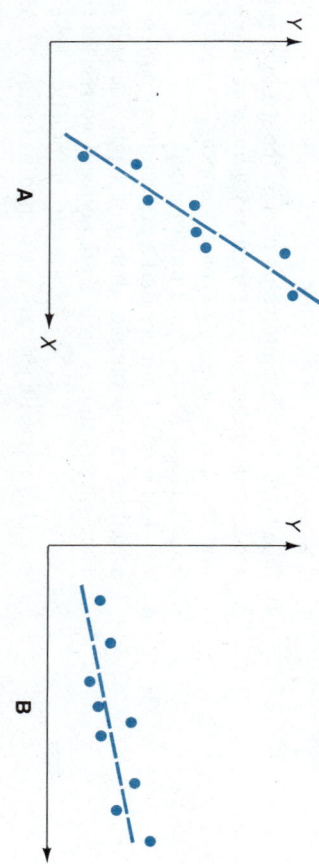

FIGURE 13.3

Although **A** has a large slope and **B** has a small slope, both are scatter diagrams for $r = .9$.

EXAMPLE 13.1

Determine the sample correlation coefficient for the real-estate data in Figure 13.1.

SOLUTION

Your calculations can be organized as follows:

FAMILY	X (INCOME)	Y (SQUARE FOOTAGE)	XY	X²	Y²
1	22	16	352	484	256
2	26	17	442	676	289
3	45	26	1,170	2,025	676
4	37	24	888	1,369	576
5	28	22	616	784	484
6	50	21	1,050	2,500	441
7	56	32	1,792	3,136	1,024
8	34	18	612	1,156	324
9	60	30	1,800	3,600	900
10	40	20	800	1,600	400
	398	226	9,522	17,330	5,370

Using the totals from this table,

$$SS_X = 17,330 - (398)^2/10 = 1489.6$$
$$SS_Y = 5370 - (226)^2/10 = 262.4$$
$$SS_{XY} = 9522 - (398)(226)/10 = 527.2$$

This value of the sample correlation coefficient is

$$r = \frac{SS_{XY}}{\sqrt{SS_X}\sqrt{SS_Y}}$$
$$= \frac{527.2}{\sqrt{1489.6}\sqrt{262.4}} = \frac{527.2}{625.2}$$
$$= .843$$

■

Using the Computer For large data sets, the only reasonable way to obtain a scatter diagram and calculate r is to use a computer. At the end of the chapter, we will show you how to do this using SAS and SPSS. A computer-generated scatter diagram of the real-estate data using MINITAB is contained in Figure 13.4.

Covariance

Another commonly used measure of the association between two variables, X and Y, is the sample covariance, written cov(X, Y). It is similar to the correlation between these two variables. For one thing, the covariance and correlation always have the *same sign*. Consequently, if large values of X are associated with large values of Y, then both the covariance and correlation are positive. Similarly, both values are negative whenever large values of X are associated with small values of Y. For any two variables, X and Y, the sample **covariance** between these variables is

$$\text{cov}(X, Y) = \frac{1}{n-1}\Sigma(x - \bar{x})(y - \bar{y}) \qquad (13.7)$$

$$\text{cov}(X, Y) = \frac{1}{n-1}SS_{XY} \qquad (13.8)$$

FIGURE 13.4

Computer-generated correlation coefficient and scatter diagram for real-estate data using MINITAB.

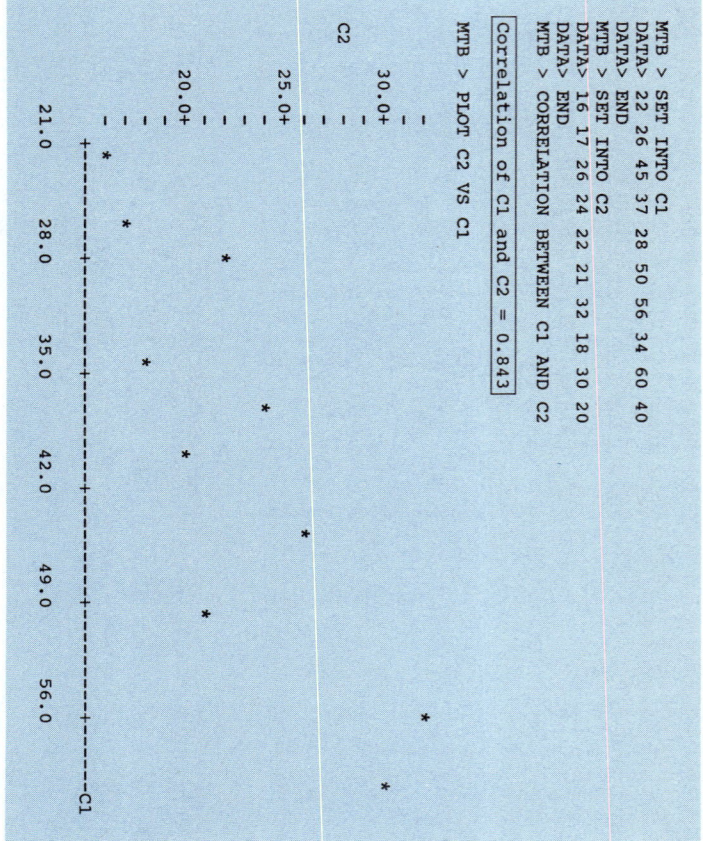

```
MTB > SET INTO C1
DATA> 22 26 45 37 28 50 56 34 60 40
DATA> END
MTB > SET INTO C2
DATA> 16 17 26 24 22 21 32 18 30 20
DATA> END
MTB > CORRELATION BETWEEN C1 AND C2

Correlation of C1 and C2 = 0.843

MTB > PLOT C2 VS C1

C2
     -
30.0+                                        -
     -                              *
     -                        *
     -                     *
25.0+
     -                  *
     -              *
     -           *
20.0+
     -        *
     -     *                                        *
     -
21.0   28.0   35.0   42.0   49.0   56.0 ---------C1
```

In Example 13.1, the covariance between income (X) and home size (Y) is

$$\text{cov}(X, Y) = \frac{1}{n-1} \text{SS}_{xy} = \frac{1}{9}(527.2) = 58.58$$

To see how the sample covariance and sample correlation (r) are related, let

$$s_x = \text{standard deviation of the } X \text{ values} = \sqrt{\frac{\text{SS}_x}{n-1}}$$

and

$$s_Y = \text{standard deviation of the } Y \text{ values} = \sqrt{\frac{\text{SS}_Y}{n-1}}$$

Then

$$r = \text{correlation between } X \text{ and } Y$$
$$= \frac{\text{cov}(X, Y)}{s_X s_Y} \tag{13.9}$$

In Example 13.1,

$$s_x = \sqrt{\frac{1489.6}{9}} = 12.865$$

$$s_Y = \sqrt{\frac{262.4}{9}} = 5.400$$

and so

$$r = \frac{58.58}{(12.865)(5.400)} = .843 \quad \text{(as before)}$$

The correlation between two variables is used more often than the covariance because r always ranges from -1 to 1. The covariance, on the other hand, has no limits and can assume any value. Furthermore, the units of measurement for a covariance are difficult to interpret. For example, the previously calculated covariance is 58.58 (thousands of dollars) · (hundreds of square feet)—a somewhat meaningless unit of measurement. So, in a sense, the correlation is a scaled version of the covariance and has no units of measurement (a nice feature). To illustrate, the sample correlation between body weight and height will be the same whether you use the metric or the English systems to obtain the sample data. The covariance, however, will *not* be the same for these two situations. The covariance does have its applications, however, particularly in financial analyses, such as determining the risk associated with a number of interrelated investment opportunities.

As a final look at these two measures, you can consider the correlation between two variables to be the covariance between the **standardized** variables. By defining

$$X' = \frac{X - \bar{X}}{s_X}$$

and

$$Y' = \frac{Y - \bar{Y}}{s_Y}$$

then

$$\text{cov}(X', Y') = \text{correlation between } X \text{ and } Y = r$$

Least Squares Line

If we believe that two variables do exhibit an underlying linear pattern, how can we determine a straight line that best passes through these points? So far, we have demonstrated only the calculations necessary to compute a correlation coefficient. We next illustrate how to construct a line through a set of points exhibiting a linear pattern; we look at the assumptions behind this procedure in the next section.

Look at the scatter diagram in Figure 13.1b, which shows one possible line through these points. This diagram and the vertical distances from each point to the line $(d_1, d_2, \ldots)$ are contained in Figure 13.5.

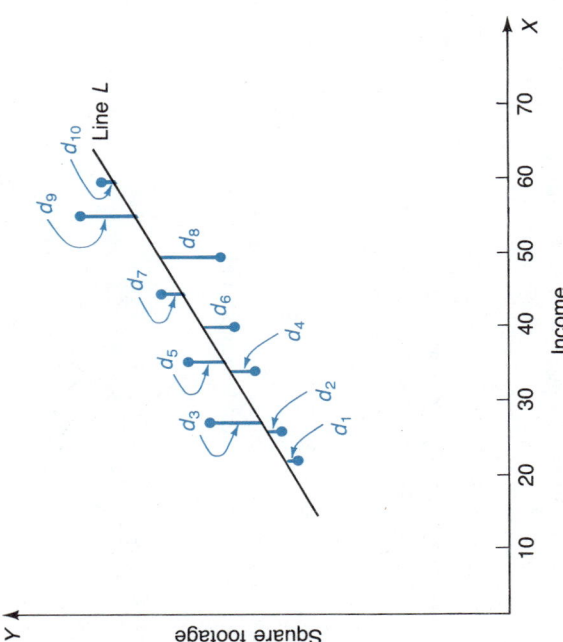

FIGURE 13.5

Vertical distances from line L to real-estate data (Example 13.1), represented by $d_1, d_2, \ldots, d_{10}$.

Is line L the best line through these points? Because we would like the distances $d_1, d_2, \ldots, d_{10}$ to be *small*, we define the best line to be the one that minimizes

$$\Sigma d^2 = d_1^2 + d_2^2 + d_3^2 + \cdots + d_{10}^2 \qquad \text{(13.10)}$$

We square each distance because some of these distances are positive (the point lies *above* line L) and some are negative (the point lies *below* line L). If we did not square each distance, d, the positive d's might cancel out the negative ones. This means that using $(d_1 + d_2 + \cdots + d_{10})$ as a *measure of fit* is *not* a good idea. A better method is to determine which line makes equation 13.10 as small as possible; this is called the **least squares line**. Deriving this line in general requires the use of calculus (derivatives, in particular).*

Because we intend to use this line to predict Y for a particular value of X, we use the notation $\hat{Y}$ (Y-hat) to describe the equation of the line. We can now define, for the least squares line, the b_0 and b_1 that minimize $(d_1^2 + d_2^2 + \cdots + d_n^2)$, given by

$$b_1 = \frac{SS_{XY}}{SS_X} \qquad \text{(13.11)}$$
$$b_0 = \bar{y} - b_1 \bar{x} \qquad \text{(13.12)}$$

where SS_X and SS_{XY} are as defined in equations 13.3 and 13.5. Also, $\bar{x} = \Sigma x/n$ and $\bar{y} = \Sigma y/n$. The resulting least squares line is

$$\hat{Y} = b_0 + b_1 X$$

In Figure 13.6, notice that each distance, d, is actually $Y - \hat{Y}$ and consists of the **residual**, encountered by using the straight line to estimate the value of Y at this point. So

$$\Sigma d^2 = \Sigma (y - \hat{y})^2$$

This term is the **sum of squares of error** (or *residual sum of squares*) and is written **SSE**. Consequently, the least squares line is the one that makes SSE as small as possible.

$$\text{SSE} = \Sigma d^2 = \Sigma (y - \hat{y})^2 \qquad \text{(13.13)}$$

* For the mathematically curious, we provide a condensed derivation of these coefficients. To minimize Σd^2, first write this expression as

$$f(b_0, b_1) = \Sigma d^2 = \Sigma (y - \hat{y})^2$$
$$= \Sigma (y - b_0 - b_1 x)^2$$

To minimize this function, determine the partial derivatives with respect to b_0 (written as f_{b_0} and with respect to b_1 (written as f_{b_1}). These are

$$f_{b_0} = 2\Sigma(y - b_0 - b_1 x)(-1) = -2[\Sigma y - nb_0 - b_1 \Sigma x]$$
$$f_{b_1} = 2\Sigma(y - b_0 - b_1 x)(-x) = -2[\Sigma xy - b_0 \Sigma x - b_1 \Sigma x^2]$$

because $\hat{y} = b_0 + b_1 x$.

Setting $f_{b_0} = f_{b_1} = 0$ and solving for b_0 and b_1 results in equations 13.11 and 13.12.

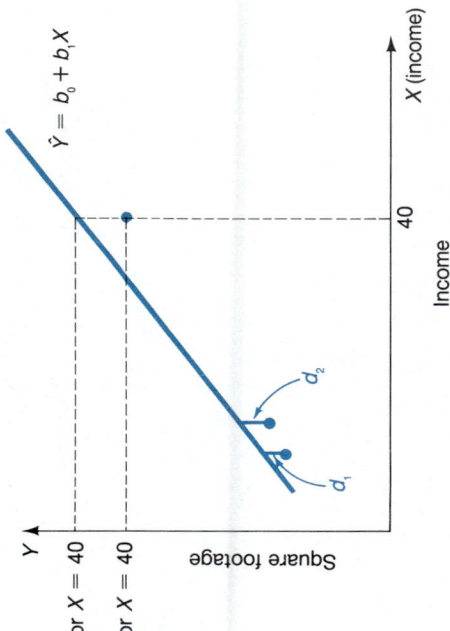

There is another method of determining SSE when using the least squares line, which avoids having to determine the value of $\hat{Y}$ at each point.

$$SSE = SS_Y - \frac{(SS_{XY})^2}{SS_X} \qquad (13.14)$$

EXAMPLE 13.2

Determine the least squares line for the real-estate data we used in Example 13.1. What is the SSE?

SOLUTION

Using the calculations from Example 13.1, $SS_{XY} = 527.2$, $SS_X = 1489.6$, and $SS_Y = 262.4$. This leads to

$$b_1 = \frac{SS_{XY}}{SS_X}$$

$$= \frac{527.2}{1489.6} = .354$$

and

$$b_0 = \bar{y} - b_1\bar{x}$$

$$= 22.6 - (.354)(39.8) = 8.51$$

because

$$\bar{y} = \Sigma y/n$$

$$= 226/10 = 22.6$$

and

$$\bar{x} = \Sigma x/n$$

$$= 398/10 = 39.8$$

So the equation of the best (least squares) line through these points is

$$\hat{Y} = 8.51 + .354X$$

This equation tells us that in the sample data an increase of \$1000 in income (X increases by 1) is accompanied by an increase of 35.4 square feet in home size (Y increases by .354), on the average. For this illustration (and many others in practice), the *intercept*, b_0, has no real meaning because it corresponds to an income of zero dollars. Furthermore, an income of zero is considerably outside the range of the incomes in the sample. It is unsafe to assume that the linear

FIGURE 13.7
Least squares line for
real-estate data
(example 13.2).

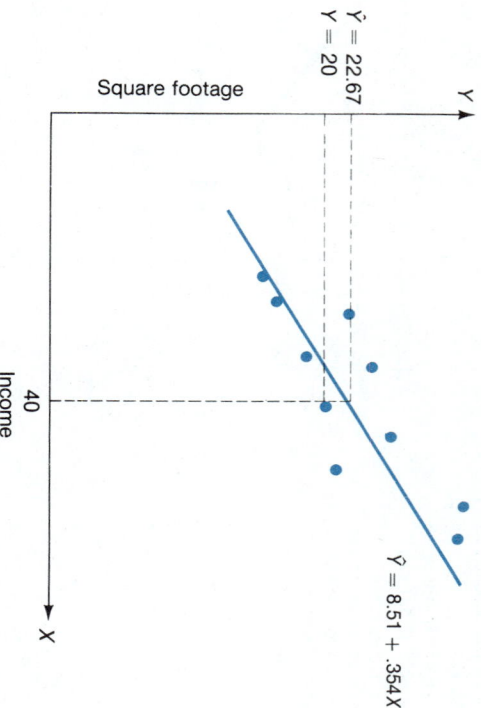

relationship between X and Y present over the range of sample incomes ($22,000
to $60,000) exists outside this range—in particular, all the way to an income
of zero. The slope, b_1, generally is the more informative value.

In Figure 13.7, the actual value of Y (in the sample data) for X = 40 is
Y = 20 (the last pair of X, Y values). The predicted value of Y using the least
squares line is

$$\hat{Y} = 8.51 + .354(40) = 22.67$$

The residual at this point is

$$\text{residual} = Y - \hat{Y} = 20 - 22.67 = -2.67$$

Repeating this for the other nine points leads to the following results. Notice
that the sum of the residuals when using the least squares line is zero. This is
always true.

X	Y	$\hat{Y}$	$Y - \hat{Y}$	$(Y - \hat{Y})^2$
22	16	16.30	– .30	.090
26	17	17.71	– .71	.504
45	26	24.44	1.56	2.434
37	24	21.61	2.39	5.712
28	22	18.42	3.58	12.816
50	21	26.21	–5.21	27.144
56	32	28.33	3.67	13.469
34	18	20.54	–2.54	6.452
60	30	29.75	.25	.063
40	20	22.67	–2.67	7.129
			0	75.81

As you can see, calculating the SSE (=75.81) using the table and equation
13.13 is tedious. Using equation 13.14 instead leads to

$$SSE = 262.4 - \frac{(527.2)^2}{1489.6}$$

$$= 262.4 - 186.59$$

$$= 75.81 \qquad \text{(the same as before)}$$

Remember, however, that equation 13.13 applies to *any* line that you choose
to construct through these points, whereas equation 13.14 applies only to the
SSE for the least squares line. ∎

In Example 13.2, we attempted to predict the size of a home (Y) using the corresponding income (X). The variable Y is the **dependent variable**, and X is the **independent variable**. By passing a straight line through the sample points with Y as the dependent variable, we are **regressing Y on X**. In linear regression, you regress the dependent variable, Y, which you are trying to predict, on the independent (or predictor or explanatory) variable, X.

EXERCISES

13.1 The manager of Hot and Crusty Pizza would like to establish a timetable to give the customers an idea of how long it will take to deliver a pizza. Twelve randomly selected deliveries were used to record the number of miles to the delivery site from Hot and Crusty Pizza and the times from the order to the delivery.

X (DISTANCE, MILES)	Y (DELIVERY TIME, MINUTES)	X (DISTANCE, MILES)	Y (DELIVERY TIME, MINUTES)
2.3	5	8.7	15
6.7	13	9.8	20
7.5	10	10.1	18
3.1	5	6.5	13
4.6	9	7.3	12
3.9	8	5.2	9

a. Draw a scatter diagram of the X and Y values. What would you estimate the coefficient of correlation to be (without calculating it)?

b. Calculate the coefficient of correlation, r.

13.2 Mr. Smart Fellow, the president of Well-Run Car, is examining the nature of the relationship between new car sales and annual advertising expenditure. His administrative assistant gathered the following information from the company's records.

YEAR	Y (NEW CAR SALES)	X (ADVERTISING DOLLARS)	YEAR	Y (NEW CAR SALES)	X (ADVERTISING DOLLARS)
1971	4,000	120,000	1977	6,000	144,000
1972	4,500	127,000	1978	6,100	147,000
1973	4,200	131,000	1979	6,800	152,000
1974	4,800	134,000	1980	7,200	160,000
1975	5,400	140,000	1981	7,800	165,000
1976	5,750	139,000	1982	9,100	170,000

a. Calculate the coefficient of correlation.

b. Find the least squares line.

c. Graph the data and the least squares line, as a check on your calculations.

d. Verify that the sum of the deviations from the least squares line is zero.

e. Interpret the coefficients of the least squares line.

13.3 Tony's used-car lot has been paying car salespeople the highest commission in town. Tony decides to compile data to substantiate his belief that yearly net earnings increase when the car salespeople are highly paid. Fifteen months are chosen:

Y: NET EARNINGS	X: TOTAL COMMISSIONS PAID	Y: NET EARNINGS	X: TOTAL COMMISSIONS PAID
10,780	3,680	11,915	3,161
15,120	5,160	25,160	7,540
18,195	5,180	26,151	8,216
21,690	7,150	18,630	6,051
14,691	5,030	15,551	4,980
16,151	5,210	16,980	5,801
11,015	2,991	24,130	7,160
10,151	3,151		

a. Graph the data and draw a line through them, using the "eyeball" method.

b. Calculate the least squares line. How does it compare to the line in part (a)?

13.4 The supervisor of a group of assembly-line workers wanted to compare last year's productivity (X) to this year's productivity (Y) for each of the 20 employees that she supervises. In the past, an approximate linear relationship has existed between these two variables. The average productivity last year per worker was 9.5 items per hour. This year, the average productivity per worker is 12.1 items per hour. The supervisor found the following sums for her 20 employees:

$$SS_{xy} = 0.4$$
$$SS_x = 0.3$$
$$SS_y = 0.8$$

a. Calculate the correlation coefficient.

b. Calculate the least squares line.

c. Calculate the sum of squares for error.

13.5 Because $b_0 = \bar{y} - b_1\bar{x}$ we can replace b_0 in $\hat{Y} = b_0 + b_1X$ by $\bar{y} - b_1\bar{x}$. Hence, we have $\hat{Y} = \bar{y} + b_1(X - \bar{x})$. From this equation, show that the point $(\bar{x}, \bar{y})$ falls on the least squares line.

13.6 Compare the formulas for the sample correlation, r, and the slope of the least squares line, b_1, and verify that $b_1 = r\sqrt{SS_y/SS_x}$. What can we say about the sign of r and b_1?

13.7 Two hundred ten restaurants were selected from a population of 500 units of a nationally franchised restaurant chain to study the relationship between certain management operations variables and sales volume. A rating for each unit in the study was assigned to the variables management, live entertainment, service hours, advertising, and cleanliness rating. The following table shows the correlation between sales volume and each of these variables.

VARIABLE	CORRELATION WITH SALES VOLUME
Management rating	.461
Live entertainment	.676
Service hours	.325
Advertising rating	.448
Cleanliness rating	.015

(*Source:* Mary Gilly and Ricky Griffen, "Correlates of Success in Franchised Restaurants," *Proceedings of the 1986 Meeting of the Decision Sciences Institute* (1986): 808–10.)

a. In your opinion, which variables are the most important in increasing sales volume?

b. Which variable would perhaps be the best predictor of sales volume?

13.8 The owner of Grandmother's Cake Shop would like to predict the quantity of cakes sold when they are marked at low prices. There are no restrictions on the quantity, because the shop can easily bake several cakes in an hour if the demand is stronger than predicted. Past data show the following results.

Y: NUMBER OF CAKES SOLD	X: PRICE OF CAKE	Y: NUMBER OF CAKES SOLD	X: PRICE OF CAKE
14	2.30	16	1.99
16	2.10	17	1.90
17	1.80	15	2.25
17	1.89	14	2.39
13	2.50	13	2.70
12	2.80		

a. Find the least squares line for X and Y.
b. Graph the data and the least squares line.
c. Suppose that the manager believes that there was a strong linear relationship between Y and X^2. Find the prediction equation for Y using X^2 only.
d. Compare the SSE for the least squares line found in question (a) with the least squares line found in part (c).

13.9 The owner of an ice-cream stand believes that there is a linear relationship between the temperature (X) and the number of ice creams sold (Y). Data are collected during the noon hour every day for 20 days. The average number of ice creams sold during this hour is 35.6 and the average temperature over the 20 days at noon is 87.4. The following sample statistics were collected.

$$SS_{xy} = 8.4$$
$$SS_x = 28.1$$
$$SS_y = 3.9$$

a. Calculate the correlation coefficient.
b. Calculate the least squares line.
c. Calculate the error sum of squares.

13.10 The manager of a city zoo would like to use his staff more efficiently to accommodate large crowds. Fifteen days were randomly selected on which attendance and high temperature for the day were recorded. Do the data indicate a significant correlation between attendance and daily high temperature?

Y (ATTENDANCE, 1000s)	X (HIGH TEMPERATURE, °F)	Y (ATTENDANCE, 1000s)	X (HIGH TEMPERATURE, °F)
1.9	82	2.6	76
0.8	104	0.7	105
1.2	90	1.3	90
1.4	92	1.6	85
2.4	75	2.1	78
2.8	70	1.8	83
1.5	86	2.3	77
1.4	87		

a. Calculate the least squares line. Interpret the coefficients of the least squares line in the context of the problem.
b. Calculate the error sum of squares.
c. Graph the data and draw the least squares line through them.

13.11 It is well known that the federal funds rate influences the yield on 13-week treasury bills. The federal funds rate is the rate at which reserves are traded among commercial banks for overnight use. Treasury bills are short-term government bills sold at an auction at a discount from the face value. The following data were collected.

Y: TREASURY BILL YIELD	X: FEDERAL FUNDS RATE	Y: TREASURY BILL YIELD	X: FEDERAL FUNDS RATE
12.89	14.23	8.79	9.43
12.36	14.51	9.39	9.56
9.71	11.01	9.05	9.45
7.93	9.29	8.71	9.48
8.08	8.65	8.71	9.34
8.42	8.80	8.96	9.47
9.19	9.46		

Find the least squares line and predict what the treasury bill rate would be if the federal funds rate was 9.67. Interpret the coefficient of the X variable in the least squares line.

13.12 The average monthly yield for new issues of 3-month treasury bills (T-Bills) and the average monthly yield on 3-month certificates of deposit (CDs) are recorded from January 1986 to October 1987.

TIME PERIOD	INTEREST RATE FOR 3-MONTH T-BILLS	INTEREST RATE FOR 3-MONTH CDs	TIME PERIOD	INTEREST RATE FOR 3-MONTH T-BILLS	INTEREST RATE FOR 3-MONTH CDs
1986 Jan.	7.04	7.82	1987 Jan.	5.49	5.87
Feb.	7.03	7.69	Feb.	5.59	6.10
Mar.	6.59	7.24	Mar.	5.56	6.17
Apr.	6.06	6.60	Apr.	5.76	6.52
May	6.12	6.65	May	5.75	6.99
Jun.	6.21	6.73	Jun.	5.69	6.94
Jul.	5.84	6.37	Jul.	5.78	6.70
Aug.	5.57	5.92	Aug.	6.00	6.75
Sep.	5.19	5.71	Sep.	6.32	7.37
Oct.	5.18	5.69	Oct.	6.40	8.02
Nov.	5.35	5.76			
Dec.	5.49	6.04			

(*Source: Standard and Poor's Statistical Service*, February 1988, p. 4.)

a. Find the correlation coefficient between the interest rate for 3-month treasury bills and the interest rate for 3-month certificates of deposit.

b. Calculate the least squares line that regresses the interest rate for 3-month treasury bills on the interest rate for 3-month certificates of deposit.

c. Calculate the error sum of squares for the least squares line in part (b).

13.2

THE SIMPLE LINEAR REGRESSION MODEL

When we construct a straight line through a set of data points, we are attempting to predict the behavior of a dependent variable, Y, using a straight line equation with one predictor (independent) variable, X. Examples 13.1 and 13.2 examined the relationship in a particular community between the square footage (Y) of a particular home and the income of the owner (X).

Another application is in attempting to predict the sales (Y) of a certain brand of shampoo using the amount of advertising expenditure (X) as the independent variable. We expect that, as more advertising dollars are spent, the sales will increase. In other words, we expect a *positive* relationship for this situation.

Regression analysis is a method of studying the relationship between two (or more) variables, one purpose being to arrive at a method for predicting a value of the dependent variable. We use only one predictor variable, X, to describe the behavior of the dependent variable, Y. Also, the relationship between X and Y is assumed to be basically linear. In **simple linear regression**, we use only *one* predictor variable, X, to describe the behavior of the dependent variable, Y.

We have learned the mechanics of constructing a line through a set of bivariate sample values. We are now ready to introduce the concept of a statistical model.

Defining the Model

Return to Example 13.2 and Figure 13.4. This set of sample data contained a value of X = 40 and Y = 20. Consider the population of *all* houses in this community where the owner's income is 40 (that is, $40,000). Will they all have the same square footage? Unless this is a very boring-looking neighborhood, certainly not. Does this mean that the straight line predictor is of no use? The answer, again, is no, because very few things in this world are that perfectly predictable. When you use a straight line to predict the square footage, you should be aware that there will be a certain amount of *error* present in this estimate. This is similar to the situation dealing with estimating the mean, μ, of a population where the sample mean, $\bar{X}$, always estimates this parameter with a certain amount of inherent error.

When we elect to use a straight-line predictor, we employ a **statistical model** of the form

$$Y = \beta_0 + \beta_1 X + e \qquad (13.15)$$

where (1) $\beta_0 + \beta_1 X$ is the *assumed* line about which *all* values of X and Y will fall, called the **deterministic** portion of the model, and (2) e is the error component, referred to as the **random** part of the model.

In other words, there exists some (unknown) line about which all X, Y values can be expected to fall. Notice that we said "about which," not "on which"— hence the necessity of the error term, e, which is the unexplained error that is part of the simple linear model. Because this model considers only one independent variable, the effect of other predictor variables (perhaps unknown to the analyst) is contained in this error term.

We emphasize that the deterministic portion, $\beta_0 + \beta_1 X$, refers to the straight line for the *population* and will remain unknown. However, by obtaining a random sample of bivariate data from this population, we are able to estimate the unknown parameters, β_0 and β_1. Thus b_0 is the **intercept** of the sample regression line and is the estimate of the population intercept, β_0. The value of b_0 can be calculated using equation 13.12. Similarly, b_1 is the **slope** of the sample regression line and is the estimate of the population slope, β_1. The value of b_1 can be calculated using equation 13.11.

Assumptions for the Simple Linear Regression Model

We can construct a least squares line through *any* set of sample points, whether or not the pattern is linear. We could construct a least squares line through a set of sample data exhibiting no linear pattern at all. However, to have an effective predictor and a model that will enable us to make statistical decisions, certain assumptions are necessary.

We treat the values of X as fixed (nonrandom) quantities when using the simple linear regression model. For any given value of X, the only source of variation comes from the error component, e, which is a random variable. In fact, there are many random variables here, one for each possible value of X. The assumptions used with this model are concerned with the nature of these random variables.

The first three assumptions are concerned with the behavior of the error component for a fixed value of X. The fourth assumption deals with the manner in which the error components (random variables) affect each other.

Assumption 1 *The mean of each error component is zero.* This is the key assumption behind simple linear regression. Look at Figure 13.8, where we once again examine a value of $X = 40$. If we consider all homes in this community whose owners have an income of $40,000 ($X = 40$), we have already decided the homes do not all have the same square footage, Y. In fact, the square-footage values will be scattered about the (unknown) line $Y = \beta_0 + \beta_1 X$, with some values lying above the line (e is positive) and some falling below it (e is negative). Consider the average of *all* Y values with $X = 40$. This is written as

$$\mu_{Y|40}$$

FIGURE 13.8
Illustration of assumption
1; see text.

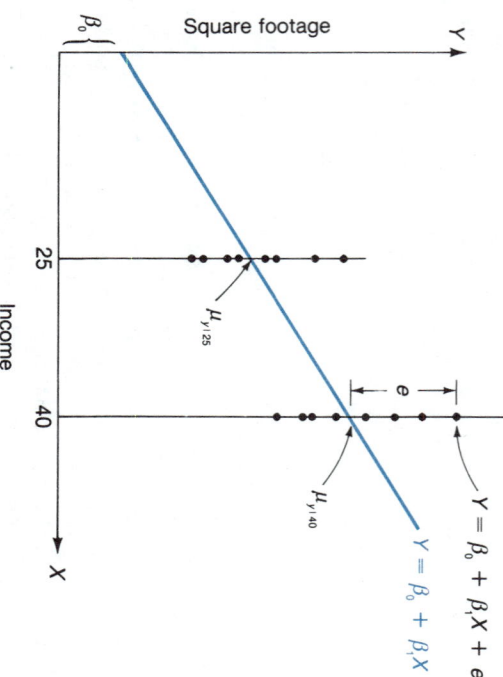

which is the mean of Y *given* X = 40. Our assumption here is that $\mu_{Y|40}$ *lies on this line*; that is, for *any* value of X, $\mu_{Y|X}$ lies on the line $Y = \beta_0 + \beta_1 X$ (such as $\mu_{Y|25}$ in Figure 13.8). Put another way, the error is zero, *on the average.*

Assumption 2   *Each error component (random variable) follows an approximate normal distribution.* In our sample of ten homes and incomes, we had one family with X = 40 and Y = 20. Figure 13.8 illustrates what we might expect if we *were* to examine other homes whose owners had an income of $40,000. We assume here that if we were to obtain 100 homes, for example, whose owners had this income, a histogram of the resulting errors (e) would be bell-shaped in appearance. So we would expect a concentration of errors near zero (from assumption 1), with one-half of them positive and one-half of them negative.

Assumption 3   *The variance of the error component,* σ_e^2, *is the same for each value of X.* For each value of X, the errors illustrated in Figure 13.8 have so far been assumed to follow a normal distribution, with a mean of zero. So each error, e, is from such a normal population. The variance of this population is σ_e^2. The assumption here is that σ_e^2 *does not change* as the value of X changes. This is the assumption of **homoscedasticity.** A situation where this assumption is violated is illustrated in Figure 13.9, where we once again consider what might occur if we *were* to obtain (we will not, actually) many values of Y for X = 25 and also for X = 50. If Figure 13.9 were the result, assumption 3 would be violated because the errors would be much larger (in absolute value) for the $50,000-income homes than they would for the $25,000-income homes. This is **heteroscedasticity,** which does pose a problem when we try to infer results from a linear regression equation.

You might argue that, proportionally, the errors for X = 50 seem about the same as those for X = 25, which means that you would expect larger errors for larger values of X here. If this is the case, the confidence intervals and tests of hypothesis that we are about to develop for the simple linear regression model are *not appropriate.* There are methods of "repairing" this situation, by applying a *transformation* to the dependent variable, Y, such as $\sqrt{Y}$ or log(Y). By using this "new" dependent variable rather than the original Y, the resulting errors often will exhibit a nearly constant variance. Such transformations, however, are beyond the scope of this text.

FIGURE 13.9

A violation of assumption 3; see text.

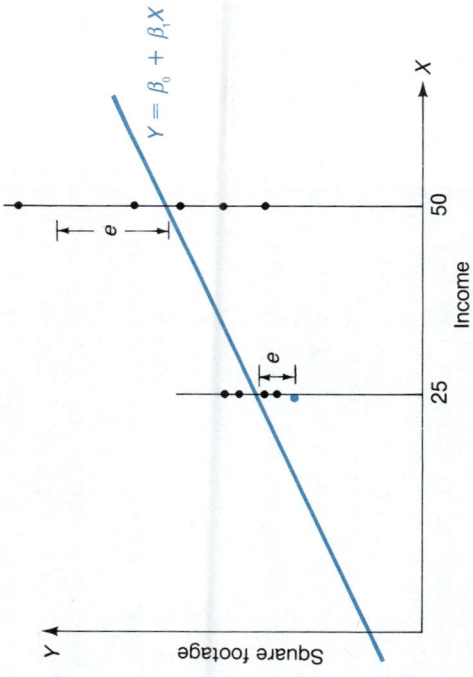

A summary of the first three assumptions is shown in Figure 13.10. Note that the distribution of errors is *identical* for each illustrated value of X; namely, it is a normal distribution with mean = zero and variance σ_e^2.

Assumption 4   *The errors are independent of each other.* This implies that the error encountered for one value of Y is unaffected by the error for any other value of Y. To illustrate, consider the real-estate data and suppose that the sample is *not* random but that instead the sampled houses are all located on a certain street. The first house has a positive error when predicting the square footage. If the probability is greater than .5 that the next house in the sample also has a positive error (that is, if its location makes it probable that it will be a certain size), then the assumption of independence is violated. In other words, the sample was poorly chosen because the houses on one street are likely to be more or less the same size and their owners are likely to have similar incomes. The nonrandom sample led to a violation of assumption 4.

We can draw two conclusions from these assumptions. First, each value of the dependent variable, Y, is a normal random variable with mean = $\beta_0 + \beta_1 X$ and variance σ_e^2. Second, the error components come from the same normal population, *regardless of the value of X*. In other words, it makes sense to examine the residuals resulting from each value of X in the sample, to construct a histogram of these residuals, and to determine whether its appearance is bell-shaped (normal), centered at zero. A key assumption when using simple

FIGURE 13.10

Illustration of assumptions 1, 2, 3; see text.

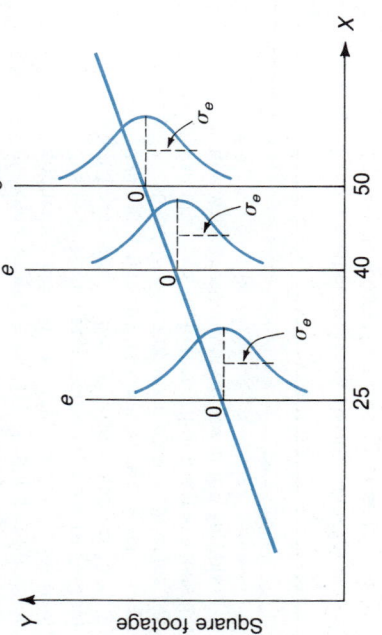

linear regression is that the errors follow a normal distribution with a mean of zero. Constructing a histogram of the sample residuals provides a convenient method of determining whether this assumption is reasonable for a particular application.

We further discuss methods of analyzing the validity of each of these assumptions in Chapter 14, where we learn how to use more than one independent variable in a linear regression equation.

Estimating the Error Variance, σ_e^2

The variance of the error components, σ_e^2, measures the variation of the error terms resulting from the simple linear regression model. The value of σ_e^2 severely affects our ability to use this model as an effective predictor for a given situation. Suppose, for example, that σ_e^2 is very large in Figure 13.10. This means that if we were to obtain many observations (square footage values, Y) for a *fixed* value of X (say, income = $40,000), these Y values would vary a great deal. This decreases the accuracy of our model; we would prefer that these values were grouped closely about the mean, $\mu_{Y|40}$.

In practice, σ_e^2 typically is unknown and must be estimated from the sample. To estimate this variance, we first determine the sum of squares of error, SSE, using SSE = $\Sigma(y - \hat{y})^2$ or equation 13.14. Estimating β_0 and β_1 for the simple regression model results in a loss of 2 df, leaving $n - 2$ df for estimating the error variance. Consequently,

$$s^2 = \hat{\sigma}_e^2 = \text{estimate of } \sigma_e^2 = \frac{\text{SSE}}{n - 2} \qquad (13.16)$$

where

$$\text{SSE} = \Sigma(y - \hat{y})^2 = SS_Y - \frac{(SS_{XY})^2}{SS_X}$$

We can determine the estimate of σ_e^2 and σ_e for the real-estate data in Example 13.2, where we calculated the value of SSE to be 75.81. Our estimate of σ_e^2 is

$$s^2 = \frac{\text{SSE}}{n - 2} = \frac{75.81}{8} = 9.476$$

and so $s = \sqrt{9.476} = 3.078$ provides an estimate of σ_e. The values of s^2 and s are a measure of the variation of the Y values about the least squares line.

We know from the empirical rule that approximately 95% of the data from a normal population should lie within two standard deviations of the mean. For this example, this implies that approximately 95% of the residuals should lie within 2(3.078) = 6.16 of the mean. In the table in Example 13.2, the sample residuals are in the fourth column. Their sum is *always* zero, when using the least squares line; therefore, their mean is zero. So, approximately 95% of the residuals should be no larger (in absolute value) than 6.16. In fact, all of them are less than 6.16—not a surprising result, given that we had only ten values in the sample.

EXERCISES

13.13 A stock broker collected data on company XYZ's quarterly earnings (X) and also on the company's closing price (Y) on the day that the quarterly earnings were reported.

X: QUARTERLY EARNINGS	Y: CLOSING PRICE	X: QUARTERLY EARNINGS	Y: CLOSING PRICE
1.09	10.125	2.0	14.0
1.10	10.0	2.10	14.25
1.12	10.25	2.50	14.37
1.80	10.75	2.85	15.0
1.95	10.5	2.65	14.55

a. Find the least squares line. Then graph the data and the least squares line.

b. Find the residual $(Y - \hat{Y})$ for each value of Y.

c. Is there any indication that the error terms may be correlated?

13.14 The following data were collected for labor hours (X, in hundreds) spent on maintenance and total cost (Y in thousands of dollars) of maintenance.

X: LABOR HOURS	Y: TOTAL COST	X: LABOR HOURS	Y: TOTAL COST
2.1	5.5	4.1	9.4
2.9	6.4	2.3	4.7
4.9	11.2	6.7	14.9
3.8	7.9	7.2	13.3
2.8	6.3	4.8	13.0
1.4	6.2	5.3	12.9
6.1	12.9	5.2	12.2
5.0	13.5	1.2	2.5
6.2	12.8	4.5	8.6
4.3	10.7	3.8	8.4

a. Calculate the least squares line. Interpret the coefficients of the least squares line in the context of the problem.

b. Find the residuals $(Y - \hat{Y})$ for each value of Y.

c. Construct a histogram for the residuals. Do they appear to follow a normal distribution?

13.15 The following data show the number of total annual bankruptcy petitions filed in the northern district of a southern state (in thousands) and the size of the permanent staff at the U.S. Bankruptcy Court for that district.

X (BANKRUPTCIES, IN THOUSANDS)	Y (PERMANENT STAFF AT U.S. BANKRUPTCY COURT)
2.1	15
3.8	18
4.1	18
10.0	59
3.2	14
3.9	18
6.1	24

a. Compute the least squares line.

b. Identify the values of the slope and the intercept for the simple linear regression model.

c. Estimate the variance of the error for the model.

d. Find the residuals $(Y - \hat{Y})$ for all the Y values.

13.16 What assumptions need to be made about the error component of a linear model in order that statistical inference can be used?

13.17 The following is a list of sample errors $(Y - \hat{Y})$ from a linear regression application:

2.1, −.3, 1.4, −2.8, −3.9, 4.2, 3.6, 4.3, 1.8, −2.7, −.8, 1.2, .9, −1.1, −4.5, −5.2, −1.3, .5, .9, −.6, 1.5, 2.1, −2.2, .9

Do the data appear to conform to the empirical rule that approximately 95% of the errors should lie within two standard deviations of the mean? Construct a histogram for the residuals.

13.18   Let X be the distance an employee lives from his or her job. Let Y be the average time that it takes the employee to drive to work. Data from 30 employees gave the following sample statistics.

$$SS_{xy} = 8.4 \quad SS_x = 9.4 \quad SS_y = 12.2$$

a.   Find the estimate of the error variance.

b.   Find the interval in which approximately 68% of the error values should fall.

13.19   The following are residuals resulting from a regression analysis:

$Y - \hat{Y}$	X	$Y - \hat{Y}$	X
.2	1	1.0	4.00
−.2	1.5	−1.5	5.00
−.5	1.75	−1.7	6.00
.6	2.00	2.1	7.00
−.5	2.50	−2.5	8.00
−.8	3.00	3.8	9.00

From these data, where Y is the dependent variable and X is the independent variable, does it appear that any of the standard assumptions of regression analysis are violated?

13.20   Why is $\Sigma(y - \hat{y})^2$ used in estimating the variance of the error term instead of $\Sigma(y - \bar{y})^2$?

13.21   In the statistical model $Y = \beta_0 + \beta_1 X + e$, is X a random variable? Comment on your answer.

13.22   Let X be a person's income. Let Y be the amount of life insurance that this person has. Data from 20 people were collected (in thousands of dollars):

X: INCOME	Y: LIFE INSURANCE	X: INCOME	Y: LIFE INSURANCE
15.4	33.2	28.6	53.7
19.8	39.5	38.7	67.6
20.6	42.2	41.5	75.4
29.4	52.5	40.1	68.3
22.3	44.3	36.5	65.2
19.5	42.3	27.4	51.2
30.8	57.6	28.6	54.9
25.5	49.2	21.4	41.0
20.4	41.6	19.8	40.9
18.4	36.7	20.1	40.2

a.   Calculate the least squares line.

b.   Find the residual $Y - \hat{Y}$ for each value of Y.

c.   Construct a histogram of the residuals.

d.   Do the residuals appear to follow a normal distribution?

13.3

INFERENCE ON THE SLOPE, β_1

Performing a Test of Hypothesis on the Slope of the Regression Line

Under the assumptions of the simple linear regression model outlined in the previous section, we are now in a position to determine whether a linear relationship exists between the variables X and Y. Examining the estimate of the slope, b_1, will provide information as to the nature of this relationship.

Consider the *population* slope, β_1. Three possible situations are demonstrated in Figure 13.11. What can you say about using X as a predictor of Y in Figure 13.11a? When $\beta_1 = 0$, the population line is perfectly horizontal. As

FIGURE 13.11

Three possible population slopes (β_1).

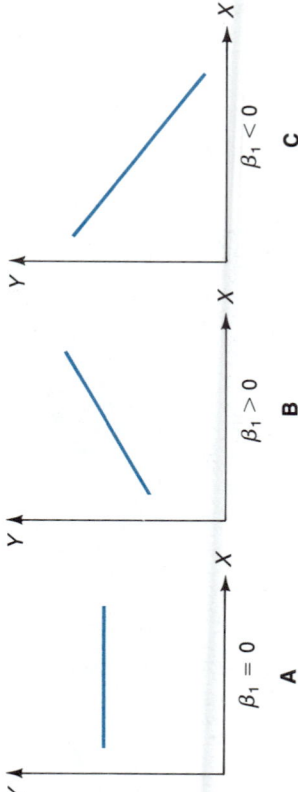

$\beta_1 = 0$

A

$\beta_1 > 0$

B

$\beta_1 < 0$

C

a result, the value of Y is the *same* for each value of X, and so X is not a good predictor of Y; the value of X provides no information regarding the value of Y. In the event $\beta_1 = 0$, the best predictor of Y is given by $\hat{Y} = \bar{y}$, and so $\beta_1 \neq 0$ is equivalent to saying that $\hat{Y}$ (using X as a predictor) is superior to using the sample mean ($\hat{Y} = \bar{y}$) as a predictor.

To determine whether X provides information in predicting Y, the hypotheses are

$H_0: \beta_1 = 0$ (X provides no information)

$H_a: \beta_1 \neq 0$ (X does provide information)

Other Alternative Hypotheses If we are attempting to demonstrate that a significant *positive* linear relationship exists between X and Y, the appropriate alternative hypothesis would be $H_a: \beta_1 > 0$. For example, do the data in Example 13.1 support the hypothesis that owners with large incomes have larger homes?

When the purpose of the analysis is to determine whether a *negative* linear relationship exists between X and Y, the alternative hypothesis should be $H_a: \beta_1 < 0$. For example, you would expect such a relationship between the number of new housing starts (Y) and the interest rate (X). As the interest rate increases, you would expect the number of new houses under construction to decrease.

The Test Statistic We use the point estimate of β_1 (that is, b_1) in the test statistic to determine the nature of β_1. What is b_1? A constant? A variable? Suppose that we obtained a different set of data and recalculated b_1. The new value would not be exactly the same as the previous value, which implies that b_1 is actually a variable. To be more precise, under the assumptions of the previous section, b_1 is a *normal* random variable with mean $= \beta_1$ and variance $= \sigma_{b_1}^2 = (\sigma_e^2)/(\text{SS}_X)$. Notice that b_1 is, on the average, equal to β_1; that is, b_1 is an *unbiased* estimator of β_1. The variance $\sigma_{b_1}^2$ is a parameter describing the variation in the b_1 values if we were to obtain random samples of n observations indefinitely.

If we replace the unknown σ_e^2 by its estimate, s^2, then the *estimated* variance of b_1 is $s_{b_1}^2 = s^2/\text{SS}_X$. As a result

$$t = \frac{b_1 - \beta_1}{s/\sqrt{\text{SS}_X}} = \frac{b_1 - \beta_1}{s_{b_1}} \tag{13.17}$$

has a t-distribution with $n - 2$ df. If the null hypothesis is $H_0: \beta_1 = 0$, the test statistic becomes

$$t = \frac{b_1}{s/\sqrt{SS_x}} \tag{13.18}$$

A summary of the testing procedure is shown in the accompanying box. As usual, for a one-tailed test, the null hypothesis can be written as an inequality (≤ 0 or ≥ 0), or as an equality using the boundary condition ($= 0$).

TEST OF HYPOTHESIS ON THE SLOPE OF THE REGRESSION LINE

TWO-TAILED TEST

$H_0: \beta_1 = 0$
$H_a: \beta_1 \neq 0$

Test statistic:

$$t = \frac{b_1}{s_{b_1}}$$

where $s_{b_1} = s/\sqrt{SS_x}$ and df $= n - 2$.
Test:

reject H_0 if $|t| > t_{\alpha/2, n-2}$

ONE-TAILED TEST

| $H_0: \beta_1 = 0$ (≤ 0) | $H_0: \beta_1 = 0$ (≥ 0) |
| $H_a: \beta_1 > 0$ | $H_a: \beta_1 < 0$ |

Test statistic:

$$t = \frac{b_1}{s_{b_1}}$$

Test statistic:

$$t = \frac{b_1}{s_{b_1}}$$

where $s_{b_1} = s/\sqrt{SS_x}$ and df $= n - 2$.
Test:

reject H_0 if $t > t_{\alpha, n-2}$

where $s_{b_1} = s/\sqrt{SS_x}$ and df $= n - 2$.
Test:

reject H_0 if $t < -t_{\alpha, n-2}$

EXAMPLE 13.3

Is there sufficient evidence, using the real-estate data in Example 13.1, to conclude that a significant positive relationship exists between income (X) and home size (Y)? Use $\alpha = .05$.

SOLUTION

Step 1. The hypotheses indicated here are

$H_0: \beta_1 = 0$
$H_a: \beta_1 > 0$

Step 2. The test statistic is

$$t = \frac{b_1}{s_{b_1}}$$

which has a t distribution with $n - 2 = 8$ df.

Step 3. The testing procedure is to

reject H_0 if $t > t_{.05, 8} = 1.860$

This is illustrated in Figure 13.12.

FIGURE 13.12

FIGURE 13.12

t curve with 8 df showing rejection region (shaded) for Example 13.3.

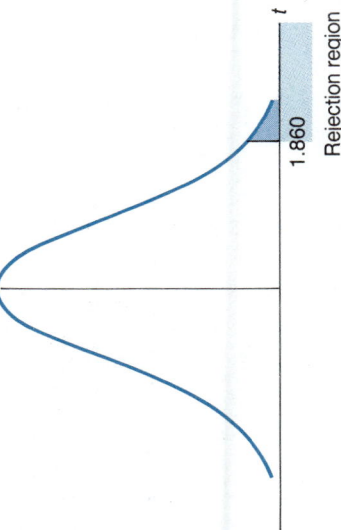

Rejection region

Step 4. We previously determined that $SS_X = 1489.6$, $b_1 = .354$, and $s = 3.078$. The calculated test statistic is then

$$t^* = \frac{.354}{3.078/\sqrt{1489.6}} = \frac{.354}{.0797} = 4.44$$

where $s_{b_1} = .0797$. Because $4.44 > 1.86$, we reject H_0.

Step 5. Based on these ten observations, we conclude that a positive linear relationship does exist between income and home size. ∎

A MINITAB solution using the real-estate data is shown in Figure 13.13. This output contains nearly all the calculations performed so far. In particular,

FIGURE 13.13

MINITAB solution to Example 13.3.

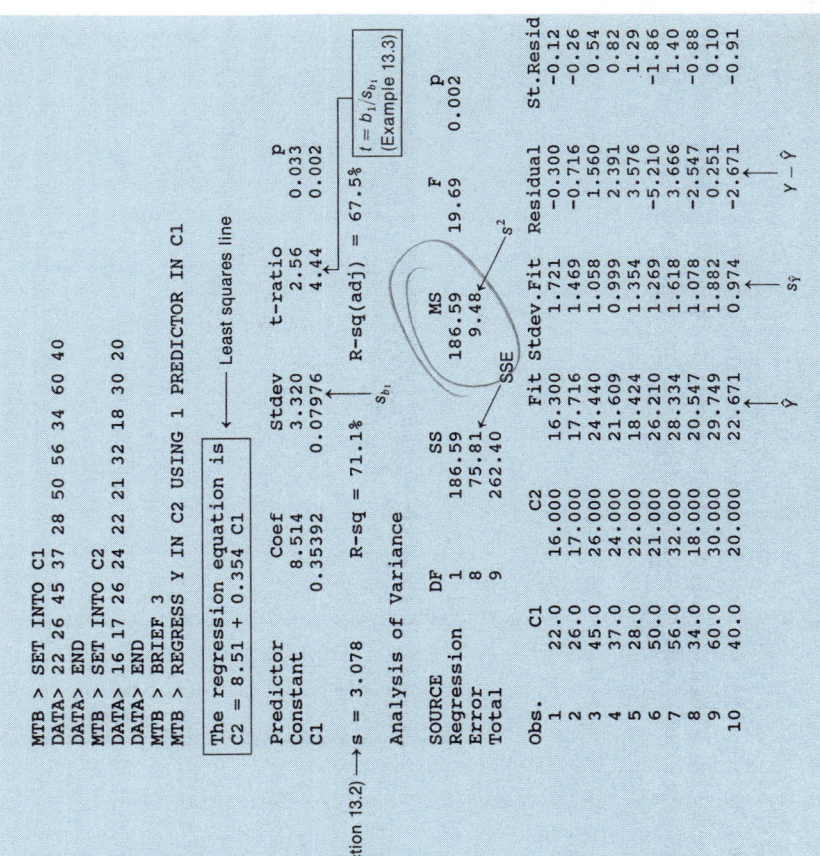

note that:

1. The least squares equation is $\hat{Y} = 8.51 + .354X$.
2. The standard deviation of b_1 is $s_{b_1} = .07976$.
3. The value of the test statistic is $t^* = b_1/s_{b_1} = 4.44$.
4. The standard deviation of the error components is $s = 3.078$.
5. The value of SSE is 75.81, contained in the ANOVA table (construction of this table is discussed in Chapter 14).
6. The column of estimated Y's $(\hat{Y}$'s) and the corresponding residuals are in the column labeled $Y - \hat{Y}$.

EXAMPLE 13.4

The firm of Smithson Financial Consultants has been hired by Blackburn Industries to determine whether a relationship exists between the age of unmarried male Blackburn employees (that is, never married, divorced, or widowed male employees) and the amount of individual liquid assets. The main question of interest is whether a linear relationship exists between these two variables, where X is defined as the age of the employee and Y is the *percentage* of annual income allocated to liquid assets (such as cash, savings accounts, and tradable stocks and bonds). A random sample of 12 unmarried male employees is selected, and the following data are obtained.

AGE (X)	LIQUID ASSETS (Y, PERCENTAGE OF ANNUAL INCOME)	AGE (X)	LIQUID ASSETS (Y, PERCENTAGE OF ANNUAL INCOME)
38	16	58	13
48	12	31	13
38	10	42	20
28	7	35	10
40	9	54	18
50	22	62	25

A scatter diagram of these 12 observations is provided in Figure 13.14, with a summary of the calculations. Using $\alpha = .10$, do you think that an employee's age provides useful information for predicting the percentage of total income allocated to liquid assets?

SOLUTION

To derive the least squares regression line, we determine

$$b_1 = \frac{SS_{XY}}{SS_X} = \frac{447.33}{1268.67} = .3526$$

and

$$b_0 = \bar{y} - b_1\bar{x}$$
$$= 14.583 - (.3526)(43.667) = -.814$$

Consequently, the least squares line is

$$\hat{Y} = -.814 + .3526X$$

Notice that the slope of this line is positive. As the following test of hypothesis will conclude, this slope is significant. Consequently, higher percentage invested in liquid assets is associated with the *older* employees. According to these data, each additional year of age is accompanied by an increase of .35 percent of income allocated to liquid assets, on the average, for the unmarried male population at Blackburn.

To carry out a test of hypothesis, we follow the usual five-step procedure.

FIGURE 13.14

Scatter diagram of annual income (X) and percentage of annual income invested in liquid assets (Y).

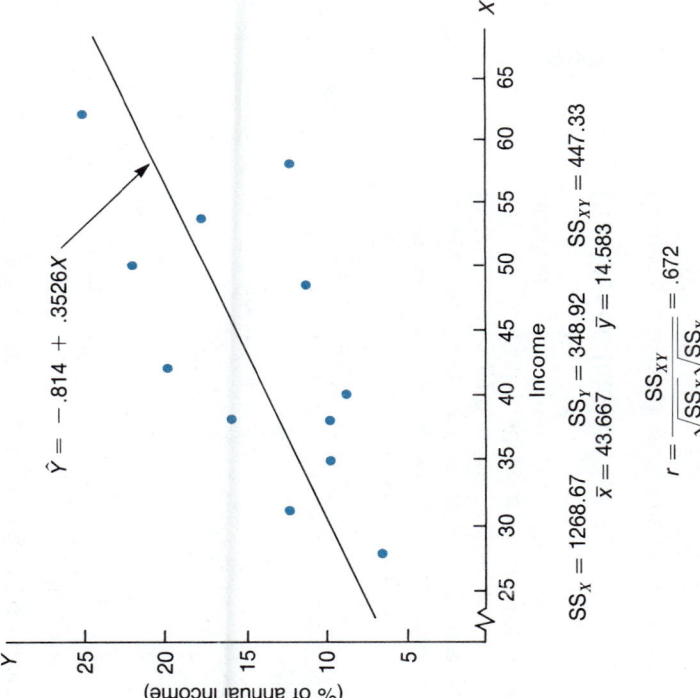

$$\hat{Y} = -.814 + .3526X$$

$SS_X = 1268.67 \qquad SS_Y = 348.92 \qquad SS_{XY} = 447.33$

$\bar{x} = 43.667 \qquad \bar{y} = 14.583$

$$r = \frac{SS_{XY}}{\sqrt{SS_X}\sqrt{SS_Y}} = .672$$

Step 1. Because the suspected direction of the relationship between these two variables (positive or negative) is unknown before the data are obtained, a two-tailed test is appropriate. The hypotheses are

$$H_0: \beta_1 = 0$$
$$H_a: \beta_1 \neq 0$$

Step 2. The test statistic is $t = b_1/s_{b_1}$, which has $n - 2 = 10$ df.

Step 3. The test procedure is to

reject H_0 if $|t| > t_{.10/2,10} = t_{.05,10} = 1.812$

Step 4. Based on the data summary in Figure 13.14 and using equation 13.14,

$$SSE = SS_Y - \frac{(SS_{XY})^2}{SS_X}$$

$$= 348.92 - \frac{(447.33)^2}{1268.67}$$

$$= 348.92 - 157.73$$

$$= 191.19$$

Consequently,

$$s^2 = \frac{SSE}{n-2} = \frac{191.19}{10} = 19.12$$

and so

$$s_{b_1} = \frac{s}{\sqrt{SS_X}} = \frac{\sqrt{19.12}}{\sqrt{1268.67}} = .1228$$

This means that the computed value of the test statistic is

$$t* = \frac{b_1}{s_{b_1}}$$

$$= \frac{.3526}{.1228} = 2.87$$

Because $t* = 2.87$ exceeds the table value of 1.812, we reject H_0 in support of H_a.

Step 5. Our conclusion is that age is a useful (although imperfect) predictor of percentage of income invested in liquid assets for this particular population.

One thing to keep in mind is that *statistical* significance does not always imply *practical* significance. In other words, rejection of $H_0: \beta_1 = 0$ (statistical significance) does not mean that precise prediction (practical significance) follows. It *does* demonstrate to the researcher that, within the sample data at least, this particular independent variable has an association with the dependent variable. ■

Confidence Interval for β_1

Following our usual procedure of providing a confidence interval with a point estimate, we use the t distribution of the previous test statistic and equation 13.17 to define a confidence interval for β_1. The narrower this confidence interval is, the more faith we have in our estimate of β_1, and in our model as an accurate, reliable predictor of the dependent variable. A $(1 - \alpha) \cdot 100\%$ confidence interval for β_1 is

$$b_1 - t_{\alpha/2, n-2} s_{b_1} \quad \text{to} \quad b_1 + t_{\alpha/2, n-2} s_{b_1}$$

EXAMPLE 13.5   Construct a 90% confidence interval for the population slope, β_1, using the real-estate data in Example 13.1.

SOLUTION   All the necessary calculations have been completed; $b_1 = .354$ and $s_{b_1} = .0797$ (from Example 13.3). Using $t_{.05, 8} = 1.860$, the resulting confidence interval is

$$.354 - (1.860)(.0797) \quad \text{to} \quad .354 + (1.860)(.0797)$$

$$= .354 - .148 \quad \text{to} \quad .354 + .148$$

$$= .206 \quad \text{to} \quad .502$$

So we are 90% confident that the value of the estimated slope ($b_1 = .354$) is within .148 of the actual slope, β_1. The large width of this interval is due in part to the lack of information (small sample size) used to derive the estimates; a larger sample would decrease the width of this confidence interval. ■

COMMENTS   A failure to reject H_0 when performing a hypothesis test on β_1 does not always indicate that no relationship exists between the two variables. Some form of nonlinear relationship may exist between these variables. For example, in Figure 13.15, there is clearly a strong curved (**curvilinear**) relationship between X and Y. However, the least squares line through these points is flat, leading to a small t-value and a failure to reject H_0. Furthermore, the sample correlation coefficient, r, for these data is zero.

Of course, you may fail to reject H_0 as the result of a type II error. In other words, you failed to reject H_0 when in fact a significant linear relationship does exist. This situation is more apt to occur when using a small sample to test the null hypothesis.

FIGURE 13.15

Curvilinear relationship.
The horizontal line is the
least squares line.

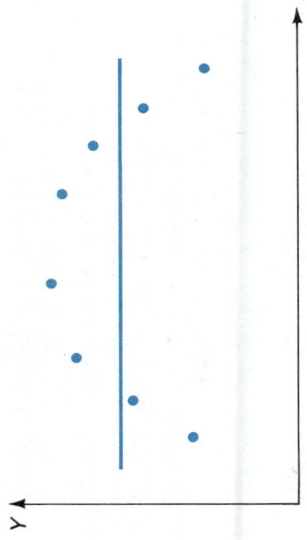

More often, a failure to reject H_0 occurs when there is no visible relationship between the two variables within the sample data. To determine whether there is no relationship or that there is a nonlinear one, you should inspect either a scatter diagram of the data, a scatter diagram of the residuals, or, better yet, both. The latter diagram is a picture of the residuals $(Y - \hat{Y})$ plotted against the independent variable, X. Residual plots are discussed further in Chapter 14.

In many situations, a business analyst has the opportunity to select the values of the independent variable, X, *before* the sample is obtained. At first glance, it might appear that the accuracy of our model is unaffected by the X values. This is partially but not completely true. Because a narrow confidence interval for β_1 lends credibility to our model, we may choose to decrease the width of this confidence interval by decreasing s_{b_1}. Now, $s_{b_1} = s/\sqrt{SS_X}$, so, if we make SS_X large, the resulting s_{b_1} will be small. Therefore, given the opportunity, select a set of X values having a *large variance*. You can accomplish this by choosing a great many X values on the lower end of your range of interest, a large number of values at the upper end, and some values in between to detect any curvature that exists (as in Figure 13.15).

EXERCISES

13.23 A banker is interested in the relationship between a person's income and the amount of money the person has in tax-free investment instruments (such as municipal bonds or IRAs). Data on 20 working individuals were collected. The results are as follows (in thousands of dollars):

X: INCOME	Y: MONEY IN TAX-FREE INVESTMENTS	X: INCOME	Y: MONEY IN TAX-FREE INVESTMENTS
20.2	5.1	24.1	5.2
33.2	7.5	25.1	4.4
35.1	8.1	34.2	7.1
29.4	6.7	33.0	5.1
33.0	7.4	45.1	12.4
40.1	9.4	41.0	9.8
41.0	9.7	45.1	8.9
45.1	10.1	40.1	8.8
42.3	9.1	31.2	6.9
45.3	11.4	24.0	4.2

a. Is there sufficient evidence using the observed data to conclude that a positive relationship exists between X and Y? Use a 5% significance level.

b. Find the p-value for the test statistic in question a. What is your conclusion based on this value?

13.24 The regression equation $\hat{Y} = 2.3 + 1.5X$ was arrived at by fitting a least squares line to 25 data points. The standard deviation (error) of the estimate of the slope was found to be 0.812. Test the null hypothesis at the .01 level of significance that the slope of the line is equal to zero.

13.25 It is believed that the size of the U.S. population (X) is a variable that influences personal consumption expenditure for housing (Y). However, the relationship historically does not appear to be linear. Therefore, a log transformation of housing expenditure is used. Fifteen observations are taken over previous years. The units of Y are millions and the units of X are billions.

X	LOG Y	X	LOG Y	X	LOG Y
183.69	3.935	196.56	4.241	207.66	4.631
186.54	4.001	198.71	4.305	209.90	4.722
189.24	4.060	200.71	4.379	211.91	4.818
191.89	4.117	202.68	4.465	213.85	4.923
194.30	4.182	205.05	4.542	215.97	5.009

From the data, does there appear to be a significant positive relationship between X and log Y? Use a significance level of .05.

13.26 The life of a lawn-mower engine can be extended by frequent oil changes. An experiment was conducted in which 20 lawn mowers were used over many years with different time intervals between oil changes. Let X be the number of hours of operation between oil changes. Let Y be the number of years that the engine was able to perform adequately.

X	Y	X	Y	X	Y
11.25	12.1	22.0	9.5	25.5	6.1
15.5	11.8	22.5	9.2	26.0	5.4
17.5	11.5	23.0	8.4	26.5	4.8
20.5	10.1	23.5	8.8	27.0	4.6
19.5	9.9	24.0	7.1	28.0	4.8
18.5	9.7	24.5	7.2	30.0	4.1
21.5	10.1	25.0	5.8		

a. Graph the data and the least squares line.

b. Is there sufficient evidence to conclude, at the 10% significance level, that a negative relationship exists between Y and X? What is the critical region?

13.27 Disposable personal income has risen steadily over the past two decades. The data given below from 1977 to 1986 is in units of billions of dollars.

YEAR	DISPOSABLE PERSONAL INCOME	YEAR	DISPOSABLE PERSONAL INCOME
1977	1314.0	1982	2180.5
1978	1474.0	1983	2428.1
1979	1650.2	1984	2668.6
1980	1828.9	1985	2841.1
1981	2041.7	1986	3022.1

(*Source: The World Almanac and Book of Facts*, 1988, pp. 111–12)

a. Plot the data, letting the variable time be on the X axis and be equal to 1, 2, . . . , 10 for years 1977, 1978, . . . , 1986.

b. Calculate the least squares line, letting time be the independent variable.

c. Is there sufficient evidence to conclude at the .01 significance level that a positive relationship exists between the independent variable and the dependent variable?

d. What assumptions about the data are necessary to ensure that the test procedure in part (c) is valid?

13.28 A medical researcher was interested in the amount of weight loss caused by a particular diuretic. In a controlled experiment with 18 rats, the amount of weight loss was recorded after 1 month of daily dose of the diuretic. Let X be the amount, in milligrams, of diuretic given. Let Y be the weight loss in pounds.

X	Y	X	Y	X	Y
.10	.05	.25	.35	.40	.44
.10	.08	.25	.31	.40	.47
.15	.11	.30	.41	.45	.51
.15	.13	.30	.42	.45	.52
.20	.19	.35	.43	.50	.54
.20	.21	.35	.42	.50	.53

Is there sufficient evidence to conclude that a significant positive relationship exists between the amount of diuretic given and the amount of weight loss? Use a significance level of 10%. Find a 90% confidence interval for the slope of the regression equation used to predict Y.

13.29 Using the data in Exercise 13.4, find a 95% confidence interval for the slope of the regression equation used to predict the current year's productivity from the previous year's productivity for each employee.

13.30 An investment counselor wanted to know the relationship between the price: earnings ratio (Y) and the yield (X) for high-yield stocks. If a stock yielded over 5.5%, it was considered to be a high-yield stock. Twenty-five high-yield stocks were randomly selected. The following sample statistics were found:

$$SS_{xy} = -10.4$$
$$SS_x = 11.4$$
$$SS_y = 21.4$$

a. Test that the slope of the regression equation used to predict the price: earnings ratio from the yield of a stock is negative. Use a 10% significance level.

b. Find a 95% confidence interval for the slope in question a.

13.31 In Exercise 13.9, find the 90% confidence interval for the slope of the regression equation used to predict the number of ice creams sold by using the independent variable, temperature.

13.32 A survey of the students of Highpoint College gathered the following information with regard to their study time (hours per week) and grade point averages.

Y (STUDY TIME)	X (GRADE-POINT AVERAGE)	Y (STUDY TIME)	X (GRADE-POINT AVERAGE)
16	4.0	8	2.2
15	3.8	6	1.5
14	3.5	4	1.0
12	3.0	2	0.5
10	2.8	0	0.2

Find the 95% confidence interval for the slope of the regression equation used to predict study time.

13.4

MEASURING THE STRENGTH OF THE MODEL

We have already used the sample coefficient of correlation, r, as a measure of the amount of linear association within a sample of bivariate data. The value of r is given by

$$r = \frac{SS_{XY}}{\sqrt{SS_X}\sqrt{SS_Y}} \qquad (13.19)$$

The possible range for r is -1 to 1.

Comparing the equations for r and b_1, we see that

$$r = b_1 \sqrt{\frac{SS_X}{SS_Y}}$$

Because SS_X and SS_Y are *always greater than zero*, r and b_1 have the same sign. Thus, if a positive relationship exists between X and Y, then both r and b_1 will be greater than zero. Similarly, they are both less than zero if the relationship is negative.

When you determine r, you use a sample of observations; r is a *statistic*. What does r estimate? It is actually an estimate of ρ (rho, pronounced "roe"), the **population correlation coefficient**. To grasp what ρ is, imagine obtaining *all possible X, Y* values and using equation 13.19 to determine a correlation. The resulting value is ρ.

The population slope, β_1, and ρ are closely related. In particular, $\beta_1 = 0$ if and only if $\rho = 0$. This leads to another method of determining whether the simple linear regression model (using X to predict Y) is satisfactory. The hypotheses are

$$H_0: \rho = 0 \qquad \text{(no linear relationship exists between } X \text{ and } Y\text{)}$$

$$H_a: \rho \neq 0 \qquad \text{(linear relationship does exist)}$$

The test statistic uses the point estimate of ρ (that is, r) and is defined by

$$t = \frac{r}{\sqrt{\dfrac{1 - r^2}{n - 2}}} \qquad\qquad (13.20)$$

where $n =$ the number of observations in the sample. This is also a t-statistic with $n - 2$ df. Although equations 13.18 and 13.20 appear to be unrelated, the two are algebraically equivalent and *their values are always the same*.

Thus, the t-test for $H_0: \beta_1 = 0$ and $H_0: \rho = 0$ produce identical results, provided both tests use the same level of significance. These tests are therefore redundant; they both produce the same conclusion. Remember, if you have already computed the sample correlation coefficient, r, equation 13.20 offers a much easier method of determining whether the simple linear model is statistically significant. Notice also in equation 13.20 that the significance of the t-value depends on the sample size, n. As a result, if the sample size is large enough, then virtually any value of r can produce a "significantly large" value of t.

EXAMPLE 13.6

Use equation 13.20 to determine whether a positive linear relationship exists between $X =$ income and $Y =$ home square footage, based on the real-estate data from Example 13.1. Use $\alpha = .05$.

SOLUTION

The hypotheses to be used here are $H_0: \rho \leq 0$ versus $H_a: \rho > 0$. In Example 13.1, we found that $r = .843$. This leads to a computed test statistic value of

$$t^* = \frac{r}{\sqrt{\dfrac{1 - r^2}{n - 2}}} = \frac{.843}{\sqrt{\dfrac{1 - (.843)^2}{8}}}$$

$$= \frac{.843}{.190} = 4.44$$

Because this value is the same as the one obtained in Example 13.3 (testing $H_0: \beta_1 \leq 0$ versus $H_a: \beta_1 > 0$), we draw the same conclusion. A positive linear relationship *does* exist between these two variables. In other words, r is large enough to justify this conclusion. ∎

Remember, there is no harm in using equation 13.20 as a substitute for equation 13.18 with $H_0: \beta_1 = 0$ (or ≤ 0, or ≥ 0), particularly if you have already determined the value of r.

Danger of Assuming Causality

A word of warning is in order here—namely, that high statistical correlation does not imply *causality*. Even if the correlation between X and Y is extremely high (say, $r = .95$), a unit increase in X does not necessarily *cause* an increase in Y. All we know is that in the sample data, as X increased, so did Y. As a simple example, consider X = percentage of gray hairs and Y = blood pressure. One might expect to observe a high correlation between these two variables, but it is probably absurd to say that an additional gray hair will *cause* a person's blood pressure to increase. What is actually happening is that there is another variable, in this case age, that is causing both percentage of gray hair and blood pressure to increase.

In many business and economics applications, we observe highly correlated variables when each pair of observations corresponds to a particular time period. For example, we would expect a high correlation between average annual wages (X) and the U.S. gross national product (GNP; Y) when measured over time. Even though wages may be a good predictor of GNP, this does not imply that an increase in wages *causes* an increase in GNP. It is much more likely that a third factor—inflation—caused both wages and GNP to increase.

Coefficient of Determination

In our earlier discussion of ANOVA techniques, we used the expression $SS(\text{total}) = \Sigma(y - \bar{y})^2$ to measure the tendency of a set of observations to group about the mean. If this value was large, then the observations (data) contained much variation and were *not* all clustered about the mean, $\bar{y}$.

In the simple linear regression model, $SS_Y = \Sigma(y - \bar{y})^2$ is computed in the same way and (as before) measures the total variation in the values of the dependent variable.

$$SS_Y = \text{total variation of the dependent variable observations}$$

When comparing the sum of squares of error, SSE, to the total variation, SS_Y, we use the ratio SSE/SS_Y. If all $\hat{Y}$ values are equal to their respective Y values, there is a perfect fit, with SSE = 0 and $r = 1$ or -1. Our model explains 100% of this total variation, and the unexplained variation is zero.

In general, SSE/SS_Y (expressed as a percentage) is the **percentage of unexplained variation**. Recall from equations 13.14 and 13.19 that

$$SSE = SS_Y - \frac{(SS_{XY})^2}{SS_X} \quad \text{and} \quad r^2 = \frac{SS_{XY}^2}{SS_X SS_Y}$$

Thus

$$r^2 = 1 - \frac{SSE}{SS_Y}$$

As a result, r^2 may be interpreted as a measure of the *explained variation* in the dependent variable using the simple linear model; r^2 is the **coefficient of determination**.

r^2 = coefficient of determination

$$= 1 - \frac{SSE}{SSY}$$

= the percentage of explained variation in the dependent
variable using the simple linear regression model   **(13.21)**

For this model, we can determine r^2 simply by squaring the coefficient of
correlation. In Chapter 14, we will predict the dependent variable, Y, using *more
than one* predictor (independent) variable. To derive the coefficient of determination for this case, we must first calculate SSE and then use equation 13.21. So,
although this definition may appear to be unnecessary, it will enable us to compute this value when we use a multiple linear regression model.

EXAMPLE 13.7

What percentage of the total variation of the home sizes is explained by means
of the single predictor, income, using the real-estate data from Example 13.1?

We previously calculated r to be .843, so the coefficient of determination is

$$r^2 = (.843)^2 = .71$$

Therefore, we have accounted for 71% of the total variation in the home sizes
by using income as a predictor of home size.

Notice that we could have determined this value by using the calculations
from Examples 13.1 and 13.2, where

$$r^2 = 1 - \frac{SSE}{SSY}$$
$$= 1 - \frac{75.81}{262.4} = .71$$

∎

Total Variation, SS_Y

In Chapter 11, when discussing the ANOVA procedure, the total variation in
the observations, measured by SS(total), was partitioned into two sums of
squares—namely, SS(factor) and SS(error). The resulting equation was

$$SS(total) = SS(factor) + SS(error)$$

In a similar fashion, we can partition the total variation of the Y values in
linear regression, measured by SS_Y, into two other sums of squares. In Figure
13.16, notice that the value of $y - \bar{y}$ can be written as the sum of two deviations,
namely,

$$y - \bar{y} = (\hat{y} - \bar{y}) + (y - \hat{y})$$

By squaring and summing over *all* the data points in the sample, we can
show that*

$$\Sigma(y - \bar{y})^2 = \Sigma(\hat{y} - \bar{y})^2 + \Sigma(y - \hat{y})^2$$

The summation on the left of the equal sign is SS_Y. The second summation
on the right is the sum of squares of error, SSE. The first summation on the
right is defined to be the **sum of squares of regression, SSR**.

* This result follows since it can be shown that $\Sigma(\hat{y} - \bar{y})(y - \hat{y}) = 0$ when using the least
squares line.

FIGURE 13.16

Splitting $(y - \bar{y})$ into two deviations, $(\hat{y} - \bar{y})$ + $(y - \hat{y})$.

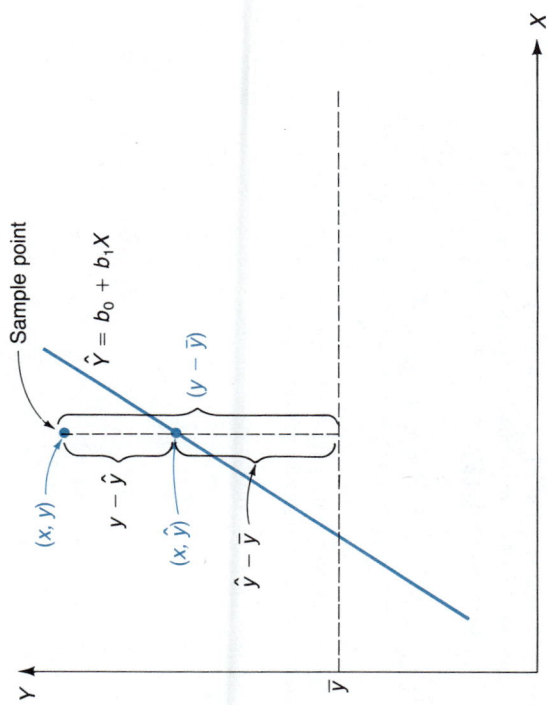

As a result, we have

$$\Sigma (\hat{y} - \bar{y})^2 = SSR$$

$$SS_Y = SSR + SSE \qquad (13.22)$$

The regression sum of squares, SSR, measures the variation in the Y values that would exist if differences in X were the *only* cause of differences among the Y's. If this were the case, then all the (X, Y) points would lie exactly on the regression line. In practice, this does not happen when using a simple linear regression model. Otherwise, we would have a deterministic phenomenon, not an object of statistical investigation. Consequently, the sample points can be assumed to lie about the regression line rather than on this line. This variation *about* the regression line is measured by the error sum of squares, SSE.

13.33 The sales manager of a real-estate firm believes that experience is the best predictor for determining the yearly sales of the various salespeople in the real-estate industry. Data were collected from 15 salespeople. Let X be the number of years of prior experience. Let Y be the annual sales (in thousands).

Y: SALES	X: EXPERIENCE	Y: SALES	X: EXPERIENCE
50	1.3	78	2.2
161	5.1	124	3.4
195	6.2	131	7.1
172	5.4	64	2.1
132	3.9	80	4.5
133	4.1	110	3.8
181	6.1	127	4.4
69	1.9		

Using a 10% significance level, test whether the population correlation coefficient between the variables X and Y is zero.

13.34 The manager of a company that relies on traveling salespersons to sell the company's products wants to examine the relationship between sales and the amount of time a salesperson spends with each established customer who regularly orders the company's products. The manager collects data on 12 salespersons. Let Y represent sales per month and X represent hours spent with customers per month.

X	Y	X	Y	X	Y
3.2	412	6.1	715	5.1	570
4.6	500	4.2	500	7.1	800
3.9	450	5.6	610	6.5	725
5.3	610	5.3	600	7.8	850

Can one conclude that the population correlation coefficient between X and Y is positive? Use a 10% significance level. Can one conclude that spending more time with customers increases sales?

13.35 Using the data in Exercise 13.3, test that there is no linear relationship between the total commissions paid and the net earnings of Tony's used-car lot. Use a 5% significance level.

13.36 Refer to Exercise 13.23. Use equation 13.20 to test whether there is a positive relationship between a person's investment in tax-free investments and a person's income. Use a significance level of .05. Is the conclusion the same as that in Exercise 13.23?

13.37 Ten cards numbered 1 through 10 are shuffled and a person is asked to pick one card. The card is replaced and the deck is reshuffled. Then the person is asked to draw a second card. If the second card is higher than the first, the dealer gives $.85 to the player. If the second card is not higher than the first, the player pays $1.15 to the dealer. A sample of 15 pairs of draws is taken to see whether there is any correlation between the first and the second cards.

a. Would you expect to observe significant correlation here? Why or why not?

b. Find the coefficient of determination for the following data and test using a 5% significance level that there is no correlation between the first and second cards. Interpret the value of the coefficient of determination.

X: FIRST CARD	Y: SECOND CARD	X: FIRST CARD	Y: SECOND CARD
7	3	10	5
3	10	3	6
8	2	4	3
5	8	6	1
2	7	7	8
7	9	8	4
9	4	2	6
1	1		

13.38 Refer to Exercise 13.25. Use equation 13.20 to test that there is no linear relationship between the size of the U.S. population and the logarithm of personal consumption expenditure on housing. Use a significance level of 5%.

13.39 For the data in Exercise 13.26, test that there is no linear relationship between the number of hours of operation between oil changes and the number of years that the engine was able to perform adequately. Use equation 13.20 and test with a 10% significance level. Is the result the same as in Exercise 13.26?

13.40 A sample of 35 pairs of observations is taken and a sample correlation coefficient is computed to be $r = .48$. Do the data provide sufficient evidence to reject the null hypothesis of no correlation? Use a 1% significance level.

13.41 Fifty people were asked to record their expenditure on vacation during the year and their yearly income. A correlation value of .39 was found. Do the data provide sufficient evidence to reject the null hypothesis of no correlation between the two variables? Use a 5% significance level.

13.42 The following pairs of observations represent the scores of a test given by a psychologist to a group before an experiment (X) and then to the same group after the experiment (Y).

X	Y	X	Y
2.1	9.4	2.4	12.9
3.4	35.6	1.9	5.2
1.6	3.5	1.3	3.4
2.7	15.4	0.2	0.1
3.2	30.1	1.5	4.6
4.5	52.7	2.1	9.3
1.8	17.4		

a. Calculate the correlation coefficient, r, between X and Y.

b. Take the log to the base 10 of Y. Calculate the correlation coefficient between X and log Y.

c. In parts (a) and (b), is there sufficient evidence to conclude that nonzero correlation exists? Use a 5% significance level.

13.5

ESTIMATION AND PREDICTION USING THE SIMPLE LINEAR MODEL

We have concentrated on predicting a value of the dependent variable (Y) for a given value of X. In the previous examples, we used a person's income, X, to predict the size of that person's home (Y). Notice in Figure 13.8 that we can also use the least squares line to estimate the *average* value of Y for a specified value of X. So we can use this line to handle two different situations.

Situation 1   The regression equation $\hat{Y} = b_0 + b_1 X$ estimates the **average** value of Y for a specified value of the independent variable, X. For $X = x_0$, this would be written $\mu_{Y|x_0}$ (the mean of Y given $X = x_0$).

For example, the least squares line passing through the real-estate data in Example 13.1 is $\hat{Y} = 8.51 + .354X$. The average square footage for *all* homes in the population with an income of \$40,000 ($X = 40$) is $\mu_{Y|40}$. Its estimate is provided by the corresponding value on the least squares line, namely,

$$\hat{Y} = 8.51 + .354(40) = 22.67$$

So the estimate of the average square footage of all such homes is 2267 square feet (Figure 13.7).

Situation 2   An **individual** predicted value of Y also uses the regression equation $\hat{Y} = b_0 + b_1 X$ for a specified value of X. This is denoted by Y_{x_0} for $X = x_0$. This is the more common application in business because a regression equation is generally used for individual forecasts.

For example, assume the Jenkins family resides in our sample community and has an income of \$40,000. A prediction of their home size (Y_{40}) is also

$$\hat{Y} = 8.51 + .354(40) = 22.67$$

which is 2267 square feet (Figure 13.7).

We see that the least squares line can be used to estimate *average values* (situation 1) or predict *individual values* (situation 2). Since $\mu_{Y|40}$ is a parameter, we use $\hat{Y}$ to *estimate* this value. On the other hand, Y_{40} represents a particular value of a dependent (random) variable, and so $\hat{Y}$ is used to *predict* this value. In the first situation, we can determine a *confidence* interval for $\mu_{Y|40}$; in the second situation, a *prediction* interval for Y_{40}.

Confidence Interval for $\mu_{Y|x_0}$ (Situation 1)

We have already established that the point estimate of $\mu_{Y|x_0}$ is the corresponding value of $\hat{Y}$. The reliability of this estimate depends on (1) the number of observations in the sample, (2) the amount of variation in the sample, and (3) the value of $X = x_0$. A confidence interval for $\mu_{Y|x_0}$ takes all three considerations into consideration.

A $(1 - \alpha) \cdot 100\%$ confidence interval for $\mu_{Y|x_0}$ is

$$\hat{Y} - t_{\alpha/2, n-2} \, s \sqrt{\frac{1}{n} + \frac{(x_0 - \bar{x})^2}{SS_x}} \text{ to } \hat{Y} + t_{\alpha/2, n-2} \, s \sqrt{\frac{1}{n} + \frac{(x_0 - \bar{x})^2}{SS_x}} \quad (13.23)$$

EXAMPLE 13.8   Determine a 95% confidence interval for the average home size of families with an income of $35,000, using the real-estate data from Example 13.1.

SOLUTION   We previously determined that $n = 10$, $\bar{x} = 39.8$, $SS_x = 1489.6$, and $s = 3.078$. The point estimate for the average square footage, $\mu_{Y|35}$, is

$$\hat{Y} = 8.51 + .354(35)$$
$$= 20.90(2090 \text{ square feet})$$

Obtaining $t_{.025, 8} = 2.306$ from Table A.5, the 95% confidence interval for $\mu_{Y|35}$ is

$$20.90 - (2.306)(3.078) \sqrt{\frac{1}{10} + \frac{(35 - 39.8)^2}{1489.6}} \text{ to } 20.90 + (2.306)(3.078) \sqrt{\frac{1}{10} + \frac{(35 - 39.8)^2}{1489.6}}$$

$$= 20.90 - (2.306)(3.078)(.340) \text{ to } 20.90 + (2.306)(3.078)(.340)$$

$$= 20.90 - 2.41 \text{ to } 20.90 + 2.41$$

$$= 18.49 \text{ to } 23.31$$

We are thus 95% confident that the average home size for families earning $35,000 is between 1849 and 2331 square feet.   ■

Using MINITAB to Construct Confidence Intervals   The MINITAB solution for the real-estate problem is contained in Figure 13.13. To construct confidence intervals, the column of interest is labeled as Stdev. Fit, which, when translated, means the standard deviation of the predicted Y. Writing this as $s_{\hat{Y}}$,

$$s_{\hat{Y}} = s \sqrt{\frac{1}{n} + \frac{(x_0 - \bar{x})^2}{SS_x}}$$

For each value of X *in the sample* (say, x_0), the corresponding confidence interval for $\mu_{Y|x_0}$ is

$$\hat{Y} - t \cdot s_{\hat{Y}} \text{ to } \hat{Y} + t \cdot s_{\hat{Y}}$$

where $t = t_{\alpha/2, n-2}$, as before, and $\hat{Y}$ is contained in the column to the left of the standard deviations. For values of X not in the sample, you can (1) approximate this confidence interval by using the value of $s_{\hat{Y}}$ corresponding to an X value *near* this particular value or (2) use the computer procedure that will be discussed in Chapter 14, which will provide an exact value for $s_{\hat{Y}}$ belonging to this particular X value.

Using the MINITAB output in Figure 13.13, we can find the confidence intervals corresponding to X values of 22, 40, and 60. The remaining seven confidence intervals are constructed in a similar manner.

For $X = 22$, the confidence interval is

$$16.300 - (2.306)(1.721) \text{ to } 16.300 + (2.306)(1.721) = 12.33 \text{ to } 20.27$$

FIGURE 13.17

95 % confidence intervals for the real-estate data derived from MINITAB output shown in Figure 13.13.

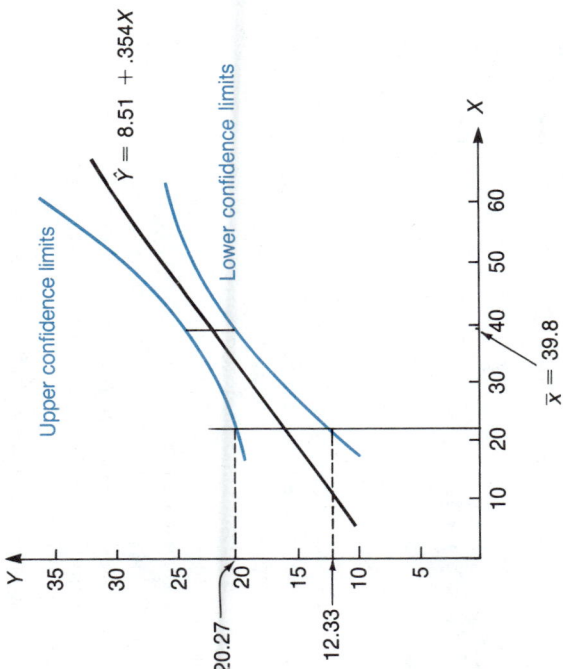

For $X = 40$, the confidence interval is

$$22.671 - (2.306)(.974) \quad \text{to} \quad 22.671 + (2.306)(.974) = 20.42 \quad \text{to} \quad 24.92$$

For $X = 60$, the confidence interval is

$$29.749 - (2.306)(1.882) \quad \text{to} \quad 29.749 + (2.306)(1.882) = 25.41 \quad \text{to} \quad 34.09$$

Notice that the confidence intervals are much wider for $X = 22$ and $X = 60$ than for $X = 40$.

By connecting the upper end of the confidence intervals for all ten data points and connecting the lower limits, we obtain Figure 13.17. Equation 13.23 indicates that the confidence interval is narrowest when $(x_0 - \bar{x})^2 = 0$, that is, at $X = x_0 = \bar{x}$. For values of X to the left or right of $\bar{x}$, the confidence interval is wider. In other words, the farther x_0 is from $\bar{x}$, the less reliable is the estimate.

The Danger of Extrapolation   Extrapolation is calculating an estimate corresponding to a value of X outside the range of the data used to derive the prediction equation (the least squares line). For example, in Figure 13.17, the least squares line could be used to estimate the average home size for families with an income of $100,000. Although we *can* estimate $\mu_{Y|100}$, the corresponding confidence interval for this parameter will be extremely wide, which means that the point estimate, $\hat{Y}$, has little practical value.

To use the simple regression model effectively for estimation, you need to stay within the range of the sampled values for the independent variable, X. This is **interpolation.** If you use values outside this range, you need to be aware that given *another* set of data, you would quite likely obtain a considerably different estimate. Furthermore, you have no assurance that the linear relationship still holds outside the range of your sample data.

Prediction Interval for Y_{x_0} (Situation 2)

The procedure of predicting individual values is most often used in business applications. The regression equation is generally used to **forecast** (predict) a future value of the dependent variable for a particular value of the independent variable. When attempting to predict a single value of the dependent variable,

Y, using the simple linear regression model, we begin, as before, with $\hat{Y}$. Substituting $X = x_0$ into the regression equation provides the best prediction of Y_{x_0}. For example, if the Johnson family has an income of $35,000, our best guess as to their home size (using this particular model) is $\hat{Y}$ for $X = 35$. From the results of Example 13.8, this is 20.90, or 2090 square feet.

We do not use the term *confidence interval* for this procedure because what we are estimating (Y_{x_0}) is not a parameter. It is a value of a random variable, so we use the term **prediction interval**.

The variability of the error in predicting a single value of Y is more than that for estimating the average value of Y (situation 1). It can be shown that an estimate of the variance of the error $(Y - \hat{Y})$, when using $\hat{Y}$ to predict an individual Y for $X = x_0$, is

$$s^2\left(1 + \frac{1}{n} + \frac{(x_0 - \bar{x})^2}{SS_x}\right) \qquad (13.24)$$

This result can be used to construct a $(1 - \alpha) \cdot 100\%$ prediction interval for Y_{x_0}, as follows:

$$\hat{Y} - t_{\alpha/2,\, n-2}\, s\, \sqrt{1 + \frac{1}{n} + \frac{(x_0 - \bar{x})^2}{SS_x}} \quad \text{to}$$
$$\hat{Y} + t_{\alpha/2,\, n-2}\, s\, \sqrt{1 + \frac{1}{n} + \frac{(x_0 - \bar{x})^2}{SS_x}} \qquad (13.25)$$

Notice that the only difference between this prediction interval and the confidence interval in equation 13.23 is the inclusion of "1 +" under the square root sign. The other two terms under the square root are usually quite small, so this "1 +" has a large effect on the width of the resulting interval. Be aware that our warning about extrapolating outside the range of the data applies here as well. In equations 13.24 and 13.25, the distance from the mean $(x_0 - \bar{x})$ is squared, which increases the risk of predicting beyond the range of the sampled data.

EXAMPLE 13.9

We previously determined that the Johnson family has an income of $35,000, and so the best prediction of their home size is $\hat{Y} = 20.90$. Determine a 95% prediction interval for this situation.

SOLUTION

We can use the calculations from Example 13.8 to derive the prediction interval for Y_{35}. The result is

$$20.90 - (2.306)(3.078)\sqrt{1 + \frac{1}{10} + \frac{(35 - 39.8)^2}{1489.6}} \quad \text{to} \quad 20.90 + (2.306)(3.078)\sqrt{1 + \frac{1}{10} + \frac{(35 - 39.8)^2}{1489.6}}$$

$$= 20.90 - (2.306)(3.078)(1.056) \quad \text{to} \quad 20.90 + (2.306)(3.078)(1.056)$$

$$= 20.90 - 7.49 \quad \text{to} \quad 20.90 + 7.49$$

$$= 13.41 \quad \text{to} \quad 28.39$$

Comparing this interval to the confidence interval for $\mu_{Y|35}$ in Example 13.8, we see that individual predictions are considerably less accurate than estimations for the mean home size. Of course, we could reduce the width of this interval by obtaining additional data. Expecting accurate results from a sample of ten observations is being a bit optimistic. ∎

Using MINITAB for Constructing Prediction Intervals We can use the MINITAB output in Figure 13.9 for Example 13.13. The values in the column labeled Stdev. Fit assume that $\hat{Y}$ is estimating $\mu_{Y|x}$; we previously used $s_{\hat{Y}}$ as a symbol for this standard deviation.

A prediction interval for a value of $X = x_0$ in the sample is provided by

$$\hat{Y} - t\sqrt{s_{\hat{Y}}^2 + s^2} \quad \text{to} \quad \hat{Y} + t\sqrt{s_{\hat{Y}}^2 + s^2}$$

where $t = t_{\alpha/2, n-2}$ is obtained from Table A.5 and $\hat{Y}$ is obtained from the column labeled FIT. For values of X *not* in the sample, you can derive a prediction interval by using the computer procedure to be discussed in Chapter 14. The following calculations determine the prediction intervals for three of the sample X values, namely, $X = 22, 40,$ and 60.

For $X = 22$, the 95% prediction interval is

$16.300 - 2.306\sqrt{(1.721)^2 + (3.078)^2} \quad$ to $\quad 16.300 + 2.306\sqrt{(1.721)^2 + (3.078)^2}$

$= 16.300 - 8.132 \quad$ to $\quad 16.300 + 8.132$

$= 8.17 \quad$ to $\quad 24.43$

For $X = 40$, the 95% prediction interval is

$22.671 - 2.306\sqrt{(.974)^2 + (3.078)^2} \quad$ to $\quad 22.671 + 2.306\sqrt{(.974)^2 + (3.078)^2}$

$= 22.671 - 7.445 \quad$ to $\quad 22.671 + 7.445$

$= 15.23 \quad$ to $\quad 30.12$

For $X = 60$, the 95% prediction interval is

$29.749 - 2.306\sqrt{(1.882)^2 + (3.078)^2} \quad$ to $\quad 29.749 + 2.306\sqrt{(1.882)^2 + (3.078)^2}$

$= 29.749 - 8.320 \quad$ to $\quad 29.749 + 8.320$

$= 21.43 \quad$ to $\quad 38.07$

Figure 13.18 shows the prediction intervals for all ten data points; the upper and lower limits have been connected. The increased width of a prediction interval versus a confidence interval is quite apparent from this graph. Also, as with that of a confidence interval, the width of a prediction interval increases as the value of X strays from $\bar{x}$.

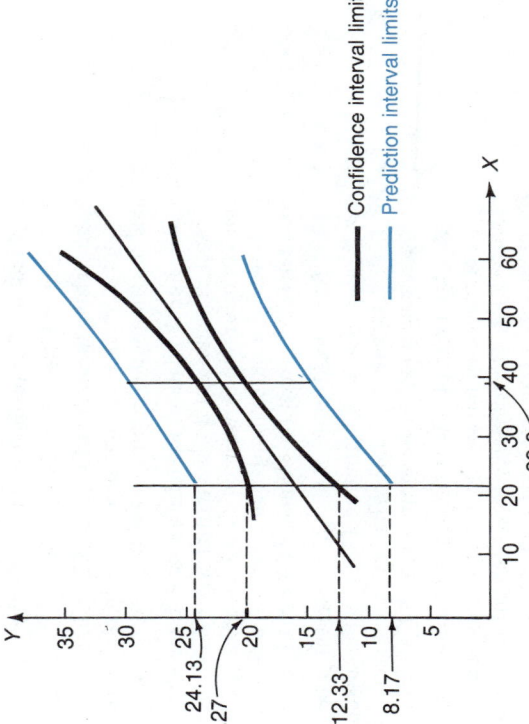

FIGURE 13.18

The 95% prediction and confidence intervals for the real-estate data.

EXERCISES

13.43 The marketing division of Astral Airlines wants to determine the relationship between the amount of money a person spends yearly on air transportation and the yearly income of the person. They randomly selected 20 airline passengers. The data (in thousands of dollars) are:

Y: YEARLY EXPENDITURE	X: YEARLY INCOME	Y: YEARLY EXPENDITURE	X: YEARLY INCOME
423	22.5	640	29.8
396	31.4	675	33.4
120	18.1	745	38.4
140	19.1	425	21.8
550	26.4	380	22.5
690	44.5	725	30.4
740	37.1	950	38.7
320	16.8	925	40.6
1200	50.5	210	18.8
470	21.8	425	21.7

a. Determine a 95% confidence interval for the average yearly airline expenditure for people with an annual income of $30,000.

b. Find a 95% prediction interval for the yearly airline expenditure of a person with an annual income of $30,000.

c. Interpret and compare the intervals in parts (a) and (b).

13.44 A statistics instructor believes that there is a strong correlation between a student's grade in college algebra and a student's grade in introductory statistics. The following statistics were collected from 20 randomly selected students. Let X be the student's grade in college algebra. Let Y be the student's grade in introductory statistics (A = 4.0, B = 3.0, and so on).

$$SS_{xy} = 31.8$$
$$SS_x = 32.3$$
$$SS_y = 45.4$$
$$\bar{x} = 2.3$$
$$\bar{y} = 2.5$$
$$n = 20$$

a. Determine a 95% confidence interval for the average grade in introductory statistics for students who make a **B** (equal to 3.0) in college algebra.

b. Find the 95% prediction interval for the grade in introductory statistics for a student who obtained a **B** in college algebra.

13.45 Using the data in Exercise 13.23, find a 90% prediction interval for the predicted amount of money an individual places in tax-free investment instruments if the individual has an annual income of $40,000.

13.46 For the data in Exercise 13.26, find the 99% confidence interval for the average number of years that a lawn mower will be able to function properly if the number of hours of operation between oil changes is 23 hours. What is the standard deviation for a predicted value of the number of years that a single lawn mower will be able to perform adequately, for which the number of hours of operation between oil changes is 23?

13.47 For a fixed value of X, which interval is larger, the confidence interval for the mean value of Y at X (equation 13.23) or the prediction interval for a predicted value of Y at X (equation 13.25)? What value can you assign to X to achieve the smallest confidence interval for the mean value of Y at X or for the predicted value of Y at X?

13.48 A sample of 200 executives who work in Chicago was taken to find out how much of their own money the executives invest each year in stock of the company that they work for. The following regression equation was developed, where X is the income (in thousands) of an executive and Y is the amount of money (in thousands) he or she invests each year in the company. The prediction equation is

$$\hat{Y} = 9.5 + 0.05X$$

Based on the regression equation, can the following statement be made? A Chicago-area executive who earns $15,000 a year would invest about $10,250 in company stock. Comment.

13.49 The manager of an engineering firm believes that an employee's overall performance is related to the employee's score on Randall and Cantrell's job-aptitude test. Fourteen employees were selected randomly and the following data were collected. Evaluation of job performance was rated by the manager on a one to ten scale, with ten being the score of a perfect employee.

Y: EVALUATION OF JOB PERFORMANCE	X: SCORE ON APTITUDE TEST	Y: EVALUATION OF JOB PERFORMANCE	X: SCORE ON APTITUDE TEST
9.5	90	4.5	59
6.0	71	5.5	63
2.0	33	7.0	84
6.5	49	9.0	94
7.5	82	8.5	83
5.0	61	6.0	74
7.5	73	7.5	51

a. Determine a 95% confidence interval for the average job performance evaluation for employees with a score of 75.

b. Find a 95% prediction interval for the job performance of an employee with a score of 75.

c. What score would give the smallest 95% prediction interval for an employee?

13.50 Using the following sample statistics, find a 90% confidence interval for the mean value of Y at $X = .5$.

$$SS_{xy} = 31.4$$
$$SS_x = 40$$
$$SS_y = 45$$
$$\bar{x} = 7.5$$
$$\bar{y} = 10.0$$
$$n = 31$$

13.51 For the data in Exercise 13.28, find a 99% prediction interval for the monthly weight loss of an individual rat that has a daily dose of .30 milligrams of the diuretic.

13.52 The owner of a used car lot would like to explain to potential car buyers the relationship between the horsepower rating of a car and the gasoline mileage. The owner collects the following data on twenty 1985 automobiles with the objective of showing that horsepower rating gives a good indication of gasoline mileage.

Y: GASOLINE MILEAGE	X: HORSEPOWER RATING	Y: GASOLINE MILEAGE	X: HORSEPOWER RATING
16.0	180	16.8	195
14.9	195	20.0	290
14.1	160	14.8	150
18.5	235	13.0	100
15.3	175	20.3	290
21.9	285	18.2	255
15.2	210	17.1	220
18.2	230	19.8	235
17.0	235	17.5	260
16.9	200	21.4	275

a. Construct a scatter diagram of the data.

b. Find the least squares line for the data.

c. Test the hypothesis that there is no linear relationship between the variables Y and X at the .10 significance level.

SUMMARY

When dealing with a pair of variables (say, X and Y), we generally are interested in determining whether the variables are related in some manner. If a relationship does exist, perhaps the **independent** variable (X) can be used to predict values of the **dependent** variable (Y). If a significant linear relationship exists within the sample data, both the direction (positive or negative) and the strength of this linear relationship can be measured using the sample **coefficient of correlation**, r. The sample correlation coefficient is an estimate of the population coefficient of correlation, ρ. Another commonly used measure of association between two variables is the sample **covariance.**

Whenever a sample of bivariate data contains a significant linear pattern, we determine the **least squares line** through the data points. This generates an equation that can be used to predict values of the dependent variable. To describe accurately the assumptions behind this procedure, we introduced the concept of a **statistical model** consisting of a **deterministic** portion (the straight line) and a **random error** component. This model can be written as $Y = \beta_0 + \beta_1 X + e$, where $\beta_0 + \beta_1 X$ is the deterministic component and e represents the error component. When we perform any test of hypothesis regarding the underlying bivariate population, we must be careful to satisfy the necessary assumptions behind this procedure. These assumptions will be examined more closely in Chapter 14.

By regressing Y on X we are able to determine the least squares line, $\hat{Y} = b_0 + b_1 X$. The value of b_0 is the **intercept** of the least squares line and estimates the population intercept β_0. The **slope** of the least squares line, b_1, estimates the population slope β_1. One question of interest is: If we regress X on Y (that is, switch the independent and dependent variables), can we rearrange the previous equation and say that $\hat{X} = (-b_0/b_1) + (1/b_1)Y$? The answer is no, although the coefficient of correlation, r, is the same in either case. Consequently, constructing a least squares line is not a good idea in the event it is not obvious which variable is the dependent variable.

Various methods for determining the utility of the model as a predictor of the dependent variable include: (1) a t-test for detecting a significant slope, b_1—a value of $\beta_1 = 0$ indicates that X has no predictive ability; (2) a t-test for determining whether the sample correlation, r, is significantly large—a value of $\rho = 0$ indicates that there is no linear relationship between the two variables; and (3) a confidence interval for the slope, β_1. The two t-tests appear to be quite different, but their computed values (and df) are *identical*; there is no point in performing both tests.

Another measure of how well the model provides estimates that fit the sample data is given by the **coefficient of determination**, r^2. For simple linear regression (one independent variable), this is the square of the correlation coefficient. Another definition of the coefficient of determination also can be used to examine more than one independent variable (called multiple linear regression), namely, $r^2 = 1 - (\text{SSE}/\text{SS}_y)$. Here, SSE is the sum of squared errors and SS_y represents the total variation in the sample Y values. For example, if $r^2 = .85$, then 85% of the variation in the sample Y values has been explained using this model.

The value of $\hat{Y}$ from the least squares regression line at a specific value of X (say, $X = x_0$), can be used to estimate an *average* value of Y, given this value of X (written $\mu_{Y|x_0}$). The value of $\hat{Y}$ centers a *confidence interval* for $\mu_{Y|x_0}$. Similarly, we can use the $\hat{Y}$ value to center a *prediction interval* for an individual value of the dependent variable, given this specific value of X (written Y_{x_0}). The value of $\hat{Y}$ can be used to *estimate* the value of $\mu_{Y|x_0}$ or *predict* the value of Y_{x_0}.

Summary of Linear Regression and Correlation Formulas

DESCRIPTION	FORMULA
Correlation between two variables	$r = \dfrac{SS_{XY}}{\sqrt{SS_X}\sqrt{SS_Y}}$ where $SS_{XY} = \Sigma xy - (\Sigma x)(\Sigma y)/n$ $SS_X = \Sigma x^2 - (\Sigma x)^2/n$ $SS_Y = \Sigma y^2 - (\Sigma y)^2/n$
Least squares line	$\hat{Y} = b_0 + b_1 X$ where $b_1 = SS_{XY}/SS_X$ and $b_0 = \bar{y} - b_1\bar{x}$
Estimate of the residual variance	$\hat{\sigma}_e^2 = s^2 = \dfrac{SSE}{n-2}$ where $SSE = \Sigma(y - \hat{y})^2$ $\quad = SS_Y - \dfrac{(SS_{XY})^2}{SS_X}$
t statistic for detecting a significant slope	$t = \dfrac{b_1}{s_{b_1}}$ (df = n − 2) where $s_{b_1} = s/\sqrt{SS_X}$
Confidence interval for the slope, β_1	$b_1 - t_{\alpha/2,\,n-2}\, s_{b_1} \quad \text{to} \quad b_1 + t_{\alpha/2,\,n-2}\, s_{b_1}$
t statistic for detecting a significant correlation	$t = \dfrac{r}{\sqrt{\dfrac{1-r^2}{n-2}}}$ (df = n − 2)
Coefficient of determination	r^2 = square of correlation coefficient = $1 - \dfrac{SSE}{SS_Y}$
Confidence interval for the average value of Y at a specific value of X (say, x_0)	$\hat{Y} \mp t_{\alpha/2,\,n-2}\, s\, \sqrt{\dfrac{1}{n} + \dfrac{(x_0 - \bar{x})^2}{SS_X}}$ or, for MINITAB, $\hat{Y} \pm t_{\alpha/2,\,n-2}\, s_{\hat{Y}}$
Prediction interval for a particular value of Y at a specific value of X (say, x_0)	$\hat{Y} \mp t_{\alpha/2,\,n-2}\, s\, \sqrt{1 + \dfrac{1}{n} + \dfrac{(x_0 - \bar{x})^2}{SS_X}}$ or, for MINITAB, $\hat{Y} \pm t_{\alpha/2,\,n-2}\, \sqrt{s_{\hat{Y}}^2 + s^2}$

REVIEW EXERCISES

13.53 An industrial engineer collected data to study the relationship between the intensity of illumination on the shop floor and the output of the workers. Ten levels of illumination, coded from 1 to 10, were studied and the output of workers was noted at each level. Output was measured as the number of items produced per hour.

ILLUMINATION INTENSITY	OUTPUT OF WORKERS	ILLUMINATION INTENSITY	OUTPUT OF WORKERS
1	70	6	94
2	70	7	100
3	75	8	92
4	88	9	90
5	91	10	85

a. Obtain the least squares line for the above data by regressing output of workers on illumination intensity.

b. Plot the data in a scatter diagram and comment on any pattern you observe.

13.54 Fans Unlimited finds that competition in the fan business had increased over the past year. The manager decides to perform an experiment by pricing the company's most popular 52-inch ceiling fan at various prices (in dollars) each week and then observing the demand. After 8 weeks, the following data had been recorded:

WEEK	X: FAN COST	Y: NUMBER SOLD	WEEK	X: FAN COST	Y: NUMBER SOLD
1	175	13	5	115	20
2	160	15	6	99	24
3	145	18	7	110	20
4	129	18	8	89	29

a. Find the least squares line for the data, with X as the independent variable and Y as the dependent variable.

b. Find the coefficient of determination.

c. Find a 90% confidence interval for the slope of the regression line.

d. Find a 90% confidence interval for the mean number of fans that will be sold if the price is $120 per fan.

e. Find a 90% prediction interval for the number of fans sold if the price is $120 per fan.

13.55 Dolls-R-Us believes that television advertising is the most effective way to market their new line of dolls. The sales manager recorded the amount of money spent on advertising and the amount of sales for 20 randomly selected months. The average cost for television advertising for the 20 months was $110,000. The average sales volume for the 20 months was $675,000. The following sample statistics were found from the data for the 20 months:

$$SS_{XY} = 198.4$$
$$SS_X = 205.3$$
$$SS_Y = 341.6$$

where Y represents the sales volume (in thousands) and X represents the television advertising costs (in thousands of dollars).

a. Calculate the least squares line.

b. Calculate the coefficient of determination.

c. Calculate the sum of squares of error.

d. What is the estimate of the variance of the error component for the model?

e. Is there sufficient evidence from the data to conclude at the .01 significance level that a positive relationship exists between X and Y?

f. Find a 95% prediction interval for the monthly sales volume if the television advertising expenditure during one particular month is $120,000.

13.56 A car rental agency has a fleet of 200 cars available for rent at Kennedy airport in New York City. The owner of the agency uses a regression equation for estimating the company's daily revenue based on the number of incoming flights that day. The regression equation is $\hat{Y} = 2500 + 21.4X$, where X is the number of daily incoming flights and Y is the daily revenue in dollars. The data used to find the least squares line are based on a sample of 100 randomly selected days in 1984. Can the following statement be made based on regression analysis? If Kennedy airport increases its daily incoming flights by 50 flights next year, then the car agency can expect to make an additional daily revenue of $1,070. Comment.

13.57 Each week, a realtor advertises the houses she manages that are available for rent. The number of telephone calls from people inquiring about the advertisement were recorded for several weeks, during which various sizes of the advertisement were used. Is there sufficient evidence from the data below to conclude, at the .10 significance level, that a nonzero correlation exists?

X (HEIGHT OF AD, INCHES)	Y (NUMBER OF INQUIRIES)	X (HEIGHT OF AD, INCHES)	Y (NUMBER OF INQUIRIES)
0.5	3	2.5	10
1.0	4	3.0	14
1.5	6	3.5	12
2.0	5	4.0	18

13.58 The manager of a firm, which specializes in assisting individuals in filling out federal income tax forms, obtained data from the Internal Revenue Service pertaining to deductions for charitable contributions. The following table provides a distribution of charitable contributions for eight groups with different adjusted gross incomes.

(X) MEDIAN ADJUSTED GROSS INCOME (IN THOUSANDS OF DOLLARS)	(Y) PERCENTAGE IN GROUP MAKING CHARITABLE CONTRIBUTIONS (CLAIMING ITEMIZED DEDUCTIONS)
5.0	17.0
7.5	36.0
12.5	40.5
17.5	38.5
25.0	29.2
40.0	14.0
75.0	4.2
100.0	1.5

a. Obtain the least squares line for these data.

b. Identify the values of the intercept, the slope, and the variance of the error for the simple linear regression model.

c. Find the residuals for all the Y values.

d. If the correlation between X and Y above was very strong, would it then be correct to conclude that an increase in income causes people to become less charitable?

13.59 The following data were collected for a certain regression analysis.

$$SS_{XY} = -138.6$$
$$SS_X = 112.3$$
$$SS_Y = 325.2$$
$$\bar{x} = 86.2$$
$$\bar{y} = 112.9$$
$$n = 41$$

a. Find the least squares line.

b. Test, using a 1% significance level, that the population correlation coefficient between the variables X and Y is negative.

c. Find a 90% confidence interval for the slope of the regression equation.

13.60 One management policy is based on the hypothesis that, the more productive a worker is, the more satisfied the worker will be. A scale from one to ten is used to measure productivity, with ten being assigned to an extremely productive worker. A second scale from one to ten is used to measure satisfaction. The worker assigns him- or herself a ten if he or she is satisfied in every aspect of the job. Twenty employees were selected randomly from the production-and-research department of Tellon Oil. The results of the data collection are as follows:

Y: SATISFACTION	X: PRODUCTIVITY	Y: SATISFACTION	X: PRODUCTIVITY
5	4	9	7
2	3	7	5
9	8	4	4
9	9	8	7
5	6	9	8
3	5	10	9
5	4	5	6
7	7	1	2
9	8	7	8
2	3	9	9

a. Draw a scatter diagram of the data.

b. Test the hypothesis, at a 5% significance level, that productivity does not positively influence a worker's satisfaction.

c. Find a 99% confidence interval for the slope of the regression equation.

d. Calculate the coefficient of determination.

e. Find a 99% prediction interval for the satisfaction of a particular worker if the measure of productivity of this worker is 7.

13.61 A certain risk-averse investor calculates the beta for a stock before investing in the stock. By regressing the weekly percent return of, say, stock XYZ on the weekly percent return of the Standard and Poor's 500 Index (S&P 500), the investor can determine the stock's beta, which is equal to the slope of the regression line. The following data represent the weekly return over 20 weeks for both the S&P 500 and stock XYZ.

WEEK	(X) S&P 500 (IN PERCENT)	(Y) STOCK XYZ (IN PERCENT)	WEEK	(X) S&P 500 (IN PERCENT)	(Y) STOCK XYZ (IN PERCENT)
1	.51	.95	11	−1.12	−1.13
2	.22	.66	12	−.80	−.74
3	−.43	−.21	13	1.55	2.32
4	−2.51	−3.00	14	2.34	3.34
5	3.05	4.11	15	−.50	−.40
6	.40	.75	16	2.81	4.00
7	−.21	.01	17	3.33	4.63
8	1.80	2.64	18	−1.64	−1.83
9	2.55	4.51	19	1.75	2.58
10	3.80	6.12	20	2.20	3.11

a. Calculate the least squares line.

b. A slope greater than one indicates that the stock is more volatile than the S&P market index. Interpret the coefficients of the regression equation.

c. Graph the data and the least squares line. Comment on the fit.

13.62 Diane's Beauty Salon is currently hiring beauticians at its new location in a popular mall. Diane wants to know what percentage of commission to pay the beauticians based on experience. A survey of 12 licensed beauticians was taken with the following results.

Y: PERCENTAGE OF COMMISSION	X: YEARS OF EXPERIENCE	Y: PERCENTAGE OF COMMISSION	X: YEARS OF EXPERIENCE
24	2	25	4
18	1	44	12
30	5	33	8
41	10	24	3
35	8	20	1
35	7	40	10

a. Find the least squares line.

b. Calculate the sum of squares due to error.

c. Test the null hypothesis that there is no linear relationship between years of experience and percentage of commissions paid. Use a significance level of .05.

d. Find a 90% confidence interval for the slope of the least squares line.

13.63 A regression line is fitted to a set of data and the values of $Y - \hat{Y}$ are calculated. Does the following set of sample errors appear to conform to the empirical rule that 95% of the data should lie within two standard deviations of the mean? Construct a histogram for the residuals. Comment on the slope of the histogram.

1.1, −0.8, 2.6, 1.5, 0.2, −0.4, 0.8, −1.8, −2.3, 0.9, −2.7, 3.1, −1.0, 0.9, 4.5, −3.4, −0.1, −0.2, 2.4, −1.7, −0.7

13.64 The foreign exchange rates of U.S. dollars for the Canadian dollar and for the British pound sterling are given from January 1986 to October 1987.

TIME PERIOD	(X) U.S. DOLLARS FOR ONE CANADIAN DOLLAR	(Y) U.S. DOLLARS FOR ONE BRITISH POUND
1986 Jan.	1.43	.7140
Feb.	1.43	.7140
Mar.	1.47	.7138
Apr.	1.50	.7213
May	1.53	.7255
Jun.	1.51	.7196
Jul.	1.52	.7252
Aug.	1.49	.7209
Sep.	1.48	.7207
Oct.	1.43	.7204
Nov.	1.42	.7215
Dec.	1.44	.7248
1987 Jan.	1.51	.7372
Feb.	1.53	.7505
Mar.	1.59	.7574
Apr.	1.63	.7568
May	1.65	.7454
Jun.	1.64	.7468
Jul.	1.61	.7531
Aug.	1.59	.7524
Sep.	1.64	.7600
Oct.	1.67	.7626

(*Source: Standard and Poor's Statistical Service*, February 1988, p. 27.)

a. Plot the values of X and Y. Does a least squares line appear to be a good fit, from eyeballing the scatter plot?

b. Calculate the correlation coefficient.

c. Find the least squares line.

13.65 The number of work stoppages (strikes) involving 1000 workers or more in the United States and the total number of workers involved in these work stoppages are given below for the years 1976–1986. The workers involved are in units of thousands:

YEAR	NUMBER OF STOPPAGES (X)	WORKERS INVOLVED (Y)	YEAR	NUMBER OF STOPPAGES (X)	WORKERS INVOLVED (Y)
1976	231	1519	1982	96	656
1977	298	1212	1983	81	909
1978	219	1006	1984	62	376
1979	235	1021	1985	54	324
1980	187	795	1986	69	533
1981	145	729			

(*Source: The World Almanac and Book of Facts*, 1988, p. 20.)

a. Find the least squares equation for predicting the number of workers involved from the number of stoppages.

b. Is there sufficient evidence to conclude that the number of stoppages is positively correlated with the number of workers involved? Use a .10 significance level.

c. Find a 90% prediction interval for the number of workers involved when the number of stoppages is equal to 140.

d. Calculate the coefficient of determination and interpret it.

13.66   Twenty employees were asked to rate on a continuous scale from 1 to 7 (1 = strongly disagree to 7 = strongly agree) their feelings about several statements to measure job satisfaction, variety, autonomy, and task significance. Variety pertains to the number of different and new tasks required of an employee. Autonomy pertains to the freedom to make decisions. Task significance pertains to the meaningfulness and importance of the tasks required of an employee.

PERSONS	JOB SATISFACTION	VARIETY	AUTONOMY	TASK SIGNIFICANCE
1	4.6	3.5	4.7	5.2
2	3.3	4.8	3.1	4.7
3	6.8	5.0	7.0	7.0
4	6.2	4.5	6.7	7.0
5	5.5	6.5	6.0	6.4
6	5.0	4.5	5.4	7.0
7	3.4	2.5	3.2	3.4
8	5.7	6.8	6.5	6.4
9	5.5	4.0	6.0	6.3
10	4.0	5.5	4.0	4.3
11	5.0	4.0	5.5	5.3
12	3.5	4.8	3.5	5.0
13	6.0	5.0	6.5	7.0
14	4.0	5.6	3.9	4.2
15	5.0	3.0	5.4	5.5
16	4.5	5.5	4.8	5.0
17	5.5	4.0	5.8	6.4
18	5.5	6.5	6.5	6.0
19	5.7	4.0	6.0	6.7
20	4.6	4.5	5.0	6.2

```
MTB > name c1 'person'
MTB > name c2 'satisfac'
MTB > name c3 'variety'
MTB > name c4 'autonomy'
MTB > name c5 'task sig'

MTB > correlations for c2-c5

           satisfac   variety   autonomy
variety     0.208
autonomy    0.983     0.276
task sig    0.873     0.211     0.874
```

a. Interpret the values of the correlation coefficients.

b. Which single variable would be the best predictor of job satisfaction?

c. Which correlations are significantly different from zero? Use a .05 significance level.

13.67 A realtor is interested in the relationship between the initial listing price of a single-family house in a large residential subdivision and the final selling price. A random selection of 16 houses is chosen from a list of homes that have been sold in the past 6 months. The units are in thousands of dollars. C1 contains the final selling price, and C2 contains the initial listing price in the accompanying printout.

```
MTB > name c1 'final'
MTB > name c2 'initial'
MTB > print c1 c2
ROW   final  initial

 1    125.5   130.1
 2    115.8   121.5
 3    105.0   110.3
 4    122.9   127.5
 5     92.0    95.0
 6    101.5   105.9
 7     88.1    99.5
 8     92.1    93.0
 9    117.0   121.3
10    112.5   118.0
11    101.0   105.0
12    107.5   110.5
13    103.0   106.9
14    113.4   117.0
15    121.0   125.5
16    115.0   119.0

MTB > regress y in c1 using 1 predictor in c2

The regression equation is
final = - 3.04 + 0.987 initial

Predictor     Coef     Stdev    t-ratio
Constant    -3.038     5.725      -0.53
initial    0.98666   0.05048      19.54

s = 2.228   R-sq = 96.5%   R-sq(adj) = 96.2%

Analysis of Variance

SOURCE      DF        SS        MS
Regression   1    1896.6    1896.6
Error       14      69.5       5.0
Total       15    1966.1

Unusual Observations
Obs. initial   final      Fit  Stdev.Fit  Residual  St.Resid
 7      100    88.100   95.135      0.875    -7.035     -3.43R

R denotes an obs. with a large st. resid.
```

a. What is the predicted selling price of a house that is listed initially for $115 thousand?

b. Does the initial listing price contribute to the prediction of the final selling price? Use a .05 significance level.

c. Find a 95% confidence interval for the slope of the regression equation.

d. If observation 7 were removed from the data, do you think the coefficient of determination would improve? Try it.

13.68 An economist was analyzing the relationship between the price (Y) and the supply (X) of a certain product. As the supply diminished, the price increased rapidly. Twelve data points were collected from historical data. A plot of Y and X is given, followed by a plot of Y and $1/X$. The variable Y is in dollars and the variable X has been coded. Comment on the adequacy of the relationship between $1/X$ and Y. Interpret the 95% prediction interval for the price at $X = 3.0$ and the 95% confidence interval for the mean price at $X = 3.0$.

```
MTB > name c1 'supply'
MTB > name c2 'price'
MTB > plot c2 c1
```

```
price -
      -
      -            *
 3.00+
      -              *
      -               **
 2.40+                  *
      -                    *
      -                2
 1.80+                  *
      -                    *
      -                      *
 1.20+
      ------+---------+---------+---------+---------+---------+------supply
           1.5       3.0       4.5       6.0       7.5       9.0
```

```
MTB > let c3 = 1/c1
MTB > name c3 '1/supply'
MTB > plot c2 c3
```

```
price -
      -
      -                                                    *
 3.00+
      -                                              *
      -                                            **
 2.40+                                         *
      -                                      *
      -                                  *  *
 1.80+                              *
      -                          **
      -                       *
 1.20+              *
      -           *
      -        *
-0.00+
      ------+---------+---------+---------+---------+---------+------1/supply
          0.15      0.30      0.45      0.60      0.75
```

```
MTB > regress c2 on 1 predictor in c3;
SUBC> predict y value for reciprocal of x equal to .333.
```

The regression equation is
price = 1.05 + 3.12 1/supply

Predictor	Coef	Stdev	t-ratio
Constant	1.0494	0.1052	9.98
1/supply	3.1207	0.2585	12.07

s = 0.1808   R-sq = 93.6%   R-sq(adj) = 92.9%

Analysis of Variance

SOURCE	DF	SS	MS
Regression	1	4.7645	4.7645
Error	10	0.3269	0.0327
Total	11	5.0914	

```
obs.1/supply    price   'Fit   Stdev.Fit  Residual  St.Resid
 1    0.769     3.3000  3.4499   0.1195    -0.1499    -1.11
 2    0.667     3.0000  3.1299   0.0964    -0.1299    -0.85
 3    0.526     2.9000  2.6919   0.0687     0.2081     1.24
 4    0.417     2.8000  2.3497   0.0547     0.4503     2.61R
 5    0.357     2.0700  2.1639   0.0522    -0.0939    -0.54
 6    0.345     2.1000  2.1255   0.0522    -0.0255    -0.15
 7    0.303     1.9000  1.9951   0.0538    -0.0951    -0.55
 8    0.263     1.8000  1.8706   0.0571    -0.0706    -0.41
 9    0.200     1.7500  1.6735   0.0655     0.0765     0.45
10    0.154     1.5000  1.5295   0.0733    -0.0295    -0.18
11    0.127     1.4000  1.4444   0.0785    -0.0444    -0.27
12    0.111     1.3000  1.3961   0.0815    -0.0961    -0.60
```

R denotes an obs. with a large st. resid.

```
   Fit   Stdev.Fit        95% C.I.            95% P.I.
 2.0886    0.0525    ( 1.9717, 2.2055)   ( 1.6690, 2.5081)
```

COMPUTER EXERCISES USING THE DATABASE

Exercise 1—Appendix H

From the database, select 50 random observations. Compute the sample correlation between the variable HPAYRENT (house payment or rent) and the variable UTILITY (monthly utility expenditure). Is there sufficient evidence to conclude that a positive correlation exists? Use a .05 significance level.

Exercise 2—Appendix H

Select 50 random observations from the database. Plot the values of total indebtedness (TOTLDEBT) and total income (INCOME1 + INCOME2). Also plot the value of total indebtedness with house payment or apartment/house rent (HPAYRENT). Choose the graph which appears to have a more linear relationship between the variables graphed. Fit a least squares line to the data. Test if the predictor variable significantly contributes to the prediction of total indebtedness. Use a .10 significance level.

Exercise 3—Appendix I

Select 50 random observations from

the database. Compute the sample correlation between the variable NETINC (net income) and each of the variables SALES (gross sales) and COSTSALE (cost of sales). Select the variable from these two that has the higher correlation with net income. Regress net income on this variable and test if this variable significantly contributes to the prediction of net income. Use a .05 significance level.

Exercise 4—Appendix I

Select 50 random observations from the database. Compute the regression line for predicting total assets (TOTAL) from current assets (ASSETS). Also compute the regression line for predicting total assets from current liabilities (LIABIL). Test for the adequacy of the fit of these two regression lines to the data. Use a .10 significance level. Which of these two regression lines has a higher value for the coefficient of determination, and what does the higher value indicate?

CASE STUDY

THE 1987 EPA MILEAGE ESTIMATES

The United States Environmental Protection Agency (EPA) releases an annual ranking of fuel efficiency in automobiles. The EPA estimates these fuel economy ratings by testing the exhaust emissions of new cars

with computerized equipment. The EPA is charged with this responsibility primarily to enable the federal government to monitor auto manufacturers' compliance with legislation mandating a minimum average

The following table was adapted from EPA data by *Consumers' Research* magazine. It shows selected models, along with their annual fuel cost in dollars based on 15,000 miles of driving, an estimated MPG rating based on a combination of city and highway driving, and engine type (capacity in cubic inches).

fuel economy that a manufacturer's fleet of cars must reach. These are known as the CAFE standards, which stands for Corporate Average Fuel Economy. For the year 1987, the CAFE standards required each automaker to meet an average fuel-efficiency rating of 26.0 miles per gallon (mpg).

CAR MODEL	ANNUAL FUEL $	EST MPG	ENGINE TYPE
Bertone X1/9	570	25	91
Ford EXP	509	28	113
Mazda RX-7	750	19	80
Nissan 300ZX	678	21	181
Pontiac Fiero	527	27	151
Subaru XT-DL	509	28	109
Alfa Romeo GTV	678	21	152
Porsche 944	678	21	151
Volkswagen Cabriolet	594	24	109
Audi Coupe GT	712	20	136
Chevrolet Camaro	648	22	173
Chevrolet Sprint	365	39	61
Chrysler Conquest	825	20	156
Chrysler LeBaron Convertible	785	21	135
Ford Mustang	527	27	140
Honda Civic	356	40	82
Honda Prelude	549	26	112
Isuzu I-Mark	445	32	90
Mitsubishi Cordia	660	25	110
Nissan Sentra Coupe	509	28	98
Plymouth Colt	460	31	90
Pontiac Firebird	648	22	173
Renault Alliance	570	25	105
Subaru Hatchback	432	33	97
Suzuki Forsa	365	39	61

(*Source:* "The 1987 EPA Mileage Estimates," *Consumers' Research* 69, no. 11 (1986): 11.)

Case Study Questions

1. Obtain a correlation matrix for the three variables, annual fuel cost, estimated MPG and engine type. Which two variables have the strongest relationship? Is this surprising?

2. **a.** Find the least squares regression equation to predict average annual fuel cost (Y) using estimated MPG as the independent variable, X. Plot the two variables.

   b. Interpret what the r^2 for the model tells us.

   c. Determine if this regression model represents a significant linear relationship between Y and X, at a 5% significance level.

   d. State the p-value for the test, and interpret what it means.

   e. What is the variance of the error in the model?

3. Repeat all five parts of question

2, but this time use the engine type as the predictor of average annual fuel cost.

4. **a.** Which of the two models do you think is better at predicting annual fuel cost?

   b. Which of the following criteria would indicate (if at all) a better model:
   i. Lower vs. higher MSE?
   ii. Lower vs. higher MSR?
   iii. Lower vs. higher correlation?
   iv. Lower vs. higher coefficient of determination?
   v. Lower vs. higher regression coefficient (b_1)?

5. Suppose you had an automobile with an engine type of 135 cubic inches and an EPA rating of 27 miles per gallon. Using the better of the two models from the above analysis, find:

   a. The point estimate for the predicted annual fuel cost for this car.

   b. The 95% interval estimate of annual fuel cost for any individual car of this type.

   c. The 95% interval estimate of mean annual fuel cost for all cars of this type.

■ ■ ■ ■ ■ ■ ■ ■ ■ S P S S ■ ■ ■ ■ ■ ■ ■ ■ ■ ■

MAINFRAME AND MICRO

SOLUTION

EXAMPLE 13.2

Example 13.2 was concerned with computing the regression equation $\hat{Y} = b_0 + b_1 X$. The purpose of this problem was to determine the relationship between family income of local residents and the square footage of their homes. The SPSS program listing in Figure 13.19 was used to request the SSE, b_0, and b_1 values as well as other statistics. The regression analysis should always include a scatter diagram of the sample data.

In this problem the SPSS commands are the same for both the mainframe and PC versions. **(Remember to end each command line with a period when using the PC version.)**

The TITLE command names the SPSS run.

The DATA LIST command gives each variable a name, and describes the data as being in free form.

The BEGIN DATA command indicates to SPSS that the input data immediately follow.

The next 10 lines contain the data values, which represent square footage in hundreds and income in thousands of dollars. For example, the first line (16 22) represents a family living in a 1600-square-foot home with an income of $22,000 per year.

The END DATA statement indicates the end of the data.

FIGURE 13.19

Input for SPSSX or SPSS/PC+. All command lines should end with a period when the PC version is used.

```
TITLE    REAL ESTATE EXAMPLE USING ONE PREDICTOR
DATA LIST FREE/SQFOOT INCOME
BEGIN DATA
16 22
17 26
26 45
24 37
22 28
21 50
32 56
18 34
30 60
20 40
END DATA
REGRESSION VARIABLES=SQFOOT,INCOME/DEPENDENT=SQFOOT/ENTER/
  SCATTERPLOT (SQFOOT,INCOME)/
```

The REGRESSION statement defines the variables SQFOOT and IN-COME as the regression variables and specifies that SQFOOT is to be the dependent variable. The ENTER subcommand means that all independent variables are to be entered into the regression analysis. In this example, however, there is only one independent variable, INCOME.

The SCATTERPLOT subcommand requests a scatter diagram of the sample data with SQFOOT on the vertical axis and INCOME on the horizontal axis. Each variable in this plot is standardized; that is, the mean is subtracted from each value and then divided by the standard deviation. Standardized values (referred to as Z scores in Chapter 3) generally range from -3 to 3.

Figure 13.20 shows the SPSSX output obtained by executing the listing in Figure 13.19, whereas Figure 13.21 shows the SPSS/PC+ output.

FIGURE 13.20 SPSSX output.

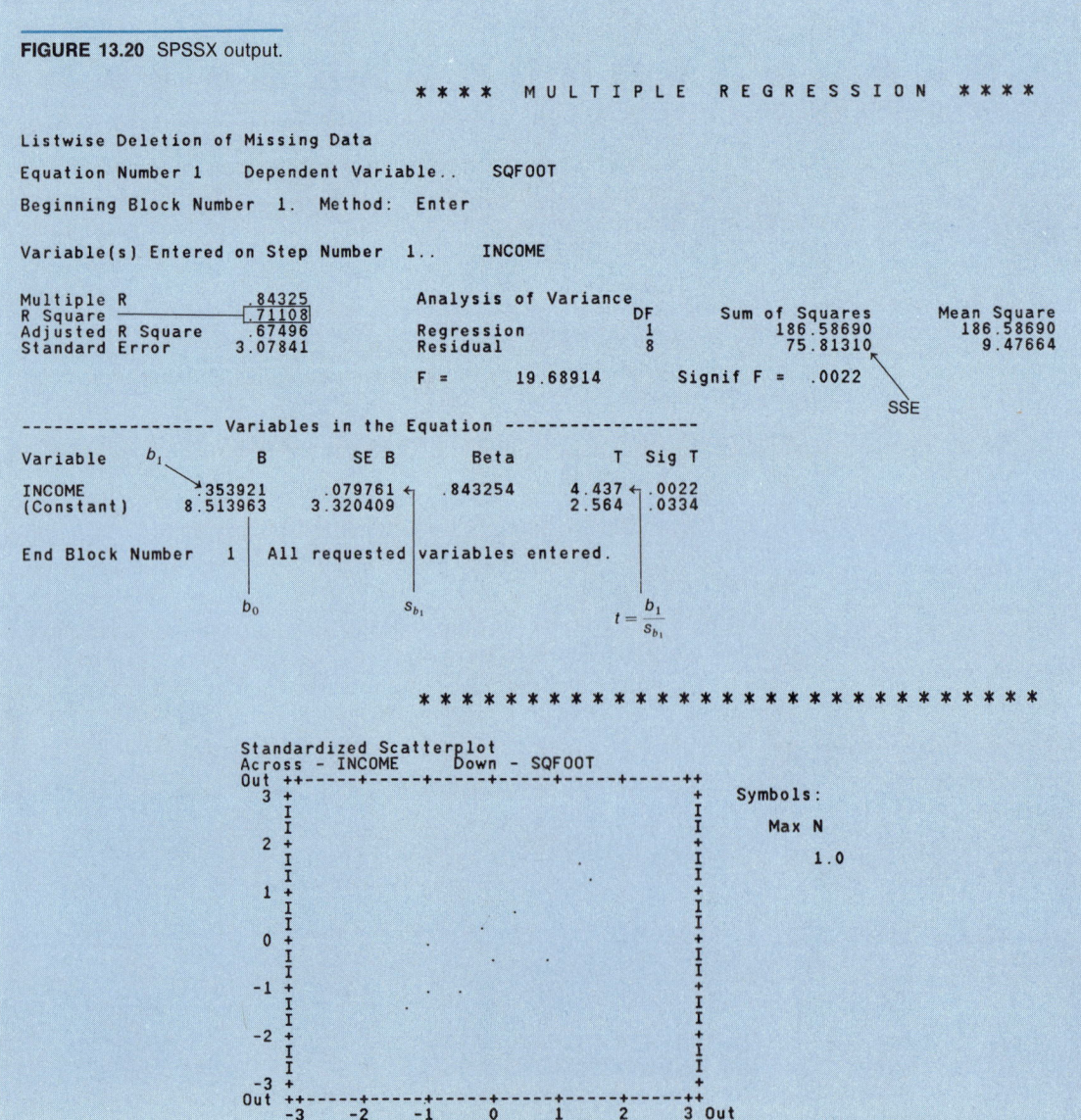

FIGURE 13.21 SPSS/PC+ output.

```
--------------------------------------------------------------------------------
                  REAL ESTATE EXAMPLE USING ONE PREDICTOR

            * * * *   M U L T I P L E   R E G R E S S I O N   * * * *

Equation Number 1    Dependent Variable..   SQFOOT


Variable(s) Entered on Step Number
    1..    INCOME


Multiple R              .84325
R Square                .71108  ←——— r²
Adjusted R Square       .67496
Standard Error         3.07841                        SSE

Analysis of Variance
                    DF        Sum of Squares          Mean Square
Regression           1            186.58690            186.58690
Residual             8             75.81310 ←           9.47664

F =       19.68914      Signif F =  .0022
--------------------------------------------------------------------------------
Page  13    REAL ESTATE EXAMPLE USING ONE PREDICTOR                    8/10/87

            * * * *   M U L T I P L E   R E G R E S S I O N   * * * *

Equation Number 1    Dependent Variable..   SQFOOT


------------------ Variables in the Equation ------------------

Variable      b₁       B         SE B        Beta          T   Sig T

INCOME            .35392      .07976 ←     .84325      4.437 ←  .0022
(Constant)      8.51396     3.32041                    2.564    .0334
              ↑ b₀                     s_b₁                 t = b₁/s_b₁

End Block Number    1    All requested variables entered.
```

b_1

b_0

s_{b_1}

$t = \dfrac{b_1}{s_{b_1}}$

```
                  REAL ESTATE EXAMPLE USING ONE PREDICTOR

Standardized Scatterplot
Across - INCOME    Down - SQFOOT
Out ++-----+-----+-----+-----+-----+-----++
  3 +                                      +      Symbols:
    |                                      |
    |                                      |        Max N
  2 +                                      +
    |                      .               |        .    1.0
    |                                      |
  1 +                                      |
    |                  .                   |
    |                .                     |
  0 +         .                            +
    |              .    .                  |
    |                                      |
 -1 +        .    .                        +
    |           .                          |
    |                                      |
 -2 +                                      +
    |                                      |
    |                                      |
 -3 +                                      +
Out ++-----+-----+-----+-----+-----+-----++
     -3    -2    -1    0     1     2     3 Out
```

■ ■ ■ ■ ■ ■ ■ ■ ■ ■ ■ ■ ■ ■ **S A S** ■ ■ ■ ■ ■ ■ ■ ■ ■ ■ ■ ■ ■ ■

MAINFRAME AND MICRO

SOLUTION

EXAMPLE 13.2

Example 13.2 was concerned with computing the regression equation $\hat{Y} = b_0 + b_1X$. The purpose of this problem was to determine the relationship between family income of local residents and the square footage of their homes. The SAS program listing in Figure 13.22 was used to request the SSE, b_0, and b_1 values as well as other statistics. The regression analysis should always include a scatter diagram of the sample data.

In this problem the SAS commands are the same for both the mainframe and PC versions.

FIGURE 13.22

Input for SAS (mainframe or micro version).

```
TITLE    'REAL ESTATE EXAMPLE USING ONE PREDICTOR';
DATA REALEST;
 INPUT SQFOOT INCOME;
CARDS;
16 22
17 26
26 45
24 37
22 28
21 50
32 56
18 34
30 60
20 40
PROC PLOT DATA = REALEST;
  PLOT SQFOOT*INCOME;
PROC REG;
 MODEL SQFOOT=INCOME;
```

The TITLE command names the SAS run.

The DATA command gives the data a name.

The INPUT command names and gives the correct order for the different fields on the data lines.

The CARDS command indicates to SAS that the input data immediately follow.

The next 10 lines contain the data values, which represent square footage in hundreds and income in thousands of dollars. For example, the first line (16 22) represents a family living in a 1600-square-foot home with an income of $22,000 per year.

The PROC PLOT command requests a scatter diagram of the sample data named in the previous DATA statement. The PLOT SQFOOT* INCOME statement plots the SQFOOT values on the vertical axis and the INCOME values on the horizontal axis.

The PROC REG command and MODEL subcommand indicate that SQFOOT and INCOME are the regression variables, with SQFOOT being the dependent variable and INCOME the independent variable.

Figure 13.23 shows the SAS output obtained by executing the listing in Figure 13.22, whereas Figure 13.24 shows the SAS/PC output.

FIGURE 13.23 SAS output.

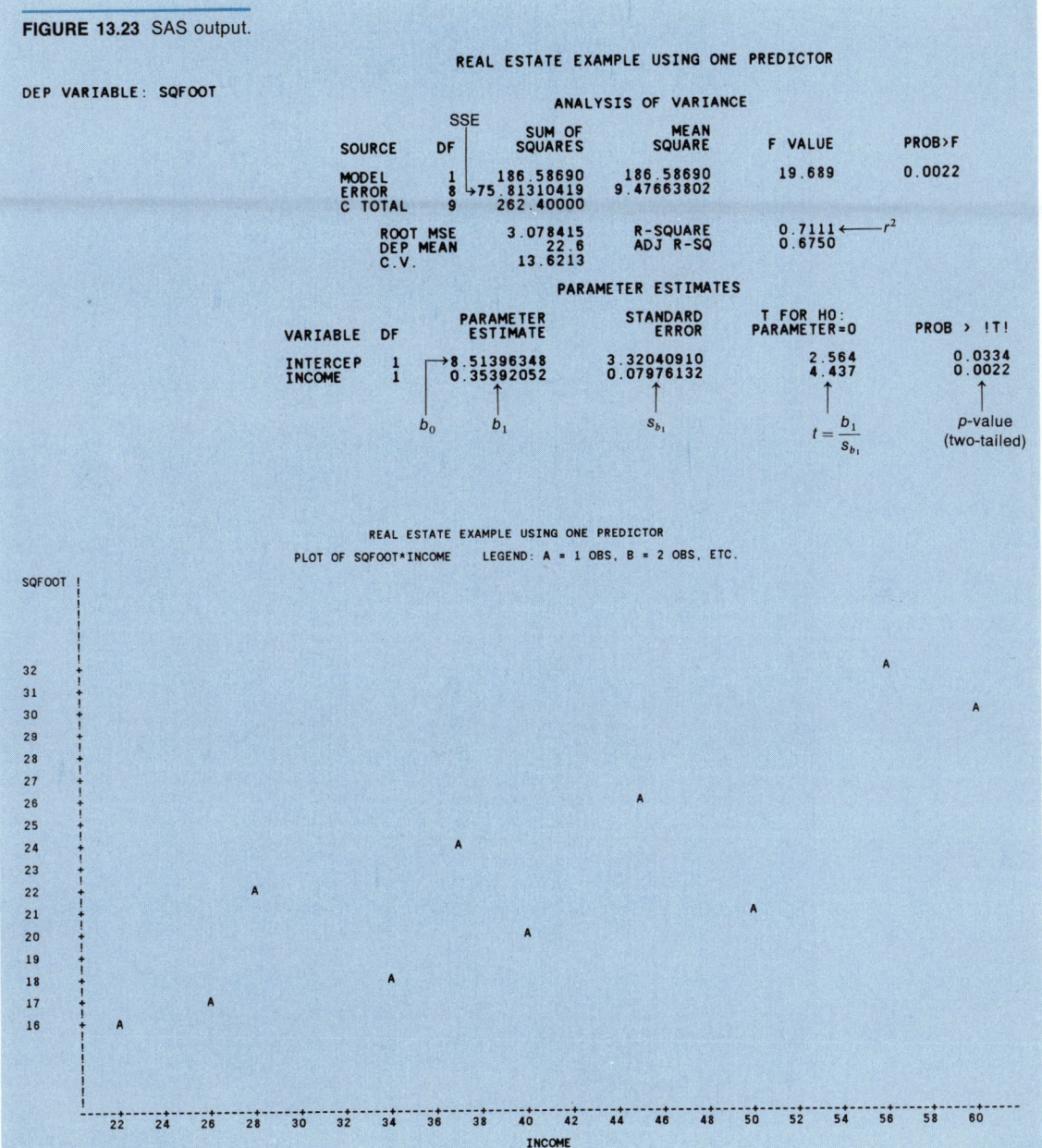

REAL ESTATE EXAMPLE USING ONE PREDICTOR

DEP VARIABLE: SQFOOT

ANALYSIS OF VARIANCE

		SSE SUM OF	MEAN		
SOURCE	DF	SQUARES	SQUARE	F VALUE	PROB>F
MODEL	1	186.58690	186.58690	19.689	0.0022
ERROR	8	75.81310419	9.47663802		
C TOTAL	9	262.40000			

ROOT MSE	3.078415	R-SQUARE	0.7111 ← r^2
DEP MEAN	22.6	ADJ R-SQ	0.6750
C.V.	13.6213		

PARAMETER ESTIMATES

VARIABLE	DF	PARAMETER ESTIMATE	STANDARD ERROR	T FOR H0: PARAMETER=0	PROB > !T!
INTERCEP	1	8.51396348	3.32040910	2.564	0.0334
INCOME	1	0.35392052	0.07976132	4.437	0.0022

b_0 b_1 s_{b_1} $t = \dfrac{b_1}{s_{b_1}}$ p-value (two-tailed)

REAL ESTATE EXAMPLE USING ONE PREDICTOR
PLOT OF SQFOOT*INCOME   LEGEND: A = 1 OBS, B = 2 OBS, ETC.

```
SQFOOT !
       !
       !
       !
   32  +                                                              A
   31  +
   30  +                                                                         A
   29  +
   28  +
   27  +
   26  +                              A
   25  +
   24  +               A
   23  +
   22  +        A
   21  +                                                    A
   20  +                     A
   19  +
   18  +          A
   17  +    A
   16  + A
       !
       !
       !
       !
       +--+--+--+--+--+--+--+--+--+--+--+--+--+--+--+--+--+--+--+--+--
         22 24 26 28 30 32 34 36 38 40 42 44 46 48 50 52 54 56 58 60
                                  INCOME
```

FIGURE 13.24 SAS/PC output.

REAL ESTATE EXAMPLE USING ONE PREDICTOR

Plot of SQFOOT*INCOME. Legend: A = 1 obs, B = 2 obs, etc.

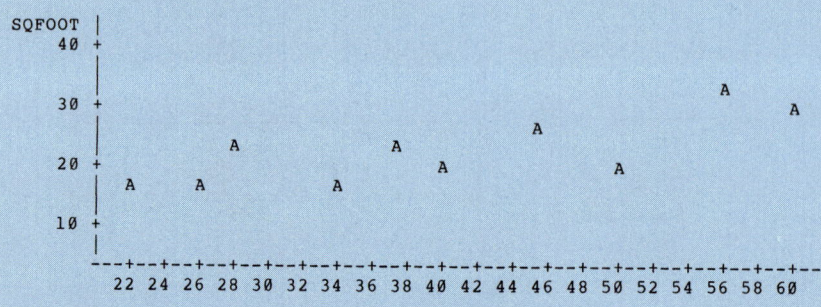

```
SQFOOT |
    40 +
       |
    30 +                                                          A
       |                                                                  A
       |                                              A
    20 +           A              A                       A
       |    A    A           A         A           A
       |
    10 +
       |
       ---+--+--+--+--+--+--+--+--+--+--+--+--+--+--+--+--+--+--+--+--+--
          22 24 26 28 30 32 34 36 38 40 42 44 46 48 50 52 54 56 58 60

                                    INCOME
```

Model: MODEL1
Dependent Variable: SQFOOT

Analysis of Variance

Source	DF	Sum of Squares	Mean Square	F Value	Prob>F
Model	1	186.58690	186.58690	19.689	0.0022
Error	8	75.81310	9.47664		
C Total	9	262.40000			

SSE → (pointing to Sum of Squares column for Model 186.58690 and Error 75.81310)

Root MSE	3.07841	R-square	0.7111 ← r^2
Dep Mean	22.60000	Adj R-sq	0.6750
C.V.	13.62130		

Parameter Estimates

Variable	DF	Parameter Estimate	Standard Error	T for H0: Parameter=0	Prob > |T|
INTERCEP	1	8.513963	3.32040910	2.564	0.0334
INCOME	1	0.353921	0.07976132	4.437	0.0022

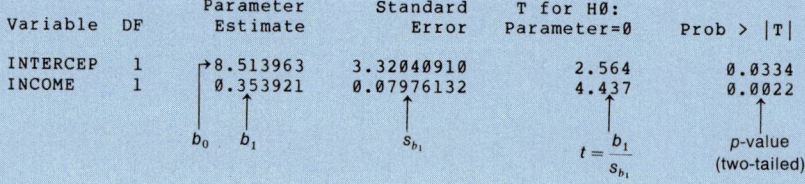

b_0 b_1 s_{b_1} $t = \dfrac{b_1}{s_{b_1}}$ *p*-value (two-tailed)

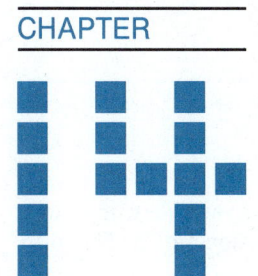

Multiple Linear Regression

Multiple regression equation for population

$$E(y|x\colon) = \beta_0 + \beta_1 X_1 + \beta_2 X_2 + \beta_3 X_3$$

A Look Back/Introduction

We used the technique of simple linear regression in Chapter 13 to explain the behavior of a dependent variable using a single predictor (independent) variable. For example, we can attempt to explain the amount of new housing construction using the interest rate as a predictor variable.

To define this procedure in statistical terms, we introduced the concept of a statistical model. This model consists of two parts. The first is the deterministic component. This is assumed to be $Y = \beta_0 + \beta_1 X$ (a straight line), implying that the underlying pattern for the X and Y variables is linear. If a simple linear regression model is appropriate for the construction illustration, a scatter diagram of the new housing starts (Y) and the corresponding interest rates (X) should reveal a basic linear pattern. We never expect all the sample data to lie *exactly* on a straight line; we realize that with any statistical model there is error involved. This makes up the random component. The actual model used for simple linear regression is $Y = \beta_0 + \beta_1 X + e$, where e represents the distance from the actual Y value to the line passing through all X, Y values. The value of e is the error and represents the error component of the model. The assumptions behind the use of this model are concerned about the behavior of these error terms—are they normally distributed, centered at zero, with the same variance? Are they independent?

In the construction example, it seems reasonable to assume that the volume of housing construction is affected not only by the interest rate but also by many other factors (variables) as well, including cost of materials, geographic location, and unemployment rate in the area. We next look at statistical models used to predict the dependent variable (such as $Y =$ the number of new housing starts) as a function of *more than one* independent variable. The concept and assumptions are the same as before—now we are merely concerned with more than one predictor variable. When we include these additional variables, the predictive ability of the model should be significantly improved. This procedure is called multiple linear regression and is a very useful statistical technique.

THE MULTIPLE LINEAR REGRESSION MODEL

Prediction Using More Than One Variable

To explain or predict the behavior of a certain dependent variable using more than one predictor variable, we use a **multiple linear regression** model. The form of this model is

$$Y = \beta_0 + \beta_1 X_1 + \beta_2 X_2 + \cdots + \beta_k X_k + e \qquad \textbf{(14.1)}$$

where $X_1, X_2, \ldots, X_k$ are the k independent (predictor) variables and e is the error associated with this model.

Notice that equation 14.1 is similar to that used in the simple linear regression model, except that the *deterministic component* is now

$$\beta_0 + \beta_1 X_1 + \cdots + \beta_k X_k \qquad \textbf{(14.2)}$$

rather than $\beta_0 + \beta_1 X$. Once again the error term, e, is included to provide for deviations about this component.

What is the appearance of the deterministic portion in equation 14.2? In Chapter 13, where we discussed simple linear regression, this was a straight line. For the multiple case, this is more difficult (usually impossible) to represent graphically. If your model contains two predictor variables, X_1 and X_2, the deterministic component becomes a plane, as shown in Figure 14.1. Consequently, the key assumption behind the use of this particular model is that the Y values will lie in this plane, *on the average*, for any particular values of X_1 and X_2.

In Chapter 13, we examined the relationship between the square footage of a home (Y) and the corresponding household income (X). The results were:

least squares line: $\hat{Y} = 8.51 + .354X$

correlation between X and Y: $r = .843$

coefficient of determination: $r^2 = .711$

significant linear relationship exists

FIGURE 14.1

The multiple linear regression model (two independent variables).

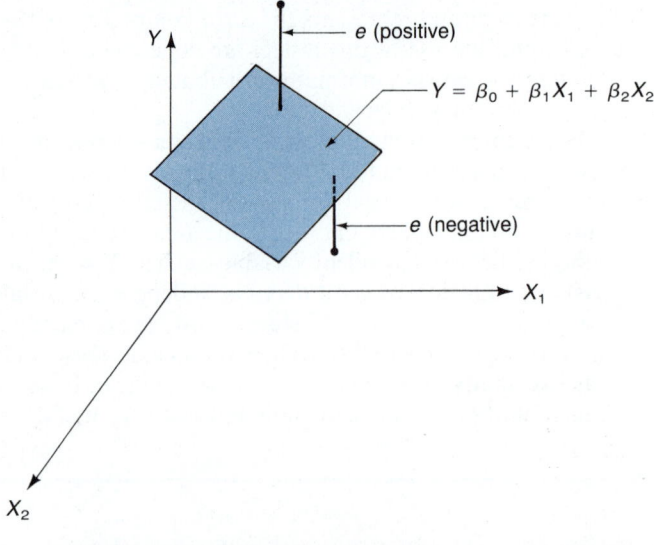

We now want to include two additional variables in the model. The real-estate developer performing the study suspects that (1) larger families have larger homes and (2) the size of the home is affected by the amount of formal education (years of college) of the wage earner(s) in the home. This results in three independent variables.

X_1 = annual income (thousands of dollars)

X_2 = family size

X_3 = combined years of formal education (beyond high school) for all household wage earners

The same ten families were used in the study, but data were collected on the two additional variables, X_2 and X_3.*

FAMILY	Y (HOME SQUARE FOOTAGE)	X_1 (INCOME)	X_2 (FAMILY SIZE)	X_3 (YEARS OF FORMAL EDUCATION)
1	16	22	2	4
2	17	26	2	8
3	26	45	3	7
4	24	37	4	0
5	22	28	4	2
6	21	50	3	10
7	32	56	6	8
8	18	34	3	8
9	30	60	5	2
10	20	40	3	6

The data configuration now has four columns (including Y) and ten rows (called *observations*). Our task is to use the data on all *three* variables (X_1, X_2, and X_3) to provide a better estimate of home size (Y).

The Least Squares Estimate   Using Figure 14.1, we proceed as we did for simple regression and determine an estimate of the β's that makes the sum of squares of the residuals as small as possible. A **residual** is defined as the difference between the actual Y value and its estimate; that is, $Y - \hat{Y}$. In other words, we attempt to find the $b_0, b_1, \ldots, b_k$ that minimize the sum of squares of error,

$$SSE = \Sigma(Y - \hat{Y})^2 \qquad (14.3)$$

where now $\hat{Y} = b_0 + b_1X_1 + b_2X_2 + \cdots + b_kX_k$ and $b_0, b_1, \ldots, b_k$ are called the **least squares estimates** of $\beta_0, \beta_1, \ldots, \beta_k$.

By determining the estimated *regression coefficients* ($b_0, b_1, \ldots, b_k$) that minimize SSE rather than $\Sigma(Y - \hat{Y})$, we once again avoid the problem of positive errors canceling out negative ones. Another advantage of this procedure is that, by means of a little calculus, we can show that a fairly simple expression exists for these sample regression coefficients. Because this expression involves the use of *matrix notation*, we omit this result.[†]

---

* A sample of size ten is unrealistically small in practice.

† Information on this expression is presented in W. Mendenhall and T. Sincich, *A Second Course in Business Statistics: Regression Analysis* 2d ed. (San Francisco: Dellen, 1986); J. Neter, W. Wasserman, and M. Kutner, *Applied Linear Regression Models* (Homewood, Ill.: Richard D. Irwin, 1983).

There is only one way to solve a multiple regression problem in practice, and that is with the help of a computer. All computer packages determine the values of $b_0, b_1, \ldots, b_k$ in the same way—namely, by minimizing SSE. As a result, these values will be identical (except for numerical rounding errors) regardless of which computer program you use.

In the example attempting to predict home size using the three predictor variables, the prediction equation is

$$\hat{Y} = b_0 + b_1 X_1 + b_2 X_2 + b_3 X_3$$

where

$$\hat{Y} = \text{predicted home size}$$
$$X_1 = \text{income}$$
$$X_2 = \text{family size}$$
$$X_3 = \text{years of education}$$

and b_0, b_1, b_2 and b_3 are the least squares estimates of $\beta_0, \beta_1, \beta_2,$ and β_3.

Figure 14.2 contains the MINITAB solution using the data we presented. According to this output, the best prediction equation (in the least squares sense) of home size is

$$\hat{Y} = 7.60 + .194 X_1 + 2.34 X_2 - .163 X_3$$

So this solution minimizes SSE. But what is the SSE here? We need to determine how well this equation "fits" the ten observations in the data set. Consider the first family, where $X_1 = 22$ (income = \$22,000), $X_2 = 2$ (family size = 2, such as an adult couple with no children), and $X_3 = 4$ (combined years of college = 4). The predicted home size here is

$$\hat{Y} = 7.60 + .194(22) + 2.34(2) - .163(4) = 15.89$$

FIGURE 14.2

MINITAB multiple regression solution to house size using three predictor variables. See text.

```
MTB > READ INTO C1-C4
DATA> 16 22 2 4
DATA> 17 26 2 8
DATA> 26 45 3 7
DATA> 24 37 4 0
DATA> 22 28 4 2
DATA> 21 50 3 10
DATA> 32 56 6 8
DATA> 18 34 3 8
DATA> 30 60 5 2
DATA> 20 40 3 6
DATA> END
       10 ROWS READ
MTB > REGRESS Y IN C1 USING 3 PREDICTORS IN C2-C4

The regression equation is
C1 = 7.60 + 0.194 C2 + 2.34 C3 - 0.163 C4

Predictor      Coef   b₀    Stdev    t-ratio       p
Constant      7.596   b₁   2.595      2.93     0.026
C2          0.19388       0.08770     2.21     0.069
C3           2.3381   b₂  0.9078      2.58     0.042
C4          -0.1628   b₃  0.2441     -0.67     0.530

s = 2.035  ⟵ s   R-sq = 90.5%     R-sq(adj) = 85.8%

Analysis of Variance
                              SSE
SOURCE        DF        SS          MS        F       p
Regression     3     237.542     79.181     19.11   0.002
Error          6      24.858      4.143
Total          9     262.400
```

Consequently, the predicted home size is 1589 square feet. The actual square footage for this observation is 1600 ($Y = 16$), so the sample residual here is $Y - \hat{Y} = 16 - 15.89 = .11$.

Using this procedure on the remaining nine observations, we find the following results:

Y	$\hat{Y}$	$Y - \hat{Y}$	$(Y - \hat{Y})^2$
16	15.89	0.11	.0121
17	16.01	0.99	.9801
26	22.19	3.81	14.5161
24	24.12	−0.12	.0144
22	22.05	−0.05	.0025
21	22.68	−1.68	2.8224
32	31.18	0.82	.6724
18	19.90	−1.90	3.6100
30	30.59	−0.59	.3481
20	21.39	−1.39	1.9321
		0	24.91 ≈ SSE

The computed value for the error sum of squares is SSE = 24.91. This value also is contained in the MINITAB output in Figure 14.2. The MINITAB value for SSE is 24.86, which differs slightly from the previous result because the computer uses much more accurate calculations than those in the table. SSE = 24.86 is more accurate, so we will use this value in the remaining discussion.

This implies that for *any* other values of b_0, b_1, b_2, and b_3, if we were to find the corresponding $\hat{Y}$'s and the resulting SSE = $\Sigma(Y - \hat{Y})^2$ using these values, this new SSE would be *larger* than 24.86. Thus, $b_0 = 7.60$, $b_1 = .194$, $b_2 = 2.34$, and $b_3 = -.163$ minimize the error sum of squares, SSE. Put still another way, these values of b_0, b_1, b_2, and b_3 provide the **best fit** to our data.

Using only income (X_1) as a predictor, in Chapter 13 we found the SSE to be 75.81 in our table. By including the additional two variables, the SSE has been reduced from 75.81 to 24.86 (a 67% reduction). It appears that either family size (X_2), years of education (X_3), or both contribute, perhaps significantly, to the prediction of Y.

Interpreting the Regression Coefficients   When using a multiple linear regression equation, such as $Y = \beta_0 + \beta_1 X_1 + \beta_2 X_2 + \beta_3 X_3 + e$, what does β_2 represent? Very simply, it reflects the change in Y that can be expected to accompany a change of one unit in X_2 provided *all other variables* (namely, X_1 and X_3) *are held constant*.

In the previous example, the sample estimate of β_2 was $b_2 = 2.34$. Can we expect an increase of 2.34 every time X_2 (the family size) increases by one if X_1 and X_3 are held constant? This type of argument is filled with problems, as we demonstrate later. The primary problem is that a change in one of the predictor variables (such as X_2) always (or almost always) is accompanied by a change in one of the other predictors (say, X_1) in the sample observations. Consequently, variables X_1 and X_2 are related in some manner, such as $X_1 \cong 1 + 5X_2$. In other words, a situation in which X_2, for instance, changed and the others remained constant would not be observed within the sample data.

In the other case (typically not observed in business applications), the predictor variables *are* totally unrelated. In this situation, a unit change in X_2, for example, can be expected to be accompanied by a change of β_2 in the dependent variable.

In general, it is *not* safe to assume that the predictor variables are unrelated. As a result, the b's usually do not reflect the true "partial effects" of the predictor variables, and you should avoid such conclusions. Section 14.4 discusses methods of dealing with this type of situation.

The Assumptions Behind the Multiple Linear Regression Model

The form of the multiple linear regression model is given by equation 14.1, which contains a linear combination of the k predictor (independent) variables as well as the error component, e. The assumptions for the case of $k > 1$ predictors are exactly the same as for $k = 1$ independent variable (simple linear regression). These assumptions, discussed in Chapter 13, are:

1. The errors follow a normal distribution, centered at zero, with common variance, σ_e^2.
2. The errors are (statistically) independent.

For the case of $k = 2$ predictor variables, this can be represented graphically, as shown in Figure 14.1 and 14.3. Using Figure 14.3, consider the situation in which $X_1 = 20$ and $X_2 = 15$. If you *were* to obtain repeated values of Y having these values for X_1 and X_2, you would obtain some Y's above the plane and some below. The assumption is that the *average* value of Y with $X_1 = 20$ and $X_2 = 15$ lies *on* the plane. Moreover, these errors are normally distributed.

The final part of assumption 1 is that the variation about this plane does not depend on the values of X_1 and X_2. You should see roughly the same amount of variation if you obtain repeated values of Y corresponding to $X_1 = 30$ and $X_2 = 5$ as you observed for $X_1 = 20$ and $X_2 = 15$. The variance of these errors, if you could observe Y indefinitely, is σ_e^2.

Finally, assumption 2 means that the error encountered at $X_1 = 30$ and $X_2 = 5$, for instance, is not affected by a known error at any other point, such as $X_1 = 20$ and $X_2 = 15$. The error associated with one pair of X_1, X_2 values has no effect on any other error.

An Estimate of σ_e^2   When using a straight line to model a relationship between Y and a single predictor, the estimate of σ_e^2 was given by equation 13.16, where

$$s^2 = \hat{\sigma}_e^2 = \frac{\text{SSE}}{n - 2}$$

FIGURE 14.3

The errors in multiple linear regression ($k = 2$).

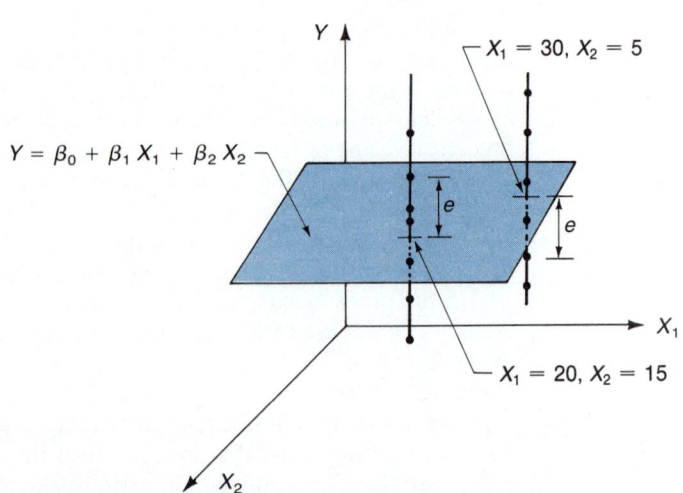

$Y = \beta_0 + \beta_1 X_1 + \beta_2 X_2$

$X_1 = 30, X_2 = 5$

$X_1 = 20, X_2 = 15$

In general, for k predictors and n observations, the estimate of this variance is

$$s^2 = \hat{\sigma}_e^2 = \frac{\text{SSE}}{n - (k + 1)} = \frac{\text{SSE}}{n - k - 1} \qquad (14.4)$$

The value of s^2 is critical in determining the reliability and usefulness of the model as a predictor. If $s^2 = 0$, then SSE $= 0$, which implies that $Y = \hat{Y}$ for each of the observations in the sample data. This rarely happens in practice, but it does point out that a small s^2 is desirable. As s^2 increases, you can expect more error when predicting a value of Y for specified values of $X_1, X_2, \ldots, X_k$. In the next section, we use s^2 as a key to determining whether the model is satisfactory and which of the independent variables are useful in the prediction of the dependent variable.

The square root of this estimated variance is the **residual standard deviation**.

$$s = \sqrt{\frac{\text{SSE}}{n - k - 1}} \qquad (14.5)$$

In the MINITAB solution in Figure 14.2, the value of s is shown in the box containing $s = 2.035$.

EXAMPLE 14.1   Determine the estimate of σ_e^2 and the residual standard deviation for the real-estate data on page 563.

SOLUTION   This example contained $n = 10$ observations and $k = 3$ predictor variables. The resulting error sum of squares was SSE $= 24.86$ (from Figure 14.2). Therefore,

$$s^2 = \hat{\sigma}_e^2$$

$$= \frac{24.86}{10 - 3 - 1} = \frac{24.86}{6} = 4.14$$

Also,

$$s = \sqrt{4.14} = 2.03$$

That is, the residual standard deviation is 203 square feet.   ∎

If a particular regression model meets all the required assumptions, then the next question of interest is whether this set of independent variables provides an accurate method of predicting the dependent variable, Y. The next section shows how to calculate the predictive ability of your model and determine which variables contribute significantly to an accurate prediction of Y.

EXERCISES

14.1   A management-consulting firm uses a regression model where variable X_1 stands for previous experience, variable X_2 for number of years at current job, and variable X_3 for score on a job-aptitude test. These variables are used in a regression model to predict job satisfaction. Job satisfaction ranges from one to 20, with 20 indicating an employee is satisfied with every aspect of his or her job. The prediction equation is:

$$\hat{Y} = 1.7 - 0.15X_1 + 0.25X_2 + 0.14X_3$$

a. What would the consulting firm predict for the job satisfaction of an employee who has 15 years of prior experience, 10 years of employment at the present job, and an aptitude test score of 85?

b. If an employee's score on the aptitude test is increased by 20, with the years of prior experience and years of employment at the current job remaining constant, what would be the net change in the employee's predicted satisfaction score?

14.2 The marketing department at Computeron would like to predict its sales volume of software for its personal computers. They believe that the sales volume of software (given in thousands) increases when the number of units of personal computers increases and when advertising expenditure (given in thousands of dollars) increases. The following data were collected for 14 months:

Y: SALES OF SOFTWARE	X_1: UNITS OF PERSONAL COMPUTERS SOLD	X_2: ADVERTISING EXPENDITURE	Y: SALES OF SOFTWARE	X_1: UNITS OF PERSONAL COMPUTERS SOLD	X_2: ADVERTISING EXPENDITURES
7.2	12	4.2	3.4	10	2.1
5.4	11	3.1	7.0	12	3.8
7.7	14	5.1	12.1	22	5.8
5.6	11	3.5	8.4	14	4.9
9.1	17	5.4	9.7	19	5.5
8.8	17	4.4	8.3	15	4.6
6.2	11	3.5	7.1	13	4.0

a. Using the least squares line $\hat{Y} = -1.21768 + 0.3141X_1 + 1.016X_2$, find the estimate of the variance of the error component.

b. Interpret the coefficients of the regression equation in the context of the problem. Would it be reasonable to expect sales of software to increase by 1.016 thousand if X_2 is increased by one?

14.3 An oil-service company decided to fit a least squares equation to a set of data to predict the total cost of building a well. The independent variables are $X_1 =$ drilling days, $X_2 =$ total depth, and $X_3 =$ intermediate casing depth. After calculating the least squares equation, the residuals were calculated to find out whether the assumptions of regression analysis are satisfied. The following are the residuals from 20 observations:

$$-0.8,\ 1.5,\ -3.7,\ 4.1,\ -3.1,\ -5.2,\ 4.3,\ -2.1,\ -1.6,\ 4.1,\ 0.9,\ -0.3,$$
$$4.5,\ -4.2,\ 3.2,\ -2.7,\ 1.7,\ -2.2,\ 3.4,\ -1.8$$

Do the residuals $Y - \hat{Y}$ appear to conform to the empirical rule that approximately 95% of the data should lie within two standard deviations of the mean?

14.4 What assumptions need to be made about the error component of a multiple linear regression model in order that the results of statistical inference can be used?

14.5 Tony owns a used-car lot. He would like to predict monthly sales volume. Tony believes that sales volume (given in thousands) is directly related to the number of salespeople employed and the number of cars on the lot for sale. The following data were collected over a period of 10 months:

Y: MONTHLY SALES VOLUME	X_1: SALESPEOPLE	X_2: CARS	Y: MONTHLY SALES VOLUME	X_1: SALESPEOPLE	X_2: CARS
5.8	4	20	8.1	4	25
7.5	5	15	13.3	8	30
11.4	7	25	15.0	9	35
7.0	3	17	8.3	5	20
5.1	2	18	6.8	4	23

a. Using a computerized statistical package, determine the least squares prediction equation here.

b. Find the value of SSE.

14.6 Using the multiple regression model $\hat{Y} = b_0 + b_1X_1 + b_2X_2 + b_3X_3$, where do you expect the average value of Y to fall for $X_1 = 3$, $X_2 = 4.1$, and $X_3 = 5.6$?

14.7 A regression analysis was performed for data with three independent variables. The following residuals were found for the 20 observations of the dependent variable:

$$5.4, \ 8.1, \ -7.4, \ 2.5, \ -3.5, \ -4.1, \ -8.1, \ 6.5, \ 4.3, \ -7.8, \ 2.8, \ 7.1, \ -6.2,$$
$$5.6, \ -5.1, \ 2.9, \ 2.8, \ -7.2, \ 8.3, \ -6.9$$

Find the residual standard deviation.

14.8 A real-estate broker uses four independent variables to predict the appraised value of homes in a certain subdivision. The variables are $X_1 = $ lot size in square feet, $X_2 = $ house size in square feet, $X_3 = $ age of the house in years, and $X_4 = X_2 X_3$. Twenty-seven observations were used to find the least squares prediction equation. The real-estate broker found the value of the residual standard deviation to be 650. What is the value of SSE?

14.2

HYPOTHESIS TESTING AND CONFIDENCE INTERVALS FOR THE β PARAMETERS

Multiple linear regression is a popular tool in the application of statistical techniques to business decisions. However, this modeling procedure does not always result in an accurate and reliable predictor. When the independent variables that you have selected account for very little of the variation in the values of the dependent variable, the model (as is) serves no useful purpose.

The first thing we demonstrate is how to determine whether your overall model is satisfactory. We begin by summarizing a regression analysis in an ANOVA table, much as we did in Chapter 11.

The ANOVA Table

The summary ANOVA table contains the usual headings.

SOURCE	df	SS	MS	F
Regression	k	SSR	MSR	MSR/MSE
Residual	$n - k - 1$	SSE	MSE	
Total	$n - 1$	SST		

where $n = $ number of observations and $k = $ number of independent variables.

$$
\begin{aligned}
\text{SST} &= \text{total sum of squares} \\
&= \text{SS}_Y \\
&= \Sigma(Y - \bar{Y})^2 = \Sigma Y^2 - (\Sigma Y)^2/n \qquad \textbf{(14.6)} \\
\text{SSE} &= \text{sum of squares for error} \\
&= \Sigma(Y - \hat{Y})^2 \qquad \textbf{(14.7)} \\
\text{SSR} &= \text{sum of squares for regression} \\
&= \Sigma(\hat{Y} - \bar{Y})^2 \\
&= \text{SST} - \text{SSE} \qquad \textbf{(14.8)} \\
\text{MSR} &= \text{mean square for regression} \\
&= \text{SSR}/k \qquad \textbf{(14.9)} \\
\text{MSE} &= \text{mean square for error} \\
&= \text{SSE}/(n - k - 1) \qquad \textbf{(14.10)}
\end{aligned}
$$

Practically all computer packages provide you with this ANOVA summary as part of the standard output. The ANOVA section of the MINITAB solution for the real-estate model is highlighted in Figure 14.4.

FIGURE 14.4

MINITAB output (see Figure 14.2). **A**: Prediction equation and ANOVA table using X_1, X_2, and X_3. **B**: Prediction equation and ANOVA table using X_1 and X_2.

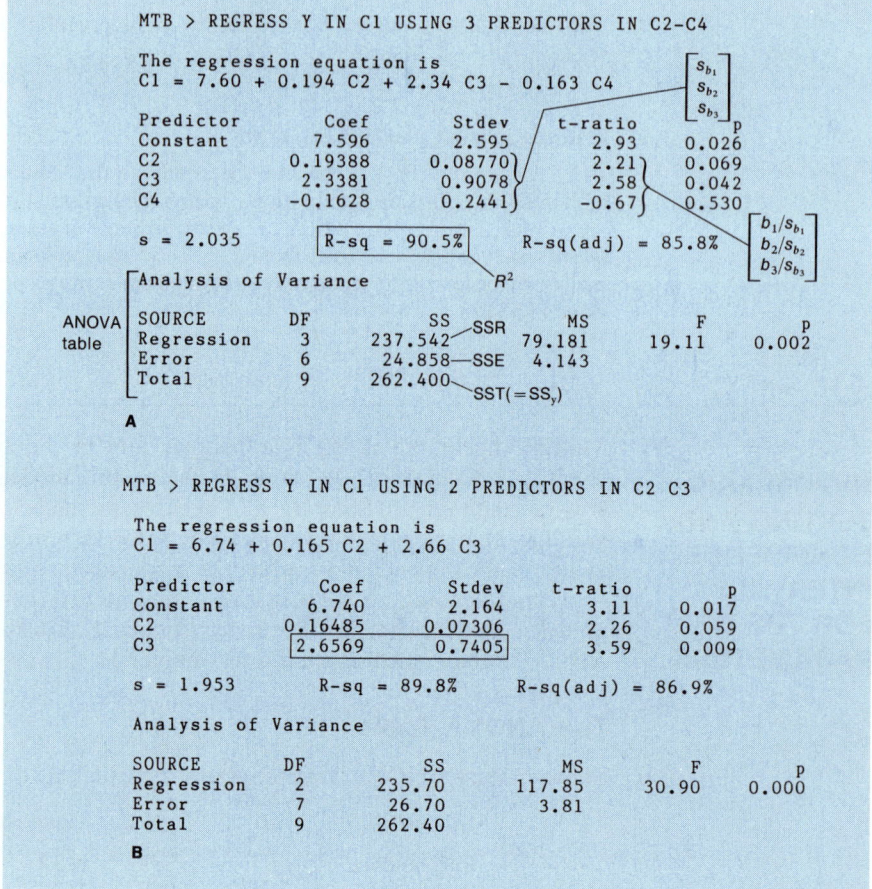

Notice that

$$SST = SS_Y$$

$$= (16^2 + 17^2 + \cdots + 20^2) - (16 + 17 + \cdots + 20)^2/10$$

$$= 262.4$$

This is the same value of SS_Y we obtained for the same example in Chapter 13 when we used only income (X_1) as the predictor variable. This is hardly surprising because *this value is strictly a function of the Y values* and is unaffected by the model that we are using to predict Y. The total sum of squares (SST) measures the total variation in the values of the dependent variable. Its value is the same, regardless of which predictor variables are included in the model.

The df for the regression source of variation are $k =$ the number of predictor variables in the analysis. The df for the error sum of squares are $n - k - 1$, where $n =$ the number of observations in the sample data.

As in the case of simple linear regression, the sum of squares of regression (SSR) measures the variation *explained* by the model—the variation in the Y values that would exist if differences in the values of the predictor variables were the only cause of differences among the Y's. On the other hand, the sum of squares of error (SSE) represents the variation *unexplained* by the model. The easiest way to determine the sum of squares of regression is to subtract $SSR = SST - SSE$.

The error mean square is $\text{MSE} = \text{SSE}/(n - k - 1) = 24.858/(10 - 3 - 1) = 4.14$. This is the same as the *estimate* of σ_e^2 determined in Example 14.1. So

$$s^2 = \hat{\sigma}_e^2 = \text{MSE}$$

A Test for H_0: all β's $= 0$

We have yet to make use of the F-value calculated in the ANOVA table, where

$$F = \frac{\text{MSR}}{\text{MSE}} \qquad\qquad (14.11)$$

When using the simple regression model, we previously argued that one way to determine whether X is a significant predictor of Y is to test $H_0\colon \beta_1 = 0$, where β_1 is the coefficient of X in the model $Y = \beta_0 + \beta_1 X + e$. If you reject H_0, the conclusion is that the independent variable X *does* contribute significantly to the prediction of Y. For example, in Example 13.3, by rejecting $H_0\colon \beta_1 = 0$, we concluded that income (X_1) was a useful predictor of home size (Y) using the simple linear model.

We use a similar test as the first step in the multiple regression analysis, where we examine the hypotheses

$$H_0\colon \beta_1 = \beta_2 = \cdots = \beta_k = 0$$
$$H_a\colon \text{at least one of the } \beta\text{'s} \neq 0$$

If we *reject* H_0, we can conclude that at least one (but maybe not all) of the independent variables contributes significantly to the prediction of Y. If we *fail to reject* H_0, we are unable to demonstrate that any of the independent variables (or combination of them) helps explain the behavior of the dependent variable, Y. For example, in our housing example, if we were to fail to reject H_0, this would imply that we are unable to demonstrate that the variation in the home sizes (Y) can be explained by the effect of income, family size, and years of education.

Test Statistic for H_0 Versus H_a The test statistic used to determine whether our multiple regression model contains at least one explanatory variable is the F statistic from the preceding ANOVA table.

When testing H_0: all β's $= 0$ (this is a poor set of predictor variables) versus H_a: at least one $\beta \neq 0$ (at least one of these variables is a good predictor), the test statistic is

$$F = \frac{\text{MSR}}{\text{MSE}}$$

which has an F distribution with k and $n - k - 1$ df. The expression $n - k - 1$ can be written as $n - (k + 1)$, where $k + 1$ is the number of coefficients (β's) estimated including the constant term.

Notice that the df for the F statistic come directly from the ANOVA table. The testing procedure is to

$$\text{reject } H_0 \text{ if } F > F_{\alpha, v_1, v_2}$$

where (1) $v_1 = k$, $v_2 = n - k - 1$ and (2) F_{α, v_1, v_2} is the corresponding F-value in Table A.7, having a *right-tail area* $= \alpha$ (Figure 14.5).

FIGURE 14.5

F curve with *k* and
n − *k* − 1 df. The
lightly shaded area is
the rejection region.

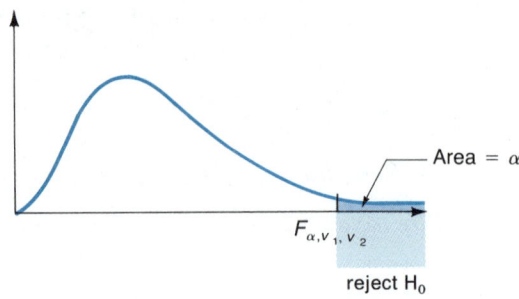

EXAMPLE 14.2

Using the real-estate data and the model we developed, what can you say about the predictive ability of the independent variables, income (X_1), family size (X_2), and years of education, (X_3)? Use $\alpha = .10$.

SOLUTION

Step 1. The hypotheses are

$$H_0: \beta_1 = \beta_2 = \beta_3 = 0$$
$$H_a: \text{at least one } \beta \neq 0$$

Remember that our hope here is to reject H_0. If you are unable to demonstrate that any of your independent variables have any predictive ability, then you will fail to reject H_0.

Step 2. The test statistic is

$$F = \frac{\text{MSR}}{\text{MSE}}$$

The mean squares are obtained from the ANOVA summary of the regression analysis (see Figure 14.4).

Step 3. The df for the F statistic are $k = 3$ and $n - k - 1 = 10 - 3 - 1 = 6$. So we will

$$\text{reject } H_0 \text{ if } F > F_{.10, 3, 6} = 3.29$$

Step 4. Using the results in Figure 14.4, the computed F-value is

$$F^* = \frac{79.18}{4.14} = 19.1$$

Because $F^* > 3.29$, we reject H_0.

Step 5. The three independent variables *as a group* constitute a good predictor of home size. This does *not* imply that all three variables have significant predictive ability; however, at least one of them does. The next section shows how you can tell *which* of these predictor variables significantly contributes to the prediction of home size.

A Test for $H_0: \beta_i = 0$

Assuming that you rejected the null hypothesis that all of the β's are zero, the next logical question would be, Which of the independent variables contributes to the prediction of Y?

In Example 14.2, we rejected the null hypothesis, so at least one of these three independent variables affects the variation of the ten home sizes in the sample. To determine the contribution of each variable, we perform three separate t-tests:

$$H_0: \beta_1 = 0 \; (X_1 \text{ does not contribute})$$
$$H_a: \beta_1 \neq 0 \; (X_1 \text{ does contribute})$$
$$H_0: \beta_2 = 0 \; (X_2 \text{ does not contribute})$$
$$H_a: \beta_2 \neq 0 \; (X_2 \text{ does contribute})$$
$$H_0: \beta_3 = 0 \; (X_3 \text{ does not contribute})$$
$$H_a: \beta_3 \neq 0 \; (X_3 \text{ does contribute})$$

One-tailed tests also can be used here, but we demonstrate this procedure using two-tailed tests. This means that we are testing to see whether this particular X contributes to the prediction of Y, but we are not concerned about the direction (positive or negative) of this relationship.

When income (X_1) was the only predictor of home size (Y), we used a t-test to determine whether the simple linear regression model was adequate. In Example 13.3, the value of the test statistic was derived, where

$$t = \frac{b_1}{s_{b_1}} \tag{14.12}$$

Also, b_1 is the estimate of β_1 in the simple regression model, and s_{b_1} is the (estimated) standard deviation of b_1.

All computer packages provide both the estimated coefficient (b_1) and its standard deviation (s_{b_1}). In Example 13.3, the computed value of this t statistic was $t^* = 4.44$. This led us to conclude that income was a good predictor of home size because a significant positive relationship existed between these two variables.

We use the same t-statistic procedure to test the effect of the individual variables in a multiple regression model. When examining the effect of an individual independent variable, X_i, on the prediction of a dependent variable, the hypotheses are

$$H_0: \beta_i = 0$$
$$H_a: \beta_i \neq 0$$

The test statistic is

$$t = \frac{b_i}{s_{b_i}}$$

where (1) b_i is the estimate of β_i, (2) s_{b_i} is the (estimated) standard deviation of b_i, and (3) the df for the t-statistic are $n - k - 1$.

The test of H_0 versus H_a is to

$$\text{reject } H_0 \text{ if } |t| > t_{\alpha/2, n-k-1}$$

where $t_{\alpha/2, n-k-1}$ is obtained from Table A.5.

We can now reexamine the real-estate data in Example 14.2.

X_1 = Income Consider the hypotheses

$$H_0: \beta_1 = 0$$
$$H_a: \beta_1 \neq 0$$

As in Example 14.2, we use $\alpha = .10$.

According to Figure 14.4, $b_1 = .194$ and $s_{b_1} = .0877$. Also contained in the output is the computed value of

$$t^* = \frac{b_1}{s_{b_1}} = \frac{.194}{.0877} = 2.21$$

Why is this value of t^* *not* the same as the value of t calculated previously in Chapter 13 for this variable, when income was the only predictor of Y? When there are three predictors in the model, t^* for income is 2.21. When income is the only predictor in the model, $t^* = 4.44$. The difference in the two values is simply that $t^* = 2.21$ provides a measure of the contribution of $X_1 = $ income, *given that X_2 and X_3 already have been included in the model.* A large value of t^* indicates that X_1 contributes significantly to the prediction of Y, even if X_2 and X_3 have been included previously as predictors.

The hypotheses can better be stated as

H_0: income *does not* contribute to the prediction of home size, *given* that family size and years of education already have been included in the model

H_a: income *does* contribute to this prediction, given that family size and years of education already have been included in the model

or as

H_0: $\beta_1 = 0$ (if X_2 and X_3 are included)
H_a: $\beta_1 \neq 0$

Because $t^* = 2.21$ exceeds the table value of $t_{\alpha/2, n-k-1} = t_{.05, 10-3-1} = t_{.05, 6} = 1.943$, we conclude that income contributes significantly to the prediction of home size and should be kept in the model.

$X_2 = $ **Family Size**   Using a similar argument, the following test of hypothesis will determine the contribution of family size, X_2, as a predictor of the home square footage, given that X_1 and X_3 already have been included. The hypotheses here are

H_0: $\beta_2 = 0$ (if X_1 and X_3 are included)
H_a: $\beta_2 \neq 0$

According to Figure 14.4, the computed t-statistic here is

$$t^* = \frac{b_2}{s_{b_2}} = \frac{2.34}{.9078} = 2.58$$

This also exceeds $t_{.05, 6} = 1.943$, and so family size provides useful information in predicting the square footage of a home. We conclude that we should keep X_2 in the model.

$X_3 = $ **Years of Education**   To test

H_0: $\beta_3 = 0$ (if X_1 and X_2 are included)
H_a: $\beta_3 \neq 0$

we once again use the t-statistic.

$$t = \frac{b_3}{s_{b_3}}$$

Using Figure 14.4, the computed value of this statistic is

$$t^* = \frac{-.163}{.2441} = -.67$$

Because $|t^*| = .67$, which does *not* exceed $t_{.05,6} = 1.943$, we fail to reject H_0. We conclude that, given the values of $X_1 = $ income and $X_2 = $ family size, the level of a family's education appears not to contribute to the prediction of the size of their home. This means that X_3 can be ignored in the final prediction equation, leaving only X_1 and X_2. As a word of warning, you should *not* simply remove this term from the equation containing all three variables. Since the predictor variables are typically related in some manner, the sample regression coefficients $(b_0, b_1, \ldots)$ change as variables are added to or deleted from the model. Referring to Figure 14.4(a), the final prediction equation is not $\hat{Y} = 7.60 + .194X_1 + 2.34X_2$. Instead, the coefficients of X_1 and X_2 should be derived by repeating the analysis using only these two variables. According to Figure 14.4(b), this prediction equation is $\hat{Y} = 6.74 + .165X_1 + 2.66X_2$.

A Confidence Interval for β_i

Using what you believe to be the "best" model, you can easily construct a $(1 - \alpha) \cdot 100\%$ confidence interval for β_i based upon the previous t-statistic:

$$b_i - t_{\alpha/2, n-k-1}s_{b_i} \quad \text{to} \quad b_i + t_{\alpha/2, n-k-1}s_{b_i} \qquad \textbf{(14.13)}$$

Once again, k represents the number of predictor variables used to estimate β_i.

EXAMPLE 14.3

Suppose you decide to retain only $X_1 = $ income and $X_2 = $ family size in the prediction equation. Referring to Figure 14.4(b), construct a 90% confidence interval for β_2, the coefficient for X_2.

SOLUTION

Since this model contains $k = 2$ predictor variables, we first find $t_{\alpha/2, n-k-1} = t_{.05,7} = 1.895$. Based upon the MINITAB results in Figure 14.4(b), the confidence interval for β_2 is

$$2.6569 - (1.895)(.7405) \quad \text{to} \quad 2.6569 + (1.895)(.7405)$$
$$= 2.6569 - 1.4032 \quad \text{to} \quad 2.6569 + 1.4032$$
$$= 1.25 \quad \text{to} \quad 4.06$$

Therefore, we are 90% confident that the estimate of β_2 (that is, $b_2 = 2.6569$) is within 1.4032 of the actual value of β_2. Notice that this is an extremely wide confidence interval. As usual, increasing the sample size would help to reduce the width of this confidence interval. ∎

EXAMPLE 14.4

The management of BB Investments decided to develop a model to predict the amount of money invested by various clients in their portfolio of high-risk securities. It was generally agreed that the income of the investor should be a major factor in predicting his or her annual investment and would explain a major portion of the variability in the amount invested. In addition, the investor's willingness to assume risk also was influenced by the investor's view of present and future economic conditions. On the assumption that the investors would use economic forecasts and economists' indices of future expectations, the financial group at BB Investments constructed an economic index that ranged from 0 to 100. When applied to any particular point in time, this index was tied to the expected increase in interest rates and borrowing levels, the expected increase in manufacturing costs because of the rate of inflation, and the expected level of price inflation at the retail level. This meant that the *lower* the index, the *better* the future economic conditions were expected to be.

Data were obtained by randomly selecting 50 high-risk portfolio customers and recording their incomes and the amounts of their investments. The income figures represent annual incomes and the economic index values are the index values at the time the investment was made.

INVESTOR	Y: (INVESTMENT)	X_1: (ECONOMIC INDEX)	X_2: (INCOME)	INVESTOR	Y: (INVESTMENT)	X_1: (ECONOMIC INDEX)	X_2: (INCOME)
1	2,500	86	55,800	26	3,600	40	61,600
2	3,700	54	60,400	27	2,800	81	60,000
3	3,900	21	72,700	28	2,200	44	50,600
4	1,700	91	41,700	29	3,800	36	66,300
5	1,000	72	35,200	30	4,300	50	70,900
6	1,700	16	41,800	31	3,300	95	66,600
7	2,500	81	43,700	32	3,300	47	64,400
8	3,400	32	67,900	33	2,100	3	52,900
9	2,500	37	53,700	34	3,800	55	68,500
10	2,900	89	57,400	35	1,700	65	36,400
11	2,100	48	47,100	36	3,800	28	69,100
12	2,600	61	55,300	37	2,000	47	44,400
13	1,700	33	40,000	38	1,900	50	47,900
14	2,100	82	40,200	39	2,400	15	57,100
15	1,500	95	36,900	40	3,800	84	67,500
16	1,700	73	40,700	41	3,700	44	61,900
17	1,400	9	35,100	42	1,500	28	36,100
18	2,400	42	50,900	43	3,000	36	58,900
19	1,000	74	36,300	44	2,200	69	50,700
20	3,200	31	63,700	45	2,400	69	49,800
21	2,500	12	46,800	46	3,200	3	62,500
22	4,500	25	75,200	47	2,600	70	51,100
23	2,400	24	42,400	48	3,000	17	52,600
24	2,000	88	42,000	49	1,900	15	43,700
25	2,900	53	54,600	50	1,700	63	40,100

Determine the predicted investment for an investor with an income of $48,500 at a time when the economic index has a value of 72.

SOLUTION

The least squares equation from Figure 14.6 is

$$\hat{Y} = -1183.3 - .127X_1 + .072X_2$$

The predicted investment is

$$\hat{Y} = -1183.3 - .127(72) + .072(48,500)$$
$$= -1183.3 - 9.1 + 3492.0$$
$$= 2299.6$$

that is, approximately $2300. Note that, similar to the argument in Chapter 13, $2300 also serves as an estimate of the average investment whenever $X_1 = 48,500$ and $X_2 = 72$. This is explored further in Section 14.5, where we discuss the construction of a confidence interval for an *average* investment or a prediction interval for an *individual* investment.

The first test of hypothesis determines whether these two variables *as a group* provide a useful model for predicting the amount of an investment.

$$H_0: \beta_1 = \beta_2 = 0$$
$$H_a: \beta_1 \neq 0, \beta_2 \neq 0, \text{ or both} \neq 0$$

FIGURE 14.6

MINITAB output for Example 14.4.

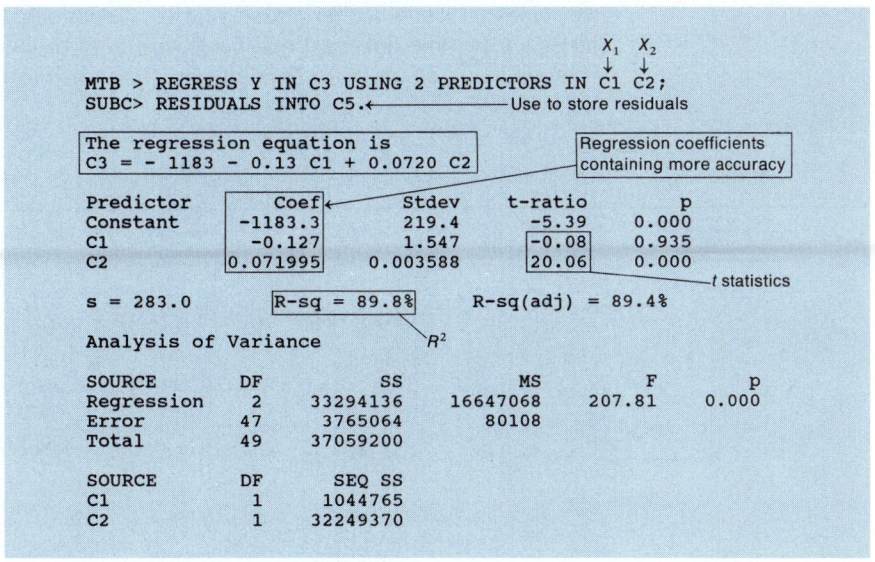

Using the ANOVA table in Figure 14.6, the value of the F statistic is

$$F^* = \frac{\text{MSR}}{\text{MSE}} = \frac{16,647,068}{81,108} = 207.8$$

The df here are $v_1 = k = 2$ and $v_2 = n - k - 1 = 50 - 2 - 1 = 47$. Because $F_{.10, 2, 47}$ is not in Table A.7, we use the nearest value, $F_{.10, 2, 40} = 2.44$. The computed F^* exceeds this value, so we reject H_0 and conclude that at least one of these two independent variables is a significant predictor of investment amounts. ■

The t-Tests   Because we rejected H_0 in Example 14.4, the next step is to examine the t-tests to determine which of the two independent variables are useful predictors. The t-value from Table A.5 is $t_{\alpha/2, n-k-1} = t_{.05, 47} \cong 1.684$. The computed t-values in Figure 14.6 lead to the following conclusions. First, the t-value for $X_1 =$ economic index is $t^* = -.08$. The absolute value of t^* is *less than* 1.684, which means that, given the presence of X_2 in the model, X_1 does not contribute useful information to the prediction of the amount of an investment. It can be removed from the model without seriously affecting the accuracy of the resulting prediction. Second, for $X_2 =$ income, the computed t-value is $t^* = 20.06$. Since this value exceeds 1.684, the investor's income *is* an excellent predictor of the amount of an investment. It was the contribution of this variable and not of X_1 that produced the extremely large F-value we obtained.

   As we have seen, a quick glance at the computer output allows you to determine whether your model is useful as a whole and, furthermore, which variables are useful predictors. But beware—the analysis is not over! Before you form your conclusions from this analysis and make critical decisions based on several tests of hypotheses, you need be sure that none of the assumptions of the multiple linear regression model (discussed earlier) has been violated. We will discuss this problem in the final section of this chapter, where we conclude the analysis by examining the sample *residuals*, $Y - \hat{Y}$.

   The use of t-tests allows you to determine the predictive contribution of each independent variable, provided you want to examine the contribution of *one* such variable while assuming that the remaining variables are included in the equation. The next section shows you how to extend this procedure to a situation in which you wish to determine the contribution of any *set* of predictor variables by using a single test.

EXERCISES

14.9   Given is selected information from a computer printout of a multiple regression analysis:

PREDICTOR	COEFFICIENT	S.D.	t-RATIO
Constant	− .50	.198	−2.525
X_1	2.40	.256	9.375
X_2	2.95	.210	14.048

R Square = 0.9381

ANALYSIS OF VARIANCE

SOURCE	df	SS	MS	F
Model	2	295.30	147.65	128.39
Residual	17	19.50	1.15	
Total	19	314.80		

Answer the following questions:

a.   Write the multiple regression equation.

b.   What percentage of variation in Y is explained by the model?

c.   What is the sample size n used for the above regression analysis?

d.   Find the value of:

i. The total sum of squares
ii. The error sum of squares
iii. The regression sum of squares
iv. The F statistic to test $H_0: \beta_1 = \beta_2 = 0$
v. The rejection region for the test in part (iv) at the .05 significance level
vi. The t-statistic to test $H_0: \beta_2 = 0$
vii. The rejection region for the test in part (vi) at the .05 significance level
viii. The estimated variance of the error in the model

14.10   Complete the following ANOVA table to test the usefulness of a model with five independent variables that attempted to explain the variation in the dependent variable:

SOURCE	df	SS	MS	F
Regression				
Error		180		
Total	50	215		

14.11   Many chief executives (CEOs) have been under serious criticism from organized labor for the fat pay checks CEOs take home. Many of the CEOs have advocated sacrifice and leaner checks for large groups of employees of their companies. An experiment was set up in which 15 observations were taken on the variables:

$$Y = \text{CEOs pay (in thousands of dollars)}$$
$$X_1 = \text{company's net profit (in millions of dollars)}$$
$$X_2 = \text{number of employees (in thousands)}$$

A computer package gave the following sample statistics:

$$b_1 = .1336 \qquad b_2 = -.86$$
$$s_{b_1} = .0424 \qquad s_{b_2} = \quad .39$$

a. Given that X_2 is in the model, does X_1 contribute to predicting the dependent variable at the .05 significance level?

b. Given that X_1 is in the model, does X_2 contribute to predicting the dependent variable at the .05 significance level?

14.12 Brown and Gilbert's law firm would like to predict the salary for a legal secretary based on years of college education (X_1), typing speed in words per minute (X_2), and years of experience (X_3). The following data were collected:

Y	X_1	X_2	X_3	Y	X_1	X_2	X_3
15,120	2	65	2	12,500	0	45	.5
12,500	1	45	2	15,800	2.5	60	2
26,000	3.5	85	9	19,600	1	70	3
19,000	0	55	11	21,800	3	75	6
16,000	4	85	1	12,400	0	60	.5
15,000	0	65	1	22,500	2	75	7

a. Using a computerized statistical package, determine the least squares prediction equation.

b. What is the value of the residual standard deviation?

c. Do the variables X_1, X_2, and X_3 contribute to predicting salaries at the .10 significance level?

d. Find a 90% confidence interval for β_1.

e. Test the null hypothesis that $\beta_1 = 0$ at the 10% significance level.

f. Interpret the results of the hypothesis test in question e.

14.13 The following sample statistics were computed for a regression analysis:

$$b_0 = 10.2 \qquad b_1 = 5.6 \qquad b_2 = 100.4 \qquad s_{b_1} = 1.04 \qquad s_{b_2} = 17.95$$

Assume that twenty observations were taken.

a. Test that X_2 significantly contributes to the prediction of Y given that X_1 is in the model. Use a 5% significance level.

b. Find a 95% confidence interval for β_1.

14.14 The model $\hat{Y} = 3.2 + 6.1X_1 + 5.2X_2$ was calculated to fit 20 data points pertaining to the growth rate of a hog. The variable X_1 represents the daily food consumption of the hog and X_2 represents the age of the hog. If the standard deviation of the estimate of β_1 is 2.5, what is a 95% confidence interval for the parameter β_1?

14.15 Complete the following ANOVA table to test the null hypothesis that the independent variables are not useful predictors of the dependent variable.

SOURCE	df	SS	MS	F
Regression				
Error		55	2.75	
Total	27	255		

14.16 Datamatics Equipment, a Seattle-based electronics firm, is interested in identifying variables in the manufacturing environment that have a linear relationship with the number of line shortages on the manufacturing floor. The sample data used in a regression analysis are as follows:

WEEK	Y	X_1	X_2	X_3	WEEK	Y	X_1	X_2	X_3
1	293	205	5.936	343	9	420	365	4.780	453
2	348	215	5.815	259	10	407	329	4.905	460
3	416	227	4.983	250	11	397	345	5.009	426
4	445	301	4.841	236	12	430	249	4.869	408
5	453	362	4.755	243	13	497	356	4.791	324
6	392	358	4.775	303	14	534	424	4.754	330
7	382	302	4.813	411	15	547	430	4.598	283
8	365	246	4.909	420					

where

$$Y = \text{number of line shortages with back-order status for a given week}$$
$$X_1 = \text{number of delinquent purchase orders for a given week}$$
$$X_2 = \text{inventory level (in millions of dollars) for prior weeks}$$
$$X_3 = \text{number of purchased items for prior weeks}$$

The least squares regression equation was found to be:

$$\hat{Y} = 710.9 + 0.4767X_1 - 70.90X_2 - 0.2525X_3$$

a. Does the complete model significantly contribute to predicting the dependent variable? Use a 10% significance level.

b. If s_{b_2} is 36.886, find a 95% confidence interval for β_2.

c. Interpret the results of the hypothesis test in question a and interpret the confidence interval in part b.

14.17 The least squares line of $\hat{Y} = 3.4 + 1.2X_1 + 4.3X_2$ was obtained. The sample residuals of the 20 observations used in fitting the regression line are:

4.1, −3.2, 1.5, 6.7, 6.4, 3.8, −4.2, −2.4, 1.6, −8.7, −3.1, 1.2, −5.1, 2.1, 0.6, 5.4, 3.4, −7.1, −6.2, 3.2

Given that the value of SST is 510, test the null hypothesis that the variables X_1 and X_2 do not contribute to predicting the variation in the dependent variable. Use a 5% significance level.

14.18 Do the number of units of personal computers sold and the advertising expenditures contribute significantly to predicting the variation in the sales of software in Exercise 14.2? Use a 5% level of significance.

14.19 The job placement center at Ozark Technological University would like to predict the starting salaries (given in thousands of dollars) for the college graduates in the engineering department. Two variables are used. The variable X_1 represents the student's overall grade point average (GPA). The variable X_2 represents the number of years of prior job-related experience. Data for fifteen randomly selected graduating students are:

Y: STARTING SALARY	X₁: OVERALL GPA	X₂: YEARS OF JOB-RELATED EXPERIENCE	Y: STARTING SALARY	X₁: OVERALL GPA	X₂: YEARS OF JOB-RELATED EXPERIENCE
27.1	3.7	0	32.3	3.8	2.5
23.3	2.9	1.1	18.1	2.1	1.4
21.4	2.4	1.5	22.5	3.0	0.3
24.2	3.2	0.5	23.8	3.4	0.5
26.1	3.6	0.8	20.9	2.8	0
19.8	2.7	0	20.0	2.5	1.0
22.8	3.1	0	27.8	3.3	2.1
20.5	2.2	2.1			

The least square equation for the data is:

$$\hat{Y} = 2.189 + 6.5144X_1 + 1.9259X_2$$

a. Find the F-value using an ANOVA table to test the hypothesis that $\beta_1 = 0$ and $\beta_2 = 0$ at the 5% level of significance.

b. Interpret the coefficients in the context of the problem.

14.20 If the residual standard deviation for a set of data is 3.82 and the total sum of squares is 269, what is the F-test for testing that a model with five independent variables does not contribute to predicting the variation in the dependent variable? Assume that 15 observations of the dependent variable were taken.

14.3

DETERMINING THE PREDICTIVE ABILITY OF CERTAIN INDEPENDENT VARIABLES

We can extend the procedure we used to examine the contribution of each independent variable, one at a time, using a t-test.

Assume that the personnel director of an accounting firm has developed a regression model to predict an individual's performance on the CPA exam. The multiple linear regression model contains eight independent variables, three of which (say, X_6, X_7, X_8) describe the physical attributes of each individual (say, height, weight, and age). Can all three of these variables be removed from the analysis without seriously affecting the predictive ability of the model?

To answer this question, we return to a statistic described in Chapter 13 that measures how well the model captures the variation in the values of your dependent variable.

Coefficient of Determination

The total variation of the sampled dependent variable is determined by

$$\text{SST} = \text{total sum of squares}$$
$$= \text{SS}_Y$$
$$= \Sigma(Y - \bar{Y})^2$$
$$= \Sigma Y^2 - (\Sigma Y)^2/n$$

where n = number of observations. To determine what percentage of this variation has been explained by the predictor variables in the regression equation, we determine the **coefficient of determination, R^2**:

$$R^2 = 1 - \frac{\text{SSE}}{\text{SST}} \qquad (14.14)$$

The range for R^2 is 0 to 1. If $R^2 = 1$, then 100% of the total variation has been explained because, in this case, $\text{SSE} = \Sigma(Y - \hat{Y})^2 = 0$, and so $Y = \hat{Y}$ for each observation in the sample; that is, the model provides a *perfect predictor*. This does not occur in practice, but the main point is that a large value of R^2 is generally desirable for a regression application. It should be mentioned that $R^2 = 1$ whenever the number of observations (n) is equal to the number of estimated coefficients ($k + 1$). This does not mean that you have a "wonderful" model; rather, you have inadequate data. As a result, you need to guard against using too small a sample in your regression analysis. A general rule of thumb is to use a sample containing at least three times as many (unique) observations as the number of predictor variables (k) in the model.

H_0: all β's = 0 A test statistic for testing H_0: all β's $= 0$ was introduced in equation 14.11, which used the ratio of two mean squares from the ANOVA table. Another way to calculate this F-value is to use

$$F = \frac{R^2/k}{(1 - R^2)/(n - k - 1)} \qquad (14.15)$$

This version of the F statistic is used to answer the question, Is the value of R^2 significantly large? If H_0 is rejected, then the answer is yes, and so this group of predictor variables has at least some predictive ability for predicting Y.

The F-value computed in this way will be exactly the *same* as the one computed using $F = MSR/MSE$, except for possible rounding error. This is illustrated in Example 14.5.

Once again, remember that *statistical* significance does not always imply *practical* significance. A large value of R^2 (rejecting H_0) does not imply that precise prediction (practical significance) will follow. However, it does inform the researcher that these predictor variables, as a group, are associated with the dependent variable.

EXAMPLE 14.5   In Chapter 13, we determined that $X =$ income explained 71% of the total variation of the home sizes (Y) in the sample, since the computed value of r^2 was .711. What percentage is explained using all three predictors (income, family size, and years of education)?

SOLUTION   The coefficient of determination using X_1 only is .711. Using the MINITAB solution in Figure 14.4(a), the coefficient of determination using X_1, X_2, and X_3 is

$$R^2 = 1 - \frac{SSE}{SST}$$

$$= 1 - \frac{24.858}{262.4} = .905$$

Consequently, 90.5% of this variation has been explained using the three independent variables.

The F-value determined in Example 14.2 for testing $H_0: \beta_1 = \beta_2 = \beta_3 = 0$ can be duplicated using equation 14.15 because

$$F = \frac{.905/3}{(1 - .905)/(10 - 3 - 1)} = \frac{.905/3}{.095/6}$$

$$= 19.1 \text{ (as before)} \qquad \blacksquare$$

COMMENTS   In Example 14.5, notice that the value of R^2 *increased* when we went from using one independent variable to using three. As you add variables to your regression model, R^2 *never decreases*. However, the increase may not be a significant one. If adding ten more predictor variables to your model causes R^2 to increase from .91 to .92, this is not a *significant* increase. Therefore, do not include these ten variables; they clutter up your model and are likely to add spurious predictive ability to it.

How can we tell if adding (or removing) a certain set of X variables causes a *significant* increase (or decrease) in R^2?

The Partial F-Test

Consider the situation in which the personnel director is trying to determine whether to retain three variables ($X_6 =$ height, $X_7 =$ weight, $X_8 =$ age) as predictors of a person's performance on a CPA exam. We know one thing—R^2 *will* be higher with these three variables included in the model. If we do not observe a *significant* increase, however, our advice would be to *remove* these variables from the analysis. To determine the extent of this increase, we use another F-test.

We define two models—one contains X_6, X_7, X_8, and one does not.

complete model: uses all predictor variables, including X_6, X_7, and X_8

reduced model: uses the same predictor variables as the complete model except X_6, X_7, and X_8

Also, let

$$R_c^2 = \text{the value of } R^2 \text{ for the complete model}$$
$$R_r^2 = \text{the value of } R^2 \text{ for the reduced model}$$

Do X_6, X_7, and X_8 contribute to the prediction of Y? We will test

$$H_0: \beta_6 = \beta_7 = \beta_8 = 0 \text{ (they do not contribute)}$$
$$H_a: \text{at least one of the } \beta\text{'s} \neq 0 \text{ (at least one of them does contribute)}$$

The test statistic here is

$$F = \frac{(R_c^2 - R_r^2)/v_1}{(1 - R_c^2)/v_2} \qquad \textbf{(14.16)}$$

where v_1 = number of β's in H_0 and $v_2 = n - 1 - $ (number of X's in the complete model).

For this illustration, $v_1 = 3$ because there are three β's contained in H_0. Assuming that there are eight variables in the complete model, then $v_2 = n - 1 - 8 = n - 9$. Here, n is the total number of observations (rows) in your data. This F statistic measures the *partial* effect of these three variables; it is a **partial F statistic**.

Equation 14.16 resembles the F statistic given in equation 14.15, which we used to test H_0: all β's = 0. If all the β's are zero, then the reduced model consists of only a constant term, and the resulting R^2 will be zero; that is, $R_r^2 = 0$. Setting $R_r^2 = 0$ in equation 14.16 produces equation 14.15, where $v_1 = k$ and $v_2 = n - k - 1$.

These variables (as a group) contribute significantly if the computed partial F-value in equation 14.16 exceeds F_{α, v_1, v_2} from Table A.7.

EXAMPLE 14.6

The personnel director gathered data from 30 individuals using all eight independent variables. These data were entered into a computer, and a multiple linear regression analysis was performed. The resulting R^2 was .857.

Next, variables X_6, X_7, and X_8 were omitted, and a second regression analysis was performed. The resulting R^2 was .824. Do the variables X_6, X_7, and X_8 (height, weight, and age) appear to have any predictive ability? Use $\alpha = .10$.

SOLUTION

Here, $n = 30$ and

$$R_c^2 = .857 \text{ (complete model)}$$
$$R_r^2 = .824 \text{ (reduced model)}$$

Based on the previous discussion, the value of the partial F statistic is

$$F^* = \frac{(.857 - .824)/3}{(1 - .857)/(30 - 1 - 8)} = \frac{.033/3}{.143/21} = 1.61$$

The procedure is to reject $H_0: \beta_6 = \beta_7 = \beta_8 = 0$ if $F^* > F_{.10, 3, 21} = 2.36$. The computed F-value does not exceed the table value, so we fail to reject H_0. We conclude that these variables should be removed from the analysis because including them in the model fails to produce a significantly larger R^2. ■

The partial F-test also can be used to determine the effect of adding a *single* variable to the model.

FIGURE 14.7

MINITAB output using $X_1 =$ income and $X_3 =$ years of education as predictors.

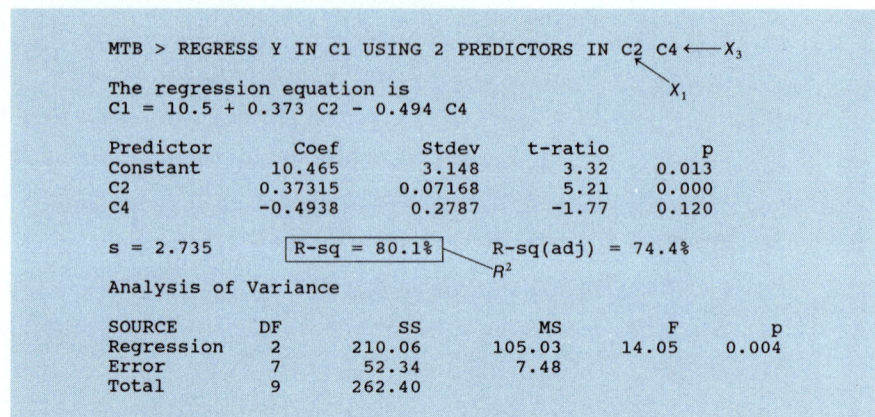

```
MTB > REGRESS Y IN C1 USING 2 PREDICTORS IN C2 C4 ←——X₃
                                                  ↑
The regression equation is                        X₁
C1 = 10.5 + 0.373 C2 - 0.494 C4

Predictor       Coef       Stdev      t-ratio        p
Constant      10.465       3.148         3.32    0.013
C2           0.37315     0.07168         5.21    0.000
C4           -0.4938      0.2787        -1.77    0.120

s = 2.735      R-sq = 80.1%      R-sq(adj) = 74.4%
                      R²
Analysis of Variance

SOURCE         DF        SS          MS        F        p
Regression      2    210.06      105.03    14.05    0.004
Error           7     52.34        7.48
Total           9    262.40
```

EXAMPLE 14.7

Using the real-estate data analyzed in Example 14.2, determine whether $X_2 =$ family size contributes to the prediction of home size, given that $X_1 =$ income and $X_3 =$ years of education are included in the model. Use a significance level of $\alpha = .10$.

SOLUTION

We test the hypotheses

$$H_0\colon \beta_2 = 0 \quad \text{(if } X_1 \text{ and } X_3 \text{ are included)}$$
$$H_a\colon \beta_2 \neq 0.$$

The complete model uses X_1, X_2, and X_3. Using Example 14.5,

$$R_c^2 = .905$$

The reduced model uses only X_1 and X_3. Figure 14.7 shows the MINITAB output for this, and

$$R_r^2 = .801$$

The value of the partial F-statistic is

$$F^* = \frac{(.905 - .801)/1}{(1 - .905)/(10 - 1 - 3)} = \frac{.104/1}{.095/6} = 6.6$$

The 1 in the numerator indicates that there is one β in H_0; subtracting the 3 in the denominator 6 indicates that there are three X's in the complete model.

This value does exceed $F_{.10, 1, 6} = 3.78$, and so $X_2 =$ family size does (as suspected from the earlier t-test) significantly improve the model's predictive ability when included with X_1 and X_3. In other words, there is a significant increase in R^2 (from .801 to .905) when X_2 is added to the model, and, as a result, our conclusion is to retain this variable in the model. ■

COMMENTS

Both Example 14.7 and the t-test for X_2 discussed on page 574 dealt with testing H_0: $\beta_2 = 0$ versus H_a: $\beta_2 \neq 0$. Both tests attempted to determine whether X_2 should be included as a predictor given that X_1 and X_3 were already included as predictor variables. Note two things: (1) the partial F-value $= 6.6 = (2.578)^2 = (t$-value$)^2$, and (2) the p-value using the t-test (.042) = the p-value using the F-test (not shown).

We can see that these tests are *identical*: They result in exactly the same p-value and the same conclusion. This means that to determine the predictive ability of an individual independent variable, we can compute the partial F statistic or the somewhat simpler t-statistic. Some computer packages use the F statistics to summarize the individual predictors, whereas others (such as MINITAB) use the t-values to measure the influence of each predictor. You should use whatever is provided (the *F* statistic or *t*-statistic) to measure the partial effect of each variable; both sets of statistics accomplish the same thing.

Using Curvilinear Models: Polynomial Regression

Mr. Bentley owns several furniture stores in a large metropolitan area. He is interested in the relationship between his monthly advertising expenditures (*X*) and the corresponding monthly sales (*Y*). He suspects that sales will increase as the amount spent on advertising increases, but after a certain point the increase will slow down; that is, sales will continue to increase but at a slower rate. In other words, after spending a certain amount on advertising, he will reach a point where there will be little gain in sales, even though he spends a much larger amount on advertising.

Data were gathered from the company records covering 15 (nonconsecutive) months of sales (in tens of thousands of dollars) and advertising expenditures (in hundreds of dollars).

MONTH	Y(SALES)	X (ADVERTISING EXPENDITURES)	MONTH	Y(SALES)	X (ADVERTISING EXPENDITURES)
1	6.9	18.1	9	8.8	29.2
2	8.5	27.3	10	10.2	50.5
3	1.2	10.1	11	9.8	37.3
4	9.4	34.8	12	9.3	40.2
5	3.2	11.8	13	9.8	43.1
6	5.2	15.0	14	9.8	45.0
7	8.0	22.9	15	9.2	31.5
8	7.2	20.4			

The scatter diagram is shown in Figure 14.8. Mr. Bentley seems to have a point—the sales do appear to level off after a certain amount of advertising expense.

FIGURE 14.8

MINITAB scatter diagram of data for advertising example.

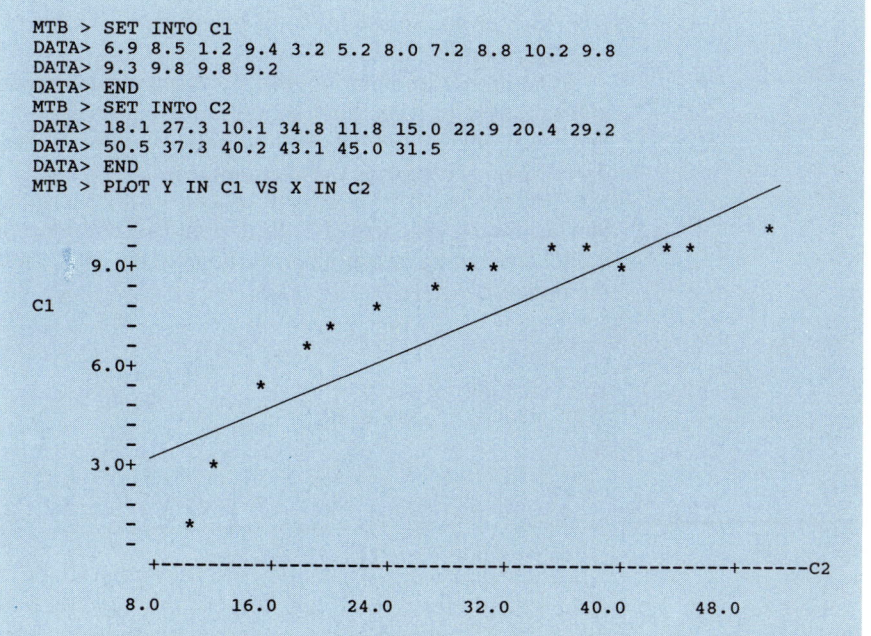

```
MTB > SET INTO C1
DATA> 6.9 8.5 1.2 9.4 3.2 5.2 8.0 7.2 8.8 10.2 9.8
DATA> 9.3 9.8 9.8 9.2
DATA> END
MTB > SET INTO C2
DATA> 18.1 27.3 10.1 34.8 11.8 15.0 22.9 20.4 29.2
DATA> 50.5 37.3 40.2 43.1 45.0 31.5
DATA> END
MTB > PLOT Y IN C1 VS X IN C2
```

FIGURE 14.9 Quadratic curves. **A**: Graph of $Y = 34 - 12X + 2X^2$. In general, this is the shape of $Y = \beta_0 + \beta_1 X + \beta_2 X^2$, where $\beta_2 > 0$. **B**: Graph of $Y = 12X - 2X^2$. In general, this is the shape of $Y = \beta_0 + \beta_1 X + \beta_2 X^2$, where $\beta_2 < 0$.

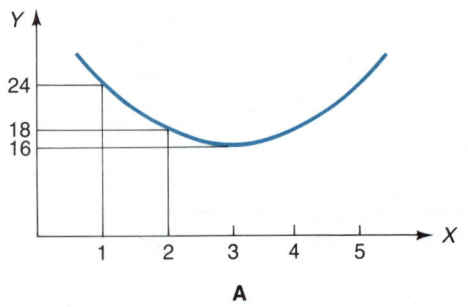

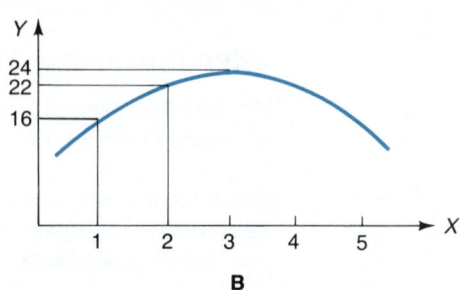

Does the simple linear model $Y = \beta_0 + \beta_1 X + e$ capture the relationship between advertising expense (X) and sales (Y)? Although Y does increase as X increases here, the linear model does not capture the "slowing down" of Y for larger values of X. The least squares line (sketched in Figure 14.8) underpredicts sales for the middle range of X but overpredicts sales for small or large values of X.

Figure 14.9 shows **quadratic curves** rather than straight lines. If we include X^2 in the model, we can describe the curved relationship that seems to exist between sales and advertising. More specifically, the left half of Figure 14.9(b) closely resembles the shape of the scatter diagram in Figure 14.8. Consider the model

$$Y = \beta_0 + \beta_1 X + \beta_2 X^2 + e \qquad (14.17)$$

Is this a linear regression model? At first glance, it would appear not to be. However, by the word *linear* we really mean that the model is **linear in the unknown β's**, not X. In equation 14.17, there are no terms such as β_1^2, $\beta_1 \beta_2$, $\sqrt{\beta_0}$, and so on. So the model is linear in the β's, and this is a (multiple) linear regression application.

The model in equation 14.17 is a **curvilinear model** and is an example of **polynomial regression**. Such models are very useful when a particular independent variable and dependent variable exhibit a definite increasing and/or decreasing relationship that is nonlinear.

Solving for β_0, β_1, and β_2   Equation 14.17 represents a multiple regression model containing two predictors, namely, $X_1 = X$ and $X_2 = X^2$. The data for the model then are

Y	X_1	X_2
6.9	18.1	327.61 ($= 18.1^2$)
8.5	27.3	745.29 ($= 27.3^2$)
1.2	10.1	102.01
$\vdots$	$\vdots$	$\vdots$
9.8	45.0	2025.0
9.2	31.5	992.25

These data for Y, X_1, and X_2 are your input to the multiple linear regression computer program. You can simplify this task by letting the computer build the $X_2 = X^2$ column of data by squaring the entries in the $X_1 = X$ column.

FIGURE 14.10

MINITAB solution using $Y = \beta_0 + \beta_1 X + \beta_2 X^2$.

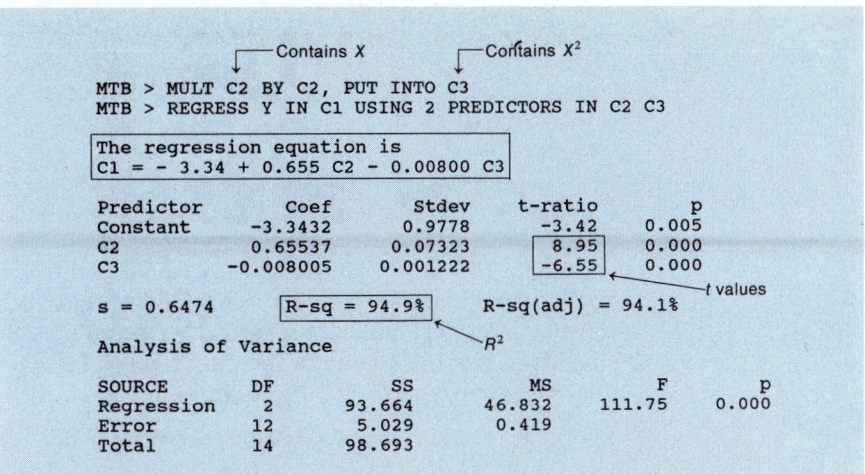

```
                    ┌──Contains X         ┌──Contains X²
MTB > MULT C2 BY C2, PUT INTO C3
MTB > REGRESS Y IN C1 USING 2 PREDICTORS IN C2 C3

┌─────────────────────────────────────────────┐
│The regression equation is                    │
│C1 = - 3.34 + 0.655 C2 - 0.00800 C3           │
└─────────────────────────────────────────────┘
Predictor         Coef       Stdev     t-ratio         p
Constant       -3.3432      0.9778       -3.42     0.005
C2             0.65537      0.07323        8.95     0.000
C3            -0.008005     0.001222      -6.55     0.000
                                                      ──t values
s = 0.6474       R-sq = 94.9%       R-sq(adj) = 94.1%

Analysis of Variance              R²

SOURCE          DF         SS         MS         F         p
Regression       2     93.664     46.832    111.75     0.000
Error           12      5.029      0.419
Total           14     98.693
```

EXAMPLE 14.8

Look at the MINITAB solution using the model $Y = \beta_0 + \beta_1 X + \beta_2 X^2 + e$ for the sales and advertising expenditure, as shown in Figure 14.10.

1. Predict the sales for a month in which Mr. Bentley spends $3000 on advertising.
2. What do the F- and t-tests tell you about this model? Use $\alpha = .10$.

SOLUTION 1

The predicted sales for $X_1 = 30$ (hundred) is

$$\hat{Y} = -3.34 + .655(30) - .008(30)^2 = 9.1$$

That is, it is approximately $91,000.

SOLUTION 2

We first examine the F-test. Our first test of hypothesis determines whether the overall model has predictive ability.

$$H_0: \beta_1 = \beta_2 = 0$$
$$H_a: \text{at least one of the } \beta\text{'s} \neq 0$$

Using the R^2 value from Figure 14.10 and equation 14.15,

$$F^* = \frac{.949/2}{.051/(15 - 2 - 1)} = \frac{.949/2}{.051/12} = 111.6$$

As we might have suspected, this model does have significant predictive ability; $F^* = 111.6$ exceeds $F_{\alpha, k, n-k-1} = F_{.10, 2, 12} = 2.81$ from Table A.7.

Now we want to look at the t-tests (same as partial F-tests). Here, we examine each variable in the model, namely, X and X^2. The t-value from Table A.5 is $t_{.10, 12} = 1.356$ for a one-tailed test. We want to determine first whether $X_1 =$ advertising expenditure should be included in the model. Increased advertising should be associated with increased sales, so β_1 should be greater than zero. As a result, we will use a one-tailed procedure to test $H_0: \beta_1 \leq 0$ versus $H_a: \beta_1 > 0$.

According to Figure 14.10, the computed t-statistic is $t^* = b_1/(\text{standard deviation of } b_1) = 8.95$. Now, $t^* = 8.95 > 1.356$, which means that the advertising variable should be retained as a predictor of sales.

Next, we want to determine whether $X_2 = (\text{advertising expenditures})^2$ contributes significantly to the prediction of sales. We are asking whether including the *quadratic term* was necessary. If this model is the correct one, then, according

FIGURE 14.11

Error resulting from extrapolation. See text and Figure 14.9b.

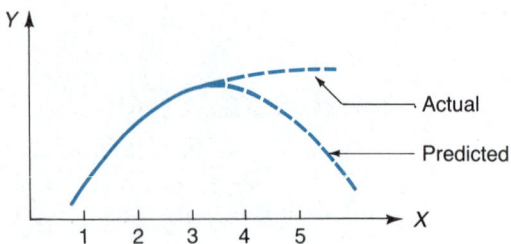

to Figure 14.9(b), β_2 should not only be unequal to zero but also, more specifically, should be less than zero. This follows since if the sales do, in fact, level off after a certain amount of advertising expenditures, the curve should resemble the left half of the quadratic curve in Figure 14.9(b).

The appropriate hypotheses are

$$H_0: \beta_2 = 0 \quad (\text{or} \geq 0)$$
$$H_a: \beta_2 < 0.$$

We reject H_0 if $t < -t_{.10, 12} = -1.356$.

From Figure 14.10, we see that $t^* = b_2/(\text{standard deviation of } b_2) = -6.55$. This lies in the rejection region, so we conclude that $\beta_2 < 0$, which means that the quadratic term, X^2, contributes significantly and in the correct direction.   ■

COMMENTS

There are three things you should note about the curvilinear model.

1. Curvilinear models often are used for situations in which the rate of increase or decrease in the dependent variable is not constant when plotted against a particular independent variable. The use of X^2 (and in some cases, X^3) in your model allows you to capture this nonlinear relationship between your variables.

2. There are other methods available for modeling a nonlinear relationship, including

$$Y = \beta_0 + \beta_1(1/X) + \text{error}$$

and

$$Y = \beta_0 + \beta_1 e^{-X} + \text{error}$$

These models also are (simple, here) linear regressions; they are linear in the unknown parameters. These models, unlike the quadratic one discussed previously, involve a **transformation** of the independent variable, X. When replacing X by the transformed X (such as $1/X$ or e^{-X}) in the model, one has many other curvilinear models that may better fit a set of sample data displaying a nonlinear pattern.

3. Avoid using the model $Y = \beta_0 + \beta_1 X + \beta_2 X^2 + e$ for values of X outside the range of data used in the analysis. Extrapolation is extremely dangerous when using this modeling technique. Consider Figure 14.9b and suppose values of X between 1 and 3 were used to derive the estimate of β_0, β_1, and β_2. Figure 14.11 shows the results. For values of X larger than 3, the predicting equation will turn down, whereas the actual relationship will probably continue to level off. So this model works for interpolation (for values of X between 1 and 3, here) but is extremely unreliable for extrapolation.

EXERCISES

14.21   If all the values of the dependent variable Y fell on the plane $\hat{Y} = 2 + 5.2X_1 + 10X_2$ and the variables X_1 and X_2 were used in a least squares fit to the data, what would be the value of the coefficient of determination?

14.22   If the residual standard deviation is 1.5 for a regression model with five independent variables and 26 observations, what is the value of R^2, assuming that SST is 231.95? Interpret the value of R^2.

14.23 The manager of the personnel department of a computer firm is interested in knowing the relationship between the pay raise (Y) given to an employee of the firm and the variables: yearly performance evaluation (X_1), years with the company (X_2), and number of credit hours of computer courses that the employee has taken in college (X_3). After observing 50 employees under different values of X_1, X_2, and X_3, the manager wishes to test that X_2 and X_3 contribute to predicting the variation in pay raises. The coefficient of determination for the model involving just Y and X_1 is .71. The coefficient of determination for the model with X_1, X_2, and X_3 is .82. Do the additional independent variables contribute significantly to the model? Use a 5% significance level.

14.24 The cognitive styles of individual workers are thought to be a valuable tool in selecting and placing workers in a job in which they can be most productive. An experiment in an industrial setting involving ten subjects was conducted to explore the relationships between task performance and cognitive style. The dependent variable was a task performance index to measure a worker's performance in visually inspecting circuit boards. A score from the widely used Embedded Figures Test (EFT) was used as an independent variable. Two other independent variables were two indexes to measure (1) extroversion or introversion (EI) and (2) perception or judgment (PI). Complete the following ANOVA table and determine the significance of the contribution of the model in predicting task performance. Use the p-value to determine your conclusions. Find the coefficient of determination.

SOURCE	df	SS	MS	F
Regression		33.462		
Error				
Total		39.058		

(*Source:* Ralph Janaro, William Shearer, and Mark Trittle, "Cognitive Style and Visual Task Performance," *Proceedings of the 1986 Annual Meeting of Decision Science Institute*, 1986, pp. 1171–73.)

14.25 The dean of the college of business at Fargo University would like to see whether several variables affect a student's grade point average. Thirty first-year students were randomly selected and data were collected on the following variables:

Y = grade point average for the first year

X_1 = average time spent per month at fraternity or sorority functions

X_2 = average time spent per month working part time

X_3 = total number of hours of coursework attempted

The SSE for the least squares line involving only Y and X_1 was found to be 5.21. The SSE for the complete model was found to be 4.31. The SST is 24.1 At the 5% significance level, test the null hypothesis that the independent variables X_2 and X_3 do not contribute to predicting the variation in Y given that X_1 is already in the model.

14.26 Data from a questionnaire survey of 166 managers was gathered to measure the effect of demographic and psychological variables on computer anxiety. A total of nine demographic and personality variables were used. The independent variables in the model were: sex, age, education, programming level, trait anxiety, focus of control, math anxiety, feeling-thinking index, and intuitive-sensing index. The coefficient of determination was found to be .25 for the full model. The model with only the four demographic variables was found to have a coefficient of determination equal to .10. At the .05 significance level, do the personality variables contribute to the prediction of computer anxiety?

(*Source:* Magid Igbaria and Saroj Parasuramon, "The Role of Individual Difference Variables In Predicting Computer Anxiety and Attitudes Toward Micro Computers," *Proceedings of the 1987 Annual Meeting of the Decision Sciences Institute*, 1987, pp. 440–42.)

14.27 In an effort to control the costs of the operations of a plant, a supervisor wishes to know the relationship between the time that it takes an employee to perform a task and the employee's mechanical aptitude and years of experience. The variables used in fitting a regression line were as follows:

$$y = \text{time it takes to perform the task}$$
$$X_1 = \text{mechanical aptitude level}$$
$$X_2 = \text{number of years of experience}$$
$$X_3 = X_1 X_2$$
$$X_4 = X_1^2$$
$$X_5 = X_2^2$$

Data were collected on 15 randomly selected employees. The coefficient of determination for the model involving only the independent variables X_1 and X_2 is .53. The coefficient of determination for the model involving all five independent variables is .75. Do the variables X_3, X_4, and X_5 contribute to predicting the variation in the time that it takes to perform the task?

14.28 An economist would like to examine the relationship between personal savings and the following independent variables:

$$X_1 = \text{total personal income}$$
$$X_2 = \text{yield on U.S. Government securities}$$
$$X_3 = \text{consumer price index}$$

The following data were collected for 14 randomly selected months:

Y	X_1	X_2	X_3	Y	X_1	X_2	X_3
80.2	2077.2	12.036	233.2	107.4	2179.4	9.259	249.4
91.6	2086.4	12.814	236.4	116.8	2205.7	10.321	252.7
87.4	2101.0	15.526	239.8	102.1	2234.3	11.580	253.9
104.9	2102.1	14.003	242.5	97.9	2257.6	13.888	256.2
116.2	2114.1	9.150	244.9	93.3	2276.6	15.661	258.4
109.1	2127.1	6.995	247.6	83.6	2300.7	14.724	260.5
110.1	2161.2	8.126	247.8	91.0	2318.2	14.905	263.2

a. Using a computerized statistical package, determine the least squares equation for these data.

b. Use only the variables X_1 and X_2. What is the new prediction equation?

c. Does the variable X_3 contribute to predicting the variation in personal savings, given that X_1 and X_2 are in the model? Use a 10% significance level.

14.29 In a study of 89 managers in an international restaurant company, data were collected for each manager on the following variables: the profits of each manager (Y), the length of service as an assistant manager (X_1), education (X_2), age (X_3), and length of experience as a restaurant manager with the current restaurant (X_4). Typically, length of employee's experience is a key factor in decisions regarding retention and promotion. This study wanted to examine the effects of all of the variables X_1 to X_4 on actual profits achieved. The profits realized by managers were regressed on these four independent variables. The coefficient of determination was found to be .35. Determine if the model with these four independent variables significantly contributes to the prediction of profits of each manager. Use a .05 significance level.

(*Source:* Mary Pat McEnrue, "Length of Experience and the Performance of Managers In The Establishment Phase of their Careers," *Academy of Management Journal* 31, no. 1 (1988): 175–85.)

14.30 The amount of money that a family spends monthly on food is believed to be related to the number of family members (X_1), the joint income of the husband and wife (X_2), and the age of the oldest child (X_3). A regression procedure was run with 50 observations on the model with X_1, X_2, and X_3. The SSE was 121,580 and the SST was 486,321. Another computer run also included the variables $\sqrt{X_1}$ and $\sqrt{X_3}$. The SSE for this complete model was 77,811. Do the variables $\sqrt{X_1}$ and $\sqrt{X_3}$ contribute to predicting the amount that a family spends on food? Use a 5% level of significance.

14.4

THE PROBLEM OF MULTICOL- LINEARITY

Another possible title for this section is, What do the individual b_i's tell you? We discuss one of the common problems in the use (or misuse) of multiple linear regression—namely, trying to extract more information from the results than they actually contain.

We examine the validity of such statements as, Because $b_1 = 10$, increasing X_1 by one while *holding X_2 constant* will result in an increase of 10 in Y.

Assume that a sample of ten employees at Bellaire Industries was examined in an effort to determine the ability of age (X_1) and years of experience (X_2) to predict an employee's salary (Y). The following data were obtained:

EMPLOYEE	Y (SALARY)	X_1 (AGE)	X_2 (YEARS OF EXPERIENCE)	EMPLOYEE	Y (SALARY)	X_1 (AGE)	X_2 (YEARS OF EXPERIENCE)
1	52	52	33	6	60	55	30
2	35	47	21	7	31	36	8
3	45	38	14	8	38	40	15
4	28	25	3	9	33	32	7
5	42	44	18	10	48	50	27

First, we can ask how well X_1 (age) predicts Y (salary)?

A MINITAB solution using the model $Y = \beta_0 + \beta_1 \cdot (\text{age}) + e$ is shown in Figure 14.12. Notice the computed t-value. Now, $k = 1$ because this model considers only one independent variable, so the tabulated value for comparison (using $\alpha = .10$) is $t_{\alpha/2, n-k-1} = t_{.05, 10-1-1} = t_{.05, 8} = 1.860$. The value of $t^* = 4.60$ is considerably larger than 1.86, so X_1 (age) is an excellent predictor of Y (salary).

FIGURE 14.12

MINITAB solution to Y(salary) = $b_0 + b_1 \cdot$ (age).

```
MTB > READ INTO C1-C3
DATA> 52 52 33
DATA> 35 47 21
DATA> 45 38 14
DATA> 28 25 3
DATA> 42 44 18
DATA> 60 55 30
DATA> 31 36 8
DATA> 38 40 15
DATA> 33 32 7
DATA> 48 50 27
DATA> END
      10 ROWS READ
MTB > REGRESS Y IN C1 USING 1 PREDICTOR IN C2

The regression equation is
C1 = 2.97 + 0.912 C2

Predictor      Coef      Stdev     t-ratio        p
Constant       2.971     8.499      0.35       0.736
C2             0.9124    0.1983     4.60       0.000

s = 5.634      R-sq = 72.6%      R-sq(adj) = 69.1%

Analysis of Variance

SOURCE      DF        SS          MS         F        p
Regression   1      671.69      671.69     21.16    0.000
Error        8      253.91       31.74
Total        9      925.60

MTB > CORRELATION BETWEEN C1 AND C2

Correlation of C1 and C2 = 0.852
```

t value

Instructions for obtaining the sample correlation between any two variables

Y X_1

FIGURE 14.13

MINITAB solution to
$Y(\text{salary}) = b_0 + b_1 \cdot$
(years of experience).

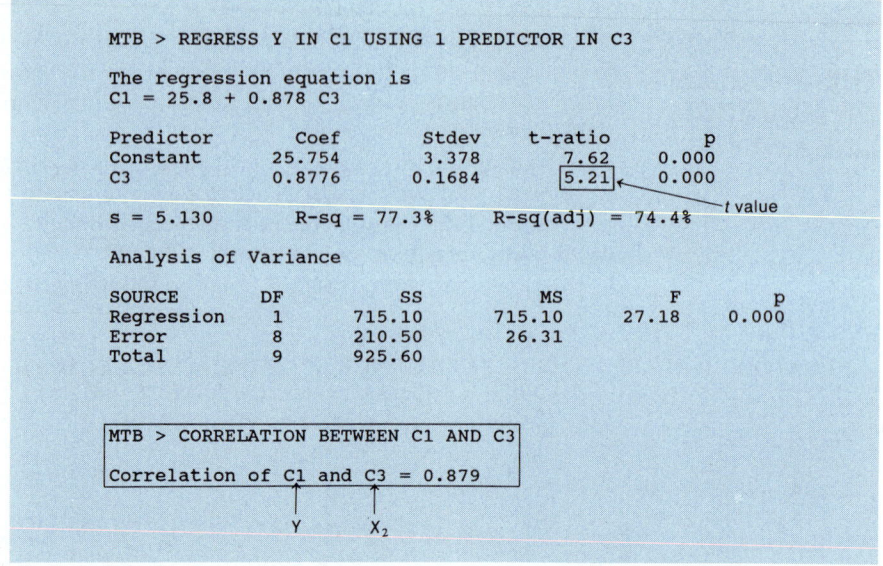

```
MTB > REGRESS Y IN C1 USING 1 PREDICTOR IN C3

The regression equation is
C1 = 25.8 + 0.878 C3

Predictor      Coef       Stdev     t-ratio         p
Constant     25.754       3.378        7.62     0.000
C3            0.8776      0.1684        5.21     0.000
                                                          ← t value
s = 5.130          R-sq = 77.3%      R-sq(adj) = 74.4%

Analysis of Variance

SOURCE        DF          SS          MS          F         p
Regression     1      715.10      715.10      27.18     0.000
Error          8      210.50       26.31
Total          9      925.60
```

```
MTB > CORRELATION BETWEEN C1 AND C3

Correlation of C1 and C3 = 0.879
                               ↑            ↑
                               Y           X₂
```

What is the *correlation* between X_1 and Y? It seems reasonable that this would be quite large because age has been shown to be a good predictor. In fact, according to Figure 14.12, this value is .852. So there is a *positive* relationship between age and salary, as one would expect.

Next, we determine how well X_2 (years of experience) predicts Y (salary). The solution using $Y = \beta_0 + \beta_1 \cdot$ (years of experience) $+ e$ is shown in Figure 14.13. Once again, the computed *t*-value $= t^* = 5.21$ is much larger than $t_{.05, 8} = 1.860$. Also, the correlation between these two variables is .879. This is not surprising; we might expect people with more years of experience to have higher salaries. Consequently, a significant positive relationship appears to exist between these two variables:

Finally, we turn to the question, how well do both X_1 (age) and X_2 (years of experience) predict salary? The model here is $Y = \beta_0 + \beta_1 X_1 + \beta_2 X_2 + e$. The least squares solution is shown in Figure 14.14.

$$\hat{Y} = 26.1 - .014X_1 + .890X_2$$

A few seemingly bizarre things show up here.

The coefficient of X_1 is $b_1 = -.014$. This would appear to indicate that larger values of X_1 (older people) produce smaller salaries. But we know from our first analysis that the *opposite* is true. We would have expected a *positive* value of b_1 here, and so the coefficient of X_1 appears to have the wrong sign.

The small *t*-values also are puzzling. The value of the F statistic (using Figure 14.14) is

$$F^* = \frac{R^2/2}{(1 - R^2)/(10 - 1 - 2)} = \frac{.773/2}{.227/7} = 11.9$$

As before, you can compute this value using the ANOVA table, where

$$F^* = \frac{\text{MSR}}{\text{MSE}} = \frac{357.55}{30.07} = 11.9$$

Using $\alpha = .10$, this is much larger than $F_{.10, 2, 7} = 3.26$, and so the model does provide a very good predictor of Y. The coefficient of determination is $R^2 = .77$; it explains 77% of the total variation in the ten salary values.

FIGURE 14.14

MINITAB solution to Y (salary) $= b_0 + b_1 \cdot$ (age) $+ b_2 \cdot$ (years of experience).

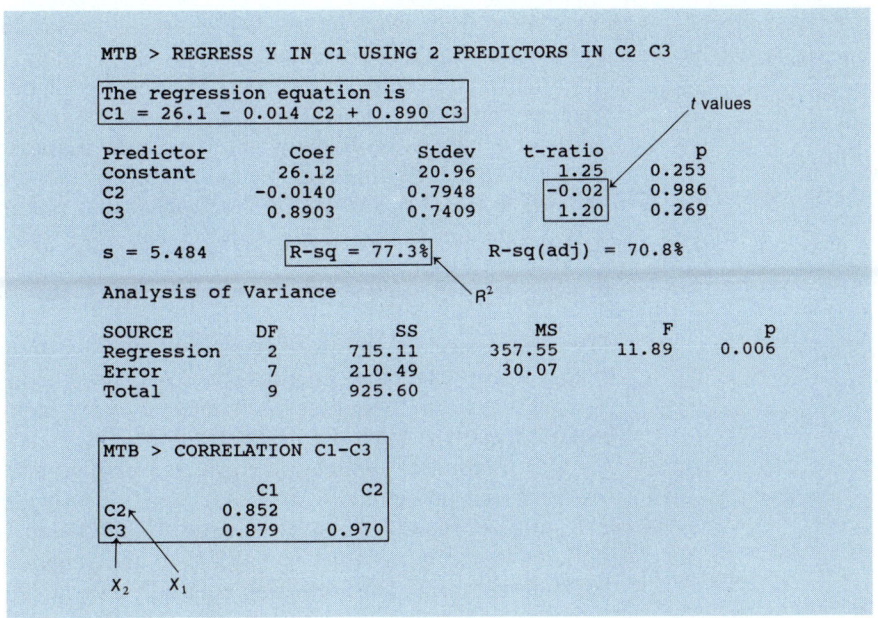

The t-values are very small; both are smaller in absolute value than $t_{\alpha/2, n-k-1} = t_{.05, 10-2-1} = t_{.05, 7} = 1.895$. Does this imply that both predictors are weak and should be removed from the model? Certainly not, as our previous analyses made clear.

This is the problem of **multicollinearity**. In multiple regression models, it is desirable for each independent variable, X, to be highly correlated with Y, but it is not *desirable* for the X's to be highly correlated *with each other*. In business applications of multiple linear regression, the independent variables typically have a certain amount of pairwise correlation (usually positive). Extremely high correlation between any pair of variables can cause a variety of problems, as we will show.

The (sample) correlation between X_1 and X_2 is

$$r = \frac{\Sigma X_1 X_2 - (\Sigma X_1)(\Sigma X_2)/n}{\sqrt{\Sigma X_1^2 - (\Sigma X_1)^2/n}\sqrt{\Sigma X_2^2 - (\Sigma X_2)^2/n}}$$

This value, using Figure 14.14, is $r = .970$. Notice in the data set that nearly every time X_1 increases, so does X_2; X_1 and X_2 are highly correlated. As a result, these data contain a great deal of multicollinearity.

Implications

First of all, the correlation of X_1 and X_2 explains the small t-values. Remember that each t-value describes the contribution of that particular independent variable *after* all other independent variables have been included in the model. X_1 is very nearly a linear function of X_2 (as evidenced by $r = .970$), so it contributes very little to the prediction of Y, given that X_2 is in the model. The same argument applies to X_2. This means that neither X_1 nor X_2 is a strong predictor given that the other variable is included—not that each one is a weak predictor by itself.

The second implication of the multicollinearity is that the situation "increasing X_1 by one while holding X_2 constant" never occurred in the sample data. In the past, as X_1 increased by one, X_2 always changed also, because X_1 and X_2 are so highly correlated.

Finally, the sample coefficients (b_1 and b_2) of our independent variables have very large variances. If we took another sample from this population, the values of b_1 and b_2 probably would change dramatically—which is not a good situation. In fact, as this example has demonstrated, these coefficients can even have the "wrong" sign, a sign different from that obtained when regressing X_1 or X_2 alone on Y.

Eliminating the Effects of Multicollinearity

The easiest way out of this dilemma is to remove some of the correlated predictors from the model. For this illustration, we should remove either X_1 or X_2 (but not both). Our best bet would be to retain X_2 = years of experience because it has the highest correlation with Y.

One method of eliminating correlated predictor variables is to use a **stepwise** selection procedure. This technique of selecting the variables to be used in a multiple linear regression equation is discussed in the next section. Essentially, it selects variables one at a time and generally (although not always) does not insert into the regression equation a variable that is highly correlated with a variable already in the equation. In the previous example, a stepwise procedure would have selected variable X_2 (the single best predictor of Y) and then informed the user that X_1 did not significantly improve the prediction of Y, given that X_2 is already included in the prediction equation.

Whenever you perform a multiple regression analysis, it is always a good idea to examine the pairwise correlations between all your variables, including the dependent variable. In this way, you often can easily detect the two independent variables that are contributing to the multicollinearity problem. These correlations can be obtained using a single command with most computer packages. The MINITAB command to generate a table (often called a **correlation matrix**) of pairwise correlations is shown in the bottom box in Figure 14.14. This output indicates that the correlation between Y and X_1 is .852, between Y and X_2 is .879, and between X_1 and X_2 is .970. Since the correlation of any variable with itself is 1, this particular correlation matrix is generally written as

$$\begin{array}{c} \\ Y \\ X_1 \\ X_2 \end{array} \begin{array}{ccc} Y & X_1 & X_2 \\ \begin{bmatrix} 1.0 & .852 & .879 \\ .852 & 1.0 & .970 \\ .879 & .970 & 1.0 \end{bmatrix} \end{array}$$

Other, more advanced methods of detecting and treating the multicollinearity problem are beyond the scope of this text. One of the more popular procedures is *ridge regression.**

We have seen that the problem of multicollinearity enters into our regression analysis when an independent variable is highly correlated with one or more other independent variables. Multicollinearity produces inflated regression coefficients that can even have the wrong sign. Also, the resulting t-statistics can be small, making it difficult to determine the predictive ability of an individual variable. Therefore, $b_1, b_2, \ldots$ tell us nothing about the partial effect of each variable, unless we can demonstrate that there is no correlation among our predictor variables. In business applications, correlation (in particular, *positive* correlation) among the independent variables is far from unusual.

---

* For an excellent discussion of this topic, see J. Neter, W. Wasserman, and M. Kutner, *Applied Linear Regression Models* (Homewood, Ill.: Richard D. Irwin, 1983).

EXERCISES

14.31 What might cause the following situation to occur for a regression model with two independent variables? The t-values for both β_1 and β_2 are nonsignificant. However, the F-test for both $\beta_1 = \beta_2 = 0$ is highly significant.

14.32 If it is known that multicollinearity exists between three independent variables, how would you choose the independent variables that should remain in the model?

14.33 Refer to Exercise 14.2. Find the correlation between the number of units of personal computers sold and the advertising expenditure. Is multicollinearity a concern?

14.34 A least squares equation was fit to a set of data for an experiment and was found to be $\hat{Y} = 30 - 501X_1 + 300X_2$. The experiment was repeated and a new set of data from the same population was fit with the least squares line $\hat{Y} = -20 + 309X_1 - 151X_2$. Is there any explanation for these two different prediction equations?

14.35 The following set of data was collected:

Y	X_1	X_2	Y	X_1	X_2
2.02	1.01	.97	4.20	1.61	2.62
7.95	2.34	5.50	2.62	1.19	1.42
2.61	1.21	1.49	.07	.07	.01
.31	.23	.05	1.53	.80	.67
1.63	.85	.72	6.19	2.03	4.17

a. Construct the correlation matrix for the variables. Does multicollinearity appear to be a problem?

b. Find the coefficient of determination for the model using only X_1. Then find it using only X_2.

c. The coefficient of determination for the complete model is .9996. Does it appear that both variables, X_1 and X_2, should stay in the model?

14.36 Consider the following set of data of 12 emerging growth-oriented companies. Y represents the growth rate of a company for the current year, X_1 represents the growth rate of the company for the previous year, and X_2 represents the percent of the market that does not use the company's product or a similar product. All values are percentages.

Y	X_1	X_2	Y	X_1	X_2
20	10	30	30	15	60
24	12	35	36	42	38
18	15	25	47	45	40
33	30	40	35	32	32
27	19	32	28	24	31
20	24	20	32	20	50

a. Construct the correlation matrix for the variables. Does multicollinearity appear to be a problem?

b. Find the coefficient of determination for the model with only X_1 included in the model.

c. Find the coefficient of determination for the model with only X_2 in the model.

d. The coefficient of determination for the complete model is .896. Does it appear from observing the values of the coefficient of determination in questions a and b that both variables X_1 and X_2 should stay in the model?

14.37 The marketing department of a local industry used a regression equation to predict monthly sales based on total advertising expenditure and television advertising expenditure (both in thousands of dollars). The least squares equation used to predict monthly sales is $\hat{Y} = 103.2 - .20X_1 + 3.4X_2$, where X_1 = total advertising expenditure and X_2 = television advertising expenditure. Can you assume that if television advertising expenditure stays constant and total advertising increases, monthly sales will decrease?

14.38 Looking at the regression equation,

$$\hat{Y} = 25 + 3.5X_1 - 1.8X_2$$

A business student interpreted the values as follows: $b_1 = 3.5$ means that Y increases by 3.5 for each unit increase in X_1 while X_2 is held constant. Similarly, $b_2 = -1.8$ means that Y decreases by 1.8 for each unit increase in X_2 while X_1 is held constant. Although this sounds very good in theory, what is the practical problem in trying to interpret the regression coefficients in this fashion?

14.5

ADDITIONAL TOPICS IN MULTIPLE LINEAR REGRESSION

The Use of Dummy Variables

The use of **dummy**, or **indicator**, variables in regression analysis allows you to include *qualitative* variables in the model. For example, if you wanted to include an employee's sex as a predictor variable in a regression model, define

$$X_1 = \begin{cases} 1 & \text{if female} \\ 0 & \text{if male} \end{cases}$$

Note that the choice of which sex is assigned the value of 1, male or female, is arbitrary. The estimated value of Y will be the same, regardless of which coding procedure is used.

Returning to the data we used in Example 14.2, the real-estate developer noticed that all the houses in the population were from three neighborhoods, A, B, and C. Taking note of which neighborhood each of the sampled houses was from led to the following data (in the discussion following Example 14.2, $X_3 = $ years of education was shown to be a weak predictor, and so is removed from the model here):

FAMILY	Y (HOME SQUARE FOOTAGE)	X_1 (INCOME)	X_2 (FAMILY SIZE)	NEIGHBORHOOD
1	16	22	2	B
2	17	26	2	C
3	26	45	3	A
4	24	37	4	C
5	22	28	4	B
6	21	50	3	C
7	32	56	6	B
8	18	34	3	B
9	30	60	5	A
10	20	40	3	A

Using these data, we can construct the necessary dummy variables and determine whether they contribute significantly to the prediction of home size (Y).

One way to code neighborhoods would be to define

$$X_3 = \begin{cases} 0 & \text{if neighborhood A} \\ 1 & \text{if neighborhood B} \\ 2 & \text{if neighborhood C} \end{cases}$$

However, this type of coding has many problems. First, because $0 < 1 < 2$, the codes imply that neighborhood A is smaller than neighborhood B, which is smaller than neighborhood C. Furthermore, any difference between neighborhoods A and C receives twice the weight (because $2 - 0 = 2$) of any difference between neighborhoods A and B or B and C. So this coding transforms data that are actually *nominal* to data that are *interval*, a much stronger type. A better procedure is to use the necessary number of dummy variables (coded 0 or 1) to represent the neighborhoods.

We needed one dummy variable with two categories (male and female) to specify a person's sex. To represent the three neighborhoods, we use two dummy variables by letting

$$X_3 = \begin{cases} 1 & \text{if house is in A} \\ 0 & \text{otherwise} \end{cases}$$

and

$$X_4 = \begin{cases} 1 & \text{if house is in B} \\ 0 & \text{otherwise} \end{cases}$$

Note that, as with the male/female dummy variable, this coding is arbitrary as far as the prediction, $\hat{Y}$, is concerned. We could have assigned $X_3 = 0$ and $X_4 = 0$ to neighborhood A, with $X_3 = 1$ for B and $X_4 = 1$ for C.

What happened to neighborhood C? It is not necessary to develop a third dummy variable here because we have the following scheme:

HOUSE IS IN NEIGHBORHOOD	X_3	X_4
A	1	0
B	0	1
C	0	0

In fact, it can be shown that a third dummy variable is not only unnecessary, it is very important that you not include it. If you attempted to use three such dummy variables in your model, you would receive a message in your computer output informing you that "no solution exists" for this model. Suppose we had introduced a third dummy variable (say, X_5) that was equal to 1 if the house was in neighborhood C. For each observation in the sample, we would have

$$X_5 = 1 - X_3 - X_4$$

Whenever any one predictor variable is a linear function (including a constant term) of one or more other predictors, then mathematically *no solution exists* for the least squares coefficients, since you have multicollinearity at its worst. To arrive at a usable equation, any such predictor variable must not be included.

The resulting model here is*

$$Y = \beta_0 + \beta_1 X_1 + \beta_2 X_2 + \beta_3 X_3 + \beta_4 X_4 + e$$

The final array of data (ready for input into a computer program) is

ROW	Y	X_1	X_2	X_3	X_4	ROW	Y	X_1	X_2	X_3	X_4
1	16	22	2	0	1	6	21	50	3	0	0
2	17	26	2	0	0	7	32	56	6	0	1
3	26	45	3	1	0	8	18	34	3	0	1
4	24	37	4	0	0	9	30	60	5	1	0
5	22	28	4	0	1	10	20	40	3	1	0

where Y = square footage of home, X_1 = income, X_2 = family size, $X_3 = 1$ if neighborhood A, and $X_4 = 1$ if neighborhood B.

* Models that include dummy variables typically contain terms that reflect any interaction between the dummy variables and the other quantitative variables. For this model, this would amount to adding four additional terms to the model, namely, $X_1 X_3$, $X_1 X_4$, $X_2 X_3$, and $X_2 X_4$. Such a model would require a larger sample size (n) than that used in this illustration, since the model would then contain $k = 8$ predictor variables.

```
MTB > READ INTO C1-C5
DATA> 16 22 2 0 1
DATA> 17 26 2 0 0
DATA> 26 45 3 1 0
DATA> 24 37 4 0 0
DATA> 22 28 4 0 1
DATA> 21 50 3 0 0
DATA> 32 56 6 0 1
DATA> 18 34 3 0 1
DATA> 30 60 5 1 0
DATA> 20 40 3 1 0
DATA> END
        10 ROWS READ
MTB > REGRESS Y IN C1 USING 4 PREDICTORS IN C2-C5

The regression equation is
C1 = 7.77 + 0.082 C2 + 3.27 C3 + 1.61 C4 - 0.90 C5

Predictor        Coef        Stdev      t-ratio          p
Constant        7.772        2.557         3.04      0.029
C2             0.0819       0.1059         0.77      0.474
C3             3.2696       0.9870         3.31      0.021
C4              1.613        1.801         0.90      0.411
C5             -0.900        1.841        -0.49      0.646

s = 2.036        R-sq = 92.1%     R-sq(adj) = 85.8%
```

A R_c^2

```
MTB > REGRESS Y IN C1 USING 2 PREDICTORS IN C2 C3

The regression equation is
C1 = 6.74 + 0.165 C2 + 2.66 C3

Predictor        Coef        Stdev      t-ratio          p
Constant        6.740        2.164         3.11      0.017
C2            0.16485      0.07306         2.26      0.059
C3            2.6569        0.7405         3.59      0.009

s = 1.953        R-sq = 89.8%     R-sq(adj) = 86.9%
```

B R_r^2

A MINITAB solution is shown in Figure 14.15. To determine whether the particular neighborhood has any effect on the prediction of home size, we test

$$H_0: \beta_3 = \beta_4 = 0 \text{ (if } X_1 \text{ and } X_2 \text{ are included)}$$
$$H_a: \text{ at least one } \beta \neq 0$$

In the complete model, the variables are X_1, X_2, X_3, and X_4, and, from Figure 14.15(a),

$$R_c^2 = .921$$

In the reduced model, the variables are X_1 and X_2 only, and, from Figure 14.15(b),

$$R_r^2 = .898$$

At first glance, it does not appear that X_3 and X_4 produced a significant increase in R^2. The partial F-test will determine whether this is true.

$$F = \frac{(R_c^2 - R_r^2)/(\text{number of } \beta\text{'s in } H_0)}{(1 - R_c^2)/[n - 1 - (\text{number of } X\text{'s in the complete model})]}$$

$$= \frac{(.921 - .898)/2}{(1 - .921)/(10 - 1 - 4)} = \frac{.023/2}{.079/5} = .73$$

Using $\alpha = .10$, this is considerably less than $F_{.10, 2, 5} = 3.78$, so there is no evidence that the neighborhood dummy variables significantly improve the prediction of home size.

In this example, the dummy variables were not significant predictors in the model. However, do not let this mislead you. In many business applications, dummy variables representing location, weather conditions, yes/no situations, time, and many other variables can have a tremendous effect on improving the results of a multiple regression model.

Stepwise Procedures

Assume you wish to predict annual divisional profits for a large corporation using, among other techniques, a multiple linear regression model. Your strategy is to consider any variable that you think *could* have an effect on these profits. You have identified twelve such variables.

One possibility is to include all these in your model and to use the t-tests to decide which variables are significant predictors. However, this procedure invites multicollinearity because your model is more apt to include correlated predictors, which severely hinders the interpretation of your model. In particular, two independent variables that are very highly correlated may both have small t-values (as we saw in the employee example), causing you possibly to discard both of them from the model. This is *not* the right thing to do because you possibly should have retained one of them.

A better way to proceed here is to use one of the several stepwise selection procedures. These techniques either choose or eliminate variables, one at a time, in an effort not to include those variables that either have no predictive ability or are highly correlated with other predictor variables. A word of caution—these procedures do not provide a guarantee against multicollinearity; however, they greatly reduce the chances of including a large set of correlated independent variables.

These procedures consist of three different selection techniques: (1) forward regression, (2) backward regression, and (3) stepwise regression.

Forward Regression   The forward regression method of model selection puts variables into the equation, one at a time, beginning with that variable having the highest correlation (or R^2) with Y. For sake of argument, call this variable X_1.

Next, it examines the remaining variables for the variable that, when included with X_1, has the highest R^2. That predictor (with X_1) is inserted into the model. This procedure continues until adding the "best" remaining variable at that stage results in an insignificant increase in R^2, according to the partial F-test.

Backward Regression   Backward regression is the opposite of forward regression: It begins with *all* variables in the model and, one by one, removes them. It begins by finding the "worst" variable—the one that causes the smallest decrease in R^2 when removed from the complete model. If the decrease is insignificant, this variable is removed, and the process continues.

The variable among those remaining in the model that causes the smallest decrease in the new R^2 is considered next. You continue this procedure of removing variables until a significant drop in R^2 is obtained, at which point you replace this significant predictor and terminate the selection.

Will the model resulting from a backward regression be the same as that obtained using forward regression? Not necessarily; usually, however, the resulting models are very similar. Of course, if two variables are highly correlated,

FIGURE 14.16

Possible solution using stepwise regression on divisional profits data.

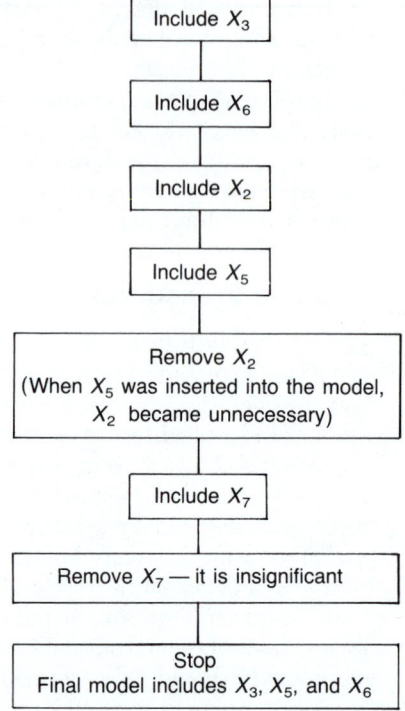

the forward procedure could choose one of the correlated predictors, whereas the backward procedure could choose the other.

Stepwise Regression Stepwise regression is a modification of forward regression. It is the most popular and flexible of the three selection techniques. It proceeds exactly as does forward regression, except that at each stage it can *remove* any variable whose partial F-value indicates that this variable does not contribute, given the present set of independent variables in the model. As with forward regression, it stops when the "best" variable among those remaining produces an insignificant increase in R^2.

Figure 14.16 illustrates this procedure using a sample of data (not shown) for the illustration concerned with predicting divisional profits. Data from all 12 independent variables, as well as from Y, are used as input to a stepwise regression program. One possible outcome from this analysis is shown by Figure 14.16.

The stepwise solution for the data we used to predict home size is contained in the end-of-chapter MINITAB appendix. As we previously determined, X_3 = educational level does not contribute significantly, and so the resulting prediction equation includes only X_1 = income and X_2 = family size. This equation is

$$\hat{Y} = 6.74 + .165X_1 + 2.66X_2$$

Using Dummy Variables in Forward or Stepwise Regression We emphasized that $C - 1$ dummy variables should be used to represent C categories if *all* the dummy variables were to be inserted into the regression equation. When using a forward or stepwise regression procedure, this may not be the best way to proceed, as the following illustration shows.

Suppose you are using nine dummy (indicator) variables to represent ten cities. The dependent variable is monthly sales, and the purpose is to determine which city (or cities) exhibits very large or very small sales. If a forward or stepwise selection procedure is used, then including one of these dummy variables indicates that specifying this particular city significantly improves the prediction of sales. In other words, this is an indication that sales for this city are not just average but are much higher (its coefficient will be positive) or lower (its coefficient will be negative).

When you use the forward or stepwise techniques, you probably will not include all nine dummy variables in the model. Your ability to predict sales (Y) is unaffected by not defining a tenth dummy variable and, in fact, as pointed out earlier, the regression analysis will not accept all ten dummy variables.

For this situation, however, there is the danger of not detecting extremely high or low sales in the tenth city that did not receive a dummy variable. When including these variables one at a time in the regression equation using a forward or stepwise procedure, we can allow the regression model to examine the effect of all ten cities. We do this defining ten such dummy variables, one for each city.*

Because a forward regression procedure generally will not attempt to include all ten dummy variables, you are able to investigate the existence of high or low sales in each of the ten cities. When using dummy variables in a forward or stepwise regression procedure, it is perfectly acceptable to use C such variables to represent C categories.

Checking the Assumptions: Examination of the Residuals

When you use a multiple linear regression model, you should keep two things in mind. First, no assumptions are necessary to derive the least squares estimates of β_1, β_2, β_3, ... The regression coefficients b_1, b_2, b_3, . . . determined by a computer solution are the "best" estimates, in the least squares sense.

Second, several key assumptions *are* required to construct confidence intervals and perform any test of hypothesis. If these assumptions are violated, you may still have an accurate prediction, $\hat{Y}$, but the validity of these inference procedures will be very questionable.

Your final step in any regression analysis should be to verify your assumptions.

Assumption 1   The errors are normally distributed, with a mean of zero.

An easy method to determine whether the errors follow a normal distribution, centered at zero, is to let the computer construct a histogram of the sample residuals ($Y - \hat{Y}$). Since the residuals *always* sum to zero, the residual histogram is typically centered at zero. This plot should reveal whether the distribution of residuals is severely skewed.

Consider the 50 residuals resulting from the analysis in Example 14.4. The computer solution for this problem is shown in Figure 14.6. Notice that the RESIDUALS subcommand following the REGRESS command can be used to store the residuals in any column (such as column C5 in Figure 14.6).

A MINITAB histogram of these values is shown in Figure 14.17. The distribution of these residuals appears to be centered at zero and, except for a slight left skew, is bell-shaped (normally distributed) in appearance. Remember that an exact normal distribution is not necessary here; problems arise only when the distribution is severely skewed and does not resemble a normal distribution.

---

* This problem is discussed in D. Dorsett and J. T. Webster, "Guidelines for Variable Selection Problems When Dummy Variables Are Used," *The American Statistician* 37, no. 4 (1983): 337.

FIGURE 14.17

Histogram of residuals
for investment data (see
Figure 14.6).

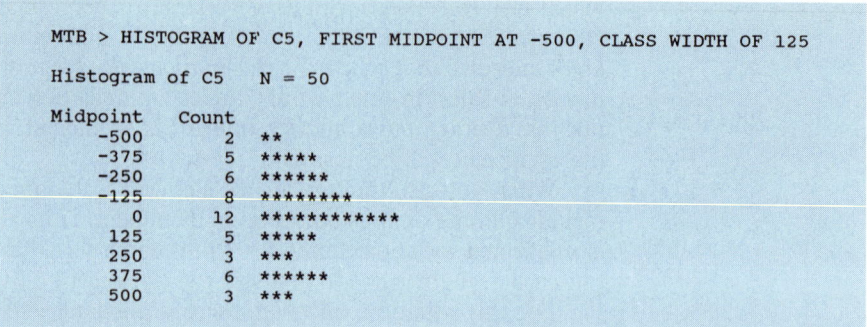

```
MTB > HISTOGRAM OF C5, FIRST MIDPOINT AT -500, CLASS WIDTH OF 125

Histogram of C5    N = 50

Midpoint    Count
   -500        2    **
   -375        5    *****
   -250        6    ******
   -125        8    ********
      0       12    ************
    125        5    *****
    250        3    ***
    375        6    ******
    500        3    ***
```

More sophisticated methods of checking this assumption involve the use of a *probability plot*, or a *chi-square goodness-of-fit test*. We do not discuss the probability plot technique here, except to say that you plot the residuals in a specialized type of graph. If the resulting graph is basically linear in appearance, the normality assumption has been verified. The goodness-of-fit test was discussed in Chapter 12, where we used a chi-square statistic to test the hypothesis that a particular set of data (in this case, the regression residuals) came from a specific distribution. The end-of-chapter exercises in Chapter 12 discuss how to use the chi-square test for a suspected *normal* population.

If you have reason to believe that this assumption of your model has been violated, then you need to search for another model. This model may include additional predictor variables that have been overlooked. Another possibility is to transform the dependent variable (such as using $\sqrt{Y}$ rather than Y) or transforming one or more of the predictor variables. As your model tends to "improve," you should observe the residuals tending toward a normal distribution.

Assumption 2   The variance of the errors remains constant. For example, you should not observe larger errors associated with larger values of $\hat{Y}$.

When the residuals $(Y - \hat{Y})$ are plotted against the predicted values $(\hat{Y})$, we hope to observe *no pattern* (a "shotgun blast" appearance) in this graph, as in Figure 14.18(a). Remember—the assumptions are essentially that the errors consist of what engineers call *noise*, with no observable pattern.

A common violation of the assumption of equal variances occurs when the value of the residual increases as $\hat{Y}$, or an individual predictor, increases. This is illustrated in Figure 14.18(b). In this figure, the variance of the residual is increasing with $\hat{Y}$. This has a serious effect on the validity of the hypothesis tests developed in this chapter, which determine the strength of the regression model and the individual predictors.

When you encounter a violation of this type, you need to resort to more

FIGURE 14.18

Examination of the
residuals. **A**: The shotgun
effect (no violation of
assumptions 1 and 2). **B**:
A violation of the equal
variance assumption
(assumption 2).

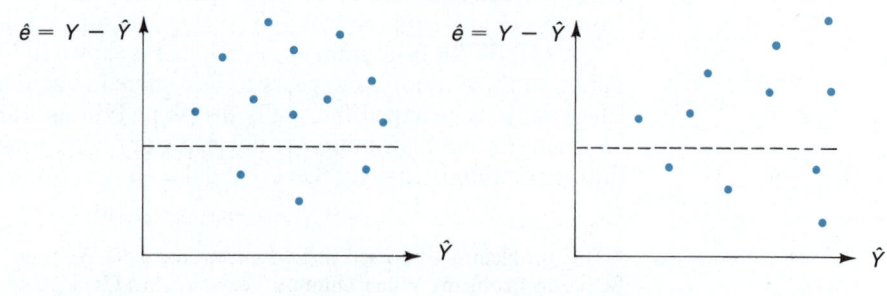

A   B

FIGURE 14.19
Autocorrelated errors.

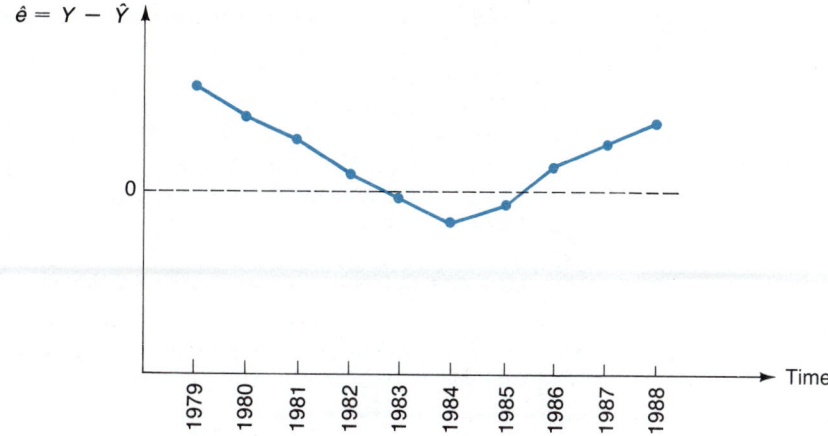

advanced modeling techniques, including *weighted least squares* or *transformations* of your dependent variable.*

Assumption 3 The errors are independent.

To examine these assumptions after the regression equation has been determined involves using the residual from each of the sample observations. For given values of $X_1, X_2, \ldots, X_k$, the actual error is

$$e = Y - (\beta_0 + \beta_1 X_1 + \beta_2 X_2 + \cdots + \beta_k X_k)$$

The β's are unknown, so we estimate the error by using the residual for this particular observation,

$$Y - \hat{Y} = Y - (b_0 + b_1 X_1 + b_2 X_2 + \cdots + b_k X_k)$$

The residuals for the real-estate data are shown in the third column, labeled $Y - \hat{Y}$, of the table on page 565. In general, a close examination of these values will reveal any departures from the regression assumptions.

When your regression data consist of *time series* data, your errors often are not independent. This type of data has the appearance

TIME	Y	X_1	X_2	$\cdots$	X_k
1979	x	x	x		x
1980	x	x	x		x
1981	x	x	x		x
⋮	⋮	⋮	⋮		⋮

Also remember that the error component *includes the effect of missing variables* in your model. In many business applications, there is a positive relationship between time-related predictor variables, such as prices and wages, because they both increase over time. This can produce a set of residuals in your regression analysis that are not independent of one another but, instead, display a pattern similar to that illustrated in Figure 14.19. This plot contains the sample residuals on the vertical axis and time on the horizontal axis. If this assumption were *not* violated here, we should observe the shotgun appearance. Instead we notice that adjacent residuals have roughly the same value and so are correlated with each other. This is **autocorrelation**. To be more specific, the pattern in

* See J. Neter, W. Wasserman, and M. Kutner, *Applied Linear Regression Models* (Homewood, Ill.: Richard D. Irwin, 1983).

Figure 14.19 is one of *positive* autocorrelation. Negative autocorrelation exists when most of the neighboring residuals are very unequal in size.

The amount of autocorrelation that exists in residuals is measured by the **Durbin-Watson statistic**. This statistic ranges from zero to four, with a value near zero indicating strong *positive* autocorrelation and a value close to four meaning that there is a significant *negative* autocorrelation. A value near two indicates that there is no (or very little) autocorrelation—the ideal situation. Chapter 16 discusses the calculation of this statistic and its use in detecting autocorrelated residuals.

The problem of autocorrelated errors is the most difficult of the three assumptions to correct. The error term is not noise, as we originally assumed, but instead has a definite pattern (as in Figure 14.19). Several ways of treating this problem are discussed in Chapter 16.

Prediction Using Multiple Regression

Once a regression equation has been derived, its primary application generally is to derive predicted values of the dependent variable. Computer packages provide an easy method of deriving such an estimate. To illustrate, consider the regression equation we developed for the real-estate data. For this illustration we include $X_3 =$ years of formal education, although, as we demonstrated in Example 14.2, this variable could be dropped without any significant loss in the prediction of home size. The resulting prediction equation was

$$\hat{Y} = 7.60 + .194X_1 + 2.34X_2 - .163X_3$$

Consider a situation in which

$$X_1 = \text{income} = 36 \text{ (thousands of dollars)}$$
$$X_2 = \text{family size} = 4$$
$$X_3 = \text{years of formal education} = 8 \text{ (years)}$$

The predicted home size (Y) here is

$$\hat{Y} = 7.60 + .194(36) + 2.34(4) - .163(8)$$
$$= 22.64 \qquad \text{(2264 square feet)}$$

We can derive this predicted value using MINITAB, SAS, or SPSS. When using the latter two packages, we can derive a predicted value by adding one additional row to our input data containing these specific values of the predictor variables and a *missing value* for the dependent variable value. The computer routine ignores this row when deriving the regression equation (and all subsequent tests of hypotheses) but attaches this row when listing the predicted values. The procedure when using MINITAB is slightly different, as described next.

Using MINITAB   The MINITAB solution for the preceding illustration is shown in Figure 14.20. By using the subcommand PREDICT following the REGRESS command, you can easily derive a predicted value for the input values of the independent variables. The output will contain the predicted (fitted) Y value, the standard deviation of the predicted Y, a 95% confidence interval, and a 95% prediction interval. The resulting predicted value for $X_1 = 36$, $X_2 = 4$, and $X_3 = 8$ is $\hat{Y} = 22.625$, which is more accurate than the previously derived value of 22.64.

FIGURE 14.20

Prediction for new data using MINITAB. For the input data, See Figure 14.2.

```
MTB > BRIEF 3
MTB > REGRESS Y IN C1 USING 3 PREDICTORS IN C2-C4;
SUBC> PREDICT FOR 36 4 8.

The regression equation is
C1 = 7.60 + 0.194 C2 + 2.34 C3 - 0.163 C4

Predictor        Coef       Stdev      t-ratio          p
Constant        7.596       2.595         2.93      0.026
C2            0.19388     0.08770         2.21      0.069
C3             2.3381      0.9078         2.58      0.042
C4            -0.1628      0.2441        -0.67      0.530

s = 2.035       R-sq = 90.5%      R-sq(adj) = 85.8%

Analysis of Variance

SOURCE         DF          SS          MS          F          p
Regression      3     237.542      79.181      19.11      0.002
Error           6      24.858       4.143
Total           9     262.400


Obs.      C2          C1      Fit Stdev.Fit  Residual  St.Resid
  1     22.0      16.000    15.886     1.193     0.114      0.07
  2     26.0      17.000    16.010     1.161     0.990      0.59
  3     45.0      26.000    22.195     0.975     3.805      2.13R
  4     37.0      24.000    24.121     1.300    -0.121     -0.08
  5     28.0      22.000    22.051     1.371    -0.051     -0.03
  6     50.0      21.000    22.676     1.331    -1.676     -1.09
  7     56.0      32.000    31.179     1.854     0.821      0.98
  8     34.0      18.000    19.900     0.949    -1.900     -1.05
  9     60.0      30.000    30.593     1.608    -0.593     -0.48
 10     40.0      20.000    21.388     0.767    -1.388     -0.74

R denotes an obs. with a large st. resid.

   Fit   Stdev.Fit          95% C.I.            95% P.I.
22.625        1.355    ( 19.308, 25.943)   ( 16.640, 28.611)
```

$\hat{Y}$ using $X_1 = 36$, $X_2 = 4$, $X_3 = 8$

Using SPSS   A solution for this illustration using SPSS is presented at the end of the chapter. A numeric value (such as -9) is used for the missing Y value, and then SPSS is informed that such a value represents a missing value using the MISSING VALUES command. This command instructs SPSS that a value of -9 appearing as a value for SQFOOT (the Y variable) should be interpreted as a missing value. The resulting predicted value of $\hat{Y} = 22.625$ appears at the end of the list of predicted values.

Using SAS   Similar to SPSS, SAS predicts values by including an additional row containing a missing value for the dependent variable. Any single *letter* (A to Z) can be used to represent the missing value; SAS is informed that such a character represents a missing value by using the MISSING statement after the DATA statement. This row is then automatically ignored during subsequent calculations, but, once again, the predicted value of $\hat{Y} = 22.625$ is generated for this set of X values at the end of the list of predicted values.

Confidence and Prediction Intervals   In the preceding illustration, what does $\hat{Y} = 22.625$ estimate? For ease of notation, let X_0 represent the set of X values used for this estimate; that is, $X_0 = (36, 4, 8)$, where $X_1 = 36$, $X_2 = 4$, and $X_3 = 8$. This value of $\hat{Y}$ estimates (1) the *average* home size of all families with this specific set of X values, written $\mu_{Y|X_0}$ and (2) the home size for an *individual* family having this specific set of X values, written Y_{X_0}.

Using the notation from Chapter 13, let

$$s_{\hat{Y}} = \text{the standard deviation of the predicted } Y \text{ mean}$$

These values can be computed and included in the output by each of the computer packages. To determine the reliability of this particular point estimate, $\hat{Y}$, you can (1) derive a *confidence interval* for $\mu_{Y|X_0}$ if your intent is to estimate the *average* value of Y given X_0 (not the usual situation) or (2) derive a *prediction interval* for Y_{X_0} if the purpose is to forecast an *individual* value of Y given this specific set of values for the predictor variables. In business applications, deriving a specific forecast is, by far, the more popular use of linear regression.

These intervals are summarized as follows. A $(1 - \alpha) \cdot 100\%$ confidence interval for $\mu_{Y|X_0}$ is

$$\hat{Y} - t_{\alpha/2, n-k-1} s_{\hat{Y}} \quad \text{to} \quad \hat{Y} + t_{\alpha/2, n-k-1} s_{\hat{Y}} \qquad \textbf{(14.18)}$$

A $(1 - \alpha) \cdot 100\%$ prediction interval for Y_{X_0} is

$$\hat{Y} - t_{\alpha/2, n-k-1} \sqrt{s^2 + s_{\hat{Y}}^2} \quad \text{to} \quad \hat{Y} + t_{\alpha/2, n-k-1} \sqrt{s^2 + s_{\hat{Y}}^2} \quad \textbf{(14.19)}$$

where $s^2 = \text{MSE}$.

Using MINITAB $[X_0 = (36, 4, 8)]$ Prediction and confidence intervals are easy to derive using MINITAB by using the PREDICT subcommand, as illustrated in Figure 14.20. The resulting confidence interval for the average home size, given X_0, is (19.308, 25.943). As usual, the 95% prediction interval for an individual family with this set of X values, namely, (16.640, 28.611), is wider than the corresponding confidence interval.

Using SPSS $[X_0 = (36, 4, 8)]$ These intervals are not directly available on this package but can be obtained easily using equations 14.18 and 14.19 and the regression output. SPSS provides the estimated value, $\hat{Y}$, and the standard deviation of the predicted value, $s_{\hat{Y}}$. Referring to the discussion at the end of the chapter, these values are $\hat{Y} = 22.625$ and $s_{\hat{Y}} = 1.355$.

Using equation 14.18, a 95% confidence interval for $\mu_{Y|X_0}$ is derived by first using Table A.5 to obtain $t_{\alpha/2, n-k-1} = t_{.025, 10-3-1} = t_{.025, 6} = 2.447$. The resulting confidence interval is

$$22.625 - (2.447)(1.355) \quad \text{to} \quad 22.625 + (2.447)(1.355)$$

$$= 22.625 - 3.316 \quad \text{to} \quad 22.625 + 3.316$$

$$= 19.309 \quad \text{to} \quad 25.941$$

Consequently, we have estimated the average home size for families with $X_1 = 36$, $X_2 = 4$, $X_3 = 8$ to within 331.6 square feet of the actual mean with 95% confidence. This confidence interval is slightly different from that obtained with MINITAB due to rounding error.

The prediction interval from equation 14.19 is derived by using MSE = 4.143 from the computer solution to obtain

$$22.625 - 2.447\sqrt{4.143 + (1.355)^2} \quad \text{to} \quad 22.625 + 2.447\sqrt{4.143 + (1.355)^2}$$

$$= 22.625 - 5.983 \quad \text{to} \quad 22.625 + 5.983$$

$$= 16.642 \quad \text{to} \quad 28.608$$

This means that we have predicted the home size of an individual family with $X_1 = 36$, $X_2 = 4$, and $X_3 = 8$ to within 598.3 square feet of the actual value with

95% confidence. Once again, this result differs slightly from that obtained with MINITAB due to rounding error.

Using SAS $[X_0 = (36, 4, 8)]$ SAS provides an easy method of determining these intervals by including CLI (for an individual Y_{X_0}), CLM (for a mean $\mu_{Y|X_0}$), or both in the final SAS statement. According to the output at the end of the chapter, the 95% confidence interval for the average home size, given X_0, is (19.309, 25.942), with a corresponding prediction interval of (16.641, 28.609).

EXERCISES

14.39 A real-estate agency was interested in the amount of rent paid (monthly) for commercial buildings near downtown Houston. The broker of the real-estate agency used the following independent variables:

$$X_1 = \text{age of building}$$

$$X_2 = \begin{cases} 1 & \text{if low-rise building} \\ 0 & \text{if not} \end{cases}$$

$$X_3 = \begin{cases} 1 & \text{if mid-rise building} \\ 0 & \text{if not} \end{cases}$$

$$X_4 = \text{square footage of building}$$

The data from 20 commercial buildings yielded the least squares equation $\hat{Y} = 130 - 50X_1 + 70X_2 - 20X_3 + 0.31X_4$.

a. What is the predicted rent for a 5-year-old high-rise commercial building that has 25,000 square feet of space?

b. Test the hypothesis that the dummy variables significantly improve the prediction of monthly rent for commercial buildings. Assume that R^2 for the model including only X_1 and X_4 is .65 and that R^2 for the complete model including X_1, X_2, X_3, and X_4 is .85.

14.40 The executive committee of Mini-Mart convenience-food-store chain would like to examine the yearly profit from its large number of stores. The following independent variables were used in analyzing the profits (Y) of each store:

$$X_1 = \text{average yearly household income of nearby area}$$

$$X_2 = \begin{cases} 1 & \text{if there is a supermarket within 1 mile} \\ 0 & \text{otherwise} \end{cases}$$

$$X_3 = \begin{cases} 1 & \text{if weekly auto traffic volume is high} \\ 0 & \text{otherwise} \end{cases}$$

The following least squares equation was obtained from observing 18 randomly selected stores:

$$\hat{Y} = 7508 + 1.5X_1 - 6235X_2 + 5987X_3$$

a. Interpret the coefficients of this equation.

b. Given that the standard deviations of the estimate of the coefficients of X_2 and X_3 are 2521 and 1873, respectively, test the hypothesis that the variable X_2 contributes to the prediction of Y, given that X_1 and X_3 are in the model. Also test that X_3 contributes to the prediction of Y given that X_1 and X_2 are in the model. Use a .01 significance level.

14.41 Nebraska Associated Insurance handles workers compensation insurance for three large manufacturing firms. The insurance company believes that the following independent variables are important in determining the total amount of compensation paid for each claim from the three manufacturers:

Age

Sex

Marital status

Length of employment

Type of injury (to a limb, to the head, or to other parts of the body)

Manufacturer employing the worker

Set up an appropriate regression model to predict total amount of compensation paid based on the independent variables. Define your variables.

14.42 Data are collected for the variables Y, X_1, and X_2. A computer printout of the correlation matrix is:

	Y	X_1	X_2
Y	1	.49	.30
X_1	.49	1	.12
X_2	.30	.12	1

a. Which independent variable, X_1 or X_2, would be selected first in a forward regression procedure?

b. Which independent variable, X_1 or X_2, would be a better predictor of Y? Why?

14.43 The following is a correlation matrix for three independent variables and one dependent variable:

	Y	X_1	X_2	X_3
Y	1	.25	.36	.59
X_1	.25	1	.54	.22
X_2	.36	.54	1	.31
X_3	.59	.22	.31	1

a. Which independent variable would be chosen for the first stage of a forward regression procedure?

b. Which independent variable would be chosen for the first step of a stepwise regression procedure?

14.44 The least squares regression equation

$$\hat{Y} = 1.5 + 3.5X_1 + 7.5X_2 - 150X_3$$

has the following t values for the independent variables:

NULL HYPOTHESIS	t-STATISTIC
$\beta_1 = 0$	4.5
$\beta_2 = 0$	1.89
$\beta_3 = 0$	1.52

Twenty observations were used in calculating the least squares equation. In the first stage of a backward selection procedure, which independent variable would be eliminated first? Use a 5% level of significance.

14.45 Describe the main difference between the forward selection procedure and the stepwise selection procedure in regression analysis.

14.46 If a statistician would like to include dummy variables to indicate one of three cities and also one of four salespeople in a regression model, how many dummy variables would be needed in the model?

14.47 Which of the standard assumptions of regression appear to have been violated from the following table, which lists the dependent variable and the residual values?

$\hat{Y}$	$Y - \hat{Y}$	$\hat{Y}$	$Y - \hat{Y}$
1.5	0.12	5.0	-1.45
2.1	$-.70$	5.5	1.61
3.5	$-.91$	6.0	1.79
4.0	1.02	7.0	-2.40
4.5	-1.18	7.5	2.10

14.48 How should a plot of the residuals $(Y - \hat{Y})$ plotted against the predicted values $\hat{Y}$ look if the standard assumptions of regression are satisfied?

14.49 A set of 20 observations is used to obtain the least squares line

$$\hat{Y} = 1.5 + 3.6X_1 + 4.9X_2$$

a. Given that the estimated standard deviation of Y at $X_1 = 1.0$ and $X_2 = 2.0$ is 3.4, find a 90% confidence interval for the mean value of Y at $X_1 = 1.0$ and $X_2 = 2.0$.

b. Given that the MSE from this analysis is 21.5, then, using the information in part(a), find a 90% prediction interval for an individual value of Y at $X_1 = 1.0$ and $X_2 = 2.0$.

14.50 Fifteen months are randomly selected to estimate the monthly sales, Y (in thousands of dollars), of a retail store based on monthly advertising expenditure, X (in thousands of dollars). The prediction equation is found to be

$$\hat{Y} = .2 + 1.5X + .4X^2$$

a. Find a 95% confidence interval for the mean monthly sales with a monthly advertising expenditure of $1.4 thousand if the estimated standard deviation of the monthly sales for $X = 1.4$ is .8.

b. Using the information in part (a), find a 95% prediction interval for the monthly sales of a month that has an advertising expenditure of $1.4 thousand if the MSE from the analysis is 1.55.

14.51 Explain the difference between a confidence interval for the mean value of Y at particular values of the independent variables and a prediction interval for a future value of Y at particular values of the independent variables. Will the prediction interval for Y always be larger than the corresponding confidence interval for particular values of the independent variables?

14.52 A real-estate firm would like to determine the monthly income (Y) of homeowners in a certain section of town by using the monthly mortgage payment (X_1), the market value of the homeowner's car(s) (X_2), and the age of the homeowner (X_3). The following data are collected from 15 randomly selected households (Y, X_1, and X_2 are in dollars; X_3 in years):

Y	X_1	X_2	X_3	Y	X_1	X_2	X_3	Y	X_1	X_2	X_3
2963	820	7,800	32	2225	725	4,380	30	3180	635	9,450	36
2100	710	5,100	33	1630	538	3,760	27	3350	758	12,600	31
2820	520	10,500	26	3070	679	7,350	37	3267	810	10,630	29
3350	630	9,500	30	2950	975	6,580	34	2120	710	5,340	28
2640	925	6,260	35	3460	1120	7,900	33	2280	504	4,690	32

Use a computerized statistical package to answer the following questions:

a. What is the mean income for a homeowner with $X_1 = 800$, $X_2 = 7000$, and $X_3 = 30$?

b. Find a 95% confidence interval for the mean monthly income of a homeowner with $X_1 = 800$, $X_2 = 7000$, and $X_3 = 30$.

c. Find a 95% prediction interval for the income of a homeowner with $X_1 = 800$, $X_2 = 7000$, and $X_3 = 30$.

14.53 The operations manager in charge of a production process is interested in the amount of time in minutes, Y, that it takes an assembly-line worker to perform a certain task relative to his or her score, X, on an aptitude test. The proposed model is

$$Y = \beta_0 + \beta_1 X + \beta_2 X^2 + e$$

Twelve assembly line workers were randomly selected with the following results:

Y	X	Y	X
49	58	19	85
37	67	17	89
12	95	50	52
60	43	67	41
33	72	39	67
22	83	35	70

a. Using a computerized statistical package and the proposed model, construct a 99% confidence interval for the mean time that it takes assembly-line workers to complete the task if the aptitude score is 80.

b. Using a computerized statistical package and the proposed model, construct a 99% prediction interval for the time it takes a worker to complete the task if the aptitude score of that worker is 80.

c. Compare the answers in parts (a) and (b).

14.54 The vice president of a computer firm is interested in funding research proposals by graduate students who wish to perform experiments in the firm's advanced technology laboratory during the summer months. The vice president receives 18 proposals and sends these proposals to the director of the laboratory for evaluation. The director rates the proposals on two criteria and gives a score between zero and ten for each criterion, with 10 representing the best score possible. The variables X_1 and X_2 are used to represent these two scores. The dependent variable Y is the level of funding that the director of the laboratory would be willing to grant for the proposal. The variable Y is given in units of 1000's. The collected data are given below:

Y	X_1	X_2	Y	X_1	X_2	Y	X_1	X_2
9.5	8.7	9.2	7.0	7.9	7.9	8.5	8.6	8.8
7.3	8.1	8.0	7.4	8.2	8.0	7.2	8.3	7.8
6.5	7.4	7.7	8.3	8.5	8.4	5.8	6.7	7.0
8.4	8.4	8.6	8.2	8.6	7.9	6.3	7.3	7.5
8.0	8.3	8.0	5.3	6.6	6.9	9.0	8.6	9.0
6.1	7.0	7.3	6.7	7.8	7.5	6.4	7.7	7.5

a. Find a 90% confidence interval for the mean value of Y at $X_1 = 8.0$ and $X_2 = 7.8$.

b. Interpret the confidence interval in part (a) in the context of this exercise.

c. Find a 90% prediction interval for the value of Y at $X_1 = 8.0$ and $X_2 = 7.8$.

d. Interpret the prediction interval in part (c) in the context of this exercise.

SUMMARY

Multiple linear regression offers a method of predicting (or modeling) the behavior of a particular **dependent** variable (Y) using two or more **independent (predictor)** variables. As in the case of simple linear regression, which uses one predictor variable, the regression coefficients are those that minimize

$$SSE = \text{sum of squares of error} = \Sigma(Y - \hat{Y})^2$$

To use this technique properly, you must pay special attention to the assumptions behind it. These are that (1) the regression errors follow a normal distribution, centered at zero, with a common variance and (2) the errors are statistically independent. An estimate of this common variance is

$$\hat{\sigma}_e^2 = s^2 = MSE = \frac{SSE}{n - k - 1}$$

To determine the adequacy of the regression model, you can test the entire set of predictor variables using an F-test with k and $n - k - 1$ degrees of freedom:

$$F = \frac{\text{MSR}}{\text{MSE}} = \frac{R^2/k}{(1 - R^2)/(n - k - 1)}$$

The contribution of an individual predictor variable (say, X_i) can be tested using a t-statistic with $n - k - 1$ df:

$$t = \frac{b_i}{s_{b_i}}$$

where s_{b_i} represents the estimated standard deviation of b_i. Here b_i is the least squares estimate of the population parameter, β_i, and centers the confidence interval for this parameter.

The **coefficient of determination, R^2**, describes the percentage of the total variation in the sample Y values explained by this set of predictor variables. To determine the contribution of a particular subset of the predictor variables—such as X_2 and X_4—R^2 is computed with X_2 and X_4 included and then with X_2 and X_4 excluded from the regression equation. A **partial F-test** is then used to determine whether the resulting decrease in R^2 is significant.

When a **curvilinear** pattern exists between two variables, X and Y, this non-linear relationship often can be modeled by including an X^2 term in the regression equation. The resulting equation is

$$\hat{Y} = b_0 + b_1 X + b_2 X^2$$

This type of model often works well in situations where Y (for example, sales) appears to increase more slowly as the independent variable, X (for example, the amount of shelf space devoted to this product) continues to increase.

The problem of **multicollinearity** arises in the application of multiple linear regression when two or more independent variables are highly correlated. The resulting regression equation contains coefficients that are highly inflated (have a large variance) with t-statistics that are extremely small, despite the fact that one or more of these seemingly insignificant variables are very useful predictors. An easy means of correcting this problem is to remove certain variables from the regression equation or to use a stepwise regression procedure.

Stepwise techniques allow you to insert variables one at a time into the equation (**forward regression**), remove them one at a time after initially including all variables in the equation (**backward regression**), or perform a combination of the two by inserting variables one at a time but removing a variable that has become redundant at any stage (**stepwise regression**). Once the variables for the model have been selected, **residual plots** should be obtained to examine the underlying assumptions that are necessary in a regression analysis.

Dummy variables can be used in a regression application to represent the categories of a qualitative variable (such as city). If all dummy variables are to be inserted into the equation, then $C - 1$ such variables should be defined to represent C categories. If a forward or stepwise selection procedure is used to define the final regression equation, then a better procedure is to define C dummy variables to represent this situation.

Use of a computer package is essential in the derivation of a multiple regression equation. In this chapter, we used MINITAB, SPSS, and SAS. They provide the sampling coefficients ($b_0, b_1, b_2, \ldots$), the statistics necessary to perform any test of hypothesis, and those needed for the prediction and confidence intervals for any specific set of predictor variable values. A **confidence interval** is derived whenever the predicted Y value is used to estimate the average value of the dependent variable for a specific set of X values. When the predicted Y value is used to predict an individual value of Y for a specific set of X values, a **prediction interval** can be used to place bounds on the actual Y value.

FORMULAS USED IN MULTIPLE LINEAR REGRESSION

H_0: all β's $= 0$

H_a: at least one $\beta \neq 0$

$$F = \frac{MSR}{MSE} = \frac{R^2/k}{(1 - R^2)/(n - k - 1)}$$

(df $= k$ and $n - k - 1$)

H_0: $\beta_i = 0$

H_a: $\beta_i \neq 0$

(or H_a: $\beta_i > 0$)

(or H_a: $\beta_i < 0$)

$$t = \frac{b_i}{s_{b_i}}$$

(df $= n - k - 1$)

Confidence interval for β_i

$$b_i - t_{\alpha/2, n-k-1} s_{b_i}$$
$$\text{to} \quad b_i + t_{\alpha/2, n-k-1} s_{b_i}$$

Coefficient of determination

$$R^2 = 1 - (SSE/SST)$$

where

$$SST = \Sigma (Y - \bar{Y})^2 = \Sigma Y^2 - (\Sigma Y)^2/n$$

and

$$SSE = \Sigma (Y - \hat{Y})^2$$

H_0: $X_i, X_{i+1}, \cdots, X_j$ do not contribute

H_a: at least one of them contributes

$$F = \frac{(R_c^2 - R_r^2)/v_1}{(1 - R_c^2)/v_2}$$

where (1) R_c^2 is the R^2 including the variables in H_0 (the complete model), (2) R_r^2 is the R^2 excluding the variables in H_0 (the reduced model), (3) $v_1 =$ the number of β's in H_0, (4) $v_2 = n - 1 -$ (the number of X's in the complete model), and (5) the degrees of freedom for the F statistic and v_1 and v_2

REVIEW EXERCISES

14.55   Given below is selected information from a computer printout of a multiple regression analysis:

ANALYSIS OF VARIANCE

	df	SUM OF SQUARES	MEAN SQUARES	F RATIO
Regression	2	86.091	43.0455	8.649
Error	17	84.609	4.9770	

VARIABLE	COEFFICIENT	ST. DEV. COEFF.	T RATIO
Intercept	.10255		
X_1	.68808	.195	3.529
X_2	.39844	.295	1.351

Answer the following questions.

a. Write down the multiple regression equation.

b. What percentage of variation in Y is explained by the model?

c. What is the size of the sample (n) used in the above regression analysis?

d. Find the value of

   i. The total sum of squares
   ii. The error sum of squares
   iii. The estimated variance of the error in the model

e. Is the model as a whole significant? Use a .05 significance level.

f. Does X_1 contribute to this model given that X_2 is in the model? Use a .05 significance level.

g. Does X_2 contribute to this model given that X_1 is in the model? Use a .05 significance level.

14.56 A company has opened several outdoor ice-skating rinks and would like to know what factors affect the attendance at the rinks. The manager believes that the following variables affect attendance:

$$X_1 = \text{temperature (forecasted high)}$$
$$X_2 = \text{wind speed (forecasted high)}$$
$$X_3 = 1 \text{ if weekend and 0 otherwise}$$
$$X_4 = X_1 X_2$$

The following least squares model was found from 30 days of data:

$$\hat{Y} = 250 + 4.8X_1 - 30X_2 + 1.3X_3 + 35X_4$$

a. What is the predicted attendance on a weekend if the forecasted high temperature is 28°F and the forecasted high wind speed is 12 miles per hour?

b. If the coefficient of determination for the model is .67, test that the overall model contributes to predicting the attendance at the ice-skating rinks. Use a 5% significance level.

c. If the standard deviation of the estimate of the coefficient of X_2 is 2.01, does the variable wind speed contribute to predicting the variation in attendance, assuming that the variables X_1, X_3, and X_4 are in the model?

14.57 An automobile dealer decided to collect data to predict the demand for automobiles using regression analysis. Using historical data, the multiple regression method gave the least squares equation:

$$\hat{Y} = -307.2 + 1.994X_1 + 0.0207X_2 + 0.00876X_3 - 10.48X_4$$

where

$$Y = \text{amount spent on new automobiles (in billions of dollars)}$$
$$X_1 = \text{U.S. population (in millions)}$$
$$X_2 = \text{disposable personal income (in billions of dollars)}$$
$$X_3 = \text{number of marriages (in thousands)}$$
$$X_4 = \text{financial interest rate (in percent) for automobile loans}$$

a. Would you expect any multicollinearity to be present in these variables? Discuss.

b. If disposable personal income increased by $200 billion and the value of the other independent variables remained constant, how much would you expect the demand for automobiles to increase? Is your conclusion valid if multicollinearity exists?

14.58 A real estate agent wanted to explore the feasibility of using multiple regression analysis in appraising the value of single-family homes within a certain community. The following variables were used:

$$Y = \text{selling price of a house (in dollars)}$$
$$X_1 = \text{total living area (in square feet)}$$
$$X_2 = \begin{cases} 1 & \text{if in neighborhood 1} \\ 0 & \text{if not} \end{cases}$$
$$X_3 = \begin{cases} 1 & \text{if in neighborhood 2} \\ 0 & \text{if not} \end{cases}$$
$$X_4 = \begin{cases} 1 & \text{if lot size is larger than the typical house lot} \\ 0 & \text{if not} \end{cases}$$

The data are as follows:

Y	X_1	X_2	X_3	X_4	Y	X_1	X_2	X_3	X_4
63,000	2020	1	0	1	31,350	640	0	1	0
36,000	980	1	0	0	49,400	1910	0	0	1
44,000	1230	0	0	1	31,000	900	1	0	0
37,000	980	0	1	0	56,000	1890	1	0	0
28,000	640	0	1	0	63,500	1900	0	0	1
28,000	720	0	1	0	49,000	2080	1	0	1
56,000	2400	1	0	1	63,000	1900	0	0	1
28,600	670	0	1	0					

Using a computerized statistical package, find the following:

a. The least squares equation

b. The 95% confidence interval for the coefficient of total living area

c. The 95% prediction interval for selling price given that $X_1 = 1800$, $X_2 = 1$, $X_3 = 0$, and $X_4 = 0$

d. The overall F-test for the model and the resulting conclusion using a 5% significance level

14.59 To predict the asking price of a used Chevrolet Camaro, the following data were collected on the car's age, condition, and mileage and on whether the seller is an individual or a dealer. The data are as follows:

ASKING PRICE (Y)	AGE (IN YEARS) (X_1)	MILEAGE (IN THOUSANDS) (X_2)	CONDITION (EXCELLENT, AVERAGE, POOR) (X_3)	(X_4)	DEALER OR INDIVIDUAL (X_5)
3,000	9	70	1	0	0
2,700	9	99	0	1	0
2,995	8	120	0	1	0
5,500	7	56	1	0	1
3,988	7	50	0	1	0
3,900	7	83	0	1	0
2,800	7	106	0	1	0
6,800	6	70	0	0	1
6,295	6	66	1	0	1
3,700	6	60	0	1	0
7,450	5	55	1	0	1
6,800	5	67	0	1	0
6,795	5	62	1	0	0
6,476	5	60	0	1	0
6,450	5	55	0	1	0
4,800	5	75	0	1	0
9,695	4	44	1	0	1
9,675	4	37	0	0	1
9,595	4	44	1	0	0
8,500	4	55	1	0	0
7,995	4	46	0	1	0
6,995	4	56	0	1	0
6,450	4	65	0	1	0
14,350	3	29	0	0	1
11,965	3	23	0	1	1
11,850	3	27	0	0	1
11,000	3	31	1	0	1
7,600	3	45	0	1	0
19,888	2	18	0	0	1
16,000	2	19	0	0	1
17,650	1	9	0	0	1

The dummy variable X_3 is equal to 1 if the car is in average condition, 0 if not. The variable X_4 is equal to 1 if the car is in poor condition, 0 if not. The dummy variable X_5 is equal to 1 if the seller is a dealer and is equal to zero if the seller is an individual. Use a computerized statistical package to answer the following questions.

a. Find the least squares equation.

b. Does the overall model contribute significantly to predicting the asking price of a used Chevrolet Camaro? Use a .01 significance level.

c. Find a 95% confidence interval for the mean asking price of a 5-year-old Camaro in average condition with 70,000 miles that is sold by an individual.

d. Calculate the correlation matrix of all the variables. Would you suspect any multicollinearity problems by observing the correlations in this matrix?

e. Do a forward regression analysis using a significance level of .10.

14.60 The owner of a photographic laboratory would like to explore the relationship between her weekly profits (Y) and

$X_1 =$ number of rolls of film sold

$X_2 =$ number of enlargements given out free for advertising purposes

$X_3 =$ number of prints

$X_4 =$ number of reprints

Several weeks were selected randomly to collect the following data:

Y	X_1	X_2	X_3	X_4	Y	X_1	X_2	X_3	X_4
350	50	15	130	50	358	62	17	125	35
414	61	18	150	39	392	55	19	150	36
385	71	12	125	45	415	59	24	157	44
429	86	21	141	36	380	63	28	140	38
415	90	22	133	40					

Use a computerized statistical package.

a. Find the least squares prediction equation.

b. Test the null hypothesis that X_4 does not contribute to predicting the variation in Y given that X_1, X_2, and X_3 are already in the model. Use a .05 significance level.

c. Find the 90% confidence interval for the mean value of Y given $X_1 = 85$, $X_2 = 20$, $X_3 = 135$, and $X_4 = 37$.

d. Find the coefficient of determination for the complete model and interpret its value.

14.61 Use a computerized statistical package to analyze the following data:

Y	X_1	X_2	X_3	Y	X_1	X_2	X_3
154	30	1	1	220	34	5	25
223	41	3	9	210	38	4	16
201	33	5	25	230	44	3	9
177	31	4	16	265	51	2	4
143	25	3	9	306	55	5	25
155	29	2	4	170	31	4	16

a. Find the least squares prediction equation.

b. Find the coefficient of determination for the model.

c. Test at the .10 significance level that X_2 and X_3 contribute to the prediction of Y, given that X_1 is in the model.

d. Test at the .10 significance level that X_1 contributes to the prediction of Y, given that X_2 and X_3 are in the model.

e. Plot the residuals of the complete model versus the predicted values. Do the residuals appear to be random?

14.62 Complete the following ANOVA table for testing whether a model with five independent variables contributes significantly to the prediction of the dependent variable:

SOURCE	df	SS	MS	F
Regression		95.6		
Error	20	159.0		
Total				

14.63   The manager of Stay Trim Health Studios would like to determine the average number of times per month a member attends the health studio (Y). The following independent variables were used in the analysis:

$$X_1 = \text{weight at initial visit (in pounds)}$$
$$X_2 = X_1^2$$
$$X_3 = \text{age at initial visit}$$
$$X_4 = \text{length of membership (in years)}$$
$$X_5 = \begin{cases} 1 & \text{if employed} \\ 0 & \text{if not} \end{cases}$$

The manager collected the following data:

Y	X_1	X_3	X_4	X_5	Y	X_1	X_3	X_4	X_5
11	202	30	1	1	13	245	35	1	0
9	180	22	2	1	15	215	24	3	0
7	130	19	1	0	11	185	43	2	1
14	175	32	4	1	12	165	27	3	1
12	225	41	2	1	12	195	38	1	0
19	191	52	5	1	11	217	42	1	1
7	142	40	1	1	10	205	40	1	1
11	208	33	2	1					

The least squares equation was found to be:

$$\hat{Y} = -11.218 + 0.15178X_1 - 0.0003X_2 + 0.08286X_3 + 1.9138X_4 - 2.299X_5$$

a. Does the overall model contribute significantly to predicting the monthly attendance at Stay Trim Health Studios? Use a 10% significance level.

b. Does weight squared contribute significantly to predicting the monthly attendance, assuming that the variables X_1, X_3, X_4, and X_5 are in the model? Use a 10% significance level.

c. Find a 95% confidence interval for the coefficient of age.

d. Find a 95% confidence interval for the coefficient of length of membership.

e. Use the model to predict the monthly attendance of a 35-year-old member who weighs 200 pounds, has a 2-year membership, and is currently employed.

f. Construct a histogram for the residuals of the complete model.

14.64   Refer to Exercise 14.12. The least squares line to fit the data involving only Y and X_1 is found to be

$$\hat{Y} = 14812.135 + 1603.915X_1$$

a. Find the coefficient of determination for the least squares equation

$$\hat{Y} = 1557.551 - 112.10X_1 + 244.152X_2$$

b. Find the coefficient of determination for the complete model involving X_1, X_2, and X_3.

c. Show that the F statistic for testing that X_1 does not contribute to predicting the variation in the dependent variable, given that X_2 and X_3 are in the model, is equal to the square of the t test used in question e of Exercise 14.12.

14.65   A study examining coupon usage suggested that three variables—price consciousness (PRICECON), time value (TIMEVAL), and satisfaction/pride (SAT/PRIDE)—are the major motivational factors in determining coupon usage by consumers. The dependent variable was the dollar amount saved using coupons (SAVING). A total of 290 consumers responded to a questionnaire designed to measure each of three variables. From the following statistics, determine at what significance level each of the independent variables significantly contributes to the prediction of SAVING assuming that the other two independent variables are included in the model. At what significance level would the overall F-test indicate that the model significantly contributes to the prediction of SAVING?

VARIABLE	t-VALUE
PRICECON	2.994
TIMEVAL	-1.749
SAT/PRIDE	3.481

Coefficient of determination is .20375.

(*Source:* Peter Tat, "Attitudinal Dimensions of Coupon Usage," William Cunningham, and Emin Bababus, *Proceedings of the 1986 Annual Meeting of the Decision Science Institute*, 1986, pp. 861–63.)

14.66 Forty of the largest companies traded on the U.S. stock exchange are listed with each company's earnings per share (EPS), sales, and profit for the 12-month period ending April 1, 1988. The sales and profits are represented in terms of millions of dollars. The data are as follows:

COMPANY	EPS (Y)	SALES (X_1)	PROFITS (X_2)
1. IBM	8.72	54,217	5,253
2. EXXON	3.43	84,116	4,841
3. GE	3.38	39,310	3,089
4. AT&T	1.88	33,598	2,044
5. FORD MOTOR	9.05	71,640	4,625
6. GM	10.06	101,782	3,551
7. AMOCO	5.77	22,390	1,360
8. MOBIL	3.08	56,446	1,264
9. DOW CHEMICAL	6.50	13,377	1,245
10. HEWLETT-PACKARD	2.76	8,540	644
11. WAL-MART	1.16	16,064	656
12. CHEVRON	2.27	29,100	773
13. SEARS	4.35	48,440	1,650
14. EASTMAN KODAK	3.52	13,305	1,178
15. COCA-COLA	2.43	7,660	916
16. JOHNSON & JOHNSON	4.83	8,010	833
17. PROCTER & GAMBLE	5.48	17,892	786
18. 3M	4.02	9,429	918
19. ARCO	6.68	16,829	1,224
20. NABISCO	4.19	15,766	1,081
21. BRISTOL-MYERS	2.47	5,400	710
22. GTE	3.29	15,421	1,119
23. SOUTHWESTERN BELL	3.48	8,003	1,047
24. AMERICAN EXPRESS	1.20	16,141	533
25. TEXACO	2.11	35,300	493
26. NISSAN	.07	33,461	55
27. ANHEUSER-BUSCH	2.04	8,258	615
28. PEPSI CO	2.30	11,485	605
29. McDONALD'S	2.89	4,894	549
30. WALT DISNEY	3.05	2,951	392
31. KRAFT	2.87	9,876	390
32. NEC	.42	17,338	103
33. PHILIP MORRIS	7.75	27,694	1,840
34. BELL SOUTH	3.46	12,269	1,665
35. BELL ATLANTIC	6.24	10,303	1,240
36. AMERITECH	8.47	9,536	1,188
37. FUJI PHOTO FILM	2.22	7,138	404
38. ABBOTT LABORATORIES	2.78	4,390	633
39. DUN & BRADSTREET	2.58	3,359	393
40. AMER. INTERNAT'L	5.37	10,679	657

(*Source: Financial World*, April 5, 1988.)

Using a computerized statistical package, find the following:

a. The least squares equation for predicting EPS.

b. The overall F-test for the model and the resulting conclusion using a 10% significance level.

c. The individual t-tests for testing the significance of each of the two independent variables and the resulting conclusions using a 10% significance level.

d. The coefficient of determination and its interpretation.

14.67   Twelve employees each take two written tests, which are used to indicate how well the employee will perform on the job. In the MINITAB printout, C2 and C3 are the scores from each of the tests. In C1, an evaluation is given of the employee's performance on the job. Using a significance level of .01, what conclusion can be drawn from the ANOVA table with regard to the prediction of the job performance by the prediction variables? Is there a conflict between the significance of the F statistic and significance of the two t-tests? How can this inconsistency be corrected?

```
MTB > NAME C1 'EVAL'
MTB > NAME C2 'SCORE1'
MTB > NAME C3 'SCORE2'
MTB > PRINT C1-C3
 ROW   EVAL   SCORE1   SCORE2

   1     99       99       66
   2     72       93       42
   3     90       75       29
   4     55       50        6
   5     95      100       54
   6     65       64       14
   7     75       80       35
   8     60       72       16
   9     83       88       36
  10     88       75       28
  11     59       73       18
  12    100      100       67

MTB > REGRESS Y IN C1 USING 2 PREDICTORS IN C2 , C3

The regression equation is
EVAL = 71.8 - 0.317 SCORE1 + 0.939 SCORE2

Predictor        Coef        Stdev      t-ratio
Constant        71.82        28.48         2.52
SCORE1        -0.3168       0.5052        -0.63
SCORE2         0.9394       0.3959         2.37

s = 8.737      R-sq = 76.4%      R-sq(adj) = 71.1%

Analysis of Variance

SOURCE         DF          SS           MS
Regression      2       2221.9       1111.0
Error           9        687.0         76.3
Total          11       2908.9

SOURCE         DF       SEQ SS
SCORE1          1       1792.1
SCORE2          1        429.8
```

14.68   An independent research firm is investigating sex discrimination in the salaries of managers of a small firm. Fourteen supervisors and midlevel managers were chosen at random and the salary, years of experience, and sex of each were recorded. The independent research firm believes that the salary compensation may increase with years of

```
MTB > LET C4 = C2*C3
MTB > NAME C1 'SALARY'
MTB > NAME C2 'EXPER'
MTB > NAME C3 'SEX'
MTB > NAME C4 'INTERACT'
MTB > PRINT C1-C4
 ROW   SALARY   EXPER   SEX   INTERACT

   1     17.1     2.3     0       0.0
   2     26.0     3.7     0       0.0
   3     30.8     4.1     1       4.1
   4     19.1     2.5     1       2.5
   5     71.2    10.3     1      10.3
   6     47.3     6.4     1       6.4
   7     25.5     3.4     1       3.4
   8     33.6     4.7     0       0.0
   9     60.3     8.3     0       0.0
  10     36.7     5.1     1       5.1
  11     26.5     3.7     0       0.0
  12     37.3     5.0     1       5.0
  13     27.7     3.8     1       3.8
  14     32.5     4.6     0       0.0
```

```
MTB > REGRESS Y IN C1 USING 3 PREDICTORS IN C2-C4

The regression equation is
SALARY = - 0.404 + 7.27 EXPER + 3.31 SEX - 0.559 INTERACT

Predictor        Coef           Stdev          t-ratio
Constant        -0.4040         0.8386          -0.48
EXPER            7.2683         0.1707          42.58
SEX              3.306          1.075            3.08
INTERACT        -0.5593         0.2092          -2.67

s = 0.7747        R-sq = 99.8%      R-sq(adj) = 99.7%

Analysis of Variance

SOURCE         DF           SS            MS
Regression      3        2999.68        999.89
Error          10           6.00          0.60
Total          13        3005.68

SOURCE         DF         SEQ SS
EXPER           1        2993.90
SEX             1           1.48
INTERACT        1           4.29

MTB > REGRESS Y IN C1 USING 1 PREDICTOR IN C2

The regression equation is
SALARY = 1.58 + 6.91 EXPER

Predictor        Coef           Stdev          t-ratio
Constant        1.5775         0.6624           2.38
EXPER           6.9148         0.1252          55.23

s = 0.9907        R-sq = 99.6%      R-sq(adj) = 99.6%

Analysis of Variance

SOURCE         DF           SS            MS
Regression      1        2993.9        2993.9
Error          12          11.8           1.0
Total          13        3005.7
```

experience differently for males and females. Data are shown in the MINITAB printout. Test that sex significantly contributes to the prediction of salaries in the model

$$Y = \beta_0 + \beta_1 X_1 + \beta_2 X_2 + \beta_3 X_3 + \epsilon$$

where X_1 is years of experiences, X_2 is 1 for male and 0 for females, and $X_3 = X_1 X_2$. Use the MINITAB printout and let the significance level be .05. Note that the variable $X_3 = X_1 X_2$ (called the interaction of sex and experience) allows for the slope of the regression line for males to be different from the slope of the regression line for females. The units of salary are in 1000s (C1). Years of experience are in C2. A 1 in C3 represents a male and a 0 represents a female. (*Hint:* The reduced model contains only the predictor variable X_1.)

14.69   Nine similar machines are used in a manufacturing process at an assembly plant. The operations manager believes that the repair costs for the machines are influenced by both the age of the machines and the operator of the machine. The MINITAB printout shows the costs of repair of the machines, over the past six months (C1), the age of

```
MTB > NAME C1 'COST'
MTB > NAME C2 'AGE'
MTB > NAME C3 'DUMMY1'
MTB > NAME C4 'DUMMY2'
MTB > PRINT C1-C4
 ROW    COST    AGE   DUMMY1   DUMMY2

   1    310      4      1        0
   2    300      8      0        1
   3    175      3      0        0
   4    200      2      1        0
   5    620     15      0        1
   6    365      9      0        0
   7    370      6      1        0
   8    175      5      0        1
   9    365      8      0        0
```

```
MTB > BRIEF 3
MTB > REGRESS Y IN C1 USING 3 PREDICTORS IN C2-C4, RES IN C5, YHATS IN C6;
SUBC> PREDICT FOR 6 1 0;
SUBC> PREDICT FOR 6 0 1;
SUBC> PREDICT FOR 6 0 0.

The regression equation is
COST = 23.7 + 41.7 AGE + 103 DUMMY1 - 47.9 DUMMY2

Predictor        Coef        Stdev      t-ratio
Constant         23.66       22.39        1.06
AGE              41.701       2.646      15.76
DUMMY1          102.87       20.73        4.96
DUMMY2          -47.87       20.73       -2.31

s = 23.87       R-sq = 98.1%    R-sq(adj) = 97.0%

Analysis of Variance

SOURCE        DF          SS          MS
Regression     3       150652       50217
Error          5         2848         570
Total          8       153500

SOURCE        DF       SEQ SS
AGE            1       126784
DUMMY1         1        20829
DUMMY2         1         3039

Obs.     AGE      COST      Fit  Stdev.Fit  Residual  St.Resid
 1       4.0    310.00   293.33    13.78      16.67      0.86
 2       8.0    300.00   309.40    14.22      -9.40     -0.49
 3       3.0    175.00   148.76    16.85      26.24      1.55
 4       2.0    200.00   209.93    14.76      -9.93     -0.53
 5      15.0    620.00   601.30    20.37      18.70      1.50
 6       9.0    365.00   398.97    15.10     -33.97     -1.84
 7       6.0    370.00   376.73    14.76      -6.73     -0.36
 8       5.0    175.00   184.30    17.93      -9.30     -0.59
 9       8.0    365.00   357.27    14.22       7.73      0.40

    Fit  Stdev.Fit      95% C.I.            95% P.I.
 376.73     14.76   ( 338.78, 414.69)   ( 304.58, 448.89)

 226.00     16.36   ( 183.93, 268.07)   ( 151.59, 300.40)

 273.87     13.89   ( 238.15, 309.59)   ( 202.86, 344.87)


MTB > NAME C5 'RESID'
MTB > NAME C6 'PRED'
MTB > PLOT C5 C6

RESID    -       *                                        *
         -
         -
    1.0+
         -                  *
         -
         -                       *
         -
  -0.0+
         -
         -                  *     *
         -       * *
         -
  -1.0+
         -
         -
         -
         -                    *
  -2.0+
         +---------+---------+---------+---------+---------+------PRED
        100       200       300       400       500       600
```

each machine at the beginning of the six month period (C2) and which of three operators used the machines (C3) and (C4). Each machine was used by only one operator during this time. What conclusions can be drawn from analyzing the MINITAB printout? Use a .05 significance level. Comment on the residual plot. Also interpret the three prediction intervals for the repair cost if a machine is 6 years old and is operated by one of the three operators.

14.70 The MINITAB computer printout shows the execution of the stepwise procedure with FENTER = 0 and FREMOVE = 0. For this situation, all independent variables are put in the model, starting with the most significant independent variable. Using a .05 significance level, determine the model which the forward regression method would choose.

```
MTB > NAME C1 'Y'
MTB > PRINT 'Y' C2-C7
ROW       Y       C2       C3        C4      C5     C6       C7

  1    1778.47   114   468.000   113.3   57.6   26.5   15.7663
  2    1628.18   114   341.759    87.4   57.6   56.0   11.7664
  3    1465.00    26   266.034    72.6   57.6   79.6   10.4331
  4    1864.35    81   547.724   128.1   72.1   73.7    2.4333
  5     610.47   125    84.172    28.2  104.0   14.7    7.7665
  6    2031.82    81   575.827   124.4   51.8   56.0   10.4331
  7     863.82   114   123.966    43.0  101.1   14.7    9.0998
  8     447.29    26    65.655    17.1  104.0   14.7    2.4333
  9    1752.71    15   532.172   131.8  104.0   14.7    9.0998
 10    1387.71    59   243.759    68.9   69.2   67.8    6.4332
 11     245.47    15    56.931     9.7   83.7   38.3    3.7666
 12    1022.71    26   128.552    39.3   98.2   20.6   14.4330
 13    1237.41   125   262.827    83.7   92.4   32.4    2.4333
 14     777.94    37   110.862    39.3   80.8   26.5   11.7664
 15     876.71    15   202.000    83.7  101.1   20.6    2.4333
 16    1336.18    26   215.345    61.5   63.4   44.2   15.7663
 17    1726.94   125   483.655   120.7   80.8   56.0    3.7666
 18     404.35    26    58.207     9.7   98.2   14.7   10.4331

MTB > STEPWISE 'Y' C2-C7;
SUBC> FENTER = 0.0;
SUBC> FREMOVE = 0.0.

 STEPWISE REGRESSION OF  Y ON 6 PREDICTORS, WITH N =   18

      STEP      1        2        3        4        5        6
  CONSTANT   437.28   255.65   129.29   -17.69   -59.78  -532.32

  C3           2.85     2.84     2.54     0.51     0.48     0.63
  T-RATIO     11.02    12.66    12.08     0.88     0.88     1.16

  C7                    22.2     24.4     28.4     28.9     35.8
  T-RATIO               2.52     3.37     5.24     5.67     4.89

  C6                              5.0      5.5      5.6      8.0
  T-RATIO                        2.88     4.33     4.67     3.63

  C4                                      9.0      8.8      8.3
  T-RATIO                                 3.62     3.76     3.60

  C2                                               0.90     1.04
  T-RATIO                                          1.65     1.91

  C5                                                        3.8
  T-RATIO                                                   1.28

  S            195      169      139      101     95.4     92.9
  R-SQ       88.35    91.82    94.87    97.45    97.92    98.19
```

COMPUTER EXERCISES USING THE DATABASE

Exercise 1—Appendix H

From the database, randomly select 40 observations. Regress the variable HPAYRENT (house payment or apartment/house rent) on the prediction variables INCOME1 (primary income), INCOME2 (secondary income), and FAMLSIZE (size of family). Find the coefficient of determination for the complete model. Find 95% confidence interval on the mean value of HPAYRENT for a family that has a principal income of $33,000, a secondary income of 18,000 and a family size equal to three.

Exercise 2—Appendix H

Using the data from the previous problem along with dummy variables representing the LOCATION of the residences, do both a forward regression analysis and a backward regression analysis with a significance level of .10. Compare the two resulting models.

Exercise 3—Appendix I

From the database, randomly select 50 observations. Consider a multiple regression model, where the dependent variable is SALES and predictor variables are COSTSALE (sales cost), EMPLOYEE (number of employees), NETINC (net income), ASSETS, and TOTAL. Using these predictor variables, what percentage of the variation in the SALES has been explained? Construct a histogram of the residuals. Do the regression assumptions appear to be satisfied?

Exercise 4—Appendix I

Using the data from the previous problem, perform both a forward regression analysis and backward regression analysis with a significance level of .10. Compare the resulting models.

CASE STUDY

MEGABUCKS FOR THE TOP EXECUTIVE WHO CAN'T PASS THE BUCK

Compensation paid to chief executive officers (CEOs) of large corporations has always been a controversial topic. The salaries and bonuses of high-profile CEOs like Lee Iacocca of Chrysler Corporation and Jim Manzi of Lotus Development run into several million dollars each year. CEO compensation has historically been high in comparison with other company salaries. It is therefore important to know how CEO compensation is determined and also what factors influence compensation.

An issue of *Forbes* magazine (May 30, 1988) profiled the CEOs of America's top 800 companies in terms of annual sales. The CEO profiles contained, among other things, the annual compensation, the actual size of the company, tenure—i.e., the number of years the CEO spent with the company—and other variables. The following table represents 35 well-known companies selected from the 800 reported by Forbes, with data on the following variables for each company:

1. Sales: the annual sales (in millions of dollars) of the corporation
2. Profit: the annual after-tax profit of the corporation, in millions of dollars
3. Tenure: the number of years the CEO has worked for the corporation
4. Age: the chronological age of the CEO in years
5. Shares owned: the market value of shares of stock owned by the CEO in the corporation, in millions of dollars
6. Compensation: the total compensation of the CEO, including salary, bonus, and stock options, in units of $100 thousand

Case Study Questions

1. Perform a multiple regression analysis on the preceding data, using the five independent variables in columns (1) through (5) to predict compensation of CEO (column 6).

   a. What percentage of variation in CEO compensation is explained by these predictors?

   b. Is the model as a whole significant at a .05 significance level?

COMPANY	(1) SALES	(2) PROFIT	(3) TENURE	(4) AGE	(5) SHARES OWNED	(6) COMPENSATION
Aetna	22,114	871.1	10	61	0.9	2,603
Albertsons	5,869	125.4	37	62	10.6	4,723
Amdahl	1,505	142.0	11	52	4.7	4,477
Am. Express	17,768	533.3	17	52	8.7	3,033
AT&T	33,598	2,044.0	31	53	0.6	1,080
Apple Computer	3,041	280.4	5	49	6.6	2,140
Bank America	9,753	−955.0	39	65	0.6	808
Bausch & Lomb	840	85.3	10	51	0.8	960
Black & Decker	2,018	64.3	3	44	0.1	793
H&R Block	737	72.8	33	65	100.2	548
Boeing	15,355	480	30	56	0.1	754
Bristol-Myers	5,401	709.6	38	63	42.6	1,365
Campbell Soup	4,642	265.2	32	61	0.3	1,007
CBS	2,762	452.8	2	65	—	1,189
Chase Manhattan	10,745	−894.8	41	61	0.7	2,011
Chevron	26,015	1,007.0	40	64	3.2	1,565
Chrysler	26,277	1,289.7	10	63	8.2	17,656
Circle K	2,564	51.6	5	59	37.7	431
Citicorp	27,519	−1,138.0	23	49	7.4	972
Coca-Cola	7,658	916.1	34	56	34.3	3,107
Colgate-Palmolive	5,647	54.0	25	49	2.4	1,436
Delta Airlines	6,351	232.6	25	46	0.2	605
Digital Equip	10,391	1,284.3	31	62	260.6	906
Dillards	2,206	91.2	49	73	1.9	1,140
Disney	2,951	455.3	4	46	11.2	6,742
Dow Chemical	13,377	1,245.0	29	52	2.8	1,176
Du Pont	30,224	1,786.0	39	64	5.4	5,959
Eastman Kodak	13,305	1,178.0	38	63	1.0	1,067
Exxon	76,416	4,840.0	36	60	6.4	5,523
Federal Express	3,522	170.9	17	43	149.3	429
Firestone	3,997	111.0	8	61	4.8	1,171
Ford	73,145	4,625.2	39	61	6.8	3,779
General Motors	101,782	3,550.9	59	62	3.8	1,436
Gillette	3,167	229.9	31	58	9.1	4,423
General Mills	5,549	249.8	30	57	4.6	1,322.0

(*Source: Forbes*, May 30, 1988)

c. Comment on how significantly each predictor variable contributes individually, given the other variables in the model.

2. Obtain a correlation matrix of all the variables. If you wished to use only *one* independent variable to predict CEO compensation, which variable would you choose? Conduct a simple regression procedure using just this one variable to predict CEO compensation.

3. Examine the correlation matrix to detect any potential for multicollinearity. What would you look for? If you detect multicollinearity, what action would you take?

4. Using the partial F-test, determine if the five-predictor model (Question 1) is better at explaining CEO compensation than the simple regression model (Question 2). Which is the "better" model of the two? State the p-value for the test.

5. Obtain a plot of the residuals $(Y - \hat{Y})$ against the predicted Y values $(\hat{Y})$. What do you look for? How would you detect

heteroscedasticity (a violation of the assumption of constant variance)?

6. Print a histogram of the residuals. What shape would you like to see? If it is not this shape, what is the problem?

7. Suppose you have a large corporation with sales of $750 million, profits of $62 million, and a 54-year-old CEO with a tenure of 15 years who owns half a million dollars worth of shares in the company.

   a. What is the predicted value (point estimate) of the CEO's compensation?

   b. Find the 95% prediction interval and the 95% confidence interval of CEO compensation for this situation.

   c. What is the difference between the prediction and confidence intervals, i.e., what different kinds of information do they provide?

8. Would the preceding regression model be applicable to all U.S. companies in general? Justify your answer. Also, comment on how the preceding model might be adversely influenced by those companies that suffered a loss.

S P S S

MAINFRAME AND MICRO

SOLUTION

EXAMPLE 14.2

At the beginning of the chapter, we computed the regression equation for predicting the estimate of home size, based on the three predictor variables income, family size, and level of education. The SPSS program listing in Figure 14.21 requests a multiple regression solution. In this problem the SPSS commands are the same for both the mainframe and PC versions. **(Remember to end each command line with a period when using the PC version.)**

The TITLE command names the SPSS run.

The DATA LIST command gives each variable a name and describes the data as being in free form.

The BEGIN DATA command indicates to SPSS that the input data immediately follow.

The next 10 lines contain the data values, which represent the four variables to be considered in the regression analysis. The first card image represents a home with 1600 square feet, an income of $22,000, a family of 2 people, and 4 years of educational experience at the college level.

The END DATA statement indicates the end of the data.

FIGURE 14.21

Input for SPSSX or SPSS/PC+. All command lines should end with a period when the PC version is being used.

```
TITLE    REAL ESTATE EXAMPLE USING TWO PREDICTORS
DATA LIST FREE/SQFOOT INCOME SIZE EDUC
BEGIN DATA
16 22 2 4
17 26 2 8
26 45 3 7
24 37 4 0
22 28 4 2
21 50 3 10
32 56 6 8
18 34 3 8
30 60 5 2
20 40 3 6
END DATA
CORRELATION SQFOOT,INCOME,SIZE,EDUC
REGRESSION VARIABLES=SQFOOT,INCOME,SIZE,EDUC/
          DEPENDENT=SQFOOT/ENTER/
```

The CORRELATION statement requests a correlation matrix of the four variables SQFOOT, INCOME, SIZE, and EDUC.

The REGRESSION statement requests that the independent variables INCOME, SIZE, and EDUC be entered in the regression equation to predict the dependent variable SQFOOT.

Figure 14.22 shows the SPSSX output obtained by executing the listing in Figure 14.21, and Figure 14.23 shows the SPSS/PC+ output.

FIGURE 14.22 SPSSX output.

```
- - - - - - - - - - - - - - P E A R S O N   C O R R E L A T I O N   C O E F F I C I E N T S - - - -

              SQFOOT      INCOME       SIZE       EDUC

SQFOOT        1.0000       .8433       .9079      -.1679
             (    0)     (   10)     (   10)     (   10)
             P= .       P= .001     P= .000     P= .321

INCOME         .8433      1.0000       .7213       .1514
             (   10)     (    0)     (   10)     (   10)
             P= .001     P= .       P= .009     P= .338

SIZE           .9079       .7213      1.0000      -.2514
             (   10)     (   10)     (    0)     (   10)
             P= .000     P= .009     P= .       P= .242

EDUC          -.1679       .1514      -.2514      1.0000
             (   10)     (   10)     (   10)     (    0)
             P= .321     P= .338     P= .242     P= .


(COEFFICIENT / (CASES) / 1-TAILED SIG)          " . " IS PRINTED IF A COEFFICIENT CANNOT BE COMPUTED



                    * * * *   M U L T I P L E   R E G R E S S I O N   * * * *

   Listwise Deletion of Missing Data

   Equation Number 1    Dependent Variable..   SQFOOT

   Beginning Block Number  1.  Method: Enter


   Variable(s) Entered on Step Number   1..     EDUC
                                        2..     INCOME
                                        3..     SIZE         p value for F test              SSE


   Multiple R          .95145          Analysis of Variance
   R Square            .90527                                DF    Sum of Squares    Mean Square
   Adjusted R Square   .85790          Regression             3        237.54155       79.18052
   Standard Error     2.03545          Residual               6         24.85845        4.14307

                                       F =      19.11154      Signif F =   .0018


   ----------------- Variables in the Equation -----------------

   Variable         B         SE B        Beta       T    Sig T
                                                                             MSE = s²
   EDUC        -.162771     .244071    -.099727   -.667   .5296
   INCOME       .193878     .087700     .461936   2.211   .0691
   SIZE        2.338108     .907791     .549625   2.576   .0420
   (Constant)  7.595501    2.594776                2.927   .0264


   End Block Number  1   All requested variables entered.
```

Regression coefficients

FIGURE 14.23 SPSS/PC+ output.

```
                 REAL ESTATE EXAMPLE USING TWO PREDICTORS

       Correlations:   SQFOOT     INCOME       SIZE       EDUC

           SQFOOT      1.0000      .8433*     .9079**    -.1679
           INCOME       .8433*    1.0000      .7213*      .1514
           SIZE         .9079**    .7213*    1.0000      -.2514
           EDUC        -.1679      .1514     -.2514      1.0000

       N of cases:    10        1-tailed Signif:  * - .01   ** - .001

       " . " is printed if a coefficient cannot be computed
```

FIGURE 14.23 Continued

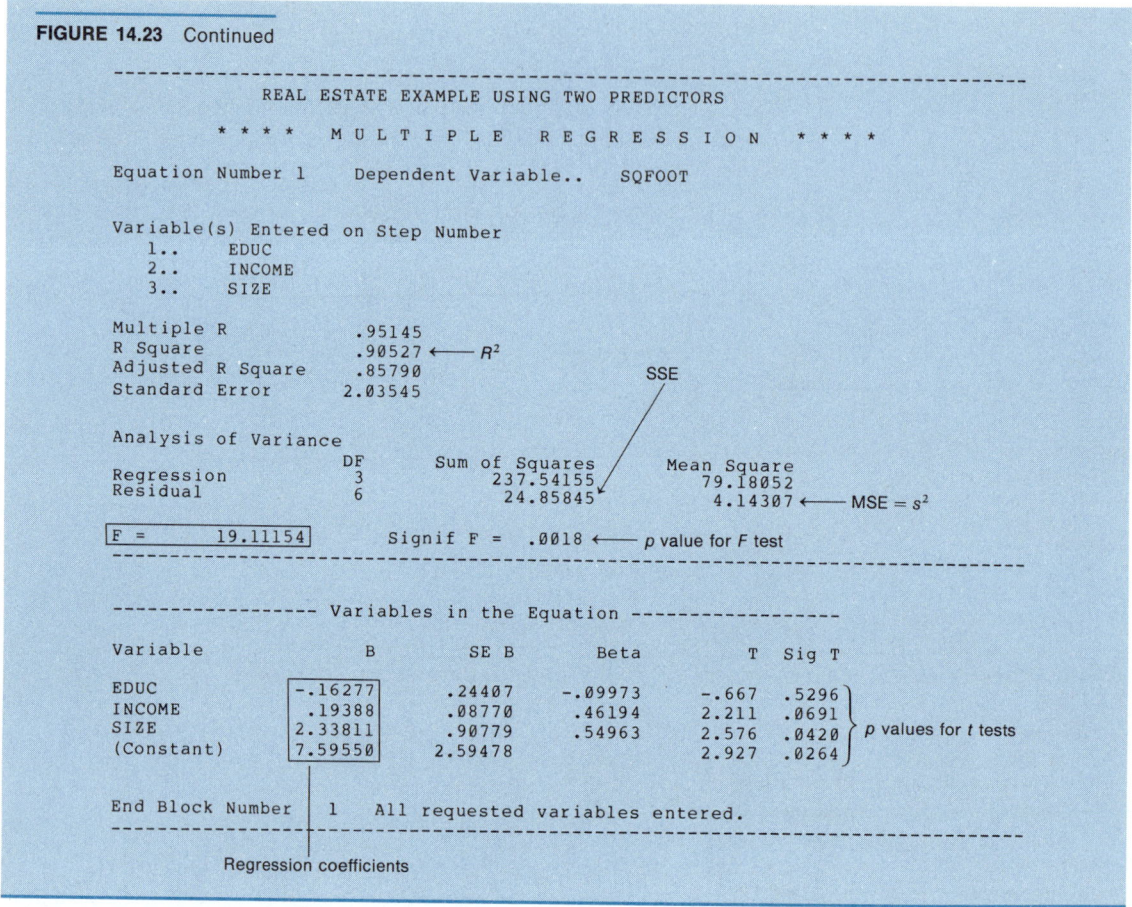

```
----------------------------------------------------------------------
                 REAL ESTATE EXAMPLE USING TWO PREDICTORS

            * * * *   M U L T I P L E   R E G R E S S I O N   * * * *

   Equation Number 1    Dependent Variable..   SQFOOT


      Variable(s) Entered on Step Number
         1..     EDUC
         2..     INCOME
         3..     SIZE


   Multiple R            .95145
   R Square              .90527  ←── R²
   Adjusted R Square     .85790
   Standard Error       2.03545                      SSE

   Analysis of Variance
                       DF      Sum of Squares        Mean Square
   Regression           3         237.54155            79.18052
   Residual             6          24.85845             4.14307  ←── MSE = s²

   F =      19.11154       Signif F =   .0018  ←── p value for F test
----------------------------------------------------------------------


   ---------------- Variables in the Equation ------------------

   Variable             B          SE B        Beta        T     Sig T

   EDUC             -.16277       .24407      -.09973     -.667   .5296 ⎫
   INCOME            .19388       .08770       .46194     2.211   .0691 ⎪
   SIZE             2.33811       .90779       .54963     2.576   .0420 ⎬ p values for t tests
   (Constant)       7.59550      2.59478                  2.927   .0264 ⎭


   End Block Number    1   All requested variables entered.
----------------------------------------------------------------------
```
Regression coefficients

SPSS

MAINFRAME AND MICRO

EXAMPLE 14.2

STEPWISE METHOD

SPSS can be used for determining the stepwise solution of predicting the estimate of home size, based on the three predictor variables income, family size, and level of education. The SPSS program listing in Figure 14.24 requests a stepwise regression solution. Notice that the word STEPWISE is substituted for ENTER and that a new line has been added. Instead of forcing all variables into the equation with an ENTER command, STEPWISE selects

FIGURE 14.24
Input for SPSSX and SPSS/PC+. All command lines should end with a period when the PC version is being used.

```
TITLE    REAL ESTATE EXAMPLE USING STEPWISE.
DATA LIST FREE/SQFOOT INCOME SIZE EDUC.
BEGIN DATA.
16 22 2 4
17 26 2 8
26 45 3 7
24 37 4 0
22 28 4 2
21 50 3 10
32 56 6 8
18 34 3 8
30 60 5 2
20 40 3 6
END DATA.
REGRESSION VARIABLES=SQFOOT,INCOME,SIZE,EDUC/
           CRITERIA=PIN(0.1)/
           DEPENDENT=SQFOOT/STEPWISE.
```

the variables which meet the entry criteria. The CRITERIA = PIN(0.1) statement specifies that each independent variable must produce a significant increase in R^2 at a significance level of .1 before it is allowed to enter into the regression equation. The POUT (.15) statement dictates that if a significant decrease in R^2 is not produced at a significance level of .15 when a variable is dropped from the model, the variable should be removed. The value contained in POUT must exceed the value contained in PIN. In this problem the SPSS commands are the same for both the mainframe and PC versions. **(Remember to end each command line with a period when using the PC version.)**

The TITLE command names the SPSS run.

The DATA LIST command gives each variable a name and describes the data as being in free form.

The BEGIN DATA command indicates to SPSS that the input data immediately follow.

The next 10 lines contain the data values, which represent the four variables to be considered in the regression analysis. The first card image represents a home with 1600 square feet, an income of $22,000, a family of 2 people, and 4 years of educational experience at the college level.

The END DATA statement indicates the end of the data.

The REGRESSION statement requests that the independent variables INCOME, SIZE, and EDUC be entered in the regression equation to predict the dependent variable SQFOOT. The regression subcommands were discussed above.

Figure 14.25 shows the SPSSX output obtained by executing the listing in Figure 14.24, and Figure 14.26 shows the SPSS/PC+ output.

FIGURE 14.25 SPSSX output.

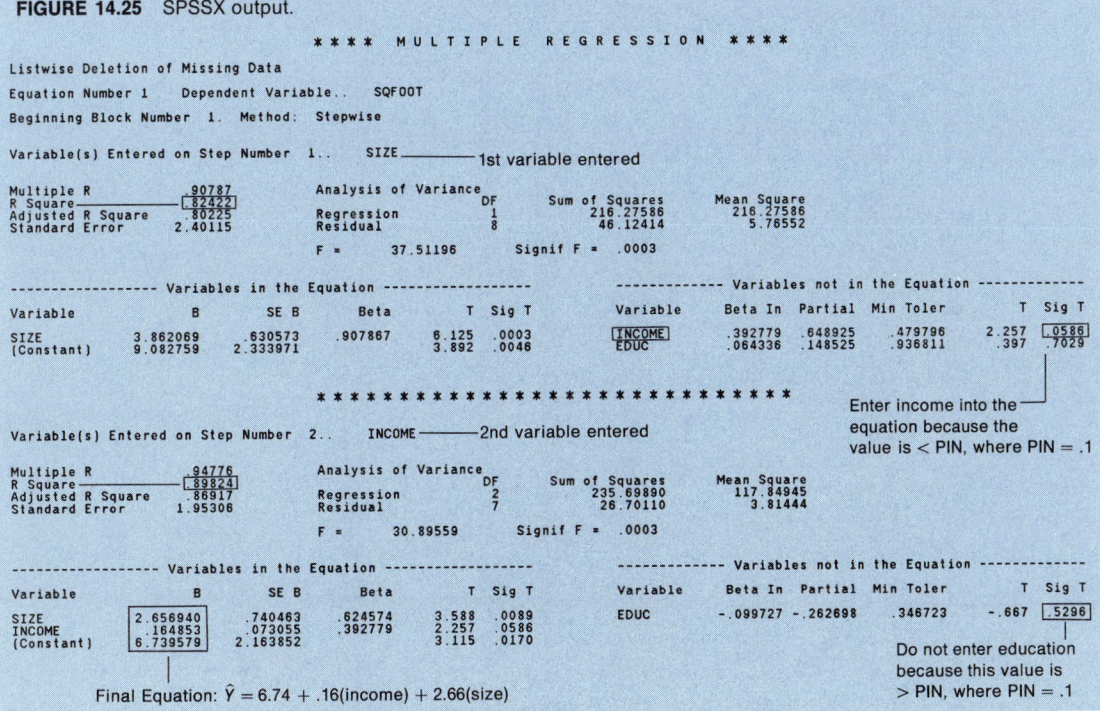

FIGURE 14.26   SPSS/PC+ output.

```
------------------------------------------------------------------------
                    REAL ESTATE EXAMPLE USING STEPWISE

            * * * *   M U L T I P L E   R E G R E S S I O N   * * * *

    Equation Number 1    Dependent Variable..   SQFOOT


    Variable(s) Entered on Step Number
        1..    SIZE ←————————————— 1st variable entered


    Multiple R              .90787
    R Square                .82422
    Adjusted R Square       .80225
    Standard Error         2.40115

    Analysis of Variance
                            DF        Sum of Squares       Mean Square
    Regression               1           216.27586         216.27586
    Residual                 8            46.12414           5.76552

    F =        37.51196      Signif F =   .0003



------------------------------------------------------------------------
                    REAL ESTATE EXAMPLE USING STEPWISE

            * * * *   M U L T I P L E   R E G R E S S I O N   * * * *

    Equation Number 1    Dependent Variable..   SQFOOT


    ----------------- Variables in the Equation ------------------

    Variable              B            SE B         Beta          T     Sig T

    SIZE              3.86207        .63057       .90787       6.125    .0003
    (Constant)        9.08276       2.33397                    3.892    .0046


    ------------- Variables not in the Equation -------------

    Variable      Beta In   Partial   Min Toler        T    Sig T

    INCOME        .39278    .64892     .47980        2.257   .0586
    EDUC          .06434    .14853     .93681         .397   .7029
```
Enter income into the equation because
the value is < PIN, where PIN = .1

```
------------------------------------------------------------------------
                    REAL ESTATE EXAMPLE USING STEPWISE

            * * * *   M U L T I P L E   R E G R E S S I O N   * * * *

    Equation Number 1    Dependent Variable..   SQFOOT


    Variable(s) Entered on Step Number
        2..    INCOME ←————————————— 2nd variable entered


    Multiple R              .94776
    R Square                .89824
    Adjusted R Square       .86917
    Standard Error         1.95306

    Analysis of Variance
                            DF        Sum of Squares       Mean Square
    Regression               2           235.69890         117.84945
    Residual                 7            26.70110           3.81444

    F =        30.89559      Signif F =   .0003
```

FIGURE 14.26 Continued

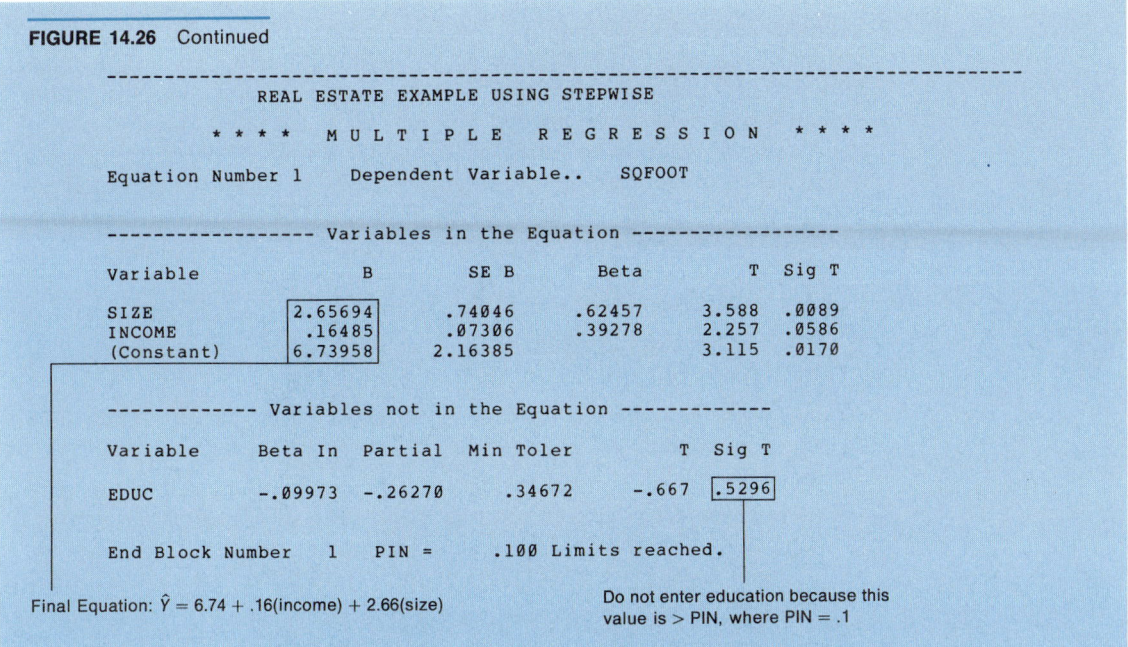

```
------------------------------------------------------------------------
                   REAL ESTATE EXAMPLE USING STEPWISE

          * * * *   M U L T I P L E   R E G R E S S I O N   * * * *

      Equation Number 1    Dependent Variable..   SQFOOT

      ----------------- Variables in the Equation ------------------

         Variable            B        SE B        Beta         T    Sig T

         SIZE            2.65694     .74046      .62457       3.588  .0089
         INCOME           .16485     .07306      .39278       2.257  .0586
         (Constant)      6.73958    2.16385                   3.115  .0170


      ------------ Variables not in the Equation -------------

         Variable     Beta In   Partial   Min Toler        T   Sig T

         EDUC         -.09973   -.26270     .34672      -.667   .5296


      End Block Number   1   PIN =      .100 Limits reached.
```

Final Equation: $\hat{Y} = 6.74 + .16(\text{income}) + 2.66(\text{size})$

Do not enter education because this value is > PIN, where PIN = .1

MAINFRAME AND MICRO

$S \; P \; S \; S$

EXAMPLE

SECTION 14.5

In the real-estate example in Section 14.5, we wished to determine the predicted square footage ($\hat{Y}$) for values of $X_1 = 36$, $X_2 = 4$, and $X_3 = 8$. Of course, one way to do this is to insert them manually into the regression equation resulting from the ten observations in the previous SPSS example. An easier way is to attach these values at the end of the input data along with a numeric value for Y (we used -9 here), which will be identified as a *missing value* using the MISSING VALUES SQFOOT(-9) statement. SPSS ignores this row of data, computes the regression equation, and then includes this row when it summarizes the predicted values. The predicted

FIGURE 14.27

Input for SPSSX or SPSS/PC+. All command lines should end with a period when the PC version is being used.

```
TITLE   PREDICTION FOR REAL ESTATE EXAMPLE
DATA LIST FREE/SQFOOT INCOME SIZE EDUC
BEGIN DATA
16 22 2 4
17 26 2 8
26 45 3 7
24 37 4 0
22 28 4 2
21 50 3 10
32 56 6 8
18 34 3 8
30 60 5 2
20 40 3 6
-9 36 4 8
END DATA
MISSING VALUES SQFOOT(-9)
REGRESSION VARIABLES=SQFOOT,INCOME,SIZE,EDUC/
          DEPENDENT=SQFOOT/ENTER/
          CASEWISE=ALL PRED SEPRED/
```

values as well as their standard errors ($s_{\hat{Y}}$) are calculated and included in the output using the CASEWISE = ALL PRED SEPRED statement. This informs SPSS that you would like to see the predicted values (and their standard errors) for *all* of the cases in the input data, including the row(s) with the missing *Y* values. The listing in Figure 14.27 is used to obtain the predicted values and corresponding standard errors. In this problem the SPSS commands are the same for both the mainframe and PC versions. **(Remember to end each command line with a period when using the PC version.)**

The TITLE command names the SPSS run.

The DATA LIST command gives each variable a name and describes the data as being in free form.

The BEGIN DATA command indicates to SPSS that the input data immediately follow.

The next 10 lines contain the data values, which represent the four variables to be considered in the regression analysis. The first card image represents a home with 1600 square feet, an income of $22,000, a family of 2 people, and 4 years of educational experience at the college level.

The END DATA statement indicates the end of the data.

The REGRESSION statement requests that the independent variables INCOME, SIZE, and EDUC be entered in the regression equation to predict the dependent variable SQFOOT.

Figure 14.28 shows the SPSSX output obtained by executing the listing in Figure 14.27, and Figure 14.29 shows the SPSS/PC+ output.

FIGURE 14.28 SPSSX output.

```
* * * *   M U L T I P L E   R E G R E S S I O N   * * * *

Listwise Deletion of Missing Data

Equation Number 1    Dependent Variable..    SQFOOT

Beginning Block Number  1.  Method:  Enter

Variable(s) Entered on Step Number    1..    EDUC
                                      2..    INCOME
                                      3..    SIZE


Multiple R            .95145        Analysis of Variance
R Square              .90527                          DF    Sum of Squares    Mean Square
Adjusted R Square     .85790        Regression         3        237.54155       79.18052
Standard Error       2.03545        Residual           6         24.85845        4.14307

                                    F =      19.11154     Signif F =   .0018


----------------- Variables in the Equation ------------------

Variable              B          SE B        Beta          T    Sig T

EDUC            -.162771       .244071    -.099727       -.667   .5296
INCOME           .193878       .087700     .461936      2.211   .0691
SIZE            2.338108       .907791     .549625      2.576   .0420
(Constant)      7.595501      2.594776                  2.927   .0264


End Block Number   1   All requested variables entered.
```

└── Regression coefficients

FIGURE 14.28 Continued

```
                                    * * * *  M U L T I P L E   R E G R E S S I O N  * * * *
Equation Number 1    Dependent Variable..   SQFOOT

Casewise Plot of Standardized Residual
*: Selected   M: Missing

            -3.0          0.0          3.0
   Case #    0:.............:.............:0          *PRED      *SEPRED
      1       .            *        .              15.8860     1.1930
      2       .            .  *     .              16.0104     1.1608
      3       .            .        *    .         22.1950      .9748
      4       .            .      *       .        24.1214     1.3002
      5       .            .      *       .        22.0510     1.3714
      6       .        *   .              .        22.6760     1.3308
      7       .            .  *           .        31.1792     1.8536
      8       .        *   .              .        19.8995      .9493
      9       .          *.              .         30.5932     1.6076
     10       .          * .              .        21.3883      .7666
     11       .            :              .        22.6254     1.3555

   Case #    0:.............:.............:0          *PRED      *SEPRED
            -3.0          0.0          3.0

            If x₁ = 36, x₂ = 4, x₃ = 8
            then Ŷ = 22.625 and sŶ = 1.355
```

If $x_1 = 36$, $x_2 = 4$, $x_3 = 8$
then $\hat{Y} = 22.625$ and $s_{\hat{Y}} = 1.355$

FIGURE 14.29 SPSS/PC+ output.

```
                    PREDICTION FOR REAL ESTATE EXAMPLE

                * * * *  M U L T I P L E   R E G R E S S I O N  * * * *

     Equation Number 1    Dependent Variable..   SQFOOT


     Variable(s) Entered on Step Number
         1..    EDUC
         2..    INCOME
         3..    SIZE


     Multiple R            .95145
     R Square              .90527
     Adjusted R Square     .85790
     Standard Error       2.03545

     Analysis of Variance
                        DF      Sum of Squares      Mean Square
     Regression          3          237.54155         79.18052
     Residual            6           24.85845          4.14307

     F =      19.11154      Signif F =  .0018
     ------------------------------------------------------------------------
                    PREDICTION FOR REAL ESTATE EXAMPLE

                * * * *  M U L T I P L E   R E G R E S S I O N  * * * *

     Equation Number 1    Dependent Variable..   SQFOOT


     ----------------- Variables in the Equation -----------------

     Variable          B          SE B        Beta        T     Sig T

     EDUC          -.16277       .24407     -.09973     -.667   .5296
     INCOME         .19388       .08770      .46194     2.211   .0691
     SIZE          2.33811       .90779      .54963     2.576   .0420
     (Constant)    7.59550      2.59478                 2.927   .0264


     End Block Number   1   All requested variables entered.
```

Regression coefficients

FIGURE 14.29 Continued

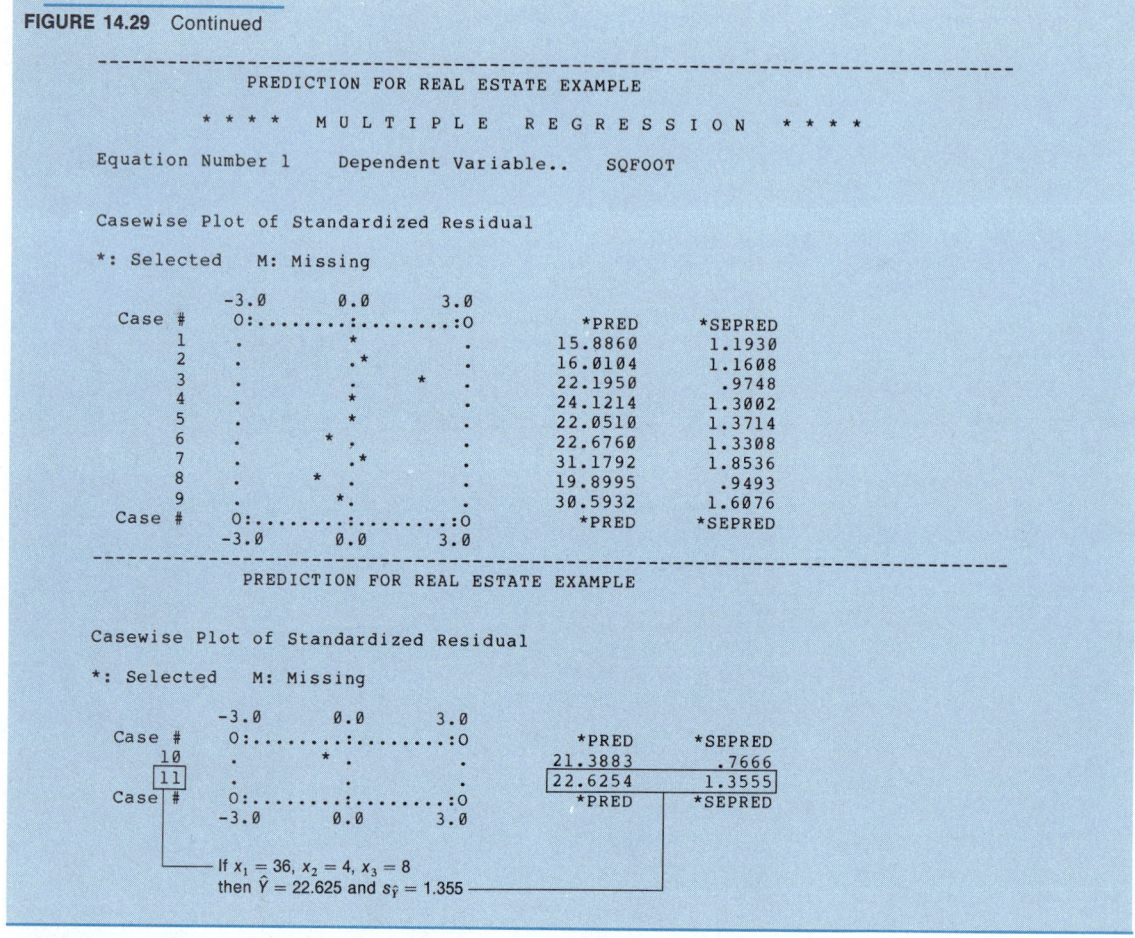

```
-----------------------------------------------------------------------
              PREDICTION FOR REAL ESTATE EXAMPLE

          * * * *   M U L T I P L E   R E G R E S S I O N   * * * *

   Equation Number 1    Dependent Variable..   SQFOOT


Casewise Plot of Standardized Residual

*: Selected   M: Missing

             -3.0       0.0        3.0
   Case #    O:.......:........:O            *PRED        *SEPRED
      1      .          *         .          15.8860       1.1930
      2      .           .*       .          16.0104       1.1608
      3      .           .     *  .          22.1950        .9748
      4      .           *        .          24.1214       1.3002
      5      .           *        .          22.0510       1.3714
      6      .      *    .        .          22.6760       1.3308
      7      .           .*       .          31.1792       1.8536
      8      .       *   .        .          19.8995        .9493
      9      .          *.        .          30.5932       1.6076
   Case #    O:.......:........:O            *PRED        *SEPRED
             -3.0       0.0        3.0
-----------------------------------------------------------------------
              PREDICTION FOR REAL ESTATE EXAMPLE


Casewise Plot of Standardized Residual

*: Selected   M: Missing

             -3.0       0.0        3.0
   Case #    O:.......:........:O            *PRED        *SEPRED
     10      .          *  .       .         21.3883        .7666
     11      .           .         .         22.6254       1.3555
   Case #    O:.......:........:O            *PRED        *SEPRED
             -3.0       0.0        3.0
```

If $x_1 = 36$, $x_2 = 4$, $x_3 = 8$
then $\hat{Y} = 22.625$ and $s_{\hat{Y}} = 1.355$

S A S

MAINFRAME AND MICRO

SOLUTION

EXAMPLE 14.2

At the beginning of the chapter, we computed the regression equation for predicting the estimate of home size, based on the three predictor variables income, family size, and level of education. The SAS program listing in Figure 14.30 requests a multiple regression solution. In this problem the SAS commands are the same for both the mainframe and PC versions.

FIGURE 14.30

Input for SAS (mainframe or micro version).

```
TITLE  'REAL ESTATE EXAMPLE USING TWO PREDICTORS';
DATA REAL ESTATE;
  INPUT SQFOOT INCOME SIZE EDUC;
CARDS;
16 22 2 4
17 26 2 8
26 45 3 7
24 37 4 0
22 28 4 2
21 50 3 10
32 56 6 8
18 34 3 8
30 60 5 2
20 40 3 6
PROC CORR;
  VAR SQFOOT INCOME SIZE EDUC;
PROC REG;
  MODEL SQFOOT=INCOME SIZE EDUC;
```

The TITLE command names the SAS run.

The DATA command gives the data a name.

The INPUT command names and gives the correct order for the different fields on the data lines.

The CARDS command indicates to SAS that the input data immediately follow.

The next 10 lines contain the data values and represent the four variables to be considered in the regression analysis. The first card image represents a home with 1600 square feet, an income of $22,000, a family of 2 people, and 4 years of educational experience at the college level.

The PROC CORR command and VAR subcommand requests a correlation matrix of the variables SQFOOT, INCOME, SIZE, and EDUC.

The PROC REG command and MODEL subcommand indicate that the independent variables INCOME, SIZE, and EDUC be entered in the regression equation to predict the dependent variable SQFOOT.

Figure 14.31 shows the SAS output obtained by executing the listing in Figure 14.30, and Figure 14.32 shows the SAS/PC output.

FIGURE 14.31 SAS output.

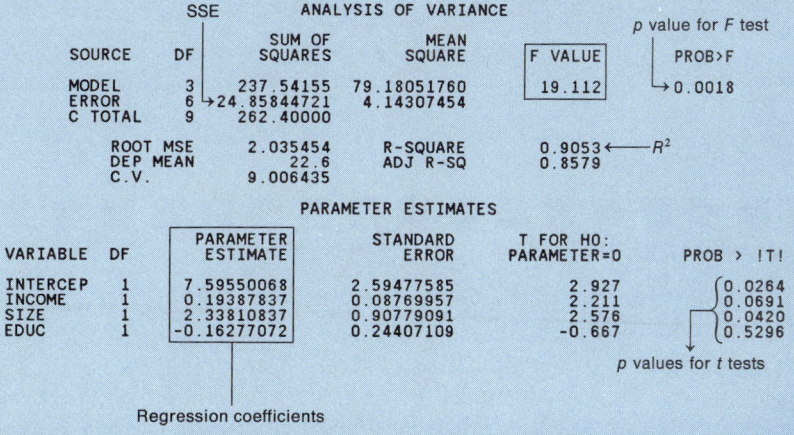

REAL ESTATE EXAMPLE USING TWO PREDICTORS

	SQFOOT	INCOME	SIZE	EDUC
SQFOOT	1.00000	0.84325	0.90787	-0.16794
	0.0000	0.0022	0.0003	0.6428
INCOME	0.84325	1.00000	0.72125	0.15142
	0.0022	0.0000	0.0186	0.6763
SIZE	0.90787	0.72125	1.00000	-0.25137
	0.0003	0.0186	0.0000	0.4836
EDUC	-0.16794	0.15142	-0.25137	1.00000
	0.6428	0.6763	0.4836	0.0000

— Correlation coefficients

REAL ESTATE EXAMPLE USING TWO PREDICTORS

DEP VARIABLE: SQFOOT

ANALYSIS OF VARIANCE

p value for F test

SOURCE	DF	SUM OF SQUARES	MEAN SQUARE	F VALUE	PROB>F
MODEL	3	237.54155	79.18051760	19.112	0.0018
ERROR	6	24.85844721	4.14307454		
C TOTAL	9	262.40000			

SSE → 24.85844721

ROOT MSE	2.035454	R-SQUARE	0.9053 ← R^2
DEP MEAN	22.6	ADJ R-SQ	0.8579
C.V.	9.006435		

PARAMETER ESTIMATES

VARIABLE	DF	PARAMETER ESTIMATE	STANDARD ERROR	T FOR H0: PARAMETER=0	PROB > !T!
INTERCEP	1	7.59550068	2.59477585	2.927	0.0264
INCOME	1	0.19387837	0.08769957	2.211	0.0691
SIZE	1	2.33810837	0.90779091	2.576	0.0420
EDUC	1	-0.16277072	0.24407109	-0.667	0.5296

p values for t tests

Regression coefficients

FIGURE 14.32   SAS/PC output.

REAL ESTATE EXAMPLE USING TWO PREDICTORS

CORRELATION ANALYSIS

Pearson Correlation Coefficients / Prob > |R| under Ho: Rho=0 / N = 10

	SQFOOT	INCOME	SIZE	EDUC
SQFOOT	1.00000	0.84325	0.90787	-0.16794
	0.0	0.0022	0.0003	0.6428
INCOME	0.84325	1.00000	0.72125	0.15142
	0.0022	0.0	0.0186	0.6763
SIZE	0.90787	0.72125	1.00000	-0.25137
	0.0003	0.0186	0.0	0.4836
EDUC	-0.16794	0.15142	-0.25137	1.00000
	0.6428	0.6763	0.4836	0.0

REAL ESTATE EXAMPLE USING TWO PREDICTORS

Model: MODEL1
Dependent Variable: SQFOOT

Analysis of Variance

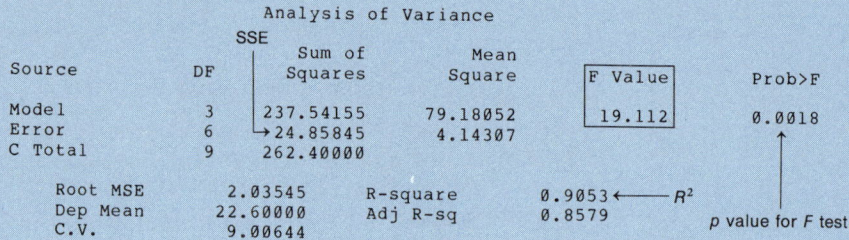

Source	DF	Sum of Squares	Mean Square	F Value	Prob>F
Model	3	237.54155	79.18052	19.112	0.0018
Error	6	24.85845	4.14307		
C Total	9	262.40000			

SSE (pointing to 24.85845)

Root MSE	2.03545	R-square	0.9053 ← R^2
Dep Mean	22.60000	Adj R-sq	0.8579
C.V.	9.00644		

p value for F test

REAL ESTATE EXAMPLE USING TWO PREDICTORS

Parameter Estimates

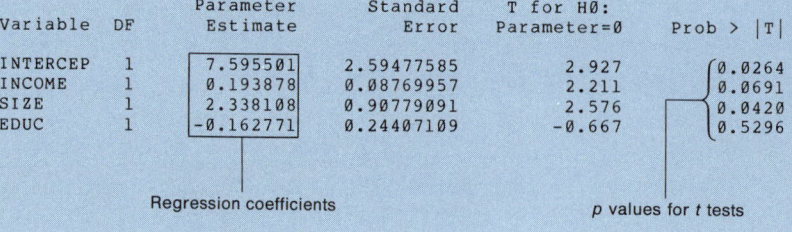

Variable	DF	Parameter Estimate	Standard Error	T for H0: Parameter=0	Prob > |T|
INTERCEP	1	7.595501	2.59477585	2.927	0.0264
INCOME	1	0.193878	0.08769957	2.211	0.0691
SIZE	1	2.338108	0.90779091	2.576	0.0420
EDUC	1	-0.162771	0.24407109	-0.667	0.5296

Regression coefficients

p values for t tests

██ ██ ██ ██ ██ ██ ██ ██ ██ S A S ██ ██ ██ ██ ██ ██ ██ ██

MAINFRAME AND MICRO

EXAMPLE 14.2

STEPWISE METHOD

SAS can be used for determining the stepwise solution used in predicting the estimate of home size, based on the three predictor variables income, family size, and level of education. Notice that the word STEPWISE is substituted for REG and that a new line has been added. Instead of forcing all variables into the equation with the REG command, STEPWISE selects

FIGURE 14.33

Input for SAS
(mainframe or micro
version).

```
TITLE  'REAL ESTATE EXAMPLE USING STEPWISE';
DATA REAL ESTATE;
 INPUT SQFOOT INCOME SIZE EDUC;
CARDS;
16 22 2 4
17 26 2 8
26 45 3 7
24 37 4 0
22 28 4 2
21 50 3 10
32 56 6 8
18 34 3 8
30 60 5 2
20 40 3 6
PROC STEPWISE;
 MODEL SQFOOT=INCOME SIZE EDUC/
        SLENTRY=0.1;
```

the variables that meet the entry criteria. The SLENTRY = 0.1 statement specifies the significance level for entering a variable into the regression equation. The SAS program listing in Figure 14.33 requests a stepwise regression solution. In this problem the SAS commands are the same for both the mainframe and PC versions.

The TITLE command names the SAS run.

The DATA command gives the data a name.

The INPUT command names and gives the correct order for the different fields on the data lines.

The CARDS command indicates to SAS that the input data immediately follow.

FIGURE 14.34 SAS output.

REAL ESTATE EXAMPLE USING STEPWISE

STEPWISE REGRESSION PROCEDURE FOR DEPENDENT VARIABLE SQFOOT

NOTE: SLSTAY HAS BEEN SET TO .15 FOR THE STEPWISE TECHNIQUE.

STEP 1 VARIABLE SIZE ENTERED		R SQUARE = 0.82422204	C(P) =	5.13282842	
	DF	SUM OF SQUARES	MEAN SQUARE	F	PROB>F
REGRESSION	1	216.27586207	216.27586207	37.51	0.0003
ERROR	8	46.12413793	5.76551724		
TOTAL	9	262.40000000			
	B VALUE	STD ERROR	TYPE II SS	F	PROB>F
INTERCEPT	9.08275862				
SIZE	3.86206897	0.63057266	216.27586207	37.51	0.0003

BOUNDS ON CONDITION NUMBER: 1,   1

STEP 2 VARIABLE INCOME ENTERED		R SQUARE = 0.89824277	C(P) =	2.44475400	
	DF	SUM OF SQUARES	MEAN SQUARE	F	PROB>F
REGRESSION	2	235.69890381	117.84945191	30.90	0.0003
ERROR	7	26.70109619	3.81444231		
TOTAL	9	262.40000000			
	B VALUE	STD ERROR	TYPE II SS	F	PROB>F
INTERCEPT	6.73957851				
INCOME	0.16485256	0.07305544	19.42304174	5.09	0.0586
SIZE	2.65693994	0.74046309	49.11200800	12.88	0.0089

BOUNDS ON CONDITION NUMBER: 2.084221,   8.336884

NO OTHER VARIABLES MET THE 0.1000 SIGNIFICANCE LEVEL FOR ENTRY INTO THE MODEL.

SUMMARY OF STEPWISE REGRESSION PROCEDURE FOR DEPENDENT VARIABLE SQFOOT

STEP	VARIABLE ENTERED	REMOVED	NUMBER IN	PARTIAL R**2	MODEL R**2	C(P)	F	PROB>F
1	SIZE		1	0.8242	0.8242	5.13283	37.5120	0.0003
2	INCOME		2	0.0740	0.8982	2.44475	5.0920	0.0586

Final equation: $\hat{Y} = 6.74 + .16(\text{income}) + 2.66(\text{size})$

The next 10 lines contain the data values and represent the four variables to be considered in the regression analysis. The first card image represents a home with 1600 square feet, an income of $22,000, a family of 2 people, and 4 years of educational experience at the college level.

The PROC STEPWISE command and MODEL subcommand indicate that the independent variables INCOME, SIZE, and EDUC be entered in the regression equation to predict the dependent variable SQFOOT. The SLENTRY subcommand was discussed earlier.

Figure 14.34 shows the SAS output obtained by executing the listing in Figure 14.33, and Figure 14.35 shows the SAS/PC output.

FIGURE 14.35 SAS/PC output.

REAL ESTATE EXAMPLE USING STEPWISE

Stepwise Procedure for Dependent Variable SQFOOT

Step 1   Variable SIZE Entered   R-square = 0.82422204   C(p) = 5.13282842

	DF	Sum of Squares	Mean Square	F	Prob>F
Regression	1	216.27586207	216.27586207	37.51	0.0003
Error	8	46.12413793	5.76551724		
Total	9	262.40000000			

Variable	Parameter Estimate	Standard Error	Type II Sum of Squares	F	Prob>F
INTERCEP	9.08275862	2.33397081	87.31381827	15.14	0.0046
SIZE	3.86206897	0.63057266	216.27586207	37.51	0.0003

Bounds on condition number: 1,   1

Step 2   Variable INCOME Entered   R-square = 0.89824277   C(p) = 2.44475400

	DF	Sum of Squares	Mean Square	F	Prob>F
Regression	2	235.69890381	117.84945191	30.90	0.0003
Error	7	26.70109619	3.81444231		
Total	9	262.40000000			

Variable	Parameter Estimate	Standard Error	Type II Sum of Squares	F	Prob>F
INTERCEP	6.73957851	2.16385176	37.00339013	9.70	0.0170
INCOME	0.16485256	0.07305544	19.42304174	5.09	0.0586
SIZE	2.65693994	0.74046309	49.11200800	12.88	0.0089

Bounds on condition number: 2.084221,   8.336884

Final equation: $\hat{Y} = 6.74 + .16(\text{income}) + 2.66(\text{size})$

REAL ESTATE EXAMPLE USING STEPWISE

All variables in the model are significant at the 0.1500 level.
No other variable met the 0.1000 significance level for entry into the model.

Summary of Stepwise Procedure for Dependent Variable SQFOOT

Step	Variable Entered Removed	Number In	Partial R**2	Model R**2	C(p)	F	Prob>F
1	SIZE	1	0.8242	0.8242	5.1328	37.5120	0.0003
2	INCOME	2	0.0740	0.8982	2.4448	5.0920	0.0586

EXAMPLE

SECTION 14.5

In the real-estate example in Section 14.5, we wished to determine the predicted square footage ($\hat{Y}$) for values of $X_1 = 36$, $X_2 = 4$, and $X_3 = 8$. Of course, one way to do this is to insert them manually into the regression equation resulting from the ten observations in the previous SAS example. An easier way is to attach these values at the end of the input data along with any single character from A to Z for the value of the dependent variable, indicating to SAS that this value is missing. Which character you use is arbitrary, but it should be specified in the MISSING statement immediately following the DATA statement. SAS ignores this row of data when it computes the regression equation but includes this row when it summarizes the predicted values.

The predicted values as well as the corresponding confidence intervals and prediction intervals are calculated and included in the output by inserting /CLI and CLM in the final MODEL statement. The CLI command generates the prediction intervals for an *individual*, whereas CLM produces confidence intervals for the *mean*. The row(s) containing the missing value(s) are included in this summary.

The SAS program listing in Figure 14.36 is used to generate the predicted value, confidence intervals, and prediction intervals. In this problem the SAS commands are the same for both the mainframe and PC versions.

The TITLE command names the SAS run.

The DATA command gives the data a name.

The INPUT command names and gives the correct order for the different fields on the data lines.

The CARDS command indicates to SAS that the input data immediately follow.

The next 10 lines contain the data values and represent the four variables to be considered in the regression analysis. The first card image represents a home with 1600 square feet, an income of $22,000, a family of 2 people, and 4 years of educational experience at the college level.

The PROC REG command and MODEL subcommand indicate that the independent variables INCOME, SIZE, and EDUC be entered in the regression equation to predict the dependent variable SQFOOT. The CLI and CLM subcommands were discussed earlier.

FIGURE 14.36

Input for SAS (mainframe or micro version).

```
TITLE  'PREDICTION FOR REAL ESTATE EXAMPLE';
DATA REAL ESTATE;
   MISSING A;
 INPUT SQFOOT INCOME SIZE EDUC;
CARDS;
16 22 2 4
17 26 2 8
26 45 3 7
24 37 4 0
22 28 4 2
21 50 3 10
32 56 6 8
18 34 3 8
30 60 5 2
20 40 3 6
A 36 4 8
PROC REG;
 MODEL SQFOOT=INCOME SIZE EDUC/ CLI CLM;
```

Figure 14.37 shows the SAS output obtained by executing the listing in Figure 14.36, and Figure 14.38 shows the SAS/PC output.

FIGURE 14.37   SAS output.

PREDICTION FOR REAL ESTATE EXAMPLE

DEP VARIABLE: SQFOOT

ANALYSIS OF VARIANCE

SOURCE	DF	SUM OF SQUARES	MEAN SQUARE	F VALUE	PROB>F
MODEL	3	237.54155	79.18051760	19.112	0.0018
ERROR	6	24.85844721	4.14307454		
C TOTAL	9	262.40000			

ROOT MSE	2.035454	R-SQUARE	0.9053	
DEP MEAN	22.6	ADJ R-SQ	0.8579	
C.V.	9.006435			

PARAMETER ESTIMATES

VARIABLE	DF	PARAMETER ESTIMATE	STANDARD ERROR	T FOR H0: PARAMETER=0	PROB > !T!
INTERCEP	1	7.59550068	2.59477585	2.927	0.0264
INCOME	1	0.19387837	0.08769957	2.211	0.0691
SIZE	1	2.33810837	0.90779091	2.576	0.0420
EDUC	1	-0.16277072	0.24407109	-0.667	0.5296

$Y - \hat{Y}$

OBS	ACTUAL	$\hat{Y}$ PREDICT VALUE	$s_{\hat{Y}}$ STD ERR PREDICT	LOWER95% MEAN	UPPER95% MEAN	LOWER95% PREDICT	UPPER95% PREDICT	RESIDUAL
1	16.0000	15.8860	1.1930	12.9669	18.8050	10.1130	21.6589	0.1140
2	17.0000	16.0104	1.1608	13.1700	18.8508	10.2768	21.7440	0.9896
3	26.0000	22.1950	0.9748	19.8098	24.5801	16.6727	27.7172	3.8050
4	24.0000	24.1214	1.3002	20.9399	27.3030	18.2114	30.0315	-0.1214
5	22.0000	22.0510	1.3714	18.6953	25.4067	16.0454	28.0566	-0.0510
6	21.0000	22.6760	1.3308	19.4197	25.9324	16.7254	28.6267	-1.6760
7	32.0000	31.1792	1.8536	26.6436	35.7147	24.4429	37.9154	0.8208
8	18.0000	19.8995	0.9493	17.5766	22.2225	14.4039	25.3952	-1.8995
9	30.0000	30.5932	1.6076	26.6595	34.5269	24.2465	36.9399	-0.5932
10	20.0000	21.3883	0.7666	19.5125	23.2642	16.0662	26.7105	-1.3883
11	A	22.6254	1.3555	19.3086	25.9422	16.6415	28.6093	

SUM OF RESIDUALS	5.39568E-14
SUM OF SQUARED RESIDUALS	24.85845
PREDICTED RESID SS (PRESS)	69.29215

CI for $\mu_{Y|X}$

Prediction interval for Y_x

$\hat{Y}$ and $s_{\hat{Y}}$ for $x_1 = 36$, $x_2 = 4$, $x_3 = 8$

FIGURE 14.38   SAS/PC output.

PREDICTION FOR REAL ESTATE EXAMPLE

Model: MODEL1
Dependent Variable: SQFOOT

Analysis of Variance

Source	DF	Sum of Squares	Mean Square	F Value	Prob>F
Model	3	237.54155	79.18052	19.112	0.0018
Error	6	24.85845	4.14307		
C Total	9	262.40000			

Root MSE	2.03545	R-square	0.9053	
Dep Mean	22.60000	Adj R-sq	0.8579	
C.V.	9.00644			

PREDICTION FOR REAL ESTATE EXAMPLE

Parameter Estimates

Variable	DF	Parameter Estimate	Standard Error	T for H0: Parameter=0	Prob > |T|
INTERCEP	1	7.595501	2.59477585	2.927	0.0264
INCOME	1	0.193878	0.08769957	2.211	0.0691
SIZE	1	2.338108	0.90779091	2.576	0.0420
EDUC	1	-0.162771	0.24407109	-0.667	0.5296

FIGURE 14.38 Continued

Obs	Dep Var SQFOOT	$\hat{Y}$ Predict Value	$s_{\hat{Y}}$ Std Err Predict	Lower95% Mean	Upper95% Mean	Lower95% Predict	Upper95% Predict
1	16.0000	15.8860	1.193	12.9669	18.8050	10.1130	21.6589
2	17.0000	16.0104	1.161	13.1700	18.8508	10.2768	21.7440
3	26.0000	22.1950	0.975	19.8098	24.5801	16.6727	27.7172
4	24.0000	24.1214	1.300	20.9399	27.3030	18.2114	30.0315
5	22.0000	22.0510	1.371	18.6953	25.4067	16.0454	28.0566
6	21.0000	22.6760	1.331	19.4197	25.9324	16.7254	28.6267
7	32.0000	31.1792	1.854	26.6436	35.7147	24.4429	37.9154
8	18.0000	19.8995	0.949	17.5766	22.2225	14.4039	25.3952
9	30.0000	30.5932	1.608	26.6595	34.5269	24.2465	36.9399
10	20.0000	21.3883	0.767	19.5125	23.2642	16.0662	26.7105
11	A	22.6254	1.355	19.3086	25.9422	16.6415	28.6093

CI for $\mu_{Y|x}$ Prediction interval for Y_x

Sum of Residuals   5.329071E-15
Sum of Squared Residuals   24.8584
Predicted Resid SS (Press)   69.2922

$\hat{Y}$ for $x_1 = 36$, $x_2 = 4$, $x_3 = 8$

■ M I N I T A B ■

MAINFRAME AND MICRO

INSTRUCTIONS FOR MULTIPLE AND STEPWISE REGRESSION

THE REGRESS COMMAND

The MINITAB REGRESS command is illustrated in many of the examples contained in this chapter. By using various subcommands you can store the residuals and the values of $b_0, b_1, \ldots, b_k$.

For example, consider the sequence

REGRESS Y IN C1 USING 2 PREDICTORS IN C2 C3;

RESIDUALS IN C4;

COEF IN C5.

This sequence stores the residuals in column C4 and the estimated regression coefficients in column C5. Also, the abbreviated form of the REGRESS command here would be

REGRESS C1 2 C2 C3;

If you wish merely to examine the residuals but not to store them, type the command BRIEF 3 at some point before performing the regression analysis. This will provide additional output (including the residuals) when the REGRESS command is used. This is illustrated in Figure 14.39.

STEPWISE REGRESSION

MINITAB also performs stepwise regression using the command

STEPWISE REGRESSION OF Y IN C1 USING PREDICTORS C2 C3 C4...

Abbreviated statement:

STEP C1 C2 C3 C4...

where C2, C3, C4, ... are the predictor variables.

This procedure enters a variable if its corresponding (partial) F-value exceeds FENTER = 4 and removes any variable whose (partial) F-value falls below FREMOVE = 4. An illustration of this procedure using the real-estate data is shown in Figure 14.40. Only two of the three variables being considered were selected (using the default values of FREMOVE = FENTER = 4). The resulting

FIGURE 14.39

Multiple regression analysis of the real-estate example.

```
MTB > BRIEF 3
MTB > REGRESS Y IN C1 USING 3 PREDICTORS IN C2-C4

The regression equation is
C1 = 7.60 + 0.194 C2 + 2.34 C3 - 0.163 C4

Predictor        Coef        Stdev      t-ratio          p
Constant        7.596        2.595         2.93      0.026
C2            0.19388      0.08770         2.21      0.069
C3             2.3381       0.9078         2.58      0.042
C4            -0.1628       0.2441        -0.67      0.530

s = 2.035        R-sq = 90.5%      R-sq(adj) = 85.8%

Analysis of Variance

SOURCE          DF          SS          MS          F          p
Regression       3     237.542      79.181      19.11      0.002
Error            6      24.858      [4.143]—MSE
Total            9     262.400

SOURCE          DF      SEQ SS
C2               1     186.587
C3               1      49.112
C4               1       1.843

Obs.      C2          C1        Fit  Stdev.Fit  Residual   St.Resid
   1    22.0      16.000     15.886      1.193     0.114        0.07
   2    26.0      17.000     16.010      1.161     0.990        0.59
   3    45.0      26.000     22.195      0.975     3.805        2.13R
   4    37.0      24.000     24.121      1.300    -0.121       -0.08
   5    28.0      22.000     22.051      1.371    -0.051       -0.03
   6    50.0      21.000     22.676      1.331    -1.676       -1.09
   7    56.0      32.000     31.179      1.854     0.821        0.98
   8    34.0      18.000     19.900      0.949    -1.900       -1.05
   9    60.0      30.000     30.593      1.608    -0.593       -0.48
  10    40.0      20.000     21.388      0.767    -1.388       -0.74

R denotes an obs. with a large st. resid.
```

FIGURE 14.40

MINITAB stepwise regression procedure.

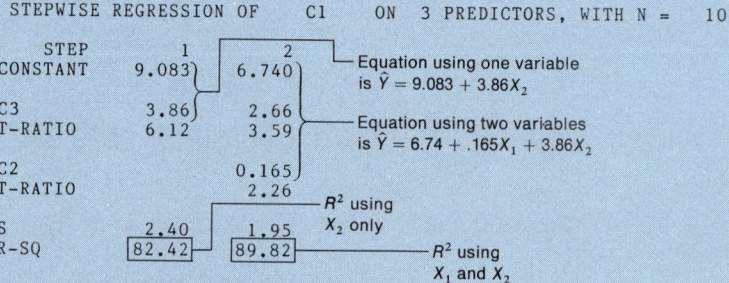

```
MTB > STEPWISE OF Y IN C1 USING PREDICTORS IN C2-C4

STEPWISE REGRESSION OF      C1     ON  3 PREDICTORS, WITH N =    10

       STEP        1        2
   CONSTANT     9.083    6.740

        C3       3.86     2.66
   T-RATIO       6.12     3.59

        C2                0.165
   T-RATIO                 2.26

         S       2.40     1.95
      R-SQ      82.42    89.82
```

Equation using one variable is $\hat{Y} = 9.083 + 3.86X_2$

Equation using two variables is $\hat{Y} = 6.74 + .165X_1 + 3.86X_2$

R^2 using X_2 only

R^2 using X_1 and X_2

equation is

$$\hat{Y} = 6.74 + .165X_1 + 2.66X_2$$

To change the values of FENTER and/or FREMOVE, use the following sequence of commands. The semicolon (;) at the end of the REGRESS command informs MINITAB that subcommands are needed. The period following the final subcommand indicates that there are no further subcommands.

```
MTB > STEPWISE REGRESSION OF Y IN C1 USING C2, C3, C4, C5, C6;

SUBC > FENTER = 3.5;

SUBC > FREMOVE = 3.5.
```

To perform a forward selection, you simply do not allow any variable to be removed once it is included in the model. Setting FREMOVE = 0 will accomplish this. The procedure ends when the (partial) F statistic for an entering variable is below FENTER.

Similarly, you can perform backwards regression by first using the subcommand ENTER:

SUBC > ENTER C2-C6;

where C2-C6 are all of your predictor variables. This enters all of your predictor variables into the model. Next, use

SUBC > FENTER = 10000.

or any large value. This procedure stops when no variable in the model has an F value less than FREMOVE.

Time Series Analysis and Index Numbers

A Look Back/Introduction

The previous two chapters introduced you to a method of predicting the value of a dependent variable using the technique of linear regression. You determined a set of one or more predictor (independent) variables $(X_1, X_2, \ldots)$ that could be used to model the behavior of the dependent variable, Y.

When the dependent variable is measured over *time*, there is another method of describing the behavior of this variable—**time series analysis**. For example, consider the following data, where Y is the amount of electrical power consumed in Pine Bluff over a 10-year period.

YEAR	POWER CONSUMPTION (MILLION kWh)
1979	95
1980	145
1981	174
1982	200
1983	224
1984	245
1985	263
1986	275
1987	283
1988	288

This is an example of a (very short) time series. Typically, a time series covers many more periods, especially when measured for each month, week, or even day. To describe the behavior of the variable Y, we examine the past data and, rather than searching for a number of predictor variables, we try to capture the patterns that exist only in the Y observations over a period of time. In other words, we assume that *time-related patterns can serve as predictors*. In this illustration, one pattern is clear—the power consumption values increase from one year to the next.

The process of using the patterns contained in the past data to predict future values is referred to as **forecasting**. Forecasting using time series data has both advantages and disadvantages. The primary advantage of using time series analysis is that often you can describe your variable of interest, Y, by using only a sample of past observations. Inherent to this type of forecasting procedure is the assumption that past patterns will continue into the future. The disadvantage of time series forecasting is that the past observations often contain patterns that are difficult to extract and, as a result, the models can become very complex.

In this chapter, we will concentrate on methods of *describing* a time series by isolating the various *components* that make up a time series (such as sales in December are always much higher than the yearly average). In the next chapter, methods of forecasting are discussed. We should note at this point that, in general, there is no single best forecasting technique. Instead, the forecaster should attempt to match the forecasting technique to patterns observed in the time series data. Consequently, this chapter and the next chapter are highly intertwined, since by describing the nature of the time series (Chapter 15), you will have a better idea as to which forecasting technique to employ (Chapter 16).

15.1

COMPONENTS OF A TIME SERIES

A **time series** represents a variable observed across time. The time increment can be years, quarters, months, or even days. The values of the time series can be presented in a table or illustrated using a scatter diagram. Usually, the points in the graph are connected by straight lines, making it easier to detect any existing patterns; this is a **line graph**.

The time series for the power-consumption data is shown in Figure 15.1. As we noted, the power-consumption values increase steadily from one year to the next. This long-term movement in the time series is called a *trend*. These values exhibit a definite increasing trend (or growth). Trend is only one of several components that describe the behavior of any time series. The **components** of a time series are:

Trend (*TR*)

Seasonal variation (*S*)

Cyclical variation (*C*)

Irregular activity (*I*)

FIGURE 15.1

Power consumption in Pine Bluff.

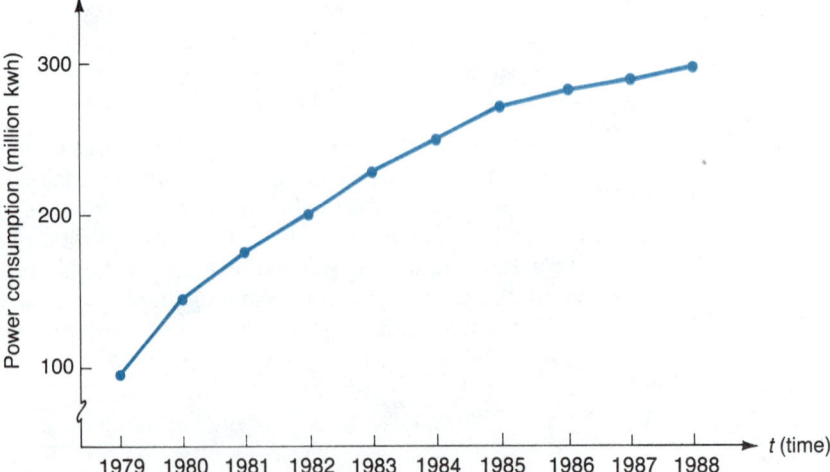

The purpose of time series analysis is to describe a particular data set by estimating the various components that make up this time series. We examine each of these components individually, although time series data usually contain a mixture of all four.

Trend (TR)

The **trend** is a steady increase or decrease in the time series. If a particular time series is neither increasing nor decreasing over its range of time, it contains *no trend*. The trend reflects any long-term growth or decline in the observations. For example, this pattern may be due to inflation, increases in the population, increases in personal income, market growth or decline, or changes in technology. Each of these could have a long-term effect on the variable of interest and would be reflected in the trend in the corresponding time series.

This long-term growth or decay pattern can take a variety of shapes. If the rate of change in Y from one time period to the next is relatively constant, this is a **linear trend**:

$$TR = b_0 + b_1 t$$

(for some b_0 and b_1), where the predictor variable is time t.

When the time series appears to be slowing down or accelerating as time increases, then a nonlinear trend may be present. It may be a **quadratic trend**,

$$TR = b_0 + b_1 t + b_2 t^2$$

or a **decaying trend**,

$$TR = b_0 + b_1 (1/t)$$

or

$$TR = b_0 + b_1 e^{-t}$$

These trend equations can be derived from the linear regression equations developed in Chapter 13 (for linear trend) and Chapter 14 (for quadratic trend). The linear trend equation is an application of *simple* linear regression, whereas the quadratic trend uses a *multiple* regression equation using two predictors, t and t^2. Simple linear regression techniques also can be used to derive b_0 and b_1 for the decay trend equations, where values of t are replaced by the values of $1/t$ or e^{-t} in the data input.

The numbers of employees from 1981 to 1988 at Video-Comp, an expanding microcomputer-software firm, are recorded in the following table and illustrated in Figure 15.2.

YEAR	NUMBER OF EMPLOYEES (thousands)
1981	1.1
1982	2.4
1983	4.6
1984	5.4
1985	5.9
1986	8.0
1987	9.7
1988	11.2

The underlying long-term growth in this time series appears to be nearly *linear*, as represented by the dotted line in Figure 15.2. To determine the equation of this line, we use the technique of simple linear regression, where X = the predictor variable = time and Y = the number of employees. We can estimate

FIGURE 15.2

Number of employees at Video-Comp (an example of linear trend).

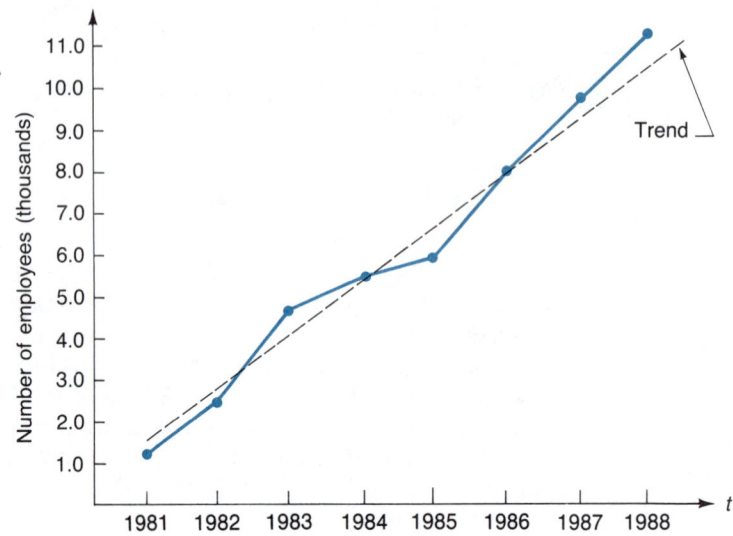

the existing trend using

$$\hat{y}_t = b_0 + b_1 t$$

where t represents the time variable and y_t is the value of Y at time period t. Here b_0 and b_1 are the least squares regression coefficients for a straight line predictor. The procedure of deriving these least squares estimates is developed later in the chapter. Figure 15.3 shows an *increasing* linear trend (y_t increases over time) and a decreasing linear trend (y_t decreases over time).

EXAMPLE 15.1

What type of trend exists in the power-consumption data (Figure 15.1)?

SOLUTION

Although this time series increases steadily, *it increases at a decreasing rate*: it starts off with large increases from one time period to the next, but these increments gradually become smaller. When the growth is linear, the values increase at a nearly constant rate. Figure 15.1 is an illustration of **quadratic trend**, where the time series randomly fluctuates about a quadratic (or curvilinear) level over time. This trend is captured by the equation

$$\hat{y}_t = b_0 + b_1 t + b_2 t^2$$

To derive these estimates, we use the multiple linear regression approach discussed in Chapter 14 (curvilinear models). Section 15.2 demonstrates this technique.

The four types of quadratic trend are summarized in Figure 15.4. ∎

FIGURE 15.3

A: Increasing linear trend: $TR = b_0 + b_1 t$ ($b_1 > 0$). **B**: Decreasing linear trend: $TR = b_0 + b_1 t$ ($b_1 < 0$).

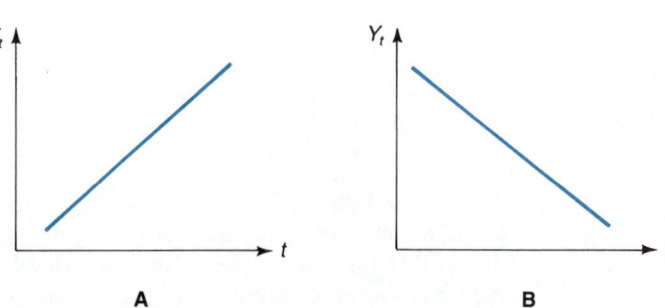

FIGURE 15.4

Quadratic trend. **A**: Y increases at a decreasing rate. **B**: Y decreases at an increasing rate. **C**: Y decreases at a decreasing rate. **D**: Y increases at an increasing rate.

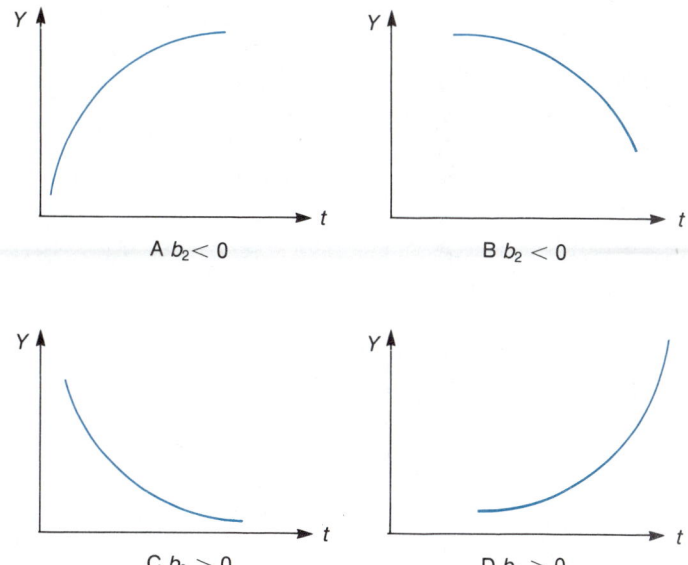

A $b_2 < 0$

B $b_2 < 0$

C $b_2 > 0$

D $b_2 > 0$

Seasonality (S)

Seasonal variation, or **seasonality**, refers to periodic increases or decreases in a time series that occur *within a calendar year*. They are very predictable because they occur every year. When a time series consists of annual data (as in Figure 15.1), you cannot see what is going on within each year. Data reported in annual increments therefore cannot be used to examine seasonality. Seasonality may or may not exist; the data are not in a form that will show whether it does.

When time series data are quarterly or monthly, seasonal variation may be evident. For example, if the power-consumption data were available for each month over these 10 years, then the resulting time series would contain $12 \cdot 10 = 120$ observations. A plot of monthly data for the last 3 years (36 observations) is shown in Figure 15.5. The seasonal effects here consist of

Extremely high power consumption during the hot summer months (July and August)

Very high consumption during the coldest part of the winter (December and January)

Gradually declining consumption during the spring, reaching a low level in April and then increasing until July

Gradually declining power consumption during the fall, but beginning to increase in November

The key is that these movements in the time series *follow the same pattern each year* and so probably are due to seasonality. An analysis of seasonal variation is often a crucial step in planning sales and production. Just because your sales drop from one month to the next does not necessarily mean that it is time to panic. If a review of the past observations indicates that sales *always* drop between these 2 months, then quite likely there is no cause for concern. On the production side, if sales always are extremely high in December, then you will need to increase production in the months prior to December so that you will have the necessary inventory level for this peak month. Measurement of this seasonal component is discussed later.

FIGURE 15.5

Illustration of seasonal
variation. These are
monthly observations;
compare with annual
data in Figure 15.1.

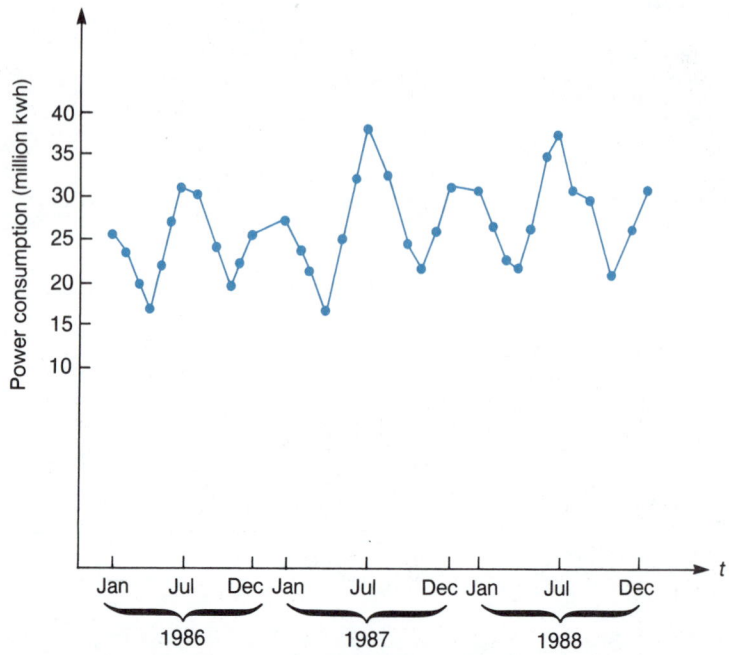

As mentioned earlier, a time series often contains the effect of trend and
seasonality (as well as cyclical and irregular activity). The sales of Wildcat sailboats, illustrated in Figure 15.6, contain a strong linear trend as well as definite
seasonal variation. In particular, the highest sales occur in the summer months
of each year.

As manager of Wildcat Enterprises, would you be concerned that the sales
of these boats in December 1987 were lower than those in July 1987? There may
or may not be a problem; due to seasonal effects, this pattern exists in Figure
15.6 despite an overall growth. More data would be required to determine

FIGURE 15.6

A time series containing
trend and seasonal
variation.

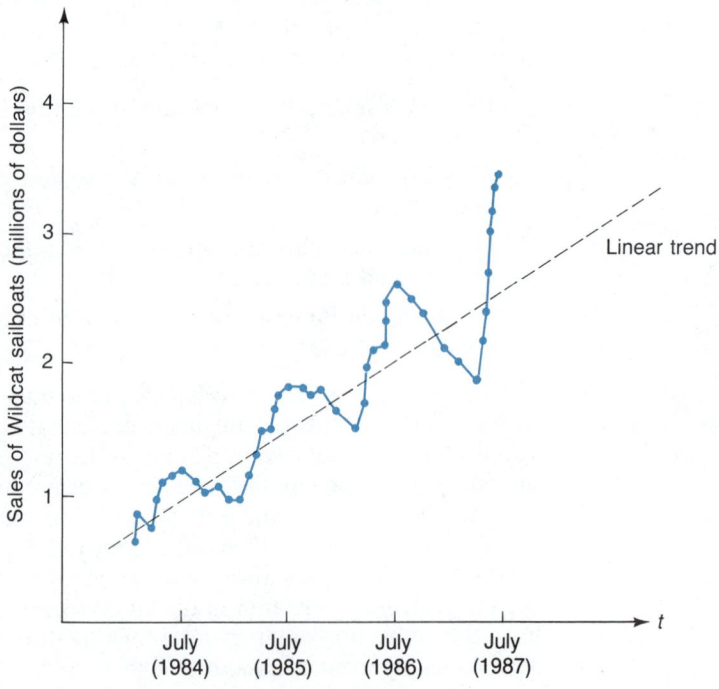

FIGURE 15.7

The cycle can be measured from P_1 to P_2, from V_1 to V_2, or from Z_1 to Z_2.

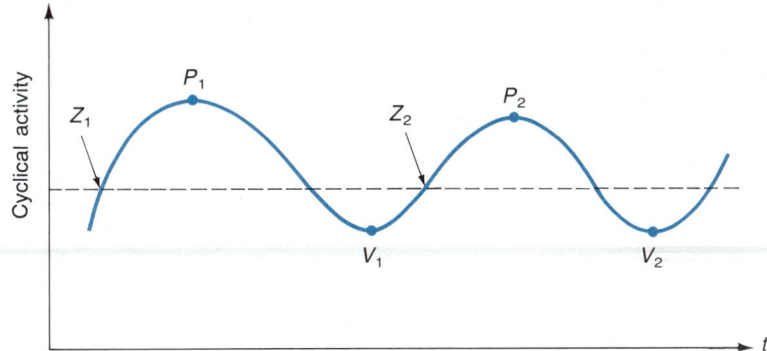

whether the December sales were lower than expected for that month. What would you think if sales in July 1988 were lower than those in July 1987? This should definitely concern you. This is a year-to-year comparison, and seasonal variation or not, we would expect the sales for July 1988 to be larger than for July 1987 if the long-term growth trend in Figure 15.6 is still present. Lower sales in July 1988 would indicate a possible leveling off or a drop in boat sales in 1988.

Cyclical Variation (C)

Cyclical variation describes a gradual cyclical movement about the trend; it is generally attributable to business and economic conditions. The length of a cycle is the **period** of that cycle. The period of a cycle can be measured from one **peak** to the next, one **trough** (valley) to the next, or from the time value at which the cycle crosses the horizontal line where no cyclic activity exists to the value where it completes the cycle and returns to this point. Figure 15.7 shows that the cycle length can be measured from P_1 to P_2, from V_1 to V_2, or from Z_1 to Z_2. In the illustrations to follow, we use the Z_1 to Z_2 approach.

In business applications, cycles typically are long-term movements, with a period ranging from 2 to 10 years. The primary difference between the cyclical and seasonal factors is the period length. Seasonal effects take place *within* one year, whereas the period for cyclical activity is *more than* one year.

Cyclical activity need not follow a definite, recurrent pattern. The cycles generally represent conditions within the economy, where a peak occurs at the height of an expansion (prosperity) period. This is generally followed by a period of contraction in economic activity. The low point (trough) of each cycle usually takes place at the low point of an economic recession or depression. This low point is then followed by a gradual increase during the recovery period.

EXAMPLE 15.2   The annual corporate taxes paid by Lindale (a clothing manufacturer) over a 25-year period are shown in Figure 15.8. How many cycles do you observe?

SOLUTION   The year 1962 began a cycle lasting approximately 8 years. There are three cycles contained within the time series, ending in the midst of an "up cycle." Notice that the cycle lengths are not the same. ■

Irregular Activity (I)

Irregular activity consists of what is "left over" after accounting for the effect of any trend, seasonality, or cyclical activity. These values should consist of noise, much like the error term in the linear regression models discussed in the previous chapters. The irregular activity should contain no observable, or predictable,

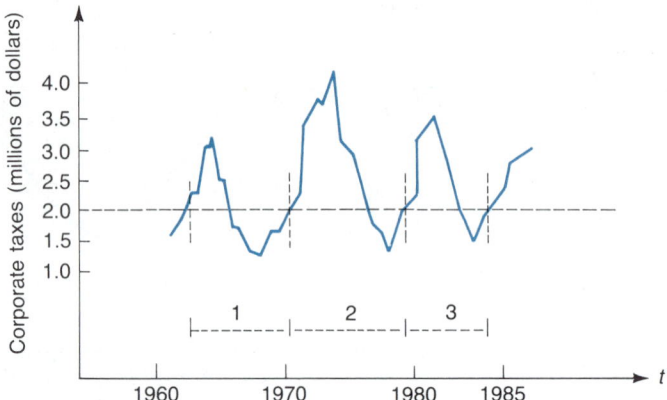

pattern. An extremely large irregular component can be caused by a measure-ment error in the variable. Such an outlier should always be checked to ensure its accuracy.

The irregular component (1) measures the random movement in your time series and (2) represents the effect introduced by unpredictable rare events, such as earthquakes, oil embargoes, or strikes.

If a noticeable jump in the resulting irregular components (when plotted across time) can be attributed to a particular rare event, you may wish to eliminate such data from the time series. You can then examine the remaining data to measure more accurately the other time series components.

Combining the Components

The time series components can be combined in various ways to describe the behavior of a particular time series. One method is to describe the time series variable, y_t, as a *sum* of these four components

$$y_t = TR_t + S_t + C_t + I_t$$

This is called the **additive structure**. The implication here is that any seasonal effects are additive from one year to the next. For example, if the seasonal effect of December for a time series representing sales is an increase of 250 units over the average yearly sales, then this same increase will occur each year regardless of the sales volume. Whether the average yearly sales are 1000 units or 10,000 units, December should show a sales volume of approximately 1250 (the first case) or 10,250 (the latter case).

Better success has been achieved by describing a time series using the **multiplicative structure**, where

$$y_t = TR_t \cdot S_t \cdot C_t \cdot I_t$$

Here, the seasonal effect increases or decreases according to the underlying trend and cyclical effect. Using the previous illustration, the difference between the December sales and the yearly average will be *higher* for the latter case where the yearly average is 10,000 units. For example, for the first case, the December sales might be 1250 (a 25% increase over the yearly average) and, for the latter situation, 12,500 (also a 25% increase). This follows from the implication in the multiplicative structure that, as the sales increase from one year to the next, the changes in volume due to seasonality also increase. For our illustration, this shift was 250 units for the first case and 2500 units for the second case.

EXERCISES

15.1 The management of a pharmaceutical firm would like to predict the effects of a technological breakthrough in an antiulcer drug in order to plan for company growth and capital expenditure. Would a time series analysis be appropriate? Why or why not?

15.2 Describe in words the trends for the quarterly sales figures (in thousands of dollars) of the companies A, B, and C. Graph the data over time.

YEAR	QUARTER	COMPANY A	COMPANY B	COMPANY C
1986	1	13.1	8.3	5.1
	2	10.2	7.3	7.1
	3	11.1	8.0	6.4
	4	16.5	7.3	5.3
1987	1	14.1	7.1	5.4
	2	11.3	6.4	7.2
	3	12.5	7.0	6.3
	4	17.8	6.4	5.1
1988	1	15.1	6.2	5.0
	2	12.2	5.3	7.0
	3	13.4	6.0	6.1
	4	18.3	5.2	5.3

15.3 Construction in the housing industry usually appears to peak in the middle of the summer and to bottom out around January. If the number of new housing starts are the same for the month of March and the month of July in a particular year, of what concern would these figures be to housing construction companies? Would they be pleased, worried, or indifferent? Why?

15.4 The end-of-year inventory levels, in dollars, of West Coast Distributing are given in the following table. Estimate the period of the cyclical component by graphing the data.

YEAR	INVENTORY	YEAR	INVENTORY
1976	80	1983	80
1977	75	1984	83
1978	71	1985	80
1979	73	1986	77
1980	82	1987	79
1981	76	1988	84
1982	78		

15.5 To which of the four components of a time series would each of the following influences on housing starts contribute?

a. Presidential election year

b. Start of the school year in September

c. Long-term growth of the housing industry

d. Shortage of lumber because of a strike

15.6 Describe in words both the trend and the seasonal components for the following sales (in thousands of dollars). (*Hint:* draw a graph for each.)

MONTH	1987	1988	MONTH	1987	1988
Jan.	1.2	2.2	Jul.	2.9	3.8
Feb.	1.4	2.4	Aug.	3.2	4.0
Mar.	1.3	2.3	Sep.	2.5	3.5
Apr.	1.5	2.4	Oct.	2.4	3.0
May	1.5	2.5	Nov.	2.3	2.8
Jun.	2.3	3.5	Dec.	2.1	2.5

15.2

MEASURING TREND: NO SEASONALITY

Suppose that you have a time series containing trend and cyclical activity but no seasonality. For example, the employment data in Figure 15.2 are annual and so contain no seasonality. The same is true for the annual power-consumption data in Figure 15.1. When data are collected on a yearly basis, we are not concerned with any seasonality in the data; we need data from quarterly or shorter intervals to identify any seasonality. Yearly data may have trend (TR), cyclical activity (C), or irregular activity (I). If we observe a strong linear trend (as in Figure 15.2) or a quadratic trend (Figure 15.4), we can estimate it using the least squares technique developed in Chapters 13 and 14. We use simple linear regression for linear trends and multiple linear regression for quadratic trends.

Linear Trend

We begin by **coding** the time variable to make the calculations (or computer input) easier.

In Figure 15.2, we can find an equation for the trend line passing through the eight observations in the time series. The least squares trend line through these eight values is sketched in Figure 15.9. The equation of the trend line is

$$\hat{y}_t = TR_t = b_0 + b_1 t$$

where t represents the time variable. For this equation, TR_t represents the trend component of the sample observation at time period t and is simply a new name for the trend effect that this equation allows us to estimate.

We could use $t = 1981, 1982, \ldots$ to represent time, but a much simpler method is to *code* the data, as illustrated in Figure 15.9. By using $t = 1, 2, \ldots$, the estimate, $\hat{y}_t$, is not affected and the calculations are easier. You are able to code the predictor variable, t, because the sample values are equally spaced—they are 1 year apart. As we saw in Chapter 13, this is not the case in all simple regression applications. Continuing the scheme in Figure 15.9, $t = 9$ represents the year 1989 and the estimated number of employees for 1989 (using trend only) is

$$\hat{y}_9 = b_0 + b_1(9)$$

To derive the "best" line through the time series data, we use the least squares estimates discussed in Chapter 13; the independent variable here is time, t. The data are

t	y_t	
1	1.1	$(=y_1)$
2	2.4	$(=y_2)$
3	4.6	
4	5.4	
5	5.9	
6	8.0	
7	9.7	
8	11.2	

The calculations are

$$\Sigma t = 1 + 2 + \cdots + 8 = 36$$
$$\Sigma t^2 = 1 + 4 + \cdots + 64 = 204$$
$$\Sigma y_t = 1.1 + 2.4 + \cdots + 11.2 = 48.3$$
$$\Sigma y_t^2 = (1.1)^2 + (2.4)^2 + \cdots + (11.2)^2 = 375.63$$
$$\Sigma t y_t = (1)(1.1) + (2)(2.4) + \cdots + (8)(11.2) = 276.3$$

FIGURE 15.9
Least squares trend line
using coded time data
(compare with Figure
15.2). $\hat{y}_t = b_0 + b_1 t$.

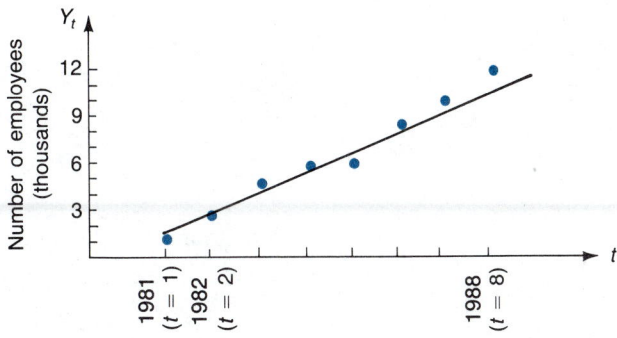

In Chapter 13, when we regressed the variable Y on a single variable X, the estimate for the slope of the least squares line (from equation 13.11) was given by

$$b_1 = \frac{\text{SS}_{XY}}{\text{SS}_X} = \frac{\Sigma\, xy - (\Sigma\, x)(\Sigma\, y)/n}{\Sigma\, x^2 - (\Sigma\, x)^2/n}$$

where n was the number of sample observations. To determine a linear trend line for a time series, this equation becomes

$$b_1 = \frac{\Sigma\, ty_t - (\Sigma\, t)(\Sigma\, y_t)/T}{\Sigma\, t^2 - (\Sigma\, t)^2/T} \qquad \text{(15.1)}$$

where $T =$ the number of observations in the time series.

The sample estimate of the intercept is

$$b_0 = \bar{y} - b_1 \bar{x} = \bar{y}_t - b_1 \bar{t} \qquad \text{(15.2)}$$

where $\bar{y}_t = (y_1 + y_2 + \cdots + y_t)/T$.

Because the time variable, t, *always* is $1, 2, \ldots, T$, there is an easier way to calculate $\Sigma\, t$, $\Sigma\, t^2$, and $\bar{t}$.

$$\Sigma\, t = 1 + 2 + \cdots + T$$
$$= \frac{T(T + 1)}{2} \qquad \text{(15.3)}$$

$$\Sigma\, t^2 = 1 + 4 + \cdots + T^2$$
$$= \frac{T(T + 1)(2T + 1)}{6} \qquad \text{(15.4)}$$

$$\bar{t} = \frac{\Sigma\, t}{T} = \frac{T + 1}{2} \qquad \text{(15.5)}$$

We use these equations to derive the least squares line in Figure 15.9. Using equations 15.3 and 15.4,

$$\Sigma\, t = \frac{T(T + 1)}{2} = \frac{(8)(9)}{2} = 36$$

and

$$\Sigma t^2 = \frac{T(T+1)(2T+1)}{6} = \frac{(8)(9)(17)}{6} = 204$$

Also,

$$\bar{t} = \frac{\Sigma t}{T} = \frac{36}{8} = 4.5$$

This value can also be found using equation 15.5:

$$\bar{t} = \frac{T+1}{2} = \frac{9}{2} = 4.5$$

So we can now calculate

$$b_1 = \frac{\Sigma t y_t - (\Sigma t)(\Sigma y_t)/T}{\Sigma t^2 - (\Sigma t)^2/T}$$

$$= \frac{276.3 - (36)(48.3)/8}{204 - (36)^2/8} = \frac{58.95}{42} = 1.4036$$

and

$$b_0 = \bar{y}_t - b_1\bar{t}$$

$$= \frac{48.3}{8} - (1.4036)(4.5) = 6.0375 - 6.3162 = -.279$$

The trend line for this time series is

$$\hat{y}_t = -.279 + 1.404t$$

We conclude that the number of employees appears to increase at the rate of 1404 per year, on the average.

The trend line is derived using the same least squares procedure as discussed in Chapter 13—you can use the computer instructions contained in the simple linear regression illustrations. A computer solution (using MINITAB) is shown in Figure 15.10. It produces the same results we obtained.

FIGURE 15.10

MINITAB solution of least squares trend line.

```
MTB > SET INTO C1
DATA> 1:8    ◄─────────────────────  This command generates
DATA> END                             integers 1 through 8
MTB > SET INTO C2
DATA> 1.1 2.4 4.6 5.4 5.9 8.0 9.7 11.2
DATA> END
MTB > REGRESS Y IN C2 USING 1 PREDICTOR IN C1;
SUBC> DW.

The regression equation is
C2 = - 0.279 + 1.40 C1                                    t value

Predictor      Coef        Stdev       t-ratio        p
Constant      -0.2786      0.3596       -0.77        0.468
C1             1.40357     0.07122      19.71        0.000

s = 0.4616      R-sq = 98.5%      R-sq(adj) = 98.2%

Analysis of Variance

SOURCE        DF          SS           MS          F          p
Regression     1        82.741       82.741     388.39      0.000
Error          6         1.278        0.213
Total          7        84.019


Durbin-Watson statistic = 1.88
```

Figure 15.10 contains the t-statistic; you may be tempted to use it to determine whether time is a significant predictor of Y = number of employees. However, to use this statistic, you must assume that the errors about the trend line are completely *independent* of one another and contain *no observable pattern*. Do not forget that there may well be considerable cyclical activity about the trend line, and this will be contained in the residuals of the regression analysis. This means that there probably will be a cyclical pattern to these residuals, so the assumption of complete independence is not met. This implies that the errors are *autocorrelated* and any test of hypothesis is invalid.

This poses no serious problems at this point, however, because our intent is simply to describe the time series by measuring the various components, and not to perform a statistical test of hypothesis. If, however, the residuals about the trend line appear to be extremely large, then this suggests that a linear trend component is not appropriate.

Quadratic Trend

The nature of a quadratic trend is illustrated in Figure 15.4. This type of trend is common for a time series that increases or decreases rapidly and then gradually levels off over the observed values. We discussed a similar situation in Chapter 14, where a quadratic model of the form

$$\hat{Y} = b_0 + b_1 X + b_2 X^2$$

was used to capture a curvilinear relationship between two variables. We use exactly the same technique to describe a quadratic trend; now X is replaced by time, t.

The power-consumption time series in Figure 15.1 indicates that as time increases, the amount of power consumption (y_t) also increases, but at a decreasing rate. More specifically, the increase for 1984 to 1985 is 18; for 1985 to 1986 is 12 (12 < 18); for 1986 to 1987 is 8 (8 < 12); and for 1987 to 1988 is 5 (5 < 8).

When you observe a series where the *changes* from one year to the next are not (approximately) constant but seem to be either increasing or decreasing with time, this indicates a quadratic trend.

The equation of this curvilinear (quadratic) trend is

$$\hat{y}_t = b_0 + b_1 t + b_2 t^2$$

To derive the least squares estimates b_0, b_1, and b_2, we use the multiple linear regression procedure of Chapter 14.

What would be the input to a computer program (such as MINITAB, SAS or SPSS) for the power-consumption data? For the regression program, you have two predictor variables, $X_1 = t$ and $X_2 = t^2$. The resulting data configuration is

y_t	t	t^2	
95	1	1	(for 1979)
145	2	4	(for 1980)
174	3	9	(for 1981)
200	4	16	(and so on)
224	5	25	
245	6	36	
263	7	49	
275	8	64	
283	9	81	
288	10	100	

[Note that here the time series data (y_t) are put in the first column of the input data. This placement is arbitrary.]

FIGURE 15.11

MINITAB solution for quadratic trend (power-consumption data).

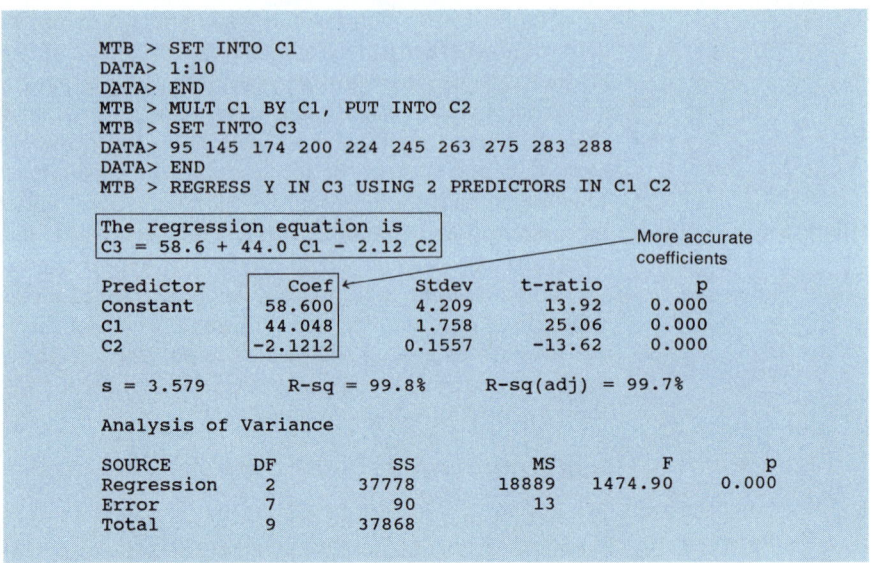

```
MTB > SET INTO C1
DATA> 1:10
DATA> END
MTB > MULT C1 BY C1, PUT INTO C2
MTB > SET INTO C3
DATA> 95 145 174 200 224 245 263 275 283 288
DATA> END
MTB > REGRESS Y IN C3 USING 2 PREDICTORS IN C1 C2

The regression equation is
C3 = 58.6 + 44.0 C1 - 2.12 C2                          More accurate
                                                       coefficients
Predictor       Coef         Stdev      t-ratio        p
Constant       58.600        4.209       13.92       0.000
C1             44.048        1.758       25.06       0.000
C2             -2.1212       0.1557     -13.62       0.000

s = 3.579        R-sq = 99.8%      R-sq(adj) = 99.7%

Analysis of Variance

SOURCE         DF          SS          MS         F         p
Regression      2        37778       18889     1474.90   0.000
Error           7          90          13
Total           9        37868
```

The solution for these data (shown in Figure 15.11) is

$$\hat{y}_t = 58.6 + 44.048t - 2.1212t^2$$

To illustrate this equation, for the second time period the actual value is $y_2 = 145$ and the predicted value is

$$\hat{y}_2 = 58.6 + 44.048(2) - 2.1212(2)^2 = 138.21$$

A First Look at Forecasting: Extending the Trend

Whenever a time series contains very little seasonality (such as *annual* data, which have *no* seasonality) and a strong trend, an easy method of providing future forecasts is to project the observed growth pattern, as measured by the trend equation, into the future. For example, if a city's tax revenues have increased steadily by approximately $15,000 per year over the past 10 years, it seems reasonable to expect that this pattern will continue, at least for a short time. (Of course, assuming that such a growth will continue indefinitely is a hazardous gamble at best!)

The process of extending a trend equation is called **forecasting**, or **extrapolation**. The following examples illustrate that extending a straight-line trend equation can provide useful estimates of future values. A quadratic trend equation is, however, useful only *within* the range of the sample data; that is, for **interpolation**.

This method of forecasting is but one of many possible ways of predicting the future time series values by capturing patterns present in the past observations. Chapter 16 examines other methods of using the past observations to forecast future values.

EXAMPLE 15.3   Using the trend line from Figure 15.9, estimate the number of employees in 1989.

SOLUTION $t = 9$ corresponds to the year 1989, so the *forecast* for this year is

$$\hat{y}_9 = -.279 + 1.404(9) = 12.357$$

that is, 12,357 employees. ∎

As mentioned earlier, the basic assumption when using the trend line to determine a forecast is that this same pattern *will continue* into the future. This may or may not be true. Very often a time series will increase at a more-or-less constant rate and then begin to level off. One example is the sales of an innovative product. Such a time series will grow from one year to the next as people think that they just have to have this product, but eventually a saturation point is reached and the sales grow at a much smaller rate. If the historical data used to determine the trend line are collected during the growth stage, then you will stop short of and miss the "slowing down" of the time series and severely over-estimate the sales. This is not a flaw in the technique; any time series model makes predictions by capturing the pattern(s) in the past observations and extending this pattern beyond the last year of the data. It does, however, place a great deal of responsibility on the person who uses the data to predict beyond the data range. If you do not know what underlying factors are driving the trend, serious errors can result.

Very often, a nonlinear growth rate can be described accurately by including a quadratic term in the trend equation. However, using such an equation to forecast *future* values is not a reliable procedure, as the following section demonstrates.

EXAMPLE 15.4

Using the trend equation from Figure 15.11

$$\hat{y}_t = TR_t = 58.6 + 44.048t - 2.1212t^2$$

what is your forecast for the power consumption during 1989? During 1990? Use only the trend equation.

SOLUTION

1989 corresponds to $t = 11$ (the last year of your data is $t = 10$ for 1988). Your forecast for 1989 is

$$\hat{y}_{11} = 58.6 + 44.048(11) - 2.1212(11)^2$$
$$= 286.46$$

that is, 2,864,600 kilowatt-hours. For 1990, your forecast is

$$\hat{y}_{12} = 58.6 + 44.048(12) - 2.1212(12)^2$$
$$= 281.72$$
∎

The sermon we delivered about projecting a trend line beyond the range of the data applies to a quadratic trend as well: By forecasting with such an equation, you assume that this quadratic (curved) pattern observed in the time series observations will continue.

In addition, there is another danger when forecasting with a quadratic trend equation. Every such equation looks like

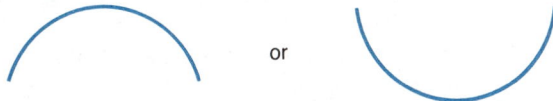

or

In other words, the curve reaches a peak (or trough) and then reverses. The problem is that you really do not know what your equation will do outside the range of your data.

The forecasts for power consumption (Example 15.4) for 1989 and 1990 provide a good illustration of this problem. Notice that the predicted value for 1989 is less than the actual value for 1988, despite a steadily increasing pattern in the time series data. Even worse, the 1990 estimate is less than that for 1989.

These values imply that the trend equation is decreasing during the years after 1988. This simply means that the trend equation forecasts appear to be poor estimates—we have no reason to suspect a downturn in the amount of power consumption for these future years. The trend is appropriately described by the quadratic curve, but only within the range of the data. Beyond this range, the curve turns down. Because we have no reason to believe that the demand for electrical power will decrease, the quadratic equation is no longer appropriate.

To describe the trend *within* the years of your time series data, the quadratic trend equation may work well. However, remember that, as a forecasting procedure, it is very dangerous; do not use it for this purpose.

This section has demonstrated how you can derive linear or quadratic trend equations by using linear regression techniques. Extending such a trend equation into the future is a method of statistical forecasting, a subject that is discussed at length in Chapter 16.

EXERCISES

15.7   A company that supplies quotation machines to stockbroker firms has had increased sales volume every year for the past 10 years. Given in thousands of dollars, the sales figures are:

YEAR	SALES	YEAR	SALES
1979	60	1984	170
1980	80	1985	190
1981	110	1986	230
1982	130	1987	240
1983	160	1988	270

a.   Using the simple regression formula, find the least squares line to describe sales from time, t, where t is equal to 1979, 1980, ... , 1988.

b.   Do the calculations in part (a) with $t = 1, 2, ... , 10$.

c.   Compare the prediction equations given in parts (a) and (b). Do these equations give the same predicted values?

15.8   The community of Farlington has seen its population grow dramatically over the past 8 years. Data on the community's population is given in the following table. Units are in thousands

YEAR	POPULATION	YEAR	POPULATION
1981	3.1	1985	22.4
1982	6.3	1986	33.5
1983	10.1	1987	49.3
1984	17.3	1988	52.1

a.   Is the time series increasing at an increasing rate?

b.   Using a computerized statistical package, find the multiple regression equation to predict the population such that a quadratic curvature is taken into account. Do you think this community can continue to grow at this rate?

15.9   Explain why a prediction equation with a quadratic trend may be dangerous to use in forecasting even though a quadratic trend fits the historic data very well.

15.10   The amount of money deposited into savings accounts at a local bank has grown steadily over the years, as the following data indicate (money = in savings accounts at the end of the year times $100,000$):

YEAR	MONEY	YEAR	MONEY
1981	2.1	1985	10.3
1982	4.2	1986	13.3
1983	6.4	1987	14.9
1984	8.5	1988	16.7

a. Does it appear that a quadratic trend exists in the data?

b. Calculate the equation you would use to describe the trend.

15.11 An insurance company would like to find the trend line for the amount of insurance sold annually (in millions of dollars) across time. The variable time is represented by t and is equal to $1, 2, \ldots, 8$ for the past 8 years. The following statistics were collected:

$$\Sigma\, ty_t = 394.5$$
$$\Sigma\, y_t = 29.4$$

Find the trend line for these time series data.

15.12 Due to rising competition from overseas, an electronics firm has been losing its share of the market. The following data show the percent of the market that the firm has captured for the past 7 years.

YEAR	SHARE OF MARKET
1982	4.7
1983	4.3
1984	3.9
1985	3.8
1986	3.6
1987	3.0
1988	2.9

a. Does the trend appear to be linear?

b. Find the equation to estimate the trend for the time series data.

c. What would be your estimate of the electronics firm's share of the market in 1989?

15.13 The interest paid on the public debt of the United States is given in units of billions of dollars from 1977 to 1986.

YEAR	INTEREST PAID ON DEBT	YEAR	INTEREST PAID ON DEBT
1977	41.9	1982	117.4
1978	48.7	1983	128.8
1979	59.8	1984	153.8
1980	74.9	1985	178.9
1981	95.6	1986	187.1

(*Source: The World Almanac and Book of Facts*, 1988, p. 96.)

Find the least squares prediction equation that best describes the trend. What is your estimate of interest paid on the debt in 1986?

15.14 The production of gold by the United States in troy ounces follows for the years 1972 to 1986.

YEAR	GOLD PRODUCTION	YEAR	GOLD PRODUCTION
1972	1,449,943	1980	969,782
1973	1,175,750	1981	1,379,161
1974	1,126,886	1982	1,465,686
1975	1,052,252	1983	2,002,526
1976	1,048,037	1984	2,084,615
1977	1,100,347	1985	2,427,232
1978	998,832	1986	3,733,190
1979	964,390		

(*Source: The World Almanac and Book of Facts*, 1985, p. 157, and 1988, p. 104.)

a. Does it appear that a quadratic trend exists in the data?

b. Calculate the equation you would use to describe the trend. What is the estimate of the gold production in the United States in 1986?

15.15 The GNP (gross national product) is an important statistic in measuring the economic growth of a country. The GNP is the value of all goods and services sold on the market during a particular time interval. The GNP (in billions of dollars) of the United States for the years 1970 through 1985 is:

YEAR	GNP	YEAR	GNP
1970	992.7	1978	2163.9
1971	1077.6	1979	2417.8
1972	1185.9	1980	2631.8
1973	1326.4	1981	2957.9
1974	1434.2	1982	3069.3
1975	1549.2	1983	3304.8
1976	1718.0	1984	3662.8
1977	1918.3	1985	3988.5

(Source: Statistical Abstract of the United States 1987, 107th ed., U.S. Department of Commerce, Bureau of the Census, p. 416.)

a. Estimate the trend line equation for the data assuming that a linear trend is present.

b. Estimate the trend line equation for the data assuming that a quadratic trend is present. Use a computerized statistical package.

c. Using the equations obtained in parts (a) and (b), estimate the GNP for the year 1986.

d. Look up the actual GNP values for 1986 in a reference book. How good were the estimates? What might account for the differences?

15.16 There is a variety of indexes of stock prices currently available. The Dow Jones Industrial average (DJIA) is probably the most widely quoted and used stock price measure. The DJIA is an unweighted arithmetic average of the prices of 30 blue-chip stocks. The DJIAs (high) for the years 1955 through 1987 are:

YEAR	DJIA	YEAR	DJIA	YEAR	DJIA
1955	488.40	1966	995.15	1977	999.75
1956	521.05	1967	943.08	1978	907.74
1957	520.77	1968	985.21	1979	897.81
1958	583.65	1969	968.85	1980	1000.17
1959	679.36	1970	842.00	1981	1024.05
1960	685.47	1971	950.82	1982	1070.55
1961	734.91	1972	1036.27	1983	1267.20
1962	726.01	1973	1051.70	1984	1286.64
1963	767.21	1974	891.66	1985	1553.10
1964	891.71	1975	881.81	1986	1955.57
1965	969.26	1976	1014.79	1987	2722.42

(Source: The World Almanac and Book of Facts, 1988, p. 98.)

a. Estimate the trend line equation for the data assuming that a linear trend is present.

b. Assuming that a quadratic trend is present, estimate the trend. Use a computerized statistical package.

c. Use the equation found in part (b) to estimate the DJIA (high) for 1987. Compare the actual DJIA figure for 1987 with the estimate. Comment on the difference.

15.3

MEASURING CYCLICAL ACTIVITY: NO SEASONALITY

Practically every time series in a business setting contains a certain amount of cyclical activity. This is a gradual movement about the trend. It is generally due to economic or other long-term conditions. The overall U.S. economy tends to fluctuate through "good times" and "bad times," producing (rather unpredictable) upward and downward variation about the long-term growth or decline in a time series.

One way of describing the cyclical activity component is to represent it as a fraction of the trend. This procedure provides accurate measures of the cyclical activity provided the time series contains *little irregular activity*. Assuming that

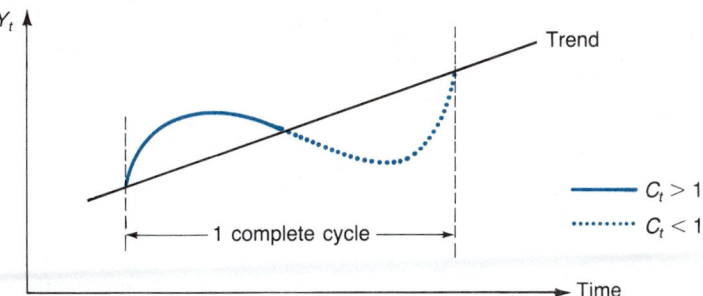

FIGURE 15.12

A complete cycle within a time series.

each time series observation is the *product* of its components, then

$$y_t = TR_t \cdot C_t \cdot I_t$$

because we are dealing with data containing no seasonality.

If we represent a small irregular activity component as i_t (rather than I_t), then a time series containing little irregular variation (noise) can be written as

$$y_t = TR_t \cdot C_t \cdot i_t$$

The cyclical components are then obtained by dividing each observation, y_t, by its corresponding estimate using trend only, $\hat{y}_t$.

$$\text{ratio of data to trend} = \frac{y_t}{\hat{y}_t} = \frac{\cancel{TR_t} \cdot C_t \cdot i_t}{\cancel{TR_t}} = C_t \cdot i_t$$

where y_t = actual time series observation at time period t and $\hat{y}_t = TR_t$ = the estimate of y_t using trend only.

Notice that the resulting ratios still contain some irregular activity. A method of reducing the irregular activity within these values is illustrated later.

An estimate of the cyclical components can be obtained by ignoring the irregular activity components in these ratios and defining

$$C_t \cong \frac{y_t}{\hat{y}_t} \tag{15.6}$$

Assuming that we are dealing with data containing no seasonality (such as annual data), equation 15.6 provides a convenient method of determining the cycles present in the data. If $C_t > 1$, the actual y_t is larger than that predicted by trend alone. Consequently, this value is somewhere in a cycle *above* the trend line. A similar argument indicates a cycle below the trend line whenever $C_t < 1$ (Figure 15.12).

EXAMPLE 15.5

For the data in Figure 15.9, we determined a least squares trend line for the number of employees (y_t) over an 8-year period at Video-Comp. We observed a linear trend with the corresponding equation

$$\hat{y}_t = -.279 + 1.404t$$

where $t = 1$ represents 1981, $t = 2$ is for 1982, and so on. Determine and graph the cyclical activity over this period.

SOLUTION

We can obtain Table 15.1 based on the preceding trend line. Here $\hat{y}_1 = -.279 + 1.404(1) = 1.125$, $\hat{y}_2 = -.279 + 1.404(2) = 2.529$, and so on.

t	y_t	$\hat{y}_t$	$C_t \cong y_t/\hat{y}_t$
1	1.1	1.125	.977
2	2.4	2.529	.949
3	4.6	3.933	1.169
4	5.4	5.337	1.012
5	5.9	6.741	.875
6	8.0	8.145	.982
7	9.7	9.549	1.016
8	11.2	10.953	1.022

The third column is the trend component, and the fourth column is the cyclical component as a fraction of the trend.

To examine the cyclical activity, we can describe each component as a percentage of the trend. For example, in Table 15.1, during the first time period, the actual number of employees is 97.7% of the trend value: C_1 is .977. An illustration of the trend and cyclical activity is shown in Figure 15.13. The cycles fluctuate about the trend line. Between the years $t = 2$ (1982) and $t = 3$ (1983), $y_t = \hat{y}_t$ and a cycle begins. This cycle is completed somewhere between $t = 6$ (1986) and $t = 7$ (1987), where, once again, $y_t = \hat{y}_t$. As discussed earlier, you can also measure cycles from peak to peak or from trough to trough.

A summary of the cyclical variation (components) over the 8 years is contained in Table 15.1 and Figure 15.14. The 4-year cycle we described is more evident in this graph. The graph clearly indicates the beginning of the cycle, where $C_t = 1$. The cycle's peak occurs at the beginning of $t = 3$ (1983), the trough is at the beginning of $t = 5$ (1985), and the cycle is finally complete when C_t is again equal to 1, toward the end of 1986.   ■

In summary, cyclical variation represents an upward or downward movement about the overall growth or decline (that is, the trend) in the time series data. Such cycles last more than 1 year. For annual data, these components can be estimated by dividing each observation (y_t) by its corresponding estimate using the trend equation ($\hat{y}_t$).

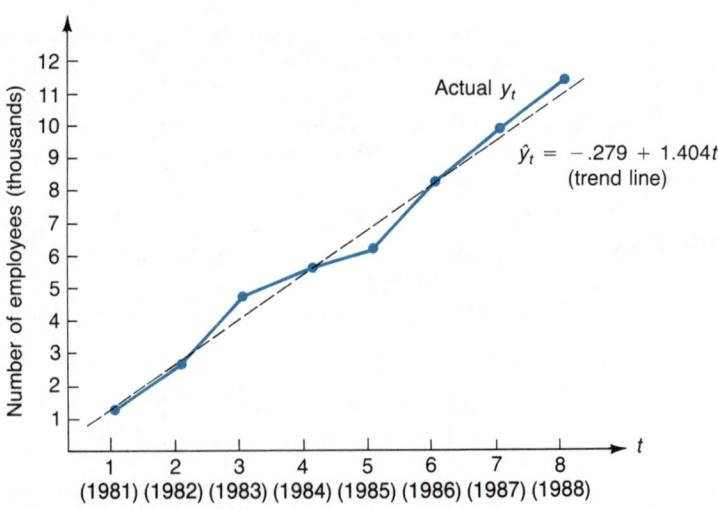

FIGURE 15.14

Cyclical components
(Example 15.5).

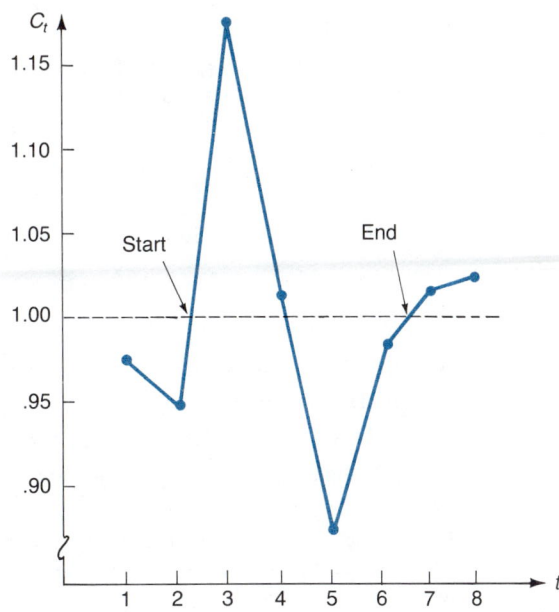

15.17 Using 15 years of data, a forecaster for an oil company found a trend line for the amount of the company's yearly contracts for oil service projects. For the past 5 years, the table gives the predicted value (in thousands of dollars) of the amount of annual contracts as well as the actual amount of annual contracts. The predicted value was arrived at by using the trend line based on 15 years of past data. Find the cyclical component for each of the past 5 years. Does the period of the cycle appear to be longer than 5 years or less than 5 years? Why?

t	y_t	$\hat{y}_t$
11	52	53
12	54	59
13	57	61
14	59	60
15	62	60

15.18 Explain the relationship between the actual value y_t and the value $\hat{y}_t$, from the trend line when $C_t > 1$ and $C_t < 1$, where C_t is defined in equation 15.6.

15.19 A food-store chain has the following record over the past 9 years of yearly sales volume (in hundreds of thousands of dollars):

YEAR	SALES VOLUME	YEAR	SALES VOLUME
1980	7	1985	17
1981	15	1986	12
1982	10	1987	8
1983	5	1988	17
1984	11		

a. Find the trend line.

b. Find the cyclical component.

c. Estimate the period of the cycle.

15.20 Residential Construction of America has been growing over the long term. Because the construction company is sensitive to cyclical variations in the economy, the level of employment for the company changes from year to year, as can be seen by the following data:

YEAR	FULL-TIME EMPLOYEES (IN HUNDREDS)	YEAR	FULL-TIME EMPLOYEES (IN HUNDREDS)
1976	2.4	1983	11.7
1977	9.2	1984	17.3
1978	11.1	1985	23.1
1979	8.5	1986	28.7
1980	10.5	1987	29.3
1981	6.8	1988	25.2
1982	5.4		

a. Find the trend line.

b. Find the cyclical components.

c. Estimate the period of the cycle.

15.21   For the data in Exercise 15.4, estimate the cyclical components.

15.22   Using the data from Exercise 15.15, determine and graph the cyclical activity for the years 1970 through 1985.

15.23   The president of Techronics is concerned about changes in the wholesale price of raw materials. The president gathers the following data on the wholesale price index for raw materials (WPI):

YEAR	WPI	YEAR	WPI
1979	105.0	1984	108.5
1980	106.0	1985	107.7
1981	105.0	1986	107.7
1982	104.9	1987	106.0
1983	105.6	1988	105.8

a. Estimate the trend line equation.

b. Determine the cyclical activity, C_t, where $C_t = y_t/\hat{y}_t$.

15.24   Using the data from Exercise 15.16, determine and graph the cyclical activity for the years 1955 through 1987.

15.25   Sales (in millions of dollars) of Konoco for the years 1979 through 1988 are as follows:

YEAR	SALES	YEAR	SALES
1979	151	1984	163
1980	194	1985	171
1981	177	1986	199
1982	157	1987	214
1983	188	1988	169

Determine and graph the cyclical activity, $C_t = y_t/\hat{y}_t$.

15.26   The percent of total U.S. petroleum products supplied by the total net petroleum imports follow for the years 1971 through 1986.

YEAR	TOTAL NET IMPORTS PERCENT OF CONSUMPTION	YEAR	TOTAL NET IMPORTS PERCENT OF CONSUMPTION
1971	24.3	1979	43.1
1972	27.6	1980	37.3
1973	34.8	1981	33.6
1974	35.4	1982	28.1
1975	35.8	1983	28.3
1976	40.6	1984	30.0
1977	46.5	1985	27.3
1978	42.5	1986	32.8

(*Source: The World Almanac and Book of Facts*, 1988, p. 128.)

Determine the cyclical component for each year.

15.4

TYPES OF SEASONAL VARIATION

Another type of variation about the trend in a time series is due to seasonality. Seasonality generally is present when the data are quarterly or monthly. It can also occur for weekly or even daily data. For example, recurrent daily effects can be expected to occur in the check-processing volume in a bank. Seasonality is any recurrent, constant source of variation caused by events at the particular time of year rather than by any long-term influence (as in cyclical activity). For example, one would expect to sell more snowmobiles in January than in July. In a sense, the seasonal variation appears as a cycle within a year; we do not refer to this as cyclical variation, however, due to its recurrent nature.

We will discuss two types of seasonal variation: additive and multiplicative.

Additive Seasonal Variation

One encounters **additive seasonal variation** when the amount of the variation due to seasonality *does not depend on the level* y_t. This type of seasonal variation is illustrated in Figure 15.15, which shows the sales of snowmobiles over a 3-year period at The Outdoor Shop. Notice that the amount of variation for each of the winter quarters remains the same (100 units), even as the unit sales increase over the 3 years. For an actual application, we assume an additive effect of seasonality if these increments are of nearly the same magnitude over the observed time series.

Assume that the sales data for Jetski snowmobiles from sales area 1 were recorded quarterly over a 5-year period from 1983 through 1987. The following trend line was derived:

$$TR_t = \hat{y}_t = 100 + 20t$$

The seasonal indexes for a seasonal time series represent the incremental effect of the seasons alone, apart from any trend or cyclical activity. For the Jetski data in sales area 1, these indexes were found to be

$$S_1 = +60 \quad \text{(winter quarter)} \qquad S_3 = -40 \quad \text{(summer quarter)}$$
$$S_2 = +30 \quad \text{(spring quarter)} \qquad S_4 = -20 \quad \text{(fall quarter)}$$

In a time series decomposition (where we actually derive these seasonal indexes), additive seasonal variation assumes that the seasonal index for, say, the winter quarter is the *same* for each year. Using the additive model, this

FIGURE 15.15

Snowmobile sales at The Outdoor Shop (additive seasonal variation).

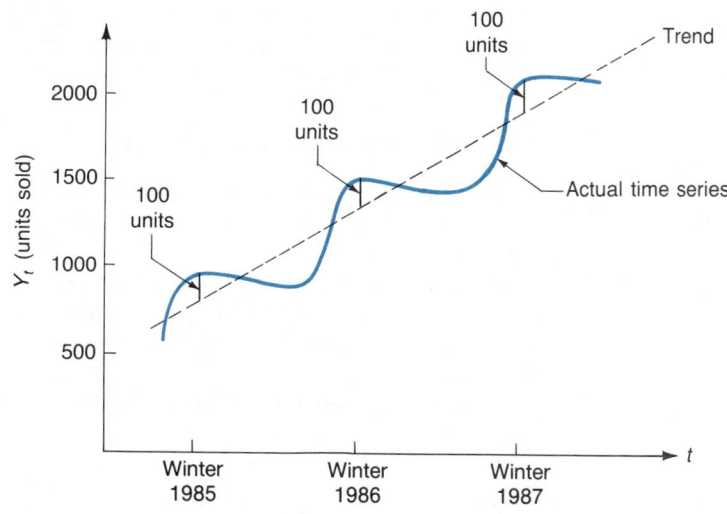

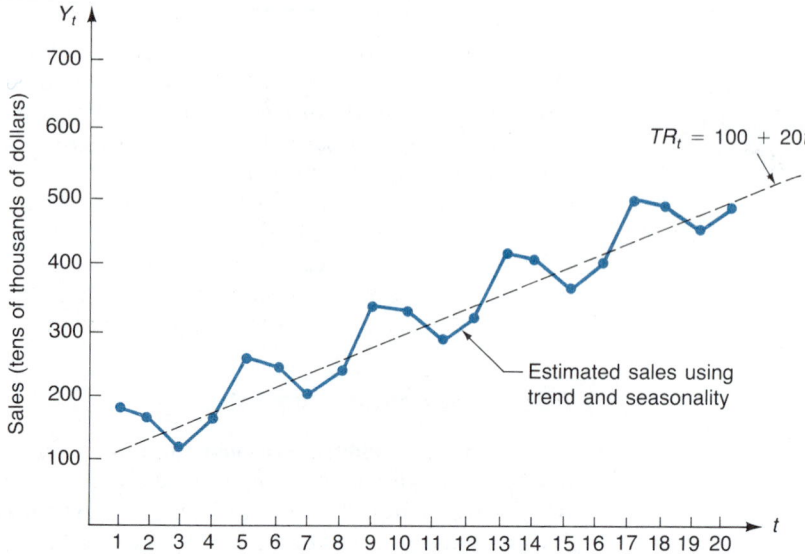

implies that the store will sell 60 more Jetski units in the winter quarter than
would be predicted by trend alone during any year. This implies that $S_1 =
S_5 = S_9 = \cdots = +60$.

To estimate y_t using only the trend and seasonality, we *add* the two corresponding components.

t (TIME)		$TR_t + S_t$ (SALES ESTIMATE)
1	(winter, 1983)	$[100 + 20(1)] + 60 = 180$
2	(spring, 1983)	$[100 + 20(2)] + 30 = 170$
3	(summer, 1983)	$[100 + 20(3)] - 40 = 120$
4	(autumn, 1983)	$[100 + 20(4)] - 20 = 160$
5	(winter, 1984)	$[100 + 20(5)] + 60 = 260$
6	(spring, 1984)	$[100 + 20(6)] + 30 = 250$
7	(summer, 1984)	$[100 + 20(7)] - 40 = 200$
8	(autumn, 1984)	$[100 + 20(8)] - 20 = 240$
⋮		⋮

A graph of the estimated sales is shown in Figure 15.16. Notice that as the
overall level of sales increases, the deviation from the trend line (due to seasonality) remains the same. If the past observations in the time series indicate
that higher levels of sales produce wider seasonal fluctuations, this is an indication of multiplicative seasonal variation.

Multiplicative Seasonal Variation

Figure 15.6 shows **multiplicative seasonal variation** in the time series for the sale
of Wildcat sailboats. Notice that in each successive year, the difference between
the actual value and the trend value for July is larger. In multiplicative seasonal variation, the seasonal fluctuation is *proportional* to the trend level for
each observation. Figure 15.17 is a general illustration of multiplicative seasonality; it shows the sales of heat pumps over a 3-year period at Handy
Home Center.

Considering only the effects of trend and seasonality, an estimate for a time
series observation is given by

$$\text{estimate of } y_t = TR_t \cdot S_t$$

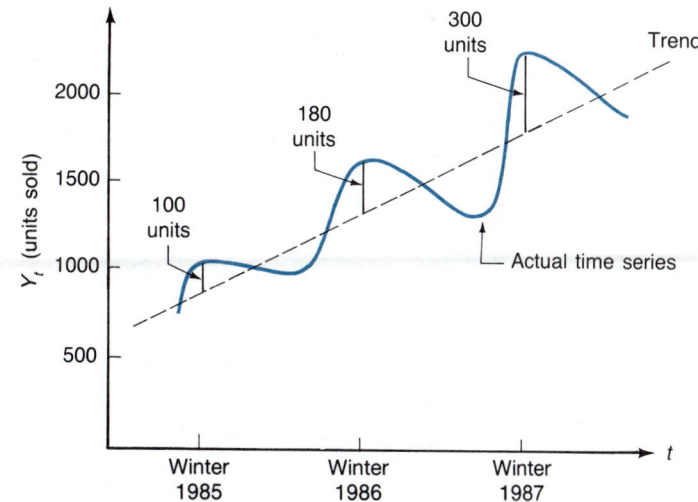

As with additive seasonal variation, the seasonal indexes, S_t, remain constant from one year to the next. When dealing with quarterly data, this means that $S_1 = S_5 = S_9 = \cdots$, $S_2 = S_6 = S_{10} = \cdots$, and so on. The next section discusses a method for determining these indexes for the case of multiplicative seasonality.

EXAMPLE 15.6

Suppose that the sales of Jetski snowmobiles from sales area 2 contain multiplicative seasonal effects with trend $= TR_t = 100 + 20t$ (as before) and seasonal indexes

$$S_1 = 1.4 \quad \text{(winter quarter)}$$
$$S_2 = 1.2 \quad \text{(spring quarter)}$$
$$S_3 = \ \ .6 \quad \text{(summer quarter)}$$
$$S_4 = \ \ .8 \quad \text{(autumn quarter)}$$

Determine the estimated sales using the trend and seasonal components.

SOLUTION

The calculations for the first 2 years are

t (TIME)		$TR_t \cdot S_t$ (ESTIMATE)
1	(winter, 1983)	$[100 + 20(1)](1.4) = 168$
2	(spring, 1983)	$[100 + 20(2)](1.2) = 168$
3	(summer, 1983)	$[100 + 20(3)](\ .6) = \ \ 96$
4	(autumn, 1983)	$[100 + 20(4)](\ .8) = 144$
5	(winter, 1984)	$[100 + 20(5)](1.4) = 280$
6	(spring, 1984)	$[100 + 20(6)](1.2) = 264$
7	(summer, 1984)	$[100 + 20(7)](\ .6) = 144$
8	(autumn, 1984)	$[100 + 20(8)](\ .8) = 208$
⋮		⋮

A graph of the estimated sales over a 5-year period is shown in Figure 15.18. Notice that seasonal patterns do exist, but (unlike additive variation) these fluctuations increase as the sales level rises. For a time series representing sales, this type of variation seems to make sense. If the volume of sales doubles, it is reasonable to expect a larger effect due to seasonality than occurred previously.

Remember that, in practice, few time series exhibit exact additive or multiplicative seasonal effects. However, you can classify a great many time series as essentially belonging to one or the other of these two classes. ■

FIGURE 15.18

Jetski sales from sales area 2 (multiplicative seasonal variation).

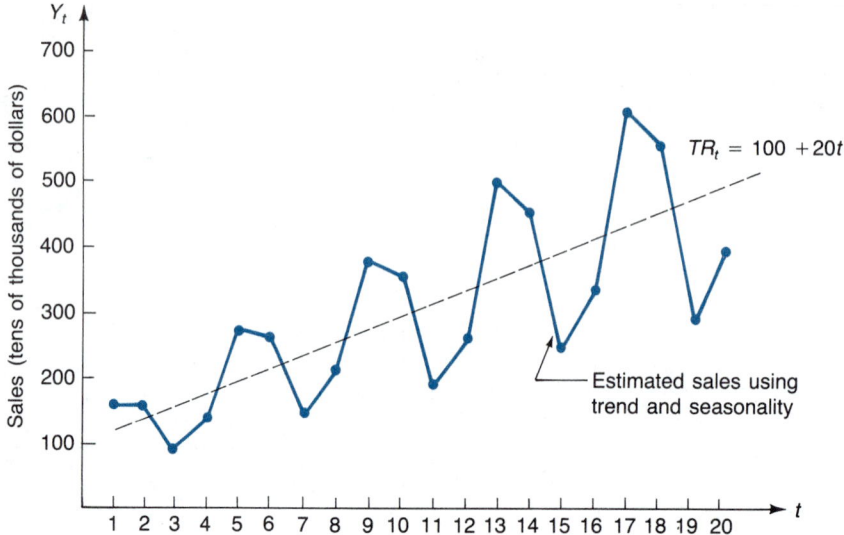

In the discussion to follow, we assume that any seasonality in the time series is *multiplicative*. Most analysts (including those in the U.S. Census Bureau) have had better success describing a time series in this manner. The decomposition method to be discussed assumes that each observation is the *product* of its various components. So, the *component structure* is assumed to be

$$y_t = TR_t \cdot S_t \cdot C_t \cdot l_t \qquad (15.7)$$

where the components representing seasonality, trend, cyclical variation, and noise are multiplied by one another.*

Four-Step Procedure (Multiplicative Components)

Based on the multiplicative component structure in equation 15.7, the following four-step procedure can be used to decompose a time series containing the effects of all four components.

Step 1. *Determine a seasonal index, S_t, for each time period.* For quarterly data, this involves determining four such indexes, S_1, S_2, S_3, and S_4. When the time series contains monthly observations, 12 seasonal indexes (S_1 through S_{12}) must be calculated, one for each month.

Step 2. *Deseasonalize the data.* This is often referred to as *adjusting for seasonality*; the seasonal component is eliminated. Because we are using a multiplicative structure, we divide each observation by its corresponding seasonal index. So

$$\text{deseasonalized observation} = d_t$$
$$= y_t/S_t$$

* Similar methods for determining the components of a time series containing additive seasonality also exist. Chapter 16 contains a small discussion of this topic. For a complete discussion of such techniques, see B. L. Bowerman and R. T. O'Connell, *Forecasting and Time Series*, 2d ed., (North Scituate, Mass.: Duxbury, 1987).

where

$$S_t = \begin{cases} S_1, S_2, S_3, \text{ or } S_4 & \text{(quarterly data)} \\ S_1, S_2, \ldots, \text{ or } S_{12} & \text{(monthly data)} \end{cases}$$

Because $y_t = TR_t \cdot S_t \cdot C_t \cdot I_t$,

$$d_t = \frac{y_t}{S_t} = \frac{TR_t \cdot \cancel{S_t} \cdot C_t \cdot I_t}{\cancel{S_t}} = TR_t \cdot C_t \cdot I_t$$

Step 3. *Determine the trend component, TR_t.* The trend is estimated by passing a least squares line through the *deseasonalized* data. The technique is identical to that discussed in Section 15.2 (which assumed no seasonality) except that we use the d_t values rather than the original time series. This is illustrated in the next section.

Step 4. *Determine the cyclical component, C_t.* You obtain C_t by first dividing each deseasonalized observation, d_t, by the corresponding trend value from step 3. So the cyclical estimates are derived by first calculating (for each time period)

$$\frac{d_t}{\hat{d}_t} = \frac{d_t}{TR_t} = \frac{\cancel{TR_t} \cdot C_t \cdot I_t}{\cancel{TR_t}} = C_t \cdot I_t$$

Notice that the resulting series contains cycles and irregular activity (but no trend or seasonality). A method for reducing the irregular component in these ratios is demonstrated later. The resulting values are the cyclical components, C_t.

We do not use the cyclical components to attempt to forecast future values of the time series because their behavior (and period) generally cannot be predicted. The cyclical components can be used in forecasting if one is willing to assume a particular phase in the business cycle. If one assumes, for example, that the cycle is in the midst of an upturn, a value of C_t (such as $C_t = 1.2$) can be assigned to this particular time period. In the discussion to follow, the cyclical components are obtained strictly as a means of *describing* the cyclical activity within a recorded time series.

EXERCISES

15.27 Explain the effect that seasonality has on the trend assuming additive and multiplicative seasonality. Would seasonality changes have a larger effect on an additive model or on a multiplicative model?

15.28 Riney, owner of Riney's Shoe Store, usually has a rush on shoe sales around September, when children are going back to school. Sales data (in thousands) for Riney's Shoe Store were recorded quarterly over a 4-year period (1985 through 1988). The following trend line was calculated from the data:

$$TR_t = 4 + 1.6t$$

where $t = 1, 2, 3, \ldots, 16$. Seasonal indexes were found to be the following:

quarter 1 $S_1 = \quad 1.3$
quarter 2 $S_2 = -1.4$
quarter 3 $S_3 = \quad 4.2$
quarter 4 $S_4 = \quad 2.6$

Assuming an additive model, estimate the quarterly sales y_t using only trend and seasonality for the four quarters of 1986.

15.29   For a 6-year period (1983 to 1988) quarterly sales data (in thousands) were used to arrive at the following trend line and seasonal indexes.

$$TR_t = 35 + 2.3t \qquad \text{for } t = 1, 2, \ldots, 24$$
$$S_1 = -8.7$$
$$S_2 = 2.5$$
$$S_3 = 8.4$$
$$S_4 = 3.1$$

Estimate the sales figures for the four quarters in 1987 using an additive equation containing only the trend and seasonality components.

15.30   Advanced Digital Components has experienced rapid growth during the past several years. The quarterly data for the past 4 years give the following trend line and seasonal indexes. Sales units are in tens of thousands.

$$TR_t = 0.85 + 0.8t \qquad \text{for } t = 1, 2, \ldots, 16$$
$$S_1 = 0.82$$
$$S_2 = 1.36$$
$$S_3 = 1.20$$
$$S_4 = 0.62$$

Estimate the sales figures for the four quarters in the most recent year using a multiplicative equation containing only the trend and seasonality components.

15.31   Monthly data from the years 1984 through 1988 were used to find the following trend line and seasonal indexes:

$$TR_t = 1.3 + 0.5t \qquad \text{for } t = 1, 2, \ldots, 60$$

$S_1 = 0.5$	$S_7 = 2.4$
$S_2 = 0.8$	$S_8 = 3.1$
$S_3 = 0.6$	$S_9 = 0.3$
$S_4 = 1.3$	$S_{10} = 0.2$
$S_5 = 1.1$	$S_{11} = 0.2$
$S_6 = 1.4$	$S_{12} = 0.1$

Assuming a multiplicative model containing only the trend and seasonality components, estimate the data for the 12 months of 1987.

15.32   Refer to Exercise 15.65. Assuming that the sales of Luz Chemicals are subject to multiplicative seasonal variation, determine the deseasonalized sales for the 12 months of the year 1988. The seasonal indexes are as follows:

$S_1 = 0.8$	$S_7 = 0.9$
$S_2 = 0.8$	$S_8 = 0.9$
$S_3 = 1.2$	$S_9 = 0.7$
$S_4 = 1.2$	$S_{10} = 1.2$
$S_5 = 1.1$	$S_{11} = 0.9$
$S_6 = 1.0$	$S_{12} = 1.3$

15.33   Rework Exercise 15.32 assuming that the sales are subject to additive seasonal variation. The seasonal indexes are as follows:

$S_1 = -25$	$S_7 = 25$
$S_2 = -50$	$S_8 = 25$
$S_3 = -60$	$S_9 = -30$
$S_4 = 25$	$S_{10} = -40$
$S_5 = 25$	$S_{11} = -40$
$S_6 = 28$	$S_{12} = -30$

15.34 The nominal GNP quarterly estimates (in billions of dollars) for a certain third world nation for the years 1986 to 1988 are:

YEAR	QUARTER	NOMINAL GNP
1986	1	206.4
	2	226.3
	3	238.5
	4	254.2
1987	1	268.6
	2	274.8
	3	287.9
	4	297.9
1988	1	313.6
	2	312.8
	3	326.7
	4	346.8

Assuming multiplicative seasonality, calculate the deseasonalized data, given the following seasonal indexes:

$$S_1 = 1.30$$
$$S_2 = 0.80$$
$$S_3 = 1.20$$
$$S_4 = 0.70$$

15.5

MEASURING SEASONALITY

Seasonality often is present in time series data collected over months or quarters. This effect is observed when, for example, some months are always higher than the average for the year. If, during the recorded values of the time series, July sales are 25% higher than the average for the year, the July index should be 1.25 using the multiplicative structure.

We derive a seasonal index for each period during the year (four for quarterly data, 12 for monthly data). We begin by developing a new series that contains *no seasonality*. This new series is obtained from the original time series and consists of the **centered moving averages**. This provides an excellent way of isolating the seasonal components from the original time series. In addition to containing no seasonality, the centered moving averages are *smoother* (contain less irregular activity) than the original time series. Consequently, the moving averages give you a clearer picture of any existing trend within a time series containing significant seasonality and irregular activity. Other methods of smoothing a time series will be discussed in Chapter 16.

Centered Moving Averages

To illustrate the calculation of a moving average, consider a time series containing quarterly observations, as shown in Table 15.2.* Here,

$$(1) = \text{sum of } y_1 \text{ through } y_4$$
$$= 85 + 41 + 92 + 45 = 263$$
$$(2) = \text{sum of } y_2 \text{ through } y_5$$
$$= 41 + 92 + 45 + 90 = 268$$
$$(3) = \text{sum of } y_3 \text{ through } y_6$$
$$= 92 + 45 + 90 + 43 = 270$$

and so on.

* An example using monthly data is contained in Section 15.6.

TIME	QUARTER	t	y_t	MOVING TOTALS
1981	1	1	85	(1) 263
	2	2	41	(2) 268
	3	3	92	(3) 270
	4	4	45	and so on
1982	1	5	90	
	2	6	43	
	3	7	95	
	4	8	47	
1983	1	9	92	
	⋮	⋮	⋮	

Because each total contains four observations (one from each quarter), any quarterly seasonal effects have been removed. Consequently, there is no seasonality within the moving totals 263, 268, 270, and so on in Table 15.2.

The first moving total in Table 15.2 is equal to $(y_1 + y_2 + y_3 + y_4)$. If we were to position this total in the center of these values, it would lie between $t = 2$ and $t = 3$, at $t = 2.5$. The second moving total is equal to $(y_2 + y_3 + y_4 + y_5)$; again, we position this total in the center between $t = 3$ and $t = 4$, at $t = 3.5$.

We then add the first two moving totals. Notice that four values went into each of these totals, so that a total of *eight* values makes up this sum. The sum of first two moving totals is $263 + 268 = 531$. The average for the 8 months in the first two moving totals is $531/8 = 66.38$. This is a **centered moving average**. The position of this moving average is midway between $t = 2.5$ and $t = 3.5$, at $t = 3$. We therefore conclude that 66.38 is the centered moving average corresponding to $t = 3$.

EXAMPLE 15.7

Continue the procedure using Table 15.2 and determine the centered moving average for (1) $t = 4$ and (2) $t = 5$.

SOLUTION 1

Here we obtain

$$268 = y_2 + y_3 + y_4 + y_5$$

(positioned at $t = 3.5$) and

$$270 = y_3 + y_4 + y_5 + y_6$$

(positioned at $t = 4.5$) So the average of the eight numbers making up $268 + 270 = 538$ would be positioned midway between 3.5 and 4.5, at $t = 4$. Consequently, the centered moving average for $t = 4$ is

$$\frac{268 + 270}{8} = 67.25$$

SOLUTION 2

Proceeding as before,

$$270 = y_3 + y_4 + y_5 + y_6$$

(positioned at $t = 4.5$) and

$$273 = y_4 + y_5 + y_6 + y_7$$

(positioned at $t = 5.5$) Therefore, the centered moving average for $t = 5$ is

$$\frac{270 + 273}{8} = 67.88$$

■

TABLE 15.3

Sales Data for
Video-Comp (Millions of
Dollars)

YEAR	QUARTER 1	QUARTER 2	QUARTER 3	QUARTER 4
1985	20	12	47	60
1986	40	32	65	76
1987	56	50	85	100
1988	75	70	101	123

Assume quarterly sales data at Video-Comp were recorded over a 4-year period. We now want to determine the centered moving averages for these data, shown in Table 15.3. There appears to be a definite seasonal effect within this time series; the highest sales occur in the fourth quarter of each year. Table 15.4 shows the centered moving averages for these data. The first *moving total* is

$$139 = y_1 + y_2 + y_3 + y_4$$
$$= 20 + 12 + 47 + 60$$

Its actual location is $t = 2.5$; it is positioned between $t = 2$ and $t = 3$. Similarly, the next moving total is centered at $t = 3.5$ and so appears between $t = 3$ and $t = 4$ in the table. This total is

$$159 = y_2 + y_3 + y_4 + y_5$$
$$= 12 + 47 + 60 + 40$$

Each moving total is centered midway between the values making up this total. For example, the last moving total, 369, is centered between $t = 14$ and $t = 15$ at $t = 14.5$. Here,

$$369 = y_{13} + y_{14} + y_{15} + y_{16}$$
$$= 75 + 70 + 101 + 123$$

The *centered moving average* at time t is the average of the moving total immediately preceding this time value and the total immediately following it. This means that, for $t = 3$,

$$37.25 = \frac{139 + 159}{8}$$

For $t = 4$,

$$42.25 = \frac{159 + 179}{8}$$

and so on. Consequently, for $t = 3$, $y_3 = 47$ and the centered moving average is 37.25.

This procedure produces 12 centered moving averages; we are unable to compute this value for $t = 1, 2, 15,$ or 16. Notice that the first two values of t are for quarters 1 and 2, whereas the remaining two correspond to quarters 3 and 4. In general, if our time series contains T observations, we can derive $T - 4$ centered moving averages using quarterly data or $T - 12$ averages for monthly data.

The moving totals and centered moving averages are formed by summing over the four quarters (seasons), so there is no seasonality present in these values. Furthermore, the irregular component has been reduced because averages always contain less random variation (noise) than do the individual values making up these averages. Representing this reduced irregular activity component as i_t (rather than I_t), we can represent a centered moving average at

TABLE 15.4

Moving Averages for
Video-Comp Sales Data

YEAR	QUARTER	t	y_t	MOVING TOTAL	CENTERED MOVING AVERAGE	RATIO TO MOVING AVERAGE
1985	1	1	20	—	—	—
	2	2	12		—	—
				139		
	3	3	47		37.25	1.26
				159		
	4	4	60		42.25	1.42
				179		
1986	1	5	40		47.00	.85
				197		
	2	6	32		51.25	.62
				213		
	3	7	65		55.25	1.18
				229		
	4	8	76		59.50	1.28
				247		
1987	1	9	56		64.25	.87
				267		
	2	10	50		69.75	.72
				291		
	3	11	85		75.13	1.13
				310		
	4	12	100		80.00	1.25
				330		
1988	1	13	75		84.50	.89
				346		
	2	14	70		89.38	.78
				369		
	3	15	101		—	—
	4	16	123	—	—	—

time t as

$$\text{centered moving average at time } t = TR_t \cdot C_t \cdot i_t$$

Because of this averaging procedure, the moving averages contain much less irregular activity and so are much "smoother" than the original time series. This procedure thus is referred to as **smoothing** the time series to get a clearer picture of any existing trend as well as of its shape (straight line or curve).

The centered moving averages in Table 15.4 show a steadily increasing trend. Because the differences between any two adjacent moving averages are nearly the same, this trend is very *linear*. This is more apparent in Figure 15.19, which contains the original data with the moving averages.

To determine the four quarterly seasonal indexes, the first step is to divide each observation, y_t, by its corresponding moving average (last column in the table).

FIGURE 15.19

Smoothing a time series
using moving averages
(Video-Comp sales
data).

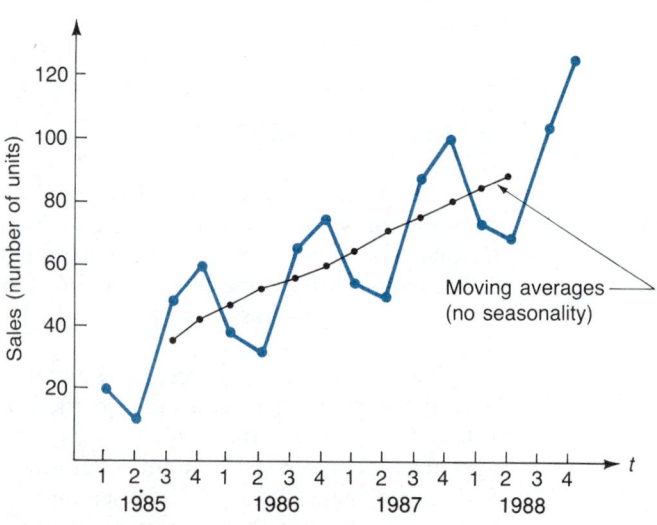

for $t = 3$: ratio $= 47/37.25 = 1.26$ (belongs to quarter 3, 1985)

for $t = 4$: ratio $= 60/42.25 = 1.42$ (belongs to quarter 4, 1985)

$\vdots$

for $t = 14$: ratio $= 70/89.38 = .78$ (belongs to quarter 2, 1988)

When we divide y_t by its corresponding centered moving average, we obtain

$$\text{ratio} = \frac{y_t}{\text{centered moving average}} = \frac{\cancel{TR_t} \cdot S_t \cdot \cancel{C_t} \cdot I_t}{\cancel{TR_t} \cdot \cancel{C_t} \cdot i_t}$$
$$= S_t \cdot I_t$$

Consequently, these ratios contain the seasonal effects as well as the irregular activity (noise) components. The following discussion illustrates how you can reduce the effect of the irregular activity factor by combining these ratios into a set of four seasonal indexes, one for each quarter.

Computing a Seasonal Index

The purpose of a seasonal index is to indicate how the time series value for each quarter (or month) compares with the average for the year. The following discussion will assume that we are dealing with a time series containing quarterly data. In the next section, we illustrate this procedure using monthly data.

We begin by collecting the ratios to moving average, placing each of them in its respective quarter. In Table 15.4, we see that 1.26 belongs to quarter 3, 1.42 to quarter 4, .85 to quarter 1, and so on. Table 15.5 is the result. Notice that there are three ratios for each quarter. In general, you always will obtain (total number of years -1) ratios under each quarter (or month). The time series in this example contains 4 years; therefore, it has three ratios. To obtain a "typical" ratio for each quarter, you have several options, including

1. Determine an average of these ratios.
2. Find the median of these values.
3. Eliminate the largest and smallest ratio within each quarter and compute a mean of the remaining ratios; this is called a **trimmed mean**.

We will follow the first procedure and calculate a mean ratio for each quarter, as illustrated in Table 15.5. When the time series contains 5 or more years of data, a trimmed mean offers you protection against an outlier ratio dominating the index for this quarter. Using the median ratios also helps guard against this type of situation.

TABLE 15.5
Ratios for Each Quarter

	QUARTER 1	QUARTER 2	QUARTER 3	QUARTER 4
	—	—	1.26	1.42
	.85	.62	1.18	1.28
	.87	.72	1.13	1.25
	.89	.78	—	—
TOTAL	2.61	2.12	3.57	3.95
AVERAGE	.870	.707	1.190	1.317

A Final Adjustment

The last step in computing the seasonal indexes is to make sure that the four computed ratio averages **sum to 4** (or 12, for monthly indexes). This is accomplished by (1) adding the four averages computed in the table (call this SUM) and (2) multiplying each average by 4/SUM. The modified average is the seasonal index for that quarter.

EXAMPLE 15.8   Using Table 15.5, determine the four seasonal indexes.

SOLUTION   First,

$$SUM = .870 + .707 + 1.190 + 1.317$$
$$= 4.084$$

This means that we need to multiply each of the four averages in Table 15.5 by $4/4.084 = .9794$.

QUARTER	SEASONAL INDEX
1	$(\ .870)(.9794) = \ \ .852$
2	$(\ .707)(.9794) = \ \ .692$
3	$(1.190)(.9794) = 1.166$
4	$(1.317)(.9794) = \underline{1.290}$
	4.0

The indexes for quarters 1 and 2 are below 1.0, so the sales during these quarters typically are below the yearly average. On the other hand, quarters 3 and 4 have seasonal indexes of 1.166 and 1.290, so the sales for these quarters are higher than the average for the year. ■

This procedure for determining seasonal effects works well, provided the ratios in Table 15.5 are reasonably *stable*. In Example 15.8, all the ratios for quarter 2 are small (near .7) and all the ratios for quarter 4 are large (near 1.3). If strong seasonality is present, such will be the case.

Seasonal indexes can be updated as you obtain an additional year's observations on the variable of interest. You have the option of deleting the most distant year's observations prior to recalculating these values. This procedure leads to seasonal indexes that change slowly over the years.

In summary, to calculate the seasonal indexes:

1. Derive the *moving* totals by summing the observations for four (quarterly data) or 12 (monthly data) consecutive time periods.
2. Average and center the totals by finding the *centered moving averages*.
3. Divide each observation by its corresponding centered moving average.
4. Place the ratios from step 3 in a table headed by the 4 quarters or 12 months.
5. For each column in this table, determine the mean of these ratios; these are the unadjusted seasonal indexes.
6. Make a final adjustment to guarantee that the final seasonal indexes sum to 4 (quarterly data) or 12 (monthly data); these adjusted means are the seasonal indexes.

Deseasonalizing the Data

To remove the seasonality from the data, we deseasonalize the time series. The resulting series contains no seasonal effects and consists of the trend, cyclical activity, and, of course, irregular activity. We write deseasonalized data as d_t.

TABLE 15.6

Deseasonalized Sales

YEAR	t	y_t	SEASONAL INDEX (S_t)	DESEASONALIZED VALUES $d_t = y_t/S_t$
1985	1	20	.852	23.47
	2	12	.692	17.34
	3	47	1.166	40.31
	4	60	1.290	46.51
1986	5	40	.852	46.95
	6	32	.692	46.24
	7	65	1.166	55.75
	8	76	1.290	58.91
1987	9	56	.852	65.73
	10	50	.692	72.25
	11	85	1.166	72.90
	12	100	1.290	77.52
1988	13	75	.852	88.03
	14	70	.692	101.16
	15	101	1.166	86.62
	16	123	1.290	95.35

$$d_t = \frac{y_t}{\text{corresponding seasonal index}}$$
$$= \frac{TR_t \cdot \cancel{S_t} \cdot C_t \cdot I_t}{\cancel{S_t}} = TR_t \cdot C_t \cdot I_t$$

The deseasonalized sales values from Table 15.3 are contained in Table 15.6. These values contain trend, cyclical effects, and irregular activity. Notice how the trend is much more apparent in the deseasonalized values than in the original time series.

In Table 15.6, we obtained deseasonalized values for all 16 of the original observations, including the two quarters on each end. We will use the "new" deseasonalized series to determine the trend and cyclical components of the original time series. This will be illustrated in the next section, where we apply the four-step procedure by (1) computing seasonal indexes, (2) deseasonalizing the data, (3) computing the trend components from the deseasonalized time series (d_t), and, finally, (4) calculating the cyclical activity.

EXERCISES

15.35 Mid-Cities Appliance store is interested in determining an approximate inventory level in dollars to control its overhead. Quarterly inventory levels (in ten thousands) for 5 years are:

YEAR	QUARTER 1	QUARTER 2	QUARTER 3	QUARTER 4
1984	1	7	12	6
1985	4	11	17	9
1986	9	14	20	14
1987	12	16	24	18
1988	16	21	27	22

Find the four-quarter centered moving averages and ratios to moving average.

15.36 Several counties in Oregon and Washington depend heavily on the lumber industry. When there is little demand for lumber, the softness in the industry causes unemployment to increase. The following data represent the unemployment percentage of workers for certain counties in Oregon and Washington.

YEAR	JAN.	FEB.	MAR.	APR.	MAY	JUN.	JUL.	AUG.	SEP.	OCT.	NOV.	DEC.
1985	10.8	9.6	8.7	7.5	6.4	5.4	6.1	7.3	8.5	8.9	10.1	10.9
1986	9.6	8.5	7.5	6.3	5.2	4.1	5.9	6.1	7.4	8.1	9.8	9.8
1987	6.9	7.4	6.3	7.5	4.9	3.7	4.2	5.4	6.5	7.7	8.0	9.1
1988	7.0	6.7	5.7	4.2	3.3	2.8	3.4	4.0	4.4	5.3	6.7	7.4

Find the 12-month centered moving averages and ratios to moving average.

15.37 Explain why a moving average is a smoothing technique.

15.38 The following table presents the ratio to moving average figure for sales at Zano Systems, a supplier of photocopy machines. Find the seasonal indexes.

YEAR	QUARTER 1	QUARTER 2	QUARTER 3	QUARTER 4
1983			.88	.87
1984	1.14	1.25	.83	.86
1985	1.19	1.22	.94	.88
1986	1.23	1.35	.90	.72
1987	1.16	1.32	.94	.81
1988	1.10	1.21		

15.39 The following table presents the ratio to moving average figure for the cost of a bushel of grapefruit in a certain county in Florida. Find the seasonal indexes.

YEAR	JAN.	FEB.	MAR.	APR.	MAY	JUN.	JUL.	AUG.	SEP.	OCT.	NOV.	DEC.
1984							1.06	1.10	1.12	1.02	1.03	.99
1985	.90	.87	.95	.93	1.00	1.04	1.08	1.14	1.15	1.06	1.04	.97
1986	.87	.84	.81	.88	1.01	1.02	1.01	1.15	1.07	1.03	1.00	.90
1987	.81	.75	.82	.89	1.05	1.04	1.10	1.21	1.18	1.10	1.07	.97
1988	.87	.81	.77	.98	1.01	1.06						

15.40 The sale of grass sod is a seasonal business. Green Garden Supplies does most of its business in May, June, July, and August. The following table presents the monthly sales (in thousands of dollars) of Green. Find the seasonal indexes. For what month is the seasonal index the largest?

YEAR	JAN.	FEB.	MAR.	APR.	MAY	JUN.	JUL.	AUG.	SEP.	OCT.	NOV.	DEC.
1984	.1	.1	1.2	2.2	4.1	4.5	5.5	5.3	3.5	1.1	.2	.1
1985	.1	.2	1.4	2.0	4.0	4.2	5.3	5.0	3.2	1.0	.1	.1
1986	.1	.2	1.3	2.2	4.3	4.4	5.6	5.3	3.5	1.1	.2	.1
1987	.1	.3	1.4	2.3	4.4	4.6	5.8	5.5	3.7	1.3	.3	.1
1988	.1	.3	1.5	2.3	4.6	4.8	6.0	5.6	3.7	1.4	.4	.1

15.41 The following table represents the ratio to moving average figure for the number of people below the poverty level in a certain county. Find the seasonal indexes.

YEAR	QUARTER 1	QUARTER 2	QUARTER 3	QUARTER 4
1984			.84	.83
1985	1.12	1.29	.91	.89
1986	1.17	1.24	.92	.90
1987	1.15	1.30	.92	.88
1988	1.13	1.26		

15.42 Seaside University has four quarterly semesters during the school year. Enrollment for each of these quarters is as follows. Find the seasonal indexes.

YEAR	QUARTER 1	QUARTER 2	QUARTER 3	QUARTER 4
1984	9,385	9,020	9,350	9,060
1985	9,970	9,671	9,928	9,701
1986	10,328	9,950	10,121	9,922
1987	10,411	9,995	10,250	9,998
1988	10,535	10,240	10,506	10,279

15.43 A major department store usually has a strong fourth quarter because of the holiday season. Earnings per share of the company are as follows. Find the seasonal indexes.

YEAR	QUARTER 1	QUARTER 2	QUARTER 3	QUARTER 4
1984	.75	.60	.80	1.40
1985	.80	.55	.82	1.51
1986	.83	.59	.81	1.63
1987	.84	.62	.83	1.75
1988	.84	.61	.82	1.79

15.44 Moody's average of yields on Aa corporate bonds, in percent, are given below, from 1981 through 1984. Find the seasonal indexes.

MOODY'S AVERAGE OF YIELDS ON Aa CORPORATE BONDS (IN PERCENT)

YEAR	AVER.	JAN.	FEB.	MAR.	APR.	MAY	JUNE	JULY	AUG.	SEPT.	OCT.	NOV.	DEC.
1984	13.31	12.71	12.70	13.22	13.48	14.10	14.33	14.12	13.47	13.27	13.11	12.66	12.50
1983	12.42	12.35	12.58	12.32	12.06	11.95	12.15	12.39	12.72	12.62	12.49	12.61	12.76
1982	14.41	15.75	15.72	15.21	14.90	14.77	15.26	15.21	14.18	13.72	12.97	12.51	12.44
1981	14.75	13.52	13.89	13.90	14.39	14.88	14.41	14.79	15.42	15.95	15.82	14.97	15.00

(*Source: Moody's Industrial Manual* 1 (1987): a33.)

15.6

A TIME SERIES CONTAINING SEASONALITY, TREND, AND CYCLES

During the summer of 1988, the owner of an import/export company decided to investigate the past behavior of U.S. retail trade figures for the years 1984 through 1987. He collected the data in Table 15.7 using monthly figures released by the U.S. Department of Commerce. He suspected that these data would indicate high retail trade during December (due to holiday sales) with much lower activity during January and, possibly, February. For the remaining months, he had no idea whether seasonal effects would be present or not. He also suspected there would be a steadily increasing trend, due to inflation and population growth.

We will perform a decomposition of the data in Table 15.7 and discuss the

TABLE 15.7
Total U.S. Retail Trade
(Sales and Inventories)
(Billions of Dollars)

	1984	1985	1986	1987
J	93.09	98.82	105.64	106.39
F	93.69	95.59	99.66	105.80
M	104.29	110.17	114.24	120.44
A	104.34	113.11	115.71	125.37
M	111.31	120.34	125.42	129.07
J	111.98	114.96	120.35	128.98
J	106.55	115.49	120.74	128.95
A	110.65	121.12	124.06	131.02
S	103.93	114.17	124.65	123.77
O	109.23	116.14	123.06	127.21
N	113.28	118.56	120.79	125.38
D	131.65	139.40	151.26	154.75

Source: *Current Business Reports* (Vols. BR-84-12, BR-85-12, BR-86-12, BR-87-12), U.S. Department of Commerce, Bureau of the Census.

results. The four-step procedure for decomposing (a gruesome term, we'll admit) a time series into the seasonal, trend, and cyclical components was introduced in Section 15.4. We demonstrate this method of describing a time series using the monthly retail trade data.

TABLE 15.8

Moving Averages for Monthly Retail Trade

YEAR	MONTH	(1) t	(2) y_t	(3) MOVING TOTAL	(4) CENTERED MOVING AVERAGE	(5) RATIO TO MOVING AVERAGE
1984	JAN.	1	93.09			
	FEB.	2	93.69			
	MAR.	3	104.29			
	APR.	4	104.34			
	MAY	5	111.31			
	JUN.	6	111.98	1293.99		
	JUL.	7	106.55	1299.72	108.07	0.986
	AUG.	8	110.65	1301.62	108.39	1.021
	SEP.	9	103.93	1307.50	108.71	0.956
	OCT.	10	109.23	1316.27	109.32	0.999
	NOV.	11	113.28	1325.30	110.07	1.029
	DEC.	12	131.65	1328.28	110.57	1.191
1985	JAN.	13	98.82	1337.22	111.06	0.890
	FEB.	14	95.59	1347.69	111.87	0.854
	MAR.	15	110.17	1357.93	112.73	0.977
	APR.	16	113.11	1364.84	113.45	0.997
	MAY	17	120.34	1370.12	113.96	1.056
	JUN.	18	114.96	1377.87	114.50	1.004
	JUL.	19	115.49	1384.69	115.11	1.003
	AUG.	20	121.12	1388.76	115.56	1.048
	SEP.	21	114.17	1392.83	115.90	0.985
	OCT.	22	116.14	1395.43	116.18	1.000
	NOV.	23	118.56	1400.51	116.50	1.018
	DEC.	24	139.40	1405.90	116.93	1.192
1986	JAN.	25	105.64	1411.15	117.38	0.900
	FEB.	26	99.66	1414.09	117.72	0.847
	MAR.	27	114.24	1424.57	118.28	0.966
	APR.	28	115.71	1431.49	119.00	0.972
	MAY	29	125.42	1433.72	119.38	1.051
	JUN.	30	120.35	1445.58	119.97	1.003
	JUL.	31	120.74	1446.33	120.50	1.002
	AUG.	32	124.06	1452.47	120.78	1.027
	SEP.	33	124.65	1458.67	121.30	1.028
	OCT.	34	123.06	1468.33	121.96	1.009
	NOV.	35	120.79	1471.98	122.51	0.986
	DEC.	36	151.26	1480.61	123.02	1.230
1987	JAN.	37	106.39	1488.82	123.73	0.860
	FEB.	38	105.80	1495.78	124.36	0.851
	MAR.	39	120.44	1494.90	124.61	0.967
	APR.	40	125.37	1499.05	124.75	1.005
	MAY	41	129.07	1503.64	125.11	1.032
	JUN.	42	128.98	1507.13	125.45	1.028
	JUL.	43	128.95			
	AUG.	44	131.02			
	SEP.	45	123.77			
	OCT.	46	127.21			
	NOV.	47	125.38			
	DEC.	48	154.75			

Step 1. *Determine the seasonal indexes.* The first step is to determine the moving totals and centered moving averages for the 48 observations in Table 15.7. These calculations are summarized in Table 15.8. Notice that, for the monthly data, there is no moving average for $t = 1$ through $t = 6$ (months 1 through 6, 1984) and for $t = 43$ through $t = 48$ (months 7 through 12, 1987). The first moving total is

$$1293.99 = y_1 + y_2 + \cdots + y_{12}$$
$$= 93.09 + 93.69 + \cdots + 131.65$$

This value is positioned midway between $t = 1$ and $t = 12$, at $t = 6.5$. The next moving total is

$$1299.72 = y_2 + y_3 + \cdots + y_{13}$$
$$= 93.69 + 104.29 + \cdots + 98.82$$

which is centered at $t = 7.5$. So the first moving *average* is centered midway between $t = 6.5$ and $t = 7.5$, at $t = 7$. This is

$$108.07 = \frac{1293.99 + 1299.72}{24}$$

Notice that we divide by 24 because 24 observations went into the sum of these two moving totals.

The final moving average is

$$125.45 = \frac{1503.64 + 1507.13}{24}$$

and corresponds to $t = 42$.

Table 15.8 also contains each ratio to moving average (column 4 divided by column 2). To illustrate,

$$.986 = 106.55/108.07$$
$$1.021 = 110.65/108.39$$

and so on. These ratios are summarized in Table 15.9, which also shows the average of the three values for each time period.

The final step is to adjust each of the averages in Table 15.9 so that they sum to 12 (because there are 12 time periods per year). Here,

$$\text{SUM} = .883 + .851 + \cdots + 1.204 = 11.990$$

and so

$$S_1 = \text{seasonal index for January}$$
$$= (.883) \cdot (12/11.99) = .88$$
$$S_2 = \text{seasonal index for February}$$
$$= (.851) \cdot (12/11.99) = .85$$
$$\vdots$$
$$S_{12} = \text{seasonal index for December}$$
$$= (1.204) \cdot (12/11.99) = 1.21$$

The final collection of seasonal indexes follows Table 15.9.

TABLE 15.9   Summary of Ratios

	MONTH											
Period	1	2	3	4	5	6	7	8	9	10	11	12
Year												
1							0.986	1.021	0.956	0.999	1.029	1.191
2	0.890	0.854	0.977	0.997	1.056	1.004	1.003	1.048	0.985	1.000	1.018	1.192
3	0.900	0.847	0.966	0.972	1.051	1.003	1.002	1.027	1.028	1.009	0.986	1.230
4	0.860	0.851	0.967	1.005	1.032	1.028						
Average	0.883	0.851	0.970	0.991	1.046	1.012	0.997	1.032	0.990	1.003	1.011	1.204

MONTH	SEASONAL INDEX	MONTH	SEASONAL INDEX
Jan.	.88	Jul.	1.00
Feb.	.85	Aug.	1.03
Mar.	.97	Sep.	.99
Apr.	.99	Oct.	1.00
May	1.05	Nov.	1.01
Jun.	1.01	Dec.	1.21

The sum of the seasonal indexes $(S_1 + S_2 + \cdots + S_{12})$ is still 11.99, which, due to the rounding of these values, is perfectly acceptable.

We observe (1) a large seasonal index for December $(S_{12} = 1.21)$ indicating large retail trade for this month, (2) low indexes for January and February, and (3) very little seasonality for any of the remaining months.

Step 2. *Deseasonalize the data.* We obtain the deseasonalized values (which contain no seasonality) by dividing each observation by its corresponding seasonal index. These values are shown in Table 15.10. The trend is more apparent now because the deseasonalized values tend to increase over time.

Step 3. *Determine the trend components.* A common method (and the one we use) of estimating trend is to construct a least squares trend line (or curve) through the deseasonalized data. From the moving averages in Table 15.8, it appears that a straight line trend equation will be appropriate; these values tend to increase at a fairly steady rate.

The calculations for the trend line are identical to those discussed in Section 15.2, using the d_t values in place of the original observations, y_t. A summary of this procedure is given in Table 15.11. The least squares line through the deseasonalized data is given by

$$TR_t = \hat{d}_t = b_0 + b_1 t$$

where

$$b_1 = \frac{\Sigma \, td_t - (\Sigma \, t)(\Sigma \, d_t)/T}{\Sigma \, t^2 - (\Sigma \, t)^2/T}$$

$$= \frac{142{,}148.19 - (1176)(5620.18)/48}{38{,}024 - (1176)^2/48} = \frac{4453.78}{9212} = .4835$$

and

$$b_0 = \bar{d}_t - b_1 \bar{t}$$

$$= \frac{5620.18}{48} - (.4835)\left(\frac{1176}{48}\right)$$

$$= 117.09 - 11.85 = 105.24$$

TABLE 15.10

Deseasonalized Monthly
Retail Trade Values

YEAR	MONTH	t	y_t	S_t	$d_t = y_t/S_t$
1984	JAN.	1	93.09	0.88	105.30
	FEB.	2	93.69	0.85	110.05
	MAR.	3	104.29	0.97	107.43
	APR.	4	104.34	0.99	105.15
	MAY	5	111.31	1.05	106.31
	JUN.	6	111.98	1.01	110.58
	JUL.	7	106.55	1.00	106.77
	AUG.	8	110.65	1.03	107.12
	SEP.	9	103.93	0.99	104.93
	OCT.	10	109.23	1.00	108.85
	NOV.	11	113.28	1.01	111.95
	DEC.	12	131.65	1.21	109.24
1985	JAN.	13	98.82	0.88	111.79
	FEB.	14	95.59	0.85	112.28
	MAR.	15	110.17	0.97	113.49
	APR.	16	113.11	0.99	113.98
	MAY	17	120.34	1.05	114.94
	JUN.	18	114.96	1.01	113.52
	JUL.	19	115.49	1.00	115.72
	AUG.	20	121.12	1.03	117.26
	SEP.	21	114.17	0.99	115.27
	OCT.	22	116.14	1.00	115.73
	NOV.	23	118.56	1.01	117.17
	DEC.	24	139.40	1.21	115.67
1986	JAN.	25	105.64	0.88	119.50
	FEB.	26	99.66	0.85	117.06
	MAR.	27	114.24	0.97	117.68
	APR.	28	115.71	0.99	116.60
	MAY	29	125.42	1.05	119.79
	JUN.	30	120.35	1.01	118.84
	JUL.	31	120.74	1.00	120.98
	AUG.	32	124.06	1.03	120.10
	SEP.	33	124.65	0.99	125.85
	OCT.	34	123.06	1.00	122.63
	NOV.	35	120.79	1.01	119.38
	DEC.	36	151.26	1.21	125.51
1987	JAN.	37	106.39	0.88	120.35
	FEB.	38	105.80	0.85	124.27
	MAR.	39	120.44	0.97	124.07
	APR.	40	125.37	0.99	126.34
	MAY	41	129.07	1.05	123.28
	JUN.	42	128.98	1.01	127.37
	JUL.	43	128.95	1.00	129.21
	AUG.	44	131.02	1.03	126.84
	SEP.	45	123.77	0.99	124.96
	OCT.	46	127.21	1.00	126.77
	NOV.	47	125.38	1.01	123.91
	DEC.	48	154.75	1.21	128.40

Consequently, the trend equation is given by

$$TR_t = \hat{d}_t$$
$$= 105.24 + .4835t$$

This equation implies that, apart from seasonal fluctuations, the U.S. retail trade is increasing at an average rate of $483.5 million each

TABLE 15.11

Calculations for Trend Line (U.S. Monthly Retail Trade Data)

t	d_t	$t \cdot d_t$	t^2
1	105.30	105.30	1
2	110.05	220.10	4
3	107.43	322.29	9
4	105.15	420.60	16
⋮	⋮	⋮	⋮
45	124.96	5,623.20	2,025
46	126.77	5,831.42	2,116
47	123.91	5,823.77	2,209
48	128.40	6,163.20	2,304
1176	5620.18	142,148.19	38,024

month. A graph of the deseasonalized data and corresponding trend line is shown in Figure 15.20. Also, a MINITAB solution for the trend line using the deseasonalized data as input is contained in Figure 15.21.

Step 4. *Determine the cyclical activity.* We begin by following the procedure outlined in Section 15.3. We divide each deseasonalized observation by the corresponding trend value,

$$\frac{d_t}{TR_t} = \frac{d_t}{\hat{d}_t} = \frac{\cancel{TR_t} \cdot C_t \cdot I_t}{\cancel{TR_t}} = C_t \cdot I_t$$

The resulting values contain cyclical effects as well as an irregular activity component. One method of reducing the irregular activity effect is to compute a series of *three-period* moving averages on the $C_t \cdot I_t$ values. This procedure greatly reduces the irregular activity effect, and the moving averages provide a much better estimate of the cyclical movement. The choice of using a three-period moving average is somewhat arbitrary, but when we use an odd number of terms, the moving averages need not be centered.

A partial solution is shown in Table 15.12. We see that the cyclical component for $t = 2$ is C_2, where

$$C_2 = \frac{.9960 + 1.0361 + 1.0069}{3} = 1.01$$

FIGURE 15.20

Deseasonalized data and trend line (monthly U.S. retail trade).

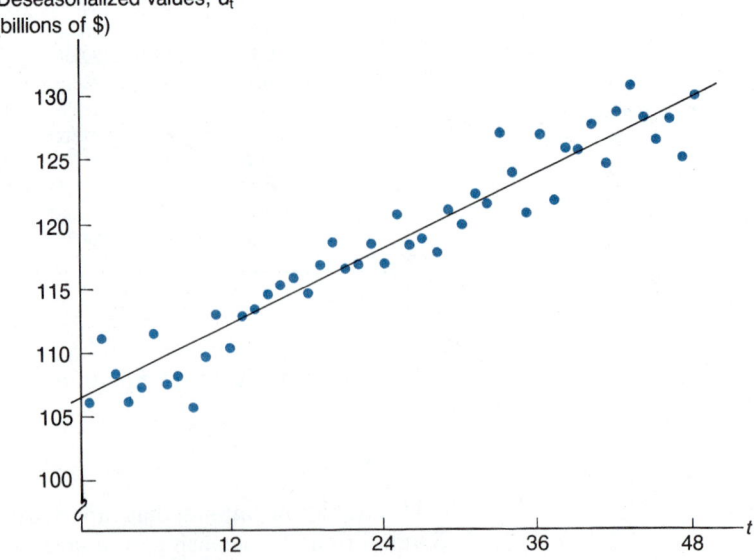

Deseasonalized values, d_t (billions of $)

FIGURE 15.21

MINITAB solution for trend line (deseasonalized monthly U.S. retail trade data).

```
MTB > SET INTO C1
DATA> 1:48  ──────────────┌─────────────────────┐
DATA> END                 │ This command generates │
MTB > SET INTO C2         │ integers 1 through 48  │
DATA> 105.30 110.05 107.43 105.15 106.31 110.58
DATA> 106.77 107.12 104.93 108.85 111.95 109.24
DATA> 111.79 112.28 113.49 113.98 114.94 113.52
DATA> 115.72 117.26 115.27 115.73 117.17 115.67
DATA> 119.50 117.06 117.68 116.60 119.79 118.84
DATA> 120.98 120.10 125.85 122.63 119.38 125.51
DATA> 120.35 124.27 124.07 126.34 123.28 127.37
DATA> 129.21 126.84 124.96 126.77 123.91 128.40
DATA> END
MTB > REGRESS Y IN C2 USING 1 PRED IN C1, RES IN C3, DHATS IN C4

The regression equation is
C2 = 105 + 0.483 C1

Predictor       Coef       Stdev      t-ratio        p
Constant     105.243       0.575       182.97    0.000
C1           0.48345     0.02044        23.66    0.000

s = 1.961       R-sq = 92.4%     R-sq(adj) = 92.2%

Analysis of Variance

SOURCE         DF          SS          MS         F         p
Regression      1       2153.1      2153.1    559.64    0.000
Error          46        177.0         3.8
Total          47       2330.1
```

and the cyclical component for $t = 3$ is

$$C_3 = \frac{1.0361 + 1.0069 + .9811}{3} = 1.01$$

Similarly,

$$C_4 = (1.0069 + .9811 + .9875)/3 = .99$$

and

$$C_5 = (.9811 + .9875 + 1.0225)/3 = 1.00$$

The complete set of cyclical components is contained in Table 15.13 and is plotted in Figure 15.22. Observe that for August, September, and October of 1984 and again in the latter part of 1987, the retail trade is in a bit of a downturn, as evidenced by the below-normal cyclical components.*

TABLE 15.12

Calculating the Cyclical Components for the U.S. Monthly Retail Trade Data

t	d_t	$\hat{d}_t$	$d_t/\hat{d}_t(C_t \cdot I_t)$	THREE-MONTH MOVING AVERAGE (C_i)
1	105.30	105.24 + .4835(1) = 105.72	.9960	—
2	110.05	105.24 + .4835(2) = 106.21	1.0361	1.01
3	107.43	105.24 + .4835(3) = 106.69	1.0069	1.01
4	105.15	105.24 + .4835(4) = 107.18	.9811	.99
5	106.31	105.24 + .4835(5) = 107.66	.9875	1.00
6	110.58	105.24 + .4835(6) = 108.14	1.0225	1.00
⋮	⋮	⋮	⋮	⋮

* In the first edition of this text, the U.S. retail trade data was examined from 1980 through 1983. For these data, a very clear cycle was observed between September 1980 and April 1983.

TABLE 15.13

Cyclical Components (Monthly U.S. Retail Trade Data)

YEAR	MONTH	d_t	$\hat{d}_t(TR_t)$	$d_t/\hat{d}_t(C_t \cdot I_t)$	THREE-MONTH MOVING AVERAGE (C_t)
1984	JAN.	105.30	105.72	0.9960	—
	FEB.	110.05	106.21	1.0361	1.01
	MAR.	107.43	106.69	1.0069	1.01
	APR.	105.15	107.18	0.9811	0.99
	MAY	106.31	107.66	0.9875	1.00
	JUN.	110.58	108.14	1.0225	1.00
	JUL.	106.77	108.63	0.9829	1.00
	AUG.	107.12	109.11	0.9818	0.97
	SEP.	104.93	109.59	0.9575	0.98
	OCT.	108.85	110.08	0.9888	0.99
	NOV.	111.95	110.56	1.0126	1.00
	DEC.	109.24	111.04	0.9837	1.00
1985	JAN.	111.79	111.53	1.0023	1.00
	FEB.	112.28	112.01	1.0024	1.00
	MAR.	113.49	112.49	1.0089	1.01
	APR.	113.98	112.98	1.0089	1.01
	MAY	114.94	113.46	1.0130	1.01
	JUN.	113.52	113.94	0.9963	1.01
	JUL.	115.72	114.43	1.0113	1.01
	AUG.	117.26	114.91	1.0204	1.01
	SEP.	115.27	115.39	0.9989	1.01
	OCT.	115.73	115.88	0.9988	1.00
	NOV.	117.17	116.36	1.0070	1.00
	DEC.	115.67	116.85	0.9899	1.01
1986	JAN.	119.50	117.33	1.0185	1.00
	FEB.	117.06	117.81	0.9936	1.00
	MAR.	117.68	118.30	0.9948	0.99
	APR.	116.60	118.78	0.9817	0.99
	MAY	119.79	119.26	1.0044	0.99
	JUN.	118.84	119.75	0.9925	1.00
	JUL.	120.98	120.23	1.0063	1.00
	AUG.	120.10	120.71	0.9949	1.01
	SEP.	125.85	121.20	1.0384	1.01
	OCT.	122.63	121.68	1.0078	1.01
	NOV.	119.38	122.16	0.9772	1.00
	DEC.	125.51	122.65	1.0233	0.99
1987	JAN.	120.35	123.13	0.9774	1.00
	FEB.	124.27	123.61	1.0053	0.99
	MAR.	124.07	124.10	0.9998	1.01
	APR.	126.34	124.58	1.0141	1.00
	MAY	123.28	125.06	0.9857	1.00
	JUN.	127.37	125.55	1.0145	1.01
	JUL.	129.21	126.03	1.0252	1.01
	AUG.	126.84	126.52	1.0026	1.00
	SEP.	124.96	127.00	0.9840	0.99
	OCT.	126.77	127.48	0.9944	0.98
	NOV.	123.91	127.97	0.9683	0.99
	DEC.	128.40	128.45	0.9997	—

A MINITAB solution for determining the cyclical components is shown in Figure 15.23. MINITAB computes the $C_t \cdot I_t$ components (cycles and irregular activity) for you. You can obtain the 3-month moving averages by using your calculator.

EXAMPLE 15.9   Once the steps in the previous sections have been completed, the various com-

FIGURE 15.22   Plot of cyclical activity (monthly U.S. retail trade data).

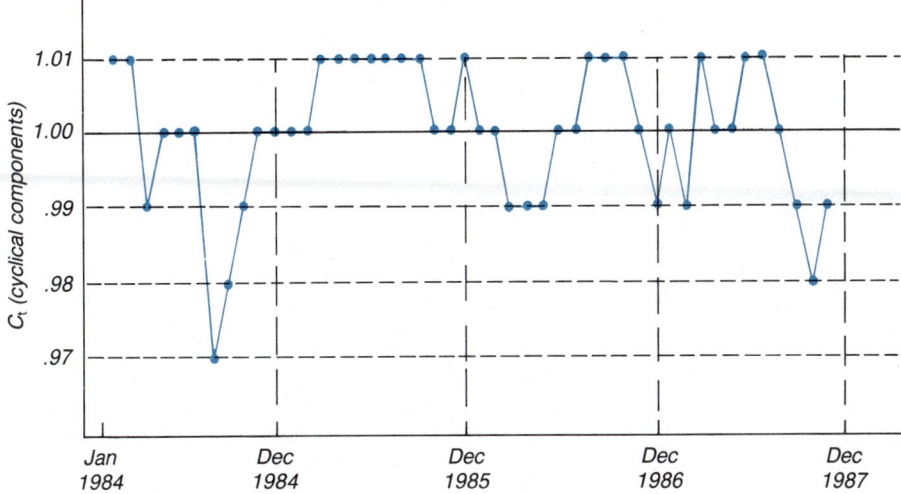

ponents can be combined for any specified value of t. Determine the four components for (1) September 1984 and (2) April 1987.

SOLUTION 1

The value of t for September 1984 is $t = 9$. The seasonal index for September is $S_9 = .99$. The trend component (from Table 15.13) is $TR_9 = 109.59$. The cyclical component (from Table 15.13) is $C_9 = .98$. The product of $S_9 \cdot TR_9 \cdot C_9 = (.99)(109.59)(.98) = 106.32$.

The actual observation during September 1984 is $y_9 = 103.93$. Since $y_9 = S_9 \cdot TR_9 \cdot C_9 \cdot I_9$,

$$I_9 = \frac{y_9}{S_9 \cdot TR_9 \cdot C_9} = \frac{103.93}{106.32} = .9775$$

and the final decomposition is

$$y_9 = 103.93 = S_9 \cdot TR_9 \cdot C_9 \cdot I_9 = (.99)(109.59)(.98)(.9775)$$

SOLUTION 2

For $t = 40$ (April 1987), we have S_{40} = seasonal index for April = $S_4 = .99$. Also, $TR_{40} = 124.58$ and $C_{40} = 1.00$ from Table 15.13. Consequently,

$$I_{40} = \frac{y_{40}}{S_{40} \cdot TR_{40} \cdot C_{40}} = \frac{125.37}{(.99)(124.58)(1.00)} = 1.0165$$

The combined decomposition for this observation is

$$y_{40} = 125.37 = S_{40} \cdot TR_{40} \cdot C_{40} \cdot I_{40} = (.99)(124.58)(1.00)(1.0165)$$ ■

FIGURE 15.23

MINITAB solution for $C_t \cdot I_t$ components (monthly U.S. retail trade data).

```
MTB > DIVIDE C2 BY C4, PUT INTO C6          C2 contains the d_t values
MTB > PRINT C6                              C4 contains the d̂_t values
                                            See Figure 15.21
C6
   0.99597   1.03616   1.00691   0.98109   0.98746   1.02253   0.98291
   0.98176   0.95744   0.98885   1.01257   0.98375   1.00235   1.00240
   1.00885   1.00887   1.01303   0.99627   1.01129   1.02044   0.99891
   0.99872   1.00694   0.98994   1.01850   0.99361   0.99479   0.98165
   1.00442   0.99243   1.00624   0.99492   1.03840   1.00781   0.97721
   1.02334   0.97742   1.00531   0.99978   1.01412   0.98573   1.01451
   1.02522   1.00257   0.98395   0.99442   0.96831   0.99962
```

Summary of Time Series Decomposition

The time series decomposition procedure allows you to examine the presence of

Trend (a long-term growth or decline)

Seasonality (a within-year recurrent pattern)

Cyclical activity (upward and downward variation about the trend)

TABLE 15.14

Time Series Components for U.S. Retail Trade Data

YEAR	MONTH	y_t	TR_t	S_t	C_t	I_t
1984	JAN.	93.09	105.72	0.88	—	—
	FEB.	93.69	106.21	0.85	1.01	1.03
	MAR.	104.29	106.69	0.97	1.01	1.00
	APR.	104.34	107.18	0.99	0.99	0.99
	MAY	111.31	107.66	1.05	1.00	0.98
	JUN.	111.98	108.14	1.01	1.00	1.03
	JUL.	106.55	108.63	1.00	1.00	0.98
	AUG.	110.65	109.11	1.03	0.97	1.02
	SEPT.	103.93	109.59	0.99	0.98	0.98
	OCT.	109.23	110.08	1.00	0.99	1.00
	NOV.	113.28	110.56	1.01	1.00	1.01
	DEC.	131.65	111.04	1.21	1.00	0.98
1985	JAN.	98.82	111.53	0.88	1.00	1.01
	FEB.	95.59	112.01	0.85	1.00	1.00
	MAR.	110.17	112.49	0.97	1.01	1.00
	APR.	113.11	112.98	0.99	1.01	1.00
	MAY	120.34	113.46	1.05	1.01	1.00
	JUN.	114.96	113.94	1.01	1.01	0.99
	JUL.	115.49	114.43	1.00	1.01	1.00
	AUG.	121.12	114.91	1.03	1.01	1.01
	SEPT.	114.17	115.39	0.99	1.01	0.99
	OCT.	116.14	115.88	1.00	1.00	1.00
	NOV.	118.56	116.36	1.01	1.00	1.01
	DEC.	139.40	116.85	1.21	1.01	0.98
1986	JAN.	105.64	117.33	0.88	1.00	1.02
	FEB.	99.66	117.81	0.85	1.00	1.00
	MAR.	114.24	118.30	0.97	0.99	1.01
	APR.	115.71	118.78	0.99	0.99	0.99
	MAY	125.42	119.26	1.05	0.99	1.01
	JUN.	120.35	119.75	1.01	1.00	1.00
	JUL.	120.74	120.23	1.00	1.00	1.00
	AUG.	124.06	120.71	1.03	1.01	0.99
	SEP.	124.65	121.20	0.99	1.01	1.03
	OCT.	123.06	121.68	1.00	1.01	1.00
	NOV.	120.79	122.16	1.01	1.00	0.98
	DEC.	151.26	122.65	1.21	0.99	1.03
1987	JAN.	106.39	123.13	0.88	1.00	0.98
	FEB.	105.80	123.61	0.85	0.99	1.02
	MAR.	120.44	124.10	0.97	1.01	0.99
	APR.	125.37	124.58	0.99	1.00	1.02
	MAY	129.07	125.06	1.05	1.00	0.98
	JUN.	128.98	125.55	1.01	1.01	1.01
	JUL.	128.95	126.03	1.00	1.01	1.03
	AUG.	131.02	126.52	1.03	1.00	1.01
	SEP.	123.77	127.00	0.99	0.99	0.99
	OCT.	127.21	127.48	1.00	0.98	1.02
	NOV.	125.38	127.97	1.01	0.99	0.98
	DEC.	154.75	128.45	1.21	—	—

The remaining component (what is left after removing the effect of these three factors) is irregular activity. Having determined these components, you are able to describe a particular time series by carefully examining and plotting the calculated components.

A summary of the components for the U.S. retail trade time series is contained in Table 15.14. The irregular activity components (I_t) are determined by continuing the procedure in Example 15.9. Graphs of these components are shown in Figure 15.24. Notice that the graph of the irregular activity components contains no obvious pattern, as we would expect. By combining the various graphs of the time series components into a single set of graphs (Figure

FIGURE 15.24   Illustration of time series components (monthly U.S. retail trade data).

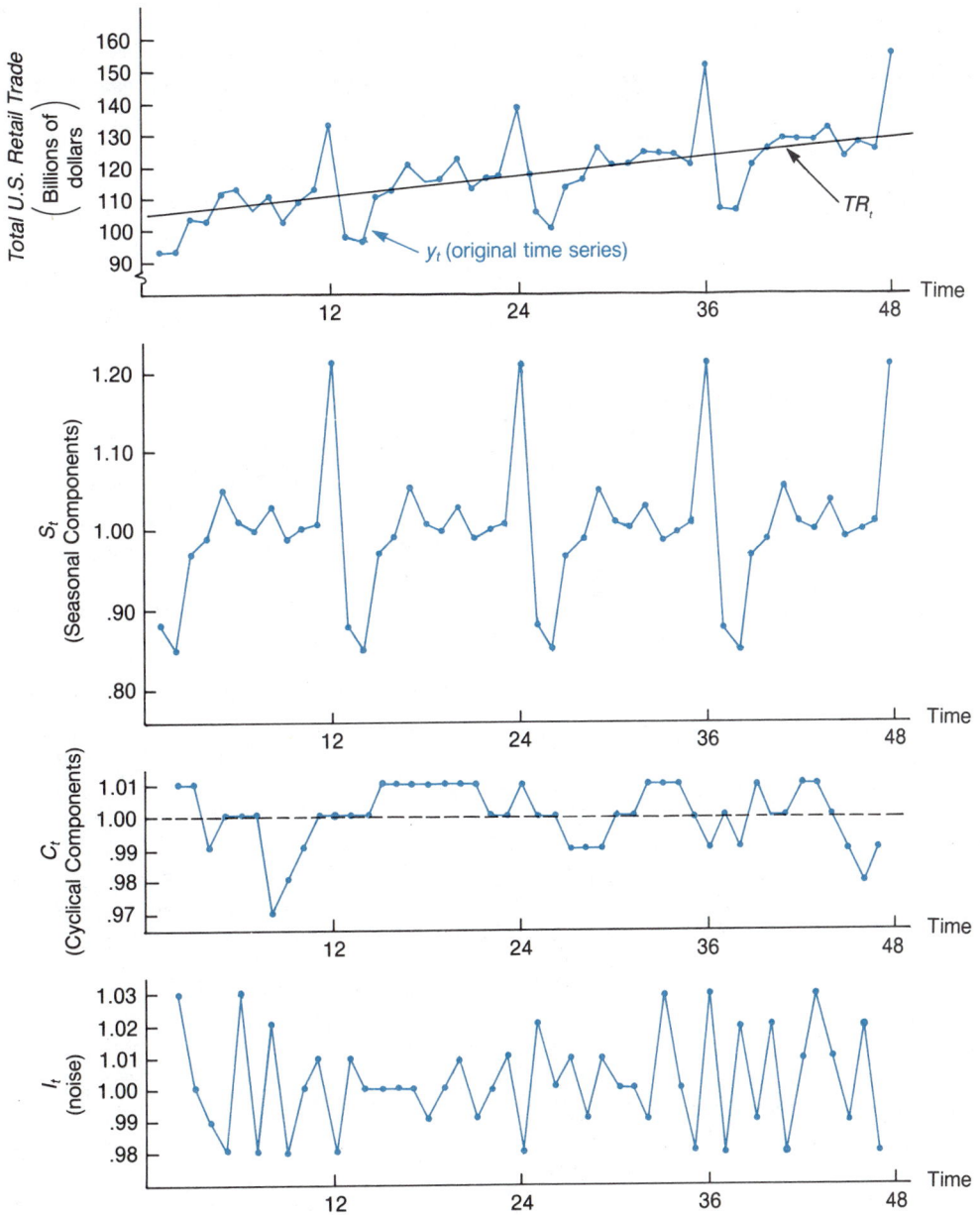

15.24), we can tell at a glance what is the nature of this series. The conclusions we can reach from this figure include

1.   There is a strong linear trend that increases over the 4-year period.
2.   There is a strong retail trade peak each December, followed by weak trading in January and February.
3.   There is a downturn in the retail trade during the latter months of 1984 and 1987.

As we discussed in Chapter 2, graphs offer an easy-to-comprehend method of summarizing data. It is time-consuming to construct this particular graph using pen and paper, but practically all microcomputer software packages have graph capabilities. Other methods of time series analysis are discussed in Chapter 16, where we examine time series forecasting.

EXERCISES

15.45   A real-estate broker, in order to understand the nature of the real estate market in a growing suburb of New Orleans, collected data from the past four years on the price per square foot for the houses that had three bedrooms, two baths, and a two-car garage. The data are as follows:

YEAR	JAN.	FEB.	MAR.	APR.	MAY	JUN.	JUL.	AUG.	SEP.	OCT.	NOV.	DEC.
1985	27.5	28.3	29.1	29.4	30.4	30.5	29.5	29.0	28.1	28.3	27.1	27.3
1986	27.1	28.6	29.5	30.4	31.4	31.8	31.3	30.7	30.4	29.5	29.6	29.0
1987	29.7	30.7	30.9	31.5	34.3	34.1	33.6	33.4	32.7	32.5	32.1	31.9
1988	31.8	32.7	35.6	36.1	36.8	36.7	35.4	36.5	35.0	34.8	34.7	34.7

a.   Determine the seasonal indexes.

b.   Determine the trend.

c.   Determine the cyclical components for 1987 using a three-period moving average.

d.   Determine the irregular components for the first 3 months of 1987.

15.46   For the month of July 1988, a researcher finds that the seasonal index is 1.14, the trend value is 65.4, and the cyclical component is .86. What is the irregular component if the actual observation for July 1988 is 80.42?

15.47   The following values represent deseasonalized observations. Find the corresponding trend values and cyclical components for each quarter.

YEAR	QUARTER 1	QUARTER 2	QUARTER 3	QUARTER 4
1984	1.79	1.77	1.75	1.68
1985	1.67	1.66	1.60	1.62
1986	1.63	1.71	1.72	1.70
1987	1.83	1.84	1.95	2.10
1988	2.13	2.11	2.15	2.16

15.48   Halston, a supplier of institutional food, sells to a restaurant in the metropolitan area of Memphis, Tennessee. Monthly data on sales (in ten thousands) are gathered over the past 5 years to help describe the company's growth pattern.

YEAR	JAN.	FEB.	MAR.	APR.	MAY	JUN.	JUL.	AUG.	SEP.	OCT.	NOV.	DEC.
1984	4.1	4.3	4.4	4.2	4.5	4.8	4.7	4.6	4.5	4.4	4.5	4.4
1985	4.3	3.6	4.7	4.9	5.3	5.5	5.9	5.7	5.5	5.3	5.2	5.0
1986	5.0	5.4	5.8	6.1	6.7	6.8	7.2	6.3	6.1	6.0	6.1	5.9
1987	5.8	6.3	6.7	6.9	7.4	7.9	8.5	7.9	7.8	7.7	7.4	7.3
1988	7.2	7.4	8.9	8.4	8.7	8.9	9.4	9.1	8.5	8.4	8.1	8.0

a.   Compute the seasonal indexes.

b. Calculate the cyclical components for 1987 using a 3-month moving average.

c. Calculate the irregular components for the last 3 months of 1987.

15.49 The manager of a large private golf course in southern California would like to examine the growth pattern of the number of golfers (given in hundreds) who would use the golf course. Monthly data were collected over a 4-year period.

YEAR	JAN.	FEB.	MAR.	APR.	MAY	JUN.	JUL.	AUG.	SEP.	OCT.	NOV.	DEC.
1985	4.3	4.4	4.7	4.5	4.8	5.1	5.4	5.6	5.7	5.0	4.7	4.7
1986	4.8	4.9	4.8	6.1	5.0	5.4	5.9	6.0	6.1	6.0	5.8	5.5
1987	5.0	5.1	5.2	5.3	5.5	5.7	6.1	6.3	6.1	5.9	5.7	5.3
1988	5.1	5.7	5.8	6.1	6.3	6.4	4.7	6.8	6.3	6.2	6.1	5.7

a. Determine the seasonal indexes.

b. Determine the trend.

c. Determine the cyclical components for 1987, using a five-period moving average.

d. Determine the irregular components for the first 3 months of 1987.

15.50 Monthly bond averages on high grade bonds for 1984 through 1986 follow.

YEAR	JAN.	FEB.	MAR.	APR.	MAY	JUN.	JUL.	AUG.	SEPT.	OCT.	NOV.	DEC.
1984	12.49	12.34	12.82	13.16	13.87	14.22	14.18	13.69	13.19	13.09	12.53	12.27
1985	12.17	12.03	12.29	12.04	11.77	11.05	11.01	11.12	11.13	11.03	10.75	10.69
1986	10.52	10.29	9.57	9.47	9.80	9.35	9.13	9.15	9.17	9.10	8.93	8.81

(*Source: Moody's Bank and Finance Manual* 2 (1987): a13.)

a. Determine the seasonal indexes.

b. Determine the trend component.

c. Determine the cyclical components using a three-period moving average.

d. Determine the irregular components for the first 3 months of 1986.

15.51 Find the cyclical components using a three-quarter moving average and the irregular components for the four quarters of 1985, using the data in Exercise 15.35.

15.52 Deseasonalize the data in Exercise 15.36. Find the cyclical components using a 3-month moving average for 1986. Find the irregular component for February of 1986.

15.7

INDEX NUMBERS

How many times have you heard a remark such as, "I can remember that 20 years ago we could have bought that house for $20,000. Now it's worth $120,000." Or, "My weekly grocery bill used to be $25. Today, it's almost $100." Many people like to talk about the prices back in the "good old days," but were goods and services actually less expensive in those days?

Perhaps a particular item consumed a greater proportion of the typical consumer's consumable income (purchasing power) in years past. To compare effectively the change in the price or value of a certain item (or group of items) between any two time periods, we use an index number. An **index number** (or index) measures the change in a particular item (typically a product or service) or a collection of items between two time periods.

The average hourly wage for production employees at Kessler Toy Company during 1970, 1975, 1980, and 1985 is shown in Table 15.15. Suppose that we wish to compare the average wages for 1975, 1980, and 1985 with those for 1970. By computing a ratio for each pair of wages (expressed as a *percentage*

TABLE 15.15

Average Hourly Wage of Production Employees at Kessler Toy Company

	1970	1975	1980	1985
WAGE	6.40	7.05	8.50	10.90
INDEX (base = 1970)	100	110.2	132.8	170.3

of the 1970 wage), we obtain the following set of index numbers:

$$\text{index number for 1975:} \quad \left(\frac{7.05}{6.40}\right) \cdot 100 = 110.2$$

$$\text{index number for 1980:} \quad \left(\frac{8.50}{6.40}\right) \cdot 100 = 132.8$$

$$\text{index number for 1985:} \quad \left(\frac{10.90}{6.40}\right) \cdot 100 = 170.3$$

When calculating an index number, we follow standard practice and round to the nearest tenth (as in Table 15.15) and omit the percent sign. For this application, all wages were compared to those in 1970, which is the **base year**. The index number for the base year is always 100.

When each index number uses the same base year, the resulting set of values is an **index time series**. An index time series is a set of index numbers determined from the same base year. The purpose of such a time series is to measure the yearly values in *constant* units (dollars, people, and so on). Because these values define a time series, they can be analyzed and decomposed by using the methods described previously. Our purpose in this section is simply to describe how to *construct* a time series of this type.

Price Indexes

Index numbers are derived for a variety of products (goods or services) as well as locations. For example, you may wish to compare the relative costs of consumer items in Los Angeles and Minneapolis if you are considering a move. Such information is readily available or can be determined from a number of business publications or government reports. The Department of Labor and the Bureau of Labor Statistics release reports (many of them monthly) on the price and quantity of many consumer items and agricultural commodities. Often, these are recorded for specific U.S. cities, providing geographical comparisons.

We focus our attention on a comparison of *prices* from one year to the next; these are **price indexes**. The most popular of these indexes is the Consumer Price Index (CPI), which combines a large number (over 400) of prices for consumer goods (such as food and housing) and family services (such as health care and recreation) into a single index. It is often called the cost-of-living index.

An index that includes more than one item is an **aggregate index**. We examine two methods of calculating an aggregate price index.

Say that we wish to measure the change in the prices of several items from 1975 to 1985, using a single price index. Table 15.16 shows four items; 1975 is the base year. Let P_0 denote the price for a particular item in the base year (1975) and P_1 represent this price during the reference year (1985). So

$$\Sigma P_0 = \text{sum of sampled prices for 1975}$$
$$= .75 + .95 + .89 + 31$$
$$= \$33.59$$

TABLE 15.16
Prices of Four Items in
1975 and 1985

ITEM	1975	1985
Eggs	.75 (doz)	1.35 (doz)
Chicken	.95 (lb)	1.79 (lb)
Cheese	.89 (lb)	1.85 (lb)
Auto battery	$31 (each)	$55 (each)

and

$$\Sigma P_1 = \text{sum of sampled prices for 1985}$$
$$= 1.35 + 1.79 + 1.85 + 55$$
$$= \$59.99$$

The ratio of these sums represents the **simple aggregate price index** for this application.

$$\text{simple aggregate price index} = \left(\frac{\Sigma P_1}{\Sigma P_0}\right) \cdot 100 \qquad \textbf{(15.8)}$$

For our example,

$$\text{index} = \left(\frac{59.99}{33.59}\right) \cdot 100 = 178.6$$

It might be tempting to conclude that, based on the prices of these four items, all prices increased by 78.6% between 1975 and 1985. Two problems arise here. The first is whether or not these sampled items are *representative* of the population of all price changes over this 10-year period. This is not a new problem—the same concern arose when we first introduced statistical sampling.

The second problem is that this index does not take into account the amounts of these items that are typically purchased by consumers. A significant change in the price for any single item will have a dramatic effect on the simple aggregate index, regardless of the demand for this product. The increase of $24 in the price of an automobile battery dominated the computed value of the aggregate price index; however, a typical consumer will spend much more annually on chicken than on car batteries. The simple aggregate price index assumes that equal amounts of each item are purchased.

For this reason, the next step is to include a measure of the quantity (Q) of each item in the price index. (We discuss methods of selecting the item quantities later.) The resulting index is known as a **weighted aggregate price index**.

$$\text{weighted aggregate price index} = \left(\frac{\Sigma P_1 Q}{\Sigma P_0 Q}\right) \cdot 100 \qquad \textbf{(15.9)}$$

EXAMPLE 15.10 Assume that a representative family each year purchases 1 automobile battery and each month consumes 6 dozen eggs, 15 pounds of chicken, and 8 pounds of cheese. Using 1975 as the base year and equation 15.9, determine the weighted aggregate price index for 1985. Use the data in Table 15.16.

SOLUTION The choice of time units on the quantities, Q, is arbitrary, but it is essential that you be consistent across all items. Converting the family purchases to

TABLE 15.17

Calculated Aggregate Price Index

	1975			1985		
ITEM	P_0	Q	P_0Q	P_1	Q	P_1Q
Eggs	.75	72	$ 54.00	1.35	72	97.20
Chicken	.95	180	171.00	1.79	180	322.20
Cheese	.89	96	85.44	1.85	96	177.60
Auto battery	31	1	31.00	55	1	55.00
			$\Sigma P_0Q = 341.44$			$\Sigma P_1Q = 652.00$

annual units, we have $6 \cdot 12 = 72$ dozen eggs, $15 \cdot 12 = 180$ pounds of chicken, $8 \cdot 12 = 96$ pounds of cheese, and 1 car battery (Table 15.17).

The weighted aggregate price index for 1985 (using 1975 as the base year) is

$$\text{index} = \left(\frac{\Sigma P_1Q}{\Sigma P_0Q}\right) \cdot 100 = \left(\frac{652}{341.44}\right) \cdot 100 = 191.0$$

In this index, the increase of 91% between 1975 and 1985 is not as severely affected by the price change for the car battery as was the simple aggregate price index, which ignored annual demand for each item. All widely used business price indexes are based on some variation of the weighted aggregate price index in equation 15.9. ◼

Selection of the Quantity, Q   Because the weights in a weighted aggregate price index usually reflect the quantities consumed, a problem arises when these quantities cannot be assumed to remain constant over the time span of the index. In Example 15.10, the same quantities, Q, were applied to both time periods, which means we are assuming an equal demand for the 2 years.

We have two options here: (1) use the quantities for the base year (1975, here) or (2) use the quantities for the reference year (1985, here). The first method is the **Laspeyres index**; the second is the **Paasche index**.

$$\text{Laspeyres index} = \left(\frac{\Sigma P_1Q_0}{\Sigma P_0Q_0}\right) \cdot 100 \qquad \textbf{(15.10)}$$

where Q_0 represents a base-year quantity.

$$\text{Paasche index} = \left(\frac{\Sigma P_1Q_1}{\Sigma P_0Q_1}\right) \cdot 100 \qquad \textbf{(15.11)}$$

where Q_1 represents a reference-year quantity.

Each of these indexes has strengths and weaknesses. The main advantage of the Laspeyres index is that the same base-year quantities apply to all future reference years. This greatly simplifies updating of this index, particularly given that most aggregate business indexes contain a large number of items. Its main disadvantage is that it tends to give more weight to those items that show a dramatic price increase. When a particular commodity's price increases sharply, this is typically accompanied by a decrease in the demand (measured by Q) for this item, or perhaps another item may be substituted by the consumer. The Laspeyres index fails to adjust for this situation. The advantages of this index outweigh its disadvantages, however, and it is more popular than the Paasche index.

The complexity of updating the reference-year quantities for the Paasche index make it difficult (and often impossible) to apply. Furthermore, because it reflects *both* price and quantity changes, we cannot use it to reflect price changes between two time periods. Its obvious advantage is that it uses current-year quantities, which provide a more realistic and up-to-date estimate of total expense.

We have seen that there is no completely reliable and accurate method of describing aggregate price changes. All such indexes include inaccuracies introduced by using a sample of items in the index as well as by the quantities to be used for weighting. Nevertheless, we treat such an index like any other sample estimate: We use the index as an estimate of relative price changes and realize that it is subject to a certain amount of error.

EXERCISES

15.53 Lemer's Clothing Store has been selling the same style of men's slacks for 6 years. The average retail price (in dollars) for the years 1983 to 1988 are as follows:

YEAR	PRICE
1983	12.75
1984	12.95
1985	13.95
1986	16.95
1987	19.95
1988	23.95

Compare the average prices for the years 1983, 1984, 1985, 1986, 1987, and 1988, using index numbers with 1983 as a base year.

15.54 The total annual profits (in millions of dollars) of car dealers in a large suburb of Chicago over a 7-year period are summarized as follows:

YEAR	TOTAL ANNUAL PROFITS
1982	2.13
1983	2.59
1984	3.60
1985	3.12
1986	3.33
1987	4.15
1988	4.54

Each total annual profit is an aggregate of profits. Find the simple aggregate price index for the years 1984, 1987, and 1988 using 1982 as a base year.

15.55 A typical family in Jackson, Mississippi, had the following weekly buying patterns in 1982 and 1987. Use 1982 as a base year. Price is in dollars.

ITEM	1982 UNIT PRICE	1982 QUANTITY	1987 UNIT PRICE	1987 QUANTITY
Meat	1.03	2	1.25	2
Milk	.97	3	1.19	2
Fish	.98	2	1.05	3
Oranges	.65	3	.75	4
Bread	.40	1	.62	2

a. Find the simple aggregate price index.

b. Construct the Laspeyres index.

c. Construct the Paasche index.

15.56 Explain the meaning, including the advantages and disadvantages, of the Paasche and Laspeyres weighted indexes. Comment on whether the indexes can be used as a representation of buying pattern.

15.57   The following table reflects the typical family's buying habits per 6 months on repairs for the family car. Use 1982 as a base year.

ITEM	1982 PRICE	1982 QUANTITY	1988 PRICE	1988 QUANTITY
Lube job	3.50	2	5.00	1
Oil change	9.50	3	13.00	2
Tune up	29.95	1	39.95	1
New tires	35.95	2	49.00	2

a. Find the simple aggregate price index.

b. Construct the Laspeyres index.

c. Construct the Paasche index.

15.58   A conglomerate is considering buying one or more of three companies. The closing prices of the stocks of these three companies for the years 1979 to 1987 are:

YEAR	BETTER FOODS	FRIENDLY INSURANCE	CHOCK FULL OF COMPUTER CHIPS
1979	13.500	20.125	39.25
1980	13.750	20.250	35.50
1981	14.250	20.500	31.75
1982	15.125	21.750	34.25
1983	15.500	21.500	37.75
1984	16.000	21.750	39.75
1985	16.125	22.500	40.00
1986	16.250	23.750	39.50
1987	16.750	23.500	42.25

Find an appropriate index to measure the change in the price of these three stocks for the years 1982, 1983, 1985, and 1987 using 1979 as a base year.

15.59   Suppose that, for a certain basket of goods, the Paasche index for 1988 is 115 and the Laspeyres index is 97. Assuming that the base year is 1982, interpret the meaning of the value of the two indexes.

15.60   The number of housing starts for four counties for the years 1986, 1987, and 1988 is:

COUNTY	1986	1987	1988
Brooks	1304	1505	1580
Litton	1264	1759	1987
Riverbed	1135	1443	1565
Tannon	1401	1605	1615

a. Compare the housing starts for Litton county for the years 1987 and 1988 using 1986 as a base year.

b. Compare the aggregate of housing starts for the years 1987 and 1988 for the four counties using 1986 as a base year.

15.61   The total revenue (in millions of dollars) of institutions of higher education for four southern states is:

STATE	1982–83	1983–1984
Alabama	1139	1222
Mississippi	650	706
Georgia	1487	1669
Louisiana	1159	1253

(*Source:* U.S. Department of Education, National Center of Education Statistics.)

a. For 1983–1984, find the simple aggregate index for the total revenue of institutions of higher education for the four states.

b. What can you conclude from the index calculated in part (a)?

15.62 A nursery purchases four different chemical ingredients to blend a certain popular fertilizer mixture. The data indicate the price per unit (PPU) paid for each ingredient and the quantity bought in 1986, 1987, and 1988.

INGREDIENTS	1986 PPU	1987 PPU	1988 PPU	1986 QUANTITY	1987 QUANTITY	1988 QUANTITY
A	.80	.81	.85	385	375	380
B	.51	.55	.60	345	360	379
C	.45	.50	.53	200	250	280
D	.37	.39	.40	150	180	195

a. Calculate the Laspeyres index for 1987 and 1988 using 1986 as a base year.

b. Calculate the Paasche index for 1987 and 1988 using 1986 as a base year.

c. Compare the two indexes in parts (a) and (b).

SUMMARY

A variable recorded over time is a **time series**. You obtain a sample of values for this variable by recording its past observations. Because this is not a random sample, it is extremely difficult (if not impossible) to obtain any tests of hypothesis or confidence intervals. Consequently, we resort to describing the past observations by deriving the components of the time series. This is **time series decomposition**. The components of a time series are (1) **trend** (a long-term growth or decline in the observations), (2) **seasonality** (within-year recurrent fluctuations), (3) **cyclical activity** (upward and downward movements of various lengths about the trend), and (4) **irregular activity** (what remains after the other three components have been removed).

We described methods for estimating these components for a time series. We first specify how we believe the components interact with one another, thus describing the time series variable, y_t. The **additive** structure assumes that each observation is the *sum* of its components. In particular, this implies that seasonal fluctuations during a particular year are not affected by the base volume for that year. In the **multiplicative** structure, each value of y_t is the *product* of the four components. Within this framework, the seasonal fluctuation for a specific month (or quarter) is more apt to be a constant *percentage* of the base volume for that year; for example, sales in December might be 35% higher than the average (base) sales for that particular year. The multiplicative structure was assumed for practically all of the illustrations in this chapter and is used more commonly in practice. The Bureau of the Census uses a variation of this procedure for their time series decomposition analyses.*

We described a four-step procedure for deriving these components for a particular time series, based on the multiplicative structure. The steps were: (1) determine a **seasonal index** for each month (monthly data) or quarter (quarterly data); (2) **deseasonalize** the data by dividing each observation by its corresponding seasonal index; (3) determine the **trend components** by deriving a least squares line or quadratic curve through the deseasonalized values; and (4) determine the **cyclical components** by, for each time period, dividing each deseasonalized value by its estimate using the trend equation and smoothing these values by computing three-period moving averages.

An **index time series**, often used by business analysts, is a time-related sequence of index numbers, where each value is a measure of the change in a

* The Bureau of the Census procedure is called the X11 program and is available on SAS/ETS. Consult the Econometric Time Series (ETS) user's guide (available from SAS) and the end-of-chapter appendix for a description of this procedure.

particular item (or group of items) from one year to the next. **Price indexes** are used to compare prices over time.

An **aggregate price index** is used to compare the relative price of a set of items for any year to the price during the base year. The index for the base year always is 100. The prices for the items can be averaged (**simple** aggregate price index) or weighted by the corresponding quantity of each item (**weighted** aggregate price index). Methods of selecting these quantities include using base-year quantities (the **Laspeyres index**) or using the reference-year quantities (the **Paasche index**).

FURTHER READING

Bowerman, B. L., and R. T. O'Connell. *Forecasting and Time Series*. 2d ed. Boston: PWS-KENT, 1987.

Makridakis, S., S. C. Wheelwright, and V. E. McGee. *Forecasting: Methods and Applications*. 2d ed. New York: John Wiley, 1983.

Mendenhall, W., and J. E. Reinmuth. *Statistics for Management and Economics*. 5th ed. Boston: PWS-KENT, 1986.

REVIEW EXERCISES

15.63   Each of the following influences on the variation in profits of a national chain of department stores would contribute to which of the four components of a time series?

a.   The long-term growth of the economy.

b.   The resignation of top managers in the company.

c.   Annual demand in spring and summer for garden equipment.

d.   The closing of several other department stores.

15.64   A manufacturer of tractors has built a record number of tractors for every year for the past 7 years. Given in thousands, the figures show the number of tractors built from 1982 to 1988.

YEAR	TRACTORS BUILT
1982	10.75
1983	11.78
1984	12.59
1985	13.4
1986	14.3
1987	15.7
1988	16.8

Find the least squares prediction equation that you would use to forecast the trend. What would you estimate the number tractors built in 1988 to be?

15.65   Luz Chemicals, which manufactures a special-purpose baking soda, is interested in estimating the equation of the trend line for their monthly sales data (in tons) for the year 1988.

MONTH	BAKING SODA SALES	MONTH	BAKING SODA SALES
Jan.	28	Jul.	34
Feb.	33	Aug.	34
Mar.	39	Sep.	35
Apr.	33	Oct.	36
May	38	Nov.	31
Jun.	31	Dec.	37

a. Without considering the seasonality present in the monthly sales, estimate the trend line equation.

b. Using the equation obtained in part (a), estimate the sales (in tons) for the month of February 1989.

15.66 Telemex, a supplier of telephone systems, has experienced moderate to rapid growth over a 12-year period. The data show the annual sales figures (in tens of thousands of dollars).

YEAR	SALES	YEAR	SALES
1977	3.1	1983	18.8
1978	6.3	1984	18.4
1979	10.5	1985	20.0
1980	10.2	1986	21.3
1981	11.5	1987	29.0
1982	14.7	1988	28.3

a. Find the trend line.

b. Find the cyclical components.

c. Graph the data and estimate the period of the cycle.

15.67 The number of retail sales of import passenger cars in the United States is given from 1976 to 1985 in units of 1000s.

YEAR	IMPORT PASSENGER CARS	YEAR	IMPORT PASSENGER CARS
1976	1498	1981	2327
1977	2076	1982	2223
1978	2000	1983	2387
1979	2329	1984	2439
1980	2398	1985	2838

(*Source: U.S. Statistical Abstracts*, 1987, p. 586.)

Graph the data.

a. Does the trend appear to be linear?

b. Find the equation to estimate the trend for the time series data.

c. What would be your estimate of the number of retail sales of import passenger cars sold in the United States in 1986?

15.68 Suppose that for the month of January 1988, the marketing department of a firm finds that the seasonal index is 1.20, the trend line value is $17,000 in sales, and the cyclical component is .79. What is the irregular component if the actual sales figure for January 1988 is $16,500?

15.69 Sales figures (in tens of thousands of dollars) for Dataphonics for a 10-year period follow. Find the corresponding trend values and cyclical components for each quarter of 1986 and 1987.

YEAR	QUARTER 1	QUARTER 2	QUARTER 3	QUARTER 4
1979	2.48	4.39	5.68	2.49
1980	2.76	4.86	5.69	2.73
1981	2.80	4.91	5.75	2.91
1982	2.90	5.10	5.85	2.95
1983	3.10	5.20	5.96	3.01
1984	3.15	5.21	6.04	3.10
1985	3.18	5.24	6.10	3.15
1986	3.20	5.30	6.14	3.19
1987	3.22	5.35	6.20	3.24
1988	3.25	5.36	6.23	3.25

15.70 The following table lists the number of building permits per month for non-residential construction during the 4-year period 1985 through 1988 in Parkins, Nebraska.

YEAR	JAN.	FEB.	MAR.	APR.	MAY	JUN.	JUL.	AUG.	SEP.	OCT.	NOV.	DEC.
1985	21	22	23	24	25	28	29	30	27	26	20	20
1986	21	24	23	25	26	25	29	32	32	27	20	18
1987	17	18	21	24	22	28	29	30	27	26	22	20
1988	17	21	23	23	24	29	31	22	28	22	21	29

a. Determine the seasonal indexes.

b. Determine the cyclical components for 1987 using a 3-month moving average.

c. Determine the irregular component for July of 1987.

15.71 The average monthly utility bill for the residents of the small community of Ridgecrest for the years 1985 to 1988 is:

YEAR	JAN.	FEB.	MAR.	APR.	MAY	JUN.	JUL.	AUG.	SEP.	OCT.	NOV.	DEC.
1985	190	180	179	130	135	145	148	153	145	153	170	185
1986	197	193	185	150	151	159	163	165	160	159	180	185
1987	215	205	193	175	171	179	185	184	180	180	173	190
1988	235	225	205	180	182	190	195	198	188	185	195	201

a. Determine the seasonal indexes.

b. Determine the trend.

c. Determine the cyclical components for 1986 using a three-period moving average.

d. Determine the irregular components for June and July 1986.

15.72 The weekly buying pattern of a typical family in a suburb in Atlanta, Georgia, for 1982 and 1988 follows. Use 1982 as a base year.

ITEM	1982 UNIT PRICE	1982 QUANTITY	1988 UNIT PRICE	1988 QUANTITY
Chicken	2.40	1	2.75	2
Milk	1.02	3	1.19	2
Bread	.39	2	.45	2
Ground beef	1.59	3	1.89	2
Tomatoes	.39	2	.78	2

a. Find the simple aggregate price index.

b. Calculate the Laspeyres index.

c. Calculate the Paasche index.

d. Compare the indexes in questions b and c.

15.73 The president of R & B Home Builders uses a housing index to obtain information about the direction of the housing market. The index for the four quarters of 1985, 1986, 1987, and 1988 yields these data:

YEAR	QUARTER 1	QUARTER 2	QUARTER 3	QUARTER 4
1985	157	155	154	147
1986	142	145	140	142
1987	143	153	152	150
1988	163	165	162	160

a. Determine the seasonal indexes for each quarter.

b. Determine the trend line.

c. Determine the cyclical components for 1986 using a three-period moving average.

d. Determine the irregular components for the first and second quarter of 1986.

15.74 Ranton House, Inc., has been building a certain style of house for the past 4 years. This house has sold for various prices over the years 1985 to 1988, as shown (in thousands of dollars).

YEAR	JAN.	FEB.	MAR.	APR.	MAY	JUN.	JUL.	AUG.	SEP.	OCT.	NOV.	DEC.
1985	49.5	51.3	51.3	51.5	52.0	57.3	57.4	58.3	57.2	56.3	55.4	58.6
1986	55.3	55.6	55.7	56.3	57.4	62.7	62.8	63.8	62.3	61.4	60.5	60.0
1987	60.5	61.3	62.4	62.7	63.0	68.6	68.9	70.1	68.6	68.4	67.4	67.0
1988	67.4	67.5	67.6	68.1	68.3	72.1	72.4	72.4	72.1	71.3	71.0	70.8

a. Determine the seasonal indexes.

b. Determine the trend.

c. Determine the cyclical components for 1987 using a three-period moving average.

d. Determine the irregular components for the months of September and October 1987.

15.75 The number of defaults per month of business loans at First State Bank are given over a 5-year period. Find the seasonal indexes.

YEAR	JAN.	FEB.	MAR.	APR.	MAY	JUN.	JUL.	AUG.	SEP.	OCT.	NOV.	DEC.
1984	54	53	52	50	48	46	48	50	52	56	58	60
1985	58	54	53	50	50	45	46	49	51	55	57	62
1986	53	52	48	47	47	44	45	48	49	52	55	60
1987	58	51	50	49	45	43	44	49	50	51	58	63
1988	59	58	56	52	54	49	50	51	54	58	60	64

15.76 The following data are the total real-estate assets of U.S. life insurance companies for 1970, 1975, 1980, and 1986 (in millions of dollars).

YEAR	TOTAL ASSETS
1970	6,320
1975	9,621
1980	15,063
1986	30,794

(*Source: Moody's Bank and Finance Manual* (1987); a15.)

The total amount of assets is an aggregate. Find the simple aggregate price index for the years 1975, 1980, and 1986, using 1970 as the base year.

15.77 Monthly redemptions of mutual funds from 1982 through 1986 are given in unit of millions of dollars.

YEAR	JAN.	FEB.	MAR.	APR.	MAY	JUN.	JUL.	AUG.	SEP.	OCT.	NOV.	DEC.
1986	4555	3511	4770	5571	5252	5567	5561	5156	6423	5901	5772	8923
1985	2158	2433	2454	2763	2624	2674	3229	3008	2469	2702	2712	4425
1984	1338	1680	1518	1570	1727	1532	1350	1604	1662	2039	1828	2369
1983	1018	956	1209	1204	1290	1491	1363	1395	992	1244	963	1561
1982	411	426	492	512	530	541	516	594	844	854	935	914

(*Source: Standard and Poor's Statistical Service, Current Statistics*, April 1988, p. 5, and *Standard and Poor's Statistical Service, Basic Statistics, Banking and Finance*, 1985, p. 27.)

a. Compute the seasonal indexes.

b. Determine the trend.

c. Calculate the cyclical components for 1985 using a 3-month moving average.

d. Calculate the irregular components for the last 3 months of 1985.

15.78 The MINITAB computer printout shows two time series plots of the values in C1 and of the values in C2. Would you say that these time series have multiplicative variation or additive variation? Draw a trend line for each time series. Estimate the length of the cycle.

```
MTB > PRINT C1-C3
 ROW     C1      C2      C3

  1      4.45   17.22     1
  2      4.66   20.36     2
  3      3.00   18.75     3
  4      2.50   18.46     4
  5      7.00   27.00     5
  6     11.70   38.36     6
  7     12.30   38.77     7
  8     10.40   28.35     8
  9      9.20   25.36     9
 10     14.50   40.00    10

MTB > MPLOT C1 VS C3, AND C2 VS C3

      -
      -
      -                                       B     B                    B
   36+
      -
      -
      -                                                      B
      -                           B                                B
   24+
      -
      -            B     B     B
      -        B                                                         A
      -
   12+                                              A     A
      -                                                         A     A
      -                                       A
      -        A     A
      -                     A     A
  -0+
        +---------+---------+---------+---------+---------+---------+------
       -0.0      2.0       4.0       6.0       8.0      10.0

          A = C1 vs. C3              B = C2 vs. C3
```

15.79 The following MINITAB computer printout shows a regression analysis of a time series with nine observations. From observing the plot of the residuals, comment on the validity of the statistical tests used in the regression analysis.

```
MTB > PRINT C1 C2
 ROW      C1     C2

  1     5.0062    1
  2     5.1638    2
  3     3.2737    3
  4     3.4650    4
  5     8.4375    5
  6    13.4437    6
  7    13.6125    7
  8    12.9825    8
  9    14.6250    9

MTB > REGRESS C1 ON 1 PREDICTOR IN C2, RESID IN C3

The regression equation is
C1 = 1.17 + 1.54 C2

Predictor        Coef         Stdev      t-ratio
Constant        1.174        1.736        0.68
C2              1.5431       0.3084        5.00

s = 2.389       R-sq = 78.1%     R-sq(adj) = 75.0%

Analysis of Variance

SOURCE          DF          SS            MS
Regression       1        142.87        142.87
Error            7         39.95          5.71
Total            8        182.82
```

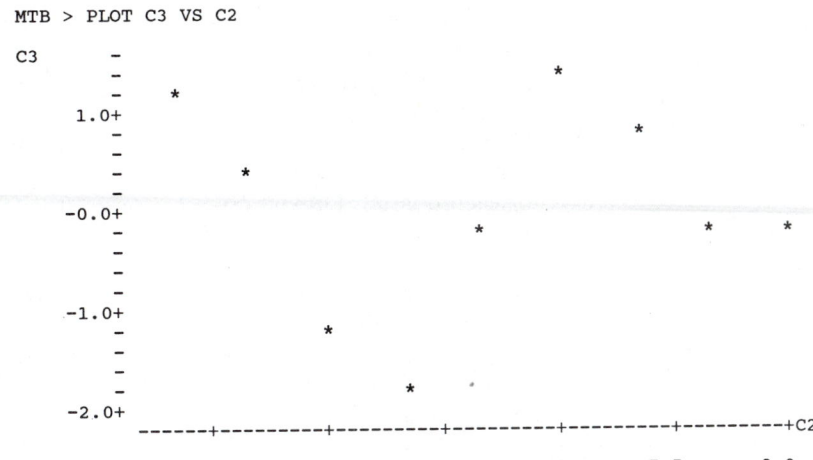

```
Exercise 15.79 (continued)

MTB > PLOT C3 VS C2

   C3     -
          -
          -                                                        *
     1.0+      *
          -
          -
          -             *
          -
    -0.0+
          -                                        *              *      *
          -
          -
          -
    -1.0+
          -                   *
          -
          -                        *
    -2.0+
          ------+---------+---------+---------+---------+--------+C2
             1.5       3.0       4.5       6.0       7.5       9.0
```

CASE STUDY

FORECASTING DEMAND FOR COMPUTER AND DATA PROCESSING SERVICE PERSONNEL

The computer revolution has created an industry that is predicted to become one of the biggest industries in the United States. Researchers and market analysts believe that the information technology industry, as they call it, can be segmented into five sectors: computer and data processing (DP) services, manifold business forms, office and computing machines, telephone and telegraph apparatus, and telephone communication.

According to *Software Industry Report*, a trade newsletter, the computer and DP services sector is the second largest employer among the five areas. Telephone communication is the largest employer but has experienced a slowly decreasing work force since 1982, whereas the computer and DP services sector has expanded its work force every quarter since the beginning of 1980.

The accompanying table of data gives a quarter-by-quarter comparison of employment in the computer and data processing services sector from 1980 to 1987, as reported by the U.S. Bureau of Labor Statistics

Quarterly Comparison
1980–1987
(In Thousands)
Computer and Data
Processing Services

YEAR	1st Q	2nd Q	3rd Q	4th Q	ANNUAL AVERAGE
1980	295.133	301.333	306.900	313.800	304.300
1981	323.467	329.366	341.600	351.933	336.600
1982	358.633	359.100	364.066	376.900	364.700
1983	391.667	410.933	425.066	436.066	415.900
1984	453.300	465.866	483.500	497.667	475.100
1985	512.066	532.866	553.933	570.633	542.400
1986	580.133	585.100	596.033	603.366	591.200
1987	617.833	630.467	644.000	656.880	637.295

(*Source:* U.S. Bureau of Labor Statistics, Washington, D.C., and *Software Industry Report* 20, no. 1 (January 15, 1988, 8–9. Published by Computer Age, a division of EDP News Services, Inc., 7043 Wimsatt Road, Springfield, VA 22151-4080. Figures for the fourth quarter of 1987 have been estimated.)

Case Study Questions

1. Using the preceding data, compute the centered moving averages. What is the benefit derived from obtaining these moving averages?

2. Compute the seasonal indexes for each quarter. Do you observe any strong seasonal patterns?

3. Deseasonalize the data, and determine the linear trend equation. How far into the future do you think the trend could be projected?

4. Assuming a multiplicative model, obtain the cyclical and irregular components.

5. Prepare a forecast of employment in the computer and data processing services sector for the four quarters of 1988.

6. Obtain the latest data published by the Bureau of Labor Statistics, or other sources, and compare these with your forecasted figures. How good were your forecasts?

■ ■ ■ ■ ■ ■ ■ ■ ■ ■ ■ ■   S A S   ■ ■ ■ ■ ■ ■ ■ ■ ■ ■ ■ ■ ■ ■

SOLUTION

SECTION 15.6

We used four-step decomposition on time series data to analyze the past behavior of U.S. trade from 1984 to 1987. (Table 15.7) The SAS program listing in Figure 15.25 was used to perform the analysis. This example will run only on the mainframe version of SAS, using the optional ETS package.

The TITLE command names the SAS run.

The DATA command gives the data a name.

The first DO statement sets a year loop from 1984 to 1987.

The second DO statement sets a month loop from 1 to 12.

SALES are to be entered as INPUT.

DATE means the first data item entered will be the first month of the first year (January 1984).

The OUTPUT statement requests that the input be printed.

FIGURE 15.25
Input for SAS.

```
TITLE 'X11 DECOMPOSITION';
DATA RETAIL;
   DO YEAR=1984 TO 1987;
      DO MONTH=1 TO 12;
         INPUT SALES @@;
         DATE=MDY(MONTH,1,YEAR);
         N+1;
         OUTPUT;
      END;
   END;
   KEEP DATE SALES N;
   FORMAT DATE MONYY5.;
   LABEL SALES=' RETAIL TRADE';
   CARDS;
93.09 93.69 104.29 104.34 111.31 111.98 106.55 110.65 103.93 109.23
113.28 131.65 98.82 95.59 110.17 113.11 120.34 114.96 115.49 121.12
114.17 116.14 118.56 139.40 105.64 99.66 114.24 115.71 125.42 120.35
120.74 124.06 124.65 123.06 120.79 151.26 106.39 105.80 120.44 125.37
129.07 128.98 128.95 131.02 123.77 127.21 125.38 154.75
PROC X11 DATA=RETAIL;
   MONTHLY DATE=DATE CHARTS=STANDARD TDREGR=TEST PRINTOUT=STANDARD;
   VAR SALES;
   TITLE2 'MONTHLY EXAMPLE SHOWING STANDARD TABLES AND CHARTS';
```

The CARDS command indicates to SAS that the input data immediately follow.

The next five lines contain the data values, which represent time series data over 48 time periods.

The PROC X11 commands requests the X11 time series decomposition procedure be used on the dataset RETAIL.

The MONTHLY statement means the data is monthly. DATE = DATE specifies when the time series starts and ends. CHARTS specifies charts to be produced by the X11 procedure. TDREGR = TEST statement is used to adjust the time series. PRINTOUT = STANDARD requests seasonal charts and trend cycle charts. VAR specifies that SALES is the input that will be analyzed by the procedure.

Figure 15.26 shows a portion of the SAS output obtained by executing the listing in Figure 15.25. The results differ slightly from the results in the text, since SAS uses a more complicated time series decomposition procedure than the one outlined in this chapter.

FIGURE 15.26   SAS output.

```
B 1 ORIGINAL SERIES
YEAR   JAN    FEB     MAR     APR     MAY     JUN     JUL     AUG     SEP     OCT     NOV     DEC     TOTAL
1984    93     94     104     104     111     112     107     111     104     109     113     132      1294
1985    99     96     110     113     120     115     115     121     114     116     119     139      1378
1986   106    100     114     116     125     120     121     124     125     123     121     151      1446
1987   106    106     120     125     129     129     129     131     124     127     125     155      1507

AVG    101     99     112     115     122     119     118     122     117     119     120     144

TOTAL -       5625 MEAN       117 S.D. -        13         Original Time Series (rounded to nearest integer)


D10 FINAL SEASONAL FACTORS
YEAR    JAN     FEB     MAR     APR      MAY      JUN      JUL      AUG     SEP     OCT      NOV      DEC     AVG
1984  88.413  85.842  97.777  98.807  104.157  101.546  100.092  103.438  98.481  99.587  100.808  120.935  99.990
1985  88.410  85.871  97.767  98.907  104.124  101.529  100.117  103.510  98.472  99.611  100.750  121.111  100.015
1986  88.360  85.833  97.734  98.881  104.102  101.518  100.122  103.526  98.493  99.638  100.788  121.168  100.014
1987  88.295  85.763  97.767  98.850  104.051  101.532  100.066  103.570  98.523  99.573  100.656  121.244  99.991
1988  88.262  85.728  97.784  98.834  104.026  101.539  100.039  103.593  98.538  99.540  100.590  121.282  99.980

AVG   88.348  85.807  97.766  98.856  104.092  101.533  100.087  103.527  98.502  99.590  100.719  121.148◄─────┐

TOTAL -      5999.870                                                                              Seasonal Indexes


D11 FINAL SEASONALLY ADJUSTED SERIES
YEAR   JAN    FEB     MAR     APR     MAY     JUN     JUL     AUG     SEP     OCT     NOV     DEC     TOTAL
1984   106    106     105     106     107     110     108     106     108     109     111     110      1292
1985   112    112     114     114     114     115     115     116     117     116     117     116      1378
1986   118    117     118     118     119     119     120     121     126     122     122     124      1445
1987   119    124     124     126     126     126     127     128     126     126     125     127      1506

AVG    114    115     115     116     116     118     117     118     119     118     119     119

TOTAL -       5621 MEAN       117 S.D. -         7         Deseasonalized values (rounded to nearest integer)
```

Quantitative Business Forecasting

A Look Back/Introduction

We have introduced you to methods of capturing the behavior of a dependent variable, Y. The first procedure was linear regression, which used a set of predictor (independent) variables to explain the observed values of this variable. In simple linear regression, a single predictor is used. When we had two or more predictor variables, we used a multiple linear regression model to attempt to account for the variation within the observed values of the dependent variable. The calculations were considerably more complex, and a computer solution was used to estimate the linear relationship between the dependent variable (Y) and the predictor variables ($X_1, X_2, \ldots$).

The success or failure of this technique lies in your ability to arrive at a set of predictor variables that can accurately predict past (and future) values of the dependent variable. Suppose your model fails to fit adequately the observed values of Y, with a resulting large sum of squares for error (SSE) and a low value of R^2 (coefficient of determination). Does this imply that multiple linear regression is not a reliable method of prediction for this situation? This could be the case, but it is just as likely that you omitted one or more key variables that would have significantly improved your prediction accuracy.

The time series decomposition technique, presented in the previous chapter, uses a different approach. This procedure attempts to explain each observed value by means of the various components that make up this observation. These include trend (long-term growth or decline in the time series), seasonality (predictable variation within each year), and cyclical activity (generally due to unpredictable swings in the national or international economy).

The key distinction between these two procedures is that the time series approach does not search for explanatory (predictor) variables. Rather, it seeks to capture the past behavior of the time series by analyzing the various components. More complex time series techniques, which were not discussed, use past observations to predict the value for the future. You can use a time series approach to forecast future values by "extending" the pattern into the future.

For example, if your company sales have been increasing approximately 150,000 units each year over the past 6 years, a reasonable forecast for next year would be a sales volume of 150,000 more than the present year's value.

Statistical forecasting is, in one sense, an extension of the prediction of a dependent variable. However, we now enter a more uncertain world—that of extrapolation. In previous chapters, we warned you of the dangers of this procedure, because outside the range of your data, the predicted values become less reliable. We can only hope that tomorrow's world will be similar to today's and that patterns observed over the past will continue. This makes forecasting fascinating. We live in an uncertain world, and a reasonably accurate forecast can be extremely valuable for a marketing or production strategy.

This chapter introduces many (certainly not all) methods of using quantitative techniques for predicting future values for the variable of interest. We demonstrate how to forecast future values by using the past observations (the time series approach) as well as by using the multiple linear regression method. By applying the proper forecast method, you often can make the future considerably less uncertain.

16.1

METHODS OF FORECASTING

Forecasting procedures come in a variety of shapes and colors. You can arrive at a sales forecast by simply assembling a panel of experts and arriving at a collective "guess" or constructing a highly complex statistical model that attempts to predict the future using past data. In the broadest sense, forecasting methods can be classified as **qualitative** (the panel of experts procedure) or **quantitative** (the statistical forecasting procedure). Also, quantitative forecasting can be carried out using two different approaches; namely, **regression** models (with several predictor variables) or a **time series** model, which utilizes past observations of the dependent variable to arrive at forecasted values.

Qualitative Forecasting

There are many instances when a qualitative approach to forecasting is appropriate. When no past data are available, it is impossible to construct a quantitative model to predict future values. This can occur when you intend to introduce a new product and no past sales data exist. Furthermore, when you introduce this product, it becomes a guessing game as to what the response will be from competitors in the field. Will they respond to your entry into the market? When will they respond? Will they lower their price to increase the demand for their product? Will they attempt to "copy" your product, and how soon can this be accomplished? Such questions do require expert opinion.

One popular method of qualitative forecasting is the **Delphi method**. To utilize this procedure, you would assemble individuals from the sales force and the market research staff and ask them to supply their predictions based upon their knowledge of the area. This can be accomplished through a questionnaire or any other written set of specific questions. After this is completed, members of the team are informed as to the responses of the entire group and asked to reevaluate their opinions based upon this new information. In this way, members of the team may be able to arrive at a "best educated" prediction of competitor response to their market entry. Of course, it is also entirely possible that no collective agreement will be reached after several rounds of this process.

We do not pursue qualitative forecasting methods in this chapter. The interested reader is referred to the text by Bowerman and O'Connell (contained in the Further Readings section at the end of the chapter) for additional qualitative procedures. The remainder of the chapter focuses on the use of quantitative forecasting techniques.

Quantitative Forecastings

To utilize a quantitative forecasting procedure, you predict future behavior of a dependent variable using information from previous time periods. This can be accomplished in one of two ways: using a *regression* model or a *time series* model.

Regression Models   The use of regression models is the multiple regression approach, where variation of the dependent variable is explained using several independent (predictor) variables. One main advantage of this approach is that you can measure the effect of changes within one or more of the predictor variables. Furthermore, this type of model is generally easily understood by those individuals responsible for making the final forecast decision, since it is clear which variables are assumed to have an effect on the value of the dependent variable. A drawback to this type of model as a forecasting instrument is that to predict future values of the dependent variable, it is necessary to predict future values of the predictor variables, which, in many instances, may be as uncertain as future values of the dependent variable. This forecasting procedure is discussed in Section 16.9.

Time Series Models   A time series forecast is made by capturing the patterns that exist in the past observations and extending them into the future. Consider the annual data reflecting the sales of the Clayton Corporation between 1974 and 1988, represented in Figure 16.1. The data reflect a strong linear trend, as shown by the line passing through the points. To estimate the sales for 1989, one simple method would be to extend this line into 1989, as illustrated in Figure 16.1. By graphically extending this line and observing the estimated value, we obtain

$$\hat{y}_{16} = \text{forecast for 1989 is approximately 350}$$

that is, 350,000 units. This procedure, along with methods of dealing with trend and seasonality, is discussed later in the chapter.

At first glance, it might appear that time series forecasting is easier to apply than are multiple regression models. After all, there is no need to search for a

FIGURE 16.1

Sales for Clayton Corporation.

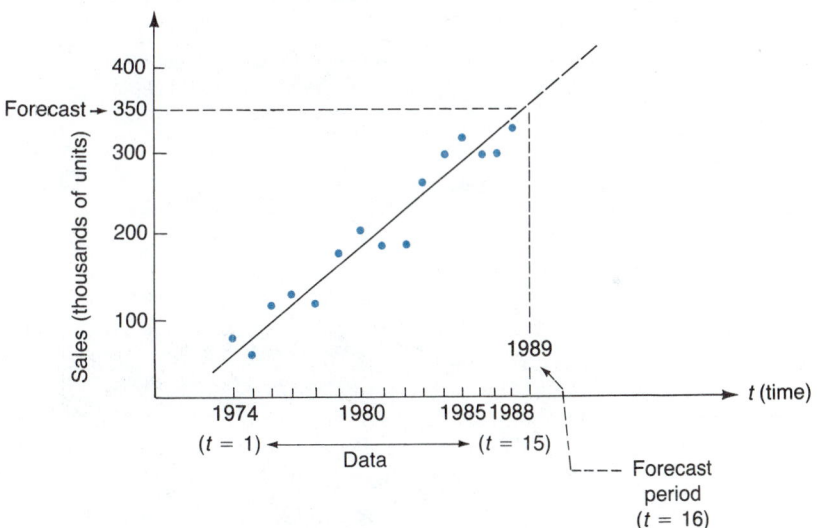

reliable set of predictor variables. It is true that time series predictors can be simple and straightforward, as is the so-called naive forecast discussed in the next section. Frequently, however, extracting the complex and interrelated structure of an observed time series requires sophisticated and complex prediction equations.

As in Chapter 15, we do not put any statistical bounds (such as a 95% confidence interval) on the predicted values. Rather, we suggest alternative methods of forecasting and demonstrate a way of determining the "best" forecasting procedure for a particular set of data. Of course, all forecasts are subject to error and are based on the assumption that the past historical patterns (such as the straight line in Figure 16.1) continue into the future.

In Sections 16.2 through 16.8, we examine several time series models and methods of evaluating the predictive ability of each procedure when applied to a particular set of time series data.

The procedure for selecting a forecasting model is summarized in Figure 16.2. Steps 1 through 5 are the model selection and forecasting stage. Steps 6 and 7 are the model review phase, during which you reevaluate your forecasting procedure. This allows you to update your model using the latest observations or to consider changing your forecasting model by returning to step 2. Any forecasting technique should be reviewed; you must reexamine the forecast errors from the previous observations.

Remember that any quantitative forecasting technique can never replace the forecast of an individual (or team of people) who uses his or her expertise and knowledge of unpredictable future events (such as strikes, wars, or market shifts) to make forecasts. Rather, the quantitative forecast is a tool the forecaster uses. A forecast offers an excellent baseline, which can be modified by informed judgment.

FIGURE 16.2

A step-by-step procedure for forecasting with time series data.

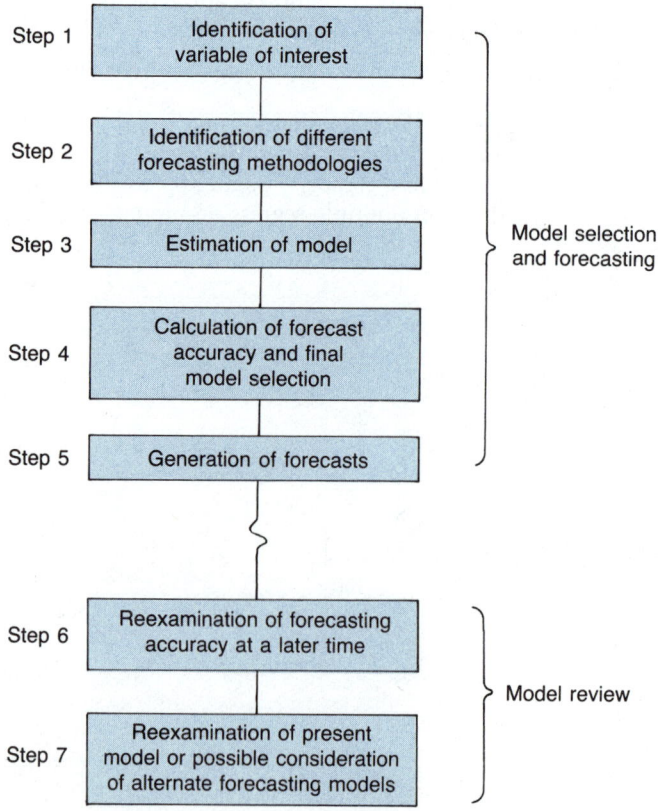

16.2
THE NAIVE FORECAST

Simply put, the naive forecast procedure states that the estimate of Y for tomorrow is the actual value from today. In general,

$$\hat{y}_{t+1} = y_t \qquad (16.1)$$

for any time period, t.

Once again, the "hat" notation is used to denote an estimate. Equation 16.1 reads: "y hat for time period $t + 1$ is y for time, t." Here, $\hat{y}_{t+1}$ represents the *forecast* for time period $t + 1$.

This method of forecasting often works well for data that are recorded for smaller time intervals (such as daily or weekly) and contain no apparent upward or downward trend among the observed values. Data of this type are not apt to shift direction suddenly from one day to the next, and the naive forecast can provide a simple, yet fairly reliable, estimate of the next day's value. On more than one occasion, this predictor has outperformed much more complex forecasting equations—particularly when applied to a difficult-to-predict time series, such as an individual stock market price. It provides an inexpensive, easy method of forecasting.

EXAMPLE 16.1

The weekly closing price for a share of Keller Toy Company stock was recorded over a 12-week period. Using the following data, determine a forecast for week 13.

WEEK	PRICE	WEEK	PRICE	WEEK	PRICE
1	60	5	$64\frac{1}{2}$	9	$63\frac{1}{4}$
2	$62\frac{1}{4}$	6	62	10	$62\frac{1}{2}$
3	$61\frac{3}{4}$	7	$63\frac{1}{2}$	11	61
4	63	8	64	12	$61\frac{1}{2}$

SOLUTION

The observed value for the last time period is $y_{12} = 61\frac{1}{2}$, so your forecast for the next time period is

$$\hat{y}_{13} = y_{12} = 61\frac{1}{2}$$

Notice that we are careful to distinguish between a *forecast*, such as $\hat{y}_{13}$, and an *observed* value, such as y_{12}.

One method of checking to see whether a particular forecasting technique is appropriate for your time series involves applying this procedure to each period of the observed data. For example, in Example 16.1, what would we have predicted for the fifth week using the naive forecasting equation 16.1? In other words, suppose we are at the end of the fourth week and need a forecast for $t = 5$. Using the naive predictor,

$$\hat{y}_5 = y_4 = 63$$

The actual value turned out to be $y_5 = 64\frac{1}{2}$, providing a **residual** of

$$\text{residual} = y_5 - \hat{y}_5 = 64\frac{1}{2} - 63 = 1\frac{1}{2} \qquad ■$$

EXAMPLE 16.2

Apply the naive forecasting procedure to the 12 time periods in Example 16.1 and determine the residual for each week.

SOLUTION

The procedure cannot be applied during the first time period ($t = 1$) because $\hat{y}_1 = y_0$, where y_0 is the closing price for the week preceding the observations in the table. If this value is available, then the forecast value for $t = 1$ can be determined; it is equal to this value. Otherwise, the forecast for this time period is left blank. (Refer to Table 16.1.) ■

TABLE 16.1

Residuals for Naive
Forecasts

WEEK	y_t	$\hat{y}_t$	RESIDUAL $(y_t - \hat{y}_t)$
1	60	—	—
2	$62\frac{1}{4}$	60	$2.25 \ (= 62\frac{1}{4} - 60)$
3	$61\frac{3}{4}$	$62\frac{1}{4}$	$-.5 \ (= 61\frac{3}{4} - 62\frac{1}{4})$
4	63	$61\frac{3}{4}$	1.25 (and so on)
5	$64\frac{1}{2}$	63	1.5
6	62	$64\frac{1}{2}$	-2.5
7	$63\frac{1}{2}$	62	1.5
8	64	$63\frac{1}{2}$	.5
9	$63\frac{1}{4}$	64	$-.75$
10	$62\frac{1}{2}$	$63\frac{1}{4}$	$-.75$
11	61	$62\frac{1}{2}$	-1.5
12	$61\frac{1}{2}$	61	.5

When we first introduced the concept of a residual (or error) in the chapters dealing with linear regression, we stressed that small residuals were desirable. When the residuals were near zero for regression applications, this meant that the model did a good job of "fitting" the sample observations.

The same idea applies to evaluating the effectiveness of a forecasting procedure. Small residuals indicate that this particular forecast technique would have done a good job of predicting the past values of this time series. A method of combining these residuals into a single measure (much like the SSE in linear regression) will be introduced in a later section.

EXERCISES

16.1 Explain the distinction between the technique of time series analysis and that of multiple regression analysis.

16.2 If a regression or time series model fits a set of data well, would the model necessarily provide small forecasting errors for future observations?

16.3 The price of the stock of Intersecond Bank has been cyclical over the years. From the following data, calculate the forecasted price of the stock using the naive model for the years 1978 to 1988. Also, calculate the residual for each forecast.

YEAR	PRICE OF STOCK	YEAR	PRICE OF STOCK
1978	28.50	1984	28.25
1979	29.25	1985	29.75
1980	31.75	1986	32.50
1981	29.50	1987	31.50
1982	28.00	1988	30.00
1983	27.50		

16.4 Sullivan's Mutual Fund invests primarily in technology stocks. The net asset value of the fund at the end of each month for the 12 months of 1988 is given. Find the forecasted value of the mutual fund for each month, starting with February, by using the naive model. Calculate the residuals.

MONTH	MUTUAL FUND PRICE	MONTH	MUTUAL FUND PRICE
Jan	8.43	Jul	8.35
Feb	8.10	Aug	9.45
Mar	7.15	Sep	9.01
Apr	6.95	Oct	10.31
May	7.25	Nov	10.25
Jun	7.95	Dec	11.04

16.5 What advantages and disadvantages can you think of in using the naive model to forecast?

16.3

PROJECTING THE LEAST SQUARES TREND EQUATION

For data containing a strong linear or curvilinear trend, a method of predicting future values of the time series is to extend the trend line (or curve) into the forecast periods. This was illustrated in Figure 16.1, in which the data from 1974 to 1988 demonstrated a very strong linear growth over these 15 years.

Suppose that a simple linear regression analysis is performed on these data, using the 15 sales values as the dependent variable and $t = 1, 2, \ldots, 15$ as the predictor variable (as discussed in Chapter 15). The resulting least squares line, shown in Figure 16.1, turns out to be

$$\hat{y}_t = 32 + 20t$$

The estimated forecast for 1989 in the earlier discussion was $\hat{y}_{16} = 350$. This was determined simply by extending the least squares line into this time period and "eyeballing" the estimate for 1989. The actual forecast is

$$y_{16} = 32 + 20(16) = 352$$

So, our estimate of sales for 1989 is 352,000 units, based on the linear trend equation.

A Time Series Containing Trend and Seasonality

The previous procedure can be adapted to situations in which the time series contains significant trend *and* seasonality. Such a situation can occur when the data are monthly or quarterly, with seasonal fluctuations about a linear or curvilinear trend.

The quarterly sales for Video-Comp over a 4-year period (1985 to 1988) are contained in Table 15.3 on page 673 and are illustrated in Figure 16.3.

The deseasonalized sales figures (often called *seasonally adjusted* sales) are summarized in Table 15.6 on page 677 and also are graphed in Figure 16.3. Notice that the extreme seasonal fluctuations of the original time series were removed when these values were divided by the appropriate seasonal index. The indexes for this application were derived in Example 15.8, indicating low

FIGURE 16.3

Quarterly sales at Video-Comp.

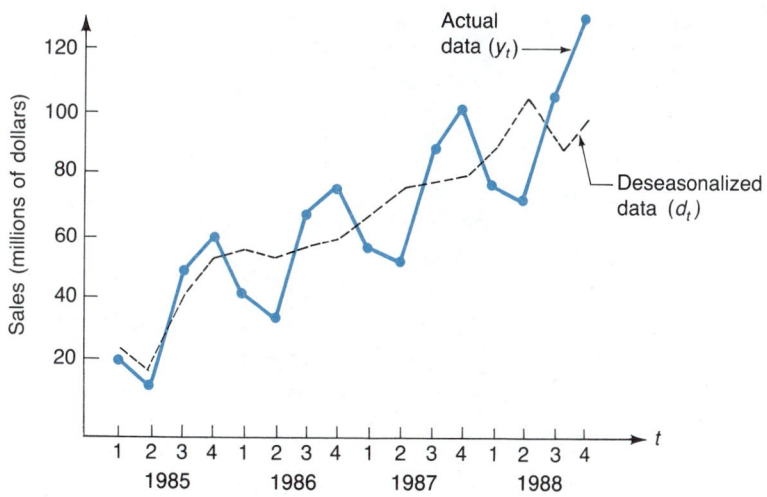

sales for the first two quarters, above-average sales for the third quarter, and extremely high sales during the fourth (holiday) quarter. The corresponding indexes were

$$S_1 = .852$$
$$S_2 = .692$$
$$S_3 = 1.166$$
$$S_4 = 1.290$$

To forecast future values when using seasonal data, you once again determine the least squares line (or curve), except that now you use the *deseasonalized data* (say, d_t) as your dependent variable. Once you have calculated the trend forecast, you obtain your final forecast by multiplying this deseasonalized estimate by the corresponding seasonal index. So the procedure for extending trend and seasonal components is:

1. Calculate the deseasonalized (seasonally adjusted) data from the original time series $(y_1, y_2, \ldots, y_T)$. Call these values $d_1, d_2, \ldots, d_T$.
2. Construct a least squares line through the deseasonalized data, where $(t = 1, 2, \ldots, T)$

$$\hat{d}_t = b_0 + b_1 t$$

3. Calculate the forecast for time period $T + 1$ using

$$\hat{y}_{T+1} = (\hat{d}_{T+1}) \cdot (\text{seasonal index for } t = T + 1)$$
$$= [b_0 + b_1(T + 1)] \cdot (\text{seasonal index for } t = T + 1)$$

EXAMPLE 16.3   Using the Video-Comp data, what would be your forecast for the first-quarter sales of 1989? Second-quarter sales?

SOLUTION   The deseasonalized data and corresponding least squares line are shown in Figure 16.4. The MINITAB solution for the least squares line is shown in Figure 16.5, where

$$\hat{d}_t = 19.372 + 5.0375t$$

This equation tells us that, apart from seasonal variation, the sales at Video-Comp are increasing by approximately $5 million each quarter. Using this

FIGURE 16.4

Trend line through deseasonalized data (quarterly sales, Video-Comp).

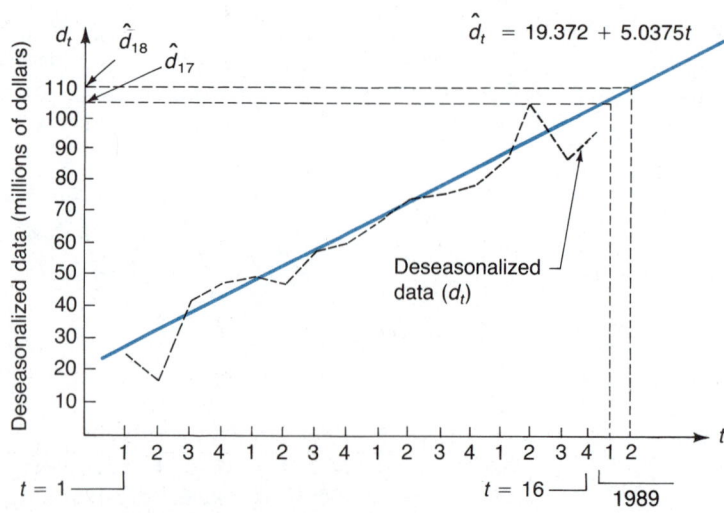

FIGURE 16.5

MINITAB solution for deseasonalized trend line.

```
MTB > SET INTO C1
DATA> 1:16
DATA> END
MTB > SET INTO C2
DATA> 23.47 17.34 40.31 46.51 46.95 46.24 55.75 58.91
DATA> 65.73 72.25 72.90 77.52 88.03 101.16 86.62 95.35
DATA> END
MTB > REGRESS Y IN C2 USING 1 PREDICTOR IN C1

The regression equation is
C2 = 19.4 + 5.04 C1

Predictor        Coef        Stdev      t-ratio         p
Constant       19.372        3.111         6.23     0.000
C1             5.0375        0.3217       15.66     0.000

s = 5.932         R-sq = 94.6%       R-sq(adj) = 94.2%

Analysis of Variance

SOURCE           DF          SS            MS         F         p
Regression        1        8627.9       8627.9    245.21     0.000
Error            14         492.6         35.2
Total            15        9120.5
```

equation and Figure 16.4, your deseasonalized forecast for the first quarter of 1989 (time period 17) is

$$\hat{d}_{17} = 19.372 + 5.0375(17)$$
$$= 105.01$$

Now, the sales for the first quarter of each year are lower than the yearly average, as reflected in the seasonal index of $S_1 = .852$ (from Example 15.8). Consequently, your actual forecast for this time period is

$$\hat{y}_{17} = \hat{d}_{17} \cdot (\text{seasonal index for quarter 1})$$
$$= 105.01 \cdot .852$$
$$= 89.5 \quad (\text{million dollars})$$

This procedure can be used to forecast any future time period. For the second quarter of 1989, the estimated sales will be

$$\hat{y}_{18} = \hat{d}_{18} \cdot (\text{seasonal index for quarter 2})$$
$$= [19.372 + 5.0375(18)] \cdot 692$$
$$= (110.05)(.692) = 76.2 \quad (\text{million dollars})$$

Do these estimates seem reasonable? Look at the values for the first and second quarters. (The forecast values are those for 1989.)

YEAR	FIRST-QUARTER SALES	SECOND-QUARTER SALES
1985	20	12
1986	40	32
1987	56	50
1988	75	70
1989	89.5	76.2

The forecast for the first quarter of 1989 seems to be about what we would expect, based on the past first-quarter sales. The predicted sales value for the second quarter of 1989 seems to be on the low side, with an increase of only 6.2 from the second quarter of 1988. Remember, however, that this forecasting technique contains the effect of *all* the quarters observed over the 4 years. By examining the past sales during the second quarter only, we are ignoring the remaining quarters, and perhaps an explanation for this seemingly low forecast lies in these values. ■

It is possible that this forecasting procedure is not a good one for the application in Example 16.3. There may be a better way to obtain a forecast for this situation. We show you several ways to forecast a time series and then determine which of these does the best job for a particular set of observed values. No one procedure always performs well for all applications.

EXERCISES

16.6   A set of quarterly data has been gathered for 3 years. The seasonal indexes are found to be $S_1 = .81$, $S_2 = .93$, $S_3 = 1.19$, $S_4 = 1.07$. The least squares line through the deseasonalized data is found to be

$$\hat{d}_t = 10.1 + 1.3t$$

from 12 quarterly periods. Find the forecast for quarterly periods 13, 14, 15, and 16.

16.7   Sands Motel, which usually has a busy summer season on the beaches of Atlantic City, New Jersey, would like to obtain a forecast of future business. Business (in thousands of dollars) for each month from 1982 to 1988 is given:

YEAR	JAN.	FEB.	MAR.	APR.	MAY	JUN.	JUL.	AUG.	SEP.	OCT.	NOV.	DEC.
1982	.4	.5	.7	.9	1.3	1.8	2.5	2.9	2.3	2.0	1.2	.7
1983	.5	.6	.8	1.1	1.2	1.9	2.8	3.1	2.7	2.1	.9	.8
1984	.7	.6	.9	1.1	1.4	2.1	2.8	3.3	2.9	2.4	1.5	.9
1985	.8	.7	.9	1.3	1.6	2.3	2.9	3.4	3.2	2.6	1.7	1.2
1986	1.0	.9	1.1	1.4	1.8	2.4	3.0	3.4	3.3	2.7	1.9	1.4
1987	1.1	.8	.9	1.3	1.9	2.5	3.2	3.6	3.2	2.6	2.0	1.5
1988	1.3	1.1	1.2	1.5	2.2	2.7	3.4	3.8	3.4	2.8	2.1	1.6

a.   Find the seasonal indexes.

b.   Find the least squares trend line for the deseasonalized data.

c.   Find the forecast for March 1989, July 1989, and December 1989.

16.8   Slater Industries would like to cut costs on the amount of inventory it holds. Quarterly data have been gathered for 5 years from 1984 through 1988. The following table lists the dollar amount of inventory in units of 10,000.

YEAR	QUARTER 1	QUARTER 2	QUARTER 3	QUARTER 4
1984	.3	.5	.4	.2
1985	.4	.7	.5	.3
1986	.5	.9	.7	.4
1987	.7	1.1	.9	.8
1988	.8	1.5	1.0	.9

a.   Find the seasonal indexes.

b.   Find the least squares trend line for the deseasonalized data.

c.   Find the forecast for each quarter of 1989.

16.9   Is the experience of the managers of a company necessary to use in aiding the forecasting process, if the model fits the past data very well?

16.10   Refer to Exercise 15.45. What is the deseasonalized forecast for January of 1987 for the price per square foot of the typical three-bedroom, two-bath, two-car-garage house? What is the actual forecast?

16.4

SIMPLE EXPONENTIAL SMOOTHING

In Chapter 15, we introduced the concept of smoothing a time series by computing a set of centered *moving averages*. The moving averages were used to derive the various seasonal indexes, but they also provided a "new" time series with considerably less random variation (irregular activity) and no seasonality. Because the moving average series was much smoother, it provided a clearer picture of any existing trend or cyclical activity.

Another method of smoothing a time series, which also serves as a forecasting procedure, is **exponential smoothing**. Unlike the moving averages, this technique uses all the preceding observations to determine a smoothed value for a particular time period. The method described in this section is called **simple** (or single) **exponential smoothing** and works well for a time series containing *no trend* (Figure 16.6). A time series (such as the one in this figure) is said to be **stationary** if the data exhibit no trend and the variance about the mean $(\bar{y}_t)$ remains constant over time. Simple exponential smoothing generally will track the original time series well, provided this series is stationary. We extend the simple exponential smoothing procedure for a series containing trend and seasonality in later sections.

The simplest way to determine a smoothed value for time period t using exponential smoothing is to find a weighted sum of the actual observation for this time period, y_t, and the previous smoothed value, S_{t-1}.

$$S_t = \text{smoothed value for time period, } t$$
$$= Ay_t + (1 - A)S_{t-1} \qquad \textbf{(16.2)}$$

where A is any number between 0 and 1.

The value of A is the **smoothing constant**. Small values of A produce smoothed values giving less weight to the corresponding observation, y_t. You should use such values (say, $A < .1$) for a volatile time series containing considerable irregular activity (noise). In this way, you give more weight to the previous smoothed value, S_{t-1}, rather than to the original observation, y_t. You can use larger values of A for a more stable time series.

The smoothing procedure used here begins by setting the first smoothed value, S_1, equal to the first observation, y_1. So,

$$S_1 = y_1$$

Then,

$$S_2 = Ay_2 + (1 - A)S_1$$
$$= Ay_2 + (1 - A)y_1$$
$$S_3 = Ay_3 + (1 - A)S_2$$
$$S_4 = Ay_4 + (1 - A)S_3$$

and so on.

FIGURE 16.6

Illustration of a stationary time series.

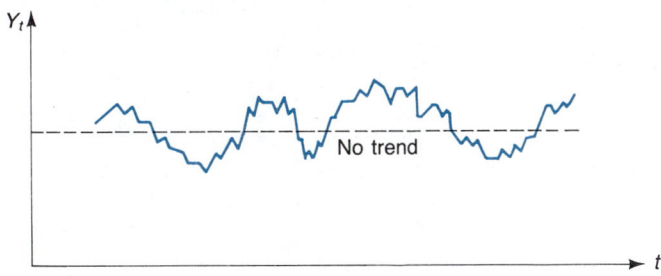

TABLE 16.2

Actual and Smoothed Values for Attendance at Jefferson Civic Center

YEAR	t	y_t	$S_t(A = .1)$	$S_t(A = .5)$	$S_t(A = .9)$
1976	1	5.0	5.0	5.0	5.0
1977	2	8.0	5.3	6.5	7.7
1978	3	2.1	4.98	4.3	2.66
1979	4	7.1	5.19	5.7	6.66
1980	5	4.8	5.15	5.25	4.99
1981	6	2.0	4.84	3.62	2.30
1982	7	7.8	5.13	5.71	7.25
1983	8	5.0	5.12	5.36	5.23
1984	9	14.1	6.02	9.73	13.21
1985	10	13.0	6.72	11.36	13.02
1986	11	13.5	7.39	12.43	13.45
1987	12	14.2	8.07	13.32	14.12
1988	13	14.0	8.67	13.66	14.01

The average attendance (in thousands—y_t) for major events held at the Jefferson County Civic Center for the past 13 years is contained in Table 16.2. We determine the exponentially smoothed values using three smoothing constants, $A = .1$, $A = .5$, and $A = .9$.

The actual time series and the three smoothed series are shown in Figure 16.7. For $A = .1$,

$$S_1 = y_1 = 5.0$$
$$S_2 = (.1)y_2 + (.9)S_1$$
$$= (.1)(8.0) + (.9)(5.0) = 5.3$$
$$S_3 = (.1)y_3 + (.9)S_2$$
$$= (.1)(2.1) + (.9)(5.3) = 4.98$$

and so on.

FIGURE 16.7

Smoothed values for attendance data (Table 16.2).

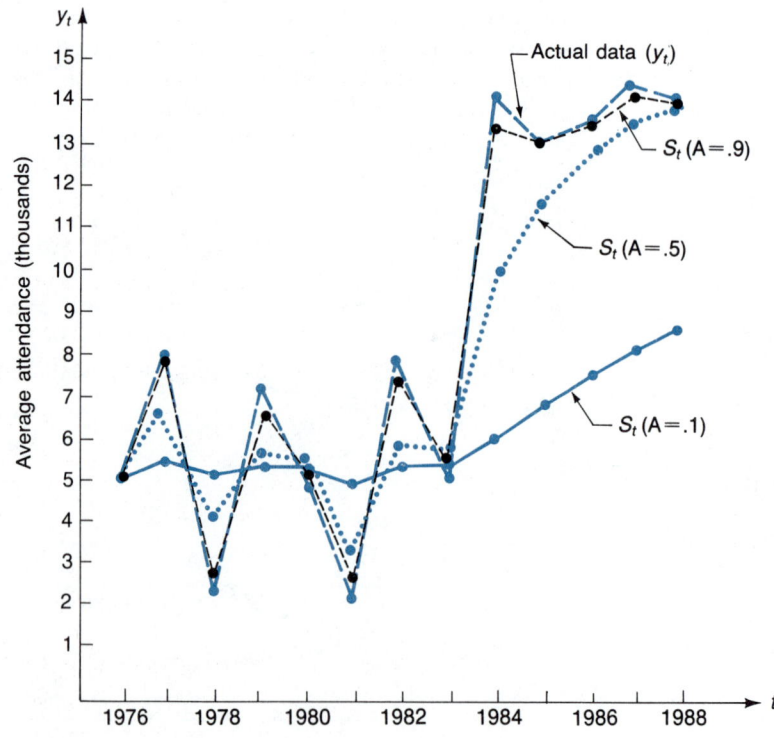

Notice that the average attendance, y_t, had a significant jump in 1984 when (it turns out) the facility was completely refurnished, providing better seating and more accessible snack booths. With the small value of $A = .1$, the smoothed values did not "track" the original series very well after this point. In general, when you use exponential smoothing with a small smoothing constant, the resulting series will be slow to detect any turning points or shifts in the observed values. However, such values of A provide considerable smoothing, as is evident from the values between the years 1976 and 1983.

The large value of $A = .9$ provides much better tracking (see Figure 16.7) but not much smoothing. Larger smoothing constants are more useful for a time series that does not contain a great deal of random fluctuation. Using $A = .5$ offers a compromise between these two extreme smoothing constants. Later we discuss methods of comparing the tracking ability for different values of A, in an effort to determine the best smoothing constant for a particular series.

To see why this procedure is called exponential smoothing, we look at how each smoothed value is obtained. First, $S_1 = y_1$. Then,

$$\begin{aligned} S_2 &= Ay_2 + (1 - A)S_1 \\ &= Ay_2 + (1 - A)y_1 \\ S_3 &= Ay_3 + (1 - A)S_2 \\ &= Ay_3 + (1 - A)[Ay_2 + (1 - A)y_1] \\ &= Ay_3 + A(1 - A)y_2 + (1 - A)^2 y_1 \\ S_4 &= Ay_4 + (1 - A)S_3 \\ &= Ay_4 + (1 - A)[Ay_3 + A(1 - A)y_2 + (1 - A)^2 y_1] \\ &= Ay_4 + A(1 - A)y_3 + A(1 - A)^2 y_2 + (1 - A)^3 y_1 \end{aligned}$$

In general,

$$\begin{aligned} S_t &= Ay_t + A(1 - A)y_{t-1} + A(1 - A)^2 y_{t-2} + \cdots + A(1 - A)^{t-2} y_2 \\ &\quad + (1 - A)^{t-1} y_1 \end{aligned}$$

For example, if $A = .5$, then

$$S_t = .5y_t + .25y_{t-1} + .125y_{t-2} + .062y_{t-3} + \cdots$$

Therefore, each smoothed value is actually a weighted sum of *all the previous observations*. Because the more recent observations have the largest weight, they have a larger effect on the smoothed value. Notice that the weights on the observations are decreasing exponentially. That is, the weight given to a particular observation is some constant (namely, $1 - A$) *times* the weight given to the preceding observation. That is why this procedure is called exponential smoothing.

Forecasting Using Simple Exponential Smoothing

The naive forecasting procedure introduced earlier predicts the time series value for tomorrow using the actual value for today. In other words, $y_{t+1} = y_t$. The exponential smoothing process is similar, except now the forecast for tomorrow is the smoothed value from today. In general,

$$\hat{y}_{t+1} = S_t \tag{16.3}$$

For the special case where $A = 1$, we have

$$\hat{y}_{t+1} = S_t = 1y_t + (1 - 1)S_{t-1} = y_t$$

and the exponential smoothing forecast is the same as that provided by the naive predictor. Because A is considerably less than 1 in practice, the smoothed forecast makes use of all the past observations, rather than only the most recent measurement.

EXAMPLE 16.4

Using simple exponential smoothing with $A = .1$, what are the predicted values and residuals for the attendance data in Table 16.2?

SOLUTION

Suppose the year is 1976 ($t = 1$) and you want a forecast for 1977 ($t = 2$). This would be the smoothed value for 1976, so $\hat{y}_2 = S_1 = 5.0$. Next, the year is 1977, and you need a forecast for 1978. Here, $\hat{y}_3 = S_2 = 5.3$ (from Table 16.2). Continuing in this way, we obtain Table 16.3.

How well does this forecasting procedure perform here? We cannot use Figure 16.7 to compare the $\hat{y}$'s and the y's because, for each time period t, we have plotted y_t and S_t. The predicted value at t, however, is $\hat{y}_t = S_{t-1}$, not S_t. So we need to shift the smoothed values in Figure 16.7 one period to the right. This is shown in Figure 16.8, which contains a plot of the values in Table 16.3 for $A = .1$.

As we might expect from Figure 16.8, the residuals using this method are quite large from 1984 on because this value of A produces smoothed values that fail to adapt to the shift that occurred in 1984. This was not a good time series for simple exponential smoothing because of this sudden shift. However, between the years 1976 and 1983 (a relatively stationary set of observations), the smoothed time series contains much less noise using $A = .1$ and gives a clear indication of the lack of any trend. ∎

This is a popular method of forecasting, particularly when there are hundreds or perhaps thousands of forecasts to be updated for each time period. Such is the case for many inventory-control systems, which are used to predict future demand levels for each item in inventory by means of a computerized forecasting procedure. Simple exponential smoothing often is used for such situations because each forecast, $\hat{y}_{t+1}$, requires only two values: the current observation, y_t, and the previous smoothed value, S_{t-1}. There is no need to store all the previous observations. Computationally, the procedure is very simple and requires less computer time than do more sophisticated forecasting techniques.

TABLE 16.3

Forecasts and Residuals Using Simple Exponential Smoothing on Attendance Data ($A = .1$)

YEAR	t	y_t	$\hat{y}_t$	RESIDUAL ($y_t - \hat{y}_t$)
1976	1	5.0	—	—
1977	2	8.0	5.0	3.00
1978	3	2.1	5.3	−3.20
1979	4	7.1	4.98	2.12
1980	5	4.8	5.19	−.39
1981	6	2.0	5.15	−3.15
1982	7	7.8	4.84	2.96
1983	8	5.0	5.13	−.13
1984	9	14.1	5.12	8.98
1985	10	13.0	6.02	6.98
1986	11	13.5	6.72	6.78
1987	12	14.2	7.39	6.81
1988	13	14.0	8.07	5.93

FIGURE 16.8

Predicted versus actual values for attendance data (Table 16.3).

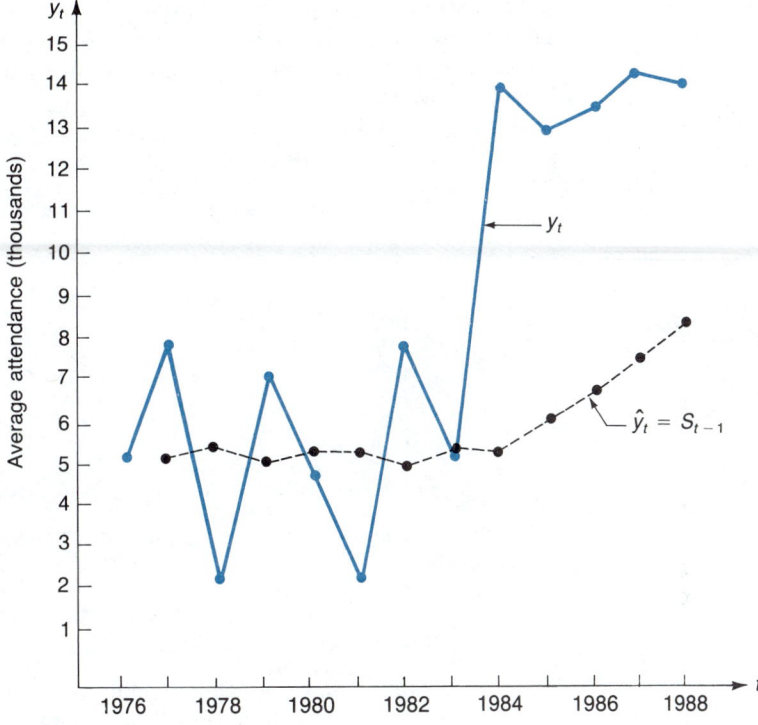

EXERCISES

16.11 If the smoothed value for time period $t = 5$ is $S_5 = 10$, find the forecast using simple exponential smoothing for time period $t = 7$ assuming that $y_6 = 14$ and $A = 0.2$.

16.12 The Fitness and Health Center has been increasing its membership over the years and is considering opening a new center. The following table lists the quarterly membership for the past 4 years.

YEAR	QUARTER 1	QUARTER 2	QUARTER 3	QUARTER 4
1985	105	117	120	115
1986	110	125	130	126
1987	120	135	140	132
1988	131	141	145	141

Using simple exponential smoothing with $A = .3$, find the forecasted membership for the first quarter of 1989.

16.13 The yield on a general obligation bond for Harrisville county fluctuates with the market. The following are the monthly quotations for the past year.

MONTH	YIELD	MONTH	YIELD
Jan.	10.17	Jul.	12.45
Feb.	10.75	Aug.	12.10
Mar.	11.03	Sep.	11.75
Apr.	11.31	Oct.	11.60
May	11.57	Nov.	11.75
Jun.	12.10	Dec.	11.03

Using simple exponential smoothing with $A = .1$, find the forecasted yield for the months of October, November, and December using the data from the previous months. Calculate the residuals.

16.14   Refer to Exercise 16.3. Using simple exponential smoothing with $A = .3$, calculate the forecasted price of the stock of Intersecond Bank for the years 1978 to 1988. Calculate the residuals. Compare the residuals from simple exponential smoothing with those from the naive model.

16.15   The gold reserves in units of millions of fine troy ounces are given for Canada for the year-end of the years 1970–1986.

YEAR	GOLD RESERVE	YEAR	GOLD RESERVE
1970	22.59	1979	22.18
1971	22.69	1980	20.98
1972	21.95	1981	20.46
1973	21.95	1982	20.26
1974	21.95	1983	20.17
1975	21.95	1984	20.14
1976	21.62	1985	20.11
1977	22.01	1986	19.72
1978	22.13		

(Source: The World Almanac and Book of Facts, 1988, p. 103.)

Using simple exponential smoothing with $A = .1$, find the forecasted yield for 1985, 1986, and 1987.

16.5

EXPONENTIAL SMOOTHING FOR A TIME SERIES CONTAINING TREND

The simple exponential smoothing technique discussed in the previous section always will lag behind a time series that contains a steadily increasing or decreasing trend. A procedure known as *Holt's two-parameter linear exponential smoothing* allows you to estimate separately the smoothed value of the time series as well as the average trend gain at each point in time. The resulting smoothed values track the past time series observations more accurately. We refer to this procedure as **linear exponential smoothing**. There are two equations for this method. The first, for smoothing the observations, is

$$S_t = Ay_t + (1 - A)(S_{t-1} + b_{t-1}) \qquad \textbf{(16.4)}$$

The second, for smoothing the trend, is

$$b_t = B(S_t - S_{t-1}) + (1 - B)b_{t-1} \qquad \textbf{(16.5)}$$

where (1) S_t is the smoothed value for time period t, (2) b_t is the smoothed *trend* estimate for this time period, and (3) A and B are smoothing constants between 0 and 1.

Smoothing the Observations (Equation 16.4)

Equation 16.4 is similar to the equation used for simple exponential smoothing, except that S_{t-1} is replaced by $(S_{t-1} + b_{t-1})$ to include the effect of the trend. The smoothing constant for this equation is $0 < A < 1$; typically, $A \le .3$.

Smoothing the Trend (Equation 16.5)

Equation 16.5 is a new addition to the smoothing process and represents the smoothed trend. It uses a separate smoothing constant, B, to smooth the trend values. This constant also is generally less than or equal to .3. This smoothed trend estimate is updated by using a weighted sum of (1) the difference between the last two smoothed values (an estimate of the current "trend") and (2) the previous smoothed trend estimate. Such a procedure significantly reduces any randomness (irregular activity) in the trend values across time.

Forecasting Using Linear Exponential Smoothing

Linear exponential forecasting uses both the smoothed observations and the smoothed trend estimates. The forecast for time period $t + 1$ is the current smoothed value plus the current smoothed trend value.

$$\hat{y}_{t+1} = S_t + b_t \qquad (16.6)$$

We also can use this procedure to forecast any number of time periods into the future, say, m periods. Here,

$$\hat{y}_{t+m} = S_t + mb_t \qquad (16.7)$$

The forecast using equation 16.6 is the **one-step ahead forecast** and the value from equation 16.7 is the **m-step ahead forecast**.

Summarizing the Results

To summarize the necessary calculations for linear exponential smoothing, you can use the format in Table 16.4. The initial year for this time series is 1976. As we did for simple exponential smoothing, we continue to set the first smoothed value, S_1, equal to the first observation, y_1. A new problem arises here, and this is an initial estimate of the trend, b_1. We examine two procedures for estimating this value.

Procedure 1 Let $b_1 = 0$. Provided you have a large number of years in your observed time series, this procedure provides an adequate initial estimate for the trend. The smoothed trend value soon "catches up" with the actual trend contained within the series.

Procedure 2 You can obtain a more accurate estimate of b_1 by using the first five (or so) time periods to estimate the initial trend. A least squares line is constructed through these five observations (exactly as discussed in Chapter 15), with the resulting equation $\hat{y} = a + bt$. The value of b provides an initial trend estimate.

We demonstrate this technique in Example 16.5, which uses both procedures to obtain the initial trend estimate, b_1.

		ACTUAL OBSERVATION	SMOOTHED OBSERVATION	SMOOTHED TREND	FORECAST	RESIDUAL
TABLE 16.4 Summary for Linear Exponential Smoothing						
YEAR	t	(y_t)	(S_t)	(b_t)	$(\hat{y}_t)$	$(y_t - \hat{y}_t)$
1976	1	y_1	$S_1 = y_1$	b_1	—	—
1977	2	y_2	S_2	b_2	$\hat{y}_2$	$y_2 - \hat{y}_2$
1978	3	y_3	S_3	b_3	$\hat{y}_3$	$y_3 - \hat{y}_3$
1979	4	y_4	S_4	b_4	$\hat{y}_4$	$y_4 - \hat{y}_4$
⋮						

EXAMPLE 16.5

The time series contained in the following table contains the city taxes (in thousands of dollars) collected in Jackson City over the past 20 quarters. Using procedures 1 and 2 to calculate an initial trend estimate, obtain the smoothed values, S_t, for each time period. Also determine the predicted values, $\hat{y}_t$, using smoothing constants $A = .1$ and $B = .3$.

YEAR	QUARTER	TAXES COLLECTED	YEAR	QUARTER	TAXES COLLECTED
1984	1	76	1987	1	403
	2	93		2	282
	3	108		3	288
	4	128		4	387
1985	1	196	1988	1	484
	2	175		2	384
	3	141		3	330
	4	236		4	497
1986	1	256			
	2	190			
	3	227			
	4	299			

SOLUTION

A summary of the results using both initial trend estimates is shown in Table 16.5. For procedure 1, $b_1 = 0$. For procedure 2, the first 5 years were used to obtain an initial trend estimate. The least squares line through these five observations is $\hat{y}_t = 37.7 + 27.5t$, and so we use $b_1 = 27.5$ for procedure 2.

To illustrate the necessary calculations here, consider $t = 10$, using procedure 1.

1. $y_{10} = 190$
2. $S_{10} = .1y_{10} + .9(S_9 + b_9)$
    $= .1(190) + .9(164.21 + 14.78) = 180.09$

TABLE 16.5

Solution to Example 16.5 Using Linear Exponential Smoothing ($A = .1$, $B = .3$)

			PROCEDURE 1				PROCEDURE 2		
t	y_t	S_t	b_t	$\hat{y}_t$	$y_t - \hat{y}_t$	S_t	b_t	$\hat{y}_t$	$y_t - \hat{y}_t$
1	76.0	76.00	0.0	—	—	76.00	27.50	—	—
2	93.0	77.70	0.51	76.00	17.00	102.45	27.18	103.50	−10.50
3	108.0	81.19	1.40	78.21	29.79	127.47	26.54	129.63	−21.63
4	128.0	87.13	2.77	82.59	45.41	151.41	25.76	154.01	−26.01
5	196.0	100.51	5.95	89.90	106.10	179.05	26.32	177.16	18.84
6	175.0	113.31	8.01	106.46	68.54	202.33	25.41	205.37	−30.37
7	141.0	123.29	8.60	121.32	19.68	219.07	22.81	227.74	−86.74
8	236.0	142.29	11.72	131.88	104.12	241.29	22.63	241.87	−5.87
9	256.0	164.21	14.78	154.01	101.99	263.13	22.39	263.92	−7.92
10	190.0	180.09	15.11	178.99	11.01	275.97	19.53	285.52	−95.52
11	227.0	198.38	16.06	195.20	31.80	288.65	17.47	295.49	−68.49
12	299.0	222.90	18.60	214.44	84.56	305.41	17.26	306.12	−7.12
13	403.0	257.65	23.44	241.50	161.50	330.70	19.67	322.67	80.33
14	282.0	281.18	23.47	281.09	0.91	343.53	17.62	350.37	−68.37
15	288.0	302.99	22.97	304.66	−16.66	353.83	15.42	361.15	−73.15
16	387.0	332.07	24.80	325.96	61.04	371.03	15.96	369.26	17.74
17	484.0	369.58	28.62	356.87	127.13	396.69	18.87	386.99	97.01
18	384.0	396.78	28.19	398.20	−14.20	412.40	17.92	415.55	−31.55
19	330.0	415.47	25.34	424.97	−94.97	420.29	14.91	430.32	−100.32
20	497.0	446.43	27.03	440.81	56.19	441.38	16.76	435.20	61.80

FIGURE 16.9

Predicted values using linear exponential smoothing ($A = .1$, $B = .3$).

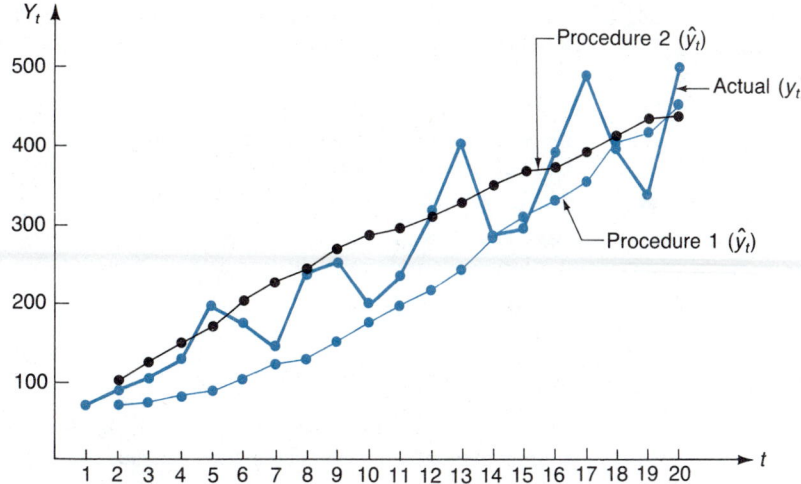

3. $b_{10} = .3(S_{10} - S_9) + .7(b_9)$

   $\qquad = .3(180.09 - 164.21) + .7(14.78) = 15.11$

4. $\hat{y}_{10} = S_9 + b_9$ (from equation 16-6) $= 164.21 + 14.78 = 178.99$

5. Residual for $t = 10$ is

   $y_{10} - \hat{y}_{10} = 190 - 178.99 = 11.01.$

   The values of y_t and $\hat{y}_t$ (the predicted value for that time period) are shown in Figure 16.9. For this particular example, procedure 2, which used the first 5 years to obtain the initial trend estimate, estimated (and smoothed) the past values more accurately. ∎

EXERCISES

16.16 Using Holt's two-parameter linear exponential smoothing technique, find the smoothed value at time $t = 9$, using these values with $A = .1$ and $B = .2$.

$$Y_9 = 10.5$$
$$S_8 = 10.0$$
$$S_7 = 9.5$$
$$b_8 = 0.5$$

16.17 The total number of cars, trucks, and buses (in millions) on the highways of the United States in each year from 1967 to 1985 is given in the following table.

YEAR	CARS, TRUCKS, AND BUSES	YEAR	CARS, TRUCKS, AND BUSES
1967	96.9	1977	142.4
1968	100.9	1978	148.4
1969	105.1	1979	151.8
1970	108.4	1980	155.8
1971	113.0	1981	158.5
1972	118.8	1982	159.5
1973	125.7	1983	161.9
1974	129.9	1984	166.5
1975	132.9	1985	170.2
1976	138.5		

(*Source: United States Statistical Abstracts*, 1987, p. 586.)

a.   Forecast the total number of cars, trucks, and buses for the year 1986. Use linear exponential smoothing with an initial estimate of zero for the slope with $A = .1$ and $B = .2$.

b.   Rework part (a) using the least squares estimate of the slope from the first 4 years.

16.18   The amount of money in the money-market account at First Louisiana State Bank has fluctuated with the interest rate over the years. The bank would like to make a forecast using linear exponential smoothing.

YEAR	AMOUNT IN MONEY-MARKET ACCOUNT	YEAR	AMOUNT IN MONEY-MARKET ACCOUNT
1979	390,121	1984	435,495
1980	395,310	1985	440,370
1981	416,432	1986	444,184
1982	427,489	1987	458,543
1983	443,560	1988	451,967

a.   Using an initial estimate of the slope to be zero, find the predicted value for the amount of money in the money-market account at the end of 1989. Let $A = .1$ and $B = .2$.

b.   Using the least squares estimate for the slope from the first 3 years, redo part (a). What is the predicted value for the amount of money in the money-market account at the end of 1989?

16.19   Using the data in Exercise 16.3, find the predicted value for the stock price of Intersecond Bank for 1989 using linear exponential smoothing. Let the initial value of the slope be zero, with $A = .09$ and $B = .15$.

16.20   Using the data in Exercise 16.12, find the forecasted membership values for the four quarters of 1989. Use the least squares estimate of the slope of the first five observations for the initial value of the slope. Let $A = .3$ and $B = .2$. Compare these forecasted values with those obtained in Exercise 16.12.

16.6

EXPONENTIAL SMOOTHING METHOD FOR TREND AND SEASONALITY

As we discussed earlier, seasonality is present in a time series whenever certain months or quarters are consistently higher or lower than the yearly average. In such cases, an extension of Holt's method, **Winter's linear and seasonal exponential smoothing**, offers additional flexibility. This three-parameter technique (that is, there are three smoothing constants) not only smooths the past observation and trend estimates (as does linear exponential smoothing) but also provides smoothed seasonality factors for each time period.

The smoothing equations for Winter's method are, for smoothing the observations,

$$S_t = A\left(\frac{y_t}{F_{t-L}}\right) + (1 - A)(S_{t-1} + b_{t-1}) \tag{16.8}$$

for smoothing the seasonality factors,

$$F_t = B\left(\frac{y_t}{S_t}\right) + (1 - B)F_{t-L} \tag{16.9}$$

and for smoothing the trend estimates,

$$b_t = C(S_t - S_{t-1}) + (1 - C)b_{t-1} \tag{16.10}$$

Here, (1) S_t is the smoothed observation for time period t; (2) F_t is the smoothed seasonality factor for this time period; (3) b_t is the smoothed estimate of trend; (4) L is the number of periods per year ($L = 4$ for quarterly data and $L = 12$ for monthly data); and (5) A, B, and C are the three smoothing constants.

Equations 16.8 and 16.10 are similar to the corresponding equations from the linear exponential smoothing procedure, except that S_t now consists of deseasonalized smoothed values. These are obtained by dividing each observation, y_t, by the smoothed seasonal factor of one year previous to that observation, F_{t-L}.

Forecasting Using Linear and Seasonal Exponential Smoothing

The procedure for forecasting using Winter's exponential smoothing method is similar to that used for Holt's. Here, the forecast for a particular quarter (month) includes the effect of all three smoothing equations. The forecast for m periods ahead is

$$\hat{y}_{t+m} = (S_t + mb_t) \cdot F_{t+m-L} \qquad \text{(16.11)}$$

The term $(S_t + mb_t)$ represents the smoothed *deseasonalized* estimate and includes the smoothed trend effect. The seasonality is included in the final estimate by multiplying by the smoothed seasonality factor belonging to the quarter (or month) one year previous to the forecast time period, namely, F_{t+m-L}. This procedure is much like that used in Section 16.3, where the deseasonalized estimate was multiplied by the corresponding seasonal index to arrive at the final forecast.

When using this procedure on the past observations, you would, for example, determine $\hat{y}_{10}$ by assuming observations $y_1, y_2, \ldots, y_9$ are available. You would do a one-step ahead forecast ($m = 1$) using the smoothed seasonal value from the previous year; that is, $F_{10-4} = F_6$, assuming quarterly data. As a result,

$$\hat{y}_{10} = [S_9 + (1)b_9] \cdot F_6$$

Similarly,

$$\hat{y}_{11} = [S_{10} + (1)b_{10}] \cdot F_7$$
$$\hat{y}_{12} = [S_{11} + (1)b_{11}] \cdot F_8$$

and so on.

Forecasting *beyond* the range of your observations (extrapolating) is illustrated in the next section.

When dealing with quarterly data, your first set of predicted values will be

$$\hat{y}_5 = [S_4 + (1)b_4] \cdot F_1$$
$$\hat{y}_6 = [S_5 + (1)b_5] \cdot F_2$$
$$\hat{y}_7 = [S_6 + (1)b_6] \cdot F_3$$

and so on. If the time series consists of monthly observations ($L = 12$), then you begin your predicted values with

$$\hat{y}_{13} = [S_{12} + (1)b_{12}] \cdot F_1$$
$$\hat{y}_{14} = [S_{13} + (1)b_{13}] \cdot F_2$$
$$\hat{y}_{15} = [S_{14} + (1)b_{14}] \cdot F_3$$

and so on.

TABLE 16.6   Summary of Linear and Seasonal Exponential Smoothing (Using Procedure 1)

YEAR	QTR.	t	ACTUAL OBSERVATIONS (y_t)	SMOOTHED OBSERVATIONS (S_t)	SMOOTHED SEASONAL FACTORS (F_t)	SMOOTHED TREND (b_t)	FORECAST $(\hat{y}_t)$	RESIDUAL $(y_t - \hat{y}_t)$
1983	1				(1) 1.0			
(year 0)	2				1.0			
	3			(3)	1.0	(2)		
	4			$S_0 = y_4$	1.0	$b_0 = 0$		
1984	1	1	y_1	S_1	F_1	b_1	$\hat{y}_1 = S_0$	$y_1 - \hat{y}_1$
(year 1)	2	2	y_2	S_2	F_2	b_2	$\hat{y}_2 = S_0$	$y_2 - \hat{y}_2$
	3	3	y_3	S_3	F_3	b_3	$\hat{y}_3 = S_0$	$y_3 - \hat{y}_3$
	4	4	y_4	S_4	F_4	b_4	$\hat{y}_4 = S_0$	$y_4 - \hat{y}_4$
1985	1	5	y_5	S_5	F_5	b_5	$\hat{y}_5$	$y_5 - \hat{y}_5$
(year 2)	2	6	y_6	S_6	F_6	b_6	$\hat{y}_6$	$y_6 - \hat{y}_6$
	⋮							

Summarizing the Results

A method of summarizing the necessary calculations is shown in Table 16.6. Suppose that the original year of the observed time series is 1984, with quarterly observations. Initial estimates must be supplied for (1) the seasonal factors for each quarter of 1983, (2) the trend estimate for quarter 4, 1983, and (3) the smoothed value corresponding to quarter 4, 1983.

Once again, we will examine two procedures for this situation—one is quick and easy, and the other is more accurate but requires additional calculations. Procedure 1 is used in Table 16.6. Both procedures are demonstrated in Example 16.6. These are not the only procedures—can you think of one or two others?

Procedure 1

1. Set the initial seasonal factors equal to 1.
2. Set the initial trend estimate (b_0) equal to 0.
3. Set the initial smoothed value for quarter 4, 1983 (S_0), equal to the actual value for quarter 4, 1984 (y_4). This is also the *forecast value* ($\hat{y}_t$) for each of the four quarters in 1984.

Procedure 2

1. Use the first 2 years of data to determine the seasonal indexes. These are the four values for F_t in 1983. Actually, any number of years of data can be used here.
2. Deseasonalize the data for the first 2 years (or any number of years), and calculate the least squares line through these deseasonalized values, d_t. Call this line $\hat{d}_t = a + bt$. The initial trend estimate (b_0) is b.
3. The initial smoothed value for quarter 4, 1983, is $S_0 = [a + b(0)] \cdot$ (seasonal index for quarter 4 in step 1) $= a \cdot$ (seasonal index), where a is the intercept of the least squares line in step 2. Also, S_0 is the *forecast value* ($\hat{y}_t$) for each of the 4 quarters in 1984.

EXAMPLE 16.6

The quarterly taxes from Jackson City in Example 16.5 indicated significant seasonality. In particular, the first-quarter taxes appeared to be considerably larger than those for the yearly average. Using the linear and seasonal expo-

TABLE 16.7

Solution Using Linear and Seasonal Exponential Smoothing, Procedure 1 ($A = .1$, $B = .3$, $C = .2$)

	t	y_t	S_t	F_t	b_t	$\hat{y}_t$	$y_t - \hat{y}_t$
				1.0			
				1.0			
				1.0			
			128	1.0	0.0		
1984	1	76.0	122.80	0.89	−1.04	128.00	−52.00
	2	93.0	118.88	0.93	−1.62	128.00	−35.00
	3	108.0	116.34	0.98	−1.80	128.00	−20.00
	4	128.0	115.89	1.03	−1.53	128.00	0.0
1985	5	196.0	125.05	1.09	0.61	101.28	94.72
	6	175.0	131.81	1.05	1.84	117.45	57.55
	7	141.0	134.70	1.00	2.05	130.78	10.22
	8	236.0	145.95	1.21	3.89	141.03	94.97
1986	9	256.0	158.34	1.25	5.59	163.36	92.64
	10	190.0	165.59	1.08	5.92	172.55	17.45
	11	227.0	177.08	1.08	7.04	171.33	55.67
	12	299.0	190.48	1.32	8.31	222.23	76.77
1987	13	403.0	211.19	1.45	10.79	248.11	154.89
	14	282.0	225.87	1.13	11.57	239.97	42.03
	15	288.0	240.26	1.12	12.13	257.35	30.65
	16	387.0	256.57	1.37	12.97	332.12	54.88
1988	17	484.0	276.05	1.54	14.27	389.79	94.21
	18	384.0	295.23	1.18	15.25	328.43	55.57
	19	330.0	308.94	1.10	14.94	347.21	−17.21
	20	497.0	327.68	1.42	15.70	444.89	52.11

nential smoothing procedures, determine the smoothed value, S_t, and predicted value, $\hat{y}_t$, for each time period. Use smoothing constants $A = .1$, $B = .3$, and $C = .2$.

SOLUTION

The computed results using procedure 1 are summarized in Table 16.7, where $b_0 = 0$, $S_0 = y_4$, and the initial seasonal factors are each 1.

The procedure 2 solution is contained in Table 16.8. The first two years were used to obtain the initial seasonal factors by finding the four seasonal indexes as described in Chapter 15. These are

$$\text{quarter } 1 = 1.23 \qquad \text{quarter } 3 = .91$$
$$\text{quarter } 2 = \ .98 \qquad \text{quarter } 4 = .88$$

Next, the data from the first 2 years were deseasonalized by dividing by the corresponding seasonal index to obtain the deseasonalized values, d_t. A least squares line through these eight values using the simple linear regression procedure from Chapter 15 produced:

$$\hat{d}_t = 44.03 + 23.02t$$

The value of 23.02 (rounded to 23) became the initial slope estimate, b_0. Finally, the initial smoothed value for quarter 4, 1983, is

$$S_0 = (44.03)(\text{initial seasonal index for quarter 4})$$
$$= (44.03)(.88) = 38.7 \text{ (rounded to 39)}$$

Also, $S_0 = 39$ becomes the forecast value ($\hat{y}_t$) for each of the quarters in 1984.

The calculations required here can be illustrated using Table 16.7 and $t = 10$ for procedure 1.

TABLE 16.8

Solution Using Linear and Seasonal Exponential Smoothing, Procedure 2 ($A = .1$, $B = .3$, $C = .2$)

	t	y_t	S_t	F_t	b_t	$\hat{y}_t$	$y_t - \hat{y}_t$
				1.23			
				.98			
				.91			
			39	.88	23		
1984	1	76.0	61.98	1.23	23.00	39.00	37.00
	2	93.0	85.97	1.01	23.19	39.00	54.00
	3	108.0	110.11	0.93	23.38	39.00	69.00
	4	128.0	134.69	0.90	23.62	39.00	89.00
1985	5	196.0	158.44	1.23	23.65	194.55	1.45
	6	175.0	181.19	1.00	23.47	184.00	−9.00
	7	141.0	199.34	0.86	22.40	190.59	−49.59
	8	236.0	225.76	0.94	23.21	199.81	36.19
1986	9	256.0	244.86	1.18	22.39	306.56	−50.56
	10	190.0	259.57	0.92	20.85	266.48	−76.48
	11	227.0	278.65	0.85	20.50	242.31	−15.31
	12	299.0	300.90	0.96	20.85	282.51	16.49
1987	13	403.0	323.85	1.20	21.27	378.24	24.76
	14	282.0	341.34	0.89	20.51	316.67	−34.67
	15	288.0	359.58	0.83	20.06	307.30	−19.30
	16	387.0	382.02	0.98	20.53	364.14	22.86
1988	17	484.0	402.76	1.20	20.58	481.55	2.45
	18	384.0	424.14	0.89	20.74	376.83	7.17
	19	330.0	439.92	0.81	19.75	371.36	−41.36
	20	497.0	464.65	1.00	20.74	448.32	48.68

1. $y_{10} = 190$

2. $S_{10} = .1\left(\dfrac{y_{10}}{F_{10-4}}\right) + .9(S_9 + b_9)$

$= .1\left(\dfrac{y_{10}}{F_6}\right) + .9(S_9 + b_9)$

$= .1\left(\dfrac{190}{1.05}\right) + .9(158.34 + 5.59)$

$= 165.59$

3. $F_{10} = .3\left(\dfrac{y_{10}}{S_{10}}\right) + .7F_6$

$= .3\left(\dfrac{190}{165.59}\right) + .7(1.05)$

$= 1.08$

4. $b_{10} = .2(S_{10} - S_9) + .8b_9$

$= .2(165.59 - 158.34) + .8(5.59)$

$= 5.92$

5. $\hat{y}_{10} = [(S_9 + (1)(b_9)]F_6$ (from equation 16.11)

$= (158.34 + 5.59)1.05$ (computer-stored value is 1.0526)

$= 172.55$

6. Residual for $t = 10$ is

$y_{10} - \hat{y}_{10} = 190 - 172.55$

$= 17.45$

FIGURE 16.10

Forecasted values using linear and seasonal exponential smoothing ($A = .1, B = .3, C = .2$).

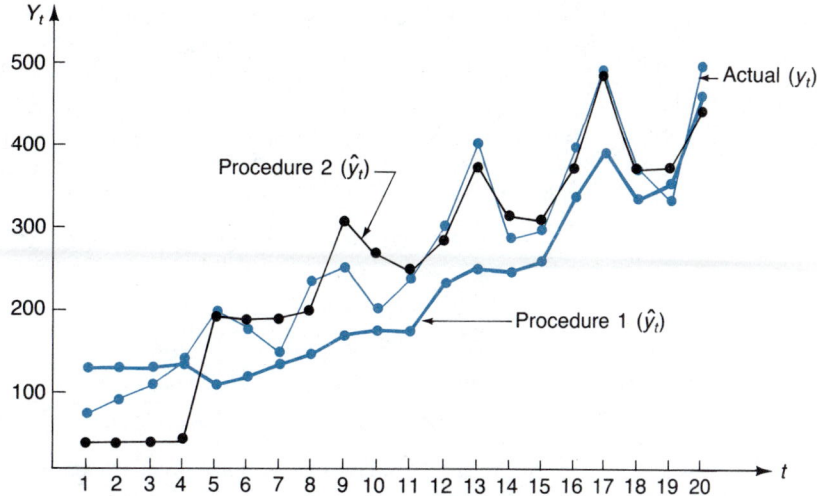

A graphical illustration of the actual observations, y_t, and the predicted value for each time period, $\hat{y}_t$, is shown in Figure 16.10. Once again, the more complex procedure 2 performed better than did procedure 1; for the last ten quarters, procedure 2 tracked the actual time series extremely well. ∎

EXERCISES

16.21 Using Winter's linear and seasonal smoothing technique, answer the following questions using these values:

$$
\begin{array}{lll}
b_6 = 0.7 & F_8 = 0.9 & B = 0.1 \\
S_7 = 15.7 & F_4 = 0.75 & C = 0.1 \\
S_6 = 14.3 & Y_8 = 15.0 & L = 4 \\
S_5 = 14.0 & A = 0.2 &
\end{array}
$$

a. Find the smoothed observation for time $t = 8$.

b. Find the forecasted value for time period 8.

16.22 Ektronics manufactures electronic testing and measuring instruments. The company has managed to capture a large share of the market over the past 4 years. Sales of its equipment (in ten thousands) is recorded monthly for 1985 to 1988.

YEAR	JAN.	FEB.	MAR.	APR.	MAY	JUN.	JUL.	AUG.	SEP.	OCT.	NOV.	DEC.
1985	1.0	1.1	1.2	1.7	1.9	2.3	2.7	3.1	2.5	2.3	2.0	1.9
1986	1.7	1.4	1.5	1.7	2.4	2.7	3.3	3.9	3.4	3.0	2.6	2.0
1987	1.9	2.0	2.1	2.3	3.1	3.5	4.1	4.7	4.3	3.4	2.9	2.8
1988	2.9	2.8	2.7	3.4	3.7	4.1	4.6	5.0	4.7	4.0	3.7	3.6

Use Winter's linear and seasonal smoothing technique to find the smoothed values for the first 4 months of 1988. Let $A = .2$, $B = .1$, and $C = .1$. Using procedure 1, set the initial estimates of the seasonal factors to 1.0 and let $b_0 = 0$.

16.23 The earnings per share of Mecta Mining, a large producer of silver, are:

YEAR	QUARTER 1	QUARTER 2	QUARTER 3	QUARTER 4
1985	.25	.20	.27	.30
1986	.26	.24	.34	.37
1987	.30	.27	.38	.45
1988	.36	.32	.47	.50

Using the linear and seasonal exponential smoothing procedure, determine the predicted value for each quarter of 1989. Use procedure 2 and the first 2 years of data to obtain b_0, S_0, and the four initial seasonal factors (F). Let $A = .3$, $B = .2$, and $C = .1$.

16.24 The average yields on 90-day Treasury bills follow for 1982 through 1986. Use Winter's linear and seasonal smoothing technique to find the smoothed values for the four quarters of 1986. Let $A = .2$, $B = .1$, and $C = .1$. Using procedure 1, set the initial estimates of the seasonal factors to 1.0 and let $b_0 = 0$.

YEAR	QUARTER 1	QUARTER 2	QUARTER 3	QUARTER 4
1982	14.23	14.51	11.01	9.29
1983	8.65	8.80	9.46	9.43
1984	9.69	10.56	11.39	9.27
1985	8.48	7.92	7.90	8.11
1986	7.83	6.92	6.21	6.27

(*Source: Moody's Bank and Finance Manual*, Vol. 1, 1987, p. a14.)

16.25 Refer to Exercise 16.8. Find the predicted amount of inventory for the 4 quarters of 1989 using the linear and seasonal exponential smoothing procedure. Use procedure 2 and the first two years of data to obtain b_0, S_0, and the four initial seasonal factors (F). Let $A = .05$, $B = .1$, and $C = .1$.

16.26 Refer to the Exercise 16.12. Find the forecasted memberships for the four quarters of 1989 using procedure 1 of the linear and seasonal exponential smoothing techniques. Compare these values to those obtained in Exercise 16.12. Let $A = .05$, $B = .1$, and $C = .1$.

16.7

CHOOSING THE APPROPRIATE FORECASTING PROCEDURE

Our purpose in showing you several different forecasting techniques is to point out that, unfortunately, no one procedure works well all the time. One method may work well on a particular steadily increasing time series that has little random fluctuation but perform poorly on a series that has considerable seasonality or random fluctuation.

As you gain more experience in time series applications, you will be better able to choose an appropriate forecasting technique. One factor to consider is the length of your forecast. We classify the forecast period as

Short-term forecast: 1 to 3 months

Medium-range forecast: greater than 3 months but less than 2 years

Long-range forecast: 2 years or more

The exponential smoothing procedures are excellent for *short-term* forecasts, whereas the component decomposition method (in Section 16.3) is useful in medium- and long-range forecasting. The latter also is a popular procedure for many short-term applications, including inventory control and production planning.

One method of deciding whether a certain forecast technique is appropriate in a particular situation is to determine how well the procedure "fits" the observed time series. You accomplish this by pretending that, in each time period, the next observation is unknown and letting the forecasting procedure "predict" the next value (the $\hat{y}_t$ values in the previous examples). Next, you compare the predicted ($\hat{y}_t$) values with the observed values (y_t).

The three most popular methods of comparing the predicted and observed values use measures involving the residuals. They are the mean absolute deviation, the predictive mean squared error, and the mean absolute percentage error.

The **mean absolute deviation (MAD)** is the average of the absolute values of each residual. Let

$$e_t = \text{residual at time } t$$
$$= y_t - \hat{y}_t$$

The mean absolute deviation is defined as

$$\text{MAD} = \frac{\Sigma |e_t|}{n} \tag{16.12}$$

where n is the number of *predicted* values obtained from the past data. For example, when using linear exponential smoothing on 20 data values, you obtain 19 predicted values because $\hat{y}_1$ is unavailable. So n is 19 and not 20.

The **predictive mean squared error (MSE)** is similar to the MAD, except we find the average of the *squared* residuals.

$$(\text{predictive}) \text{ MSE} = \frac{\Sigma e_t^2}{n} \tag{16.13}$$

where, again, n is the number of predicted values.*

The **mean absolute percentage error (MAPE)** considers the *relative* error of each forecast. The relative error at time period t is defined as e_t/y_t. The mean absolute percentage error is defined to be

$$\cdot \text{ MAPE} = \frac{\Sigma \left|\frac{e_t}{y_t}\right|}{n} \tag{16.14}$$

where n is the number of predicted values.

If, during a particular time period, the actual value is $y_t = 50$ and the forecast value is $\hat{y}_t = 60$, the absolute percentage error is

$$\left|\frac{50 - 60}{50}\right| = .2$$

So, the error at this time period is 20% of the actual value. Consequently, for a particular time series, the MAPE is the sum of the absolute percentage error for each predicted value divided by the number of predicted values.

The MSE severely penalizes large residuals because it *squares* each value. Consequently, you use the MSE for situations in which you prefer several small residuals to one large value and wish to be warned if there is one larger residual. The primary advantage of using the MAPE is that it can be used to compare the predictive ability of a certain forecasting technique on two different time

---

* The MSE that we compute as a measure of how well a forecasting procedure fits the observed data is not the same as the MSE computed in a normal ANOVA table. The ANOVA MSE is equal to SSE/(degrees of freedom for residual). In contrast, the predictive MSE is not used in any test of hypothesis and is merely the average of the squared deviations.

TABLE 16.9

Comparison of the Mean Absolute Deviation (MAD), the Mean Squared Error (MSE), and the Mean Absolute Percentage Error (MAPE)

FORECAST	y_t	$\hat{y}_t$	$e_t = y_t - \hat{y}_t$	$|e_t|$	e_t^2	$|e_t/y_t|$
Method 1	36	32	4	4	16	.111
	42	46	−4	4	16	.095
	45	49	−4	4	16	.089
				12	48	.295

MAD = 12/3 = 4.0
MSE = 48/3 = 16.0
MAPE = .295/3 = .098

Method 2	36	34	2	2	4	.056
	42	40	2	2	4	.048
	45	52	−7	7	49	.156
				11	57	.260

MAD = 11/3 = 3.67
MSE = 57/3 = 19.0
MAPE = .260/3 = .087

series. By using relative error, rather than actual error, the effect of the magnitude of the time series observations has been removed from the predictive measure.

To illustrate these measures, consider Table 16.9. For forecasting method 1, there are no large residuals, whereas method 2 results in one large residual. So the MSE is smaller for method 1, but the MAD is smaller for method 2. When using any of these measures, the *smaller* this value, the *more accurate* your forecast procedure.

There is no consensus among statisticians as to which measure is preferable. Instead, it depends on the results of having large forecast residuals. If a large error is disastrous (such as in predicting the inventory level of an expensive product), then using the MSE is preferable. On the other hand, if you can afford to overlook a single severe miss provided the general tracking is close, then the MAD serves better. When comparing the prediction accuracy for two different time series, the MAPE is the appropriate measure.

EXAMPLE 16.7

We used two types of exponential smoothing to smooth (and predict) the city taxes collected in Jackson City over the past 5 years. Data from the past 20 quarters are contained in the table in Example 16.5, in which we used linear exponential smoothing (with smoothing constants $A = .1$ and $B = .3$) to reduce randomness within the observations and trend values. The results are summarized in Table 16.5, using the two procedures for providing initial estimates.

Example 16.6 examined the same data using linear and seasonal exponential smoothing, with smoothing constants $A = .1$, $B = .3$, and $C = .2$. A much better fit was obtained using the more sophisticated method of providing initial smoothed estimates (procedure 2). These results are summarized in Tables 16.7 and 16.8 and are presented graphically in Figure 16.10.

Determine the predictive MSE for each of these four methods. Using the appropriate procedure, determine the forecasted tax revenue for each quarter of 1989.

SOLUTION

1. *Linear exponential smoothing (procedure 1).* The residuals from this forecasting procedure are contained in Table 16.5. The computed predictive mean squared error is

$$MSE = \frac{(17.00)^2 + (29.79)^2 + \cdots + (56.19)^2}{19} = 5670.11$$

2. *Linear exponential smoothing (procedure 2).* Based on Figure 16.9, we would expect a much smaller predictive MSE here. There are no surprises, because

$$MSE = \frac{(-10.50)^2 + (-21.63)^2 + \cdots + (61.80)^2}{19} = 3426.46$$

3. *Linear and seasonal exponential smoothing (procedure 1).* These residuals are listed in Table 16.7, with a corresponding predictive mean squared error of

$$MSE = \frac{(-52.00)^2 + (-35.00)^2 + \cdots + (52.11)^2}{20} = 4414.95$$

(Note that we divide by 20 here because 20 predicted values are available using this procedure.)

A warning: It is not valid to conclude, based on the large MSE value, that this forecasting method is less appropriate than linear exponential smoothing. Remember that we are at the mercy of the particular values of the smoothing constants, A, B, and C. Perhaps a different set of constants would have resulted in a significantly smaller MSE. Finding the best set of constants for any one application involves finding the set of values for A, B, and C that *minimize* the resulting predictive MSE. This (not insignificant) computational burden is one of the drawbacks to using Holt's and Winter's exponential smoothing techniques.

4. *Linear and seasonal exponential smoothing (procedure 2).* Based on Figure 16.10, we observe excellent agreement between the actual time series, y_t, and the predicted series, $\hat{y}_t$, using the smoothed estimates. A very small predictive MSE value would be expected here, and such is the case.

$$MSE = \frac{(37.00)^2 + (54.00)^2 + \cdots + (48.68)^2}{20} = 1828.89$$

We conclude that the best choice of these four alternatives is the linear and seasonal exponential smoothing method using procedure 2 to derive the original estimates.

5. *Forecasted tax revenue.* Using equation 16.11 and the results in Table 16.8, the forecasts for 1989 would be as follows. For the first quarter (one step ahead): $t = 20$, $L = 4$, $m = 1$, and

$$\hat{y}_{21} = [S_{20} + (1)b_{20}] \cdot F_{17}$$
$$= [464.65 + (1)(20.74)](1.20) = 582$$

For the second quarter (two steps ahead): $t = 20$, $L = 4$, $m = 2$, and

$$\hat{y}_{22} = [S_{20} + (2)b_{20}] \cdot F_{18}$$
$$= [464.65 + (2)(20.74)] \cdot (.89) = 450$$

For the third quarter (three steps ahead): $t = 20$, $L = 4$, $m = 3$, and

$$\hat{y}_{23} = [S_{20} + (3)b_{20}] \cdot F_{19}$$
$$= [464.65 + (3)(20.74)](.81) = 427$$

For the fourth quarter (four steps ahead): $t = 20$, $L = 4$, $m = 4$, and

$$\hat{y}_{24} = [S_{20} + (4)b_{20}] \cdot F_{20}$$
$$= [464.65 + (4)(20.74)] \cdot (1.00) = 548$$

■

Selecting the Smoothing Constants

As mentioned earlier, the computed MSE (or MAD) value for any exponential smoothing procedure is determined not only by the procedure itself but also by the value of the necessary smoothing constants. In Example 16.6, the smoothing constants were $A = .1$, $B = .3$, and $C = .2$, with a corresponding MSE value of 1828.89, using procedure 2. By changing these constants, you might improve the fit (lower the MSE), or you might obtain a less desirable solution (a larger MSE).

To illustrate this point, using Example 16.6 and procedure 2, for

$$A = .1, B = .4, C = .3: \quad \text{MSE} = 1748.10 \quad \text{(an improvement)}$$
$$A = .2, B = .2, C = .2: \quad \text{MSE} = 2127.67$$

To arrive at the smallest possible predictive MSE, you must examine a variety of values, compute the MSE for each combination, and select the set of values that provides the smallest MSE. For example, if you consider all non-zero values of A, B, and C between 0 and .4, in increments of .05, this results in $(.4/.05)^3 = 8^3 = 512$ different passes through the procedure to determine the corresponding 512 MSE values. The set of A, B, and C values that provides the smallest MSE is the one you should use in forecasting future values of the time series.

This procedure is not extremely difficult to perform with the help of a computer, but it takes away one main advantage of exponential smoothing—namely, the computational simplicity of this procedure in calculating and updating smoothed estimates. If you are using this method to perform a small number of forecasts, then this poses no problem. On the other hand, if the technique is being used to forecast future demand levels continuously for thousands of inventory items, then this added complexity is a cause for concern. You will have to consider complexity versus cost on an individual application basis.

As a final note here, you should be made aware that computer packages exist for determining the optimal values of the smoothing constants for a given time series.* Such procedures fall under the heading of **automated forecasting procedures**, whereby the computer finds the best fit (using a specified measure of fit, such as MAPE, MSE, or MAD) for a particular forecasting model. By utilizing such procedures, you can drastically simplify the calculations necessary to fit a model to your time series, but, on the negative side, such a "black-box" procedure implies that you sacrifice some control and knowledge of the fitting process.

We can increase the computational burden (but also improve the accuracy) even more by using different values of the smoothing constant(s) at different times in the analysis of a time series. Such techniques are computer controlled. The constant(s) are changed automatically to adapt the process to shifts in the structure of the time series, using **adaptive control procedures**.

We showed you several forecasting procedures and methods for comparing the predictive accuracy of these techniques. Our purpose is to give you an arsenal of methodologies that will allow you to apply each procedure to a particular time series and then summarize and compare the resulting residuals.

---

* Two forecasting packages available for the microcomputer are FORECAST MASTER (Scientific Systems, Inc., Cambridge, Mass.) and SIBYL-RUNNER (Applied Decision Systems, Lexington, Mass).

In this way, you can determine the most accurate procedure for a particular time series and use this method to arrive at a forecast.

Next we turn to another forecasting model, the autoregressive model. With this procedure, we again use the past observations to predict future values but in a slightly different way: We use the past values as variables in a regression equation.

EXERCISES

16.27 Consider the following forecasts for the yearly sales of Dentroff Wholesale Plumbing Supplies. Sales are given in units of 100,000.

YEAR	ACTUAL	FORECAST	YEAR	ACTUAL	FORECAST
1979	1.1	—	1984	2.1	2.0
1980	1.2	1.0	1985	2.0	2.2
1981	1.5	1.3	1986	2.5	2.3
1982	1.9	1.6	1987	2.7	2.5
1983	2.3	1.8	1988	3.4	2.9

a. Compute the MAD, MAPE, and predictive MSE for the forecasts.

b. Compute the MAD, MAPE, and predictive MSE using the naive forecasts of sales.

c. Compare the forecasts in parts (a) and (b).

16.28 The advertising expenditures for a local supermarket (in thousands of dollars) are as follows.

YEAR	JAN.	FEB.	MAR.	APR.	MAY	JUN.	JUL.	AUG.	SEP.	OCT.	NOV.	DEC.
1985	.3	.4	.4	.5	.5	.6	.7	.7	.8	.7	.7	.6
1986	.5	.6	.6	.7	.6	.8	.9	1.0	1.2	1.1	1.1	1.0
1987	.7	.8	.9	.9	.9	.8	.9	1.0	1.1	1.4	1.3	1.2
1988	.9	1.0	1.0	1.2	1.1	1.3	1.4	1.6	1.6	1.4	1.5	1.3

a. Using the naive model, obtain a forecast for the 47 time periods, omitting the first time period. Find the MAD, MAPE, and the predictive MSE.

b. Obtain a forecast for the 47 time periods, omitting the first time period, by using a least squares line that represents just trend. Find the MAD, MAPE, and the predictive MSE.

c. Obtain the forecast for the 47 time periods, omitting the first time period, using only the trend and seasonal components to forecast. Find the MAD, MAPE, and the predictive MSE.

d. Compare the forecasts obtained in parts (a), (b) and (c).

16.29 Two forecasting procedures produce the following set of forecast errors.

YEAR	MONTH	PROCEDURE 1	PROCEDURE 2	YEAR	MONTH	PROCEDURE 1	PROCEDURE 2
1987	Jan.	+5	+1		Nov.	+6	+2
	Feb.	+7	−2		Dec.	+1	+1
	Mar.	+6	+3	1988	Jan.	−3	−3
	Apr.	+2	−1		Feb.	−5	+1
	May	−1	+2		Mar.	−4	0
	Jun.	−2	0		Apr.	+3	−2
	Jul.	−3	+1		May	+2	−19
	Aug.	+2	+1		Jun.	−1	−20
	Sep.	+4	0				
	Oct.	+7	−1				

Compute the MAD and predictive MSE for each forecasting procedure. Comment on the adequacy of the forecasting procedure.

16.30 The following data represent the number of single-family housing starts in a certain sector of the state of California. The units are in 10,000s.

YEAR	QUARTER 1	QUARTER 2	QUARTER 3	QUARTER 4
1985	.6	.8	1.4	.8
1986	.9	1.1	1.7	1.3
1987	1.2	1.4	2.1	1.6
1988	1.4	1.7	2.6	1.9

a. Using the simple exponential procedure with $A = 0.3$, find the predicted number of housing starts for each time period, omitting the first time period. Find the predictive MSE and MAD.

b. Use Holt's two-parameter linear exponential smoothing technique to obtain a forecast for each time period, omitting the first time period. Let $A = .3$ and $B = .2$. Use the least squares estimate of the slope for the initial value of the slope from the first 5 periods. Find the predictive MSE and MAD.

c. Compare the forecasts obtained in parts (a) and (b).

16.31 If an investor invested $10,000 into the T. Krow long-term-growth mutual fund, the investor would have realized a gain of 93% after 5 years. The following table shows the performance over this period of time.

YEAR	QUARTER 1	QUARTER 2	QUARTER 3	QUARTER 4
1984	10,031	9,638	12,591	12,480
1985	12,691	11,745	13,721	13,980
1986	13,043	12,680	15,376	15,860
1987	14,932	14,280	17,035	17,210
1988	16,830	15,923	18,671	19,300

a. Use Winter's linear and seasonal smoothing technique to find the forecasted value of the original $10,000 invested for each of the quarters of 1986, 1987, and 1988. Let $A = .2$, $B = .1$, and $C = .1$. Using procedure 1 and Winter's technique, set the initial estimates of the seasonal factors to 1.0 and $b_0 = 0$. Find the predictive MSE.

b. Redo question (a) with $A = .1$, $B = .2$, and $C = .2$. Find the predictive MSE.

c. Compare the forecasts found in parts (a) and (b) using the predictive MSEs.

16.32 Explain how the MAD, MAPE, and the predictive MSE differ in what they measure. Why should the sum of the forecast errors divided by the number of forecasts not be used to compare two forecasting procedures?

16.33 Find the predictive MSE and MAPE for the forecasts of the amount of money in the money-market account at First Louisiana State Bank in Exercise 16.18(b) for the years 1980 to 1988 using the linear exponential smoothing technique specified in that exercise.

16.34 Find the MAPE and MAD for the forecast of monthly sales of Ektronics in Exercise 16.22 for the months of 1987 and 1988 using the linear and seasonal technique specified in that exercise.

16.8

AUTO-REGRESSIVE FORECASTING TECHNIQUES

So far, the forecasting procedures have used either a member of the exponential smoothing family or the method of time series decomposition. The exponential smoothing technique greatly reduces the randomness (irregular activity) within the observed time series, as well as smoothing any existing trend or seasonal effects.

For the case of simple exponential smoothing, the forecast for the next time period $(\hat{y}_{t+1})$ is the smoothed value for the current period (S_t). When you use the other exponential smoothing procedures, your forecast includes the effect of the smoothed seasonality or trend.

The time series decomposition method determines the various components in each observation, including seasonality, trend, cycles, and random activity. Forecasts are derived by extending the trend and seasonal components into the future. Unlike exponential smoothing, this method can provide reliable long-range forecasts. Naturally, the longer this forecast period is, the less reliable your forecasted value becomes.

This section examines yet another method of forecasting, which can be used when the time series variable is related to past values of itself. By regressing y_t on some combination of its past values, we are able to derive a forecasting equation. So we return to multiple linear regression, except now the dependent variable is y_t, and the predictor variables are the past values, y_{t-1}, y_{t-2}, ... This forecasting technique is **autoregression**; we are essentially regressing the time series variable on itself.

We can expect the autoregressive forecast technique to perform reasonably well for a time series that (1) is not extremely volatile and does not contain extreme amounts of random movement and (2) requires a short-term or medium-range forecast (that is, less than 2 years). The fact that the autoregressive procedure does not perform well on a time series containing a great deal of irregular activity is not a serious disadvantage; practically all forecasting techniques perform poorly in this situation.

Suppose we attempt to predict the values of y_t using the previous two observations. The prediction equation is

$$\hat{y}_t = b_0 + b_1 y_{t-1} + b_2 y_{t-2} \qquad (16.15)$$

The values of b_0, b_1, and b_2 are the least squares regression estimates, obtained from any multiple linear regression computer package. There are two predictor variables here: the **lagged variables**, y_{t-1} and y_{t-2}. Equation 16.15 is a **second-order** autoregressive equation because it uses the first two lagged terms. In general, a pth-order autoregressive equation is written

$$\hat{y}_t = b_0 + b_1 y_{t-1} + b_2 y_{t-2} + \cdots + b_p y_{t-p} \qquad (16.16)$$

We illustrate the computer-input procedure for the second-order equation with an example. Earlier, we used the naive forecasting procedure (forecast for tomorrow is the observed value for today) to predict the closing price of Keller Toy Company stock. The closing prices for a 12-week period are shown on page 711; the predicted values are summarized in Table 16.1.

Suppose we use the second-order autoregressive equation (16.15) to predict these values. The input data required by the linear regression routine consist of the actual time series data and the two columns of lagged data, as illustrated in Table 16.10. The ten input rows are below the line. Notice that we lose the first two observations due to the missing values for the lagged variables. If these data are available from the 2 weeks prior to week 1, they can be used to fill in the missing values, providing 12 rows of data.

A computer solution to this problem using MINITAB is in Figure 16.11. The prediction equation is

$$\hat{y}_t = 45.5 + .278 y_{t-1} - .004 y_{t-2}$$

Also, $R^2 = .068$, indicating that the two lagged variables account for only 7% of the total variation in the ten time series values used as input (y_3 through y_{12}).

To determine whether this is the best way to forecast a particular time series,

TABLE 16.10

Input for the Second
Order Autoregressive
Predictor

y_t	y_{t-1}	y_{t-2}
60	—	—
62.25	60	—
61.75	62.25	60
63	61.75	62.25
64.5	63	61.75
62	64.5	63
63.5	62	64.5
64	63.5	62
63.25	64	63.5
62.5	63.25	64
61	62.5	63.25
61.5	61	62.5

we can use the procedure discussed in the previous section. This involves calculating an MSE (or MAPE or MAD), using the autoregressive technique on the past observations and comparing this MSE with the MSE using other forecasting methods. For example, we obtain an improvement over the naive forecasting procedure here because, from Figure 16.11,

$$\text{(predictive) MSE} = \frac{10.926}{10} = 1.09 \text{ (for autoregressive forecaster)}$$

and from Table 16.1,

$$\text{(predictive) MSE} = \frac{(2.25)^2 + (-.5)^2 + \cdots + (.5)^2}{11} = \frac{21.5}{11}$$

$$= 1.95 \text{ (for naive forecaster)}$$

So, despite the low value of R^2, we obtain a better fit to the observed data using the second-order autoregressive technique. This value of R^2, however,

FIGURE 16.11

MINITAB procedure for
second-order
autoregression.

```
MTB > SET INTO C1
DATA> 60 62.25 61.75 63 64.5 62 63.5 64 63.25 62.5 61 61.5
DATA> END
MTB > LAG BY 1 OF C1, PUT INTO C2
MTB > LAG BY 2 OF C1, PUT INTO C3
MTB > REGRESS Y IN C1 USING 2 PREDICTORS IN C2 C3

The regression equation is
C1 = 45.5 + 0.278 C2 - 0.004 C3

10 cases used 2 cases contain missing values

Predictor       Coef        Stdev      t-ratio        p
Constant       45.50        29.35        1.55       0.165
C2            0.2775       0.3923        0.71       0.502
C3           -0.0035       0.3285       -0.01       0.992

s = 1.249       R-sq = 6.8%       R-sq(adj) = 0.0%

Analysis of Variance
                              SSE
SOURCE        DF        SS          MS         F          p
Regression     2      0.799       0.400      0.26       0.781
Error          7     10.926       1.561
Total          9     11.725
```

Predictive MSE
$$= \frac{10.926}{10}$$
$$= 1.09$$

does indicate that the search for a more accurate forecasting procedure should continue.

Determining Autocorrelations

There are several methods of calculating the correlation between a time series, y_t, and its past values. For example, in Table 16.10,

y_t	y_{t-1}	y_{t-2}
61.75	62.25	60
63	61.75	62.25
⋮	⋮	⋮
61.5	61	62.5

To find the correlation between y_t and y_{t-1}, we could use the equation for a sample correlation coefficient, r, defined in Chapter 13. We also could find the correlation between y_t and y_{t-2} using the same procedure.

There is, however, a computationally more efficient way to find each of these correlations, which also helps us to identify a time series that is not stationary. This equation for the correlation between y_t and y_{t-k} (for any lag k) is

$$r_k = \frac{\sum_{t=1}^{T-k} (y_t - \bar{y})(y_{t+k} - \bar{y})}{\sum_{t=1}^{T} (y_t - \bar{y})^2} \qquad (16.17)$$

where (1) k is the lag under consideration, (2) r_k is the **autocorrelation** for lag k, (3) $\bar{y}$ is the average of the observed time series—that is,

$$\bar{y} = \frac{1}{T} \sum_{t=1}^{T} y_t$$

and (4) T is the number of observations in the time series.

EXAMPLE 16.8   Determine r_1 and r_2 using the following time series:

t	y_t
1	5
2	12
3	20
4	15
5	13

SOLUTION   Here $T = 5$ and

$$\bar{y} = \frac{1}{5}(5 + 12 + 20 + 15 + 13) = 13$$

Consequently,

r_1 correlation between y_t and y_{t-1}

$$= \frac{(y_1 - \bar{y})(y_2 - \bar{y}) + (y_2 - \bar{y})(y_3 - \bar{y}) + (y_3 - \bar{y})(y_4 - \bar{y}) + (y_4 - \bar{y})(y_5 - \bar{y})}{(y_1 - \bar{y})^2 + (y_2 - \bar{y})^2 + \cdots + (y_5 - \bar{y})^2}$$

$$= \frac{(-8)(-1) + (-1)(7) + (7)(2) + (2)(0)}{(-8)^2 + (-1)^2 + (7)^2 + (2)^2 + (0)^2} = \frac{15}{118} = .13$$

FIGURE 16.12

Correlogram for Example
16.8.

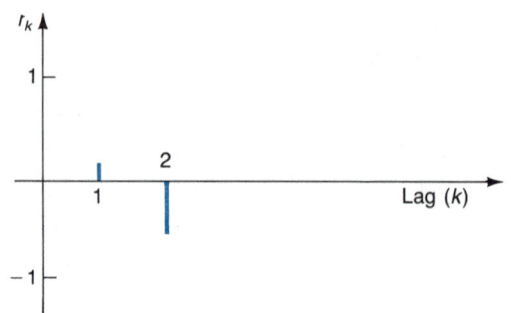

Also,

$$r_2 = \frac{(y_1 - \bar{y})(y_3 - \bar{y}) + (y_2 - \bar{y})(y_4 - \bar{y}) + (y_3 - \bar{y})(y_5 - \bar{y})}{118}$$

$$= \frac{(-8)(7) + (-1)(2) + (7)(0)}{118} = \frac{-58}{118} = -.49$$ ∎

A graphical representation of these autocorrelations is a **correlogram**. The correlogram for Example 16.8 contains the values of r_1 and r_2, as illustrated in Figure 16.12. By inspecting a correlogram, you can determine which lagged variables appear to contribute to the prediction of your time series variable. The autoregressive equation includes those lagged variables for which the corresponding autocorrelation is large. There are statistical procedures for identifying significantly large autocorrelations. These are discussed in Bowerman and O'Connell, and also Makridakis, Wheelright, and McGee (see the end-of-chapter Further Reading).

Detecting Seasonality

The autoregressive forecasting approach does allow you to detect seasonality in your time series data. If seasonality is present in quarterly data, we expect a significant positive correlation between y_t and y_{t-4}; that is, r_4 will be large. This implies that the y value of 1 year ago is a good predictor of the y value today. Similarly, for monthly data, we can expect r_{12} to be large if there is significant seasonality present.

EXAMPLE 16.9

Table 16.11 contains data for the quarterly profits from Ken's Auto Paint Shop. Which lagged variables appear to be correlated with $y_t =$ profit during time period t?

SOLUTION

The MINITAB output for the autocorrelation equation (16.17) is shown in Figure 16.13. Here the *autocorrelation function* (called ACF by MINITAB) computes the first $10 + \sqrt{T}$ autocorrelations, where T (20, here) is the number of observations in the time series. If $10 + \sqrt{T}$ is not an integer, it is rounded *down* to the nearest integer, which is 14 in this case.

TABLE 16.11

Quarterly Profits of
Ken's Auto Paint Shop
(in Thousands of Dollars)

QUARTER	1984	1985	1986	1987	1988
Spring	5.56	5.11	4.12	6.31	4.81
Summer	16.36	15.21	14.33	15.02	16.82
Fall	2.12	5.72	5.25	2.83	4.75
Winter	3.15	2.65	6.75	4.56	8.54

FIGURE 16.13

Autocorrelations for Example 16.9 using MINITAB.

```
MTB > SET INTO C1
DATA> 5.56 16.36 2.12 3.15 5.11 15.21 5.72 2.65 4.12 14.33
DATA> 5.25 6.75 6.31 15.02 2.83 4.56 4.81 16.82 4.75 8.54
DATA> END
MTB > ACF OF C1 ◄─────────── Generates the autocorrelations, r₁, r₂, . . .

ACF of C1

         -1.0 -0.8 -0.6 -0.4 -0.2  0.0  0.2  0.4  0.6  0.8  1.0
           +----+----+----+----+----+----+----+----+----+----+
    1 -0.315                      XXXXXXXXX
    2 -0.260                      XXXXXXX
    3 -0.286                      XXXXXXXX
    4  0.678                         XXXXXXXXXXXXXXXXXX ◄────┐
    5 -0.213                      XXXXX                      │
    6 -0.168                      XXXX                       │
    7 -0.156                      XXXX                       │
    8  0.512                         XXXXXXXXXXXXXX ◄────────┼── Large
    9 -0.235                      XXXXXXX                    │
   10 -0.134                      XXXX                       │
   11 -0.104                      XXXX                       │
   12  0.388                         XXXXXXXXXXX ◄───────────┘
   13 -0.155                      XXXXX
   14 -0.060                      XX
```

The seasonality pattern of four-quarter duration can be seen from the resulting autocorrelations in Figure 16.13. The large r_k values are $r_4 = .678$, $r_8 = .512$, and $r_{12} = .388$. Notice that $r_4 > r_8 > r_{12}$, which is typical in the presence of strong four-period seasonality. The large value of r_4 is your clue that such seasonality exists, and the large values of r_8 and r_{12} confirm this suspicion.

Using y_{t-4} as the predictor variable, we find from Figure 16.14 that the autoregression equation is

$$\hat{y}_t = 1.42 + .87 y_{t-4}$$

The corresponding predictive MSE using this model is

$$(\text{predictive}) \text{ MSE} = \frac{\text{SSE}}{(\text{number of observations used})}$$

$$= \frac{64.40}{16} = 4.025$$

FIGURE 16.14

MINITAB solution for Example 16.9.

```
                                       ┌──────── See Figure 16.13
                                       │
    MTB > LAG BY 4 OF C1, PUT INTO C2
    MTB > REGRESS Y IN C1 USING 1 PREDICTOR IN C2

    The regression equation is
    C1 = 1.42 + 0.870 C2

    16 cases used 4 cases contain missing values

    Predictor      Coef       Stdev     t-ratio        p
    Constant     1.4198      0.9623        1.48    0.162
    C2           0.8697      0.1111        7.83    0.000

    s = 2.145      R-sq = 81.4%     R-sq(adj) = 80.1%

    Analysis of Variance

    SOURCE         DF        SS          MS          F        p
    Regression      1     281.82      281.82      61.26    0.000
    Error          14      64.40        4.60
    Total          15     346.22
                           └──── SSE
```

To decide whether this procedure performs well for this time series, we need to compare this MSE with the MSE obtained using other forecasting techniques.   ◼

Removing Nonstationarity

The autoregressive procedures discussed so far are effective if the time series is *stationary*—that is, if it contains no trend and has constant variance about the mean, $\bar{y}$. One method of detecting nonstationarity in a time series is to examine the correlogram. If you notice that the autocorrelations in the correlogram do not die down rapidly (say after the second or third lag), then your time series is *not stationary*.

With such a time series, an autoregressive procedure is *not appropriate*, unless you modify the time series to make it more stationary. That is, the data should be transformed to a stationary series before attempting to determine seasonality. This can be achieved by using the **differencing** method, which replaces y_t by the first difference, defined by

$$y'_t = y_t - y_{t-1}$$

To illustrate this technique, consider the series

$$2, 5, 8, 11, 14, \ldots$$

This series clearly contains a linear trend; each value is three more than the preceding value. The first differences are

$$y'_2 = 5 - 2 = 3$$
$$y'_3 = 8 - 5 = 3$$
$$y'_4 = 11 - 8 = 3$$
$$y'_5 = 14 - 11 = 3$$

and so on. These values contain no trend, and so the resulting series of first differences is stationary.

If this procedure has been successful in producing a stationary time series, the resulting correlogram (using the y'_t values) should die out quickly. If such is not the case, using the *second* differences of the original time series values $(y_1, y_2, \ldots, y_T)$ often will produce a stationary series. Here, the second differences can be found by deriving first differences of the y'_t values, namely,

$$y''_t = y'_t - y'_{t-1} = y_t - 2y_{t-1} + y_{t-2}$$

You generally can achieve stationarity in your original time series by continuing to take differences until the autocorrelations of the "new" series drop to near zero after two or three time lags (except for possible large values, or spikes, due to seasonal effects). It usually is necessary to determine only first- or second-order differences when dealing with a nonstationary time series.

EXAMPLE 16.10   In Chapter 15, we examined the quarterly sales data of Video-Comp, contained in Table 15.3 on page 673. These data contained a strong linear trend and a definite seasonal pattern, with low sales in the first two quarters and high sales in the final two quarters. A graph of the data is contained in Figure 16.3. Is the seasonal effect more apparent using the original data or the first differences?

SOLUTION   The MINITAB autocorrelations using the original data are summarized and plotted in Figure 16.15. You can see the nonstationarity of this series—the autocorrelations fail to die out after two or three periods. Because of the strong trend component, the seasonal effect is not apparent.

FIGURE 16.15

MINITAB
autocorrelations using
sales data (Example
16.10).

```
MTB > SET INTO C1
DATA> 20 12 47 60 40 32 65 76 56 50 85 100 75 70 101 123
DATA> END
MTB > ACF C1

ACF of C1

          -1.0 -0.8 -0.6 -0.4 -0.2  0.0  0.2  0.4  0.6  0.8  1.0
           +----+----+----+----+----+----+----+----+----+----+
   1  0.541                          XXXXXXXXXXXXXX
   2  0.105                          XXXX
   3  0.280                          XXXXXXXX
   4  0.473                          XXXXXXXXXXXXX
   5  0.122                          XXXX
   6 -0.232                   XXXXXXX
   7 -0.087                      XXX
   8  0.063                          XXX
   9 -0.176                      XXXXX
  10 -0.422              XXXXXXXXXXXX
  11 -0.279                   XXXXXXXX
  12 -0.121                      XXXX
  13 -0.234                   XXXXXXX
  14 -0.343                XXXXXXXXXX
```

The first differences are formed by subtracting adjacent y_t values. These are as follows:

FIRST DIFFERENCES

y_t	$y'_t = y_t - y_{t-1}$
20	*
12	-8
47	35
60	13
40	-20
$\vdots$	$\vdots$

The MINITAB autocorrelations using the first differences are computed in Figure 16.16. Now the seasonality pattern is much clearer, with large negative values for r_2, r_6, and r_{10} as well as large positive values for r_4, r_8, and r_{12}. The negative values are a result of a high (low) sales value in time period t followed by a low (high) sales figure two quarters later. Similarly, the large value of r_4 indicates a strong four-quarter seasonal effect; the large values of r_8 and r_{12} support this conclusion.

FIGURE 16.16

MINITAB
autocorrelations using
first differences
(Example 16.10).

```
MTB > SET INTO C1
DATA> 20 12 47 60 40 32 65 76 56 50 85 100 75 70 101 123
DATA> END
MTB > DIFFERENCES OF LAG 1 FOR C1, PUT INTO C2
MTB > ACF C2

ACF of C2

          -1.0 -0.8 -0.6 -0.4 -0.2  0.0  0.2  0.4  0.6  0.8  1.0
           +----+----+----+----+----+----+----+----+----+----+
   1  0.001                          X
   2 -0.821     XXXXXXXXXXXXXXXXXXXXXX
   3 -0.034                         XX
   4  0.698                          XXXXXXXXXXXXXXXXXX
   5  0.001                          X
   6 -0.546            XXXXXXXXXXXXXX
   7 -0.024                         XX
   8  0.431                          XXXXXXXXXXX
   9 -0.005                          X
  10 -0.307                  XXXXXXXXX
  11 -0.019                          X
  12  0.148                          XXXXX
  13  0.010                          X
```

If you wished to use an autoregressive model for this application, you would use y_t' (not y_t) as your dependent variable because the y_t' values *are* stationary. An excellent set of predictor variables (using Figure 16.16) would be y_{t-2}' and y_{t-4}'. ■

EXERCISES

16.35 Malcom Chemicals manufactures 12 different speciality chemicals. The company's net profit from these speciality chemicals has been stable over the past 5 years, as the data indicate. The figures in the tables are net profit per quarter in units of $10,000.

YEAR	QUARTER 1	QUARTER 2	QUARTER 3	QUARTER 4
1984	4.3	5.7	8.3	2.4
1985	4.7	5.9	8.0	2.7
1986	4.4	5.5	7.6	2.5
1987	4.9	5.9	8.4	2.8
1988	5.3	6.2	8.6	2.9

Fit the following two autoregressive processes to the data:

$$y_t = b_0 + b_1 y_{t-1} + b_2 y_{t-2}$$
$$y_t = b_0 + b_1 y_{t-1} + b_2 y_{t-2} + b_3 y_{t-3} + b_4 y_{t-4}$$

Find the predictive MSE for each autoregressive process and compare the values. Use a computerized statistical package to find the coefficients of the autoregressive equations.

16.36 The debt-to-equity capitalization ratio for Dooper Industries, a maker of machinery parts, has never been above 20% for the past 6 years as a result of excellent management. These ratios (given as a percentage) for the past 6 years are as follows:

YEAR	QUARTER 1	QUARTER 2	QUARTER 3	QUARTER 4
1983	12	15	18	13
1984	11	14	16	12
1985	10	16	19	12
1986	11	17	20	14
1987	12	15	18	12
1988	11	14	17	13

a. Find the autocorrelations for lags of $k = 1, 2, 3$, and 4.

b. Using a computer package, find the coefficients for the autoregressive process

$$\hat{y}_t = b_0 + b_1 y_{t-1} + b_2 y_{t-2} + b_3 y_{t-3} + b_4 y_{t-4}$$

16.37 The percent of Medical International's total revenue derived from freestanding centers for cardiac rehabilitation is given for the past 16 years.

YEAR	PERCENT OF REVENUE	YEAR	PERCENT OF REVENUE
1973	45	1981	42
1974	47	1982	43
1975	46	1983	40
1976	43	1984	37
1977	40	1985	35
1978	36	1986	37
1979	35	1987	40
1980	39	1988	43

a. Find the autocorrelations for lags $k = 1, 2, 3$, and 4.

b. Fit a first-order autoregressive equation to the data.

16.38 How should the graph of a correlogram look if the time series is stationary?

16.39 Refer to Exercise 16.7. Find the autocorrelations for the seasonal data for Sands Motel, using lags of $k = 1, 2, 3, \ldots, 12$. Do the data appear to be stationary?

16.40 The manager of a children's clothing store has tabulated a sales index of children's clothes sold each quarter. The sales of children's clothes has been in an upward trend over the past 5 years, but the sales are also affected by seasonal variation.

YEAR	QUARTER 1	QUARTER 2	QUARTER 3	QUARTER 4
1984	108	256	201	190
1985	185	380	290	320
1986	280	421	360	331
1987	300	504	432	450
1988	400	862	510	480

a. Calculate the autocorrelations through lag 12.

b. Calculate the first differences for the time series.

c. Calculate the autocorrelations for the data obtained by differencing in part (b).

d. Compare the autocorrelations in parts (a) and (c). Comment on whether the autocorrelations describe a stationary time series.

16.41 Determine whether the following data are stationary. If the data are not stationary, take differences until it becomes clear that the resulting time series is stationary. The data represent quarterly interest (in thousands of dollars) paid by a local savings and loan association to those of its depositors who have regular saving accounts.

YEAR	QUARTER 1	QUARTER 2	QUARTER 3	QUARTER 4
1984	25	38	15	21
1985	46	57	36	32
1986	65	80	54	49
1987	84	97	77	70
1988	104	109	96	90

16.42 The total retail sales in billions of dollars for the automotive industry is given for the years 1981 to 1986.

YEAR	JAN.	FEB.	MAR.	APR.	MAY	JUN.	JUL.	AUG.	SEPT.	OCT.	NOV.	DEC.
1986	23.84	23.26	26.04	28.05	30.38	29.51	28.80	29.06	35.04	28.07	24.22	29.54
1985	21.91	22.47	26.51	27.64	29.65	27.56	27.95	28.41	28.48	24.99	23.14	23.16
1984	19.76	21.68	24.03	24.10	26.13	26.25	24.27	24.09	21.20	23.87	22.47	21.01
1983	14.55	15.35	20.31	19.35	20.62	22.36	20.62	20.16	19.61	20.21	20.03	19.91
1982	12.37	14.20	17.50	16.95	17.81	16.69	16.43	16.27	16.19	16.11	17.19	15.82
1981	13.03	13.99	16.87	15.83	15.37	16.36	16.48	16.94	15.62	15.04	13.63	13.54

(*Source: Standard and Poor's Statistical Service, Basic Statistics, Income and Trade*, 1987, p. 109.)

a. Calculate the autocorrelations through lag 15.

b. Calculate the first differences for the time series.

c. Calculate the autocorrelations for the difference data obtained in (b).

d. Comment on the autocorrelations obtained in parts (a) and (c).

16.43 The total number of stamps and stamped paper that the U.S. Postal Service sold in the years from 1970 to 1985 follows. Fit a second-order autoregressive process to the data. Sales are in units of millions of dollars.

YEAR	SALES	YEAR	SALES
1970	1936	1978	3943
1971	1999	1979	4382
1972	2371	1980	4287
1973	2399	1981	4625
1974	2504	1982	5559
1975	2819	1983	5709
1976	3155	1984	6023
1977	3658	1985	6520

(*Source: United States Statistical Abstracts*, 1987, p. 528.)

We have already used linear regression procedures in many of the time series forecasting techniques we have discussed. For instance, simple linear regression was used to describe the *trend* present in time series data. Multiple linear regression was used in the previous section to predict a future time series value using the past one or more observations. This was the autoregressive forecasting method, where, for example,

$$\hat{y}_t = b_0 + b_1 y_{t-1} + b_2 y_{t-2}$$

Here the predictor variables are the *lagged* time series variables, y_{t-1} and y_{t-2}.

Often, you will wish to combine one or more autoregressive terms and several time-related predictor variables into the regression equation, such as

$$\hat{y}_t = b_0 + b_1 y_{t-1} + b_2 y_{t-2} + b_3 X_{1,t} + b_4 X_{2,t} + b_5 X_{3,t}$$

Or you can omit the autoregressive terms and use an equation such as

$$\hat{y}_t = b_0 + b_1 X_{1,t} + b_2 X_{2,t} + b_3 X_{3,t}$$

The notation $X_{1,t}, X_{2,t}, \ldots$ is used (rather than $X_1, X_2, \ldots$) to denote values of the predictor variables during time period t. For example, $X_{2,5}$ represents the observed value of X_2 during the fifth time period. In this way, we can refer to *specific* values of these variables. There is nothing different about this model and the multiple regression model in Chapter 14, except for a slight change in notation because we are dealing with time series data.

Linear regression techniques on time series data offer a variety of opportunities for better forecasting precision, including (1) the use of dummy variables to capture *additive* seasonality and (2) the use of lagged independent variables to allow for time delay effects between the dependent and predictor variables. On the other hand, when this technique is used on time series data, it becomes increasingly difficult to satisfy the linear regression assumptions discussed in Chapter 14. In particular, the error term from one time period often is seriously affected by the previous errors, violating the assumption of independent errors. This means that we can expect to find that the *residuals* are *autocorrelated*. The degree of autocorrelation in the residuals can be measured and tested for significance by calculating the **Durbin-Watson statistic**, which is obtained from the residuals.

These matters are discussed in the remaining sections of this chapter. We look at how regression techniques expand the area of time series analysis but also present a new set of problems.

Use of Dummy Variables for Seasonality

We introduced the concept of additive seasonality previously. Essentially, this type of seasonal effect is present whenever the amount of the seasonal variation is unaffected by the underlying trend in the time series. This was illustrated in Figure 15.15 on page 665. Notice that even as the sales grow over time, the seasonal effect remains the same.

Ignoring any cyclical activity, each observation, y_t, can be described by

$$y_t = TR_t + S_t + I_t$$

where (1) TR_t is the trend component described by a straight line ($TR_t = \beta_0 + \beta_1 t$) or a quadratic curve ($TR_t = \beta_0 + \beta_1 t + \beta_2 t^2$), (2) S_t is the seasonal effect, and (3) I_t is the irregular activity component.

Both the trend and seasonal components can be obtained by using multiple linear regression. The seasonal effects are captured by including a set of *dummy variables* in the regression equation. This type of variable was first introduced in Chapter 14, where we used a set of dummy variables to represent the categories of a *qualitative* variable—such as seasons of the year, in this application.

We use the same procedure for defining dummy variables here—we define one less dummy variable than the number of seasons (categories), L. Because $L = 4$ for quarterly data, we need $L - 1 = 3$ dummy variables. One possible scheme is to define

$$Q_1 = \begin{cases} 1 & \text{if quarter 1} \\ 0 & \text{otherwise} \end{cases} \qquad Q_2 = \begin{cases} 1 & \text{if quarter 2} \\ 0 & \text{otherwise} \end{cases} \qquad Q_3 = \begin{cases} 1 & \text{if quarter 3} \\ 0 & \text{otherwise} \end{cases}$$

$$(16.18)$$

With this procedure, no dummy variable is defined for the fourth quarter. The resulting coefficient of Q_1, Q_2, or Q_3 in the prediction equation will compare the effect of that quarter *against the fourth quarter*. For example, if the coefficient of Q_2 from your computer solution is -5, then, apart from any changes due to trend, quarter 2 produces a value of the dependent variable *five* less than during quarter 4. Quarter 4 is called the **base** quarter. Remember that the base period you select has absolutely *no effect* on the predicted values.

For monthly data, you define $L - 1 = 12 - 1 = 11$ dummy variabl[es] before, you can omit the dummy variable for any one month, which comes the base month for all the computed dummy variable coe[fficients] you omitted a variable for December, then December would be th[e base] and your corresponding set of dummy variables would be

$$M_1 = \begin{cases} 1 \text{ if January} \\ 0 \text{ otherwise} \end{cases}$$

$$M_2 = \begin{cases} 1 \text{ if February} \\ 0 \text{ otherwise} \end{cases} \quad \cdots \quad M_{11}$$

The quarterly sales at Video-[...]
16.3, assuming *multiplicative* se[asonal...]
in reviewing this solution, th[...]
to be increasing along wi[th...]
actually is present. We[...]
ality and calculate [...]
know whether th[...]
observed sales [...]

The predict[ion...]

where (1) $\widehat{TR}_t = b_0 + b_1$.
$b_2 Q_1 + b_3 Q_2 + b_4 Q_3$. Her[e,]
in equation 16.18.

The input configuration use[d...]

and five columns:

y_t	t	Q_1	Q_2	Q_3
20	1	1	0	0
12	2	0	1	0
47	3	0	0	1
60	4	0	0	0
40	5	1	0	0
32	6	0	1	0
65	7	0	0	1
76	8	0	0	0
56	9	1	0	0
50	10	0	1	0
85	11	0	0	1
100	12	0	0	0
75	13	1	0	0
70	14	0	1	0
101	15	0	0	1
123	16	0	0	0

The MINITAB solution for $b_0, b_1, \ldots, b_4$ is provided in Figure 16.17. The resulting equation is

$$\hat{y}_t = 41.75 + 4.8t - 27.6Q_1 - 39.15Q_2 - 10.45Q_3 \qquad (16.20)$$

Notice that $R^2 = .996$, which indicates a strong fit to the 16 observations.

FIGURE 16.17

MINITAB solution using dummy variables to represent quarterly sales Video-Comp.

```
MTB > READ INTO C1-C5
DATA> 20 1 1 0 0
DATA> 12 2 0 1 0
DATA> 47 3 0 0 1
DATA> 60 4 0 0 0
DATA> 40 5 1 0 0
DATA> 32 6 0 1 0
DATA> 65 7 0 0 1
DATA> 76 8 0 0 0
DATA> 56 9 1 0 0
DATA> 50 10 0 1 0
DATA> 85 11 0 0 1
DATA> 100 12 0 0 0
DATA> 75 13 1 0 0
DATA> 70 14 0 1 0
DATA> 101 15 0 0 1
DATA> 123 16 0 0 0
DATA> END
      16 ROWS READ
MTB > REGRESS Y IN C1 USING 4 PREDICTORS IN C2-C5

The regression equation is
C1 = 41.7 + 4.80 C2 - 27.6 C3 - 39.2 C4 - 10.4 C5

Predictor      Coef       Stdev    t-ratio        p
Constant     41.750       1.688      24.74    0.000
C2           4.8000      0.1258      38.16    0.000
C3          -27.600       1.635     -16.88    0.000
C4          -39.150       1.611     -24.30    0.000
C5          -10.450       1.596      -6.55    0.000

s = 2.250      R-sq = 99.6%      R-sq(adj) = 99.4%

Analysis of Variance

SOURCE       DF          SS          MS        F        p
Regression    4     13629.3      3407.3   672.90    0.000
Error        11        55.7         5.1
Total        15     13685.0
```

TABLE 16.12

Summary of Multiplicative Versus Additive Seasonal Forecasting (Quarterly Sales of Video-Comp)

MULTIPLICATIVE SEASONALITY				ADDITIVE SEASONALITY			
t	y_t	$\hat{y}_t$	$y_t - \hat{y}_t$	t	y_t	$\hat{y}_t$	$y_t - \hat{y}_t$
1	20	20.80	−.80	1	20	18.95	1.05
2	12	20.38	−8.38	2	12	12.20	−.20
3	47	40.21	6.79	3	47	45.70	1.30
4	60	50.98	9.02	4	60	60.95	−.95
5	40	37.96	2.04	5	40	38.15	1.85
6	32	34.32	−2.32	6	32	31.40	.60
7	65	63.70	1.30	7	65	64.90	.10
8	76	76.98	−.98	8	76	80.15	−4.15
9	56	55.13	.87	9	56	57.35	−1.35
10	50	48.26	1.74	10	50	50.60	−.60
11	85	87.20	−2.20	11	85	84.10	.90
12	100	102.97	−2.97	12	100	99.35	.65
13	75	72.30	2.70	13	75	76.55	−1.55
14	70	62.21	7.79	14	70	69.80	.20
15	101	110.69	−9.69	15	101	103.30	−2.30
16	123	128.96	−5.96	16	123	118.55	4.45

Forecasting equation: $\hat{y}_t = (19.372 + 5.0375t) \times S_t$

$S_1 = S_5 = S_9 = \cdots = .852$
$S_2 = S_6 = S_{10} = \cdots = .692$

$S_3 = S_7 = S_{11} = \cdots = 1.166$

$S_4 = S_8 = S_{12} = \cdots = 1.290$

predictive MSE $= \dfrac{\Sigma (y_t - \hat{y}_t)^2}{16}$

$= \dfrac{425.36}{16} = 26.58$

Forecasting equation: $\hat{y}_t = 41.75 + 4.8t - 27.6Q_1 - 39.15Q_2 - 10.45Q_3$

predictive MSE $= \dfrac{\Sigma (y_t - \hat{y}_t)^2}{16}$

$= \dfrac{55.7}{16} = 3.48$

How does this method of forecasting compare to the one used in Example 16.3, where we assumed *multiplicative* seasonality in the quarterly sales data? Comparing the two MSE's in Table 16.12, it appears that the seasonality effect is in fact additive, and the marketing vice president was correct.

From equation 16.20, we find that

1. The sales are increasing at an average rate of 4.8 (million dollars) per quarter.
2. Apart from trend effects, sales for the first quarter are 27.6 (million dollars) less than the sales for the fourth quarter.
3. Apart from trend effects, sales for the second quarter are 39.15 less than for the fourth quarter.
4. Sales for the third quarter are 10.45 lower than those during the fourth quarter, ignoring trend effects.

After you have decided to use the additive seasonality equation, to determine the 1989 forecasts you use the appropriate value for t with the 0 or 1 values for the dummy variables.

Forecast for first quarter of 1989: Here, $Q_1 = 1$, $Q_2 = 0$, $Q_3 = 0$, and

$$\hat{y}_{17} = 41.75 + 4.8(17) - 27.6(1) - 39.15(0) - 10.45(0)$$
$$= 95.75 \quad \text{(million dollars)}$$

Forecast for second quarter of 1989: Here, $Q_2 = 1$ (all other Q's $= 0$) and

$$\hat{y}_{18} = 41.75 + 4.8(18) - 27.6(0) - 39.15(1) - 10.45(0)$$
$$= 89$$

Forecast for third quarter of 1989: Now, $Q_3 = 1$ (all other Q's $= 0$) and

$$\hat{y}_{19} = 41.75 + 4.8(19) - 27.6(0) - 39.15(0) - 10.45(1)$$
$$= 122.5$$

Forecast for the fourth quarter of 1989: Since $Q_1 = Q_2 = Q_3 = 0$,

$$\hat{y}_{20} = 41.75 + 4.8(20) - 27.6(0) - 39.15(0) - 10.45(0)$$
$$= 137.75$$

EXERCISES

16.44   The following multiple regression equation was used to fit quarterly sales data. The data are in units of 10,000.

$$\hat{y}_t = 0.5 + 1.8t + 3Q_1 - 0.6Q_2 + 1.1Q_3$$

where

$$Q_1 = \begin{cases} 1 & \text{if Quarter 1} \\ 0 & \text{otherwise} \end{cases}$$

$$Q_2 = \begin{cases} 1 & \text{if Quarter 2} \\ 0 & \text{otherwise} \end{cases}$$

$$Q_3 = \begin{cases} 1 & \text{if Quarter 3} \\ 0 & \text{otherwise} \end{cases}$$

a.   Apart from seasonality, how fast are sales increasing each quarter?

b.   Apart from trend, how much are sales for the first quarter ahead of sales for the fourth quarter?

c.   Apart from trend, how much lower are sales for the second quarter than for the fourth quarter?

d.   What is the forecasted sales for time period 12, which is a fourth quarter?

16.45   Activity in the federal funds market consists of short-term (usually 1 day) loans, of perhaps several millions, by one commercial bank with surplus reserve funds to another bank, which is short of reserves. The interest rate charged for these loans is the federal funds rate. The federal funds rate is given below from 1981 through 1986. Determine the multiple regression equation which takes into account trend and seasonality.

YEAR	QUARTER 1	QUARTER 2	QUARTER 3	QUARTER 4
1981	16.57	17.78	17.58	13.59
1982	14.23	14.51	11.01	9.29
1983	8.65	8.80	9.46	9.43
1984	9.69	10.56	11.39	9.27
1985	8.48	7.92	7.90	8.11
1986	7.83	6.92	6.21	6.27

(*Source: Moody's Bank and Finance Manual*, Vol. 1, 1987, p. a14.)

16.46   Quality Homes Inc. builds single-family houses in several large cities. The number of carpenters that it hires fluctuates with the demand for housing. For the 6 years shown, find a multiple regression equation to predict the number of carpenters on the payroll at Quality Homes. The multiple regression equation should take into account the trend and the monthly seasonality. What percentage of the total variation has been explained using these variables?

YEAR	JAN.	FEB.	MAR.	APR.	MAY	JUN.	JUL.	AUG.	SEP.	OCT.	NOV.	DEC.
1983	145	148	150	169	197	250	267	290	280	230	180	160
1984	155	150	166	178	220	290	320	325	300	270	200	190
1985	180	195	210	213	255	308	350	368	345	320	280	250
1986	230	245	258	290	330	342	394	405	380	350	310	290
1987	285	298	310	345	396	408	451	465	441	430	390	370
1988	350	361	372	395	420	439	480	495	483	450	420	410

16.47 Refer to Exercise 16.7. Using a computerized statistical package, find the coefficients of a multiple regression model for the monthly data for Sands Motel that takes into account both the trend and the effect due to the particular month.

16.48 Explain the difference between multiplicative seasonality and additive seasonality. Which forecasting techniques are best suited for each of these situations?

16.49 Refer to Exercise 16.36. Find the coefficients of the multiple regression model for the quarterly data from Dooper Industries that takes into account both the trend and seasonality.

Use of Lagged Independent Variables

When using multiple linear regression on time series data, we can represent the model as

$$y_t = \beta_0 + \beta_1 X_{1,t} + \beta_2 X_{2,t} + \cdots + \beta_k X_{k,t} + e_t$$

where each $X_{i,t}$ represents the value of predictor variable X_i in time period t, and e_t is the error component for this period.

Look at the data in Table 16.13. The object is to predict the number of home loans financed by Liberty Savings and Loan. The data in this table consist of semiannual figures from the past 8 years. The predictor variables were chosen to be average interest rate (X_1), advertising expenditure (X_2), an election-year dummy variable $(X_3 = 1$ for an election year and 0 otherwise), and a seasonal

TABLE 16.13

Housing Data for Liberty Savings and Loan

YEAR	t	NUMBER OF HOME LOANS y_t	AVERAGE INTEREST RATE $(X_{1,t})$	ADVERTISING EXPENDITURE (THOUSANDS OF $) $(X_{2,t})$	ELECTION YEAR VARIABLE $(X_{3,t})$	SEASONAL VARIABLE $(X_{4,t})$
1981	1	115	13.0	9.1	0	1
	2	84	14.7	9.5	0	0
1982	3	76	14.1	6.1	0	1
	4	81	12.0	8.2	0	0
1983	5	122	11.8	10.4	0	1
	6	118	12.4	6.7	0	0
1984	7	106	11.0	7.5	1	1
	8	140	14.5	7.8	1	0
1985	9	86	11.0	5.1	0	1
	10	96	10.1	6.8	0	0
1986	11	110	12.9	6.8	0	1
	12	76	14.8	9.1	0	0
1987	13	62	10.5	7.5	0	1
	14	104	9.8	5.1	0	0
1988	15	135	10.1	8.8	1	1
	16	120	10.8	4.3	1	0

dummy variable, where $X_4 = 1$ for the first 6 months and $X_4 = 0$ for the final 6 months.

A financial analyst at Liberty Savings saw two problems with the data in Table 16.13. First, she thought that the number of home loans during a particular 6-month period should be more affected by the *previous* 6-month interest rate, due to the time delay between loan application and actual funding. The value of y_t increases by 1 each time a loan is funded, not when the application is turned in. This time delay generally ran between 3 and 6 months. For the same reason, she believed that the effect of any increased (or decreased) advertising would be reflected in the loan amounts of the next period.

The procedure to follow in this situation is to lag the predictor variable by the corresponding time lag. For this example, we can lag X_1 and X_2 by one time period, which results in the following regression model:

$$y_t = \beta_0 + \beta_1 X_{1,t-1} + \beta_2 X_{2,t-1} + \beta_3 X_{3,t} + \beta_4 X_{4,t} + e_t$$

FIGURE 16.18

MINITAB solution using lagged interest and advertising variables (data from Table 16.13).

```
MTB > READ INTO C1-C5
DATA> 115 13.0 9.1 0 1
DATA> 84 14.7 9.5 0 0
DATA> 76 14.1 6.1 0 1
DATA> 81 12.0 8.2 0 0
DATA> 122 11.8 10.4 0 1
DATA> 118 12.4 6.7 0 0
DATA> 106 11.0 7.5 1 1
DATA> 140 14.5 7.8 1 0
DATA> 86 11.0 5.1 0 1
DATA> 96 10.1 6.8 0 0
DATA> 110 12.9 6.8 0 1
DATA> 76 14.8 9.1 0 0
DATA> 62 10.5 7.5 0 1
DATA> 104 9.8 5.1 0 0
DATA> 135 10.1 8.8 1 1
DATA> 120 10.8 4.3 1 0
DATA> END
      16 ROWS READ
MTB > LAG BY 1 OF C2, PUT INTO C12
MTB > LAG BY 1 OF C3, PUT INTO C13
MTB > REGRESS Y IN C1 USING 4 PREDICTORS IN C12,C13,C4,C5;
SUBC> DW.

The regression equation is
C1 = 193 - 9.64 C12 + 2.51 C13 + 17.0 C4 + 4.59 C5

Predictor        Coef        Stdev      t-ratio
Constant        192.57       31.31        6.15
C12              -9.635       2.503       -3.85
C13               2.510       2.429        1.03
C4               17.048       8.964        1.90
C5                4.590       7.205        0.64

s = 13.23       R-sq = 77.0%      R-sq(adj) = 67.8%

Analysis of Variance
                                     R²
SOURCE        DF         SS           MS
Regression     4       5843.9       1461.0
Error         10       1749.0        174.9
Total         14       7592.9
                                  SSE

SOURCE        DF       SEQ SS
C12            1       4930.6
C13            1        112.5
C4             1        729.8
C5             1         71.0


Durbin-Watson statistic = 1.10
```

The other problem she foresaw with using the regression variables in Table 16.13 is a common difficulty in applying regression techniques to a forecasting situation. If we had not lagged X_1 and X_2, then any forecast for 1989 would involve *specifying values for X_1 and X_2 for this future time period*. Because these values may be just as difficult to predict as the dependent variable, our model has little potential as a forecaster. By lagging these variables, we have removed this problem for a one-period-ahead forecast because now the lagged values for tomorrow are the actual values for today.

A portion of the input data using the lagged predictors is:

y_t	$X_{1,t-1}$	$X_{2,t-1}$	$X_{3,t}$	$X_{4,t}$
84	13.0	9.1	0	0
76	14.7	9.5	0	1
81	14.1	6.1	0	0
122	12.0	8.2	0	1
118	11.8	10.4	0	0
106	12.4	6.7	1	1
140	11.0	7.5	1	0
⋮				

The resulting prediction equation, shown in Figure 16.18, is

$$\hat{y}_t = 192.57 - 9.635X_{1,t-1} + 2.510X_{2,t-1} + 17.048X_{3,t} + 4.590X_{4,t}$$

$$\text{(16.21)}$$

We can draw several conclusions from Figure 16.18:

1. Based on the t values, the lagged interest rate variable ($X_{1,t-1}$) and the election year variable ($X_{3,t}$) are the only significant predictors of the number of home loans financed by Liberty.
2. Because $R^2 = .770$, 77% of the total variation of the y_t values has been explained using these four predictors.
3. The predictive mean squared error (for comparison purposes) is

$$(\text{predictive}) \text{ MSE} = \frac{\text{SSE}}{15} = \frac{1749}{15} = 116.6$$

To illustrate the effect of lagging the independent variables, Figure 16.19 contains the solution to this example, where neither X_1 nor X_2 are lagged. Two things are striking. First, the R^2 value drops from .770 to .398. Second, the interest variable, X_1, is *no longer significant*, based on its small t-value.

For each application, try lagging the independent variables that could possibly have a delayed action on the dependent variable. You also should vary the lag period to account for predictor effects that show up several time periods later.

What would be the forecast for the first half of 1989 using the prediction equation 16.21? Here, $X_{1,t-1} = X_{1,16}$ = interest rate for the last half of 1988 = 10.8, and $X_{2,t-1} = X_{2,16} = 4.3$. Also, $X_{3,17} = 0$ (1989 is not an election year) and $X_{4,17} = 1$ (this applies to the first half of 1989). So,

$$\hat{y}_{17} = 192.57 - 9.635(10.8) + 2.510(4.3) + 17.048(0) + 4.590(1)$$

$$= 103.9$$

This results in a drop from the value for the last half of 1988 (120). This is primarily due to the small amount spent on advertising in the final half of 1988 ($4300) and the fact that 1989 is not an election year.

FIGURE 16.19

MINITAB solution without lagging the interest and advertising variables (data from Table 16.13).

```
MTB > REGRESS Y IN C1 USING 4 PREDICTORS IN C2-C5;
SUBC> DW.

The regression equation is
C1 = 91.5 - 1.11 C2 + 2.33 C3 + 31.4 C4 - 2.6 C5

Predictor        Coef       Stdev     t-ratio
Constant        91.53       40.89        2.24        No longer
C2              -1.109       3.493       -0.32   ←   significant
C3               2.330       3.459        0.67
C4              31.36       12.10         2.59
C5              -2.63       10.78        -0.24

s = 20.63        R-sq = 39.8%      R-sq(adj) = 17.9%

Analysis of Variance

SOURCE        DF          SS           MS
Regression     4        3095.0       773.8
Error         11        4679.9       425.4
Total         15        7774.9

SOURCE        DF       SEQ SS
C2             1         89.1
C3             1        110.3
C4             1       2870.2
C5             1         25.4


Durbin-Watson statistic = 2.18
```

EXERCISES

16.50 What is the importance of using lagged independent variables in a regression equation?

16.51 The following table lists the food price index (FPI) and the per-capita income (PCI) index for a certain third-world nation. The indexes are listed in 6-month increments.

YEAR	MONTH	Y: FPI	X: PCI
1980	Jun.	109	104
	Dec.	101	116
1981	Jun.	104	124
	Dec.	107	120
1982	Jun.	105	167
	Dec.	114	133
1983	Jun.	108	188
	Dec.	126	148
1984	Jun.	112	101
	Dec.	100	158
1985	Jun.	114	115
	Dec.	104	127
1986	Jun.	107	112
	Dec.	102	141
1987	Jun.	110	124
	Dec.	105	162
1988	Jun.	113	184
	Dec.	120	140

a. Find the simple regression equation

$$\hat{y}_t = b_0 + b_1 X_t$$

b. Find the simple regression equation

$$\hat{y}_t = b_0 + b_1 X_{t-1}$$

c. Compare the R^2 for the two equations found in parts (a) and (b).

16.52 Credit Corp finances small home-improvement projects for 1 year or less. Usually, the company does not screen clients rigorously, because a mechanics lien is placed on the home. Credit Corp has found that a significant correlation exists between the interest rate on loans and the number of defaults. The following data give the average interest rate charged on a loan for that quarter and the number of times a loan holder has been more than 30 days behind on a payment.

YEAR	QUARTER	Y: AVERAGE INTEREST RATE	X: TIMES BEHIND ON A LOAN PAYMENT
1985	1	8.5	53
	2	8.0	45
	3	9.0	44
	4	9.5	47
1986	1	9.0	49
	2	10.0	46
	3	10.5	49
	4	10.0	52
1987	1	10.5	50
	2	11.25	51
	3	11.50	55
	4	11.00	57
1988	1	12.25	52
	2	12.50	60
	3	12.00	63
	4	11.50	58

a. Find the simple regression equation

$$\hat{y}_t = b_0 + b_1 X_t$$

b. Find the simple regression equation

$$\hat{y}_t = b_0 + b_1 X_{t-1}$$

c. Compare the R^2 for the two equations.

16.53 Refer to Exercise 16.35. The following is the number of salespeople working for Malcom Chemical Company over the 5-year period 1984 through 1988.

YEAR	QUARTER 1	QUARTER 2	QUARTER 3	QUARTER 4
1984	30	41	13	23
1985	31	39	15	22
1986	27	37	12	25
1987	30	42	14	26
1988	32	43	15	25

Find the multiple regression equation to predict the net profit per quarter for Malcom Chemical. Use dummy variables to represent seasonality and the variable, number of salespeople, lagged by one time period. Also find the predictive MSE.

16.54 Refer to Exercise 16.36. The following is the level of inventory for Dooper Industries in units of 10,000 for the years 1983 to 1988.

YEAR	QUARTER 1	QUARTER 2	QUARTER 3	QUARTER 4
1983	4.8	6.1	4.2	3.8
1984	4.2	5.2	4.1	3.3
1985	5.0	6.4	4.2	3.2
1986	5.5	6.8	4.2	3.1
1987	5.1	6.1	4.0	3.8
1988	4.2	5.9	4.1	3.7

Find the multiple regression equation to predict the debt-to-equity capitalization ratio for Dooper Industries using a variable to represent seasonality and the variable, level of inventory lagged by 1 quarter. Find the R^2.

16.55 A local used-car dealer believes that advertising has greatly increased sales at the used car lot. The sales (in hundreds of dollars) of cars for each month and also the corresponding advertising expenditure are given at the top of page 758.

YEAR	MONTH	Y: SALES	X: ADVERTISING EXPENDITURE	YEAR	MONTH	Y: SALES	X: ADVERTISING EXPENDITURE
1987	Jan.	19	250	1988	Jan.	16	300
	Feb.	16	405		Feb.	17	350
	Mar.	20	308		Mar.	19	401
	Apr.	17	425		Apr.	20	560
	May	21	550		May	23	630
	Jun.	24	300		Jun.	25	725
	Jul.	16	450		Jul.	28	630
	Aug.	22	522		Aug.	26	550
	Sep.	23	630		Sep.	23	430
	Oct.	26	510		Oct.	21	400
	Nov.	23	320		Nov.	21	350
	Dec.	17	250		Dec.	19	260

a. Find the simple regression equation

$$\hat{y}_t = b_0 + b_1 X_t$$

b. Find the simple regression equation

$$\hat{y}_t = b_0 + b_1 X_{t-1}$$

c. Compare the R^2 for the regression equations in parts (a) and (b).

16.10

THE PROBLEM OF AUTO-CORRELATION: THE DURBIN-WATSON STATISTIC

A problem you will encounter frequently when using multiple linear regression on time series data is that the residual terms (e_t) are not independent. We discussed autoregressive forecasting, in which an observation (y_t) is related to its past values. For this situation, we said that significant autocorrelation was present in the *observations*.

When we have an autocorrelated time series, we simply regress y_t on the past values. However, when a particular model dealing with least squares estimates of the unknown parameters (such as multiple linear regression) results in **autocorrelated residuals**, problems do arise. In particular, all tests of hypothesis, including the *t*-tests for individual predictors, become extremely suspect.

Detecting Autocorrelated Residuals

The Durbin-Watson statistic frequently is used to test for significant autocorrelation in the residuals. If its value is very small, significant *positive* autocorrelation exists. This means that each value of e_t is very close to its neighbors, e_{t-1} and e_{t+1}. A large value indicates high *negative* autocorrelation, where each e_t value is very different from the adjacent residuals.

The value of the Durbin-Watson statistic (DW) is determined using each residual value, e_t, and its previous value, e_{t-1}.*

$$DW = \frac{\sum_{t=2}^{T} (e_t - e_{t-1})^2}{\sum_{t=1}^{T} e_t^2} \qquad (16.22)$$

where T is the number of observations in the time series.

---

* The Durbin-Watson statistic can be approximated using the autocorrelation of lag one (called r_1) discussed in Section 16.8. This approximation is $DW \cong (1 - r_1)$.

The range of possible values for the Durbin-Watson statistic is from 0 to 4. The **ideal value of DW** is 2. For this situation, the errors are completely uncorrelated, and there is no violation of the independent errors assumption. As DW decreases from 2, positive autocorrelation of the errors increases. Values between 2 and 4 indicate various degrees of negative autocorrelation.

The common problem of autocorrelated errors has to do with *positive* correlation between neighboring errors. When this occurs, the errors are not independent of one another; instead, each error is largely determined by its previous value. This implies that a similar behavior will exist for the estimated residuals in the regression model—that is, we can expect the estimated residuals to be positively correlated. The test for autocorrelation using the DW statistic is unique, in that there is a certain range of DW values for which we can neither reject H_0: no autocorrelation exists, nor fail to reject it. The testing procedure uses Table A.9; the value of k in Table A.9 represents the number of predictor variables in the regression equation. The hypotheses are

$$H_0: \text{no autocorrelation exists}$$

$$H_a: \text{positive autocorrelation exists}$$

The testing procedure, using the values of d_L and d_U from Table A.9, is

reject H_0 if $DW < d_L$

fail to reject H_0 if $DW > d_U$

the test is inconclusive if $d_L \leq DW \leq d_U$

The assumption is that the errors follow a normal distribution.

EXAMPLE 16.11

Determine the value of the Durbin-Watson statistic if equation 16.21 is used on the home-loan data in Table 16.13. Use $\alpha = .05$.

SOLUTION

Using Figure 16.18 to obtain the estimated values from equation 16.21, Table 16.14 shows the necessary calculations that make up the Durbin-Watson statistic. This is a standard portion of the MINITAB output, as indicated in Figure 16.18.

TABLE 16.14

Calculating the Durbin-Watson Statistic for Example 16.11

t	y_t	$\hat{y}_t$	$e_t = y_t - \hat{y}_t$	$e_t - e_{t-1}$	$(e_t - e_{t-1})^2$	e_t^2
1	115	—	—	—	—	—
2	84	90.16	−6.1600	—	—	37.946
3	76	79.37	−3.3700	2.79	7.784	11.357
4	81	72.03	8.9700	12.34	152.276	80.461
5	122	102.13	19.8700	10.90	118.810	394.817
6	118	104.99	13.0100	−6.86	47.060	169.260
7	106	111.55	−5.5500	−18.56	344.474	30.803
8	140	122.46	17.5400	23.09	533.148	307.652
9	86	77.03	8.9700	−8.57	73.445	80.461
10	96	99.39	−3.3900	−12.36	152.770	11.492
11	110	116.92	−6.9200	−3.53	12.461	47.886
12	76	85.35	−9.3500	−2.43	5.905	87.422
13	62	77.41	−15.4100	−6.06	36.724	237.468
14	104	110.23	−6.2300	9.18	84.272	38.813
15	135	132.59	2.4100	8.64	74.650	5.808
16	120	134.40	−14.4000	−16.81	282.576	207.360
					1926.4*	1749.0**

* Numerator for DW
** Denominator for DW ($=$SSE in Figure 16.18)

Using Table 16.14, the Durbin-Watson statistic for this situation is

$$DW = \frac{1926.4}{1749} = 1.10$$

Using $\alpha = .05$, $n = 15$, $k = 4$, and Table A.9, $d_L = .69$ and $d_U = 1.97$. Notice that the value of n is 15 because only 15 values of y_t were estimated. Also, equation 16.21 uses four variables to predict y_t, and so $k = 4$. The test of hypothesis will be to

reject H_0 if $DW < .69$

fail to reject H_0 if $DW > 1.97$,

the test is inconclusive if $.69 \leq DW \leq 1.97$

Because $DW = 1.10$ falls in the gray area between .69 and 1.97, positive autocorrelation *could* exist, but this test is inconclusive. We need additional information before we can draw any conclusion concerning possible error autocorrelation. ∎

Procedures for Correcting Autocorrelated Errors

All is not lost if the Durbin-Watson test concludes that significant autocorrelation is present in the residuals. We do not attempt to describe fully all the remedies for this situation. (For discussions of these methods, consult Bowerman and O'Connell, *Forecasting and Time Series*, and Makridakis, Wheelright, and McGee, *Forecasting: Methods and Applications*, in the Further Reading at the end of this chapter.) However, the following procedures are often used to modify the model such that the "new" residuals are uncorrelated.

1. Replace y_t by the *first difference*, as discussed in Section 16.8. The new dependent variable is

$$y'_t = y_t - y_{t-1}$$

2. Replace y_t by the *percentage change* during year t,

$$z_t = \left(\frac{y_t - y_{t-1}}{y_{t-1}}\right)100$$

3. Include the lagged dependent variables, $y_{t-1}, y_{t-2}, \ldots$ as predictors of y_t in the regression equation. This is a modification of the autoregressive technique in Section 16.8; now the lagged dependent variables are used with $X_{1,t}, X_{2,t}, \ldots$ (which might also be lagged) to predict the time series variable, y_t.

4. Improve the existing model by attempting to discover other significant predictor variables. Because the residuals include the effect of these missing variables, residual autocorrelation often can be improved by including these additional variables. This procedure offers the best solution to the autocorrelation problem but, unfortunately, is easier said than done.

5. Because the errors are autocorrelated, you can *model the error term* in much the same way as we handled the situation of autocorrelated observations. This involves describing each residual, e_t, by its previous values, such as

$$e_t = \phi_1 e_{t-1} + \phi_2 e_{t-2} + \cdots + \phi_j e_{t-j} + u_t$$

The value of j is arbitrary and represents the maximum period over which errors are correlated. Now the problem becomes to estimate not only the coefficients of the predictor variables ($\beta_0, \beta_1, \beta_2, \ldots$) but also those of $\phi_1, \phi_2, \ldots$.

The hope is that the new error term, u_t, will contain mostly noise with little autocorrelation.

COMMENT

Practically all computer packages automatically print out the Durbin-Watson statistic when performing multiple linear regression. When your data are *not* collected over time (but rather from different families, cities, companies, and so on), *this statistic is meaningless* and should be ignored.

EXERCISES

16.56 If autocorrelation is present in a data set, what procedure used in multiple regression analysis would possibly become invalid?

16.57 What type of autocorrelation is indicated by a value of the Durbin-Watson statistic equal to zero? To two? To four?

16.58 Find the value of the Durbin-Watson statistic for the following yearly data.

YEAR	DATA	YEAR	DATA
1973	32	1981	42
1974	40	1982	48
1975	48	1983	54
1976	52	1984	64
1977	41	1985	42
1978	31	1986	35
1979	28	1987	32
1980	27	1988	27

a. Use the model $\hat{y}_t = b_0 + b_1 y_{t-1}$

b. Test for positive autocorrelation. Use a significance level of .05.

16.59 Test for positive autocorrelation in the residuals of the second-order process in Exercise 16.35. Use a significance level of .05.

16.60 Test for positive autocorrelation in the residuals of the fourth-order autoregressive process in Exercise 16.36. Use a significance level of .05.

16.61 Test for positive autocorrelation in the residuals of the multiple regression equation in Exercise 16.46. Use a significance level of .05.

16.62 If significant correlations are present in the residuals, what procedures can be used to modify the model so that the new residuals are uncorrelated?

SUMMARY

In this chapter, we have looked briefly at several popular **forecasting** techniques. To cover all aspects of time series forecasting would fill an entire textbook. It is a fascinating side of statistics because anyone having a reliable "crystal ball" technique for predicting the future definitely is one step ahead of the game. We hope that this chapter has whetted your appetite to pursue further reading in this area.

Forecasting methods can be divided into two broad categories: qualitative procedures and quantitative techniques. When arriving at a **qualitative** forecast, expert opinion is used to arrive at a "best educated" estimate of future behavior. One such method is the **Delphi method**, which requires input from a team of experts. Each team member is then informed as to the responses from all other members and asked to reevaluate his or her opinion in light of this information. This is continued for several rounds until each member of the team feels confident in his or her final decision.

Quantitative forecasting was the main emphasis of this chapter and dealt with two (sometimes overlapping) sets of procedures: time series techniques and multiple linear regression on time series data. **Time series procedures** attempt to capture the past behavior of the time series and use this information to predict future values. No external predictors are considered; only the past observations are used to describe and predict the future value of the time series variable. Time series methods include (1) the **decomposition** procedure, which extracts and extends the trend and seasonal components, (2) **exponential smoothing**, which reduces randomness and forecasts future values by using the smoothed values, and (3) **autoregressive** forecasting, which predicts future values by using a linear combination of past values.

There are various exponential smoothing procedures; the proper one to use depends on the nature of the time series. **Simple exponential smoothing** works best when the time series contains neither trend nor seasonality. **Linear exponential smoothing** is better for a time series that does contain trend but has no seasonality, and **linear and seasonal exponential smoothing** should be used for a time series that has both components.

Exponential Smoothing

TYPE	STRUCTURE OF TIME SERIES	
	Contains Trend	Contains Seasonality
Simple Exponential Smoothing	NO	NO
Linear Exponential Smoothing	YES	NO
Linear and Seasonal Exponential Smoothing	YES	YES

There are many factors that determine the strengths of any forecasting procedure. These include the (1) time horizon of the forecast, (2) stationarity of the data, and (3) presence of trend, seasonality, or cyclical activity. To measure the forecast accuracy of a particular method, you can calculate the predictive mean squared error (MSE), the mean absolute deviation (MAD), or the mean absolute percentage error (MAPE). The **MSE** is found by squaring each of the residuals obtained by applying this technique to the past observations and then deriving the average of these squared residuals. This measure is very sensitive to one or two very large residuals. The **MAD** is calculated by averaging the absolute values of the residuals and is less sensitive to a single large residual. The **MAPE** uses the *relative* error of each forecast value to arrive at a measure of prediction accuracy. It is very useful for comparing the accuracy of a particular forecasting technique on two different time series, since the effect of the magnitude of the observations has been removed.

The advantage of the time series methods is that there is no need to search for external predictors to explain the behavior of the dependent variable. One disadvantage is that the patterns within the observed values can be extremely complex and difficult to determine. Such methods often are hard to "sell" to managers, who may not be able to understand the technique.

Multiple linear regression forecasting requires additional input data; for each time period, data are recorded for each predictor (independent) variable as well as for the dependent variable, y_t. The predictor variables can include **lagged** dependent or independent variables or **dummy** variables to represent seasonality or the occurrence (or nonoccurrence) of a particular event (such as an election year).

When you use multiple linear regression techniques on time series data, you often will violate the assumption of independent errors. The **Durbin-Watson statistic** can be used to test for significant autocorrelation in the regression residuals. If significant autocorrelation is present, the tests of hypothesis and confidence intervals contained in the regression output are unreliable.

The advantages of multiple linear regression on time series data include: (1) it is a very flexible approach, in that a wide variety of explanatory variables can be included in the model; (2) it allows for lagging the predictor variables, including lagged values of the dependent variable; and (3) it is generally easier to explain to managers, who can see easily which variables are predicting the behavior of the dependent variable. On the other hand, residual autocorrelation often is a problem. This may be caused by missing variables in the prediction equation, which typically are extremely difficult to determine. Also, a very complex pattern within the observed time series may be difficult to capture using a linear combination of predictor variables. Finally, forecasting with this technique becomes extremely difficult unless lagged variables are used. Dummy variables can be included, provided that they can be predicted with certainty. A dummy variable representing an election year would be acceptable, whereas one representing the occurrence (or nonoccurrence) of an earthquake would not be.

FURTHER READING

Bowerman, B. L., and R. T. O'Connell. *Forecasting and Time Series.* 2d ed., North Scituate, Mass.: Duxbury, 1987.

Hanke, J., A. Reitsch, and J. P. Dickson. *Statistical Decision Models for Management.* Newton, Mass.: Allyn and Bacon, 1984.

Makridakis, S., S. C. Wheelright, and V. E. McGee, *Forecasting: Methods and Applications.* 2d. ed. New York: John Wiley, 1983.

Makridakis, S., and S. C. Wheelright, eds. *Handbook of Forecasting: A Manager's Guide.* New York: John Wiley, 1987.

Mendenhall, W. and T. Sincich. *A Second Course in Business Statistics: Regression Analysis.* 2d ed. San Francisco: Dellen, 1986.

REVIEW EXERCISES

16.63 A set of monthly data has been gathered over 3 years from January 1983 to December 1985. From these data, the seasonal indexes for the 12 months are found to be

$$S_1 = 0.75 \quad S_2 = 0.85 \quad S_3 = 0.95 \quad S_4 = 0.99 \quad S_5 = 0.90 \quad S_6 = 1.01$$
$$S_7 = 1.20 \quad S_8 = 1.10 \quad S_9 = 1.15 \quad S_{10} = 1.05 \quad S_{11} = 1.11 \quad S_{12} = 0.94$$

The least squares line through the deseasonalized data is found to be

$$\hat{d}_t = 2.73 + 0.62t$$

for the 36 monthly periods. Find the forecast for monthly periods 37, 38, 39, 40, and 41.

16.64 The total monthly volume of trade on the stock of Xcon Corp is given for a 4-year period. Units are in millions of shares.

YEAR	JAN.	FEB.	MAR.	APR.	MAY	JUN.	JUL.	AUG.	SEP.	OCT.	NOV.	DEC.
1985	1.1	1.2	1.4	1.3	1.2	1.5	1.9	1.8	1.3	1.1	0.9	0.8
1986	1.3	1.4	1.5	1.4	1.4	1.7	2.1	1.9	1.6	1.4	1.1	1.1
1987	1.4	1.4	1.6	1.7	1.6	1.9	2.3	2.2	1.8	1.7	1.4	1.2
1988	1.3	1.5	1.5	1.6	1.7	2.0	2.2	2.3	1.9	1.7	1.5	1.4

Using simple exponential smoothing with the parameter $A = .3$, find the forecasted membership for the first month of 1989.

16.65 The number of employees at Computeron has fluctuated over the past 9 years. The table lists the number of employees on the payroll at Computeron at the end of each year for the years 1980 to 1988. The company would like you to forecast employment in 1989 by using the linear exponential smoothing technique.

YEAR	EMPLOYEES	YEAR	EMPLOYEES
1980	1030	1985	1075
1981	1020	1986	1130
1982	1041	1987	1135
1983	1050	1988	1175
1984	1062		

a. Using the initial estimate of the slope to be zero with $A = .2$ and $B = .2$, determine the predicted value for the employment at the end of 1989.

b. Using the least squares estimate for the slope, from the first 3 years, redo part (a).

c. Compare the predicted values in parts (a) and (b).

16.66 Two forecasting procedures were used to forecast the 12 quarters from 1986 to 1988. The forecast errors for each quarter are given below. Compute the predictive MSE and MAD for each forecasting procedure. Interpret the results.

YEAR	QUARTER	PROCEDURE 1 FORECAST ERROR	PROCEDURE 2 FORECAST ERROR
1986	1	-1.0	.7
	2	.5	$-.5$
	3	-2.1	$-.2$
	4	2.5	.9
1987	1	.9	.1
	2	2.1	.2
	3	-1.3	$-.3$
	4	1.6	.9
1988	1	2.7	11.2
	2	-1.9	$-.1$
	3	2.4	.2
	4	-1.2	$-.2$

16.67 The amount of money spent on research and development by Energy Today in finding economical uses of alternative fuels for energy is given over a 4-year period. Units are in $10,000.

YEAR	QUARTER 1	QUARTER 2	QUARTER 3	QUARTER 4
1985	4.2	4.5	4.8	4.0
1986	4.3	4.7	5.6	4.4
1987	4.6	4.9	5.7	4.5
1988	4.7	5.0	5.8	4.7

Use Holt's two-parameter linear exponential smoothing technique to obtain a forecast for each of the quarters, omitting the first time period. Let $A = .3$, and $B = .2$. Use the least squares estimate of the trend from the first 5 periods for the initial value of the slope. Calculate the predictive MSE.

16.68 The following table represents the number of homes sold monthly in a growing community in California over the past 5 years:

MONTH	1984	1985	1986	1987	1988
Jan.	48	83	117	156	192
Feb.	51	85	121	158	195
Mar.	65	98	133	170	208
Apr.	67	102	134	173	209
May	69	105	139	176	214
Jun.	78	114	149	184	221
Jul.	81	118	153	187	224
Aug.	85	123	155	191	227
Sep.	67	105	132	173	210
Oct.	71	109	142	175	213
Nov.	72	113	145	179	217
Dec.	75	115	148	184	220

a. Calculate the autocorrelations through lag 15.

b. Calculate the first differences for the time series.

c. Calculate the autocorrelations for the difference data obtained in question b.

d. Comment on the autocorrelations in parts (a) and (c).

16.69 National Finance Company provides short-term loans to consumers to finance household goods. The amount of interest received quarterly is given below in units of $10,000.

YEAR	QUARTER 1	QUARTER 2	QUARTER 3	QUARTER 4
1984	20	31	39	42
1985	28	35	43	45
1986	31	38	45	49
1987	35	40	43	52
1988	38	44	48	56

Determine the multiple regression equation, which takes into account trend and seasonality.

16.70 An independent gas station allows its customers to buy gasoline on credit. The amount of credit on the books for the 20 quarters of the years 1984 through 1988 follows. Find the multiple regression equation that takes into account trend and seasonality. The figures in the table are in units of $10,000.

YEAR	QUARTER 1	QUARTER 2	QUARTER 3	QUARTER 4
1984	2.3	2.7	3.4	3.0
1985	2.4	3.0	3.6	3.2
1986	2.6	3.1	3.8	3.4
1987	3.0	3.3	4.0	3.2
1988	3.2	3.4	4.4	3.5

16.71 Using Holt's two-parameter linear exponential smoothing technique, find the smoothed value at time $t = 12$, where the observed value at $t = 12$ is 13.6 and the observed value at $t = 11$ is 12.1. Also let $S_{11} = 7.4$, $S_{10} = 10.4$, $b_{11} = 0.6$, $A = .1$ and $B = .2$.

16.72 In Holt's two-parameter linear exponential smoothing technique, why do you think the values of A and B are typically less than or equal to .3?

16.73 Explain what one should look for in determining the appropriate forecasting procedure.

16.74 The total number of retail sales for the years 1967 to 1985 for the United States are given in the following table. Units are in billions of dollars. Find the first- and second-order autoregressive models. Calculate the predictive MSE for each model.

YEAR	SALES	YEAR	SALES
1967	293.0	1977	725.2
1968	324.4	1978	806.9
1969	346.7	1979	899.4
1970	368.4	1980	960.8
1971	406.2	1981	1043.5
1972	449.1	1982	1074.6
1973	509.5	1983	1174.0
1974	541.0	1984	1293.1
1975	588.1	1985	1373.9
1976	657.4		

(*Source: United States Statistical Abstracts,* 1987.)

16.75 The interest paid on savings deposits at FSLIC-insured savings institutions in the United States is given for 1970 to 1986 in millions of dollars. Using the linear exponential smoothing technique with an initial estimate of the slope to be zero with $A = .2$ and

$B = .2$, determine the predictive value for the interest paid on savings deposits at the end of 1987.

YEAR	INTEREST ON SAVINGS DEPOSITS	YEAR	INTEREST ON SAVINGS DEPOSITS
1970	6,895	1979	32,151
1971	8,274	1980	41,627
1972	9,967	1981	54,075
1973	11,688	1982	58,600
1974	13,646	1983	58,275
1975	16,020	1984	71,790
1976	19,090	1985	73,471
1977	22,547	1986	68,045
1978	26,024		

(*Source: Moody's Bank and Finance Manual*, Vol. 1, 1987, p. a8.)

16.76 The manager of an assembly plant is interested in predicting the time spent on maintenance of machines at the plant. The data in C1 of the MINITAB computer printout represent the number of hours that was spent on maintenance of machines at the assembly plant per quarter. The data in C2 represent the time starting with period one, and incrementing by one for each quarter. The dummy variables in C3 to C5 represent the quarter of the year. Test for positive autocorrelation in the residuals for the model in

```
MTB > NAME C1 = 'Y'
MTB > NAME C2 = 'TIME'
MTB > NAME C3 = 'DUMMY1'
MTB > NAME C4 = 'DUMMY2'
MTB > NAME C5 = 'DUMMY3'
MTB > NAME C6 = 'RESID'
MTB > PRINT C1-C5
 ROW        Y    TIME   DUMMY1   DUMMY2   DUMMY3

   1    77.10      1        1        0        0
   2    46.78      2        0        1        0
   3   179.43      3        0        0        1
   4   228.70      4        0        0        0
   5   152.90      5        1        0        0
   6   122.58      6        0        1        0
   7   247.65      7        0        0        1
   8   289.34      8        0        0        0
   9   213.54      9        1        0        0
  10   190.80     10        0        1        0
  11   323.45     11        0        0        1
  12   387.88     12        0        0        0
  13   285.55     13        1        0        0
  14   255.23     14        0        1        0
  15   384.09     15        0        0        1
  16   467.47     16        0        0        0

MTB > REGRESS Y IN C1 USING 4 PREDICTORS IN C2-C5;
SUBC> RESID IN C6;
SUBC> DW.

The regression equation is
Y = 163 + 18.0 TIME - 107 DUMMY1 - 153 DUMMY2 - 41.7 DUMMY3

Predictor        Coef        Stdev      t-ratio
Constant      163.086        6.975        23.38
TIME          18.0262        0.5199       34.67
DUMMY1       -106.996        6.759       -15.83
DUMMY2       -153.447        6.658       -23.05
DUMMY3        -41.666        6.597        -6.32

s = 9.300        R-sq = 99.5%        R-sq(adj) = 99.3%

Analysis of Variance

SOURCE        DF           SS           MS
Regression     4       197336        49334
Error         11          951           86
Total         15       198288
```

```
SOURCE        DF      SEQ SS
TIME          1       143889
DUMMY1        1         4227
DUMMY2        1        45770
DUMMY3        1         3451

Durbin-Watson statistic = 1.71

MTB > PRINT C6
RESID
    2.9846    1.0895    3.9322   -6.4902    6.6799    4.7849    0.0475
  -17.9551   -4.7847    0.9002    3.7427    8.4802   -4.8796   -6.7744
   -7.7222   15.9656
```

the MINITAB computer printout given above. Use a significance level of .05. Determine the MAD, MAPE, and predictive MSE for the forecasts.

16.77 The data in Exercise 16.76 is analyzed in the following MINITAB computer printout with a model that contains the independent variables time and the Y variable lagged by four time periods. Test for positive autocorrelation in the residuals for the model in the MINITAB computer printout. Determine the MAD, MAPE, and predictive MSE for the forecasts and compare these values to those of the model in Exercise 16.76.

```
MTB > LAG BY 4 OF C1, PUT INTO C7
MTB > NAME C7 = 'LAG4'
MTB > PRINT C1 C2 C7
 ROW      Y     TIME      LAG4

   1    77.10     1        *
   2    46.78     2        *
   3   179.43     3        *
   4   228.70     4        *
   5   152.90     5       77.10
   6   122.58     6       46.78
   7   247.65     7      179.43
   8   289.34     8      228.70
   9   213.54     9      152.90
  10   190.80    10      122.58
  11   323.45    11      247.65
  12   387.88    12      289.34
  13   285.55    13      213.54
  14   255.23    14      190.80
  15   384.09    15      323.45
  16   467.47    16      387.88

MTB > REGRESS Y IN C1 USING 2 PREDICTORS IN C2, C7;
SUBC> RESIDUALS IN C6;
SUBC> DW.

The regression equation is
Y = 71.0 - 0.77 TIME + 1.04 LAG4

12 cases used 4 cases contain missing values

Predictor      Coef       Stdev     t-ratio
Constant      70.97       10.87        6.53
TIME          -0.770       1.679      -0.46
LAG4           1.04298     0.06086    17.14

s = 11.57      R-sq = 98.9%      R-sq(adj) = 98.7%

Analysis of Variance

SOURCE        DF        SS          MS
Regression     2      113258       56629
Error          9        1205         134
Total         11      114462

SOURCE        DF      SEQ SS
TIME           1       73946
LAG4           1       39311

Durbin-Watson statistic = 1.44

MTB > PRINT C6
RESID
      *         *         *         *      5.3684    7.4418   -5.0699
  -13.9978   -9.9693   -0.3159    2.6582   24.3765    1.8745   -3.9575
  -12.6794    4.2715
```

**FORECASTING
WATER DEMAND
FOR A CITY (JUST
DON'T EXCEED 15
MILLION GALLONS)**

The Water Department of the city of Fort Worth, Texas, has to purchase untreated lake water from several administrative entities around the city in order to fulfill the water requirements of its population. Most of these entities have a fixed price per thousand gallons of water, except for the entity in charge of water from Clear

OBSERVATION	YEAR	MONTH	CLEAR FORK LAKE WATER DEMAND (THOUSANDS OF GALLONS)
1	1983	Sep.	1292.00
2		Oct.	900.00
3		Nov.	360.00
4		Dec.	1456.00
5	1984	Jan.	352.00
6		Feb.	1037.00
7		Mar.	1564.00
8		Apr.	379.00
9		May	1013.00
10		Jun.	1227.35
11		Jul.	2204.76
12		Aug.	2352.87
13		Sep.	2334.73
14		Oct.	887.05
15		Nov.	1027.49
16		Dec.	1500.01
17	1985	Jan.	1758.02
18		Feb.	2171.23
19		Mar.	1883.52
20		Apr.	1937.84
21		May	2068.83
22		Jun.	2175.52
23		Jul.	2027.15
24		Aug.	1860.79
25		Sep.	702.81
26		Oct.	890.18
27		Nov.	1956.44
28		Dec.	2335.61
29	1986	Jan.	3119.55
30		Feb.	2091.26
31		Mar.	1271.86
32		Apr.	3448.84
33		May	3592.00
34		Jun.	4029.18
35		Jul.	4043.03
36		Aug.	3330.48
37		Sep.	3372.23
38		Oct.	3015.32
39		Nov.	2226.68
40		Dec.	2538.55
41	1987	Jan.	3064.31
42		Feb.	2981.15
43		Mar.	2835.10
44		Apr.	2900.00
45		May	3700.22
46		Jun.	4950.11
47		Jul.	5000.22
48		Aug.	4800.30

(*Source:* City of Fort Worth Annual Reports, Water Department, 1983–1987.)

Fork Lake. If the total amount of water obtained from Clear Fork Lake in any month is above 15 million gallons, their price increases by 30%.

Therefore, it is very important for Fort Worth city officials to be able to make good forecasts of their city's raw water needs as budgeting periods approach. In order to make such projections, monthly data pertaining to Clear Fork Lake have been provided by the Water Department for the period September 1983 to August 1987.

Suppose you are consulted by city officials who request your assistance in developing a good forecast of their requirements of water from Clear Fork Lake. You are informed that water consumption varies from fall to spring and summer to winter. Each season has a peak period and the summer time tends to have greater water consumption, which is not unusual, considering the extremely hot Texas summer weather. Population also has been steadily increasing annually in Fort Worth, and this is expected to be reflected in a positive linear trend in demand for water.

Case Study Questions

1. Discuss the relevance of the three exponential techniques of time series analysis discussed in this chapter to the preceding situation, namely, simple exponential smoothing, Holt's two-parameter technique, and Winter's three-parameter technique. Which is appropriate and why?

2. Using Winter's technique, develop a time series forecasting model to forecast water needs from September 1987 to August 1988. Try different weights for the smoothing constants, and compute the predictive MSE for each model. Pick the best model having the lowest predictive MSE and prepare a forecast for water demand from September 1987 to August 1988.

3. Using the autoregressive technique, develop a second-order autoregressive forecasting model for the above data, and prepare a forecast for water demand from September 1987 to August 1988.

4. Compute the predictive MSE for the autoregressive model, and compare it with the predictive MSE for the Winter's model developed earlier. Comment on any differences.

5. Suppose city officials made available data on population growth and suggested using this as a predictor variable, in a multiple regression model. What potential problems should you alert them to about using such a technique with time series data? Mention the relevance of autocorrelation and the Durbin-Watson statistic in this context.

| SOLUTION | Example 16.9 is divided into two computer runs. In the first run we used |

S P S S X

SOLUTION

EXAMPLE 16.9

Example 16.9 is divided into two computer runs. In the first run we used the autocorrelation function to determine the significant autocorrelations among the given lag periods. In the second run, the significant lag period (lag 4) was included in the autoregression equation. The SPSSX program listings in Figures 16.20 and 16.21 were used to perform the two procedures. This example will only run on the mainframe version of SPSS.

FIGURE 16.20

Input for SPSSX to determine significant autocorrelations.

```
TITLE    KENS AUTO PAINT SHOP - AUTOCORRELATION FUNCTION
DATA LIST FREE / T YT
BEGIN DATA
1 5.56
2 16.36
3 2.12
4 3.15
    .
    .
    .
17 4.81
18 16.82
19 4.75
20 8.54
END DATA
PRINT / T YT
VARIABLE LABELS T 'TIME'
               YT 'OBSERVED VALUES'
BOX-JENKINS VARIABLE=YT/IDENTIFY
```

FIGURE 16.21

Input for SPSSX to determine the autocorrelation equation using one lagged variable.

```
TITLE    KENS AUTO PAINT SHOP - AUTOREGRESSIVE MODEL
DATA LIST FREE / T YT
BEGIN DATA
1 5.56
2 16.36
3 2.12
4 3.15
    .
    .
    .
17 4.81
18 16.82
19 4.75
20 8.54
END DATA
VARIABLE LABELS T 'TIME'
               YT 'OBSERVED VALUES'
COMPUTE YT4=LAG(YT,4)
PRINT / T YT YT4
REGRESSION VARIABLES=YT YT4/DEP=YT/ENTER/SAVE PRED(PREDX)
PRINT / YT PREDX
EXECUTE
```

The TITLE command names the SPSSX run.

The DATA LIST command gives each variable a name, and describes the data as being in free form.

The BEGIN DATA command indicates to SPSSX that the input data immediately follow.

The next 20 lines contain the data values, which represent time series data over 20 time periods. For example, 1 5.56 represents the OBSERVED VALUE of 5.56 for time period 1.

The END DATA statement indicates the end of the data.

The PRINT command causes the printing of the variables T and YT.

The VARIABLE LABELS statement assigns a descriptive label to the variables. This descriptive label is substituted for the variable name when output is printed.

The BOX-JENKINS VARIABLE = YT/IDENTIFY statement means that the YT time series is to be analyzed, and we wish to produce statistics for model identification.

The commands in Figure 16.21 are the same as those in Figure 16.20, with the following exceptions:

The COMPUTE YT4 = LAG(YT, 4) command provides YT4 with the value of YT for the fourth case before the current one. For example, the value of YT in case 16 is 4.56. Therefore, the value of YT4 in case 20 is also 4.56.

The REGRESSION VARIABLES statement specifies that variables YT and YT4 are to be used in the regression equation, and that YT is the dependent variable.

The ENTER option specifies that all independent variables are to be applied in the equation simultaneously.

The SAVE option saves the predicted values under the name PREDX.

FIGURE 16.22 SPSSX output from Figure 16.20.

```
* * * * * * * * * * * * * * * * * * * A R I M A      A N A L Y S I S * * * * * * * * * * * * * * * * * * * * * *
VARIABLE YT      CONTAINS THE TIME SERIES
 DEGREE OF NONSEASONAL DIFFERENCING -   0
 DEGREE OF SEASONAL DIFFERENCING -   0
 SEASONAL SPAN -   1
 MEAN VALUE OF THE PROCESS
   0.74985E+01
 STANDARD DEVIATION OF THE PROCESS
   0.48828E+01
 AUTOCORRELATION FUNCTION FOR VARIABLE YT
 AUTOCORRELATIONS *
 TWO STANDARD ERROR LIMITS

     AUTO. STAND.
LAG  CORR.  ERR.  -1  -.75  -.5  -.25   0   .25   .5   .75   1
                  :----:----:----:----:----:----:----:----:
  1  -0.315  0.202              .  *****:
  2  -0.260  0.197              .  ***** :
  3  -0.286  0.191              .  *****:
  4   0.678  0.185              :  ***** .*******
  5  -0.213  0.178              .  ****:
  6  -0.168  0.172              .  ***:
  7  -0.156  0.165              .  ***:
  8   0.512  0.158              :  ***** .****
  9  -0.235  0.151              .  *****:
 10  -0.134  0.143              .  ***:
 11  -0.104  0.135              .  **:
 12   0.388  0.126              :  ****.***
 13  -0.155  0.117              .  ***:
 14  -0.060  0.107              .  *:
 15  -0.123  0.095              .  **:
 16   0.206  0.083              :  **.*
 17  -0.101  0.067              .  **:
 18   0.031  0.048              :  .*.
```

r_k

— Use this lagged variable in the autoregressive equation

FIGURE 16.23 SPSSX output from Figure 16.21.

```
                        * * * *  M U L T I P L E   R E G R E S S I O N  * * * *

Listwise Deletion of Missing Data

Equation Number 1    Dependent Variable..    YT    OBSERVED VALUES

Beginning Block Number  1.  Method: Enter


Variable(s) Entered on Step Number  1..    YT4
                                                                        SSE
Multiple R          .90221      Analysis of Variance
R Square            .81398                        DF    Sum of Squares    Mean Square
Adjusted R Square   .80069      Regression         1       281.81898       281.81898
Standard Error     2.14484      Residual          14        64.40499         4.60036

                                F =     61.26025      Signif F =  .0000


------------------ Variables in the Equation ------------------

Variable           B          SE B      Beta        T    Sig T

YT4              .869732     .111121    .902208    7.827  .0000
(Constant)      1.419837     .962273              1.476  .1622


End Block Number     1  | All requested variables entered.
```

$\hat{y}_t = 1.42 + .87y_{t-4}$

FIGURE 16.23
(Continued)

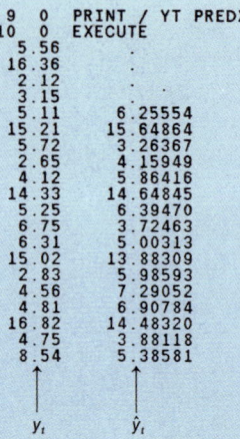

```
  9  0   PRINT / YT PREDX
 10  0   EXECUTE
   5.56         .
  16.36         .
   2.12         .
   3.15         .
   5.11       6.25554
  15.21      15.64864
   5.72       3.26367
   2.65       4.15949
   4.12       5.86416
  14.33      14.64845
   5.25       6.39470
   6.75       3.72463
   6.31       5.00313
  15.02      13.88309
   2.83       5.98593
   4.56       7.29052
   4.81       6.90784
  16.82      14.48320
   4.75       3.88118
   8.54       5.38581
```

y_t $\hat{y}_t$

Figure 16.22 shows the SPSSX output obtained by executing the listing in Figure 16.20, and Figure 16.23 shows the SPSSX output obtained by executing the listing in Figure 16.21.

S P S S X

SOLUTION

SECTION 16.9

We used multiple linear regression on time series data to predict the number of home loans financed by Liberty Savings. The predictor variables used were average interest rate, advertising expenditure, an election year dummy variable, and a seasonal dummy variable. The SPSSX program listing in Figure 16.24 below was used to compute the regression equation from the time series data. This example will only run on the mainframe version of SPSS.

FIGURE 16.24
Input for SPSSX.

```
TITLE    LIBERTY SAVINGS AND LOAN
DATA LIST FREE / YEAR TIME YT X1T X2T X3T X4T
BEGIN DATA
1981 1 115 13.0 9.1 0 1
1981 2 84 14.7 9.5 0 0
1982 3 76 14.1 6.1  0 1
1982 4 81 12.0 8.2 0 0
1983 5 122 11.8 10.4 0 1
1983 6 118 12.4 6.7 0 0
1984 7 106 11.0 7.5 1 1
1984 8 140 14.5 7.8 1 0
1985 9 86 11.0 5.1  0 1
1985 10 96 10.1 6.8 0 0
1986 11 110 12.9 6.8 0 1
1986 12 76 14.8 9.1 0 0
1987 13 62 10.5 7.5 0 1
1987 14 104 09.8 5.1 0 0
1988 15 135 10.1 8.8 1 1
1988 16 120 10.8 4.3 1 0
END DATA
VARIABLE LABELS    YT 'NUMBER OF HOME LOANS'
                   X1T 'AVERAGE INTEREST RATE'
                   X2T 'ADVERTISING EXPENDITURE'
                   X3T 'ELECTION YEAR VARIABLE'
                   X4T 'SEASONAL VARIABLE'
COMPUTE X1TLAG=LAG(X1T,1)
COMPUTE X2TLAG=LAG(X2T,1)
PRINT / YEAR TIME YT X1T X2T X3T X4T X1TLAG X2TLAG
REGRESSION VARIABLES=YT X1TLAG X2TLAG X3T X4T/
          DEPENDENT=YT/ENTER/RESID=DURBIN
```

The TITLE command names the SPSSX run.

The DATA LIST command gives each variable a name, and describes the data as being in free form.

The BEGIN DATA command indicates to SPSSX that the input data immediately follow.

The next 16 lines contain the data values, which represent time series data over 16 time periods. The first line, for example, represents the year 1981, the first time period, 115 home loans, average interest rate of 13%, and so on.

The END DATA statement indicates the end of the data.

The VARIABLE LABELS statement assigns a descriptive label to the variables. This descriptive label is substituted for the variable name when output is printed.

The COMPUTE X1TLAG = LAG (X1T, 1) command provides X1TLAG with the value of X1T for the case before the current one. For example, in 1981, the average interest rate (X1T) was 13.0%. Therefore, the value of X1TLAG for 1982 is also 13.0%.

The COMPUTE X2TLAG = LAG (X2T, 1) command provides X2TLAG with the value of X2T for the case before the current one.

The REGRESSION VARIABLES statement specifies that variables YT, X1TLAG, X2TLAG, X3T, and X4T are to be used in the regression equation, and that YT is the dependent variable.

The ENTER option specifies that all independent variables are to be applied in the equation simultaneously.

The RESID = DURBIN option specifies that the DURBIN-WATSON statistic is to be computed.

FIGURE 16.25 SPSSX output.

```
                         * * * *   M U L T I P L E   R E G R E S S I O N   * * * *

Listwise Deletion of Missing Data

Equation Number 1     Dependent Variable..   YT   NUMBER OF HOME LOANS

Beginning Block Number  1.  Method: Enter

Variable(s) Entered on Step Number   1..    X4T        SEASONAL VARIABLE
                                     2..    X2TLAG
                                     3..    X3T        ELECTION YEAR VARIABLE
                                     4..    X1TLAG


Multiple R            .87730        Analysis of Variance
R Square              .76965                             DF     Sum of Squares      Mean Square
Adjusted R Square     .67751        Regression            4        5843.91660       1460.97915
Standard Error      13.22504        Residual             10        1749.01674        174.90167

                                    F =      8.35315     Signif F =   .0031


----------------- Variables in the Equation ------------------

Variable             B         SE B        Beta          T    Sig T

X4T            4.590323     7.204781     .101786       .637   .5384
X2TLAG         2.510174     2.429178     .169530      1.033   .3258
X3T           17.047599     8.964411     .335073      1.902   .0864
X1TLAG        -9.635256     2.503154    -.730334     -3.849   .0032
(Constant)   192.574914    31.307754                  6.151   .0001


End Block Number   1   All requested variables entered.
```
$$\hat{y}_t = 192.57 - 9.64 X_{1,t-1} + 2.51 X_{2,t-1} + 17.05 X_{3,t} + 4.59 X_{4,t}$$

FIGURE 16.25   (Continued)

* * * *   M U L T I P L E   R E G R E S S I O N   * * * *

Equation Number 1   Dependent Variable..   YT   NUMBER OF HOME LOANS

Residuals Statistics:

	Min	Max	Mean	Std Dev	N
*PRED	72.0299	134.3960	101.0667	20.4309	15
*RESID	-15.4060	19.8744	.0000	11.1772	15
*ZPRED	-1.4212	1.6313	.0000	1.0000	15
*ZRESID	-1.1649	1.5028	.0000	.8452	15

Total Cases =   16

Durbin-Watson Test =   1.10159

Figure 16.25 shows the SPSSX output obtained by executing the listing in Figure 16.24.

S A S

SOLUTION

EXAMPLE 16.4

We used simple exponential smoothing on time series data to predict the average attendance at the Jefferson County Civic Center (Table 16.3). The SAS program listing in Figure 16.26 was used to compute the predicted values and residuals for the attendance data using a smoothing constant of $A = .1$. This example will only run on the mainframe version of SAS.

The TITLE command names the SAS run.

The DATA command gives the data a name.

The INPUT command names and gives the correct order for the different fields on the data lines.

FIGURE 16.26
SAS input.

```
TITLE 'SIMPLE EXPONENTIAL SMOOTHING';
DATA ATTEND;
   INPUT TIME A;
   LABEL TIME='YEAR'
            A='ATTENDANCE';
   CARDS;
1 5.0
2 8.0
3 2.1
4 7.1
5 4.8
6 2.0
7 7.8
8 5.0
9 14.1
10 13.0
11 13.5
12 14.2
13 14.0
;
PROC FORECAST DATA=ATTEND OUT=B OUTEST=C METHOD=EXPO
     TREND=1 WEIGHT=.1 OUTALL LEAD=1;
   ID TIME;
   VAR A;
PROC PRINT DATA=B;
   TITLE 'SMOOTHING FORECAST';
```

The LABEL statement assigns labels for variables TIME and A. These labels are substituted for the variable names when output is printed.

The CARDS command indicates to SAS that the input data immediately follow.

The next 13 lines contain the data values, which represent time series data over 13 time periods. The first line, for example, represents the year 1976 when the attendance was 5 (thousand).

The PROC FORECAST command requests the smoothed forecast.

OUT = B names an output dataset to hold the forecast. OUTEST = C names an output dataset to hold the parameter estimates. METHOD specifies the exponential smoothing method be used. TREND = 1 selects the constant trend model. WEIGHT = .1 specifies the smoothing weight. OUTALL requests all output. LEAD = 1 specifies the number of periods ahead for the forecast. ID and VAR mean identify by time and that A is the input data variable to be analyzed.

The PROC PRINT command requests that the generated data and estimated values be printed.

Figure 16.27 shows a portion of the SAS output obtained by executing the listing in Figure 16.26.

FIGURE 16.27
SAS output.

SMOOTHING FORECAST

OBS	TIME	_TYPE_	_LEAD_	A
1	1	ACTUAL	0	5.0000
2	1	FORECAST	0	.
3	1	RESIDUAL	0	.
4	2	ACTUAL	0	8.0000 ← y_2
5	2	FORECAST	0	5.0000 ← $\hat{y}_2$
6	2	RESIDUAL	0	3.0000 ← $y_2 - \hat{y}_2$
7	3	ACTUAL	0	2.1000
8	3	FORECAST	0	5.3000
9	3	RESIDUAL	0	-3.2000
10	4	ACTUAL	0	7.1000
11	4	FORECAST	0	4.9800
12	4	RESIDUAL	0	2.1200
13	5	ACTUAL	0	4.8000
14	5	FORECAST	0	5.1920
15	5	RESIDUAL	0	-0.3920
16	6	ACTUAL	0	2.0000
17	6	FORECAST	0	5.1528
18	6	RESIDUAL	0	-3.1528
19	7	ACTUAL	0	7.8000
20	7	FORECAST	0	4.8375
21	7	RESIDUAL	0	2.9625
22	8	ACTUAL	0	5.0000
23	8	FORECAST	0	5.1338
24	8	RESIDUAL	0	-0.1338
25	9	ACTUAL	0	14.1000
26	9	FORECAST	0	5.1204
27	9	RESIDUAL	0	8.9796
28	10	ACTUAL	0	13.0000
29	10	FORECAST	0	6.0184
30	10	RESIDUAL	0	6.9816
31	11	ACTUAL	0	13.5000
32	11	FORECAST	0	6.7165
33	11	RESIDUAL	0	6.7835
34	12	ACTUAL	0	14.2000
35	12	FORECAST	0	7.3949
36	12	RESIDUAL	0	6.8051
37	13	ACTUAL	0	14.0000 ← y_{13}
38	13	FORECAST	0	8.0754 ← $\hat{y}_{13}$
39	13	RESIDUAL	0	5.9246 ← $y_{13} - \hat{y}_{13}$
40	14	FORECAST	1	8.6678
41	14	L95	1	-2.0570
42	14	STD	1	5.4720
43	14	U95	1	19.3927

■ ■ ■ ■ ■ ■ ■ ■ ■ ■ ■ ■ 🅂 🅰 🅂 ■ ■ ■ ■ ■ ■ ■ ■ ■ ■ ■ ■

SOLUTION

EXAMPLE 16.5

We used linear exponential smoothing on time series data to predict the city taxes collected in Jackson City over the past 20 quarters (Table 16.5). The SAS program listing in Figure 16.28 below was used to compute an initial trend estimate, and to obtain the smoothed values S_t, for each time period. It was also used to determine the predicted values, $\hat{y}_t$, using smoothing constants $A = .1$ and $B = .3$. This example will only run on the mainframe version of SAS.

FIGURE 16.28
Input for SAS.

```
TITLE 'LINEAR EXPONENTIAL SMOOTHING';
DATA TAXES;
  INPUT TIME T;
  LABEL T='TAXES';
  CARDS;
1 76
2 93
3 108
4 128
  :
  :
17 484
18 384
19 330
20 497
;
PROC FORECAST DATA=TAXES OUT=B OUTEST=C METHOD=WINTERS TREND=2
  WEIGHT=(.1,.3,0.001) OUTDATA OUT1STEP OUTLIMIT LEAD=1;
  ID TIME;
  VAR T;
PROC PRINT DATA=B;
  TITLE 'SMOOTHING FORECAST';
```

The TITLE command names the SAS run.

The DATA command gives the data a name.

The INPUT command names and gives the correct order for the different fields on the data lines.

The LABEL statement assigns a label for the variable TAXES.

The CARDS command indicates to SAS that the input data immediately follow.

The next 20 lines contain the data values, which represent time series data over 20 time periods. The first line, for example, represents the year 1984 when the taxes collected were 76 (thousand).

The PROC FORECAST command requests the smoothed forecast.

OUT = B names an output dataset to hold the forecast. OUTEST = C names an output dataset to hold the parameter estimates. METHOD specifies the WINTERS smoothing method be used. TREND = 2 selects the linear trend model. WEIGHT = (.1, .3, .001) specifies the smoothing weights. OUTDATA requests that the observations used to fit the model be included in the dataset. OUT1STEP requests that one step ahead forecasts be included in the dataset. OUTLIMIT requests that the forecast confidence limits be included in the dataset. LEAD = 1 specifies the number of periods ahead for the forecast. ID and VAR mean identify by time and that T is the input data set to be analyzed.

The PROC PRINT command requests that the generated data and estimated values be printed.

FIGURE 16.29

SAS output.

SMOOTHING FORECAST

OBS	TIME	_TYPE_	_LEAD_	T
1	1	ACTUAL	0	76.000
2	1	FORECAST	0	.
3	2	ACTUAL	0	93.000 ← y_2
4	2	FORECAST	0	76.000 ← $\hat{y}_2$
5	3	ACTUAL	0	108.000
6	3	FORECAST	0	78.210
7	4	ACTUAL	0	128.000
8	4	FORECAST	0	82.593
9	5	ACTUAL	0	196.000
10	5	FORECAST	0	89.899
11	6	ACTUAL	0	175.000
12	6	FORECAST	0	106.458
13	7	ACTUAL	0	141.000
14	7	FORECAST	0	121.318
15	8	ACTUAL	0	236.000
16	8	FORECAST	0	131.882
17	9	ACTUAL	0	256.000
18	9	FORECAST	0	154.013
19	10	ACTUAL	0	190.000
20	10	FORECAST	0	178.990
21	11	ACTUAL	0	227.000
22	11	FORECAST	0	195.200
23	12	ACTUAL	0	299.000
24	12	FORECAST	0	214.443
25	13	ACTUAL	0	403.000
26	13	FORECAST	0	241.499
27	14	ACTUAL	0	282.000
28	14	FORECAST	0	281.094
29	15	ACTUAL	0	288.000
30	15	FORECAST	0	304.656
31	16	ACTUAL	0	387.000
32	16	FORECAST	0	325.963
33	17	ACTUAL	0	484.000
34	17	FORECAST	0	356.870
35	18	ACTUAL	0	384.000
36	18	FORECAST	0	398.201
37	19	ACTUAL	0	330.000
38	19	FORECAST	0	424.972
39	20	ACTUAL	0	497.000 ← y_{20}
40	20	FORECAST	0	440.817 ← $\hat{y}_{20}$
41	21	FORECAST	1	473.463
42	21	L95	1	312.851
43	21	U95	1	634.074

Figure 16.29 shows a portion of the SAS output obtained by executing the listing in Figure 16.28.

S A S

SOLUTION

EXAMPLE 16.6

We used Winter's linear and seasonal exponential smoothing on time series data to predict the city taxes collected in Jackson City over the past 20 quarters (Table 16.7). The SAS program listing in Figure 16.30 below was used

FIGURE 16.30

Input for SAS.

```
TITLE 'WINTERS METHOD OF EXPONENTIAL SMOOTHING';
DATA A;
   INPUT TIME T:
   LABEL T='TAXES';
   CARDS;
1 76
2 93
3 108
4 128
  .
  .
17 484
18 384
19 330
20 497
;
PROC FORECAST DATA=A OUT=B OUTEST=C METHOD=WINTERS
      TREND=2 WEIGHT=(.1 .3 .2) OUTDATA OUT1STEP OUTLIMIT
      LEAD=1;
   ID TIME;
   VAR T;
PROC PRINT DATA=B;
   TITLE 'THE OUTPUT FROM PROC FORECAST: WINTERS METHOD';
```

to compute an initial trend estimate, and to obtain the smoothed values S_t, for each time period. It was also used to determine the predicted values, $\hat{y}_t$, using smoothing constants $A = .1$, $B = .3$, and $C = .2$. This example will only run on the mainframe version of SAS.

The TITLE command names the SAS run.

The DATA command gives the data a name.

The INPUT command names and gives the correct order for the different fields on the data lines.

The LABEL statement assigns a label for the variable TAXES.

The CARDS command indicates to SAS that the input data immediately follow.

The next 20 lines contain the data values, which represent time series data over 20 time periods. The first line, for example, represents the year 1984 when the taxes collected were 76 (thousand).

The PROC FORECAST command requests the smoothed forecast.

OUT = B names an output dataset to hold the forecast. OUTEST = C names an output dataset to hold the parameter estimates. METHOD specifies the WINTERS smoothing method be used. TREND = 2 selects the linear trend and seasonal model. WEIGHT = (.1, .3, .2) specifies the smoothing weights. OUTDATA requests that the observations used to fit the model be included in the dataset. OUT1STEP requests that one step ahead forecasts be included in the dataset. OUTLIMIT requests that the forecast confidence limits be included in the dataset. LEAD = 1

FIGURE 16.31

SAS output.

```
                    THE OUTPUT FROM PROC FORECAST: WINTERS METHOD

      OBS      TIME      _TYPE_        _LEAD_          T

       1         1      ACTUAL           0          76.000
       2         1      FORECAST         0             .
       3         2      ACTUAL           0          93.000        y₂
       4         2      FORECAST         0          76.000
       5         3      ACTUAL           0         108.000
       6         3      FORECAST         0          78.210        ŷ₂
       7         4      ACTUAL           0         128.000
       8         4      FORECAST         0          82.593
       9         5      ACTUAL           0         196.000
      10         5      FORECAST         0          89.899
      11         6      ACTUAL           0         175.000
      12         6      FORECAST         0         106.458
      13         7      ACTUAL           0         141.000
      14         7      FORECAST         0         121.318
      15         8      ACTUAL           0         236.000
      16         8      FORECAST         0         131.882
      17         9      ACTUAL           0         256.000
      18         9      FORECAST         0         154.013
      19        10      ACTUAL           0         190.000
      20        10      FORECAST         0         178.990
      21        11      ACTUAL           0         227.000
      22        11      FORECAST         0         195.200
      23        12      ACTUAL           0         299.000
      24        12      FORECAST         0         214.443
      25        13      ACTUAL           0         403.000
      26        13      FORECAST         0         241.499
      27        14      ACTUAL           0         282.000
      28        14      FORECAST         0         281.094
      29        15      ACTUAL           0         288.000
      30        15      FORECAST         0         304.656
      31        16      ACTUAL           0         387.000
      32        16      FORECAST         0         325.963
      33        17      ACTUAL           0         484.000
      34        17      FORECAST         0         356.870
      35        18      ACTUAL           0         384.000
      36        18      FORECAST         0         398.201
      37        19      ACTUAL           0         330.000
      38        19      FORECAST         0         424.972        y₂₀
      39        20      ACTUAL           0         497.000
      40        20      FORECAST         0         440.817
      41        21      FORECAST         1         473.463        ŷ₂₀
      42        21      L95              1         312.851
      43        21      U95              1         634.074
```

specifies the number of periods ahead for the forecast. ID and VAR mean identify by time and that T is the input data set to be analyzed.

The **PROC PRINT** command requests that the generated data and estimated values be printed.

Figure 16.31 shows a portion of the SAS output obtained by executing the listing in Figure 16.30. These results differ from those in Table 16.7 because of the use of a different initialization procedure.

■ ■ ■ ■ ■ ■ ■ ■ ■ ■ ■ ■ ■ S A S ■ ■ ■ ■ ■ ■ ■ ■ ■ ■ ■ ■ ■ ■

SOLUTION

EXAMPLE 16.9

EXAMPLE 16.9 is divided into two computer runs. In the first run we used the autocorrelation function to determine the significant autocorrelations among the given lag periods. In the second run, the significant lag period (lag 4) was included in the autoregression equation. The SAS program listings in Figures 16.32 and 16.33 below were used to perform the two procedures. This example will only run on the mainframe version of SAS.

The TITLE command names the SAS run.

The DATA command gives the data a name.

FIGURE 16.32

Input for SAS to determine significant autocorrelations.

```
TITLE  'KENS AUTO PAINT SHOP - AUTOCORRELATION FUNCTION';
DATA AUTO;
 INPUT T YT;
 LABEL T=TIME
       YT=OBSERVED VALUES;
CARDS;
1 5.56
2 16.36
3 2.12
4 3.15
   :
17 4.81
18 16.82
19 4.75
20 8.54
PROC ARIMA;
 IDENTIFY VAR=YT;
```

FIGURE 16.33

Input for SAS to determine the autocorrelation equation using one lagged variable.

```
TITLE  'KENS AUTO PAINT SHOP - AUTOREGRESSIVE MODEL';
DATA AUTO;
 INPUT T YT;
 LABEL T=TIME
       YT=OBSERVED VALUES;
YT4=LAG4(YT);
CARDS;
1 5.56
2 16.36
3 2.12
4 3.15
   :
17 4.81
18 16.82
19 4.75
20 8.54
PROC REG;
 MODEL YT=YT4;
 OUTPUT OUT=C PRED=P;
PROC PRINT;
 VAR YT P;
```

FIGURE 16.34 SAS output from Figure 16.32.

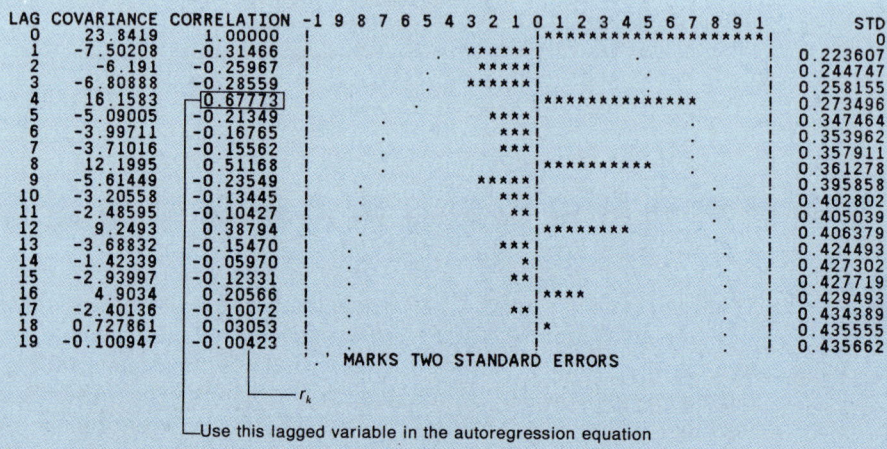

```
               KENS AUTO PAINT SHOP - AUTOCORRELATION FUNCTION

                              ARIMA PROCEDURE

                    NAME OF VARIABLE       =      YT
                    MEAN OF WORKING SERIES=    7.4985
                    STANDARD DEVIATION     =   4.88281
                    NUMBER OF OBSERVATIONS=       20
                              AUTOCORRELATIONS

LAG  COVARIANCE  CORRELATION  -1 9 8 7 6 5 4 3 2 1 0 1 2 3 4 5 6 7 8 9 1      STD
 0    23.8419     1.00000      !                 !********************!       0
 1    -7.50208   -0.31466      !         .    ******!          .     !     0.223607
 2    -6.191     -0.25967      !              *****!                 !     0.244747
 3    -6.80888   -0.28559      !              ******!                !     0.258155
 4    16.1583     0.67773      !              .     !***************  !     0.273496
 5    -5.09005   -0.21349      !                ****!          .     !     0.347464
 6    -3.99711   -0.16765      !                 ***!          .     !     0.353962
 7    -3.71016   -0.15562      !                 ***!          .     !     0.357911
 8    12.1995     0.51168      !              .     !**********       !     0.361278
 9    -5.61449   -0.23549      !                *****!          .     !     0.395858
10    -3.20558   -0.13445      !                 ***!          .     !     0.402802
11    -2.48595   -0.10427      !                  **!          .     !     0.405039
12     9.2493     0.38794      !              .     !********         !     0.406379
13    -3.68832   -0.15470      !                 ***!          .     !     0.424493
14    -1.42339   -0.05970      !                   *!          .     !     0.427302
15    -2.93997   -0.12331      !                  **!          .     !     0.427719
16     4.9034     0.20566      !              .     !****            !     0.429493
17    -2.40136   -0.10072      !                  **!          .     !     0.434389
18    0.727861    0.03053      !                    !*                !     0.435555
19    -0.100947  -0.00423      !              .     !          .     !     0.435662
                          '.' MARKS TWO STANDARD ERRORS
```

r_k

Use this lagged variable in the autoregression equation

FIGURE 16.35 SAS output from Figure 16.33.

```
                    KENS AUTO PAINT SHOP - AUTOREGRESSIVE MODEL

DEP VARIABLE: YT      OBSERVED VALUES
                                          ANALYSIS OF VARIANCE

                                  SUM OF         MEAN
              SOURCE     DF       SQUARES        SQUARE      F VALUE    PROB>F

              MODEL       1     281.81898     281.81898     61.260     0.0001
              ERROR      14      64.40499305    4.60035665
              C TOTAL    15     346.22397

                 ROOT MSE      2.144844     R-SQUARE      0.8140
                 DEP. MEAN     7.67375      ADJ R-SQ      0.8007
                 C.V.         27.9504

                              PARAMETER ESTIMATES

                      PARAMETER     STANDARD      T FOR H0:                   VARIABLE
      VARIABLE   DF    ESTIMATE      ERROR      PARAMETER=0   PROB > !T!        LABEL

      INTERCEP    1   1.41983685   0.96227306      1.476        0.1622       INTERCEPT
      YT4         1   0.86973151   0.11112092      7.827        0.0001
```

$\hat{y}_t = 1.42 + .87 y_{t-4}$

```
              KENS AUTO PAINT SHOP - AUTOREGRESSIVE MODEL

                    OBS      YT         P

                     1      5.56       .
                     2     16.36       .
                     3      2.12       .
                     4      3.15       .
                     5      5.11      6.2555
                     6     15.21     15.6486
                     7      5.72      3.2637
                     8      2.65      4.1595
                     9      4.12      5.8642
                    10     14.33     14.6485
                    11      5.25      6.3947
                    12      6.75      3.7246
                    13      6.31      5.0031
                    14     15.02     13.8831
                    15      2.83      5.9859
                    16      4.56      7.2905
                    17      4.81      6.9078
                    18     16.82     14.4832
                    19      4.75      3.8812
                    20      8.54      5.3858
```

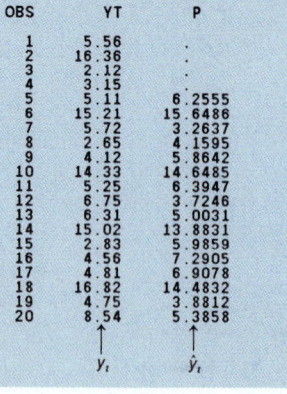

y_t $\hat{y}_t$

The INPUT command names and gives the correct order for the different fields on the data lines.

The LABEL statement assigns the labels TIME for variable T and OB-SERVED VALUES for variable YT. These labels are substituted for the variable names when output is printed.

The CARDS command indicates to SAS that the input data immediately follow.

The next 20 lines contain the data values, which represent time series data over 20 time periods. For example, 1 5.56 represents the OB-SERVED VALUE of 5.56 housing starts for the first time period.

The PROC ARIMA and IDENTIFY commands are used to analyze the time series YT and identify the significant autocorrelations.

The commands in Figure 16.33 are the same as those in Figure 16.32, with the following exceptions:

The YT4 = LAG4(YT) command provides YT4 with the value of YT for the fourth case before the current one. For example, the value of YT in case 16 is 4.56. Therefore, the value of YT4 in case 20 is also 4.56.

The PROC REG command and MODEL subcommand specify that variables YT and YT4 are to be used in the regression equation, and that YT is the dependent variable.

The OUTPUT command specifies the output from PROC REG is to be stored in an output data set named C, and that the predicting values in the data set are named PRED.

Figure 16.34 shows the SAS output obtained by executing the listing in Figure 16.32, and Figure 16.35 shows the SAS output obtained by executing the listing in Figure 16.33.

■ ■ ■ ■ ■ ■ ■ ■ ■ ■   ⧈Ⓐ⧈   ■ ■ ■ ■ ■ ■ ■ ■ ■ ■

MAINFRAME AND MICRO

SOLUTION
─────────
SECTION 16.9

We used multiple linear regression on time series data to predict the number of home loans financed by Liberty Savings. The predictor variables used were average interest rate, advertising expenditure, an election year dummy variable, and a seasonal dummy variable. The SAS program listing in Figure 16.36 below was used to compute the regression equation from the time series data. This example will only run on the mainframe version of SAS.

The TITLE command names the SAS run.

The DATA command gives the data a name.

The INPUT command names and gives the correct order for the different fields on the data lines.

The LABEL statement assigns labels for variables YT, X1T, X2T, X3T, and X4T. These labels are substituted for the variable names when output is printed.

The X1TLAG = LAG1(X1T) command provides X1TLAG with the value of X1T for the case before the current one. For example, the average interest rate (X1T) in 1981 was 13.0%. Therefore, the value of X1TLAG for 1982 is also 13.0%.

FIGURE 16.36
Input for SAS.

```
TITLE   'LIBERTY SAVINGS AND LOAN MODEL';
DATA AUTO;
  INPUT YEAR TIME YT X1T X2T X3T X4T;
  LABEL YT=NUMBER OF HOME LOANS
        X1T=AVERAGE INTEREST RATE
        X2T=ADVERTISING EXPENDITURE
        X3T=ELECTION YEAR VARIABLE
        X4T=SEASONAL VARIABLE;
  X1TLAG=LAG1(X1T);
  X2TLAG=LAG1(X2T);
CARDS;
1981 1 115 13.0 9.1 0 1
1981 2 84 14.7 9.5 0 0
1982 3 76 14.1 6.1  0 1
1982 4 81 12.0 8.2 0 0
1983 5 122 11.8 10.4 0 1
1983 6 118 12.4 6.7 0 0
1984 7 106 11.0 7.5 1 1
1984 8 140 14.5 7.8 1 0
1985 9 86 11.0 5.1  0 1
1985 10 96 10.1 6.8 0 0
1986 11 110 12.9 6.8 0 1
1986 12 76 14.8 9.1 0 0
1987 13 62 10.5 7.5 0 1
1987 14 104 9.8 5.1  0 0
1988 15 135 10.1 8.8 1 1
1988 16 120 10.8 4.3 1 0
PROC REG;
  MODEL YT=X1TLAG X2TLAG X3T X4T/ DW;
```

The X2TLAG = LAG1(X2T) command provides X2TLAG with the value of X2T for the case before the current one.

The CARDS command indicates to SAS that the input data immediately follow.

The next 16 lines contain the data values, which represent time series data over 16 time periods. The first line, for example, represents the year 1981, the first time period, 115 home loans, average interest rate of 13.0%, and so on.

The PROC REG command and MODEL subcommand specify that variables YT, X1TLAG, X2TLAG, X3T, and X4T are to be used in the regression equation, and that YT is the dependent variable.

The DW option requests that the Durbin-Watson statistic be computed.

Figure 16.37 shows the SAS output obtained by executing the listing in Figure 16.36.

FIGURE 16.37   SAS output.

LIBERTY SAVINGS AND LOAN MODEL

DEP VARIABLE: YT   NUMBER OF HOME LOANS

ANALYSIS OF VARIANCE

SOURCE	DF	SUM OF SQUARES	MEAN SQUARE	F VALUE	PROB>F
MODEL	4	5843.91660	1460.97915	8.353	0.0031
ERROR	10	1749.01674	174.90167		
C TOTAL	14	7592.93333			

ROOT MSE	13.22504	R-SQUARE	0.7697 ← — R^2
DEP MEAN	101.0667	ADJ R-SQ	0.6775
C.V.	13.08546		

PARAMETER ESTIMATES

VARIABLE	DF	PARAMETER ESTIMATE	STANDARD ERROR	T FOR H0: PARAMETER=0	PROB > !T!	VARIABLE LABEL
INTERCEP	1	192.57491	31.30775358	6.151	0.0001	INTERCEPT
X1TLAG	1	-9.63525641	2.50315402	-3.849	0.0032	
X2TLAG	1	2.51017445	2.42917767	1.033	0.3258	
X3T	1	17.04759942	8.96441113	1.902	0.0864	ELECTION YEAR VARIABLE
X4T	1	4.59032329	7.20478095	0.637	0.5384	SEASONAL VARIABLE

DURBIN-WATSON D   1.102
(FOR NUMBER OF OBS.)   15
1ST ORDER AUTOCORRELATION   0.379

$$\hat{y}_t = 192.57 - 9.64X_{1,t-1} + 2.51X_{2,t-1} + 17.05X_{3,t} + 4.59X_{4,t}$$

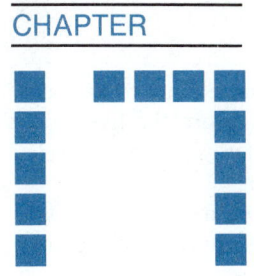

Decision Making Under Uncertainty

A Look Back/Introduction

You have been exposed to decision making in which you use a statistical test of hypothesis. The final step for any test of hypothesis is to reject or fail to reject the null hypothesis, H_0. The previous chapters examined a number of such tests, ranging from a decision regarding two means (for example, is $\mu_1 > \mu_2$?) to selecting predictor variables for a multiple regression equation.

The previous tests of hypothesis concentrated on the *probabilistic* aspects of decision making. For example, if the probability of observing a t-value as large as the computed sample value was less than a predetermined significance level α, we rejected H_0 and decided to keep a particular predictor variable in the regression equation. Such procedures are intended to help you make statistical decisions regarding certain population **parameters**, such as the mean (μ) or standard deviation (σ) of a normal population, the population coefficient (β) for a certain independent variable in a regression equation, or the proportion of successes (p) for a binomial situation.

We now examine a different side of decision making, which is particularly useful when money is involved. This area of statistics allows the decision-maker to consider the various benefits and losses associated with each possible alternative in an effort to find the best decision. For example, if your aging car one day rolls over and goes to that great garage in the sky, you may be faced with the decision of buying a new car or leasing one. If you are lucky enough to have the option of buying a new car that runs well and will incur few repair bills, perhaps your best bet would be to purchase it. Other factors to consider would be the length of time that you intend to keep the car and the tax advantages for each alternative. The main problem, however, is that the future reliability of the new car is uncertain.

The problem of decision making in such a situation would be greatly simplified if you had a crystal ball. A perfect predictor would tell you whether the new car you are thinking of buying is a lemon. Unfortunately, no such device exists, so you need to develop various decision strategies to deal with future uncertainties. Certain strategies are conservative; you stand neither to gain nor to lose a large amount. Other procedures attempt to measure the likelihood of future events by using probabilities. If you are a gambler at heart, strategies exist that allow you to defy odds in hopes of a big payoff.

If your life savings are at stake, you may elect to use a more conservative strategy in your investment decisions; gambling with these funds could be too dangerous. On the other hand, you may have certain reserve funds that you would be willing to invest in more speculative ventures, hoping for a large return at the risk of losing your investment. The purchase of a lottery ticket is a small-scale example of such an investment.

This chapter examines the various strategies available to the decision-maker and illustrates these techniques for different business situations.

17.1

DEFINING THE DECISION PROBLEM

Essentially, when confronted with making a decision in the face of uncertainty, you need to be concerned with two basic questions:

1. What are my possible **actions** (alternatives) for this problem?
2. What is it about the future that affects the desirability of each action?

For the buy versus lease example, you are faced with two possible actions,

Action	Description
A_1	Purchase the car
A_2	Lease the car

When you are trying to decide between these two options, what would you like to know? One possibility is to describe the future (as it applies to this decision) by means of three events, or **states of nature**.

S_1:   the new car will have less-than-average repair costs

S_2:   the new car will have average repair costs

S_3:   the new car will have above-average repair costs

These future states of nature are **outcomes**. The key distinction between an action and a state of nature is that the action taken is *under your control*, whereas the state of nature that occurs is strictly a matter of chance. We will assume that *one and only one* of the states of nature will occur in the future; that is, they are *mutually exclusive*.

Other questions that require the decision-maker to specify corresponding states of nature are ones such as:

What will be the future demand for a new computer software package?

How long will it be until the newly purchased electrical pump breaks down and needs to be replaced?

Will the stock market turn up or down in the next 3 months?

Will this year's winter be milder or colder than the average for the past 20 years?

Associated with each action (A) and state of nature (S) is a corresponding **payoff** or **profit**. We will assume that the payoff associated with each particular state of nature and action is known with certainty. These payoffs can be summarized in a **payoff table**, shown in Table 17.1. The entry corresponding to action A_2 (row 2) and state of nature S_3 (column 3), for example, is denoted by π_{23} and is the payoff associated with action A_2 should state of nature S_3 occur.

TABLE 17.1
Payoff Table

Action	STATES OF NATURE				
	S_1	S_2	S_3	...	S_n
A_1	π_{11}	π_{12}	π_{13}		π_{1n}
A_2	π_{21}	π_{22}	π_{23}		π_{2n}
A_3	π_{31}	π_{32}	π_{33}		π_{3n}
$\vdots$					
A_k	π_{k1}	π_{k2}	π_{k3}		π_{kn}

Mr. Larson is owner of Sailtown, a store in southern Minnesota specializing in the sale of small sailboats. Each spring he is forced to place an order for his entire stock of Bluefin sailboats to be sold during the summer months because the Bluefin manufacturer is unable to supply any additional boats once the summer has begun.

Mr. Larson's main concern when ordering his summer inventory is the demand for his product during the next 5 months. He has discovered that this demand seems to be largely dependent on economic conditions, in particular on the prevailing interest rate. He has four possible actions (order quantities)

A_1: purchase 50 boats

A_2: purchase 75 boats

A_3: purchase 100 boats

A_4: purchase 150 boats

The states of nature for this problem are

S_1: the interest rate increases significantly (more than 1.5%) from the current rate

S_2: the interest rate holds steady

S_3: the interest rate drops significantly (more than 1.5%)

Based on his expected sales under each condition, the payoff table in Table 17.2 was constructed. To demonstrate how the payoffs were determined, consider $\pi_{42} = 5$, which is the resulting profit if he orders 150 sailboats (action A_4) and if the interest rate remains basically unchanged (state of nature S_2). Mr. Larson believes that, for S_2, the resulting demand will be 100 sailboats. His profit per sale is $300, and his cost for holding and returning an unsold boat at the end of the fall season is $500. Consequently, if the demand is for 100 boats and he decides to stock 150, he ends up selling 100 boats and returning 50 of them to the manufacturer. The resulting dollar amounts are

profit for selling 100 boats = 100 · $300 = $30,000
loss for returning 50 boats =   50 · $500 = $25,000
net payoff = π_{42} = $ 5,000

TABLE 17.2
Profit Table for Sailtown
(Thousands of Dollars)

Amount Ordered	AVERAGE INTEREST RATE		
	Increases (S_1)	Steady (S_2)	Decreases (S_3)
50 (A_1)	15	15	15
75 (A_2)	2.5	22.5	22.5
100 (A_3)	−10	30	30
150 (A_4)	−35	5	45

TABLE 17.3

Demand for Sailboats for Each State of Nature

STATE OF NATURE	INTEREST RATE	CORRESPONDING DEMAND
S_1	Increases	50
S_2	Holds steady	100
S_3	Decreases	150

TABLE 17.4

Payoff for Sailtown Under S_1 (Example 17.1)

ACTION	REVENUE FOR BOATS SOLD	LOSS DUE TO RETURNED BOATS	NET PAYOFF
A_1	$50 \cdot 300 = \$15,000$	$0 \cdot 500 = \$0$	$\$15,000$
A_2	$50 \cdot 300 = \$15,000$	$25 \cdot 500 = \$12,500$	$\$ 2,500$
A_3	$50 \cdot 300 = \$15,000$	$50 \cdot 500 = \$25,000$	$-\$10,000$
A_4	$50 \cdot 300 = \$15,000$	$100 \cdot 500 = \$50,000$	$-\$35,000$

So Mr. Larson calculates that the interest rate holding steady is equivalent to a demand for 100 sailboats. Similarly, he determines that an increase in the interest rate will result in a demand for 50 boats, whereas a decrease in the interest rate will produce a demand for 150 boats (Table 17.3).

EXAMPLE 17.1

Using Table 17.3, determine the payoff for each of the four alternatives if the interest rate increases over the summer months. What would be the best action to take if you knew that this particular state of nature would occur?

SOLUTION

We are given that state of nature S_1 will occur. Mr. Larson thinks that under this state of nature, 50 people will walk in his front door wanting to purchase a Bluefin sailboat. Under this assumption, the payoffs in Table 17.4 can be derived.

If we know that the interest rate will increase during the summer months, Table 17.4 tells us that the ideal action is to purchase 50 boats (action A_1).

Is action A_1 the ideal action for each state of nature? Given state of nature S_2 and using Table 17.2, the payoffs are $15,000 (for A_1), $22,500 (for A_2), $30,000 (for A_3), and $5000 (for A_4). In this case, you would achieve maximum profit by purchasing 100 boats (action A_3). The third column of Table 17.2 shows that action A_4 (purchasing 150 boats) provides the maximum profit in the event that the interest rate declines. ∎

This is an example of **decision making under certainty**. Although the decision-maker will never have this luxury, the technique at least enables him or her to determine the maximum profit under each state of nature. Also, if the *same action* provides the maximum payoff regardless of the state of nature, then this particular action is the obvious choice for this situation. Such was not the case in Table 17.2. There is no obviously superior action here, so we need to consider more elaborate decision strategies.

EXERCISES

17.1 A stockbroker is trying to decide which of three possible actions to recommend to a client. One action (A_1) would be to invest 100% in a mutual fund that invests in stocks. Another action (A_2) would be to invest 50% in a fixed account yielding 10% per year and 50% in a stock market mutual fund. The third action (A_3) would be to invest 100% in the fixed account, yielding 10% per year. The stockbroker believes the following returns can be made in a particular mutual fund in 1 year with regard to the stock market's direction.

S_1: STOCK MARKET GOES UP	S_2: STOCK MARKET HAS NO DIRECTION	S_3: STOCK MARKET GOES DOWN
20%	5%	-10%

Using the three states of nature, S_1, S_2, and S_3 and the three actions A_1, A_2, and A_3, construct a payoff table if a client has $10,000 to invest for 1 year.

17.2 Would the following decisions be made under certainty or uncertainty?

a. Whether to buy or lease a new car.

b. Whether to accept a client who wants to pay you $40 per visit for five visits.

c. Whether to borrow $10,000 at 12% interest per year for 5 years.

d. Whether to switch jobs or stay at the same job.

17.3 A seminar on sales motivation is being given at a local hotel. The cost of handouts and materials per attendee is $10. The total cost of the hotel arrangements is $75. Each attendee pays $25 for the seminar. The coordinator of the seminar must plan the number of handouts and materials. The coordinator must plan for 7, 8, 9, 10, 11, 12, or 13 attendees. Construct the payoff table if the demand is 5, 6, 7, 8, 9, 10, 11, 12, or 13 attendees.

17.4 The Jones family has moved out of their house and now has it up for sale. They need to decide whether to price their house at the top of market, $88,000, at the market, $85,000, or below the market at $82,000. Each month that it takes to sell the house, the Jones family loses $700 from making the monthly payment. Construct a payoff table that shows the selling price of the home minus $700 for each month the house remains unsold. Use S_1 = sold in 1 month, S_2 = sold in 2 months, . . . , S_6 = sold in 6 months.

17.5 Mini-Super, a convenience food store, orders milk each week. The store pays $1.10 per gallon and sells the milk for $2.00 per gallon. Unsold milk at the end of the week is given to an orphanage. The manager must decide on ordering 100 gallons, 125 gallons, or 150 gallons per week. Construct the payoff table for this problem if demand is 100 gallons, 125 gallons, or 150 gallons a week.

17.2

DECISION STRATEGIES

When the action providing a maximum payoff depends on an uncertain state of nature, the decision-maker is forced to consider all the values in the payoff table to choose the most attractive action. The various strategies discussed here allow you to choose a procedure that best fits your style of making decisions. We begin with a conservative strategy, the minimax procedure.

The Conservative (Minimax) Strategy

A conservative strategy is basically one that, when choosing between a savings account and an extremely risky venture, selects the savings account. It does this because, under the *worst* conditions, your loss is smaller with a savings account than with the high-risk venture.

We examined the ideal action for Mr. Larson to take in the event that he knew that the interest rate was going to increase (S_1 in Table 17.2). This action was to order 50 boats for a payoff of $\pi_{11} = 15$ (thousand). If he had taken action A_2 instead, his profit would be only $2500. For this situation, we say that the opportunity loss is $15,000 - $2500 = $12,500.

The **opportunity loss**, L_{ij}, is the difference between the payoff for action i and the payoff for the action that would have the largest payoff under state of nature j.

The opportunity loss is not a loss in the accounting sense; rather, it describes how much more profit you *would have made* had you chosen the best action for this state of nature. The opportunity loss for action A_3 (assuming state of nature S_1) is

$$\text{opportunity loss} = L_{31} = 15 - (-10)$$
$$= 25 \quad \text{(thousand dollars)}$$

and this value for action A_4 (under S_1) is

$$\text{opportunity loss} = L_{41} = 15 - (-35)$$
$$= 50 \quad \text{(thousand dollars)}$$

EXAMPLE 17.2

Construct the remaining opportunity losses for Mr. Larson, and summarize them in an opportunity loss table.

SOLUTION

Keep in mind that opportunity losses are determined one *column* at a time in Table 17.2 by assuming that each individual state of nature occurs and then looking for the best action under this condition. This is exactly the same procedure that we used when discussing the unrealistic situation of decision making under certainty.

If the interest rate holds steady (S_2 occurs), the best action is to stock 100 boats (action A_3) with a payoff of $\pi_{32} = 30$. Table 17.5 shows the opportunity loss for this situation.

Similarly, action A_4 (stock 150 boats) is the ideal action in the event the interest rate decreases (S_3) and sales increase, as shown in Table 17.6.

We format this as an **opportunity loss table**, as shown in Table 17.7. Notice that each column of an opportunity loss table contains a zero, and all values in this table are nonnegative (≥ 0). ∎

TABLE 17.5

Opportunity Loss for Assumption: S_2 Occurs; Best Action: A_3; Maximum Payoff: 30

ACTION	PAYOFF	OPPORTUNITY LOSS
A_1	15	$L_{12} = 30 - 15 = 15$
A_2	22.5	$L_{22} = 30 - 22.5 = 7.5$
A_3	30	$L_{32} = 30 - 30 = 0$
A_4	5	$L_{42} = 30 - 5 = 25$

TABLE 17.6

Opportunity Loss for Assumption: S_3 Occurs; Best Action: A_4; Maximum Payoff: 45

ACTION	PAYOFF	OPPORTUNITY LOSS
A_1	15	$L_{13} = 45 - 15 = 30$
A_2	22.5	$L_{23} = 45 - 22.5 = 22.5$
A_3	30	$L_{33} = 45 - 30 = 15$
A_4	45	$L_{43} = 45 - 45 = 0$

TABLE 17.7

Opportunity Loss Table for Sailtown Decision Problem (Thousands of Dollars)

	STATE OF NATURE		
Action	S_1	S_2	S_3
A_1	0	15	30
A_2	12.5	7.5	22.5
A_3	25	0	15
A_4	50	25	0

The Minimax Strategy The minimax strategy is to:

1. Construct an opportunity loss table by using the maximum payoff for each state of nature.
2. Determine the maximum opportunity loss for each action.
3. Find the minimum value of those found in step 2; the corresponding action is the one selected by the minimax strategy.

This is a very conservative approach that does not search for large payoffs; rather, it selects the action that has the smallest "worst case" opportunity loss.

The minimax procedure begins by examining the worst possible situation for each action. So you examine Table 17.7 one *row* at a time and determine the largest opportunity loss for each action. Thus, we have

ACTION	MAXIMUM OPPORTUNITY LOSS
A_1	30
A_2	22.5
A_3	25
A_4	50

This is the *max* part of the minimax strategy. The *mini* side is finding the *minimum* of these four values. In this way, you attempt to offset the worst possible situation scenario. For this example, of 30, 22.5, 25, and 50, the minimum is 22.5, which belongs to action A_2, so the **minimax decision** is to order 75 sailboats.

In actual practice, a decision analysis rarely begins in the form of a payoff table. Instead, this table perhaps can be constructed using the information available and then the appropriate decision strategy can be applied to arrive at the corresponding action. This is illustrated in the next example.

EXAMPLE 17.3

Tastee Yogurt Factory is a family-run, franchised store selling frozen yogurt. The operator of the store has two options available during the first year of operation: (1) lease the premises at a fixed rate of $1300 per month or (2) pay a monthly lease rate of $1000 per month (based on a 9% annual interest rate) and pay an adjustment at the end of the year, which adjusts the total annual lease payment based upon the prevailing interest rate at year end. For example, if the interest rate at the end of the year is 11%, then had this interest rate been applied for the entire year, the lease payment would have been $1550 per month (or $18,600 annually). Thus, the operator of Tastee would have due

$$(1550 \times 12) - (1000 \times 12) = \$6600$$

Since the operator of Tastee has put down a sizeable deposit to secure the franchise, there is no risk to the company owning the operation in offering the variable payment option. After a careful review of present and forecasted economic conditions, it appears that the possible interest rates at the end of the following year are 9%, 10%, 11%, and 12%, with the following total lease payments:

$$9\%: \quad \$1000 \times 12 = \$12,000$$
$$10\%: \quad \$1250 \times 12 = \$15,000$$
$$11\%: \quad \$1550 \times 12 = \$18,600$$
$$12\%: \quad \$1900 \times 12 = \$22,800$$

In addition, the operator of the store estimates revenues for the following year to be $74,000 with fixed overhead and payroll costs estimated to be $12,000.

Construct a payoff table for this situation and decide whether to take the fixed or variable leasing option using the minimax strategy.

SOLUTION

This illustration contains two possible actions:

A_1: use the fixed lease price of $1300 per month

A_2: use the variable lease price, adjusted at year end

There are four states of nature defined by the prevailing interest rate at the end of the year:

S_1: interest rate at end of year is 9%

S_2: interest rate at end of year is 10%

S_3: interest rate at end of year is 11%

S_4: interest rate at end of year is 12%

The annual payoff (regardless of the state of nature) for action A_1 is

$$74,000 - 12,000 - (1300 \times 12) = \$46,400$$

When employing action A_2, the resulting annual payoff depends upon the state of nature:

for S_1: payoff $= 74,000 - 12,000 - (1000 \times 12)$
$$= 50,000$$

for S_2: payoff $= 74,000 - 12,000 - (1250 \times 12)$
$$= 47,000$$

for S_3: payoff $= 74,000 - 12,000 - (1550 \times 12)$
$$= 43,400$$

for S_4: payoff $= 74,000 - 12,000 - (1900 \times 12)$
$$= 39,200$$

The resulting payoff table is shown in Table 17.8.

If state of nature S_1 should occur, the best action is A_2 with a payoff of $50,000 and an opportunity loss of zero. Consequently, A_1 has an opportunity loss of $50,000 - 46,400 = \$3600$. Continuing in this way, we can construct Table 17.9.

TABLE 17.8
Payoff Table for Tastee Yogurt Factory (Dollars)

		STATE OF NATURE			
Action		$9\%(S_1)$	$10\%(S_2)$	$11\%(S_3)$	$12\%(S_4)$
fixed (A_1)		46,400	46,400	46,400	46,400
variable (A_2)		50,000	47,000	43,400	39,200

TABLE 17.9
Opportunity Loss Table for Tastee Yogurt Factory (Dollars)

		STATE OF NATURE				
Action		$9\%(S_1)$	$10\%(S_2)$	$11\%(S_3)$	$12\%(S_4)$	**Maximum**
fixed (A_1)		3600	600	0	0	3600
variable (A_2)		0	0	3000	7200	7200

Next we find the maximum opportunity loss for each action.

ACTION	MAXIMUM OPPORTUNITY LOSS
A_1	Maximum of 3600, 600, 0, 0 = \$3600
A_2	Maximum of 0, 0, 3000, 7200 = \$7200

Finally, we select the minimum of these values. This is \$3600, corresponding to action A_1. The minimax decision is to use the fixed lease price of \$1300 per month. This action is the conservative strategy and is the one that minimizes the maximum difference between the profit received and the profit that could have been received if the state of nature (interest rate at year end) had been known in advance. ■

The Gambler (Maximax) Strategy

The maximax strategy is the opposite of the minimax procedure and appeals to those who are gamblers at heart. The **maximax strategy** is to choose that action having the largest possible payoff. It is not a recommended procedure for most business decisions because, by choosing that action with the largest payoff, it fails to consider the possibility of large accounting losses or opportunity losses.

EXAMPLE 17.4

Using the information in Table 17.2, which action should Sailtown select using the maximax strategy?

SOLUTION

Of the 12 payoffs in Table 17.2, the largest is 45, which corresponds to action A_4. If Sailtown is desperate for a large payoff, the appropriate action using this strategy would be to order 150 sailboats for the summer months. Of course, the company also stands to lose the most using this action; the loss will be \$35,000 if the interest rate increases. ■

EXAMPLE 17.5

Based on the payoff table in Table 17.8 and the maximax strategy, which leasing option (fixed or variable) should the operator of Tastee Yogurt Factory use?

SOLUTION

The maximum payoff is \$50,000 in Table 17.8, and this corresponds to action A_2. So, if the operator wants to gamble for a large payoff, the corresponding action would be to use the variable lease price. This procedure ignores the effect of the ideal state of nature for this action (S_1) *not* occurring. ■

EXERCISES

17.6 Why is the following table not an opportunity loss table? Give two reasons.

ACTION	S_1	S_2	S_3
A_1	4	6	80
A_2	29	0	150
A_3	57	−1	0

17.7 The owner of a bookstand orders the local daily newspapers and charges 25¢ per copy. The owner pays 10¢ for each copy. Each day, the owner orders either 70, 80, 90 or 100 copies. At the end of the day, the leftover newspapers are discarded. Construct the payoff table and opportunity loss table if the demand is 50, 60, 70, 80, 90, or 100 copies. What is the minimax decision? What is the maximax decision?

17.8 The owner of a small commercial building can either pay \$1000 per year to insure the \$200,000 building or not insure and save a \$1000 per year. If the states of nature are complete loss of the commercial building or no loss at all, what would the payoff table and opportunity loss table be for this situation? What is the minimax decision?

17.9   What would the opportunity loss table be using the following payoff table?

ACTION	S_1	S_2	S_3	S_4
A_1	150	2	89	5
A_2	-20	10	76	-10
A_3	0	15	94	-20

17.10   Refer to Exercise 17.1. Construct the opportunity loss table for the actions of the stockbroker and the states of nature of the stock market. What is the minimax decision? What is the maximax decision?

The Strategist (Maximizing Expected Payoff)

In many respects, a more sensible approach to any decision problem is to con-sider the likelihood that each state of nature will occur. In this way, you can use any information you have to help evaluate the possibilities of each of the states of nature. If you believe strongly that the chance of the interest rate declining is small, a decision strategy that uses this information would be useful. The strategy discussed here differs from previous procedures, in that we begin by determining the *probability* asssociated with each state of nature.

Selecting the Probabilities   The probability for each state of nature measures to what degree you believe this state of nature will occur in the future. One way to obtain these probabilities is from past experience—referred to as **empirical evidence**. For example, if, under similar conditions in the past, the stock market declined 15% of the time, we would set

$$P(\text{stock market declines}) = .15$$

In this way, you can determine the probability for each state of nature. Because we are assuming that one (and only one) of these states *must* occur, these probabilities must sum to 1.

Another method of selecting these probabilities is the **subjective approach**. With this procedure, an individual or group of individuals will select each prob-ability such that (1) each value represents their confidence that each state of nature will occur and (2) the probabilities sum to 1.

To someone unfamiliar with the concept of a probability, you can pose the question, Given this set of circumstances 100 different times, how often do you think the stock market will decline? If the answer is about 15 times, then once again you have

$$P(\text{stock market declines}) = .15$$

If the resulting probabilities do not sum to 1 on the first pass, you can state that these probabilities are *all* a little too small (or large) and try again. By continuing in this manner, you eventually will arrive at a set of probabilities for this situation.

The strengths and weaknesses of using these probabilities in the decision process lie in the accuracy of their values. If they are inaccurate, you may well choose an action that incurs a small (or negative) payoff. As a result, this strategy can lead to poor decisions, particularly if the action chosen was based on un-reliable subjective probabilities. Nevertheless, it continues to be a popular deci-sion strategy because it allows the decision-maker to place probabilities on the unknown future and consider the alternatives.

TABLE 17.10
Probabilities for
S_1, S_2, and
S_3 from Table 17.2

STATE OF NATURE		PROBABILITY
S_1	Interest rate increases	$P(S_1) = .3$
S_2	Interest rate remains unchanged	$P(S_2) = .2$
S_3	Interest rate decreases	$P(S_3) = .5$

The Decision Strategy   When using probabilities for each of the states of nature, you determine the *average* payoff for each action in the long run—the average payoff if you repeatedly took this action. This is the **expected payoff** for each action. The strategy in this case is to choose that action having the *largest* expected payoff.

Consider Table 17.2. Suppose that the owner of Sailtown believes that there is a 30% chance that the interest rate will increase over the summer months. This can be written as

$$P(S_1) = .3$$

The chance of the interest rate holding steady is believed to be 20%, whereas the value corresponding to a drop in the rate is 50% (Table 17.10).

One of the alternatives for this problem was to stock 150 sailboats (action A_4). Using Table 17.2, the respective payoffs are a loss of $35,000 should the interest rate increase (S_1), a profit of $5000 if it holds steady (S_2), and a profit of $45,000 if it decreases ($S_3$). So, if you repeatedly took this action (under the same conditions facing the owner of Sailtown), then you would

lose $35,000, 30% of the time (S_1 occurs)

make $5000, 20% of the time ($S_2$ occurs)

make $45,000, 50% of the time (S_3 occurs)

This is the situation discussed in Chapter 5, where we examined *discrete random variables*. The random variable for this situation is

$$X = \text{payoff under action } A_4$$

Based on the preceding discussion, we have

$$X = \begin{cases} -35 \text{ with probability} & .3 \\ 5 \text{ with probability} & .2 \\ 45 \text{ with probability} & \underline{.5} \\ & 1.0 \end{cases} \qquad \textbf{(17.1)}$$

The expected payoff for this action is simply the *mean of the random variable,* X. In Chapter 5, this was defined to be

$$\text{expected payoff for } A_4 = \text{mean of } X$$
$$= \Sigma(\text{each value of } X)(\text{its probability})$$
$$= (-35)(.3) + (5)(.2) + (45)(.5)$$
$$= 13$$

This implies that, if the owner of Sailtown repeatedly ordered 150 sailboats (under similar conditions), he would make a profit of $13,000 on the average.

We next use these expected payoffs to form a decision strategy.

TABLE 17.11

Expected Payoffs for Sailtown (Thousands of Dollars)

ACTION		EXPECTED PAYOFF
A_1	Order 50 sailboats	$(15)(.3) + (15)(.2) + (15)(.5) = 15$
A_2	Order 75 sailboats	$(2.5)(.3) + (22.5)(.2) + (22.5)(.5) = 16.5$
A_3	Order 100 sailboats	$(-10)(.3) + (30)(.2) + (30)(.5) = 18$
A_4	Order 150 sailboats	$(-35)(.3) + (5)(.2) + (45)(.5) = 13$

EXAMPLE 17.6

Determine the expected payoff for each of the actions in Table 17.2. Using this procedure, how many sailboats should Mr. Larson order?

SOLUTION

Based on the four expected payoffs, the appropriate action is to order 100 sailboats (A_3) with an expected (average) payoff of $18,000 (Table 17.11). If Mr. Larson chooses this alternative, his payoff for a one-time decision will be not a profit of $18,000 but, rather, a loss of $10,000 [with probability $P(S_1) = .3$] or a gain of $30,000 [with probability $P(S_2) + P(S_3) = .7$]. Mr. Larson will select this action if he believes that his long-term gain under this alternative has been maximized. In a sense, he has measured the uncertainty of the future in order to select the best action.   ■

EXAMPLE 17.7

In Example 17.3, the operator of Tastee Yogurt Factory used the minimax strategy and opted for the "safe" fixed lease price option. When assessing the economic conditions for the coming year, he arrived at the following set of probabilities for the four states of nature:

$$P(S_1) = .2 \qquad P(S_2) = .4 \qquad P(S_3) = .3 \qquad P(S_4) = .1$$

If he elects to use the expected payoff strategy, which leasing option offers the largest expected payoff?

SOLUTION

The expected payoffs for the two actions are:

A_1:   expected payoff $= (.2)(46,400) + (.4)(46,400) + (.3)(46,400) + (.1)(46,400)$
$$= \$46,400$$

A_2:   expected payoff $= (.2)(50,000) + (.4)(47,000) + (.3)(43,400) + (.1)(39,200)$
$$= \$45,740$$

Consequently, A_1 has the largest expected payoff and both the minimax and expected payoff strategies indicate that the best option is to use the fixed lease price.   ■

EXAMPLE 17.8

Omega is about to introduce a new line of microcomputers. Their main concern is what selling price they should charge for their computers. The managers can estimate accurately the demand at each price; they are primarily concerned about the time it will take their competitors to catch up and introduce a similar product. They intend to determine a selling price and then not change it for the next 2 years. They decide to structure the decision problem using four possible alternatives (actions):

A_1:   set selling price at $1500

A_2:   set selling price at $1750

A_3:   set selling price at $2000

A_4:   set selling price at $2500

TABLE 17.12

Profit Table for
Omega Computer-Price
Problem (Millions
of Dollars)

SELLING PRICE	<6 MONTHS (S_1)	6–12 MONTHS (S_2)	12–18 MONTHS (S_3)	>18 MONTHS (S_4)
A_1 $1500	250	320	350	400
A_2 $1750	150	260	300	370
A_3 $2000	120	290	380	450
A_4 $2500	80	280	410	550

The states of nature specify the amount of time until a similar product is introduced by one of their competitors. These are

$$S_1 = \text{less than 6 months}$$
$$S_2 = \text{6 to 12 months}$$
$$S_3 = \text{12 to 18 months}$$
$$S_4 = \text{longer than 18 months}$$

The next step for this decision problem is to construct a payoff table. This is *not* an easy step because the managers must consider price-demand, cost-volume, and consumer-preference information in order to specify a payoff for each action under each state of nature. After many meetings between the production and marketing staffs, Table 17.12 was derived, showing projected profits over the next 2 years.

What is the appropriate action (selling price) if:

1. The minimax strategy is used?
2. Omega decides to maximize the expected payoff?

Use

$$P(S_1) = .1 \qquad P(S_2) = .5 \qquad P(S_3) = .3 \qquad P(S_4) = .1$$

SOLUTION 1

Using the minimax strategy, we first construct an opportunity loss table for this situation (Table 17.13). We do this by considering each state of nature and finding the action with the largest payoff under each state. The opportunity

TABLE 17.13

Construction of
Opportunity Loss
Table for Omega
(Example 17.8)

STATE OF NATURE	ACTION WITH LARGEST PAYOFF	OPPORTUNITY LOSS
S_1	A_1	for A_1: 250 − 250 = 0 for A_2: 250 − 150 = 100 for A_3: 250 − 120 = 130 for A_4: 250 − 80 = 170
S_2	A_1	for A_1: 320 − 320 = 0 for A_2: 320 − 260 = 60 for A_3: 320 − 290 = 30 for A_4: 320 − 280 = 40
S_3	A_4	for A_1: 410 − 350 = 60 for A_2: 410 − 300 = 110 for A_3: 410 − 380 = 30 for A_4: 410 − 410 = 0
S_4	A_4	for A_1: 550 − 400 = 150 for A_2: 550 − 370 = 180 for A_3: 550 − 450 = 100 for A_4: 550 − 550 = 0

TABLE 17.14

Opportunity Loss Table for Omega Computer-Price Problem (Millions of Dollars)

SELLING PRICE	<6 MONTHS (S_1)	6–12 MONTHS (S_2)	12–18 MONTHS (S_3)	>18 MONTHS (S_4)
A_1 $1500	0	0	60	150
A_2 $1750	100	60	110	180
A_3 $2000	130	30	30	100
A_4 $2500	170	40	0	0

TABLE 17.15

Expected Profits for Omega (Example 17.8)

ACTION	EXPECTED PROFIT
A_1	$(.1)(250) + (.5)(320) + (.3)(350) + (.1)(400) = 330$
A_2	$(.1)(150) + (.5)(260) + (.3)(300) + (.1)(370) = 272$
A_3	$(.1)(120) + (.5)(290) + (.3)(380) + (.1)(450) = 316$
A_4	$(.1)(80) + (.5)(280) + (.3)(410) + (.1)(550) = 326$

loss for each action is the maximum payoff under this state of nature minus the payoff for this particular action.

Next, we find the maximum opportunity loss *for each action* (row in Table 17.14).

ACTION	MAXIMUM OPPORTUNITY LOSS
A_1	150
A_2	180
A_3	130
A_4	170

The minimum of these values is 130, belonging to A_3. The minimax strategy would be to select a selling price of $2000 for the next two years.*

SOLUTION 2

The expected profit for each action is summarized in Table 17.15.

The maximum expected profit is 330, for action A_1. So the strategy here is to set the selling price at $1500, with an expected payoff of $330 million. Notice, however, that the three largest expected values are quite close to each other. The implications of this are that one of the other alternatives might surpass A_1 if the state of nature probabilities are adjusted slightly. The preference for A_1 may be very sensitive to these probabilities. This should concern a decision-maker, especially if the probabilities are determined subjectively. ■

Example 17.8 suggests another important element of the decision process—a sensitivity analysis.

Sensitivity Analysis

Typically, there is no way to determine a state of nature probability with certainty. You can consider past observations and derive an empirical estimate or merely make up a value that measures your belief that this event will occur

* The minimax procedure is often confused with the *maximin* strategy, which examines the minimum payoff for each action and selects that action having the maximum of these minimum payoffs. For this application, the minimum payoffs for each action are A_1: 250, A_2: 150, A_3: 120, and A_4: 80. The maximum value here is 250 (belonging to A_1), and the maximin strategy is to select action A_1. The resulting conclusions using minimax and maximin are not the same here because the minimax strategy (which selects that action minimizing the maximum opportunity loss) is to use action A_3. Both strategies are typically very conservative.

TABLE 17.16
Summary of Sensitivity
Analysis (Values in
Color Represent the
Action with the Largest
Expected Payoff)

			EXPECTED PAYOFF			
$P(S_1)$	$P(S_2)$	$P(S_3)$	A_1	A_2	A_3	A_4
.4	.2	.4	15	14.5	14	5
.4	.3	.3	15	14.5	14	1
.4	.1	.5	15	14.5	14	9
.5	.2	.3	15	12.5	10	−3
.5	.1	.4	15	12.5	10	1
.3	.3	.4	15	16.5	18	9
.3	.2	.5	15	16.5	18	13

(the subjective approach). The next step when using the maximum expected payoff strategy is to examine what happens to this solution under other sets of realistic probabilities. This is a **sensitivity analysis**.

In Example 17.6, the expected payoff procedure selected action A_3. By ordering 100 sailboats, Sailtown achieved a maximum expected profit of $18,000. The state of nature probabilities used here were

$$P(S_1) = .3 \quad \text{(interest rate increases)}$$
$$P(S_2) = .2 \quad \text{(interest rate remains unchanged)}$$
$$P(S_3) = .5 \quad \text{(interest rate decreases)}$$

Although Mr. Larson and his financial advisor are uncertain as to the precise values of these probabilities, they believe that:

1. There is no more than a 50% chance that the interest rate will increase $(P(S_1) \leq .5)$.
2. There is no more than a 30% chance that the rate will remain unchanged $(P(S_2) \leq .3)$.
3. The probability that the rate will decrease is between .3 and .5.

They decide to examine the expected payoffs under the probability conditions listed in Table 17.16. The expected payoffs under each set of probabilities are determined as in Example 17.6. As an illustration, using $P(S_1) = .4$ and $P(S_2) = P(S_3) = .3$, the expected payoff for action A_4 is $(.4)(-35) + (.3)(5) + (.3)(45) = 1$ (that is, $1000).

The sensitivity summary in Table 17.16 indicates that action A_1 (ordering 50 sailboats) may be much more attractive than we thought. In fact, if there is more than a 30% chance that the interest rate will increase $(P(S_1) > .3)$, this action produces the largest expected payoff. Under this decision, Mr. Larson can expect to sell all his inventory, resulting in a profit of $15,000 *regardless of the state of nature*. Consequently, Sailtown would be seeking an expected gain of $3000 ($18,000 for A_3 minus $15,000 for A_1) by speculating on the uncertain future.

Without such a sensitivity analysis, Mr. Larson would not have noticed these results. In five of the six cases using probabilities other than those used in Example 17.6, action A_1 produced the maximum expected payoff. If Mr. Larson is uncertain in his original determination of these probabilities, this action is a better solution to his decision problem.

Evaluating Risk

Using the preceding sensitivity analysis and the payoffs in Table 17.2, we noticed that the payoff for action A_1 was 15 (thousand dollars), regardless of the state of nature. Action A_4, on the other hand, has possible payoffs of -35, 5, and 45.

This implies that you will encounter a higher risk using A_4 rather than A_1. In fact, action A_1 has *no risk* because its payoff is known with certainty.

Take a closer look at action A_4. In discussing equation 17.1, we remarked that the payoff for this action, X, is a random variable, where

$$X = \begin{cases} -35 & \text{with probability } .3 \\ 5 & \text{with probability } .2 \\ 45 & \text{with probability } .5 \end{cases}$$

The expected payoff for this action is the *mean of X*.

A risky alternative (action) is one that has larger probabilities attached to extremely large or small payoffs. A good measure of this risk is simply the *variance* of X; the variance of the possible payoffs for each action is a measure of the risk associated with this alternative. The larger the variance is, the more risk will be incurred using this action. The variance of this *discrete* random variable is found in the same way as it was in Chapter 5 and is summarized here.

Let X_i be the payoff associated with action A_i. Then

$$X_i = \begin{cases} x_1 & \text{with probability } p_1 \\ x_2 & \text{with probability } p_2 \\ \vdots \\ x_n & \text{with probability } p_n \end{cases}$$

where n represents the number of states of nature.

The **expected payoff** for this action is the mean of X_i, where

$$\text{expected payoff} = \mu_i = \Sigma\, xp \tag{17.2}$$

The **risk** associated with action A_i is the variance of X_i. So,

$$\text{risk} = \Sigma\, x^2 p - \mu_i^2 \tag{17.3}$$

EXAMPLE 17.9

Compute the risk for each of the actions in Table 17.2, using the state of nature probabilities from Example 17.6. Based on these results and the sensitivity analysis in Table 17.16, which action appears to be the best one for this situation?

SOLUTION

Using equation 17.3, we can find the risk associated with each of the four alternatives, as shown in Table 17.17. The purpose of examining the risk for each action is that often the decision maker will prefer a less risky alternative over a riskier action with a larger expected payoff. In this example, action A_1 has no risk and also has a maximum expected payoff for most of the situations examined in the sensitivity analysis. On the other hand, action A_3 (the other suggested approach) carries the second-largest risk, as measured in Table 17.17.

TABLE 17.17

Risk Calculations for Sailtown Decision Problem

ACTION	EXPECTED PAYOFF (μ_i)	RISK (USING EQUATION 17.3)
A_1	15	$[(15)^2(.3) + (15)^2(.2) + (15)^2(.5)] - 15^2 = 0$
A_2	16.5	$[(2.5)^2(.3) + (22.5)^2(.2) + (22.5)^2(.5)] - 16.5^2 = 84$
A_3	18	$[(-10)^2(.3) + (30)^2(.2) + (30)^2(.5)] - 18^2 = 336$
A_4	13	$[(-35)^2(.3) + (5)^2(.2) + (45)^2(.5)] - 13^2 = 1216$

TABLE 17.18

Risk Calculations for
Omega Decision
Problem

ACTION	EXPECTED PAYOFF (μ_i)	RISK (USING EQUATION 18.3)
A_1	330	$[(250)^2(.1) + (320)^2(.5) + (350)^2(.3) + (400)^2(.1)] - 330^2 = 1{,}300$
A_2	272	$[(150)^2(.1) + (260)^2(.5) + (300)^2(.3) + (370)^2(.1)] - 272^2 = 2{,}756$
A_3	316	$[(120)^2(.1) + (290)^2(.5) + (380)^2(.3) + (450)^2(.1)] - 316^2 = 7{,}204$
A_4	326	$[(80)^2(.1) + (280)^2(.5) + (410)^2(.3) + (550)^2(.1)] - 326^2 = 14{,}244$

For these reasons, the soundest alternative appears to be action A_1, with a known payoff of $15,000. ∎

EXAMPLE 17.10 Which of the actions facing the operator of Tastee Yogurt Factory (Example 17.7) has the least amount of risk?

SOLUTION The fixed lease price option contains a known payoff ($46,400) regardless of the state of nature, and consequently has zero risk. This action appears to offer the best alternative since it has minimum risk and the largest expected payoff. ∎

EXAMPLE 17.11 Which of the alternatives in Example 17.8 has the least amount of risk?

SOLUTION As before, the risk associated with each action is the variance of the corresponding random variable, shown in Table 17.18. The most desirable action for Omega is A_1 because it wins on two counts. This action (selling price = $1500) not only has the largest expected profit, it also has the smallest risk. For a great many decision problems, this will not be the case, and so the decision-maker will have to decide how much risk he or she is willing to assume in an effort to gain a higher expected profit. If a heavy loss would be devastating to a company, it may be forced into adopting strategies that select alternatives with reasonably attractive profits but considerably less risk. ∎

EXERCISES

17.11 The following table gives the payoff for four different states of nature and three different actions. If each state of nature is equally likely, find the decision resulting from maximizing expected payoff. What is the risk for each action?

ACTION	S_1	S_2	S_3	S_4
A_1	180	150	10	50
A_2	55	55	55	55
A_3	80	160	100	40

17.12 The manager of a hardware store orders several cords of split logs to be sold to customers for firewood. More wood typically is sold when a winter is colder than usual. The manager figures that the chance of an extremely cold winter (S_1) is 0.25, the chance of a normal winter (S_2) is 0.50, and the chance of a relatively mild winter (S_3) is 0.25. The manager must decide on ordering either 50, 40, 30, or 20 cords of wood. The payoff table follows.

ACTION	S_1	S_2	S_3
A_1 (50)	5000	3000	1000
A_2 (40)	4400	3200	1200
A_3 (30)	3200	2800	1400
A_4 (20)	3000	2500	2000

a. What is the minimax decision?

b. What is the decision based on the maximum expected payoff?

c. What is the risk of each action?

d. What is the decision based on minimum risk?

17.13 Programs need to be printed for a theatrical performance. The programs are sold at the entrance of the theater before the performance starts. The director of the theater believes that there is a 35% chance that there will be a heavy turnout (S_1). The director also believes that there is a 50% chance for a normal turnout (S_2) and a 15% chance for a low turnout (S_3). The director must decide to have 200 copies (A_1), 300 copies (A_2), 400 copies (A_3), or 500 copies (A_4) of the program printed. The payoff table follows. Unsold programs would result in a loss.

ACTION	S_1	S_2	S_3
A_1	100	100	100
A_2	150	140	110
A_3	200	160	75
A_4	250	120	-50

a. Find the minimax decision.

b. Find the decision based on the maximum expected payoff.

c. Find the risk associated with each decision.

17.14 In Exercise 17.1, what is the decision based on the maximum expected payoff if the probability that the stock market goes up is 10%, the probability that the stock market has no direction is 50%, and the probability that the stock market goes down is 40%?

17.15 In Exercise 17.7, what is the maximum expected payoff if the demand for newspapers has the following probabilities?

DEMAND	PROBABILITY	DEMAND	PROBABILITY
50	0.10	80	0.25
60	0.10	90	0.20
70	0.25	100	0.10

17.16 In Exercise 17.8, what is the maximum expected payoff if the probability of a complete loss of the commercial building is .05? Assume that the probability of no loss is 0.95. Find the risk associated with each action.

Dominated Actions and the Value of a Crystal Ball

In a decision problem, we often can eliminate an action from consideration if another action in the problem has a larger payoff, regardless of the state of nature. Consider actions A_1 and A_2 from Table 17.12. Notice that the payoff for A_1 exceeds that for A_2 for all four states of nature. In this case, we say that A_1 **dominates** A_2. Action A_i dominates A_j if the payoff for A_i is greater than or equal to that for A_j under each state of nature. For at least one state of nature, the payoff for A_i must exceed that for A_j. In our example, there is no reason to consider A_2 for any decision strategy because A_1 produces a larger profit, regardless of what happens in the future. We say that action A_2 is inadmissible; it will not be included in the group of actions to be considered in the problem solution. Action A_i is **inadmissible** if it is dominated by any other action. Consequently, A_i is **admissible** if no other action under consideration dominates it.

EXAMPLE 17.12 Which of the actions in Table 17.2 are admissible?

SOLUTION The procedure here is to determine whether any of these actions are dominated by any other action. For example, for actions A_2 and A_3, A_2 has a larger payoff, given state of nature S_1, but A_3 produces a bigger profit, given S_3. Therefore, neither of these two actions dominates the other. Comparing A_1 with A_2 and

A_1 with A_3 produces a similar argument; no one action produces a larger pay-off for all states of nature. This implies that *all three* actions are admissible, and so they will all be considered in the search for the best action under the selected decision strategy.

Note that there is no serious harm in considering a dominated action; such an action never is selected by any of the decision strategies. By eliminating a dominated action from consideration, however, we simplify the decision process, since we have fewer actions to consider.

Expected Value of Perfect Information   When using the strategy of maximizing expected profit, one value of interest is how much you would be willing to pay for a predictor that could tell you the future state of nature correctly 100% of the time. For example, you might have the (very unrealistic) situation of a con-sulting firm that predicts the future correctly all the time or, just as farfetched, a crystal ball. Because the future is in fact never perfectly predictable, any infor-mation about the future will be imperfect. Consequently, you use the value of a perfect predictor to evaluate any cost that you might incur for such imperfect information.

In Example 17.6, what would Mr. Larson, the owner of Sailtown, expect to make if a perfect predictor existed? Referring to Table 17.2, because $P(S_1) = .3$, the crystal ball will predict state of nature S_1 30% of the time. In this case, Mr. Larson inspects this column of Table 17.2, realizes that action A_1 has the largest payoff, and so orders 50 sailboats. His profit for this decision will be 15 (thousand dollars).

Now suppose the crystal ball predicts that the interest rate will remain unchanged (S_2 occurs). In this event, Mr. Larson will order 100 sailboats (A_3), because this is the largest payoff in the column under S_2 in Table 17.2. His profit will be 30. Finally, if he is informed that S_3 will occur, he selects action A_4, with a payoff of 45, because this produces the largest profit given that the interest rate will decrease.

In the long run, with a perfect predictor, Mr. Larson would make

$15,000 30% of the time   (when S_1 occurs)

$30,000 20% of the time   (when S_2 occurs)

$45,000 50% of the time   (when S_3 occurs)

This means that his *expected payoff with a perfect predictor* is

$$(15,000)(.3) + (30,000)(.2) + (45,000)(.5) = \$33,000$$

Finally, recall that the action that maximized the expected payoff (from Example 17.6) was A_3, with a value of $18,000. So, Mr. Larson would make $33,000 on the average *with* a crystal ball. Conversely, he would earn $18,000 on the average *without* it by taking action A_3 each time. This means that the maximum price he should be willing to pay for a perfect predictor is

$$33,000 - 18,000 = \$15,000$$

This is the expected value of perfect information.

When you use expected payoffs in your decision strategy, you select that action (say, A') having the largest expected payoff. The **expected value of perfect information (EVPI)** is

$$\text{EVPI} = \frac{\text{(average payoff using a perfect predictor)} -}{\text{(average payoff for } A')} \qquad (17.4)$$

EXAMPLE 17.13   In Example 17.8, the managers of Omega attempted to choose a selling price for their new computer. The states of nature for this problem were concerned with the amount of time until a major competitor introduced a similar product. The payoffs for this situation are summarized in Table 17.12. Assuming that $P(S_1) = .1$, $P(S_2) = .5$, $P(S_3) = .3$, and $P(S_4) = .1$, determine the expected value of perfect information.

SOLUTION   The action having the largest expected profit (according to Example 17.8) is A_1, with an expected value of 330. Consequently, the payoff with a selling price of $1500 would be $330 million on the average.

Given a perfect predictor, the following payoffs are available.

STATE OF NATURE (S_i)	MAXIMUM PAYOFF	PROBABILITY $P(S_i)$
S_1	250 (for A_1)	.1
S_2	320 (for A_1)	.5
S_3	410 (for A_4)	.3
S_4	550 (for A_4)	.1

Consequently, the expected payoff using a perfect predictor is

$$(.1)(250) + (.5)(320) + (.3)(410) + (.1)(550) = 363 \quad \text{(million dollars)}$$

From these results, the expected value of perfect information (from equation 17.4) is

$$\text{EVPI} = 363 - 330 = 33 \quad \text{(million dollars)}$$

So what is the maximum amount that Omega Corporation should be willing to pay an outside consulting firm for information regarding the time until a competitor introduces a similar model into the market? This is what the EVPI represents—an upper limit for the price of *any* information regarding the future. If Omega elects to pay an outside firm for information, they realize that the predicted state of nature could be wrong (that is, this information will be imperfect). For this reason, this information is worth *considerably less* than the EVPI of $33 million. Its value will depend in part on the reliability of the consulting firm, as measured by the latter's past performance in similar situations. This topic is pursued further in Section 17.4.   ∎

EXERCISES

17.17   Consider the following payoff table, in which each state of nature is equally likely. Find the EVPI. Are all actions admissible?

ACTION	S_1	S_2	S_3	S_4
A_1	40	10	4	15
A_2	30	15	0	35
A_3	20	10	−5	30
A_4	35	20	10	10

17.18   A builder usually builds 20-, 50-, or 100-unit apartments. The builder is concerned with three states of nature: low demand (S_1), medium demand (S_2), and high demand (S_3). The builder believes that the probability of S_1 is .40, the probability of S_2 is .30, and the probability of S_3 is .30. From the following payoff table, what is the maximum amount that the builder would be willing to pay a consultant for advice regarding the market demand for apartments? Are all the actions admissible?

ACTION	S_1	S_2	S_3
A_1 (20 units)	10,000	13,000	16,000
A_2 (50 units)	8,000	23,000	25,000
A_3 (100 units)	8,000	20,000	40,000

17.19   Greetings card shop must decide on whether to order 1500, 2000, or 2500 holiday cards before the holiday season. The card shop makes a profit of 50¢ on each card it sells and loses 30¢ on each card that remains unsold. If the demand for holiday cards is strong (S_1), 2500 cards should sell. If the demand is average (S_2), 2000 cards should sell. If the demand is weak (S_3), 1,500 cards should sell. The manager of Greetings believes that the following probabilities are representative of past sales: $P(S_1) = .40$, $P(S_2) = .40$, and $P(S_3) = .20$. What is the maximum amount that the manager of Greetings would be willing to pay for perfect information regarding the demand for holiday cards?

17.20   If the EVPI for a certain company is $40,000 and the average payoff based on the decision from the maximum expected payoff is $25,000, what is the maximum payoff using a perfect market predictor?

17.21   Refer to Exercise 17.1. What is the maximum that the stockbroker would be willing to pay a consultant regarding the direction of the stock market, assuming that all of the three states of nature are equally likely?

17.22   Refer to Exercise 17.12. What is the maximum that the manager of the hardware store would be willing to pay to obtain perfect information regarding the type of winter for the current year?

17.23   Refer to Exercise 17.13. What is the EVPI for the manager of the theater, regarding attendance turnout? Are any of the actions inadmissible?

17.3

THE CONCEPT OF UTILITY

We have concentrated on choosing the best action under various decision strategies by using the values contained in the payoff table. For example, one strategy determines that action having the largest expected payoff. Another (the minimax procedure) examines opportunity losses derived from the payoff table. Still another strategy examines expected payoffs. In other words, each action is evaluated by the corresponding *dollar amount* resulting from a particular strategy.

There are many instances in which it is more advantageous *not* to use expected payoffs, particularly when large amounts of money are at stake. Anyone who purchases an insurance policy or buys a lottery ticket generally is trying neither to minimize expected losses nor maximize expected gains. Rather, such a person *gambles* his or her money, trying to guard against a heavy loss (insurance) or hoping to strike it rich (lottery).

There is something else besides money involved in the decision to purchase an insurance policy or lottery ticket. In the case of insurance, suppose you have a $100,000 home insured for the full amount. For most people, a gift of $100,000 would be nice—in fact, *very* nice—but a $100,000 loss would be totally devastating. This is the underlying concept behind the insurance philosophy. A gain of $100,000 does not have the same effect on the positive side as does a $100,000 loss on the negative one.

When you fail to purchase insurance, you are betting that your house will not go up in smoke. Your *risk* here is that the house may burn. When we look at expected payoffs only, we ignore risk. On the other hand, we also discussed a method of examining the risk of each action by finding the variance of the respective payoffs. What we need is a method that combines the decision-maker's attitude toward the payoff with the corresponding risk of each alternative. An action with a possible higher payoff (or loss, in the case of insurance) often contains more risk. We measure the attractiveness of each outcome using utility values, which we will now develop.

The **utility value** of a particular outcome is used to measure both the attractiveness and the risk associated with this dollar amount.

Constructing a Utility Value

Suppose you have $10,000 saved up for college expenses 1 year before you begin your freshman year. A friend of yours has offered you part interest in an oil-drilling venture for your $10,000. If the venture fails, you lose your entire investment, but if it succeeds, you stand to gain $40,000. According to the latest geological survey, the probability of hitting oil is .3. Also, if oil exists, the expected life of the venture is 1 year, with the payoff of $40,000.

Your other option is to invest the money for one year in a money-market account at an expected interest rate of 12%. If you choose to maximize your expected payoff (dollars on hand at the end of the year), which action should you select?

The decision problem involves two actions,

A_1: put the $10,000 into the money market   (interest $= 12\%$)

A_2: invest $10,000 in the oil venture

with two states of nature

S_1: oil does not exist on the site

S_2: oil does exist on the site

Your corresponding payoffs, should you select the oil investment, are

$0   if S_1 occurs (and you lose your investment)

$40,000   if S_2 occurs

The payoff table is shown next with the corresponding state of nature probabilities in parentheses.

The expected payoffs here are

	$S_1(.7)$	$S_2(.3)$
A_1	11,200	11,200
A_2	0	40,000

for action A_1: $11,200(.7) + 11,200(.3) = 11,200$

for action A_2: $(0)(.7) + (40,000)(.3) = 12,000$

The oil venture (A_2) has a larger expected payoff, so, using this decision strategy, you would elect to gamble your money in hopes of a large payoff. But is this a realistic strategy? Assume that the loss of your $10,000 would result in your not going to college. All things considered, this would be disastrous. Although the large payoff would be terrific, the high probability of a heavy loss might make you wonder if the gamble is a good idea.

The problem in this illustration is that a large payoff often is very attractive, but it is offset by a risk associated with it. We say that the **utility** associated with a gain of, say, $100,000 without risk is *higher* than the utility of this amount with a high risk. For each decision problem, we ask the decision-maker to determine the utility value associated with the various payoffs in the problem.

In this way, the person can build in his or her attitudes regarding avoiding risk or preferring a gamble with a big payoff (a *risk taker*).

To illustrate the construction of a utility value, consider the payoff table we just constructed. There are many ways to proceed here, although all the various ways of assigning utility values produce the *same* decision when using these values to arrive at the best alternative. We use a two-step procedure.

Step 1. Assign a utility value of zero to the smallest payoff amount (π_{min}) and a value of 100 to the largest (π_{max}). For this example, this would be written as $U(0) = 0$ and $U(40,000) = 100$, because $\pi_{min} = 0$ and $\pi_{max} = 40,000$. *All utility values range from* 0 *to* 100. Whether you assign utility values from 0 to 1, 0 to 100, 1 to 5, or any range does not matter. It is not the actual value of the utility that is important but rather its value *relative* to the range of all values.

There is one other payoff in the table to consider, namely, $11,200. What is the utility of this dollar amount to the decision-maker—you, in this case? We consider both the attractiveness and the risk involved with this payoff in the following situation.

Consider the largest payoff of $40,000 and the smallest of $0. You need to decide what the probability, P, would have to be before you would consider

$11,200 with certainty

to be as attractive as

$40,000 with probability P and $0 with probability $1 - P$

Suppose you decide that you would need at least a 50% chance of striking oil. So, $P = .5$. We next define the utility of the $11,200 payoff by using

$$\text{utility value} = P \cdot 100$$
$$= .5 \cdot 100$$
$$= 50$$

That is,

$$U(11,200) = 50$$

A graphical illustration of these utilities is shown in Figure 17.1. An easy way to measure your attitude toward risk is to connect the lower left (utility = 0) and upper right (utility = 100) corners. If the utility values you have assigned fall *above* this line, you tend to avoid risk. If they

FIGURE 17.1

Illustration of utility
values for oil venture.

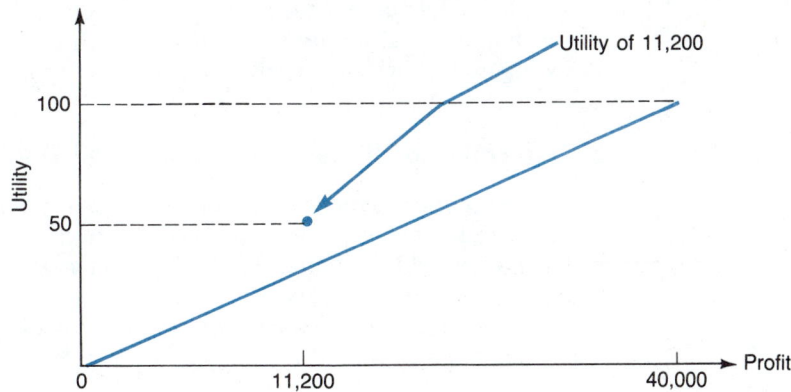

fall *below* the line, you are a risk taker. For this example, $U(11,200) = 50$ lies above the diagonal line, which indicates that, for this situation, you are a risk avoider. A summary of this procedure is step 2 of the utility value assignment.

Step 2. The utility value for any payoff (say, π_{ij}) under consideration is found by using

$$U(\pi_{ij}) = P \cdot 100$$

where P is the probability such that

$$\pi_{ij} \text{ with certainty}$$

is equally as attractive as

$$\pi_{max} \text{ with probability } P \text{ and } \pi_{min} \text{ with probability } 1 - P$$

and π_{max} and π_{min} are determined in step 1.

The resulting table of utility values for this example is as shown.

ACTION	S_1: OIL VENTURE IS UNSUCCESSFUL (.7)	S_2: OIL VENTURE IS SUCCESSFUL (.3)
A_1 money-market account	50	50
A_2 oil venture	0	100

Notice that we return to the *original* probabilities for the states of nature (.7 and .3) when constructing this table. The values of .5 and .5 were used only to measure your willingness to take a risk. This does not change the fact that S_1 will occur 70% of the time (even though you may wish it would occur only 50% of the time).

To use the utility values in choosing one of the alternative actions, we proceed exactly as before, except that we select that action having the *largest expected utility* rather than the largest expected payoff.

> expected utility for each action
> $= \Sigma \,(\text{each utility value}) \cdot (\text{its probability})$ **(17.5)**

EXAMPLE 17.14   Using the preceding table, which action, based on expected utility, is the more attractive of the two?

SOLUTION

$$A_1: \quad \text{expected utility} = (50)(.7) + (50)(.3) = 50$$
$$A_2: \quad \text{expected utility} = (0)(.7) + (100)(.3) = 30$$

Because action A_1 (the money-market account) has a larger expected utility, we choose this action over the riskier oil venture, A_2. This was also suggested by Figure 17.1, which indicated that, for this application, you are a risk avoider.

■

Determining Utility Values for Large Decision Problems

Whenever your payoff table contains a large number of values, there is an alternative to the two-step procedure just described. The main problem here is the second step, which requires that the decision maker determine the utility of *each* payoff contained in the payoff table. This can be quite difficult, because there is a requirement. The requirement for step 2 is that if

$$\text{payoff } \pi_{ij} < \text{payoff } \pi_{st}$$

then it is necessary that

$$U(\pi_{ij}) < U(\pi_{st})$$

This means that if one payoff is larger than another, the corresponding utility values must be in the *same order*.

When forced to determine the utility of many payoffs, some of which are nearly the same, the decision-maker may rate one payoff lower than another, but the utility values may be in the opposite order. One way to avoid this situation is *not* to use step 2 on every payoff involved in the problem; rather, you use this step on *between five and ten payoff values over the range of payoffs* for this problem. Consequently, you would examine π_{min} and π_{max} from step 1 and select, say, six payoffs between these values. These *need not* be actual payoff values from the payoff table. You then use the step 2 procedure to determine the utility value (U) for each of these six payoffs.

Your next step is to plot these values in a graph and connect them to form a **utility curve**. The utility of each value within the payoff table can be obtained by approximating it from the resulting graph. Because of the requirement for step 2, you need to make sure that the utility curve always *increases as the payoff increases*. We will demonstrate this technique in Example 17.15.

EXAMPLE 17.15

Table 17.12 contains the various payoffs for the selling-price decision facing the Omega Corporation. The minimum payoff is $\pi_{min} = 80$ (million dollars), and the maximum is $\pi_{max} = 550$. So, for step 1, we have

$$U(80) = 0$$

and

$$U(550) = 100$$

Describe a procedure for determining the utility of the remaining 14 values.

SOLUTION

One method, of course, is to have the decision-maker choose 14 corresponding utility values using step 2 on each payoff. An easier procedure is to request this information for payoffs of, say, 100, 150, 200, 300, 400, and 500. Notice that these payoffs are not necessarily contained in Table 17.12, but they do cover the range from 80 to 550. For a payoff of 200, we ask for that value of P such that a payoff of 200 (million dollars) with certainty is equally as attractive as a payoff of 500 with probability P and a payoff of 80 with probability $1 - P$. Suppose the decision-maker's response is $P = .55$. Then the utility value of this payoff is

$$U(200) = .55 \cdot 100 = 55$$

Consider the set of probabilities (P) and corresponding utilities in Table 17.19. It will be much easier for the decision-maker to supply these six values than to choose the 14 values remaining in Table 17.12. A key ingredient to making any quantitative procedure useable is to keep it reasonably simple!

The utilities for this problem are plotted in Figure 17.2; the curve through these points represents the decision-maker's utility curve. Notice that the utility

TABLE 17.19
Utilities for Example 17.15

	PAYOFF					
	100	**150**	**200**	**300**	**400**	**500**
Probability (P)	.20	.40	.55	.75	.90	.97
Utility [$P(100)$]	20	40	55	75	90	97

FIGURE 17.2

Utility curve for Omega computer-price decision (Example 17.15).

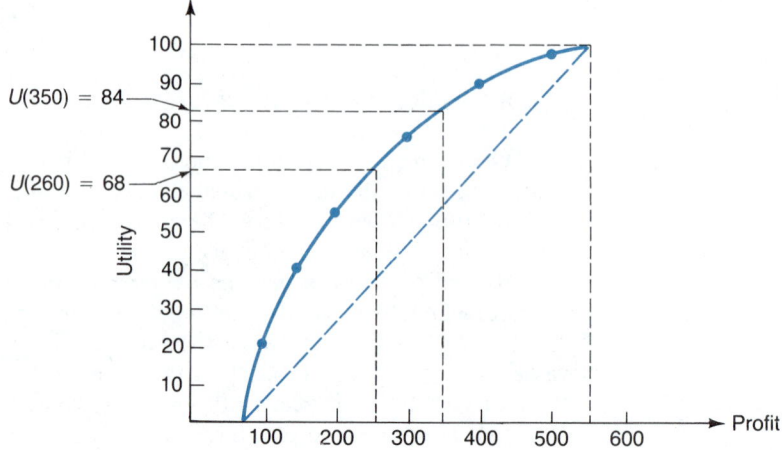

values *do* increase as the payoffs increase, so the requirement for step 2 is satisfied. As with Figure 17.1, the utility values lie *above* the line connecting the corners, indicating that this individual is a risk avoider.   ■

EXAMPLE 17.16

Using the utility curve in Figure 17.2, determine the utility for each payoff in Table 17.12. Which action (selling price) has the largest expected utility?

SOLUTION

From step 1, we have $U(80) = 0$ and $U(550) = 100$. The remaining utilities can be estimated from the utility curve constructed in Example 17.15. This is illustrated for payoffs of 260 (action A_2, state of nature S_2) and 350 (action A_1, state of nature S_3) in Figure 17.2. Consequently,

$$U(260) = 68$$
$$U(350) = 84$$

Continuing this procedure results in Table 17.20. The expected utilities are, for example,

$$\text{expected utility for } A_2 = (.1)(40) + (.5)(68) + (.3)(75) + (.1)(87) = 69.2$$

If we choose that action with the largest expected utility, our decision is to select action A_1 (selling price $1500). For this application, A_1 maximizes both expected payoff (see Example 17.8) and expected utility.   ■

Shape of Utility Curves

The shape of a decision-maker's utility curve indicates his or her preference for or aversion to risk. There are essentially three categories of people in regard to risk: (1) the risk avoider, (2) the risk neutral, and (3) the risk taker.

TABLE 17.20

Utility Table for Omega Computer-Price Decision Problem (Example 17.16)

ACTION	S_1 (.1)	S_2 (.5)	S_3 (.3)	S_4 (.1)	EXPECTED UTILITY
A_1: price = $1500	67	79	84	90	80.4
A_2: price = $1750	40	68	75	87	69.2
A_3: price = $2000	30	74	88	94	75.8
A_4: price = $2500	0	72	91	100	73.3

FIGURE 17.3

Three classes of utility curves. **A**: The risk avoider. **B**: The risk neutral. **C**: The risk taker.

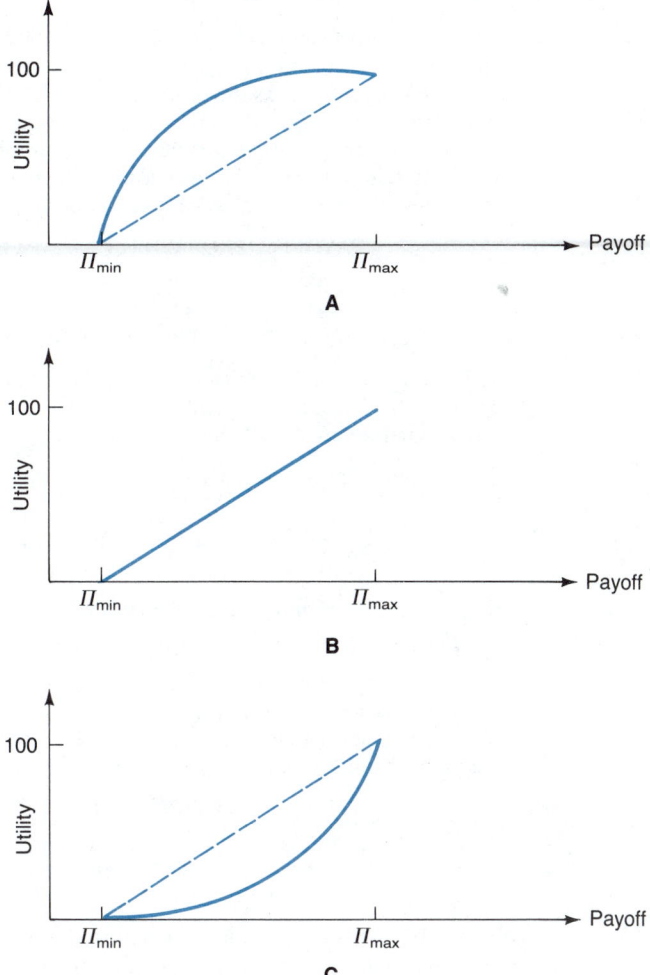

The basic shape of the utility curve for each of these classifications is contained in Figure 17.3. Notice that, in all three situations, the utility curves increase as the payoff increases. Variations of these curves also can occur; for example, a person may prefer a risk for small payoffs but then avoid a risk for large payoffs. A utility curve for such a person is S-shaped.

An individual who is **risk neutral** will have resulting utility values that lie close to the line connecting the corners. It makes no difference whether you maximize expected payoff or expected utility—the resulting best action for this person is the *same* in both cases. Very wealthy people often demonstrate this behavior because, for them, the utility of each dollar remains nearly constant.

Most people are **risk avoiders**, particularly when large payoffs or losses are involved. For two actions with equal expected payoffs, the risk avoider prefers the one with the smaller risk. This person also prefers a smaller expected payoff with a small risk over a larger expected payoff with a large risk. The **risk taker**, on the other hand, is the gambler; he or she prefers an action with a possible large payoff, even if the risk is more severe.

In summary, utility values allow the decision-maker to combine both payoff and risk into a single measure. However, the assignment of these values is subjective and special care must be taken in their determination.

EXERCISES

17.24 A utility function is assigned the value 100 for the largest payoff, which is $20,000. Also, the utility function is assigned the value zero for the smallest payoff, which is $0. The following table gives the probability, P, that, if the amount in the left hand column were certain, it would be equivalent to having a payoff of $20,000 with probability P and a payoff of $0 with probability $1 - P$. Graph the utility function and determine the attitude toward risk of a person with this utility curve. From the graph, determine the approximate utility value for a payoff of $17,000.

PAYOFF	PROBABILITY	PAYOFF	PROBABILITY
2,000	0.125	12,000	0.60
4,000	0.230	16,000	0.85
8,000	0.40	18,000	0.90

17.25 Suppose that a person has a utility function $U(x) = 2\sqrt{x}$, where x is assumed to be any value between 1 and 100. Consider the following payoff table in which all four states of nature are equally likely.

ACTION	S_1	S_2	S_3	S_4
A_1	1	100	50	10
A_2	80	30	40	25
A_3	90	20	30	10

a. Find the decision based on the maximum expected payoff.

b. Find the decision based on the maximum expected utility of the payoff.

17.26 Suppose that a person is risk neutral and has the utility function $U(x) = 3x$. In the following payoff table, $P(S_1) = .20$, $P(S_2) = .40$, $P(S_3) = 0.30$, and $P(S_4) = .10$. Show that the decision based on the maximum expected payoff is equivalent to the decision based on the maximum expected utility of the payoff.

ACTION	S_1	S_2	S_3	S_4
A_1	50	10	30	10
A_2	20	20	30	60
A_3	10	50	10	20

17.27 If a person has a utility function $U(x) = x$, what probability, P, would the person have to assign to a maximum payoff of $10,000 and a $1 - P$ probability to a minimum payoff of $0, assuming that the person could receive $8100 with certainty?

17.28 The utilities of the payoffs given in the table in Exercise 17.11 are as follows:

ACTION	S_1	S_2	S_3	S_4
A_1	2.83	2.72	1.59	2.19
A_2	2.23	2.23	2.23	2.23
A_3	2.40	2.76	2.51	2.09

Find the decision based on the maximum expected utility of the payoff.

17.29 If the utility curve of the manager of the hardware store in exercise 17.12 is $U(x) = \log_{10}(x)$, find the decision based on the maximum expected utility. Is the manager a risk neutral, a risk taker, or a risk avoider?

17.30 Suppose that a money manager is a risk avoider with the utility function $U(X) = 3 \cdot x\sqrt{x}$, where x can range from 0 to 200. In the following payoff table, $P(S_1) = .25$, $P(S_2) = .30$, $P(S_3) = .40$, $P(S_4) = .05$.

ACTION	S_1	S_2	S_3	S_4
A_1	151	33	95	40
A_2	75	75	97	180
A_3	29	162	30	50

a. Find the decision based on the maximum expected payoff.

b. Find the decision based on the maximum expected utility of the payoff.

17.4

DECISION TREES AND BAYES' RULE

This section describes a device useful for structuring and illustrating the uncertain outcomes associated with any decision problem. This is a decision tree, which graphically represents and offers you a "picture" of the entire decision problem, including a representation of:

1. The possible actions facing the decision-maker.
2. The outcomes (states of nature) that can occur.
3. The relationships between these actions and outcomes.

The decision tree makes it easier for you to compute the expected values and to understand the process of making a decision. We will demonstrate how to construct a tree diagram and discuss a procedure for using the diagram to examine the alternatives and arrive at a decision.

Constructing Decision Trees

A convenient way of representing a set of alternatives and states of nature is by means of a decision tree. A **decision tree** is a picture of the actions under consideration as well as of the states of nature that affect the profitability of each action. It is a convenient way of illustrating the entire decision problem, because you can tell at a glance exactly which alternatives are being considered and what the payoff is under each state of nature.

In Example 17.8, Omega needed to make a decision regarding a selling price for their new computer. A decision tree for this situation is shown in Figure 17.4.

A decision tree represents a sequence of *decisions*, represented by boxes, and *outcomes* left strictly to chance, represented by circles. The boxes are decision nodes, and the circles are chance nodes.

When you reach a **decision node**, you need to make a decision at this point in the decision tree. The path you select reflects your choice of the best action

FIGURE 17.4

Decision tree for the Omega computer-price problem.

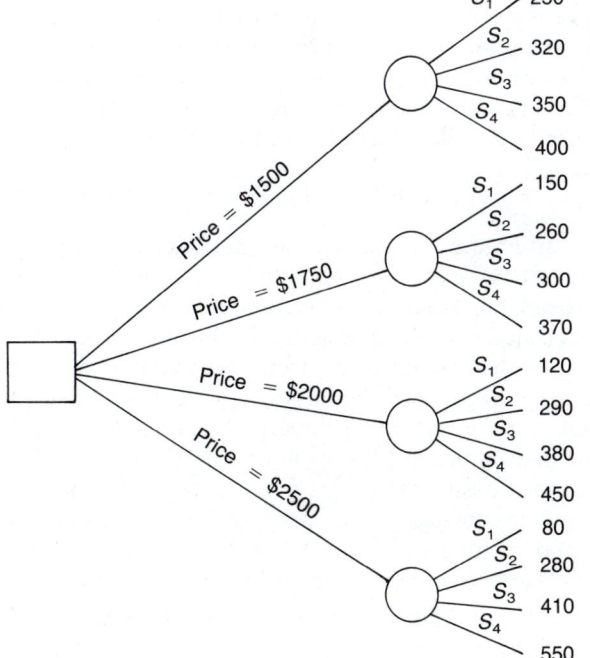

Time until competitor response
S_1: < 6 months
S_2: 6–12 months
S_3: 12–18 months
S_4: > 18 months

FIGURE 17.5

Completed decision tree
for the Omega computer-
price problem.

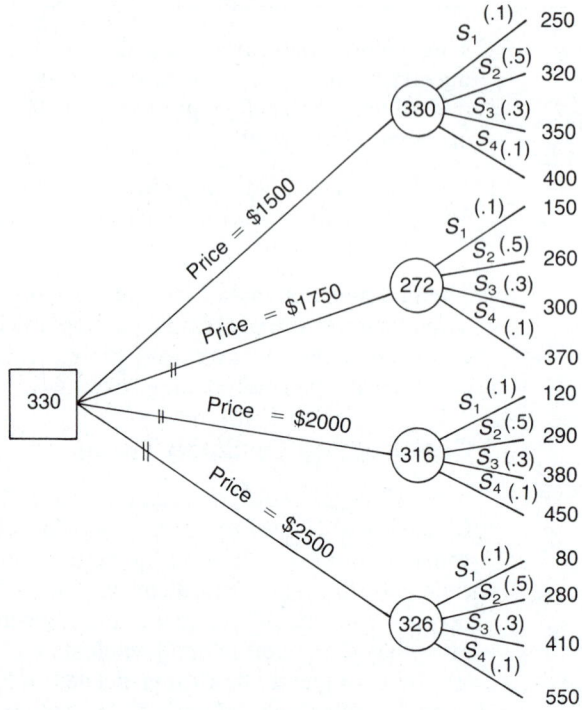

to take at this point. This decision is under your control. The paths away from
a **chance node** represent states of nature $(S_1, S_2, \ldots)$. There is no choice for you
to make here; rather, each of these paths will occur with a certain probability,
written as $P(S_1), P(S_2), \ldots$.

The final step in completing a decision tree is to determine a dollar amount
(or utility amount, if you are using utility values) within each chance node and
decision node. The amount placed inside a chance node is the *expected payoff*
at this point, using the probability for each state of nature. Consider the top
chance node in Figure 17.4. Using the state-of-nature probabilities from example
17.8, the completed tree (Figure 17.5) contains the expected payoff for each
selling price. To illustrate, the expected payoff for a selling price of $1500 is

$$(.1)(250) + (.5)(320) + (.3)(350) + (.1)(400) = 330$$

In Figure 17.5, the amount in each *decision* node is not an expected value,
because there are no probabilities with the paths leading away from this point.
Instead, the dollar (or utility) amount, or the expected dollar (or utility) amount,
associated with the best action at this point is contained within the box. Of the
four paths leading away from the decision node in Figure 17.5, action A_1
(price = $1500) has the largest expected payoff, so this amount goes into the
box. On the remaining three paths at this node, a double vertical bar across
the path indicates that we have struck out these alternatives because they are
not the ones to use at this point in the decision path.

Our conclusion from reading this tree would be to select a selling price of
$1500 for an expected payoff of 330 (million dollars).

EXAMPLE 17.17   Structure the decision problem with the Sailtown data from Table 17.2 as a
decision tree.

FIGURE 17.6   Decision tree for Sailtown decision problem (Example 17.17).

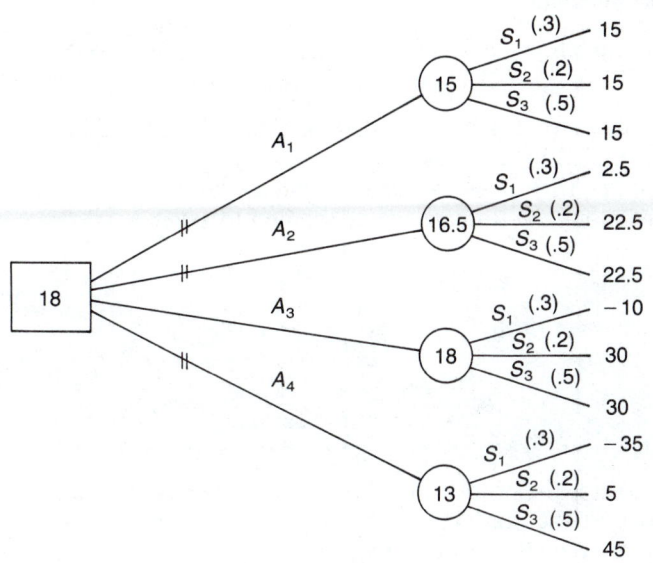

Actions
A_1: order 50 sailboats
A_2: order 75 sailboats
A_3: order 100 sailboats
A_4: order 150 sailboats

States of nature
S_1: interest rate increases
S_2: interest rate holds steady
S_3: interest rate decreases

SOLUTION

Once again, the decision path begins with a decision node (how many sailboats to purchase) followed by a sequence of chance nodes reflecting the change in the interest rate. Figure 17.6 contains the completed tree for this problem. As in the previous analysis, when you are maximizing expected payoffs, your best alternative is to order 100 sailboats (action A_3) with an expected payoff of 18 (thousand dollars). ∎

Once again you should perform a follow-up *sensitivity analysis* to determine how sensitive this solution is to the state-of-nature probabilities. When you summarize the results of a sensitivity analysis, you can construct a decision tree for each set of probabilities under consideration, which will indicate the optimum path under this condition.

An Application of Decision Trees: Bayes' Rule

Thomas Bayes was an English clergyman who lived in the 1700s. He is credited with developing a procedure that allows you to revise the probabilities for each state of nature in light of sample results or outside information. For example, in Table 17.2, the probability that the interest rate will increase (S_1) was $P(S_1) = .3$. What if a reliable consulting firm using various economic indicators predicted that the interest rate would increase? The value of $P(S_1) = .3$ was a subjective estimate, measuring Mr. Larson's belief that the rate would increase. In light of the *new information* obtained from the consulting firm, we would expect the probability of S_1 to increase—if, in fact, the firm is reliable.

The initial probability of .3 was obtained prior to receiving the new information and is thus a **prior probability**.

prior probability: $P(S_1) = .3$

This was obtained by examining the existing conditions and did not take into account the new information provided by the consulting firm. The revised probability uses the consulting firm's information; therefore, it is a conditional probability. Because it is obtained after receiving the new information, it is called a **posterior probability**.

> posterior probability: $P(S_1|\text{firm predicts a rate increase})$ = (to be determined using Bayes' Rule)

This section discusses a method of determining these probabilities using the procedure developed by Thomas Bayes.

Bayes' rule states that given the final event, B, the probability that this event was reached along the ith path corresponding to event E_i is

$$P(E_i|B) = \frac{P(E_i \text{ and } B)}{P(B)}$$
$$= \frac{i\text{th path}}{\text{sum of the paths}} \qquad (17.6)$$

where (1) the ith path is the product of all probabilities along this path and (2) the sum of the paths is the sum of all such products for the tree diagram. This is the probability of event B, that is, $P(B)$.

We illustrate Bayes' procedure for revising probabilities in the following example. The Pine Bluff Credit Union obtains a credit rating for each person that applies for an automobile loan. This rating is either type A (the best), type B, or type C (the worst). The latest statistics released by the local credit bureau indicate that 50% of the residents are rated type A, 30% are type B, and 20% are type C.

The credit union also has determined from past experience that 5% of type A people will default on their car loan. The corresponding percentages for type B and type C are 15% and 35%, respectively. We wish to determine (1) the probability that an individual applying for a car loan at the credit union will default and (2) the probability that a person has a type A credit rating if it is known that this person has just defaulted on a car loan.

Although this is not a decision problem, we can construct a tree diagram for this situation. It consists of all chance nodes, as shown in Figure 17.7. Notice that we intentionally have left off a branch on each of the final chance nodes. The complete picture for the top node is shown in Figure 17.8. When you use a tree diagram to revise probabilities, we recommend that you omit these branches and end the tree with the *same event listed down the right side*. This event should be the new information provided in the problem. For this example, this information would be that a certain person defaulted on a car loan.

FIGURE 17.7

Tree diagram for Pine Bluff Credit Union.

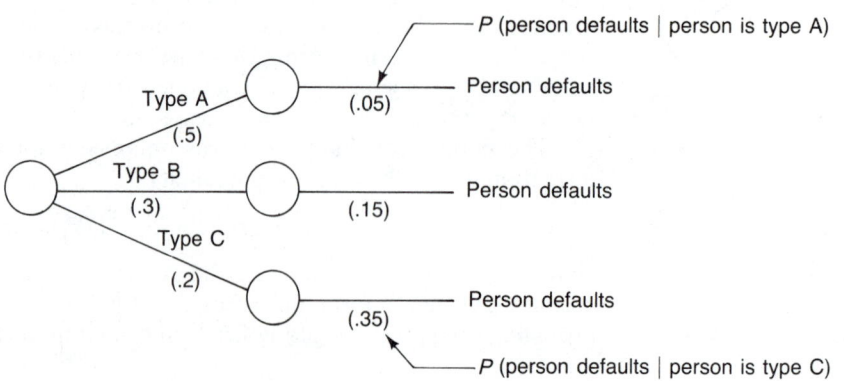

FIGURE 17.8
Top node for Figure 17.7.

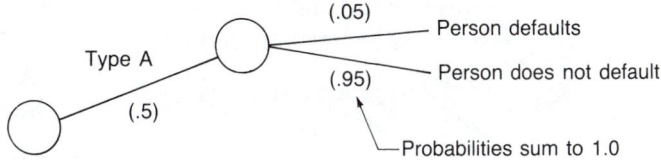

The probabilities on the first three branches represent the *prior* probabilities of having a type A, type B, or type C credit rating. The final three probabilities are conditional probabilities, representing the chances of each of these credit types defaulting. According to the information provided,

$$.05 = P(\text{person defaults} \mid \text{person is type A})$$
$$.15 = P(\text{person defaults} \mid \text{person is type B})$$
$$.35 = P(\text{person defaults} \mid \text{person is type C})$$

The advantage of constructing the tree as in Figure 17.7 is that the probability of the final event (person defaults) is simply the **sum of the paths**. So, when we use an incomplete tree,

the probability of the final event = sum of the paths (sum of the product of all probabilities along each path)

To determine the probability that a person applying for a car loan will default, refer to Figure 17.7. Here,

$$\text{sum of the paths} = (.5)(.05) + (.3)(.15) + (.2)(.35)$$
$$= .14$$

Consequently,

$$P(\text{person defaults}) = .14$$

which means that 14% of all people applying for a car loan will default.

Now you are given the information that the person has defaulted on a car loan and are asked to find the probability that he or she has a type A credit rating, in light of this new information. That is, you are given that the final event occurred and are asked to find the probability that it occurred *along the first path*, belonging to type A. This can be written as

$$P(\text{type A} \mid \text{person defaults})$$

Recalling from Chapter 4 that for any events A and B,

$$P(A \mid B) = \frac{P(A \text{ and } B)}{P(B)} \tag{17.7}$$

it follows that

$$P(\text{type A} \mid \text{person defaults}) = \frac{P(\text{type A and person defaults})}{P(\text{person defaults})}$$

Also from Chapter 4,

$$P(A \text{ and } B) = P(B)P(A \mid B)$$
$$= P(A)P(B \mid A)$$

for any events A and B. The first equation is simply a rearrangement of equation 17.7. Using the second equation,

$$P(\text{type A}|\text{person defaults}) = \frac{P(\text{type A and person defaults})}{P(\text{person defaults})}$$

$$= \frac{P(\text{type A})P(\text{person defaults}|\text{type A})}{P(\text{person defaults})}$$

$$= \frac{(.5)(.05)}{.14} = .18$$

The values .5 and .05 are from the tree diagram in Figure 17.7, and .14 was derived earlier in this discussion. We see that 18% of the people who default have a type A credit rating. The probabilities .5 and .05 lie along the first path, so this can be written as

$$P(\text{type A}|\text{person defaults}) = \frac{\text{first path}}{\text{sum of the paths}}$$

This example thus provided an illustration of Bayes' rule. You apply this rule whenever you are given that the final event in the tree diagram has occurred and you are asked to find the probability that you traveled along a particular path.

Using Bayes' rule (equation 17.6) in our example, the probability that a person has a type C credit rating, given the information that the person defaulted, is

$$P(\text{type C}|\text{person defaults}) = \frac{\text{3rd path}}{\text{sum of the paths}}$$

$$= \frac{(.2)(.35)}{.14} = .50$$

Consequently, once we know that a person has defaulted, the *revised probability* of this person having a type C rating goes from .2, which is the *prior* probability (before this information), to .5, which is the *posterior* or *revised* probability (in light of this information).

Using Bayes' Rule to Maximize Profits

An excellent opportunity to use Bayes' rule arises when you want to update your prior probabilities based on recent information regarding the states of nature in your decision problem. This information can come from such sources as an outside consulting firm or a questionnaire developed by your company's marketing staff. Based on the new information, you can maximize your expected payoff (or utility) by replacing the prior probabilities with their corresponding posterior probabilities.

EXAMPLE 17.18   Now take another look at the Sailtown example. Mr. Larson, the owner, has decided to purchase the services of an outside consultant in an effort to determine more accurately the movement of the interest rate over the summer months. The information supplied by the consultant will be one of the following:

I_1: consultant predicts an increase in the interest rate

I_2: consultant predicts no change in this rate

I_3: consultant predicts a drop in this rate

The information in Table 17.21 also was provided; it describes the past performance of this consultant when predicting interest rates. The values in the

TABLE 17.21
Conditional Probabilities
for Consultant [= $P(I|S)$]
(Example 17.18)

		ACTUALLY OCCURRED		
Consultant Predicted		**An Increase (S_1)**	**No Change (S_2)**	**A Decrease (S_3)**
I_1	an increase	.7	.4	.2
I_2	no change	.2	.5	.2
I_3	a decrease	.1	.1	.6
		1.0	1.0	1.0

FIGURE 17.9 Decision tree, given information I_1.

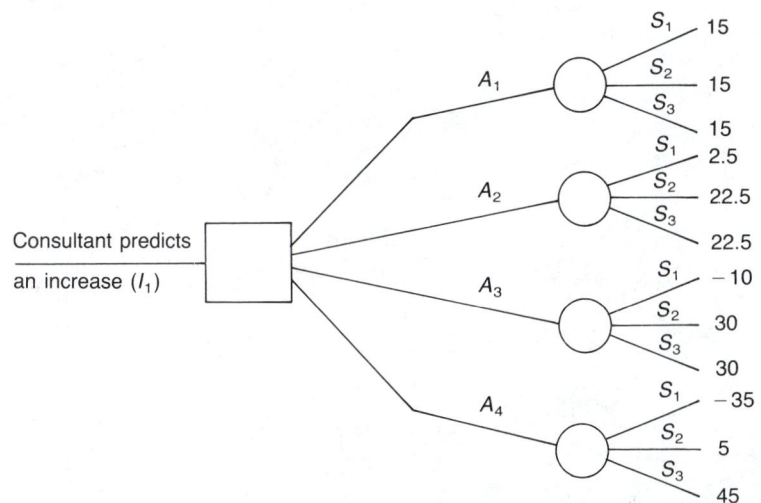

table contain conditional probabilities for the consultant's prediction under each state of nature. For example, $.7 = P(I_1|S_1)$, $.4 = P(I_1|S_2)$, and so forth. This means that 70% of the time she predicted an increase when, in fact, it actually did increase, and 40% of the time she predicted an increase when there was no change in the interest rate. If the consultant is extremely reliable, the numbers from the upper left to the lower right (.7, .5, and .6) should be near 1. The remaining values should be small.

The consultant predicts an increase (I_1) in the interest rate. What is the best action for Mr. Larson to take in light of this information? What is his expected profit?

SOLUTION
Figure 17.9 shows the new decision tree. Notice that Figures 17.6 and 17.9 are very similar, including the payoff amounts. The big difference is that the prior probabilities of $P(S_1) = .3$, $P(S_2) = .2$, and $P(S_3) = .5$ have been revised in light of the new information: $P(S_1)$ is replaced by $P(S_1|I_1)$, $P(S_2)$ by $P(S_2|I_1)$, and $P(S_3)$ by $P(S_3|I_1)$. ∎

Deriving the Posterior Probabilities

To derive the posterior probabilities, we begin by constructing a tree diagram with the new information as the event on the *far right*, as shown in Figure 17.10. We can then obtain the probabilities along the various branches from the prior probabilities and the information in Table 17.21.

FIGURE 17.10

Partial tree diagram for deriving posterior probabilities; new information is the event on the far right.

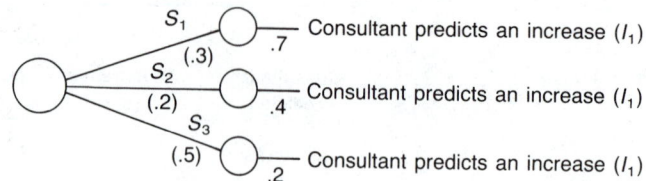

Using Bayes' rule,

$$P(I_1) = \text{sum of the paths}$$
$$= (.3)(.7) + (.2)(.4) + (.5)(.2)$$
$$= .39$$

The posterior probabilities are given by

$$P(S_1 | I_1) = \frac{\text{1st path}}{\text{sum of the paths}}$$
$$= \frac{.21}{.39} = .54$$

$$P(S_2 | I_1) = \frac{\text{2nd path}}{\text{sum of the paths}}$$
$$= \frac{.08}{.39} = .20$$

$$P(S_3 | I_1) = \frac{\text{3rd path}}{\text{sum of the paths}}$$
$$= \frac{.10}{.39} = .26$$

Placing these values in the decision tree results in Figure 17.11. The expected payoffs using the posterior probabilities are found in the usual manner. For

FIGURE 17.11

Completed decision tree using posterior probabilities.

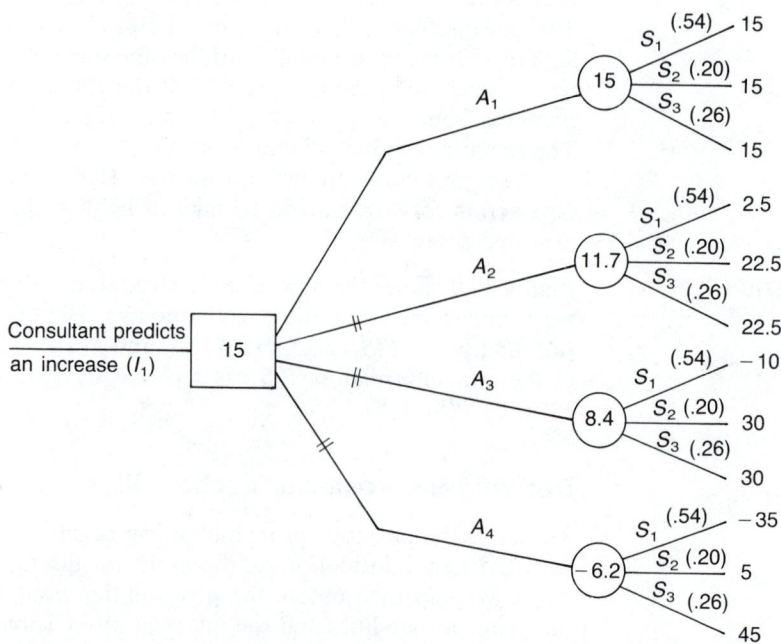

action A_2,

$$\text{expected payoff} = (.54)(2.5) + (.20)(22.5) + (.26)(22.5) = 11.7$$

Our conclusion is that, given the information that the consultant has predicted a rise in the interest rate, Mr. Larson's best alternative is to order 50 sailboats (action A_1) with an expected payoff of 15 (thousand dollars). Remember that, given no information at all, we use the *prior* probabilities to select that action having the largest expected payoff. These expected values were summarized in Table 17.11, where action A_3 (ordering 100 sailboats) provided the largest expected value. Notice also that in Table 17.11, the expected payoff for action A_2, given no information from the consultant, is 16.5. In other words, our revised expected payoff for this action, given the consultant's forecast, drops from 16.5 to 11.7.

Evaluating Sample Information

By combining a decision tree with Bayes' rule for calculating posterior probabilities, the decision-maker is able to determine whether purchasing new information is a good idea. We will refer to this new information as **sample information**. This may be collected from one of many sources, including a sample of questionnaires, a recently released government report, or, as in the previous example, an outside consultant. Typically, such information costs money; by using a decision-tree analysis, you will be able to decide between:

1. Not purchasing any additional information and using the prior probabilities to determine that action with the maximum expected payoff (or utility).
2. Purchasing this information because the expected payoff (or utility) for this decision is larger than that obtained using prior probabilities only.

In Example 17.18, the owner of Sailtown used information provided by a consultant to revise his prior probabilities regarding a possible change in the interest rate. Based on the information provided (the interest rate will increase), Mr. Larson derived the posterior probabilities and decided to purchase 50 sailboats (action A_1).

Was it a good idea for Mr. Larson to purchase the consultant's services in the first place? The cost of this information was $2500. We previously found the expected value of perfect information (EVPI) for this situation to be $15,000. The consultant's fee is considerably less than this amount, so Mr. Larson was willing to evaluate the alternative of purchasing this information.

To construct a decision tree for the full problem, we begin exactly as we did in Figure 17.6. Our next step is to include an additional branch for purchasing information from the consultant. To complete this branch, we can use a two-step procedure (refer to Figure 17.12).

Step 1. The next node will be a *chance* node, representing the possible information to be provided. In Figure 17.12, this is

I_1: consultant predicts an increase in the interest rate

I_2: consultant predicts no change in the interest rate

I_3: consultant predicts a decrease in the interest rate

Step 2. For each branch representing $I_1, I_2, \ldots$ in step 1, we reconstruct the decision tree in Figure 17.6 because the possible actions and states of nature from this point on are the *same as before*. However, the probabilities for $S_1, S_2, \ldots$ will be the *posterior* probabilities rather than the prior probabilities in Figure 17.6.

FIGURE 17.12   Completed decision tree for Sailtown decision problem.

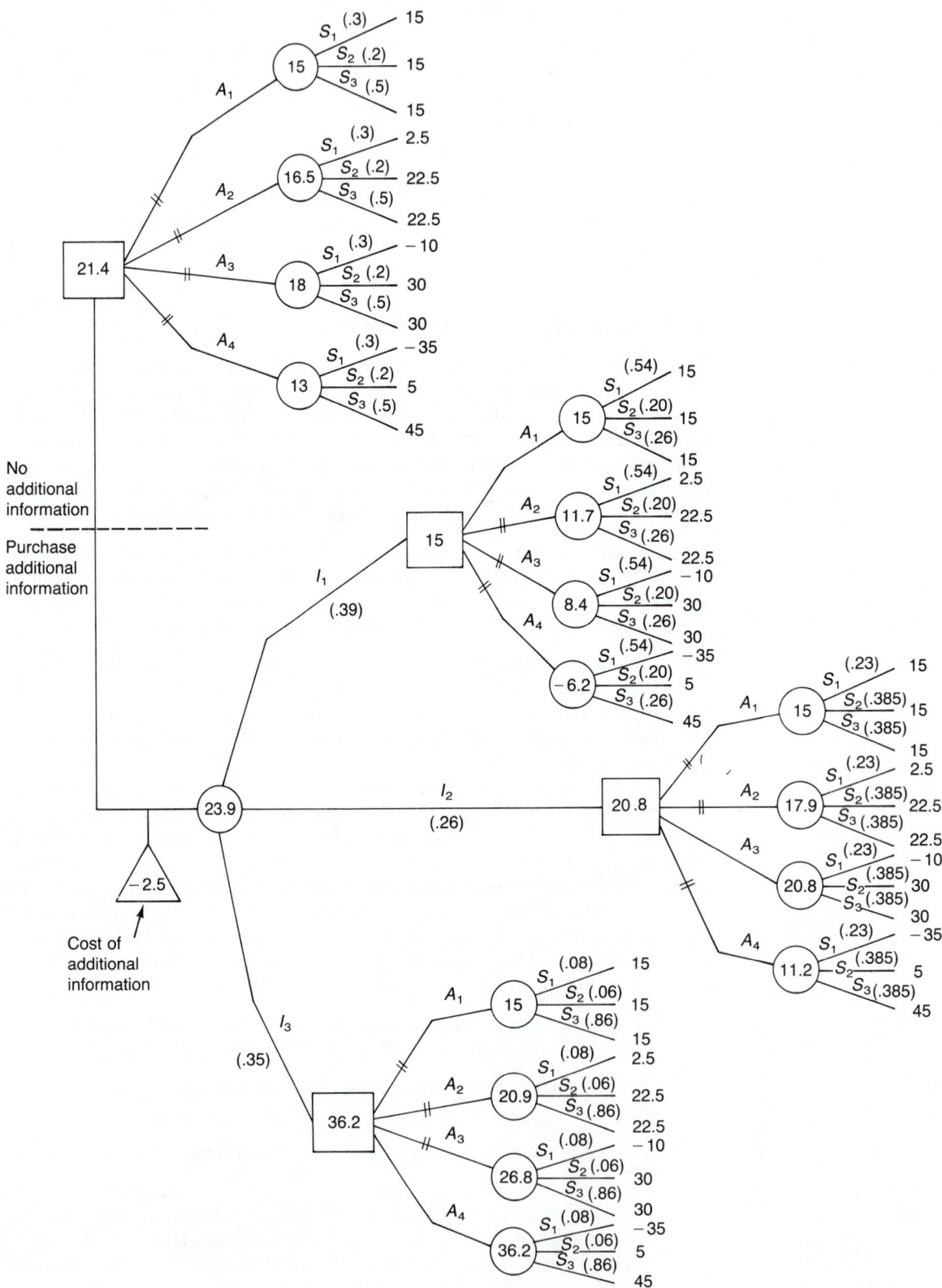

FIGURE 17.13

Deriving the posterior probabilities for the Sailtown decision problem.

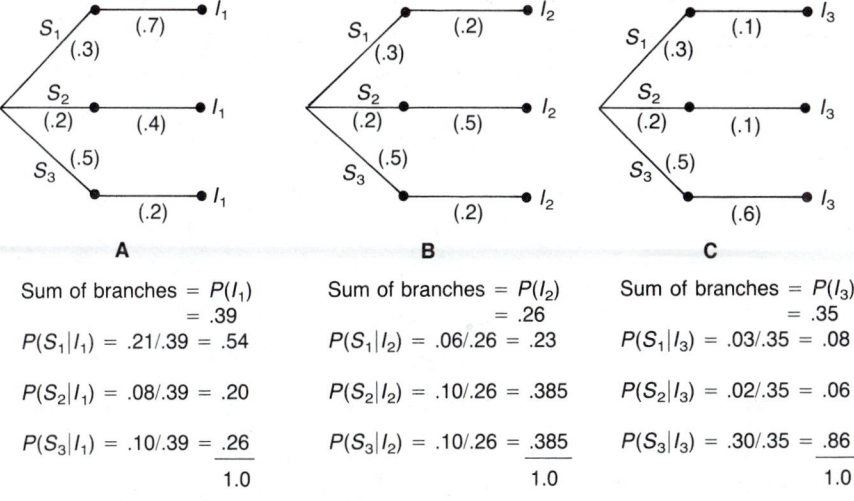

Sum of branches = $P(I_1)$
= .39
$P(S_1|I_1) = .21/.39 = .54$

$P(S_2|I_1) = .08/.39 = .20$

$P(S_3|I_1) = .10/.39 = \underline{.26}$
1.0

Sum of branches = $P(I_2)$
= .26
$P(S_1|I_2) = .06/.26 = .23$

$P(S_2|I_2) = .10/.26 = .385$

$P(S_3|I_2) = .10/.26 = \underline{.385}$
1.0

Sum of branches = $P(I_3)$
= .35
$P(S_1|I_3) = .03/.35 = .08$

$P(S_2|I_3) = .02/.35 = .06$

$P(S_3|I_3) = .30/.35 = \underline{.86}$
1.0

Having constructed the decision tree, you next need to calculate the posterior probabilities. This was illustrated in Example 17.18, where I_1 occurred (the consultant predicted an increase in the interest rate). Notice that the tree for this situation in Figure 17.11 becomes a portion of the large tree in Figure 17.12. A summary of the posterior probabilities is contained in Figure 17.13.

Also contained in Figure 17.13 are the probabilities for each of the possible predictions by the consultant. Here, $P(I_1) = .39$, $P(I_2) = .26$, and $P(I_3) = .35$. Because this prediction is *not* under your control, step 1 constructs a chance node at this point, including these three probabilities.

We find the expected payoff given each consultant's prediction (15, 20.8, and 36.2) by using the posterior probabilities, as in Example 17.18 and Figure 17.13. We then calculate the expected payoff when using the consultant, where

$$\text{expected payoff with consultant} = (.39)(15) + (.26)(20.8) + (.35)(36.2) = 23.9$$

From this amount you need to subtract the cost of this information (2.5 thousand dollars), providing a net expected payoff of 21.4 (thousand dollars). Because this exceeds the four expected payoffs where no additional information is purchased, this action maximizes the expected payoff and provides the best alternative.

We thus conclude that the owner of Sailtown was right to purchase the services of the external consultant. The best action for him to take for each prediction is summarized in Table 17.22. The net profit is obtained for each case by subtracting the cost of information, 2.5 (thousand dollars).

TABLE 17.22

Best Actions for Mr. Larson, Given the Consultant's Advice

CONSULTANT PREDICTS	BEST ACTION
A rise in the interest rate (I_1)	Order 60 sailboats (A_1) Expected payoff: 15 Net profit: 12.5
No change in the interest rate (I_2)	Order 100 sailboats (A_3) Expected payoff: 20.8 Net profit: 18.3
A drop in the interest rate (I_3)	Order 150 sailboats (A_4) Expected payoff: 36.2 Net profit: 33.7

EXERCISES

17.31 Complete the following tree diagram and determine the decision based on the maximum expected payoff. Let $P(S_1) = .4$, $P(S_2) = .2$, $P(S_3) = .1$, and $P(S_4) = .3$.

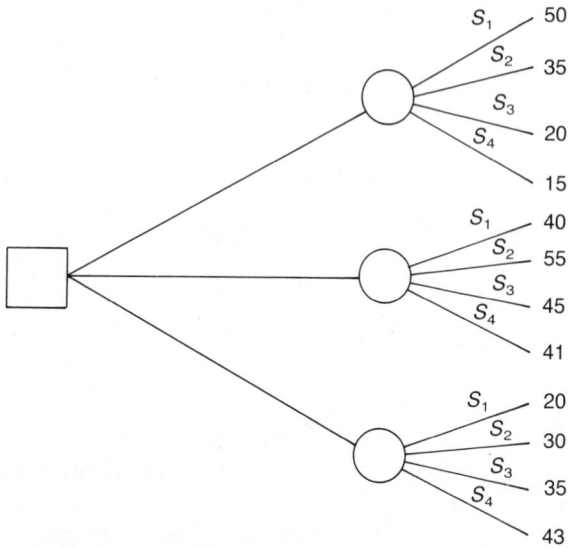

17.32 Security Designs sells alarm systems to businesses for protection against burglary and fire. If the number of small businesses continues to increase at a medium or fast rate, then Security Designs believes it should advertise more to increase its business. The manager at Security Designs believes that there is a 15% chance that the number of small businesses will decrease (S_1). He believes that there is a 25% chance that the number will stay the same (S_2), and a 25% chance and 35% chance that the number of small businesses will increase moderately (S_3) and rapidly (S_4), respectively. Construct a decision tree using the following payoff table to find the decision based on the maximum expected payoff. The payoff is the amount of monthly profit. A_1: no advertising; A_2: keep advertising at current period; A_3: increase advertising 15%; A_4: increase advertising 30%.

ADVERTISING	S_1	S_2	S_3	S_4
A_1	1451	1840	2050	2300
A_2	−1091	1685	2430	2900
A_3	−2015	1100	3060	3561
A_4	−3460	−1350	3340	4300

17.33 Consider the following payoff table, which lists the utilities of the payoff to an investor who needs to decide on one of three different investment strategies: A_1, A_2, and A_3. The investor believes that there are four different states of nature that can affect the return on the investment. States S_1, S_2, S_3, and S_4 are believed to occur with probabilities .1, .3, .4, and .2, respectively.

ACTION	S_1	S_2	S_3	S_4
A_1	30	25	32	45
A_2	35	15	40	42
A_3	44	27	20	18

Construct a decision tree to find the decision based on the maximum expected utility of the payoff.

17.34 Sullivan and Orr, in an article in *Industrial Engineering* using simulation analysis to determine optional decisions, describe a company that was contemplating making an investment of $150,000 in a new product line. Because of the uncertainty of the market, the following probabilities were assigned to the events that the product will be discontinued in either 8, 10, 12, or 14 years: .1, .2, .5, and .2, respectively. Suppose that a manager of the company could invest in one of 3 products: A_1, A_2, and A_3. Consider the following payoff table, which lists the utilities of the payoff to a manager who must decide on one of three actions. States S_1, S_2, S_3, and S_4 refer to a market life of 8, 10, 12, or 14 years respectively.

ACTION	S_1	S_2	S_3	S_4
A_1	480	750	2950	4900
A_2	−360	1240	3800	6210
A_3	−1180	1500	5300	7360

(*Source:* W. Sullivan and R. G. Orr, "Monte Carlo Simulation Analyzes Alternatives in Uncertain Economy," *Industrial Engineering* (November 1982): 43–49.)

Construct a decision tree to find the decision based on the maximum expected utility of the payoff.

17.35 Refer to Exercise 17.25. Construct the decision tree to find the decision based on the maximum expected utility of the payoff.

17.36 Refer to Exercise 17.26. Construct the decision tree to find the decision based on the maximum expected utility of the payoff.

17.37 Let A_1, A_2, A_3, and A_4 be a set of outcomes with $P(A_1) = .2$, $P(A_2) = .3$, $P(A_3) = .3$, and $P(A_4) = .2$. If it is given that $P(B|A_1) = .4$, $P(B|A_2) = .1$, $P(B|A_3) = .2$, and $P(B|A_4) = .1$, what are the probabilities $P(A_1|B)$, $P(A_2|B)$, $P(A_3|B)$, and $P(A_4|B)$?

17.38 A survey shows that 30% of the fashions that were found to be unprofitable were marketed by the major fashion clothes stores; 60% of the fashions found to be profitable were marketed by the major fashion clothes stores. If 70% of all fashions are profitable to market, find the probability that a fashion will be profitable if the major fashion clothes stores market it. What is the probability that the major fashion clothes stores market a particular fashion?

17.39 David, Harold, and Daniel are three salespeople at Southeast Insurance. David sells 40% of all insurance policies, Harold sells 33%, and Daniel sells 27%. The percent of policies sold by David that are whole-life insurance policies is 5%. That sold by Harold is 8% and that sold by Daniel is 10%. If a whole-life insurance policy at Southeast Insurance is selected at random, what is the probability that the insurance policy was sold by Harold?

17.40 Refer to Exercise 17.32. Assume that the manager at Security Designs has obtained some additional information from a consultant who predicts that the number of small businesses will stay the same; that is, state of nature S_2 is predicted to occur. The consultant has the following record, letting I be the event that the consultant predicts the number of small businesses will stay the same: $P(I|S_1) = .2$, $P(I|S_2) = .7$, $P(I|S_3) = .2$, and $P(I|S_4) = .1$. Find the decision based on the maximum expected payoff using revised probabilities with the additional information.

17.41 The investor in Exercise 17.33 subscribes to the stock market newsletter *Prudent Investor*. The newsletter forecasts state of nature S_1. Let I be the event that the newsletter forecasts S_1. The stock market newsletter has the following record: $P(I|S_1) = .6$, $P(I|S_2) = .2$, $P(I|S_3) = .3$, and $P(I|S_4) = .1$. Using revised probabilities based on this additional information, find the decision based on the maximum expected utility of the payoff.

17.42 Four legal secretaries type legal documents at a certain law firm. Secretary A types 15% of the work load, secretary B types 25%, secretary C types 20%, and secretary D types 40% of the work load. The secretaries produce the following proportions of the typographical errors:

SECRETARY	PERFORMANCE
A	.04
B	.06
C	.08
D	.03

If a typographical error is found on a legal document, which secretary would have the highest probability of having typed the error?

17.43 A decision-maker must determine whether to conduct an experiment that will give one of three predictions, I_1, I_2, or I_3. If the experiment indicates I_1 and the decision-maker uses this information, the expected profit is $15,000. If I_2 is indicated, the expected profit is $5000. But if I_3 is indicated, the expected profit is only $1000. If the experiment is not performed, the maximum expected profit would be $4000. Assuming that each of the predictions I_1, I_2, and I_3 is equally likely, is it worthwhile to conduct an experiment that costs $2000?

17.44 Nutritious Cereals would like to market a new multigrain cereal. The manager is trying to decide whether to produce the cereal in large quantities (A_1), moderate quantities (A_2), or small quantities (A_3). The manager believes that the probability of strong demand (S_1) is .4, of moderate demand (S_2) is .4, and of weak demand (S_3) is .2. A survey that can be conducted would predict strong demand (I_1), moderate demand (I_2), or weak demand (I_3). Historical data show the following conditional probabilities with regard to the predictions of the survey $[P(I|S)]$:

PREDICTION	S_1	S_2	S_3
I_1	0.8	0.3	0.3
I_2	0.1	0.5	0.1
I_3	0.1	0.2	0.6

The profit resulting from the different actions of Nutritious Cereal with regard to marketing the product is given in the following payoff table in thousands of dollars.

ACTION	S_1	S_2	S_3
A_1	88	53	20
A_2	75	66	32
A_3	57	50	39

If the survey costs $20,000 to conduct, should the management of Nutritious Cereals undertake it?

17.45 The following table lists conditional probabilities for certain predictions that are made by the consultant in Exercise 17.32, given the four states of nature of S_1, S_2, S_3, and S_4. Assume that I_1 represents the event that the consultant predicts the number of small businesses will decrease, I_2 that they will stay the same, I_3 that they will increase moderately, and I_4 that they will increase rapidly.

PREDICTION	S_1	S_2	S_3	S_4
I_1	0.80	0.10	0.20	0.10
I_2	0.10	0.70	0.20	0.20
I_3	0.05	0.10	0.50	0.30
I_4	0.05	0.10	0.10	0.40

Would the consultant's fee of $1200 be so high that Security Designs would not consider using the consultant's service?

17.46 Refer to Exercise 17.18. Let I_1, I_2, and I_3 be the events that an economist forecasts states of nature S_1, S_2, and S_3, respectively. The following table lists the conditional probabilities that the consultant makes one of these predictions given a particular state of nature. Would it be worth paying $2000 for the economist's services?

PREDICTION	S_1	S_2	S_3
I_1	.8	.4	.2
I_2	.1	.4	.2
I_3	.1	.2	.6

SUMMARY

This chapter presented a different approach to using probabilities—arriving at a **decision** when the future is uncertain. For example, should you lease a building or incur the extra expense of building one? Should your recently acquired inheritance be put in a money-market account or should you take advantage of a reliable (in the past, at least) stock market report and invest in a newly formed corporation?

When facing such a problem, the decision-maker must define the possible **actions** or **alternatives** (such as lease versus purchase or invest in the money market versus invest in stocks) and **states of nature** that describe the uncertain future (such as company sales will be below expected, equal to expected, or greater than expected). For each action and state of nature, the decision-maker must determine the corresponding **payoff** amount. These values can be summarized in a **payoff table**. This is certainly the most difficult and crucial step in the decision process because each payoff value must reflect such factors as future costs to the company and responses of competitors. Any action whose payoff is less than that belonging to another action *regardless of the state of nature* is said to be **dominated** and can be removed from consideration.

Different strategies exist for any decision problem. If you elect to describe the uncertain future by assigning a probability to each state of nature, then a popular strategy is to select the action that **maximizes the expected payoff**. Typically, these probabilities are subjective, so any decision based on this method always should be followed up by a **sensitivity analysis** that repeats the decision procedure under various sets of probabilities. In other words, it is a "what-if" process that says, If the future is described by the following set of probabilities, then the best action using this strategy is . . . .

The minimax and maximax procedures do not require state-of-nature probabilities. The **minimax** strategy is very conservative. It begins by constructing an opportunity loss table that summarizes, for each state of nature, the loss the decision-maker incurs by failing to take the most profitable action, given that this state of nature occurs. The action to take using this strategy is the one that minimizes the maximum opportunity loss for each of the actions under consideration. The **maximax** strategy is suited to the gambler; it selects that action having the largest possible payoff. Because it fails to take into consideration any heavy losses, it is not appropriate for most business decisions.

When using the expected payoff strategy, you should examine not only the payoffs that you can expect in the long run from each action but also the **risk** associated with each action. Here you measure the variation in the possible payoffs corresponding to each alternative. You often will select a less risky alternative and sacrifice a small amount of expected payoff. When you use the expected payoff strategy, a useful piece of information is the **expected value of perfect information** (EVPI), which is how much a decision-maker should be willing to pay for a perfect prediction of tomorrow's state of nature—for a crystal ball. Because any information about the future probably will be imperfect (for example, the consultant might be wrong), such information should cost considerably less than the EVPI.

It is not necessary to set up a decision problem by defining a payoff table in financial units. An alternative is to use **utility values**, which measure both the attractiveness and the risk associated with each dollar amount. For example, a $100,000 gain might be attractive, but a $100,000 loss may well be disastrous to a struggling company. The utility value for each dollar amount can be summarized in a **utility curve**. You use the shape of this curve to identify a decision-maker as a **risk avoider**, a **risk neutral**, or a **risk taker**.

A complex decision problem can be summarized best using a **decision tree**.

The tree identifies clearly the actions under consideration, the states of nature for the problem, and the expected payoffs for various segments of the decision analysis. **Bayes' rule** puts such a tree to good use by allowing the decision maker to revise the subjective probability for each state of nature (the **prior probabilities**) in light of new information about the future. This new information could be a recent stock-market analysis or predictions made by a consulting firm. The revised probabilities are **posterior probabilities**. Bayes' rule allows you to analyze a decision problem by determining the expected payoff of (1) not purchasing this information and using the prior probabilities, or (2) purchasing this information and basing your decision on the results of this prediction.

REVIEW EXERCISES

17.47   Pay-Lo drive-in grocery must decide how many loaves of bread to order each day. The demand per day is 29, 30, 31, 32, 33, 34, 35, or 36 loaves of bread. Given that 60¢ profit is made on each loaf of bread sold and a loss of 20¢ is incurred on each loaf not sold, construct the payoff table and the opportunity loss table, if any number of loaves between 29 and 36 are ordered for a particular day. What is the minimax decision? What is the maximax decision?

17.48   A computer company is considering marketing software that will take daily financial data and compute various statistics and give a complete financial analysis as well as an up-to-date forecast of financial conditions. The introduction of the software will cost approximately $200,000 in fixed cost. A profit of $20 is expected from the sale of each financial software package. The vice president of the company believes that sales will amount to 5,000, 10,000, 15,000, 20,000 or 25,000 packages of the software. Construct the opportunity loss table. What is the minimax decision?

17.49   Refer to Example 17.7. The managers at Omega have decided that a sensitivity analysis should be made before making a decision. Rework Example 17.7 using the following sets of probabilities. How sensitive is the decision based on the maximum expected payoff?

$P(S_1)$	$P(S_2)$	$P(S_3)$	$P(S_4)$
.1	.5	.2	.2
.1	.5	.1	.3
.1	.5	.3	.1
.1	.5	.2	.2
.2	.5	.2	.1

17.50   S & W Bookstore competes with the bookstore on a university campus for selling textbooks to students. *Introductory Statistics* is one of the textbooks that sells in large quantities. The manager of S & W Bookstore believes that there is a 30% chance that there will be a heavy enrollment (S_1) in this course. The probabilities for a normal enrollment (S_2) and a low enrollment (S_3) are .55 and .15, respectively. The manager must decide to order either 300, 400, or 500 copies of the textbooks. The payoff table follows.

ACTION	S_1	S_2	S_3
A_1 (300)	830	750	710
A_2 (400)	1230	1125	620
A_3 (500)	1850	910	330

a.   Are all the actions admissible?

b.   What is the minimax decision?

c.   What is the decision based on the maximum expected payoff?

d.   What is the risk of each action?

e. What is the decision based on minimum risk?

f. What is the EVPI?

17.51 Suppose you have $5000 that you would like to invest in either a no-load mutual fund that invests completely in stocks (A_1) or a fixed money-market account that yields 12% for 1 year (A_2). Assume that there are two states of nature: the stock market goes up (S_1), or the stock market goes down (S_2). An investment advisor gives you the following payoff table. Determine what value your personal utility function would have for the payoffs.

ACTION	S_1	S_2
A_1	1200	−575
A_2	600	600

17.52 You want to find several values of your utility for money from $0 to $2000. Find the value of your utility function at $0, $500, $1000, $1500, and $2000. Then, by graphing, approximate the value of your utility function at $700 and $1200.

17.53 Complete the tree diagram in Figure 17.14 and determine the decision based on the maximum expected payoff. Assume $P(S_1) = .3$, $P(S_2) = .3$, $P(S_3) = .2$, and $P(S_4) = .2$. What is the EVPI?

17.54 Refer to Exercise 17.50. Assume that the manager at S & W Bookstore has obtained additional information from a consultant that the enrollment in the introductory statistics course will be heavy (I_1). Evidence from the consultant's previous performance indicates the following probabilities: $P(I_1|S_1) = .6$, $P(I_1|S_2) = .2$, and $P(I_1|S_3) = .2$. Determine the decision based on the maximum expected payoff, using revised probabilities.

17.55 Refer to Exercise 17.54. Suppose that the consultant's predictions I_2 and I_3, which represent a normal enrollment and a low enrollment, respectively, have the following conditional probabilities: $P(I_2|S_1) = .1$, $P(I_2|S_2) = .5$, $P(I_2|S_3) = .3$, $P(I_3|S_1) = .3$, $P(I_3|S_2) = .3$, $P(I_3|S_3) = .5$. Would the consultant's fee of $350 make it worthwhile for S & W Bookstore to hire this consultant?

17.56 At JBM, a computer company, 30% of the employees are females. Of the female employees, 20% are in top management positions. Also, 25% of the male employees are in top management. What is the probability that a randomly selected person from top management is a female?

17.57 Thirty percent of the clients of an investment broker invest in only long-term–growth mutual funds. Of this group of clients, 60% are under the age of 40 years. Of

FIGURE 17.14
Tree diagram for
Exercise 17.53.

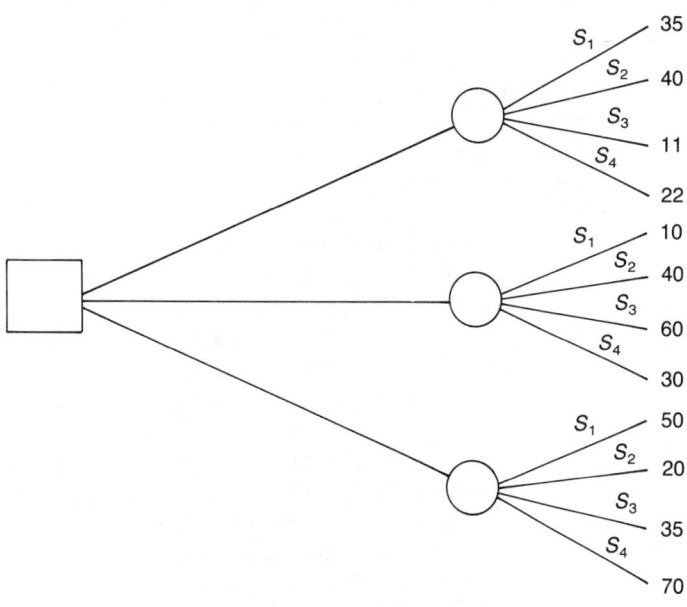

those clients who do not invest solely in only long-term mutual funds, 15% are under the age of 40 years. What is the probability that a randomly selected client under 40 years of age invests only in long-term-growth mutual funds?

17.58   A brokerage firm is conducting a seminar to explain the use of technical indicators in adjusting one's portfolio. The seminar director can decide to have either one (A_1), two (A_2), or three (A_3) guest speakers present at the seminar. Usually the more guest speakers, the more inclined the attendees will be to invest their money with the brokerage firm. The director believes that there is a 40% chance for a heavy turnout (S_1), a 35% chance for a normal turnout (S_2), and a 25% chance for a low turnout (S_3). The cost incurred of having the three guest speakers with a normal or low turnout would cause a loss. The payoff table is as follows.

ACTION	S_1	S_2	S_3
A_1	450	400	350
A_2	700	500	200
A_3	800	-100	-300

a.   Find the minimax decision.

b.   Find the decision based on the maximum expected payoff.

c.   Find the risk associated with each action.

17.59   At an investment firm, 62% of the buy recommendations on stocks are made by advisor A. The remaining 38% of the buy recommendations are made by advisor B. Of the recommendations made by advisor A, 55% are good buys. Of the recommendations made by advisor B, 70% are good buys. What is the probability that a randomly selected stock recommendation, which is a good buy, was recommended by advisor B?

CASE STUDY

WHETHER TO USE 100% INSPECTION: A DECISION THEORY ANALYSIS

A great deal of the success of the Japanese in world markets has been attributed to the implementation of quality-control procedures developed by W. E. Deming. Deming has derived formulas for the total cost of inspection, which include the cost of performing the initial inspection (say, k_1 dollars per unit) and the detrimental cost if faulty material is processed further (say, k_2 dollars per unit). The latter cost may include a cost due to damaged customer loyalty and reputation.

If we denote the proportion of nonconforming parts being produced by the process (while under control) as p, then a key element of the decision strategy is the value of $(k_2/k_1)p$. We will demonstrate this in the analysis that follows. Deming also advocates the use of control charts (a simple graph containing limits within which the process must fall) to detect the time when a process goes out of control, stop and repair the process, and quarantine and sort the parts made during the out-of-control period.

The total cost for incoming material is

$$TC = (Nk_1/q)\{1 + Qq[(k_2/k_1)p' - 1][1 - n/N]\}$$

where

N = number of parts in the lot

n = sample size drawn from each lot for inspection

k_1 = cost to inspect one unit at the beginning of the process

k_2 = cost to the firm when one nonconforming item goes farther into production

p = average fraction of rejectable parts

$q = 1 - p$

Q = fraction of lots accepted at initial inspection

p' = the average fraction nonconforming in lots that are accepted and put straight into the production line.

For sake of illustration, let us assume that p and p' are equal. In fact, p' can be smaller but only somewhat smaller than p. Consider a situation where $N = 1000$, $Q = .95$, $k_1 = \$2.00$, and $k_2 = \$100$. Con-

sequently, the total cost per lot of incoming material is

$$TC = (2000/q)\{1 + .95q[50p - 1][1 - n/1000]\}$$

(*Source:* E. P. Papadakis, The Deming Inspection Criterion for Choosing Zero or 100 Percent Inspection, *Journal of Quality Technology* 17, no. 3 (July 1985).)

Case Study Questions

1. Suppose that the value of p is believed to be .01, .03, or .05. Consequently we have three states of nature defined by these three proportions. The actions under consideration are to use sample sizes (n) of 0 (that is, inspection can be eliminated provided the process is assumed to be under control), 100, or 1000 (that is, 100% inspection). Construct a total cost table defined by these actions and states of nature. For example, the total cost (rounded to the nearest dollar) for a sample size of $n = 100$ and a proportion of $p = .03$ is

$$TC = (2000/.97)\{1 + (.95)(.97)[1.5 - 1][1 - .1]\}$$
$$= \$2917$$

2. Construct an opportunity loss table for this situation by de-

fining the opportunity loss for each action and state of nature to be the TC for this situation minus the minimum TC given this state of nature. As a result, each opportunity loss is nonnegative.

3. Are any of the three actions inadmissible?

4. Using the minimax criterion, which action (sample size) would you select?

5. If it is believed that each of the values of p are equally likely, which action minimizes the expected total cost?

6. The main point of Deming's argument is that one should use 100% inspection whenever the value of $(k_2/k_1)p$ is larger than one and use 0% inspection otherwise. Is this consistent with the total cost table constructed in Exercise 1?

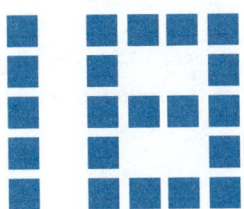

Nonparametric Statistics

A Look Back/Introduction

Although last, this chapter is far from the least in importance. **Nonparametric** statistical techniques are used extensively for a variety of business-related applications.

In the previous chapters, we introduced a large assortment of tests of hypothesis. These tests generally were concerned with things such as the mean or variance of a population. A mean, variance, or proportion is referred to as a *parameter* in statistics, and so these tests are called *parametric* tests of hypothesis. The common underlying assumption in testing a parameter from a continuous population is that this population has a *normal* distribution. Any time you use a *t*-statistic, you assume a normal population distribution. When you test more than two means using the ANOVA procedure, you also assume that the shape of the populations is normal.

What can you do if you have reason to believe that the populations under study are not normally distributed? For example, suppose that data collected previously from these populations have been extremely skewed (not symmetric). One option when dealing with means or proportions is to collect large samples. In such a situation, the central limit theorem assures us that the distribution of the sample estimators is approximately normal *regardless* of the population distribution. The other alternative, particularly for small or moderate sample sizes, is to use a nonparametric statistical procedure that deals with cases in which the assumptions of normality are not true.

Many of the nonparametric statistical tests try to answer the same sorts of questions as do those tests discussed previously. With these tests, however, the assumptions can be relaxed considerably. In earlier chapters, means and medians were referred to as *measures of central tendency*. A nonparametric test concerning such a measure does not assume an underlying normal population—unlike its parametric counterpart. Consequently, nonparametric methods are used for situations that violate the assumptions of the parametric procedures.

A common method for practically all nonparametric techniques is the use of **ranks**. Given a set of data, we obtain a set of ranks by replacing each data value by its relative *position*. To illustrate this idea, consider the following eight observations.

10.8, 6.4, 11.7, 5.3, 9.5, 2.5, 15.1, 10.4

Arranged in order, these are

Position	1	2	3	4	5	6	7	8
Value	2.5	5.3	6.4	9.5	10.4	10.8	11.7	15.1

We say that the rank of the value 2.5 is 1, the rank of the value 5.3 is 2, and so forth. Replacing each value by its rank and maintaining the original position produces

6, 3, 7, 2, 4, 1, 8, 5

Most nonparametric procedures use the eight *ranks* rather than the original data values. By using ranks, we are able to relax the assumptions regarding the underlying populations and develop tests that apply to a wider variety of situations.

You often will encounter an application in which a numeric measurement is extremely difficult to obtain, but a rank value is not. One example of this is a consumer taste test; each participant finds it much easier to rank several different brands of soft drinks than to assign a numeric value to each one. The data for analysis consist of the rank assigned to each brand.

Such data are said to be *ordinal* because only the relative position of each value has any meaning (see Chapter 1). This is a "weaker" form of data than is the *interval* form, for which not only the positions but also the *differences* between data values are meaningful. In this text, we have dealt mostly with interval data. Data consisting of temperatures, for example, are interval data; the difference between 60°F and 70°F is the same as that between 65°F and 75°F (10°F). When dealing with ranks, this is not the case; there is no reason to assume that the difference between ranks 1 and 3 is the same as that between ranks 3 and 5.

In Chapter 12, we introduced one nonparametric procedure, which used the chi-square statistic to test for goodness-of-fit or independence between two classifications. In this chapter, we examine other popular nonparametric methods used in a business setting. These tests of hypothesis by no means constitute all the nonparametric techniques used in practice, but they should provide you with a basis for knowing how and when to apply such a method to a particular set of sample data.

18.1

A TEST FOR RANDOMNESS: THE RUNS TEST

A crucial assumption behind a great many statistical procedures is based on the concept of *randomness*. In the earlier chapters, all samples were assumed to be random. The reliability of any statistical test—even if run on a high-powered computer—is suspect if the sample was not obtained in a random manner. Similarly, the t- and F-tests in linear regression contain the assumption that the resulting sample residuals are *independent*, with no observable pattern. This means that the *signs* of these errors should be random.

When you examine a sequence of observations or residuals, one method of detecting a lack of randomness is to observe the number of runs contained in the sequence. For a sequence containing two possible values (A and B, + and −, and so on) a **run** consists of a string of identical values.

Suppose we flip a coin ten times, where each flip results in a head (H) or a tail (T). Consider the following three outcomes, each containing five heads and five tails.

Sequence 1	H	H	H	H	H	T	T	T	T	T
Sequence 2	H	T	H	T	H	T	H	T	H	T
Sequence 3	H	H	T	H	T	T	H	T	T	H

Only sequence 3 exhibits a random pattern. To see why, we will examine each sequence.

Sequence 1   These ten observations contain only two runs

$$\underline{\text{H H H H H}} \qquad \underline{\text{T T T T T}}$$
$$\text{Run 1} \qquad\qquad \text{Run 2}$$

The small number of runs indicate that this sequence was not generated in a random manner.

Sequence 2   At first glance, this pattern may appear to be random, but there are an excessive number (ten) of runs

$$\underline{\text{H}}\ \underline{\text{T}}\ \underline{\text{H}}\ \underline{\text{T}}\ \underline{\text{H}}\ \underline{\text{T}}\ \underline{\text{H}}\ \underline{\text{T}}\ \underline{\text{H}}\ \underline{\text{T}}$$
$$\text{Run 1} \qquad\qquad\qquad\qquad \text{Run 10}$$

Once again, the process that generated this sequence is not random, as indicated by the large number of runs.

Sequence 3   This sequence seems to be a compromise between the first two, exhibiting neither too few runs nor too many.

$$\underline{\text{H H}}\ \underline{\text{T}}\ \underline{\text{H}}\ \underline{\text{T T}}\ \underline{\text{H}}\ \underline{\text{T T}}\ \underline{\text{H}}$$
$$\text{Run 1} \qquad\qquad\qquad\qquad \text{Run 7}$$

It appears that the sequence was generated in a random manner.

In this section, we use the runs test statistical procedure to test for randomness using the number of observed runs.

The Runs Test (Small Samples)

Consider a sequence of n observations, containing n_1 symbols of the first type (H, in our example) and n_2 symbols of the second type (T). So, $n = n_1 + n_2$. Let

$$R = \text{number of runs within these } n \text{ observations}$$

The situation we consider here is for small samples, where $n_1 \le 10$ and $n_2 \le 10$. This provides a good method of demonstrating how this particular nonparametric technique is indeed distribution free; that is, it makes no assumptions about the population of H's and T's.

Consider the case where $n_1 = 2$ and $n_2 = 3$. In this section, we assume (without any loss of generality) that $n_1 \le n_2$. So we have two H's and three T's. How many such arrangements (permutations) of these five symbols are there?* There are ten, provided in Table 18.1. This value in general can be found using

---

* The formulas for the number of permutations given in Chapter 4 do not apply here because the n objects (symbols) are not all different (distinct).

TABLE 18.1

Arrangements When
$n_1 = 2$, $n_2 = 3$

ARRANGEMENT	NUMBER OF RUNS (R)
H H T T T	2
H T H T T	4
H T T H T	4
H T T T H	3
T H H T T	3
T H T H T	5
T H T T H	4
T T H H T	3
T T H T H	4
T T T H H	2

$$\text{number of arrangements} = A = \frac{n!}{n_1! n_2!} \qquad \textbf{(18.1)}$$

where $n = n_1 + n_2$. For this illustration,

$$A = \frac{5!}{2!3!} = \frac{(5)(4)(3)(2)}{(2)(3)(2)} = 10$$

Each of the ten arrangements in Table 18.1 is equally likely to occur *providing* the process generating this sequence is *random*, so each has probability .1. Some conclusions we can draw from Table 18.1 include:

1. For two of these sequences, there are two runs, and so $P(R = 2) = .2$.
2. For three of the sequences, there are three runs, and so $P(R = 3) = .3$.
3. For four of the sequences, there are four runs, and so $P(R = 4) = .4$.
4. For one of the sequences, there are five runs, and so $P(R = 5) = .1$.

Notice that these probabilities sum to 1, as they should.

Consequently, for this situation, we can make the following statements:

$$P(R \leq 2) = P(R = 2) = .2$$
$$P(R \leq 3) = P(R = 2) + P(R = 3) = .2 + .3 = .5$$
$$P(R \leq 4) = .2 + .3 + .4 = .9$$
$$P(R \leq 5) = .2 + .3 + .4 + .1 = 1.0$$

This means that, for example, the probability of observing three or less runs if the sequence has been produced in a random manner, is .5. What we are seeing here is that these probabilities are obtained *without* assuming any probability distribution for the underlying population (process) that generated a sequence of $n_1 = 2$ values of H and $n_2 = 3$ values of T. This is the beauty of nonparametric methods.

Probabilities such as those just discussed are summarized in Table A.15. The top portion of this table is reproduced in Table 18.2. The table entries contain the probability that $R \leq a$ for the possible values of a. Notice that the first row of this table is identical to the $\leq$ probabilities (called *cumulative* probabilities) that we just derived.

The hypotheses under investigation here are

H_0: the sequence was generated in a random manner

H_a: the sequence was not generated in a random manner

TABLE 18.2
A Portion of Table A.15 for the Runs Test. Each Entry is $P(R \leq a)$ Where the Values of a Run Across the Table

$(n_1 n_2)$	2	3	4	5	6	7	8	9	10
(2, 3)	.200	.500	.900	1.000					
(2, 4)	.133	.400	.800	1.000					
(2, 5)	.095	.333	.714	1.000					
(2, 6)	.071	.286	.643	1.000					
(2, 7)	.056	.250	.583	1.000					
(2, 8)	.044	.222	.533	1.000					
(2, 9)	.036	.200	.491	1.000					
(2, 10)	.030	.182	.455	1.000					
(3,3)	.100	.300	.700	.900	1.000				
(3, 4)	.057	.200	.543	.800	.971	1.000			
(3, 5)	.036	.143	.429	.714	.929	1.000			
(3, 6)	.024	.107	.345	.643	.881	1.000			
(3, 7)	.017	.083	.283	.583	.833	1.000			
(3, 8)	.012	.067	.236	.533	.788	1.000			
(3, 9)	.009	.055	.200	.491	.745	1.000			
(3, 10)	.007	.045	.171	.455	.706	1.000			
(4, 4)	.029	.114	.371	.629	.886	.971	1.000		
(4, 5)	.016	.071	.262	.500	.786	.929	.992	1.000	
(4, 6)	.010	.048	.190	.405	.690	.881	.976	1.000	
(4, 7)	.006	.033	.142	.333	.606	.833	.954	1.000	
(4, 8)	.004	.024	.109	.279	.533	.788	.929	1.000	
(4, 9)	.003	.018	.085	.236	.471	.745	.902	1.000	
(4, 10)	.002	.014	.068	.203	.419	.706	.874	1.000	
(5,5)	.008	**.040**	.167	.357	.643	.833	**.960**	.992	1.000
(5,6)	.004	.024	.110	.262	.522	.738	.911	.976	.998

As we mentioned earlier, we reject H_0 whenever the number of runs is too *small* (say, whenever $R \leq k_1$) or too *large* (say, whenever $R \geq k_2$).

To illustrate the testing procedure here, consider our original sequences for the ten coins with $n_1 = 5$ and $n_2 = 5$. According to equation 18.1, there are $A = 10!/5!5! = 252$ possible arrangements here—sequences 1, 2, and 3 are three of these. We are looking for some "cutoff number" of runs, k_1, where we are fairly sure that fewer than k_1 are "too few" and more than k_2 are "too many." Using Table 18.2 with $n_1 = 5$ and $n_2 = 5$, we find, for example, that (assuming H_0 is true)

$$P(R \leq 3) = .04$$
$$P(R \leq 8) = .96$$

Consequently, $P(R > 8) = P(R \geq 9) = 1 - .96 = .04$. In other words, the event of observing three or less runs is very unlikely (with probability .04) if H_0 is true, so a value of $R \leq 3$ indicates that H_0 is not true and should be rejected. The same reasoning applies to $R \geq 9$.

This means that, with a significance level of $\alpha = .04 + .04 = .08$, the values of k_1 and k_2 are $k_1 = 3$ and $k_2 = 9$. The corresponding testing procedure is to

reject H_0 if $R \leq 3$ or $R \geq 9$

We will formalize the procedure for using this information.

The overall significance level of this test is $.04 + .04 = .08$. One disadvantage of small-sample nonparametric procedures is that you cannot derive a test for any specified significance level (such as $\alpha = .05$ here). Rather, you are at the mercy of the available values in this table. We can summarize this testing procedure as follows.

Hypotheses:

H_0: pattern was generated in a random manner

H_a: pattern was not generated in a random manner

Test statistic (for small samples): R, where R denotes the number of runs in the sequence.

Procedure:

reject H_0 if $R \le k_1$ or $R \ge k_2$

where (1) k_1 is the value from Table A.15 such that $P(R \le k_1) = \alpha/2$, and (2) k_2 is the value from Table A.15 such that $P(R \ge k_2) = \alpha/2$.

EXAMPLE 18.1

Using a significance level between .05 and .10 and as close to .05 as possible, determine which of the three sequences of five H's and five T's in the earlier discussion were generated in a random manner.

SOLUTION

Using Table 18.2 (or Table A.15), the three smallest available significance levels for this test are

$$.008 + (1 - .992) = .008 + .008 = .016 \ (k_1 = 2, k_2 = 10)$$
$$.040 + (1 - .960) = .040 + .040 = .08 \ (k_1 = 3, k_2 = 9)$$
$$.167 + (1 - .833) = .167 + .167 = .334 \ (k_1 = 4, k_2 = 8)$$

Because $\alpha = .08$ comes closest to satisfying our desired significance level, we will

reject H_0 if $R \le 3$ or $R \ge 9$

The results are

for sequence 1: $R = 2$, so reject H_0

for sequence 2: $R = 10$, so reject H_0

for sequence 3: $R = 7$, so fail to reject H_0

For the first two sequences we conclude that these arrangements were not the result of a random process. For the third sequence, we have no reason to suspect the presence of a nonrandom process. ∎

The Runs Test (Large Samples)

For large samples ($n_1 > 10$ and $n_2 > 10$), the approximate distribution for R if the generating process is random will be *normal* with mean

$$\mu_R = 1 + \frac{2n_1 n_2}{n_1 + n_2} \tag{18.2}$$

and standard deviation

$$\sigma_R = \sqrt{\frac{2n_1 n_2(2n_1 n_2 - n_1 - n_2)}{(n_1 + n_2)^2(n_1 + n_2 - 1)}} \tag{18.3}$$

By standardizing R in the usual way, we obtain the following summary.

Hypotheses:

H_0: pattern was generated in a random manner

H_a: pattern was not generated in a random manner

Test statistic (for large samples):

$$Z = \frac{R - \mu_R}{\sigma_R} \qquad (18.4)$$

where (1) R denotes the number of runs in the data sequence and (2) μ_R and σ_R are the mean and standard deviation of this random variable, defined in equations 18.2 and 18.3.

The testing procedure using the standard normal random variable is the same as in previous tests using Z. For the randomness test, a nonrandom pattern is indicated by a Z value in the right tail (too many runs) or in the left tail (too few runs).

EXAMPLE 18.2

The president of Northside National Bank requested the savings-account balance for 45 randomly selected accounts of nonmarried customers. When she examined the data, she began to question the randomness of the procedure used to select the accounts. Letting M denote a male account and F a female account, the following sequence was obtained, listed in the order in which they were selected for the supposedly random sample.

M M F F F F M F F M M M M M M F F F F M M F

M M F F M F F F F M M M M M F F F F M M M

Based upon this sequence, would you conclude that this sample consists of 45 randomly selected males and females? Use $\alpha = .05$.

SOLUTION

The preceding sequence contains $R = 15$ runs. Also,

$$n_1 = \text{number of males} = 22$$
$$n_2 = \text{number of females} = 23$$

For these values of n_1 and n_2, the mean number of runs if H_0 is true is

$$\mu_R = 1 + \frac{(2)(22)(23)}{45} = 23.49$$

This implies that, on the average, whenever $n_1 = 22$ and $n_2 = 23$, you will obtain 23.49 runs.

The sample contains only 15 runs, so it could be that this sequence exhibits a nonrandom pattern, due to insufficient runs. However, this depends heavily on the standard deviation of R; therefore, to complete the analysis, we next find

$$\sigma_R = \sqrt{\frac{(2)(22)(23)[(2)(22)(23) - 45]}{(45)^2(44)}} = \sqrt{10.9832} = 3.314$$

To determine whether $R = 15$ is sufficiently small to reject the random sequence hypothesis, we calculate the test statistic.

$$Z^* = \frac{15 - 23.49}{3.314} = -2.56$$

The test procedure here (using $\alpha = .05$) is to

$$\text{reject } H_0 \text{ if } |Z| > 1.96$$

The computed Z value does have an absolute value larger than 1.96, and so we reject H_0. There is evidence that the male-female sequence is nonrandom, indicating a lack of randomness in the sampling procedure used in selecting the individual accounts from the bank records. ∎

We encounter another application of the runs test when we examine the residuals from a linear regression analysis. A key assumption when using linear regression is that the residuals are *independent*. Consequently, you should observe a random pattern in the sample residuals. If the observations in your data set are recorded across time (say, 24 consecutive months), this often results in residuals that are *not* independent. In this case, we would say that the errors are correlated—more precisely, they are *autocorrelated*: they are correlated with each other. In Chapter 16, we computed the Durbin-Watson (DW) statistic to measure the degree of autocorrelation.

The DW statistic assumes that the errors follow a normal distribution, as do all the tests of hypothesis when using a linear regression equation. The nonparametric runs test also can be used to examine the residuals, by recording the *sign* ($+$ or $-$) of each residual and counting the number of runs. This test is valid regardless of the distribution of the residuals and can be used for any model that assumes the residuals are uncorrelated.

EXAMPLE 18.3

In Example 14.4, a multiple linear regression model was used to predict $Y =$ the amount of investment in a portfolio of high-risk securities using $X_1 =$ value of economic index (a measure of present and future economic conditions) and $X_2 =$ annual income of investor. Figure 14.6 on page 577 contains the MINITAB solution for this problem, which includes a calculation of the $n = 50$ residuals. These values are repeated in Table 18.3. Using $\alpha = .05$, is there any evidence that the residuals are not random?

SOLUTION

We begin by forming a sequence containing the sign of each residual.

$$\underline{-\ +}\ \ \underline{-\ -\ -\ -}\ \ \underline{+}\ \ \underline{-\ -\ -\ -\ -}\ \ \underline{+\ +\ +}\ \ \underline{-}\ \ \underline{+}$$

$$\underline{-\ -\ -}\ \ \underline{+\ +\ +\ +\ +\ +}\ \ \underline{-\ -}\ \ \underline{+\ +}\ \ \underline{-\ -\ -\ -}$$

$$\underline{+\ +\ +}\ \ \underline{-\ -\ -}\ \ \underline{+\ +\ +}\ \ \underline{-\ -}\ \ \underline{+}\ \ \underline{-}\ \ \underline{+\ +}\ \ \underline{-}\ \ \underline{+}$$

The number of runs here is $R = 22$. Also, there are $n_1 = 24$ pluses and $n_2 = 26$ minuses. So the expected number of runs for this situation, if H_0 is true, is

$$\mu_R = 1 + \frac{(2)(24)(26)}{50} = 25.96$$

Also,

$$\sigma_R = \sqrt{\frac{(2)(24)(26)(1248 - 50)}{(50)^2(49)}} = 3.49$$

Using $\alpha = .05$, the test procedure here is to

$$\text{reject } H_0 \text{ if } |Z| > 1.96$$

The computed test statistic for this example is

$$Z^* = \frac{22 - 25.96}{3.49} = -1.13$$

TABLE 18.3

Residuals for Multiple Regression in Example 18.3

−323.1	541.7	−148.0	−107.3	−341.7	−124.0	547.4	−301.1
−178.1	−37.9	−101.5	−190.2	7.7	399.6	38.8	−37.6
57.4	−75.9	−420.7	−198.8	315.5	272.5	533.8	170.7
159.1	353.5	−326.1	−254.0	214.6	385.2	−299.5	−147.2
−524.8	58.7	271.0	12.0	−7.3	−358.9	−525.7	134.3
432.4	87.9	−52.6	−258.0	6.7	−116.0	113.3	398.6
−60.9	4.3						

This value does not lie in the rejection region, so we fail to reject H_0. There is not enough evidence to indicate that the residuals are autocorrelated. Incidentally, the value of the Durbin-Watson statistic for this example is $DW = 1.77$, which is *not* significant using the Durbin-Watson procedure discussed in Chapter 16. Here, both procedures agree that the assumption of independent errors is met. ■

A computer solution using MINITAB is contained in Figure 18.1. This

FIGURE 18.1

MINITAB solution for runs test.

```
MTB > REGRESS Y IN C3 USING 2 PREDICTORS IN C1 C2;
SUBC> RESIDUALS INTO C5;                              ↑  ↑
SUBC> DW.                                            X₁ X₂

The regression equation is
C3 = - 1183 - 0.13 C1 + 0.0720 C2

Predictor        Coef        Stdev      t-ratio          p
Constant       -1183.3       219.4        -5.39      0.000
C1               -0.127       1.547        -0.08      0.935
C2             0.071995    0.003588        20.06      0.000

s = 283.0          R-sq = 89.8%      R-sq(adj) = 89.4%

Analysis of Variance

SOURCE         DF          SS          MS          F          p
Regression      2    33294136    16647068     207.81      0.000
Error          47     3765064       80108
Total          49    37059200

SOURCE         DF      SEQ SS
C1              1     1044765
C2              1    32249370
```

Durbin-Watson statistic = 1.77

```
MTB > PRINT C5

C5
 -323.059    541.696   -148.036   -107.296   -341.745   -124.031    547.443
 -301.062   -178.100    -37.869   -101.535   -190.240      7.721    399.552
   38.788    -37.590     57.444    -75.879   -420.685   -198.811    315.486
  272.485    533.789    170.724    159.139    353.522   -326.073   -254.026
  214.638    385.242   -299.459   -147.173   -524.827     58.665    270.971
   12.035     -7.276   -358.877   -525.680    134.347    432.432     87.865
  -52.600   -258.047      6.749   -115.978    113.282    398.552    -60.949
    4.336

MTB > RUNS ABOVE AND BELOW 0, USING C5      (Abbreviated: RUNS 0 C5)

    C5

    K =        0.0000
                                               R
    THE OBSERVED NO. OF RUNS =    22 ◄
    THE EXPECTED NO. OF RUNS =  25.9600  ◄── μ_R
    24 OBSERVATIONS ABOVE K    26 BELOW
              THE TEST IS SIGNIFICANT AT    0.2573  ◄── p-value
              CANNOT REJECT AT ALPHA = 0.05
      Here, K is set equal to 0 ─┘
```

solution contains the number of runs as well as the mean of the runs statistic if H_0 is true. The MINITAB procedure also could be used for the male-female sequence in Example 18.2 by using -1 for male and $+1$ for female in the computer input. Also contained in Figure 18.1 is the p-value for this test, $p = .2573$. Because this value is larger than the significance level of $\alpha = .05$, we once again fail to reject H_0 and conclude that there is insufficient evidence to indicate that the errors are autocorrelated. When you perform a one-tailed runs test, you should divide this value by two and then compare it to the level of significance (α) before making the decision. Also be sure that the sign of the Z value is compatible with your one-tailed alternative hypothesis; that is, it should be positive when testing H_a: too many runs and negative when testing H_a: too few runs.

EXERCISES

18.1   What assumptions need to be made about the data when using the runs test?

18.2   A jar contains two balls, a red one and a blue one. A person is asked to draw a ball at random. The ball is then replaced in the jar and the experiment is repeated. The results of repeating the experiment 17 times follow, where R represents the red ball and B represents the blue ball. At the .05 level of significance is there any evidence that the sequence is not randomly generated?

R B B R R B R R B B B R B R R B R

18.3   Are the negative and positive numbers randomly ordered in the following sequence? Use a .10 level of significance.

-1, 2, -5, 4, -10, 3, -1, 4, 6, 9, -7, 8, -3, 5

18.4   Ozark County Bank is taking applications for the position of loan officer. The following sequence lists the order in which either a male (M) or a female (F) applied for the position. Is there evidence to indicate that the sequence is not randomly generated? Use a .05 significance level.

M M M F M M F M M F M F F M M F M M M F F F F M M F F F

18.5   After a television debate between two political candidates, a telephone line is open to viewers wishing to express their opinion on whether the democratic (D) or the republican (R) candidate won the debate. The following sequence represents 19 opinions of viewers in the order in which they telephoned. Using a runs test and a significance level of 5%, does the sequence indicate a nonrandom order?

R R D D R D D R R R R D D R D R D D D

18.6   Conduct a runs test on the following sequence of 3s and 4s to see if there is evidence that the sequence is not randomly generated. Use a 5% significance level.

3 3 3 4 4 3 4 3 4 3 3 3 4 4 3 4 4 3 3 3 3 4 3

18.7   A certain computer program generates a sequence of random digits. Test whether there is any evidence that the following sequence of numbers is nonrandom by considering the sequence of odd and even numbers. Use a 10% significance level.

4 8 7 9 3 2 1 6 7 9 4 1 8 3 2 5

18.8   The amounts of government securities, in units of millions of dollars, held by life insurance companies in the United States are given below for 1976–1986. A regression equation is fit through the data with time as the independent variable. The residuals for this regression are also given. Does the sequence of positive and negative residuals appear to be random with respect to time? Use a 5% significance level. Use the "RUNS" command in MINITAB.

YEAR	GOVERNMENT SECURITIES	RESIDUALS	YEAR	GOVERNMENT SECURITIES	RESIDUALS
1976	20,260	9,377.0	1982	29,291	−14,569.0
1977	23,555	7,175.9	1983	43,290	−6,066.2
1978	26,552	4,676.7	1984	55,810	957.7
1979	29,719	2,347.5	1985	69,282	8,933.5
1980	20,731	−12,136.6	1986	79,899	14,054.4
1981	23,613	−14,750.8			

(*Source: Moody's Bank and Finance Manual*, Vol. 1, 1987, p. a13.)

18.9   Thirty-five true-or-false questions were given on a history test. The following sequence contains the answers to the questions in the order in which they appeared. At the 5% level, is there evidence that the true and false answers are not randomly assigned?

F F F F T T F T T F F F F T F T F F T T T T F F T T F F T F T F T T T F

18.10   The hourly earnings, in dollars per hour, of workers in the petroleum refining industry are listed for January 1986 to October 1987. Letting $t = 1$ for January 1986 and incrementing t by one for each month, the regression equation for predicting hourly earnings is found to be $\hat{y} = 15.1623 + .27781t$. Find the residuals for each time t. At the 5% significance level, is there evidence that the positive and negative residuals are not random with respect to time?

YEAR	JAN.	FEB.	MAR.	APR.	MAY	JUN.	JUL.	AUG.	SEP.	OCT.	NOV.	DEC.
1987	15.46	15.47	15.62	15.68	15.75	15.63	15.65	15.71	15.93	15.79	—	—
1986	15.33	15.30	15.36	15.35	15.20	15.30	15.36	15.19	15.41	15.36	15.33	15.42

(*Source: Standard and Poor's Statistical Service, Current Statistics*, February 1988, p. 10.)

18.2

NONPARA-METRIC TESTS OF CENTRAL TENDENCY

Chapter 3 introduced you to measures of central tendency. The more commonly used measures are the mean and median, which attempt to identify the middle of a set of sample data. In Chapter 9, we introduced two populations, where the question of interest was whether the two means were the same (a two-tailed test) or whether one mean exceeded the other (a one-tailed test). This is illustrated in Figure 18.2, where the variable of interest is height.

The main assumption in Figure 18.2 is that the two populations are normally distributed. When you sample from these populations, if both sample sizes (n_1 and n_2) are *large*, you can remove this assumption. However, there is a need for a nonparametric technique for this two-population situation, when (1) you have small samples and you suspect that one or both populations do not follow a normal distribution or (2) your data are such that only the relative ranks are available within each sample, such as in a consumer taste test. In other words, you are dealing with *ordinal data*. The t-tests from Chapter 9 assumed that the measurement scale of the data was at least interval, and so these tests are inappropriate for data consisting of ranks.

FIGURE 18.2

Two-population test of hypothesis for means.

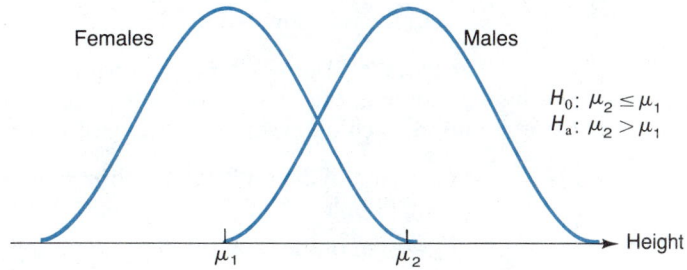

FIGURE 18.3

Illustration of dependent
(paired) samples.

Data

Sample 1
(women)

Sample 2
(men)

X		X	—— Couple 1
X		X	—— Couple 2
X		X	—— Couple 3
⋮		⋮	(etc.)

Also in Chapter 9, when dealing with samples from two populations, we looked at two situations:

1. The two samples are *independent*. In Figure 18.2, this would mean that a sample of n_1 female heights is obtained independently of the n_2 male heights. There is no reason to match up the first male height with the first female height, the second male height with the second female height, and so on in your two samples.

2. The two samples are *dependent* or *paired*. This might occur in Figure 18.3 if the question were, Are husbands taller than wives? The data then consist of n_1 wives and $n_2 = n_1$ husbands.

This portion of the chapter discusses the nonparametric counterparts to these two parametric tests of hypothesis. The assumptions behind the application of these methods are considerably weaker than those for the t-tests in Chapter 9. These nonparametric techniques, named after the people responsible for their development, are the Mann-Whitney U test—a nonparametric procedure for situation 1 (two independent samples)—and the Wilcoxon signed rank test—a nonparametric procedure for situation 2 (two paired samples).

The parametric tests in Chapter 9 were concerned with population means. If two populations have different means, we say that these populations differ in **location**. This implies that population 1 is shifted to the left or right of population 2. When defining the corresponding nonparametric test, the hypotheses will be stated in terms of differing location rather than differing means.

The Mann-Whitney U Test for Independent Samples

The **Mann-Whitney U test** for two means is named after H. B. Mann and R. Whitney, who developed this test in the 1940s. The purpose of this procedure is to provide a test for differing location that does not require the assumption of normal populations. The test is an alternative to the t-tests from Chapter 9, which do contain this assumption.

The two-tailed hypotheses for the Mann-Whitney test can be written

H_0: the two populations have identical probability distributions

H_a: the two populations differ in location

To use the Mann-Whitney nonparametric technique, we begin by combining (pooling) the two samples into one large sample and then determining the rank of each observation in the pooled sample. Next, let

T_1 = sum of the ranks of the observations from the first sample in this pooled sample

T_2 = sum of the ranks of the observations from the second sample

The procedure is (if $n_1 = n_2$)

reject H_0 if T_1 is "too much different" than T_2

To illustrate this, consider the following pooled sample, where the pooled observations have been arranged in order from smallest to largest. Here A represents a value from population A and B is a value from population B.

Value	A	A	A	A	A	B	B	B	B	B
Rank	1	2	3	4	5	6	7	8	9	10

For this pooled sample, we have

$n_1 = 5$

T_1 = sum of ranks of the five A observations in the pooled sample

$= 1 + 2 + 3 + 4 + 5 = 15$

$n_2 = 5$

T_2 = sum of the ranks of the B observations

$= 6 + 7 + 8 + 9 + 10 = 40$

Now consider another pooled sample:

Value	A	B	B	A	B	A	A	A	B	B
Rank	1	2	3	4	5	6	7	8	9	10

For this situation, the values are $T_1 = 1 + 4 + 6 + 7 + 8 = 26$ and $T_2 = 2 + 3 + 5 + 9 + 10 = 29$.

In the first pooled sample, there is clear evidence that the second population is shifted to the right of the first population, as indicated by the large difference between $T_1 = 15$ and $T_2 = 40$. This is also evident when you examine this pooled sample because the values from population A are all less than those from population B. From a parametric view, this implies that $\mu_B > \mu_A$. The Mann-Whitney procedure will result in rejecting H_0: the two populations have identical probability distributions in favor of H_a: the two populations differ in location (or H_a: population A is shifted to the left of population B had we used a one-tailed test). For the second set of ten pooled observations, there is no indication of a difference in population location; the A and B values are fairly well mixed in the combined sample. This is evidenced by the values of $T_1 = 26$ and $T_2 = 29$, which are nearly equal here. The Mann-Whitney test will lead to a failure to reject the null hypothesis.

Mann-Whitney Test for Small Samples For this test of hypothesis, small samples are defined as both $n_1 \leq 10$ and $n_2 \leq 10$. *Regardless* of the sample sizes, the procedure begins by finding T_1 and T_2 as described previously and then letting

$$U_1 = n_1 n_2 + \frac{n_1(n_1 + 1)}{2} - T_1 \qquad (18.5)$$

and

$$U_2 = n_1 n_2 + \frac{n_2(n_2 + 1)}{2} - T_2 \qquad (18.6)$$

The Mann-Whitney test is summarized in the accompanying box.

THE MANN-WHITNEY TEST FOR SMALL SAMPLES

Null Hypothesis:

H_0: the two populations have identical probability distributions

Assumptions:

1. Random samples are obtained from each population.
2. The two samples are independent of one another—respective observations are not paired.
3. The sample data are at least ordinal.

Procedure:

1. Assume that $n_1 \leq n_2$ (if this is not the case, reverse your populations, so that n_1 is the smaller sample size).
2. Determine U_1 and U_2 from equations 18.5 and 18.6.
3. Use the value from Table A.10 to test H_0 versus H_a where, once again, small p-values lead to rejecting H_0.

TWO-SIDED TEST	ONE-SIDED TEST	
H_a: the two populations differ in location	H_a: population 1 is shifted to the right of population 2	H_a: population 1 is shifted to the left of population 2
Reject H_0 if Table A.10 value for U is less than $\alpha/2$, where $U =$ minimum of U_1 and U_2	Reject H_0 if Table A.10 value for U is less than α, where $U = U_1$.	Reject H_0 if Table A.10 value for U is less than α, where $U = U_2$.

EXAMPLE 18.4

A local auto dealer wants to know whether single male buyers purchase the same amount of "extras" (such as air conditioning, power steering, exterior trim) as do single females when ordering a new car. The researcher in charge of the analysis has no reason to believe the amounts are normally distributed and so elects to test the hypotheses using the Mann-Whitney procedure. A sample of eight males and nine females was obtained; the data consist of the dollar amounts of the ordered extras.

Male purchases	2450	1436	850	1240	3645	1766	1226	2840	
Female purchases	1742	3146	2740	2160	3436	2750	562	1290	2060

Use $\alpha = .05$ to test for a difference between the amounts purchased by the male and female buyers.

SOLUTION

The hypotheses are

H_0: the two populations have identical probability distributions

H_a: the two populations differ in location

The pooled sample here is

562, 850, 1226, 1240, 1290, 1436, 1742, 1766, 2060, 2160, 2450, 2740, 2750, 2840, 3146, 3436, 3645

Next, we indicate from which sample each value came in the pooled sample.

RANK	MALE SAMPLE	FEMALE SAMPLE	RANKS FOR MALE SAMPLE	RANKS FOR FEMALE SAMPLE
1		562		1
2	850		2	
3	1226		3	
4	1240		4	
5		1290		5
6	1436		6	
7		1742		7
8	1766		8	
9		2060		9
10		2160		10
11	2450		11	
12		2740		12
13		2750		13
14	2840		14	
15		3146		15
16		3436		16
17	3645		17	
			$T_1 = 65$	$T_2 = 88$

Using equations 18.5 and 18.6,

$$U_1 = (8)(9) + \frac{(8)(9)}{2} - 65 = 43$$

$$U_2 = (8)(9) + \frac{(9)(10)}{2} - 88 = 29$$

Because this is a two-sided alternative, we let U = the minimum of 29 and 43, so $U = 29$.

For $n_1 = 8$, $n_2 = 9$, and $U = 29$, the value in Table A.10 is .2707. Because this is greater than $\alpha/2 = .025$, we fail to reject H_0. Based on these data, there is insufficient evidence to indicate a difference between male and female purchase amounts.

The p-value for this test is $(2)(.2707) = .5414$, which is extremely large. For a one-sided test, the p-value would be obtained by finding the value from Table A.10 and *not* doubling it. ∎

Ties When the pooled sample contains two or more identical observations, each is assigned a rank equal to the *average* of the ranks of the tied observations. For example, if there are two observations tied for sixth and seventh place, each is assigned a rank of 6.5. The rank of the next largest sample value is 8. We illustrate this procedure in the next section.

Mann-Whitney Test for Large Samples Whenever n_1 or n_2 is greater than ten, a large sample approximation can be used for the distribution of the Mann-Whitney U statistic. For this case, we can use either U_1 or U_2 in the test statistic for both one-sided *and* two-sided tests. The following discussion uses U_2.

In the event that the two populations have identical probability distributions (that is, H_0 is true), the U_2 statistic is approximately *normally* distributed with mean

$$\mu_{U_2} = \frac{n_1 n_2}{2} \tag{18.7}$$

and standard deviation

$$\sigma_{U_2} = \sqrt{\frac{n_1 n_2 (n_1 + n_2 + 1)}{12}} \tag{18.8}$$

The rejection region for the various alternative hypotheses are defined in the accompanying box.

The corresponding test statistic here is

$$Z = \frac{U_2 - \mu_{U_2}}{\sigma_{U_2}} \tag{18.9}$$

THE MANN-WHITNEY TEST FOR LARGE SAMPLES

Null hypothesis:

H_0: the two populations have identical probability distributions

Assumptions: Same as for small samples
Procedure: Determine

$$U_2 = n_1 n_2 + \frac{n_2(n_2 + 1)}{2} - T_2$$

where $T_2 =$ sum of the ranks for the second sample in the pooled sample.

TWO-SIDED TEST	ONE-SIDED TEST	
H_a: the two populations differ in location	H_a: population 1 is shifted to the right of population 2	H_a: population 1 is shifted to the left of population 2
Reject H_0 if $|Z| > Z_{\alpha/2}$	Reject H_0 if $Z > Z_\alpha$.	Reject H_0 if $Z < -Z_\alpha$.

where (1) Z is defined in equation 18.9 and (2) $Z_{\alpha/2}$ is the value from Table A.4 having a right-tail area of $\alpha/2$.

EXAMPLE 18.5　Food World operates two supermarkets in a large metropolitan area. One of their services to customers is to cash personal checks at no charge. The owner of Food World is concerned that one of the stores (store A), situated in a low-income neighborhood, may have a greater number of checks returned due to insufficient funds in the customers' checking accounts than does store B, which is located in a higher-income area. Data were collected for 12 randomly selected 6-month periods from store A, consisting of the number of returned

TABLE 18.4

Pooled Sample for Example 18.5

RANK	STORE A SAMPLE	STORE B SAMPLE	RANKS FOR STORE A	RANKS FOR STORE B
1		8		1
2		10		2
3		14		3
4		15		4
5		17		5.5
6		17		5.5
7		19		7
8		20		8
9		22		9
10		24		10
11		28		11
12		35		12
13	38		13	
14	42		15	
15		42		15
16	42		15	
17		45		17
18	47		18	
19		50		19
20	55		20	
21	57		21	
22	59		22	
23	60		23	
24	65		24	
25	68		25	
26	71		26	
27	76		27	
			$T_1 = 249$	$T_2 = 129$

checks over this period. This was repeated for 15 randomly selected 6-month periods for store B.

Store A   42,   65,   38,   55,   71,   60,   47,   59,   68,   57,   76,   42
Store B   22,   17,   35,   19,   8,   24,   42,   14,   28,   17,   10,   15,   20,   45,   50

The pooled sample and corresponding ranks are summarized in Table 18.4. Notice that there are two values of 17, which are tied for fifth and sixth place. Consequently, each is given a rank of $(5 + 6)/2 = 5.5$. Similarly, there is a three-way tie for fourteenth, fifteenth, and sixteenth place, so a rank of $(14 + 15 + 16)/3 = 15$ is given to each.

Using $\alpha = .05$, is there sufficient evidence to indicate that store A has a larger number of returned checks than does store B?

SOLUTION

The hypotheses for this situation are

> H_0: the two populations have identical probability distributions
>
> H_a: population A is shifted to the right of population B

The test procedure is to

$$\text{reject } H_0 \text{ if } Z > 1.645$$

where $1.645 = Z_{.05}$ is obtained from Table A.4. From Table 18.4, we find that

$T_2 = $ sum of ranks for store B

$= 1 + 2 + 3 + 4 + 5.5 + 5.5 + \cdots + 15 + 17 + 19 = 129$

and so

$$U_2 = (12)(15) + \frac{(15)(16)}{2} - 129 = 171$$

Also, the mean and standard deviation of the U_2 statistic are

$$\mu_{U_2} = \frac{(12)(15)}{2} = 90$$

and

$$\sigma_{U_2} = \sqrt{\frac{(12)(15)(28)}{12}} = 20.49$$

The value of the resulting test statistic is

$$Z = \frac{U_2 - \mu_{U_2}}{\sigma_{U_2}} = \frac{171 - 90}{20.49} = 3.95$$

This exceeds 1.645, and so we reject H_0 and conclude that store A does in fact have a larger volume of returned checks than store B. ◼

A MINITAB solution for Example 18.5 is contained in Figure 18.4. The Mann-Whitney statistic is denoted by W, which is actually the sum of the first sample ranks, T_1. T_2 can be obtained by using the identity

$$T_1 + T_2 = \frac{n(n + 1)}{2}$$

where $n =$ pooled sample size $= n_1 + n_2$.

Wilcoxon Signed Rank Test for Paired Samples

When your sample data consist of *paired* observations from two populations, the Mann-Whitney procedure from the previous section does not apply; it assumes *independent* samples. By *paired observations*, we mean that respective

FIGURE 18.4

MINITAB procedure for Mann-Whitney test (Example 18.5).

```
                                                    Alternative:
                                                    1 for Hₐ: pop. A is shifted to the
                                                        right of pop. B
                                                    −1 for Hₐ: pop. A is shifted to
MTB > SET INTO C1                                        the left of pop. B
DATA> 42 65 38 55 71 60 47 59 68 57 76 42           Omit for Hₐ: pop. A and pop. B
DATA> END                                               differ in location
MTB > SET INTO C2
DATA> 22 17 35 19 8 24 42 14 28 17 10 15 20 45 50
DATA> END
MTB > MANN-WHITNEY (ALTERNATIVE=1) USING C1 AND C2 ←

Mann-Whitney Confidence Interval and Test

   C1          N =  12      MEDIAN =        58.00
   C2          N =  15      MEDIAN =        20.00
→POINT ESTIMATE FOR ETA1-ETA2 IS             33.00
   95.2  PCT C.I. FOR ETA1-ETA2 IS (    21.99,    44.00)
→W =    249.0
   TEST OF ETA1 = ETA2  VS.  ETA1 G.T. ETA2 IS SIGNIFICANT AT   0.0000
   = T₁. So, T₂ = (27)(28)/2 − 249 = 129
```

ETA1 − ETA2 represents the difference in population locations

p-value for *Z* statistic is ≅ 0

observations from each sample are matched with one another. Examples include husband-wife, brother-sister, and before-after combinations.

A method of testing population means under this type of sampling procedure was introduced in Chapter 9, where we used a t-test on the sample differences. However, as with all t-tests, a key assumption using this method of testing two means is that the differences are *normally distributed*. When small samples from suspected nonnormal populations are used, a nonparametric technique is required. The **Wilcoxon signed rank test** is used for such situations.

The Wilcoxon test begins like its parametric counterpart, the paired-sample t-test, by subtracting the data pairs and using the differences to perform the test. As with the paired-sample t-test, the hypotheses are written in terms of the location of the probability distribution for the population differences.

The steps involved in applying the Wilcoxon test are:

1. Determine the difference for each sample pair.
2. Arrange the *absolute value* of these differences in order, assigning a rank to each.
3. Let T_+ = sum of the ranks having a positive value and T_- = sum of the ranks for the negative values.
4. T_+, T_-, or T = the minimum of T_+ and T_- is used to define a test of H_0 versus H_a.

To demonstrate the test, suppose we are interested in determining the effects of a vigorous 6-month advertising campaign. Sales figures are collected before and after the campaign from ten different cities. The results are shown in Table 18.5.

We determine the paired differences and rank the corresponding absolute values in order. Ties are handled as before by assigning a rank equal to the average of the tied positions. Also, if a pair of observations has a difference equal to zero, then this pair should be *deleted* from the sample, and n is reduced by one. Other methods exist for handling zero differences, but this is the simplest procedure and works well provided there are not many zero differences.

According to Table 18.6, the negative differences are -6 and -8. Their corresponding ranks are 2 and 4. Therefore,

$$T_- = 2 + 4 = 6$$

A rule that can simplify the calculations here and serve as a check for arithmetic is that

$$T_+ + T_- = \frac{n(n + 1)}{2}$$

TABLE 18.5

Sales (Thousands of $)
for Ten Cities

CITY	SALES BEFORE	SALES AFTER
Denver	61	63
Boston	50	57
Salt Lake City	18	34
Seattle	56	48
Miami	29	44
Dallas	25	38
Atlanta	34	28
Baltimore	48	68
Topeka	37	57
Minneapolis	14	26

TABLE 18.6

Illustration of Wilcoxon
Signed Rank Procedure
(Example 18.6)

SALES BEFORE	SALES AFTER	DIFFERENCE (AFTER–BEFORE)	|DIFFERENCE|	RANK
61	63	2	2	1
50	57	7	7	3
18	34	16	16	8
56	48	−8	8	4(−)
29	44	15	15	7
25	38	13	13	6
34	28	−6	6	2(−)
48	68	20	20	9.5
37	57	20	20	9.5
14	26	12	12	5

where n = the number of sample pairs. In our example, $n = 10$, so

$$T_+ + T_- = \frac{(10)(11)}{2} = 55$$

which means that $T_+ = 55 - T_- = 55 - 6 = 49$

The Wilcoxon Signed Rank Test for Small Samples (Paired)   Once T_+ and T_- have been obtained, you can use the Wilcoxon signed rank test for testing hypotheses about the location of the population differences.

THE WILCOXON SIGNED RANK TEST FOR SMALL SAMPLES (PAIRED)

Null Hypothesis:

H_0: the population differences are centered at 0

Assumptions:

1. Each data pair is randomly selected.
2. The absolute values of the differences can be ranked.

Procedure:

1. Determine the n differences using each sample pair, where each difference is defined to be sample 1 − sample 2.
2. Assign a rank to the absolute value of each difference; define T_+ = sum of the ranks of the positive values and T_- = sum of the ranks of the negative values.

Table A.11 is used to define the rejection region for the following tests.

TWO-SIDED TEST	ONE-SIDED TEST	
H_a: the population differences are not centered at 0	H_a: the population differences are centered at a value >0	H_a: the population differences are centered at a value <0
Using the two-sided value from Table A.11, reject H_0 if $T \le$ table value, where T = minimum of T_+ and T_-.	Using the one-sided value from Table A.11, reject H_0 if $T_- \le$ table value.	Using the one-sided value from Table A.11, reject H_0 if $T_+ \le$ table value.

EXAMPLE 18.6

Table 18.5 contains the sales results from ten cities before and after the 6-month advertising campaign. Using $\alpha = .05$, are we able to conclude that there was a significant increase in sales after the advertising campaign?

SOLUTION

The hypotheses here can be stated as (A = after, B = before)

H_0: the population differences are centered at 0.

H_a: the population differences are centered at a value > 0.

We refer to the "after" population as population 1 and the "before" population as population 2. This agrees with the difference column in Table 18.6 because our procedure assumes that each difference is sample 1 (A, here) $-$ sample 2 (B).
The values of T_+ and T_- also are derived following Table 18.6, where

$$T_- = 6$$

and

$$T_+ = 49$$

The one-sided value in Table A.11 corresponding to $n = 10$ and $\alpha = .05$ is 11. Consequently, the test is to

reject H_0 if $T_- \leq 11$

Because the value of T_- is smaller than 11, we reject H_0 and conclude that there is sufficient evidence of a sales increase after the advertising campaign.

■

The Wilcoxon Signed Rank Test for Large Samples (Paired)   For samples consisting of $n > 15$ pairs, a large-sample approximation to the Wilcoxon test statistic can be used. An advantage to using this procedure is that p-values are much easier to determine. (A p-value is once again a measure of the strength of your conclusion.)
   When using the large sample procedure, we can define a test using either T_+ or T_-. The following hypothesis tests use T_+, the sum of the ranks for the positive differences. If the population differences are centered at zero (that is, H_0 is true), then T_+ is approximately a normal random variable with mean

$$\mu_{T_+} = \frac{n(n + 1)}{4} \tag{18.10}$$

and standard deviation

$$\sigma_{T_+} = \sqrt{\frac{n(n + 1)(2n + 1)}{24}} \tag{18.11}$$

The corresponding test statistic is

$$Z = \frac{T_+ - \mu_{T_+}}{\sigma_{T_+}} \tag{18.12}$$

The one- and two-sided large sample procedures are summarized in the accompanying box.

THE WILCOXON SIGNED RANK TEST
FOR LARGE SAMPLES (PAIRED)

Null Hypothesis:

H_0: the population differences are centered at 0.

Assumptions: Same as for small samples
Procedure: (1) and (2) are the same as for small samples. Each paired difference is defined to be sample 1 − sample 2.

TWO-SIDED TEST	ONE-SIDED TEST	
H_a: the population differences are not centered at 0	H_a: the population differences are centered at a value >0	H_a: the population differences are centered at a value <0
Reject H_0 if $|Z| > Z_{\alpha/2}$, where Z is defined as in equation 18.12 and $Z_{\alpha/2}$ is the value from Table A.4 having a right-tail area of $\alpha/2$.	Reject H_0 if $Z > Z_\alpha$.	Reject H_0 if $Z < -Z_\alpha$.

EXAMPLE 18.7

The owner of Worldwide Travel Agency is interested in seeing whether her customers flew more miles in 1988 than in 1978. She randomly selects 20 customers who used her agency during 1978 and 1988, and she records the total number of passenger miles for each of the 2 years.

CUSTOMER	1978	1988	CUSTOMER	1978	1988
1	55	45	11	121	171
2	101	79	12	46	21
3	62	77	13	112	78
4	93	138	14	70	106
5	40	68	15	97	87
6	120	80	16	47	91
7	110	138	17	106	80
8	77	81	18	84	104
9	49	49	19	50	88
10	90	80	20	75	117

Use the Wilcoxon signed rank test to determine whether the owner's belief—that her customers flew more miles in 1988—is correct. Let $\alpha = .05$.

SOLUTION

If we let the 1978 population be population 1, then the correct hypotheses are

H_0: the population differences are centered at 0

H_a: the population differences are centered at a value <0.

This agrees with Table 18.7, in which each difference is calculated using the 1978 value (sample 1) minus the 1988 value (sample 2). Because customer 9 had no change in the passenger miles for these 2 years, a difference of zero is the result, and so this customer is removed from the sample. This leaves $n = 19$ pairs in the sample.

Based on a significance level of $\alpha = .05$, the proper test is to

reject H_0 if $Z < -1.645$

TABLE 18.7
Paired Samples for
Example 18.7

MILES (THOUSANDS)		DIFFERENCE:	RANK OF	+RANKS	−RANKS
1978	1988	1978–1988	ABSOLUTE VALUE		
55	45	10	3	3*	
101	79	22	7	7	
62	77	−15	5(−)		5
93	138	−45	18(−)		18
40	68	−28	10.5(−)		10.5**
120	80	40	15	15	
110	138	−28	10.5(−)		10.5**
77	81	−4	1 (−)		1
[49	49	0	—	removed, so use $n = 19$ pairs]	
90	80	10	3	3*	
121	171	−50	19(−)		19
46	21	−25	8	8	
112	78	34	12	12	
70	106	−36	13(−)		13
97	87	10	3	3*	
47	91	−44	17(−)		17
106	80	26	9	9	
84	104	−20	6(−)		6
50	88	−38	14(−)		14
75	111	−42	16(−)		16
				$T_+ = 60$	$T_- = 130$

* three-way tie; assigned rank = $(2 + 3 + 4)/3 = 3$.
** two-way tie; assigned rank = $(10 + 11)/2 = 10.5$.

For a value of $n = 19$, the mean of T_+ (assuming H_0 is true) is

$$\mu_{T_+} = \frac{(19)(20)}{4} = 95$$

with a standard deviation of

$$\sigma_{T_+} = \sqrt{\frac{(19)(20)(39)}{24}} = 24.85$$

Table 18.7 informs us that $T_+ = 60$, and so the value of the test statistic here is

$$Z^* = \frac{60 - 95}{24.85} = -1.41$$

This value is not less than -1.645, so there is insufficient evidence to conclude that her long-term customers flew more passenger miles in 1988 than in 1978.

The p-value for this test is obtained in the usual manner by finding, in this case, the area under a standard normal curve to the left of the calculated test statistic of $Z^* = -1.41$. According to Figure 18.5 and Table A.4, the p-value

FIGURE 18.5
p-value for Example 18.7.

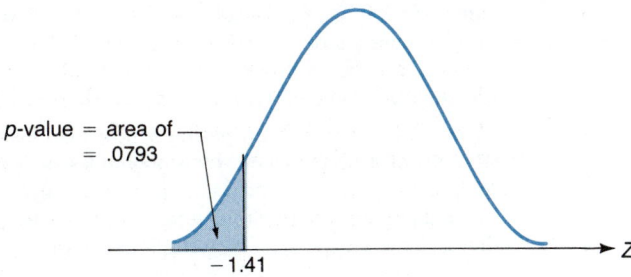

p-value = area of
= .0793

-1.41

Z

FIGURE 18.6

MINITAB commands for
Wilcoxon signed rank
test (Example 18.7).

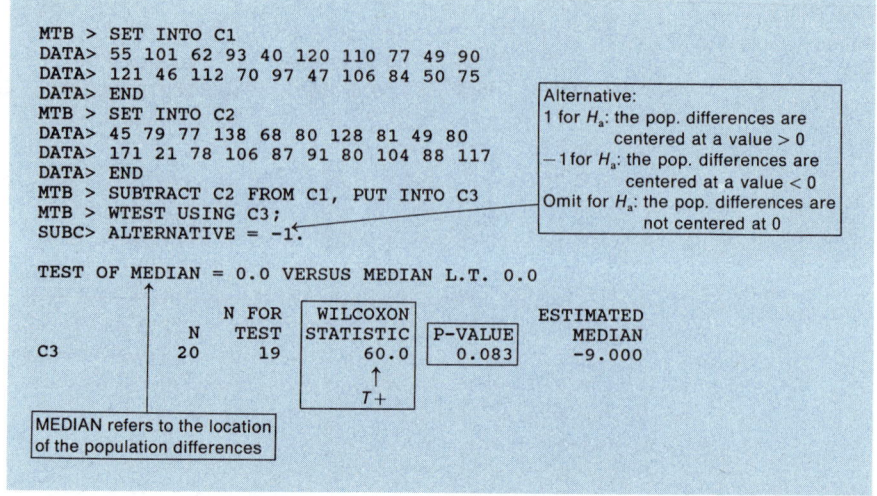

```
MTB > SET INTO C1
DATA> 55 101 62 93 40 120 110 77 49 90
DATA> 121 46 112 70 97 47 106 84 50 75
DATA> END
MTB > SET INTO C2
DATA> 45 79 77 138 68 80 128 81 49 80
DATA> 171 21 78 106 87 91 80 104 88 117
DATA> END
MTB > SUBTRACT C2 FROM C1, PUT INTO C3
MTB > WTEST USING C3;
SUBC> ALTERNATIVE = -1.
```

Alternative:
1 for H_a: the pop. differences are
    centered at a value > 0
−1 for H_a: the pop. differences are
    centered at a value < 0
Omit for H_a: the pop. differences are
    not centered at 0

```
TEST OF MEDIAN = 0.0 VERSUS MEDIAN L.T. 0.0

                        N FOR   WILCOXON            ESTIMATED
                 N      TEST    STATISTIC  P-VALUE    MEDIAN
C3              20       19        60.0     0.083     -9.000
                                     ↑
                                    T+
```

MEDIAN refers to the location
of the population differences

is .0793. Using our rule-of-thumb procedure from before, this p-value is neither large ($>.1$) nor small ($<.01$), but it *is* greater than $\alpha = .05$, which leads us to fail to reject H_0. ∎

To use MINITAB for the Wilcoxon procedure, begin by subtracting the two samples and then using the WTEST command on the differences, as illustrated in Figure 18.6. Note that the p-value in this figure is slightly different from our previous result because the MINITAB procedure includes a continuity adjustment (similar to that used in Chapter 6) and is slightly more accurate. If you have the MINITAB package available, use it to obtain more accurate p-values for this nonparametric test.

EXERCISES

18.11 What assumptions need to be made about the type and distribution of the data when the Mann-Whitney test is used?

18.12 If 15 observations are randomly selected from each of two populations and the sum of the ranks for the sample from population 1 (T_1) is 230, what is the sum of the ranks (T_2) for the sample from population 2?

18.13 Two groups of randomly selected students are given an aptitude test on understanding the financial markets. The first group has eight students, selected from second-semester freshmen, and the sum of the ranks (T_1) of these eight students is 65. If the second group has nine students selected from first-semester freshmen, is there sufficient evidence to conclude that the aptitude of the second-semester freshmen is higher than the aptitude of the first-semester freshmen on understanding the financial markets? Use a 5% significance level.

18.14 A real-estate agent claims that the homes in two neighborhoods have the same average value. A random sample from neighborhood A contains the following dollar values: 85,000, 70,000, 74,000, 69,000, 88,000, 89,000. A random sample from neighborhood B consists of the following values: 71,000, 64,000, 68,000, 73,000, 81,000, 69,000, 72,000. Test the alternative hypothesis that the value of homes in neighborhood B is less than the value of homes in neighborhood A. Use a 10% significance level.

18.15 An engineer is proposing a new manufacturing process to increase the tensile strength of a certain wire. Eleven samples of wire manufactured under the proposed process are collected, and 14 samples are collected from wire manufactured under the existing process. Use the following data and a 10% significance level, test the hypothesis that there is greater tensile strength in the proposed technique.

PROPOSED TECHNIQUE (PSI)	EXISTING TECHNIQUE (PSI)	PROPOSED TECHNIQUE (PSI)	EXISTING TECHNIQUE (PSI)
1.4	1.2	2.2	1.8
2.1	1.6	2.0	1.7
1.8	1.7	1.9	2.0
1.7	1.7	2.0	1.8
1.6	2.0		1.9
1.9	1.3		1.6
1.4	1.4		2.1

18.16 The head lawyer of the Brown and Smith firm would like to know whether there is a difference in the number of errors made by the two secretaries employed in the firm. Five randomly selected documents are given to secretary A to type, and five are given to secretary B. The number of errors per document is shown in the following table. Using a 5% level of significance, test the hypothesis that there is no difference in the number of errors made by each secretary.

SECRETARY A	SECRETARY B
3	2
5	0
4	4
2	3
0	1

18.17 A nursery is experimenting with two blends of fertilizer for fertilizing lawns in a certain area. Twenty-four randomly selected patches of grass are selected for experimenting with the fertilizer. Twelve patches are randomly assigned to fertilizer A and another 12 patches are randomly assigned to fertilizer B. The increase in the height of the grass after 2 weeks is given in the following table. Using a 10% level of significance, test the hypothesis that fertilizer B is more effective than fertilizer A.

FERTILIZER A	FERTILIZER B	FERTILIZER A	FERTILIZER B
1.2	1.0	0.8	1.2
0.9	1.1	1.4	1.3
1.3	1.0	0.7	0.8
0.5	0.9	0.9	1.0
0.3	0.7	0.8	0.8
0.9	0.8	0.7	1.1

18.18 Two samples of light bulbs are taken from the brands Everglo and Britelite. The following table gives the life of the bulbs for the collected sample. At the 1% level, is there sufficient evidence to indicate that Britelite bulbs last longer than do Everglo bulbs?

EVERGLO (HOURS)	BRITELITE (HOURS)	EVERGLO (HOURS)	BRITELITE (HOURS)
1134	1405	1107	1290
1255	1251	1095	1210
1313	1106	1401	1198
1012	1384	1109	1203
1265	1193	1150	1295
1375	1208		1102
1102	1110		1185

18.19 An economist wishes to compare the percent increase in personal income for two suburbs of Chicago. Using the data in the following table, test at the 5% significance level the hypothesis that there is no difference in the percent increase in personal income.

SUBURB A (%)	SUBURB B (%)	SUBURB A (%)	SUBURB B (%)
5.2	2.6	1.3	11.3
3.1	9.7	8.4	8.1
10.6	1.2	9.1	4.2
11.4	1.4	11.3	1.6
1.2	5.0	12.1	2.7
0.0	9.8	9.8	7.9

18.20 A supermarket manager was curious as to which of the two vending machines located at opposite ends of the store was used the most during peak hours. On 12 randomly selected days during the peak hours, the number of users were counted for machine A. On another randomly selected 12 days during the peak hours, the number of users were counted for machine B. From the following data, test at the 5% level of significance the hypothesis that there is no difference in the use of the two vending machines.

VENDING MACHINE A	VENDING MACHINE B	VENDING MACHINE A	VENDING MACHINE B
10	9	13	9
12	11	19	8
13	14	11	12
11	10	10	13
10	13	15	14
15	14	12	11

18.21 At the 5% significance level, test the hypothesis that there is no significant difference in the time for machine 1 and machine 2 to produce an item in Exercise 9.34, using the Mann-Whitney test. Compare to the results obtained using the t-test.

18.22 For the data in Exercise 9.38, use the Mann-Whitney test in testing the hypothesis that the two parent populations differ in location. Use a significance level of .05. When would you prefer the Mann-Whitney test to the t-test?

18.23 What assumption needs to be made about the sample data used in the Wilcoxon signed rank test? What assumption needs to be made about the distributions of the population?

18.24 From 12 paired observations, it is found that by ranking the magnitude of the differences of the observations in each pair, T_+ (the sum of the ranks of the positive differences) is 27. Using a .05 significance level, can it be concluded that there is a difference in the location of the two populations?

18.25 A psychologist conducts a seminar to increase a person's self-esteem. A before-and-after test that measures each person's self-esteem is given to nine individuals. Using the following scores and a 5% significance level, is there evidence to conclude that the scores after the seminar are greater than the scores before the seminar?

BEFORE	AFTER	BEFORE	AFTER
70	74	55	58
72	88	43	41
75	71	51	63
61	62	84	80
82	89		

18.26 An insurance company believes that employees who have a college degree when hired progress faster in the company than those who do not. Pairs of employees are randomly selected; each pair consists of two people hired at the same time, one person with a college degree and the other without a college degree. The percent increase in pay for the employees after 3 years is recorded below. At the 10% level of significance, can you conclude that employees who have a college degree when hired progress faster than those who do not?

WITHOUT COLLEGE DEGREE (%)	WITH COLLEGE DEGREE (%)	WITHOUT COLLEGE DEGREE (%)	WITH COLLEGE DEGREE (%)
10	13	12	13
9	10	9	8
8	6	18	16
13	13	9	12
14	18	15	17
7	10	10	9
12	11	11	13
11	15	10	9
16	20		

18.27 Seven randomly selected faculty members were asked to evaluate two research project proposals on a scale from 0 to 10, with a higher score indicating a more acceptable proposal. The scores follow. Using a 5% significance level, can you conclude that the proposal for research project 2 is more acceptable than the proposal for research project 1?

RESEARCH PROJECT 1	RESEARCH PROJECT 2
5	7
3	5
6	9
7	6
8	9
4	6
7	10

18.28 Martin's Weight Control Center claims that if a woman maintains her same diet but attends aerobic classes three times per week, she will definitely lose weight. Seventeen women who attended the program for 3 months were randomly selected. From the table, which gives the participants' weights before and after, test the claim of Martin's Weight Control Center. Use a 5% significance level.

BEFORE	AFTER	BEFORE	AFTER	BEFORE	AFTER
119	117	148	140	114	108
131	130	152	138	122	114
135	125	180	171	125	111
125	121	110	114	120	118
140	143	130	132	112	113
119	114	118	110		

18.29 The manager of a calculator-assembly plant wanted to know whether machine operators with little experience produced more defective calculators than did the experienced machine operators. The number of defective calculators produced by 20 randomly selected experienced machine operators in 1 week was recorded. Then, these 20 experienced operators were replaced by inexperienced machine operators and the number of defective calculators produced at these positions was recorded for 1 week. If the operators at each position can be considered to be a pair, use the Wilcoxon test to test that the experienced operators produced fewer defective calculators. Use a 5% significance level.

EXPERIENCED EMPLOYEES	INEXPERIENCED EMPLOYEES	EXPERIENCED EMPLOYEES	INEXPERIENCED EMPLOYEES
10	14	13	19
13	14	18	21
15	12	19	18
18	25	10	13
14	13	19	26
10	15	25	26
30	21	15	17
14	18	21	20
22	23	20	28
15	13	12	19

18.30 The manager of an insurance company sent ten randomly selected salespeople to a sales-motivation lecture given by several top selling insurance salespeople. The manager recorded the dollar amount (in hundreds of thousands) of insurance sold by the

BEFORE LECTURE	AFTER LECTURE	BEFORE LECTURE	AFTER LECTURE
1.2	1.9	6.2	6.3
1.8	3.4	1.5	1.4
3.8	3.1	3.3	4.9
1.9	4.5	2.4	3.5
5.8	5.0	3.1	3.0

ten salespeople during the 4 months prior to attending the lecture and the 4 months after the lecture. At the 5% significance level, is there evidence to suggest that the sales-motivation lecture improved sales?

18.31 A paired-difference experiment yielded a value of 280 for the sum of the ranks of the positive differences from 30 observations. Using a 5% significance level, is there evidence to suggest that the population differences are not centered at zero?

18.32 A large discount department-store chain decided to rearrange the layout of its merchandise at six of its stores to encourage customers to buy more on impulse. Sales from the 6 months prior to the rearrangement and sales from the 6 months after the rearrangement are given below. Using a 5% significance level, can it be concluded that there is an increase in sales after the rearrangement.

SALES BEFORE THE REARRANGEMENT ($\times$ \$100,000)	SALES AFTER THE REARRANGEMENT ($\times$ \$100,000)
3.4	3.9
2.8	2.9
4.1	4.0
3.6	3.8
4.8	5.6
5.1	5.4

18.33 The following values are the differences from 17 pairs of observations. At the 10% level, determine whether there is sufficient evidence to indicate that the population differences are not centered at zero?

$-1.1, \; -2.3, \; 4.5, \; 1.6, \; 2.3, \; -4.3, \; 1.9, \; -2.6, \; 1.8, \; 1.6, \; -2.7, \; 1.8, \; 2.1, \; -3.8, \; -1.0, \; 1.4, \; 2.5$

18.34 For Exercise 9.47, test at the 5% significance level that the blood pressure is less after the drug is administered. Use the Wilcoxon signed rank test. Compare the results using the paired t-test and a 5% significance level.

18.35 In Exercise 9.50, use the Wilcoxon signed rank test to test the hypothesis that there is no difference in the weekly sales of the two restaurants. Use a 5% significance level. When would you prefer the Wilcoxon test to the paired t-test?

18.3

COMPARING MORE THAN TWO POPULATIONS: THE KRUSKAL-WALLIS TEST AND THE FRIEDMAN TEST

The Kruskal-Wallis Test

When comparing the means of more than two populations, a popular technique is the ANOVA procedure discussed in Chapter 11. The assumptions behind this technique include that you are dealing with normally distributed populations; the F-test used in the ANOVA table is invalid unless all of the populations are nearly normally distributed with equal variances.

The nonparametric counterpart to the one-way ANOVA method is the **Kruskal-Wallis test**. It is named after W. H. Kruskal and W. A. Wallis, who published their results in 1952. This test, like many other nonparametric procedures, is relatively new, unlike most of the parametric hypothesis tests, which were developed much earlier. The assumption of normal populations is *not necessary* for the Kruskal-Wallis test, which makes it an ideal technique for samples exhibiting a nonsymmetric (skewed) pattern. It is also less sensitive than the ANOVA procedure to the assumption of equal variances. The test also is useful when the data consist of rankings (ordinal data) within each sample.

The assumption of normal populations becomes quite critical when dealing with small samples. As we've seen in many of the earlier tests of hypothesis, this assumption can be relaxed when larger samples are used, due to the Central Limit Theorem. However, many experiments of a business nature dealing with product comparisons result in the destruction of the product being tested. Consequently, small samples are often a necessity for such experiments, and nonparametric techniques are widely used to analyze the resulting data.

The Kruskal-Wallis test is actually an extension of the Mann-Whitney U test discussed earlier for *two* independent samples. Both procedures require that the sample values have a measurement scale that is at least ordinal, which means that each sample can be ranked from smallest to largest.

The hypotheses for this situation are similar to the Mann-Whitney hypotheses in that they are stated in terms of differing population location. The Kruskal-Wallis hypotheses are

H_0: the k populations have identical probability distributions

H_a: at least two of the populations differ in location

Procedure You first obtain random samples of size $n_1, n_2, \ldots, n_k$ from each of the k populations. The total sample size is $n = n_1 + n_2 + \cdots + n_k$. As with the Mann-Whitney procedure, you next pool the samples and arrange them in order, assigning a rank to each. For ties, you assign the average rank to the tied positions.

Let T_i = the total of the ranks from the ith sample. The Kruskal-Wallis test statistic (KW) is

$$KW = \frac{12}{n(n+1)} \sum_{i=1}^{k} \frac{T_i^2}{n_i} - 3(n+1) \qquad (18.13)$$

The distribution of the KW statistic approximately follows a chi-square distribution with $k - 1$ df. This approximation is good even if the sample sizes are small. To test H_0 versus H_a, the procedure is to

reject H_0 if KW is "large"

that is, if KW is in the right tail of the chi-square curve. This right-tail critical value is obtained from Table A.6, using a significance level = α and df = $k - 1$.

THE KRUSKAL-WALLIS TEST

Hypotheses:

H_0: the k populations have identical probability distributions

H_a: at least two of the populations differ in location

Assumptions:

1. Random samples are obtained from each of the k populations.
2. The individual samples are obtained independently.
3. Values within each sample can be ranked.

Procedure: The individual samples are pooled and then ranked from smallest to largest. Letting T_i = the sum of the ranks of the ith sample, the KW statistic is determined using equation 18.13. The null hypothesis, H_0, is rejected if

$$KW > \chi^2_{\alpha, \, df}$$

where $\chi^2_{\alpha, \, df}$ is the value from Table A.6 corresponding to df = $k - 1$, with a right-tail area = α.

TABLE 18.8

Amount of Down Time for Copying Machines (Example 18.8)

BRAND 1	RANK	BRAND 2	RANK	BRAND 3	RANK	BRAND 4	RANK
28	12	5	1	10	3	45	18
41	17	16	6	8	2	30	13
34	15	20	8	18	7	49	19
52	20	24	9	14	4	32	14
25	10	15	5	26	11	36	16
	$T_1 = 74$		$T_2 = 29$		$T_3 = 27$		$T_4 = 80$

EXAMPLE 18.8

Drexton Industries has a number of different brands of copying machines at their main facility. A critical factor in the attractiveness of each brand is the amount of time that a machine is not working and is waiting for repair (downtime). Management requested a study to be made on four different brands of machines to determine whether there is a difference in the amount of down time for these brands. Data were collected by finding the total downtime per month for 20 randomly selected months. In this way, the downtimes for five randomly selected months were obtained for each of the four brands of machine. These results are shown in Table 18.8.

Do these data indicate a difference in the amount of downtime for the four brands? Use $\alpha = .05$.

SOLUTION

There are $k = 4$ populations here, so we need the $\chi^2_{.05, 3}$ value from Table A.6. Based on this value, the testing procedure is to

$$\text{reject } H_0 \text{ if } KW > \chi^2_{.05, 3} = 7.81$$

From Table 18.8, we are able to compute the value of the KW statistic using the ranks of the observations in the pooled sample:

$$KW = \frac{12}{(20)(21)} \left[\frac{74^2}{5} + \frac{29^2}{5} + \frac{27^2}{5} + \frac{80^2}{5} \right] - 3(21)$$

$$= 13.83$$

As a check of your calculations at this point, make sure the ranks sum to $n(n + 1)/2$. For this example, $n = $ total number of observations $= 20$, and so the total of the ranks should be $(20)(21)/2 = 210$; here, $74 + 29 + 27 + 80 = 210$.

The calculated KW value exceeds 7.81, and so our conclusion is that there is a difference in downtime among the four brands. From the small values of T_2 and T_3, it appears that these two brands have much less downtime and are superior in this respect to brands 1 and 4.

Finally, the p-value here is $<.005$, indicating a very strong conclusion. In other words, these data indicate a clear difference in location for the four brands. This p-value is illustrated in Figure 18.7. ∎

FIGURE 18.7

p-value for KW statistic; χ^2 curve with 3 df (Example 18.8).

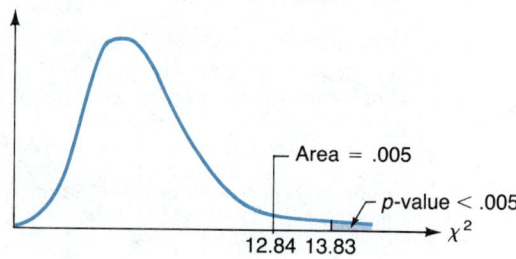

FIGURE 18.8

MINITAB procedure for Kruskal-Wallis test (Example 18.8).

```
MTB > SET INTO C1
DATA> 28 41 34 52 25 5 16 20 24 15
DATA> 10 8 18 14 26 45 30 49 32 36
DATA> END
MTB > SET INTO C2
DATA> 1 1 1 1 1 2 2 2 2 2 3 3 3 3 3 4 4 4 4 4
DATA> END
MTB > KRUSKAL-WALLIS USING DATA IN C1, LEVELS IN C2

    LEVEL     NOBS    MEDIAN   AVE. RANK    Z VALUE
        1        5     34.00        14.8       1.88
        2        5     16.00         5.8      -2.05
        3        5     14.00         5.4      -2.23
        4        5     36.00        16.0       2.40
  OVERALL       20                  10.5

 H = 13.83  ←——— KW
```

The Kruskal-Wallis test statistic is computed using MINITAB in Figure 18.8. Notice that the downtime values are stored in column C1, whereas column C2 contains the sample number of each observation (1, 2, 3, or 4). The value of the Kruskal-Wallis statistic is called H and agrees with the previous result.

The Friedman Test

The topic of comparing more than two population means using dependent samples was introduced in Chapter 11, where we examined the randomized block design. Such a design consists of a single factor of interest with k levels, where the sample data are organized into blocks. For this situation, the samples are not independently obtained, but data within the same block may be gathered from the same city or person or at the same point in time.

The key assumption behind the use of the randomized block design is that the variable being measured (the *dependent* variable) is normally distributed within each factor level/block combination. There are two situations for a blocked design where the use of the parametric randomized block technique is inappropriate, requiring the use of a nonparametric procedure. These are:

1. You have no evidence to support the assumption of normality.

or

2. The sample data are ordinal (that is, consist of rankings).

When confronted with either of these two situations, the nonparametric **Friedman** test provides a correct method of testing for differences in location for the k factor level populations. The corresponding hypotheses are

H_0: the k populations have identical probability distributions

H_a: at least two of the populations differ in location

Procedure The k observations *within each block* are rank ordered, using the usual procedure of assigning the average of the tied positions in the event of ties within a block. Of course, this step is omitted if ordinal data are obtained initially, such as asking 50 people to rank order four particular products (or brands) according to a specified set of criteria. For this illustration, you would have $b = 50$ blocks and $k = 4$ populations defined by the four products. In addition, to remove potential bias introduced by fatigue, familiarity, and so on, the order in which the four brands are evaluated should be *randomly* determined for each person.

Define T_i = the total of the ranks for the ith population. The test statistic for the Friedman test is defined as

$$FR = \frac{12}{bk(k+1)} \sum_{i=1}^{k} T_i^2 - 3b(k+1) \qquad (18.14)$$

where b = number of blocks and k = number of factor levels (populations).

The distribution of the FR statistic approximately follows a chi-square distribution with $k - 1$ df. This approximation works well provided the number of blocks (b) or the number of factor levels (k) exceeds five. Similar to the Kruskal-Wallis test, the Friedman test procedure is to reject H_0 if FR lies in the right-tail of the chi-square curve; that is,

$$\text{reject } H_0 \text{ if } FR > \chi^2_{\alpha, \text{df}}$$

where $\chi^2_{\alpha, \text{df}}$ is the value from Table A.6 corresponding to df $= k - 1$ and right-tail area $= \alpha$. This procedure is summarized in the accompanying box.

THE FRIEDMAN TEST

Hypotheses:

H_0: the k populations have identical probability distributions

H_a: at least two of the populations differ in location

Assumptions:

1. The factor levels are applied in a random manner within each block.
2. The number of blocks (b) or the number of factor levels (k) exceeds five.
3. Values within each block can be ranked.

Procedure: Values are ranked within each block. Letting T_i = sum of the ranks within the ith factor level, the FR statistic is calculated using equation 18.14. The null hypothesis is rejected if

$$FR > \chi^2_{\alpha, \text{df}}$$

where $\chi^2_{\alpha, \text{df}}$ is the chi-square value (Table A.6) with df $= k - 1$ and right-tail area $= \alpha$.

EXAMPLE 18.9

In Section 11.4 we used the randomized block design to compare the perceived new-car warranty among three competing brands of automobiles; Henry, GA, and Roadster. The editors of ten different automotive magazines were asked to evaluate the warranties and assign a composite score to each warranty, ranging from 0 (worst possible) to 100 (ideal). Here the ten editors represent the $b = 10$ blocks and the factor of interest here is brand of automobile, consisting of $k = 3$ levels.

Suppose that 5 years after this study was completed you wished to repeat the analysis using the current warranties for these three automobiles and the editors (perhaps, new) of these ten magazines. However, there was some controversy regarding asking each editor to come up with a score to represent the attractiveness of the warranty. This was due to the belief that each person had his or her own set of criteria used to judge a warranty, not to mention different weights assigned to these various criteria.

TABLE 18.9

Results of New-Car
Warranty Ranking
(Example 18.9)

HENRY	GA	ROADSTER
3	2	1
2	1	3
3	1	2
2	1	3
3	2	1
2	1	3
1	2	3
2	1	3
3	1	2
1	2	3
$T_1 = 22$	$T_2 = 14$	$T_3 = 24$

Consequently, it was decided simply to ask each editor to rank the three warranties, rather than determine a score for each one. The warranties were assigned in a random manner for each editor. The results are summarized in Table 18.9. Use the Friedman test and a significance level of .10 to determine if there is a difference in the perceived quality of the three warranties.

SOLUTION   The computed value of the Friedman statistic is

$$FR = \frac{12}{(10)(3)(4)}\left[22^2 + 14^2 + 24^2\right] - 3(10)(4)$$

$$= 125.6 - 120$$

$$= 5.6$$

Referring to the chi-square table (A.6), using $k - 1 = 2$ df and right-tail area $= .10$, the test procedure is to

$$\text{reject } H_0 \text{ if } FR > 4.60517$$

Since $5.6 > 4.60517$, we reject H_0 and conclude that there *is* a difference in the perceived quality of the three warranties. The apparent reason for this result is the fact that the GA warranty was ranked first or second by all ten editors and far outranked the warranties for Henry and Roadster. This agrees with the conclusion reached in the earlier study (discussed in Section 11.4), which demonstrated the superior perception of the GA warranty using the randomized block design. Finally, the p-value here is between .05 and .10, indicating a fairly weak result, yet one that is statistically significant at a level of $\alpha = .10$.   ■

EXERCISES

18.36 Samples of nine observations were taken from each of three populations, making a total of 27 observations. When the observations were pooled and then ranked from smallest to largest, the sum of the ranks for the first, second, and third samples was 120, 148, and 110, respectively. At a significance level of .05, do the data indicate a difference in location for the three populations?

18.37 Ron, Ted, and James are three sales representatives covering three separate territories for a company that sells farm equipment at the wholesale level. To test the hypothesis that there is no difference in the monthly commission of the three salespeople, 8 months were chosen at random for each and their commissions were recorded in thousand-dollar units. Use the Kruskal-Wallis test on the data to test this hypothesis. Use a 1% significance level.

MONTH	RON	TED	JAMES	MONTH	RON	TED	JAMES
1	2.3	2.8	2.0	5	4.6	1.7	4.4
2	2.1	2.9	1.9	6	3.8	3.4	2.2
3	1.5	3.1	2.6	7	3.1	2.3	3.8
4	2.3	1.8	3.9	8	2.6	2.1	2.5

18.38   The following table lists observations from a sample of four populations. Is there reason to believe that the four populations differ in location? Use a 1% significance level.

POPULATION 1	POPULATION 2	POPULATION 3	POPULATION 4
101	104	104	105
110	99	102	110
120	86	100	120
105	105	111	121
100	110	103	127
107	120	102	118
106	114		112
	110		

18.39   A survey was taken of the starting salaries of students who had completed a degree in business administration at either Oceanspray College, Stanton University, or Hillside College. Do the following data indicate, at the .05 level, a difference in the starting salaries of students with a degree in business administration from one of the three schools?

OCEANSPRAY COLLEGE	STANTON UNIVERSITY	HILLSIDE COLLEGE
27,100	24,500	22,500
25,300	27,250	23,000
22,450	26,700	26,000
26,800	27,000	25,750
25,100	26,500	21,630
21,350	22,300	22,500
22,500	27,600	23,650
25,000	28,150	24,180
		22,750

18.40   Four machines are used to package 16-ounce bags of puffed wheat. Each machine is designed to package the bags so that the average bag has 16 ounces of cereal in it. From the data, in which samples of eight bags were randomly selected from each machine, is there an indication that the amount of puffed wheat packaged is not the same for all four machines? Use a .05 significance level.

MACHINE 1	MACHINE 2	MACHINE 3	MACHINE 4
15.9	16.1	15.8	16.4
15.8	16.3	15.9	16.5
16.0	16.0	16.0	16.0
15.7	15.9	16.1	16.1
16.1	16.4	16.0	16.4
16.2	15.8	16.4	16.3
15.6	16.2	16.1	16.1
15.8	16.1	15.7	16.4

18.41   Thirty new employees were selected to test two training programs. Ten of them (group A) were randomly selected for a self-paced training program. Another ten (group B) were randomly selected for a classroom training program. The remaining ten employees (group C) were not given any training. After the completion of the experiment, the manager evaluated the 30 employees on their productivity over a 2-week span. The following ranks were given by the manager, with the highest rankings being given to those who were not productive.

PROGRAM A	PROGRAM B	PROGRAM C	PROGRAM A	PROGRAM B	PROGRAM C
6	5	1	30	12	3
22	21	10	16	18	7
25	15	11	23	24	14
26	4	13	28	27	17
20	8	2	9	29	19

From the data, is there a significant difference in the productivity of the three groups? Use a .05 significance level.

18.42 Joe's Delicatessen sells cheese sandwiches, ham sandwiches, and roast-beef sandwiches. Joe would like to know if there is a significant difference in the number of sandwiches of each type sold. On 30 randomly selected days, the following number of sandwiches of each type were sold. Ten days were therefore selected for each type of sandwich sold. Is there a significant difference at the 10% level?

CHEESE	HAM	ROAST BEEF	CHEESE	HAM	ROAST BEEF
27	30	33	26	22	34
24	21	25	25	24	31
23	20	23	24	21	30
29	31	22	25	22	30
21	32	27	28	20	27

18.43 The management of a company that markets Soft and Fresh Detergent would like to increase sales of detergent by including a free drinking glass in the box, including a coupon worth 50 cents toward the next purchase, or using a colorful see-through plastic container for the detergent. Ten stores in different cities were randomly selected to market the detergent in one of the three ways, providing a total of 30 stores in the sample. The number of boxes sold over a 1-month period at the stores using each of the three marketing strategies follows. Do the data indicate a difference in the number of boxes sold for each of the marketing strategies? Use a 5% significance level.

FREE GLASS	COUPON	SEE-THROUGH PLASTIC CONTAINER	FREE GLASS	COUPON	SEE-THROUGH PLASTIC CONTAINER
350	320	374	270	311	349
310	315	371	340	318	331
250	300	332	310	330	322
380	315	361	290	340	368
290	390	356	375	314	351

18.44 The number of hours it takes three workers to complete a task is given in the following table. The task is assigned to each worker four times. Do the data indicate a significant difference in the time it takes each worker to complete the task? Use a significance level of .05.

WORKER 1	WORKER 2	WORKER 3
3.1	3.4	3.5
3.4	3.3	3.2
3.0	3.4	3.3
3.1	3.2	3.1

18.45 The number of cars passing each of three different intersections in Crossroads City between 5:00 P.M. and 5:30 P.M. is given in the following table for randomly selected

INTERSECTION 1	INTERSECTION 2	INTERSECTION 3
440	480	433
420	392	406
530	386	427
401	456	338
454	427	397

days. Fifteen days were randomly selected and then the amount of traffic was recorded for 5 of the 15 days at each intersection. Test the null hypothesis that there is no difference in the amount of traffic at each intersection between 5:00 P.M. and 5:30 P.M. Use a 10% significance level.

18.46 A manager believes that the higher-salaried employees in a certain company are more satisfied with their job than are the lower-salaried employees. A sample of ten employees from each of the salary levels indicated by the following table was taken. Is there a significant difference in the satisfaction level, measured on a scale of 1 to 10 (10 being a perfectly satisfied employee), for the three groups? Use a significance level of 5%.

$25,000 TO $40,000	$40,001 TO $60,000	OVER $60,000	$25,000 TO $40,000	$40,001 TO $60,000	OVER $60,000
4	7	8	9	3	9
3	8	7	1	4	10
7	6	6	8	9	3
6	7	7	7	6	8
5	9	5	6	7	7

18.47 For the data in Exercise 11.1, use the Kruskal-Wallis test statistic to test that there is no difference in the amount of wear on three different designs of rubber soles. Use a significance level of .05. Compare with the results obtained from the ANOVA procedure.

18.48 In Exercise 11.7, can you conclude that there is a significant difference in the monthly sales of the three salespeople using the Kruskal-Wallis test statistic? Use a 5% significance level. When would you prefer the Kruskal-Wallis test to the usual ANOVA procedure?

18.49 The vice president of quality assurance at an airline company is interested in whether its three quality engineers are usually in agreement on the ratings they give to different airplane seating designs. Seven different designs are chosen at random and the quality engineers are asked to rate the comfort to the passengers on a scale from one to ten, with ten representing the highest level of comfort possible. Do the given data indicate that there is a difference in the ratings of the three quality engineers? Use a .10 significance level.

DESIGN	QUALITY ENGINEER 1	QUALITY ENGINEER 2	QUALITY ENGINEER 3
1	5	7	8
2	4	3	5
3	6	5	4
4	9	7	8
5	5	7	4
6	8	7	6
7	9	6	8

18.50 The manager at a manufacturing plant is interested in whether there is a significant difference in the number of times four machines need to be readjusted after going out of control. Eight months are randomly selected and the number of times the machines are readjusted per month is recorded. Do the data support the conclusion that some machines need more adjusting than others? Use a .05 significance level.

MONTH	MACHINE 1	MACHINE 2	MACHINE 3	MACHINE 4
1	5	6	3	6
2	4	3	7	5
3	4	5	6	3
4	5	3	4	7
5	4	4	5	6
6	10	11	9	11
7	15	18	13	16
8	3	4	5	4

18.51 The manager of a management information systems department wishes to know if there is a preference in the level of color and graphics used in the computer-aided instruction courses that are available for the programmer. Eight programmers are asked to rank the three different levels of color and graphics, with the rank of 1 representing the most desirable level. Do the data support the conclusion that not all three levels of color and graphics are equally preferred? Use a .05 significance level.

PROGRAMMER	LEVEL 1	LEVEL 2	LEVEL 3
1	1	3	2
2	2	1	3
3	1	2	3
4	1	3	2
5	1	3	2
6	1	2	3
7	2	1	3
8	1	3	2

18.52 The vice president of a chain of convenience stores is interested in whether the sale of three brands of a certain product differ significantly. Convenience stores are chosen at random and a week is randomly selected to record the sales of each of the three brands of the products. Do the data indicate that not all brands are equally preferred? Use a .10 significance level.

STORE	BRAND A	BRAND B	BRAND C
1	25	38	41
2	20	31	35
3	15	19	13
4	10	8	12
5	49	51	60
6	26	23	29
7	37	33	35
8	7	10	15
9	12	10	14

18.53 A property manager wished to know if there was a significant difference in the bids of three independent contractors. The property manager recorded the bids by the contractors on seven randomly selected jobs. Do the data support the conclusion that the bids are significantly different for the contractors? Use a .05 significance level.

JOB	CONTRACTOR 1	CONTRACTOR 2	CONTRACTOR 3
1	6500	7310	7400
2	5130	4950	5800
3	2500	2300	2650
4	2900	2800	2675
5	7800	8000	8300
6	4650	4500	4725
7	1250	1050	1400

18.54 Seven financial advisors were randomly selected and asked to rank their choice of three investments for the next year: bonds, blue-chip stocks, and small company stocks. Do the data given below support the conclusion that not all three investments are expected to perform equally as well? Use a .05 significance level.

FINANCIAL ADVISOR	BONDS	BLUE-CHIP STOCKS	SMALL COMPANY STOCKS
1	2	1	3
2	3	1	2
3	1	2	3
4	2	3	1
5	2	1	3
6	3	1	2
7	2	1	3

18.4

A MEASURE OF ASSOCIATION: SPEARMAN'S RANK CORRELATION

Whenever you encounter data describing two variables (say, X and Y), one measure of interest is the degree of **association** between X and Y. Are large values of X associated with large values of Y (a *positive* relationship)? Or do you observe smaller values of Y with larger values of X (a *negative* relationship)? Another possibility is that no relationship is observed between these two variables.

Consider the following data, in which a sample of ten families is used to determine the relationship (if any) that exists between X = market value of the family's home and Y = their total indebtedness (excluding the home mortgage; in thousands of dollars). Included in Y are any charge accounts, automobile loans, and other current liabilities.

FAMILY	X (MARKET VALUE OF HOME)	Y (TOTAL INDEBTEDNESS)
1	85	12
2	147	27
3	340	45
4	94	10
5	120	17
6	105	4
7	135	20
8	162	25
9	480	35
10	88	14

The president of Metro Savings and Loan believes that larger home values are associated with larger indebtedness; that is, a positive relationship exists between these two variables. This is confirmed in the scatter plot contained in Figure 18.9.

One method of measuring the association between two variables is the Pearson product moment correlation, r, introduced in Chapter 13. The equation

FIGURE 18.9

MINITAB scatter plot of home value versus debt.

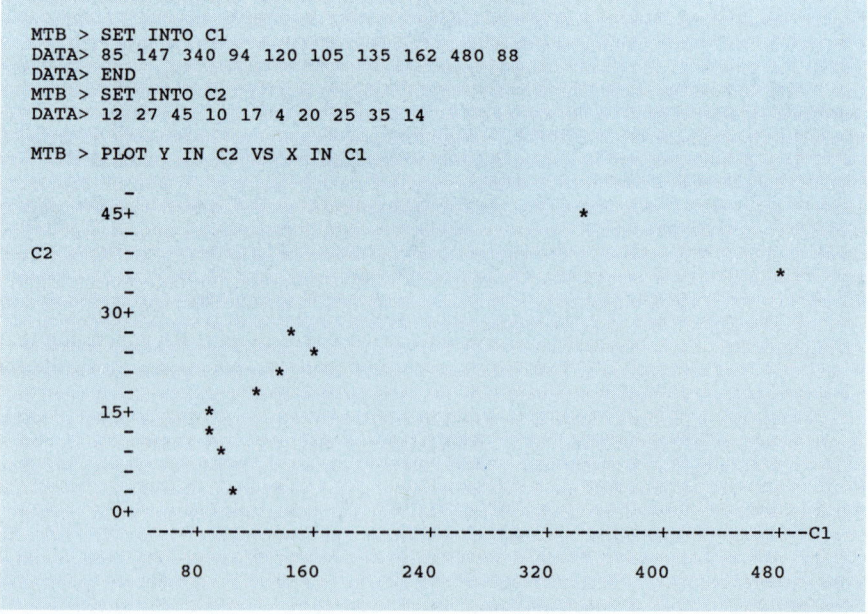

used to determine this measure is

$$r = \frac{\Sigma\, xy - (\Sigma\, x)(\Sigma\, y)/n}{\sqrt{\Sigma\, x^2 - (\Sigma\, x)^2/n}\sqrt{\Sigma\, y^2 - (\Sigma\, y)^2/n}} \qquad (18.15)$$

where n represents the number of observations (pairs).

The value of r, often called the *sample correlation coefficient*, measures the amount of linearity that exists between the sample values of X and Y. It is used to measure ρ (rho), the *population* correlation coefficient. The value of ρ can be thought of as the correlation between *all* possible X, Y pairs, not just those contained in the sample.

In the previous discussions, a significant relationship between X and Y existed if we were able to reject H_0: $\rho = 0$. The test statistic here was a t-statistic, so this test assumes a normal distribution for the X, Y variables.

An alternative to this procedure is a measure of association derived from the *ranks* of the X and Y variables. This nonparametric measure does *not* assume a normal distribution; it assumes only that the values within the X and Y samples can be ranked. For data such as the home price versus debt values, each of the X and Y values also are replaced by their ranks. If we use these ranks in place of the actual data in equation 18.15, we obtain another measure of association, called the **Spearman rank correlation coefficient, r_s**:

$$r_s = \frac{\Sigma\, R(x)R(y) - [\Sigma\, R(x)][\Sigma\, R(y)]/n}{\sqrt{\Sigma\, R^2(x) - [\Sigma\, R(x)]^2/n}\sqrt{\Sigma\, R^2(y) - [\Sigma\, R(y)]^2/n}} \qquad (18.16)$$

where $R(x)$ = rank of the X observation and $R(y)$ = rank of the Y observation.

If there are no ties, a second formula provides a much easier method of finding r_s. If there are a few ties, this still serves as a very good approximation to r_s. The shortcut method of finding r_s is

$$r_s = 1 - \frac{6\Sigma\, d^2}{n(n^2 - 1)} \qquad (18.17)$$

where, for each observation, d is the difference between the X and Y ranks; that is, $d = R(x) - R(y)$.

The Pearson product moment correlation, r, measures the amount of the *linear* relationship between X and Y and ranges from -1 to 1. Also, $r = 1$ or -1 only if all of the points fall exactly on a straight line. Similarly, the range for the Spearman rank correlation is

$$-1 \leq r_s \leq 1$$

One difference here is that r_s will equal 1 provided Y increases every time X does in the sample observations. This rate of increase need not be linear. This is illustrated in Figure 18.10 for a sample of five observations that do not lie on a straight line. Consequently, r is less than 1 but, as the following table shows, r_s does equal 1.

FIGURE 18.10

Measure of association.
Pearson product
moment correlation, r,
and Spearman rank
correlation, r_s.

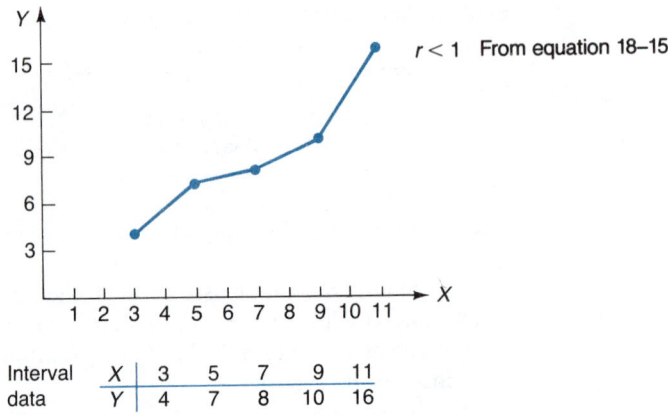

Interval data	X	3	5	7	9	11
	Y	4	7	8	10	16

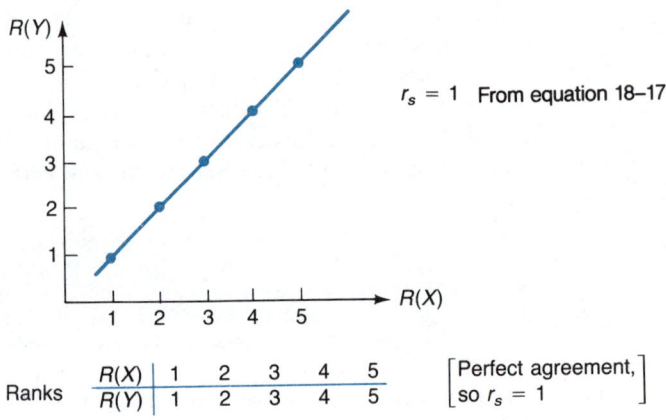

Ranks	$R(X)$	1	2	3	4	5
	$R(Y)$	1	2	3	4	5

$\begin{bmatrix} \text{Perfect agreement,} \\ \text{so } r_s = 1 \end{bmatrix}$

X	RANK $R(X)$	Y	RANK $R(Y)$	DIFFERENCE OF RANKS (d)	d^2
3	1	4	1	$1 - 1 = 0$	0
5	2	7	2	$2 - 2 = 0$	0
7	3	8	3	$3 - 3 = 0$	0
9	4	10	4	$4 - 4 = 0$	0
11	5	16	5	$5 - 5 = 0$	0
					$\Sigma d^2 = \overline{0}$

You can also see this in Figure 18.10, where the pairwise ranks *are* perfectly linear, which is why $r_s = 1$.

$$r_s = 1 - \frac{6\Sigma d^2}{n(n^2 - 1)}$$

$$= 1 - \frac{(6)(0)}{(5)(24)} = 1$$

When it is possible to calculate both, the values of r and r_s are generally not the same, although they are usually quite close. They will have the same sign. A positive value indicates a *positive* relationship between the two variables. Similarly, if r_s and r are negative, then a *negative* relationship exists (Y decreases as X increases). If the values of r and r_s are nearly the same, this usually indicates high linearity between the two variables.

As we discussed earlier, nonparametric methods are well suited for situations in which (1) the data are *ordinal* or (2) the distribution of the population(s) from which the data are obtained is suspected to be nonnormal. When using data of the ordinal type, it is no longer appropriate to use the Pearson coefficient of correlation (r) for performing a t-test to determine whether a significant linear relationship exists. This procedure requires interval or ratio data. Similarly, when dealing with nonnormal populations, the t-test is no longer valid because the normality assumption is a vital part of this procedure.

Suppose that Mr. Roberts and Mr. Clauson each evaluated eight brands of television sets by inspecting eight different sets. After inspecting the sets, rather than assigning each of them a score of some kind, they merely ranked them from 1 (best) to 8 (worst). The results were

BRAND	X (ROBERTS)	Y (CLAUSON)	DIFFERENCE
A	1	2	−1
B	4	3	1
C	2	1	1
D	6	6	0
E	8	7	1
F	3	5	−2
G	7	8	−1
H	5	4	1

Notice that each set of rankings consists of ordinal data because the only meaningful information contained within these values is the *order*, not the difference between them. For example, Mr. Robert's three favorite brands were brands A (first), C (second), and F (third). We have no way of knowing whether brand A was much better than brand C or only slightly better; the same is true for brand C versus brand F. Consequently, there is no way of knowing if the distances between ranks 1 and 2 and between ranks 2 and 3 are the same, and so these differences are meaningless.

A measure of how well the two people agree is the Spearman rank correlation, r_s. The larger this value is, the more agreement there is between the two sets of rankings. A value of $r_s = 1$ would indicate perfect agreement, whereas perfect disagreement would result in a value of $r_s = -1$. Here the rank correlation is

$$r_s = 1 - \frac{6\Sigma\, d^2}{n(n^2 - 1)}$$

$$= 1 - \frac{6[(-1)^2 + (1)^2 + \cdots + (-1)^2 + (1)^2]}{8(64 - 1)}$$

$$= 1 - \frac{(6)(10)}{(8)(63)} = .881$$

This value appears to be quite large, although we will need a formal testing procedure to determine whether there is significant agreement between the two testers (discussed next).

EXAMPLE 18.10   Refer to the table for the home values and debt data. The president of Metro Savings and Loan would like a measure of association between these two variables. Based on past experience, he is reluctant to assume that the home values are normally distributed; they usually are skewed right. There generally are enough homes with an extremely large market value (two, here) to produce a

FIGURE 18.11

MINITAB procedure for
finding the Spearman
rank correlation
(Example 18.10).

```
MTB > SET INTO C1
DATA> 85 147 340 94 120 105 135 162 480 88
DATA> END
MTB > SET INTO C2
DATA> 12 27 45 10 17 4 20 25 35 14
MTB > RANK C1, PUT INTO C11
MTB > RANK C2, PUT INTO C12
MTB > CORRELATION OF C11 AND C12          $r_s$

Correlation of C11 and C12 = 0.867
```

skewed distribution (see Figure 18.9). He asks you to use the Spearman measure
of correlation and determine the value of r_s.

SOLUTION

The ranks are calculated as follows:

FAMILY	X	RANK $R(X)$	Y	RANK $R(Y)$	DIFFERENCE (d)	d^2
1	85	(1)	12	(3)	-2	4
2	147	(7)	27	(8)	-1	1
3	340	(9)	45	(10)	-1	1
4	94	(3)	10	(2)	1	1
5	120	(5)	17	(5)	0	0
6	105	(4)	4	(1)	3	9
7	135	(6)	20	(6)	0	0
8	162	(8)	25	(7)	1	1
9	480	(10)	35	(9)	1	1
10	88	(2)	14	(4)	-2	4

$$\Sigma d^2 = 22$$

Then,

$$r_s = 1 - \frac{(6)(22)}{(10)(99)} = .867$$

Based on this value, it appears that a significant positive relationship exists
between the market value of a family's home and their total debts.

To determine whether a derived value of r_s is "large enough" to support
a conclusion, as in Example 18.10, we develop a test of hypothesis that uses
the rank correlation, r_s, as the test statistic.

To obtain a computer solution for Spearman's rank correlation, you ask
MINITAB to rank the X and Y values and to compute r for the ranks. Recall
that equation 18.17 was a shortcut for determining the rank correlation. This
is illustrated in Figure 18.11, where, as before, $r_s = .867$.   ■

A Test of Hypothesis Using the Rank Correlation

In Chapter 13, we used the Pearson correlation coefficient, r, to determine
whether a linear relationship existed between two variables, X and Y. This
relationship could be either positive (Y increases as X increases) or negative
(Y decreases as X increases). When using Spearman's rank correlation, r_s, we
drop the "linear" term in the hypotheses and test

H_0: no association exists between the X and Y variables

H_a: association does exist between the X and Y variables

You can also perform one-sided tests, as summarized in the accompanying
box. In this way, you can test for a significant positive or negative relationship
between the two variables.

THE SPEARMAN TEST FOR RANK CORRELATION

TWO-SIDED TEST	ONE-SIDED TEST	
H_0: no association exists between X and Y	H_0: no association exists between X and Y	H_0: no association exists between X and Y
H_a: association does exist between X and Y	H_a: a positive relationship exists between X and Y	H_a: a negative relationship exists between X and Y
Assumption: Sample values for each variable can be ranked.		
Procedure: Determine the value from Table A.12 using the sample size, n, and the column corresponding to $\alpha/2$.	Use the column in Table A.12 corresponding to α.	Use the column in Table A.12 corresponding to α.
Reject H_0 if $|r_s| >$ (table value).	Reject H_0 if $r_s >$ (table value).	Reject H_0 if $r_s <$ $-$(table value).

EXAMPLE 18.11

In the television-ranking example, is there sufficient evidence to indicate that there was general agreement between the rankings made by Mr. Roberts and those made by Mr. Clauson? Use $\alpha = .05$.

SOLUTION

The appropriate hypotheses here are

H_0: no association exists between the two ranks

H_a: a positive association exists between the two ranks

According to Table A.12, using $\alpha = .05$ and $n = 8$, the testing procedure is to

reject H_0 if $r_s > .643$

Because the computed value of r_s is .881 (as we derived previously), we reject H_0 and conclude that there *was* significant agreement between the two sets of rankings.

The p-value for this result can be obtained by looking across the row in Table A.12 corresponding to $n = 8$. Here we find that .881 corresponds to $\alpha = .005$. Consequently, the p-value here is .005. ■

EXAMPLE 18.12

The president of Metro Savings and Loan is attempting to demonstrate that a positive relationship exists between $X =$ market value of a family's home and $Y =$ their total indebtedness (excluding the home mortgage). Using the results of Example 18.10, is there sufficient evidence of a positive relationship between these two variables? Use $\alpha = .05$.

SOLUTION

In Example 18.10, the sample rank correlation was found to be $r_s = .867$. Using Table A.12 for $\alpha = .05$ and $n = 10$, we test

H_0: no association exists between the home market value and total indebtedness

H_a: a positive relationship exists

using

$$\text{reject } H_0 \text{ if } r_s > .564$$

The computed value of .867 exceeds the table value, so we reject H_0 and conclude that there is a tendency for larger values of $X = $ home value and $Y = $ family indebtedness to be paired together. This large value of r_s also is off the right side of Table A.12, indicating that for this test the p-value is $< .005$. This extremely small value is strong evidence of a positive relationship between these two variables. ■

EXERCISES

18.55 A market stand sells watermelons each week at different prices, depending on the supply of watermelons. Calculate the Pearson product moment correlation and the Spearman rank correlation of the quantity sold and the price.

PRICE	QUANTITY SOLD	PRICE	QUANTITY SOLD
1.80	53	2.50	46
2.00	45	2.10	46
1.50	60	1.50	70
1.25	75	1.75	65
1.75	50	1.90	47
2.25	48		

18.56 The rank correlation coefficient between 15 pairs of observations is .31. Using a significance level of .05, test the null hypothesis that there is no positive relationship between the two variables sampled.

18.57 The following data represent the high temperature of the day (°F) and the number of sno-cones sold at Dairy Freeze. Using a nonparametric procedure and a 5% significance level, test the hypothesis that there is a positive relationship between the two variables.

DAILY HIGH TEMPERATURE	NUMBER OF SNO-CONES SOLD	DAILY HIGH TEMPERATURE	NUMBER OF SNO-CONES SOLD
90	49	93	55
91	48	92	51
86	40	90	52
85	38	88	46
84	40		

18.58 A factory wants to know what the relationship is between the age of its machines and the number of breakdowns per year. Ten machines were selected at random and the following table was constructed. Using the Spearman test for rank correlation, is there a relationship between the age of the machine and the number of breakdowns per year? Use a 10% significance level.

AGE (YEARS)	BREAKDOWNS PER YEAR	AGE (YEARS)	BREAKDOWNS PER YEAR
2	6	7	20
4	10	5	16
5	12	6	15
8	24	3	10
6	17	4	12

18.59 A physician would like to know the relationship between a person's diastolic blood pressure and the average number of hours spent exercising each week. Twenty people of age 30 years were selected randomly. Is there an indication from the data that there may be a negative relationship between exercise and blood pressure? Test with the Spearman test for rank correlation and use a 1% significance level.

DIASTOLIC BLOOD PRESSURE	HOURS EXERCISED	DIASTOLIC BLOOD PRESSURE	HOURS EXERCISED
74	9	80	8
70	10	64	18
62	16	56	20
58	15	68	10
82	6	72	11
84	3	78	8
90	0	84	7
84	4	88	4
72	12	70	12
70	9	66	15

18.60 Two taste testers were asked to rank ten beers in order of taste preference, with the best tasting beers receiving the highest ranks.

BEER	TASTE TESTER 1	TASTE TESTER 2	BEER	TASTE TESTER 1	TASTE TESTER 2
1	5	7	6	7	5
2	3	3	7	2	4
3	6	6	8	10	9
4	1	2	9	4	1
5	8	10	10	9	8

Using a .05 significance level, is there a significant agreement between the first beer tester's rankings and the second beer tester's ranking?

18.61 The manager of Sales Unlimited wanted to know whether there was a significant positive relationship between a salesperson's travel expenses (in thousands of dollars) and his or her sales (in tens of thousands of dollars). Using the following data, test that there is a positive relationship at the .05 significance level.

TRAVEL EXPENSES	SALES	TRAVEL EXPENSES	SALES
1.5	3.4	2.5	4.1
2.0	3.9	2.2	3.6
3.5	6.1	3.8	5.7
1.6	2.5	4.1	4.3
1.8	3.4	2.6	4.6

18.62 A supervisor at a computer firm wanted to determine whether a relationship existed between the level of satisfaction (measured from 1 to 10) of an employee with his or her job at the firm and the number of years the employee has been employed with the firm. Test the null hypothesis that there is no relationship and use a 10% significance level.

LEVEL OF SATISFACTION	NUMBER OF YEARS EMPLOYED AT THE FIRM	LEVEL OF SATISFACTION	NUMBER OF YEARS EMPLOYED AT THE FIRM
6	1	3	2
3	5	9	10
7	10	7	7
5	12	8	12
8	6	7	16
4	5		

18.63 The management of a firm wishes to know whether there is a positive relationship

TIME ON MARKET (YEARS)	PERCENT OF MARKET	TIME ON MARKET (YEARS)	PERCENT OF MARKET
1	1.2	6	3.9
2	2.6	7	3.2
3	1.8	8	4.1
4	2.7	9	3.8
5	2.9	10	4.6

between the length of time a certain product has been on the market and the percent of market that the product has captured. Do the data indicate that a positive relationship exists? Use a 5% significance level.

18.64   The owner of a used car lot believes that there is a negative relationship between the number of cars sold monthly and the average monthly interest rate used for financing the cars. Twelve months were randomly selected to obtain the following data. Use a 5% significance level to determine whether the owner's belief is correct.

NUMBER OF CARS SOLD MONTHLY	AVERAGE MONTHLY INTEREST RATE (%)	NUMBER OF CARS SOLD MONTHLY	AVERAGE MONTHLY INTEREST RATE (%)
35	12	50	11
25	15	20	15
28	16	35	11.5
31	12	42	12.5
40	11.5	46	12
48	11	28	15

18.65   The following ranks are given for the most popular television commercials of 1986 and 1987. At the .10 significance level, do the data indicate that a relationship exists between the rankings?

1987 RANK	1986 RANK	TELEVISION COMMERCIALS
1	3	California Raisin
2	7	Bud Light
3	5	Pepsi/Diet Pepsi
4	4	Miller Lite
5	2	McDonald's
6	6	Bartles & Jaymes
7	1	Coca-Cola

(*Source: Wall Street Journal*, March 3, 1988, p. 19.)

SUMMARY

A key step in applying any statistical technique correctly is to make sure that it is appropriate for the type of data that is involved. For example, performing a *t*-test using a small sample containing *ordinal* data (such as a set of consumer rankings) is never correct. For situations in which your data are ordinal or from populations that you suspect are nonnormally distributed, a **nonparametric technique** is often preferable. This chapter has introduced some (certainly not all) of the more popular nonparametric procedures.

The **runs test** examines a sequence containing an arrangement of two symbols (M or F, yes or no, + or −, and so on) to determine whether the sequence was generated in a random manner. We defined tests for both small samples (using Table A.15) and large samples (using a test statistic having an approximate normal distribution and Table A.4).

The **Mann-Whitney *U* test** is a nonparametric procedure for determining if two populations differ in location using two independent samples. Unlike its counterpart, the *t*-test, this test does not require that the populations be normally distributed. By combining (pooling) the samples and finding the ranks of the combined sample, you can calculate a value of the test statistic. This method can be applied to both small samples (using Table A.10) and large samples (using an approximate normal distribution and Table A.4).

The **Wilcoxon signed rank test** is a nonparametric procedure used for determining if the population of differences is centered at zero when dealing with two dependent (paired) samples. The Wilcoxon technique determines the differences

FIGURE 18.12
Summary of
nonparametric tests
of central tendency.

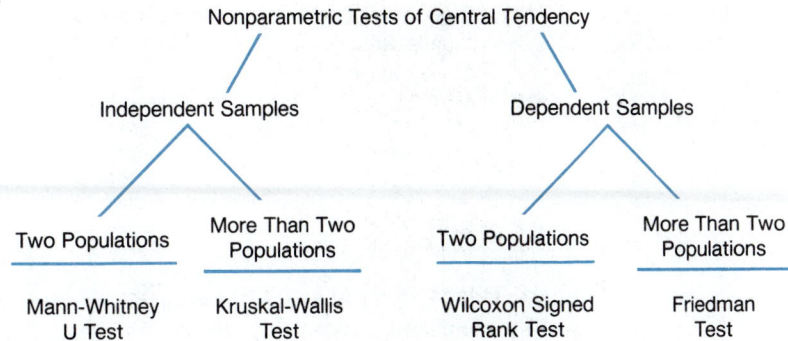

of the paired observations and then calculates a value of the test statistic using the ranks of these differences. Both small samples (using Table A.11) and large samples (using an approximate normal distribution and Table A.4) can be used.

The **Kruskal-Wallis test** is an extension of the Mann-Whitney test. It is used to test if two or more populations differ in location when using independent samples. As in the Mann-Whitney procedure, the samples are pooled and then ranked from smallest to largest. The resulting ranks are then used to define a test statistic. This statistic has an approximate chi-square distribution (tabulated in Table A.6), even for fairly small sample sizes.

A nonparametric alternative to the randomized block technique (discussed in Chapter 11) is the **Friedman test**. The Friedman test does not require normal populations with equal variances, unlike the randomized block procedure. Values are ranked within each block and then summed for the various populations under consideration. The Friedman test statistic is calculated using these sums and has an approximate chi-square distribution.

These tests are summarized in Figure 18.12.

The final nonparametric technique, the **Spearman rank correlation**, allows you to measure the *association* between sample values on two variables that consist of ordinal data. If the data are of the interval or ratio type, they can be converted to ordinal data by using the ranks of these values. The Spearman rank correlation coefficient, denoted as r_s, is a nonparametric measure of association between the observations of these two variables. This statistic is computed using the ranks of the observations and can be used to test for a significant relationship between the two variables.

REVIEW EXERCISES

18.66 Using the runs test, is there evidence that the following sequence of A's and B's is not randomly generated? Use a 10% significance level.

A A B A B B A A B B B A A B A B A A B B A A B A B A A A A B

18.67 A radio station requests that people telephone the station to express their opinion on the new property tax that the city is levying. An F represents a person telephoning in who is for the tax, and an A represents one against it. Test whether the following sequence of people telephoning the radio station is nonrandomly generated. Use a 10% significance level.

F A F A A A F A A F A A A F F A F A F A F A A A F A A

18.68 The manager of a small-town savings and loan association is interested in finding out whether there is a relationship between the average monthly balance of a savings account and the age of the savings account. Fifteen accounts were selected at random. Do the data indicate a relationship at the .05 significance level?

AVERAGE MONTHLY BALANCE	AGE OF ACCOUNT (YEARS)	AVERAGE MONTHLY BALANCE	AGE OF ACCOUNT (YEARS)
2510	1.5	6148	3.8
3612	2.6	5134	4.7
5634	3.5	2614	1.1
3698	1.8	2581	1.9
3978	2.1	2501	0.5
6751	4.3	3986	4.2
5869	10.1	6645	4.1
		3582	2.3

18.69 Explain the differences in the assumptions necessary to perform parametric tests and nonparametric tests on two population means.

18.70 A statistician wants to determine whether two populations differ in location. After collecting a sample of size eight from each population, the statistician finds that the sum of the ranks of the observations in population 1 is 60. Is there evidence to indicate that there is a difference in location for the two populations? Use a 5% significance level.

18.71 An economist believes that the cost of a typical basket of goods bought by a family of four costs more in Atlanta, Georgia, than it does in Houston, Texas. Seven grocery stores were randomly selected in each of the two cities to collect the following data (cost of basket of goods, in dollars). Test that there is no difference in the costs of the basket of goods for the two cities. Use a 10% significance level.

ATLANTA	HOUSTON
30	37
33	36
41	39
43	40
37	32
39	36
41	32

18.72 Jeff's Auto Parts store special orders parts from either warehouse A or warehouse B. Warehouse A claims that it delivers parts faster than any other warehouse. Jeff special ordered 18 auto parts—9 from warehouse A and 9 from warehouse B—and noted the number of days it took to receive them. Using the following data, test warehouse A's claim. Use a 5% significance level.

WAREHOUSE A	WAREHOUSE B	WAREHOUSE A	WAREHOUSE B
2.50	3.00	2.25	2.00
2.25	3.25	3.00	2.25
3.50	2.50	2.50	2.75
2.00	2.75	2.50	2.25
2.75	3.00		

18.73 Vulcan Construction believes that a minicourse in safety to increase the employees' awareness of potential accidents would reduce the incidence of on-the-job accidents. The following data represent the number of accidents reported at 15 construction sites for the month before and the month after the workers attended the minicourse. Do the data indicate that the minicourse reduced accidents? Use a 5% significance level.

BEFORE	AFTER	BEFORE	AFTER
4	3	11	6
5	6	7	7
7	6	8	5
8	5	5	3
7	8	10	11
10	3	9	4
12	2	11	9
9	8		

18.74 A cooking contest was conducted; two top chefs baked chicken and then asked 12 tasters to judge the quality of the cooking on a scale from 1 to 10, 10 being the highest score. Test the null hypothesis that there was no difference between the taster's judgment of the quality of cooking for the two chefs at a significance level of .05.

TASTER	CHEF A	CHEF B	TASTER	CHEF A	CHEF B
1	8	7	7	9	9
2	6	10	8	10	7
3	5	9	9	9	10
4	10	4	10	8	6
5	5	8	11	4	9
6	3	7	12	7	3

18.75 Paint A, paint B, and paint C were painted on metallic surfaces and then subjected to high temperatures. Nine replications of the experiment were made. A measure of the cohesiveness of the paint was then taken. The following coded data represent the cohesiveness of the individual paints. Test the hypothesis that there was no difference in the cohesiveness for the three paints. Use a 10% significance level.

PAINT A	PAINT B	PAINT C	PAINT A	PAINT B	PAINT C
1.3	2.1	3.4	1.6	1.9	2.5
1.6	2.7	2.8	2.6	2.3	2.4
3.1	1.6	1.9	2.7	1.8	1.9
2.6	1.9	2.0	1.9	1.6	1.7
4.3	1.6	2.8			

18.76 A chemist was interested in knowing whether three different drugs used for insomnia were equally effective. Three groups of mice, with ten mice to a group, were used. Each group of mice was given the adult-equivalent dosage for one of the three drugs. The time in minutes it took for the mice to fall asleep was recorded. Using the following data, test the null hypothesis that each drug is equally effective in reducing sleep latency. Use a significance level of .05.

DRUG 1	DRUG 2	DRUG 3	DRUG 1	DRUG 2	DRUG 3
32	38	31	41	35	30
35	37	33	28	39	28
40	42	29	34	40	33
30	44	34	39	41	37
33	37	31	28	42	35

18.77 An automobile worker would like to know whether there is any difference in the comfort of three different cars. Three groups of five drivers were selected to judge the riding comfort of the three cars; one of the three cars was assigned to each group. The drivers rated the comfort of each car on a scale from 1 to 5, 5 being the most comfortable. Test that there is no difference in the comfort of the three cars from these data. Use a .05 significance level.

CAR 1	CAR 2	CAR 3
2	5	4
4	3	1
4	4	2
3	3	5
5	2	4

18.78 A microbiologist wants to know whether there is any difference in the time in hours it takes to make yogurt from three different starters. Seven batches of yogurt were made with each of the starters. The following data give the times it took to make each batch. At the .05 significance level, do the data indicate a difference in the time it takes to make yogurt from the three starters?

LACTOBACILLUS ACIDOPHILUS	BULGARIUS	MIXTURE OF ACIDOPHILUS AND BULGARIUS
6.8	6.1	7.3
6.3	6.4	6.1
7.4	5.7	6.4
6.1	6.5	7.2
8.2	6.9	7.4
7.3	6.3	6.5
6.9	6.7	6.8

18.79   A company uses four advertising methods to increase sales: newspaper, mailers, television, and radio. Four 6-month periods were randomly selected, and sales (in thousands of dollars) were recorded monthly for one of the 6-month periods for each advertising method. Do the following data indicate a difference in sales promotion for the four advertising methods? Use a .05 significance level.

NEWSPAPER	MAILERS	TELEVISION	RADIO
2.1	4.1	5.1	4.2
5.6	2.6	4.2	4.3
6.2	2.1	3.7	4.0
3.4	3.3	2.0	3.1
3.1	3.0	3.6	3.8
4.1	3.1	3.1	3.7

18.80   The following data represent the years of job-related experience and last year's salary (in thousands of dollars) for ten randomly selected realtors. Do the data indicate a positive relationship? Use a 5% significance level.

YEARS OF EXPERIENCE	ANNUAL SALARY	YEARS OF EXPERIENCE	ANNUAL SALARY
13.2	42.5	7.8	38.4
10.1	36.8	5.4	29.6
4.6	15.9	3.8	21.6
5.7	27.6	9.6	33.7
6.7	34.3	8.4	35.3

18.81   The owner of a convenience food store is interested in whether there is a positive relationship between the number of cars that pass the store weekly (given in thousands) and the weekly sales of the store (in thousands of dollars). Test the hypothesis that there is a positive relationship using a 5% significance level.

WEEK	NUMBER OF CARS	SALES	WEEK	NUMBER OF CARS	SALES
1	24.6	1.3	6	24.6	1.2
2	29.7	2.6	7	32.7	2.9
3	22.6	1.4	8	35.1	3.4
4	30.4	2.8	9	29.9	3.9
5	20.1	1.1			

18.82   A cable television company is interested in whether there is a difference in the preference of three movie channels that households would like to have. Listed are rankings from 1 to 3 that each household rated three movie channels—channel 22, channel 29, and channel 32. A ranking of one indicates the most desirable channel. Do the

HOUSEHOLD	CHANNEL 22	CHANNEL 29	CHANNEL 32	HOUSEHOLD	CHANNEL 22	CHANNEL 29	CHANNEL 32
1	1	3	2	6	1	3	2
2	3	2	1	7	3	1	2
3	1	2	3	8	2	3	1
4	2	3	1	9	1	3	2
5	3	2	1	10	3	2	1

data provide sufficient evidence to indicate that there is a difference in the preference of these channels by household? Use a .10 significance level.

18.83   Lifetime Exterior sells three types of siding: vinyl, aluminum, and steel. The manager of Lifetime Exterior is interested in whether one type of siding is sold more than another. Twelve salespersons are randomly selected and the number of sales of each type of siding to homes is recorded over a 6-month period. From the data, can the manager conclude that not all types of siding sell equally as well? Use a .05 significance level.

SALESPERSON	VINYL	ALUMINUM	STEEL	SALESPERSON	VINYL	ALUMINUM	STEEL
1	9	5	8	7	8	14	10
2	6	10	11	8	9	8	3
3	13	9	7	9	6	9	5
4	11	10	6	10	11	15	17
5	15	12	11	11	25	21	15
6	21	22	14	12	10	19	20

18.84   The composite index of 12 leading economic indicators is given for each month from January 1985 to November 1987. A regression equation, with the time variable t as the independent variable ($t = 1, 2, \ldots, 35$), is found to be $\hat{y} = 163.011 + .88098t$. Is there sufficient evidence that the sequence of positive and negative residuals is not random with respect to time? Use a .01 significance level.

YEAR	JAN.	FEB.	MAR.	APR.	MAY	JUN.	JUL.	AUG.	SEP.	OCT.	NOV.	DEC.
1987	185.4	185.8	187.5	187.8	188.9	190.8	191.6	192.7	192.7	193.3	190.0	—
1986	174.1	175.0	176.4	178.2	178.6	178.4	179.7	180.0	179.7	180.9	182.5	186.8
1985	165.5	166.5	167.2	165.9	166.9	167.3	168.5	169.3	170.2	171.2	171.1	174.0

(*Source: Standard and Poor's Statistical Service, Current Statistics,* April 1988, p. 7.)

18.85   The 1987 recruiting fees, in units of millions, are given for the biggest U.S.-based recruiting firms. The 1987 recruiting fee is given for both worldwide recruiting and recruiting only in the United States. The ranks both worldwide and in the United States are given. Is there sufficient evidence of a positive relationship between the fees worldwide and in the United States for the biggest U.S.-based recruiting firms? Use a .01 significance level.

COMPANY	1987 RECRUITING FEES			
	WORLDWIDE		U.S.	
	in millions	rank	in millions	rank
Korn/Ferry International[2]	$68.4	1	$45.2	1
Russell Reynolds Associates	$67.7	2	$44.0	2
Spencer Stuart	$56.0	3	$26.2	4
Heidrick & Struggles[3]	$38.9	4	$27.8	3
Ward Howell International[4]	$35.4	5	$12.0	6
Boyden International[5]	$26.8	6	$13.1	5
Peat Marwick Main & Co.	$18.2[6]	7	$7.4	9
Paul R. Ray & Co.	$12.5	8	$11.8	7
Nordeman Grimm	N.A.[7]	9	$9.6	8
Handy Associates	$7.1	10	$7.1	10
Lamalie Associates	$6.4	11	$6.4	11
Kearney Executive Search	$5.2	12	$4.1	14
Witt Associates	$4.9	13	$4.9	12
Higdon Joys & Mingle	$4.4	14	$4.4	13

(*Source:* "How the Top Recruiters Really Rank," *Fortune,* May 9, 1988, p. 112.)

18.86   The vice president of a credit union wishes to determine the relationship between the number of automobile loan applications per month (Y) and the loan interest rate (X). Twenty months were randomly selected. The data over these months is printed in the MINITAB computer printout. Comment on the models used in the regression analysis given in the MINITAB printout. Which model appears to be more appropriate? Why? Comment on the significance of the rank correlation of the Y and X variables.

```
MTB > name c1 = 'Y'
MTB > NAME C2 = 'X'
MTB > PRINT C1 C2
 ROW      Y       X
   1     143     7.0
   2     144     7.1
   3     136     8.0
   4     135     8.5
   5     137     8.8
   6     131     9.0
   7     126     9.5
   8     127    10.0
   9     125    10.6
  10     116    11.8
  11     117    12.0
  12     113    12.4
  13     110    13.0
  14     104    13.8
  15      96    14.8
  16      95    15.0
  17      91    16.0
  18      88    16.2
  19      79    16.7
  20      75    17.0
MTB > REGRESS Y IN C1 ON 1 PREDICTOR IN C2, RESIDUALS IN C5

The regression equation is
y = 190 - 6.37 X

Predictor        Coef        Stdev       t-ratio
Constant      189.919        2.470         76.88
X             -6.3676        0.2009        -31.69

s = 2.914        R-sq = 98.2%      R-sq(adj) = 98.1%

Analysis of Variance

SOURCE        DF          SS            MS
Regression     1        8528.0        8528.0
Error         18         152.8           8.5
Total         19        8680.8

MTB > NAME C5 = 'RESID'
MTB > PRINT C5
RESID
 -0.87983   -0.26538   -1.09030   -0.28820    1.12352   -0.57931   -1.22401
  0.26863    0.91094    0.42885    1.22942    0.71828    1.00997    0.69426
  0.11544    0.21458    1.09068    0.45702   -1.71673   -2.52136

MTB > RUNS ABOVE AND BELOW 0 USING C5

     RESID
     K =       0.0000

     THE OBSERVED NO. OF RUNS =    5
     THE EXPECTED NO. OF RUNS =   10.6000
     12 OBSERVATIONS ABOVE K     8 BELOW
               THE TEST IS SIGNIFICANT AT   0.0073

MTB > LET C3 = C2 * C2
MTB > NAME C3 = 'X*X'
MTB > REGRESS Y IN C1 USING 1 PREDICTOR IN C3, RESIDUALS IN C6

The regression equation is
y = 154 - 0.264 X*X

Predictor        Coef        Stdev       t-ratio
Constant      154.284        1.091        141.44
X*X         -0.263823        0.006402     -41.21

s = 2.249        R-sq = 99.0%      R-sq(adj) = 98.9%
```

```
Analysis of Variance

SOURCE        DF           SS           MS
Regression     1         8589.7       8589.7
Error         18           91.1          5.1
Total         19         8680.8

MTB > NAME C6 = 'RESID2'
MTB > PRINT C6
RESID2
   0.78555    1.43934   -0.65991   -0.10429    1.46990   -0.89202   -2.07377
  -0.41584    0.16503   -0.70701    0.32249   -0.32766    0.13812   -0.01895
  -0.23080    0.03571    2.03880    1.42479   -0.83880   -1.51430

MTB > RUNS ABOVE AND BELOW 0, USING C6

    RESID2
    K =      0.0000

    THE OBSERVED NO. OF RUNS =   12
    THE EXPECTED NO. OF RUNS =   10.9000
      9 OBSERVATIONS ABOVE K     11 BELOW
            THE TEST IS SIGNIFICANT AT   0.6096
            CANNOT REJECT AT ALPHA = 0.05

MTB > RANK C1, PUT INTO C10
MTB > RANK C2, PUT INTO C11
MTB > CORRELATION BETWEEN C10 AND C11

Correlation of C10 and C11 = -0.991
```

18.87 A real-estate agent was interested in whether there was a significance difference in the size of a family for three neighborhoods. The data collected on family size is given in C1 below and the neighborhoods are represented by 1, 2, or 3 in C2. Using the results from the Kruskal-Wallis test given in the following analysis, what can the real-estate agent conclude? Find the p-value.

```
MTB > PRINT C1 C2
ROW    C1   C2

  1     2    1
  2     4    1
  3     5    1
  4     3    1
  5     2    1
  6     4    1
  7     2    2
  8     1    2
  9     2    2
 10     3    2
 11     2    2
 12     2    2
 13     3    2
 14     1    2
 15     4    3
 16     5    3
 17     3    3
 18     2    3
 19     3    3
 20     5    3
 21     7    3
 22     3    3
 23     4    3
 24     2    3

MTB > KRUSKAL-WALLIS USING DATA IN C1, LEVELS IN C2

LEVEL     NOBS     MEDIAN    AVE. RANK    Z VALUE
    1        6      3.500       14.3        0.70
    2        8      2.000        7.0       -2.69
    3       10      3.500       15.9        1.96
OVERALL     24                  12.5

H = 7.452
H(ADJ. FOR TIES) = 7.913
```

COMPUTER EXERCISES USING THE DATABASE

Exercise 1—Appendix H

Choose ten observations at random from the database of households that own their home and another ten observations at random from households that rent their home. (Refer to variable OWNORENT.) Using the Mann-Whitney test, is there sufficient evidence to conclude that the income of the principal wage earner is significantly different for the households that own their home and for the households that rent their home? Use a .05 significance level.

Exercise 2—Appendix H

Choose 12 observations at random from the database of households that have a secondary income (INCOME 2) above $20,000. Choose another 12 observations from households that have a secondary income (INCOME 2) that is positive and less than $20,000. Also, choose a random sample of 12 observations from households with no secondary income. Can one conclude that there is a difference in the house payment or apartment/house rent (HPAYRENT) for the three groups? Use the Kruskal-Wallis test with a .05 significance level.

Exercise 3—Appendix I

Randomly select 12 observations from the database of companies with an A bond rating, 12 from companies with a B bond rating, and 12 from companies with a C bond rating. Can one conclude that there is a difference in the TOTAL (long term assets) for the three groups? Use the Kruskal-Wallis test with a .05 significance level.

Exercise 4—Appendix I

Choose 25 observations at random from the database. Is there evidence that a positive relationship exists between EMPLOYEE (the number of employees) and SALES (the amount of sales)? Use Spearman's rank correlation coefficient and test at the .01 significance level.

CASE STUDY

FLEXTIME, WORKTIME, ANYTIME

Productivity in the workplace is a perennial issue in the management of any enterprise, public or private. According to past studies, the employee work schedule (i.e., the time the employee gets to work, the time he or she gets for lunch and the time the employee gets off work) has an effect on productivity. Almost all the studies that have investigated this matter have said that flexible work schedules, or *flextime*, improves worker satisfaction and boosts productivity. Flextime is different from fixed work hours in that it allows employees to vary their work schedules to suit their particular needs better.

Some old-fashioned managers might regard this as a "touchy-feely" kind of issue, but organizational behavior theorists argue that flextime benefits both the employee and the organization. They offer several reasons. Firstly, giving employees more control over their hours of work enhances their sense of responsibility and increases opportunities for job enrichment. Secondly, flexible scheduling allows workers to adjust their schedules to their own biological clock to work when they are most productive. For example, early risers can work early mornings, whereas those who like to sleep late can work later in the day. Finally, flextime allows the employees to make better arrangements for transportation to and from their place of work. They can thus avoid the rush hours and not spend too much time commuting.

Following on numerous previous studies, McGuire and Liro (1986) investigated two forms of flexible work scheduling: flextime and staggered fixed hours. Since they felt that previous research had been not statistically rigorous, they decided to include a control group in their experimental design. Independent samples from three different New York State government agencies were taken. One sample ($n = 115$) was drawn from a group that had a *staggered fixed schedule*: Employees could schedule their arrival anytime between 8:00 A.M. and 9:30 A.M., and could leave after putting in 8 hours of work. Work schedules could be changed every quarter. The second sample ($n = 130$) was drawn from a group that had true flextime work schedules: Employees could arrive anytime between 7:00 and 9:00 A.M. and leave anytime between 3:00 and 5:00 P.M. after putting in $7\frac{1}{2}$ hours of work. This schedule could be

changed daily. The third sample ($n = 29$) was drawn from an agency that had *standard fixed hours* (9 to 5); this was the control group.

The survey instrument asked the employees to rank their preferences in two areas: whether they were satisfied with their job and whether they were satisfied with their work environment. The rankings were 1 = greatly dislike through 7 = greatly enjoy. The researchers also investigated other issues like whether the workers wished to continue their present schedules and whether their schedules increased or decreased their commuting times to and from work. However, we shall only be concerned with the two questions dealing with ranked responses, since for the other issues the researchers used the chi-square test of proportions, which has already been discussed in an earlier chapter. The following table summarizes the results.

Work Scheduling and Employee Attitudes

SURVEY QUESTION	FLEXTIME % FREQUENCY		STAGGERED FIXED HOURS % FREQUENCY		STANDARD FIXED HOURS % FREQUENCY	
(1) Satisfaction with work day						
Attitude and Rank						
Dislike (1–3)	6.9%	9	8.7%	10	13.8	4
Indifference (4)	9.2	12	10.4	12	13.8	4
Enjoy (5–7)	83.9	109	80.9	93	72.4	21
$KW = 3.13$	100.0	130	100.0	115	100.0	29
(2) Satisfaction with work environment						
Attitude and Rank						
Dislike (1–3)	7.7%	10	8.7%	10	13.8	4
Indifference (4)	32.3	42	53.0	61	48.3	14
Enjoy (5–7)	60.0	78	38.3	44	37.9	11
$KW = 6.63$	100.0	130	100.0	115	100.0	29

(*Source:* Adapted and modified from Jean B. McGuire and Joseph R. Liro, "Flexible Work Schedules, Work Attitudes, and Perceptions of Productivity," *Public Personnel Management* 15, no. 1 (Spring 1986): 65–73.)

Case Study Questions

1. The researchers used the Kruskal-Wallis test to detect differences in the three groups: flextime, staggered, and standard

hours. Explain why the Mann-Whitney test and Wilcoxon test would not be appropriate for these data.

2. Employing a nonparametric procedure avoids the necessity of assuming normally distributed populations. However, because of the nature of the data, even if the responses could be assumed to be normally distributed, the equivalent parametric procedure (ANOVA) might still not be appropriate. Can you explain why?

3. Explain the purpose of using standard fixed hours as a control group in the above research design.

4. At a .05 significance level, test the null hypothesis that these groups do not differ significantly in their satisfaction with their work. What is your conclusion?

5. At a .05 significance level, is there sufficient evidence to reject the null hypothesis that no significant difference exists between the three groups in their feelings about satisfaction with the work environment?

6. The preceding table does not provide the original data but merely reports the Kruskal-Wallis test statistic, KW. If, for some reason, you wish to simulate recreating (approximately) the original data, you can do so by using the middle rank of each class. For example, in the control group's responses to the question of work satisfaction, we could pretend that we had the rank 2 appear 4 times, rank 4 appear 4 times, and the rank 6 appear 21 times. As an optional exercise, regenerate the individual rankings in this way, and perform the Kruskal-Wallis test procedure yourself. Comment on the value of the KW statistic obtained in this way, and the one provided by the researchers in the table. Are the KW statistics close? Do you come to the same conclusions for the hypothesis test?

▪▪▪▪▪▪▪▪▪▪ S P S S ▪▪▪▪▪▪▪▪▪▪▪

MAINFRAME AND MICRO

SOLUTION

EXAMPLE 18.2

Example 18.2 used the nonparametric runs test to determine whether a sample of savings accounts at the Northside National Bank was determined in a random manner. The runs test was used to see if the sample consisted of 45 randomly selected males and females. The SPSS program listing in Figure 18.13 was used to compute the number of runs and the Z value. In this problem the SPSS commands are the same for both the mainframe and PC versions. **(Remember to end each command line with a period when using the PC version.)**

The TITLE command names the SPSS run.

The DATA LIST command gives each variable a name and describes the data as being in free form.

The BEGIN DATA command indicates the SPSS that the input data immediately follow.

The next 45 lines contain the data values, with each line representing either a −1 for a female or a +1 for a male.

The END DATA statement indicates the end of the data.

FIGURE 18.13

FIGURE 18.13

Input for SPSSX or
SPSS/PC+. All
command lines
should end with a
period when the PC
version is being used.

```
TITLE    RUNS TEST FOR NORTHSIDE NATIONAL BANK
DATA LIST FREE/SEX
BEGIN DATA
1
1
-1
-1
-1
-1
-1
1
-1
-1
1
1
1
1
1
1
-1
-1
-1
-1
1
1
-1
1
1
-1
-1
1
-1
-1
-1
-1
-1
1
1
1
1
-1
-1
-1
1
1
1
END DATA
VALUE LABELS SEX -1 'FEMALE' 1 'MALE'/
NPAR TESTS RUNS(1)=SEX/
```

The VALUE LABELS statement assigns labels to the values of the variable SEX. The value -1 is assigned the label FEMALE, whereas the value $+1$ is assigned the label MALE.

The NPAR TESTS command indicates that we wish to do a nonparametric test. In this case it is the runs test. We are testing for runs that are greater than 0 and less than 0. In other words, runs of females (-1) or males ($+1$).

Figure 18.14 shows the SPSSX output obtained by executing the listing in Figure 18.13, and Figure 18.15 shows the SPSS/PC+ output.

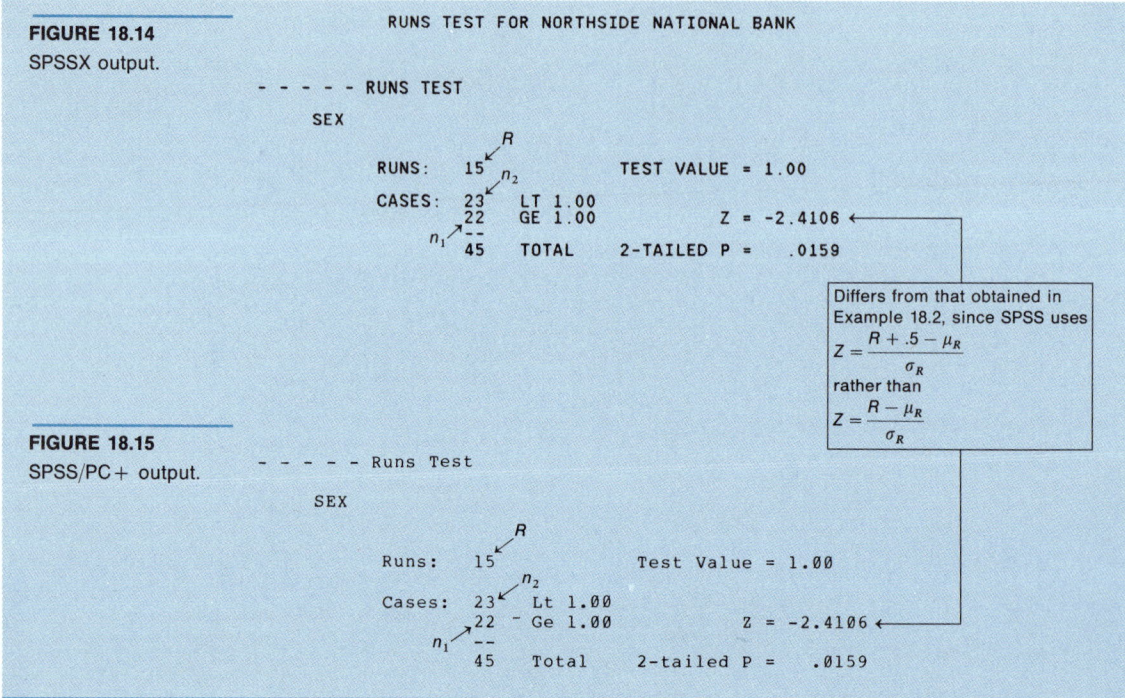

FIGURE 18.14
SPSSX output.

RUNS TEST FOR NORTHSIDE NATIONAL BANK

- - - - - RUNS TEST

        SEX

                 R
        RUNS: 15   TEST VALUE = 1.00
                    n_2
        CASES: 23   LT 1.00
              22   GE 1.00   Z = -2.4106 ←
       n_1 --
              45   TOTAL   2-TAILED P =   .0159

> Differs from that obtained in Example 18.2, since SPSS uses
> $$Z = \frac{R + .5 - \mu_R}{\sigma_R}$$
> rather than
> $$Z = \frac{R - \mu_R}{\sigma_R}$$

FIGURE 18.15
SPSS/PC+ output.

- - - - - Runs Test

        SEX

                 R
        Runs: 15   Test Value = 1.00
                    n_2
        Cases: 23   Lt 1.00
              22   - Ge 1.00   Z = -2.4106 ←
       n_1 --
              45   Total   2-tailed P =   .0159

⬛⬛⬛⬛ ⬛ ⬛ ⬛ ⬛ ⬛ ⬛ S P S S ⬛ ⬛ ⬛ ⬛⬛⬛⬛

MAINFRAME AND MICRO

SOLUTION

EXAMPLE 18.5

Example 18.5 used the nonparametric Mann-Whitney U-test to determine whether there was sufficient evidence to indicate that Food World's store A had a larger number of returned checks than did their store B. The SPSS program listing in Figure 18.16 below was used to compute the Mann-

FIGURE 18.16
Input for SPSSX or SPSS/PC+. All command lines should end with a period when the PC version is being used.

```
TITLE    FOOD WORLD SUPERMARKETS
DATA LIST FREE/STORE GROUP
BEGIN DATA
42 0
65 0
38 0
55 0
71 0
60 0
47 0
59 0
68 0
57 0
76 0
42 0
22 1
17 1
35 1
19 1
8  1
24 1
42 1
14 1
28 1
17 1
10 1
15 1
20 1
45 1
50 1
END DATA
NPAR TESTS M-W = STORE BY GROUP (0,1)
```

Whitney test statistic. In this problem the SPSS commands are the same for both the mainframe and PC versions. **(Remember to end each command line with a period when using the PC version.)**

The TITLE command names the SPSS run.

The DATA LIST command gives each variable a name and describes the data as being in free form.

The BEGIN DATA command indicates to SPSS that the input data immediately follow.

The next 27 lines contain the data values with each line representing the number of returned checks per month per store and the group in which the specific store belongs (0 = Store A, 1 = Store B).

The END DATA statement indicates the end of the data.

The NPAR TESTS command indicates that we wish to do a nonparametric Mann-Whitney test, comparing the number of returned checks (variable STORE) by the store category (variable GROUP).

Figure 18.17 shows the SPSSX output obtained by executing the listing in Figure 18.16, and Figure 18.18 shows the SPSS/PC+ output.

FIGURE 18.17
SPSSX output.

```
- - - - - MANN-WHITNEY U - WILCOXON RANK SUM W TEST

      STORE
 BY GROUP

   MEAN RANK      CASES        n₁
                            /
      20.75          12   GROUP =  .00
       8.60          15   GROUP = 1.00
                       --
                 n₂  27   TOTAL
```

n_1, n_2

```
                                  EXACT              CORRECTED FOR TIES
        U              W         2-TAILED P            Z      2-TAILED P
       9.0           249.0        0.0000           -3.9554      0.0001
```

U_1 T_1

Test statistic using U_1.
The value using U_2 is $Z = 3.9554$

$$T_2 = \frac{n(n+1)}{2} - T_1$$

$$= \frac{(27)(28)}{2} - 249 = 129$$

and

$$U_2 = n_1 n_2 + \frac{n_2(n_2+1)}{2} - T_2$$

$$= (12)(15) + \frac{(15)(16)}{2} - 129 = 171$$

FIGURE 18.18
SPSS/PC+ output.

```
- - - - - Mann-Whitney U - Wilcoxon Rank Sum W Test

      STORE
 by GROUP

   Mean Rank      Cases        n₁
                            /
      20.75          12   GROUP =  .00
       8.60          15   GROUP = 1.00
                       --
                 n₂  27   Total
```

n_1, n_2

```
                                  EXACT              Corrected for Ties
        U              W         2-tailed P            Z      2-tailed P
       9.0           249.0        .0000            -3.9554      .0001
```

U_1 T_1

Test statistic using U_1.
The value using U_2 is $Z = 3.9554$

▪ ▪ ▪ ▪ ▪ ▪ ▪ ▪ ▪ ▪ ▪ S P S S ▪ ▪ ▪ ▪ ▪ ▪ ▪ ▪ ▪ ▪ ▪

MAINFRAME AND MICRO

SOLUTION

EXAMPLE 18.7

Example 18.7 used the nonparametric Wilcoxon signed ranks test to determine whether customers at Worldwide Travel flew more miles in 1988 than they did in 1978. The SPSS program listing in Figure 18.19 was used to compute the Wilcoxon Z statistic for the two time periods. In this problem the SPSS commands are the same for both the mainframe and PC versions. **(Remember to end each command line with a period when using the PC version.)**

FIGURE 18.19

Input for SPSSX or SPSS/PC+. All command lines should end with a period when the PC version is being used.

```
TITLE    WORLDWIDE TRAVEL CUSTOMER MILEAGE COMPARISON
DATA LIST FREE/YEAR78 YEAR88
BEGIN DATA
55   45
101  79
62   77
93   138
40   68
120  80
110  138
77   81
49   49
90   80
121  171
46   21
112  78
70   106
97   87
47   91
106  80
84   104
50   88
75   117
END DATA
VARIABLE LABELS
        YEAR78  'MILES FLOWN IN 1978'
        YEAR88  'MILES FLOWN IN 1988'
NPAR TESTS  WILCOXON=ALL
```

The TITLE command names the SPSS run.

The DATA LIST command gives each variable a name and describes the data as being in free form.

The BEGIN DATA command indicates to SPSS that the input data immediately follow.

The next 20 lines contain the data values with each line representing the number of miles flown by continuing customers in both 1978 and 1988.

The END DATA statement indicates the end of the data.

The VARIABLE LABELS statement assigns a descriptive label to the variables. This descriptive label is substituted for the variable name when output is printed.

The NPAR TESTS command indicates that we wish to do the nonparametric Wilcoxon test as well as generate both the Z value and two-tailed p-value.

Figure 18.20 shows the SPSSX output obtained by executing the listing in Figure 18.19, and Figure 18.21 shows the SPSS/PC+ output.

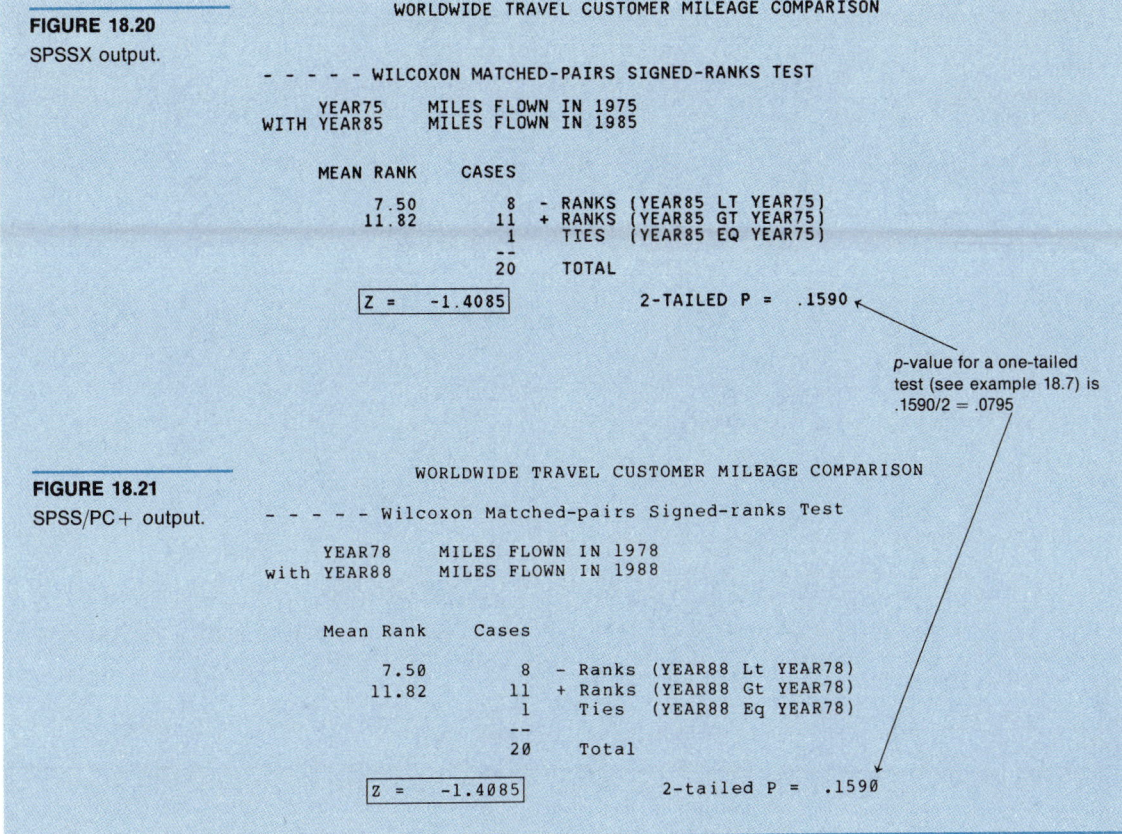

FIGURE 18.20
SPSSX output.

FIGURE 18.21
SPSS/PC+ output.

─ ─ ─ ─ ─ WILCOXON MATCHED-PAIRS SIGNED-RANKS TEST

WORLDWIDE TRAVEL CUSTOMER MILEAGE COMPARISON

p-value for a one-tailed test (see example 18.7) is .1590/2 = .0795

MAINFRAME AND MICRO

SOLUTION

EXAMPLE 18.8

Example 18.8 used the nonparametric Kruskal-Wallis one-way test to determine whether there was a difference in the amount of downtime among four different copying machines. Data were collected by observing the downtime for 20 randomly selected months. In this way, five randomly selected months were observed for each of the four brands of machines. The SPSS program listing in Figure 18.22 was used to compute the Kruskal-Wallis statistic and the *p*-value. In this problem the SPSS commands are the same for both the mainframe and PC versions. **(Remember to end each command line with a period when using the PC version.)**

The TITLE command names the SPSS run.

The DATA LIST command gives each variable a name, and describes the data as being in free form.

The BEGIN DATA command indicates to SPSS that the input data immediately follow.

The next 20 lines contain the data values, with each line representing the length of downtime, in hours, of each machine as well as the machine type (1, 2, 3, or 4).

The END DATA statement indicates the end of the data.

FIGURE 18.22

Input for SPSSX or
SPSS/PC+. All
command lines
should end with a
period when the PC
version is being used.

```
TITLE    DREXTON INDUSTRIES COPYING MACHINES
DATA LIST FREE/DOWNTIME GROUP
BEGIN DATA
28 1
41 1
34 1
52 1
25 1
5   2
16 2
20 2
24 2
15 2
10 3
8  3
18 3
14 3
26 3
45 4
30 4
49 4
32 4
36 4
END DATA
NPAR TESTS   K-W = DOWNTIME BY GROUP (1,4)
```

The **NPAR TESTS** command indicates that we wish to do a Kruskal-Wallis test, analyzing DOWNTIME by GROUP.

Figure 18.23 shows the SPSSX output obtained by executing the listing in Figure 18.22, and Figure 18.24 shows the SPSS/PC+ output.

FIGURE 18.23

SPSSX output.

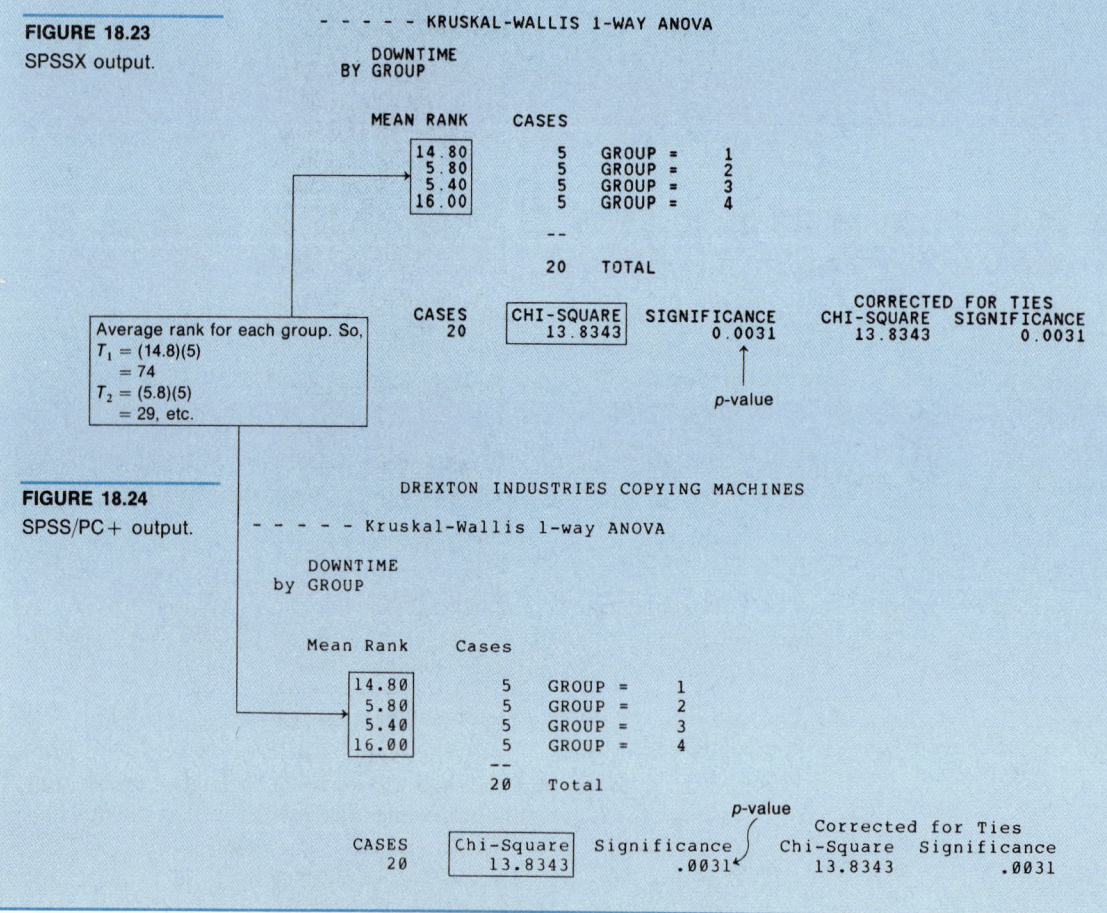

FIGURE 18.24

SPSS/PC+ output.

SOLUTION

EXAMPLE 18.9

Example 18.9 used the nonparametric FRIEDMAN test to determine whether there was a difference in the perceived quality of the three automobile manufacturers' warranties. The SPSS program listing in Figure 18.25 was used to compute the Friedman test statistic. In this problem the SPSS commands are the same for both the mainframe and PC versions. **(Remember to end each command line with a period when using the PC version.)**

FIGURE 18.25

Input for SPSSX or SPSS/PC+. All command lines should end with a period when the PC version is being used.

```
TITLE    WARRANTY ASSESSMENTS
DATA LIST FREE/HENRY GA ROADSTER
BEGIN DATA
3 2 1
2 1 3
3 1 2
2 1 3
3 2 1
2 1 3
1 2 3
2 1 3
3 1 2
1 2 3
END DATA
NPAR TESTS FRIEDMAN = HENRY GA ROADSTER
```

The TITLE command names the SPSS run.

The DATA LIST command gives each variable a name and describes the data as being in free form.

The BEGIN DATA command indicates to SPSS that the input data immediately follow.

The next 10 lines contain the data values, with each line representing the rankings of one editor for the three automobile warranties. The first line indicates that editor 1 rated the Roadster first, the GA second, and the Henry third.

The END DATA statement indicates the end of the data.

The NPAR TESTS command indicates that we wish to do a nonparametric FRIEDMAN test, analyzing the rankings of the warranties of the three types of automobiles.

FIGURE 18.26

SPSSX output.

```
- - - - - FRIEDMAN TWO-WAY ANOVA

    MEAN RANK     VARIABLE

       2.20       HENRY
       1.40       GA
       2.40       ROADSTER

       CASES         CHI-SQUARE        D.F.    SIGNIFICANCE
        10            5.6000            2          .0608
```

FIGURE 18.27

SPSS/PC+ output.

```
                WARRANTY ASSESSMENTS

- - - - - Friedman Two-way ANOVA


    Mean Rank     Variable

       2.20       HENRY
       1.40       GA
       2.40       ROADSTER

       Cases         Chi-Square        D.F.    Significance
        10            5.6000            2          .0608
```

Figure 18.26 shows the SPSSX output obtained by executing the listing in Figure 18.25, and Figure 18.27 shows the SPSS/PC+ output.

■ ■ ■ ■ ■ ■ ■ ■ ■ S P S S X ■ ■ ■ ■ ■ ■ ■ ■ ■

SOLUTION

EXAMPLES 18.10 AND 18.12

Examples 18.10 and 18.12 used the nonparametric Spearman rank correlation coefficient to determine the relationship that exists between the market value of homes and the personal indebtedness of the homeowner. The SPSSX program listing in Figure 18.28 was used to compute the Spearman rank correlation coefficient r. This program will run only on the mainframe version of SPSSX.

FIGURE 18.28
Input for SPSSX

```
TITLE  HOME VALUE AND FAMILY INDEBTEDNESS ANALYSIS
DATA LIST FREE/MKTVALUE DEBT
BEGIN DATA
85   12
147  27
340  45
94   10
120  17
105  4
135  20
162  25
480  35
88   14
END DATA
VARIABLE LABELS
       MKTVALUE 'MARKET VALUE OF HOME'
       DEBT     'TOTAL INDEBTEDNESS'
NONPAR CORR MKTVALUE WITH DEBT
```

The TITLE command names the SPSSX run.

The DATA LIST command gives each variable a name and describes the data as being in free form.

The BEGIN DATA command indicates to SPSSX that the input data immediately follow.

The next 10 lines contain the data values, with each line representing the total market value of the home and the total indebtedness of the homeowner.

The END DATA statement indicates the end of the data.

The VARIABLE LABELS statement assigns a descriptive label to the variables. This descriptive label is substituted for the variable name when the output is printed.

The NONPAR CORR statement requests that the rank correlation be computed for variables MKTVALUE and DEBT.

FIGURE 18.29 SPSSX output.

- - - - - - - - - - - - - S P E A R M A N   C O R R E L A T I O N   C O E F F I C I E N T S - - -

```
                DEBT

MKTVALUE       .8667  ←—— rs
      N(   10)
      SIG .001  ←—— p-value for H0: no association exists between X and Y
```

" . " IS PRINTED IF A COEFFICIENT CANNOT BE COMPUTED.

Figure 18.29 shows the SPSSX output obtained by executing the listing in Figure 18.28.

■ ■ ■ ■ ■ ■ ■ ■ ■ ■ ■ S A S ■ ■ ■ ■ ■ ■ ■ ■ ■ ■ ■ ■

MAINFRAME AND MICRO

SOLUTION

EXAMPLE 18.5

Example 18.5 used the nonparametric Mann-Whitney U test to determine whether there was sufficient evidence to indicate that Food World's store A had a larger number of returned checks than did their store B. The SAS program listing in Figure 18.30 was used to compute the Mann-Whitney test statistic. In this problem the SAS commands are the same for both the mainframe and PC versions.

FIGURE 18.30

Input for SAS (mainframe and micro version).

```
TITLE  'FOOD WORLD SUPERMARKETS';
DATA STORES;
 INPUT STORE GROUP;
CARDS;
42 0
65 0
38 0
55 0
71 0
60 0
47 0
59 0
68 0
57 0
76 0
42 0
22 1
17 1
35 1
19 1
8  1
24 1
42 1
14 1
28 1
17 1
10 1
15 1
20 1
45 1
50 1
PROC NPAR1WAY WILCOXON;
 VAR STORE;
 CLASS GROUP;
```

The TITLE command names the SAS run.

The DATA command gives the data a name.

The INPUT command names and gives the correct order for the different fields on the data lines.

The CARDS command indicates to SAS that the input data immediately follow.

The next 27 lines contain the data values with each line representing the number of returned checks per month per store and the group in which the store belongs (0 = Store A, 1 = Store B).

The Mann-Whitney test is sometimes called the Wilcoxon test. The following command set was used to obtain the statistics:

PROC NPAR1WAY WILCOXON
VAR STORE
CLASS GROUP

Using this command set, we are able to compare the number of returned checks (variable STORE) by the store category (variable GROUP).

Figure 18.31 shows the SAS output obtained by executing the listing in Figure 18.30, and Figure 18.32 shows the SAS/PC output.

FIGURE 18.31
SAS output.

```
                      FOOD WORLD SUPERMARKETS

       ANALYSIS FOR VARIABLE STORE CLASSIFIED BY VARIABLE  GROUP
                  AVERAGE SCORES WERE USED FOR TIES


                  WILCOXON SCORES (RANK SUMS)


                              SUM OF     EXPECTED     STD DEV      MEAN
        LEVEL         N       SCORES     UNDER HO     UNDER HO     SCORE
     T₁
                0     12      249.00     168.00       20.48        20.75
     T₂         1     15      129.00     210.00       20.48        8.60
             WILCOXON 2-SAMPLE TEST (NORMAL APPROXIMATION)
             (WITH CONTINUITY CORRECTION OF .5)
     T₁ ──→ S=  249.00     Z= 3.9310      PROB >!Z!=0.0001 ←──── p-value

        T-TEST APPROX. SIGNIFICANCE=0.0006

        KRUSKAL-WALLIS TEST (CHI-SQUARE APPROXIMATION)
        CHISQ=  15.65      DF= 1    PROB > CHISQ=0.0001
```

> This value is slightly different from the Z value in Example 18.5, since SAS adds .5 to the numerator of Z as well as adjusting this Z value for any ties obtained in the pooled sample.

FIGURE 18.32
SAS/PC output.

```
                      FOOD WORLD SUPERMARKETS


              N P A R 1 W A Y   P R O C E D U R E

       Wilcoxon Scores (Rank Sums) for Variable STORE
                 Classified by Variable GROUP


                            Sum of      Expected      Std Dev          Mean
        GROUP       N       Scores      Under H0      Under H0         Score
              T₁
        0           12      249.0       168.0         20.4782561       20.7500000
        1           15      129.0       210.0         20.4782561       8.6000000
              T₂      Average Scores were used for Ties
              Wilcoxon 2-Sample Test (Normal Approximation)
              (with Continuity Correction of .5)

        T₁ ──→ S=  249.000     Z=  3.93100     Prob > |Z| =   0.0001 ←── p-value

        T-Test approx. Significance =      0.0006


        Kruskal-Wallis Test (Chi-Square Approximation)
        CHISQ=  15.645     DF= 1     Prob > CHISQ=      0.0001
```

■■■ ■ ■ ■ ■ ■ ■ ■ ■ ■ S A S ■ ■ ■ ■ ■ ■ ■ ■ ■ ■ ■

MAINFRAME AND MICRO

SOLUTION

EXAMPLE 18.8

Example 18.8 used the nonparametric Kruskal-Wallis one-way test to determine whether there was a difference in the amount of downtime among four different copying machines. Data were collected by observing the downtime for 20 randomly selected months. In this way, five randomly selected months were observed for each of the four brands of machines. The SAS program listing in Figure 18.33 was used to compute the Kruskal-Wallis statistic and the *p*-value. In this problem the SAS commands are the same for both the mainframe and PC versions.

FIGURE 18.33

Input for SAS (mainframe or micro version).

```
TITLE   'DREXTON INDUSTRIES COPYING MACHINES';
DATA MACHINES;
 INPUT DOWNTIME BRAND;
CARDS;
28 1
41 1
34 1
52 1
25 1
5  2
16 2
20 2
24 2
15 2
10 3
8  3
18 3
14 3
26 3
45 4
30 4
49 4
32 4
36 4
PROC NPAR1WAY WILCOXON;
 VAR DOWNTIME;
 CLASS BRAND;
```

The TITLE command names the SAS run.

The DATA command gives the data a name.

The INPUT command names and gives the correct order for the different fields on the data lines.

The CARDS command indicates to SAS that the input data immediately follow.

The next 20 lines contain the data values with each line representing the downtime, in hours, of a copier and the brand (1, 2, 3, or 4) of the copier.

The following command set was used to obtain the Kruskal-Wallis statistic:

PROC NPAR1WAY WILCOXON
VAR DOWNTIME
CLASS BRAND

Using this command set, we are analyzing the difference in average downtime for each of the four brands of copiers.

Figure 18.34 shows the SAS output obtained by executing the listing in Figure 18.33, and Figure 18.35 shows the SAS/PC output.

FIGURE 18.34
SAS output.

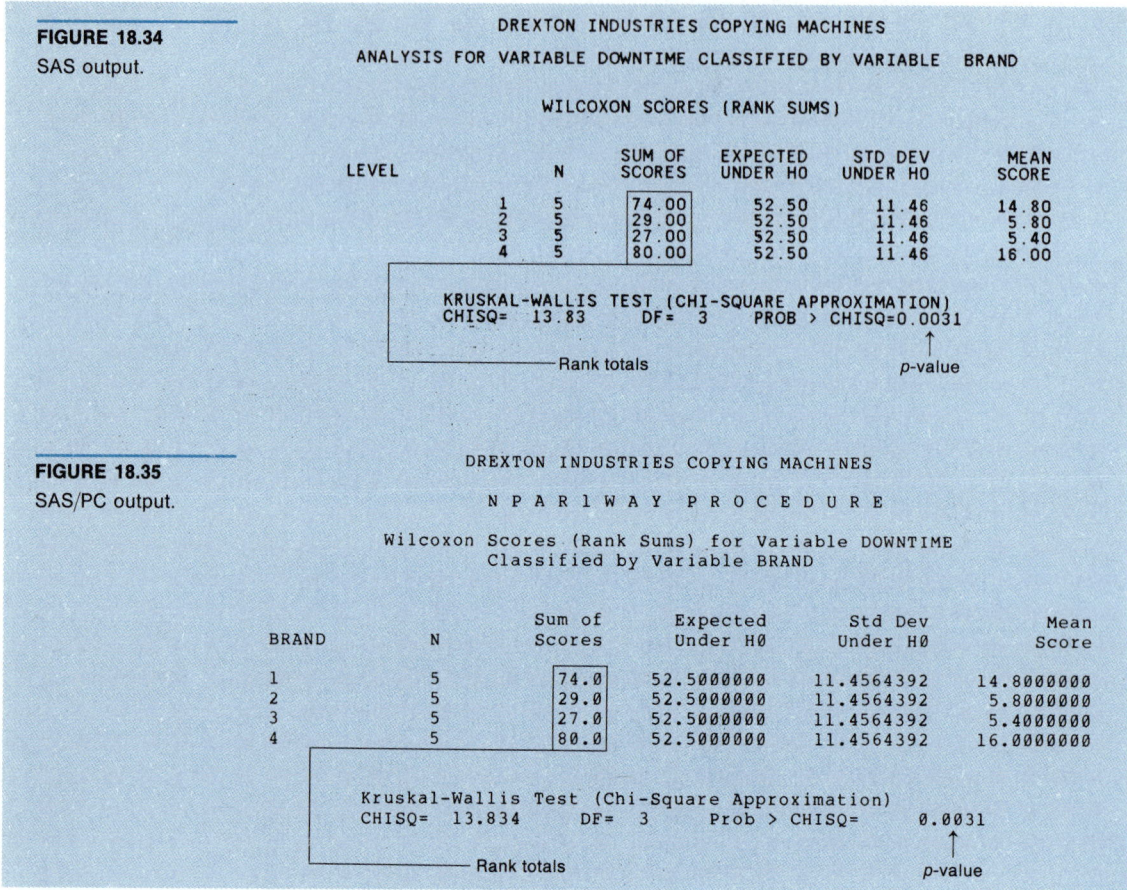

DREXTON INDUSTRIES COPYING MACHINES

ANALYSIS FOR VARIABLE DOWNTIME CLASSIFIED BY VARIABLE BRAND

WILCOXON SCORES (RANK SUMS)

LEVEL	N	SUM OF SCORES	EXPECTED UNDER H0	STD DEV UNDER H0	MEAN SCORE
1	5	74.00	52.50	11.46	14.80
2	5	29.00	52.50	11.46	5.80
3	5	27.00	52.50	11.46	5.40
4	5	80.00	52.50	11.46	16.00

KRUSKAL-WALLIS TEST (CHI-SQUARE APPROXIMATION)
CHISQ= 13.83   DF= 3   PROB > CHISQ=0.0031

└── Rank totals   *p*-value

FIGURE 18.35
SAS/PC output.

DREXTON INDUSTRIES COPYING MACHINES

N P A R 1 W A Y   P R O C E D U R E

Wilcoxon Scores (Rank Sums) for Variable DOWNTIME
Classified by Variable BRAND

BRAND	N	Sum of Scores	Expected Under H0	Std Dev Under H0	Mean Score
1	5	74.0	52.5000000	11.4564392	14.8000000
2	5	29.0	52.5000000	11.4564392	5.8000000
3	5	27.0	52.5000000	11.4564392	5.4000000
4	5	80.0	52.5000000	11.4564392	16.0000000

Kruskal-Wallis Test (Chi-Square Approximation)
CHISQ= 13.834   DF= 3   Prob > CHISQ=   0.0031

└── Rank totals   *p*-value

■ ■ ■ ■ ■ ■ ■ ■ ■ ■ ■ ■ **S A S** ■ ■ ■ ■ ■ ■ ■ ■ ■ ■ ■ ■

MAINFRAME AND MICRO

SOLUTION

EXAMPLES 18.10
AND 18.12

Examples 18.10 and 18.12 used the nonparametric Spearman rank correlation coefficient to determine the relationship between the market value of homes and the personal indebtedness of the homeowner. The SAS program listing in Figure 18.36 was used to compute the Spearman rank correlation

FIGURE 18.36
Input for SAS
(mainframe or micro
version).

```
TITLE  'HOME VALUE AND FAMILY INDEBTEDNESS ANALYSIS';
DATA HOMES;
 INPUT MKTVALUE DEBT;
 LABEL MKTVALUE='MARKET VALUE OF HOMES'
       DEBT='TOTAL INDEBTEDNESS';
CARDS;
85  12
147 27
340 45
94  10
120 17
105 4
135 20
162 25
480 35
88  14
PROC CORR NOSIMPLE SPEARMAN;
 VAR MKTVALUE DEBT;
```

coefficient r_s. In this problem the SAS commands are the same for both the mainframe and PC versions.

The TITLE command names the SAS run.

The DATA command gives the data a name.

The INPUT command names and gives the correct order for the different fields on the data lines.

The LABEL statements assign a descriptive label to the variables. This descriptive label is substituted for the variable name when output is printed.

The CARDS command indicates to SAS that the input data immediately follow.

The next 10 lines contain the data values, with each line representing the market value of a home and the personal indebtedness of the homeowner.

The following command set was used to calculate the Spearman rank correlation coefficient, r_s, between the variables MKTVALUE and DEBT:

PROC CORR NOSIMPLE SPEARMAN
VAR MKTVALUE DEBT

Figure 18.37 shows the SAS output obtained by executing the listing in Figure 18.36, and Figure 18.38 shows the SAS/PC output.

FIGURE 18.37
SAS output.

```
                 HOME VALUE AND FAMILY INDEBTEDNESS ANALYSIS
SPEARMAN CORRELATION COEFFICIENTS / PROB > !R! UNDER HO:RHO=0 / N = 10
                                   MKTVALUE      DEBT

          MKTVALUE                  1.00000    0.86667
          MARKET VALUE OF HOMES      0.0000     0.0012

          DEBT                      0.86667    1.00000
          TOTAL INDEBTEDNESS         0.0012     0.0000
```

r_s

p-value for H_0: no association exists between *X* and *Y*

FIGURE 18.38
SAS/PC output.

```
              HOME VALUE AND FAMILY INDEBTEDNESS ANALYSIS

                      CORRELATION ANALYSIS

                2 'VAR' Variables:  MKTVALUE DEBT

Spearman Correlation Coefficients / Prob > |R| under Ho: Rho=0    N = 10

                                  MKTVALUE               DEBT

        MKTVALUE                   1.00000             0.86667
        MARKET VALUE OF HOMES       0.0                 0.0012

        DEBT                       0.86667             1.00000
        TOTAL INDEBTEDNESS          0.0012              0.0
```

r_s

p-value for H_0: no association exists between *X* and *Y*

Appendixes

APPENDIX A Tables

TABLE A-1 Binomial Probabilities $[{}_nC_x p^x(1-p)^{n-x}]$

n	x	0.01	0.05	0.10	0.20	0.30	0.40	0.50	0.60	0.70	0.80	0.90	0.95	0.99	x
2	0	980	902	810	640	490	360	250	160	090	040	010	002	0+	0
	1	020	095	180	320	420	480	500	480	420	320	180	095	020	1
	2	0+	002	010	040	090	160	250	360	490	640	810	902	980	2
3	0	970	857	729	512	343	216	125	064	027	008	001	0+	0+	0
	1	029	135	243	384	441	432	375	288	189	096	027	007	0+	1
	2	0+	007	027	096	189	288	375	432	441	384	243	135	029	2
	3	0+	0+	001	008	027	064	125	216	343	512	729	857	970	3
4	0	961	815	656	410	240	130	062	026	008	002	0+	0+	0+	0
	1	039	171	292	410	412	346	250	154	076	026	004	0+	0+	1
	2	001	014	049	154	265	346	375	346	265	154	049	014	001	2
	3	0+	0+	004	026	076	154	250	346	412	410	292	171	039	3
	4	0+	0+	0+	002	008	026	062	130	240	410	656	815	961	4
5	0	951	774	590	328	168	078	031	010	002	0+	0+	0+	0+	0
	1	048	204	328	410	360	259	156	077	028	006	0+	0+	0+	1
	2	001	021	073	205	309	346	312	230	132	051	008	001	0+	2
	3	0+	001	008	051	132	230	312	346	309	205	073	021	001	3
	4	0+	0+	0+	006	028	077	156	259	360	410	328	204	048	4
	5	0+	0+	0+	0+	002	010	031	078	168	328	590	774	951	5
6	0	941	735	531	262	118	047	016	004	001	0+	0+	0+	0+	0
	1	057	232	354	393	303	187	094	037	010	002	0+	0+	0+	1
	2	001	031	098	246	324	311	234	138	060	015	001	0+	0+	2
	3	0+	002	015	082	185	276	312	276	185	082	015	002	0+	3
	4	0+	0+	001	015	060	138	234	311	324	246	098	031	001	4
	5	0+	0+	0+	002	010	037	094	187	303	393	354	232	057	5
	6	0+	0+	0+	0+	001	004	016	047	118	262	531	735	941	6
7	0	932	698	478	210	082	028	008	002	0+	0+	0+	0+	0+	0
	1	066	257	372	367	247	131	055	017	004	0+	0+	0+	0+	1
	2	002	041	124	275	318	261	164	077	025	004	0+	0+	0+	2
	3	0+	004	023	115	227	290	273	194	097	029	003	0+	0+	3
	4	0+	0+	003	029	097	194	273	290	227	115	023	004	0+	4
	5	0+	0+	0+	004	025	077	164	261	318	275	124	041	002	5
	6	0+	0+	0+	0+	004	017	055	131	247	367	372	257	066	6
	7	0+	0+	0+	0+	0+	002	008	028	082	210	478	698	932	7
8	0	923	663	430	168	058	017	004	001	0+	0+	0+	0+	0+	0
	1	075	279	383	336	198	090	031	008	001	0+	0+	0+	0+	1
	2	003	051	149	294	296	209	109	041	010	001	0+	0+	0+	2
	3	0+	005	033	147	254	279	219	124	047	009	0+	0+	0+	3
	4	0+	0+	005	046	136	232	273	232	136	046	005	0+	0+	4
	5	0+	0+	0+	009	047	124	219	279	254	147	033	005	0+	5
	6	0+	0+	0+	001	010	041	109	209	296	294	149	051	003	6
	7	0+	0+	0+	0+	001	008	031	090	198	336	383	279	075	7
	8	0+	0+	0+	0+	0+	001	004	017	058	168	430	663	923	8
9	0	914	630	387	134	040	010	002	0+	0+	0+	0+	0+	0+	0
	1	083	299	387	302	156	060	018	004	0+	0+	0+	0+	0+	1
	2	003	063	172	302	267	161	070	021	004	0+	0+	0+	0+	2
	3	0+	008	045	176	267	251	164	074	021	003	0+	0+	0+	3
	4	0+	001	007	066	172	251	246	167	074	017	001	0+	0+	4
9	5	0+	0+	001	017	074	167	246	251	172	066	007	001	0+	5
	6	0+	0+	0+	003	021	074	164	251	267	176	045	008	0+	6
	7	0+	0+	0+	0+	004	021	070	161	267	302	172	063	003	7
	8	0+	0+	0+	0+	0+	004	018	060	156	302	387	299	083	8
	9	0+	0+	0+	0+	0+	0+	002	010	040	134	387	630	914	9
10	0	904	599	349	107	028	006	001	0+	0+	0+	0+	0+	0+	0
	1	091	315	387	268	121	040	010	002	0+	0+	0+	0+	0+	1
	2	004	075	194	302	233	121	044	011	001	0+	0+	0+	0+	2
	3	0+	010	057	201	267	215	117	042	009	001	0+	0+	0+	3
	4	0+	001	011	088	200	251	205	111	037	006	0+	0+	0+	4
	5	0+	0+	001	026	103	201	246	201	103	026	001	0+	0+	5
	6	0+	0+	0+	006	037	111	205	251	200	088	011	001	0+	6
	7	0+	0+	0+	001	009	042	117	215	267	201	057	010	0+	7
	8	0+	0+	0+	0+	001	011	044	121	233	302	194	075	004	8
	9	0+	0+	0+	0+	0+	002	010	040	121	268	387	315	091	9
	10	0+	0+	0+	0+	0+	0+	001	006	028	107	349	599	904	10

From Mosteller, Rourke, & Thomas, *Probability with Statistical Applications*, 2d ed. © 1970, Addison-Wesley, Reading, Mass. Table on pp. 475–477. Reprinted with permission.

TABLE A-1 (continued)

n	x	0.01	0.05	0.10	0.20	0.30	0.40	P 0.50	0.60	0.70	0.80	0.90	0.95	0.99	x
11	0	895	569	314	086	020	004	0+	0+	0+	0+	0+	0+	0+	0
	1	099	329	384	236	093	027	005	001	0+	0+	0+	0+	0+	1
	2	005	087	213	295	200	089	027	005	001	0+	0+	0+	0+	2
	3	0+	014	071	221	257	177	081	023	004	0+	0+	0+	0+	3
	4	0+	001	016	111	220	236	161	070	017	002	0+	0+	0+	4
	5	0+	0+	002	039	132	221	226	147	057	010	0+	0+	0+	5
	6	0+	0+	0+	010	057	147	226	221	132	039	002	0+	0+	6
	7	0+	0+	0+	002	017	070	161	236	220	111	016	001	0+	7
	8	0+	0+	0+	0+	004	023	081	177	257	221	071	014	0+	8
	9	0+	0+	0+	0+	001	005	027	089	200	295	213	087	005	9
	10	0+	0+	0+	0+	0+	001	005	027	093	236	384	329	099	10
	11	0+	0+	0+	0+	0+	0+	0+	004	020	086	314	569	895	11
12	0	886	540	282	069	014	002	0+	0+	0+	0+	0+	0+	0+	0
	1	107	341	377	206	071	017	003	0+	0+	0+	0+	0+	0+	1
	2	006	099	230	283	168	064	016	002	0+	0+	0+	0+	0+	2
	3	0+	017	085	236	240	142	054	012	001	0+	0+	0+	0+	3
	4	0+	002	021	133	231	213	121	042	008	001	0+	0+	0+	4
	5	0+	0+	004	053	158	227	193	101	029	003	0+	0+	0+	5
	6	0+	0+	0+	016	079	177	226	177	079	016	0+	0+	0+	6
	7	0+	0+	0+	003	029	101	193	227	158	053	004	0+	0+	7
	8	0+	0+	0+	001	008	042	121	213	231	133	021	002	0+	8
	9	0+	0+	0+	0+	001	012	054	142	240	236	085	017	0+	9
	10	0+	0+	0+	0+	0+	002	016	064	168	283	230	099	006	10
	11	0+	0+	0+	0+	0+	0+	003	017	071	206	377	341	107	11
	12	0+	0+	0+	0+	0+	0+	0+	002	014	069	282	540	886	12
13	0	878	513	254	055	010	001	0+	0+	0+	0+	0+	0+	0+	0
	1	115	351	367	179	054	011	002	0+	0+	0+	0+	0+	0+	1
	2	007	111	245	268	139	045	010	001	0+	0+	0+	0+	0+	2
	3	0+	021	100	246	218	111	035	006	001	0+	0+	0+	0+	3
	4	0+	003	028	154	234	184	087	024	003	0+	0+	0+	0+	4
	5	0+	0+	006	069	180	221	157	066	014	001	0+	0+	0+	5
	6	0+	0+	001	023	103	197	209	131	044	006	0+	0+	0+	6
	7	0+	0+	0+	006	044	131	209	197	103	023	001	0+	0+	7
	8	0+	0+	0+	001	014	066	157	221	180	069	006	0+	0+	8
	9	0+	0+	0+	0+	003	024	087	184	234	154	028	003	0+	9
	10	0+	0+	0+	0+	001	006	035	111	218	246	100	021	0+	10
	11	0+	0+	0+	0+	0+	001	010	045	139	268	245	111	007	11
	12	0+	0+	0+	0+	0+	0+	002	011	054	179	367	351	115	12
	13	0+	0+	0+	0+	0+	0+	0+	001	010	055	254	513	878	13
14	0	869	488	229	044	007	001	0+	0+	0+	0+	0+	0+	0+	0
	1	123	359	356	154	041	007	001	0+	0+	0+	0+	0+	0+	1
	2	008	123	257	250	113	032	006	001	0+	0+	0+	0+	0+	2
	3	0+	026	114	250	194	085	022	003	0+	0+	0+	0+	0+	3
	4	0+	004	035	172	229	155	061	014	001	0+	0+	0+	0+	4
	5	0+	0+	008	086	196	207	122	041	007	0+	0+	0+	0+	5
	6	0+	0+	001	032	126	207	183	092	023	002	0+	0+	0+	6
	7	0+	0+	0+	009	062	157	209	157	062	009	0+	0+	0+	7
	8	0+	0+	0+	002	023	092	183	207	126	032	001	0+	0+	8
	9	0+	0+	0+	0+	007	041	122	207	196	086	008	0+	0+	9
	10	0+	0+	0+	0+	001	014	061	155	229	172	035	004	0+	10
	11	0+	0+	0+	0+	0+	003	022	085	194	250	114	026	0+	11
	12	0+	0+	0+	0+	0+	001	006	032	113	250	257	123	008	12
	13	0+	0+	0+	0+	0+	0+	001	007	041	154	356	359	123	13
	14	0+	0+	0+	0+	0+	0+	0+	001	007	044	229	488	869	14
15	0	860	463	206	035	005	0+	0+	0+	0+	0+	0+	0+	0+	0
	1	130	366	343	132	031	005	0+	0+	0+	0+	0+	0+	0+	1
	2	009	135	267	231	092	022	003	0+	0+	0+	0+	0+	0+	2
	3	0+	031	129	250	170	063	014	002	0+	0+	0+	0+	0+	3
	4	0+	005	043	188	219	127	042	007	001	0+	0+	0+	0+	4
	5	0+	001	010	103	206	186	092	024	003	0+	0+	0+	0+	5
	6	0+	0+	002	043	147	207	153	061	012	001	0+	0+	0+	6
	7	0+	0+	0+	014	081	177	196	118	035	003	0+	0+	0+	7
	8	0+	0+	0+	003	035	118	196	177	081	014	0+	0+	0+	8
	9	0+	0+	0+	001	012	061	153	207	147	043	002	0+	0+	9
	10	0+	0+	0+	0+	003	024	092	186	206	103	010	001	0+	10
	11	0+	0+	0+	0+	001	007	042	127	219	188	043	005	0+	11
	12	0+	0+	0+	0+	0+	002	014	063	170	250	129	031	0+	12
	13	0+	0+	0+	0+	0+	0+	003	022	092	231	267	135	009	13

16 cert. bus drivers
12 bus routes
P = .10 (won't show up)

.981

n = 16

16 ρ = .10

TABLE A-1 (continued)

n	x	0.01	0.05	0.10	0.20	0.30	0.40	P 0.50	0.60	0.70	0.80	0.90	0.95	0.99	x
	14	0+	0+	0+	0+	0+	0+	001	005	031	132	343	366	130	14
	15	0+	0+	0+	0+	0+	0+	0+	0+	005	035	206	463	860	15
16	0	852	440	185	028	003	0+	0+	0+	0+	0+	0+	0+	0+	0
	1	138	371	329	113	023	003	0+	0+	0+	0+	0+	0+	0+	1
	2	010	146	274	211	073	015	002	0+	0+	0+	0+	0+	0+	2
	3	0+	036	142	246	146	047	008	001	0+	0+	0+	0+	0+	3
	4	0+	006	051	200	204	101	028	004	0+	0+	0+	0+	0+	4
	5	0+	001	014	120	210	162	067	014	001	0+	0+	0+	0+	5
	6	0+	0+	003	055	165	198	122	039	006	0+	0+	0+	0+	6
	7	0+	0+	0+	020	101	189	175	084	018	001	0+	0+	0+	7
	8	0+	0+	0+	006	049	142	196	142	049	006	0+	0+	0+	8
	9	0+	0+	0+	001	018	084	175	189	101	020	0+	0+	0+	9
	10	0+	0+	0+	0+	006	039	122	198	165	055	003	0+	0+	10
	11	0+	0+	0+	0+	001	014	067	162	210	120	014	001	0+	11
	12	0+	0+	0+	0+	0+	004	028	101	204	200	051	006	0+	12
	13	0+	0+	0+	0+	0+	001	008	047	146	246	142	036	0+	13
	14	0+	0+	0+	0+	0+	0+	002	015	073	211	274	146	010	14
	15	0+	0+	0+	0+	0+	0+	0+	003	023	113	329	371	138	15
	16	0+	0+	0+	0+	0+	0+	0+	0+	003	028	185	440	852	16
17	0	843	418	167	022	002	0+	0+	0+	0+	0+	0+	0+	0+	0
	1	145	374	315	096	017	002	0+	0+	0+	0+	0+	0+	0+	1
	2	012	158	280	191	058	010	001	0+	0+	0+	0+	0+	0+	2
	3	001	042	156	239	124	034	005	0+	0+	0+	0+	0+	0+	3
	4	0+	008	060	209	187	080	018	002	0+	0+	0+	0+	0+	4
	5	0+	001	018	136	208	138	047	008	001	0+	0+	0+	0+	5
	6	0+	0+	004	068	178	184	094	024	003	0+	0+	0+	0+	6
	7	0+	0+	001	027	120	193	148	057	010	0+	0+	0+	0+	7
	8	0+	0+	0+	008	064	161	186	107	028	002	0+	0+	0+	8
	9	0+	0+	0+	002	028	107	186	161	064	008	0+	0+	0+	9
	10	0+	0+	0+	0+	010	057	148	193	120	027	001	0+	0+	10
	11	0+	0+	0+	0+	003	024	094	184	178	068	004	0+	0+	11
	12	0+	0+	0+	0+	001	008	047	138	208	136	018	001	0+	12
	13	0+	0+	0+	0+	0+	002	018	080	187	209	060	008	0+	13
	14	0+	0+	0+	0+	0+	0+	005	034	124	239	156	042	001	14
	15	0+	0+	0+	0+	0+	0+	001	010	058	191	280	158	012	15
	16	0+	0+	0+	0+	0+	0+	0+	002	017	096	315	374	145	16
	17	0+	0+	0+	0+	0+	0+	0+	0+	002	022	167	418	843	17
18	0	834	397	150	018	002	0+	0+	0+	0+	0+	0+	0+	0+	0
	1	152	376	300	081	013	001	0+	0+	0+	0+	0+	0+	0+	1
	2	013	168	284	172	046	007	001	0+	0+	0+	0+	0+	0+	2
	3	001	047	168	230	105	025	003	0+	0+	0+	0+	0+	0+	3
	4	0+	009	070	215	168	061	012	001	0+	0+	0+	0+	0+	4
	5	0+	001	022	151	202	115	033	004	0+	0+	0+	0+	0+	5
	6	0+	0+	005	082	187	166	071	014	001	0+	0+	0+	0+	6
	7	0+	0+	001	035	138	189	121	037	005	0+	0+	0+	0+	7
	8	0+	0+	0+	012	081	173	167	077	015	001	0+	0+	0+	8
	9	0+	0+	0+	003	039	128	186	128	039	003	0+	0+	0+	9
	10	0+	0+	0+	001	015	077	167	173	081	012	0+	0+	0+	10
	11	0+	0+	0+	0+	005	037	121	189	138	035	001	0+	0+	11
	12	0+	0+	0+	0+	001	014	071	166	187	082	005	0+	0+	1
	13	0+	0+	0+	0+	0+	004	033	115	202	151	022	001	0+	13
	14	0+	0+	0+	0+	0+	001	012	061	168	215	070	009	0+	14
	15	0+	0+	0+	0+	0+	0+	003	025	105	230	168	047	001	15
	16	0+	0+	0+	0+	0+	0+	001	007	046	172-	284	168	013	16
	17	0+	0+	0+	0+	0+	0+	0+	001	013	081	300	376	152	17
	18	0+	0+	0+	0+	0+	0+	0+	0+	002	018	150	397	834	18
19	0	826	377	135	014	001	0+	0+	0+	0+	0+	0+	0+	0+	0
	1	159	377	285	068	009	001	0+	0+	0+	0+	0+	0+	0+	1
	2	014	179	285	154	036	005	0+	0+	0+	0+	0+	0+	0+	2
	3	001	053	180	218	087	018	002	0+	0+	0+	0+	0+	0+	3
	4	0+	011	080	218	149	047	007	0+	0+	0+	0+	0+	0+	4
	5	0+	002	027	164	192	093	022	002	0+	0+	0+	0+	0+	5
	6	0+	0+	007	096	192	145	052	008	0+	0+	0+	0+	0+	6
	7	0+	0+	001	044	152	180	096	024	002	0+	0+	0+	0+	7
	8	0+	0+	0+	017	098	180	144	053	008	0+	0+	0+	0+	8
	9	0+	0+	0+	005	051	146	176	098	022	001	0+	0+	0+	9
	10	0+	0+	0+	001	022	098	176	146	051	005	0+	0+	0+	10
	11	0+	0+	0+	0+	008	053	144	180	098	017	0+	0+	0+	11

TABLE A-1   (continued)

n	x	0.01	0.05	0.10	0.20	0.30	0.40	P 0.50	0.60	0.70	0.80	0.90	0.95	0.99	x
	12	0+	0+	0+	0+	002	024	096	180	152	044	001	0+	0+	12
	13	0+	0+	0+	0+	0+	008	052	145	192	096	007	0+	0+	13
	14	0+	0+	0+	0+	0+	002	022	093	192	164	027	002	0+	14
	15	0+	0+	0+	0+	0+	0+	007	047	149	218	080	011	0+	15
	16	0+	0+	0+	0+	0+	0+	002	018	087	218	180	053	001	16
	17	0+	0+	0+	0+	0+	0+	0+	005	036	154	285	179	014	17
	18	0+	0+	0+	0+	0+	0+	0+	001	009	068	285	377	159	18
	19	0+	0+	0+	0+	0+	0+	0+	0+	001	014	135	377	826	19
20	0	818	358	122	012	001	0+	0+	0+	0+	0+	0+	0+	0+	0
	1	165	377	270	058	007	0+	0+	0+	0+	0+	0+	0+	0+	1
	2	016	189	285	137	028	003	0+	0+	0+	0+	0+	0+	0+	2
	3	001	060	190	205	072	012	001	0+	0+	0+	0+	0+	0+	3
	4	0+	013	090	218	130	035	005	0+	0+	0+	0+	0+	0+	4
	5	0+	002	032	175	179	075	015	001	0+	0+	0+	0+	0+	5
	6	0+	0+	009	109	192	124	037	005	0+	0+	0+	0+	0+	6
	7	0+	0+	002	054	164	166	074	015	001	0+	0+	0+	0+	7
	8	0+	0+	0+	022	114	180	120	036	004	0+	0+	0+	0+	8
	9	0+	0+	0+	007	065	160	160	071	012	0+	0+	0+	0+	9
	10	0+	0+	0+	002	031	117	176	117	031	002	0+	0+	0+	10
	11	0+	0+	0+	0+	012	071	160	160	065	007	0+	0+	0+	11
	12	0+	0+	0+	0+	004	036	120	180	114	022	0+	0+	0+	12
	13	0+	0+	0+	0+	001	015	074	166	164	054	002	0+	0+	13
	14	0+	0+	0+	0+	0+	005	037	124	192	109	009	0+	0+	14
	15	0+	0+	0+	0+	0+	001	015	075	179	175	032	002	0+	15
	16	0+	0+	0+	0+	0+	0+	005	035	130	218	090	013	0+	16
	17	0+	0+	0+	0+	0+	0+	001	012	072	205	190	060	001	17
	18	0+	0+	0+	0+	0+	0+	0+	003	028	137	285	189	016	18
	19	0+	0+	0+	0+	0+	0+	0+	0+	007	058	270	377	165	19
	20	0+	0+	0+	0+	0+	0+	0+	0+	001	012	122	358	818	20

TABLE A-2   Values of e^{-a}

a	e^{-a}	a	e^{-a}	a	e^{-a}	a	e^{-a}
0.00	1.000000	2.60	.074274	5.10	.006097	7.60	.000501
0.10	.904837	2.70	.067206	5.20	.005517	7.70	.000453
0.20	.818731	2.80	.060810	5.30	.004992	7.80	.000410
0.30	.740818	2.90	.055023	5.40	.004517	7.90	.000371
0.40	.670320	3.00	.049787	5.50	.004087	8.00	.000336
0.50	.606531	3.10	.045049	5.60	.003698	8.10	.000304
0.60	.548812	3.20	.040762	5.70	.003346	8.20	.000275
0.70	.496585	3.30	.036883	5.80	.003028	8.30	.000249
0.80	.449329	3.40	0.33373	5.90	.002739	8.40	.000225
0.90	.406570	3.50	.030197	6.00	.002479	8.50	.000204
1.00	.367879	3.60	.027324	6.10	.002243	8.60	.000184
1.10	.332871	3.70	.024724	6.20	.002029	8.70	.000167
1.20	.301194	3.80	.022371	6.30	.001836	8.80	.000151
1.30	.272532	3.90	.020242	6.40	.001661	8.90	.000136
1.40	.246597	4.00	.018316	6.50	.001503	9.00	.000123
1.50	.223130	4.10	.016573	6.60	.001360	9.10	.000112
1.60	.201897	4.20	.014996	6.70	.001231	9.20	.000101
1.70	.182684	4.30	.013569	6.80	.001114	9.30	.000091
1.80	.165299	4.40	.012277	6.90	.001008	9.40	.000083
1.90	.149569	4.50	.011109	7.00	.000912	9.50	.000075
2.00	.135335	4.60	.010052	7.10	.000825	9.60	.000068
2.10	.122456	4.70	.009095	7.20	.000747	9.70	.000061
2.20	.110803	4.80	.008230	7.30	.000676	9.80	.000056
2.30	.100259	4.90	.007447	7.40	.000611	9.90	.000050
2.40	.090718	5.00	.006738	7.50	.000553	10.00	.000045
2.50	.082085						

TABLE A-3 Poisson Probabilities $\left[\dfrac{e^{-\mu}\mu^x}{x!}\right]$

watch decimals of μ   μ

x	0.005	0.01	0.02	0.03	0.04	0.05	0.06	0.07	0.08	0.09
0	0.9950	0.9900	0.9802	0.9704	0.9608	0.9512	0.9418	0.9324	0.9231	0.9139
1	0.0050	0.0099	0.0192	0.0291	0.0384	0.0476	0.0565	0.0653	0.0738	0.0823
2	0.0000	0.0000	0.0002	0.0004	0.0008	0.0012	0.0017	0.0023	0.0030	0.0037
3	0.0000	0.0000	0.0000	0.0000	0.0000	0.0000	0.0000	0.0001	0.0001	0.0001

x	0.1	0.2	0.3	0.4	0.5	0.6	0.7	0.8	0.9	1.0
0	0.9048	0.8187	0.7408	0.6703	0.6065	0.5488	0.4966	0.4493	0.4066	0.3679
1	0.0905	0.1637	0.2222	0.2681	0.3033	0.3293	0.3476	0.3595	0.3659	0.3679
2	0.0045	0.0164	0.0333	0.0536	0.0758	0.0988	0.1217	0.1438	0.1647	0.1839
3	0.0002	0.0011	0.0033	0.0072	0.0126	0.0198	0.0284	0.0383	0.0494	0.0613
4	0.0000	0.0001	0.0002	0.0007	0.0016	0.0030	0.0050	0.0077	0.0111	0.0153
5	0.0000	0.0000	0.0000	0.0001	0.0002	0.0004	0.0007	0.0012	0.0020	0.0031
6	0.0000	0.0000	0.0000	0.0000	0.0000	0.0000	0.0001	0.0002	0.0003	0.0005
7	0.0000	0.0000	0.0000	0.0000	0.0000	0.0000	0.0000	0.0000	0.0000	0.0001

x	1.1	1.2	1.3	1.4	1.5	1.6	1.7	1.8	1.9	2.0
0	0.3329	0.3012	0.2725	0.2466	0.2231	0.2019	0.1827	0.1653	0.1496	0.1353
1	0.3662	0.3614	0.3543	0.3452	0.3347	0.3230	0.3106	0.2975	0.2842	0.2707
2	0.2014	0.2169	0.2303	0.2417	0.2510	0.2584	0.2640	0.2678	0.2700	0.2707
3	0.0738	0.0867	0.0998	0.1128	0.1255	0.1378	0.1496	0.1607	0.1710	0.1804
4	0.0203	0.0260	0.0324	0.0395	0.0471	0.0551	0.0636	0.0723	0.0812	0.0902
5	0.0045	0.0062	0.0084	0.0111	0.0141	0.0176	0.0216	0.0260	0.0309	0.0361
6	0.0008	0.0012	0.0018	0.0026	0.0035	0.0047	0.0061	0.0078	0.0098	0.0120
7	0.0001	0.0002	0.0003	0.0005	0.0008	0.0011	0.0015	0.0020	0.0027	0.0034
8	0.0000	0.0000	0.0001	0.0001	0.0001	0.0002	0.0003	0.0005	0.0006	0.0009
9	0.0000	0.0000	0.0000	0.0000	0.0000	0.0000	0.0001	0.0001	0.0001	0.0002

x	2.1	2.2	2.3	2.4	2.5	2.6	2.7	2.8	2.9	3.0
0	0.1225	0.1108	0.1003	0.0907	0.0821	0.0743	0.0672	0.0608	0.0050	0.0498
1	0.2572	0.2438	0.2306	0.2177	0.2052	0.1931	0.1815	0.1703	0.1596	0.1494
2	0.2700	0.2681	0.2652	0.2613	0.2565	0.2510	0.2450	0.2384	0.2314	0.2240
3	0.1890	0.1966	0.2033	0.2090	0.2138	0.2176	0.2205	0.2225	0.2237	0.2240
4	0.0992	0.1082	0.1169	0.1254	0.1336	0.1414	0.1488	0.1557	0.1622	0.1680
5	0.0417	0.0476	0.0538	0.0602	0.0668	0.0735	0.0804	0.0872	0.0940	0.1008
6	0.0146	0.0174	0.0206	0.0241	0.0278	0.0319	0.0362	0.0407	0.0455	0.0504
7	0.0044	0.0055	0.0068	0.0083	0.0099	0.0118	0.0139	0.0163	0.0188	0.0216
8	0.0011	0.0015	0.0019	0.0025	0.0031	0.0038	0.0047	0.0057	0.0068	0.0081
9	0.0003	0.0004	0.0005	0.0007	0.0009	0.0011	0.0014	0.0018	0.0022	0.0027
10	0.0001	0.0001	0.0001	0.0002	0.0002	0.0003	0.0004	0.0005	0.0006	0.0008
11	0.0000	0.0000	0.0000	0.0000	0.0000	0.0001	0.0001	0.0001	0.0002	0.0002
12	0.0000	0.0000	0.0000	0.0000	0.0000	0.0000	0.0000	0.0000	0.0000	0.0001

x	3.1	3.2	3.3	3.4	3.5	3.6	3.7	3.8	3.9	4.0
0	0.0450	0.0408	0.0369	0.0334	0.0302	0.0273	0.0247	0.0224	0.0202	0.0183
1	0.1397	0.1304	0.1217	0.1135	0.1057	0.0984	0.0915	0.0850	0.0789	0.0733
2	0.2165	0.2087	0.2008	0.1929	0.1850	0.1771	0.1692	0.1615	0.1539	0.1465
3	0.2237	0.2226	0.2209	0.2186	0.2158	0.2125	0.2087	0.2046	0.2001	0.1954
4	0.1734	0.1781	0.1823	0.1858	0.1888	0.1912	0.1931	0.1944	0.1951	0.1954
5	0.1075	0.1140	0.1203	0.1264	0.1322	0.1377	0.1429	0.1477	0.1522	0.1563
6	0.0555	0.0608	0.0662	0.0716	0.0771	0.0826	0.0881	0.0936	0.0989	0.1042
7	0.0246	0.0278	0.0312	0.0348	0.0385	0.0425	0.0466	0.0508	0.0551	0.0595
8	0.0095	0.0111	0.0129	0.0148	0.0169	0.0191	0.0215	0.0241	0.0269	0.0298
9	0.0033	0.0040	0.0047	0.0056	0.0066	0.0076	0.0089	0.0102	0.0116	0.0132
10	0.0010	0.0013	0.0016	0.0019	0.0023	0.0028	0.0033	0.0039	0.0045	0.0053
11	0.0003	0.0004	0.0005	0.0006	0.0007	0.0009	0.0011	0.0013	0.0016	0.0019
12	0.0001	0.0001	0.0001	0.0002	0.0002	0.0003	0.0003	0.0004	0.0005	0.0006
13	0.0000	0.0000	0.0000	0.0000	0.0001	0.0001	0.0001	0.0001	0.0002	0.0002
14	0.0000	0.0000	0.0000	0.0000	0.0000	0.0000	0.0000	0.0000	0.0000	0.0001

x	4.1	4.2	4.3	4.4	4.5	4.6	4.7	4.8	4.9	5.0
0	0.0166	0.0150	0.0136	0.0123	0.0111	0.0101	0.0091	0.0082	0.0074	0.0067
1	0.0679	0.0630	0.0583	0.0540	0.0500	0.0462	0.0427	0.0395	0.0365	0.0337
2	0.1393	0.1323	0.1254	0.1188	0.1125	0.1063	0.1005	0.0948	0.0894	0.0842
3	0.1904	0.1852	0.1798	0.1743	0.1687	0.1631	0.1574	0.1517	0.1460	0.1404
4	0.1951	0.1944	0.1933	0.1917	0.1898	0.1875	0.1849	0.1820	0.1789	0.1755

TABLE A-3 (continued)

					μ					
x	4.1	4.2	4.3	4.4	4.5	4.6	4.7	4.8	4.9	5.0
5	0.1600	0.1633	0.1662	0.1687	0.1708	0.1725	0.1738	0.1747	0.1753	0.1755
6	0.1093	0.1143	0.1191	0.1237	0.1281	0.1323	0.1362	0.1398	0.1432	0.1462
7	0.0640	0.0686	0.0732	0.0778	0.0824	0.0869	0.0914	0.0959	0.1002	0.1044
8	0.0328	0.0360	0.0393	0.0428	0.0463	0.0500	0.0537	0.0575	0.0614	0.0653
9	0.0150	0.0168	0.0188	0.0209	0.0232	0.0255	0.0280	0.0307	0.0334	0.0363
10	0.0061	0.0071	0.0081	0.0092	0.0104	0.0118	0.0132	0.0147	0.0164	0.0181
11	0.0023	0.0027	0.0032	0.0037	0.0043	0.0049	0.0056	0.0064	0.0073	0.0082
12	0.0008	0.0009	0.0011	0.0014	0.0016	0.0019	0.0022	0.0026	0.0030	0.0034
13	0.0002	0.0003	0.0004	0.0005	0.0006	0.0007	0.0008	0.0009	0.0011	0.0013
14	0.0001	0.0001	0.0001	0.0001	0.0002	0.0002	0.0003	0.0003	0.0004	0.0005
15	0.0000	0.0000	0.0000	0.0000	0.0001	0.0001	0.0001	0.0001	0.0001	0.0002

x	5.1	5.2	5.3	5.4	5.5	5.6	5.7	5.8	5.9	6.0
0	0.0061	0.0055	0.0050	0.0045	0.0041	0.0037	0.0033	0.0030	0.0027	0.0025
1	0.0311	0.0287	0.0265	0.0244	0.0225	0.0207	0.0191	0.0176	0.0162	0.0149
2	0.0793	0.0746	0.0701	0.0659	0.0618	0.0580	0.0544	0.0509	0.0477	0.0446
3	0.1348	0.1293	0.1239	0.1185	0.1133	0.1082	0.1033	0.0985	0.0938	0.0892
4	0.1719	0.1681	0.1641	0.1600	0.1558	0.1515	0.1472	0.1428	0.1383	0.1339
5	0.1753	0.1748	0.1740	0.1728	0.1714	0.1697	0.1678	0.1656	0.1632	0.1606
6	0.1490	0.1515	0.1537	0.1555	0.1571	0.1584	0.1594	0.1601	0.1605	0.1606
7	0.1086	0.1125	0.1163	0.1200	0.1234	0.1267	0.1298	0.1326	0.1353	0.1377
8	0.0692	0.0731	0.0771	0.0810	0.0849	0.0887	0.0925	0.0962	0.0998	0.1033
9	0.0392	0.0423	0.0454	0.0486	0.0519	0.0552	0.0586	0.0620	0.0654	0.0688
10	0.0200	0.0220	0.0241	0.0262	0.0285	0.0309	0.0334	0.0359	0.0386	0.0413
11	0.0093	0.0104	0.0116	0.0129	0.0143	0.0157	0.0173	0.0190	0.0207	0.0225
12	0.0039	0.0045	0.0051	0.0058	0.0065	0.0073	0.0082	0.0092	0.0102	0.0113
13	0.0015	0.0018	0.0021	0.0024	0.0028	0.0032	0.0036	0.0041	0.0046	0.0052
14	0.0006	0.0007	0.0008	0.0009	0.0011	0.0013	0.0015	0.0017	0.0019	0.0022
15	0.0002	0.0002	0.0003	0.0003	0.0004	0.0005	0.0006	0.0007	0.0008	0.0009
16	0.0001	0.0001	0.0001	0.0001	0.0001	0.0002	0.0002	0.0002	0.0003	0.0003
17	0.0000	0.0000	0.0000	0.0000	0.0000	0.0001	0.0001	0.0001	0.0001	0.0001

x	6.1	6.2	6.3	6.4	6.5	6.6	6.7	6.8	6.9	7.0
0	0.0022	0.0020	0.0018	0.0017	0.0015	0.0014	0.0012	0.0011	0.0010	0.0009
1	0.0137	0.0126	0.0116	0.0106	0.0098	0.0090	0.0082	0.0076	0.0070	0.0064
2	0.0417	0.0390	0.0364	0.0340	0.0318	0.0296	0.0276	0.0258	0.0240	0.0223
3	0.0848	0.0806	0.0765	0.0726	0.0688	0.0652	0.0617	0.0584	0.0552	0.0521
4	0.1294	0.1269	0.1205	0.1162	0.1118	0.1076	0.1034	0.0992	0.0952	0.0912
5	0.1579	0.1549	0.1519	0.1487	0.1454	0.1420	0.1385	0.1349	0.1314	0.1277
6	0.1605	0.1601	0.1595	0.1586	0.1575	0.1562	0.1546	0.1529	0.1511	0.1490
7	0.1399	0.1418	0.1435	0.1450	0.1462	0.1472	0.1480	0.1486	0.1489	0.1490
8	0.1066	0.1099	0.1130	0.1160	0.1188	0.1215	0.1240	0.1263	0.1284	0.1304
9	0.0723	0.0757	0.0791	0.0825	0.0858	0.0891	0.0923	0.0954	0.0985	0.1014
10	0.0441	0.0469	0.0498	0.0528	0.0558	0.0588	0.0618	0.0649	0.0679	0.0710
11	0.0245	0.0265	0.0285	0.0307	0.0330	0.0353	0.0377	0.0401	0.0426	0.0452
12	0.0124	0.0137	0.0150	0.0164	0.0179	0.0194	0.0210	0.0227	0.0245	0.0264
13	0.0058	0.0065	0.0073	0.0081	0.0089	0.0098	0.0108	0.0119	0.0130	0.0142
14	0.0025	0.0029	0.0033	0.0037	0.0041	0.0046	0.0052	0.0058	0.0064	0.0071
15	0.0010	0.0012	0.0014	0.0016	0.0018	0.0020	0.0023	0.0026	0.0029	0.0033
16	0.0004	0.0005	0.0005	0.0006	0.0007	0.0008	0.0010	0.0011	0.0013	0.0014
17	0.0001	0.0002	0.0002	0.0002	0.0003	0.0003	0.0004	0.0004	0.0005	0.0006
18	0.0000	0.0001	0.0001	0.0001	0.0001	0.0001	0.0001	0.0002	0.0002	0.0002
19	0.0000	0.0000	0.0000	0.0000	0.0000	0.0000	0.0000	0.0001	0.0001	0.0001

x	7.1	7.2	7.3	7.4	7.5	7.6	7.7	7.8	7.9	8.0
0	0.0008	0.0007	0.0007	0.0006	0.0006	0.0005	0.0005	0.0004	0.0004	0.0003
1	0.0059	0.0054	0.0049	0.0045	0.0041	0.0038	0.0035	0.0032	0.0029	0.0027
2	0.0208	0.0194	0.0180	0.0167	0.0156	0.0145	0.0134	0.0125	0.0116	0.0107
3	0.0492	0.0464	0.0438	0.0413	0.0389	0.0366	0.0345	0.0324	0.0305	0.0286
4	0.0874	0.0836	0.0799	0.0764	0.0729	0.0696	0.0663	0.0632	0.0602	0.0573
5	0.1241	0.1204	0.1167	0.1130	0.1094	0.1057	0.1021	0.0986	0.0951	0.0916
6	0.1468	0.1445	0.1420	0.1394	0.1367	0.1339	0.1311	0.1282	0.1252	0.1221
7	0.1489	0.1486	0.1481	0.1474	0.1465	0.1454	0.1442	0.1428	0.1413	0.1396
8	0.1321	0.1337	0.1351	0.1363	0.1373	0.1382	0.1388	0.1392	0.1395	0.1396
9	0.1042	0.1070	0.1096	0.1121	0.1144	0.1167	0.1187	0.1207	0.1224	0.1241
10	0.0740	0.0770	0.0800	0.0829	0.0858	0.0887	0.0914	0.0941	0.0967	0.0993
11	0.0478	0.0504	0.0531	0.0558	0.0585	0.0613	0.0640	0.0667	0.0695	0.0722

TABLE A-3 (continued)

					μ					
x	7.1	7.2	7.3	7.4	7.5	7.6	7.7	7.8	7.9	8.0
12	0.0283	0.0303	0.0323	0.0344	0.0366	0.0388	0.0411	0.0434	0.0457	0.0481
13	0.0154	0.0168	0.0181	0.0196	0.0211	0.0227	0.0243	0.0260	0.0278	0.0296
14	0.0078	0.0086	0.0095	0.0104	0.0113	0.0123	0.0134	0.0145	0.0157	0.0169
15	0.0037	0.0041	0.0046	0.0051	0.0057	0.0062	0.0069	0.0075	0.0083	0.0090
16	0.0016	0.0019	0.0021	0.0024	0.0026	0.0030	0.0033	0.0037	0.0041	0.0045
17	0.0007	0.0008	0.0009	0.0010	0.0012	0.0013	0.0015	0.0017	0.0019	0.0021
18	0.0003	0.0003	0.0004	0.0004	0.0005	0.0006	0.0006	0.0007	0.0008	0.0009
19	0.0001	0.0001	0.0001	0.0002	0.0002	0.0002	0.0003	0.0003	0.0003	0.0004
20	0.0000	0.0000	0.0001	0.0001	0.0001	0.0001	0.0001	0.0001	0.0001	0.0002
21	0.0000	0.0000	0.0000	0.0000	0.0000	0.0000	0.0000	0.0000	0.0001	0.0001

x	8.1	8.2	8.3	8.4	8.5	8.6	8.7	8.8	8.9	9.0
0	0.0003	0.0003	0.0002	0.0002	0.0002	0.0002	0.0002	0.0002	0.0001	0.0001
1	0.0025	0.0023	0.0021	0.0019	0.0017	0.0016	0.0014	0.0013	0.0012	0.0011
2	0.0100	0.0092	0.0086	0.0079	0.0074	0.0068	0.0063	0.0058	0.0054	0.0050
3	0.0269	0.0252	0.0237	0.0222	0.0208	0.0195	0.0183	0.0171	0.0160	0.0150
4	0.0544	0.0517	0.0491	0.0466	0.0443	0.0420	0.0398	0.0377	0.0357	0.0337
5	0.0882	0.0849	0.0816	0.0784	0.0752	0.0722	0.0692	0.0663	0.0635	0.0607
6	0.1191	0.1160	0.1128	0.1097	0.1066	0.1034	0.1003	0.0972	0.0941	0.0911
7	0.1378	0.1358	0.1338	0.1317	0.1294	0.1271	0.1247	0.1222	0.1197	0.1171
8	0.1395	0.1392	0.1388	0.1382	0.1375	0.1366	0.1356	0.1344	0.1332	0.1318
9	0.1256	0.1269	0.1280	0.1290	0.1299	0.1306	0.1311	0.1315	0.1317	0.1318
10	0.1017	0.1040	0.1063	0.1084	0.1104	0.1123	0.1140	0.1157	0.1172	0.1186
11	0.0749	0.0776	0.0802	0.0828	0.0853	0.0878	0.0902	0.0925	0.0948	0.0970
12	0.0505	0.0530	0.0555	0.0579	0.0604	0.0629	0.0654	0.0679	0.0703	0.0728
13	0.0315	0.0334	0.0354	0.0374	0.0395	0.0416	0.0438	0.0459	0.0481	0.0504
14	0.0182	0.0196	0.0210	0.0225	0.0240	0.0256	0.0272	0.0289	0.0306	0.0324
15	0.0098	0.0107	0.0116	0.0126	0.0136	0.0147	0.0158	0.0169	0.0182	0.0194
16	0.0050	0.0055	0.0060	0.0066	0.0072	0.0079	0.0086	0.0093	0.0101	0.0109
17	0.0024	0.0026	0.0029	0.0033	0.0036	0.0040	0.0044	0.0048	0.0053	0.0058
18	0.0011	0.0012	0.0014	0.0015	0.0017	0.0019	0.0021	0.0024	0.0026	0.0029
19	0.0005	0.0005	0.0006	0.0007	0.0008	0.0009	0.0010	0.0011	0.0012	0.0014
20	0.0002	0.0002	0.0002	0.0003	0.0003	0.0004	0.0004	0.0005	0.0005	0.0006
21	0.0001	0.0001	0.0001	0.0001	0.0001	0.0002	0.0002	0.0002	0.0002	0.0003
22	0.0000	0.0000	0.0000	0.0000	0.0001	0.0001	0.0001	0.0001	0.0001	0.0001

x	9.1	9.2	9.3	9.4	9.5	9.6	9.7	9.8	9.9	10.0
0	0.0001	0.0001	0.0001	0.0001	0.0001	0.0001	0.0001	0.0001	0.0001	0.0000
1	0.0010	0.0009	0.0009	0.0008	0.0007	0.0007	0.0006	0.0005	0.0005	0.0005
2	0.0046	0.0043	0.0040	0.0037	0.0034	0.0031	0.0029	0.0027	0.0025	0.0023
3	0.0140	0.0131	0.0123	0.0115	0.0107	0.0100	0.0093	0.0087	0.0081	0.0076
4	0.0319	0.0302	0.0285	0.0269	0.0254	0.0240	0.0226	0.0213	0.0201	0.0189
5	0.0581	0.0555	0.0530	0.0506	0.0483	0.0460	0.0439	0.0418	0.0398	0.0378
6	0.0881	0.0851	0.0822	0.0793	0.0764	0.0736	0.0709	0.0682	0.0656	0.0631
7	0.1145	0.1118	0.1091	0.1064	0.1037	0.1010	0.0982	0.0955	0.0928	0.0901
8	0.1302	0.1286	0.1269	0.1251	0.1232	0.1212	0.1191	0.1170	0.1148	0.1126
9	0.1317	0.1315	0.1311	0.1306	0.1300	0.1293	0.1284	0.1274	0.1263	0.1251
10	0.1198	0.1210	0.1219	0.1228	0.1235	0.1241	0.1245	0.1249	0.1250	0.1251
11	0.0991	0.1012	0.1031	0.1049	0.1067	0.1083	0.1098	0.1112	0.1125	0.1137
12	0.0752	0.0776	0.0799	0.0822	0.0844	0.0866	0.0888	0.0908	0.0928	0.0948
13	0.0526	0.0549	0.0572	0.0594	0.0617	0.0640	0.0662	0.0685	0.0707	0.0729
14	0.0342	0.0361	0.0380	0.0399	0.0419	0.0439	0.0459	0.0479	0.0500	0.0521
15	0.0208	0.0221	0.0235	0.0250	0.0265	0.0281	0.0297	0.0313	0.0330	0.0347
16	0.0118	0.0127	0.0137	0.0147	0.0157	0.0168	0.0180	0.0192	0.0204	0.0217
17	0.0063	0.0069	0.0075	0.0081	0.0088	0.0095	0.0103	0.0111	0.0119	0.0128
18	0.0032	0.0035	0.0039	0.0042	0.0046	0.0051	0.0055	0.0060	0.0065	0.0071
19	0.0015	0.0017	0.0019	0.0021	0.0023	0.0026	0.0028	0.0031	0.0034	0.0037
20	0.0007	0.0008	0.0009	0.0010	0.0011	0.0012	0.0014	0.0015	0.0017	0.0019
21	0.0003	0.0003	0.0004	0.0004	0.0005	0.0006	0.0006	0.0007	0.0008	0.0009
22	0.0001	0.0001	0.0002	0.0002	0.0002	0.0002	0.0003	0.0003	0.0004	0.0004
23	0.0000	0.0001	0.0001	0.0001	0.0001	0.0001	0.0001	0.0001	0.0002	0.0002
24	0.0000	0.0000	0.0000	0.0000	0.0000	0.0000	0.0000	0.0001	0.0001	0.0001

TABLE A-4 Areas of the Standard Normal Distribution

The entries in this table are the probabilities that a standard normal random variable is between 0 and *z* (the shaded area).

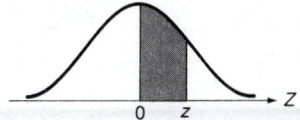

z	0.00	0.01	0.02	0.03	0.04	0.05	0.06	0.07	0.08	0.09
0.0	0.0000	0.0040	0.0080	0.0120	0.0160	0.0199	0.0239	0.0279	0.0319	0.0359
0.1	0.0398	0.0438	0.0478	0.0517	0.0557	0.0596	0.0636	0.0675	0.0714	0.0753
0.2	0.0793	0.0832	0.0871	0.0910	0.0948	0.0987	0.1026	0.1064	0.1103	0.1141
0.3	0.1179	0.1217	0.1255	0.1293	0.1331	0.1368	0.1406	0.1443	0.1480	0.1517
0.4	0.1554	0.1591	0.1628	0.1664	0.1700	0.1736	0.1772	0.1808	0.1844	0.1879
0.5	0.1915	0.1950	0.1985	0.2019	0.2054	0.2088	0.2123	0.2157	0.2190	0.2224
0.6	0.2257	0.2291	0.2324	0.2357	0.2389	0.2422	0.2454	0.2486	0.2517	0.2549
0.7	0.2580	0.2611	0.2642	0.2673	0.2704	0.2734	0.2764	0.2794	0.2823	0.2852
0.8	0.2881	0.2910	0.2939	0.2967	0.2995	0.3023	0.3051	0.3078	0.3106	0.3133
0.9	0.3159	0.3186	0.3212	0.3238	0.3264	0.3289	0.3315	0.3340	0.3365	0.3389
1.0	0.3413	0.3438	0.3461	0.3485	0.3508	0.3531	0.3554	0.3577	0.3599	0.3621
1.1	0.3643	0.3665	0.3686	0.3708	0.3729	0.3749	0.3770	0.3790	0.3810	0.3830
1.2	0.3849	0.3869	0.3888	0.3907	0.3925	0.3944	0.3962	0.3980	0.3997	0.4015
1.3	0.4032	0.4049	0.4066	0.4082	0.4099	0.4115	0.4131	0.4147	0.4162	0.4177
1.4	0.4192	0.4207	0.4222	0.4236	0.4251	0.4265	0.4279	0.4292	0.4306	0.4319
1.5	0.4332	0.4345	0.4357	0.4370	0.4382	0.4394	0.4406	0.4418	0.4429	0.4441
1.6	0.4452	0.4463	0.4474	0.4484	0.4495	0.4505	0.4515	0.4525	0.4535	0.4545
1.7	0.4554	0.4564	0.4573	0.4582	0.4591	0.4599	0.4608	0.4616	0.4625	0.4633
1.8	0.4641	0.4649	0.4656	0.4664	0.4671	0.4678	0.4686	0.4693	0.4699	0.4706
1.9	0.4713	0.4719	0.4726	0.4732	0.4738	0.4744	0.4750	0.4756	0.4761	0.4767
2.0	0.4772	0.4778	0.4783	0.4788	0.4793	0.4798	0.4803	0.4808	0.4812	0.4817
2.1	0.4821	0.4826	0.4830	0.4834	0.4838	0.4842	0.4846	0.4850	0.4854	0.4857
2.2	0.4861	0.4864	0.4868	0.4871	0.4875	0.4878	0.4881	0.4884	0.4887	0.4890
2.3	0.4893	0.4896	0.4898	0.4901	0.4904	0.4906	0.4909	0.4911	0.4913	0.4916
2.4	0.4918	0.4920	0.4922	0.4925	0.4927	0.4929	0.4931	0.4932	0.4934	0.4936
2.5	0.4938	0.4940	0.4941	0.4943	0.4945	0.4946	0.4948	0.4949	0.4951	0.4952
2.6	0.4953	0.4955	0.4956	0.4957	0.4959	0.4960	0.4961	0.4962	0.4963	0.4974
2.7	0.4965	0.4966	0.4967	0.4968	0.4969	0.4970	0.4971	0.4972	0.4973	0.4974
2.8	0.4974	0.4975	0.4976	0.4977	0.4977	0.4978	0.4979	0.4979	0.4980	0.4981
2.9	0.4981	0.4982	0.4982	0.4983	0.4984	0.4984	0.4985	0.4985	0.4986	0.4986
3.0	0.4987	0.4987	0.4987	0.4988	0.4988	0.4989	0.4989	0.4989	0.4990	0.4990
3.1	0.4990	0.4991	0.4991	0.4991	0.4992	0.4992	0.4992	0.4992	0.4993	0.4993
3.2	0.4993	0.4993	0.4994	0.4994	0.4994	0.4994	0.4994	0.4995	0.4995	0.4995
3.3	0.4995	0.4995	0.4995	0.4996	0.4996	0.4996	0.4996	0.4996	0.4996	0.4997
3.4	0.4997	0.4997	0.4997	0.4997	0.4997	0.4997	0.4997	0.4997	0.4997	0.4998
3.5	0.4998									
4.0	0.49997									
4.5	0.499997									
5.0	0.4999997									

Reprinted with permission from *Standard Mathematical Tables*, 15th ed., © CRC Press, Inc., Boca Raton, FL.

TABLE A-5   Critical Values of *t*

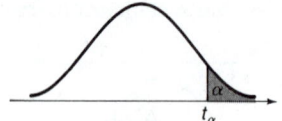

DEGREES OF FREEDOM	$t_{.100}$	$t_{.050}$	$t_{.025}$	$t_{.010}$	$t_{.005}$
1	3.078	6.314	12.706	31.821	63.657
2	1.886	2.920	4.303	6.965	9.925
3	1.638	2.353	3.182	4.541	5.841
4	1.533	2.132	2.776	3.747	4.604
5	1.476	2.015	2.571	3.365	4.032
6	1.440	1.943	2.447	3.143	3.707
7	1.415	1.895	2.365	2.998	3.499
8	1.397	1.860	2.306	2.896	3.355
9	1.383	1.833	2.262	2.821	3.250
10	1.372	1.812	2.228	2.764	3.169
11	1.363	1.796	2.201	2.718	3.106
12	1.356	1.782	2.179	2.681	3.055
13	1.350	1.771	2.160	2.650	3.012
14	1.345	1.761	2.145	2.624	2.977
15	1.341	1.753	2.131	2.602	2.947
16	1.337	1.746	2.120	2.583	2.921
17	1.333	1.740	2.110	2.567	2.898
18	1.330	1.734	2.101	2.552	2.878
19	1.328	1.729	2.093	2.539	2.861
20	1.325	1.725	2.086	2.528	2.845
21	1.323	1.721	2.080	2.518	2.831
22	1.321	1.717	2.074	2.508	2.819
23	1.319	1.714	2.069	2.500	2.808
24	1.318	1.711	2.064	2.492	2.797
25	1.316	1.708	2.060	2.485	2.787
26	1.315	1.706	2.056	2.479	2.779
27	1.314	1.703	2.052	2.473	2.771
28	1.313	1.701	2.048	2.467	2.763
29	1.311	1.699	2.045	2.462	2.756
30	1.310	1.697	2.042	2.457	2.750
40	1.303	1.684	2.021	2.423	2.704
60	1.296	1.671	2.000	2.390	2.660
120	1.289	1.658	1.980	2.358	2.617
∞	1.282	1.645	1.960	2.326	2.576

From M. Merrington, ''Table of Percentage Points of the *t*-Distribution,'' *Biometrika*, 1941, 32, 300. Reproduced by permission of the *Biometrika* trustees.

TABLE A-6 Critical Values of χ^2

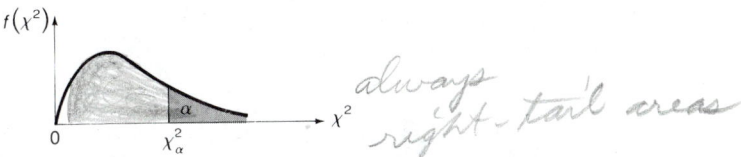

DEGREES OF FREEDOM	$\chi^2_{.995}$	$\chi^2_{.990}$	$\chi^2_{.975}$	$\chi^2_{.950}$	$\chi^2_{.900}$
1	0.0000393	0.0001571	0.0009821	0.0039321	0.0157908
2	0.0100251	0.0201007	0.0506356	0.102587	0.210720
3	0.0717212	0.114832	0.215795	0.351846	0.584375
4	0.206990	0.297110	0.484419	0.710721	1.063623
5	0.411740	0.554300	0.831211	1.145476	1.61031
6	0.675727	0.872085	1.237347	1.63539	2.20413
7	0.989265	1.239043	1.68987	2.16735	2.83311
8	1.344419	1.646482	2.17973	2.73264	3.48954
9	1.734926	2.087912	2.70039	3.32511	4.16816
10	2.15585	2.55821	3.24697	3.94030	4.86518
11	2.60321	3.05347	3.81575	4.57481	5.57779
12	3.07382	3.57056	4.40379	5.22603	6.30380
13	3.56503	4.10691	5.00874	5.89186	7.04150
14	4.07468	4.66043	5.62872	6.57063	7.78953
15	4.60094	5.22935	6.26214	7.26094	8.54675
16	5.14224	5.81221	6.90766	7.96164	9.31223
17	5.69724	6.40776	7.56418	8.67176	10.0852
18	6.26481	7.01491	8.23075	9.39046	10.8649
19	6.84398	7.63273	8.90655	10.1170	11.6509
20	7.43386	8.26040	9.59083	10.8508	12.4426
21	8.03366	8.89720	10.28293	11.5913	13.2396
22	8.64272	9.54249	10.9823	12.3380	14.0415
23	9.26042	10.19567	11.6885	13.0905	14.8479
24	9.88623	10.8564	12.4011	13.8484	15.6587
25	10.5197	11.5240	13.1197	14.6114	16.4734
26	11.1603	12.1981	13.8439	15.3791	17.2919
27	11.8076	12.8786	14.5733	16.1513	18.1138
28	12.4613	13.5648	15.3079	16.9279	18.9392
29	13.1211	14.2565	16.0471	17.7083	19.7677
30	13.7867	14.9535	16.7908	18.4926	20.5992
40	20.7065	22.1643	24.4331	26.5093	29.0505
50	27.9907	29.7067	32.3574	34.7642	37.6886
60	35.5346	37.4848	40.4817	43.1879	46.4589
70	43.2752	45.4418	48.7576	51.7393	55.3290
80	51.1720	53.5400	57.1532	60.3915	64.2778
90	59.1963	61.7541	65.6466	69.1260	73.2912
100	67.3276	70.0648	74.2219	77.9295	82.3581

TABLE A-6   (continued)

DEGREES OF FREEDOM	$\chi^2_{.100}$	$\chi^2_{.050}$	$\chi^2_{.025}$	$\chi^2_{.010}$	$\chi^2_{.005}$
1	2.70554	3.84146	5.02389	6.63490	7.87944
2	4.60517	5.99147	7.37776	9.21034	10.5966
3	6.25139	7.81473	9.34840	11.3449	12.8381
4	7.77944	9.48773	11.1433	13.2767	14.8602
5	9.23635	11.0705	12.8325	15.0863	16.7496
6	10.6446	12.5916	14.4494	16.8119	18.5476
7	12.0170	14.0671	16.0128	18.4753	20.2777
8	13.3616	15.5073	17.5346	20.0902	21.9550
9	14.6837	16.9190	19.0228	21.6660	23.5893
10	15.9871	18.3070	20.4831	23.2093	25.1882
11	17.2750	19.6751	21.9200	24.7250	26.7569
12	18.5494	21.0261	23.3367	26.2170	28.2995
13	19.8119	22.3621	24.7356	27.6883	29.8194
14	21.0642	23.6848	26.1190	29.1413	31.3193
15	22.3072	24.9958	27.4884	30.5779	32.8013
16	23.5418	26.2962	28.8454	31.9999	34.2672
17	24.7690	27.5871	30.1910	33.4087	35.7185
18	25.9894	28.8693	31.5264	34.8053	37.1564
19	27.2036	30.1435	32.8523	36.1908	38.5822
20	28.4120	31.4104	34.1696	37.5662	39.9968
21	29.6151	32.6705	35.4789	38.9321	41.4010
22	30.8133	33.9244	36.7807	40.2894	42.7956
23	32.0069	35.1725	38.0757	41.6384	44.1813
24	33.1963	36.4151	39.3641	42.9798	45.5585
25	34.3816	37.6525	40.6465	44.3141	46.9278
26	35.5631	38.8852	41.9232	45.6417	48.2899
27	36.7412	40.1133	43.1944	46.9630	49.6449
28	37.9159	41.3372	44.4607	48.2782	50.9933
29	39.0875	42.5569	45.7222	49.5879	52.3356
30	40.2560	43.7729	46.9792	50.8922	53.6720
40	51.8050	55.7585	59.3417	63.6907	66.7659
50	63.1671	67.5048	71.4202	76.1539	79.4900
60	74.3970	79.0819	83.2976	88.3794	91.9517
70	85.5271	90.5312	95.0231	100.425	104.215
80	96.5782	101.879	106.629	112.329	116.321
90	107.565	113.145	118.136	124.116	128.229
100	118.498	124.342	129.561	135.807	140.169

TABLE A-7 Percentage Points of the F Distribution

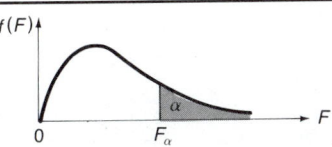

v_2	v_1 NUMERATOR DEGREES OF FREEDOM								
	1	**2**	**3**	**4**	**5**	**6**	**7**	**8**	**9**
1	39.86	49.50	53.59	55.83	57.24	58.20	58.91	59.44	59.86
2	8.53	9.00	9.16	9.24	9.29	9.33	9.35	9.37	9.38
3	5.54	5.46	5.39	5.34	5.31	5.28	5.27	5.25	5.24
4	4.54	4.32	4.19	4.11	4.05	4.01	3.98	3.95	3.94
5	4.06	3.78	3.62	3.52	3.45	3.40	3.37	3.34	3.32
6	3.78	3.46	3.29	3.18	3.11	3.05	3.01	2.98	2.96
7	3.59	3.26	3.07	2.96	2.88	2.83	2.78	2.75	2.72
8	3.46	3.11	2.92	2.81	2.73	2.67	2.62	2.59	2.56
9	3.36	3.01	2.81	2.69	2.61	2.55	2.51	2.47	2.44
10	3.29	2.92	2.73	2.61	2.52	2.46	2.41	2.38	2.35
11	3.23	2.86	2.66	2.54	2.45	2.39	2.34	2.30	2.27
12	3.18	2.81	2.61	2.48	2.39	2.33	2.28	2.24	2.21
13	3.14	2.76	2.56	2.43	2.35	2.28	2.23	2.20	2.16
14	3.10	2.73	2.52	2.39	2.31	2.24	2.19	2.15	2.12
15	3.07	2.70	2.49	2.36	2.27	2.21	2.16	2.12	2.09
16	3.05	2.67	2.46	2.33	2.24	2.18	2.13	2.09	2.06
17	3.03	2.64	2.44	2.31	2.22	2.15	2.10	2.06	2.03
18	3.01	2.62	2.42	2.29	2.20	2.13	2.08	2.04	2.00
19	2.99	2.61	2.40	2.27	2.18	2.11	2.06	2.02	1.98
20	2.97	2.59	2.38	2.25	2.16	2.09	2.04	2.00	1.96
21	2.96	2.57	2.36	2.23	2.14	2.08	2.02	1.98	1.95
22	2.95	2.56	2.35	2.22	2.13	2.06	2.01	1.97	1.93
23	2.94	2.55	2.34	2.21	2.11	2.05	1.99	1.95	1.92
24	2.93	2.54	2.33	2.19	2.10	2.04	1.98	1.94	1.91
25	2.92	2.53	2.32	2.18	2.09	2.02	1.97	1.93	1.89
26	2.91	2.52	2.31	2.17	2.08	2.01	1.96	1.92	1.88
27	2.90	2.51	2.30	2.17	2.07	2.00	1.95	1.91	1.87
28	2.89	2.50	2.29	2.16	2.06	2.00	1.94	1.90	1.87
29	2.89	2.50	2.28	2.15	2.06	1.99	1.93	1.89	1.86
30	2.88	2.49	2.28	2.14	2.05	1.98	1.93	1.88	1.85
40	2.84	2.44	2.23	2.09	2.00	1.93	1.87	1.83	1.79
60	2.79	2.39	2.18	2.04	1.95	1.87	1.82	1.77	1.74
120	2.75	2.35	2.13	1.99	1.90	1.82	1.77	1.72	1.68
∞	2.71	2.30	2.08	1.94	1.85	1.77	1.72	1.67	1.63

DENOMINATOR DEGREES OF FREEDOM

(a) (continued)

v_2 \ v_1	NUMERATOR DEGREES OF FREEDOM									
	10	12	15	20	24	30	40	60	120	∞
1	60.19	60.71	61.22	61.74	62.00	62.26	62.53	62.79	63.06	63.33
2	9.39	9.41	9.42	9.44	9.45	9.46	9.47	9.47	9.48	9.49
3	5.23	5.22	5.20	5.18	5.18	5.17	5.16	5.15	5.14	5.13
4	3.92	3.90	3.87	3.84	3.83	3.82	3.80	3.79	3.78	3.76
5	3.30	3.27	3.24	3.21	3.19	3.17	3.16	3.14	3.12	3.10
6	2.94	2.90	2.87	2.84	2.82	2.80	2.78	2.76	2.74	2.72
7	2.70	2.67	2.63	2.59	2.58	2.56	2.54	2.51	2.49	2.47
8	2.54	2.50	2.46	2.42	2.40	2.38	2.36	2.34	2.32	2.29
9	2.42	2.38	2.34	2.30	2.28	2.25	2.23	2.21	2.18	2.16
10	2.32	2.28	2.24	2.20	2.18	2.16	2.13	2.11	2.08	2.06
11	2.25	2.21	2.17	2.12	2.10	2.08	2.05	2.03	2.00	1.97
12	2.19	2.15	2.10	2.06	2.04	2.01	1.99	1.96	1.93	1.90
13	2.14	2.10	2.05	2.01	1.98	1.96	1.93	1.90	1.88	1.85
14	2.10	2.05	2.01	1.96	1.94	1.91	1.89	1.86	1.83	1.80
15	2.06	2.02	1.97	1.92	1.90	1.87	1.85	1.82	1.79	1.76
16	2.03	1.99	1.94	1.89	1.87	1.84	1.81	1.78	1.75	1.72
17	2.00	1.96	1.91	1.86	1.84	1.81	1.78	1.75	1.72	1.69
18	1.98	1.93	1.89	1.84	1.81	1.78	1.75	1.72	1.69	1.66
19	1.96	1.91	1.86	1.81	1.79	1.76	1.73	1.70	1.67	1.63
20	1.94	1.89	1.84	1.79	1.77	1.74	1.71	1.68	1.64	1.61
21	1.92	1.87	1.83	1.78	1.75	1.72	1.69	1.66	1.62	1.59
22	1.90	1.86	1.81	1.76	1.73	1.70	1.67	1.64	1.60	1.57
23	1.89	1.84	1.80	1.74	1.72	1.69	1.66	1.62	1.59	1.55
24	1.88	1.83	1.78	1.73	1.70	1.67	1.64	1.61	1.57	1.53
25	1.87	1.82	1.77	1.72	1.69	1.66	1.63	1.59	1.56	1.52
26	1.86	1.81	1.76	1.71	1.68	1.65	1.61	1.58	1.54	1.50
27	1.85	1.80	1.75	1.70	1.67	1.64	1.60	1.57	1.53	1.49
28	1.84	1.79	1.74	1.69	1.66	1.63	1.59	1.56	1.52	1.48
29	1.83	1.78	1.73	1.68	1.65	1.62	1.58	1.55	1.51	1.47
30	1.82	1.77	1.72	1.67	1.64	1.61	1.57	1.54	1.50	1.46
40	1.76	1.71	1.66	1.61	1.57	1.54	1.51	1.47	1.42	1.38
60	1.71	1.66	1.60	1.54	1.51	1.48	1.44	1.40	1.35	1.29
120	1.65	1.60	1.55	1.48	1.45	1.41	1.37	1.32	1.26	1.19
∞	1.60	1.55	1.49	1.42	1.38	1.34	1.30	1.24	1.17	1.00

DENOMINATOR DEGREES OF FREEDOM

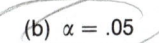 **A15**

(b) $\alpha = .05$

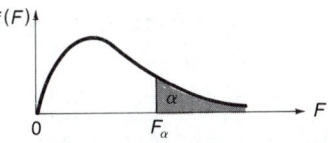

v_1				NUMERATOR DEGREES OF FREEDOM					
v_2	1	2	3	4	5	6	7	8	9
1	161.4	199.5	215.7	224.6	230.2	234.0	236.8	238.9	240.5
2	18.51	19.00	19.16	19.25	19.30	19.33	19.35	19.37	19.38
3	10.13	9.55	9.28	9.12	9.01	8.94	8.89	8.85	8.81
4	7.71	6.94	6.59	6.39	6.26	6.16	6.09	6.04	6.00
5	6.61	5.79	5.41	5.19	5.05	4.95	4.88	4.82	4.77
6	5.99	5.14	4.76	4.53	4.39	4.28	4.21	4.15	4.10
7	5.59	4.74	4.35	4.12	3.97	3.87	3.79	3.73	3.68
8	5.32	4.46	4.07	3.84	3.69	3.58	3.50	3.44	3.39
9	5.12	4.26	3.86	3.63	3.48	3.37	3.29	3.23	3.18
10	4.96	4.10	3.71	3.48	3.33	3.22	3.14	3.07	3.02
11	4.84	3.98	3.59	3.36	3.20	3.09	3.01	2.95	2.90
12	4.75	3.89	3.49	3.26	3.11	3.00	2.91	2.85	2.80
13	4.67	3.81	3.41	3.18	3.03	2.92	2.83	2.77	2.71
14	4.60	3.74	3.34	3.11	2.96	2.85	2.76	2.70	2.65
15	4.54	3.68	3.29	3.06	2.90	2.79	2.71	2.64	2.59
16	4.49	3.63	3.24	3.01	2.85	2.74	2.66	2.59	2.54
17	4.45	3.59	3.20	2.96	2.81	2.70	2.61	2.55	2.49
18	4.41	3.55	3.16	2.93	2.77	2.66	2.58	2.51	2.46
19	4.38	3.52	3.13	2.90	2.74	2.63	2.54	2.48	2.42
20	4.35	3.49	3.10	2.87	2.71	2.60	2.51	2.45	2.39
21	4.32	3.47	3.07	2.84	2.68	2.57	2.49	2.42	2.37
22	4.30	3.44	3.05	2.82	2.66	2.55	2.46	2.40	2.34
23	4.28	3.42	3.03	2.80	2.64	2.53	2.44	2.37	2.32
24	4.26	3.40	3.01	2.78	2.62	2.51	2.42	2.36	2.30
25	4.24	3.39	2.99	2.76	2.60	2.49	2.40	2.34	2.28
26	4.23	3.37	2.98	2.74	2.59	2.47	2.39	2.32	2.27
27	4.21	3.35	2.96	2.73	2.57	2.46	2.37	2.31	2.25
28	4.20	3.34	2.95	2.71	2.56	2.45	2.36	2.29	2.24
29	4.18	3.33	2.93	2.70	2.55	2.43	2.35	2.28	2.22
30	4.17	3.32	2.92	2.69	2.53	2.42	2.33	2.27	2.21
40	4.08	3.23	2.84	2.61	2.45	2.34	2.25	2.18	2.12
60	4.00	3.15	2.76	2.53	2.37	2.25	2.17	2.10	2.04
120	3.92	3.07	2.68	2.45	2.29	2.17	2.09	2.02	1.96
∞	3.84	3.00	2.60	2.37	2.21	2.10	2.01	1.94	1.88

DENOMINATOR DEGREES OF FREEDOM

(b) (continued)

v_2 \ v_1	10	12	15	20	24	30	40	60	120	∞
				NUMERATOR DEGREES OF FREEDOM						
1	241.9	243.9	245.9	248.0	249.1	250.1	251.1	252.2	253.3	254.3
2	19.40	19.41	19.43	19.45	19.45	19.46	19.47	19.48	19.49	19.50
3	8.79	8.74	8.70	8.66	8.64	8.62	8.59	8.57	8.55	8.53
4	5.96	5.91	5.86	5.80	5.77	5.75	5.72	5.69	5.66	5.63
5	4.74	4.68	4.62	4.56	4.53	4.50	4.46	4.43	4.40	4.36
6	4.06	4.00	3.94	3.87	3.84	3.81	3.77	3.74	3.70	3.67
7	3.64	3.57	3.51	3.44	3.41	3.38	3.34	3.30	3.27	3.23
8	3.35	3.28	3.22	3.15	3.12	3.08	3.04	3.01	2.97	2.93
9	3.14	3.07	3.01	2.94	2.90	2.86	2.83	2.79	2.75	2.71
10	2.98	2.91	2.85	2.77	2.74	2.70	2.66	2.62	2.58	2.54
11	2.85	2.79	2.72	2.65	2.61	2.57	2.53	2.49	2.45	2.40
12	2.75	2.69	2.62	2.54	2.51	2.47	2.43	2.38	2.34	2.30
13	2.67	2.60	2.53	2.46	2.42	2.38	2.34	2.30	2.25	2.21
14	2.60	2.53	2.46	2.39	2.35	2.31	2.27	2.22	2.18	2.13
15	2.54	2.48	2.40	2.33	2.29	2.25	2.20	2.16	2.11	2.07
16	2.49	2.42	2.35	2.28	2.24	2.19	2.15	2.11	2.06	2.01
17	2.45	2.38	2.31	2.23	2.19	2.15	2.10	2.06	2.01	1.96
18	2.41	2.34	2.27	2.19	2.15	2.11	2.06	2.02	1.97	1.92
19	2.38	2.31	2.23	2.16	2.11	2.07	2.03	1.98	1.93	1.88
20	2.35	2.28	2.20	2.12	2.08	2.04	1.99	1.95	1.90	1.84
21	2.32	2.25	2.18	2.10	2.05	2.01	1.96	1.92	1.87	1.81
22	2.30	2.23	2.15	2.07	2.03	1.98	1.94	1.89	1.84	1.78
23	2.27	2.20	2.13	2.05	2.01	1.96	1.91	1.86	1.81	1.76
24	2.25	2.18	2.11	2.03	1.98	1.94	1.89	1.84	1.79	1.73
25	2.24	2.16	2.09	2.01	1.96	1.92	1.87	1.82	1.77	1.71
26	2.22	2.15	2.07	1.99	1.95	1.90	1.85	1.80	1.75	1.69
27	2.20	2.13	2.06	1.97	1.93	1.88	1.84	1.79	1.73	1.67
28	2.19	2.12	2.04	1.96	1.91	1.87	1.82	1.77	1.71	1.65
29	2.18	2.10	2.03	1.94	1.90	1.85	1.81	1.75	1.70	1.64
30	2.16	2.09	2.01	1.93	1.89	1.84	1.79	1.74	1.68	1.62
40	2.08	2.00	1.92	1.84	1.79	1.74	1.69	1.64	1.58	1.51
60	1.99	1.92	1.84	1.75	1.70	1.65	1.59	1.53	1.47	1.39
120	1.91	1.83	1.75	1.66	1.61	1.55	1.50	1.43	1.35	1.25
∞	1.83	1.75	1.67	1.57	1.52	1.46	1.39	1.32	1.22	1.00

DENOMINATOR DEGREES OF FREEDOM

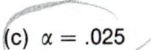

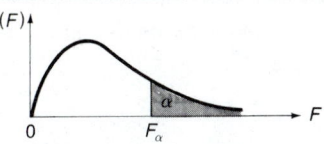

(c) $\alpha = .025$

v_2 \ v_1	NUMERATOR DEGREES OF FREEDOM								
	1	2	3	4	5	6	7	8	9
1	647.8	799.5	864.2	899.6	921.8	937.1	948.2	956.7	963.3
2	38.51	39.00	39.17	39.25	39.30	39.33	39.36	39.37	39.39
3	17.44	16.04	15.44	15.10	14.88	14.73	14.62	14.54	14.47
4	12.22	10.65	9.98	9.60	9.36	9.20	9.07	8.98	8.90
5	10.01	8.43	7.76	7.39	7.15	6.98	6.85	6.76	6.68
6	8.81	7.26	6.60	6.23	5.99	5.82	5.70	5.60	5.52
7	8.07	6.54	5.89	5.52	5.29	5.12	4.99	4.90	4.82
8	7.57	6.06	5.42	5.05	4.82	4.65	4.53	4.43	4.36
9	7.21	5.71	5.08	4.72	4.48	4.32	4.20	4.10	4.03
10	6.94	5.46	4.83	4.47	4.24	4.07	3.95	3.85	3.78
11	6.72	5.26	4.63	4.28	4.04	3.88	3.76	3.66	3.59
12	6.55	5.10	4.47	4.12	3.89	3.73	3.61	3.51	3.44
13	6.41	4.97	4.35	4.00	3.77	3.60	3.48	3.39	3.31
14	6.30	4.86	4.24	3.89	3.66	3.50	3.38	3.29	3.21
15	6.20	4.77	4.15	3.80	3.58	3.41	3.29	3.20	3.12
16	6.12	4.69	4.08	3.73	3.50	3.34	3.22	3.12	3.05
17	6.04	4.62	4.01	3.66	3.44	3.28	3.16	3.06	2.98
18	5.98	4.56	3.95	3.61	3.38	3.22	3.10	3.01	2.93
19	5.92	4.51	3.90	3.56	3.33	3.17	3.05	2.96	2.88
20	5.87	4.46	3.86	3.51	3.29	3.13	3.01	2.91	2.84
21	5.83	4.42	3.82	3.48	3.25	3.09	2.97	2.87	2.80
22	5.79	4.38	3.78	3.44	3.22	3.05	2.93	2.84	2.76
23	5.75	4.35	3.75	3.41	3.18	3.02	2.90	2.81	2.73
24	5.72	4.32	3.72	3.38	3.15	2.99	2.87	2.78	2.70
25	5.69	4.29	3.69	3.35	3.13	2.97	2.85	2.75	2.68
26	5.66	4.27	3.67	3.33	3.10	2.94	2.82	2.73	2.65
27	5.63	4.24	3.65	3.31	3.08	2.92	2.80	2.71	2.63
28	5.61	4.22	3.63	3.29	3.06	2.90	2.78	2.69	2.61
29	5.59	4.20	3.61	3.27	3.04	2.88	2.76	2.67	2.59
30	5.57	4.18	3.59	3.25	3.03	2.87	2.75	2.65	2.57
40	5.42	4.05	3.46	3.13	2.90	2.74	2.62	2.53	2.45
60	5.29	3.93	3.34	3.01	2.79	2.63	2.51	2.41	2.33
120	5.15	3.80	3.23	2.89	2.67	2.52	2.39	2.30	2.22
∞	5.02	3.69	3.12	2.79	2.57	2.41	2.29	2.19	2.11

DENOMINATOR DEGREES OF FREEDOM

(c) (continued)

v_2 \\ v_1	NUMERATOR DEGREES OF FREEDOM									
	10	12	15	20	24	30	40	60	120	∞
1	968.6	976.7	984.9	993.1	997.2	1001	1006	1010	1014	1018
2	39.40	39.41	39.43	39.45	39.46	39.46	39.47	39.48	39.49	39.50
3	14.42	14.34	14.25	14.17	14.12	14.08	14.04	13.99	13.95	13.90
4	8.84	8.75	8.66	8.56	8.51	8.46	8.41	8.36	8.31	8.26
5	6.62	6.52	6.43	6.33	6.28	6.23	6.18	6.12	6.07	6.02
6	5.46	5.37	5.27	5.17	5.12	5.07	5.01	4.96	4.90	4.85
7	4.76	4.67	4.57	4.47	4.42	4.36	4.31	4.25	4.20	4.14
8	4.30	4.20	4.10	4.00	3.95	3.89	3.84	3.78	3.73	3.67
9	3.96	3.87	3.77	3.67	3.61	3.56	3.51	3.45	3.39	3.33
10	3.72	3.62	3.52	3.42	3.37	3.31	3.26	3.20	3.14	3.08
11	3.53	3.43	3.33	3.23	3.17	3.12	3.06	3.00	2.94	2.88
12	3.37	3.28	3.18	3.07	3.02	2.96	2.91	2.85	2.79	2.72
13	3.25	3.15	3.05	2.95	2.89	2.84	2.78	2.72	2.66	2.60
14	3.15	3.05	2.95	2.84	2.79	2.73	2.67	2.61	2.55	2.49
15	3.06	2.96	2.86	2.76	2.70	2.64	2.59	2.52	2.46	2.40
16	2.99	2.89	2.79	2.68	2.63	2.57	2.51	2.45	2.38	2.32
17	2.92	2.82	2.72	2.62	2.56	2.50	2.44	2.38	2.32	2.25
18	2.87	2.77	2.67	2.56	2.50	2.44	2.38	2.32	2.26	2.19
19	2.82	2.72	2.62	2.51	2.45	2.39	2.33	2.27	2.20	2.13
20	2.77	2.68	2.57	2.46	2.41	2.35	2.29	2.22	2.16	2.09
21	2.73	2.64	2.53	2.42	2.37	2.31	2.25	2.18	2.11	2.04
22	2.70	2.60	2.50	2.39	2.33	2.27	2.21	2.14	2.08	2.00
23	2.67	2.57	2.47	2.36	2.30	2.24	2.18	2.11	2.04	1.97
24	2.64	2.54	2.44	2.33	2.27	2.21	2.15	2.08	2.01	1.94
25	2.61	2.51	2.41	2.30	2.24	2.18	2.12	2.05	1.98	1.91
26	2.59	2.49	2.39	2.28	2.22	2.16	2.09	2.03	1.95	1.88
27	2.57	2.47	2.36	2.25	2.19	2.13	2.07	2.00	1.93	1.85
28	2.55	2.45	2.34	2.23	2.17	2.11	2.05	1.98	1.91	1.83
29	2.53	2.43	2.32	2.21	2.15	2.09	2.03	1.96	1.89	1.81
30	2.51	2.41	2.31	2.20	2.14	2.07	2.01	1.94	1.87	1.79
40	2.39	2.29	2.18	2.07	2.01	1.94	1.88	1.80	1.72	1.64
60	2.27	2.17	2.06	1.94	1.88	1.82	1.74	1.67	1.58	1.48
120	2.16	2.05	1.94	1.82	1.76	1.69	1.61	1.53	1.43	1.31
∞	2.05	1.94	1.83	1.71	1.64	1.57	1.48	1.39	1.27	1.00

DENOMINATOR DEGREES OF FREEDOM

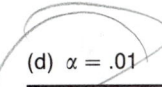

(d) $\alpha = .01$

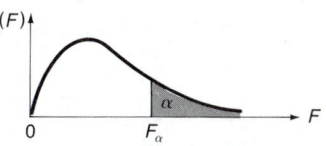

v_1 v_2	NUMERATOR DEGREES OF FREEDOM								
	1	2	3	4	5	6	7	8	9
1	4,052	4,999.5	5,403	5,625	5,764	5,859	5,928	5,982	6,022
2	98.50	99.00	99.17	99.25	99.30	99.33	99.36	99.37	99.39
3	34.12	30.82	29.46	28.71	28.24	27.91	27.67	27.49	27.35
4	21.20	18.00	16.69	15.98	15.52	15.21	14.98	14.80	14.66
5	16.26	13.27	12.06	11.39	10.97	10.67	10.46	10.29	10.16
6	13.75	10.92	9.78	9.15	8.75	8.47	8.26	8.10	7.98
7	12.25	9.55	8.45	7.85	7.46	7.19	6.99	6.84	6.72
8	11.26	8.65	7.59	7.01	6.63	6.37	6.18	6.03	5.91
9	10.56	8.02	6.99	6.42	6.06	5.80	5.61	5.47	5.35
10	10.04	7.56	6.55	5.99	5.64	5.39	5.20	5.06	4.94
11	9.65	7.21	6.22	5.67	5.32	5.07	4.89	4.74	4.63
12	9.33	6.93	5.95	5.41	5.06	4.82	4.64	4.50	4.39
13	9.07	6.70	5.74	5.21	4.86	4.62	4.44	4.30	4.19
14	8.86	6.51	5.56	5.04	4.69	4.46	4.28	4.14	4.03
15	8.68	6.36	5.42	4.89	4.56	4.32	4.14	4.00	3.89
16	8.53	6.23	5.29	4.77	4.44	4.20	4.03	3.89	3.78
17	8.40	6.11	5.18	4.67	4.34	4.10	3.93	3.79	3.68
18	8.29	6.01	5.09	4.58	4.25	4.01	3.84	3.71	3.60
19	8.18	5.93	5.01	4.50	4.17	3.94	3.77	3.63	3.52
20	8.10	5.85	4.94	4.43	4.10	3.87	3.70	3.56	3.46
21	8.02	5.78	4.87	4.37	4.04	3.81	3.64	3.51	3.40
22	7.95	5.72	4.82	4.31	3.99	3.76	3.59	3.45	3.35
23	7.88	5.66	4.76	4.26	3.94	3.71	3.54	3.41	3.30
24	7.82	5.61	4.72	4.22	3.90	3.67	3.50	3.36	3.26
25	7.77	5.57	4.68	4.18	3.85	3.63	3.46	3.32	3.22
26	7.72	5.53	4.64	4.14	3.82	3.59	3.42	3.29	3.18
27	7.68	5.49	4.60	4.11	3.78	3.56	3.39	3.26	3.15
28	7.64	5.45	4.57	4.07	3.75	3.53	3.36	3.23	3.12
29	7.60	5.42	4.54	4.04	3.73	3.50	3.33	3.20	3.09
30	7.56	5.39	4.51	4.02	3.70	3.47	3.30	3.17	3.07
40	7.31	5.18	4.31	3.83	3.51	3.29	3.12	2.99	2.89
60	7.08	4.98	4.13	3.65	3.34	3.12	2.95	2.82	2.72
120	6.85	4.79	3.95	3.48	3.17	2.96	2.79	2.66	2.56
∞	6.63	4.61	3.78	3.32	3.02	2.80	2.64	2.51	2.41

DENOMINATOR DEGREES OF FREEDOM

(d) (continued)

v_2	v_1 NUMERATOR DEGREES OF FREEDOM									
	10	12	15	20	24	30	40	60	120	∞
1	6,056	6,106	6,157	6,209	6,235	6,261	6,287	6,313	6,339	6,366
2	99.40	99.42	99.43	99.45	99.46	99.47	99.47	99.48	99.49	99.50
3	27.23	27.05	26.87	26.69	26.60	26.50	26.41	26.32	26.22	26.13
4	14.55	14.37	14.20	14.02	13.93	13.84	13.75	13.65	13.56	13.46
5	10.05	9.89	9.72	9.55	9.47	9.38	9.29	9.20	9.11	9.02
6	7.87	7.72	7.56	7.40	7.31	7.23	7.14	7.06	6.97	6.88
7	6.62	6.47	6.31	6.16	6.07	5.99	5.91	5.82	5.74	5.65
8	5.81	5.67	5.52	5.36	5.28	5.20	5.12	5.03	4.95	4.86
9	5.26	5.11	4.96	4.81	4.73	4.65	4.57	4.48	4.40	4.31
10	4.85	4.71	4.56	4.41	4.33	4.25	4.17	4.08	4.00	3.91
11	4.54	4.40	4.25	4.10	4.02	3.94	3.86	3.78	3.69	3.60
12	4.30	4.16	4.01	3.86	3.78	3.70	3.62	3.54	3.45	3.36
13	4.10	3.96	3.82	3.66	3.59	3.51	3.43	3.34	3.25	3.17
14	3.94	3.80	3.66	3.51	3.43	3.35	3.27	3.18	3.09	3.00
15	3.80	3.67	3.52	3.37	3.29	3.21	3.13	3.05	2.96	2.87
16	3.69	3.55	3.41	3.26	3.18	3.10	3.02	2.93	2.84	2.75
17	3.59	3.46	3.31	3.16	3.08	3.00	2.92	2.83	2.75	2.65
18	3.51	3.37	3.23	3.08	3.00	2.92	2.84	2.75	2.66	2.57
19	3.43	3.30	3.15	3.00	2.92	2.84	2.76	2.67	2.58	2.49
20	3.37	3.23	3.09	2.94	2.86	2.78	2.69	2.61	2.52	2.42
21	3.31	3.17	3.03	2.88	2.80	2.72	2.64	2.55	2.46	2.36
22	3.26	3.12	2.98	2.83	2.75	2.67	2.58	2.50	2.40	2.31
23	3.21	3.07	2.93	2.78	2.70	2.62	2.54	2.45	2.35	2.26
24	3.17	3.03	2.89	2.74	2.66	2.58	2.49	2.40	2.31	2.21
25	3.13	2.99	2.85	2.70	2.62	2.54	2.45	2.36	2.27	2.17
26	3.09	2.96	2.81	2.66	2.58	2.50	2.42	2.33	2.23	2.13
27	3.06	2.93	2.78	2.63	2.55	2.47	2.38	2.29	2.20	2.10
28	3.03	2.90	2.75	2.60	2.52	2.44	2.35	2.26	2.17	2.06
29	3.00	2.87	2.73	2.57	2.49	2.41	2.33	2.23	2.14	2.03
30	2.98	2.84	2.70	2.55	2.47	2.39	2.30	2.21	2.11	2.01
40	2.80	2.66	2.52	2.37	2.29	2.20	2.11	2.02	1.92	1.80
60	2.63	2.50	2.35	2.20	2.12	2.03	1.94	1.84	1.73	1.60
120	2.47	2.34	2.19	2.03	1.95	1.86	1.76	1.66	1.53	1.38
∞	2.32	2.18	2.04	1.88	1.79	1.70	1.59	1.47	1.32	1.00

DENOMINATOR DEGREES OF FREEDOM

TABLE A-8 Confidence Interval for a Population Proportion, Small Sample

$n = 5$

	$\alpha = .05$		$\alpha = .10$	
	P_L	P_U	P_L	P_U
$x = 1$	0.005	0.716	0.010	0.657
2	0.053	0.853	0.076	0.811
3	0.147	0.947	0.189	0.924
4	0.284	0.995	0.343	0.990

$n = 6$

	$\alpha = .05$		$\alpha = .10$	
	P_L	P_U	P_L	P_U
$x = 1$	0.004	0.641	0.009	0.582
2	0.043	0.777	0.063	0.729
3	0.118	0.882	0.153	0.847
4	0.223	0.957	0.271	0.937
5	0.359	0.996	0.418	0.991

$n = 7$

	$\alpha = .05$		$\alpha = .10$	
	P_L	P_U	P_L	P_U
$x = 1$	0.004	0.579	0.007	0.521
2	0.037	0.710	0.053	0.659
3	0.099	0.816	0.129	0.775
4	0.184	0.901	0.225	0.871
5	0.290	0.963	0.341	0.947
6	0.421	0.996	0.479	0.993

$n = 8$

	$\alpha = .05$		$\alpha = .10$	
	P_L	P_U	P_L	P_U
$x = 1$	0.003	0.527	0.006	0.471
2	0.032	0.651	0.046	0.600
3	0.085	0.755	0.111	0.711
4	0.157	0.843	0.193	0.807
5	0.245	0.915	0.289	0.889
6	0.349	0.968	0.400	0.954
7	0.473	0.997	0.529	0.994

$n = 9$

	$\alpha = .05$		$\alpha = .10$	
	P_L	P_U	P_L	P_U
$x = 1$	0.003	0.482	0.006	0.429
2	0.028	0.600	0.041	0.550
3	0.075	0.701	0.098	0.655
4	0.137	0.788	0.169	0.749
5	0.212	0.863	0.251	0.831
6	0.299	0.925	0.345	0.902
7	0.400	0.972	0.450	0.959
8	0.518	0.997	0.571	0.994

$n = 10$

	$\alpha = .05$		$\alpha = .10$	
	P_L	P_U	P_L	P_U
$x = 1$	0.003	0.445	0.005	0.394
2	0.025	0.556	0.037	0.507
3	0.067	0.652	0.087	0.607
4	0.122	0.738	0.150	0.696
5	0.187	0.813	0.222	0.778
6	0.262	0.878	0.304	0.850
7	0.348	0.933	0.393	0.913
8	0.444	0.975	0.493	0.963
9	0.555	0.997	0.606	0.995

$n = 11$

	$\alpha = .05$		$\alpha = .10$	
	P_L	P_U	P_L	P_U
$x = 1$	0.002	0.413	0.005	0.364
2	0.023	0.518	0.033	0.470
3	0.060	0.610	0.079	0.564
4	0.109	0.692	0.135	0.650
5	0.167	0.766	0.200	0.729
6	0.234	0.833	0.271	0.800
7	0.308	0.891	0.350	0.865
8	0.390	0.940	0.436	0.921
9	0.482	0.977	0.530	0.967
10	0.587	0.998	0.636	0.995

$n = 12$

	$\alpha = .05$		$\alpha = .10$	
	P_L	P_U	P_L	P_U
$x = 1$	0.002	0.385	0.004	0.339
2	0.021	0.484	0.030	0.438
3	0.055	0.572	0.072	0.527
4	0.099	0.651	0.123	0.609
5	0.152	0.723	0.181	0.685
6	0.211	0.789	0.245	0.755
7	0.277	0.848	0.315	0.819
8	0.349	0.901	0.391	0.877
9	0.428	0.945	0.473	0.928
10	0.516	0.979	0.562	0.970
11	0.615	0.998	0.661	0.996

$n = 13$

	$\alpha = .05$		$\alpha = .10$	
	P_L	P_U	P_L	P_U
$x = 1$	0.002	0.360	0.004	0.316
2	0.019	0.454	0.028	0.410
3	0.050	0.538	0.066	0.495
4	0.091	0.614	0.113	0.573
5	0.139	0.684	0.166	0.645
6	0.192	0.749	0.224	0.713
7	0.251	0.808	0.287	0.776
8	0.316	0.861	0.355	0.834
9	0.386	0.909	0.427	0.887
10	0.462	0.950	0.505	0.934
11	0.546	0.981	0.590	0.972
12	0.640	0.998	0.684	0.996

$n = 14$

	$\alpha = .05$		$\alpha = .10$	
	P_L	P_U	P_L	P_U
$x = 1$	0.002	0.339	0.004	0.297
2	0.018	0.428	0.026	0.385
3	0.047	0.508	0.061	0.466
4	0.084	0.581	0.104	0.540
5	0.128	0.649	0.153	0.610
6	0.177	0.711	0.206	0.675
7	0.230	0.770	0.264	0.736
8	0.289	0.823	0.325	0.794
9	0.351	0.872	0.390	0.847
10	0.419	0.916	0.460	0.896
11	0.492	0.953	0.534	0.939
12	0.572	0.982	0.615	0.974
13	0.661	0.998	0.703	0.996

$n = 15$

	$\alpha = .05$		$\alpha = .10$	
	P_L	P_U	P_L	P_U
$x = 1$	0.002	0.319	0.003	0.279
2	0.017	0.405	0.024	0.363
3	0.043	0.481	0.057	0.440
4	0.078	0.551	0.097	0.511
5	0.118	0.616	0.142	0.577
6	0.163	0.677	0.191	0.640
7	0.213	0.734	0.244	0.700
8	0.266	0.787	0.300	0.756
9	0.323	0.837	0.360	0.809
10	0.384	0.882	0.423	0.858
11	0.449	0.922	0.489	0.903
12	0.519	0.957	0.560	0.943
13	0.595	0.983	0.637	0.976
14	0.681	0.998	0.721	0.997

TABLE A-8   (continued)

n = 16	α = .05 P_L	P_U	α = .10 P_L	P_U
x = 1	0.002	0.302	0.003	0.264
2	0.016	0.383	0.023	0.344
3	0.040	0.456	0.053	0.417
4	0.073	0.524	0.090	0.484
5	0.110	0.587	0.132	0.548
6	0.152	0.646	0.178	0.609
7	0.198	0.701	0.227	0.667
8	0.247	0.753	0.279	0.721
9	0.299	0.802	0.333	0.773
10	0.354	0.848	0.391	0.822
11	0.413	0.890	0.452	0.868
12	0.476	0.927	0.516	0.910
13	0.544	0.960	0.583	0.947
14	0.617	0.984	0.656	0.977
15	0.698	0.998	0.736	0.997

n = 17	α = .05 P_L	P_U	α = .10 P_L	P_U
x = 1	0.001	0.287	0.003	0.250
2	0.015	0.364	0.021	0.326
3	0.038	0.434	0.050	0.396
4	0.068	0.499	0.085	0.461
5	0.103	0.560	0.124	0.522
6	0.142	0.617	0.166	0.580
7	0.184	0.671	0.212	0.636
8	0.230	0.722	0.260	0.689
9	0.278	0.770	0.311	0.740
10	0.329	0.816	0.364	0.788
11	0.383	0.858	0.420	0.834
12	0.440	0.897	0.478	0.876
13	0.501	0.932	0.539	0.915
14	0.566	0.962	0.604	0.950
15	0.636	0.985	0.674	0.979
16	0.713	0.999	0.750	0.997

n = 18	α = .05 P_L	P_U	α = .10 P_L	P_U
x = 1	0.001	0.273	0.003	0.238
2	0.014	0.347	0.020	0.310
3	0.036	0.414	0.047	0.377
4	0.064	0.476	0.080	0.439
5	0.097	0.535	0.116	0.498
6	0.133	0.590	0.156	0.554
7	0.173	0.643	0.199	0.608
8	0.215	0.692	0.244	0.659
9	0.260	0.740	0.291	0.709
10	0.308	0.785	0.341	0.756
11	0.357	0.827	0.392	0.801
12	0.410	0.867	0.446	0.844

n = 18	α = .05 P_L	P_U	α = .10 P_L	P_U
13	0.465	0.903	0.502	0.884
14	0.524	0.936	0.561	0.920
15	0.586	0.964	0.623	0.953
16	0.653	0.986	0.690	0.980
17	0.727	0.999	0.762	0.997

n = 19	α = .05 P_L	P_U	α = .10 P_L	P_U
x = 1	0.001	0.260	0.003	0.226
2	0.013	0.331	0.019	0.296
3	0.034	0.396	0.044	0.359
4	0.061	0.456	0.075	0.419
5	0.091	0.512	0.110	0.476
6	0.126	0.565	0.147	0.530
7	0.163	0.616	0.188	0.582
8	0.203	0.665	0.230	0.632
9	0.244	0.711	0.274	0.680
10	0.289	0.756	0.320	0.726
11	0.335	0.797	0.368	0.770
12	0.384	0.837	0.418	0.812
13	0.435	0.874	0.470	0.853
14	0.488	0.909	0.524	0.890
15	0.544	0.939	0.581	0.925
16	0.604	0.966	0.641	0.956
17	0.669	0.987	0.704	0.981
18	0.740	0.999	0.774	0.997

n = 20	α = .05 P_L	P_U	α = .10 P_L	P_U
x = 1	0.001	0.249	0.003	0.216
2	0.012	0.317	0.018	0.283
3	0.032	0.379	0.042	0.344
4	0.057	0.437	0.071	0.401
5	0.087	0.491	0.104	0.456
6	0.119	0.543	0.140	0.508
7	0.154	0.592	0.177	0.558
8	0.191	0.639	0.217	0.606
9	0.231	0.685	0.259	0.653
10	0.272	0.728	0.302	0.698
11	0.315	0.769	0.347	0.741
12	0.361	0.809	0.394	0.783
13	0.408	0.846	0.442	0.823
14	0.457	0.881	0.492	0.860
15	0.509	0.913	0.544	0.896
16	0.563	0.943	0.599	0.929
17	0.621	0.968	0.656	0.958
18	0.683	0.988	0.717	0.982
19	0.751	0.999	0.784	0.997

Construction of Table A-8   Confidence limits for the binomial parameter can be determined using the incomplete beta function. A beta distribution has two parameters, say A and B. Consider a 95% confidence interval (α = .05) for the binomial parameter, p. To find the lower confidence limit, P_L, for a given sample size (n) and observed number of successes (x), set

$$A = x \quad \text{and} \quad B = n - x + 1.$$

Then, P_L is that value of the beta random variable for which

$$\left(\begin{array}{l}\text{area under the beta distribution} \\ \text{with parameters A and B to the} \\ \text{left of } P_L\end{array}\right) = \frac{\alpha}{2} = .025.$$

This can be determined by using a computerized incomplete beta function, as shown below.

Similarly, to find the upper limit, P_U, for a given sample size (n) and observed number of successes (x), set

$$A = x + 1 \quad \text{and} \quad B = n - x.$$

Then P_U is that value of beta random variable for which

$$\left(\begin{array}{l}\text{area under the beta distribution} \\ \text{with parameters A and B to the} \\ \text{left of } P_U\end{array}\right) = 1 - \frac{\alpha}{2} = .975.$$

Again, this can be determined by using a computerized incomplete beta function.

The listing of FORTRAN statements used to generate the values in Table A-8 is shown below. To change the starting value of n ($=5$, here), the final value of n ($=20$, here), or the two significance levels to be considered (.05 and .10, here), change one or more of the first four lines of code. The incomplete beta function subroutine used here is MDBETI, available through International Mathematical and Statistical Libraries, Inc. (IMSL), Sixth Floor, NBC Building, 7500 Bellaire Boulevard, Houston, Texas 77036.

```
0001        NST=5
0002        NEND=20
0003        ALPHA1=.05
0004        ALPHA2=.10
0005        DO 10 N=NST,NEND
0006        PRINT 500 , N,ALPHA1,ALPHA2
0007        NP1 = N + 1
0008        DO 10 J=2,N
0009        JX = J - 1
0010        T2 = ALPHA1/2.0
0011        T1 = 1.0 - T2
0012        A1 = JX + 1
0013        A2 = JX
0014        B1 = N - JX
0015        B2 = N - JX + 1
0016        CALL MDBETI(T1,A1,B1,P2,IER1)
0017        CALL MDBETI(T2,A2,B2,P1,IER2)
0018        IF(IER1.GT.0) PRINT 100 , ALPHA1,IER1
0019        IF(IER2.GT.0) PRINT 150 , ALPHA1,IER2
0020        TT2 = ALPHA2/2.0
0021        TT1 = 1.0 - TT2
0022        CALL MDBETI(TT1,A1,B1,PP2,IER1)
0023        CALL MDBETI(TT2,A2,B2,PP1,IER2)
0024        IF(IER1.GT.0) PRINT 100 , ALPHA2, IER1
0025        IF(IER2.GT.0) PRINT 150 , ALPHA2, IER2
0026        IF(J.EQ.2) PRINT 200 , JX,P1,P2,PP1,PP2
0027        IF(J.NE.2) PRINT 250 , JX,P1,P2,PP1,PP2
         C
0028     10 CONTINUE
0029    100 FORMAT('0FOR ALPHA = ',F5.2,' ERROR NUMBER',I3,
               *' OCCURRED WHEN FINDING P2')
0030    150 FORMAT('0FOR ALPHA = ',F5.2,' ERROR NUMBER',I3,
               *' OCCURRED WHEN FINDING P1')
0031    200 FORMAT('0 X = ',I2,4(4X,F5.3))
0032    250 FORMAT('0',5X,I2,4(4X,F5.3))
0033    500 FORMAT('1',///' N = ',I2,8X,'ALPHA = ',F3.2,8X,'ALPHA = ',
               *F3.2,//,13X,' P1',6X,' P2',6X,' P1',6X,' P2')
0034        STOP
0035        END
```

TABLE A-9   Critical Values for the Durbin-Watson DW Statistic

(a) $\alpha = .05$

n	k = 1 d_L	k = 1 d_U	k = 2 d_L	k = 2 d_U	k = 3 d_L	k = 3 d_U	k = 4 d_L	k = 4 d_U	k = 5 d_L	k = 5 d_U
15	1.08	1.36	0.95	1.54	0.82	1.75	0.69	1.97	0.56	2.21
16	1.10	1.37	0.98	1.54	0.86	1.73	0.74	1.93	0.62	2.15
17	1.13	1.38	1.02	1.54	0.90	1.71	0.78	1.90	0.67	2.10
18	1.16	1.39	1.05	1.53	0.93	1.69	0.82	1.87	0.71	2.06
19	1.18	1.40	1.08	1.53	0.97	1.68	0.86	1.85	0.75	2.02
20	1.20	1.41	1.10	1.54	1.00	1.68	0.90	1.83	0.79	1.99
21	1.22	1.42	1.13	1.54	1.03	1.67	0.93	1.81	0.83	1.96
22	1.24	1.43	1.15	1.54	1.05	1.66	0.96	1.80	0.86	1.94
23	1.26	1.44	1.17	1.54	1.08	1.66	0.99	1.79	0.90	1.92
24	1.27	1.45	1.19	1.55	1.10	1.66	1.01	1.78	0.93	1.90
25	1.29	1.45	1.21	1.55	1.12	1.66	1.04	1.77	0.95	1.89
26	1.30	1.46	1.22	1.55	1.14	1.65	1.06	1.76	0.98	1.88
27	1.32	1.47	1.24	1.56	1.16	1.65	1.08	1.76	1.01	1.86
28	1.33	1.48	1.26	1.56	1.18	1.65	1.10	1.75	1.03	1.85
29	1.34	1.48	1.27	1.56	1.20	1.65	1.12	1.74	1.05	1.84
30	1.35	1.49	1.28	1.57	1.21	1.65	1.14	1.74	1.07	1.83
31	1.36	1.50	1.30	1.57	1.23	1.65	1.16	1.74	1.09	1.83
32	1.37	1.50	1.31	1.57	1.24	1.65	1.18	1.73	1.11	1.82
33	1.38	1.51	1.32	1.58	1.26	1.65	1.19	1.73	1.13	1.81
34	1.39	1.51	1.33	1.58	1.27	1.65	1.21	1.73	1.15	1.81
35	1.40	1.52	1.34	1.58	1.28	1.65	1.22	1.73	1.16	1.80
36	1.41	1.52	1.35	1.59	1.29	1.65	1.24	1.73	1.18	1.80
37	1.42	1.53	1.36	1.59	1.31	1.66	1.25	1.72	1.19	1.80
38	1.43	1.54	1.37	1.59	1.32	1.66	1.26	1.72	1.21	1.79
39	1.43	1.54	1.38	1.60	1.33	1.66	1.27	1.72	1.22	1.79
40	1.44	1.54	1.39	1.60	1.34	1.66	1.29	1.72	1.23	1.79
45	1.48	1.57	1.43	1.62	1.38	1.67	1.34	1.72	1.29	1.78
50	1.50	1.59	1.46	1.63	1.42	1.67	1.38	1.72	1.34	1.77
55	1.53	1.60	1.49	1.64	1.45	1.68	1.41	1.72	1.38	1.77
60	1.55	1.62	1.51	1.65	1.48	1.69	1.44	1.73	1.41	1.77
65	1.57	1.63	1.54	1.66	1.50	1.70	1.47	1.73	1.44	1.77
70	1.58	1.64	1.55	1.67	1.52	1.70	1.49	1.74	1.46	1.77
75	1.60	1.65	1.57	1.68	1.54	1.71	1.51	1.74	1.49	1.77
80	1.61	1.66	1.59	1.69	1.56	1.72	1.53	1.74	1.51	1.77
85	1.62	1.67	1.60	1.70	1.57	1.72	1.55	1.75	1.52	1.77
90	1.63	1.68	1.61	1.70	1.59	1.73	1.57	1.75	1.54	1.78
95	1.64	1.69	1.62	1.71	1.60	1.73	1.58	1.75	1.56	1.78
100	1.65	1.69	1.63	1.72	1.61	1.74	1.59	1.76	1.57	1.78

(b) $\alpha = .01$

n	k = 1 d_L	k = 1 d_U	k = 2 d_L	k = 2 d_U	k = 3 d_L	k = 3 d_U	k = 4 d_L	k = 4 d_U	k = 5 d_L	k = 5 d_U
15	0.81	1.07	0.70	1.25	0.59	1.46	0.49	1.70	0.39	1.96
16	0.84	1.09	0.74	1.25	0.63	1.44	0.53	1.66	0.44	1.90
17	0.87	1.10	0.77	1.25	0.67	1.43	0.57	1.63	0.48	1.85
18	0.90	1.12	0.80	1.26	0.71	1.42	0.61	1.60	0.52	1.80
19	0.93	1.13	0.83	1.26	0.74	1.41	0.65	1.58	0.56	1.77
20	0.95	1.15	0.86	1.27	0.77	1.41	0.68	1.57	0.60	1.74
21	0.97	1.16	0.89	1.27	0.80	1.41	0.72	1.55	0.63	1.71
22	1.00	1.17	0.91	1.28	0.83	1.40	0.75	1.54	0.66	1.69
23	1.02	1.19	0.94	1.29	0.86	1.40	0.77	1.53	0.70	1.67
24	1.04	1.20	0.96	1.30	0.88	1.41	0.80	1.53	0.72	1.66
25	1.05	1.21	0.98	1.30	0.90	1.41	0.83	1.52	0.75	1.65
26	1.07	1.22	1.00	1.31	0.93	1.41	0.85	1.52	0.78	1.64
27	1.09	1.23	1.02	1.32	0.95	1.41	0.88	1.51	0.81	1.63
28	1.10	1.24	1.04	1.32	0.97	1.41	0.90	1.51	0.83	1.62
29	1.12	1.25	1.05	1.33	0.99	1.42	0.92	1.51	0.85	1.61
30	1.13	1.26	1.07	1.34	1.01	1.42	0.94	1.51	0.88	1.61
31	1.15	1.27	1.08	1.34	1.02	1.42	0.96	1.51	0.90	1.60

(b) (continued)

n	k = 1 d_L	d_U	k = 2 d_L	d_U	k = 3 d_L	d_U	k = 4 d_L	d_U	k = 5 d_L	d_U
32	1.16	1.28	1.10	1.35	1.04	1.43	0.98	1.51	0.92	1.60
33	1.17	1.29	1.11	1.36	1.05	1.43	1.00	1.51	0.94	1.59
34	1.18	1.30	1.13	1.36	1.07	1.43	1.01	1.51	0.95	1.59
35	1.19	1.31	1.14	1.37	1.08	1.44	1.03	1.51	0.97	1.59
36	1.21	1.32	1.15	1.38	1.10	1.44	1.04	1.51	0.99	1.59
37	1.22	1.32	1.16	1.38	1.11	1.45	1.06	1.51	1.00	1.59
38	1.23	1.33	1.18	1.39	1.12	1.45	1.07	1.52	1.02	1.58
39	1.24	1.34	1.19	1.39	1.14	1.45	1.09	1.52	1.03	1.58
40	1.25	1.34	1.20	1.40	1.15	1.46	1.10	1.52	1.05	1.58
45	1.29	1.38	1.24	1.42	1.20	1.48	1.16	1.53	1.11	1.58
50	1.32	1.40	1.28	1.45	1.24	1.49	1.20	1.54	1.16	1.59
55	1.36	1.43	1.32	1.47	1.28	1.51	1.25	1.55	1.21	1.59
60	1.38	1.45	1.35	1.48	1.32	1.52	1.28	1.56	1.25	1.60
65	1.41	1.47	1.38	1.50	1.35	1.53	1.31	1.57	1.28	1.61
70	1.43	1.49	1.40	1.52	1.37	1.55	1.34	1.58	1.31	1.61
75	1.45	1.50	1.42	1.53	1.39	1.56	1.37	1.59	1.34	1.62
80	1.47	1.52	1.44	1.54	1.42	1.57	1.39	1.60	1.36	1.62
85	1.48	1.53	1.46	1.55	1.43	1.58	1.41	1.60	1.39	1.63
90	1.50	1.54	1.47	1.56	1.45	1.59	1.43	1.61	1.41	1.64
95	1.51	1.55	1.49	1.57	1.47	1.60	1.45	1.62	1.42	1.64
100	1.52	1.56	1.50	1.58	1.48	1.60	1.46	1.63	1.44	1.65

TABLE A-10 Distribution Function for the Mann-Whitney U Statistic*

This table contains the value of $P(U \leq U_0)$ where $n_1 \leq n_2$.

$n_2 = 3$ U_0	1	2	3
0	.25	.10	.05
1	.50	.20	.10
2		.40	.20
3		.60	.35
4			.50

$n_2 = 4$ U_0	1	2	3	4
0	.2000	.0667	.0286	.0143
1	.4000	.1333	.0571	.0286
2	.6000	.2667	.1143	.0571
3		.4000	.2000	.1000
4		.6000	.3143	.1714
5			.4286	.2429
6			.5714	.3429
7				.4429
8				.5571

* Computed by M. Pagano, Dept. of Statistics, University of Florida. Reprinted by permission from *Statistics for Management and Economics*, 5th ed., by William Mendenhall and James E. Reinmuth. Copyright © 1986 by PWS-KENT Publishers, Boston.

TABLE A-10   (continued)

$n_2 = 5$ U_0	1	2	n_1 3	4	5
0	.1667	.0476	.0179	.0079	.0040
1	.3333	.0952	.0357	.0159	.0079
2	.5000	.1905	.0714	.0317	.0159
3		.2857	.1250	.0556	.0278
4		.4286	.1964	.0952	.0476
5		.5714	.2857	.1429	.0754
6			.3929	.2063	.1111
7			.5000	.2778	.1548
8				.3651	.2103
9				.4524	.2738
10				.5476	.3452
11					.4206
12					.5000

$n_2 = 6$ U_0	1	2	3	n_1 4	5	6
0	.1429	.0357	.0119	.0048	.0022	.0011
1	.2857	.0714	.0238	.0095	.0043	.0022
2	.4286	.1429	.0476	.0190	.0087	.0043
3	.5714	.2143	.0833	.0333	.0152	.0076
4		.3214	.1310	.0571	.0260	.0130
5		.4286	.1905	.0857	.0411	.0206
6		.5714	.2738	.1286	.0628	.0325
7			.3571	.1762	.0887	.0465
8			.4524	.2381	.1234	.0660
9			.5476	.3048	.1645	.0898
10				.3810	.2143	.1201
11				.4571	.2684	.1548
12				.5429	.3312	.1970
13					.3961	.2424
14					.4654	.2944
15					.5346	.3496
16						.4091
17						.4686
18						.5314

$n_2 = 7$ U_0	1	2	3	n_1 4	5	6	7
0	.1250	.0278	.0083	.0030	.0013	.0006	.0003
1	.2500	.0556	.0167	.0061	.0025	.0012	.0006
2	.3750	.1111	.0333	.0121	.0051	.0023	.0012
3	.5000	.1667	.0583	.0212	.0088	.0041	.0020
4		.2500	.0917	.0364	.0152	.0070	.0035
5		.3333	.1333	.0545	.0240	.0111	.0055
6		.4444	.1917	.0818	.0366	.0175	.0087
7		.5556	.2583	.1152	.0530	.0256	.0131
8			.3333	.1576	.0745	.0367	.0189
9			.4167	.2061	.1010	.0507	.0265
10			.5000	.2636	.1338	.0688	.0364
11				.3242	.1717	.0903	.0487
12				.3939	.2159	.1171	.0641
13				.4636	.2652	.1474	.0825
14				.5364	.3194	.1830	.1043
15					.3775	.2226	.1297
16					.4381	.2669	.1588
17					.5000	.3141	.1914
18						.3654	.2279
19						.4178	.2675
20						.4726	.3100
21						.5274	.3552
22							.4024
23							.4508
24							.5000

TABLE A-10 (continued)

$n_2 = 8$	U_0	1	2	3	4	5	6	7	8
	0	.1111	.0222	.0061	.0020	.0008	.0003	.0002	.0001
	1	.2222	.0444	.0121	.0040	.0016	.0007	.0003	.0002
	2	.3333	.0889	.0242	.0081	.0031	.0013	.0006	.0003
	3	.4444	.1333	.0424	.0141	.0054	.0023	.0011	.0005
	4	.5556	.2000	.0667	.0242	.0093	.0040	.0019	.0009
	5		.2667	.0970	.0364	.0148	.0063	.0030	.0015
	6		.3556	.1394	.0545	.0225	.0100	.0047	.0023
	7		.4444	.1879	.0768	.0326	.0147	.0070	.0035
	8		.5556	.2485	.1071	.0466	.0213	.0103	.0052
	9			.3152	.1414	.0637	.0296	.0145	.0074
	10			.3879	.1838	.0855	.0406	.0200	.0103
	11			.4606	.2303	.1111	.0539	.0270	.0141
	12			.5394	.2848	.1422	.0709	.0361	.0190
	13				.3414	.1772	.0906	.0469	.0249
	14				.4040	.2176	.1142	.0603	.0325
	15				.4667	.2618	.1412	.0760	.0415
	16				.5333	.3108	.1725	.0946	.0524
	17					.3621	.2068	.1159	.0652
	18					.4165	.2454	.1405	.0803
	19					.4716	.2864	.1678	.0974
	20					.5284	.3310	.1984	.1172
	21						.3773	.2317	.1393
	22						.4259	.2679	.1641
	23						.4749	.3063	.1911
	24						.5251	.3472	.2209
	25							.3894	.2527
	26							.4333	.2869
	27							.4775	.3227
	28							.5225	.3605
	29								.3992
	30								.4392
	31								.4796
	32								.5204

$n_2 = 9$	U_0	1	2	3	4	5	6	7	8	9
	0	.1000	.0182	.0045	.0014	.0005	.0002	.0001	.0000	.0000
	1	.2000	.0364	.0091	.0028	.0010	.0004	.0002	.0001	.0000
	2	.3000	.0727	.0182	.0056	.0020	.0008	.0003	.0002	.0001
	3	.4000	.1091	.0318	.0098	.0035	.0014	.0006	.0003	.0001
	4	.5000	.1636	.0500	.0168	.0060	.0024	.0010	.0005	.0002
	5		.2182	.0727	.0252	.0095	.0038	.0017	.0008	.0004
	6		.2909	.1045	.0378	.0145	.0060	.0026	.0012	.0006
	7		.3636	.1409	.0531	.0210	.0088	.0039	.0019	.0009
	8		.4545	.1864	.0741	.0300	.0128	.0058	.0028	.0014
	9		.5455	.2409	.0993	.0415	.0180	.0082	.0039	.0020
	10			.3000	.1301	.0559	.0248	.0115	.0056	.0028
	11			.3636	.1650	.0734	.0332	.0156	.0076	.0039
	12			.4318	.2070	.0949	.0440	.0209	.0103	.0053
	13			.5000	.2517	.1199	.0567	.0274	.0137	.0071
	14				.3021	.1489	.0723	.0356	.0180	.0094
	15				.3552	.1818	.0905	.0454	.0232	.0122
	16				.4126	.2188	.1119	.0571	.0296	.0157
	17				.4699	.2592	.1361	.0708	.0372	.0200
	18				.5301	.3032	.1638	.0869	.0464	.0252
	19					.3497	.1942	.1052	.0570	.0313
	20					.3986	.2280	.1261	.0694	.0385
	21					.4491	.2643	.1496	.0836	.0470
	22					.5000	.3035	.1755	.0998	.0567
	23						.3445	.2039	.1179	.0680
	24						.3878	.2349	.1383	.0807
	25						.4320	.2680	.1606	.0951
	26						.4773	.3032	.1852	.1112
	27						.5227	.3403	.2117	.1290
	28							.3788	.2404	.1487
	29							.4185	.2707	.1701
	30							.4591	.3029	.1933

TABLE A-10   (continued)

$n_2 = 9$ U_0	1	2	3	4	5	6	7	8	9
31							.5000	.3365	.2181
32								.3715	.2447
33								.4074	.2729
34								.4442	.3024
35								.4813	.3332
36								.5187	.3652
37									.3981
38									.4317
39									.4657
40									.5000

$n_2 = 10$ U_0	1	2	3	4	5	6	7	8	9	10
0	.0909	.0152	.0035	.0010	.0003	.0001	.0001	.0000	.0000	.0000
1	.1818	.0303	.0070	.0020	.0007	.0002	.0001	.0000	.0000	.0000
2	.2727	.0606	.0140	.0040	.0013	.0005	.0002	.0001	.0000	.0000
3	.3636	.0909	.0245	.0070	.0023	.0009	.0004	.0002	.0001	.0000
4	.4545	.1364	.0385	.0120	.0040	.0015	.0006	.0003	.0001	.0001
5	.5455	.1818	.0559	.0180	.0063	.0024	.0010	.0004	.0002	.0001
6		.2424	.0804	.0270	.0097	.0037	.0015	.0007	.0003	.0002
7		.3030	.1084	.0380	.0140	.0055	.0023	.0010	.0005	.0002
8		.3788	.1434	.0529	.0200	.0080	.0034	.0015	.0007	.0004
9		.4545	.1853	.0709	.0276	.0112	.0048	.0022	.0011	.0005
10		.5455	.2343	.0939	.0376	.0156	.0068	.0031	.0015	.0008
11			.2867	.1199	.0496	.0210	.0093	.0043	.0021	.0010
12			.3462	.1518	.0646	.0280	.0125	.0058	.0028	.0014
13			.4056	.1868	.0823	.0363	.0165	.0078	.0038	.0019
14			.4685	.2268	.1032	.0467	.0215	.0103	.0051	.0026
15			.5315	.2697	.1272	.0589	.0277	.0133	.0066	.0034
16				.3177	.1548	.0736	.0351	.0171	.0086	.0045
17				.3666	.1855	.0903	.0439	.0217	.0110	.0057
18				.4196	.2198	.1099	.0544	.0273	.0140	.0073
19				.4725	.2567	.1317	.0665	.0338	.0175	.0093
20				.5275	.2970	.1566	.0806	.0416	.0217	.0116
21					.3393	.1838	.0966	.0506	.0267	.0144
22					.3839	.2139	.1148	.0610	.0326	.0177
23					.4296	.2461	.1349	.0729	.0394	.0216
24					.4765	.2811	.1574	.0864	.0474	.0262
25					.5235	.3177	.1819	.1015	.0564	.0315
26						.3564	.2087	.1185	.0667	.0376
27						.3962	.2374	.1371	.0782	.0446
28						.4374	.2681	.1577	.0912	.0526
29						.4789	.3004	.1800	.1055	.0615
30						.5211	.3345	.2041	.1214	.0716
31							.3698	.2299	.1388	.0827
32							.4063	.2574	.1577	.0952
33							.4434	.2863	.1781	.1088
34							.4811	.3167	.2001	.1237
35							.5189	.3482	.2235	.1399
36								.3809	.2483	.1575
37								.4143	.2745	.1763
38								.4484	.3019	.1965
39								.4827	.3304	.2179
40								.5173	.3598	.2406
41									.3901	.2644
42									.4211	.2894
43									.4524	.3153
44									.4841	.3421
45									.5159	.3697
46										.3980
47										.4267
48										.4559
49										.4853
50										.5147

TABLE A-11 Critical Values of the Wilcoxon Signed Rank Test ($n = 5, \ldots, 50$)

1-sided	2-sided	n = 5	n = 6	n = 7	n = 8	n = 9	n = 10
$\alpha = .05$	$\alpha = .10$	1	2	4	6	8	11
$\alpha = .025$	$\alpha = .05$		1	2	4	6	8
$\alpha = .01$	$\alpha = .02$			0	2	3	5
$\alpha = .005$	$\alpha = .01$				0	2	3

1-sided	2-sided	n = 11	n = 12	n = 13	n = 14	n = 15	n = 16
$\alpha = .05$	$\alpha = .10$	14	17	21	26	30	36
$\alpha = .025$	$\alpha = .05$	11	14	17	21	25	30
$\alpha = .01$	$\alpha = .02$	7	10	13	16	20	24
$\alpha = .005$	$\alpha = .01$	5	7	10	13	16	19

1-sided	2-sided	n = 17	n = 18	n = 19	n = 20	n = 21	n = 22
$\alpha = .05$	$\alpha = .10$	41	47	54	60	68	75
$\alpha = .025$	$\alpha = .05$	35	40	46	52	59	66
$\alpha = .01$	$\alpha = .02$	28	33	38	43	49	56
$\alpha = .005$	$\alpha = .01$	23	28	32	37	43	49

1-sided	2-sided	n = 23	n = 24	n = 25	n = 26	n = 27	n = 28
$\alpha = .05$	$\alpha = .10$	83	92	101	110	120	130
$\alpha = .025$	$\alpha = .05$	73	81	90	98	107	117
$\alpha = .01$	$\alpha = .02$	62	69	77	85	93	102
$\alpha = .005$	$\alpha = .01$	55	61	68	76	84	92

1-sided	2-sided	n = 29	n = 30	n = 31	n = 32	n = 33	n = 34
$\alpha = .05$	$\alpha = .10$	141	152	163	175	188	201
$\alpha = .025$	$\alpha = .05$	127	137	148	159	171	183
$\alpha = .01$	$\alpha = .02$	111	120	130	141	151	162
$\alpha = .005$	$\alpha = .01$	100	109	118	128	138	149

1-sided	2-sided	n = 35	n = 36	n = 37	n = 38	n = 39
$\alpha = .05$	$\alpha = .10$	214	228	242	256	271
$\alpha = .025$	$\alpha = .05$	195	208	222	235	250
$\alpha = .01$	$\alpha = .02$	174	186	198	211	224
$\alpha = .005$	$\alpha = .01$	160	171	183	195	208

1-sided	2-sided	n = 40	n = 41	n = 42	n = 43	n = 44	n = 45
$\alpha = .05$	$\alpha = .10$	287	303	319	336	353	371
$\alpha = .025$	$\alpha = .05$	264	279	295	311	327	344
$\alpha = .01$	$\alpha = .02$	238	252	267	281	297	313
$\alpha = .005$	$\alpha = .01$	221	234	248	262	277	292

1-sided	2-sided	n = 46	n = 47	n = 48	n = 49	n = 50
$\alpha = .05$	$\alpha = .10$	389	408	427	446	466
$\alpha = .025$	$\alpha = .05$	361	379	397	415	434
$\alpha = .01$	$\alpha = .02$	329	345	362	380	398
$\alpha = .005$	$\alpha = .01$	307	323	339	356	373

From F. Wilcoxon and R. A. Wilcox, "Some Rapid Approximate Statistical Procedures," 1964. Reprinted by permission of Lederle Labs, a division of the American Cyanamid Co.

TABLE A-12 Critical Values of Spearman's Rank Correlation Coefficient

n	$\alpha = .05$	$\alpha = .025$	$\alpha = .01$	$\alpha = .005$	n	$\alpha = .05$	$\alpha = .025$	$\alpha = .01$	$\alpha = .005$
5	0.900	—	—	—	8	0.643	0.738	0.833	0.881
6	0.829	0.886	0.943	—	9	0.600	0.683	0.783	0.833
7	0.714	0.786	0.893	—	10	0.564	0.648	0.745	0.794

* From E. G. Olds, "Distribution of Sums of Squares of Rank Differences for Small Samples," *Annals of Mathematical Statistics*, Vol. 9 (1938). Reprinted with permission of the Institute of Mathematical Statistics.

TABLE A-12 (continued)

n	α = .05	α = .025	α = .01	α = .005	n	α = .05	α = .025	α = .01	α = .005
11	0.523	0.623	0.736	0.818	21	0.368	0.438	0.521	0.576
12	0.497	0.591	0.703	0.780	22	0.359	0.428	0.508	0.562
13	0.475	0.566	0.673	0.745	23	0.351	0.418	0.496	0.549
14	0.457	0.545	0.646	0.716	24	0.343	0.409	0.485	0.537
15	0.441	0.525	0.623	0.689	25	0.336	0.400	0.475	0.526
16	0.425	0.507	0.601	0.666	26	0.329	0.392	0.465	0.515
17	0.412	0.490	0.582	0.645	27	0.323	0.385	0.456	0.505
18	0.399	0.476	0.564	0.625	28	0.317	0.377	0.448	0.496
19	0.388	0.462	0.549	0.608	29	0.311	0.370	0.440	0.487
20	0.377	0.450	0.534	0.591	30	0.305	0.364	0.432	0.478

TABLE A-13 Random Numbers

12651	61646	11769	75109	86996	97669	25757	32535	07122	76763
81769	74436	02630	72310	45049	18029	07469	42341	98173	79260
36737	98863	77240	76251	00654	64688	09343	70278	67331	98729
82861	54371	76610	94934	72748	44124	05610	53750	95938	01485
21325	15732	24127	37431	09723	63529	73977	95218	96074	42138
74146	47887	62463	23045	41490	07954	22597	60012	98866	90959
90759	64410	54179	66075	61051	75385	51378	08360	95946	95547
55683	98078	02238	91540	21219	17720	87817	41705	95785	12563
79686	17969	76061	83748	55920	83612	41540	86492	06447	60568
70333	00201	86201	69716	78185	62154	77930	67663	29529	75116
14042	53536	07779	04157	41172	36473	42123	43929	50533	33437
59911	08256	06596	48416	69770	68797	56080	14223	59199	30162
62368	62623	62742	14891	39247	52242	98832	69533	91174	57979
57529	97751	54976	48957	74599	08759	78494	52785	68526	64618
15469	90574	78033	66885	13936	42117	71831	22961	94225	31816
18625	23674	53850	32827	81647	80820	00420	63555	74489	80141
74626	68394	88562	70745	23701	45630	65891	58220	35442	60414
11119	16519	27384	90199	79210	76965	99546	30323	31664	22845
41101	17336	48951	53674	17880	45260	08575	49321	36191	17095
32123	91576	84221	78902	82010	30847	62329	63898	23268	74283
26091	68409	69704	82267	14751	13151	93115	01437	56945	89661
67680	79790	48462	59278	44185	29616	76531	19589	83139	28454
15184	19260	14073	07026	25264	08388	27182	22557	61501	67481
58010	45039	57181	10238	36874	28546	37444	80824	63981	39942
56425	53996	86245	32623	78858	08143	60377	42925	42815	11159
82630	84066	13592	60642	17904	99718	63432	88642	37858	25431
14927	40909	23900	48761	44860	92467	31742	87142	03607	32059
23740	22505	07489	85986	74420	21744	97711	36648	35620	97949
32990	97446	03711	63824	07953	85965	87089	11687	92414	67257
05310	24058	91946	78437	34365	82469	12430	84754	19354	72745
21839	39937	27534	88913	49055	19218	47712	67677	51889	70926
08833	42549	93981	94051	28382	83725	72643	64233	97252	17133
58336	11139	47479	00931	91560	95372	97642	33856	54825	55680
62032	91144	75478	47431	52726	30289	42411	91886	51818	78292
45171	30557	53116	04118	58301	24375	65609	85810	18620	49198
91611	62656	60128	35609	63698	78356	50682	22505	01692	36291
55472	63819	86314	49174	93582	73604	78614	78849	23096	72825
18573	09729	74091	53994	10970	86557	65661	41854	26037	53296
60866	02955	90288	82136	83644	94455	06560	78029	98768	71296
45043	55608	82767	60890	74646	79485	13619	98868	40857	19415
17831	09737	79473	75945	28394	79334	70577	38048	03607	06932
40137	03981	07585	18128	11178	32601	27994	05641	22600	86064
77776	31343	14576	97706	16039	47517	43300	59080	80392	63189
69605	44104	40103	95635	05635	81673	68657	09559	23510	95875
19916	52934	26499	09821	97331	80993	61299	36979	73599	35055
02606	58552	07678	56619	65325	30705	99582	53390	46357	13244
65183	73160	87131	35530	47946	09854	18080	02321	05809	04893
10740	98914	44916	11322	89717	88189	30143	52687	19420	60061
98642	89822	71691	51573	83666	61642	46683	33761	47542	23551
60139	25601	93663	25547	02654	94829	48672	28736	84994	13071

Reprinted with permission from *A Million Random Digits with 100,000 Normal Deviates* by the Rand Corporation (New York: The Free Press, 1955). Copyright 1955 and 1983 by the Rand Corp.

TABLE A-14 Critical Values of Hartley's *H*-statistic, $\alpha = .05$

n = number of observations in each sample
k = number of samples

						k						
n	2	3	4	5	6	7	8	9	10	11	12
3	39.0	87.5	142	202	266	333	403	475	550	626	704
4	15.4	27.8	39.2	50.7	62.0	72.9	83.5	93.9	104	114	124
5	9.60	15.5	20.6	25.2	29.5	33.6	37.5	41.1	44.6	48.0	51.4
6	7.15	10.8	13.7	16.3	18.7	20.8	22.9	24.7	26.5	28.2	29.9
7	5.82	8.38	10.4	12.1	13.7	15.0	16.3	17.5	18.6	19.7	20.7
8	4.99	6.94	8.44	9.70	10.8	11.8	12.7	13.5	14.3	15.1	15.8
9	4.43	6.00	7.18	8.12	9.03	9.78	10.5	11.1	11.7	12.2	12.7
10	4.03	5.34	6.31	7.11	7.80	8.41	8.95	9.45	9.91	10.3	10.7
11	3.72	4.85	5.67	6.34	6.92	7.42	7.87	8.28	8.66	9.01	9.34
13	3.28	4.16	4.79	5.30	5.72	6.09	6.42	6.72	7.00	7.25	7.48
16	2.86	3.54	4.01	4.37	4.68	4.95	5.19	5.40	5.59	5.77	5.93
21	2.46	2.95	3.29	3.54	3.76	3.94	4.10	4.24	4.37	4.49	4.59
61	2.07	2.40	2.61	2.78	2.91	3.02	3.12	3.21	3.29	3.36	3.39
∞	1.67	1.85	1.96	2.04	2.11	2.17	2.22	2.26	2.30	2.33	2.36
	1.00	1.00	1.00	1.00	1.00	1.00	1.00	1.00	1.00	1.00	1.00

TABLE A-15 Distribution Function for the Number of Runs *R*, in Samples of Size (n_1, n_2)

Each entry is $P(R \leqslant a)$

						a				
(n_1, n_2)	2	3	4	5	6	7	8	9	10
(2, 3)	.200	.500	.900	1.000					
(2, 4)	.133	.400	.800	1.000					
(2, 5)	.095	.333	.714	1.000					
(2, 6)	.071	.286	.643	1.000					
(2, 7)	.056	.250	.583	1.000					
(2, 8)	.044	.222	.533	1.000					
(2, 9)	.036	.200	.491	1.000					
(2, 10)	.030	.182	.455	1.000					
(3, 3)	.100	.300	.700	.900	1.000	1.000			
(3, 4)	.057	.200	.543	.800	.971	1.000			
(3, 5)	.036	.143	.429	.714	.929	1.000			
(3, 6)	.024	.107	.345	.643	.881	1.000			
(3, 7)	.017	.083	.283	.583	.833	1.000			
(3, 8)	.012	.067	.236	.533	.788	1.000			
(3, 9)	.009	.055	.200	.491	.745	1.000			
(3, 10)	.007	.045	.171	.455	.706	1.000			
(4, 4)	.029	.114	.371	.629	.886	.971	1.000		
(4, 5)	.016	.071	.262	.500	.786	.929	.992	1.000	
(4, 6)	.010	.048	.190	.405	.690	.881	.976	1.000	
(4, 7)	.006	.033	.142	.333	.606	.833	.954	1.000	
(4, 8)	.004	.024	.109	.279	.533	.788	.929	1.000	
(4, 9)	.003	.018	.085	.236	.471	.745	.902	1.000	
(4, 10)	.002	.014	.068	.203	.419	.706	.874	1.000	
(5, 5)	.008	.040	.167	.357	.643	.833	.960	.992	1.000
(5, 6)	.004	.024	.110	.262	.522	.738	.911	.976	.998
(5, 7)	.003	.015	.076	.197	.424	.652	.854	.955	.992
(5, 8)	.002	.010	.054	.152	.347	.576	.793	.929	.984
(5, 9)	.001	.007	.039	.119	.287	.510	.734	.902	.972
(5, 10)	.001	.005	.029	.095	.239	.455	.678	.874	.958

TABLE A-15 (continued)

(n_1, n_2)	2	3	4	5	6	7	8	9	10
(6, 6)	.002	.013	.067	.175	.392	.608	.825	.933	.987
(6, 7)	.001	.008	.043	.121	.296	.500	.733	.879	.966
(6, 8)	.001	.005	.028	.086	.226	.413	.646	.821	.937
(6, 9)	.000	.003	.019	.063	.175	.343	.566	.762	.902
(6, 10)	.000	.002	.013	.047	.137	.288	.497	.706	.864
(7, 7)	.001	.004	.025	.078	.209	.383	.617	.791	.922
(7, 8)	.000	.002	.015	.051	.149	.296	.514	.704	.867
(7, 9)	.000	.001	.010	.035	.108	.231	.427	.622	.806
(7, 10)	.000	.001	.006	.024	.080	.182	.355	.549	.743
(8, 8)	.000	.001	.009	.032	.100	.214	.405	.595	.786
(8, 9)	.000	.001	.005	.020	.069	.157	.319	.500	.702
(8, 10)	.000	.000	.003	.013	.048	.117	.251	.419	.621
(9, 9)	.000	.000	.003	.012	.044	.109	.238	.399	.601
(9, 10)	.000	.000	.002	.008	.029	.077	.179	.319	.510
(10, 10)	.000	.000	.001	.004	.019	.051	.128	.242	.414

(n_1, n_2)	11	12	13	14	15	16	17	18	19	20
(2, 3)										
(2, 4)										
(2, 5)										
(2, 6)										
(2, 7)										
(2, 8)										
(2, 9)										
(2, 10)										
(3, 3)										
(3, 4)										
(3, 5)										
(3, 6)										
(3, 7)										
(3, 8)										
(3, 9)										
(3, 10)										
(4, 4)										
(4, 5)										
(4, 6)										
(4, 7)										
(4, 8)										
(4, 9)										
(4, 10)										
(5, 5)										
(5, 6)	1.000									
(5, 7)	1.000									
(5, 8)	1.000									
(5, 9)	1.000									
(5, 10)	1.000									
(6, 6)	.998	1.000								
(6, 7)	.992	.999	1.000							
(6, 8)	.984	.998	1.000							
(6, 9)	.972	.994	1.000							
(6, 10)	.958	.990	1.000							
(7, 7)	.975	.996	.999	1.000						
(7, 8)	.949	.988	.998	1.000	1.000					
(7, 9)	.916	.975	.994	.999	1.000					
(7, 10)	.879	.957	.990	.998	1.000					
(8, 8)	.900	.968	.991	.999	1.000	1.000				
(8, 9)	.843	.939	.980	.996	.999	1.000	1.000			
(8, 10)	.782	.903	.964	.990	.998	1.000	1.000			
(9, 9)	.762	.891	.956	.988	.997	1.000	1.000	1.000		
(9, 10)	.681	.834	.923	.974	.992	.999	1.000	1.000	1.000	
(10, 10)	.586	.758	.872	.949	.981	.996	.999	1.000	1.000	1.000

TABLE A-16 Critical Values of the Studentized Range (Q) Distribution

The values listed in the table are the critical values of Q for $\alpha = .05$ and $.01$, as a function of degrees of freedom for MS(error) and k (the number of means)

df for MS(ERROR) (v)	α	2	3	4	5	6	7	8	9	10	11
5	.05	3.64	4.60	5.22	5.67	6.03	6.33	6.58	6.80	6.99	7.17
	.01	5.70	6.98	7.80	8.42	8.91	9.32	9.67	9.97	10.24	10.48
6	.05	3.46	4.34	4.90	5.30	5.63	5.90	6.12	6.32	6.49	6.65
	.01	5.24	6.33	7.03	7.56	7.97	8.32	8.61	8.87	9.10	9.30
7	.05	3.34	4.16	4.68	5.06	5.36	5.61	5.82	6.00	6.16	6.30
	.01	4.95	5.92	6.54	7.01	7.37	7.68	7.94	8.17	8.37	8.55
8	.05	3.26	4.04	4.53	4.89	5.17	5.40	5.60	5.77	5.92	6.05
	.01	4.75	5.64	6.20	6.62	6.96	7.24	7.47	7.68	7.86	8.03
9	.05	3.20	3.95	4.41	4.76	5.02	5.24	5.43	5.59	5.74	5.87
	.01	4.60	5.43	5.96	6.35	6.66	6.91	7.13	7.33	7.49	7.65
10	.05	3.15	3.88	4.33	4.65	4.91	5.12	5.30	5.46	5.60	5.72
	.01	4.48	5.27	5.77	6.14	6.43	6.67	6.87	7.05	7.21	7.36
11	.05	3.11	3.82	4.26	4.57	4.82	5.03	5.20	5.35	5.49	5.61
	.01	4.39	5.15	5.62	5.97	6.25	6.48	6.67	6.84	6.99	7.13
12	.05	3.08	3.77	4.20	4.51	4.75	4.95	5.12	5.27	5.39	5.51
	.01	4.32	5.05	5.50	5.84	6.10	6.32	6.51	6.67	6.81	6.94
13	.05	3.06	3.73	4.15	4.45	4.69	4.88	5.05	5.19	5.32	5.43
	.01	4.26	4.96	5.40	5.73	5.98	6.19	6.37	6.53	6.67	6.79
14	.05	3.03	3.70	4.11	4.41	4.64	4.83	4.99	5.13	5.25	5.36
	.01	4.21	4.89	5.32	5.63	5.88	6.08	6.26	6.41	6.54	6.66
15	.05	3.01	3.67	4.08	4.37	4.59	4.78	4.94	5.08	5.20	5.31
	.01	4.17	4.84	5.25	5.56	5.80	5.99	6.16	6.31	6.44	6.55
16	.05	3.00	3.65	4.05	4.33	4.56	4.74	4.90	5.03	5.15	5.26
	.01	4.13	4.79	5.19	5.49	5.72	5.92	6.08	6.22	6.35	6.46
17	.05	2.98	3.63	4.02	4.30	4.52	4.70	4.86	4.99	5.11	5.21
	.01	4.10	4.74	5.14	5.43	5.66	5.85	6.01	6.15	6.27	6.38
18	.05	2.97	3.61	4.00	4.28	4.49	4.67	4.82	4.96	5.07	5.17
	.01	4.07	4.70	5.09	5.38	5.60	5.79	5.94	6.08	6.20	6.31
19	.05	2.96	3.59	3.98	4.25	4.47	4.65	4.79	4.92	5.04	5.14
	.01	4.05	4.67	5.05	5.33	5.55	5.73	5.89	6.02	6.14	6.25
20	.05	2.95	3.58	3.96	4.23	4.45	4.62	4.77	4.90	5.01	5.11
	.01	4.02	4.64	5.02	5.29	5.51	5.69	5.84	5.97	6.09	6.19
24	.05	2.92	3.53	3.90	4.17	4.37	4.54	4.68	4.81	4.92	5.01
	.01	3.96	4.55	4.91	5.17	5.37	5.54	5.69	5.81	5.92	6.02
30	.05	2.89	3.49	3.85	4.10	4.30	4.46	4.60	4.72	4.82	4.92
	.01	3.89	4.45	4.80	5.05	5.24	5.40	5.54	5.65	5.76	5.85
40	.05	2.86	3.44	3.79	4.04	4.23	4.39	4.52	4.63	4.73	4.82
	.01	3.82	4.37	4.70	4.93	5.11	5.26	5.39	5.50	5.60	5.69
60	.05	2.83	3.40	3.74	3.98	4.16	4.31	4.44	4.55	4.65	4.73
	.01	3.76	4.28	4.59	4.82	4.99	5.13	5.25	5.36	5.45	5.53
120	.05	2.80	3.36	3.68	3.92	4.10	4.24	4.36	4.47	4.56	4.64
	.01	3.70	4.20	4.50	4.71	4.87	5.01	5.12	5.21	5.30	5.37
∞	.05	2.77	3.31	3.63	3.86	4.03	4.17	4.29	4.39	4.47	4.55
	.01	3.64	4.12	4.40	4.60	4.76	4.88	4.99	5.08	5.16	5.23

From E. S. Pearson and H. O. Hartley (eds.), *Biometrika Tables for Statisticians*, 3rd ed., 1966. Reproduced by permission of *Cambridge University Press*.

APPENDIX B   Derivation of Minimum Total Sample Size

Claim: When obtaining two independent samples, the maximum error for the difference of the two population means $\mu_1 - \mu_2$, is

$$E = Z_{\alpha/2} \sqrt{\frac{\sigma_1^2}{n_1} + \frac{\sigma_2^2}{n_2}} \qquad (\sigma_1, \sigma_2 \text{ known})$$

or estimated using

$$E = Z_{\alpha/2} \sqrt{\frac{s_1^2}{n_1} + \frac{s_2^2}{n_2}} \qquad (\sigma_1, \sigma_2 \text{ unknown})$$

For a specific value of E, the sample sizes, n_1 and n_2, that minimize the total sample size, $n = n_1 + n_2$, are given by

$$n_1 = \frac{Z_{\alpha/2}^2 s_1 (s_1 + s_2)}{E^2}$$

and

$$n_2 = \frac{Z_{\alpha/2}^2 s_2 (s_1 + s_2)}{E^2}$$

Proof: For ease of notation, define

$$Z = Z_{\alpha/2}$$
$$a = s_1$$
$$b = s_2$$
$$x = n_1$$
$$y = n_2$$

(For the case where the σ's are known, then $a = \sigma_1$ and $b = \sigma_2$.) Now,

$$E = Z \sqrt{\frac{a^2}{x} + \frac{b^2}{y}}$$

is fixed. Solving for y yields

$$y = \frac{Z^2 b^2 x}{E^2 x - Z^2 a^2}$$

The total sample size is $n = x + y$, and so

$$n = f(x) = x + \frac{Z^2 b^2 x}{E^2 x - Z^2 a^2}$$

To determine the value of x that minimizes $n = f(x)$, the procedure will be to find $f'(x)$, set it to zero, and solve for x.

$$f'(x) = 1 + \frac{(E^2 x - Z^2 a^2)(Z^2 b^2) - Z^2 b^2 x (E^2)}{(E^2 x - Z^2 a^2)^2} = 1 - \frac{Z^4 a^2 b^2}{(E^2 x - Z^2 a^2)^2}$$

Now,

$$f'(x) = 0$$
$$\text{iff } x^2(E^4) + x(-2Z^2 E^2 a^2) + (Z^4 a^4 - Z^4 a^2 b^2) = 0$$
$$\text{iff } x = \frac{2Z^2 E^2 a^2 \mp \sqrt{4Z^4 E^4 a^4 - 4Z^4 E^4 a^2 (a^2 - b^2)}}{2E^4} = \frac{Z^2 a^2 \mp Z^2 ab}{E^2}$$

Now,

$$f''(x) = \frac{2Z^4E^2a^2b^2(E^2x - Z^2a^2)}{(E^2x - Z^2a^2)^4}$$

Consequently,

$$f''(x) > 0 \quad \text{iff} \, (E^2x - Z^2a^2) > 0$$

Letting

$$x = \frac{Z^2a^2 + Z^2ab}{E^2}$$

then

$$E^2x - Z^2a^2 = Z^2ab > 0$$

because a and b are > 0. Letting

$$x = \frac{Z^2a^2 - Z^2ab}{E^2}$$

then

$$E^2x - Z^2a^2 = -Z^2ab < 0$$

Conclusion:

1. $f(x)$ has a local minimum at

$$x = \frac{Z^2a(a + b)}{E^2}$$

2. $f(x)$ has a local maximum at

$$x = \frac{Z^2a(a - b)}{E^2}$$

Because we are restricted to values of x (that is, n_1) such that $f(x) =$ total sample size is positive, and because $f(x)$ approaches ∞ as x approaches ∞, for the admissible values of x, $f(x)$ has a global minimum at

$$x = n_1 = \frac{Z^2a(a + b)}{E^2}$$

$$= \frac{Z^2s_1(s_1 + s_2)}{E^2}$$

Solving for n_2, we previously stated that

$$y = \frac{Z^2b^2x}{E^2x - Z^2a^2}$$

Substituting

$$x = \frac{Z^2s_1(s_1 + s_2)}{E^2}$$

into this expression produces

$$y = n_2 = \frac{Z^2s_2(s_1 + s_2)}{E^2}$$

APPENDIX C   Introduction to MINITAB—
Mainframe or Microcomputer Versions

MINITAB is an easy-to-use, flexible statistical package. It was originally designed for students; over the years, it has been constantly improved. It is one of the more powerful statistical systems currently available. MINITAB will allow you to "speak" to the computer using commands that are similar to English sentences. The sequence of steps in a typical problem resembles the same steps you would take if solving the problem by hand.

MINITAB consists of a worksheet containing rows and columns. The data for each variable are stored in a particular *column*. The following discussion will provide a brief introduction of data entry and use of MINITAB commands. No distinction is made between mainframe and microcomputer MINITAB commands since they can be used on either system.

Entering the Data: Data for each of the variables in your data set are stored in columns. For example, suppose you have four scores (80, 75, 43, and 91) that you want to enter as one variable. These can be entered as shown in Figure C.1 using the SET command. Here, the four test scores are stored in column C1. Notice that (1) the data values are separated by blanks (commas are OK) and (2) at the end of the data string, the word END is entered on the next line. If your data will not fit on a single line, type as many values as you wish, enter these, continuing typing on the next line, enter another line of data, and so forth until you have entered all of your data. Your last line always will be END.

FIGURE C.1

Commands to enter four test scores.

```
MTB > SET INTO C1
DATA> 80 75 43 91
DATA> END
```

Suppose that you have test scores for three people (Joe, Mary, and Al), each of whom has four scores. The previous four test scores (80, 75, 43, and 91) belong to Joe. The other method of entering data is to read in your data one *row* at a time using the READ command. Each line contains a single row of data on each of the variables (three variables, here). This is shown in Figure C.2.

These steps will input the 12 test scores one row (exam) at a time. Note that C1–C3 means C1 through C3.

FIGURE C.2

Commands to enter four test scores for each of three students.

```
MTB > READ INTO C1-C3
DATA> 80 70 100 ─────────── Scores from exam 1
DATA> 75 95 65                    ⋮
DATA> 43 83 76 ─────────── Scores from exam 2
DATA> ┌91 90 86
DATA> │ END
```
Scores for Al (variable 3)

Scores for Mary (variable 2)

Scores for Joe (variable 1)

Output of the Data: To display the data, you should use the PRINT command, as shown in Figure C.3. When using the microcomputer version of MINITAB, to send your results to the printer and screen simultaneously, type

MTB > PAPER

FIGURE C.3
Commands to display
(or print) data.

```
MTB > PRINT C1

C1
      80      75      43      91

MTB > PRINT C1-C3
     ROW      C1      C2      C3

       1      80      70     100
       2      75      95      65
       3      43      83      76
       4      91      90      86
```

To stop sending output to the printer at any time, type

MTB > NOPAPER

MINITAB Shorthand: One exceptionally nice feature of MINITAB is that most of the commands can be shortened to save time and effort. You are able to do this because the MINITAB system does not process the entire line that you enter—it reads only those pieces of information that it needs to know and ignores everything else! Consequently, you can misspell words or leave out unnecessary words and MINITAB will still execute your command. For example, in Figure C.1, you could have used the command

MTB > SET C1

That is, the word "into" was not necessary. We used it originally because it makes the statement easier to comprehend for someone who wants to know what we are doing at this step.

Many of the MINITAB commands are illustrated on the inside front cover of this book. Only those portions of each statement that are colored must be included. What you put between the colored portions is up to you—you can leave them blank (the shorthand version) or put in any words you wish to make the statement easier to understand. For example, in Figure C.1, we could have used

MTB > SET THE EXAM SCORES FOR JOE INTO C1

Here, all MINITAB needs to see is SET and C1. What you put in between is your decision.

In addition, for the commands on the inside front cover, any portion of the statement enclosed in brackets [ ] need not be included in the statement. If you want to omit this portion of the statement, you may do so. The information in the brackets generally allows you to be more specific in your input to MINITAB or informs MINITAB in which columns you would like certain information from the output to be stored.

MINITAB Subcommands: Some of the more sophisticated MINITAB commands allow you to specify further information by using one or more *subcommands*. For those commands that allow subcommands (not all of them do), you should end the main command line with a *semicolon, ;*. This informs MINITAB that subcommands will follow. Each subcommand line should end with a semicolon unless it is the last subcommand—the last one ends with a

FIGURE C.4

The REGRESS command, with subcommands.

```
MTB > REGRESS Y IN C1 USING 2 PREDICTORS IN C2 C3;
SUBC> NOCONSTANT;
SUBC> COEF INTO C4;
SUBC> RESIDS INTO C5.
```

Informs MINITAB that this
is the end of the subcommands

Informs MINITAB that one or
more subcommands will follow

period. If you forget to type the period at the end of the last subcommand, simply type a period on the next line.

An example of a command utilizing subcommands (called REGRESS, discussed in Chapters 13 and 14) is shown in Figure C.4. The MINITAB solutions contained throughout the text will clarify what commands allow the use of subcommands and what these possible subcommands are.

APPENDIX D   Introduction to SPSSX—Mainframe Version

Computer packages often are used to perform various statistical analysis procedures. When used properly, these computer packages can save time and decrease the probability of human error. The purpose of this appendix is to provide a basic overview of one such mainframe package, the Statistical Package for the Social Sciences (SPSSX).

To solve a statistical problem using SPSSX, you must define the data and the format in which it is to be interpreted, and specify the statistical procedure to be performed. For example, the data in Figure D.1 are test grades for three students. If you wish to find the mean grade for each of the students, you define each individual test score and specify the SPSSX procedure to obtain mean values (CONDESCRIPTIVE). Figure D.2 shows the statements and data required to perform this task under SPSSX.

FIGURE D.1

Test grades for three students.

80	70	100
75	95	65
43	83	76
91	90	86

FIGURE D.2

Statements and data to find mean grades for three students.

```
TITLE            SPSSX-APPENDIX
DATA LIST FREE /JOE MARY AL
BEGIN DATA
80  70  100
75  95  65
43  83  76
91  90  86
END DATA
CONDESCRIPTIVE ALL
```

Analysis of Statements: The following is an analysis of each of the statements used to obtain the mean grades for each student:

TITLE SPSSX-APPENDIX   Defines the name of the SPSSX run.

DATA LIST FREE / JOE MARY AL   Defines three variables (students) named Joe, Mary, and Al. FREE specifies the data is not in any for-

	mat, other than it is to be read sequentially, starting at column one.
BEGIN DATA	Specifies that the following lines are input data lines.
80 70 100	Gives the test-score values for the first test (Joe scored 80, Mary scored 70, and Al scored 100).
75 95 65	Gives the test-score values for the second test.
43 83 76	Gives the test-score values for the third test.
91 90 86	Gives the test-score values for the fourth test.
	As shown, the values must be separated by one or more spaces, and there must be a value for all four tests for each of the three students.
END DATA	Indicates the end of the data card images.
CONDESCRIPTIVE ALL	Specifies the SPSSX procedure to be performed. In this program, the mean, as well as other statistics, will be calculated for all variables.

Figure D.3 shows the printed results of running this procedure under SPSSX.

FIGURE D.3
Results of running listing in Figure D.2.

```
NUMBER OF VALID OBSERVATIONS (LISTWISE) =      4.00
VARIABLE        MEAN     STD DEV   MINIMUM    MAXIMUM VALID N   LABEL

JOE            72.250    20.614    43.00       91.00      4
MARY           84.500    10.847    70.00       95.00      4
AL             81.750    14.886    65.00      100.00      4
```

Most SPSSX procedures require additional statements, either to describe the problem or to request different options for solving the problem in different ways. All procedures, statements, and options necessary for solving the problems in this text are discussed in the end-of-chapter appendixes. For further information on SPSSX, refer to the *SPSSX Introductory Statistics Guide* (Marija Norusis, New York: McGraw-Hill, 1983).

Basic SPSSX Rules: As with all computer packages, you must observe certain rules when running SPSSX programs. The following is a list of the basic rules for SPSSX:

1. All commands must begin in column 1 and cannot exceed column 15.
2. Any additional information must appear in columns 16 through 80.
3. All command keywords must be separated by only one blank.
4. Multiple blanks are allowed beyond column 15.
5. Include the decimal point when entering decimal data (such as 38.95).

Additional rules concerning individual procedures are contained in the *SPSSX Introductory Statistics Guide*.

JCL Statements: Job Control Language (JCL) statements must be included with SPSSX statements in order for the SPSSX program to execute properly. These statements identify the user and the procedure to be performed. Typically, you need only a job statement, an execute statement (EXEC SPSSX), and a card image input statement (SYSIN DD*). The format and order of these statements may differ, so consult your computer center before attempting to run your first program.

APPENDIX E   Introduction to SPSS/PC+—Microcomputer Version

Computer packages often are used to perform various statistical analysis procedures. When used properly, these computer packages can save time and decrease the probability of human error. The purpose of this appendix is to provide a basic overview of one such microcomputer package, the Statistical Package for the Social Sciences (SPSS/PC+).

To solve a statistical problem using SPSS/PC+, you must define the data and the format in which it is to be interpreted, and specify the statistical procedure to be performed. For example, the data in Figure E.1 are test grades for three students. If you wish to find the mean grade for each of the students, you define each individual test score and specify the SPSS procedure to obtain mean values (CONDESCRIPTIVE). Figure E.2 shows the statements and data required to perform this task under SPSS/PC+.

FIGURE E.1

Test grades for three students.

80	70	100
75	95	65
43	83	76
91	90	86

FIGURE E.2

Statements and data to find mean grades for three students.

```
TITLE      SPSS/PC+   APPENDIX.
DATA LIST FREE /JOE MARY AL.
BEGIN DATA.
80   70   100
75   95   65
43   83   76
91   90   86
END DATA.
CONDESCRIPTIVE ALL.
```

Analysis of Statements: The following is an analysis of each of the statements used to obtain the mean grades for each student. Each statement *must* end with a period.

TITLE SPSS/PC+ APPENDIX.	Defines the name of the SPSS/PC+ run.
DATA LIST FREE / JOE MARY AL.	Defines three variables (students) named Joe, Mary, and Al. FREE specifies the data is not in any format, other than it is to be read sequentially, starting at column one.
BEGIN DATA.	Specifies that the following lines are input data lines.
80 70 100	Gives the test-score values for the first test (Joe scored 80, Mary scored 70, and Al scored 100).
75 95 65	Gives the test-score values for the second test.
43 83 76	Gives the test-score values for the third test.
91 90 86	Gives the test-score values for the fourth test. As shown, the values must be separated by one or more spaces, and there must be a value for all four tests for each of the three students.
END DATA.	Indicates the end of the data card images.
CONDESCRIPTIVE ALL.	Specifies the SPSS/PC+ procedure to be performed. In this program, the mean, as well as other statistics, will be calculated for all variables.

FIGURE E.3
Results of running
listing in Figure E.2.

```
-----------------------------------------------------------------------
              SPSS/PC+    APPENDIX

Number of Valid Observations (Listwise) =        4.00

Variable      Mean    Std Dev   Minimum   Maximum    N  Label

JOE          72.25     20.61     43.00     91.00     4
MARY         84.50     10.85     70.00     95.00     4
AL           81.75     14.89     65.00    100.00     4
-----------------------------------------------------------------------
```

Figure E.3 shows the printed results of running this procedure under SPSS/PC+.

Most SPSS/PC+ procedures require additional statements, either to describe the problem or to request different options for solving the problem in different ways. All procedures, statements, and options necessary for solving the problems in this text are discussed in the end-of-chapter appendixes. For further information on SPSS/PC+ refer to the *SPSS/PC+ for the IBM PC/XT/AT* (Marija Norusis, Chicago: SPSS Inc., 1986).

Basic SPSS/PC+ Rules: As with all computer packages, you must observe certain rules when running SPSS/PC+ programs. The following is a list of the basic rules for SPSS/PC+.

1. All SPSS/PC+ statements must end with a period.
2. Lines containing data do not end with a period.
3. All command keywords must be separated by only one blank.
4. Multiple blanks are allowed beyond column 15.
5. Include the decimal point when entering decimal data (such as 38.95).
6. All data must be defined before requesting SPSS/PC+ to perform a statistical procedure.
7. To return to DOS, type FINISH at the end of your SPSS/PC+ session.

Additional rules concerning individual procedures are contained in the *SPSS/PC+ for the IBM PC/XT/AT*.

Obtaining Paper Output: At any point in the SPSS/PC+ session, you can obtain output on your printer by typing

SET   PRINTER = ON.

All subsequent output will be sent to the screen and the printer. To terminate the printed output, type

SET   PRINTER = OFF.

APPENDIX F   Introduction to SAS— Mainframe Version

Computer packages often are used to perform various statistical analysis procedures. When used properly, these computer packages can save time and decrease the probability of human error. The purpose of this appendix is to provide a basic overview of one such mainframe package, the Statistical Analysis System (SAS).

To solve a statistical problem using SAS, you must define the data to be used and specify the statistical procedure to be performed. For example, the data in Figure F.1 lists test grades for three students. If you wish to find the mean grade for each of the students, you define each individual test score and specify the SAS procedure to obtain mean values (PROC MEANS). Figure F.2 shows the statements and data required to perform this task under SAS.

FIGURE F.1

Test grades for three students.

80	70	100
75	95	65
43	83	76
91	90	86

FIGURE F.2

Statements and data to find mean grades for three students.

```
TITLE      'SAS-APPENDIX';
DATA       GRADES;
INPUT      JOE MARY AL;
CARDS;
80   70   100
75   95   65
43   83   76
91   90   86
PROC  MEANS;
```

Analysis of Statements: The following is an analysis of each of the statements used to obtain the mean grades for each student:

TITLE 'SAS APPENDIX';	Defines the name of the SAS run. Enclose the title in single quotes.
DATA GRADES;	Defines a dataset named grades.
INPUT JOE MARY AL;	Defines three variables (students) named Joe, Mary, and Al.
CARDS;	Defines the input medium as card images.
80 70 100	Gives the test-score values for the first test (Joe scored 80, Mary scored 70, and Al scored 100).
75 95 65	Gives the test-score values for the second test.
43 83 76	Gives the test-score values for the third test.
91 90 86	Gives the test-score values for the fourth test.
	As shown, the values must be separated by one or more spaces, and there must be a value for all four tests for each of the three students.
PROC MEANS;	Specifies the SAS procedure to be performed.

Figure F.3 shows the printed results of running this procedure under SAS.

Most SAS procedures require additional statements, either to describe the problem or to request different options for solving the problem in different ways. All procedures, statements, and options necessary for solving the problems in this text are discussed in the end-of-chapter appendixes. For further information on SAS, refer to the *SAS User's Guide: Statistics Version*, 5th ed. (SAS Institute, Inc., Cary, NC, 1985).

FIGURE F.3   Results of running listing in Figure F.2.

SAS-APPENDIX

VARIABLE	N	MEAN	STANDARD DEVIATION	MINIMUM VALUE	MAXIMUM VALUE	STD ERROR OF MEAN	SUM	VARIANCE	C.V.
JOE	4	72.25000000	20.61350690	43.00000000	91.00000000	10.30675345	289.00000000	424.91666667	28.531
MARY	4	84.50000000	10.84742673	70.00000000	95.00000000	5.42371337	338.00000000	117.66666667	12.837
AL	4	81.75000000	14.88567544	65.00000000	100.00000000	7.44283772	327.00000000	221.58333333	18.209

Basic SAS Rules: As with all computer packages, you must observe certain rules when running SAS programs. The following is a list of the basic rules for SAS:

1. SAS statements must begin in column 1 and cannot exceed column 72. Any SAS statement longer than 72 columns may be continued on the next line.
2. All SAS statements must end with a semicolon (;).
3. Lines containing data do not end with a semicolon.
4. All data must be defined before a procedure can be run.
5. More than one procedure can be performed by adding more procedure statements.
6. You must include the decimal point when entering decimal data (such as 38.95).

Additional rules concerning individual procedures are contained in the *User's Guide.*

JCL Statements: Job Control Language (JCL) statements must be included with SAS statements in order for the SAS program to execute properly. These statements identify the user and the procedure to be performed. Typically, you need only a job statement, an execute statement (EXEC SAS), and a data input statement (SYSIN DD*). The format and order of these statements may differ, so consult your computer center before attempting to run your first program.

APPENDIX G Introduction to SAS—Microcomputer Version

Computer packages often are used to perform various statistical analysis procedures. When used properly, these computer packages can save time and decrease the probability of human error. The purpose of this appendix is to provide a basic overview of one such microcomputer package, the Statistical Analysis System (SAS).

To solve a statistical problem using SAS, you must define the data to be used and specify the statistical procedure to be performed. For example, the data in Figure G.1 lists test grades for three students. If you wish to find the mean grade for each of the students, you define each individual test score and specify the SAS procedure to obtain mean values (PROC MEANS). Figure G.2 shows the statements and data required to perform this task under SAS.

FIGURE G.1

Test grades for three students.

80	70	100
75	95	65
43	83	76
91	90	86

FIGURE G.2

Statements and data to find mean grades for three students.

```
TITLE     'SAS-APPENDIX';
DATA      GRADES;
INPUT     JOE MARY AL;
CARDS;
80   70  100
75   95  65
43   83  76
91   90  86
PROC  MEANS;
```

Analysis of Statements: The following is an analysis of each of the statements used to obtain the mean grades for each student:

TITLE 'SAS APPENDIX';	Defines the name of the SAS run. Enclose the title in single quotes.
DATA GRADES;	Defines a dataset named grades.
INPUT JOE MARY AL;	Defines three variables (students) named Joe, Mary, and Al.
CARDS;	Defines the input medium as card images.
80 70 100	Gives the test-score values for the first test (Joe scored 80, Mary scored 70, and Al scored 100).
75 95 65	Gives the test-score values for the second test.
43 83 76	Gives the test-score values for the third test.
91 90 86	Gives the test-score values for the fourth test.
	As shown, the values must be separated by one or more spaces, and there must be a value for all four tests for each of the three students.
PROC MEANS;	Specifies the SAS procedure to be performed.

Figure G.3 shows the printed results of running this procedure under SAS. Most SAS procedures require additional statements, either to describe the problem or to request different options for solving the problem in different ways. All procedures, statements, and options necessary for solving the problems in this text are discussed in the end-of-chapter appendixes. For further information on SAS, refer to the *SAS Introductory Guide for Personal Computers Version*, 6th ed. (SAS Institute, Inc., Cary, NC, 1985).

FIGURE G.3   Results of running listing in Figure G.2.

N Obs	Variable	N	Minimum	Maximum	Mean	Std Dev
4	JOE	4	43.0000000	91.0000000	72.2500000	20.6135069
	MARY	4	70.0000000	95.0000000	84.5000000	10.8474267
	AL	4	65.0000000	100.0000000	81.7500000	14.8856754

Basic SAS Rules: As with all computer packages, you must observe certain rules when running SAS programs. The following is a list of the basic rules for SAS:

1. All SAS statements must end with a semicolon (;).
2. Lines containing data do not end with a semicolon.
3. All data must be defined before a procedure can be run.
4. More than one procedure can be performed by adding more procedure statements.
5. You must include the decimal point when entering decimal data (such as 38.95).
6. To return to DOS, at the COMMAND = prompt, enter the single letter X.

Additional rules concerning individual procedures are contained in the *SAS Introductory Guide for Personal Computers*.

Obtaining Paper Output: To obtain printed output during interactive full screen sessions, in the OUTPUT window type FILE 'PRN:'. This should be typed after the output has appeared in the window.

If submitting the SAS job from a previously created ASCII file (say, SASRUN) containing the SAS statements and data, the SAS output will be contained in a newly created file (created by SAS), called SASRUN.LST. To submit the SAS job, at the hard disk drive prompt, type

SAS   B:SASRUN

This assumes you have saved SASRUN on a floppy diskette, contained in drive B. No output is routed to the screen; instead it will be routed to the B drive into the file SASRUN.LST. This will be an ASCII file and can be printed in the usual manner (such as the DOS command PRINT B:SASRUN.LST).

APPENDIX H   Database Using Household Financial Variables

VARIABLE	DESCRIPTION
INCOME1	Income of principal wage earner
INCOME2	Income of secondary wage earner
FAMLSIZE	Family size
OWNORENT	Own or rent (1 = own, 0 = rent)
TOTLDEBT	Total indebtedness (excluding home mortgage)
HPAYRENT	House payment or apartment/house rent
UTILITY	Average monthly utility expenditure
LOCATION	Location of residence (1 = NE sector, 2 = NW sector, 3 = SW sector, 4 = SE sector)

OBS	INCOME1	INCOME2	FAMLSIZE	OWNORENT	TOTLDEBT	HPAYRENT	UTILITY	LOCATION
1	36741	20691	4	1	14634	1138	295	1
2	27242	25454	5	1	14445	1108	260	2
3	39076	0	6	1	10383	1012	274	1
4	41633	0	3	1	9569	867	214	1
5	32980	25396	3	1	17490	1026	245	2
6	44143	24302	5	1	20354	1201	306	1
7	32082	25715	4	0	11870	1100	281	2
8	43450	0	5	1	12226	1035	284	3
9	41534	22739	5	1	15023	1494	206	1
10	40070	32128	5	0	15882	1995	346	1
11	31603	0	2	1	15883	924	95	4
12	45092	15136	5	0	17848	1312	281	1
13	36554	0	4	1	16753	1362	148	4
14	28554	23431	2	1	15479	1246	460	3
15	35066	20418	2	0	13997	965	238	1
16	26937	0	3	1	10595	908	213	1
17	46632	0	4	1	11626	1037	259	1
18	38521	23135	4	1	16336	1935	228	1
19	45319	25028	6	1	19307	757	164	3
20	55899	0	4	1	15751	1515	390	2
21	31989	0	1	1	7537	728	213	1
22	36905	0	7	0	11512	803	179	3
23	28203	17720	4	1	10293	906	223	2
24	35549	19045	2	1	14694	555	163	1
25	35210	0	5	0	7832	943	259	1
26	34432	0	4	1	13489	1018	250	2
27	43170	22311	2	1	12497	679	193	2
28	36217	16510	5	0	9966	439	102	3
29	35484	0	2	1	10285	1067	277	4
30	50576	0	5	1	14471	1139	273	2
31	29138	0	2	1	5893	775	183	1
32	32329	0	2	0	12599	505	398	2
33	41609	15323	6	1	8924	1624	160	1
34	30993	23354	2	0	15534	552	136	4
35	40365	0	2	1	14887	647	169	4
36	42082	0	3	1	3368	648	136	2
37	39816	0	2	0	6594	545	266	2
38	28878	0	2	0	18138	894	232	1
39	31882	22231	3	1	10442	1114	212	1
40	31307	0	3	1	10172	1816	293	1
41	50716	13937	5	0	6379	1370	193	2
42	41204	13865	3	1	15644	1157	245	1
43	28438	13500	3	1	16614	1077	269	2
44	37141	0	4	1	16560	1097	227	1
45	40547	23413	3	1	16611	943	296	4
46	40966	0	2	1	17357	1176	272	2
47	39926	25335	5	1	17235	1160	247	1
48	41204	16905	2	0	10730	973	353	2
49	28205	15424	4	1	10767	1442	222	1
50	44604	23670	3	1	12325	362	203	1
51	37954	24519	4	1	13903	738	134	4
52	26882	0	2	1	17021	918	226	2
53	20224	14458	1	1	16104	612	309	1
54	35449	19396	3	0	11723	704	269	2
55	37275	14486	2	1	10438	585	221	1
56	28205	0	2	1	10418	558	156	1
57	38847	0	2	0	8408	961	249	1
58	34812	15729	5	1	10421	737	202	2
59	33384	17879	2	0	15182	952	142	1
60	42104	15961	3	1	19605	520	315	1
61	27980	15729	4	1	16079	1357	229	1
62	31397	20081	2	1	20081	799	190	1
63	32081	31060	4	0	12586	894	272	4
64	29001	21761	3	1	9043	1138	116	1
65	31637	26655	2	1	18699	540	459	2
66	26583	22201	5	1	16332	1872	201	1
67	38847	0	2	1	14872	877	385	4
68	34405	24182	5	0	10421	768	207	2
69	34112	23384	2	0	12314	1657	248	1
70	27980	17739	2	1	11314	399	118	1
71	31397	15961	2	1	9834	1156	249	2
72	41397	0	2	1		722	175	2
73	31870	23477	7	1				
74	29949	24656	2	1				
75	41779	21196	6	0				
76	38300	0	2	1				
77	40674	22843	1	0				
78	39044	0	4	1				
79	32559	16284	4	1				
80	31905	0	1	0				
81								
82								

OBS	INCOME1	INCOME2	FAMLSIZE	OWNORENT	TOTLDEBT	HPAYRENT	UTILITY	LOCATION
83	15802	0	5	0	2987	311	127	3
84	28890	0	2	1	4787	417	120	4
85	21374	21302	2	1	8739	429	76	1
86	22261	18313	5	1	9938	1192	315	2
87	34402	0	2	1	10580	372	87	1
88	36236	13286	4	1	13968	708	162	2
89	36540	19533	1	1	13164	1289	295	1
90	21208	0	3	1	6942	359	134	1
91	41167	0	5	1	5596	329	108	2
92	36544	24913	2	1	16138	1058	272	2
93	32442	19149	5	1	13338	763	235	1
94	32442	23785	3	0	17795	1511	379	2
95	46144	10439	4	1	17848	929	219	1
96	29707	12731	1	1	18699	1283	311	1
97	31498	0	4	0	6940	904	122	1
98	32080	0	3	0	6352	606	136	1
99	31715	13275	1	0	12068	713	123	1
100	33182	26546	2	1	13902	349	133	1
101	22264	0	2	1	10221	1642	146	2
102	30757	21576	3	1	8854	612	466	4
103	24405	17153	3	0	7154	859	163	3
104	30467	0	3	1	7661	894	183	2
105	29922	0	4	1	16996	1457	346	3
106	37858	12108	4	1	11849	1251	336	1
107	36112	27812	4	1	11148	430	278	2
108	30782	14429	3	1	9642	312	103	2
109	40463	0	2	1	11988	1297	112	1
110	45203	12267	5	1	9805	1799	299	3
111	26553	0	4	0	9112	840	192	3
112	37220	9836	6	1	11969	827	203	2
113	32400	0	2	1	6127	872	225	2
114	27727	0	2	1	10893	716	229	1
115	47121	0	3	1	14430	785	247	3
116	36059	18619	3	1	13717	1339	192	1
117	29495	27820	2	1	15250	578	335	2
118	32998	19829	3	2	15577	542	143	2
119	35568	0	6	1	10261	1022	138	3
120	23962	0	2	1	8678	613	265	1
121	44452	26966	4	1	5465	890	166	3
122	27949	23340	5	0	15882	1244	230	2
123	29622	15559	2	1	12709	972	258	1
124	43379	0	3	0	8974	869	278	1
125	41722	0	4	0	10493	859	250	2
126	21292	16231	4	0	12146	960	211	1
127	23425	16917	3	1	15347	689	173	3
128	32964	23340	3	1	6911	945	214	2
129	34485	0	2	0	18893	857	189	1
130	30210	20091	5	1	19200	808	185	2
131	37026	17543	2	1	10313	1030	199	1
132	34325	0	2	0	11967	998	223	2
133	33462	15012	2	0	11053	724	268	2
134	31418	22205	3	1	11083	912	219	2
135	58869	0	2	1	9689	1472	142	1
136	31357	0	2	1	12847	742	114	2
137	27122	25137	1	1	11247	518	159	4
138	29511	19970	2	1	3395	564	96	1
139	33021	0	2	0	9155	972	255	4
140	36083	13960	6	1	10313	673	179	2
141	31629	2509	2	0	14976	798	176	4
142	32583	20009	4	1	14939	724	182	4
143	40260	0	2	1	11967	912	236	4
144	26134	20728	2	1	11053	604	142	4
145	35481	0	2	1	9689	531	114	2
146	21432	20080	1	1	12897	718	187	4
147	27814	14213	2	0	3395	742	96	3
148	27122	0	2	1	9155	518	176	2
149	29519	0	2	1	8482	564	116	2
150	33021	13960	2	1	12802	799	182	2
151	36083	2509	2	1	20724	957	227	2
152	44939	20009	2	0	13754	917	222	4
153	32583	0	4	1	14943	751	179	4
154	36103	16450	3	0	9364	735	236	3
155	35626	15952	4	1	9340	860	181	2
156	21167	13480	2	1	7236	703	202	1
157	21432	13807	4	0	10124	860	343	1
158	27814	18529	2	1	14207	1386	202	1
159	29995	20112	3	1	11927	820	207	1

Left table (OBS 165–246):

OBS	INCOME1	INCOME2	FAMLSIZE	OWNORENT	TOTLDEBT	HPAYRENT	UTILITY	LOCATION
165	36582	22969	4	0	14593	1250	283	1
166	54131	0	4	1	11101	1174	256	1
167	39248	32401	4	1	11019	784	241	1
168	42107	0	2	0	11628	655	187	2
169	22041	0	2	0	8888	760	161	4
170	38947	1819	3	1	15641	1040	153	4
171	24537	821	3	0	14001	821	263	4
172	47566	24430	4	1	18761	1387	362	1
173	36022	19247	6	1	13525	1635	353	1
174	36185	0	3	0	10114	643	152	1
175	25143	21342	3	1	13773	430	110	4
176	26068	22518	2	0	14504	980	225	1
177	36548	0	5	1	11271	813	178	2
178	25212	18405	2	1	14342	385	116	2
179	27972	22958	4	0	18203	1204	283	4
180	25212	0	2	1	14271	320	116	1
181	37104	0	6	0	7901	1536	369	2
182	35460	0	3	1	7627	767	167	4
183	16414	14081	2	1	2703	478	126	1
184	28707	0	6	0	10165	911	246	3
185	34495	0	3	0	9330	720	151	1
186	38171	0	1	1	7026	576	113	2
187	32418	20092	2	0	11574	382	277	2
188	46646	22870	2	1	11888	925	224	1
189	32014	20579	2	0	18847	903	261	2
190	27800	15052	3	1	10844	1533	164	1
191	27789	17687	2	1	10154	500	211	1
192	27047	19462	3	0	12675	522	184	2
193	33675	0	6	1	10690	777	144	1
194	40282	21965	2	1	14232	783	157	2
195	36118	0	3	0	10232	482	326	1
196	30080	0	7	1	10103	1240	257	2
197	35259	0	3	0	12493	746	156	1
198	41035	15224	5	1	15197	969	294	2
199	43144	18874	5	0	11646	443	113	1
200	21916	13634	2	1	9383	980	218	1
201	29937	0	2	0	8891	485	165	1
202	54284	0	2	0	17924	774	178	1
203	31934	17601	2	1	11549	320	204	1
204	29301	24023	2	1	13704	787	108	1
205	49466	22873	2	0	14201	474	143	4
206	33652	0	2	1	7416	651	164	4
207	34861	12200	2	0	14789	442	137	1
208	29301	25523	2	1	13688	1190	208	1
209	32131	23216	4	0	14903	922	210	2
210	35247	23904	2	1	15514	853	218	4
211	35247	23368	2	0	6226	556	135	1
212	33904	0	6	1	17488	625	186	1
213	33725	0	3	1	10199	418	242	1
214	14441	0	2	1	4705	740	201	1
215	33058	0	5	0	12331	1195	163	4
216	43058	21252	3	1	15389	1043	219	2
217	30669	0	2	0	16864	1899	242	1
218	35192	23904	2	1	15525	957	101	1
219	51467	0	7	0	17265	937	414	4
220	30981	9884	3	1	8510	633	336	4
221	28909	0	4	1	33300	725	173	1
222	30974	11841	5	1	11503	1261	317	1
223	30974	14419	2	0	15469	1730	229	2
224	39019	26251	2	1	15503	935	223	1
225	27041	0	6	0	17726	1873	312	3
226	40138	24547	2	1	14946	542	105	2
227	31634	0	2	1	16703	418	202	1
228	28184	0	2	0	4609	740	295	4
229	31509	0	5	1	12331	542	105	1
230	57203	0	2	0	9370	1195	219	1
231	26802	24125	4	1	15525	1043	244	4
232	35192	19669	2	0	99864	1899	336	3
233	22589	21899	7	1	17265	1453	193	1
234	38649	0	3	1	8872	858	294	4
235	33197	0	3	0	16549	1113	182	2
236	30981	24539	4	1	51399	1136	236	4
237	29909	22225	5	1	11186	1663	265	1
238	30974	12129	2	1	14436	805	246	1
239	49379	19283	6	0	14207	1056	162	4
240	26130	20260	3	1	8137	812	128	1
241	35010	21413	4	1	13975	1104	170	1
242	37812	26843	2	1	11556	375	125	1
243	26309	0	4	1	8137	812	162	4
244	38210	23149	2	0	13975	1704	112	1
245	36162	12834	2	0	11556	375	125	1

Right table (OBS 247–328):

OBS	INCOME1	INCOME2	FAMLSIZE	OWNORENT	TOTLDEBT	HPAYRENT	UTILITY	LOCATION
247	33467	22385	4	1	10755	1097	279	2
248	40159	30462	1	1	16537	805	205	1
249	41635	0	1	1	15009	883	202	3
250	42107	32401	2	1	15555	706	204	4
251	32636	12091	2	1	16282	861	202	2
252	22041	14967	2	1	15639	1137	279	1
253	38947	0	3	1	16168	1336	93	2
254	24537	19333	3	1	12731	549	104	3
255	47566	23080	5	1	24871	1078	229	1
256	36022	19247	6	1	2407	93	209	1
257	37490	0	3	0	8143	714	253	1
258	36185	0	4	0	13552	958	241	3
259	25143	16438	3	1	13862	979	186	1
260	26068	21605	3	0	13097	683	212	2
261	36548	0	1	1	14480	771	212	1
262	25212	0	2	0	17311	780	209	2
263	43322	22758	3	1	14033	849	182	4
264	21711	19323	3	0	12258	602	179	1
265	19056	0	3	1	6577	740	241	2
266	39236	16558	3	0	12450	936	253	1
267	27413	20001	3	0	6232	951	139	3
268	23203	0	2	1	16612	579	304	4
269	24514	8361	4	0	16542	1328	305	2
270	24514	23309	4	0	17994	1072	293	2
271	24880	0	3	1	22431	642	241	4
272	49514	25935	3	0	22431	457	110	2
273	32562	25880	2	1	21146	659	184	1
274	46698	0	4	1	13871	522	134	2
275	28347	16114	2	0	11901	584	198	1
276	28347	21371	3	1	13328	846	186	2
277	34271	22908	3	0	11210	1020	267	3
278	34271	20626	3	0	17550	936	252	2
279	29796	24450	2	1	11635	746	169	1
280	43599	28762	2	1	8791	682	200	4
281	30264	0	3	1	9448	1297	323	3
282	26603	18612	5	0	6946	424	327	2
283	35176	16174	5	0	5047	668	135	3
284	23237	0	3	0	13269	1301	286	1
285	31215	20905	3	0	4817	1063	279	2
286	27931	16138	5	1	13823	1107	217	4
287	35061	18840	2	1	12240	1063	289	3
288	37428	16366	4	0	16642	1080	152	2
289	40706	17934	3	1	17002	524	163	3
290	34800	13903	2	1	12488	601	172	1
291	49669	0	2	1	12485	931	229	1
292	15514	0	4	0	2429	468	125	3
293	27953	0	4	1	6554	489	144	4
294	27953	19985	7	1	11921	1060	328	4
295	33047	0	3	1	11465	1410	259	1
296	33300	0	3	1	16401	513	142	2
297	36411	0	3	0	18807	1142	108	4
298	36533	0	6	1	18807	752	204	4
299	36264	15983	2	1	6526	1233	163	2
300	27715	19536	2	1	10449	763	305	2
301	36533	0	4	1	6617	939	162	1
302	43732	17125	3	1	9650	1364	200	2
303	20385	0	5	1	9616	786	235	1
304	25878	15882	2	1	7823	317	228	2
305	35943	19701	5	1	6142	902	375	2
306	43425	0	4	1	6972	866	94	4
307	28749	22975	4	1	12024	933	271	1
308	27356	22441	1	1	11115	1154	259	4
309	46700	14630	2	1	16779	1045	100	1
310	25758	17049	1	1	9339	702	240	4
311	51109	11453	3	1	12262	570	200	3
312	44151	0	6	1	9658	1191	112	1
313	28392	23814	3	1	8555	394	111	2
314	36379	0	3	1	13014	996	252	4
315	30379	27849	2	1	4412	616	186	1
316	24281	23441	6	1	16098	610	207	1
317	24612	0	1	1	11103	777	343	1
318	24612	23696	5	1	10883	1512	112	3
328	24612	23696	5	1	9669	1705	381	2

OBS	INCOME1	INCOME2	FAMLSIZE	OWNORENT	TOTLDEBT	HPAYRENT	UTILITY	LOCATION
329	29969	0	3	0	6233	519	159	3
330	27484	26902	2	1	15794	872	209	1
331	29375		1	0	15958	458	208	3
332	38188	0	4	0	11031	636	132	2
333	37304		2	1	15794	681	209	3
334	49501	20769	3	0	15989	1185	281	2
335	23468	22058	4	1	8611	930	249	1
336	22260		4	1	7555	738	145	1
337	33949	26538	3	1	13002	595	168	3
338	34498		2	0	13035	718	191	2
339	34898	29146	2	1	10029	498	100	1
340	28763		4	0	9542	420	155	3
341	28761	28424	4	1	13713	1123	308	4
342	45234		3	0	10968	460	162	2
343	35134	19722	1	1	14862	585	123	1
344	44571	21099	3	1	18255	819	229	1
345	42837	19991	4	1	12132	1269	280	3
346	29544	17096	5	1	18154	1243	301	2
347	31257		2	0	12991	547	153	1
348	40570	13727	2	1	13712	1218	263	2
349	41795		2	1	4551	539	149	1
350	30941	21863	4	0	16220	711	108	2
351	30852	14936	3	1	16933	530	294	1
352	39867	14002	4	1	16682	755	204	2
353	33975	21124	2	1	6386	1324	209	4
354	30096	26124	6	0	15251	449	221	2
355	30227	22726	3	1	14465	934	140	2
356	40148	15120	4	0	14499	612	221	2
357	28649		4	1	5461	616	108	4
358	28720	21857	2	1	16645	344	101	4
359	28151		6	0	10652	750	180	3
360	57626	0	3	1	11116	370	101	1
361	35364		4	0	15271	662	207	4
362	30695	27583	3	0	10048	590	146	1
363	27415	15628	1	0	15592	568	147	2
364	62146		4	0	11705	1298	325	2
365	35418	23107	4	0	11152	886	257	1
366	29767	16100	4	1	14876	736	197	3
367	27191	22227	3	1	9689	802	214	4
368	34354		4	0	16418	1164	300	1
369	29783	17399	3	0	3507	643	153	1
370	25461	13346	5	1	9392	332	116	2
371	41700	21732	4	0	10542	869	260	4
372	35507		2	0	13288	1234	303	3
373	33502	21368	5	1	5282	537	98	1
374	29519		1	1	11662	858	244	2
375	29969	22336	4	0	10971	1006	209	2
376	26671	21018	5	1	19976	581	237	1
377	17339	12077	3	0	8840	471	123	1
378	36881	22336	1	1	10983	419	173	3
379	45819		5	0	11048	570	149	1
380	45819	2353	3	0	18100	1122	281	3
381	37594	21516	3	0	18100	1167	294	1
382	38269	1507	3	1	8170	1663	166	1
383	37594		3	0	9281	853	232	4
384	32146	15560	4	0	11155	1727	437	4
385	26057		8	1	2112	1009	164	4
386	32878	21368	5	0	6685	757	274	3
387	25745		1	0	7700	447	164	1
388	26195	17888	2	0	7104	349	97	3
389	20089	15579	3	1	14013	1046	225	1
390	27973	19952	7	1	11418	1580	382	2
391	21763		2	0	10040	616	185	3
392	33715		2	0	14202	800	135	3
393	33316		5	0	6100	561	94	2
394	38932		2	1	12239	698	205	2
395	35720	21433	2	1	19294	818	212	1
396	32710		3	1	4323	1306	327	3
397	35242	22743	3	0	12893	1352	177	3
398	28886		2	0	8741	488	105	2
399	36805		2	0	3776	320	108	1
400	32646	18010	2	1	8564	534	212	2
401	35828	23321	2	1	16464	700	194	1
402	35528		6	1	16532	778	224	3
403	36947		3	1	18134	927	230	3
404	47932	15888	3					

OBS	INCOME1	INCOME2	FAMLSIZE	OWNORENT	TOTLDEBT	HPAYRENT	UTILITY	LOCATION
411	32982	7882	2	1	10538	521	101	1
412	29817		3	1	9486	817	236	3
413	53279	0	2	1	14959	546	114	1
414	31788		3	0	7569	791	188	2
415	41289		3	1	10553	1251	291	3
416	24220	20948	4	0	7226	1242	237	1
417	26952	18987	5	1	14082	1420	342	1
418	39441	27834	4	1	15447	1532	381	1
419	24770	11034	3	1	9524	1077	230	2
420	42500		5	0	10449	1260	321	2
421	31648	0	2	0	6243	555	147	4
422	46508		5	1	13098	1364	309	2
423	30635		3	0	6205	696	163	3
424	38596	20171	3	0	5681	561	145	3
425	29310	16062	5	1	12500	1250	282	2
426	29151	20962	2	1	10433	1243	253	1
427	39333		1	1	13054	531	133	3
428	16502		3	0	3789	622	142	1
429	35574	23429	3	0	18854	1036	256	2
430	36347		2	1	10082	915	182	3
431	36433	22725	2	1	8878	829	191	3
432	31611		6	1	6593	470	141	2
433	44815	27183	4	1	17406	1421	358	1
434	26570	17808	4	1	17406	1095	277	1
435	36965	25967	4	1	12655	1014	249	3
436	34718	15864	6	1	12655	1324	246	1
437	32269	22882	2	1	18700	903	314	2
438	32265	22868	2	1	13796	817	209	2
439	28649	18485	6	1	13060	1084	227	4
440	25095		2	1	13060	1053	305	3
441	25085	32793	6	0	14239	529	252	2
442	27274	17883	2	1	11665	717	207	2
443	25240	25727	2	1	10712	884	203	1
444	34647		8	0	15578	1457	373	3
445	38611		7	1	8154	1534	351	3
446	34549		3	1	9442	464	101	1
447	29236		4	1	9224	785	193	2
448	22185	12357	4	1	5468	310	116	2
449	37405	21001	4	1	11911	1090	264	1
450	32031	22862	2	1	15935	623	134	3
451	38435		3	1	17005	409	189	3
452	33718	23011	5	1	15615	506	184	1
453	46672	14978	2	1	14682	683	349	1
454	46123	15228	6	1	17356	1401	88	3
455	26390	12141	2	1	11458	272	168	1
456	30550		5	1	12657	636	233	2
457	18085	15940	6	1	5817	970	259	1
458	33327		6	1	15523	1104	226	3
459	31112	21647	3	0	15045	1360	353	4
460	42887	18430	2	1	15726	513	103	1
461	26724	27125	4	1	19412	1448	364	1
462	30926	15547	5	1	12087	1141	265	2
463	33491	12215	2	1	17743	1260	284	1
464	30561		5	1	16743	777	233	2
465	28335	20711	2	0	16802	854	207	3
466	36196	10108	5	1	10529	953	266	2
467	45816	19915	2	1	10523	967	161	3
468	37662		4	1	13691	335	105	3
469	23982	21786	3	1	9932	852	210	4
470	34391		1	0	7909	570	85	2
471	19647	12847	5	0	7652	382	109	2
472	27088		6	1	4990	1159	271	3
473	28495	21154	4	1	5539	469	154	4
474	46495	17174	3	1	74495	975	239	4
475	40013	23859	2	0	18553	859	377	1
476	32602		1	1	11662	632	228	3
477	23025		5	0	15531	403	106	3
478	33962	21380	6	1	3842	700	149	4
479	20430	18115	4	0	12507	978	259	3
480	24457	21329	3	1	11888	993	270	2
481	35137		2	1	11888	1060	93	4
482	37284	21380	1	0	12103	369	253	4
483	26318	18115	4	1	13405	983	93	2
484	44632	21387	2	1	15775	1341	327	1

OBS	INCOME1	INCOME2	FAMLSIZE	OWNORENT	TOTLDEBT	HPAYRENT	UTILITY	LOCATION
575	41156	13463	4	1	12552	1094	287	1
576	39066		4	0	13674	994	257	1
577	37341	18284	3	0	13539	995	233	1
578	25287	20735	6	1	12100	1115	288	2
579	30707	0	1	1	6272		104	2
580	39883	26150	1	1	11746	321	267	1
581	27782	21048	5	1	15046	1210	372	2
582	34170		4	1	14826	1443	105	1
583	37482		2	1	9305	343	363	3
584	18729	17719	6	0	8443	1253	105	1
585	43294	19521	2	0	16098	792	212	2
586	32366	19812	2	0	11425	819	264	1
587	46759		3	1	13701	972	156	3
588	35981	18257	1	1	14508	685	162	4
589	38596		2	0	15797	1014	229	4
590	30382		4	1	10416	1203	197	3
591	43835	21831	4	0	8501	909	193	1
592	36986		3	1	11589	648	166	3
593	41690		2	1	4584	1788	313	1
594	19114		1	1	11573	767	180	3
595	54478	18632	2	1	10785	774	215	1
596	20672	15923	3	1	15928	1168	405	1
597	37813		7	0	33523	872	317	3
598	32613		5	1	8445	1259	143	2
599	42489	19674	2	1	15585	504	289	1
600	35996		4	1	9250	1346	162	2
601	21566	19389	4	1	14381	410	283	2
602	44106	13984	4	0	19984	1025	241	1
603	49294	25892	2	0	5204	797	256	4
604	22341		3	0	9234	349	187	1
605	37936		2	1	7496	639	271	4
606	27450		6	0	9245	956	121	2
607	29517		2	1	2242	569	162	2
608	45802		4	1	10513	601	319	2
609	38680		3	0	12676	1147	386	1
610	22046	23252	5	1	12250	1229	106	3
611	30159	1657	7	1	14757	386	111	4
612	32035		2	1	15502	648	136	2
613	38224	20112	4	1	14809	873	211	1
614	32063		3	1	13154	591	134	3
615	24613	16662	5	1	17242	1489	363	4
616	41234	26216	2	1	17897	609	125	2
617	26444	13623	4	1	14690	647	136	1
618	33064	17467	3	1	9264	1289	299	1
619	35257	22753	7	0	19768	1286	324	3
620	40040		2	1	9724	966	166	4
621	42574	25874	4	1	10401	1079	273	2
622	32942	21643	3	1	15573	738	296	2
623	43899	30505	2	1	10573	667	217	2
624	60047	24689	3	1	19363	1296	313	1
625	24895	8487	5	1	14623	890	202	1
626	28382	168880	2	1	12906	957	261	2
627	41883	24354	4	1	11753	1032	229	1
628	42633	19248	3	1	13272	658	197	3
629	30749	28249	3	1	12011	119	325	2
630	35692	23325	2	0	13918	1269	354	4
631	25980	27404	5	1	20777	947	240	4
632	25730	20631	3	1	12351	1029	236	2
633	32470	154427	1	0	12641	980	132	2
634	42141	23069	2	1	10881	1086	262	3
635	31529	23873	5	1	20165	636	272	4
636	20672	17380	4	0	16223	406	131	2
637	33683	13096	2	1	13201	459	143	1
638	36147		3	0	65563	874	236	4
639	29247	12295	3	1	81163	915	196	1
640	33920	8865	2	0	6898	760	161	1

OBS	INCOME1	INCOME2	FAMLSIZE	OWNORENT	TOTLDEBT	HPAYRENT	UTILITY	LOCATION
493	40364	0	6	1	11330	1024	291	2
494	29959		4	0	16507	737	145	3
495	30755	14775	3	0	14224	806	203	2
496	24476		1	1	5459	450	101	4
497	31819		2	0	6283	675	140	2
498	45305		1	1	12276	398	143	1
499	37149	22913	3	1	14276	1068	270	1
500	35717		2	1	5456	322	94	2
501	37913		4	0	13482	697	203	1
502	36893	13109	4	1	10519	1031	258	2
503	32450		1	1	15830	535	98	1
504	37293	29326	5	0	11389	1371	337	3
505	22225	19766	2	1	15306	478	95	1
506	24998	23537	3	1	14155	960	389	2
507	35260	16627	7	1	15306	1505	247	3
508	34290		3	1	11163	918	168	1
509	22271		4	0	4707	744	163	2
510	40059	19743	2	0	6828	363	166	4
511	33587	18563	5	1	17215	7584	290	2
512	32032	13252	4	1	14248	1315	279	2
513	37112	22472	2	0	17081	795	198	1
514	23151		3	1	13695	538	120	2
515	40771	18309	3	0	17313	1035	257	1
516	31453	19914	2	1	17313	766	193	4
517	36436	20107	4	1	14124	784	186	3
518	36456	26948	2	1	9770	850	198	1
519	27085	14452	4	1	14625	306	210	3
520	35167	27421	2	0	17215	1534	212	1
521	41118	20803	5	1	12466	570	382	4
522	41782	11717	3	0	11369	733	113	4
523	25266		6	0	5865	1483	340	4
524	53266		1	0	11084	403	103	2
525	24648	18477	5	0	4810	903	213	2
526	25871	27082	4	0	11760	1025	275	3
527	24809	23159	2	1	15345	1939	217	1
528	28614		3	1	11407	1130	277	3
529	35457		1	1	8003	823	122	1
530	23778	19914	4	0	16990	516	203	1
531	32328	20657	4	1	17246	1329	253	4
532	15879	18208	7	1	10684	1409	372	1
533	41958	23336	3	0	10790	971	210	3
534	49823		3	1	12278	967	203	2
535	32628	21745	3	1	8392	673	269	3
536	44549	18208	4	1	9408	1548	261	1
537	32076	23336	4	1	15971	469	353	3
538	21042	18337	3	1	10661	452	130	2
539	26080		7	1	14616	567	290	1
540	26741		2	1	85559	1016	279	1
541	31588	26312	4	1	5931	886	151	4
542	36610	16361	2	1	12205	725	168	2
543	32276	21575	4	1	90095	697	230	4
544	33224		3	0	10057	1067	234	4
545	25713	26312	3	1	14511	1060	362	3
546	34414	25748	4	1	14771	1826	308	3
547	39376	19936	4	1	18712	1283	295	1
548	28108	17875	3	1	17446	1113	257	2
549	42254	21873	4	1	10737	1915	257	3
550	18712		3	0	28614	857	184	2
551	38004		3	1	16886	1236	310	2
552	43597	18371	2	1	9134	601	115	4
553	51722	14872	4	0	11140	458	208	2
554	25913		2	1	8995	1098	252	4
555	30772		2	0	1817	1557	160	3
556	19262	21364	6	1	18860	1634	163	4
557	20109		3	0	15869	1095	163	4
558	48950		5	0	11129	872	186	2
559	32115	24517	2	1	13996	773	199	1
560	30582		4	0	13585	841	293	3
561	30362	14152	3	0	12336	1181	293	3
562	12801	17886	5	0	13252	841	199	3
563	38282		5	0	15420	1181	293	3
564	48554		4	1		714	269	2

OBS	INCOME1	INCOME2	FAMLSIZE	OWNORENT	TOTLDEBT	HPAYRENT	UTILITY	LOCATION
657	36269	0	4	1	8429	956	195	1
658	22005	16043	4	1	19901	615	183	2
659	34790	24228	2	1	13787	724	180	2
660	36860	22204	2	1	13154	670	150	1
661	22951	0	1	0	7825	366	109	4
662	36673	0	1	1	9590	785	145	2
663	22490	0	3	0	6781	504	175	4
664	24273	20166	2	0	12549	527	115	2
665	25756	17508	2	0	10448	363	107	3
666	40430	0	4	0	10044	627	168	1
667	26315	21348	2	1	14649	520	136	1
668	26879	20079	4	0	9060	738	214	1
669	36826	20093	2	1	17998	714	199	2
670	26223	0	2	1	17264	389	116	1
671	37323	22437	4	0	15507	1184	340	2
672	36623	12541	2	1	13242	389	128	2
673	35161	0	5	1	8538	1076	244	4
674	25763	0	3	1	17241	773	164	1
675	30313	19896	3	0	12120	674	137	2
676	28511	0	2	0	5710	837	204	4
677	37963	0	1	1	6363	525	111	2
678	21653	0	4	1	4228	870	233	2
679	52809	0	4	1	13569	788	191	4
680	31014	26447	4	0	14133	923	210	3
681	28880	17399	3	1	13569	863	260	1
682	49848	25953	5	1	14495	948	224	4
683	23809	18578	3	0	14166	1098	297	1
684	37402	0	2	0	9897	698	178	2
685	35881	12698	6	1	9409	489	140	3
686	40395	13561	6	0	13431	1267	298	3
687	29980	26536	3	0	12747	1260	181	1
688	35353	14433	3	1	12530	807	387	2
689	36915	19337	4	1	11998	330	139	4
690	39515	19991	2	1	15888	1091	268	1
691	20688	17146	4	1	16533	835	242	1
692	46017	0	2	0	6512	802	232	3
693	31441	25083	4	1	9974	678	435	2
694	38167	16937	4	1	15568	1558	214	3
695	31023	19884	2	1	12888	854	234	1
696	37396	20155	4	0	12888	1356	297	2
697	32048	15552	5	1	14073	791	294	2
698	29551	0	4	1	12145	899	280	2
699	27585	20979	5	0	12937	1111	307	1
700	33952	16302	3	0	8124	301	294	4
701	20768	0	2	1	13757	1032	223	2
702	35562	0	3	1	13889	910	186	1
703	36311	24078	2	1	14150	662	162	1
704	41817	17680	5	1	14431	519	107	1
705	25405	16639	5	1	11208	1239	309	1
706	42256	20289	5	0	11203	686	182	2
707	34276	0	2	1	12595	616	159	1
708	46017	17002	2	0	8817	645	135	1
709	29501	0	2	0	9974	653	172	4
710	31023	21739	3	1	4429	485	112	2
711	30461	0	7	0	8895	962	202	2
712	32426	0	3	0	13881	1681	414	1
713	52341	0	1	1	8083	842	233	1
714	39467	28272	2	1	9494	637	130	4
715	39976	0	1	1	12404	907	214	2
716	32878	11833	2	1	8810	625	192	2
717	29079	0	4	1	12893	1355	352	1
718	55868	0	7	1	10038	897	266	3
719	35065	22433	2	1	10917	764	167	1
720	31219	19663	2	1	12309	627	186	2
721	28830	14207	2	1	14560	1123	295	1
722	28810	14622	5	0	15268	1073	251	2
723	25440	17806	3	1	13542	1180	337	2
724	43876	17996	4	0	12552	1140	269	2
725	27783	0	2	1	12252	581	160	1
726	36144	20277	4	0	9432	1074	266	3
727	42129	13594	2	1	9604	1104	301	1
728	26735	13554	5	1	16149	1276	292	1
729	29313	15576	4	1	16205	962	267	1
730	44508	21908	3	1	11908	1704	384	1
731	53336	19141	7	0	10790	1548	375	4

OBS	INCOME1	INCOME2	FAMLSIZE	OWNORENT	TOTLDEBT	HPAYRENT	UTILITY	LOCATION
739	53955	0	2	1	11072	759	177	2
740	44600	14550	4	1	17267	1167	316	1
741	47925	13585	4	1	17523	1063	245	1
742	54527	0	3	0	13807	540	159	1
743	28228	9830	2	0	9318	715	182	2
744	22434	0	3	0	1157	452	149	1
745	22811	14568	3	1	10178	1275	341	3
746	20822	0	6	0	9346	974	223	1
747	20422	0	2	1	5023	330	101	1
748	21528	22554	2	1	6884	687	201	1
749	21288	19916	2	0	6812	1101	229	4
750	33138	17906	2	1	12011	468	127	1
751	47138	0	2	1	12011	700	141	4
752	37355	0	3	0	6014	882	204	3
753	39023	11439	5	1	8920	1176	293	2
754	38233	0	2	1	11631	992	174	2
755	38035	13522	3	0	9023	781	252	2
756	32311	23194	2	1	10574	659	188	2
757	32293	24414	3	1	10554	738	165	1
758	32229	0	2	0	10076	444	119	3
759	36432	0	4	1	10571	1109	198	1
760	43191	13230	2	0	9245	1566	182	1
761	21152	26422	4	1	16829	1726	202	3
762	35201	20987	2	0	8148	1468	281	5
763	44170	0	5	1	15008	1069	282	4
764	32516	13196	1	0	10401	934	212	2
765	29737	23647	3	1	10234	949	180	1
766	29156	0	4	0	15142	1363	207	2
767	29880	28218	2	1	5995	741	239	3
768	37414	0	4	1	54095	915	333	4
769	36687	0	4	0	15458	1466	179	3
770	37446	11311	4	1	11535	615	202	2
771	31563	0	2	1	9373	845	230	2
772	31563	14358	4	1	15162	1026	107	4
773	37937	26466	5	1	9471	397	231	3
774	36202	0	2	1	7934	962	86	3
775	35126	0	3	1	14523	354	217	1
776	44835	0	3	1	14523	842	108	3
777	52703	0	2	1	9161	511	126	3
778	37503	25073	4	1	7621	425	294	1
779	37250	13570	5	1	15253	1258	289	2
780	23996	17733	2	0	12253	1127	133	1
781	29864	0	4	0	14395	474	159	1
782	50996	0	7	1	16331	457	246	4
783	41574	21161	6	1	14040	1128	383	2
784	26013	15530	2	1	12064	1598	396	1
785	25189	26783	3	0	18534	1605	113	4
786	30137	0	7	1	12234	1016	278	2
787	41200	22205	5	1	10891	752	226	4
788	33535	21991	3	1	13465	978	168	4
789	31862	21527	7	1	13138	497	431	1
790	22891	17016	5	0	5938	1647	353	2
791	25812	0	3	1	11105	1497	122	2
792	28190	0	4	1	8449	573	132	1
793	53604	0	2	0	10075	895	239	1
794	28138	22357	4	0	13649	459	217	2
795	32293	16005	5	0	12033	968	92	2
796	37671	0	2	1	12033	302	211	3
797	26991	19732	3	1	10937	847	236	2
798	41529	21434	6	1	13958	991	202	3
799	45033	0	1	1	4285	691	169	2
800	39426	0	2	1	6339	1582	372	3
801	37446	24193	2	0	9341	430	103	1
802	37246	25644	3	1	15541	336	162	1
803	29770	0	3	1	12871	953	241	2
804	29770	2230	5	0	9736	631	177	4
805	23227	0	4	1	17693	634	112	1
806	23263	0	3	1	9736	1016	257	2
807	29916	24193	4	0	12149	632	155	3
808	23418	25086	5	1	17560	473	114	1
809	42873	2230	4	0	9597	1245	280	2
810	20180	0	2	1	9212	477	126	1
811	29770	24193	3	1	12149	632	155	3
812	37446	25644	3	0	12871	634	114	3
813	29770	0	5	1	9736	1016	167	2
814	23227	24193	4	0	17693	632	257	1
815	23263	25086	3	1	12149	473	155	3
816	29916	2230	5	0	9597	1245	114	2
817	23418	15863	4	1	9212	632	280	2
818	42873	22220	4	1	11002	477	126	1
819	34418	0	2	1	75509	900	220	1
820	20180	0	2	1	1339	810	157	3

OBS	INCOME1	INCOME2	FAMLSIZE	OWNORENT	TOTLDEBT	HPAYRENT	UTILITY	LOCATION
821	25752	14218	2	1	12492	844	205	3
822	38967	0	2	0	10921	641	154	3
823	43653	19954	4	0	15931	681	177	3
824	34017	15554	2	1	13256	725	193	2
825	52262	0	4	0	13735	619	209	1
826	34743	25608	3	1	16362	1042	267	2
827	25903	0	2	0	6787	255	110	1
828	26034	0	2	1	8402	596	161	2
829	26077	14048	3	1	7998	720	166	4
830	42774	17390	5	1	15434	951	234	2
831	29208	20891	2	0	14962	460	162	1
832	42302	0	5	0	11054	1140	301	1
833	31043	0	1	1	10054	687	149	3
834	31630	30026	1	0	18873	567	93	1
835	50616	0	7	0	19561	1312	367	1
836	53440	0	2	1	5244	442	170	4
837	31441	28758	2	1	12243	584	149	1
838	25441	24558	1	0	12437	644	237	3
839	33762	0	3	1	4421	412	97	4
840	24488	23151	2	1	3809	850	164	3
841	29204	0	6	1	17736	1158	244	1
842	28391	16384	3	1	16094	1601	372	2
843	35463	30055	3	0	10177	894	217	1
844	30927	0	2	1	6049	510	235	4
845	32112	20987	2	1	12617	395	144	1
846	41416	22310	5	1	14826	1073	256	1
847	46093	24734	1	1	15534	1277	352	2
848	42389	25528	2	0	14257	786	208	1
849	30557	0	5	1	11970	812	192	2
850	36557	18797	6	1	10083	388	170	3
851	29950	14798	3	1	11581	1286	370	1
852	35663	0	2	0	12858	709	195	2
853	29823	0	1	1	5143	183	84	4
854	30064	17290	4	0	11726	456	120	1
855	30068	0	2	1	8130	889	174	1
856	31143	0	4	0	16035	1089	271	4
857	41313	17055	2	1	10059	1733	333	2
858	28005	24299	2	0	10359	1214	282	1
859	33361	25626	4	1	8849	1339	235	1
860	23361	25659	6	1	12052	879	223	1
861	21408	18610	2	1	11854	919	242	1
862	31446	20036	3	1	12854	858	216	1
863	42389	24452	4	1	14696	861	321	2
864	27463	24436	2	0	13963	889	183	1
865	28851	18763	2	1	10748	1138	321	1
866	29606	0	5	0	13664	907	183	3
867	33357	13791	2	1	12724	766	90	2
868	30178	20688	2	0	8549	451	285	1
869	32034	14309	4	1	7225	1196	181	1
870	27011	0	1	0	16035	911	138	1
871	32067	17290	4	0	18130	642	269	2
872	28710	0	2	1	10224	1158	246	3
873	30416	0	3	0	13018	1129	127	2
874	38113	0	4	1	5802	438	148	1
875	45114	25434	2	1	12999	559	183	4
876	42022	0	1	0	9681	657	246	2
877	55414	17709	5	1	12300	1049	243	1
878	26405	20240	3	0	10811	808	132	2
879	27211	19202	2	1	10575	490	128	3
880	33260	0	4	0	13382	531	257	1
881	44416	21128	7	1	20612	959	386	2
882	38165	0	3	1	6376	844	208	4
883	20637	0	1	0	9107	548	119	1
884	31978	21128	2	1	9583	559	136	2
885	31795	0	4	0	16508	1480	235	1
886	25099	0	2	1	12282	1026	267	1
887	38868	12432	4	1	14734	632	198	3
888	37500	23650	3	1	10327	985	362	1
889	38320	24011	6	1	13320	1296	136	1
890	32841	0	3	0	2945	475	207	2
891	30888	19654	4	1	6111	800	111	2
892	27432	0	1	0	14990	444	249	4
893	24011	0	3	1	4760	993	99	3
894	37794	20621	6	0	8535	412	149	4
895	22551	0	3	1	8359	734	189	2
896	51699	17994	7	0	13222	1455	379	2
897	32499							
898	32190							
899	32211							
900	32650							
901	36105							
902	31968							
903	23555	13624	3	0	11138	614	160	2
904	47238	0	7	0	8943	1398	363	3
905	24237	16296	2	1	7078	616	173	4
906	28530	22096	2	1	8442	984	219	2
907	48290	14753	6	1	15116	1014	206	2
908	25288	0	1	1	10964	1128	277	1
909	43549	0	4	1	12832	1492	101	1
910	30993	28851	5	1	14895	1487	316	2
911	33020	17296	4	1	12216	983	244	4
912	28686	20998	2	1	6880	680	124	1
913	29628	0	7	0	11644	1596	429	2
914	41738	17225	4	0	14448	1076	293	1
915	29401	24944	3	0	18500	810	204	4
916	34298	0	5	1	9106	859	139	4
917	25443	19085	2	1	8919	478	200	3
918	19315	0	1	1	11317	871	251	3
919	45778	0	3	1	11739	667	164	2
920	34480	16868	3	1	9190	801	299	1
921	39587	0	2	1	8207	1191	217	4
922	43406	28236	2	1	14923	889	255	1
923	41536	24736	4	1	14606	1077	228	1
924	20610	16849	3	1	7944	832	145	3
925	36416	0	3	1	7089	703	274	2
926	31808	0	3	1	5033	867	272	2
927	36578	17592	4	1	8695	1021	231	1
928	31102	21506	5	1	2688	1082	302	1
929	33153	12441	4	1	14848	1214	231	2
930	38924	0	2	1	18126	1071	337	2
931	42265	0	7	1	14106	1075	96	1
932	41211	0	2	1	7952	1114	131	2
933	28731	23615	3	1	3332	335	325	4
934	32349	11830	6	0	6755	620	221	2
935	38806	0	5	1	12250	1089	137	3
936	28224	18183	2	1	10734	817	224	3
937	26181	0	3	1	11649	462	310	2
938	27077	29927	4	1	7636	946	317	1
939	26135	0	2	1	9198	521	192	3
940	32241	20889	3	1	17027	918	290	1
941	31840	12851	4	0	6327	1040	260	1
942	45769	12866	5	1	19919	1095	176	2
943	42435	0	4	1	11038	632	235	1
944	41211	0	4	1	16774	1334	284	3
945	40778	16214	5	1	15407	1291	204	1
946	26050	11337	3	1	12743	709	304	4
947	37080	0	3	1	18205	1111	394	3
948	46241	22881	2	1	19197	1865	204	2
949	41572	16683	4	1	14912	655	182	1
950	22064	15574	2	1	11234	747	161	3
951	29439	0	4	1	6722	806	177	2
952	32203	7204	2	1	6526	651	168	2
953	47959	20102	4	1	15804	1091	261	3
954	42486	0	4	1	9638	957	238	1
955	43817	26322	2	1	21287	594	243	1
956	26041	21053	2	0	14585	494	320	1
957	35584	20375	7	1	12978	1263	250	2
958	29191	0	4	1	14569	402	117	4
959	28452	0	3	0	9455	733	166	3
960	37698	11908	3	1	10842	709	217	1
961	30894	19500	4	0	14273	1076	266	2
962	32841	0	2	1	14281	1793	417	1
963	33588	17196	5	1	7359	483	355	3
964	30786	19922	2	0	13379	1485	191	1
965	24602	17954	7	1	12427	816	165	2
966	33897	0	2	1	12628	722	283	2
967	20632	18744	2	1	5969	1273	140	1
968	37570	22188	4	0	16627	742	308	1
969	49508	1847	7	1	13126	1612	146	3
970	28040	0	4	1	12614	942	206	2
971	53856	20628	4	0	12006	1566	209	2
972	34034	0	2	1	12930	794	199	1
973	38678	15143	3	1	14103	680	189	1

OBS	INCOME1	INCOME2	FAMLSIZE	OWNORENT	TOTLDEBT	HPAYRENT	UTILITY	LOCATION
1067	42312	33693	4	1	22918	1066	266	1
1068	44621	0	2	1	3702	548	155	1
1069	32400	17673	5	0	10971	1435	336	1
1070	37673	22379	3	1	22319	873	220	1
1071	46619	14391	3	1	13851	1124	270	1
1072	40044	22724	2	1	14965	561	124	1
1073	38060	0	5	0	7164	1050	267	4
1074	38605	0	2	0	7164	329	99	3
1075	35380	25869	2	0	13630	766	210	3
1076	64025	0	2	1	13567	500	207	1
1077	42788	0	1	1	11253	481	99	3
1078	43350	0	3	0	10415	626	129	2
1079	44601	0	3	1	9412	642	169	1
1080	34449	0	1	1	11966	402	102	2
1081	48358	0	3	1	14344	721	201	2
1082	55409	0	6	1	14160	470	128	1
1083	25804	21194	3	0	13381	1390	348	4
1084	25855	0	2	0	14078	567	128	1
1085	36447	0	1	1	2714	1225	293	1
1086	36397	0	4	1	8476	860	172	3
1087	30998	0	3	1	3957	475	131	1
1088	29928	0	1	0	5180	427	119	3
1089	50959	22766	2	1	17678	827	232	2
1090	36420	0	2	1	9603	635	106	1
1091	25886	11015	2	1	9214	455	90	3
1092	37051	23838	1	1	15505	833	210	1
1093	33335	0	4	1	19943	390	111	1
1094	30610	17371	2	1	14377	1145	296	3
1095	26604	0	3	1	10051	382	129	3
1096	38543	18524	2	1	9653	370	137	1
1097	38774	0	1	0	9943	475	116	2
1098	27873	14399	5	0	11768	802	236	2
1099	22847	22640	4	1	15067	1109	277	2
1100	44470	8100	2	0	9302	312	99	2
1101	61755	0	5	1	6944	869	216	1
1102	33643	17660	4	0	10085	1063	208	1
1103	33954	18725	5	1	11527	637	258	1
1104	25414	0	4	0	14178	1231	302	4
1105	29735	0	3	1	10591	980	250	3
1106	35803	0	3	1	9308	913	228	3
1107	22337	17187	2	1	9067	478	101	3
1108	30345	21234	5	0	16941	1175	175	1
1109	47670	19687	1	1	12284	1123	295	2
1110	28650	0	4	0	12093	910	188	1
1111	28030	13551	3	1	13309	822	244	3
1112	25332	10370	1	1	5855	799	237	1
1113	31635	22238	2	0	14380	881	214	2
1114	36497	26629	2	1	8935	554	199	2
1115	31309	0	4	1	9975	839	165	2
1116	21120	19914	2	1	9525	653	216	2
1117	28667	18617	2	0	12749	951	227	4
1118	43630	0	3	1	12137	903	165	1
1119	25806	23440	2	0	9362	803	191	1
1120	29316	12172	2	1	8644	688	241	2
1121	37401	0	3	1	17661	904	179	4
1122	23914	23227	2	1	16850	976	223	1
1123	26035	23318	2	0	16781	1188	265	2
1124	39950	11256	3	1	8237	762	363	3
1125	35555	0	1	0	12033	1010	210	2
1126	45305	0	2	1	19808	1017	226	1
1127	36615	23227	2	1	10821	900	210	2
1128	32908	17792	2	1	14498	454	123	2
1129	27762	23789	3	1	14269	827	215	2
1130	32168	21946	3	1	17005	1911	457	4
1140					7922	1121	251	3

OBS	INCOME1	INCOME2	FAMLSIZE	OWNORENT	TOTLDEBT	HPAYRENT	UTILITY	LOCATION
985	28056	0	3	0	6154	790	212	4
986	23615	23524	3	0	10577	760	188	2
987	20980	0	1	1	11167	387	89	3
988	39852	0	1	1	12141	439	101	1
989	32924	0	5	1	9002	1410	354	4
990	34380	24573	3	1	9718	1295	292	2
991	50192	2147	1	1	12346	426	256	2
992	33867	0	3	1	12938	301	94	1
993	28530	20933	2	0	13649	777	176	2
994	32052	0	3	1	8255	328	169	2
995	30996	0	3	0	9283	830	189	4
996	20383	0	5	1	7344	703	204	4
997	29372	20373	2	0	13565	810	178	3
998	24744	17761	2	0	12354	1109	235	2
999	31596	17230	4	0	14854	1100	297	2
1000	29291	0	4	0	10005	578	248	1
1001	32494	0	4	1	8496	898	214	3
1002	24539	23963	4	1	5290	858	206	4
1003	25408	23347	2	1	9601	732	214	1
1004	41166	24475	4	1	17230	1228	265	1
1005	27979	22000	2	1	13988	614	169	2
1006	27720	0	2	1	14854	825	183	3
1007	43901	0	4	1	4942	690	149	1
1008	36633	25869	2	1	6817	683	178	1
1009	25887	19453	4	1	11385	956	277	2
1010	48930	18071	2	1	17442	1574	338	2
1011	27763	15905	5	1	9904	1312	357	2
1012	42199	0	4	0	16220	1438	357	1
1013	33868	0	3	1	8864	1056	264	3
1014	39313	27509	2	1	14178	677	216	2
1015	45033	23628	4	1	16706	1137	245	4
1016	26358	25414	4	1	9605	1128	290	4
1017	28317	19960	2	0	14266	1157	150	3
1018	41226	12900	3	1	16720	723	180	1
1019	35939	0	2	0	15723	736	298	2
1020	41633	18157	5	0	15715	717	191	1
1021	34054	15100	3	1	8764	846	81	1
1022	33868	0	2	1	8935	320	144	4
1023	29671	16518	2	0	5526	315	90	4
1024	32212	0	2	1	13152	982	389	2
1025	44433	0	6	1	16285	529	153	1
1026	20854	14564	3	0	10762	1209	314	2
1027	36142	0	3	0	6643	1140	295	1
1028	30398	0	2	1	19188	620	180	1
1029	39436	23298	3	1	14460	899	205	2
1030	42715	14177	2	0	10769	887	92	1
1031	26602	15553	2	1	12860	308	154	3
1032	32317	16997	4	1	13187	610	209	2
1033	30968	15000	3	0	18841	1370	334	4
1034	27132	15718	2	1	12528	444	104	1
1035	26597	23581	1	1	12069	869	217	3
1036	38585	0	4	1	9303	610	228	2
1037	31115	22067	2	1	12167	785	121	1
1038	29339	0	4	0	6362	522	197	4
1039	20573	18150	2	1	13858	878	188	3
1040	27925	0	2	1	7758	650	184	4
1041	26367	21866	3	1	13333	992	280	3
1042	30634	15453	4	1	10931	841	326	1
1043	39480	0	4	1	5948	1096	113	4
1044	44891	0	4	0	11264	625	145	2
1045	35718	16718	2	1	11588	1470	338	4
1046	34053	14435	1	1	15421	702	164	4
1047	51582	0	4	1	15472	751	227	3
1048	37476	0	4	1	9731	759	140	1
1049	35798	18397	3	1	11020	1064	143	3
1050	41214	0	2	0	13100	392	186	4
1051	23571	20924	3	1	14272	620	172	1
1052	26438	23057	2	0	10829	440	87	3
1053	27380	20052	3	1	12976	655	185	1
1054	31308	19615	2	1	12215	710	225	4
1055	25717	0	4	1	6265	840	172	1
1056	45601	12582	3	1	14081	786	225	1
1057	31039	23761	4	1	13413	843	221	1
1058	28739	0	6	1	8318	1006	255	4

APPENDIX I Database Using Financial Variables of Companies

The observations that comprise the database are random selections of companies listed in the Moody's Investor Service Industrial Manual. Each observation includes the following variables.

NUMBER	NAME	DESCRIPTION
1	BONDRATE	Bond rating as given by Moody's where Aaa-A = 1, Baa-B = 2, Caa-C = 3
2	REGION	Region of United States where main office is located, where Northeast = 1, Southeast = 2, Southwest = 3, Northwest = 4
3	EMPLOYEE	Number of employees
4	SALES	Gross sales in thousands of dollars
5	COSTSALE	Cost of sales in thousands of dollars, i.e. the cost to the company to produce or manufacture the products sold
6	NETINC	Net income
7	ASSETS	Current assets
8	LIABIL	Current liabilities
9	TOTAL	Total assets or total liabilities

VARIABLES

OBS.	1	2	3	4	5	6	7	8	9
1	1	1	11600	1204236	932014	38378	355606	167371	650812
2	2	1	23000	2303731	1713703	-107331	947651	662653	1989945
3	3	1	4000	911002	692227	3116	389884	196724	511393
4	1	4	15592	9259100	6705200	310000	1424600	112590	5282000
5	3	1	6100	1268580	991863	40723	221308	1509417	4974267
6	2	1	37481	3516289	3237018	-229627	20296124	114844	469184
7	3	1	8300	701059	460513	31215	279361	76659	409064
8	3	1	5320	413668	266110	34905	244063	409064	12243000
9	1	1	51300	11113000	8332000	732000	4569000	2808000	26733000
10	1	1	141268	27148000	151290000	1538000	8960000	5636000	685750
11	1	1	6200	1155711	857101	19088	321878	196566	3769000
12	3	1	13200	901890	613606	48844	413346	144692	1856700
13	1	3	60700	4952900	3208100	408900	2432300	867200	248067
14	1	1	34173	2094800	1407900	-10200	930800	581200	114079
15	3	2	11082	635076	119407	25611	84052	67696	739445
16	1	1	7000	239352		1798	60369060	43298	449348
17	3	1	1401	725241		20000360	271643	181676	2593000
18	2	1	7830	100094161	200072391	24082	1296000	131605	377849
19	2	1	53000	3501000	2711000	85000	1101247	844000	2685797
20	3	2	9880	98394	2237000	29531	18458000	80256	37933000
21	1	1	28064	3002700	5866200	152500	335606	1138923	2096345
22	1	1	382274	62715800	88298000	3285100	282933	3525600	459164
23	1	1	16623	1549290	334305	146391	273500	315248	235840
24	3	1	750	543986	722000	9272	320402	133545	1608900
25	2	2	5614	629700	425400	-25700	659121	650487	719257
26	2	2	16800	692900	628523	-12100	331613	208000	1347065
27	2	1	9174	701194	496181	39575	804400	789164	691546
28	2	1	4250	753774	525101	293161	26768400	166950	2086200
29	2	1	9400	648419	2563900	4674	241193	150600	72593000
30	1	4	62056	4586600		-183500	611700	762800	290059
31	1	1	876000	102813700		2944700	631900	22848100	1819700
32	2	3	7000	514589	325545	-20426	1496400	68024	1265100
33	1	1	11914	2553300	1887300	-32700	110466	432600	4097200
34	2	1	18605	908800	497900	-101800	582649	266800	155585
35	2	1	41400	3725700	2517500	-472300	470600	1118700	819962
36	3	4	2110	2250701	109915	14296	177766	43615	2077000
37	3	1	5578	1159590	773141	86112	458596	178399	80053
38	1	1	14000	2039200	1637800	85300	637237	276100	1709495
39	2	1	3650	366205	236911	7600	131889	75305	966924
40	1	2	16000	968555	431093	70476	71073	246642	194958
41	2	4	10000	713685	566787	-1310	44776	341659	181853
42	1	3	800	154536	70185	16701	18696	30770	46474
43	3	4	265	911500	80944	-21304	66041	20676	67293
44	3	3	305	104212	45068	3126	176568	4692	92145
45	2	1	2200	54263	133083	839	45085	8216	728855
46	2	1	27477	198038	73104	8498	5460	41474	686575
47	3	4	31024	93282	40353	-10873	1117317	37358	8860
48	2	1	2422	189074	1098	-182761	7149	2060	10270
49	2	4	11	6902	2398	4336	1546	1371	128118
50	3	3	1809	3601	80551	-3440	76659	20874	14093
51	2	1	167	134183	8603	20250	343142	5975	11716
52	3	3	2839	11629	6775	-3744	1836	4147	156946
53	2	3	3100	7459	188830	-1116	3001	37529	541438
54	1	2	109	229728	6056	-7927	9395	143351	2065
55	3	1	12	2404	142	-4245	13865	614	4815
56	3	1	867	416	1377	341	1040	3316	13211
57	3	1	240	1458	13277	-7374	3001	5393	79385
58	3	2	1086	23897	31798	560	9395	11574	1576
59	3	2	130	27481	3059	-7389	13865	738	
60	1	4		5046		-20	1040		

VARIABLES

OBS.	1	2	3	4	5	6	7	8	9
61	2	3	27	2369	1459	307	1643	248	2226
62	2	3	198	784	383	-567	160	230	434
63	3	1	22000	1012451	416322	74425	561254	178721	873302
64	3	2	24	4873	2268	1509	6174	396	6353
65	1	4	298	16084	12418	1217	20635	1380	27607
66	2	1	350	52684	42682	1762	34316	27731	80770
67	1	1	11	216	183	-1367	785	304	1955
68	2	4	18	1345	376	-104	8357	3139	8357
69	1	3	575	28079	18922	1148	25677	13935	33712
70	1	4	261	25244	20085	1667	7179	5842	42725
71	3	1	1222	18874	11144	-1855	985	3057	12894
72	1	2	322	29575	22577	5470	27578	7659	38873
73	1	3	17	4935	4130	-635	3529	3642	4700
74	1	1	5	18404	232245	-218346	119139	173732	331874
75	1	1	19	311813	130158	-458900	1947351	1203353	2307941
76	2	1	311	11803	4961	800	4844	2661	9667
77	1	3	520	52888	39948	1837	13548	5749	35890
78	3	3	550	3153	2118	-363	521	668	597
79	3	2	92	25911	21515	669	8698	4014	34817
80	2	2	278	92	339	-71	93	93	186
81	2	2	1226	17733	9248	-72	15418	3197	22637
82	3	2	36	36308	18648	793	13090	3378	14586
83	3	2	750	50217	36964	5801	5788	5602	88957
84	2	3	239	734	401	-1386	2126	7296	9080
85	3	1	175	19266	13169	2601	3396	7251	14160
86	3	1	12200	23297	21762	890	7613	5714	17433
87	3	3	46	16705	13173	710	13471	9800	18709
88	3	1	900	240314	226140	-21103	70772	44819	177085
89	2	1	448	4703	2686	-680	2385	1180	6502
90	3	4	460	16895	7507	-5624	35503	5322	37860
91	1	3	4100	65412	21440	4950	37718	12166	54499
92	1	2	5	56979	49128	-29806	34047	27093	46611
93	2	1	220	37050	26020	-5394	45569	14330	68953
94	2	1	333	306	27	28	101	153	259
95	2	4	133	150118	60118	7998	76174	44215	121398
96	2	4	28	6600	1796	113	2057	802	2648
97	3	1	36	67	7085	-5980	3111	9891	12117
98	2	1	16	26048	1642	-1575	72341	10542	577697
99	3	4	22	32672	26048	-25478	248242	10100	2520964
100	3	1	606	64505	1610	901	1837	1394	1984
101	1	1	115	13575	244	-1144	403	403	2079
102	1	2	164	28536	26336	987	58478	9605	112820
103	1	1	213	9193	42653	-119237	18715	1455	400027514
104	2	1	250	2406	5165	-400	6125	12715	8836
105	3	2	14	1199	21493	1201	19095	2108	23173
106	1	1	1470	161106	7100	107	7038	3210	8934
107	2	3	22	53794	1378	1061	4511	277	16667
108	2	3	550	25406	963	137	383	99636	2253
109	1	3	600	13552	92321	-18175	25897	924882	219786
110	2	1	38	221	20000	-187850	33961	8167	235433
111	1	2	667	16453	15329	1152	15852	5046	19135
112	3	1	470	33109	10303	1079	11224	4119	13329
113	2	1	3900	1162	388	-2403	407	1850	948
114	2	1	412	47040	13360	-2711	4333	1636	18989
115	2	1	37	892	14368	3656	28010	261	34215
116	2	4	2896	17	994	2463	1552	8625	17887
117	1	3	170	12103	31478	-7941	21768	274	26998
118	3	3			947	-309	271	547	405
119	3	2			42	42	505	9519	809
120	1	4			10744	-1416	9990		17152

VARIABLES

OBS.	1	2	3	4	5	6	7	8	9
181	2	1	37500	4332000	3913100	-152700	1183900	940300	4668900
182	2	1	21700	1791194	1160379	6304	1048104	692976	1580571
183	3	4	124196	16341000	15711000	665000	8478000	5659000	11068000
184	3	1	23333	3739970	2925870	101540	860785	547902	3533647
185	1	1	4800	919690	623817	49444	254380	115535	1600665
186	3	1	25500	1400196	808453	40138	493963	159392	658318
187	3	1	2953	216336	172221	8896	70971	35223	132180
188	2	1	46976	4378714	3173491	223225	1334792	626052	2762785
189	1	1	18300	7321000	5950000	760000	3363000	2180000	6288000
190	3	1	53731	4753700	4469100	375100	2063400	1299600	3370300
191	3	1	7074	1114442	872913	57274	436601	156671	812734
192	1	1	32200	4387623	3590044	200832	801502	730474	6025690
193	2	1	132422	22586500	18635200	1403600	5364000	5121000	14463200
194	3	1	433	35635	18045	2330	31136	14012	44947
195	3	1	52	158	309	4637	1411	937	2986
196	3	1	8207	266800	300000	-41700	228700	98900	585200
197	3	1	5100	1089020	525076	95610	417237	177057	849225
198	1	2	28030	8669000	4645000	934000	3739412	2754814	8373438
199	3	4	1310572	1616267	1199454	89270	533599	325157	1145018
200	2	3	2957	885411	805473	14226	200087	94085	295559
201	1	1	2342	466320	380831	58948	256731	106031	1214177
202	3	4	42176	5911046	5150120	105285	1283476	926158	1819696
203	1	1	25100	1856300	1220400	177100	860500	486000	2360800
204	1	1	28000	4548000	3168000	219200	1283100	1469500	3650600
205	3	1	3900	294297	215147	6487	95564	48215	152350
206	3	1	273	51300	1834000	-292000	699000	559000	3090000
207	1	1	51300	8742200	5970600	4132000	4749300	1211300	2383000
208	1	1	29100	3720400	2861000	558200	2072200	1162500	4595800
209	2	4	17500	1058702	543761	-951	587517	252545	705561
210	2	3	22000	3217000	1938202	-215742	975564	1213631	2974542
211	1	1	20600	3217000	2608900	36300	820900	635100	2721200
212	3	1	7000	634627	609940	17107	143959	91500	740485
213	2	3	638	359638	219468	70605	404602	212473	2620608
214	2	1	6965	4541296	4210826	-44957	923301	634312	2098619
215	2	1	24300	3644410	2756150	15643	1247572	1310080	2194882
216	3	1	24300	2955900	2685400	154200	1701800	503100	2932500
217	1	1	40000	4476000	1720400	660000	2339900	1170600	5163700
218	1	1	111000	25409000	16437000	1478000	5914000	4482000	17642000
219	2	1	29166	1239496	643823	167924	791584	671788	2027577
220	2	4	2600	310228	297427	15815	76635	38782	207158
221	1	1	36500	4687100	2799600	316400	1615900	975700	4461400
222	2	3	124617	15978000	8383000	1064000	611200	5934	17019
223	1	1	121000	12295700	9913000	28254	3837500	3411400	7703400
224	1	1	3300	36356	167411	89301	79085	37097	212762
225	3	1	87000	7937722	5403149	223455	1779481	1151067	3503106
226	1	1	485400	44281500	4289710	135130	2104140	1425810	6599460
227	2	1	13706	1552931	916886	105952	753492	430290	1145647
228	2	3	67174	8577749	6603164	200445	746811	952541	3421088
229	1	1	39700	9218956	7592469	-344670	3454795	2445542	15955241
230	3	1	30350	9065819	8243368	89301	854565	742197	1797887
231	3	3	10700	3172260	2578671	58328	513939	252864	781291
232	1	1	72000	5023300	4011900	239200	2281200	1285200	5557100
233	2	1	21500	3762000	3454000	376000	1431000	1032000	4230000
234	1	1	6307	957796	717817	93217	315008	131912	834659
235	2	1	24485	2017775	1341293	370300	637626	473800	2795038
236	1	1	15110	1811937	1345108	108096	549808	197541	2060066
237	1	1	48000	5958000	3965200	17700	168400	88100	2850000
238	3	1	77100	7002900	2630100	-329500	3201500	2292700	5876700
239	2	4	754	113655	269234	-184224	85803	228029	453542
240	1	1	17383	3340700	1744600	318900	729600	686400	2084200

VARIABLES

OBS.	1	2	3	4	5	6	7	8	9
121	1	4	444	19821	11676	-7423	5225	10709	17116
122	2	1	116	2720	1236	-376	1645	1579	2934
123	3	3	2050	68028	77415	4795	74395	32028	125990
124	3	1	1649	125790	86436	2402	53453	20154	72129
125	2	4	800	54662	23634	5138	28333	12705	75935
126	2	1	5065	396403	150208	54149	279364	91014	417016
127	1	1	722	110719	95927	-2228	31937	12003	106879
128	2	2	4300	616463	341375	38275	168173	88248	429618
129	1	1	68143	3923220	2342926	176260	933803	441291	1492533
130	3	1	694	61376	31528	2910	34821	5720	45944
131	3	1	19000	1625989	1095989	832222	649372	211377	928426
132	1	1	36600	3246139	1263369	450855	4214698	997955	4214698
133	2	3	2844	447755	231038	33289	131829	70662	415657
134	2	3	241	9048	529	8519	4405	2624	4405
135	1	1	83	16100	10019	1043	7792	1636	14411
136	2	1	109	14874	10810	1136	13818	3139	17877
137	2	2	23	16419	13045	900	5492	2265	19308
138	2	3	4640	532754	237877	102195	251591	237027	1310572
139	2	1	92	5429	3314	658	1663	834	6418
140	1	1	3800	192032	139229	-328	125691	68463	223898
141	1	1	1350	94570	81336	6347	40461	8456	58348
142	1	2	1200	91245	16607	2601	76845	52684	102512
143	1	3	810	83923	57542	1454	33198	12525	43078
144	1	1	183	188207	175456	-379	44878	24714	107854
145	2	4	118	40530	18645	10309	17342	13617	130774
146	3	1	150	1115406	561883	35455	501915	237036	985774
147	2	1	1227	222375	129290	16471	71770	17608	114270
148	2	3	4006	336963	183587	-18674	226914	120646	300617
149	2	1	803	62375	34255	3394	36882	12477	58566
150	2	1	31	6179	3515	594	1810	3896	2536
151	1	1	85700	7039000	4163000	667000	3119000	1208000	5760000
152	1	2	492	64076	59623	-1555	21165	13344	55256
153	1	3	3400	930708	559949	72778	241162	230860	1981396
154	1	2	1012	56997	34530	2677	40891	7365	63105
155	2	3	1627	123325	68926	6725	73249	26255	117438
156	1	2	40	419	2120	-2742	865	354	1651
157	2	3	2464	63064	50913	667	24790	24815	44005
158	2	1	4252	169168	129118	3763	68369	19793	126061
159	1	2	1073	74947	54525	-23804	59407	23810	74767
160	2	1	11750	643831	488472	13096	228206	88742	369265
161	2	1	5400	216985	94206	12045	15801	31828	342772
162	2	3	375	39755	26113	-6156	7827	47365	60415
163	1	1	2700	423444	328995	11533	190262	78461	967444
164	2	1	215	11905	6840	867	6095	2388	16660
165	1	1	16600	1982134	1146089	4735	704536	471908	2705730
166	3	1	5400	423220	420583	-8553	209471	105311	334065
167	1	1	35700	4667200	3509600	254100	1551800	944300	6766700
168	1	1	34462	3816000	2086000	203000	1891200	1126500	3667000
169	2	1	38000	2997692	2227551	188467	712121	712121	2208436
170	2	3	129000	14021484	10578985	144528	1638472	1312524	3590174
171	2	1	7505	800136	640518	-37608	279131	143352	521482
172	1	3	17823	995620	821549	41207	418692	198523	803310
173	2	1	2200	195010	144619	1559	98483	56593	185193
174	1	2	20000	1204246	543875	105960	769465	273284	1313414
175	1	3	3133	1132120	40075678	40048624	398629	243313	8638136
176	1	1	35200	1885400	1034000	158700	1046700	672900	2296300
177	1	1	4640	1010307	872500	16603	157847	165300	732700
178	3	1	4697	439727	269089	16603	157847	103188	277828
179	2	4	2708	370882	292054	-13053	257724	117991	404158
180	2	4	7947	864670	674397	25631	242637	138025	495472

OBS.	1	2	3	4	VARIABLES 5	6	7	8	9
301	3	1	900	901875	345687	78808	787621	603680	1699314
302	2	1	1070	199246	154581	-6258	74946	30810	261306
303	2	1	15565	868629	738231	28980	823096	319922	1257428
304	3	1	20600	2233511	1776514	158268	663759	367130	1741251
305	3	1	58000	3113506	2359278	339990	1509141	982785	3117664
306	2	1	7900	648337	585573	38808	274425	108928	477008
307	1	1	43428	3811000	2870500	137600	1676500	646300	3025100
308	3	1	23000	1145122	951664	44897	264191	161937	445173
309	2	1	1010	205624	154826	8836	63095	43234	161558
310	3	1	26100	2683961	2270866	-60900	936188	699894	2002894
311	2	1	1900	335000	136600	28700	1386200	1161744	2234400
312	1	2	12500	997837	678598	125820	144028	161744	1042389
313	1	1	482000	28139000	20757000	2492000	14288000	11461000	34591000
314	1	2	39000	7223000	5783000	296000	1420000	837000	5114000
315	3	1	32100	2818300	1183800	15800	1489700	900700	2539500
316	2	1	7200	350587	103802	12515	248606	127784	336884
317	2	1	33400	3440125	3159806	78690	1305388	624229	1963670
318	2	1	600	103044	76499	-3676	222543	81461	4243400
319	1	3	47000	4366177	2622239	301734	1614694	910407	2837364
320	1	3	35000	740366	442746	97839	301188	127551	1302273
321	3	1	38162	1865700	225200	-136700	1394100	937500	4712900
322	1	1	29857	2799481	2105587	93974	1441196	711009	2355273
323	2	1	403508	34276000	16197000	4879000	27749000	1274300	57814000
324	2	1	13900	1177700	1047400	-217000	768100	337100	2427100
325	1	1	4400	5500000	3991000	305000	1628000	1332000	7848000
326	2	4	4004	401109	315456	66092	112571	59048	318045
327	3	1	1392	138158	89442	-873	111120	50195	140670
328	2	1	4000	422600	308100	-50700	271700	91900	317100
329	3	1	7200	321769	156183	22623	216282	93048	274956
330	3	1	70	2477	1038	-33224	233808	55975	240518
331	1	1	123	1911664	4219591	-1809591	1921227	3040119	12958397
332	3	3	1000	243542	175685	6236	54507	26882	86033
333	3	1	32000	1091675	788550	-59259	140683	273444	568493
334	3	1	5500	314429	209241	1610	86029	88019	234095
335	2	1	16300	919818	703576	42426	305882	120168	587945
336	3	1	4500	528483	357597	18922	158816	83884	291180
337	2	1	2100	587820	431961	20256	208205	98854	315642
338	1	1	110	1033560	772020	36436	184877	65540	213423
339	3	1	1729	772241	603988	6695	324689	234328	628924
340	3	1	1187	975727	637382	29739	466805	266556	849814
341	3	1	1864	884726	474275	73753	262230	69973	192557
342	3	1	2200	561858	329341	32666	124750	69244	401892
343	3	1	2100	553068	479267	13045	3614	1150	4813
344	3	3	110	532	589	-8793	54938	18428	85079
345	2	3	1729	143677	73908	11968	4889	566	8560
346	2	3	1187	221	257	-4101	52080	25880	90447
347	3	1	1864	108382	82877	3321	332149	151798	385194
348	1	1	2200	1285811	1172318	15600	65080	5750	96809
349	2	1	265	34732	31232	2137	125724	120154	278656
350	2	2	5500	587492	501072	22137	11648	2219	16052
351	1	3	156	23035	5822	3328	12046	14182	31476
352	2	2	967	30463	9047	-6460	352	764	1061
353	2	2	4	646	600	-691	16526	1411	26414
354	3	3	75	2158	5859	-5859	8539	4552	15110
355	1	3	190	8131	4394	-2915	3013	2674	10427
356	2	2	23	5975	5675	182	18468	13379	39821
357	2	3	5000	50811	39297	1125	92354	30300	186680
358	2	2	3917	200602	171578	6670000	15542	4188	21081
359	3	1	457	36605	26327	1530	15542	4188	21081
360	1	4	154	20813	12837	1967	17255	1239	22469

OBS.	1	2	3	VARIABLES 4	5	6	7	8	9
241	3	1	4800	355377	216322	571	156951	55509	300024
242	2	1	36490	4303100	2712200	269400	1032900	804800	3676000
243	2	1	10944	1396401	1029008	77480	455659	278743	1016624
244	2	4	55500	4521002	3592465	71137	2717684	1747786	4569321
245	1	1	22950	8629988	7941301	545503	19024309	16107685	19024309
246	1	1	847	826500	881400	-20600	239600	173600	1680200
247	3	4	12714	1506200	1197110	187910	309720	210970	1888970
248	3	3	44000	6440871	4917211	225940	564050	757685	1551671
249	3	1	8600	817797	388933	59526	393187	277000	870130
250	3	3	19400	1920262	1451786	81228	913677	304393	2513343
251	2	3	5105	1430036	335530	78282	386802	234967	1404615
252	3	3	2567	399267	90210	55049	156002	80835	300351
253	1	1	68500	4752537	4421794	202344	1005376	829641	2470745
254	2	1	21000	1452010	929650	203420	786850	209765	2226150
255	1	3	151700	10376000	7533000	381000	3525000	1604000	6209000
256	3	2	9200	741586	502326	19461	423097	208670	691685
257	1	2	127400	49865000	48458000	1407000	10869000	10432000	39411000
258	1	3	51703	6879000	4341000	433000	2808000	1716000	8269000
259	3	1	15660	2110348	1985022	39411	585915	369101	969393
260	2	1	12895	1729600	1413000	75660	1206800	172494	2904400
261	2	3	16000	972819	716833	5778	340000	172494	1656127
262	1	3	8465	374583	181210	17103	195609	76143	272127
263	3	3	4940	402357	283221	-24009	164384	335282	335282
264	2	1	6000	549314	531246	45764	477970	415135	1149950
265	2	1	307	129285	242993	-64258	-64258	63316	590179
266	2	4	51000	15344143	12640306	181064	3711130	3313121	17466777
267	3	1	16300	17072610	13175600	752260	6500980	390577	1544588
268	3	4	875	811292	765098	27783	238785	105881	478464
269	3	3	220	164991	158898	5004	29482	24020	64589
270	2	1	9600	1107564	717718	51947	466912	219168	996685
271	1	3	6257	1921223	995746	45443	851472	287827	3369278
272	1	1	214000	29290800	3731800	457800	2503800	2223100	8028600
273	3	3	14581	1290558	726975	72572	1032886	496958	1398613
274	2	3	21800	9786000	7914000	228000	2800000	2231000	12399000
275	1	1	6345	899065	644636	50228	200728	108165	665042
276	3	1	2044	217296	146335	-3304	156219	120288	222451
277	3	3	1484	158829	132669	-2070	56616	15924	86126
278	2	1	27100	1611281	1413389	-59562	2089903	940711	3619807
279	2	1	2445	243255	198246	11036	76552	40658	160699
280	2	2	25600	3638900	3028500	191800	1407100	814900	3708800
281	1	3	12060	2067000	1346000	138000	941000	434000	1842000
282	2	4	8000	634162	553910	47318	361758	154393	502370
283	1	3	32641	18222000	12792000	883000	4081000	3483000	26214000
284	3	4	1134	50152	39833	5769	13004	9246	185983
285	2	2	1100	104192	61776	7225	43096	9562	87166
286	2	1	28000	17725200	1464600	82600	681100	334800	1287900
287	2	1	35724	3807634	1868402	278328	1888872	1303521	3865609
288	1	1	13722	11794000	9111000	605000	4760600	365500	1126800
289	2	1	47000	2696993	2290943	-97279	604825	541357	63468
290	2	1	20900	2640500	2367800	-472000	949500	685300	264200
291	2	1	7000	1034953	1056538	9140	412333	265576	1840314
292	1	3	26000	14993000	9495000	615000	4743000	3750000	21604000
293	2	1	46500	221654	175615	23701	422872	413592	2313968
294	1	1	61000	5543000	3526000	444000	2344000	1881000	7068000
295	3	1	80185	9319800	7456900	1167100	2755600	3593800	21090900
296	3	1	34900	4835900	1515200	589500	2758600	1015200	4183000
297	3	1	2266	241428	162961	22675	157228	56782	335910
298	2	1	6600	1309405	967662	71051	565527	287037	1509744
299	3	1	51095	26245000	5272000	715000	9050000	6485000	34583000
300	1	1	24149	2551469	2163448	50886	1175044	1240412	2161408

VARIABLES

OBS.	1	2	3	4	5	6	7	8	9
361	1	2	425	24300	15172	1908	6556	5166	29941
362	1	1	75	1084	638	-1426	2510	7744	4193
363	3	1	2400	108280	2157−	390150	91781	6609	120956
364	3	2	810	124399	100770	52610	51766	8679	71806
365	3	1	265	6685	124789	-6882	9117	4981	21039
366	3	1	2485	163005	2128	7196	131685	26313	161602
367	3	1	52	4148	107508	-52270	3813	3500	7077
368	2	2	3000	45098	3727	17671	133533	63272	640576
369	3	1	742	94205	34644	2495	37188	9980	60853
370	2	1	532	35824	68254	711	23628	6503	30518
371	2	2	10336	223918	190749	4656	32504	13743	149302
372	2	2	2365	303951	183230	16951	261860	69658	406338
373	1	1	691	146800	105906	6083	95213	42935	109090
374	1	1	1545	125668	98133	5070	36267	12798	89472
375	3	1	310	36949	53992	2859	27096	16544	33580
376	1	3	910	104949	5749	7489	60545	1692	79828
377	1	3	42	5397	10022	-212	6955	4229	33930
378	3	2	320	15016	158669	700	12868	2372	20622
379	3	2	4	221258	48928	-290324	2197508	7224	9868389
380	3	2	550	56260	12544	720	1257	15956	29732
381	3	2	215	20758	8750	188	6796	4106	11509
382	2	3	500	11217	453000	740	5006	20804	23976
383	2	1	1434	514090	5039	-16942	303344	6812	1317613
384	1	2	976	11739	125	1521	6137	2719	6854
385	3	2	71	174	820	-12084	1858	6608	14980
386	3	3	33	1541	58394	2812	2766	622	130543
387	3	1	3420	80471	19581	1205	45762	8408	21197
388	2	2	480	31146	7808	188	11096	1590	34490
389	2	1	383	19612	53942	6838	25497	14264	61474
390	1	2	1700	119979	268441	45975	201938	931000	1160881
391	1	3	6	101577	35275	1694	11251	308	20997
392	2	1	270	43809	361	-2849	2704	3750	5041
393	3	3	34	1235	19300	58	14630	7169	35739
394	3	1	845	42004	3737	-430	3974	6444	16873
395	1	2	94	4822	25587	4280	26725	193	31709
396	2	3	104	52166	2770000	-426000	1578000	338	4698000
397	2	1	22400	3172000	179	-439	14432	9472	1641
398	2	3	148	181	48341	1393	10538	2671	16960
399	1	3	1000	61805	22730	208	24176	399	12902
400	1	4	330	29655	23126	6930	2538	27895	27412
401	1	3	261	46441	395	-2305	2537	652187	2892
402	2	3	29	335	105	-1949	638	7765	3197
403	2	3	25	156	3130	8661	63553	80355	9472
404	2	2	67	6197	7874	-1415	2624	72829	72661
405	3	4	111	6603	1227	-452	42095	323545	4378
406	3	1	972	1525	127156	1025	1188435	44749	53790
407	2	1	886	155327	4230629	88974	7591	3561	1879122
408	3	1	63500	5350635	20104	-1206	112349	16788	38707
409	3	3	325	27908	328248	-96405	170198	78229	405322
410	3	3	176	410345	111163	23173	609464	3699	244062
411	1	3	2664	266788	656581	-58601	59861	11170	1170548
412	2	4	1800	766217	86213	2033	7033	10452	81378
413	3	2	931	106096	18036	1130	62627	34427	11741
414	4	3	460	23279	53636	10193	180494		83804
415	3	1	1017	64483	210275	-8190	7912		292196
416	3	3	3645	288664	1021	6520	41115		14451
417	3	1	87	4007	61787	4642	34653		107448
418	2	2	1100	93285	20223	-24594	189635		53478
419	3	1	520	39357	62975				453681
420	2	1	800	87180					

VARIABLES

OBS.	1	2	3	4	5	6	7	8	9
421	3	1	8	17254	11630	2830	34163	34163	34163
422	3	1	127	16868	8781	2107	124992	2000	225622
423	2	3	7432	263952	213431	40952	185829	30749	567383
424	3	1	599	66549	45814	1696	20951	11396	31451
425	3	2	2100	96200	67497	3645	42137	8646	54848
426	2	1	3200	315357	227653	14320	194088	75744	308890
427	3	3	31	671	1504	-794	441	651	2294
428	3	2	308	15608	9186	1626	22632	1690	25975
429	2	1	1250	53829	39066	1466	24090	11798	35934
430	2	1	485	41446	30466	1308	22685	4738	25950
431	1	2	1522	207589	175901	850	45723	37831	76129
432	3	1	7500	523146	366301	9252	135982	45413	32337
433	3	1	570	50203	33446	1510	28503	16960	147866
434	3	3	1700	251815	192014	3055	112658	83482	108751
435	1	1	3273	228599	131040	2430	80396	58937	23517
436	3	1	431	29811	13243	1983	14915	8584	83110
437	3	1	1800	115078	91191	9648	56585	20528	51699
438	2	1	1000	101944	88035	-968	31321	21081	1006511
439	2	1	2794	130854	101562	1004	275867	89871	42301
440	1	1	860	32934	23365	2151	26547	4365	24353
441	3	1	1002	50303	34223	2211	16667	4370	629256
442	1	3	4560	174224	94237	-61670	329653	222512	99205
443	2	1	690	73825	13920	-5200	16269	18913	25074
444	2	1	556	20476	313430	-399	31221	2143	996267
445	3	2	365	327272	12204	-650314	752206	157444	64906
446	1	3	1018	114057	69803	-2866	10803	12257	202084
447	3	1	2880	42177	14428	-4664	98908	56933	120691
448	2	1	350	40126	27733	2794	61490	33243	21203
449	4	1	450	117	106	2785	12128	7482	1028
450	1	2	30	294818	232110	594	480	920	146714
451	3	3	3307	47747	35752	19092	59755	35321	20109
452	3	1	583	10163	7130	629	16244	9566	8538
453	2	3	150	2684	1076	-1183	6475	3751	3871
454	3	1	375	23566	16422	-440	3143	904	241560
455	3	1	1786	11512	8653	4740	49767	36723	16069
456	2	3	272	23439	17029	682	9347	4853	40562
457	3	2	6516	108767	110417	-3261	24301	19470	47268
458	3	3	898	58707	46379	-800	28275	10089	622
459	3	1	858	73	75	90000174	84	319	28741
460	3	2	2600	36299	20880	1745	28021	10493	205267
461	3	3	2000	192599	133555	16980	36932	22800	165861
462	3	1	1177	271947	165143	15728	127477	36953	75924
463	1	1	362	142568	94404	8515	44162	12364	30498
464	3	1	6823	149519	29929	5419	20819	5092	166143
465	3	3	2641	174644	129685	1743	29153	36943	260908
466	2	1	128	1-20601	93567	18741	152852	96792	9164
467	3	2	85	10241	12316	861	6895	2330	40014
468	3	2	27	13309	11266	-2710	17431	4358	79420
469	2	4	14	747	549	2155	38279	11637	53528
470	3	3	400	1366381	744609	159	17	6	1128692
471	2	3	1600	32892	13668	107657	279206	1241383	37126
472	1	4	5	95993	60796	4248	19667	4173	59385
473	1	2	205	314	190	5865	35578	11607	502
474	2	4	4303	21020	11474	-273	413	243	16223
475	2	1	1419	531640	323196	-722	10265	6730	311837
476	1	1	1300	195453	170739	37818	185089	62964	128265
477	1	1	208	31072	167113	4678	42652	25602	45601
478	2	4	2	7703	6514	-2421	32751	21898	11682
479	2	3			2790	774	5389	3096	25749

VARIABLES

OBS.	1	2	3	4	5	6	7	8	9
480	1	1	724	82097	69112	1278	36433	23863	39489
481	3	4	200	1726	574	-214	425	306	665
482	3	4	1554	95637	52441	8346	73347	15104	113539
483	3	4	300	41527	40138	-156	42263	5030	61087
484	3	1	7	663875	412882	-2035663	1189863	1068744	4733485
485	3	2	2000	68274	46906	14923	43570	20578	233376
486	3	2	77	12236	8068	213	6300	5361	28639
487	3	1	792	9654	7338	351	7163	2963	12546
488	1	1	26700	2216636	1488646	59609	1041910	549747	1793160
489	3	1	25120	2615110	1998741	226733	1079821	419683	2914319
490	1	1	15980	2169614	1453440	132764	393420	202143	1356303
491	2	2	43600	1647788	342617	103370	2278189	477067	2278189
492	3	1	6000	1960237	1893884	39079	320723	124524	205614
493	1	1	89000	4930652	4109528	174644	1239606	715694	6793372
494	1	1	13700	961077	623338	79583	597837	304262	1276230
495	2	1	22459	3173242	3073622	-19264	894956	466954	2526557
496	2	1	7800	745049	505448	33345	279903	178944	178598
497	3	1	83	34830	8395	6142	23122	5197	30045
498	3	3	45	3	1	-983	474	137	2678
499	3	3	4959	334437	151442	23406	161761	59915	369011
500	3	1	500	46391	41375	-2810	43596	9504	95791
501	2	1	5100	195359	157650	-3089	128676	32489	251433
502	3	1	390	66475	48135	8590	22435	9469	60517
503	3	4	280	31399	25941	405	22706	3777	52974
504	3	3	12400	433560	259396	71164	81007	51436	115065
505	2	2	14000	666186	502776	43655	211444	145569	924533
506	3	2	3500	87205	44056	-5323	13389	14419	40057
507	3	3	190	23950	16623	167	8741	6054	12479
508	3	3	5470	460279	203851	9000	346432	251415	851426
509	3	1	1797	184296	72009	15346	132060	63273	216741
510	3	3	1601	131175	59866	7750	105599	32369	130799
511	2	1	498	53259	47236	3434	55159	15763	69364
512	2	1	1700	68274	22637	3948	36435	4632	48261
513	3	1	1486	113184	89184	6328	124629	51982	173771
514	3	2	2530	190718	171782	10118	89694	42829	122575
515	3	3	80	114319	110855	117	15001	11558	74903
516	3	2	988	473561	372511	17558	36598	35424	566769
517	2	1	1700	931934	854117	17104	105670	84456	260126
518	3	3	250	22369	50953	-27402	8276	6385	82053
519	3	3	343	16372	11749	-3639	4789	1933	10648
520	3	2	46	9249	7569	162	3768	1451	6844
521	3	4	14	121677	75259	5529	65463	51776	119456
522	2	2	23	212886	117964	21360	104797	41371	175473
523	3	1	21800	1933055	1267918	1016571	959381	483918	1786870
524	2	3	4269	381046	305238	-46294	174763	95580	340577
525	1	1	1600	180920	91568	4073	30918	28245	96853
526	1	1	9900	757820	603568	6562	260558	121721	508098
527	1	1	1619	63225	51524	23665	40709	22813	84382
528	1	4	2150	23008	18002	-3170	15479	7212	26502
529	3	4	3300	294502	120687	-149032	442930	150400	67354
530	3	3	33747	3745400	1315200	521100	1892800	1355300	4222000
531	2	2	4730	1792632	2308205	-304300	1255525	633502	3288484
532	3	2	1600	339190	241026	-13100	116568	62846	565746
533	3	1	6277	341382	253913	14335	162864	55853	274461
534	1	1	20500	1403472	934393	98928	481225	232787	1131988
535	1	1	16915	1784629	566049	396291	1505506	750631	2408831
536	2	1	9411	2988396	2494357	26186	530775	401302	1770232
537	3	4	4000	380501	148029	25648	110259	90442	202200
538	2	2	2300	208025	152107	-896	97055	49633	131237
539	2	1	22800	1671871	1396682	53687	604797	220378	961058

VARIABLES

OBS.	1	2	3	4	5	6	7	8	9
540	1	1	4700	267978	215548	-20861	147115	35358	188817
541	3	2	15500	2032325	1564619	35415	530459	203410	1523600
542	1	1	23645	10047000	7109000	385000	3289000	2386000	11684000
543	1	1	16000	1433940	944297	45400	669418	255211	1399176
544	1	2	27941	2920310	1623351	408085	659082	458869	2929081
545	1	2	16565	1058055	875006	2736	406206	305490	1403529
546	3	1	2325	157232	103788	5664	94212	37008	130531
547	3	2	898	780278	743890	17580	114883	66976	491889
548	2	3	857	56530	38684	1352	40947	1352	56808
549	2	1	23000	2667912	2326185	12928	1304547	848835	3027518
550	1	1	78556	6035900	4678200	217700	1748900	1351700	3909300
551	1	1	10000	796351	627078	37500	387535	166797	591908
552	2	1	7500	961411	734983	11945	279713	105526	501707
553	2	3	17355	2045215	1340174	129934	568849	262070	2751482
554	1	1	50292	6343000	4343000	496000	2414000	1881000	7571000
555	2	1	98300	5115900	3681600	-43400	4536400	3342100	9408800
556	1	1	8982	633564	635564	-80071	67521	236991	725649
557	2	3	193500	15669157	12373271	72727	6940346	4626477	11091787
558	2	3	18005	8091000	5225000	176000	1835000	1175000	10133000
559	2	3	20700	2291448	795614	252646	1201840	564040	2664956
560	1	1	21700	2723664	1884596	225508	686547	617053	2006068
561	2	3	2080	1880083	1652081	-99888	312085	323055	1697322
562	3	1	42100	3660553	2550072	103137	720164	388834	1197082
563	1	1	53200	3102918	1057281	309484	1510068	969806	2515923
564	1	2	6400	215064	671199	100173	219422	185113	1145227
565	3	1	2733	235635	160217	17020	78477	42847	237405
566	3	1	117267	10731000	7771200	670800	4635300	4196400	8481800
567	3	3	30956	4008700	3595500	199700	1088200	537300	2203300
568	2	3	5500	1858600	1480700	-239700	1036600	876800	3907900
569	2	3	100367	4822000	2412000	465000	3973000	2206000	10608000
570	1	1	30	2673	153000	294	1175	986	2594
571	1	3	462	77636	46442	16052	13615	5424	64836
572	3	3	160	22285	20973	1920	8216	3472	73871
573	2	4	39	1606	1556	46	833	849	1399
574	3	1	8	1358	1106	-1242	760	2004	3443
575	1	2	135	182344	123542	2528	65917	45053	80788
576	3	3	744	113	331	-216	168	90	540
577	3	4	273	94031	71620	3210	37561	13321	38892
578	3	1	957	23321	14666	2633	14438	2777	24390
579	3	1	130	9484	6592	-833	3266	3443	9458
580	3	2	1200	143356	118253	-2161	44006	17049	63271
581	2	3	221	17957	17966	-1000	8414	2851	19651
582	3	3	28	6387	4478	-1052	4319	3809	17797
583	1	4	5	208	208	-420	298	7	332
584	2	3	1400	126502	96362	4921	35509	18463	69321
585	3	3	24	1454	1454	-909	179	1524	5790
586	3	2	1509	86052	68440	1748	26074	10488	36386
587	3	4	16	410	296	-406	120	561	473
588	1	4	238	5679	4540	-852	21652	5734	23937
589	3	1	26	1157	844	408	1299	302	2019
590	2	3	2300	209787	147536	4469	72903	26021	94204
591	1	1	103	861630	632040	13220	117592	82287	205579
592	3	4	28	149	69	-46	859	36	1641
593	1	4	5	10	40	-30	181	12	232
594	3	1	227	31290	16445	617	10286	8663	35150
595	3	1	976	626900	613919	2906	158565	67538	164697
596	3	3	300	11201	11853	879	5326	4757	12621
597	3	3	152	1257	889	-1341	1333	875	2008
598	2	2	42	5008	1095	486	3945	554	6204
599	2	1	75	4887	2264	300	3171	413	4700
600	1	1							

VARIABLES

OBS.	1	2	3	4	5	6	7	8	9
601	2	1	77	2563	2467	-3929	9808	1283	14582
602	2	1	29	1092	797	-1105	2005	608	3583
603	2	4	37	32	7	-6	10	5	195
604	3	2	526	44102	17914	5222	28373	5325	33469
605	3	2	6	380	196	-533	481	239	12072
606	1	1	600	17813	11255	-1098	6653	2603	780
607	1	1	16	767	480	-534	583	356	3882
608	2	1	270	7132	5891	-309	2306	1092	10302
609	3	1	123	6245	2912	-119	7596	839	33118
610	3	1	103	7027	4562	-4015	24757	1713	2563
611	3	1	30	1423	603	-822	2006	541	147
612	1	3	8	83	32	-3	47	25	2956
613	3	1	23	1811	928	-931	1201	1594	7379
614	1	1	1700	151828	118070	7867	49505	23838	143602
615	1	4	94	7464	4473	-72	6933	4522	32689
616	2	1	486	45974	34068	-6613	21579	18289	290083
617	3	1	2600	274191	87677	28743	132373	50890	222714
618	3	1	6438	35914	25095	5556	19060	9166	219695
619	3	1	1860	201901	187380	806	44014	24961	33738
620	2	1	149	11088	8944	-12002	13340	2646	1303
621	2	4	15	1328	508	-319	1008	125	4294
622	2	4	85	193	197	541	452	976	27772
623	2	3	68	6361	5550	-179	2480	430	20295
624	1	4	298	34520	33226	1264	15104	14662	9693
625	1	1	984	9910	4651	282	6156	969	7737
626	3	1	115	14529	7953	1515	6993	892	17743
627	2	1	684	13558	11167	-614	13734	1409	8173
628	1	4	400	6162	4028	-132	2628	1563	25794
629	3	1	350	232487	224625	588	19616	14601	94477
630	3	1	1352	109899	49385	9447	55962	15261	275663
631	1	1	432	72209	19793	-19412	89212	36717	38889
632	1	1	334	119152	98577	3899	35980	20738	6568
633	3	2	88	17800	10731	1164	6166	3834	40334
634	3	2	302	25080	22258	667	18434	1798	59886
635	2	4	473	46180	18334	5065	51680	10864	13295
636	2	1	700	54111	41395	628	8813	5801	190230
637	1	2	4400	310668	234683	12710	129085	38506	3924
638	2	4	48	4177	3566	-351	2630	509	291459
639	3	3	3347	1018431	809854	30251	133841	85753	2565
640	3	4	14	80	131	-1350	119	88	14140
641	1	1	807	44291	32986	929	11817	5448	33399
642	3	1	186	32779	20027	3234	23329	3391	36663
643	3	1	1400	43379	22976	-1224	24671	7939	16830
644	3	1	169	22536	13537	2166	14791	4253	193278
645	2	1	2331	197659	134541	9518	144852	26544	10619
646	2	4	98	12040	6698	1120	6823	1098	92886
647	2	1	10	259	81	-8906	1761	23294	8870
648	2	1	336	16483	10225	-16920	4910	3876	68824
649	2	1	1080	145308	71221	13534	56242	16187	101651
650	1	1	2737	249570	205308	233	70894	29324	1678
651	3	3	30	2268	2179	-644	1018	10045	29984
652	3	1	630	25872	17193	1655	11679	10782	90564
653	2	2	465	51748	22089	10367	80499	8547	60
654	2	2	43	1161	515	-50	293	390	2442
655	3	3	5	510	633	-425	2126	495	1973
656	3	3	175	460	263	-99	1118	522	154138
657	3	4	1415	149261	88902	-1242	104166	32384	1813
658	3	1	20	50	774	-1034	53	1568	19232
659	2	1	22	678	588	406	1483	887	
660	2	3	32	2882	1435	-4359	2773	3014	

VARIABLES

OBS.	1	2	3	4	5	6	7	8	9
661	1	4	600	78824	58592	2024	35519	10381	45688
662	1	1	124	9214	6917	93	1512	1458	3777
663	2	4	13	783	581	-375	530	239	1755
664	3	1	850	79533	58348	5243	56898	21963	72896
665	3	2	1250	4223	4135	-2398	1528	7770	5318
666	3	4	897	4253	2728	2577	5716	735	6633
667	3	1	129	10218	8154	-1980	5833	4043	7726
668	3	1	12	480	199	-1981	325	737	869
669	2	3	63	18034	15680	-199	10495	7478	11911
670	1	2	136	212	344	-141	898	82	968
671	2	4	1570	210799	38625	30100	129729	44740	174731
672	3	1	455	76019	36623	5889	88795	22969	96689
673	3	3	11	1011	463	190	693	333	3295
674	3	3	280	1898	7143	535	3946	1090	7599
675	3	1	79	1898	1179	-1021	3250	4064	29600
676	1	4	3300	81506	27445	-2913	16705	11234	54321
677	3	1	532	172299	76492	27177	95822	19436	108801
678	2	1	955	12134	7435	198	4373	1143	6987
679	2	1	200	17146	10457	969	17561	2169	30257
680	2	1	217	12212	7642	2828	42604	3574	42995
681	2	1	2224	458377	353315	-3688	177749	74452	241939
682	2	2	400	36901	26776	-2542	19553	4070	23440
683	1	1	15	14167	230520	-244356	64393	243314	1430578
684	3	3	3787	263286	198047	-14788	120620	55002	194341
685	2	1	550	35332	31575	-10489	46446	6407	124620
686	2	3	275	10021	8472	-2297	4663	1121	6147
687	2	4	112	318	19606	-19348	119853	100100	900891
688	3	1	111	994121	687580	129019	257449	100454	323540
689	1	3	1800	7935	4959	-228	5766	937	8845
690	1	1	326	10609	6332	115	2492	1066	2754
691	2	3	631	39051	34797	-7038	20808	13184	45162
692	3	4	58	4357	942	488	5818	1381	8743
693	2	2	836	22145	9557	-1044	2224	4764	6549
694	2	1	150	17545	11613	-7851	11489	3992	12859
695	2	1	181	16309	8190	-5154	14327	3212	18685
696	1	4	32	4686	1730	584	3673	2307	4422
697	2	3	68	3400	2504	-232	1661	1030	2575
698	1	3	192	10456	9424	-3935	5934	1924	17698
699	2	1	68	3818	2307	-596	1255	1610	3298
700	2	1	266	24575	15333	2771	12370	5507	22826
701	1	1	227	89528	79921	1388	26760	17033	27890
702	1	3	943	64038	60930	3815	50284	22732	62025
703	3	2	330	34685	22018	1375	9777	3641	12263
704	2	4	66	20697	10133	1347	22572	3224	44390
705	3	1	900	31024	4426	-261	3061	1112	3640
706	3	4	14	549	18122	7555	45022	8237	46966
707	1	3	118	7817	471	9	471	356	1316
708	1	1	600	37255	7930	-1947	6555	4699	14212
709	1	1	8300	421612	343610	799	21347	9079	32545
710	3	3	559	3000	19478	19478	194267	28031	277006
711	3	3	1159	10078	1719	-1948	1593	1706	1964
712	2	2	1501	6671	6671	332	3016	1939	7464
713	2	1	30	112788	57338	3736	66001	24838	113771
714	1	2	13	720	126	-1076	1729	172	2176
715	1	1	58	5415	536	-95	231	342	375
716	3	3	660	57240	3185	-294	3718	739	5165
717	1	4	123	30965	31026	715	11991	8304	15443
718	1	4	162	16169	25538	-943	14168	3911	15213
719	1	2	2058	92743	13023	-4269	2990	1866	5324
720	1	2		67301	67301	3778	24338	14161	57490

VARIABLES

OBS.	1	2	3	4	5	6	7	8	9
721	2	3	236	32585	6822	5057	31402	4694	37598
722	2	4	360	14604	17563	-5689	4522	8049	10103
723	2	4	946	104421	75734	3463	46869	18470	67138
724	2	4	200	67828	17547	2741	13821	8288	61605
725	3	1	532	43306	36219	1457	15910	10058	34458
726	3	1	898	16593	17404	-2427	3115	4724	9693
727	3	1	146	148240	117507	-52723	162944	299776	274224
728	3	1	50	2949	1604	-1554	1536	505	3282
729	1	1	696	1671	1606	-1554	1232	863	2540
730	1	1	487	197	1045	-1260	1054	1784	3046
731	1	2	281	7769	3960	-3677	2753	2492	8639
732	2	4	15	295	111	5	168	41	197
733	3	1	28	795	654	2383	907	1159	1168
734	1	1	1000	130427	94548	1150	111390	61862	132835
735	1	1	601	5878	3341	-1031	1334	1669	2129
736	1	4	895	98682	5645	5645	48793	19616	78608
737	3	1	1725	280274	237735	6249	15254	13893	72091
738	3	3	238	17938	15800	488	4365	2466	6595
739	3	3	83	9592	5870	-1102	4529	1322	6216
740	3	1	3500	137702	1268137	1797	252644	140783	427078
741	1	1	300	45955	29465	2630	29900	7459	40771
742	1	3	665	122424	93816	507	21903	8004	29263
743	1	1	85	8716	5229	-2936	6966	1905	8169
744	1	3	433	253292	237142	-3447	76670	40007	223622
745	2	3	229	19134	16869	29	34337	15122	54067
746	1	4	105	5077	4735	-412	2550	1904	3319
747	3	4	730	7365	4485	106	13643	5224	27115
748	2	1	2443	110112	79851	613	60574	25224	77971
749	1	1	256	33642	27765	-242	10644	7002	15266
750	2	2	4061	521234	338904	40469	135397	67829	237346
751	1	2	356	55004	20153	4422	26733	17137	64178
752	1	3	181	4128	3495	-61	80	1093	2364
753	1	4	454	35649	15414	6208	25385	6131	36225
754	3	4	688	23044	19385	-5057	11994	12019	28790
755	3	1	813	104001	65123	6792	43317	19802	50079
756	1	1	2834	160555	160555	2732	55506	39862	126685
757	3	4	300	32788	25554	1006	11522	8241	27184
758	3	2	16400	269265	269265	9055	96275	26011	229435
759	2	4	1924	29524	29107	6957	14015	6148	47214
760	3	3	4500	106990	29504	4799	14170	9795	63110
761	3	3	460	275424	183459	12797	105654	39955	229688
762	3	3	460	22983	17231	723	5966	4237	14620
763	2	1	1607	22093	39789	11488	40210	8852	75527
764	1	4	177	18018	7305	2081	13907	2948	15456
765	2	4	262	15360	4919	560	10982	4977	6218
766	2	4	107	21417	11790	89	26449	16025	44033
767	1	3	50	4709	230	1093	2154	1358	4103
768	3	1	1660	150648	676	14368	75457	13153	122779
769	2	2	200	35543	9924	-927	14525	6676	16460
770	2	3	1500	9098	26736	6838	2781	1205	3588
771	2	1	27	3492	6179	327	2312	1837	2467
772	1	1	4169	299524	1329	360	204831	50430	335307
773	1	4	50	345	27026	-113	468	711	1908
774	3	3	1460	1618	230	-1734	1858	790	2175
775	3	3	579	160334	676	11908	29301	14124	53869
776	3	1	2700	232282	87441	14368	65844	34843	131377
777	3	1	78	102691	126836	6838	31252	9003	41607
778	3	4	22	53557	86763	327	28772	12858	34556
779	3	3	1210	4132	39610	3510	1391	940	9859
780	3	3	1780	174012	1837	261	82950	18831	116994

VARIABLES

OBS.	1	2	3	4	5	6	7	8	9
781	3	2	1219	507670	227848	-12980	88534	64604	499292
782	3	1	78	6063	3698	-2469	6973	1935	7971
783	3	1	416	17173	20040	-201190	1058532	87312	1095518
784	2	4	200	21447	3546	1800	15449	60002059	32205
785	1	3	299	60914	46075	1201	21032	8142	23353
786	3	1	1027	11083	7433	228	4613	1615	7022
787	3	1	3300	274510	188191	10616	118673	55572	27449
788	1	3	302	42611	37324	-2308	22882	31624	24552
789	1	1	210	14813	12454	523	8465	1667	13846
790	3	1	8	420	148	-230	550	416	843
791	2	4	311	1652	709	75	757	424	991
792	2	4	241	20412	11987	-3851	10160	6442	14775
793	2	4	127	10861	5199	1402	5620	1665	7276
794	1	2	60	320	337	-336	97	200	169
795	2	3	1298	116201	65291	215	27846	12722	70439
796	1	4	60	9918	98883	-18050	38181	8190	38181
797	3	4	321	9146	7250	439	3887	3176	9287
798	3	1	1130	17387	2209	251	273	5632	8038
799	2	1	231	68872	13466	-1657	3837	16313	43880
800	3	1	935	354927	326544	1342	19517	67891	155527
801	2	1	2700	31824	142	-7338	911	176	7761
802	3	1	31	1824	1442	-557	5736	363	5915
803	2	1	47	3189	1941	182	2889	3988	28668
804	3	2	350	30132	3561	-638	5879	735	3812
805	3	3	36	3041	12226	-2292	8032	2883	7594
806	3	2	65	4265	98770	-688	22376	4846	22106
807	3	3	50	15938	3971	-9428	764	26593	38048
808	3	2	800	112361	461652	318	184984	1865	6745
809	3	1	312	4874	27273	87542	4538	67879	723181
810	3	1	4338	653040	22267	878	10231	11364	47904
811	3	1	3600	84822	23712	-5255	10233	13745	45352
812	2	1	165	77587	16484	524	11862	5051	22811
813	3	1	420	28292	102216	809	32010	10803	17543
814	3	1	31	22394	47231	3711	20004	23001	78355
815	3	1	1736	122517	2791	-6484	3125	16692	36811
816	3	1	400	84622	55593	-693	59902	1124	4952
817	3	3	79	4287	58300	18371	130407	36108	139277
818	2	1	4910	257547	8267	-21500	10639	31078	176314
819	3	2	1000	116027	20979	781	9938	5761	13246
820	3	3	187	12768	194671	-3333	104609	8103	12862
821	3	1	260	22873	20979	781	130407	8103	12862
822	3	1	1556	263268	194671	-3333	59902	74799	126950
823	3	3	56	4538	2319	13244	9938	1603	5339
824	3	1	143	11710	6432	-698	104609	2913	7900
825	3	1	3600	82047	51944	-30	5049	25754	142260
826	3	1	4300	417628	322660	23382	4819	155563	396325
827	3	1	658	36769	21594	170	68968	6623	17240
828	3	2	1320	137245	109531	6680	291587	23898	113228
829	3	2	477	87144	78472	1626	9121	11678	34021
830	3	3	1600	1036450	794759	18094	102647	57343	240447
831	3	3	1226	150747	119681	4339	17829	21316	83348
832	3	1	82	10746	4972	-642	87847	1599	3926
833	3	3	14	437	805	1481	56583	1201	890
834	3	4	280	23979	10834	1150	12877	6714	27700
835	3	3	86088	810822	586649	21125	334053	99468	1106263
836	3	2	5642	399006	284120	-3516	46464	45088	146542
837	3	1	683	107149	34524	-1081	131946	18340	174447
838	3	4	631	173476	128984	-529	34914	13917	57138
839	3	3	65	1137	493	9	961	170	1093
840	3	3	615	66464	24458	4528	59389	9991	81353

Upper table

OBS.	1	2	3	4	5	6	7	8	9
901	2	1	21	769	654	411	440	925	485
902	3	1	331	21428	20457	-4447	6974	12560	27001
903	2	2	465	24734	20979	-6029	21873	2297	30733
904	2	4	722	48865	64808	-33941	32978	9658	132343
905	2	2	1050	38234	38027	18250	106297	21681	132839
906	2	4	22	1875	1090	-299	948	631	1265
907	3	3	107	5301	87	-184	1870	34	200
908	3	2	118	2613	2613	-235	126	688	3355
909	3	1	311	24609	16686	-1145	8033	15642	22091
910	1	1	91	3505	1539	275	1138	471	1162
911	3	1	300	62004	46810	176	26716	13996	29167
912	3	1	84	4408	3485	-683	2517	659	3454
913	2	1	4868	443092	196362	39064	255495	90074	398065
914	2	1	660	68148	45137	4013	42896	12514	51831
915	3	1	300	13301	9077	855	3634	2942	4877
916	3	3	2917	184861	137936	2430	102197	46686	136677
917	3	2	346	15652	13592	127	6384	1590	9440
918	1	1	3579	349027	245360	19434	128644	62588	236630
919	1	4	36	1211	12	539	1772	241	2060
920	2	1	30	1100	533	-2078	2168	394	5287
921	1	1	4403	291913	158400	32039	185824	56318	328240
922	2	1	1149	79769	35257	4710	42594	6687	57169
923	2	4	1601	383735	312202	10076	99849	27219	131412
924	2	3	16000	524358	188933	10947	61373	39702	212460
925	3	1	112	6206	2335	-912	6250	936	8268
926	3	1	100	2303	1566	98	906	670	2877
927	2	1	451	2534	1150	581	4079	759	4592
928	3	2	841	51869	40876	-1420	38864	20535	51374
929	3	2	46	1901	1648	-306	603	99	614
930	3	2	12	524	447	-735	701	894	809
931	2	4	794	7189	4801	-809	4410	2029	6949
932	3	4	218	30237	26058	-1008	50618	27056	95002
933	2	1	12	15993	12195	439	9008	7819	22541
934	3	4	18	42	70	-22	798	92	1800
935	3	2	67	741	587	-125	404	3091	1226
936	1	3	1500	11252	3573	1616	9403	22175	11643
937	1	1	2	11530	77139	3367	49243	890	63970
938	1	2	1214	4036	2867	307	1424	890	2161
939	3	1	7	125906	83917	6373	64236	29998	94970
940	2	4	1841	54	19	-423	1754	67	2195
941	1	4	671	166086	87792	13855	112561	35252	154523
942	3	1	78	13794	7582	962	11585	4384	41176
943	1	3	24000	6293	4499	11	5256	3059	6308
944	3	1	115	4705	2673	647	4992	321	5770
945	2	2	618	944356	783649	28003	205730	161708	653369
946	3	2	450	10211	6993	405	5724	5383	8697
947	1	1	1200	59323	42976	1063	27259	7459	42478
948	3	4	370	27724	24381	-723	14610	3439	22069
949	2	4	465	236629	140785	6762	54248	28541	111511
950	2	4	4000	58717	39992	7461	12688	6910	36390
951	3	1	4214	42756	42976	426	27121	4023	40421
952	2	1	479	270325	191747	27121	135995	48588	224608
953	2	3	3000	77925	59209	5033	37198	14635	45698
954	3	2	85	47681	20050	39	25331	9298	33065
955	3	1	130	8310	5313	-2909	510	3479	7925
956	2	4	6300	2650	817	-2121	5924	1190	7571
957	2	1	1561	21363	11723	646	11997	1907	15465
958	3	1	26	16736	21523	472	140320	50767	255898
959	3	1	2032	394092	256490	-4807	140320	821	30971
960	1	4	248	82526	44008	9854	73337	20477	90177

Lower table — VARIABLES

OBS.	1	2	3	4	5	6	7	8	9
841	2	1	20	591	452	-805	1553	96	1888
842	2	1	154	14207	7839	959	6169	1586	7329
843	2	1	790	67400	30175	3028	42958	18107	56259
844	2	2	170	19212	12411	858	15511	1293	16814
845	3	1	61	10211	6987	269	5208	1015	7515
846	3	4	367	18381	12683	1449	10677	6710	24200
847	3	2	470	101354	87491	1625	29422	14381	35658
848	3	2	32	1625	695	-2885	895	1779	3305
849	3	1	1888	221634	207625	833	217708	157621	293582
850	3	1	169	13428	10777	-164	17702	4382	32642
851	3	2	675	40515	12272	3629	10226	2645	21091
852	3	4	281	21410	16277	-1775	7804	3654	19848
853	2	2	22	676	137	-854	131	195	507
854	3	4	2200	40031	18180	-1384	15188	4667	20428
855	1	1	390	20781	10250	-2487	14568	6854	20739
856	3	1	911	17452	7527	1135	23111	9169	31697
857	1	1	1172	108792	33699	18077	167615	16593	193830
858	1	1	100	2160	1622	770	1087	1099	1223
859	2	4	126	6343	5601	-335	3908	20338	6436
860	3	4	444	96023	25149	8126	35078	16199	47466
861	3	1	1172	95400	58295	-12662	72201	2379	114874
862	2	1	94	3466	3157	-926	2123	444	4499
863	3	1	16	446	210	-315	705	18163	1037
864	2	4	1306	129298	64412	-15730	81481	3660	154741
865	1	1	33	2877	2028	-13600	992	23650	1644
866	2	1	218	169767	161097	-463	16731	2265	37355
867	3	1	324	9010	6648	-210	3129	1390	4395
868	3	1	55	6275	4648	96	2354	87100	4061
869	2	1	4	59433	64830	-63060	78925	7022	135226
870	1	4	175	22034	14387	-1987	7361	4271	10444
871	1	4	231	27025	18117	2412	28276	5829	34163
872	4	4	1050	24867	10944	9487	27419	8790	32112
873	2	4	2100	75095	50935	8271	53988	13117	59294
874	3	1	4300	67009	59648	2222	18384	29228	33063
875	1	3	1300	431591	326402	6784	27875	13526	86249
876	3	1	497	76608	57634	-1079	13085	19700	34060
877	3	1	12	95730	71424	2363	45379	454	55121
878	3	4	741	171	121	-940	135	10154	179
879	3	1	210	66257	47076	2614	27049	6662	38328
880	3	2	1368	11493	6515	46	16984	23223	26743
881	1	1	120	130718	95708	7173	62354	7384	79294
882	3	1	22	17334	12718	91	15172	1037	23703
883	3	1	900	3814	1040	1365	3556	53773	4442
884	1	1	225	136063	100237	7110	168295	1900	258052
885	1	1	570	19689	9750	1785	11321	4627	18291
886	1	1	41	14770	12390	-340	6346	267	7488
887	3	3	1082	3147	1825	994	2084	2057	2584
888	3	2	52	16981	3766	2652	16162	1160	25201
889	3	3	353	5348	3705	-2458	1287	9824	3009
890	3	1	370	49377	30455	2957	9495	2259	34566
891	3	1	370	16277	10833	553	30769	25696	14553
892	3	1	3850	207493	142302	7485	135995	83	72531
893	3	3	29	860	457	-708	1240	10658	1333
894	2	4	373	46594	33696	1314	19772	7598	29531
895	2	1	13	1313	363	-621	7598	682	7598
896	3	1	48	363	3145	321	3083	21397	5129
897	2	1	400	5489	34013	10943	35427	3943	63296
898	3	1	2032	86498	13021	14	8667	20638	11114
899	3	1	2032	189538	93109	10789	109272	39152	186781
900	1	4	248	221418	200300	30112	65569	39152	68554

OBS.	1	2	3	4	5	6	7	8	9
961	1	1	139	14330	5495	1101	8142	2490	12095
962	1	1	870	74444	54034	4825	26051	5991	48720
963	1	1	3	614	348	-21	166	232	277
964	1	1	2205	9031	7722	-790	4695	1450	5468
965	3	4	376	891	581	-1362	777	2373	1944
966	2	4	3000	288160	228052	17971	69631	22634	168816
967	2	1	1607	90784	75611	1972	45421	27489	50354
968	2	4	13	482	254	-389	331	445	2061
969	3	1	1138	64771	33404	-7445	65664	13412	90318
970	2	1	5672	419991	289779	24664	186911	71576	269887
971	2	2	6720	298238	216796	22190	113435	35671	173281
972	2	1	3600	423413	317117	2607	152391	77618	250427
973	1	1	207	9281	5359	-366	4627	989	6256
974	2	1	733	16016	7783	1276	11113	4265	12773
975	2	1	46	1728	904	-759	391	830	1968
976	2	1	57	4363	2930	-75	3425	1117	4214
977	1	3	1438	8086	5755	1942	4809	2362	8160
978	1	1	467	27246	13935	1425	10417	8176	15112
979	1	3	3406	237686	174743	1970	76243	44776	158461
980	2	3	500	8510	2298	4559	2529	2324	9367
981	1	1	1043	95588	52030	5581	86205	17162	128819
982	2	1	45	3901	1767	-709	2131	589	3914
983	2	1	220	53615	34352	2247	23462	9288	26704
984	3	4	180	110890	50418	5818	55283	20356	67384
985	3	1	443	84604	74095	1877	40503	5205	58203
986	3	1	1343	263164	204696	5038	86799	31182	106804
987	2	1	234	16904	6682	436	9747	2466	14303
988	3	3	313	46006	30789	4246	29793	6491	49586
989	2	1	9	306	990	-1530	184	2900	1903
990	2	1	802	70667	49443	6042	14731	6747	36304
991	2	4	4119	543041	263681	31753	135165	110798	633558
992	2	1	7531	813497	502247	86194	294387	81216	355502
993	2	3	255	24834	17293	506	13524	4590	17277
994	3	3	75	7743	5545	795	6250	2251	7122
995	2	3	4534	391700	417100	-117100	152000	88700	480200
996	1	4	3300	487685	361921	33484	99710	75445	484975
997	3	1	16	362	44	-1155	726	59	1017
998	3	3	156	1826	1028	-324	1023	266	1559
999	2	1	135	27908	15965	2410	11519	3016	18394
1000	2	4	1540	194335	129150	3855	280546	32150	451404

APPENDIX J   Answers to Odd-Numbered Exercises

Chapter 1

1.1 The population of interest is all small businesses in Los Angeles. The sample is the group of 50 randomly selected small businesses.

1.3 a. Represents a sample of employees at General Motors if randomly selected from all employees **b.** Represents a sample of students on the university campus   **c.** Represents a sample of people listed in the telephone directory   **d.** Represents the population of all possible ways of choosing 2 cards from a deck of 52 cards

1.5 Inferential statistics uses a sample to form conclusions about a population. A census obtains information about everyone in a population.

1.7 Age is ratio. Sex is nominal. Grade point average is interval. Classification is nominal.

1.9 Answers on a marketing survey that indicate strongly agree to strongly disagree

Chapter 2

2.1 b. There is no "correct" number of classes. Consider using $K = 8$ classes.
c. $CW = (100 - 18)/8 = 10.25$ (round to 10).
e. Relative frequency is equal to the frequency divided by the total number of values in the data set.

2.3 a. Relative frequencies are .38, .31, .21, .06, .03, .01.   **b.** Lower class limits are 0, 2, 4, 6, 8 and the upper class limits are 2, 4, 6, 8, 10.   **c.** Class midpoints are 1, 3, 5, 7, 9.   **d.** No

2.5 a. Discrete data

2.7 a. $CW = (17.1 - 2.2)/5 = 2.98$ (round to 3)
b. Determine the basic shape of the data.
c. Consider $K = 6$ classes, 2 and under 5, 5 and under 8, etc.

2.13 b. The shape indicates that only one class has a low frequency and that most of the data fell in the larger class intervals   **c.** No, the shape would not change.

2.19 a. Classes are 0–3, 4–7, 8–11, 12–15, 16–19; frequencies are 12, 10, 7, 1, 1; cumulative relative frequencies are .39, .71, .94, .97, 1.00.   **b.** 29%.

2.23 c. The distribution peaks between 70 and 80, and nearly 90% of the distribution is between 40 and 100.

2.25 b. $X = 12$ (approximately)

2.27 d. A "typical" return on sales is between 3 and 6   **e.** Yes, Pfizer and Merck have much higher returns.

2.29 Proprietorships are 76.97% (277°), partnerships are 7.82% (28°), and corporations are 15.21% (55°).

2.31 The class 10,000–14,999 is 17.5% (63°); the class 15,000–19,999 is 10.0% (36°); the class 20,000–24,999 is 32.5% (117°); the class 25,000–29,999 is 15.0% (54°); and the class 30,000–34,999 is 25.0% (90°).

2.33 c. Yes, consumption varies between 8 and 17 kilowatt-hours.

2.35 c. There are two peaks in the distribution; one in the 20–24 age group and another in the 45–49 age group.

2.39 a. The data can be considered to be continuous.

2.49 c. Nearly 35% of the loan officers are between 45 and 54 years of age, with another 35% between 35 and 44.

2.51 c. 75% of the boxes contain less than 15 defective fuses.

2.53 b. For 15 days out of 30 there are at least 11 workers absent.

2.55 b. Approximately 96% of the managers received a rating of 4 or higher and over half received a rating of 6.

2.57 b. The percentage of incidents related to the theft and destruction of computer hardware is 40.18%. The percentage related to the theft and destruction of computer software is 17.35%.

2.61 There are slightly more female workers at lower salary levels and more male workers at the high end of the class intervals.

Chapter 3

3.1 Mean is 117.5, median is 118, mode is 118, midrange is 116.5.

3.3 Mean is 3.227, median is 2.96, no mode, midrange is 3.655.

3.5 The median would be close to $20,000.

3.7 Mean is 6.97, median is 6.3, modes are 5.0 and 6.3.

3.9 Mean is 74.35, median is 57.5, mode is 130.

3.11 a. Mean is 41.0, median is 28.1.   **b.** Mean is 34.29, median is 25.1.

3.13 a. Range is 87.   **b.** MAD is 24.09.   **c.** Variance is 794.08.   **d.** s is 28.18.   **e.** CV is 56.27.

3.15 s is 10.489.

3.19 a. Player A is better (on the average).
b. Player B is more consistent.

3.21 Variance is 1742.13.

3.23 Approximately the 39th percentile

3.25 a. 32   **b.** 19   **c.** 17.5

3.27 a. $Z = .78$ **b.** $X = 63$ is .78 standard deviation to the right of the sample mean.

3.29 s is 2.5.

3.31 $(x - \bar{x}) = 14.24$

3.33 a. Mean is 3023.73, median is 3115.2, variance is 144,166, s is 379.7   **b.** $-.72$

3.35 $17,710 to $19,490

3.37 Near zero

3.39 a. CVs are 20.5, 23.6, 16.7, 22.0.   **b.** 2.79–4.23, 2.94–4.76, 2.90–4.06, 2.69–4.21

3.41 15.8 to 40.6

3.43 0 to 134.62

3.45 $200 to $360

3.47 Median is approximately $26,666.67.

3.49 Median is approximately 24.99 years.

3.51 a. Mean is 75.96, variance is 162.8049.
b. s is 12.76.

3.55 b. 12 defectives are most frequently found.
c. Data are skewed right.

3.57 a. Mean is 8.99, median is 8.5.   **b.** s is 3.062, MAD is 2.26.   **e.** Data are skewed slightly to the right.

3.61 Mean is 412.125, s is 7.53.

3.63 Mean is 1760, s is 2932.

3.65 Mean is .00133, s is .0001555.

3.67 a. 6.3   **b.** 5.669   **c.** 5.60   **d.** 5.6   **e.** s is 1.752
f. 30.9   **g.** 4.3   **h.** 7.7   **i.** 2.7

3.69 b. .32   **c.** Between -1.155 and 1.655

3.71 $45 to $105

3.73 a. -4.0

3.75 Mean is 53.756, s is 2.78.

3.77 a. 167.758   **b.** 164.9   **c.** No mode   **d.** 3263.08
e. 57.123   **f.** 37.29   **g.** 280.8   **h.** 34   **i.** .15   **j.** 139.8
k. 221.5   **l.** $-.381$ **m.** 53.512 to 282.004
n. Approximately 68%

3.79 a. .5736   **b.** Range is 17.8, midrange is 10.4, interquartile range is 5.25.   **c.** Between 0 and 21.036

Chapter 4

4.1 5/4 cannot be a probability.

4.3 .286; yes, this is the relative frequency approach.

4.7 a. .16   **b.** .45   **c.** .676   **d.** .84

4.9 a. .541   **b.** .015   **c.** .5   **d.** .949   **e.** Yes

4.11 a. .776   **b.** .403   **c.** .263   **d.** They are mutually exclusive but not independent.

4.13 a. .703   **b.** .447   **c.** .638   **d.** .768

4.17 a. .3   **b.** .8   **c.** .286   **d.** .5   **e.** .714   **f.** .2,
g. .8

4.19 .75

4.21 No

4.23 .25

4.25 P(both nondefective) $= .64$, P(both defective) $= .04$

4.27 $P(M \text{ and } E) = .1$

4.29 a. .54   **b.** .86   **c.** .675

4.31 .90, assuming they are independent

4.33 .676

4.35 $P(\text{over } 15,000) = .28$, $P(A \text{ or } B) = .486$

4.39 a. Yes   **b.** .849   **c.** 0

4.41 .00148

4.43 .86

4.45 24

4.47 120

4.49 210

4.51 90

4.53 20,000

4.55 60

4.57 45

4.59 1/595

4.61 Probability is 1/495. This does constitute a random sample.

4.63 1/45

4.65 a. .5   **b.** .667   **c.** .667

4.67 a. .091   **b.** .125   **c.** .273

4.69 a. .143   **b.** .143   **c.** .2857   **d.** 1

4.71 a. .195   **b.** .755   **c.** .354   **d.** No

4.73 847,660,528

4.75 a. .8   **b.** .2

4.77 a. Yes   **b.** .8629   **c.** .9812

4.79 a. .49   **b.** No   **c.** .667   **d.** .237   **e.** .355

4.81 a. .4627   **b.** .8495   **c.** .3275

4.83 a. .75   **b.** .45   **c.** .25

Chapter 5

5.1 The number of customers or daily account size of its customers

5.3 a. Discrete   **b.** Discrete   **c.** Continuous
d. Continuous   **e.** Continuous

5.5 12 outcomes, each with probability 1/12

5.7 .25

5.9 a. $P(X = x) = 1/6$ for $x = 1, 2, 3, 4, 5, 6$

5.11 $P(X = 0) = 9/25$, $P(X = 1) = 12/25$,
$P(X = 2) = 4/25$

5.13 Yes

5.15 Rolling a die

5.17 $P(X = x) = 1/3$ for $x = 2, 4, 6$

5.19 $P(X = 1) = 1/6$, $P(X = 2) = 2/6$, $P(X = 3) = 3/6$.
Sum of probabilities is 1.

5.21 Mean is .2, variance is .18.

5.23 Mean is 3, variance is 3.

5.25 Mean is 17.4, standard deviation is 4.98.

5.27 b. Mean is 2.025, standard deviation is 1.1.

5.29 .2963

5.31 $P(x \geq 5) = .991$, $P(4 \leq x \leq 10) = .913$, $P(x \leq 7) = .275$

5.33 .506

5.35 .404

5.37 a. .001 **b.** 3 **c.** 1.55 **d.** .982

5.39 .7636

5.41 a. .835 **b.** .798

5.43 .8

5.45 .6353

5.47 .9286

5.49 a. .011 **b.** .599 **c.** .673

5.51 Mean is 3, probability is .5768

5.53 a. .9004 **b.** 2.83

5.55 a. .3528 **b.** .0335, yes

5.57 Yes

5.59 Mean is 3.5 and the variance is 18.85.

5.61 a. .029 **b.** .237 **c.** .117 **d.** .085

5.63 .623

5.65 Mean is 1.01 and the standard deviation is 1.3304.

5.67 .3658 (approximately)

5.69 Probability is .224, standard deviation is 1.73.

5.71 .2378

5.73 a. .677 **b.** .1085

5.75 .2415

5.77 b. $\bar{x}$ is 8.508, μ is 8.5. **c.** Sample variance is 4.595, population variance is 4.25.

Chapter 6

6.1 The mean and variance indicate where the curve is centered and how wide the curve is, respectively.

6.3 a. .3413 **b.** .0919 **c.** .6826 **d.** .0606

6.5 a. .9439 **b.** .0179 **c.** .1986 **d.** .9010

6.7 a. $z = .35$ **b.** $z = .67$ **c.** $z = 1.57$ **d.** $z = -1.62$

6.9 $z = .65$ and $z = 1.57$

6.11 a. .3632 **b.** .2119 **c.** .3842 **d.** .1791

6.13 8.08%

6.15 $82.5

6.17 101 is the passing score

6.19 a. .9772 **b.** Yes, because this probability is .0013

6.21 a. .2514 **b.** .1056

6.23 Mean is 9.45.

6.25 Mean is 1.29 and standard deviation is .556.

6.27 a. .688 **b.** .692

6.29 .9612

6.31 .8564

6.33 .0441 (Poisson approximation)

6.35 a. .5 **b.** Value is 1.8.

6.37 a. .524 **b.** 23.15 **c.** 1.819

6.39 .454

6.41 .2997

6.43 .3836

6.45 Answer varies for different data sets.

6.47 a. .0133 **b.** .8664 **c.** .2734

6.49 .8461

6.51 Values are -7.025 and 7.025.

6.53 Value is 87.12.

6.55 Probability is .3442, standard deviation is 30.

6.57 58.89%

6.59 .0668

6.61 .4168

6.63 a. Mean is 1.0, standard deviation is .289. **b.** .3

6.65 .5

6.67 a. .6826 **b.** .3811 **c.** .3446 **d.** .1587

6.69 Variance is 43.03.

6.71 .0392

6.73 a. .1788 **b.** .7123

Chapter 7

7.1 a. .017 **b.** .1446 **c.** .2974 **d.** .7108

7.3 c.

$\bar{X}$	26.7	43.3	53.3	56.7	176.7
P	.1	.1	.1	.1	.1

$\bar{X}$	186.7	190.0	203.3	206.7	216.7
P	.1	.1	.1	.1	.1

7.5 a. .2033 **b.** .0019

7.7 a. .8145 **b.** .701

7.9 Without replacement: .0020; with replacement: .0031

7.11 .7960, assuming the population of electric bill amounts follows a normal distribution

7.13 .0823

7.15 .0268

7.17 $n = 40$, probability is .8790; $n = 20$, probability is .7852.

7.19 19.86 to 21.34

7.21 92.41 to 107.59

7.23 70.62 to 443.67

7.25 22.24 to 23.36

7.27 12.72 to 13.68

7.29 7.86 to 8.74

7.31 a. 1.311 **b.** 2.160 **c.** 1.734 **d.** -1.325 **e.** -1.708 **f.** 1.303

7.33 27.78 to 32.22

7.35 a. 62.24 to 73.76 **b.** 63.21 to 72.79

7.37 302.37 to 317.63

7.39 9.79 to 12.11

7.41 92.54 to 99.66, assuming a normal population for the tensile strengths

7.43 89

7.45 16

7.47 6.1033

7.49 24

7.51 57

7.53 601

7.55 Estimate is .80, confidence interval is .716 to .884 (in thousands).

7.57 Estimate is 5.0422, confidence interval is 4.47 to 5.61.

7.59 Estimate is 3.3923, confidence interval is 2.54 to 4.25.

7.61 a. .2266 **b.** .0329

7.63 28.69 to 30.11

7.65 59

7.67 a. .0344 **b.** .0455

7.69 Estimate is 3.94, confidence interval is 3.46 to 4.42.

7.71 a. Variance of sample mean is .0571. **b.** .4817

7.73 Estimate is 10.174, confidence interval is 9.56 to 10.79.

Chapter 8

8.1 a. When failing to reject the null hypothesis, the possible outcomes are: (1) the loan is made and paid back and (2) the loan is made and not paid back. When rejecting the null hypothesis, the possible outcomes are: (1) the loan is not made and would not have been paid back had it been granted and (2) the loan is not made and would have been paid back had it been granted. **b.** Type II **c.** No, a Type I error may have been made.

8.3 a. False **b.** False **c.** True **d.** False

8.5 $Z^* = 4.65$, reject H_0.

8.7 $Z^* = -3.486$, reject H_0.

8.9 Reject H_0 since the 95% confidence interval does not contain 2.0.

8.11 a. 12.7274 to 12.7326 **b.** Reject H_0.

8.13 a. .9732 **b.** Approximately 1 **c.** .9131

8.15 $P(Z < -15.29) + P(Z > -11.37) = 1$

8.17 a. $Z > 1.645$ **b.** $Z < -1.645$ or $Z > 1.645$ **c.** $Z < -2.33$

8.19 $Z^* = -2.93$, reject H_0.

8.21 $Z^* = -14.14$, reject H_0.

8.23 $Z^* = 3.64$, reject H_0.

8.25 a. $Z^* = 2.59$, reject H_0. **b.** 9.84 to 10.56

8.27 a. FTR H_0 **b.** Reject H_0. **c.** Reject H_0. **d.** FTR H_0

8.31 $Z^* = 1.03$, p-value is .1515, FTR H_0

8.33 $Z^* = -1.29$, p-value is .0985, FTR H_0 for a significance level of .05 and .01.

8.35 $Z^* = 2.80$, p-value is .0026.

8.37 a. $t < -1.729$ **b.** $t > 1.318$ **c.** $t > 2.145$ or $t < -2.145$

8.39 $t^* = .59$, FTR H_0

8.41 $Z^* = -5.54$, reject H_0.

8.43 $t^* = 3.0$, p-value $< .005$, reject H_0.

8.45 $t^* = -6.56$, p-value $< .005$, reject H_0.

8.47 a. 15.987 **b.** 46.979 **c.** 7.261 **d.** 45.642

8.49 a. 7.612 to 27.439 **b.** 2.76 to 5.24 **c.** FTR H_0

8.51 Chi-square $= 19.81$, FTR H_0

8.53 Chi-square $= 21.39$, p-value $> .10$, FTR H_0

8.55 Chi-square $= 45.24$, FTR H_0

8.57 a. Increasing (decreasing) the significance level will increase (decrease) the rejection region. **b.** Increasing (decreasing) the significance level will decrease (increase) the probability of a Type II error.

8.59 $Z^* = -2.99$, p-value $= .0028$, reject H_0.

8.61 a. $P(Z < -.75) + P(Z > 3.17) = .2274$ **b.** $P(Z < -2.77) + P(Z > 1.15) = .1279$ **c.** $P(Z < -6.04) + P(Z > -2.12) = .9830$

8.63 $Z^* = -3.5$, p-value $= .0004$, reject H_0.

8.65 $t^* = -9.22$, reject H_0.

8.67 a. 14.193 to 14.807 **b.** $Z^* = 3.19$, reject H_0.

8.69 a. $t^* = 3.058$, reject H_0. **b.** p-value $< .005$

8.71 a. 36.85 to 40.35 **b.** $Z^* = -1.57$, FTR H_0 **c.** .0582

8.73 Chi-square $= 14.37$, FTR H_0

8.75 Chi-square $= 27.22$, FTR H_0

8.77 a. $Z^* = -2.345$, reject H_0 **b.** Normally distributed

8.79 a. Mean is not equal to 1 hour. **b.** Yes **c.** Large sample size

Chapter 9

9.1 a. Paired **b.** Independent **c.** Independent

9.3 Dependent samples

9.5 No, the samples are dependent.

9.7 The samples are independent.

9.9 a. $Z^* = 1.77$, p-value $= .0384$ **b.** $Z^* = -.19$, p-value $= .8494$ **c.** $Z^* = -.06$, p-value $= .4761$

9.11 7.485 to 8.171

9.13 -4.849 to -2.951

9.15 $Z^* = .697$, FTR H_0

9.17 $Z^* = .405$, FTR H_0

9.19 $Z^* = -3.2$, reject H_0.

9.21 $Z^* = -2.66$, reject H_0.

9.23 $Z^* = 2.19$, reject H_0.

9.25 $t^* = 1.53$, FTR H_0

9.27 df $= 13$, $t'^* = .4595$, FTR H_0

9.29 $t^* = .4595$, FTR H_0

9.31 -48.364 to -1.636

9.33 a. 70   **b.** Significant difference
c. Not a significant difference

9.35 df $= 14$, confidence interval is -4.026 to $-.934$.

9.37 $F^* = 1.746$, FTR H_0

9.39 .511 to 12.72

9.41 a. .23 to 1.91   **b.** $F^* = 1.49$, FTR H_0

9.43 .70 to 3.91

9.45 $F^* = 2.0$, FTR H_0

9.47 a. $t^* = 3.5355$, reject H_0 **b.** Use dependent
samples t statistic

9.49 $t^* = .707$, FTR H_0

9.51 p-value is between .10 and .20

9.53 a. $t^* = .638$, FTR H_0 **b.** The samples are
dependent.

9.55 a. $t^* = 2.67$, FTR H_0 **b.** Differences are
independent and normally distributed.

9.57 The samples are dependent since both brands of
tires are placed on the same car.

9.59 $t^* = 3.8443$, reject H_0.

9.61 $t^* = 1.038$, FTR H_0

9.63 $t^* = .6286$, FTR H_0

9.65 a. Significant difference for work, supervision,
and co-workers   **b.** Work: .8 to 8.76, supervision:
.41 to 9.41, pay: -1.25 to 1.29, promotion: $-.28$ to
4.78, co-workers: 30.72 to 40.14

9.67 $t^* = 3.90$, reject H_0.

9.69 a. Not a significant difference for electrical
engineering, industrial engineering, civil engineering,
and material science   **b.** Significant difference for
electrical engineering and industrial engineering; not
a significant difference for civil engineering and
material science   **c.** Appears to be accurate

9.71 a. As the confidence level increases, the width
of the confidence intervals increase.   **b.** Data are from
a normal population.

Chapter 10

10.1 .394 to .783

10.3 601

10.5 95% confidence interval is (.465, .903); 90%
confidence interval is (.502, .884).

10.7 655

10.9 8141

10.11 2018

10.13 .001 to .069

10.15 a. .038 to .562   **b.** .357 to .423

10.17 p-value $= .7296$, FTR H_0

10.19 Since $.10 < .123$, reject H_0.

10.21 Sample size must be at least 167.

10.23 $Z^* = .3985$, FTR H_0

10.25 p-value $= .3015$

10.27 p-value $= .121$, FTR H_0

10.29 $Z^* = 1.59$, FTR H_0

10.31 $Z^* = 2.38$, reject H_0.

10.33 The sign of Z^* is reversed.

10.35 $n_1 = 26$, $n_2 = 29$

10.37 $-.099$ to .239

10.39 $n_1 = 35$, $n_2 = 42$

10.41 a. Since $.10 < .226$, FTR H_0.   **b.** .003 to .226

10.43 .213 to .734

10.45 593

10.47 $Z^* = 1.3488$, reject H_0.

10.49 a. .23   **b.** .13   **c.** 41

10.53 p-value $= .3557$, FTR H_0

10.55 $Z^* = -.512$, FTR H_0

10.57 a. Since $8.77 > 1.96$, reject H_0.
b. Since $-3.179 < -1.96$, reject H_0.
c. Since $-1.65 > -1.96$, FTR H_0.

10.59 .694 to .851

10.61 .9833 to .9967

10.63 $-.0398$ to .1864

Chapter 11

11.1 a. $H_0: \mu_1 = \mu_2 = \mu_3$; H_a: not all three means are
equal.   **b.** The samples are taken independently from
normal populations with a common variance.
c. For μ_1: 4.533, for μ_2: 5.333, for μ_3: 4.633   **d.** 1.29

e.

Source	df	SS	MS	F
Factor	2	2.28	1.14	.88
Error	15	19.38	1.29	
Total	17	21.66		

$F^* = .88$, FTR H_0
f. Not appropriate

11.3

Source	df	SS	MS	F
Factor	1	6.5333	6.5333	2.2603
Error	28	80.9334	2.8905	
Total	29	87.4667		

$F^* = 2.2603$, FTR H_0

11.5 a.

Source	df	SS	MS	F
Factor	2	54.111	27.056	12.82
Error	15	31.667	2.111	
Total	17	85.778		

b. p-value $< .01$ **c.** The type of word-processing
software does significantly affect the performance.
d. The within-sample variation increases.   **e.** Means 1
and 3 differ, means 2 and 3 differ.

11.7 a.

Source	df	SS	MS	F
Factor	2	74026.1	37013.05	8.7670
Error	21	88659.2	4221.8667	
Total	23	162685.3		

$F^* = 8.7670$, reject H_0. **b.** p-value $< .01$
c. Means for John and Randy differ; means for John and Ted differ.

11.9

Source	df	SS	MS	F
Factor	2	11.6667	5.8335	1.232
Error	27	127.8	4.7333	
Total	29	139.4667		

$F^* = 1.232$, FTR H_0, p-value $> .10$

11.11 Means 1 and 3 differ; means 2 and 3 differ.

11.13 a.

Source	df	SS	MS	F
Factor	1	1.691	1.691	3.07
Error	36	19.829	.551	
Total	37	21.520		

b. p-value is between .05 and .10, reject at a significance level of .10.

11.15 $H^* = 4.591$, reject H_0.

11.17 a.

Source	df	SS	MS	F
Factor	2	1.2381	0.61905	0.07359
Error	18	151.4286	8.4127	
Total	20	152.6667		

$F^* = .07359$, FTR H_0, p-value $> .10$
b. Right eye: .6968 to 5.5032; left eye: 1.1254 to 5.7318; both eyes: 1.2682 to 5.8746

11.19 b. Randomized block design **d.** One dependent variable **e.** Four treatments with 24 observations **f.** 24 treatments with a minimum of 48 observations

11.21 a. Completely randomized design **b.** Randomized block design

11.23 a. $t^* = -.8355$, FTR H_0

b.

Source	df	SS	MS	F
Factor	1	517,750	517,750	.70
Blocks	10	1,730,841,990	173,084,199	
Error	10	7,349,590	734,959	
Total	21	1,738,709,330		

$F^* = .70$, FTR H_0
c. The F value is the square of the t value.

11.25

Source	df	SS	MS	F
Factor	3	1998	666	57.61
Blocks	3	734	244.67	21.17
Error	9	104	11.56	
Total	15	2836		

11.27 a.

Source	df	SS	MS	F
Factor	2	20.33	10.17	17.93
Blocks	5	112.5	22.5	39.68
Error	10	5.667	.567	
Total	17	138.5		

11.29 No change in sum of squares

11.31

Source	df	SS	MS	F
Factor	3	130	43.33	4.73
Blocks	4	280	70.0	
Error	12	110	9.167	
Total	19	520		

11.33

Source	df	SS	MS	F
Factor	2	.8184	.4092	13.46
Blocks	12	5.5225	.4602	
Error	24	.7283	.0304	
Total	38	7.0692		

$F^* = 13.46$, reject H_0.

11.35

Source	df	SS	MS	F
Factor	3	293.1	97.7	15.4988
Blocks	9	5160.2	573.3556	90.955
Error	27	170.2	6.3037	
Total	39	5623.5		

Both F values are significant.

11.37

Source	df	SS	MS	F
Factor	2	77.0556	38.5278	16.8582
Blocks	11	42.2223	3.8384	1.6795
Error	22	50.2777	2.2854	
Total	35	169.5556		

The factor F value is significant. The block F value is not significant.

11.39 a.

Source	df	SS	MS	F
A	2	242.67	121.335	68.242
B	2	8.22	4.11	2.312
Interaction	4	11.11	2.778	1.562
Error	18	32	1.778	
Total	26	294		

e. p-value for $F_1 < .01$, p-value for $F_2 > .10$, p-value for $F_3 > .10$

11.41 BSD is the only significant factor.

11.43 a. Five replicates in each treatment combination

b.

Source	df	SS	MS	F
A	1	3.2	3.2	.27
B	1	105.8	105.8	8.82
Interaction	1	16.2	16.2	1.35
Error	16	192.0	12.0	
Total	19	317.2		

d. The mean of group instruction and computer-assisted training is significantly different from the mean of computer-assisted training and self-paced programs.

11.45 a. $H_0: \mu_1 = \mu_2 = \mu_3$; H_a: The means are not equal **b.** SS(factor) $= 14.0952$ **c.** 53.1429

d.

Source	df	SS	MS	F
Factor	2	14.0952	7.0477	2.3871
Error	18	53.1429	2.9524	
Total	20	67.2381		

$F^* = 2.3871$, FTR H_0, p-value > 0.1

11.47 a.

Source	df	SS	MS	F
Factor	2	1591	795	4.55
Error	12	2098	175	
Total	14	3689		

b. $F^* = 4.55$, reject H_0 **c.** 24.71 to 50.49
d. -6.63 to 29.83 **e.** Means 1 and 3 differ

11.49 a.

Source	df	SS	MS	F
Manager level	2	780.8	390.4	5.93
Sex	1	122.7	122.7	1.86
Interaction	2	84.8	42.4	.64
Error	12	790.0	65.8	
Total	17	1778.3		

b. Reject H_0 **c.** FTR H_0 **d.** FTR H_0 **e.** For part
(a), p-value is between .01 and .025; for part
(b), p-value $> .10$; for part (c), p-value $> .10$.

11.51

Source	df	SS	MS	F
Factor	3	217.9	72.6333	706.549
Error	36	3.7	0.1028	
Total	39	221.6		

$F^* = 706.549$, reject H_0, p-value $< .01$

11.53

Source	df	SS	MS	F
Factor	9	190	21.111	2.5503
Blocks	2	250	125	15.1006
Error	18	149	8.2778	
Total	29	589		

11.55

Source	df	SS	MS	F
Factor	2	58.3333	29.1667	2.8824
Blocks	7	74	10.5714	
Error	14	141.6667	10.1191	
Total	23	274		

11.57 $H^* = 2.67$, FTR H_0
11.59 a. 9 **b.** 15 **c.** Increase the sample size.
11.61 b. $F^* = 6.78$ would be significant at the .05 level
if the error df is greater than 4. **c.** Yes, the F^* values
of .45 and .44 would have p-values greater than .10.
11.63 Since $F^* = 1.98$, FTR H_0.
11.65 $F_1 = 2.66$, $F_2 = 2.0$, $F_3 = 3.14$. Only the
interaction term is significant at the .10 level.

Chapter 12

12.1 a. $Z^* = -.8386$, FTR H_0
b. Chi-square $= .7033$, FTR H_0
c. The chi-square value is the square of the Z value.
12.5 Chi-square $= .8125$, FTR H_0
12.7 Pool the last three groups. Chi-square $= 1.2$,
FTR H_0
12.9 Using classes $\leq 3, 4, 5, 6, 7, 8, \geq 9$,
chi-square $= 6.106$, FTR H_0.
12.11 Estimate of proportion $\cong .20$. Using the classes
$\leq 1, 2, 3, \geq 4$, chi-square $= .4184$, FTR H_0.

12.13 Chi-square $= 2.24$, FTR H_0
12.15 Chi-square $= 8.9839$, reject H_0
12.17 Pool SATISFIED and NOT SATISFIED.
Chi-square $= .5641$, FTR H_0
12.19 Chi-square $= 15.085$, p-value $< .005$
12.21 a. Chi-square $= 6.004$, FTR H_0
b. p-value $> .10$
12.23 Chi-square $= 6.06514$, FTR H_0
12.25 Chi-square $= 8.8571$, reject H_0
12.27 Using $E_1 = 50$, $E_2 = 27$, $E_3 = 12$, $E_4 = 11$,
chi-square $= 2.276$, FTR H_0.
12.29 Chi-square $= 2.133$, FTR H_0
12.31 Estimate of mean $= 3$. Using the classes $\leq 1, 2$,
3, 4, ≥ 5, chi-square $= .351$, FTR H_0.
12.33 Estimate of proportion $= .70$. Using the classes
$\leq 5, 6, 7, 8, \geq 9$, chi-square $= 2.148$, FTR H_0.
12.35 Chi-square $= 4.69$, FTR H_0
12.37 Chi-square $= 15.32$, FTR H_0, p-value slightly
above .05
12.39 Chi-square $= .4144$, FTR H_0
12.41 Chi-square $= 15.92$, reject H_0
12.43 Chi-square $= 3.2511$, reject H_0
12.45 a. $\bar{x} = 28.35$, $s = 19.96$ **b.** Chi-square $=$
36.882, reject H_0 **c.** p-value $< .005$ **d.** no change
12.47 a. Chi-square $= 2.99$, FTR H_0 **b.** p-value $>$
.10 **c.** Insufficient evidence to conclude that these
qualities are not independent
12.49 Chi-square $= 3.558$, FTR H_0
12.51 Since the p-value is greater than .10, fail to
reject the null hypothesis.

Chapter 13

13.1 b. $r = .948$
13.3 b. $\hat{Y} = 544.5516 + 3.084X$
13.7 a. Management rating, live entertainment, and
advertising rating have the highest correlations. These
variables should be considered as very important.
b. Since the correlation of live entertainment with
sales volume is .676, this variable is the single best
predictor of sales.
13.9 a. $r = .8024$ **b.** $\hat{Y} = 9.4761 + .2989X$
c. SSE $= 1.3890$
13.11 $\hat{Y} = 1.5823 + .7659X$. For $X = 9.67$, $\hat{Y} = 8.9886$.
13.13 a. $\hat{Y} = 6.662 + 2.984X$

b.

Y:	10.125	10.000	10.250	10.750	10.500
$Y - \hat{Y}$:	.210	.055	.246	-1.283	-1.981

Y:	14.000	14.250	14.370	15.000	14.550
$Y - \hat{Y}$:	1.370	1.321	.248	$-.167$	$-.020$

c. The error terms do not appear to be correlated.

13.15 a. $\hat{Y} = -3.889 + 5.820X$ **b.** SSE = 129.531, $s^2 = 25.906$

13.17 SSE = 155.94, $s = 2.6624$, all the sample residuals are within two standard deviations

13.19 Yes, the variance of the error component is not constant.

13.21 No

13.23 a. SSE = 15.7775, $t^* = 9.9296$, there is a positive relationship   **b.** p-value < .005, there is a positive relationship

13.25 Using log Y as the dependent variable, SSE = .0285, $t^* = 27.1601$, there is a positive relationship between X and log Y.

13.27 b. $\hat{Y} = 1079.22 + 193.7636X$ **c.** SSE = 8232.91, $s^2 = 1029.114$, $t^* = 54.861$, reject H_0

13.29 .8665 to 1.8001

13.31 .2080 to .3898

13.33 $r = .8371$, $t^* = 5.5172$, there is a linear relationship.

13.35 $r = .9742$, $t^* = 15.5525$, there is a linear relationship.

13.37 a. No   **b.** $r = -.1225$, $r^2 = .015$, $t^* = -.4450$, FTR H_0

13.39 $r = -.93167$, $t^* = -10.8799$, reject H_0 (the same result as before)

13.41 $t^* = 2.9344$, reject H_0

13.43 a. $\hat{Y} = -201.7469 + 25.9865X$. For $X = 30$, $\hat{Y} = 577.8481$, confidence interval is 520.9389 to 634.7573.   **b.** Prediction interval is 317.9274 to 837.7688

13.45 $\hat{Y} = 9.1036$, prediction interval is 7.4260 to 10.7812

13.47 The prediction interval is wider. Both intervals are the narrowest for $X = \bar{x}$.

13.49 a. $\hat{Y} = .29255 + .0909X$. For $X = 75$, $\hat{Y} = 7.1101$, confidence interval is 6.3529 to 7.8673   **b.** Prediction interval is 4.3343 to 9.8859   **c.** 69.07

13.51 $\hat{Y} = -.0267 + 1.2233(.30) = .3403$, prediction interval is .2098 to .4708

13.53 a. $\hat{Y} = 72.201 + 2.418X$

13.55 a. $\hat{Y} = 568.969 + .9664X$ **b.** .5613   **c.** SSE = 149.8681   **d.** 8.3260   **e.** $t^* = 4.7990$, reject H_0
f. $\hat{Y} = 684.664$, prediction interval is 677.15 to 692.18

13.57 $r = .9445$, $t^* = 7.04$, reject H_0

13.59 a. $\hat{Y} = 219.2880 - 1.2342X$ **b.** $r = -.7253$ $t^* = -6.5794$, reject H_0 **c.** SSE = 154.1407, confidence interval is -1.543 to $-.926$

13.61 $\hat{Y} = .322 + 1.3602X$

13.63 SSE = 80.75, $s = 2.06155$, one of the sample residuals lies outside two standard deviations, the histogram appears to be nearly symmetric

13.65 a. $\hat{Y} = 289.44 + 3.516X$ **b.** SSE = 384013,

$s = 206.56$, $t^* = 4.598$, reject H_0 **c.** $\hat{Y} = 781.68$, prediction interval is 779.76 to 783.60   **d.** $r^2 = .7014$

13.67 a. $\hat{Y} = 110.428$ **b.** $t^* = 19.54$, reject H_0
c. .8783 to 1.0949   **d.** $r^2 = .9928$ if observation 7 is removed.

Chapter 14

14.1 a. $\hat{Y} = 13.85$ **b.** 2.8

14.3 SSE = 191.16, $s = 3.4565$, all the residuals lie within two standard deviations, empirical rule approximately holds.

14.5 a. $\hat{Y} = .0090 + 1.1102X_1 + .13855X_2$
b. SSE = 4.7289

14.7 6.6646

14.9 a. $\hat{Y} = -.50 + 2.40X_1 + 2.95X_2$ **b.** $R^2 = .9381$
c. $n = 20$ **d. i.** 314.80   **ii.** 19.50   **iii.** 295.30
iv. 128.39   **v.** $F > 3.59$ **vi.** 14.048   **vii.** $t^* > 2.110$
viii. MSE = 1.15

14.11 a. $t^* = 3.1509$, reject H_0 **b.** $t^* = -2.205$, reject H_0

14.13 a. $t^* = 5.593$, X_2 contributes
b. 3.4056 to 7.7944

14.15

Source	df	SS	MS	F
Regression	7	200	28.5714	10.3896
Residual	20	55	2.75	
Total	27	255		

14.17

Source	df	SS	MS	F
Regression	2	97.12	48.56	1.9994
Residual	17	412.88	24.287	
Total	19	510		

$F^* = 1.9994$, FTR H_0

14.19 $F^* = 196.5$, reject H_0

14.21 One

14.23 $F^* = 14.056$, X_2 and X_3 contribute

14.25 $F^* = 2.718$, X_2 and X_3 do not contribute

14.27 $F^* = 2.6367$, the three variables do not contribute

14.29 $F^* = 11.31$, reject H_0

14.31 Presence of multicollinearity

14.33 Correlation = .8625, multicollinearity may be present

14.35 a. Correlation matrix is

	Y	X_1	X_2
Y	1	.975	.9965
X_1		1	.955
X_2			1

b. Using X_1, .951; using X_2, .993   **c.** No, due to the high correlation between X_1 and X_2

14.37 No

14.39 a. $\hat{Y} = 7630$ **b.** $F^* = 10.0$, the dummy variables contribute

14.41 $Y = \beta_0 + \beta_1 X_1 + \beta_2 X_2 + \beta_3 X_3 + \beta_4 X_4 + \beta_5 X_5 + \beta_6 X_6 + \beta_7 X_7 + \beta_8 X_8 + e$
  Y = total amount of compensation paid for a claim
  X_1 = age of employee (years)

$$X_2 = \begin{cases} 1 \text{ if male} \\ 0 \text{ if female} \end{cases}$$

$$X_3 = \begin{cases} 1 \text{ if employee is single} \\ 0 \text{ if not} \end{cases}$$

X_4 = length of employment (years)

$$X_5 = \begin{cases} 1 \text{ if injury is to head} \\ 0 \text{ if not} \end{cases}$$

$$X_6 = \begin{cases} 1 \text{ if injury is to a limb} \\ 0 \text{ if not} \end{cases}$$

$$X_7 = \begin{cases} 1 \text{ if employee works for} \\ \quad \text{manufacturer \#1} \\ 0 \text{ if not} \end{cases}$$

$$X_8 = \begin{cases} 1 \text{ if employee works for} \\ \quad \text{manufacturer \#2} \\ 0 \text{ if not} \end{cases}$$

14.43 a. X_3 **b.** X_3

14.45 Stepwise regression can remove variables previously included.

14.47 Equal variance for error terms

14.49 a. $\hat{Y} = 14.9$, confidence interval is 8.984 to 20.816 **b.** 4.895 to 24.905

14.51 The prediction interval is always wider than the corresponding confidence interval.

14.53 a. $\hat{Y} = 25.1434$, confidence interval is 22.9139 to 27.3729 **b.** 18.7494 to 31.5374

14.55 a. $\hat{Y} = .1026 + .688X_1 + .398X_2$ **b.** $R^2 = .506$ **c.** $n = 20$ **d.** SST = 171.30, SSE = 84.609, MSE = 4.977 **e.** Model is significant. **f.** X_1 contributes. **g.** X_2 does not contribute.

14.57 a. Yes, since the predictor variables will quite likely be highly correlated. **b.** 4.14 billion dollars. This is not a valid conclusion if multicollinearity is present.

14.59 a. $\hat{Y} = 17,357 - 1132X_1 - 33.2X_2 - 2556X_3 - 3275X_4 + 776X_5$ **b.** $F^* = 43.29$, reject H_0 **c.** 5876 to 9304 **d.** Correlation matrix is

	Y	X_1	X_2	X_3	X_4	X_5
Y	1	$-.872$	$-.857$	$-.077$	$-.527$	.659
X_1		1	.845	.125	.309	$-.462$
X_2			1	$-.032$	.480	$-.599$
X_3				1	$-.619$	.177
X_4					1	$-.692$
X_5						1

e. Using the forward procedure at the .10 significance level, the model is $\hat{Y} = 14,510 - 1581X_1 + 2841X_5$

14.61 a. $\hat{Y} = 15.24 + 4.8676X_1 - 5.802X_2 - 2.248X_3$ **b.** .987 **c.** $F^* = 18.1538$, X_2 and X_3 contribute **d.** $t^* = 22.778$, X_1 contributes **e.** The residuals appear to be random.

14.63 a. $F^* = 37.16$, reject H_0 **b.** $F^* = 1.48$, X_2 does not contribute. **c.** .0192 to .1465 **d.** 1.4757 to 2.3519 **e.** $\hat{Y} = 11.5657$

14.65 Error df = 286; p-value for PRICECON < .01, p-value for TIMEVAL is between .05 and .1, p-value for SAT/PRIDE < .01. $F^* = 24.395$ with a corresponding p-value < .01

14.67 $F^* = 14.56$, reject H_0. Neither t value is significant, suggesting multicollinearity.

14.69 $F^* = 88.1$, reject H_0. Each t statistic is significant.

Chapter 15

15.3 Due to seasonal effects, there may be a decline in demand.

15.5 a. Cyclical **b.** Seasonal **c.** Trend **d.** Irregular

15.7 a. $\hat{y}_t = -45,516.606 + 23.0303t$ **b.** $\hat{y}_t = 37.333 + 23.0303t$ **c.** The predicted values are the same.

15.9 The nature of the quadratic curve is unknown outside the range of the time series data.

15.11 $\hat{y}_t = -24.418 + 6.2429t$

15.13 a. $y_t = 12.97 + 17.404t$ **b.** 187.01

15.15 a. $\hat{y}_t = 512.912 + 199.95t$ **b.** $\hat{y}_t = 900.06 + 70.90t + 7.59t^2$ **c.** $\hat{y}_{17} = 3911.98$ using the linear equation, and 4299.08 using the quadratic equation

15.17 $C_{11} = .9811$, $C_{12} = .9152$, $C_{13} = .9344$, $C_{14} = .9833$, $C_{15} = 1.033$. The period of the cycle appears to be longer than 5 years.

15.19 a. $\hat{y}_t = 8.417 + .5833t$, where $t = 1$ corresponds to 1980 **b.** $C_t = y_t/\hat{y}_t$. These components are .778, 1.565, .984, .465, .971, 1.427, .96, .612, 1.244. **c.** Approximately 4 years.

15.21 The trend equation is $\hat{y}_t = 74.654 + .5220t$. The cyclical components are 1.062, .991, .932, . . . , .958, .976, 1.031.

15.23 a. The trend equation is $\hat{y}_t = 105.153 + .1940t$. **b.** The cyclical components are .997, 1.004, .993, .990, .995, 1.021, 1.011, 1.009, .992, .988.

15.25 The trend equation is $\hat{y}_t = 164.0 + 2.6t$. The cyclical components are .906, 1.147, 1.030, .900, 1.062, .908, .939, 1.077, 1.142, .890.

15.29 65.4, 78.9, 87.1, 84.1

15.31 9.9, 16.24, 12.48, 27.69, 23.98, 31.22, 54.72, 72.23, 7.14, 4.86, 4.96, 2.53

15.33 53, 83, 99, 8, 13, 3, 9, 9, 65, 76, 71, 67

15.35 Centered moving averages are 6.88, 7.75, 8.88, . . . , 19.13, 20.13, 21.00.

15.39 Seasonal indexes are .8710, .8256, .8458, .9291, 1.0280, 1.0503, 1.0730, 1.1614, 1.1411, 1.0629, 1.0452, .9670.

15.41 Seasonal indexes are 1.0913, 1.2155, .8573, .8358.

15.43 Seasonal indexes are .87, .61, .86, 1.65.

15.45 a. Seasonal indexes are .94, .97, 1.01, 1.02, 1.07, 1.06, 1.04, 1.02, .99, .98, .96, .94. **b.** The deseasonalized trend line is $\hat{d}_t = 27.1648 + .1773t$. **c.** The cyclical components are .9888; .9825, .9694, .9711, .9817, .9932, .9937, .9959, .9992, 1.0019, 1.0053, 1.0048. **d.** The irregular components are 1.0113, 1.0138, .9878.

15.47 The trend equation is $\hat{y}_t = 1.552 + .0263t$. The cyclical components for $t = 2, 3, 4, \ldots, 17, 18, 19$ are 1.103, 1.063, 1.026, . . . , 1.059, 1.053, 1.043.

15.49 a. Seasonal indexes are .89, .94, .94, 1.04, .98, 1.02, 1.08, 1.10, 1.10, 1.02, .97, .93. **b.** $\hat{d}_t = 3.7214 + .0836t$ **c.** Cyclical components are 1.0358, 1.0005, .9829, .9771, .9800, .9812, .9930, .9941, .9987, .9958, .9913, 1.0047. **d.** Irregular components are .9850, .9800, 1.0122.

15.51 Cyclical components are .9465, .9879, 1.0627, 1.0356. Irregular components are .7508, 1.1694, 1.0340, .9012.

15.53 Index numbers are 100.0, 101.6, 109.4, 132.9, 156.5, 187.8.

15.55 a. 120.6 **b.** 119.0 **c.** 118.7

15.57 a. 135.6 **b.** 136.11 **c.** 135.9

15.61 a. 109.36

15.63 a. Trend **b.** Irregular **c.** Seasonal **d.** Irregular

15.65 a. $\hat{y}_t = 32.697 + .213t$ **b.** $\hat{y}_{14} = 35.68$

15.67 a. Yes **b.** $\hat{y}_t = 1713.27 + 97.861t$ **c.** 2789.73

15.69 The deseasonalized trend equation is $\hat{d}_t = 3.8255 + .0215t$. The cyclical components for 1986 and 1987 are 1.0082, .9928, .9953, .9979, .9985, .9823, .9880, .9907.

15.71 a. Seasonal indexes are 1.20, 1.15, 1.07, .92, .91, .95, .96, .96, .92, .93, .99, 1.05 **b.** $\hat{d}_t = 151.806 + 1.094t$ **d.** Irregular components are 1.00, 1.00.

15.73 a. Seasonal indexes are .99, 1.02, 1.00, .98 **b.** $\hat{d}_t = 144.7426 + .8401t$ **c.** .9719, .9444, .9435, .9426 **d.** Iregular components for the first two quarters are .9909 and 1.0050.

15.75 Seasonal indexes are 1.10, 1.03, .99, .95, .94, .87, .89, .95, .98, 1.03, 1.10, 1.18.

15.77 a. Seasonal indexes are 1.02, .99, 1.05, 1.09, 1.07, 1.06, .95, .96, .89, .96, .86, 1.10. **b.** $\hat{d}_t = -560.48 + 96.852t$ **d.** Irregular components are .96, .95, 1.04.

15.79 Residuals follow a nonrandom pattern, violating the assumptions of the regression statistical tests.

Chapter 16

16.3 Estimates for 1979–1988 are 28.50, 29.25, 31.75, 29.50, 28.00, 27.50, 28.25, 29.75, 32.50, 31.50.

16.7 a. .48, .42, .52, .68, .88, 1.21, 1.62, 1.84, 1.63. 1.33, .83, .58 **b.** $\hat{d}_t = 1.2595 + .0137t$ **c.** $\hat{y}_{87} = 1.275$, $\hat{y}_{91} = 4.060$, $\hat{y}_{96} = 1.493$

16.11 10.8

16.13 Estimates values are 11.06, 11.11, 11.18. Residuals are .54, .64, −.15

16.15 Forecasts are 21.48, 21.34, 21.18

16.17 a. 172.52 **b.** 178.51

16.19 29.79

16.21 a. 17.176 **b.** 12.35

16.23 Initial seasonal factors are .95, .83, 1.07 and 1.16. Least squares line is $.2287 + .0109t$. Forecasts for each quarter are .404, .361, .491, .538.

16.25 Initial seasonal factors are .91, 1.50, 1.09 and .50. Least squares line is $.2709 + .0342t$. Forecasts for each quarter are 1.05, 1.78, 1.29, .72.

16.27 a. MAD = .267, MAPE = .125, MSE = .089 **b.** MAD = .32, MAPE = .144, MSE = .14

16.29 Procedure 1: MAD = 3.556, MSE = 3.333; procedure 2: MAD = 16.556, MSE = 44.556. Procedure 1 is superior overall, and procedure 2 is superior if the last two observations are ignored.

16.31 a. Predicted values for 1986–1988 are 12500.97, 12564.44, 13096.19, . . . , 15940.29, 17151.46, 17656.35. MSE = 2,049,098.2. **b.** Predicted values are 12067.43, 11996.72, 13013.65, . . . , 15160.85, 17268.70, 17738.42. MSE = 2,694,686.4. **c.** The procedure in part **a** is better.

16.33 MAD = 16,931.80, MSE = 472,762,368, MAPE = .0321

16.35 Second-order equation is $\hat{y}_t = 10.919 - .4813y_{t-1} - .5393y_{t-2}$
Fourth-order equation is $\hat{y}_t = -.483 + .0171y_{t-1} + .0515y_{t-2} + .0376y_{t-3} + 1.0115y_{t-4}$
MSE for second-order equation is 2.9641. MSE for fourth-order equation is .1155

16.37 a. $r_1 = .693, r_2 = .108, r_3 = -.355, r_4 = -.428$ **b.** One possible model is $\hat{y}_t = b_0 + b_1y_{t-1} = 11.542 + .7105y_{t-1}$

16.39 $r_1 = .848, r_2 = .500, r_3 = .066, r_4 = -.334, r_5 = -.599, r_6 = -.688, r_7 = -.590, r_8 = -.320, r_9 = .046, r_{10} = .435, r_{11} = .726, r_{12} = .832$. The data do not appear to be stationary.

16.41 The series of second differences appear to be stationary with two period seasonal spikes.

16.43 $\hat{y}_t = 184.1 + .7921y_{t-1} + .2645y_{t-2}, R^2 = .972$

16.45 $\hat{y}_t = 15.25 - .422t + .313Q_1 + .909Q_2 + .842Q_3$

16.47 Let $M_1 = 1$ for Jan., $M_2 = 1$ for Feb., . . . , $M_{11} = 1$ for Nov. The regression equation is $\hat{y}_t = .6143 - .2042M_1 - .3012M_2 - .1268M_3 + .1619M_4 + .5506M_5 + 1.5355M_6 + 1.8422M_7 +$

$2.2452M_8 + 1.8768M_9 + 1.3226M_{10} + .4685M_{11} + .01131t$

16.49 $\hat{y}_t = 12.617 - 1.489Q_1 + 2.507Q_2 + 5.337Q_3 + .0036t$

16.51 a. $\hat{y}_t = 108.52 + .0031X_t$ **b.** $\hat{y}_t = 74.454 + .252X_{t-1}$ **c.** R^2 for part **a** is approximately .0, R^2 for part **b** is .957

16.53 $\hat{y}_t = .5134 + .1555X_{t-1} + .5783Q_1 + .6600Q_2 + 1.3822Q_3$, MSE = .0091

16.55 a. $\hat{y}_t = 14.532 + .01459X_t$ **b.** $\hat{y}_t = 9.5539 + .02569X_{t-1}$ **c.** R^2 for part **a** is .334, R^2 for part **b** is .972

16.57 positive autocorrelation, no autocorrelation, negative autocorrelation

16.59 DW = 2.66, FTR H_0

16.61 DW = .56, possible positive autocorrelation

16.63 $\hat{y}_{37} = 19.25$, $\hat{y}_{38} = 22.35$, $\hat{y}_{39} = 25.56$, $\hat{y}_{40} = 27.25$, $\hat{y}_{41} = 25.34$

16.65 a. 1123.2 **b.** 1135.78

16.67 MSE = .32

16.69 Let $Q_1 = 1$ for quarter 1, $Q_2 = 1$ for quarter 2, $Q_3 = 1$ for quarter 3. The regression equation is $\hat{y}_t = 39.275 - 16.019Q_1 - 9.613Q_2 - 4.406Q_3 + .79375t$

16.71 8.56

16.75 $\hat{y}_{18} = 76647.83$

16.77 MAD = 7.665, MSE = 100.40, MAPE = .029

Chapter 17

17.1

States of Nature

Action	S_1	S_2	S_3
A_1	2,000	500	−1,000
A_2	1,500	750	0
A_3	1,000	1,000	1,000

17.3

States of Nature

Action	5	6	7	8	9	10	11	12	13
7	−20	5	30	30	30	30	30	30	30
8	−30	−5	20	45	45	45	45	45	45
9	−40	−15	10	35	60	60	60	60	60
10	−50	−25	0	25	50	75	75	75	75
11	−60	−35	−10	15	40	65	90	90	90
12	−70	−45	−20	5	30	55	80	105	105
13	−80	−55	−30	−5	20	45	70	95	120

17.5

States of Nature

Action	$S_1(100)$	$S_2(125)$	$S_3(150)$
$A_1(100)$	90	90	90
$A_2(125)$	62.5	112.5	112.5
$A_3(150)$	35	85	135

17.7 a.

States of Nature

Action	50	60	70	80	90	100
70	5.5	8	10.5	10.5	10.5	10.5
80	4.5	7	9.5	12	12	12
90	3.5	6	8.5	11	13.5	13.5
100	2.5	5	7.5	10	12.5	12.5

b.

States of Nature

Action	50	60	70	80	90	100
70	0	0	0	1.5	3.5	4.5
80	1	1	1	0	1.5	3
90	2	2	2	1	0	1.5
100	3	3	3	2	1	0

Minimax decision is to order 90 copies. Maximax decision is to order 100 copies.

17.9

States of Nature

Action	S_1	S_2	S_3	S_4
A_1	0	13	5	0
A_2	170	5	18	15
A_3	150	0	0	25

17.11 a. $E(A_1) = 97.5$, $E(A_2) = 55$, $E(A_3) = 95$; expected payoff is maximized using A_1
b. risk $(A_1) = 4868.75$, risk $(A_2) = 0$, risk $(A_3) = 1875$

17.13 a. A_3 **b.** A_3 **c.** risk $(A_1) = 0$, risk $(A_2) = 169$, risk $(A_3) = 1642.1875$, risk $(A_4) = 9850$

17.15 Maximum expected payoff is 10.125.

17.17 $EVPI = 6.25$; action A_3 is inadmissible.

17.19 Expected payoff by ordering 2000 cards is $840; expected payoff with a perfect predictor is $950; maximum amount is $110.

17.21 Expected payoff using A_3 is 1000; expected payoff with a perfect predictor is 1333.33; maximum amount is 333.33.

17.23 $EVPI = 22.75$, A_1 is inadmissible.

17.25 a. A_2
b. Table of utility values:

States of Nature

Action	S_1	S_2	S_3	S_4
A_1	2	20	14.14	6.32
A_2	17.89	10.95	12.65	10
A_3	18.97	8.94	10.95	6.32

$E(A_1) = 10.615$, $E(A_2) = 12.873$, $E(A_3) = 11.295$; the decision is A_2.

17.27 $p = .81$

17.29 $E(A_1) = 3.413$, $E(A_2) = 3.433$, $E(A_3) = 3.386$, $E(A_4) = 3.394$. Because the decision is A_2, the manager is a risk avoider.

17.31

17.33

17.35

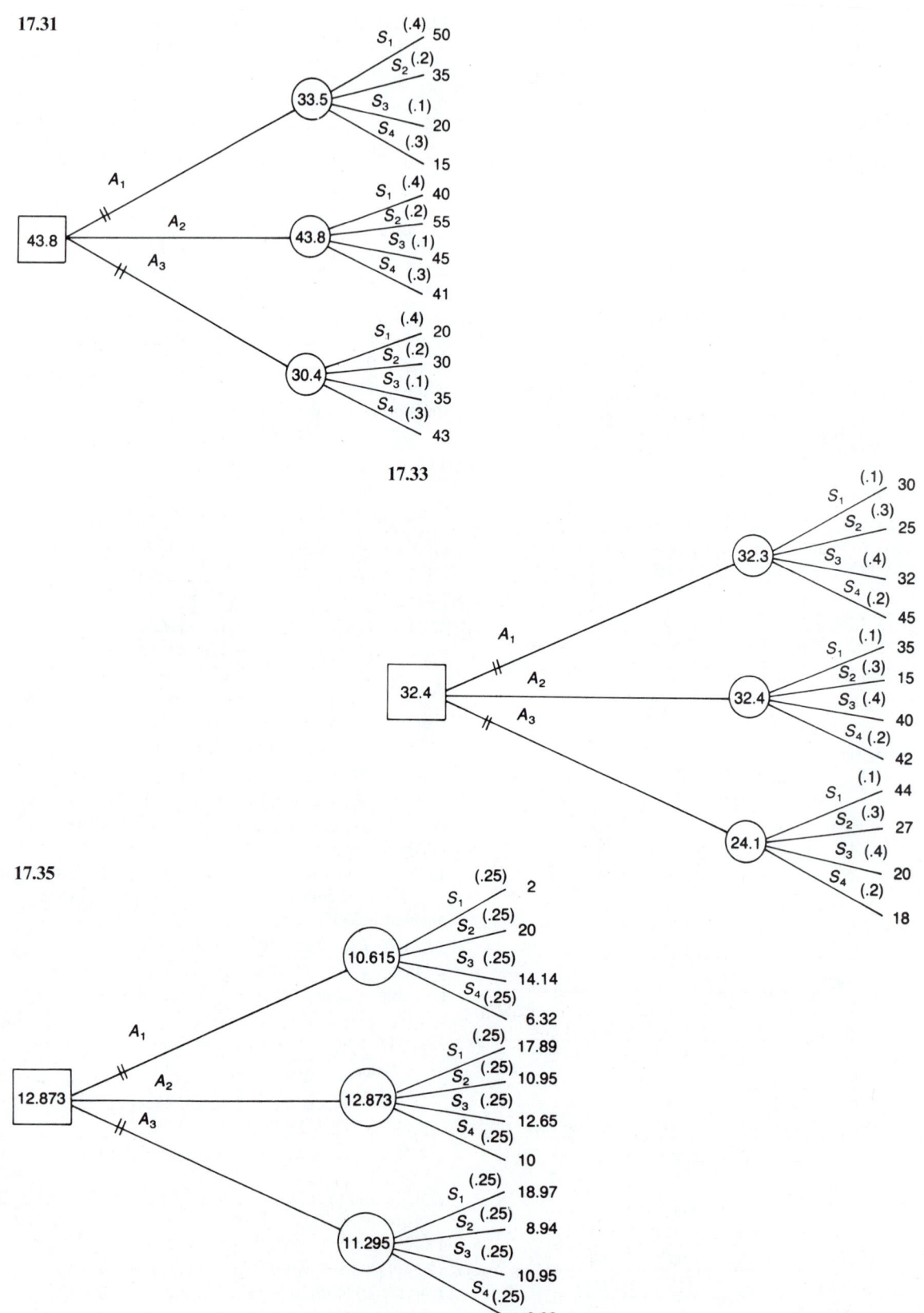

17.37 $P(A_1|B) = .421$, $P(A_2|B) = .158$, $P(A_3|B) =$ .316, $P(A_4|B) = .105$

17.39 $.0264/.0734 = .3597$

17.41 $P(B) = .26$, $P(S_1|B) = .2308$, $P(S_2|B) = .2308$, $P(S_3|B) = .4615$, $P(S_4|B) = .0769$; maximum expected utility of 33.23 occurs for A_2.

17.43 $EVPI = 3000$; yes

17.45 $EVPI = 2442.27 - 1995.15 = 447.12$. It is not worthwhile to hire the consultant.

17.47 Minimax decision is to order 34 or 35 loaves; maximax decision is to order 36 loaves.

17.49 The change of probabilities will not affect the minimax decision. Using the expected payoffs, action A_4 is the optimal action for sets 1, 2, and 4, and action A_1 is optimal for sets 3 and 5.

17.51 One example would be to let $U(x) = 0$ for $x < 0$, $U(x) = 2x$ for $0 \le x \le 1000$, and $U(x) = 2000 + (x - 1000)^2$ for $x > 1000$.

18.5 $R = 10$, FTR H_0

18.7 $R = 12$, FTR H_0

18.9 $R = 17$, $Z^* = (17 - 18.486)/2.91 = -.51$, FTR H_0

18.13 $U = U_1 = 43$, FTR H_0

18.15 $Z^* = (97 - 77)/18.267 = 1.09$, FTR H_0

18.17 $Z^* = (53.5 - 72)/17.32 = -1.068$, FTR H_0

18.19 $Z^* = (88.5 - 72)/17.32 = .953$, FTR H_0

18.21 $U = 14$, reject H_0

18.25 $T_+ = 12$, FTR H_0

18.27 $T_- = 26.5$, reject H_0

18.29 $Z^* = (48 - 105)/26.786 = -2.13$, reject H_0

18.31 $Z^* = (280 - 232.5)/48.618 = .9769$, FTR H_0

18.33 $Z^* = (81.5 - 76.5)/21.12 = .24$, FTR H_0

18.35 $T = 6$, FTR H_0

18.37 $KW = .485$, FTR H_0

17.53

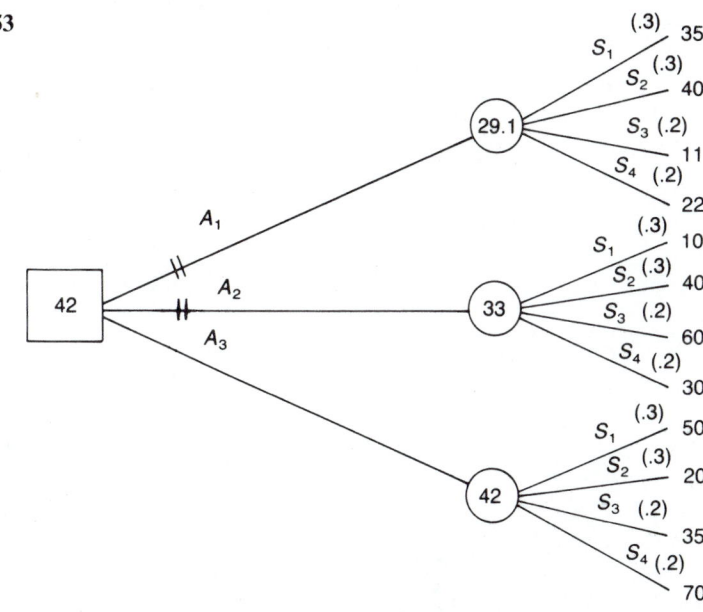

17.55 The maximum payoff with the consultant is 1157.08. The maximum payoff without the consultant is 1105 using A_3. Since $1157.08 - 1105 = 52.08$, which is less than 350, it is not worthwhile to have the consultant's service.

17.57 $P(\text{long}|\text{under 40}) = .632$

17.59 $P(A) = .62$, $P(B) = .38$, $P(G|A) = .55$, $P(G|B) = .70$, $P(G) = .607$, $P(B|G) = .344$

Chapter 18

18.1 The only assumption needed is that you have a sequence of n observations containing n_1 symbols of the first type and n_2 symbols of the second type.

18.3 $R = 12$, reject H_0

18.39 $KW = 6.167$, reject H_0

18.41 $KW = 7.649$, *reject H_0*

18.43 $KW = 6.83$, reject H_0

18.45 $KW = 2.105$, FTR H_0

18.47 $KW = 1.51$, FTR H_0

18.49 $FR = 2$, FTR H_0

18.51 $FR = 7$, reject H_0

18.53 $FR = 6$, reject H_0

18.55 $r = -.83$, rank correlation $= -.865$

18.57 rank correlation $= .874$, reject H_0

18.59 rank correlation $= -.944$, reject H_0

18.61 rank correlation $= .88$, reject H_0

18.63 rank correlation $= .93$, reject H_0

18.65 rank correlation $= .84$, reject H_0

18.67 $R = 16$, $Z^* = (16 - 12.52)/2.25 = 1.55$, FTR H_0

18.71 $U = 16$, FTR H_0

18.73 $T = 10.5$, reject H_0 (use $n = 14$)

18.75 $KW = 2.93$, FTR H_0

18.77 $KW = .14$, FTR H_0

18.79 $KW = 4.10$, FTR H_0

18.81 rank correlation $= .845$, reject H_0

18.83 $FR = 31.2$, reject H_0

18.85 Rank correlation $= .9648$, reject H_0

18.87 $KW = 7.452$, reject H_0. The p-value is between .01 and .025.

Index

$(1.61 \le z \le 1.72) = 1.72 - 1.61 = .0110$

Areas of the Standard Normal Distribution

The entries in this table are the probabilities that a standard normal random variable is between 0 and z (the shaded area).

$P(0 \le z \le \text{\# in table})$

z	0.00	0.01	0.02	0.03	0.04	0.05	0.06	0.07	0.08	0.09
					Second Decimal Place in z					
0.0	0.0000	0.0040	0.0080	0.0120	0.0160	0.0199	0.0239	0.0279	0.0319	0.0359
0.1	0.0398	0.0438	0.0478	0.0517	0.0557	0.0596	0.0636	0.0675	0.0714	0.0753
0.2	0.0793	0.0832	0.0871	0.0910	0.0948	0.0987	0.1026	0.1064	0.1103	0.1141
0.3	0.1179	0.1217	0.1255	0.1293	0.1331	0.1368	0.1406	0.1443	0.1480	0.1517
0.4	0.1554	0.1591	0.1628	0.1664	0.1700	0.1736	0.1772	0.1808	0.1844	0.1879
0.5	0.1915	0.1950	0.1985	0.2019	0.2054	0.2088	0.2123	0.2157	0.2190	0.2224
0.6	0.2257	0.2291	0.2324	0.2357	0.2389	0.2422	0.2454	0.2486	0.2517	0.2549
0.7	0.2580	0.2611	0.2642	0.2673	0.2704	0.2734	0.2764	0.2794	0.2823	0.2852
0.8	0.2881	0.2910	0.2939	0.2967	0.2995	0.3023	0.3051	0.3078	0.3106	0.3133
0.9	0.3159	0.3186	0.3212	0.3238	0.3264	0.3289	0.3315	0.3340	0.3365	0.3389
1.0	0.3413	0.3438	0.3461	0.3485	0.3508	0.3531	0.3554	0.3577	0.3599	0.3621
1.1	0.3643	0.3665	0.3686	0.3708	0.3729	0.3749	0.3770	0.3790	0.3810	0.3830
1.2	0.3849	0.3869	0.3888	0.3907	0.3925	0.3944	0.3962	0.3980	0.3997	0.4015
1.3	0.4032	0.4049	0.4066	0.4082	0.4099	0.4115	0.4131	0.4147	0.4162	0.4177
1.4	0.4192	0.4207	0.4222	0.4236	0.4251	0.4265	0.4279	0.4292	0.4306	0.4319
1.5	0.4332	0.4345	0.4357	0.4370	0.4382	0.4394	0.4406	0.4418	0.4429	0.4441
1.6	0.4452	0.4463	0.4474	0.4484	0.4495	0.4505	0.4515	0.4525	0.4535	0.4545
1.7	0.4554	0.4564	0.4573	0.4582	0.4591	0.4599	0.4608	0.4616	0.4625	0.4633
1.8	0.4641	0.4649	0.4656	0.4664	0.4671	0.4678	0.4686	0.4693	0.4699	0.4706
1.9	0.4713	0.4719	0.4726	0.4732	0.4738	0.4744	0.4750	0.4756	0.4761	0.4767
2.0	0.4772	0.4778	0.4783	0.4788	0.4793	0.4796	0.4803	0.4808	0.4812	0.4817
2.1	0.4821	0.4826	0.4830	0.4834	0.4838	0.4842	0.4846	0.4850	0.4854	0.4857
2.2	0.4861	0.4864	0.4868	0.4871	0.4875	0.4878	0.4881	0.4884	0.4887	0.4890
2.3	0.4893	0.4896	0.4898	0.4901	0.4904	0.4906	0.4909	0.4911	0.4913	0.4916
2.4	0.4918	0.4920	0.4922	0.4925	0.4927	0.4929	0.4931	0.4932	0.4934	0.4936
2.5	0.4938	0.4940	0.4941	0.4943	0.4945	0.4946	0.4948	0.4949	0.4951	0.4952
2.6	0.4953	0.4955	0.4956	0.4957	0.4959	0.4960	0.4961	0.4962	0.4963	0.4974
2.7	0.4965	0.4966	0.4967	0.4968	0.4969	0.4970	0.4971	0.4972	0.4973	0.4974
2.8	0.4974	0.4975	0.4976	0.4977	0.4977	0.4978	0.4979	0.4979	0.4980	0.4981
2.9	0.4981	0.4982	0.4982	0.4983	0.4984	0.4984	0.4985	0.4985	0.4986	0.4986
3.0	0.4987	0.4987	0.4987	0.4988	0.4988	0.4989	0.4989	0.4989	0.4990	0.4990
3.1	0.4990	0.4991	0.4991	0.4991	0.4992	0.4992	0.4992	0.4992	0.4993	0.4993
3.2	0.4993	0.4993	0.4994	0.4994	0.4994	0.4994	0.4994	0.4995	0.4995	0.4995
3.3	0.4995	0.4995	0.4995	0.4996	0.4996	0.4996	0.4996	0.4996	0.4996	0.4997
3.4	0.4997	0.4997	0.4997	0.4997	0.4997	0.4997	0.4997	0.4997	0.4997	0.4998
3.5	0.4998									
4.0	0.49997									
4.5	0.499997									
5.0	0.4999997									

$P(z \ge \text{\# in table}) = .5 - P(\text{in table})$

$(z \ge 2.15) = .5 - .4842 = .0158$

Reprinted with permission from *Standard Mathematical Tables*, 15th ed., © CRC Press. Inc., Boca Raton, FL.